Emissions Scenarios

How will the world's climate change in the coming century? The answer to this question depends on how human societies develop in terms of demographics and economic development, technological change, energy supply and demand, and land use change.

The Intergovernmental Panel on Climate Change (IPCC) Special Report on Emissions Scenarios describes new scenarios of the future, and predicts greenhouse gas emissions associated with such developments. These scenarios are based on a thorough review of the literature, the development of narrative "storylines", and the quantification of these storylines with the help of six different integrated models from different countries. The scenarios provide the basis for future assessments of climate change and possible response strategies. The report illustrates that future emissions, even in the absence of explicit climate policies, depend very much on the choices people make: how economies are structured, which energy sources are preferred, and how people use available land resources. The IPCC published previous greenhouse gas emissions scenarios in 1990 and 1992. The current scenarios introduce many innovative aspects, such as narrative descriptions of the scenarios, inclusion of information on the availability of energy technologies, and an analysis of international equity issues.

This IPCC Special Report is the most comprehensive and state-of-the-art assessment available of greenhouse gas emissions scenarios, and provides invaluable information for industry, policy-makers, environmental organizations, and researchers in global change, technology, engineering and economics.

Nebojša Nakićenović is the Leader of the Transitions to New Technologies Project at the International Institute for Applied Systems Analysis (IIASA), in Austria.

Rob Swart is Head of the Technical Support Unit of Working Group III on Mitigation of the Intergovernmental Panel on Climate Change (IPCC), in the Netherlands.

Special Report on Emissions Scenarios

Nebojša Nakićenović, Joseph Alcamo, Gerald Davis, Bert de Vries, Joergen Fenhann, Stuart Gaffin, Kenneth Gregory, Arnulf Grübler, Tae Yong Jung, Tom Kram, Emilio Lebre La Rovere, Laurie Michaelis, Shunsuke Mori, Tsuneyuki Morita, William Pepper, Hugh Pitcher, Lynn Price, Keywan Riahi, Alexander Roehrl, Hans-Holger Rogner, Alexei Sankovski, Michael Schlesinger, Priyadarshi Shukla, Steven Smith, Robert Swart, Sascha van Rooijen, Nadejda Victor, Zhou Dadi

A Special Report of Working Group III of the Intergovernmental Panel on Climate Change

Published for the Intergovernmental Panel on Climate Change

PUBLISHED BY THE PRESS SYNDICATE OF THE UNIVERSITY OF CAMBRIDGE
The Pitt Building, Trumpington Street, Cambridge, United Kingdom

CAMBRIDGE UNIVERSITY PRESS
The Edinburgh Building, Cambridge CB2 2RU, UK http://www.cup.cam.ac.uk
40 West 20th Street, New York, NY 10011-4211, USA http://www.cup.org
10 Stamford Road, Oakleigh, Melbourne 3166, Australia
Ruiz de Alarcón 13, 28014 Madrid, Spain

First published 2000

Printed in the United States of America

A catalog record for this book is available from the British Library

Library of Congress Cataloging-in-Publication Data available

ISBN 0 521 80081 1 hardback
ISBN 0 521 80493 0 paperback

Special Report on Emissions Scenarios

Coordinating Lead Author: Nebojša Nakićenović (Austria)

Lead Authors: Joseph Alcamo (Germany), Gerald Davis (United Kingdom), Bert de Vries (The Netherlands),[1] Joergen Fenhann (Denmark), Stuart Gaffin (United States), Kenneth Gregory (United Kingdom), Arnulf Grübler (IIASA), Tae Yong Jung (Republic of Korea), Tom Kram (The Netherlands), Emilio Lebre La Rovere (Brazil), Laurie Michaelis (United Kingdom), Shunsuke Mori (Japan),[1] Tsuneyuki Morita (Japan), William Pepper (United States),[1] Hugh Pitcher (United States),[1] Lynn Price (United States), Keywan Riahi (IIASA),[1] Alexander Roehrl (IIASA),[1] Hans-Holger Rogner (IAEA), Alexei Sankovski (United States),[1] Michael Schlesinger (United States), Priyadarshi Shukla (India), Steven Smith (United States), Robert Swart (The Netherlands), Sascha van Rooijen (The Netherlands), Nadejda Victor (United States), Zhou Dadi (China)

Contributing Authors: Dennis Anderson (United Kingdom), Johannes Bollen (The Netherlands),[1] Lex Bouwman (The Netherlands), Ogunlade Davidson (IPCC), Jae Edmonds (United States), Christopher Elvidge (United States), Michel den Elzen (The Netherlands),[1] Henryk Gaj (Poland), Erik Haites (Canada), William Hare (The Netherlands), Marco Janssen (The Netherlands),[1] Kejun Jiang (China),[1] Anne Johnson (United States), Eric Kreileman (The Netherlands),[1] Mathew Luhanga (United Republic of Tanzania), Nicolette Manson (United States), Toshihiko Masui (Japan),[1] Alan McDonald (IIASA), Douglas McKay (United Kingdom), Bert Metz (IPCC), Leena Srivastava (India), Cees Volkers (The Netherlands), Robert Watson (IPCC), John Weyant (United States), Ernst Worrell (United States), Xiaoshi Xing (United States)

Review Editors: Eduardo Calvo (Peru), Michael Chadwick (United Kingdom), Yukio Ishiumi (Japan), Jyoti Parikh (India)

Report Editors: Nebojša Nakićenović (Austria) and Robert Swart (The Netherlands)

Acknowledgements: Thomas Büttner (United Nations), Renate Christ (IPCC), Ewa Delpos (IIASA), Angela Dowds (IIASA), Günther Fischer (IIASA), Anne Goujon (IIASA), Andrei Gritsevskii (IIASA), Niklas Hohne (UNFCCC), Fortunat Joos (University of Bern), Frank Kasper (University of Kassel), Kathy Kienleitner (IIASA), Ger Klaassen (IIASA), Katalin Kuszko (IIASA), Matt Lloyd (Cambridge University Press), Yuzuru Matsuoka (Kyoto University), Mack McFarland (DuPont), Anita Meier (IPCC-WGIII Technical Support Unit/RIVM), Martin Middelburg (Report Layout, RIVM), Roberta Miller (CIESIN), Janina Onigkeit (University of Kassel), John Ormiston (Copy Editor), Jiahua Pan (IPCC-WGIII Technical Support Unit/RIVM), Joyce Penner (University of Michigan), Michael Prather (University of California), Paul Reuter (University of Kassel), Susan Riley (IIASA), Cynthia Rosenzweig (Goddard Institute for Space Studies), Stephen Schneider (Stanford University), Leo Schrattenholzer (IIASA), Dennis Tirpak (UNFCCC), David Victor (US Council on Foreign Relations), Patricia Wagner (IIASA)

[1] Members of the Modeling Teams
Affiliations indicate country of residency or an international organization and not necessarily citizenship.

Contents

Foreword

The Intergovernmental Panel on Climate Change (IPCC) was jointly established by the World Meteorological Organization (WMO) and the United Nations Environment Programme (UNEP) to assess the scientific, technical and socio-economic information relevant for the understanding of the risk of human-induced climate change. Since its inception the IPCC has produced a series of comprehensive Assessment Reports on the state of understanding of causes of climate change, its potential impacts and options for response strategies. It prepared also Special Reports, Technical Papers, methodologies and guidelines. These IPCC publications have become standard works of reference, widely used by policymakers, scientists and other experts.

In 1992 the IPCC released emission scenarios to be used for driving global circulation models to develop climate change scenarios. The so-called IS92 scenarios were pathbreaking. They were the first global scenarios to provide estimates for the full suite of greenhouse gases. Much has changed since then in our understanding of possible future greenhouse gas emissions and climate change. Therefore the IPCC decided in 1996 to develop a new set of emissions scenarios which will provide input to the IPCC third assessment report but can be of broader use than the IS92 scenarios. The new scenarios provide also input for evaluating climatic and environmental consequences of future greenhouse gas emissions and for assessing alternative mitigation and adaptation strategies. They include improved emission baselines and latest information on economic restructuring throughout the world, examine different rates and trends in technological change and expand the range of different economic-development pathways, including narrowing of the income gap between developed and developing countries. To achieve this a new approach was adopted to take into account a wide range of scientific perspectives, and interactions between regions and sectors. Through the so-called "open process" input and feedback from a community of experts much broader than the writing team were solicited. The results of this work show that different social, economic and technological developments have a strong impact on emission trends, without assuming explicit climate policy interventions. The new scenarios provide also important insights about the interlinkages between environmental quality and development choices and will certainly be a useful tool for experts and decision makers.

As usual in the IPCC, success in producing this report has depended first and foremost on the co-operation of scientists and other experts worldwide. In the case of this report the active contribution of a broad expert community to the open process was an important element of the success. These individuals have devoted enormous time and effort to producing this report and we are extremely grateful for their commitment to the IPCC process. We would like to highlight in particular the enthusiasm and tireless efforts of the co-ordinating lead author for this report Nebojša Nakićenović and his team at the International Institute for Applied Systems Analysis (IIASA) in Laxenburg/Austria who ensured the high quality of this report.

Further we would like to express our sincere thanks to:

- Robert T. Watson, the Chairman of the IPCC,
- The Co-chairs of Working Group III Bert Metz and Ogunlade Davidson,
- The members of the writing team,
- The staff of the Working Group III Technical Support Unit, including Rob Swart, Jiahua Pan, Tom Kram, and Anita Meier,
- N. Sundararaman, the Secretary of the IPCC, Renate Christ Deputy Secretary of the IPCC and the staff of the IPCC Secretariat Rudie Bourgeois, Chantal Ettori and Annie Courtin.

G.O.P. Obasi

Secretary-General
World Meteorological Organization

K. Töpfer

Executive Director
United Nations Environment Programme
and
Director-General
United Nations Office in Nairobi

Preface

The Intergovernmental Panel on Climate Change (IPCC) was established jointly by the World Meteorological Organisation (WMO) and the United Nations Environment Programme (UNEP) to assess periodically the science, impacts, and socio-economics of climate change and of adaptation and mitigation options. The IPCC provides, on request, scientific and technical advice to the Conference of the Parties to the United Nations Framework Convention on Climate Change (UNFCCC) and its bodies. In response to a 1994 evaluation of the earlier IPCC IS92 emissions scenarios, the 1996 Plenary of the IPCC requested this Special Report on Emissions Scenarios (SRES) (see Appendix I for the Terms of Reference). This report was accepted by the Working Group III (WGIII) plenary session in March 2000. The long-term nature and uncertainty of climate change and its driving forces require scenarios that extend to the end of the 21st century. This report describes the new scenarios and how they were developed.

The SRES scenarios cover a wide range of the main driving forces of future emissions, from demographic to technological and economic developments. As required by the Terms of Reference, none of the scenarios in the set includes any future policies that explicitly address climate change, although all scenarios necessarily encompass various policies of other types. The set of SRES emissions scenarios is based on an extensive assessment of the literature, six alternative modelling approaches, and an "open process" that solicited wide participation and feedback from many groups and individuals. The SRES scenarios include the range of emissions of all relevant species of greenhouse gases (GHGs) and sulfur and their driving forces.

The SRES writing team included more than 50 members from 18 countries who represent a broad range of scientific disciplines, regional backgrounds, and non-governmental organizations (see Appendix II). The team, led by Nebojša Nakićenović of the International Institute for Applied Systems Analysis (IIASA) in Austria, included representatives of six scenario modeling groups and lead authors from all three earlier IPCC scenario activities – the 1990 and 1992 scenarios and the 1994 scenario evaluation. The SRES preparation included six major steps:

- analysis of existing scenarios in the literature;
- analysis of major scenario characteristics, driving forces, and their relationships;
- formulation of four narrative scenario "storylines" to describe alternative futures;
- quantification of each storyline using a variety of modelling approaches;
- an "open" review process of the resultant emission scenarios and their assumptions; and
- three revisions of the scenarios and the report subsequent to the open review process, i.e., the formal IPCC Expert Review and the final combined IPCC Expert and Government Review.

As required by the Terms of Reference, the SRES preparation process was open with no single "official" model and no exclusive "expert teams." To this end, in 1997 the IPCC advertised in relevant scientific journals and other publications to solicit wide participation in the process. A web site documenting the SRES process and intermediate results was created to facilitate outside input. Members of the writing team also published much of their background research in the peer-reviewed literature and on web sites.

In June 1998, the IPCC Bureau agreed to make the unapproved, preliminary scenarios available to climate modelers, who could use the scenarios as a basis for the assessment of climatic changes in time for consideration in the IPCC's Third Assessment Report. We recommend that the new scenarios be used not only in the IPCC's future assessments of climate change, its impacts, and adaptation and mitigation options, but also as the basis for analyses by the wider research and policy community of climate change and other environmental problems.

Ogunlade Davidson and Bert Metz
Co-chairs of IPCC WGIII

Summary for Policymakers

Special Report on Emission Scenarios

A Special Report of Working Group III of the Intergovernmental Panel on Climate Change

Based on a draft prepared by:

Nebojša Nakićenović, Ogunlade Davidson, Gerald Davis, Arnulf Grübler, Tom Kram, Emilio Lebre La Rovere, Bert Metz, Tsuneyuki Morita, William Pepper, Hugh Pitcher, Alexei Sankovski, Priyadarshi Shukla, Robert Swart, Robert Watson, Zhou Dadi

CONTENTS

Why new Intergovernmental Panel on Climate Change scenarios?

The Intergovernmental Panel on Climate Change (IPCC) developed long-term emission scenarios in 1990 and 1992. These scenarios have been widely used in the analysis of possible climate change, its impacts, and options to mitigate climate change. In 1995, the IPCC 1992 scenarios were evaluated. The evaluation recommended that significant changes (since 1992) in the understanding of driving forces of emissions and methodologies should be addressed. These changes in understanding relate to, e.g., the carbon intensity of energy supply, the income gap between developed and developing countries, and to sulfur emissions. This led to a decision by the IPCC Plenary in 1996 to develop a new set of scenarios. The new set of scenarios is presented in this report.

What are scenarios and what is their purpose?

Future greenhouse gas (GHG) emissions are the product of very complex dynamic systems, determined by driving forces such as demographic development, socio-economic development, and technological change. Their future evolution is highly uncertain. Scenarios are alternative images of how the future might unfold and are an appropriate tool with which to analyze how driving forces may influence future emission outcomes and to assess the associated uncertainties. They assist in climate change analysis, including climate modeling and the assessment of impacts, adaptation, and mitigation. The possibility that any single emissions path will occur as described in scenarios is highly uncertain.

What are the main characteristics of the new scenarios?

A set of scenarios was developed to represent the range of driving forces and emissions in the scenario literature so as to reflect current understanding and knowledge about underlying uncertainties. They exclude only outlying "surprise" or "disaster" scenarios in the literature. Any scenario necessarily includes subjective elements and is open to various interpretations. Preferences for the scenarios presented here vary among users. No judgment is offered in this report as to the preference for any of the scenarios and they are not assigned probabilities of occurrence, neither must they be interpreted as policy recommendations.

The scenarios are based on an extensive assessment of driving forces and emissions in the scenario literature, alternative modeling approaches, and an "open process"[1] *that solicited wide participation and feedback.* These are all-important elements of the Terms of Reference (see Appendix I).

Four different narrative storylines were developed to describe consistently the relationships between emission driving forces and their evolution and add context for the scenario quantification. Each storyline represents different demographic, social, economic, technological, and environmental developments, which may be viewed positively by some people and negatively by others.

The scenarios cover a wide range of the main demographic, economic, and technological driving forces of GHG and sulfur emissions[2] *and are representative of the literature.* Each scenario represents a specific quantitative interpretation of one of four storylines. All the scenarios based on the same storyline constitute a scenario "family".

As required by the Terms of Reference, the scenarios in this report do not include additional climate initiatives, which means that no scenarios are included that explicitly assume implementation of the United Nations Framework Convention for Climate Change (UNFCCC) or the emissions targets of the Kyoto Protocol. However, GHG emissions are directly affected by non-climate change policies designed for a wide range of other purposes. Furthermore government policies can, to varying degrees, influence the GHG emission drivers such as demographic change, social and economic development, technological change, resource use, and pollution management. This influence is broadly reflected in the storylines and resultant scenarios.

For each storyline several different scenarios were developed using different modeling approaches to examine the range of outcomes arising from a range of models that use similar assumptions about driving forces. Six models were used which are representative of integrated assessment frameworks in the literature. One advantage of a multi-model approach is that the resultant 40 SRES scenarios together encompass the current range of uncertainties of future GHG emissions arising from different characteristics of these models, in addition to the current knowledge of and uncertainties that arise from scenario driving forces such as demographic, social and economic, and broad technological developments that drive the models, as described in the storylines. Thirteen of these 40 scenarios explore variations in energy technology assumptions.

[1] The open process defined in the Special Report on Emissions Scenarios (SRES) Terms of Reference calls for the use of multiple models, seeking inputs from a wide community as well as making scenario results widely available for comments and review. These objectives were fulfilled by the SRES multi-model approach and the open SRES website.

[2] Included are anthropogenic emissions of carbon dioxide (CO_2), methane (CH_4), nitrous oxide (N_2O), hydrofluorocarbons (HFCs), perfluorocarbons (PFCs), sulfur hexafluoride (SF_6), hydrochlorofluorocarbons (HCFCs), chlorofluorocarbons (CFCs), the aerosol precursor and the chemically active gases sulfur dioxide (SO_2), carbon monoxide (CO), nitrogen oxides (NO_x), and non-methane volatile organic compounds (NMVOCs). Emissions are provided aggregated into four world regions and global totals. In the new scenarios no feedback effect of future climate change on emissions from biosphere and energy has been assumed.

Box SPM-1: The Main Characteristics of the Four SRES Storylines and Scenario Families.

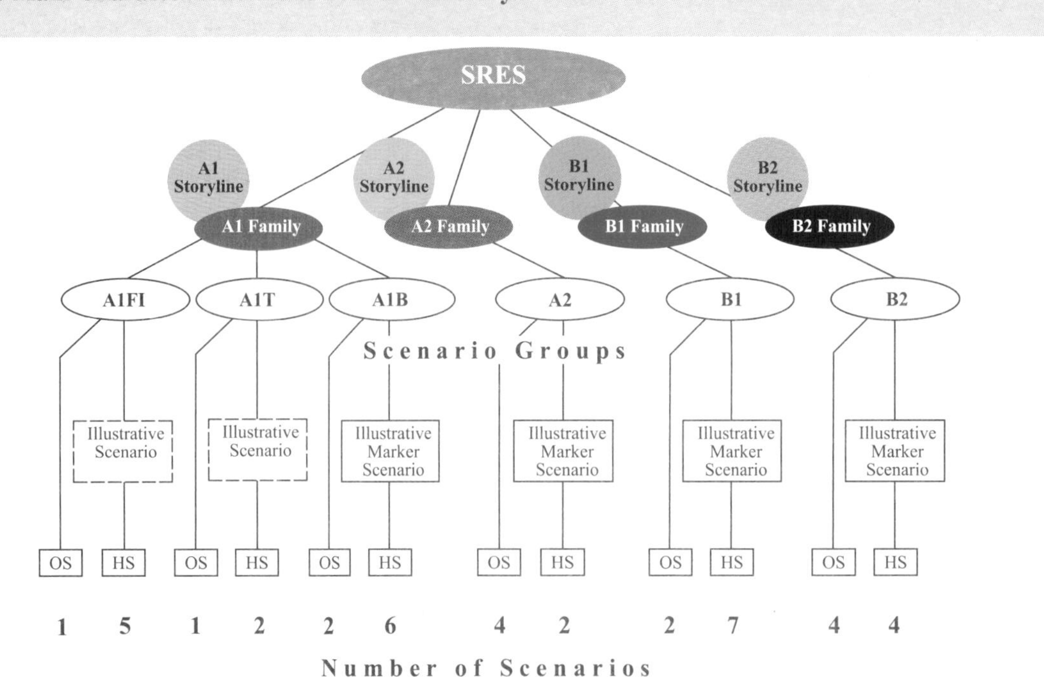

Figure SPM-1: Schematic illustration of SRES scenarios. Four qualitative storylines yield four sets of scenarios called "families": A1, A2, B1, and B2. Altogether 40 SRES scenarios have been developed by six modeling teams. All are equally valid with no assigned probabilities of occurrence. The set of scenarios consists of six scenario groups drawn from the four families: one group each in A2, B1, B2, and three groups within the A1 family, characterizing alternative developments of energy technologies: A1FI (fossil fuel intensive), A1B (balanced), and A1T (predominantly non-fossil fuel). Within each family and group of scenarios, some share "harmonized" assumptions on global population, gross world product, and final energy. These are marked as "HS" for harmonized scenarios. "OS" denotes scenarios that explore uncertainties in driving forces beyond those of the harmonized scenarios. The number of scenarios developed within each category is shown. For each of the six scenario groups an illustrative scenario (which is always harmonized) is provided. Four illustrative marker scenarios, one for each scenario family, were used in draft form in the 1998 SRES open process and are included in revised form in this report. Two additional illustrative scenarios for the groups A1FI and A1T are also provided and complete a set of six that illustrate all scenario groups. All are equally sound.

By 2100 the world will have changed in ways that are difficult to imagine – as difficult as it would have been at the end of the 19th century to imagine the changes of the 100 years since. Each storyline assumes a distinctly different direction for future developments, such that the four storylines differ in increasingly irreversible ways. Together they describe divergent futures that encompass a significant portion of the underlying uncertainties in the main driving forces. They cover a wide range of key "future" characteristics such as demographic change, economic development, and technological change. For this reason, their plausibility or feasibility should not be considered solely on the basis of an extrapolation of *current* economic, technological, and social trends.

- The A1 storyline and scenario family describes a future world of very rapid economic growth, global population that peaks in mid-century and declines thereafter, and the rapid introduction of new and more efficient technologies. Major underlying themes are convergence among regions, capacity building, and increased cultural and social interactions, with a substantial reduction in regional differences in per capita income. The A1 scenario family develops into three groups that describe alternative directions of technological change in the energy system. The three A1 groups are distinguished by their technological emphasis: fossil intensive (A1FI), non-fossil energy sources (A1T), or a balance across all sources (A1B)[3].

[3] Balanced is defined as not relying too heavily on one particular energy source, on the assumption that similar improvement rates apply to all energy supply and end use technologies.

- The A2 storyline and scenario family describes a very heterogeneous world. The underlying theme is self-reliance and preservation of local identities. Fertility patterns across regions converge very slowly, which results in continuously increasing global population. Economic development is primarily regionally oriented and per capita economic growth and technological change are more fragmented and slower than in other storylines.
- The B1 storyline and scenario family describes a convergent world with the same global population that peaks in mid-century and declines thereafter, as in the A1 storyline, but with rapid changes in economic structures toward a service and information economy, with reductions in material intensity, and the introduction of clean and resource-efficient technologies. The emphasis is on global solutions to economic, social, and environmental sustainability, including improved equity, but without additional climate initiatives.
- The B2 storyline and scenario family describes a world in which the emphasis is on local solutions to economic, social, and environmental sustainability. It is a world with continuously increasing global population at a rate lower than A2, intermediate levels of economic development, and less rapid and more diverse technological change than in the B1 and A1 storylines. While the scenario is also oriented toward environmental protection and social equity, it focuses on local and regional levels.

Within each scenario family two main types of scenarios were developed – those with harmonized assumptions about global population, economic growth, and final energy use and those with alternative quantification of the storyline. Together, 26 scenarios were harmonized by adopting common assumptions on global population and gross domestic product (GDP) development. Thus, the harmonized scenarios in each family are not independent of each other. The remaining 14 scenarios adopted alternative interpretations of the four scenario storylines to explore additional scenario uncertainties beyond differences in methodologic approaches. They are also related to each other within each family, even though they do not share common assumptions about some of the driving forces.

There are six scenario groups that should be considered equally sound that span a wide range of uncertainty, as required by the Terms of Reference. These encompass four combinations of demographic change, social and economic development, and broad technological developments, corresponding to the four families (A1, A2, B1, B2), each with an illustrative "marker" scenario. Two of the scenario groups of the A1 family (A1FI, A1T) explicitly explore alternative energy technology developments, holding the other driving forces constant, each with an illustrative scenario. Rapid growth leads to high capital turnover rates, which means that early small differences among scenarios can lead to a large divergence by 2100. Therefore the A1 family, which has the highest rates of technological change and economic development, was selected to show this effect.

In accordance with a decision of the IPCC Bureau in 1998 to release draft scenarios to climate modelers for their input in the Third Assessment Report, and subsequently to solicit comments during the open process, one marker scenario was chosen from each of four of the scenario groups based on the storylines. The choice of the markers was based on which of the initial quantifications best reflected the storyline, and features of specific models. Marker scenarios are no more or less likely than any other scenarios, but are considered by the SRES writing team as illustrative of a particular storyline. These scenarios have received the closest scrutiny of the entire writing team and via the SRES open process. Scenarios have also been selected to illustrate the other two scenario groups. Hence, this report has an illustrative scenario for each of the six scenario groups.

What are the main driving forces of the GHG emissions in the scenarios?

This Report reinforces our understanding that the main driving forces of future greenhouse gas trajectories will continue to be demographic change, social and economic development, and the rate and direction of technological change. This finding is consistent with the IPCC 1990, 1992 and 1995 scenario reports. Table SPM-1 (see later) summarizes the demographic, social, and economic driving forces across the scenarios in 2020, 2050, and 2100[4]. The intermediate energy result (shown in table SPM 2, see later) and land use results[5] reflect the influences of driving forces.

Recent global population projections are generally lower than those in the IS92 scenarios. Three different population trajectories that correspond to socio-economic developments in the storylines were chosen from recently published projections. The A1 and B1 scenario families are based on the low International Institute for Applied Systems Analysis (IIASA) 1996 projection. They share the lowest trajectory, increasing to 8.7 billion by 2050 and declining toward 7 billion by 2100, which combines low fertility with low mortality. The B2 scenario family is based on the long-term UN Medium 1998 population projection of 10.4 billion by 2100. The A2 scenario family is based on a high population growth scenario of 15 billion by 2100 that assumes a significant decline in fertility for most regions and stabilization at above replacement levels. It falls below the long-term 1998 UN High projection of 18 billion.

[4] Technological change is not quantified in table SPM-1.

[5] Because of the impossibility of including the complex way land use is changing between the various land use types, this information is not in the table.

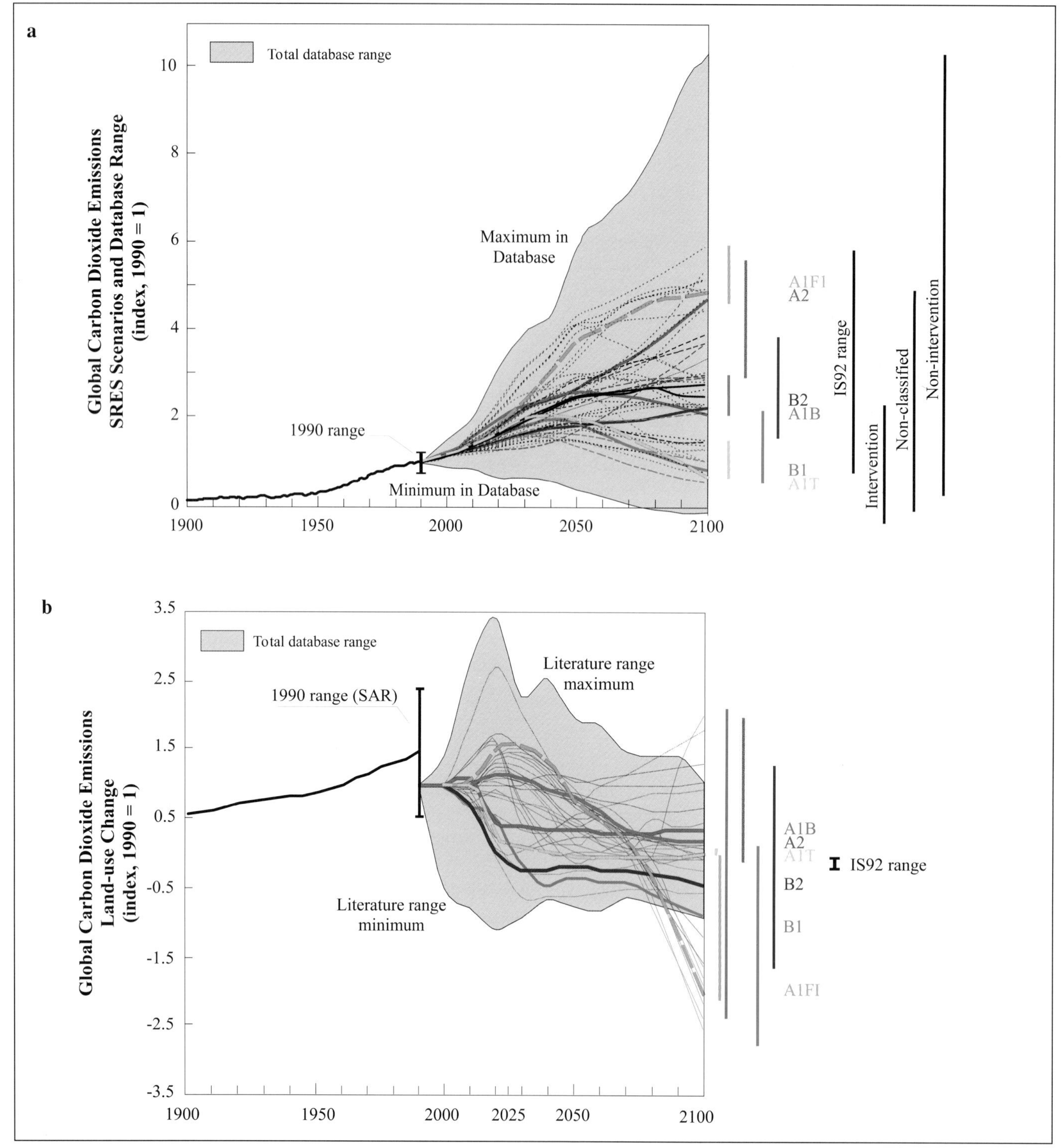

Figure SPM-2: Global CO_2 emissions related to energy and industry (Figure SPM-2a) and land-use changes (Figure SPM-2b) from 1900 to 1990, and for the 40 SRES scenarios from 1990 to 2100, shown as an index (1990 = 1). The dashed time-paths depict individual SRES scenarios and the area shaded in blue the range of scenarios from the literature as documented in the SRES database. The scenarios are classified into six scenario groups drawn from the four scenario families. Six illustrative scenarios are highlighted. The colored vertical bars indicate the range of emissions in 2100. The four black bars on the right of Figure SPM-1a indicate the emission ranges in 2100 for the IS92 scenarios and three ranges of scenarios from the literature, documented in the SRES database. These three ranges indicate those scenarios that include some additional climate initiatives (designated as "intervention" scenarios), those that do not ("non-intervention"), and those that cannot be assigned to either category ("non-classified"). This classification is based on a subjective evaluation of the scenarios in the database and was possible only for energy and industry CO_2 emissions. SAR, Second Assessment Report.

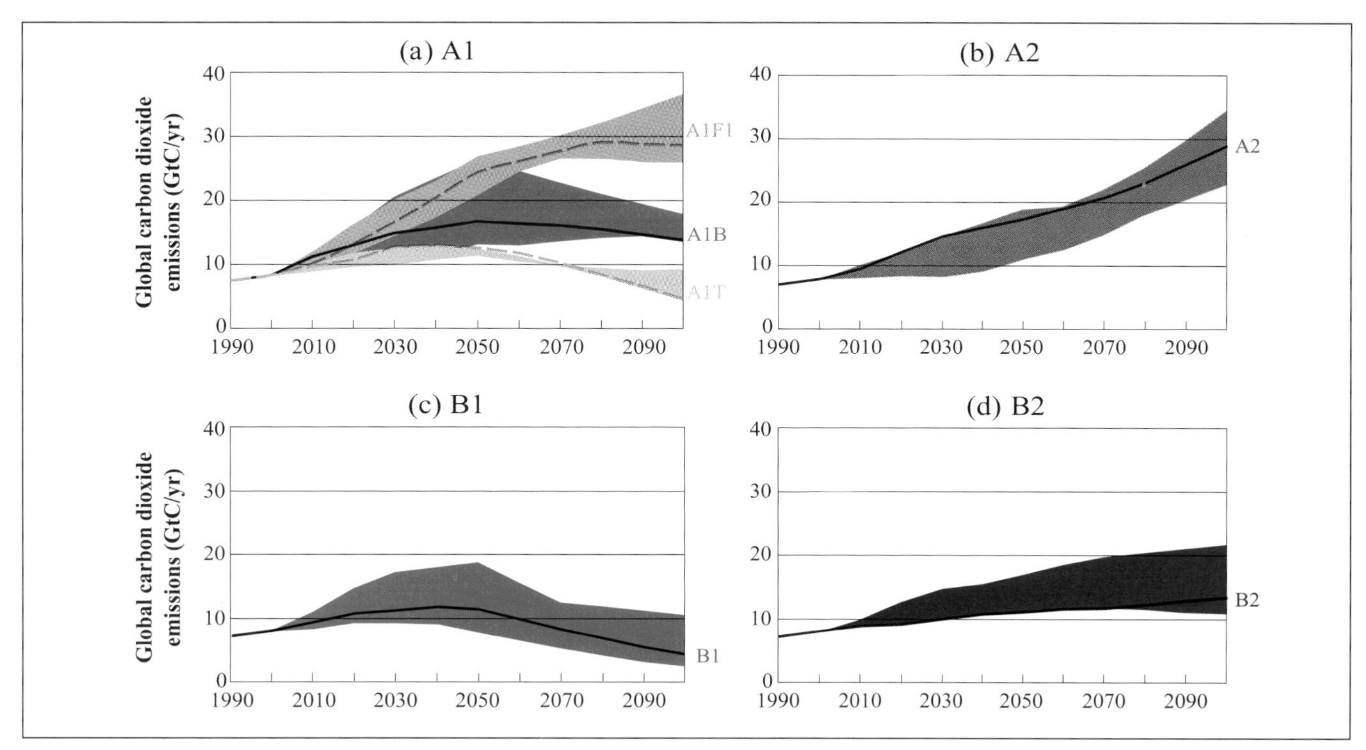

Figure SPM-3: Total global annual CO_2 emissions from all sources (energy, industry, and land-use change) from 1990 to 2100 (in gigatonnes of carbon (GtC/yr) for the families and six scenario groups. The 40 SRES scenarios are presented by the four families (A1, A2, B1, and B2) and six scenario groups: the fossil-intensive A1FI (comprising the high-coal and high-oil-and-gas scenarios), the predominantly non-fossil fuel A1T, the balanced A1B in Figure SPM-3a; A2 in Figure SPM-3b; B1 in Figure SPM-3c, and B2 in Figure SPM-3d. Each colored emission band shows the range of harmonized and non-harmonized scenarios within each group. For each of the six scenario groups an illustrative scenario is provided, including the four illustrative marker scenarios (A1, A2, B1, B2, solid lines) and two illustrative scenarios for A1FI and A1T (dashed lines).

All scenarios describe futures that are generally more affluent than today. The scenarios span a wide range of future levels of economic activity, with gross world product rising to 10 times today's values by 2100 in the lowest to 26-fold in the highest scenarios.

A narrowing of income differences among world regions is assumed in many of the SRES scenarios. Two of the scenario families, A1 and B1, explicitly explore alternative pathways that gradually close existing income gaps in relative terms.

Technology is at least as important a driving force as demographic change and economic development. These driving forces are related. Within the A1 scenario family, scenarios with common demographic and socio-economic driving forces but different assumptions about technology and resource dynamics illustrate the possibility of very divergent paths for developments in the energy system and land-use patterns.

The SRES scenarios cover a wider range of energy structures than the IS92 scenarios. This reflects uncertainties about future fossil resources and technological change. The scenarios cover virtually all the possible directions of change, from high shares of fossil fuels, oil and gas or coal, to high shares of non-fossils.

In most scenarios, global forest area continues to decrease for some decades, primarily because of increasing population and income growth. This current trend is eventually reversed in most scenarios with the greatest eventual increase in forest area by 2100 in the B1 and B2 scenario families, as compared to 1990. Associated changes in agricultural land use are driven principally by changing food demands caused by demographic and dietary shifts. Numerous other social, economic, institutional, and technological factors also affect the relative shares of agricultural lands, forests, and other types of land use. Different analytic methods lead to very different results, indicating that future land use change in the scenarios is very model specific.

All the above driving forces not only influence CO_2 emissions, but also the emissions of other GHGs. The relationships between the driving forces and non-CO_2 GHG emissions are generally more complex and less studied, and the models used for the scenarios less sophisticated. Hence, the uncertainties in the SRES emissions for non-CO_2 greenhouse gases are generally greater than those for energy CO_2[6].

[6] Therefore, the ranges of non-CO_2 GHG emissions provided in the Report may not fully reflect the level of uncertainty compared to CO_2, for example only a single model provided the sole value for halocarbon emissions.

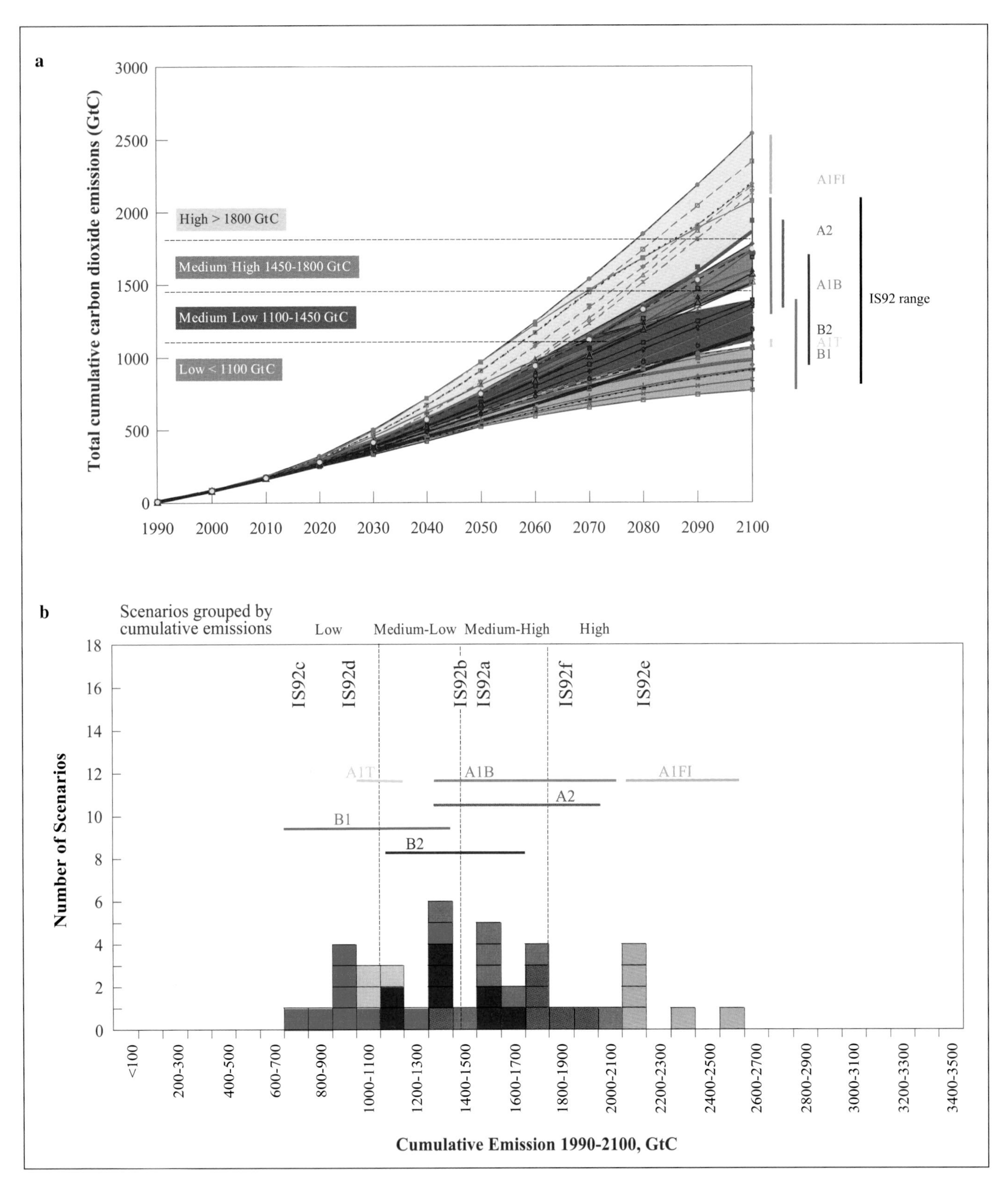

Figure SPM-4: Total global cumulative CO_2 emissions (GtC) from 1990 to 2100 (SPM-4a) and histogram of their distribution by scenario groups (SPM-4b). No probability of occurrence should be inferred from the distribution of SRES scenarios or those in the literature. Both figures show the ranges of cumulative emissions for the 40 SRES scenarios. Scenarios are also grouped into four cumulative emissions categories: low, medium–low, medium–high, and high emissions. Each category contains one illustrative marker scenario plus alternatives that lead to comparable cumulative emissions, although often through different driving forces. This categorization can guide comparisons using either scenarios with different driving forces yet similar emissions, or scenarios with similar driving forces but different emissions. The cumulative emissions of the IS92 scenarios are also shown.

What is the range of GHG emissions in the SRES scenarios and how do they relate to driving forces?

The SRES scenarios cover most of the range of carbon dioxide (CO_2; see Figures SPM-2a and SPM-2b), other GHGs, and sulfur emissions found in the recent literature and SRES scenario database. Their spread is similar to that of the IS92 scenarios for CO_2 emissions from energy and industry as well as total emissions but represents a much wider range for land-use change. The six scenario groups cover wide and overlapping emission ranges. The range of GHG emissions in the scenarios widens over time to capture the long-term uncertainties reflected in the literature for many of the driving forces, and after 2050 widens significantly as a result of different socio-economic developments. Table SPM-2b summarizes the emissions across the scenarios in 2020, 2050, and 2100. Figure SPM-3 shows in greater detail the ranges of total CO_2 emissions for the six scenario groups of scenarios that constitute the four families (the three scenario families A2, B1, and B2, plus three groups within the A1 family A1FI, A1T, and A1B).

Some SRES scenarios show trend reversals, turning points (i.e., initial emission increases followed by decreases), and crossovers (i.e., initially emissions are higher in one scenario, but later emissions are higher in another scenario). Emission trend reversals (see Figures SPM-2 and SPM-3) depart from historical emission increases. In most of these cases, the upward emissions trend due to income growth is more than compensated by productivity improvements combined with a slowly growing or declining population.

In many SRES scenarios CO_2 emissions from loss of forest cover peak after several decades and then gradually decline[7] *(Figure SPM-1b).* This pattern is consistent with scenarios in the literature and can be associated with slowing population growth, followed by a decline in some scenarios, increasing agricultural productivity, and increasing scarcity of forest land. These factors allow for a reversal of the current trend of loss of forest cover in many cases. Emissions decline fastest in the B1 family. Only in the A2 family do net anthropogenic CO_2 emissions from land use change[2] remain positive through 2100. As was the case for energy-related emissions, CO_2 emissions related to land-use change in the A1 family cover the widest range. The diversity across these scenarios is amplified through the high economic growth, increasing the range of alternatives, and through the different modeling approaches and their treatment of technology.

Total cumulative SRES carbon emissions from all sources through 2100 range from approximately 770 GtC to approximately 2540 GtC. According to the IPCC Second Assessment Report (SAR), "any eventual stabilised concentration is governed more by the accumulated anthropogenic CO_2 emissions from now until the time of stabilisation than by the way emissions change over the period." Therefore, the scenarios are also grouped in the report according to their cumulative emissions.[8] (see Figure SPM-4). The SRES scenarios extend the IS92 range toward higher emissions (SRES maximum of 2538 GtC compared to 2140 GtC for IS92), but not toward lower emissions. The lower bound for both scenario sets is approximately 770 GtC.

Total anthropogenic methane (CH_4) and nitrous oxide (N_2O) emissions span a wide range by the end of the 21st century (see Figures SPM-5 and SPM-6 derived from Figures 5.5 and 5.7). Emissions of these gases in a number of scenarios begin to decline by 2050. The range of emissions is wider than in the IS92 scenarios due to the multimodel approach, which leads to a better treatment of uncertainties and to a wide range of driving forces. These totals include emissions from land use, energy systems, industry, and waste management.

Methane and nitrous oxide emissions from land use are limited in A1 and B1 families by slower population growth followed by a decline, and increased agricultural productivity. After the initial increases, emissions related to land use peak and decline. In the B2 family, emissions continue to grow, albeit very slowly. In the A2 family, both high population growth and less rapid increases in agricultural productivity result in a continuous rapid growth in those emissions related to land use.

The range of emissions of HFCs in the SRES scenario is generally lower than in earlier IPCC scenarios. Because of new insights about the availability of alternatives to HFCs as replacements for substances controlled by the Montreal Protocol, initially HFC emissions are generally lower than in previous IPCC scenarios. In the A2 and B2 scenario families HFC emissions increase rapidly in the second half of the this century, while in the A2 and B2 scenario families the growth of emissions is significantly slowed down or reversed in that period.

Sulfur emissions in the SRES scenarios are generally below the IS92 range, because of structural changes in the energy system as well as concerns about local and regional air pollution. These reflect sulfur control legislation in Europe, North America, Japan, and (more recently) other parts of Asia and other developing regions. The timing and impact of these changes and controls vary across scenarios and regions[9]. After

[7] In the new scenarios no feedback effect of future climate change on emissions from the biosphere has been assumed.

[8] In this Report, cumulative emissions are calculated by adding annual net anthropogenic emissions in the scenarios over their time horizon. When relating these cumulative emissions to atmospheric concentrations, all natural processes that affect carbon concentrations in the atmosphere have to be taken into account.

[9] Although global emissions of SO_2 for the SRES scenarios are lower than the IS92 scenarios, uncertainty about SO_2 emissions and their effect on sulfate aerosols has increased compared to the IS92 scenarios because of very diverse regional patterns of SO_2 emissions in the scenarios.

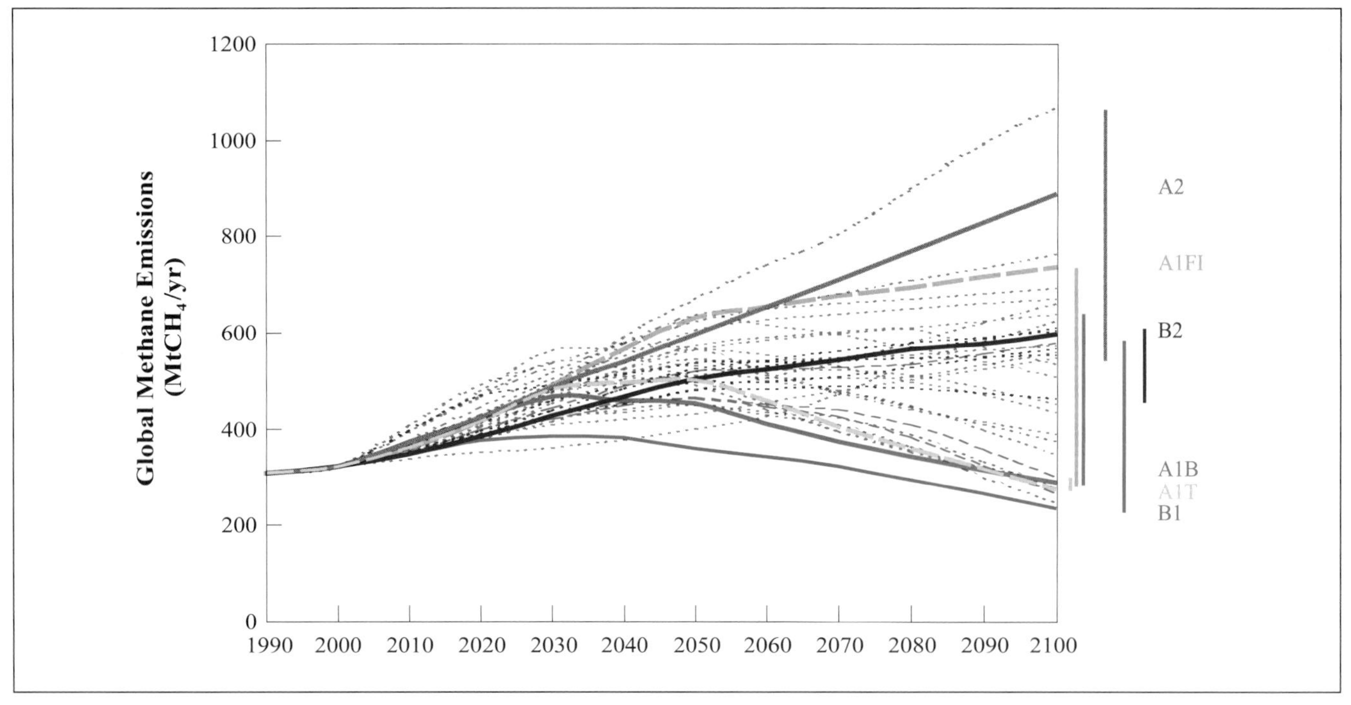

Figure SPM-5: Standardized (to common 1990 and 2000 values) global annual methane emissions for the SRES scenarios (in $MtCH_4$/yr). The range of emissions by 2100 for the six scenario groups is indicated to the right. Illustrative (including marker) scenarios are highlighted.

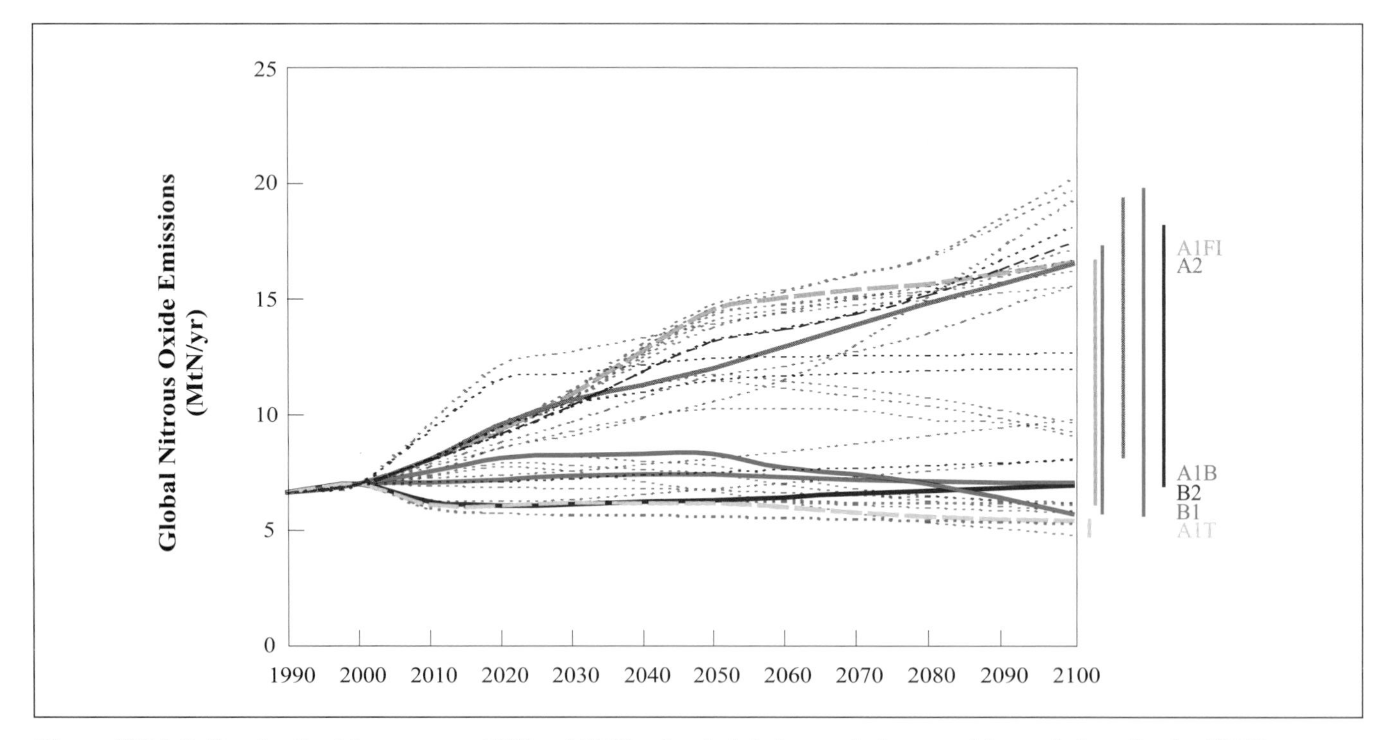

Figure SPM-6: Standardized (to common 1990 and 2000 values) global annual nitrous oxides emissions for the SRES scenarios (in MtN/yr). The range of emissions by 2100 for the six scenario groups is indicated to the right. Illustrative (marker) scenarios are highlighted.

initial increases over the next two to three decades, global sulfur emissions in the SRES scenarios decrease (see Table SPM-1b), consistent with the findings of the 1995 IPCC scenario evaluation and recent peer-reviewed literature.

Similar future GHG emissions can result from very different socio-economic developments, and similar developments of driving forces can result in different future emissions. Uncertainties in the future developments of key emission driving forces create large uncertainties in future emissions, even within the same socio-economic development paths. Therefore, emissions from each scenario family overlap substantially with emissions from other scenario families. The overlap implies that a given level of future emissions can arise from very different combinations of driving forces. Figures SPM-1, SPM-2, and SPM-3 show this for CO_2.

Convergence of regional per capita incomes can lead to either high or low GHG emissions. Tables SPM-1a and SPM-1b indicate that there are scenarios with high per capita incomes in all regions that lead to high CO_2 emissions (e.g., in the high-growth, fossil fuel intensive scenario group A1FI). They also indicate that there are scenarios with high per capita incomes that lead to low emissions (e.g., the A1T scenario group or the B1 scenario family). This suggests that in some cases other driving forces may have a greater influence on GHG emissions than income growth.

How can the SRES scenarios be used?

It is recommended that a range of SRES scenarios with a variety of assumptions regarding driving forces be used in any analysis. Thus more than one family should be used in most analyses. The six scenario groups – the three scenario families A2, B1, and B2, plus three groups within the A1 scenario family, A1B, A1FI, and A1T – and four cumulative emissions categories were developed as the smallest subsets of SRES scenarios that capture the range of uncertainties associated with driving forces and emissions.

The important uncertainties ranging from driving forces to emissions may be different in different applications – for example climate modeling; assessment of impacts, vulnerability, mitigation, and adaptation options; and policy analysis. Climate modelers may want to cover the range reflected by the cumulative emissions categories. To assess the robustness of options in terms of impacts, vulnerability, and adaptation may require scenarios with similar emissions but different socio-economic characteristics, as reflected by the six scenario groups. For mitigation analysis, variation in both emissions and socio-economic characteristics may be necessary. For analysis at the national or regional scale, the most appropriate scenarios may be those that best reflect specific circumstances and perspectives.

There is no single most likely, "central", or "best-guess" scenario, either with respect to SRES scenarios or to the underlying scenario literature. Probabilities or likelihood are not assigned to individual SRES scenarios. None of the SRES scenarios represents an estimate of a central tendency for all driving forces or emissions, such as the mean or median, and none should be interpreted as such. The distribution of the scenarios provides a useful context for understanding the relative position of a scenario but does not represent the likelihood of its occurrence.

The driving forces and emissions of each SRES scenario should be used together. To avoid internal inconsistencies, components of SRES scenarios should not be mixed. For example, the GHG emissions from one scenario and the SO_2 emissions from another scenario, or the population from one and economic development path from another, should not be combined.

While recognizing the inherent uncertainties in long-term projections[10]*, the SRES scenarios may provide policymakers with a long-term context for near-term analysis.* The modeling tools that have been used to develop these scenarios that focus on the century time scale are less suitable for analysis of near term (a decade or less) developments. When analyzing mitigation and adaptation options, the user should be aware that although no additional climate initiatives are included in the SRES scenarios, various changes have been assumed to occur that would require other interventions, such as those leading to reductions in sulfur emissions and significant penetration of new energy technologies.

What future work on emissions scenarios would be useful?

- Establishment of a program for on-going evaluations and comparisons of long-term emission scenarios, including a regularly updated scenario database;
- Capacity building, particularly in developing countries, in the area of modeling tools and emissions scenarios;
- Multiple storyline, multi-model approaches in future scenario analyses;
- New research activities to assess future developments in key GHG driving forces in greater regional, subregional, and sectoral detail which allow for a clearer link between emissions scenarios and mitigation options;
- Improved specification and data for, and integration of, the non-CO_2 GHG and non-energy sectors, such as land use, land-use change and forestry, in models, as well as model inter-comparison to improve scenarios and analyses;
- Integration into models emissions of particulate, hydrogen, or nitrate aerosol precursors, and processes,

[10] Confidence in the quantification of any scenario decreases substantially as the time horizon increases because the basis for the assumptions becomes increasingly speculative. This is why a set of scenarios was developed.

such as feedback of climate change on emissions, that may significantly influence scenario results and analyses;

- Development of additional gridded emissions for scenarios which would facilitate improved regional assessment;
- Assessment of strategies that would address multiple national, regional, or global priorities;
- Development of methods for scientifically sound aggregation of emissions data;
- More detailed information on assumptions, inputs, and the results of the 40 SRES scenarios should be made available at a web site and on a CD-ROM. Regular maintenance of the SRES web site is needed;
- Extension of the SRES web site and production of a CD-ROM to provide, if appropriate, time-dependent geographic distributions of driving forces and emissions, and concentrations of GHGs and sulfate aerosols.
- Development of a classification scheme for classifying scenarios as intervention or non-intervention scenarios.

Table SPM-1a*: Overview of main primary driving forces in 1990, 2020, 2050, and 2100. Bold numbers show the value for the illustrative scenario and the numbers between brackets show the value for the range[a] across all 40 SRES scenarios in the six scenario groups that constitute the four families. Units are given in the table. Technological change is not quantified in the table.*

Family		A1			A2	B1	B2
Scenario group	**1990**	**A1FI**	**A1B**	**A1T**	**A2**	**B1**	**B2**
Population (billion)	5.3						
2020		**7.6** (7.4-7.6)	**7.5** (7.2-7.6)	**7.6** (7.4-7.6)	**8.2** (7.5-8.2)	**7.6** (7.4-7.6)	**7.6** (7.6-7.8)
2050		**8.7**	**8.7** (8.3-8.7)	**8.7**	**11.3** (9.7-11.3)	**8.7** (8.6-8.7)	**9.3** (9.3-9.8)
2100		**7.1** (7.0-7.1)	**7.1** (7.0-7.7)	**7.0**	**15.1** (12.0-15.1)	**7.0** (6.9-7.1)	**10.4** (10.3-10.4)
World GDP (10^{12} 1990US$/yr)	21						
2020		**53** (53-57)	**56** (48-61)	**57** (52-57)	**41** (38-45)	**53** (46-57)	**51** (41-51)
2050		**164** (163-187)	**181** (120-181)	**187** (177-187)	**82** (59-111)	**136** (110-166)	**110** (76-111)
2100		**525** (522-550)	**529** (340-536)	**550** (519-550)	**243** (197-249)	**328** (328-350)	**235** (199-255)
Per capita income ratio: developed countries and economies in transition (Annex-I) to developing countries (Non-Annex-I)	16.1						
2020		**7.5** (6.2-7.5)	**6.4** (5.2-9.2)	**6.2** (5.7-6.4)	**9.4** (9.0-12.3)	**8.4** (5.3-10.7)	**7.7** (7.5-12.1)
2050		**2.8**	**2.8** (2.4-4.0)	**2.8** (2.4-2.8)	**6.6** (5.2-8.2)	**3.6** (2.7-4.9)	**4.0** (3.7-7.5)
2100		**1.5** (1.5-1.6)	**1.6** (1.5-1.7)	**1.6** (1.6-1.7)	**4.2** (2.7-6.3)	**1.8** (1.4-1.9)	**3.0** (2.0-3.6)

[a] For some driving forces, no range is indicated because all scenario runs have adopted exactly the same assumptions.

***Table SPM-1b**: Overview of main primary driving forces in 1990, 2020, 2050, and 2100. Bold numbers show the value for the illustrative scenario and the numbers between brackets show the value for the range[a] across 26 harmonized SRES scenarios in the six scenario groups that constitute the four families. Units are given in the table. Technological change is not quantified in the table.*

Family		A1			A2	B1	B2
Scenario group	**1990**	**A1FI**	**A1B**	**A1T**	**A2**	**B1**	**B2**
Population (billion)	5.3						
2020		**7.6** (7.4-7.6)	**7.4** (7.4-7.6)	**7.6** (7.4-7.6)	**8.2**	**7.6** (7.4-7.6)	**7.6**
2050		**8.7**	**8.7**	**8.7**	**11.3**	**8.7** (8.6-8.7)	**9.3**
2100		**7.1** (7.0-7.1)	**7.1** (7.0-7.1)	**7.0**	**15.1**	**7.0** (6.9-7.1)	**10.4**
World GDP (10^{12} 1990US$/yr)	21						
2020		**53** (53-57)	**56** (52-61)	**57** (56-57)	**41**	**53** (51-57)	**51** (48-51)
2050		**164** (164-187)	**181** (164-181)	**187** (182-187)	**82**	**136** (134-166)	**110** (108-111)
2100		**525** (525-550)	**529** (529-536)	**550** (529-550)	**243**	**328** (328-350)	**235** (232-237)
Per capita income ratio: developed countries and economies in transition (Annex-I) to developing countries (Non-Annex-I)	16.1						
2020		**7.5** (6.2-7.5)	**6.4** (5.2-7.5)	**6.2** (6.2-6.4)	**9.4** (9.4-9.5)	**8.4** (5.3-8.4)	**7.7** (7.5-8.0)
2050		**2.8**	**2.8** (2.4-2.8)	**2.8**	**6.6**	**3.6** (2.7-3.9)	**4.0** (3.8-4.6)
2100		**1.5** (1.5-1.6)	**1.6** (1.5-1.7)	**1.6**	**4.2**	**1.8** (1.6-1.9)	**3.0** (3.0-3.5)

[a] For some driving forces, no range is indicated because all scenario runs have adopted exactly the same assumptions.

***Table SPM-2a**: Overview of main secondary scenario driving forces in 1990, 2020, 2050, and 2100. Bold numbers show the value for the illustrative scenario and the numbers between brackets show the value for the range across all 40 SRES scenarios in the six scenario groups that constitute the four families. Units are given in the table.*

Family		A1			A2	B1	B2
Scenario group	**1990**	**A1FI**	**A1B**	**A1T**	**A2**	**B1**	**B2**
Final energy intensity (10^6J/US$)[a]	16.7						
2020		**9.4** (8.5-9.4)	**9.4** (8.1-12.0)	**8.7** (7.6-8.7)	**12.1** (9.3-12.4)	**8.8** (6.7-11.6)	**8.5** (8.5-11.8)
2050		**6.3** (5.4-6.3)	**5.5** (4.4-7.2)	**4.8** (4.2-4.8)	**9.5** (7.0-9.5)	**4.5** (3.5-6.0)	**6.0** (6.0-8.1)
2100		**3.0** (2.6-3.2)	**3.3** (1.6-3.3)	**2.3** (1.8-2.3)	**5.9** (4.4-7.3)	**1.4** (1.4-2.7)	**4.0** (3.7-4.6)
Primary energy (10^{18} J/yr)[a]	351						
2020		**669** (653-752)	**711** (573-875)	**649** (515-649)	**595** (485-677)	**606** (438-774)	**566** (506-633)
2050		**1431** (1377-1601)	**1347** (968-1611)	**1213** (913-1213)	**971** (679-1059)	**813** (642-1090)	**869** (679-966)
2100		**2073** (1988-2737)	**2226** (1002-2683)	**2021** (1255-2021)	**1717** (1304-2040)	**514** (514-1157)	**1357** (846-1625)
Share of coal in primary energy (%)[a]	24						
2020		**29** (24-42)	**23** (8-28)	**23** (8-23)	**22** (18-34)	**22** (8-27)	**17** (14-31)
2050		**33** (13-56)	**14** (3-42)	**10** (2-13)	**30** (24-47)	**21** (2-37)	**10** (10-49)
2100		**29** (3-48)	**4** (4-41)	**1** (1-3)	**53** (17-53)	**8** (0-22)	**22** (12-53)
Share of zero carbon in primary energy (%)[a]	18						
2020		**15** (10-20)	**16** (9-26)	**21** (15-22)	**8** (8-16)	**21** (7-22)	**18** (7-18)
2050		**19** (16-31)	**36** (21-40)	**43** (39-43)	**18** (14-29)	**30** (18-40)	**30** (15-30)
2100		**31** (30-47)	**65** (27-75)	**85** (64-85)	**28** (26-37)	**52** (33-70)	**49** (22-49)

[a] 1990 values include non-commercial energy consistent with IPCC WGII SAR (Energy Primer) but with SRES accounting conventions. Note that ASF, MiniCAM, and IMAGE scenarios do not consider non-commercial renewable energy. Hence, these scenarios report lower energy use.

***Table SPM-2b**: Overview of main secondary scenario driving forces in 1990, 2020, 2050, and 2100. Bold numbers show the value for the illustrative scenario and the numbers between brackets show the value for the range across 26 harmonized SRES scenarios in the six scenario groups that constitute the four families. Units are given in the table.*

Family		**A1**			**A2**	**B1**	**B2**
Scenario group	**1990**	**A1FI**	**A1B**	**A1T**	**A2**	**B1**	**B2**
Final energy intensity (10^6J/US$)[a]	16.7						
2020		**9.4** (8.5-9.4)	**9.4** (8.7-12.0)	**8.7** (7.6-8.7)	**12.1** (11.3-12.1)	**8.8** (6.7-11.6)	**8.5** (8.5-9.1)
2050		**6.3** (5.4-6.3)	**5.5** (5.0-7.2)	**4.8** (4.3-4.8)	**9.5** (9.2-9.5)	**4.5** (3.5-6.0)	**6.0** (6.0-6.6)
2100		**3.0** (3.0-3.2)	**3.3** (2.7-3.3)	**2.3**	**5.9** (5.5-5.9)	**1.4** (1.4-2.1)	**4.0** (3.9-4.1)
Primary energy (10^{18} J/yr)[a]	351						
2020		**669** (657-752)	**711** (589-875)	**649** (611-649)	**595** (595-610)	**606** (451-774)	**566** (519-590)
2050		**1431** (1377-1601)	**1347** (1113-1611)	**1213** (1086-1213)	**971** (971-1014)	**813** (642-1090)	**869** (815-941)
2100		**2073** (2073-2737)	**2226** (1002-2683)	**2021** (1632-2021)	**1717** (1717-1921)	**514** (514-1157)	**1357** (1077-1357)
Share of coal in primary energy (%)[a]	24						
2020		**29** (24-42)	**23** (8-26)	**23** (23-23)	**22** (20-22)	**22** (19-27)	**17** (14-31)
2050		**33** (13-52)	**14** (3-42)	**10** (10-13)	**30** (27-30)	**21** (4-37)	**10** (10-35)
2100		**29** (3-46)	**4** (4-41)	**1** (1-3)	**53** (45-53)	**8** (0-22)	**22** (19-37)
Share of zero carbon in primary energy (%)[a]	18						
2020		**15** (10-20)	**16** (9-26)	**21** (15-21)	**8** (8-16)	**21** (7-22)	**18** (12-18)
2050		**19** (16-31)	**36** (23-40)	**43** (41-43)	**18** (18-29)	**30** (18-40)	**30** (21-30)
2100		**31** (30-47)	**65** (39-75)	**85** (67-85)	**28** (28-37)	**52** (44-70)	**49** (22-49)

[a] 1990 values include non-commercial energy consistent with IPCC WGII SAR (Energy Primer) but with SRES accounting conventions. Note that ASF, MiniCAM, and IMAGE scenarios do not consider non-commercial renewable energy. Hence, these scenarios report lower energy use.

***Table SPM-3a**: Overview of GHG, SO_2, and ozone precursor emissions[a] in 1990, 2020, 2050, and 2100, and cumulative carbon dioxide emissions to 2100. Bold numbers show the value for the illustrative scenario and the numbers between brackets show the value for the range across all 40 SRES scenarios in the six scenario groups that constitute the four families. Units are given in the table.*

Family		A1			A2	B1	B2
Scenario group	**1990**	**A1FI**	**A1B**	**A1T**	**A2**	**B1**	**B2**
Carbon dioxide, fossil fuels (GtC/yr)	6.0						
2020		**11.2** (10.7-14.3)	**12.1** (8.7-14.7)	**10.0** (8.4-10.0)	**11.0** (7.9-11.3)	**10.0** (7.8-13.2)	**9.0** (8.5-11.5)
2050		**23.1** (20.6-26.8)	**16.0** (12.7-25.7)	**12.3** (10.8-12.3)	**16.5** (10.5-18.2)	**11.7** (8.5-17.5)	**11.2** (11.2-16.4)
2100		**30.3** (27.7-36.8)	**13.1** (12.9-18.4)	**4.3** (4.3-9.1)	**28.9** (17.6-33.4)	**5.2** (3.3-13.2)	**13.8** (9.3-23.1)
Carbon dioxide, land use (GtC/yr)	1.1						
2020		**1.5** (0.3-1.8)	**0.5** (0.3-1.6)	**0.3** (0.3-1.7)	**1.2** (0.1-3.0)	**0.6** (0.0-1.3)	**0.0** (0.0-1.9)
2050		**0.8** (0.0-0.9)	**0.4** (0.0-1.0)	**0.0** (-0.2-0.5)	**0.9** (0.6-0.9)	**-0.4** (-0.7-0.8)	**-0.2** (-0.2-1.2)
2100		**-2.1** (-2.1-0.0)	**0.4** (-2.4-2.2)	**0.0** (0.0-0.1)	**0.2** (-0.1-2.0)	**-1.0** (-2.8-0.1)	**-0.5** (-1.7-1.5)
Cumulative carbon dioxide, fossil fuels (GtC)							
1990-2100		**2128** (2079-2478)	**1437** (1220-1989)	**1038** (989-1051)	**1773** (1303-1860)	**989** (794-1306)	**1160** (1033-1627)
Cumulative carbon dioxide, land use (GtC)							
1990-2100		**61** (31-69)	**62** (31-84)	**31** (31-62)	**89** (49-181)	**-6** (-22-84)	**4** (4-153)
Cumulative carbon dioxide, total (GtC)							
1990-2100		**2189** (2127-2538)	**1499** (1301-2073)	**1068** (1049-1113)	**1862** (1352-1938)	**983** (772-1390)	**1164** (1164-1686)
Sulfur dioxide, (MtS/yr)	70.9						
2020		**87** (60-134)	**100** (62-117)	**60** (60-101)	**100** (66-105)	**75** (52-112)	**61** (48-101)
2050		**81** (64-139)	**64** (47-120)	**40** (40-64)	**105** (78-141)	**69** (29-69)	**56** (42-107)
2100		**40** (27-83)	**28** (26-71)	**20** (20-27)	**60** (60-93)	**25** (11-25)	**48** (33-48)
Methane, ($MtCH_4$/yr)	310						
2020		**416** (415-479)	**421** (400-444)	**415** (415-466)	**424** (354-493)	**377** (377-430)	**384** (384-469)
2050		**630** (511-636)	**452** (452-636)	500(492-500)	**598** (402-671)	**359** (359-546)	**505** (482-536)
2100		**735** (289-735)	**289** (289-640)	**274** (274-291)	**889** (549-1069)	**236** (236-579)	**597** (465-613)

[a] The uncertainties in the SRES emissions for non-CO_2 greenhouse gases are generally greater than those for energy CO_2. Therefore, the ranges of non-CO_2 GHG emissions provided in the Report may not fully reflect the level of uncertainty compared to CO_2, for example only a single model provided the sole value for halocarbon emissions.

Table SPM-3a (continued)

Family			**A1**		**A2**	**B1**	**B2**
Scenario group	**1990**	**A1FI**	**A1B**	**A1T**	**A2**	**B1**	**B2**
Nitrous Oxide, (MtN/yr)	6.7						
2020		**9.3** (6.1-9.3)	**7.2** (6.1-9.6)	**6.1** (6.1-7.8)	**9.6** (6.3-12.2)	**8.1** (5.8-9.5)	**6.1** (6.1-11.5)
2050		**14.5** (6.3-14.5)	**7.4** (6.3-14.3)	**6.1** (6.1-6.7)	**12.0** (6.8-13.9)	**8.3** (5.6-14.8)	**6.3** (6.3-13.2)
2100		**16.6** (5.9-16.6)	**7.0** (5.8-17.2)	**5.4** (4.8-5.4)	**16.5** (8.1-19.3)	**5.7** (5.3-20.2)	**6.9** (6.9-18.1)
CFC/HFC/HCFC (MtC equiv./yr)[b]	1672						
2020		**337**	**337**	**337**	**292**	**291**	**299**
2050		**566**	**566**	**566**	**312**	**338**	**346**
2100		**614**	**614**	**614**	**753**	**299**	**649**
PFC, (MtC equiv./yr)[b]	32.0						
2020		**42.7**	**42.7**	**42.7**	**50.9**	**31.7**	**54.8**
2050		**88.7**	**88.7**	**88.7**	**92.2**	**42.2**	**106.6**
2100		**115.3**	**115.3**	**115.3**	**178.4**	**44.9**	**121.3**
SF_6, (MtC equiv./yr)[b]	37.7						
2020		**47.8**	**47.8**	**47.8**	**63.5**	**37.4**	**54.7**
2050		**119.2**	**119.2**	**119.2**	**104.0**	**67.9**	**79.2**
2100		**94.6**	**94.6**	**94.6**	**164.6**	**42.6**	**69.0**
CO, (MtCO/yr)	879						
2020		**1204**	**1032**	**1147**	**1075**	**751**	**1022**
		(1123-1552)	(978-1248)	(1147-1160)	(748-1100)	(751-1162)	(632-1077)
2050		**2159**	**1214**	**1770**	**1428**	**471**	**1319**
		(1619-2307)	(949-1925)	(1244-1770)	(642-1585)	(471-1470)	(580-1319)
2100		**2570**	**1663**	**2077**	**2326**	**363**	**2002**
		(2298-3766)	(1080-2532)	(1520-2077)	(776-2646)	(363-1871)	(661-2002)
NMVOC, (Mt/yr)	139						
2020		**192** (178-230)	**222** (157-222)	**190** (188-190)	**179** (166-205)	**140** (140-193)	**180** (152-180)
2050		**322** (256-322)	**279** (158-301)	**241** (206-241)	**225** (161-242)	**116** (116-237)	**217** (147-217)
2100		**420** (167-484)	**194** (133-552)	**128** (114-128)	**342**(169-342)	**87** (58-349)	**170** (130-304)
NO_x, (MtN/yr)	30.9						
2020		**50** (46-51)	**46** (46-66)	**46** (46-49)	**50** (42-50)	**40** (38-59)	**43** (38-52)
2050		**95** (49-95)	**48** (48-100)	**61** (49-61)	**71** (50-82)	**39** (39-72)	**55** (42-66)
2100		**110** (40-151)	**40** (40-77)	**28** (28-40)	**109** (71-110)	**19** (16-35)	**61** (34-77)

[b] In the SPM the emissions of CFC/HFC/HCFC, PFC, and SF6 are presented as carbon-equivalent emissions. This was done by multiplying the emissions by weight of each substance (see Table 5-8) by its global warming potential (GWP; see Table 5-7) and subsequent summation. The results were then converted from CO_2-equivalents (reflected by the GWPs) into carbon-equivalents. Note that the use of GWP is less appropriate for emission profiles that span a very long period. It is used here, in the interest of readability of the SPM in preference to a more detailed breakdown by the 27 substances listed in Table 5-7. The method here is also preferred over the even less desirable option to display weighted numbers for the aggregate categories in this table.

Table SPM-3b: *Overview of GHG, SO_2, and ozone precursor emissions[a] in 1990, 2020, 2050, and 2100, and cumulative carbon dioxide emissions to 2100. Bold numbers show the value for the illustrative scenario and the numbers between brackets show the value for the range across 26 harmonized SRES scenarios in the six scenario groups that constitute the four families. Units are given in the table.*

Family		**A1**			**A2**	**B1**	**B2**
Scenario group	**1990**	**A1FI**	**A1B**	**A1T**	**A2**	**B1**	**B2**
Carbon dioxide, fossil fuels (GtC/yr)	6.0						
2020		**11.2** (10.7-14.3)	**12.1** (8.7-14.7)	**10.0** (9.8-10.0)	**11.0** (10.3-11.0)	**10.0** (8.2-13.2)	**9.0** (8.8-10.2)
2050		**23.1** (20.6-26.8)	**16.0** (12.7-25.7)	**12.3** (11.4-12.3)	**16.5** (15.1-16.5)	**11.7** (8.5-17.5)	**11.2** (11.2-15.0)
2100		**30.3** (30.3-36.8)	**13.1** (13.1-17.9)	**4.3** (4.3-8.6)	**28.9** (28.2-28.9)	**5.2** (3.3-7.9)	**13.8** (13.8-18.6)
Carbon dioxide, land use (GtC/yr)	1.1						
2020		**1.5** (0.3-1.8)	**0.5** (0.3-1.6)	**0.3** (0.3-1.7)	**1.2** (1.1-1.2)	**0.6** (0.0-1.3)	**0.0** (0.0-1.1)
2050		**0.8** (0.0-0.8)	**0.4** (0.0-1.0)	**0.0** (-0.2-0.0)	**0.9** (0.8-0.9)	**-0.4** (-0.7-0.8)	**-0.2** (-0.2-1.2)
2100		**-2.1** (-2.1-0.0)	**0.4** (-2.0-2.2)	**0.0** (0.0-0.1)	**0.2** (0.0-0.2)	**-1.0** (-2.6-0.1)	**-0.5** (-0.5-1.2)
Cumulative carbon dioxide, fossil fuels (GtC)							
1990-2100		**2128** (2096-2478)	**1437** (1220-1989)	**1038** (1038-1051)	**1773** (1651-1773)	**989** (794-1306)	**1160** (1160-1448)
Cumulative carbon dioxide, land use (GtC)							
1990-2100		**61** (31-61)	**62** (31-84)	**31** (31-62)	**89** (81-89)	**-6** (-22-84)	**4** (4-125)
Cumulative carbon dioxide, total (GtC)							
1990-2100		**2189** (2127-2538)	**1499** (1301-2073)	**1068** (1068-1113)	**1862** (1732-1862)	**983** (772-1390)	**1164** (1164-1573)
Sulfur dioxide, (MtS/yr)	70.9						
2020		**87** (60-134)	**100** (62-117)	**60** (60-101)	**100** (80-100)	**75** (52-112)	**61** (61-78)
2050		**81** (64-139)	**64** (47-64)	**40** (40-64)	**105** (104-105)	**69** (29-69)	**56** (44-56)
2100		**40** (27-83)	**28** (28-47)	**20** (20-27)	**60** (60-69)	**25** (11-25)	**48** (33-48)
Methane, ($MtCH_4$/yr)	310						
2020		**416** (416-479)	**421** (406-444)	**415** (415-466)	**424** (418-424)	**377** (377-430)	**384** (384-391)
2050		**630** (511-630)	**452** (452-636)	**500** (492-500)	**598** (598-671)	**359** (359-546)	**505** (482-505)
2100		**735** (289-735)	**289** (289-535)	**274** (274-291)	**889** (889-1069)	**236** (236-561)	**597** (465-597)

a The uncertainties in the SRES emissions for non-CO_2 greenhouse gases are generally greater than those for energy CO_2. Therefore, the ranges of non-CO_2 GHG emissions provided in the Report may not fully reflect the level of uncertainty compared to CO_2, for example only a single model provided the sole value for halocarbon emissions.

Table SPM-3b (continued)

Family			**A1**		**A2**	**B1**	**B2**
Scenario group	**1990**	**A1FI**	**A1B**	**A1T**	**A2**	**B1**	**B2**
Nitrous oxide, (MtN/yr)	6.7						
2020		**9.3** (6.1-9.3)	**7.2** (6.1-9.6)	**6.1** (6.1-7.8)	**9.6** (6.3-9.6)	**8.1** (5.8-9.5)	**6.1** (6.1-7.1)
2050		**14.5** (6.3-14.5)	**7.4** (6.3-13.8)	**6.1** (6.1-6.7)	**12.0** (6.8-12.0)	**8.3** (5.6-14.8)	**6.3** (6.3-7.5)
2100		**16.6** (5.9-16.6)	**7.0** (5.8-15.6)	**5.4** (4.8-5.4)	**16.5** (8.1-16.5)	**5.7** (5.3-20.2)	**6.9** (6.9-8.0)
CFC/HFC/HCFC (MtC equiv./y) [b]	1672						
2020		**337**	**337**	**337**	**292**	**291**	**299**
2050		**566**	**566**	**566**	**312**	**338**	**346**
2100		**614**	**614**	**614**	**753**	**299**	**649**
PFC, (MtC equiv./yr) [b]	32.0						
2020		**42.7**	**42.7**	**42.7**	**50.9**	**31.7**	**54.8**
2050		**88.7**	**88.7**	**88.7**	**92.2**	**42.2**	**106.6**
2100		**115.3**	**115.3**	**115.3**	**178.4**	**44.9**	**121.3**
SF_6 , (MtC equiv./yr) [b]	37.7						
2020		**47.8**	**47.8**	**47.8**	**63.5**	**37.4**	**54.7**
2050		**119.2**	**119.2**	**119.2**	**104.0**	**67.9**	**79.2**
2100		**94.6**	**94.6**	**94.6**	**164.6**	**42.6**	**69.0**
CO, (MtCO/yr)	879						
2020		**1204** (1123-1552)	**1032** (1032-1248)	**1147** (1147-1160)	**1075** (1075-1100)	**751** (751-1162)	**1022** (941-1022)
2050		**2159** (1619-2307)	**1214** (1214-1925)	**1770** (1244-1770)	**1428** (1428-1585)	**471** (471-1470)	**1319** (1180-1319)
2100		**2570** (2298-3766)	**1663** (1663-2532)	**2077** (1520-2077)	**2326** (2325-2646)	**363** (363-1871)	**2002** (1487-2002)
NMVOC, (Mt/yr)	139						
2020		**192** (178-230)	**222** (194-222)	**190** (188-190)	**179** (179-204)	**140** (140-193)	**180** (179-180)
2050		**322** (256-322)	**279** (259-301)	**241** (206-241)	**225** (225-242)	**116** (116-237)	**217** (197-217)
2100		**420** (167-484)	**194** (137-552)	**128** (114-128)	**342** (311-342)	**87** (58-349)	**170** (130-170)
NO_x, (MtN/yr)	30.9						
2020		**50** (46-51)	**46** (46-66)	**46** (46-49)	**50** (47-50)	**40** (38-59)	**43** (38-43)
2050		**95** (49-95)	**48** (48-100)	**61** (49-61)	**71** (66-71)	**39** (39-72)	**55** (42-55)
2100		**110** (40-151)	**40** (40-77)	**28** (28-40)	**109** (109-110)	**19** (16-35)	**61** (34-61)

[b] In the SPM the emissions of CFC/HFC/HCFC, PFC, and SF6 are presented as carbon-equivalent emissions. This was done by multiplying the emissions by weight of each substance (see Table 5-8) by its global warming potential (GWP; see Table 5-7) and subsequent summation. The results were then converted from CO_2-equivalents (reflected by the GWPs) into carbon-equivalents. Note that the use of GWP is less appropriate for emission profiles that span a very long period. It is used here, in the interest of readability of the SPM in preference to a more detailed breakdown by the 27 substances listed in Table 5-7. The method here is also preferred over the even less desirable option to display weighted numbers for the aggregate categories in this table.

Technical Summary

CONTENTS

1. Introduction and Background

The Intergovernmental Panel on Climate Change (IPCC) decided at its September 1996 plenary session in Mexico City to develop a new set of emissions scenarios (see Appendix I for the Terms of Reference). This Special Report on Emission Scenarios (SRES) describes the new scenarios and how they were developed.

The SRES writing team formulated a set of emissions scenarios. These scenarios cover a wide range of the main driving forces of future emissions, from demographic to technological and economic developments. The scenarios encompass different future developments that might influence greenhouse gas (GHG) sources and sinks, such as alternative structures of energy systems and land-use changes. As required by the Terms of Reference however, *none* of the scenarios in the set includes any future policies that explicitly address additional climate change initiatives[1], although GHG emissions are directly affected by non-climate change policies designed for a wide range of other purpose.

The set of SRES emissions scenarios is based on an extensive assessment of the literature, six alternative modeling approaches, and an "open process" that solicited wide participation and feedback from many groups and individuals. The set of scenarios includes anthropogenic emissions of all relevant GHG species, sulfur dioxide (SO_2), carbon monoxide (CO), nitrogen oxides (NO_x), and non-methane volatile organic compounds (NMVOCs), see Table 1-1 in Chapter 1. It covers most of the range of GHG emissions compared with the published scenario literature. For example, emissions of carbon dioxide (CO_2) in 2100 range from more than 40 to less than 6 giga (or billion) tons[2] of elemental carbon (GtC), that is, from almost a sevenfold increase to roughly the same emissions level as in 1990.

Future emissions and the evolution of their underlying driving forces are highly uncertain, as reflected in the very wide range of future emissions paths in the literature that is also captured by the SRES scenarios. The use of scenarios in this report addresses the uncertainties related to *known* factors. Uncertainties related to *unknown* factors can of course never be persuasively captured by any approach. As the prediction of future anthropogenic GHG emissions is impossible, alternative GHG emissions scenarios become a major tool for analyzing potential long-range developments of the socio-economic system and corresponding emission sources.

Emissions scenarios are a central component of any assessment of climate change. GHG and SO_2 emissions are the basic input for determining future climate patterns with simple climate models, as well as with complex general circulation models (GCMs). Possible climate change, together with the major driving forces of future emissions, such as demographic patterns, economic development and environmental conditions, provide the basis for the assessment of vulnerability, possible adverse impacts and adaptation strategies and policies to climate change. The major driving forces of future emissions also provide the basis for the assessment of possible mitigation strategies and policies designed to avoid climate change. The new set of emissions scenarios is intended for use in future IPCC assessments and by wider scientific and policymaking communities for analyzing the effects of future GHG emissions and for developing mitigation and adaptation measures and policies.

[1] For example, no scenarios are included that explicitly assume implementation of the emission targets in the UNFCCC and the Kyoto protocol.

[2] Metric tons are used throughout this report. Unless otherwise specified, monetary units are 1990 US dollars (see Chapter 4).

2. Emissions Scenarios and Their Purposes

Scenarios are images of the future, or alternative futures. They are neither predictions nor forecasts. Rather, each scenario is one alternative image of how the future might unfold (see Chapters 1 and 4 for more detail). As such they enhance our understanding of how systems behave, evolve and interact. They are useful tools for scientific assessments, learning about complex systems behavior and for policymaking and assist in climate change analysis, including climate modeling and the assessment of impacts, adaptation and mitigation.

Future levels of global GHG emissions are a product of very complex, ill-understood dynamic systems, driven by forces such as population growth, socio-economic development, and technological progress among others, thus making long-term predictions about emissions virtually impossible. However, near-term policies may have profound long-term climate impacts. Consequently, policy makers need a summary of what is understood about possible future GHG emissions, and given the uncertainties in both emissions models and our understanding of key driving forces, scenarios are an appropriate tool for summarizing both current understanding and current uncertainties.

GHG emissions scenarios are usually based on an internally consistent and reproducible set of assumptions about the key relationships and driving forces of change, which are derived from our understanding of both history and the current situation. Often these scenarios are formulated with the help of formal models. Sometimes GHG emissions scenarios are less quantitative and more descriptive, and in a few cases they do not involve any formal analysis and are expressed in qualitative terms. The SRES scenarios involve both qualitative and quantitative components; they have a narrative part called "storylines" and a number of corresponding quantitative scenarios for each storyline. SRES scenarios can be viewed as a linking tool that integrates qualitative narratives or stories about the future and quantitative formulations based on different formal modeling approaches. Although no scenarios are value free, the SRES scenarios are descriptive and are not intended to be desirable or undesirable in their own right. They have been built as descriptions of plausible alternative futures, rather than preferred developments.

However, developing scenarios for a period of one hundred years is a relatively new field. This is not only because of large scientific uncertainties and data inadequacies. For example, within the 21st century technological discontinuities should be expected, and possibly major shifts in societal values and in the balance of geopolitical power. The study of past trends over such long periods is hampered by the fact that most databases are incomplete if we go back much further than 50 years. Given these gaps in our data, methods, and understanding, scenarios are the best way to integrate demographic, economic, societal, and technological knowledge with our understanding of ecological systems to evaluate sources and sinks of GHG emissions. Scenarios as an integration tool in the assessment of climate change allow a role for intuition, analysis, and synthesis, and thus we turn to scenarios in this report to take advantage of those features to aid the assessment of future climate change, impacts, vulnerabilities, adaptation, and mitigation. Since the scenarios focus on the century time scale, tools have been used that have been developed for this purpose. These tools are less suitable for analysis of near-term developments and this report does not intend to provide reliable projections for the near term.

3. Review of Past IPCC Emissions Scenarios

The IPCC developed sets of emissions scenarios in 1990 and 1992. The six IS92 scenarios developed in 1992 (Leggett *et al.*, 1992; Pepper *et al.*, 1992), have been used very widely in climate change assessments. In 1995 the IPCC formally evaluated the 1992 scenarios and found that they were innovative at the time of their publication, path-breaking in their coverage of the full range of GHG emissions and useful for the purpose of driving atmospheric and climate models (Alcamo *et al.*, 1995). Specifically, their global carbon emissions spanned most of the range of other scenarios identified in the literature at that time.

The review also identified a number of weaknesses. These included the limited range of carbon intensities of energy (carbon emissions per unit energy) and the absence of any scenario with significant closure in the income gap between developed and developing countries, even after a full century (Parikh, 1992). Furthermore, rapid growth of sulfur emissions in the IS92 scenarios had been questioned on the basis that they did not reflect recent legislation in Japan, Europe, and North America and that in general regional and local air quality concerns might prompt limits on future sulfur emissions.

An important recommendation of the 1995 IPCC review was that, given the degree of uncertainty about future climate change, analysts should use the full range of IS92 emissions as input to climate models rather than a single scenario. This is in stark contrast to the actual use of one scenario from the set, the IS92a scenario, as the reference scenario in numerous studies. The review concluded that the mere fact of the IS92a being an intermediate, or central, CO_2 emissions scenario at the global level at that time does not equate it with being the most likely scenario. Indeed, the conclusion was that there was no objective basis on which to assign likelihood to any of the scenarios. Furthermore, the IS92a scenario was shown to be "central" for only a few of its salient characteristics such as global population growth, global economic development and global CO_2 emissions. In other ways, IS92a was found not to be central with respect to the published literature, particularly in some of its regional assumptions and emissions. The same is the case with the new set of SRES scenarios, as is shown below.

The new set of SRES scenarios presented here is designed to respond to the IS92 weaknesses identified in the 1995 IPCC scenario evaluation and to incorporate advances in the state of the art since 1992. As in the case of the IS92 scenario series, also in this new set of SRES scenarios there is no single central case with respect to all characteristics that are relevant for different uses of emissions scenarios and there is no objective way to assign likelihood to any of the scenarios. Hence there is no "best guess" or "business-as-usual" scenario.

4. SRES Writing Team, Approach, and Process

IPCC Working Group III (WGIII) appointed the SRES writing team in January 1997. After some adjustments, it eventually came to include more than 50 members from 18 countries. Together they represent a broad range of scientific disciplines, regional backgrounds, and non-governmental organizations. In particular, the team includes representatives of six scenario modeling groups and a number of lead authors from all three earlier IPCC scenario activities: the 1990 and 1992 scenarios and the 1995 scenario evaluation. Their expertise and familiarity with earlier IPCC emissions scenario work assured continuity and allowed the SRES effort to build efficiently upon prior work. The SRES team worked in close collaboration with colleagues on the IPCC Task Group on Climate Scenarios for Impact Assessment (TGCIA) and with colleagues from all three IPCC Working Groups (WGs) of the Third Assessment Report (TAR). Appendix II lists the members of the writing team and their affiliations and Chapter 1 gives a more detailed description of the SRES approach and process.

Taking the above audiences and purposes into account, the following more precise specifications for the new SRES scenarios were developed. The new scenarios should:

- cover the full range of radiatively important gases, which include direct and indirect GHGs and SO_2;
- have sufficient spatial resolution to allow regional assessments of climate change in the global context;
- cover a wide spectrum of alternative futures to reflect relevant uncertainties and knowledge gaps;

- use a variety of models to reflect methodological pluralism and uncertainty;
- incorporate input from a wide range of scientific disciplines and expertise from non-academic sources through an open process;
- exclude additional initiatives and policies specifically designed to reduce climate change;
- cover and describe to the extent possible a range of policies that could affect climate change although they are targeted at other issues, for example, reductions in SO_2 emissions to limit acid rain;
- cover as much as possible of the range of major underlying driving forces of emission scenarios identified in the open literature;
- be transparent with input assumptions, modeling approaches, and results open to external review;
- be reproducible – document data and methodology adequately enough to allow other researchers to reproduce the scenarios; and
- be internally consistent – the various input assumptions and data of the scenarios are internally consistent to the extent possible.

The writing team agreed that the scenario formulation process would consist of five major components:

- review of existing scenarios in the literature;
- analysis of their main characteristics and driving forces;
- formulation of narrative "storylines" to describe alternative futures;
- quantification of storylines with different modeling approaches; and
- "open" review process of emissions scenarios and their assumptions

As is evident from the components of the work program, there was agreement that the process be an open one with no single "official" model and no exclusive "expert teams." In 1997 the IPCC advertised in a number of relevant scientific journals and other publications to solicit wide participation in the process. To facilitate participation and improve the usefulness of the new scenarios, the SRES web site (www.sres.ciesin) was created. In addition, members of the writing team published much of the background work used for formulating SRES scenarios in the peer-reviewed literature[3] and on web sites (see Appendix IV). Finally, the revised set of scenarios, the web sites, and the draft of this report have been evaluated through the IPCC expert and government review processes. This process resulted in numerous changes and revisions of the report. In particular, during the approval process of the Summary for Policymakers (SPM) in March 2000 at the 5th Session of the WG III in Katmandu changes in this SPM were agreed that necessitated some changes in the underlying document, including this Technical Summary. These changes have been implemented in agreement with the Lead Authors.

5. Scenario Literature Review and Analysis

The first step in the formulation of the SRES scenarios was the review and the analysis of the published literature and the development of the database with more than 400 emissions scenarios that is accessible through the web site (www-cger.nies.go.jp/cger-e/db/ipcc.html); 190 of these extend to 2100 and are considered in the comparison with the SRES scenarios in the subsequent Figures. Chapters 2 and 3 give a more detailed description of the literature review and analysis.

Figure TS-1 shows the global energy-related and industrial CO_2 emission paths from the database as "spaghetti" curves for the period to 2100 against the background of the historical emissions from 1900 to 1990. These curves are plotted against an index on the vertical axis rather than as absolute values because of the large differences and discrepancies for the values assumed for the base year 1990. These sometimes arise from genuine differences among the scenarios (e.g., different data sources, definitions) and sometimes from different base years assumed in the analysis or from alternative calibrations.[4] The differences among the scenarios in the specification of the base year illustrate the large genuine scientific and data uncertainty that surrounds emissions and their main driving forces captured in the scenarios. The literature includes scenarios with additional climate polices, which are sometimes referred to as mitigation or intervention scenarios.

There are many ambiguities associated with the classification of emissions scenarios into those that include additional climate initiatives and those that do not. Many cannot be classified in this way on the basis of the information available from the database. Figure TS-1 indicates the ranges of emissions in 2100 from scenarios that apparently include additional climate initiatives (designated as "intervention" emissions range), those that do not ("non-intervention") and those that cannot be assigned to either of these two categories ("non-classified"). This classification is based on the subjective evaluation of the scenarios in the database by the members of the writing team and is explained in Chapter 2. The range of the whole sample of scenarios has significant

[3] Alcamo and Nakićenović, 1998; Alcamo and Swart, 1998; Anderson, 1998; Gaffin, 1998; Gregory, 1998; Gregory and Rogner, 1998; Grübler, 1998; Michaelis, 1998; Morita and Lee, 1998; Nakićenović *et al.*, 1998; Price *et al.*, 1998; de Vries *et al.*, 2000; Fenhann, 2000; Jiang *et al.*, 2000, Jung *et al.*, 2000; Kram *et al.*, 2000; Mori, 2000; Nakićenović, 2000; Riahi and Roehrl, 2000; Roehrl and Riahi, 2000; Sankovski *et al.*, 2000.

[4] The 1990 emissions from energy production and use are estimated by Marland *et al.* (1994) at 5.9 GtC excluding cement production. The 1990 base year values in the scenarios reviewed range from 4.8 (CETA/EMF14, Scenario MAGICC CO_2) to 6.4 GtC (ICAM2/EMF14), see Dowlatabadi *et al.*, 1995; Peck and Teisberg, 1995.

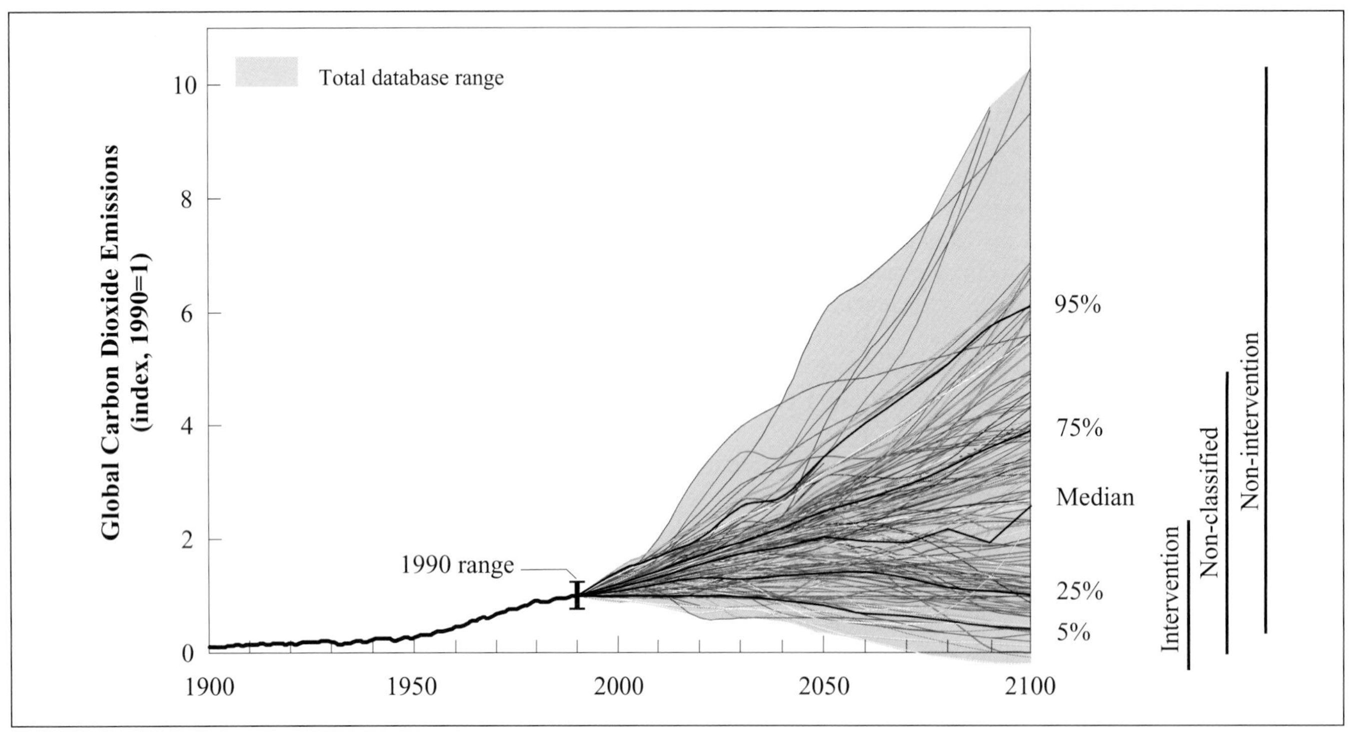

Figure TS-1: Global energy-related and industrial CO_2 emissions – historical development and future scenarios, shown as an index (1990 = 1). The median (50th), the 5th, and 95th percentiles of the frequency distribution are also shown. The statistics associated with scenarios from the literature do not imply probability of occurrence (e.g., the frequency distribution of the scenarios may be influenced by the use of IS92a as a reference for many subsequent studies). The emissions paths indicate a wide range of future emissions. The range is also large in the base year 1990 and is indicated by an "error" bar. To separate the variation due to base-year specification from different future paths, emissions are indexed for the year 1990, when actual global energy-related and industrial CO_2 emissions were about 6 GtC. The coverage of CO_2 emissions sources may vary across the 256 different scenarios from the database included in the figure. The scenario samples used vary across the time steps (for 1990 256 scenarios, for 2020 and 2030 247, for 2050 211, and for 2100 190 scenarios). Also shown, as vertical bars on the right of the figure, are the ranges of emissions in 2100 of IS92 scenarios and for scenarios from the literature that apparently include additional climate initiatives (designated as "intervention" scenarios emissions range), those that do not ("non-intervention"), and those that cannot be assigned to either of these two categories ("non-classified"). This classification is based on the subjective evaluation of the scenarios in the database by the members of the writing team and is explained in Chapter 2. Data sources: Morita and Lee, 1998a, 1998b; Nakićenović *et al.*, 1998.

overlap with the range of those that cannot be classified and they share virtually the same median (15.7 and 15.2 GtC in 2100, respectively), but the non-classified scenarios do not cover the high part of the range. Also, the range of the scenarios that apparently do not include climate polices (non-intervention) has considerable overlap with the other two ranges (the lower bound of non-intervention scenarios is higher than the lower bounds of the intervention and non-classified scenarios), but with a significantly higher median (of 21.3 GtC in 2100).

Historically, gross anthropogenic CO_2 emissions have increased at an average rate of about 1.7% per year since 1900 (Nakićenović *et al.*, 1996); if that historical trend continues global emissions would double during the next three to four decades and increase more than sixfold by 2100. Many scenarios in the database describe such a development. However, the range is very large around this historical trend so that the highest scenarios envisage about a tenfold increase of global emissions by 2100 as compared with 1990, while the lowest have emissions lower than today. The median and the average of the scenarios lead to about a threefold emissions increase over the same time period or to about 16 GtC by 2100. This is lower than the median of the IS92 set and is lower than the IS92a scenario, often considered as the "central" scenario with respect to some of its tendencies. However, the distribution of emissions is asymmetric. The thin emissions "tail" that extends above the 95th percentile (i.e., between the six- and tenfold increase of emissions by 2100 compared to 1990) includes only a few scenarios. The range of other emissions and the main scenario driving forces (such as population growth, economic development and energy production, conversion and end use) for the scenarios documented in the database is also large and comparable to the variation of CO_2 emissions. Statistics associated with scenarios from the literature do not imply probability of occurrence or likelihood of the scenarios. The frequency distribution of the database may be influenced by the use of IS92a as a reference for scenario studies and by the fact that many scenarios in the database share common assumptions prescribed for the purpose of model comparisons with similar scenario driving forces.

One of the recommendations of the writing team is that IPCC or a similar international institution should maintain such a database thereby ensuring continuity of knowledge and scientific progress in any future assessments of GHG scenarios. An equivalent database for documenting narrative and other qualitative scenarios is considered to be also very useful for future climate-change assessments. One difficulty encountered in the analysis of the emissions scenarios is that the distinction between climate policies and non-climate policy scenarios and other scenarios appeared to be to a degree arbitrary and often impossible to make. Therefore, the writing team recommends that an effort should be made in the future to develop an appropriate emissions scenario classification scheme.

6. Narrative Scenarios and Storylines

Given these large ranges of future emissions and their driving forces, there is an infinite number of possible alternative futures to explore. The SRES scenarios cover a finite, albeit a very wide range, of future emissions. The approach involved the development of a set of four alternative scenario "families" comprising 40 SRES scenarios subdivided into seven scenario groups. During the approval process of the SPM in March 2000 at the 5th Session of WGIII in Katmandu, it was decided to combine two of these groups into one, resulting in six groups. To facilitate the process of identifying and describing alternative future developments, each scenario family includes a coherent narrative part called a "storyline," and a number of alternative interpretations and quantifications of each storyline developed by six different modeling approaches. All the interpretations and quantifications of one storyline together are called a scenario family (see also Box 1-1 in Chapter 1 on terminology). Each storyline describes a demographic, social, economic, technological, environmental, and policy future for one of these scenario families. Storylines were formulated by the writing team in a process which identified driving forces, key uncertainties, possible scenario families, and their logic. Within each family different scenarios explore variations of global and regional developments and their implications for GHG and sulfur emissions. Each of these scenarios is consistent with the broad framework of that scenario family as specified by the storyline. Consequently, each scenario family and scenario group is equally sound. Chapters 4 and 5 give a more detailed description of the storylines, their quantifications, and the resultant 40 emissions scenarios.

The main reasons for formulating storylines are:

- to help the writing team to think more coherently about the complex interplay among scenario driving forces within each and across alternative scenarios;
- to make it easier to explain the scenarios to the various user communities by providing a narrative description of alternative futures that goes beyond quantitative scenario features;
- to make the scenarios more useful, in particular to analysts who contribute to IPCC WGII and WGIII; the social, political, and technological context described in the scenario storylines is all-important in analyzing the effects of policies either to adapt to climate change or to reduce GHG emissions; and
- to provide a guide for additional assumptions to be made in detailed climate impact and mitigation analyses, because at present no single model or scenario can possibly respond to the wide variety of informational and data needs of the different user communities of long-term emissions scenarios.

The SRES writing team reached broad agreement that there could be no "best guess" scenarios; that the future is inherently unpredictable and that views will differ on which of the storylines could be more likely. The writing team decided on four storylines: an even number helps to avoid the impression that there is a "central" or "most likely" case. The team wanted more than two in order to help illustrate that the future depends on many different underlying dynamics; the team did not want more than four, as they wanted to avoid complicating the process by too many alternatives. There is no "business-as-usual" scenario. Nor should the scenarios be taken as policy recommendations. The storylines represent the playing out of certain social, economic, technological, and environmental paradigms, which will be viewed positively by some people and negatively by others. The scenarios cover a wide range, but not all possible futures. In particular, it was decided that possible "surprises" would not be considered and that there would be no "disaster" scenarios that are difficult to quantify with the aid of formal models.

The titles of the storylines have been kept simple: A1, A2, B1, and B2. There is no particular order among the storylines; Box TS-1 lists them in alphabetic order. Figure TS-2 schematically illustrates the four storylines and scenario families. Each is based on a common specification of the main driving forces. They are shown, very simplistically, as branches of a two-dimensional tree. The two dimensions shown indicate the global-regional and the development-environmental orientation, respectively. In reality, the four scenario families share a space of a much higher dimensionality given the numerous driving forces and other assumptions needed to define any given scenario in a particular modeling approach. The team decided to carry out sensitivity tests within some of the storylines by considering alternative scenarios with different fossil-fuel reserves, rates of economic growth, or rates of technical change within a given scenario family. For example, four scenario "groups" within the A1 scenario family were explored. As mentioned, two of these four groups that explore fossil-intensive developments in the energy system were merged in the SPM. Together with the other three scenario families this results in seven groups of scenarios -

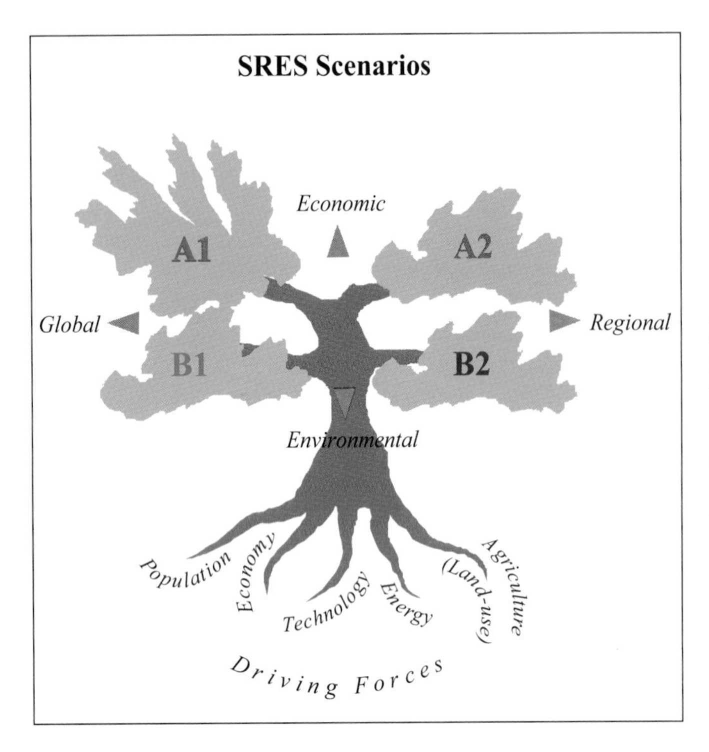

Figure TS-2: Schematic illustration of SRES scenarios. The four scenario "families" are shown, very simplistically, as branches of a two-dimensional tree. In reality, the four scenario families share a space of a much higher dimensionality given the numerous assumptions needed to define any given scenario in a particular modeling approach. The schematic diagram illustrates that the scenarios build on the main driving forces of GHG emissions. Each scenario family is based on a common specification of some of the main driving forces. The A1 storyline branches out into four groups of scenarios to illustrate that alternative development paths are possible within one scenario family. Two of these groups were merged in the SPM.

Box TS-1: The Main Characteristics of the Four SRES Storylines and Scenario Families

By 2100 the world will have changed in ways that are hard to imagine – as hard as it would have been at the end of the 19th century to imagine the changes of the 100 years since. Each storyline assumes a distinctly different direction for future developments, such that the four storylines differ in increasingly irreversible ways. Together they describe divergent futures that encompass a significant portion of the underlying uncertainties in the main driving forces. They cover a wide range of key "future" characteristics such as population growth, economic development, and technological change. For this reason, their plausibility or feasibility should not be considered solely on the basis of an extrapolation of *current* economic, technological, and social trends.

- The A1 storyline and scenario family describes a future world of very rapid economic growth, low population growth, and the rapid introduction of new and more efficient technologies. Major underlying themes are convergence among regions, capacity building, and increased cultural and social interactions, with a substantial reduction in regional differences in per capita income. The A1 scenario family develops into four groups that describe alternative directions of technological change in the energy system. Two of the fossil-intensive groups were merged in the SPM.
- The A2 storyline and scenario family describes a very heterogeneous world. The underlying theme is self-reliance and preservation of local identities. Fertility patterns across regions converge very slowly, which results in high population growth. Economic development is primarily regionally oriented and per capita economic growth and technological change are more fragmented and slower than in other storylines.
- The B1 storyline and scenario family describes a convergent world with the same low population growth as in the A1 storyline, but with rapid changes in economic structures toward a service and information economy, with reductions in material intensity, and the introduction of clean and resource-efficient technologies. The emphasis is on global solutions to economic, social, and environmental sustainability, including improved equity, but without additional climate initiatives.
- The B2 storyline and scenario family describes a world in which the emphasis is on local solutions to economic, social, and environmental sustainability. It is a world with moderate population growth, intermediate levels of economic development, and less rapid and more diverse technological change than in the B1 and A1 storylines. While the scenario is also oriented toward environmental protection and social equity, it focuses on local and regional levels.

After determining the basic features and driving forces for each of the four storylines, the team began modeling and quantifying the storylines. This resulted in 40 scenarios, each of which constitutes an alternative interpretation and quantification of a storyline. All the interpretations and quantifications associated with a single storyline are called a scenario family (see Chapter 1 for terminology and Chapter 4 for further details).

effectively six equally sound groups after the merging of the two fossil-intensive groups in the SPM - that share common assumptions of some of the key driving forces and are thus not independent of each other.

All four storylines and scenario families describe future worlds that are generally more affluent compared to the current situation. They range from very rapid economic growth and technological change to high levels of environmental protection, from low to high global populations, and from high to low GHG emissions. What is perhaps even more important is that all the storylines describe dynamic changes and transitions in generally different directions. Although they do not include additional climate initiatives, none of them are policy free. As time progresses, the storylines diverge from each other in many of their characteristic features. In this way they allow us to span the relevant range of GHG emissions and different combinations of their main sources.

7. Quantitative Scenarios and Modeling Approaches

The storylines were essentially complete by January 1998. After determining the basic features and driving forces for each of the four storylines, the six modeling groups represented on the writing team (on a voluntary basis) began quantifying them. The six modeling groups that quantified the storylines are listed in Box TS-2. Each model quantification of a storyline constitutes a scenario, and all scenarios derived from one storyline constitute a scenario family. The six models are representative of different approaches to modeling emissions scenarios and different integrated assessment (IA) frameworks in the literature and include so-called top-down and bottom-up models. The writing team recommends that IPCC or a similar international institution should assure participation of modeling groups around the world and especially from developing countries in any future scenario development and assessment efforts. Clearly, this would also require resources specifically directed at assisting modeling groups from developing countries. Indeed, a concerted effort was made to engage modeling groups and experts from developing countries in SRES as a direct response to the recommendations of the last IPCC scenario evaluation (Alcamo *et al.*, 1995).

The six models have different regional aggregations. The writing team decided to group the various global regions into four "macro-regions" common to all the different regional aggregations across the six models. Box TS-3 indicates that the four macro-regions (see Appendix III) are broadly consistent with the allocation of countries in the United Nations Framework Convention on Climate Change (UNFCCC, 1992), although the correspondence is not exact because of changes in the countries listed in Annex I of UNFCCC (1997).

All the qualitative and quantitative features of scenarios belonging to the same family were set to conform to the corresponding features of the underlying storyline. Together, 26 scenarios were "harmonized" to share agreed common assumptions about global population and GDP (gross domestic product) development (a few that also share common population, GDP, and final energy trajectories at the level of the four SRES macro-regions are called "fully harmonized," see Section 4.1. in Chapter 4). Thus, the harmonized scenarios are not independent within each of the four families. However, scenarios within each family vary quite substantially in such

Box TS-2: The Six Modeling Teams that Quantified the 40 SRES Scenarios

In all, six models were used to generate the 40 scenarios:

- Asian Pacific Integrated Model (AIM) from the National Institute of Environmental Studies in Japan (Morita *et al.*, 1994);
- Atmospheric Stabilization Framework Model (ASF) from ICF Consulting in the USA (Lashof and Tirpak, 1990; Pepper *et al.*, 1992, 1998; Sankovski *et al.*, 2000);
- Integrated Model to Assess the Greenhouse Effect (IMAGE) from the National Institute for Public Health and Environmental Hygiene (RIVM) (Alcamo *et al.*, 1998; de Vries *et al.*, 1994, 1999, 2000), used in connection with the Dutch Bureau for Economic Policy Analysis (CPB) WorldScan model (de Jong and Zalm, 1991), the Netherlands;
- Multiregional Approach for Resource and Industry Allocation (MARIA) from the Science University of Tokyo in Japan (Mori and Takahashi, 1999; Mori, 2000);
- Model for Energy Supply Strategy Alternatives and their General Environmental Impact (MESSAGE) from the International Institute of Applied Systems Analysis (IIASA) in Austria (Messner and Strubegger, 1995; Riahi and Roehrl, 2000); and
- Mini Climate Assessment Model (MiniCAM) from the Pacific Northwest National Laboratory (PNNL) in the USA (Edmonds *et al.*, 1994, 1996a, 1996b).

These six models are representative of emissions scenario modeling approaches and different IA frameworks in the literature and include so-called top-down and bottom-up models. For a more detailed description of the modeling approaches see Appendix IV.

Box TS-3: SRES World "Macro-Regions" Used by All Six Modeling Teams

The six models have different regional aggregations. The writing team decided to group the various global regions into four "macro-regions" common to all the different regional aggregations across the six models. The four macro-regions (see Appendix III) are broadly consistent with the allocation of the countries in the United Nations Framework Convention on Climate Change (UNFCCC, 1992), although the correspondence is not exact due to changes in the countries listed in Annex I of UNFCCC (1997):

- The OECD90 region groups together all countries belonging to the Organization for Economic Cooperation and Development (OECD) as of 1990, the base year of the participating models, and corresponds to Annex II countries under UNFCCC (1992).
- The REF region stands for countries undergoing economic reform and groups together the East European countries and the Newly Independent States of the former Soviet Union. It includes Annex I countries outside Annex II as defined in UNFCCC (1992).
- The ASIA region stands for all developing (non-Annex I) countries in Asia.
- The ALM region stands for rest of the world and includes all developing (non-Annex I) countries in Africa, Latin America, and the Middle East.

In other words, the OECD90 and REF regions together correspond to the developed (industrialised) countries (referred to as IND in this report) while the ASIA and ALM regions together correspond to the developing countries (referred to as DEV in this report). The OECD90 and REF regions are consistent with the Annex I countries under the Framework Convention on Climate Change, while the ASIA and ALM regions correspond to the non-Annex I countries (UNFCCC, 1992).

characteristics as the assumptions about availability of fossil-fuel resources, the rate of energy-efficiency improvements, the extent of renewable-energy development, and, hence, resultant GHG emissions. Thus, after the modeling teams had quantified the key driving forces and made an effort to harmonize them with the storylines by adjusting control parameters, there still remained diversity in the assumptions about the driving forces and in the resultant emissions (see Chapter 4).

The remaining 14 scenarios adopted alternative interpretations of the four scenario storylines to explore additional scenario uncertainties beyond differences in methodologic approaches, such as different rates of economic growth and variations in population projections. These variations reflect the "modeling teams' choice" of alternative but plausible global and regional development compared to "harmonized" scenarios and also stem from the differences in the underlying modeling approaches. Each of the 40 quantifications of one of the storylines constitutes a SRES scenario. This approach generated a large variation and richness in different scenario quantifications, often with overlapping ranges of main driving forces and GHG emissions across the four families.

In addition, the A1 scenario family branched out into four distinct scenario groups. They are based on four alternative technological developments in future energy systems, from carbon-intensive development to decarbonization. Similar storyline variations were considered for other scenario families, but they did not result in genuine scenario groupings within the respective families. This further increased richness in different GHG and SO_2 emissions paths, because this variation in the structure of the future energy systems in itself resulted in a range of emissions almost as large as that generated through the variation of other main driving forces such as population and economic development. It should be noted that future energy systems variations could be applied to the other storylines, but they may evolve differently from those in A1. They have been introduced into the A1 storyline because of its "high growth with high technology" nature, where differences in alternative technology developments translate into large differences in future GHG emission levels. Altogether the 40 SRES scenarios fall into seven groups: the three scenario families, A2, B1, and B2, plus four groups within the A1 scenario. In the SPM, two of these groups, the coal and gas and oil intensive groups, were merged into one fossil-intensive group, leading to six groups.

As in the case of the storylines, no single scenario – whether it represents a modeler's choice or harmonized assumptions – was treated as being more or less "probable" than the others belonging to the same family. Initially, for each storyline, one modeling group was given principal responsibility, and the quantification produced by that group is referred to as the "marker" scenario for that storyline. The four preliminary marker scenarios were used in 1998 to solicit comments during the "open process" and as input for climate modelers in accordance with a decision of the IPCC Bureau in 1998. The four marker scenarios were posted on the SRES web site (www.sres.ciesin) in June 1998 and were subsequently revised to account for comments and suggestions received through this open scenario review process that lasted until January 1999. In addition to many revisions, the marker scenarios were also harmonized along with the other 26 scenarios that adopted common assumptions for the main driving forces within the four respective families. The choice of the markers was based on extensive discussion of:

- range of emissions across all of marker scenarios;
- which of the initial quantifications (by the modelers) reflected the storyline best;
- preference of some of the modeling teams and features of specific models;
- use of four different models for the four markers.

As a result the markers were not intended to be the median or mean scenarios from their respective families. Indeed, in general it proved impossible to develop scenarios in which all relevant characteristics match mean or median values. Thus, marker scenarios are no more or less likely than any other scenarios, but are those scenarios considered by the SRES writing team as illustrative of a particular storyline. These scenarios have received the closest scrutiny of the entire writing team and via the SRES open process compared to other scenario quantifications. The marker scenarios are also those SRES scenarios that have been most intensively tested in terms of reproducibility. As a rule, different modeling teams have attempted to replicate the model quantification of marker scenarios. Available time and resources have not allowed a similar exercise to be conducted for all SRES scenarios, although some effort was devoted to reproduce the four scenario groups (merged into three in the SPM) that constitute different interpretations of the A1 storyline with different models.

Additional scenarios using the same harmonized assumptions as the marker scenarios developed by different modeling teams and other scenarios that give alternative quantitative interpretations of the four storylines constitute the final set of 40 SRES scenarios. This also means that the 40 scenarios are not independent of each other as they are all based on four storylines and subdivided into seven scenario groups (after merging two groups, six in the SPM) that share many common assumptions. In addition to many revisions of the marker and other harmonized scenarios, other alternative scenarios were formulated by the six modeling teams within each of the four scenario families. The result is a more complete, refined set of 40 emissions scenarios that reflects the broad spectrum of modeling approaches and regional perspectives. However, differences in modeling approaches have meant that not all of the scenarios provide estimates for all the direct and indirect GHG emissions for all the sources and sectors. In addition to the marker scenarios, two scenarios were also selected in the SPM to illustrate the alternative energy systems developments in the A1 family. Hence, this report has an illustrative scenario for each of the six scenario groups in the SPM. The four SRES marker scenarios and the two illustrative scenarios (selected in the SPM) cover all the relevant gas species and emission categories comprehensively and thus constitute the smallest set of independent and fully documented SRES scenarios.

The scenario groups and cumulative emissions categories were developed as the smallest subsets of SRES scenarios that capture the range of uncertainties associated with driving forces and emissions. Together, the four markers and the two additional illustrative scenarios selected in the SPM from the A1 scenario groups constitute the set of SRES scenarios that reflects the uncertainty ranges in the emissions and their driving forces. Furthermore, the writing team recommends that, to the extent possible, these scenarios, but at least the four markers and the two additional illustrative scenarios selected in the SPM, should always be used together, and that, in general, no individual scenario should be singled out for any purpose. Multiple baselines and overlapping emissions ranges have important implications for making policy analysis, e.g., similar policies might have different impacts in different scenarios. Combination of policies might shape the future development in the direction of certain scenarios.

8. Main Scenario Driving Forces Based on the Literature

The scenarios cover a wide range of driving forces, from demographic to social and economic developments. This section summarizes the assumptions on important scenario drivers. For simplicity, only three of these are presented separately here following the exposition in Chapters 2, 3, and 4. Nonetheless, it is important to keep in mind that the future evolution of these and other main driving forces is interrelated in the SRES scenarios (see Tables TS-2 and TS-3 for a summary of the ranges of the main driving forces across the scenario groups in 2100).

The SRES scenarios span a wide range of assumptions for the most salient scenario drivers, and thus reflect the uncertainty of the future. Evidently, views of the future are a time-specific phenomenon, and this report and its scenarios are no exception. However, it is important to emphasize that this is an explicit part of the Terms of Reference for the SRES writing team – to reflect a range of views, based on current knowledge and the most recently available literature (see Appendix I). The scenario quantification results reflect well the literature range, except for extreme scenarios.

8.1. Population Projections

Three different population trajectories were chosen for SRES scenarios to reflect future demographic uncertainties based on published population projections (Lutz, 1996; UN, 1998; see Chapter 3). The population projections are exogenous input to all the models used to develop the SRES scenarios. The models used do not develop population from other assumptions within the model. Figure TS-3 shows the three population projections in comparison with the three population projections used in the IS92 scenarios. Global population ranges between seven and 15 billion people by 2100 across the scenarios, depending on the rate and extent of the demographic transition. The insert in Figure TS-3 shows population development in the developed (industrialized) regions. The range of future populations is smaller than in the IS92 scenarios, particularly in the developed (industrialized) regions, for which the lowest scenario indicates a very modest population decline compared to IS92 scenarios. The greatest uncertainty about future growth

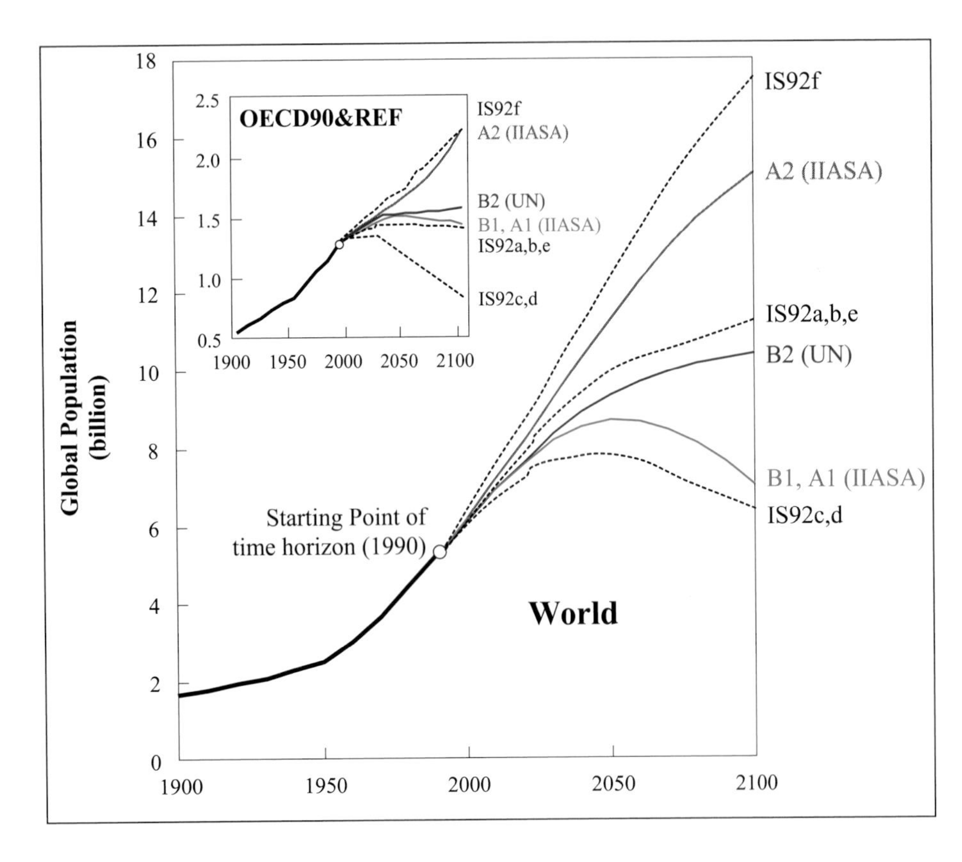

Figure TS-3: Population projections – historical data from 1900 to 1990 (based on Durand, 1967; Demeny; 1990; UN, 1998), and SRES scenarios (based on Lutz, 1996, for high and low, and UN, 1998, for medium) and IPCC IS92 scenarios (Leggett *et al.*, 1992; Pepper *et al.*, 1992) from 1990 to 2100.

lies in the developing regions across all scenarios in the literature. An equally pervasive trend across all scenarios is urbanization (see Chapter 3). Altogether, three different population projections were used in the 26 harmonized scenarios. Other scenarios explored alternative population projections consistent with the storylines.

The lowest population trajectory is assumed for the A1 and B1 scenario families and is based on the low population projection in Lutz (1996), which combines low fertility with low mortality and central migration rate assumptions. After peaking at 8.7 billion in the middle of the 21st century, world population declines to 7.1 billion by the year 2100. As discussed in Chapters 3 and 4, this population development is somewhat higher than the previous low population used in the IS92 scenarios. The B2 scenario family is based on the UN median 1998 population projection (UN, 1998). The global population increases to about 9.4 billion people by 2050 and to about 10.4 billion by 2100. This population scenario is characteristic of recent median global population projections, which describe a continuation of historical trends toward a completion of the demographic transition that would lead to a constant global population, and is consistent with recent faster fertility declines in the world together with declining mortality rates. Hence, the population is somewhat lower than previous UN median projections, as used in the IS92 scenarios. This median scenario projects very low population growth in today's industrialized countries, with stabilization of growth in Asia in the second half of the 21st century and in the rest of the world towards the end of the 21st century. The A2 scenario family is based on the high population growth of 15 billion by 2100 reported in Lutz (1996), which assumes a significant decline in fertility for most regions and a stabilization at above replacement levels. It falls below the long-term 1998 UN high projection of 18 billion. It is also lower than in the highest IS92 scenario (17.6 billion by 2100). Nevertheless, this scenario represents very high population growth compared with that in current demographic literature. Some demographers attach a probability of more than 90% that actual population will be lower than the trajectory adopted in the A2 scenario family (Lutz *et al.*, 1997). A more detailed discussion of the population projections used to quantify the four scenario families is given in Chapters 3 and 4.

8.2. Economic Development

The SRES scenarios span a wide range of future levels of economic activity (expressed in gross world product). The A1 scenario family with a "harmonized" gross world product of US$529 trillion (all values in 1990 US dollars unless otherwise indicated) in 2100 delineates the SRES upper bound, whereas B2 with "harmonized" US$235 trillion in 2100 represents its lower bound. The range of gross world product across all scenarios is even higher, from US$197 to 550 by 2100.

Although the SRES scenarios span a wide range, still lower and higher gross world product levels can be found in the literature (see Chapters 2, 3, and 4). Uncertainties in future gross world product levels are governed by the pace of future productivity growth and population growth, especially in developing regions. Different assumptions on conditions and

Table TS-1: *Income per capita in the world and by SRES region for the IS92 (Leggett et al., 1992) and four marker scenarios by 2050 and 2100, measured by GDP per capita in 1000 US dollars (at 1990 prices and exchange rates).*

Income per Capita by World and Regions
(103 1990US$ per capita)

		Regions						
Year	**Scenario**	**OECD90**	**REF**	**IND**	**ASIA**	**ALM**	**DEV**	**WORLD**
1990	SRES MESSAGE	19.1	2.7	13.7	0.5	1.6	0.9	4.0
2050	IS92a,b	49.0	23.2	39.7	3.7	4.8	4.1	9.2
	IS92c	35.2	14.6	27.4	2.2	2.9	2.5	6.3
	IS92d	54.4	25.5	43.4	4.1	5.4	4.6	10.5
	IS92e	67.4	38.3	56.9	5.9	7.7	6.6	13.8
	IS92f	43.9	21.5	35.8	3.3	4.1	3.6	8.1
	A1B*	50.1	29.3	44.2	14.9	17.5	15.9	20.8
	A2	34.6	7.1	26.1	2.6	6.0	3.9	7.2
	B1	49.8	14.3	39.1	9.0	13.6	10.9	15.6
	B2	39.2	16.3	32.5	8.9	6.9	8.1	11.7
2100	IS92a,b	85.9	40.6	69.5	15.0	14.2	14.6	21.5
	IS92c	49.2	17.6	36.5	6.4	5.8	6.1	10.1
	IS92d	113.9	51.3	88.8	20.3	17.7	19.1	28.2
	IS92e	150.6	96.6	131.0	34.6	33.0	33.8	46.0
	IS92f	69.7	31.3	54.9	11.9	10.7	11.4	16.8
	A1B*	109.2	100.9	107.3	71.9	60.9	66.5	74.9
	A2	58.5	20.2	46.6	7.8	15.2	11.0	16.1
	B1	79.7	52.2	72.8	35.7	44.9	40.2	46.6
	B2	61.0	38.3	54.4	19.5	16.1	18.0	22.6

* The two additional illustrative scenarios A1F1 and AIT share similar assumptions with A1B. See also Appendix VII for more details

possibilities for development "catch-up" and for narrowing per capita income gaps in particular explain the wide range in projected future gross world product levels. Given a qualitatively negative relationship between population growth and per capita income growth discussed in Chapters 2 and 3, uncertainties in future population growth rates tend to narrow the range of associated gross world product projections. High population growth would, *ceteris paribus*, lower per capita income growth, whereas low population growth would tend to increase it. This relationship is evident in empiric data – high per capita income countries are generally also those that have completed their demographic transition. The affluent live long and generally have few children. (Exceptions are some countries with small populations, high birth rates, and significant income from commodity exports.) This relationship between affluence and longevity again identifies development as one of the most important indicators of human well being. Yet even assuming this relationship holds for an extended time into the future, its quantification is subject to considerable theoretic and empiric uncertainties (Alcamo *et al.*, 1995).

Two of the SRES scenario families, A1 and B1, explicitly explore alternative pathways to gradually close existing income gaps. As a reflection of uncertainty, development catch-up diverges in terms of geographically distinct economic growth patterns across the four SRES scenario families. Table TS-1 summarizes per capita income for SRES and IS92 scenarios for the four SRES world regions. SRES scenarios indicate a smaller difference between the now industrialized and developing countries compared with the IS92 scenarios. This tendency toward a substantially narrower income "gap" compared with the IS92 scenarios overcomes one of the major shortcomings of the previous IPCC scenarios cited in the literature (Parikh, 1992).

8.3. *Structural and Technological Change*

In this brief summary of the SRES scenarios, structural and technological changes are illustrated by using energy and land use as examples. These examples are characteristic for the driving forces of emissions because the energy system and land use are the major sources of GHG and sulfur emission.

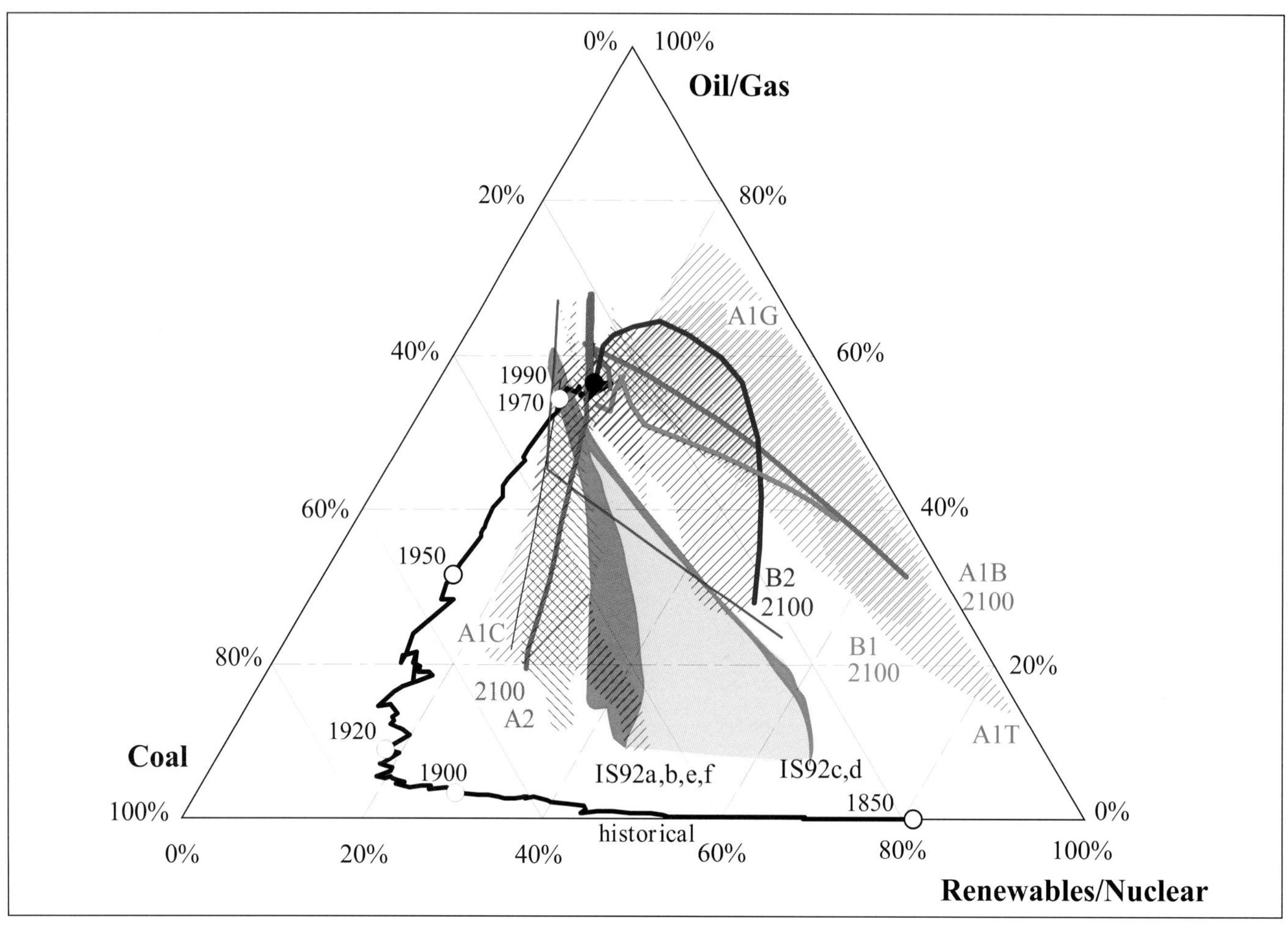

Figure TS-4: Global primary energy structure, shares (%) of oil and gas, coal, and non-fossil (zero-carbon) energy sources – historical development from 1850 to 1990 and in SRES scenarios. Each corner of the triangle corresponds to a hypothetical situation in which all primary energy is supplied by a single source – oil and gas on the top, coal to the left, and non-fossil sources (renewables and nuclear) to the right. Constant market shares of these energies are denoted by their respective isoshare lines. Historical data from 1850 to 1990 are based on Nakićenović *et al.* (1998). For 1990 to 2100, alternative trajectories show the changes in the energy systems structures across SRES scenarios. They are grouped by shaded areas for the scenario families A1B, A2, B1, and B2 with respective markers shown as lines. In addition, the four scenario groups within the A1 family A1B, A1C, A1G, and A1T, which explore different technological developments in the energy systems, are shaded individually. In the SPM, A1C and A1G are combined into one fossil-intensive group A1FI. For comparison the IS92 scenario series are also shown, clustering along two trajectories (IS92c,d and IS92a,b,e,f). For model results that do not include non-commercial energies, the corresponding estimates from the emulations of the various marker scenarios by the MESSAGE model were added to the original model outputs.

Chapter 4 gives a more detailed treatment of the full range of emissions driving forces across the SRES scenarios.

Figure TS-4 illustrates that the change of world primary energy structure diverges over time. It shows the contributions of individual primary energy sources – the percentage supplied by coal, that by oil and gas, and that by all non-fossil sources taken together (for simplicity of presentation and because not all models distinguish between renewables and nuclear energy). Each corner of the triangle corresponds to a hypothetical situation in which all primary energy is supplied by a single source – oil and gas at the top, coal to the left, and non-fossil sources (renewables and nuclear) to the right. Historically, the primary energy structure has evolved clockwise according to the two "grand transitions" (discussed in Chapter 3) that are shown by the two segments of the "thick black" curve. From 1850 to 1920 the first transition can be characterized as the substitution of traditional (non-fossil) energy sources by coal. The share of coal increased from 20% to about 70%, while the share of non-fossils declined from 80% to about 20%. The second transition, from 1920 to 1990, can be characterized as the replacement of coal by oil and gas (while the share of non-fossils remained essentially constant). The share of oil and gas increased to about 50% and the share of coal declined to about 30%.

Figure TS-4 gives an overview of the divergent evolution of global primary energy structures between 1990 and 2100, regrouped into their respective scenario families and four A1 scenarios groups (three in the SPM) that explore different technological developments in the energy systems. The SRES scenarios cover a wider range of energy structures than the previous IS92 scenario series, which reflects advances in knowledge on the uncertainty ranges of future fossil resource availability and technological change.

In a clockwise direction, A1B, A1T, and B1 scenario groups map the structural transitions toward higher shares of non-fossil energy in the future, which almost closes the historical "loop" that started in 1850. The B2 scenarios indicate a more "moderate" direction of change with about half of the energy coming from non-fossil sources and the other half shared by coal on one side and oil and gas on the other. Finally, the A2 scenario group marks a stark transition back to coal. Shares of oil and gas decline while non-fossils increase moderately. What is perhaps more significant than the diverging developments in these three marker scenarios is that the whole set of 40 scenarios covers virtually all possible directions of change, from high shares of oil and gas to high shares of coal and non-fossils. In particular, the A1 scenario family covers basically the same range of structural change as all the other scenarios together. In contrast, the IS92 scenarios cluster into two groups, one of which contains IS92c and IS92d and the other the four others. In all of these the share of oil and gas declines, and the main structural change occurs between coal on the one hand and non-fossils on the other. This divergent nature in the structural change of the energy system and in the underlying technological base of the SRES results in a wide span of future GHG and sulfur emissions.

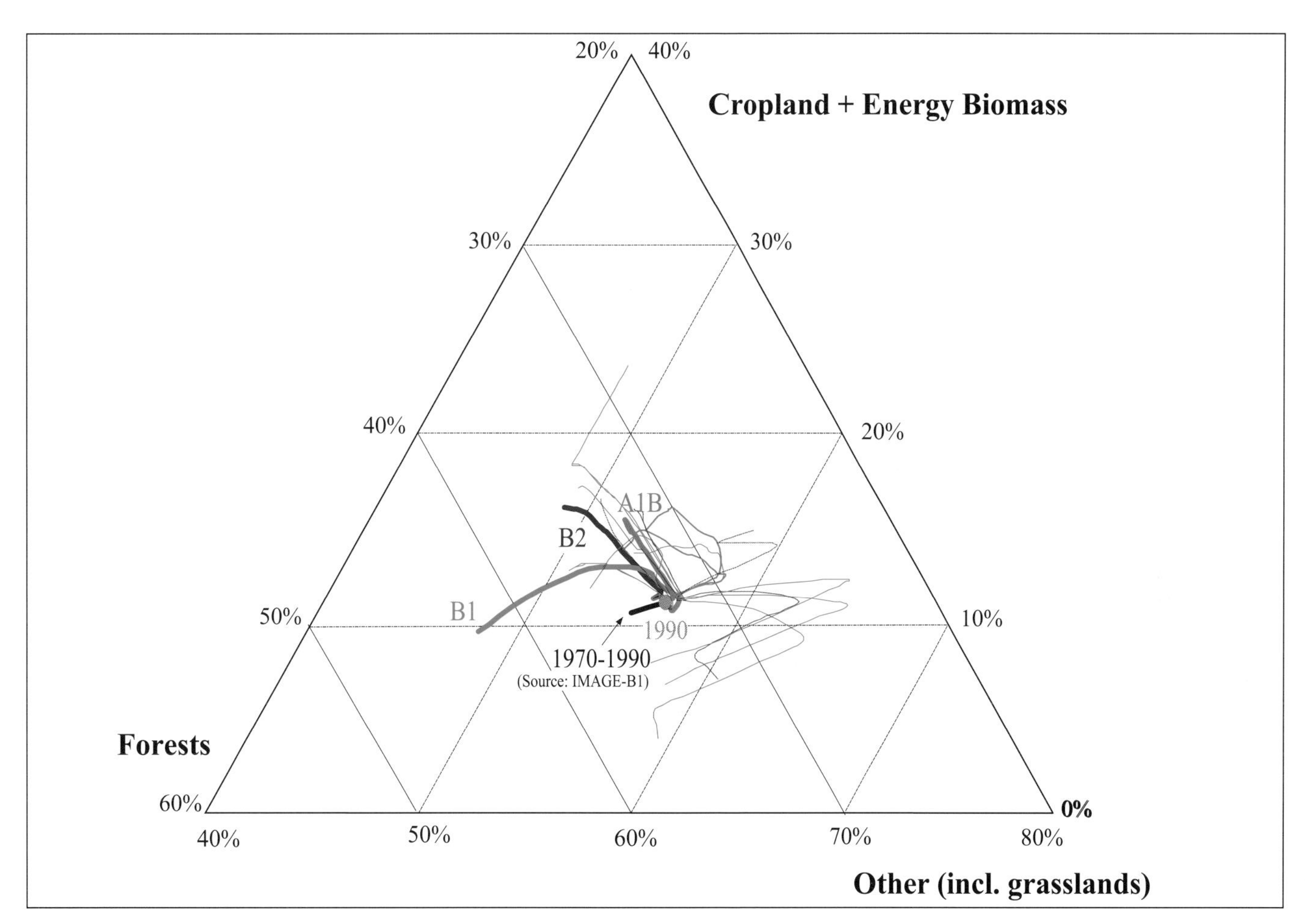

Figure TS-5: Global land-use patterns, shares (%) of croplands and energy biomass, forests, and other categories including grasslands – historical development from 1970 to 1990 (based on B1-IMAGE) and in SRES scenarios. As for the energy triangle in Figure 6-3, each corner corresponds to a hypothetical situation in which land use is dedicated to a much greater extent than today to one category – 60% to cropland and energy biomass at the top, 80% to forests to the left, and 80% to other categories (including grasslands) to the right. Constant shares in total land area of cropland and energy biomass, forests, and other categories are denoted by their respective isoshare lines. For 1990 to 2100, alternative trajectories are shown for the SRES scenarios. The three marker scenarios A1B, B1, and B2 are shown as thick colored lines, and other SRES scenarios as thin colored lines. The ASF model used to develop the A2 marker scenario projects only land-use change related GHG emissions. Comparable data on land cover changes are therefore not available.The trajectories appear to be largely model specific and illustrate the different views and interpretations of future land-use patterns across the scenarios (e.g. the scenario trajectories on the right that illustrate larger increases in grasslands and decreases in cropland are MiniCAM results).

Figure TS-5 illustrates that land-use patterns are also diverging over time. It shows the main land-use categories – the percentages of total land area use that constitute the forests, the joint shares of cropland and energy biomass, and all the other categories including grasslands. As for the energy triangle in Figure TS-4, in Figure TS-5 each corner corresponds to a hypothetical situation in which land use is dedicated to a much greater extent than today to two of the three land-use categories – 40% to cropland and energy biomass and 20% to forests at the top, 60% to forests and 40% to other categories (including grasslands) to the left, and 80% to other categories (including grasslands) to the right.

In most scenarios, the current trend of shrinking forests is eventually reversed because of slower population growth and increased agricultural productivity. Reversals of deforestation trends are strongest in the B1 and A1 families. In the B1 family pasture lands decrease significantly because of increased productivity in livestock management and dietary shifts away from meat, thus illustrating the importance of both technological and social developments.

The main driving forces for land-use changes are related to increasing demands for food because of a growing population and changing diets. In addition, numerous other social, economic, and institutional factors govern land-use changes such as deforestation, expansion of cropland areas, or their reconversion back to forest cover (see Chapter 3). Global food production can be increased, either through intensification (by multi-cropping, raising cropping intensity, applying fertilizers, new seeds, improved farming technology) or through land expansion (cultivating land, converting forests). Especially in developing countries, there are many examples of the potential to intensify food production in a more or less ecological way (e.g. multi-cropping; agroforestry) that may not lead to higher GHG emissions.

Different assumptions on these processes translate into alternative scenarios of future land-use changes and GHG emissions, most notably CO_2, methane (CH_4), and nitrous oxide (N_2O). A distinguishing characteristic of several models (e.g., AIM, IMAGE, MARIA, and MiniCAM) used in SRES is the explicit modeling of land-use changes caused by expanding biomass uses and hence exploration of possible land-use conflicts between energy and agricultural sectors. The corresponding scenarios of land-use changes are illustrated in Figure TS-5 for all SRES scenarios. In some contrast to the structural changes in energy systems shown in Figure TS-4, different land-use scenarios in Figure TS-5 appear to be rather model specific, following the general trends as indicated by the respective marker scenario developed with a particular model.

9. Greenhouse Gases and Sulfur Emissions

The SRES scenarios generally cover the full range of GHG and sulfur emissions consistent with the storylines and the underlying range of driving forces from studies in the literature, as documented in the SRES database. This section summarizes the emissions of CO_2, CH_4, and SO_2. For simplicity, only these three important gases are presented separately, following the more detailed exposition in Chapter 5 (see Table TS-4 for a summary of the ranges of emissions across the scenario groups).

9.1. Carbon Dioxide Emissions

9.1.1. Carbon Dioxide Emissions and Their Driving Forces

Figure TS-6 illustrates the CO_2 emissions across the SRES scenarios in relation to each of the three main scenario driving forces – global population, gross world product and primary energy requirements. The general tendencies across the driving forces are consistent with the underlying literature. All else being equal, the higher future global populations, higher gross world product, or higher primary energy requirements would be associated with higher emissions. However, it is important to note that the range of emissions is large across the whole range of driving forces considered in SRES, indicating the magnitude of the uncertainty associated with emission scenarios. For instance, emissions can range widely for any given level of future population (e.g. between 5 to 20 GtC in case of a low population scenario of seven billion by 2100). Conversely, emissions in the range of 20 GtC are possible with global population levels ranging from seven to 15 billion by 2100. While the SRES scenarios do not map all possibilities, they do indicate general tendencies, with an uncertainty range consistent with the underlying literature. This emphasizes an important SRES conclusion: alternative combinations of main scenario driving forces can lead to similar levels of GHGs emissions by the end of the 21st century. Alternatively, similar future worlds with respect to socio-economic developments can result in wide differences in future GHGs emissions, primarily as a result of alternative technological developments. This suggests that technology is at least as important a driving force of future GHG emissions as population and economic development across the set of 40 SRES scenarios.

9.1.2. Carbon Dioxide Emissions from Energy, Industry, and Land Use

Figure TS-7 illustrates the range of CO_2 emissions of the SRES scenarios against the background of all the IS92 scenarios and other emissions scenarios from the literature documented in the SRES scenario database (blue shaded area). The range of future emissions is very large so that the highest scenarios envisage a tenfold increase of global emissions by 2100 while the lowest have emissions lower than today.

The literature includes scenarios with additional climate initiatives and policies, which are also referred to as mitigation or intervention scenarios. As shown in Chapter 2, many ambiguities are associated with the classification of emissions scenarios into those that include additional climate initiatives and those that do not. Many cannot be classified in this way on

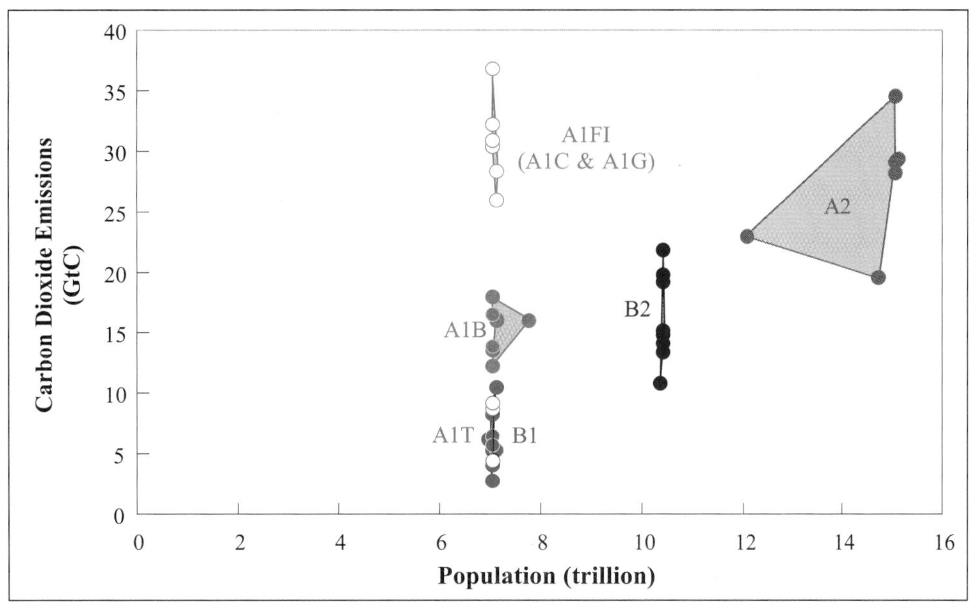

Figure TS-6a: Global carbon dioxide emissions (standardized) across SRES scenarios in relation to global population in 2100 for the four scenario families and six scenario groups. A1C and A1G have been combined into one fossil-intensive group A1F1. Shaded areas indicate scenario space for each scenario family and scenario group (in A1) (see Chapters 4 and 5).

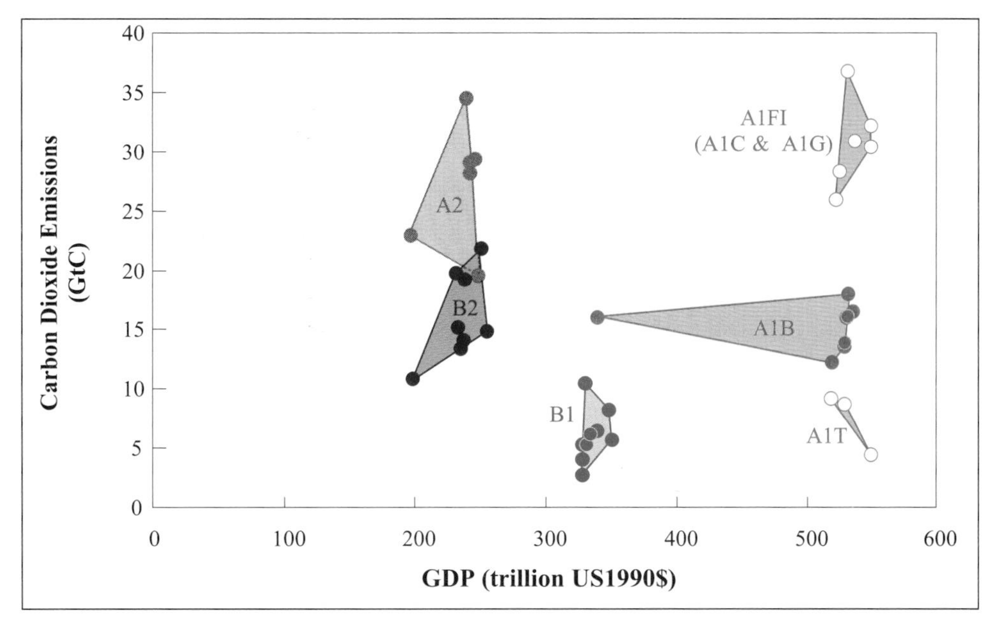

Figure TS-6b: Global carbon dioxide emissions (standardized) across SRES scenarios in relation to gross world product in 2100 for the four scenario families and six scenario groups. A1C and A1G have been combined into one fossil-intensive group A1F1. Shaded areas indicate scenario space for each scenario family and scenario group (in A1) (see Chapters 4 and 5).

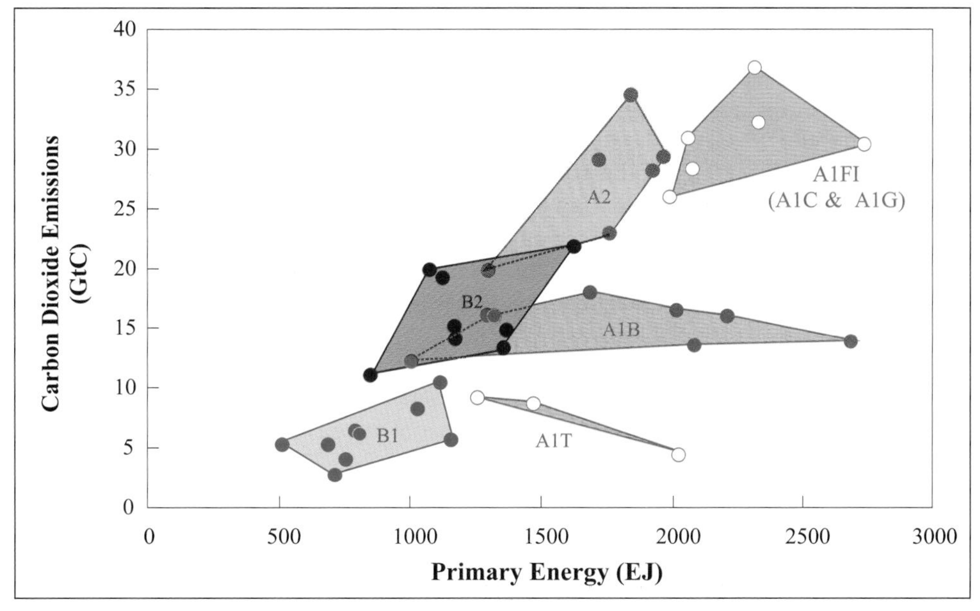

Figure TS-6c: Global carbon dioxide emissions (standardized) across SRES scenarios in relation to global primary energy requirements in 2100 for the four scenario families and six scenario groups. A1C and A1G have been combined into one fossil-intensive group A1F1. Shaded areas indicate scenario space for each scenario family and scenario group (in A1) (see Chapters 4 and 5).

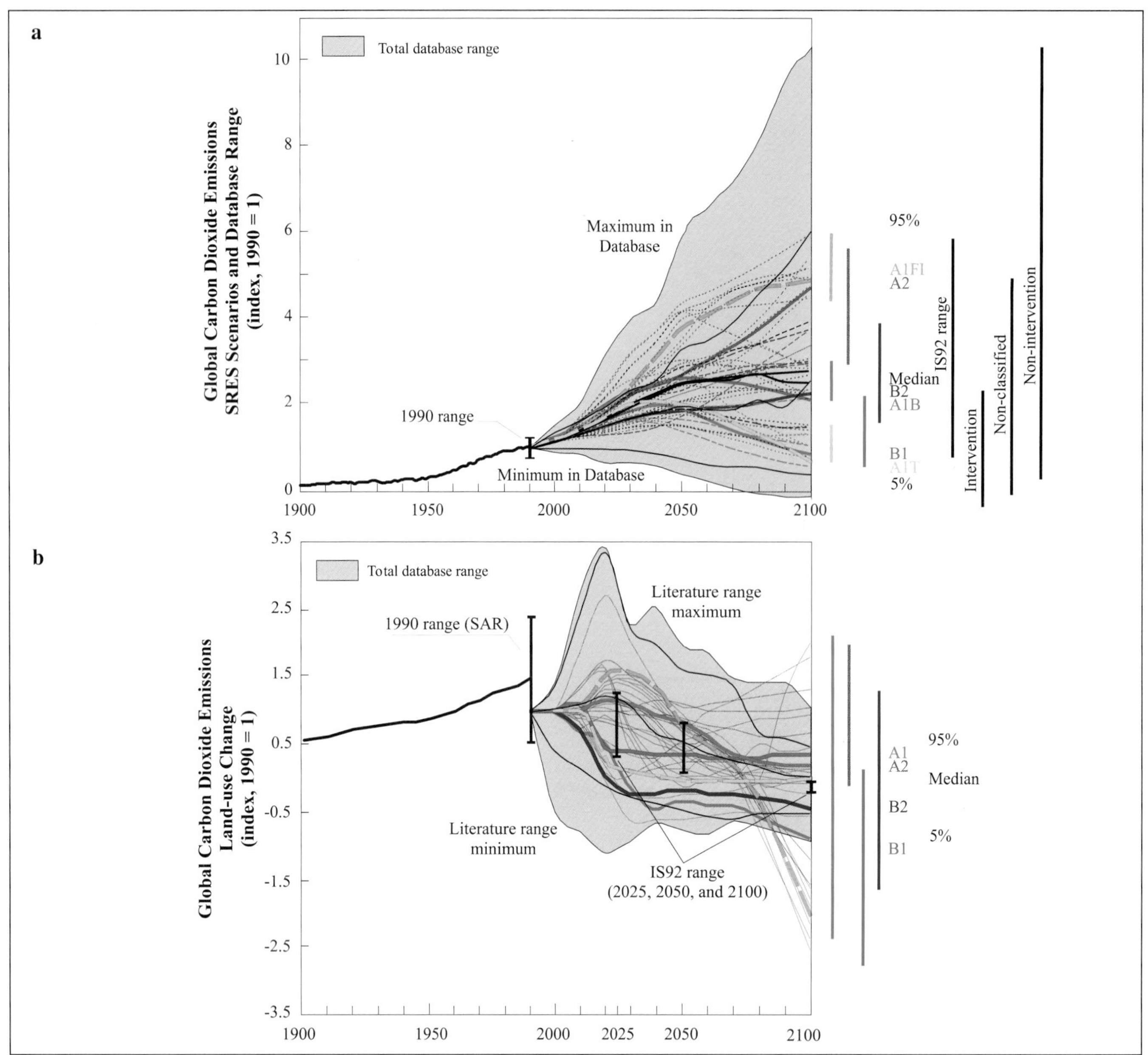

Figure TS-7: Global CO_2 emissions from energy and industry in Figure TS-7a and from land-use change in Figure TS-7b – historical development from 1900 to 1990 and in 40 SRES scenarios from 1990 to 2100, shown as an index (1990 = 1). The range is large in the base year 1990, as indicated by an "error" bar, but is excluded from the indexed future emissions paths. The dashed time-paths depict individual SRES scenarios and the blue shaded area the range of scenarios from the literature (as documented in the SRES database). The median (50th), 5th, and 95th percentiles of the frequency distribution are shown. The statistics associated with the distribution of scenarios do not imply probability of occurrence (e.g., the frequency distribution of the scenarios in the literature may be influenced by the use of IS92a as a reference for many subsequent studies). The 40 SRES scenarios are classified into six groups (that result after A1C and A1G are combined into one fossil-intensive group A1FI, as in the SPM), which constitute four scenario families and three A1 scenario groups. Jointly the scenarios span most of the range of the scenarios in the literature. The emissions profiles are dynamic, ranging from continuous increases to those that curve through a maximum and then decline. The colored vertical bars indicate the range of the four SRES scenario families in 2100. Also shown as vertical bars on the right of Figure TS-7a are the ranges of emissions in 2100 of IS92 scenarios and of scenarios from the literature that apparently include additional climate initiatives (designated as "intervention" scenarios emissions range), those that do not ("non-intervention"), and those that cannot be assigned to either of these two categories ("non-classified").[5] Three vertical bars in Figure TS-7b indicate the range of IS92 land-use emissions in 2025, 2050, and 2100.

[5] This classification is based on a subjective evaluation of the scenarios in the database by the members of the writing team and is explained in Chapter 2. It was not possible to develop an equivalent classification for land-use emissions scenarios.

basis of the information available from the SRES scenario database and the published literature.

Figure TS-7a indicates the ranges of emissions in 2100 from scenarios that apparently include additional climate initiatives (designated as "intervention" emissions range), those that do not ("non-intervention"), and those that cannot be assigned to either of these two categories ("non-classified"). This classification is based on the subjective evaluation of the scenarios in the database by the members of the writing team and is explained in Section 5 and in Chapter 2 in greater detail. It should be noted that the distributions of emissions of scenarios from the literature is asymmetric (see the emissions histogram in Figure 6-5 in Chapter 6) and that the thin tail that extends above 30 GtC by 2100 includes only a few scenarios.

Figure TS-7a shows the ranges of emissions of the four families (vertical bars next to each of the four marker scenarios), which illustrate that the scenario groups by themselves cover a large portion of the overall scenario distribution. Together, they cover much of the range of future emissions, both with respect to the scenarios in the literature and all SRES scenarios. Adding all other scenarios increases the covered range. For example, the SRES scenarios span jointly from the 95th percentile to just above the 5th percentile of the distribution of energy and industry emissions scenarios from the literature. This illustrates again that they only exclude the most extreme emissions scenarios found in the literature that are situated out in the tails of the distribution. What is perhaps more important is that each of the four scenario families covers a substantial part of this distribution. This leads to a substantial overlap in the emissions ranges of the four scenario families. In other words, a similar quantification of driving forces can lead to a wide range of future emissions and a given level of future emissions can result from different combinations of driving forces. This result is of fundamental importance for the assessments of climate change impacts and possible mitigation and adaptation strategies. Thus, it warrants some further discussion.

Another interpretation is that a given combination of the main driving forces, such as the population and economic growth, is not sufficient to determine the future emissions paths. Different modeling approaches and different specifications of other scenario assumptions overshadow the influence of the main driving forces. A particular combination of driving forces, such as specified in the A1 scenario family, is associated with a whole range of possible emission paths from energy and industry. The nature of climate change impacts and adaptation and mitigation strategies would be fundamentally different depending on whether emissions are high or low, given a particular combination of scenario driving forces. Thus, the implication is that the whole range needs to be considered in the assessments of climate change, from high emissions and driving forces to low ones.

The A1 scenario family explored variations in energy systems most explicitly and hence covers the largest part of the scenario distribution shown in Figure TS-7a, from the 95th to just above the 10th percentile. The A1 scenario family includes four groups of scenarios that explore different structures of future energy systems, from carbon-intensive development paths to high rates of decarbonization. Two of the fossil-intensive groups were merged into one group, as in SPM, resulting in three A1 groups. All A1 groups otherwise share the same assumptions about the main driving forces (see Chapter 6 and for further detail Chapters 4 and 5). This indicates that different structures of the energy system can lead to basically the same variation in future emissions as generated by different combinations of the other main driving forces – population, economic activities, and energy consumption levels. The implication is that decarbonization of energy systems – the shift from carbon-intensive to less carbon-intensive and carbon-free sources of energy – is of similar importance in determining the future emissions paths as other driving forces. Sustained decarbonization requires the development and successful diffusion of new technologies. Thus investments in new technologies during the coming decades might have the same order of influence on future emissions as population growth, economic development, and levels of energy consumption taken together.

Figure TS-7b shows that CO_2 emissions from deforestation peak in many SRES scenarios after several decades and subsequently gradually decline. This pattern is consistent with many scenarios in the literature and can be associated with slowing population growth and increasing agricultural productivity. These allow a reversal of current deforestation trends, leading to eventual CO_2 sequestration. Emissions decline fastest in the B1 family. Only in the A2 family do net anthropogenic CO_2 emissions from land use remain positive through to 2100. As was the case for energy-related emissions, CO_2 emissions related to land-use in the A1 family cover the widest range. The range of land-use emissions across the IS92 scenarios is narrower in comparison.

9.1.3. Scenario Groups and Four Categories of Cumulative Emissions

This comparison of SRES scenario characteristics implies that similar future emissions can result from very different socio-economic developments, and similar developments of driving forces can result in different future emissions. Uncertainties in the future development of key emission driving forces create large uncertainties in future emissions even within the same socio-economic development paths. Therefore, emissions from each scenario family overlap substantially with emissions from other scenario families.

To facilitate comparisons of emissions and their driving forces across the scenarios, the writing team grouped them into four categories of cumulative emissions between 1990 and 2100. However, any categorization of scenarios based on emissions of multiple gases is quite difficult. Figure TS-8 shows total CO_2 emissions from all sources (from Figures TS-7a and b). The 40 scenarios are shown aggregated into six groups, three for the A1

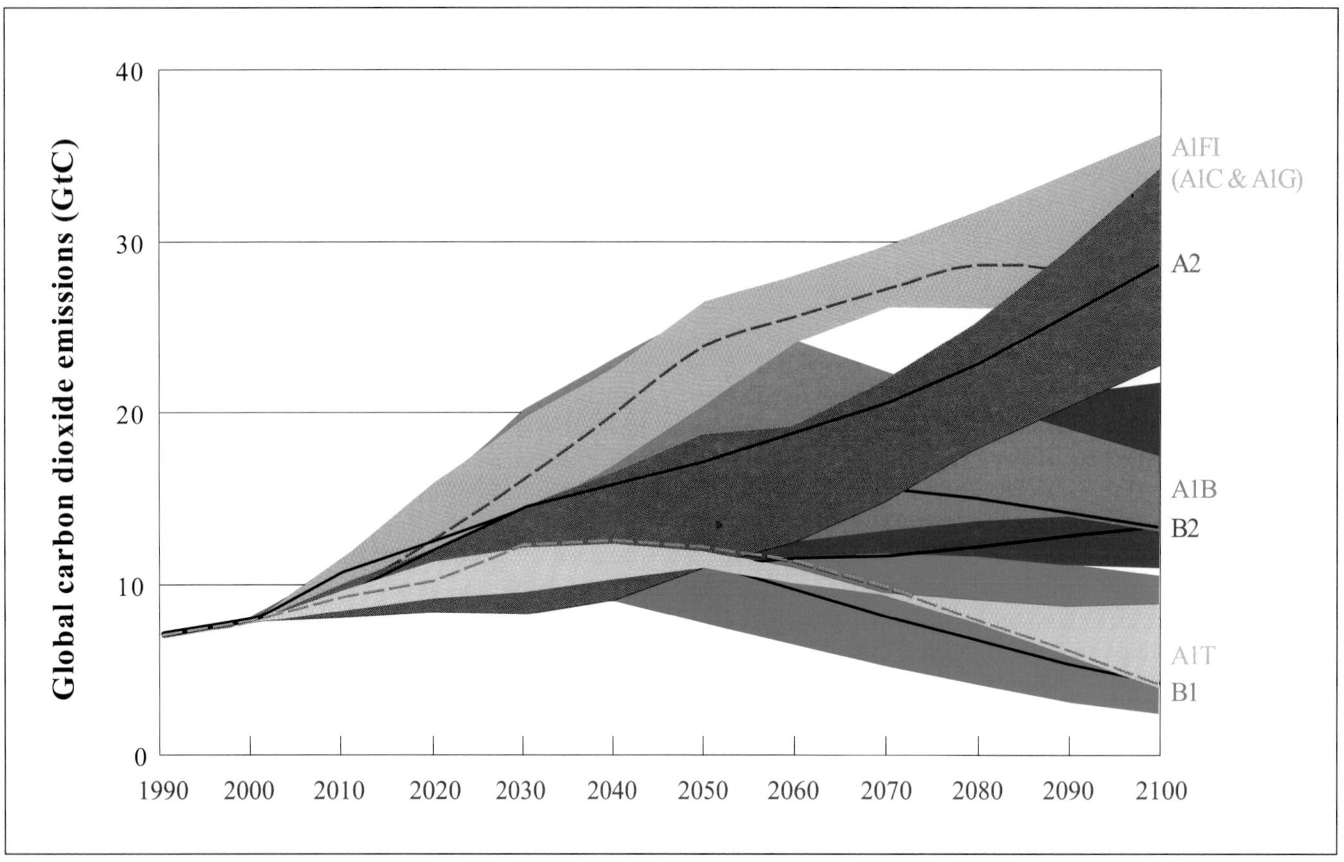

Figure TS-8: Global CO_2 emissions (GtC/yr, standardized) from all sources for the four scenario families from 1990 to 2100. Scenarios are also presented for the three constituent groups of the A1 family (fossil-intensive A1FI group, resulting by merging A1C and A1G as in the SPM, the high non-fossil fuel A1T, and the balanced A1B) and for the other three families (A2, B1, and B2), forming six scenario groups altogether. Each colored emission band shows the range of the scenarios within one group that share common global input assumptions for population and GDP. The scenarios remaining outside the six groups adopted alternative interpretations of the four scenario storylines.

family (that result by merging the two fossil-intensive groups A1C and A1G into one A1FI group as in the SPM). The scenarios that remain outside the six groups adopted alternative interpretations of the four scenario storylines. The emission trajectories ("bands") of the scenario groups display different dynamics, from monotonic increases to non-linear trajectories in which there is a subsequent decline from a maximum. The dynamics of the individual scenarios are also different across gasses, sectors, or world regions. This particularly diminishes the significance of focusing scenario categorization on any given year, such as 2100. In addition, all gases that contribute to radiative forcing should be considered, but methods of combining gases such as the use of global warming potentials (GWPs) are appropriate only for near-term GHG inventories[6]. In light of these difficulties, the classification approach presented here uses cumulative CO_2 emissions between 1990 and 2100. CO_2 is the dominant GHG and cumulative CO_2 emissions are expected to be roughly proportional to CO_2 radiative forcing over the time scale of a century. According to the IPCC SAR, "any eventual stabilized concentration is governed more by the accumulated anthropogenic CO_2 emissions from now until the time of stabilization than by the way emissions change over the period" (Houghton *et al.*, 1996). Therefore, the writing team also grouped the scenarios according to their cumulative emissions.

Cumulative SRES carbon emissions through to 2100 range from less than 800 GtC to more than 2500 GtC with a median of about 1500 GtC. To represent this range, the scenario classification uses four intervals as follows: less than 1100 GtC (low), between 1100 and 1450 GtC (medium–low), between 1450 and 1800 GtC (medium–high), and greater than 1800 GtC (high). Each CO_2 interval contains multiple scenarios and scenarios from more than one family. Each category also includes one of the four marker scenarios. Figure TS-9 shows how cumulative carbon emissions from the 40 SRES scenarios fit within the selected emission intervals (see Chapter 4 for further details). The 40 SRES scenarios extend the IS92 range

[6] In particular, in the IPCC WGI Second Assessment Report (SAR) GWPs are calculated for constant concentrations (Houghton *et al.*, 1996). In long-term scenarios, concentrations may change significantly, as do GWP values. It is unclear how to apply GWPs to long-term scenarios in a meaningful manner. In addition, the GWP approach is not applicable to gases such as SO_2 and ozone precursors.

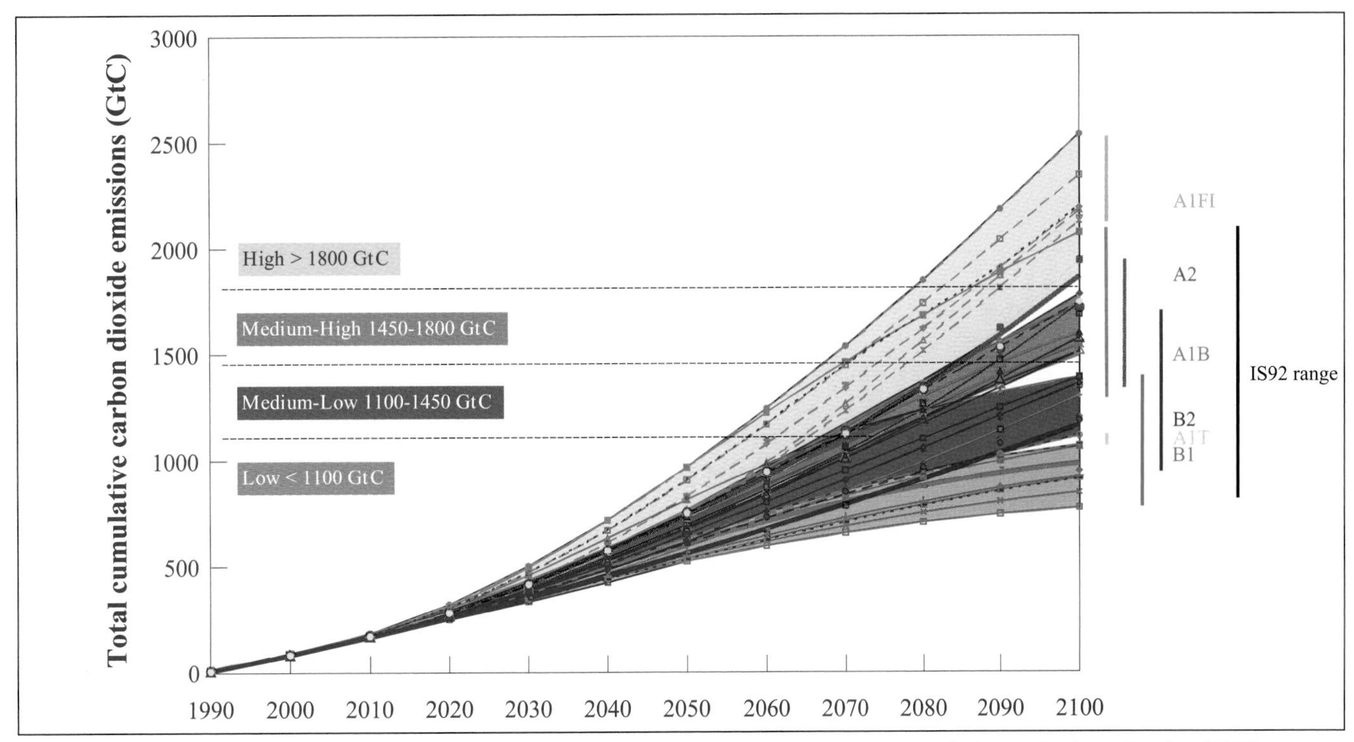

Figure TS-9: Global cumulative CO_2 emissions (GtC, standardized). The ranges of cumulative emissions for the SRES scenarios are shown. Scenarios are grouped into four categories: low, medium–low, medium–high, and high emissions. Each category contains one marker scenario plus alternatives that lead to comparable cumulative emissions, although often through different driving forces. The ranges of cumulative emissions of the six SRES scenario groups are shown as colored vertical bars and the range of the IS92 scenarios as a black vertical bar.

toward higher emissions (SRES maximum of 2570 GtC compared to 2140 GtC for IS92), but not toward lower emissions. The lower bound for both scenario sets is just below 800 GtC.

This categorization can guide comparisons using either scenarios with different driving forces yet similar emissions, or scenarios with similar driving forces but different emissions. This characteristic of SRES scenarios also has very important implications for the assessment of climate-change impacts, mitigation, and adaptation strategies. Two future worlds with fundamentally different characteristic features, such as the A1B and B2 marker scenarios, also have different cumulative CO_2 emissions and radiative forcing, but very similar CO_2 emissions in 2100. In contrast, scenarios that are in the same category of cumulative emissions can have fundamentally different driving forces and different CO_2 emissions in 2100, but very similar cumulative emissions and radiative forcing. Presumably, adverse impacts and effective adaptation measures would vary among the scenarios from different families that share similar cumulative emissions but have different demographic, socio-economic, and technological driving forces. This is another reason for considering the entire range of emissions in future assessments of climate change.

9.2. *Other Greenhouse Gases*

Of the GHGs, CO_2 is the main contributor to anthropogenic radiative forcing because of changes in concentrations from pre-industrial times. According to Houghton *et al.* (1996) well-mixed GHGs (CO_2, CH_4, N_2O, and the halocarbons) induced additional radiative forcing of around 2.5 W/m^2 on a global and annually averaged basis. CO_2 accounted for 60% of the total, which indicates that the other GHGs are significant as well. Whereas CO_2 emissions are by-and-large attributable to two major sources, energy consumption and land-use change, other emissions arise from many different sources and a large number of sectors and applications (e.g. see Table 5-3 in Chapter 5).

The SRES emissions scenarios also have different emissions for other GHGs and chemically active species such as CO, NO_x, nitrogen oxides, and NMVOCs. The uncertainties that surround the emissions sources of the other GHGs, and the more complex set of driving forces behind them are considerable and unresolved. Therefore, the models and approaches employed for the SRES analyses cannot produce unambiguous and generally approved estimates for different sources and world regions over a century. Keeping these caveats above in mind, Table TS-4 (see later) shows the emissions of all relevant direct and indirect GHGs for the four marker scenarios and, in brackets, the range of the other scenarios in the same family (or scenario groups for the A1 family). Chapter 5 gives further detail about the full range of GHG emissions across the SRES scenarios. The emissions of other gases follow dynamic patterns much like those shown in Figures TS-7 and TS-8 for carbon dioxide emissions. A summary of GHG emissions is given in Chapter 6 and further details in Chapter 5.

9.2.1. *Methane Emissions*

Anthropogenic CH_4 emissions arise from a variety of activities, dominated by biologic processes, each associated with considerable uncertainty. The future CH_4 emissions in the scenarios depend in part on the consumption of fossil fuels, adjusted for assumed changes in technology and operational practices, but more strongly on scenario-specific, regional demographic and affluence developments, together with assumptions on preferred diets and agricultural practices. The writing team recommends further research into the sources and modeling approaches to capture large uncertainties surrounding future CH_4 emissions.

The resultant CH_4 emission trajectories for the four SRES scenario families portray complex patterns (as displayed in Figure 5-5 in Chapter 5). For example, the emissions in A2 and B2 marker scenarios increase throughout the whole time horizon to the year 2100. Increases are most pronounced in the high population A2 scenarios where emissions rise to between 549 and 1069 (A2 marker: 900) $MtCH_4$ by 2100, compared to 310 $MtCH_4$ in 1990. The emissions range by 2100 in the B2 scenarios is between 465 and 613 (B2 marker: 600) $MtCH_4$. In the A1B and B1 marker scenarios, the CH_4 emissions level off and subsequently decline sooner or later in the 21st century. This phenomenon is most pronounced in the A1B marker, in which the fastest growth in the first few decades is followed by the steepest decline; the 2100 level ends up slightly below the current emission of 310 $MtCH_4$. The range of emissions in Table TS-4 indicates that alternative developments in energy technologies and resources could yield a higher range in CH_4 emissions compared to the "balanced" technology A1B scenario group that includes the A1B marker scenario discussed above. In the fossil-intensive A1FI group (combined from A1C and A1G groups, as in the SPM), CH_4 emissions could reach some 735 $MtCH_4$ by 2100, whereas in the post-fossil A1T scenario group emissions are correspondingly lower (some 300 $MtCH_4$ by 2100). Interestingly, the A1 scenarios generally have comparatively low CH_4 emissions from non-energy sources because of a combination of low population growth and rapid advances in agricultural productivity. Hence the SRES scenarios extend the uncertainty range of the IS92 scenario series somewhat toward lower emissions. However, both scenario sets indicate an upper bound of emissions of some 1000 $MtCH_4$ by 2100.

9.2.2. *Nitrous Oxide Emissions*

Even more than for CH_4, the assumed future food supply will be a key determinant of future N_2O emissions. Size, age structure, and regional spread of the global population will be reflected in the emission trajectories, together with assumptions on diets and improvements in agricultural practices. Other things being equal, N_2O emissions are generally highest in the high population scenario family A2. Importantly, as the largest anthropogenic source of N_2O (cultivated soils) is already very uncertain in the base year, all future emission trajectories are affected by large uncertainties, especially if calculated with different models as is the case in this SRES report. Therefore, the writing team recommends further research into the sources and modeling of long-term N_2O emissions. Uncertainty ranges are correspondingly large, and are sometimes asymmetric. For example, while the range in 2100 reported in all A1 scenarios is between 5 and 10 MtN (7 MtN in the A1B marker), the A2 marker reports 17 MtN in 2100. Other A2 scenarios report emissions that fall within the range reported for A1 (from 8 to 19 MtN in 2100). Thus, different model representations of processes that lead to N_2O emissions and uncertainties in source strength can outweigh easily any underlying differences between individual scenarios in terms of population growth, economic development, etc. Different assumptions with respect to future crop productivity, agricultural practices, and associated emission factors, especially in the very populous regions of the world, explain the very different global emission levels even for otherwise shared main scenario drivers. Hence, the SRES scenarios extend the uncertainty range of future emissions significantly toward higher emissions (4.8 to 20.2 MtN by 2100 in SRES compared to 5.4 to 10.8 MtN in the IS92 scenarios. (Note that natural sources are excluded in this comparison.)

9.2.3. *Halocarbons and Halogenated Compounds*

The emissions of halocarbons (chlorofluorocarbons (CFCs), hydrochlorofluorocarbons (HCFCs), halons, methylbromide, and hydrofluorocarbons (HFCs)) and other halogenated compounds (polyfluorocarbons (PFCs) and sulfur hexafluoride (SF_6)) across the SRES scenarios are described in detail on a substance-by-substance basis in Chapter 5 and Fenhann (2000). However, none of the six SRES models has its own projections for emissions of ozone depleting substances (ODSs), their detailed driving forces, and their substitutes. Hence, a different approach for scenario generation was adopted.

First, for ODSs, an external scenario, the Montreal Protocol scenario (A3, maximum allowed production) from WMO/UNEP (1998) is used as direct input to SRES. In this scenario corresponding emissions decline to zero by 2100 as a result of international environmental agreements, a development not yet anticipated in some of the IS92 scenarios (Pepper *et al.*, 1992). For the other gas species, most notably for CFC and HCFC substitutes, a simple methodology of developing different emission trajectories consistent with aggregate SRES scenario driving force assumptions (population, GDP, etc.) was developed. Scenarios are further differentiated as to assumed future technological change and control rates for these gases, varied across the scenarios consistently with the interpretation of the SRES storylines presented in Chapter 4 as well as the most recent literature.

Second, different assumptions about CFC applications as well as substitute candidates were developed. These were initially based on Kroeze and Reijnders (1992) and information given in Midgley and McCulloch (1999), but updated with the most recent information from the Joint IPCC/TEAP Expert Meeting on Options for the Limitation of Emissions of HFCs and PFCs (WMO/UNEP, 1999). An important assumption, on the basis of the latest information from the industry, is that relatively few Montreal gases will be replaced fully by HFCs. Current indications are that substitution rates of CFCs by HFCs will be less than 50% (McCulloch and Midgley, 1998). In Fenhann (2000) a further technological development is assumed that would result in about 25% of the CFCs ultimately being substituted by HFCs (see Table 5-9 in Chapter 5). This low percentage not only reflects the introduction of non-HFC substitutes, but also the notion that smaller amounts of halocarbons will be used in many applications when changing to HFCs (efficiency gains with technological change). A general assumption is that the present trend, not to substitute with high GWP substances (including PFCs and SF_6), will continue. As a result of this assumption, the emissions reported here may be underestimates. This substitution approach is used in all four scenarios, and the technological options adopted are those known at present. Further substitution away from HFCs is assumed to require a climate policy and is therefore not considered in SRES scenarios. The range of emissions of HFCs in the SRES scenario is initially generally lower than in earlier IPCC scenarios because of new insights about the availability of alternatives to HFCs as replacements for substances controlled by the Montreal Protocol. In two of the four scenarios in the report, HFC emissions increase rapidly in the second half of the next century, while in two others the growth of emissions is significantly slowed down or reversed in that period.

Aggregating all the different halocarbons (CFCs, HCFCs, HFCs) as well as halogenated compounds (PFCs and SF_6) into MtC-equivalents (using GWPs from IPCC SAR, notwithstanding the caveats given in footnote 6) indicates a range between 386 and 1096 MtC-equivalent by 2100 for the SRES scenarios. This compares with a range of 746 to 875 MtC-equivalent for IS92 (which, however, does not include PFCs and SF_6). (The comparable SRES range excluding PFCs and SF_6 is between 299 and 753 MtC-equivalent by 2100.) The scenarios presented here indicate a wider range of uncertainty compared to IS92, particularly toward lower emissions (because of the technological and substitution reasons discussed above).

The effect on climate of each of the substances aggregated to MtC-equivalents given in Table TS-4 varies greatly, because of differences in both atmospheric lifetime and the radiative effect per molecule of each gas. The net effect on climate of these substances is best determined by a calculation of their radiative forcing – which is the amount by which these gases enhance the anthropogenic greenhouse effect. The radiative forcing will be addressed in IPCC TAR and is thus not discussed in this report.

9.3. Sulfur Dioxide Emissions

Emissions of sulfur portray even more dynamic patterns in time and space than the CO_2 emissions shown in Figures TS-7 and TS-8. Factors other than climate change (namely regional and local air quality, and transformations in the structure of the energy system and end use) intervene to limit future emissions. Figure TS-10 shows the range of global sulfur emissions for all SRES scenarios and the four markers against the emissions range of the IS92 scenarios, more than 80 scenarios from the literature, and the historical development.

A detailed review of long-term global and regional sulfur emission scenarios is given in Grübler (1998) and summarized in Chapter 3. The most important new finding from the scenario literature is recognition of the significant adverse impacts of sulfur emissions on human health, food production, and ecosystems. As a result, scenarios published since 1995 generally assume various degrees of sulfur controls to be implemented in the future, and thus have projections substantially lower than previous ones, including the IS92 scenario series. Of these, only the two low-demand scenarios IS92c and IS92d fall within the range of more recent long-term sulfur emission scenarios. A related reason for lower sulfur emission projections is the recent tightening of sulfur-control policies in the OECD countries, such as the Amendments of the Clean Air Act in the USA and the implementation of the Second European Sulfur Protocol. Such legislative changes were not reflected in previous long-term emission scenarios, as noted in Alcamo *et al.* (1995) and Houghton *et al.* (1995). Similar sulfur control initiatives due to local air quality concerns are beginning to impact sulfur emissions also in a number of developing countries in Asia and Latin America (see IEA, 1999; La Rovere and Americano, 1998; Streets and Waldhoff, 2000; for a more detailed discussion see Chapter 3). As a result, even the highest range of recent sulfur-control scenarios is significantly below that of comparable, high-demand IS92 scenarios (IS92a, IS92b, IS92e, and IS92f). The scenarios with the lowest ranges project stringent sulfur-control levels that lead to a substantial decline in long-term emissions and a return to emission levels that prevailed at the beginnings of the 20th century. The SRES scenario set brackets global anthropogenic sulfur emissions of between 27 and 169 MtS by 2050 and between 11 and 93 MtS by 2100 (see Table TS-4). In contrast, the range of the IS92 scenarios (Pepper *et al.*, 1992) is substantially higher starting at 80 MtS and extending all the way to 200 MtS by 2050 and from 55 to 230 MtS by 2100.

Reflecting recent developments and the literature, it is assumed that sulfur emissions in the SRES scenarios will also be controlled increasingly outside the OECD. As a result, both long-term trends and regional patterns of sulfur emissions evolve differently from carbon emissions in the SRES scenarios. As a general pattern, global sulfur emissions do not rise substantially, and eventually decline even in absolute terms during the second half of the 21st century, as indicated by the median of all scenarios in Figure TS-10 (see also Chapters 2 and 3). The spatial distribution of emissions changes markedly.

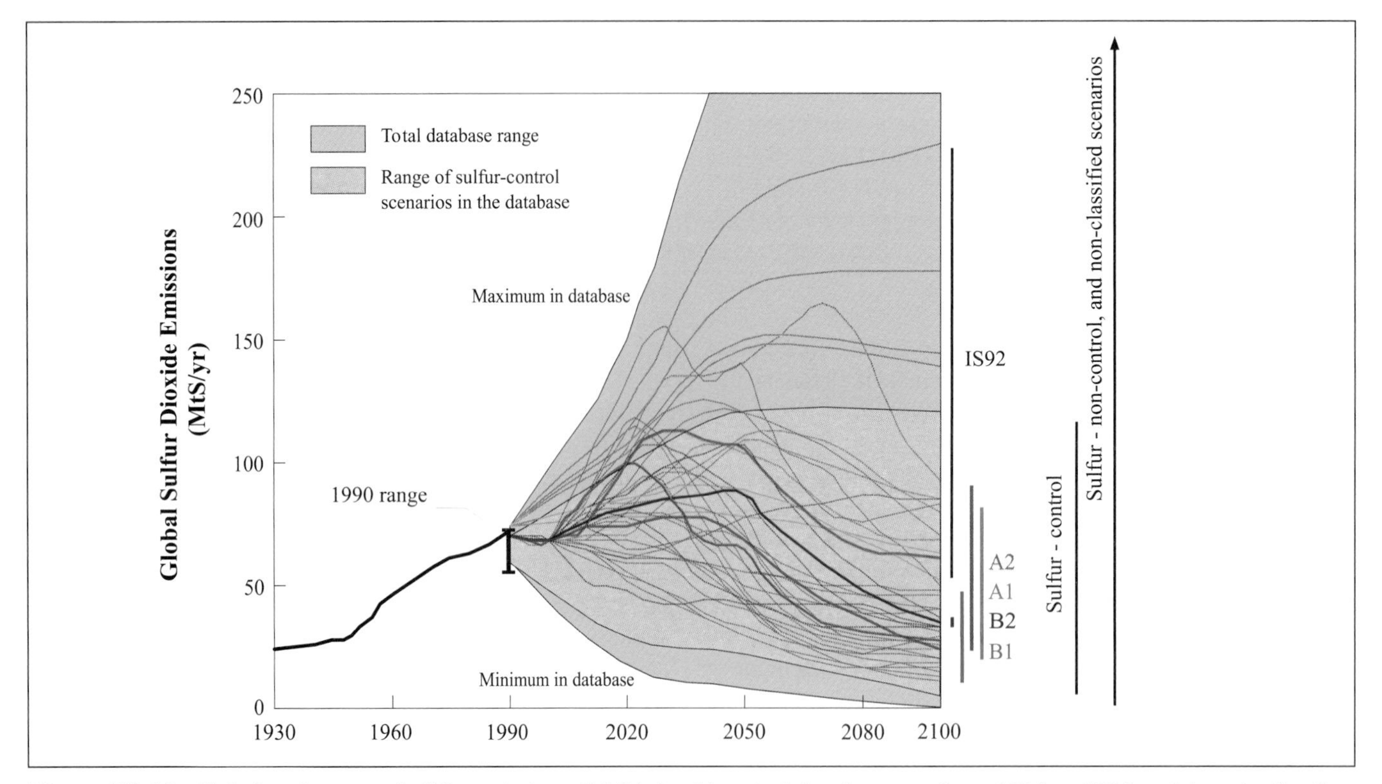

Figure TS-10: Global anthropogenic SO_2 emissions (MtS/yr) – historical development from 1930 to 1990 and (standardized) in the SRES scenarios. The dashed colored time-paths depict individual SRES scenarios, the solid colored lines the four marker scenarios, the solid thin curves the six IS92 scenarios, the shaded areas the range of 81 scenarios from the literature, the gray shaded area the sulfur-control and the blue shaded area the range of sulfur-non-control scenarios or "non-classified" scenarios from the literature that exceeds the range of sulfur control scenarios. The colored vertical bars indicate the range of the SRES scenario families in 2100. For details of the two additional illustrative A1 scenarios see Appendix VII. Database source: Grübler (1998).

Emissions in the OECD countries continue their recent declining trend (reflecting the tightening of control measures). Emissions outside the OECD rise initially, most notably in Asia, which compensates for the declining OECD emissions. Over the long term, however, sulfur emissions decline throughout the world, but the timing and magnitude vary across the scenarios. It should be noted that SRES scenarios assume sulfur controls *only* and do not assume any additional climate policy measures. Nevertheless, one important implication of this varying pattern of sulfur emissions is that the historically important, but uncertain, negative radiative forcing of sulfate aerosols may decline in the very long run. This view is also confirmed by the model calculations reported in Subak *et al.* (1997) and Nakićenović *et al.* (1998) based on recent long-term GHG and sulfur emission scenarios.

9.4. Other Chemically Active Gases

The SRES emissions scenarios also have different emissions for other GHGs and chemically active species such as CO, NO_x, and volatile organic compounds. The uncertainties that surround the emissions sources of these gases, and the more complex set of driving forces behind them are considerable and unresolved. Hence, model projections of these gases are particularly uncertain and the scenarios presented here are no exception. Improved inventories and studies linking driving forces to changing emissions in order to improve the representation of these gases in global and regional emission models remain an important future research task.

The emissions of other gases follow dynamic patterns much like those shown in Figure TS-7 for carbon dioxide emissions. Further details about GHG emissions are given in Chapter 5.

9.4.1. Nitrogen Oxides Emissions

Some models of the six SRES models do not provide a comprehensive description of NO_x emissions or include only specific sectors (e.g., energy-related sources) and have adopted other source categories from corresponding model runs derived from other models. Even with a simplified model representation, future NO_x emission levels are mainly determined by two set of variables: levels of fossil energy use (see Chapter 4), and level and timing of emission controls, inspired by local air quality concerns.

As a result the spread of NO_x emissions is largest within the A1 scenario family (28 to 151 MtN/yr by 2100), almost as large as the range across all 40 SRES scenarios (see Table TS-4). Only in the highest emission scenarios (the fossil fuel intensive scenarios within the A1 scenario family and the high population, coal intensive A2 scenario family) do emissions rise continuously throughout the 21st century. In the A1B ("balanced") scenario group and in the B2 scenario family, NO_x emission levels rise less. NO_x emissions tend to increase up to 2050 and stabilize thereafter, the result of a gradual substitution of fossil fuels by alternatives as well as of the increasing diffusion of NO_x control technologies. Low emission futures are described by various B1 family scenarios, as well as in the A1T scenario group, that describe futures in which NO_x emissions are controlled because of either local air quality concerns or rapid technological change away from conventional fossil technologies. Overall, the SRES scenarios describe a similar upper range of NO_x emissions as the previous IS92 scenarios (151 MtN versus 134 MtN, respectively, by 2100), but extend the IS92 uncertainty range toward lower emission levels (16 versus 54 MtN by 2100 in the SRES and IS92 scenarios, respectively).

9.4.2. *Volatile Organic Compounds, Excluding Methane*

NMVOCs arise from fossil fuel combustion (as with NO_x, wide ranges of emission factors are typical for internal combustion engines), and also from industrial processes, fuel storage (fugitive emissions), use of solvents (e.g., in paint and cleaners), and a variety of other activities. In this report NMVOCs are discussed as one group. As for NO_x emissions, not all models include the NMVOCs emissions category or all of its sources.

A relatively robust trend across all 40 scenarios (see Chapter 5) is a gradual increase in NMVOC emissions up to about 2050, with a range between 190 and 260 Mt. Beyond 2050, uncertainties increase with respect to both emission levels and trends. By 2100, the range is between 58 and 552 Mt, which extends the IS92 scenario range of 136 to 403 Mt by 2100 toward both higher and lower emissions (see Table TS-4). As for NO_x emissions, the upper bounds of NMVOC emissions are formed by the fossil fuel intensive scenarios within the A1 scenario family, and the lower bounds by the scenarios within the B1 scenario family. Characteristic ranges are between 60 and 90 Mt NMVOC by 2100 in the low emissions cluster and between 370 and 550 Mt NMVOC in the high emissions cluster. All other scenario families and individual scenarios fall between these two emissions clusters; the B2 marker scenario (B2-MESSAGE) closely tracks the median of global NMVOC emissions from all the SRES scenarios (see Chapter 5).

9.4.3. *Carbon Monoxide Emissions*

The same caveats as stated above for NO_x and NMVOC emissions also apply to CO emissions – the number of models that represent all the emission source categories is limited and modeling and data uncertainties, such as emission factors, are considerable. As a result, CO emission estimates across scenarios are highly model specific and future emission levels overlap considerably between the four SRES scenario families (see Table TS-4). Generally, emissions are highest in the high growth fossil fuel intensive scenarios within the A1 scenario family. Lowest emission levels are generally associated with the B1 and B2 scenario families. By 2100, emissions range between 363 and 3766 Mt CO, a considerably larger uncertainty range, particularly toward higher emissions, than in IS92, for which the 2100 emission range was between 450 and 929 Mt CO (see Table TS-4).

9.5. *Emissions Overview*

Table TS-4 (see later) summarizes the emissions of GHGs, sulfur dioxide and other radiatively active species by 2100 for the four markers and the ranges for other 36 scenarios. Combined with Tables TS-2 and TS-3, the tables provide a concise summary of the new SRES scenarios. Data are given for both the harmonized and all scenarios.

10. Summary, Conclusions, and Recommendations

In summary, the SRES scenarios lead to the following findings:

- Alternative combinations of driving forces can lead to similar levels and structure of energy and land-use patterns, as illustrated by different scenarios and groups. Hence, even for a given scenario outcome (e.g., in terms of GHG emissions) there are alternative combinations of driving forces and pathways that could lead to that outcome. For instance, significant global changes could result from a scenario of high population growth, even if per capita incomes rise only modestly, as well as from a scenario in which a rapid demographic transition (to low population levels) coincides with high rates of income growth and affluence.
- Important possibilities for further bifurcations in future development trends exist within one scenario family, even when particular values are adopted for the important scenario driving force variables to illustrate a particular development path. The technology scenario groups in the A1 family illustrate such alternative development paths with similar quantifications of the main driving forces.
- Emissions profiles are dynamic across the range of SRES scenarios. They portray trend reversals and indicate possible emissions crossover among different scenarios. They do not represent mere extensions of continuous increase of GHGs and SO_2 emissions into the future. This more complex pattern of future emissions across the range of SRES scenarios, time periods, world regions, and sectors reflects recent scenario literature.
- Describing potential future developments involves inherent ambiguities and uncertainties. One and only one possible development path (as alluded to, for instance, in concepts such as "business-as-usual scenario") simply

does not exist alone. And even for each alternative development path described by any given scenario, there are numerous combinations of driving forces and numeric values that can be consistent with a particular scenario description. The numeric precision of any model result should not distract from the basic fact that uncertainty abounds. However, the multi-model approach increases the value of the SRES scenario set, since uncertainties in the choice of model input assumptions can be separated more explicitly from the specific model behavior and related modeling uncertainties.

- Any scenario has subjective elements and is open to various interpretations. While the writing team as a whole has no preference for any of the scenarios, and has no judgment as to the probability or desirability of different scenarios, the open process and initial reactions to draft versions of this report show that individuals and interest groups do have such judgments. The writing team hopes that this will stimulate an open discussion in the policymaking arena about potential futures and choices that can be made in the context of climate change response. For the scientific community, the SRES scenario exercise has led to the identification of a number of recommendations for future research that can further increase the understanding of potential developments of socio-economic driving forces and their interactions, and the associated GHG emissions. A summary of the main findings and recommendations for potential users of the SRES scenarios is given in Boxes TS-4 and Box TS-5. The writing teams' suggestions for consideration by the IPCC are summarized in Box TS-6.
- Finally, the writing team believes that the SRES scenarios largely fulfill all specifications set out in Chapter 1. To support reproducibility, more detailed information than can be included in this report will be made available by individual modeling groups and members of the writing team through other means, such as web sites, peer-reviewed literature, or background documentation, if additional resources can be made available.

In conclusion, Tables TS-2, TS-3, and TS-4 summarize the main characteristics of the scenario groups that constitute the four families, both for the harmonized and for all scenarios. Tables TS-2 and TS-3 summarize the ranges of the primary and secondary scenario driving forces, respectively. Table TS-4 summarizes the emissions of GHGs, SO_2, and ozone precursor emissions. Together, the three tables provide a concise summary of the new SRES scenarios.

Box TS-4: Main Findings and Implications of SRES Scenarios

- The four scenario families each have a narrative storyline and consist of 40 scenarios developed by six modeling groups.
- The 40 scenarios cover the full range of GHGs and SO_2 emissions consistent with the underlying range of driving forces from scenario literature.
- The 40 SRES scenarios fall into various groups – the three scenario families A2, B1, and B2, plus different groups within the A1 scenario family. The A1 groups are distinguished by their technological emphasis – on coal (A1C), oil and gas (A1G), non-fossil energy sources (A1T), or a balance across all sources (A1B). In the SPM, the A1C and A1G scenario groups are combined into one fossil intensive group A1FI. All scenario groups are equally sound.
- The scenarios are also grouped into four categories of cumulative CO_2 emissions, which indicate that scenarios with different driving forces can lead to similar cumulative emissions and those with similar driving forces can branch out into different categories of cumulative emissions.
- Four from 40 scenarios are designated as marker scenarios that are characteristic of the four scenarios families. Together with the two additional illustrative scenarios selected from the scenario groups in the A1 family, they capture most of the emissions and driving forces spanned by the full set of the scenarios.
- There is no single central or "best guess" scenario, and probabilities or likelihood are not assigned to individual scenarios. Instead, the writing team recommends that the smallest set of scenarios used should include the four designated marker scenarios and the two additional illustrative scenarios selected from the scenario groups in the A1 family.
- Distinction between scenarios that envisage stringent environmental policies and those that include direct climate policies was very difficult to make, a difficulty associated with many definitional and other ambiguities.
- All scenarios describe futures that are generally more affluent than today. Many of the scenarios envisage a more rapid convergence in per capita income ratios in the world compared to the IS92 scenarios while, at the same time, they jointly cover a wide range of GHGs and SO_2 emissions.
- Emissions profiles are more dynamic than the IS92 scenarios, which reflects changes in future emissions trends for some scenarios and GHG species.
- The levels of GHG emissions are generally lower than the IS92 levels, especially toward the end of the 21st century, while emissions of SO_2, which have a cooling effect on the atmosphere, are significantly lower than in IS92.
- Alternative combinations of main scenario driving forces can lead to similar levels of GHG emissions by the end of the 21st century. Scenarios with different underlying assumptions can result in very similar climate changes.
- Technology is at least as important a driving force of GHG emissions as population and economic development across the set of 40 SRES scenarios.

Box TS-5: Recommendations for Consideration by the User Communities

The writing team recommends that the SRES scenarios be the main basis for the assessment of future emissions and their driving forces in the TAR. Accordingly, the SRES writing team makes the following recommendations regarding the emissions scenarios to be used in the atmosphere/ocean general circulation models (A/O GCMs) simulations for WGI, for the models that will be used in the assessment of climate change impacts by WGII, and for the mitigation and stabilization assessments by WGIII:

- *It is recommended that a range of SRES scenarios from more than one family be used in any analysis.* The scenario groups – the three scenario families A2, B1, and B2, plus the groups within the A1 scenario family, A1B, A1C & A1G (combined into A1FI in the SPM), and A1T – and four cumulative emissions categories were developed as the smallest subsets of SRES scenarios that capture the range of uncertainties associated with driving forces and emissions.
- *The important uncertainties may be different in different applications – for example climate modeling; assessment of impacts, vulnerability, mitigation, and adaptation options; and policy analysis.* Climate modelers may want to cover the range reflected by the cumulative emissions categories. To assess the robustness of options in terms of impacts, vulnerability, and adaptation may require scenarios with similar emissions but different socio-economic characteristics, as reflected by the seven groups. For mitigation analysis, variation in both emissions and socio-economic characteristics may be necessary. For analysis at the national or regional scale, the most appropriate scenarios may be those that best reflect specific circumstances and perspectives.
- *There is no single most, likely "central" or "best-guess" scenario, either with respect to other SRES scenarios or to the underlying scenario literature.* Probabilities or likelihoods are not assigned to individual SRES scenarios. None of the SRES scenarios represents an estimate of a central tendency for all driving forces and emissions, such as the mean or median, and none should be interpreted as such. The statistics associated with the frequency distributions of SRES scenarios do not represent the likelihood of their occurrence. The writing team cautions against constructing a central, "best-estimate" scenario from the SRES scenarios; instead it recommends use of the SRES scenarios as they are.
- *Concerning large-scale climate models, the writing team recommends that the minimum set of SRES scenarios should include the four designated marker scenarios and the two additional illustrative scenarios selected in the SPM from the scenario groups in the A1 family.* At the minimum (a) a simulation for one and the same SRES marker scenario should be performed by every TAR climate model for a given stabilization ceiling, and (b) the set of simulations performed by the TAR climate models and stabilization runs for a given ceiling should include all four of the SRES marker scenarios and the two additional illustrative scenarios selected in SPM from the scenario groups in the A1 family.
- *The driving forces and emissions of each SRES scenario should be used together.* To avoid internal inconsistencies, components of SRES scenarios should not be mixed. For example, the GHG emissions from one scenario and the SO_2 emissions from another scenario, or the population from one and economic development path from another, should not be combined.
- *The SRES scenarios can provide policy makers with a long-term context for near-term decisions.* This implies that they are not necessarily well suited for the analysis of near-term developments. When analyzing mitigation and adaptation options, the user should be aware that although no additional climate initiatives are included in the SRES scenarios, various changes have been assumed to occur that would require other policy interventions.
- *All 40 SRES emissions scenarios, their main driving forces, and underlying assumptions should be made widely available.* Depending on resources available the scenario documentation, should e.g., be placed on the web and made available on a CD-ROM. In addition, the time-dependent geographic distributions of the concentrations of GHGs and sulfate aerosol burden, together with their corresponding radiative forcings, should also be placed on the web.

Box TS-6: Recommendations for Consideration by the IPCC

- Assure that the SRES scenarios, their main assumptions, and modeling approaches are widely available through a web site or a CD-ROM.
- Establish a long-term facility for documentation and comparison of emissions scenarios to succeed the SRES open process. This should include a scenario database and analytic evaluation capabilities and should be regularly maintained.
- An effort should be made in the future to develop an appropriate emissions scenario classification scheme.
- Identify resources for capacity building in the area of emissions scenarios and modeling tools, with a particular emphasis to involve strong participation from developing countries.
- Promote activities within and outside the IPCC to extend the SRES multi-baseline and multi-model approach in future assessments of climate change impacts, adaptation, and mitigation.
- Initiate new programs to assess GHG emissions from land use and sources of emissions other than energy-related CO_2 emissions, to go beyond the effort of SRES, which was limited by time and resources.
- Initiate new programs to assess future developments of driving forces and GHG emissions for different regions and for different sectors (taking the set of SRES scenarios as reference for overall global and regional developments) to provide more regional and sectorial detail than time and resources allowed SRES to achieve.

***Table TS-2a**: Overview of main primary driving forces in 1990, 2020, 2050, and 2100. Bold numbers show the value for the illustrative scenario and the numbers between brackets show the value for the range[a] across all 40 SRES scenarios in the six scenario groups that constitute the four families. Units are given in the table. Technological change is not quantified in the table.*

Family		A1			A2	B1	B2
Scenario group	**1990**	**A1FI**	**A1B**	**A1T**	**A2**	**B1**	**B2**
Population (billion)	5.3						
2020		**7.6** (7.4-7.6)	**7.5** (7.2-7.6)	**7.6** (7.4-7.6)	**8.2** (7.5-8.2)	**7.6** (7.4-7.6)	**7.6** (7.6-7.8)
2050		**8.7**	**8.7** (8.3-8.7)	**8.7**	**11.3** (9.7-11.3)	**8.7** (8.6-8.7)	**9.3** (9.3-9.8)
2100		**7.1** (7.0-7.1)	**7.1** (7.0-7.7)	**7.0**	**15.1** (12.0-15.1)	**7.0** (6.9-7.1)	**10.4** (10.3-10.4)
World GDP (10^{12} 1990US$/yr)	21						
2020		**53** (53-57)	**56** (48-61)	**57** (52-57)	**41** (38-45)	**53** (46-57)	**51** (41-51)
2050		**164** (163-187)	**181** (120-181)	**187** (177-187)	**82** (59-111)	**136** (110-166)	**110** (76-111)
2100		**525** (522-550)	**529** (340-536)	**550** (519-550)	**243** (197-249)	**328** (328-350)	**235** (199-255)
Per capita income ratio: developed countries and economies in transition (Annex-I) to developing countries (Non-Annex-I)	16.1						
2020		**7.5** (6.2-7.5)	**6.4** (5.2-9.2)	**6.2** (5.7-6.4)	**9.4** (9.0-12.3)	**8.4** (5.3-10.7)	**7.7** (7.5-12.1)
2050		**2.8**	**2.8** (2.4-4.0)	**2.8** (2.4-2.8)	**6.6** (5.2-8.2)	**3.6** (2.7-4.9)	**4.0** (3.7-7.5)
2100		**1.5** (1.5-1.6)	**1.6** (1.5-1.7)	**1.6** (1.6-1.7)	**4.2** (2.7-6.3)	**1.8** (1.4-1.9)	**3.0** (2.0-3.6)

[a] For some driving forces, no range is indicated because all scenario runs have adopted exactly the same assumptions.

***Table TS-2b**: Overview of main primary driving forces in 1990, 2020, 2050, and 2100. Bold numbers show the value for the illustrative scenario and the numbers between brackets show the value for the range[a] across 26 harmonized SRES scenarios in the six scenario groups that constitute the four families. Units are given in the table. Technological change is not quantified in the table.*

Family			**A1**		**A2**	**B1**	**B2**
Scenario group	**1990**	**A1FI**	**A1B**	**A1T**	**A2**	**B1**	**B2**
Population (billion)	5.3						
2020		**7.6** (7.4-7.6)	**7.4** (7.4-7.6)	**7.6** (7.4-7.6)	**8.2**	**7.6** (7.4-7.6)	**7.6**
2050		**8.7**	**8.7**	**8.7**	**11.3**	**8.7** (8.6-8.7)	**9.3**
2100		**7.1** (7.0-7.1)	**7.1** (7.0-7.1)	**7.0**	**15.1**	**7.0** (6.9-7.1)	**10.4**
World GDP (10^{12} 1990US$/yr)	21						
2020		**53** (53-57)	**56** (52-61)	**57** (56-57)	**41**	**53** (51-57)	**51** (48-51)
2050		**164** (164-187)	**181** (164-181)	**187** (182-187)	**82**	**136** (134-166)	**110** (108-111)
2100		**525** (525-550)	**529** (529-536)	**550** (529-550)	**243**	**328** (328-350)	**235** (232-237)
Per capita income ratio: developed countries and economies in transition (Annex-I) to developing countries (Non-Annex-I)	16.1						
2020		**7.5** (6.2-7.5)	**6.4** (5.2-7.5)	**6.2** (6.2-6.4)	**9.4** (9.4-9.5)	**8.4** (5.3-8.4)	**7.7** (7.5-8.0)
2050		**2.8**	**2.8** (2.4-2.8)	**2.8**	**6.6**	**3.6** (2.7-3.9)	**4.0** (3.8-4.6)
2100		**1.5** (1.5-1.6)	**1.6** (1.5-1.7)	**1.6**	**4.2**	**1.8** (1.6-1.9)	**3.0** (3.0-3.5)

[a] For some driving forces, no range is indicated because all scenario runs have adopted exactly the same assumptions.

***Table TS-3a**: Overview of main secondary scenario driving forces in 1990, 2020, 2050, and 2100. Bold numbers show the value for the illustrative scenario and the numbers between brackets show the value for the range across all 40 SRES scenarios in the six scenario groups that constitute the four families. Units are given in the table.*

Family		A1			A2	B1	B2
Scenario group	**1990**	**A1FI**	**A1B**	**A1T**	**A2**	**B1**	**B2**
Final energy intensity (10^6J/US$)[a]	16.7						
2020		**9.4** (8.5-9.4)	**9.4** (8.1-12.0)	**8.7** (7.6-8.7)	**12.1** (9.3-12.4)	**8.8** (6.7-11.6)	**8.5** (8.5-11.8)
2050		**6.3** (5.4-6.3)	**5.5** (4.4-7.2)	**4.8** (4.2-4.8)	**9.5** (7.0-9.5)	**4.5** (3.5-6.0)	**6.0** (6.0-8.1)
2100		**3.0** (2.6-3.2)	**3.3** (1.6-3.3)	**2.3** (1.8-2.3)	**5.9** (4.4-7.3)	**1.4** (1.4-2.7)	**4.0** (3.7-4.6)
Primary energy (10^{18} J/yr)[a]	351						
2020		**669** (653-752)	**711** (573-875)	**649** (515-649)	**595** (485-677)	**606** (438-774)	**566** (506-633)
2050		**1431** (1377-1601)	**1347** (968-1611)	**1213** (913-1213)	**971** (679-1059)	**813** (642-1090)	**869** (679-966)
2100		**2073** (1988-2737)	**2226** (1002-2683)	**2021** (1255-2021)	**1717** (1304-2040)	**514** (514-1157)	**1357** (846-1625)
Share of coal in primary energy (%)[a]	24						
2020		**29** (24-42)	**23** (8-28)	**23** (8-23)	**22** (18-34)	**22** (8-27)	**17** (14-31)
2050		**33** (13-56)	**14** (3-42)	**10** (2-13)	**30** (24-47)	**21** (2-37)	**10** (10-49)
2100		**29** (3-48)	**4** (4-41)	**1** (1-3)	**53** (17-53)	**8** (0-22)	**22** (12-53)
Share of zero carbon in primary energy (%)[a]	18						
2020		**15** (10-20)	**16** (9-26)	**21** (15-22)	**8** (8-16)	**21** (7-22)	**18** (7-18)
2050		**19** (16-31)	**36** (21-40)	**43** (39-43)	**18** (14-29)	**30** (18-40)	**30** (15-30)
2100		**31** (30-47)	**65** (27-75)	**85** (64-85)	**28** (26-37)	**52** (33-70)	**49** (22-49)

[a] 1990 values include non-commercial energy consistent with IPCC WGII SAR (Energy Primer) but with SRES accounting conventions. Note that ASF, MiniCAM, and IMAGE scenarios do not consider non-commercial renewable energy. Hence, these scenarios report lower energy use.

***Table TS-3b**: Overview of main secondary scenario driving forces in 1990, 2020, 2050, and 2100. Bold numbers show the value for the illustrative scenario and the numbers between brackets show the value for the range across 26 harmonized SRES scenarios in the six scenario groups that constitute the four families. Units are given in the table.*

Family		**A1**			**A2**	**B1**	**B2**
Scenario group	**1990**	**A1FI**	**A1B**	**A1T**	**A2**	**B1**	**B2**
Final energy intensity (10^6J/US$)[a]	16.7						
2020		**9.4** (8.5-9.4)	**9.4** (8.7-12.0)	**8.7** (7.6-8.7)	**12.1** (11.3-12.1)	**8.8** (6.7-11.6)	**8.5** (8.5-9.1)
2050		**6.3** (5.4-6.3)	**5.5** (5.0-7.2)	**4.8** (4.3-4.8)	**9.5** (9.2-9.5)	**4.5** (3.5-6.0)	**6.0** (6.0-6.6)
2100		**3.0** (3.0-3.2)	**3.3** (2.7-3.3)	**2.3**	**5.9** (5.5-5.9)	**1.4** (1.4-2.1)	**4.0** (3.9-4.1)
Primary energy (10^{18} J/yr)[a]	351						
2020		**669** (657-752)	**711** (589-875)	**649** (611-649)	**595** (595-610)	**606** (451-774)	**566** (519-590)
2050		**1431** (1377-1601)	**1347** (1113-1611)	**1213** (1086-1213)	**971** (971-1014)	**813** (642-1090)	**869** (815-941)
2100		**2073** (2073-2737)	**2226** (1002-2683)	**2021** (1632-2021)	**1717** (1717-1921)	**514** (514-1157)	**1357** (1077-1357)
Share of coal in primary energy (%)[a]	24						
2020		**29** (24-42)	**23** (8-26)	**23** (23-23)	**22** (20-22)	**22** (19-27)	**17** (14-31)
2050		**33** (13-52)	**14** (3-42)	**10** (10-13)	**30** (27-30)	**21** (4-37)	**10** (10-35)
2100		**29** (3-46)	**4** (4-41)	**1** (1-3)	**53** (45-53)	**8** (0-22)	**22** (19-37)
Share of zero carbon in primary energy (%)[a]	18						
2020		**15** (10-20)	**16** (9-26)	**21** (15-21)	**8** (8-16)	**21** (7-22)	**18** (12-18)
2050		**19** (16-31)	**36** (23-40)	**43** (41-43)	**18** (18-29)	**30** (18-40)	**30** (21-30)
2100		**31** (30-47)	**65** (39-75)	**85** (67-85)	**28** (28-37)	**52** (44-70)	**49** (22-49)

[a] 1990 values include non-commercial energy consistent with IPCC WGII SAR (Energy Primer) but with SRES accounting conventions. Note that ASF, MiniCAM, and IMAGE scenarios do not consider non-commercial renewable energy. Hence, these scenarios report lower energy use.

***Table TS-4a**: Overview of GHG, SO_2 and ozone precursor emissions[a] in 1990, 2020, 2050, and 2100, and cumulative carbon dioxide emissions to 2100. Bold numbers show the value for the illustrative scenario and the numbers between brackets show the value for the range across all 40 SRES scenarios in the six scenario groups that constitute the four families. Units are given in the table.*

Family		**A1**			**A2**	**B1**	**B2**
Scenario group	**1990**	**A1FI**	**A1B**	**A1T**	**A2**	**B1**	**B2**
Carbon dioxide, fossil fuels (GtC/yr)	6.0						
2020		**11.2** (10.7-14.3)	**12.1** (8.7-14.7)	**10.0** (8.4-10.0)	**11.0** (7.9-11.3)	**10.0** (7.8-13.2)	**9.0** (8.5-11.5)
2050		**23.1** (20.6-26.8)	**16.0** (12.7-25.7)	**12.3** (10.8-12.3)	**16.5** (10.5-18.2)	**11.7** (8.5-17.5)	**11.2** (11.2-16.4)
2100		**30.3** (27.7-36.8)	**13.1** (12.9-18.4)	**4.3** (4.3-9.1)	**28.9** (17.6-33.4)	**5.2** (3.3-13.2)	**13.8** (9.3-23.1)
Carbon dioxide, land use (GtC/yr)	1.1						
2020		**1.5** (0.3-1.8)	**0.5** (0.3-1.6)	**0.3** (0.3-1.7)	**1.2** (0.1-3.0)	**0.6** (0.0-1.3)	**0.0** (0.0-1.9)
2050		**0.8** (0.0-0.9)	**0.4** (0.0-1.0)	**0.0** (-0.2-0.5)	**0.9** (0.6-0.9)	**-0.4** (-0.7-0.8)	**-0.2** (-0.2-1.2)
2100		**-2.1** (-2.1-0.0)	**0.4** (-2.4-2.2)	**0.0** (0.0-0.1)	**0.2** (-0.1-2.0)	**-1.0** (-2.8-0.1)	**-0.5** (-1.7-1.5)
Cumulative carbon dioxide, fossil fuels (GtC)							
1990-2100		**2128** (2079-2478)	**1437** (1220-1989)	**1038** (989-1051)	**1773** (1303-1860)	**989** (794-1306)	**1160** (1033-1627)
Cumulative carbon dioxide, land use (GtC)							
1990-2100		**61** (31-69)	**62** (31-84)	**31** (31-62)	**89** (49-181)	**-6** (-22-84)	**4** (4-153)
Cumulative carbon dioxide, total (GtC)							
1990-2100		**2189** (2127-2538)	**1499** (1301-2073)	**1068** (1049-1113)	**1862** (1352-1938)	**983** (772-1390)	**1164** (1164-1686)
Sulfur dioxide, (MtS/y)	70.9						
2020		**87** (60-134)	**100** (62-117)	**60** (60-101)	**100** (66-105)	**75** (52-112)	**61** (48-101)
2050		**81** (64-139)	**64** (47-120)	**40** (40-64)	**105** (78-141)	**69** (29-69)	**56** (42-107)
2100		**40** (27-83)	**28** (26-71)	**20** (20-27)	**60** (60-93)	**25** (11-25)	**48** (33-48)
Methane, ($MtCH_4$/yr)	310						
2020		**416** (415-479)	**421** (400-444)	**415** (415-466)	**424** (354-493)	**377** (377-430)	**384** (384-469)
2050		**630** (511-636)	**452** (452-636)	500(492-500)	**598** (402-671)	**359** (359-546)	**505** (482-536)
2100		**735** (289-735)	**289** (289-640)	**274** (274-291)	**889** (549-1069)	**236** (236-579)	**597** (465-613)

[a] The uncertainties in the SRES emissions for non-CO_2 greenhouse gases are generally greater than those for energy CO_2. Therefore, the ranges of non-CO_2 GHG emissions provided in the Report may not fully reflect the level of uncertainty compared to CO_2, for example only a single model provided the sole value for halocarbon emissions.

Table TS-4a (continued)

Family		A1			A2	B1	B2
Scenario group	**1990**	**A1FI**	**A1B**	**A1T**	**A2**	**B1**	**B2**
Nitrous oxide, (MtN/yr)	6.7						
2020		**9.3** (6.1-9.3)	**7.2** (6.1-9.6)	**6.1** (6.1-7.8)	**9.6** (6.3-12.2)	**8.1** (5.8-9.5)	**6.1** (6.1-11.5)
2050		**14.5** (6.3-14.5)	**7.4** (6.3-14.3)	**6.1** (6.1-6.7)	**12.0** (6.8-13.9)	**8.3** (5.6-14.8)	**6.3** (6.3-13.2)
2100		**16.6** (5.9-16.6)	**7.0** (5.8-17.2)	**5.4** (4.8-5.4)	**16.5** (8.1-19.3)	**5.7** (5.3-20.2)	**6.9** (6.9-18.1)
CFC/HFC/HCFC (MtC equiv./yr)[b]	1672						
2020		**337**	**337**	**337**	**292**	**291**	**299**
2050		**566**	**566**	**566**	**312**	**338**	**346**
2100		**614**	**614**	**614**	**753**	**299**	**64**
PFC, (MtC equiv./yr) [b]	32.0						
2020		**42.7**	**42.7**	**42.7**	**50.9**	**31.7**	**54.8**
2050		**88.7**	**88.7**	**88.7**	**92.2**	**42.2**	**106.6**
2100		**115.3**	**115.3**	**115.3**	**178.4**	**44.9**	**121.3**
SF6 , (MtC equiv./yr) [b]	37.7						
2020		**47.8**	**47.8**	**47.8**	**63.5**	**37.4**	**54.7**
2050		**119.2**	**119.2**	**119.2**	**104.0**	**67.9**	**79.2**
2100		**94.6**	**94.6**	**94.6**	**164.6**	**42.6**	**69.0**
CO, (Mt CO/y)	879						
2020		**1204** (1123-1552)	**1032** (978-1248)	**1147** (1147-1160)	**1075** (748-1100)	**751** (751-1162)	**1022** (632-1077)
2050		**2159** (1619-2307)	**1214** (949-1925)	**1770** (1244-1770)	**1428** (642-1585)	**471** (471-1470)	**1319** (580-1319)
2100		**2570** (2298-3766)	**1663** (1080-2532)	**2077** (1520-2077)	**2326** (776-2646)	**363** (363-1871)	**2002** (661-2002)
NMVOC, (Mt/yr)	139						
2020		**192** (178-230)	**222** (157-222)	**190** (188-190)	**179** (166-205)	**140** (140-193)	**180** (152-180)
2050		**322** (256-322)	**279** (158-301)	**241** (206-241)	**225** (161-242)	**116** (116-237)	**217** (147-217)
2100		**420** (167-484)	**194** (133-552)	**128** (114-128)	**342**(169-342)	**87** (58-349)	**170** (130-304)
NO_x, (MtN/yr)	30.9						
2020		**50** (46-51)	**46** (46-66)	**46** (46-49)	**50** (42-50)	**40** (38-59)	**43** (38-52)
2050		**95** (49-95)	**48** (48-100)	**61** (49-61)	**71** (50-82)	**39** (39-72)	**55** (42-66)
2100		**110** (40-151)	**40** (40-77)	**28** (28-40)	**109** (71-110)	**19** (16-35)	**61** (34-77)

[b] In the SPM the emissions of CFC/HFC/HCFC, PFC, and SF6 are presented as carbon-equivalent emissions. This was done by multiplying the emissions by weight of each substance (see Table 5-8) by its global warming potential (GWP; see Table 5-7) and subsequent summation. The results were then converted from CO_2-equivalents (reflected by the GWPs) into carbon-equivalents. Note that the use of GWP is less appropriate for emission profiles that span a very long period. It is used here, in the interest of readability of the SPM in preference to a more detailed breakdown by the 27 substances listed in Table 5-7. The method here is also preferred over the even less desirable option to display weighted numbers for the aggregate categories in this table.

***Table TS-4b**: Overview of GHG, SO_2, and ozone precursor emissions[a] in 1990, 2020, 2050, and 2100, and cumulative carbon dioxide emissions to 2100. Bold numbers show the value for the illustrative scenario and the numbers between brackets show the value for the range across 26 harmonized SRES scenarios in the six scenario groups that constitute the four families. Units are given in the table.*

Family		**A1**			**A2**	**B1**	**B2**
Scenario group	**1990**	**A1FI**	**A1B**	**A1T**	**A2**	**B1**	**B2**
Carbon dioxide, fossil fuels (GtC/yr)	6.0						
2020		**11.2** (10.7-14.3)	**12.1** (8.7-14.7)	**10.0** (9.8-10.0)	**11.0** (10.3-11.0)	**10.0** (8.2-13.2)	**9.0** (8.8-10.2)
2050		**23.1** (20.6-26.8)	**16.0** (12.7-25.7)	**12.3** (11.4-12.3)	**16.5** (15.1-16.5)	**11.7** (8.5-17.5)	**11.2** (11.2-15.0)
2100		**30.3** (30.3-36.8)	**13.1** (13.1-17.9)	**4.3** (4.3-8.6)	**28.9** (28.2-28.9)	**5.2** (3.3-7.9)	**13.8** (13.8-18.6)
Carbon dioxide, land use (GtC/yr)	1.1						
2020		**1.5** (0.3-1.8)	**0.5** (0.3-1.6)	**0.3** (0.3-1.7)	**1.2** (1.1-1.2)	**0.6** (0.0-1.3)	**0.0** (0.0-1.1)
2050		**0.8** (0.0-0.8)	**0.4** (0.0-1.0)	**0.0** (-0.2-0.0)	**0.9** (0.8-0.9)	**-0.4** (-0.7-0.8)	**-0.2** (-0.2-1.2)
2100		**-2.1** (-2.1-0.0)	**0.4** (-2.0-2.2)	**0.0** (0.0-0.1)	**0.2** (0.0-0.2)	**-1.0** (-2.6-0.1)	**-0.5** (-0.5-1.2)
Cumulative carbon dioxide, fossil fuels (GtC)							
1990-2100		**2128** (2096-2478)	**1437** (1220-1989)	**1038** (1038-1051)	**1773** (1651-1773)	**989** (794-1306)	**1160** (1160-1448)
Cumulative carbon dioxide, land use (GtC)							
1990-2100		**61** (31-61)	**62** (31-84)	**31** (31-62)	**89** (81-89)	**-6** (-22-84)	**4** (4-125)
Cumulative carbon dioxide, total (GtC)							
1990-2100		**2189** (2127-2538)	**1499** (1301-2073)	**1068** (1068-1113)	**1862** (1732-1862)	**983** (772-1390)	**1164** (1164-1573)
Sulfur dioxide (MtS/yr)	70.9						
2020		**87** (60-134)	**100** (62-117)	**60** (60-101)	**100** (80-100)	**75** (52-112)	**61** (61-78)
2050		**81** (64-139)	**64** (47-64)	**40** (40-64)	**105** (104-105)	**69** (29-69)	**56** (44-56)
2100		**40** (27-83)	**28** (28-47)	**20** (20-27)	**60** (60-69)	**25** (11-25)	**48** (33-48)
Methane ($MtCH_4$/yr)	310						
2020		**416** (416-479)	**421** (406-444)	**415** (415-466)	**424** (418-424)	**377** (377-430)	**384** (384-391)
2050		**630** (511-630)	**452** (452-636)	**500** (492-500)	**598** (598-671)	**359** (359-546)	**505** (482-505)
2100		**735** (289-735)	**289** (289-535)	**274** (274-291)	**889** (889-1069)	**236** (236-561)	**597** (465-597)

[a] The uncertainties in the SRES emissions for non-CO_2 greenhouse gases are generally greater than those for energy CO_2. Therefore, the ranges of non-CO_2 GHG emissions provided in the Report may not fully reflect the level of uncertainty compared to CO_2, for example only a single model provided the sole value for halocarbon emissions.

Table TS-4b (continued)

Family		A1			A2	B1	B2
Scenario group	**1990**	**A1FI**	**A1B**	**A1T**	**A2**	**B1**	**B2**
Nitrous Oxide, (MtN/y)	6.7						
2020		**9.3** (6.1-9.3)	**7.2** (6.1-9.6)	**6.1** (6.1-7.8)	**9.6** (6.3-9.6)	**8.1** (5.8-9.5)	**6.1** (6.1-7.1)
2050		**14.5** (6.3-14.5)	**7.4** (6.3-13.8)	**6.1** (6.1-6.7)	**12.0** (6.8-12.0)	**8.3** (5.6-14.8)	**6.3** (6.3-7.5)
2100		**16.6** (5.9-16.6)	**7.0** (5.8-15.6)	**5.4** (4.8-5.4)	**16.5** (8.1-16.5)	**5.7** (5.3-20.2)	**6.9** (6.9-8.0)
CFC/HFC/HCFC (MtC equiv./y) [b]	1672						
2020		**337**	**337**	**337**	**292**	**291**	**299**
2050		**566**	**566**	**566**	**312**	**338**	**346**
2100		**614**	**614**	**614**	**753**	**299**	**64**
PFC, (MtC equiv./yr) [b]	32.0						
2020		**42.7**	**42.7**	**42.7**	**50.9**	**31.7**	**54.8**
2050		**88.7**	**88.7**	**88.7**	**92.2**	**42.2**	**106.6**
2100		**115.3**	**115.3**	**115.3**	**178.4**	**44.9**	**121.3**
SF6 , (MtC equiv./yr) [b]	37.7						
2020		**47.8**	**47.8**	**47.8**	**63.5**	**37.4**	**54.7**
2050		**119.2**	**119.2**	**119.2**	**104.0**	**67.9**	**79.2**
2100		**94.6**	**94.6**	**94.6**	**164.6**	**42.6**	**69.0**
CO, (Mt CO/y)	879						
2020		**1204** (1123-1552)	**1032** (1032-1248)	**1147** (1147-1160)	**1075** (1075-1100)	**751** (751-1162)	**1022** (941-1022)
2050		**2159** (1619-2307)	**1214** (1214-1925)	**1770** (1244-1770)	**1428** (1428-1585)	**471** (471-1470)	**1319** (1180-1319)
2100		**2570** (2298-3766)	**1663** (1663-2532)	**2077** (1520-2077)	**2326** (2325-2646)	**363** (363-1871)	**2002** (1487-2002)
NMVOC, (Mt/y)	139						
2020		**192** (178-230)	**222** (194-222)	**190** (188-190)	**179** (179-204)	**140** (140-193)	**180** (179-180)
2050		**322** (256-322)	**279** (259-301)	**241** (206-241)	**225** (225-242)	**116** (116-237)	**217** (197-217)
2100		**420** (167-484)	**194** (137-552)	**128** (114-128)	**342** (311-342)	**87** (58-349)	**170** (130-170)
NOx, (MtN/y)	30.9						
2020		**50** (46-51)	**46** (46-66)	**46** (46-49)	**50** (47-50)	**40** (38-59)	**43** (38-43)
2050		**95** (49-95)	**48** (48-100)	**61** (49-61)	**71** (66-71)	**39** (39-72)	**55** (42-55)
2100		**110** (40-151)	**40** (40-77)	**28** (28-40)	**109** (109-110)	**19** (16-35)	**61** (34-61)

[b] In the SPM the emissions of CFC/HFC/HCFC, PFC, and SF6 are presented as carbon-equivalent emissions. This was done by multiplying the emissions by weight of each substance (see Table 5-8) by its global warming potential (GWP; see Table 5-7) and subsequent summation. The results were then converted from CO_2-equivalents (reflected by the GWPs) into carbon-equivalents. Note that the use of GWP is less appropriate for emission profiles that span a very long period. It is used here, in the interest of readability of the SPM in preference to a more detailed breakdown by the 27 substances listed in Table 5-7. The method here is also preferred over the even less desirable option to display weighted numbers for the aggregate categories in this table.

References:

Alcamo, J., A. Bouwman, J. Edmonds, A. Grübler, T. Morita, and A. Sugandhy, 1995: An evaluation of the IPCC IS92 emission scenarios. In *Climate Change 1994, Radiative Forcing of Climate Change* and *An Evaluation of the IPCC IS92 Emission Scenarios,* J.T. Houghton, L.G. Meira Filho, J. Bruce, Hoesung Lee, B.A. Callander, E. Haites, N. Harris and K. Maskell (eds.), Cambridge University Press, Cambridge, pp. 233-304.

Alcamo, J., E. Kreileman, and R. Leemans (eds.), 1998: *Global Change Scenarios of the 21st Century. Results from the IMAGE 2.1 model.* Elseviers Science, London.

Alcamo, J. and Nakićenović, N., (eds.), 1998: Long-term Greenhouse Gas Emission Scenarios and Their Driving Forces, *Mitigation and Adaptation Strategies for Global Change* **3**, Nos. 2-4.

Alcamo, J., and R. Swart, 1998: Future trends of land-use emissions of major greenhouse gases. *Mitigation and Adaptation Strategies for Global Change,* **3**(2-4), 343-381.

Anderson, D., 1998: On the effects of social and economic policies on future carbon emissions. *Mitigation and Adaptation Strategies for Global Change*, **3**(2-4), 419-453.

De Jong, A. and G. Zalm, 1991: Scanning the future: a long-term scenario study of the world economy 1990-2015. In: *Long-term Prospects of the World Economy.* OECD, Paris, France, pp. 27-74.

Demeny, P., 1990: Population. In *The Earth As Transformed by Human Action.* B.L. Turner II *et al.*, (ed.), Cambridge University Press, Cambridge, pp. 41-54.

De Vries, H.J.M., J.G.J. Olivier, R.A. van den Wijngaart, G.J.J. Kreileman, and A.M.C. Toet, 1994: Model for calculating regional energy use, industrial production and greenhouse gas emissions for evaluating global climate scenarios. *Water, Air, Soil Pollution*, **76**, 79-131.

De Vries, B., M. Janssen, and A. Beusen, 1999: Perspectives on global energy futures – simulations with the TIME model. *Energy Policy,* **27**, 477-494.

De Vries, B., J. Bollen, L. Bouwman, M. den Elzen, M. Janssen, and E. Kreileman, 2000: Greenhouse gas emissions in an equity-, environment- and service-oriented world: An IMAGE-based scenario for the next century. *Technological Forecasting & Social Change*, **63**(2-3). (In press.)

Dowlatabadi, H., and M. Kandlikar, 1995: *Key Uncertainties in Climate Change Policy: Results from ICAM-2.* Proceedings of the 6th Global Warming Conference, San Francisco, CA.

Durand, J.D., 1967: The modern expansion of world population. *Proceedings of the American Philosophical Society*, **111**(3), 136-159.

Edmonds, J., M. Wise, and C. MacCracken, 1994: *Advanced energy technologies and climate change.* An Analysis Using the Global Change Assessment Model (GCAM). PNL-9798, UC-402, Pacific Northwest Laboratory, Richland, WA, USA.

Edmonds, J., M. Wise, H. Pitcher, R. Richels, T. Wigley, and C. MacCracken, 1996a: An integrated assessment of climate change and the accelerated introduction of advanced energy technologies: An application of MiniCAM 1.0. *Mitigation and Adaptation Strategies for Global Change*, **1**(4), 311-339.

Edmonds, J., M. Wise, R. Sands, R. Brown, and H. Kheshgi, 1996b: *Agriculture, land-use, and commercial biomass energy.* A Preliminary integrated analysis of the potential role of Biomass Energy for Reducing Future Greenhouse Related Emissions. PNNL-11155, Pacific Northwest National Laboratories, Washington, DC.

Fenhann, J., 2000: Industrial non-energy, non-CO_2 greenhouse gas emissions. *Technological Forecasting & Social Change,* **63**(2-3). (In press.)

Gaffin, S.R., 1998: World population projections for greenhouse gas emissions scenarios. *Mitigation and Adaptation Strategies for Global Change*, **3**(2-4), 133-170.

Gregory, K., 1998: Factors affecting future emissions of methane from non-land-use sources. *Mitigation and Adaptation Strategies for Global Change*, **3**(2-4), 321-341.

Gregory, K., and H.-H. Rogner, 1998: Energy resources and conversion technologies for the 21st century. *Mitigation and Adaptation Strategies for Global Change*, **3**(2-4), 171-229.

Grübler, A., 1998: A review of global and regional sulfur emission scenarios. *Mitigation and Adaptation Strategies for Global Change,* **3**(2-4), 383-418.

Houghton, J.T., L.G. Meira Filho, J. Bruce, Hoesung Lee, B.A. Callander, E. Haites, N. Harris, and K. Maskell (eds.), 1995: *Climate Change 1994: Radiative Forcing of Climate Change and an Evaluation of the IPCC IS92 Emissions Scenarios.* Cambridge University Press, Cambridge, 339 pp.

Houghton, J.T., L.G. Meira Filho, B.A. Callander, N. Harris, A. Kattenberg, and K. Maskell (eds.), 1996: *Climate Change 1995. The Science of Climate Change.* Contribution of Working Group I to the Second Assessment Report of the Intergovernmental Panel on Climate Change, Cambridge University Press, Cambridge.

IEA (International Energy Agency), 1999. *Non-OECD Coal-Fired Power Generation – Trends in the 1990s.* IEA Coal Research, London.

Jiang, K., T. Masui, T. Morita, and Y. Matsuoka, 2000: Long-term GHG emission scenarios of Asia-Pacific and the world. *Technological Forecasting & Social Change*, **63**(2-3). (In press.)

Jung, T.-Y., E.L. La Rovere, H. Gaj, P.R. Shukla, and D. Zhou, 2000: Structural changes in developing countries and their implication to energy-related CO_2 emissions. *Technological Forecasting & Social Change*, **63**(2-3). (In press.)

Kram, T., K. Riahi, R.A. Roehrl, S. van Rooijen, T. Morita, and B. de Vries, 2000: Global and regional greenhouse gas emissions scenarios. *Technological Forecasting & Social Change*, **63**(2-3). (In press).

Kroeze, C. and Reijnders, L., 1992: Halocarbons and global warming III, *The Science of Total Environment*, **112**, 291-314.

La Rovere, E.L., and B. Americano, 1998: *Environmental Impacts of Privatizing the Brazilian Power Sector.* Proceedings of the International Association of Impact Assessment Annual Meeting, April 1998, Christchurch, New Zealand.

Lashof, D., and Tirpak, D.A., 1990: *Policy Options for Stabilizing Global Climate.* 21P-2003. U.S. Environmental Protection Agency, Washington, DC.

Leggett, J., W.J. Pepper, and R.J. Swart, 1992: Emissions scenarios for IPCC: An update. In *Climate Change 1992. Supplementary Report to the IPCC Scientific Assessment,* J.T. Houghton, B.A. Callander, S.K. Varney (eds.), Cambridge University Press, Cambridge, pp. 69-95.

Lutz, W. (ed.), 1996: *The Future Population of the World: What can we assume today?* 2nd Edition. Earthscan, London.

Lutz, W., W. Sanderson, and S. Scherbov, 1997: Doubling of world population unlikely. *Nature*, **387**(6635), 803-805.

Marland, G., R.J. Andres, and T.A. Boden, 1994: Global, regional, and national CO_2 emissions. In *Trends '93: A Compendium of Data on Global Change,* ORNL/CDIAC-65, T.A. Boden, D.P. Kaiser, R.J. Sepanski, and F.W. Stoss (eds).,Carbon Dioxide Information Analysis Center, Oak Ridge National Laboratory, Oak Ridge, TN, pp. 505-584.

McCulloch, A., and P.M. Midgley, 1998: Estimated historic emissions of fluorocarbons from the European Union, *Atmospheric Environment*, **32**(9), 1571-1580.

Messner, S., and M. Strubegger, 1995: *User's Guide for MESSAGE III.* WP-95-69, International Institute for Applied Systems Analysis, Laxenburg, Austria, 155 pp.

Michaelis, L., 1998: Economic and technological development in climate scenarios. *Mitigation and Adaptation Strategies for Global Change*, **3**(2-4), 231-261.

Midgley, P.M., and A. McCulloch, 1999: Properties and applications of industrial halocarbons. In *Reactive Halogen Compounds in the Atmosphere,* **4**, *Part E.* P. Fabian, and O.N. Singh (eds.), *The Handbook of Environmental Chemistry*, Springer-Verlag, Berlin/Heidelberg.

Mori, S., and M. Takahashi, 1999: An integrated assessment model for the evaluation of new energy technologies and food productivity. *International Journal of Global Energy Issues*, **11**(1-4), 1-18.

Mori, S., 2000: The development of greenhouse gas emissions scenarios using an extension of the MARIA model for the assessment of resource and energy technologies. *Technological Forecasting & Social Change*, **63**(2-3). (In press.)

Morita, T., Y. Matsuoka, I. Penna, and M. Kainuma, 1994: *Global Carbon Dioxide Emission Scenarios and their Basic Assumptions: 1994 Survey.* CGER-1011-94, Center for Global Environmental Research, National Institute for Environmental Studies, Tsukuba, Japan.

Morita, T., and H.-C. Lee, 1998a: Appendix to emissions scenarios database and review of scenarios. *Mitigation and Adaptation Strategies for Global Change*, **3**(2-4), 121-131.

Morita, T., and H.-C. Lee, 1998b: *IPCC SRES Database*, Version 0.1, Emission Scenario Database prepared for IPCC Special Report on Emissions Scenarios (http:www-cger.nies.go.jp/cger-e/db/ipcc.html).

Nakićenović, N., A. Grübler, H. Ishitani, T. Johansson, G. Marland, J.R. Moreira, and H.-H. Rogner, 1996: Energy primer. In *Climate Change 1995. Impacts, Adaptations and Mitigation of Climate Change: Scientific Analyses.* R. Watson, M.C. Zinyowera, R. Moss (eds.), Cambridge University Press, Cambridge, 75-92.

Nakićenović, N., N. Victor, and T. Morita, 1998: Emissions scenarios database and review of scenarios. *Mitigation and Adaptation Strategies for Global Change*, **3**(2-4), 95-120.

Nakićenović, N., 2000: Greenhouse gas emissions scenarios. *Technological Forecasting & Social Change*, **65**(3). (In press.)

Parikh, J.K., 1992: IPCC strategies unfair to the south. *Nature*, **360**, 507-508.

Pepper, W.J., J. Leggett, R. Swart, J. Wasson, J. Edmonds, and I. Mintzer, 1992: Emissions scenarios for the IPCC. An update: Assumptions, methodology, and results: Support Document for Chapter A3. In *Climate Change 1992: Supplementary Report to the IPCC Scientific Assessment,* J.T. Houghton, B.A. Callandar, S.K. Varney (eds.), Cambridge University Press, Cambridge.

Pepper, W.J., Barbour, W., Sankovski, A., and Braaz, B., 1998: No-policy greenhouse gas emission scenarios: revisiting IPCC 1992. *Environmental Science & Policy,* **1**, 289-312.

Price, L., L. Michaelis, E. Worell, and M. Khrushch, 1998: Sectoral trends and driving forces of global energy use and greenhouse gas emissions. *Mitigation and Adaptation Strategies for Global Change,* **3**(2-4), 263-319.

Riahi, K., and R.A. Roehrl, 2000: Greenhouse gas emissions in a dynamics-as-usual scenario of economic and energy development. *Technological Forecasting & Social Change,* **63**(2-3). (In press.)

Roehrl, R.A., and K. Riahi, 2000: Technology dynamics and greenhouse gas emissions mitigation - a cost assessment. *Technological Forecasting & Social Change*, **63**(2-3). (In press.)

Sankovski, A., W. Barbour, and W. Pepper, 2000: Quantification of the IS99 emission scenario storylines using the atmospheric stabilization framework (ASF). *Technological Forecasting & Social Change*, **63**(2-3). (In press).

Streets, D.G., and S.T. Waldhoff, 2000: Present and future emissions of air pollutants in China: SO_2, NO_x, and CO. *Atmospheric Environment,* **34**(3), 363-374. (In press).

Subak, S., M. Hulme, and L. Bohn, 1997: *The Implications of FCCC Protocol Proposals for Future Global Temperature: Results Considering Alternative Sulfur Forcing*. CSERGE Working Paper GEC-97-19, CSERGE University of East Anglia, Norwich, UK.

UN (United Nations), 1998: *World Population Projections to 2150.* UN Department of Economics and Social Affairs (Population Division), United Nations, New York, NY.

UNFCCC (United Nations Framework Convention on Climate Change), 1992: *United Nations Framework Convention on Climate Change.* Convention text, UNEP/WMO Information Unit of Climate Change (IUCC) on behalf of the Interim Secretariat of the Convention. IUCC, Geneva.

UNFCCC (United Nations Framework Convention on Climate Change), 1997: *Kyoto Protocol to the United Nations Framework Convention on Climate Change*, FCCC/CP/L7/Add.1, 10 December 1997. UN, New York , NY.

WMO/UNEP (World Meteorological Organisation and United Nations Environment Programme), 1998: *Scientific Assessment of Ozone Depletion: 1998.* WMO Global Ozone Research & Monitoring Project, December 1998, WMO, Geneva.

WMO/UNEP (World Meteorological Organisation), 1999: *Proceedings of the Joint IPCC/TEAP Expert Meeting on Options for the Limitation of Emissions of HFCs and PFCs*, Petten, the Netherlands, 26-28 May 1999.

World Bank, 1999: *1999 World Development Indicators.* World Bank, Washington, DC.

1

Background and Overview

CONTENTS

1.1. Introduction

The Intergovernmental Panel on Climate Change (IPCC) decided at its September 1996 plenary session in Mexico City to develop a new set of emissions scenarios (see Appendix I for the Terms of Reference). This Special Report on Emission Scenarios (SRES) describes the new scenarios and how they were developed.

The SRES writing team formulated a set of emissions scenarios. These scenarios cover a wide range of the main driving forces of future emissions, from demographic to technological and economic developments. The scenarios encompass different future developments that might influence greenhouse gas (GHG) sources and sinks, such as alternative structures of energy systems and land-use changes. As required by the terms of reference however, *none* of the scenarios in the set includes any future policies that explicitly address additional climate change initiatives[1] although all necessarily encompass various assumed future policies of other types that may indirectly influence GHGs sources and sinks.

The set of SRES emissions scenarios is based on an extensive assessment of the literature, six alternative modeling approaches, and an "open process" that solicited wide participation and feedback from many groups and individuals. The set of scenarios includes anthropogenic emissions of all relevant GHG species and sulfur dioxide (SO_2), carbon monoxide (CO), nitrogen oxides (NO_x) and non-methane volatile organic hydrocarbons (VOCs), as shown in Table 1-1. It covers most of the range of GHG emissions compared with the published scenario literature. For example, in the SRES scenarios, emissions of CO_2 in 2100 range from more than 40 to less than 6 giga (or billion) tons[2] of elemental carbon (GtC), that is, from almost a sevenfold increase to roughly the same emissions level as in 1990.

Emissions scenarios are a central component of any assessment of climate change. Scenarios facilitate the assessment of future developments in complex systems that are either inherently unpredictable or have high scientific uncertainties, and the assessment of future emissions is an essential component of the overall assessment of global climate change by the IPCC.

Emissions of GHGs and SO_2 are the basic input for determining future climate patterns with simple climate models, as well as with complex general circulation models (GCMs). Possible climate changes together with the major driving forces of future emissions, such as demographic patterns, economic development and environmental conditions, provide the basis for the assessment of vulnerability, possible adverse impacts and adaptation strategies and policies to climate change. The major driving forces of future emissions also provide the basis for the assessment of possible mitigation strategies and policies designed to avoid climate change.

[1] For example, no scenarios are included that explicitly assume implementation of the emission targets in the Kyoto protocol.

[2] Metric tons are used throughout this report. Unless otherwise specified, monetary units are 1990 US dollars (see Chapter 4).

Table 1-1: *Names and chemical formulae or abbreviations of anthropogenic emissions of GHGs and other gases covered in the emissions scenarios.*

Carbon Dioxide	CO_2
Carbon Monoxide	CO
Hydrochlorofluorocarbons	HCFCs
Hydrofluorocarbons	HFCs
Methane	CH_4
Nitrous Oxide	N_2O
Nitrogen Oxides	NO_x
Non-Methane Hydrocarbons	NMVOCs
Perfluorocarbons	PFCs
Sulfur Dioxide	SO_2
Sulfur Hexafluoride	SF_6

Future emissions and the evolution of their underlying driving forces are highly uncertain, as reflected in the very wide range of future emissions paths in the literature. Of the many ways that uncertainties have been classified in the literature (see Box 1-1 in Section 1.2 below), this introduction uses the three categories of Funtowicz and Ravetz (1990): "data uncertainties," "modeling uncertainties" and "completeness uncertainties." This categorization has the advantages of a small number of categories and of clear descriptive titles. Data uncertainties reflect the reality that most historical and base year data sets are neither fully complete nor fully reliable. This is certainly true for data on population, energy consumption, energy efficiency, gross world product, energy resources and reserves, and probably true for every parameter mentioned in this report. Modeling uncertainties refer both to the approximations necessary in any model of complex phenomena like GHG emissions, and to the range of plausible but different modeling approaches, each with its own strengths and weaknesses. Completeness uncertainties encompass, first, relevant factors that can be identified but are nonetheless excluded from an analysis – for example exclusion of criteria other than cost minimization in an energy model, such as energy security, the protection of domestic industries, and free trade. Second, they also include factors that may be relevant but are as essentially unknown to us as jet airplanes were to Thomas Malthus or 3-D seismic techniques in oil exploration were to John D. Rockefeller. The use of scenarios and storylines in this report partially addresses completeness uncertainties related to *known* factors. Completeness uncertainties related to *unknown* factors can, of course, never be persuasively captured by any approach.

The IPCC developed sets of emissions scenarios in 1990 (Houghton *et al.*, 1990) and 1992 (Leggett *et al.*, 1992; Pepper *et al.*, 1992). In 1994 the IPCC formally evaluated the 1992 scenario set (Alcamo *et al.*, 1995) and, in 1996, it initiated the effort described in this report. The new set of emissions scenarios is intended for use in future IPCC assessments and by wider scientific and policymaking communities who analyze the effects of future GHG emissions and develop mitigation and adaptation measures and policies. The emissions profiles of the new scenarios can provide inputs for GCMs and simplified models of climate change. The new scenarios also contain information, such as the level of economic activity, rates of technological change and demographic developments in different world regions, required to assess climate-change impacts and vulnerabilities, adaptation strategies and policies. The same kind of information, in conjunction with emissions trajectories, can serve as a benchmark for the evaluation of alternative mitigation measures and policies. Finally, the new set of scenarios may provide a common basis and an integrative element for the Third Assessment Report (TAR).

IPCC Working Group III (WGIII) appointed the SRES writing team in January 1997. After some adjustments it eventually included more than 50 scientists. Together they represent a broad range of scientific disciplines, regional backgrounds and non-governmental organizations. In particular the team includes representatives of six leading groups with extensive expertise in modeling alternative emissions scenarios. It also includes a number of members who were convening and lead authors in all three earlier IPCC scenario activities (see above). Their expertise and familiarity with earlier IPCC emissions scenario work assured continuity and allowed the SRES effort to build efficiently upon prior work. Appendix II lists the members of the writing team and their affiliations.

The writing team reached a consensus concerning the overall work program. It was agreed that the scenario development process would consist of four major components.

- First, a review of existing global and regional emissions scenarios from the published literature and development of a unique database of 416 global and regional scenarios (accessible at a web site: www-cger.nies.go.jp/cger-e/db/ipcc.html).
- Second, an analysis of the range of the scenarios' main characteristics, their relationships, and their "driving forces" (such as population, economic development, energy consumption, rates of technological change and GHG emissions) and the documentation of the results, some of which are published in the peer-reviewed literature.[3]
- Third, a formulation of narrative "storylines" to describe the main scenario characteristics, the development of quantitative prototype scenarios by six leading groups representing the main modeling approaches from around the world, and the publication of the prototype scenarios on a specially developed IPCC web site (sres.ciesin.org), on web sites of the modeling teams,[4] and in the peer-reviewed literature.[5]
- Fourth, an "open" process through the IPCC web site (sres.ciesin.org) that involves feedback from modeling groups and experts worldwide, followed by the IPCC expert and government reviews that were coordinated by four review editors.

Most of the background material and findings of the assessments conducted by the writing team have been documented in this report and in a series of publications including two special issues of international scientific journals, *Mitigation and Adaptation Strategies for Global Change*[3] and *Technological Forecasting and Social Change*.[5]

1.2. What are Scenarios?

Scenarios are images of the future, or alternative futures. They are neither predictions nor forecasts. Rather, each scenario is one alternative image of how the future might unfold. A set of scenarios assists in the understanding of possible future developments of complex systems. Some systems, those that are well understood and for which complete information is available, can be modeled with some certainty, as is frequently the case in the physical sciences, and their future states predicted. However, many physical and social systems are poorly understood, and information on the relevant variables is so incomplete that they can be appreciated only through intuition and are best communicated by images and stories. Prediction is not possible in such cases (see Box 1-1 on uncertainties inherent in scenario analysis).

Scenarios can be viewed as a linking tool that integrates qualitative narratives or stories about the future and quantitative formulations based on formal modeling. As such they enhance our understanding of how systems work, behave and evolve. Scenarios are useful tools for scientific assessments, for learning about complex systems behavior and for policy making (Jefferson, 1983; Davis, 1999). In scientific assessments, scenarios are usually based on an internally consistent and reproducible set of assumptions or theories about the key relationships and driving forces of change, which are derived from our understanding of both history and the current situation. Often scenarios are formulated with the help of numeric or analytic formal models.

Future levels of global GHG emissions are the products of a very complex, ill-understood dynamic system, driven by forces

[3] Anderson, 1998; Alcamo and Swart, 1998; Gaffin, 1998; Gregory, 1998; Gregory and Rogner, 1998; Grübler, 1998; Michaelis, 1998; Morita and Lee, 1998a; Nakićenović *et al.*, 1998; Price *et al.*, 1998.

[4] e.g., www-cger.nies.go.jp/cger-e/db/ipcc.html; www.iiasa.ac.at.

[5] de Vries *et al.*, 2000; Fenhann, 2000; Jiang *et al.*, 2000, Jung *et al.*, 2000; Kram *et al.*, 2000; Mori, 2000; Nakićenović, 2000; Riahi and Roehrl, 2000; Roehrl and Riahi, 2000; Sankovski *et al.*, 2000.

Box 1-1: Uncertainties and Scenario Analysis

In general, there are three types of uncertainty: uncertainty in quantities, uncertainty about model structure and uncertainties that arise from disagreements among experts about the value of quantities or the functional form of the model (Morgan and Henrion, 1990). Sources of uncertainty could be statistical variation, subjective judgment (systematic error), imperfect definition (linguistic imprecision), natural variability, disagreement among experts and approximation (Morgan and Henrion, 1990). Others (Funtowicz and Ravetz, 1990*)* distinguish three main sources of uncertainty: "data uncertainties," "modeling uncertainties" and "completeness uncertainties." Data uncertainties arise from the quality or appropriateness of the data used as inputs to models. Modeling uncertainties arise from an incomplete understanding of the modeled phenomena, or from approximations that are used in formal representation of the processes. Completeness uncertainties refer to all omissions due to lack of knowledge. They are, in principle, non-quantifiable and irreducible.

Scenarios help in the assessment of future developments in complex systems that are either inherently unpredictable, or that have high scientific uncertainties. In all stages of the scenario-building process, uncertainties of different nature are encountered. A large uncertainty surrounds future emissions and the possible evolution of their underlying driving forces, as reflected in a wide range of future emissions paths in the literature. The uncertainty is further compounded in going from emissions paths to climate change, from climate change to possible impacts and finally from these driving forces to formulating adaptation and mitigation measures and policies. The uncertainties range from inadequate scientific understanding of the problems, data gaps and general lack of data to inherent uncertainties of future events in general. Hence the use of alternative scenarios to describe the range of possible future emissions.

For the current SRES scenarios, the following sources of uncertainties are identified:
Choice of Storylines. Freedom in choice of qualitative scenario parameter combinations, such as low population combined with high gross domestic product (GDP), contributes to scenario uncertainty.
Authors Interpretation of Storylines. Uncertainty in the individual modeler's translation of narrative scenario storyline text in quantitative scenario drivers. Two kinds of parameters can be distinguished:

- Harmonized drivers such as population, GDP, and final energy (see Section 4.1. in Chapter 4). Inter-scenario uncertainty is reduced in the harmonized runs as the modeling teams decided to keep population and GDP within certain agreed boundaries.
- Other assumed parameters were chosen freely by the modelers, consistent with the storylines.

Translation of the Understanding of Linkages between Driving Forces into Quantitative Inputs for Scenario Analysis. Often the understanding of the linkages is incomplete or qualitative only. This makes it difficult for modelers to implement these linkages in a consistent manner.
Methodological Differences.

- Uncertainty induced by conceptual and structural differences in the way models work (model approaches) and in the ways models are parameterized.
- Uncertainty in the assumptions that underlie the relationships between scenario drivers and output, such as the relationship between average income and diet change.

Different Sources of Data. Data differ from a variety of well-acknowledged scientific studies, since "measurements" always provide ranges and not exact values. Therefore, modelers can only choose from *ranges* of input parameters for. For example:

- Base year data.
- Historical development trajectories.
- Current investment requirements.

Inherent Uncertainties. These uncertainties stem from the fact that unexpected "rare" events or events that a majority of researchers currently consider to be "rare future events" might nevertheless occur and produce outcomes that are fundamentally different from those produced by SRES model runs.

such as population growth, socio-economic development, and technological progress; thus to predict emissions accurately is virtually impossible. However, near-term policies may have profound long-term climate impacts. Consequently, policy-makers need a summary of what is understood about possible future GHG emissions, and given the uncertainties in both emissions models and our understanding of key driving forces, scenarios are an appropriate tool for summarizing both current understanding and current uncertainties. For such scenarios to be useful for climate models, impact assessments and the design of mitigation and adaptation policies, both the main outputs of the SRES scenarios (emissions) and the main inputs or driving forces (population growth, economic growth, technological, e.g., as it affects energy and land-use) are equally important.

GHG emissions scenarios are usually based on an internally consistent and reproducible set of assumptions about the key relationships and driving forces of change, which are derived from our understanding of both history and the current situation. Often these scenarios are formulated with the help of formal models. Such scenarios specify the future emissions of GHGs in quantitative terms and, if fully documented, they are also reproducible. Sometimes GHG emissions scenarios are less quantitative and more descriptive, and in a few cases they do not involve any formal analysis and are expressed in qualitative terms. The SRES scenarios involve both qualitative and quantitative components; they have a narrative part called "storylines" and a number of corresponding quantitative scenarios for each storyline. Figure 1-1 illustrates the interrelated nature of these alternative scenario formulations.

Although no scenarios are value free, it is often useful to distinguish between normative and descriptive scenarios. Normative (or prescriptive) scenarios are explicitly values-based and teleologic, exploring the routes to desired or undesired endpoints (utopias or dystopias). Descriptive scenarios are evolutionary and open-ended, exploring paths into the future. The SRES scenarios are descriptive and should not be construed as desirable or undesirable in their own right. They are built as descriptions of possible, rather than preferred, developments. They represent pertinent, plausible, alternative futures. Their pertinence is derived from the need for policy makers and climate-change modelers to have a basis for assessing the implications of future possible paths for GHG and SO_2 emissions, and the possible response strategies. Their plausibility is based on an extensive review of the emissions scenarios available in the literature, and has been tested by alternative modeling approaches, by peer review (including the "open process" through the IPCC web site), and by the IPCC review and approval processes. Good scenarios are challenging and court controversy, since not everybody is comfortable with every scenario, but used intelligently they allow policies and strategies to be designed in a more robust way.

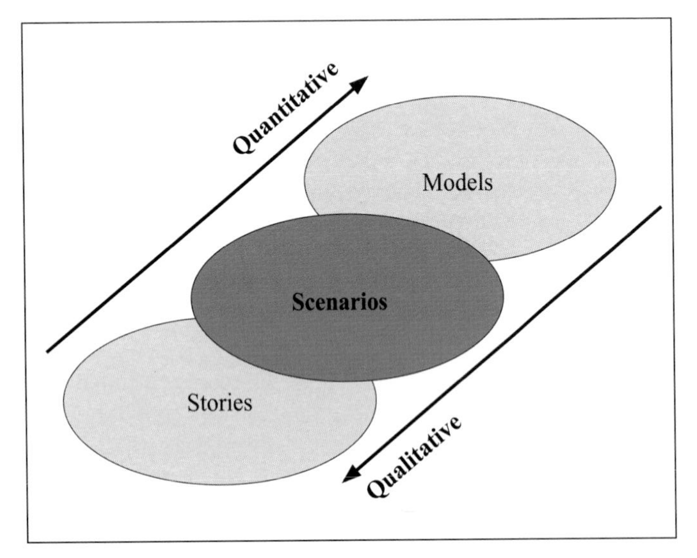

Figure 1-1: Schematic illustration of alternative scenario formulations, from narrative storylines to quantitative formal models.

1.3. Purposes and Uses of SRES Emissions Scenarios

The assessment of climate change dictates a global perspective and a very long time horizon that covers periods of at least a century. As the prediction of future anthropogenic GHG emissions is impossible, alternative GHG emissions scenarios become a major tool for the analysis of potential long-range developments of the socio-economic system and corresponding emission sources.

However, to develop scenarios for a period of 100 years is a relatively new field. Difficulties arise not only from large scientific uncertainties and data inadequacies, but also because people are not trained to think in such time-spans. We are educated in narrow disciplines, and our ability to model complex systems, at the global level, is still in its infancy. For example, within the next century technological discontinuities should be expected, and possibly major shifts in societal values and in the balance of geopolitical power. The study of past trends over such long periods is hampered because most databases are incomplete if more than 50 years old. Given these gaps in our data, methods and understanding, scenarios are the best way to integrate our demographic, economic, societal and technological knowledge with our understanding of ecologic systems to evaluate sources and sinks of GHG emissions. Scenarios as an integration tool in the assessment of climate change allow a role for intuition, analysis and synthesis, and thus we turn to scenarios to take advantage of these features and aid the assessment of future climate change, impacts, vulnerabilities, adaptation and mitigation. Since the scenarios focus on the century time scale, tools are used that have been developed for this purpose. These tools are less suitable for the analysis of near-term developments, so this report does not intend to provide reliable projections for the near term.

The IPCC's 1994 evaluation of its 1992 emissions scenarios identified four principal uses (Alcamo *et al.*, 1995):

- To provide input for evaluating climatic and environmental consequences of alternative future GHG emissions in the absence of specific measures to reduce such emissions or enhance GHG sinks.
- To provide similar input for cases with specific alternative policy interventions to reduce GHG emissions and enhance sinks.
- To provide input for assessing mitigation and adaptation possibilities, and their costs, in different regions and economic sectors.
- To provide input to negotiations of possible agreements to reduce GHG emissions.

The SRES emissions scenarios are intended for the first, third and fourth uses. They do not include any additional (explicit) policies or measures directed at reducing GHG sources and enhancing sinks. Thus, they cannot be directly applied to the second purpose of emissions scenarios. Instead, they could be used as reference cases for the introduction of specific policy interventions and measures in new model runs that share the same specifications for the other principal driving forces of future emissions. However, the SRES emissions scenarios include a host of other policies and measures that are not directed at reducing sources and increasing sinks of GHGs, but that nevertheless have an indirect effect on future emissions. For example, policies directed at achieving greater environmental protection may also lead to lower emissions of GHGs. Moreover, afforestation and reforestation measures increase CO_2 sinks, and a shift to renewable energy sources reduces the sources of emissions.

Within three of the broad objectives listed above, the new SRES emissions scenarios are also intended to meet the specific needs of three main IPCC user communities:

- Working Group I (WGI), which includes climate modelers who need future emission trajectories for GHGs and aerosol precursors as inputs for the GCMs used to develop climate change scenarios.
- Working Group II (WGII), which analyzes climate impacts and adaptation policies, first need the climate-change scenarios produced by WGI's climate modelers. Second, analysts need to know the socio-economic changes associated with specific emissions scenarios, as impacts of climate change on ecosystems and people depend on many factors. Among these are whether the people are numerous or few, rich or poor, free to move or relatively immobile, and included or excluded from world markets in technologies, food, etc.
- WGIII, which analyzes potential mitigation policies for climate change, also needs to know the socio-economic settings against which policy options are to be evaluated. Are markets open or protected? Are technological options and economic resources plentiful or scarce? Are people vulnerable or adaptable?

The interests of these three user groups create certain requirements that the SRES scenarios attempt to fulfill. For example, climate modelers and those who analyze climate impacts need scenarios on the order of 100 years because of the long response time of the climate system. At the same time adaptation-policy analysis tends to be focused more on the medium-term, around 20 to 50 years. The SRES scenarios attempt to include enough information and specific details to be useful to these groups. Spatially explicit emissions and socio-economic variables are required for slightly different reasons. Some emissions, such as the SO_2 emissions that contribute to sulfate aerosols, have impacts that vary depending on where they are emitted. Climate modelers therefore need spatially explicit emission estimates. Similarly, impacts depend on the geographic patterns of changing temperatures, rainfall, humidity and cloud cover, and how these compare to evolving socio-economic patterns in specific scenarios. Impact modelers therefore need spatially explicit estimates of, in particular, population growth, migration, and the economic variables that reflect the expected adaptability or vulnerability of different populations and regional economies to different regional climate changes.

Taking the above audiences and purposes into account, the following more precise specifications for the new SRES scenarios were developed. The new scenarios should:

- cover the full range of radiatively important gases, which include direct and indirect GHGs and SO_2;
- have sufficient spatial resolution to allow regional assessments of climate change in the global context;
- cover a wide spectrum of alternative futures to reflect relevant uncertainties and knowledge gaps;
- use a variety of models to reflect methodological pluralism and uncertainty;
- incorporate input from a wide range of scientific disciplines and expertise from non-academic sources through an open process;
- exclude additional initiatives and policies specifically designed to reduce climate change;
- cover and describe to the extent possible a range of policies that could affect climate change although they are targeted at other issues, for example, reductions in SO_2 emissions to limit acid rain;
- cover as much as possible of the range of major underlying "driving forces" of emissions scenarios identified in the open literature;
- be transparent with input assumptions, modeling approaches and results open to external review;
- be reproducible – input data and methodology are documented adequately enough to allow other researchers to reproduce the scenarios; and
- be internally consistent – the various input assumptions and data of the scenarios are internally consistent to the extent possible.

1.4. Review of Past IPCC Emissions Scenarios

The 1994 IPCC evaluation of the usefulness of the IPCC 1992 (IS92) scenarios found that for the purposes of driving atmospheric and climate models, the CO_2 emissions trajectories in these provided a reasonable reflection of variations found in the open literature. Specifically, their global CO_2 emissions spanned most of the range of other scenarios identified in the literature. Figure 1-2 shows the global, energy-related and industrial CO_2 emissions of the IS92 scenarios ranging from very high emissions of 35.8 GtC to very low emissions of 4.6 GtC by 2100 (corresponding to a sixfold increase and a decline by a third compared to 1990 levels, respectively). The shaded area in Figure 1-2 indicates the coverage of the IS92 scenarios while the "spaghetti-like" curves indicate other energy-related emissions scenarios found by the IPCC review to be representative of the scenarios available in the open literature at that time (Alcamo *et al.*, 1995). In the open literature, emissions trajectories for other gases were extremely thin in many instances, but the IS92 cases were not dissimilar to these.

Another important recommendation of the 1994 IPCC review was that, given the degree of uncertainty about future climate

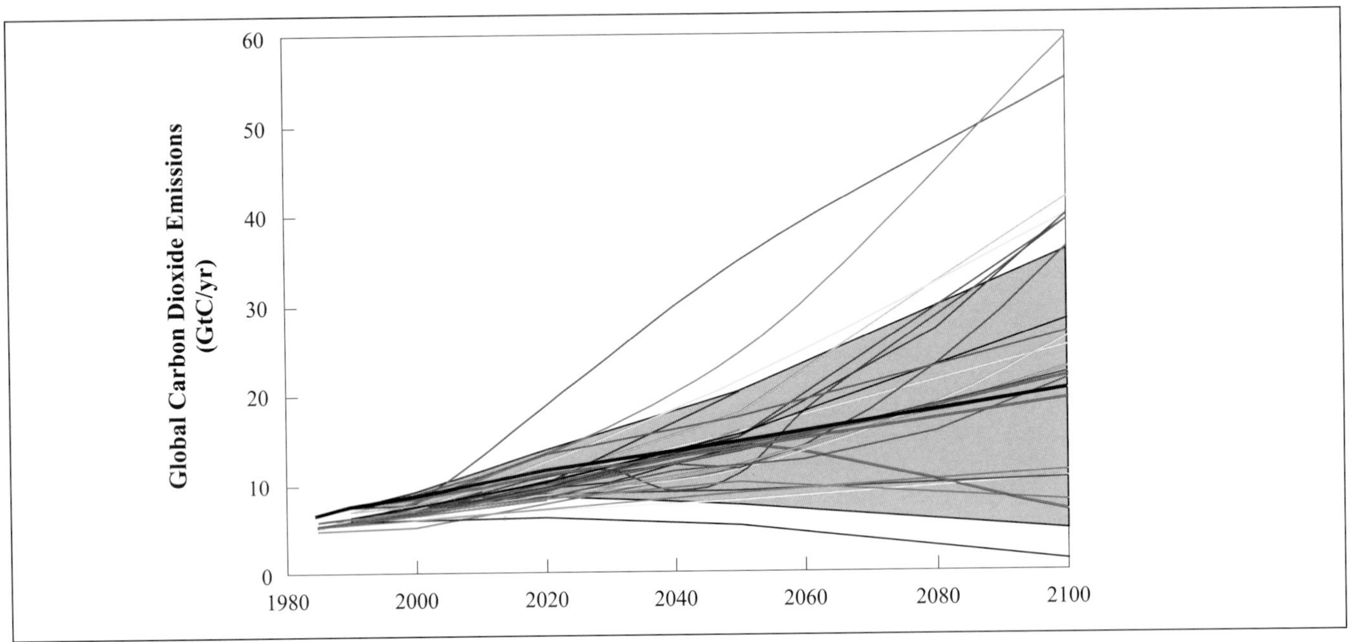

Figure 1-2: Energy-related and industrial global CO_2 emissions for scenarios reviewed in the IPCC Report Climate Change 1994 (Alcamo *et al.*, 1995). The shaded area indicates coverage of IS92 scenarios while the "spaghetti-like" curves indicate other energy-related emissions scenarios found by the IPCC review to be representative of the scenarios available in the open literature at that time. (Individual scenarios are listed in the Appendix of Alcamo *et al.*, 1995.)

change, analysts should use the full range of IS92 emissions as input to climate models rather than a single scenario. This is in stark contrast to the actual use of one particular scenario in the set, the IS92a scenario, as the reference case in numerous studies. In fact, the IS92a scenario is often referred to in climate change modeling and impact studies as the "business-as-usual" scenario and used as the only reference emissions trajectory. The review concluded that the mere fact of the IS92a being an intermediate, or central, CO_2 emissions scenario at the global level does not equate it with being the most likely scenario. Indeed, the conclusion was that there was no objective basis on which to assign likelihood to any of the scenarios. Furthermore, the IS92a scenario was shown to be "central" for only a few of its salient characteristics such as global population growth, global economic development and global CO_2 emissions. In other ways, IS92a was found not to be central with respect to the published literature, particularly in some of its regional input assumptions. The same is the case with the new set of SRES scenarios, as is shown below. No single scenario can be central with respect to all the characteristics relevant for different uses of emissions scenarios and there is no objective way to assign likelihood to any of the scenarios.

1.5. Why New IPCC Emissions Scenarios?

The 1994 IPCC evaluation of the IS92 scenarios found that the scenarios were innovative at the time of their publication and path-breaking in their coverage of the full range of GHG and SO_2 emissions, on both a global and a regional basis. The review also identified a number of weaknesses. These included the limited range of CO_2 intensities of energy (CO_2 emissions per unit energy) reflected in the six scenarios and the absence of any scenario with significant closure in the income gap between developed and developing countries, even after a full century (Parikh, 1992). Across all six scenarios the per capita income in the developing countries grows only to a share of 17% to 26% of that of the industrial countries, compared to 6% today.

Furthermore, since the development of the IS92 scenarios much has changed in our understanding of both the possible future GHG emissions and the climate impacts that might result. The most straightforward change arises because the IS92 scenarios had to estimate data for the base year 1990 (actual data were unavailable at the time); 1990 is no longer a forecast year and actual data can be used. But other changes are at least as important. It is now recognized that in some regions SO_2 emissions may be as important to climate change as all non-CO_2 greenhouse related gases combined – at least in the near-term. As a result the rapid growth of SO_2 emissions in the IS92 scenarios has been questioned. The 1994 evaluation of the IS92 scenario series (Alcamo *et al.*, 1995) concluded, for instance, that projected emissions in the Organization for Economic Cooperation and Development (OECD) countries did not reflect recent legislative changes such as the Amendments of the Clean Air Act in the US or the Second European Sulfur Protocol. Thus, factors other than climate change, namely regional and local air quality, may well prompt limits on future SO_2 emissions independent of global warming concerns. Restructuring in Eastern and Central Europe and the Newly Independent States of the former Soviet Union has powerfully affected economic activity, with reductions in CO_2 emissions unforeseen in the IS92 scenarios. The advent of integrated assessment (IA) models has also made it possible now to construct internally consistent emissions scenarios that jointly consider the interactions among energy use, the economy, and land-use change. Some IA models account for interactions in both directions between driving forces of GHG and SO_2 emissions and possible impacts of climate change. Progress has also been made in achieving greater consistency among scenario characteristics such as the rates of technological change in different sectors.

Owing to these advances, the 1994 IPCC review of the IS92 scenarios concluded that new scenarios, if developed, should include these improvements:

- estimation of emissions baselines and future non-CO_2 emissions, particularly from land use;
- incorporation of the latest information on economic restructuring throughout the world;
- expand the range of economic-development pathways, including a narrowing of the income gap between developing and industrialized countries;
- examine different trends in and rates of technological change;
- evaluate possible consequences of trade and market liberalization and privatization; and
- reflect current emission commitments in connection with the United Nations Framework Convention on Climate Change (UNFCCC, 1997).

However, the Terms of Reference for SRES explicitly preclude additional initiatives, measures and policies specifically designed to reduce climate change (see Appendix I).

1.6. SRES Approach and Process

The 1994 IPCC review also offered recommendations about the process by which a new set of scenarios might be prepared. It recommended that the IPCC or another suitable organization act as an "umbrella" under which different groups could develop comparable, comprehensive emissions scenarios. They further recommended that the process for developing scenarios should draw on increasing experience in scenario harmonization and model calculations, and that it should emphasize:

- openness to broad participation by the research (and stakeholder) community, particularly from developing countries and countries with economies in transition;
- extensive documentation of modeling assumptions, inputs and outputs;
- pluralism and diversity of groups, approaches and methods, although the final set of scenarios should be harmonized;
- comparability across the scenarios that necessitates standardized reporting conventions for model inputs and outputs; and
- harmonization of emission scenarios in the set to provide common benchmarks for scenarios from different modeling groups.

Further recommendations included wide dissemination of the scenarios to countries, international organizations, non-governmental organizations and the scientific community. As part of this effort, a central archive should be established to make available the results of new scenarios to any group. The archive should also make available some aspects of the models and input assumptions used to derive the scenarios. In addition, special efforts are needed to improve the capabilities of researchers to analyze and develop scenarios, especially in developing countries and countries with economies in transition.

As described at the beginning of this chapter, IPCC WGIII appointed the SRES writing team in January 1997, and the team reached early consensus on the four major components of an overall work program (outlined in Section 1.1). The SRES team worked in close collaboration with colleagues on the IPCC Task Group on Climate Scenarios for Impact Assessment (TGCIA) and with colleagues from all three IPCC working groups. As is evident from the four components of the work program, it was agreed that the process be an open one with no "official" model and no exclusive "expert teams." High priority was given to wide participation so that any research group capable of preparing scenarios for any region could participate. In 1997 the IPCC advertised in a number of relevant scientific journals and other publications to solicit wide participation in the process. All global modeling teams and regional modelers were invited and encouraged to participate. In this way, researchers with local expertise from both developing and developed regions could contribute to the global exercise even if their own research was exclusively regional. To facilitate participation and improve the usefulness of the new scenarios, the open-process web site mentioned above was created. The open process provided a wide access to preliminary marker (see below) SRES scenario results and greatly facilitated coordination among the writing team. It also provided feedback about the needs of those who would use the final scenarios, and suggestions for improvements. The open process also served to document all relevant results and associated assumptions for the preliminary scenarios developed by the participating modeling groups.

Four storylines were developed by the whole writing team in an iterative process that identified driving forces, key uncertainties, and quantitative scenario families. The team was fortunate to have a number of skilled practitioners in scenario building. The process of quantifying the four storylines deserves some elaboration. The storylines were essentially complete by January 1998, at which time the modeling groups represented on the writing team began to quantify them. For each storyline, one modeling group was given principal responsibility, and the quantification produced by that group is referred to as the "marker scenario" for that storyline. The four preliminary marker scenarios were posted on the web site of the open process. The choice of the markers was based on extensive discussion of:

- range of emissions across all of marker scenarios;
- which of the initial quantifications (by the modelers) reflected the storyline;
- preference of some of the modeling teams and features of specific models;
- use of different models for the four markers.

As a result the markers are not necessarily the median or mean of the scenario family, but are those scenarios considered by the SRES writing team as illustrative of a particular storyline. These scenarios have received the closest scrutiny of the entire writing team and via the SRES open process compared to other scenario quantifications. The marker scenarios are also those SRES scenarios that have been most intensively tested in terms of reproducibility. As a rule, different modeling teams have attempted to replicate the model quantification of a particular marker scenario. Available time and resources have not allowed a similar exercise to be conducted for all SRES scenarios, although some effort was devoted to reproduce the four scenario groups[6] that constitute different interpretations of one of the four storylines (see Figure 1-4) with different models. Additional versions of the preliminary marker scenarios by different modeling teams and other scenarios that give alternative quantitative interpretations of the four storylines constitute the final set of 40 SRES scenarios. This also means that the 40 scenarios are not independent of each other as they are all based on four storylines. However, differences in modeling approaches have meant that not all of the scenarios provide estimates for all the direct and indirect GHG emissions for all the sources and sectors. The four SRES marker scenarios cover all the relevant gas species and emission categories comprehensively

The four marker scenarios were posted on the IPCC web site (sres.ciesin.org) in June 1998, and the open scenario review process through the IPCC web site lasted until January 1999. The submissions invited through the open process and web site fell into three categories (see Appendix VI):

- additional scenarios published in the reviewed literature that had not been included in the scenario database (see Appendix V);
- new scenarios based on the SRES marker scenarios; and
- general suggestions to improve the work of the SRES writing team as posted on the web site (preferably based on referenced literature).

The submissions were used to revise the marker scenarios and to develop additional alternatives within each of the four scenario families. The result is a more complete, refined set of 40 new scenarios that reflects the broad spectrum of modeling approaches and regional perspectives. The preliminary marker scenarios posted on the web site were provided also to climate modelers, with the approval of the IPCC Bureau.

[6] Please note that in the Summary for Policymakers, two of these groups were merged into one. See also the endnote in Box 1-2.

1.7. SRES Emissions Scenarios

1.7.1. Literature Review and Analysis

The first step in formulation of the SRES emissions scenarios was to review both the published scenario literature and the development of the scenario database accessible through the web site (www-cger.nies.go.jp/cger-e/db/ipcc.html). Chapters 2 and 3 give a more detailed description of the literature review and analysis. Figure 1-3 shows the global energy-related CO_2 emission paths from the database as "spaghetti" curves for the period to 2100 against the background of the historical emissions from 1900 to 1990. These curves are plotted against an index on the vertical axis rather than as absolute values because of the large differences and discrepancies for the values assumed for the base year 1990. These sometimes arise from genuine differences among the scenarios (e.g., different data sources, definitions) and sometimes from different base years assumed in the analysis or in alternative calibrations.[7] The differences among the scenarios in the specification of the base year illustrate the large genuine scientific and data uncertainty that surrounds emissions and their main driving forces captured in the scenarios. The literature includes scenarios with additional climate polices, which are sometimes referred to as mitigation or intervention scenarios. There are many ambiguities associated with the classification of emissions scenarios into those that include additional climate initiatives and those that do not. Many cannot be classified in this way on basis of the information available from the database. Figure 1-3 indicates the ranges of emissions in 2100 from scenarios that apparently include additional climate initiatives (designated as intervention emissions range), those that do not (non-intervention) and those that cannot be assigned to either of these two categories (non-classified). This classification is based on the subjective evaluation of the scenarios in the database by the members of the writing team and is explained in Chapter 2. The range of the whole sample of scenarios has significant overlap with the range of those that cannot be classified and they share virtually the same median (15.7 and 15.2 GtC in 2100, respectively) but the non-classified scenarios do not cover the high part of the range. Also, the range of the scenarios that apparently do not include climate polices (non-intervention) has considerable overlap with the other two ranges (lower bound is higher) but with a significantly higher median (of 21.3 GtC in 2100).

[7] The 1990 emissions from energy production and use are estimated by Marland *et al.* (1994) at 5.9 GtC excluding cement production. The 1990 base year values in the scenarios reviewed range from 4.8 (CETA/EMF14, Scenario MAGICC CO_2) to 6.4 GtC (ICAM2/EMF14); see Dowlatabadi and Kandlikar 1995; Peck and Teisberg, 1995.

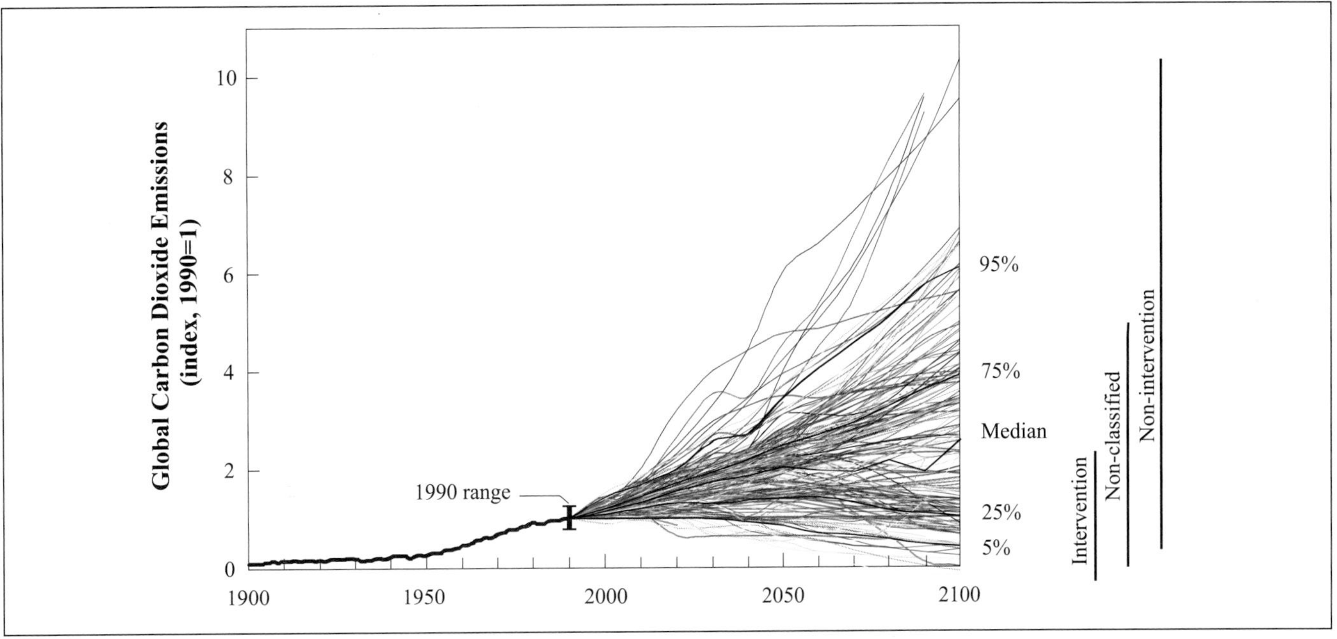

Figure 1-3: Global energy-related and industrial CO_2 emissions – historical development and future scenarios, shown as an index (1990 = 1). The median (50th), the 5th, 25th, 75th and 95th percentiles of the frequency distribution are shown. The statistics associated with scenarios from the literature do not imply probability of occurrence (e.g., the frequency distribution of the scenarios may be influenced by the use of IS92a as a reference for many subsequent studies). The emissions paths indicate a wide range of future emissions. The range is also large in the base year 1990 and is indicated by an "error" bar. To separate the variation due to base-year specification from different future paths, emissions are indexed for the year 1990, when actual global energy-related and industrial CO_2 emissions were about 6 GtC. The coverage of CO_2 emissions sources may vary across the 256 different scenarios from the database included in the figure. The scenario samples used vary across the time steps (for 1990 256 scenarios, for 2020 and 2030 247, for 2050 220, and for 2100 190 scenarios). Also shown, as vertical bars on the right of the figure, are the ranges of emissions in 2100 for scenarios from the literature that apparently include additional climate initiatives (designated as "intervention" scenarios emissions range), those that do not ("non-intervention"), and those that cannot be assigned to either of these two categories ("non-classified"). This classification is based on the subjective evaluation of the scenarios in the database by the members of the writing team and is explained in Chapter 2. Data sources: Morita and Lee, 1998a, 1998b; Nakićenović *et al.*, 1998.

Historically, gross CO_2 emissions have increased at an average rate of about 1.7% per year since 1900 (Nakićenović *et al.*, 1996); if that historical trend continues global emissions would double during the next three to four decades and increase more than sixfold by 2100. Many scenarios in the database describe such a development. However, the range is very large around this historical trend so that the highest scenarios envisage more than a sevenfold increase of global emissions by 2100 as compared with 1990, while the lowest have emissions lower than those of today. The median and the average of the scenarios lead to about a threefold emissions increase over the same time period or to about 16 GtC. This is lower than the median of the IS92 set and is lower than the IS92a scenario, often considered as the "central" scenario with respect to some of its tendencies. However, the distribution of emissions is asymmetric. The thin emissions "tail" that extends above the 95th percentile (i.e., between the six and tenfold increase of emissions by 2100 compared to 1990) includes only a few scenarios. The range of other emissions and the main scenario driving forces (such as population growth, economic development and energy production, conversion and end use) for the scenarios documented in the database is also large and comparable to the variation of CO_2 emissions. Statistics associated with scenarios from the literature do not imply probability of occurrence or likelihood of the scenarios. The frequency distribution of the database may be influenced by the use of IS92a as a reference for scenario studies.

1.7.2. Narrative Storylines and Scenario Quantifications

Given these large ranges of future emissions and their driving forces, there are an infinite number of possible alternative futures to explore. The SRES scenarios cover a finite, albeit a very wide, range of future emissions. To facilitate the process of identifying alternative future developments, the writing team decided to describe their scenarios coherently by narrative storylines. The storylines describe developments in many different economic, technical, environmental and social dimensions. The main reasons for formulating storylines are to:

- help the writing team to think more coherently about the complex interplay between scenario driving forces within each and across alternative scenarios;

- make it easier to explain the scenarios to the various user communities by providing a narrative description of alternative futures that goes beyond quantitative scenario features;
- make the scenarios more useful, in particular to analysts who contribute to IPCC WGII and WGIII; the social, political and technological context described in the scenario storylines is all-important in analyzing the effects of policies either to adapt to climate change or to reduce GHG emissions; and
- provide a guide for additional assumptions to be made in detailed climate impact and mitigation analyses, because at present no single model or scenario can possibly respond to the wide variety of informational and data needs of the different user communities of long-term emissions scenarios.

The writing team consciously applied the principle of Occam's Razor (i.e., the economy of thought, Eatwell *et al.*, 1998). They sought the minimum number of scenarios that could still serve as an adequate basis to assess climate change and that would still challenge policy makers to test possible response strategies against a significant range of plausible futures. The team decided on four storylines, as an even number helps to avoid the impression that there is a "central" or "most likely" case. The writing team wanted more than two storylines to help to illustrate that the future depends on many different underlying dynamics; the team did not want more than four, as it wanted to avoid complicating the process by too many alternatives. The scenarios would cover a wide range of – but not all possible – futures. In particular, there would be no "disaster" scenarios. None of the scenarios include new explicit climate policies. The team decided to carry out sensitivity tests within some of the storylines by considering alternative scenarios with different fossil-fuel reserves, rates of economic growth, or rates of technical change.

The storylines describe developments in many different social, economic, technological, environmental and policy dimensions. The titles of the storylines have been kept simple: A1, A2, B1 and B2. There is no particular order among the storylines; they are listed in the alphabetic and numeric order:

- The A1 storyline and scenario family describes a future world of very rapid economic growth, low population growth, and the rapid introduction of new and more efficient technologies. Major underlying themes are convergence among regions, capacity building and increased cultural and social interactions, with a substantial reduction in regional differences in per capita income. The A1 scenario family develops into four groups that describe alternative directions of technological change in the energy system.[8]
- The A2 storyline and scenario family describes a very heterogeneous world. The underlying theme is self-reliance and preservation of local identities. Fertility patterns across regions converge very slowly, which results in high population growth. Economic development is primarily regionally oriented and per capita economic growth and technological change are more fragmented and slower than in other storylines.
- The B1 storyline and scenario family describes a convergent world with the same low population growth as in the A1 storyline, but with rapid changes in economic structures toward a service and information economy, with reductions in material intensity, and the introduction of clean and resource-efficient technologies. The emphasis is on global solutions to economic, social, and environmental sustainability, including improved equity, but without additional climate initiatives.
- The B2 storyline and scenario family describes a world in which the emphasis is on local solutions to economic, social, and environmental sustainability. It is a world with moderate population growth, intermediate levels of economic development, and less rapid and more diverse technological change than in the B1 and A1 storylines. While the scenario is also oriented toward environmental protection and social equity, it focuses on local and regional levels.

Figure 1-4 schematically illustrates the SRES scenarios. It shows that the scenarios build on the main driving forces of GHG emissions. Each scenario family is based on a common specification of the main driving forces. The four scenario families are illustrated, very simplistically, as branches of a two-dimensional tree. The two dimensions indicate global and regional scenario orientation and development and environmental orientation, respectively. In reality, the four scenarios share a space of a much higher dimensionality given the numerous driving forces and other assumptions needed to define any given scenario in a particular modeling approach.

After determining the basic features and driving forces for each of the four storylines, the teams began to model and quantify them. This resulted in 40 scenarios, each constituting an alternative interpretation and quantification of a storyline. All the interpretations and quantifications associated with a single storyline are called a scenario "family" (see also Box 1-2 on terminology and Chapter 4 for further details).

In all, six models were used to generate the 40 scenarios:

- Asian Pacific Integrated Model (AIM) from the National Institute of Environmental Studies in Japan (Morita *et al.*, 1994);
- Atmospheric Stabilization Framework Model (ASF) from ICF Consulting in the USA (Lashof and Tirpak, 1990; Pepper *et al.*, 1992, 1998; Sankovski *et al.*, 2000);

[8] Please note that in the Summary for Policymakers, two of these groups were merged into one. See also the endnote in Box 1-2.

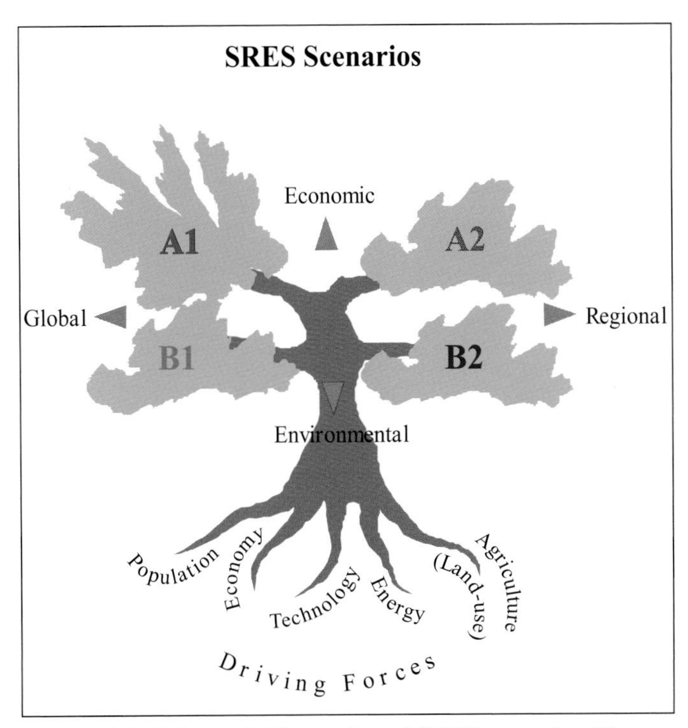

Figure 1-4: Schematic illustration of SRES scenarios. The four scenario "families" are illustrated, very simplistically, as branches of a two-dimensional tree. In reality, the four scenario families share a space of a much higher dimensionality given the numerous assumptions needed to define any given scenario in a particular modeling approach. The schematic diagram illustrates that the scenarios build on the main driving forces of GHG emissions. Each scenario family is based on a common specification of some of the main driving forces.

- Integrated Model to Assess the Greenhouse Effect (IMAGE) from the National Institute for Public Health and Environmental Hygiene (RIVM) (Alcamo *et al.*, 1998; de Vries *et al.*, 1994, 1999, 2000), used in connection with the Dutch Bureau for Economic Policy Analysis (CPB) WorldScan model (de Jong and Zalm, 1991), the Netherlands;
- Multiregional Approach for Resource and Industry Allocation (MARIA) from the Science University of Tokyo in Japan (Mori and Takahashi, 1999; Mori, 2000);
- Model for Energy Supply Strategy Alternatives and their General Environmental Impact (MESSAGE) from the International Institute of Applied Systems Analysis (IIASA) in Austria (Messner and Strubegger, 1995; Riahi and Roehrl, 2000); and the
- Mini Climate Assessment Model (MiniCAM) from the Pacific Northwest National Laboratory (PNNL) in the USA (Edmonds *et al.*, 1994, 1996a, 1996b).

These six models are representative of emissions scenario modeling approaches and different IA frameworks in the literature and include so-called top-down and bottom-up models.

The six models have different regional aggregations. The writing team decided to group the various global regions into four "macro-regions" common to all different regional aggregations across the six models. The four macro-regions (see Appendix III) are broadly consistent with the allocation of the countries in the United Nations Framework Convention on Climate Change (UNFCCC, 1997), although the correspondence is not exact because of changes in the countries listed in Annex I of the UNFCCC:

- The OECD90 region groups together all countries that belong to the OECD as of 1990, the base year of the participating models, and corresponds to Annex II countries under UNFCCC (1992).
- The REF region comprises those countries undergoing economic reform and groups together the East European countries and the Newly Independent States of the former Soviet Union. It includes Annex I countries outside Annex II as defined in UNFCCC (1992).
- The ASIA region stands for all developing (non-Annex I) countries in Asia.
- The ALM region stands for rest of the world and includes all developing (non-Annex I) countries in Africa, Latin America and the Middle East.

In other words, the OECD90 and REF regions together correspond to the developed (i.e., industrialized) countries while the ASIA and ALM regions together correspond to the developing countries. The OECD90 and REF regions are consistent with the Annex I countries in the Framework Convention on Climate Change, while the ASIA and ALM regions correspond to the non-Annex I countries.

1.7.3. The Range of SRES Emissions and their Implications

The 40 SRES scenarios cover the full range of GHG and SO_2 emissions consistent with the storylines and underlying ranges of driving forces from studies in the literature as documented in the SRES database. The four marker scenarios are characteristic of the four scenario families and jointly capture most of the ranges of emissions and driving forces spanned by the full set of scenarios. Figure 1-6 illustrates the range of global energy-related and industrial CO_2 emissions for the 40 SRES scenarios against the background of all the emissions scenarios in the SRES scenario database shown in Figure 1-3. Figure 1-6 also shows a range of emissions of the four scenario families

Figure 1-6 shows that the SRES scenarios cover most of the range of global energy-related CO_2 emissions from the literature, from the 95th percentile at the high end of the distribution down to low emissions just above the 5th percentile of the distribution. Thus, they only exclude the most extreme emissions scenarios found in the literature – those situated in the tails of the distribution. What is perhaps more important is

Box 1-2

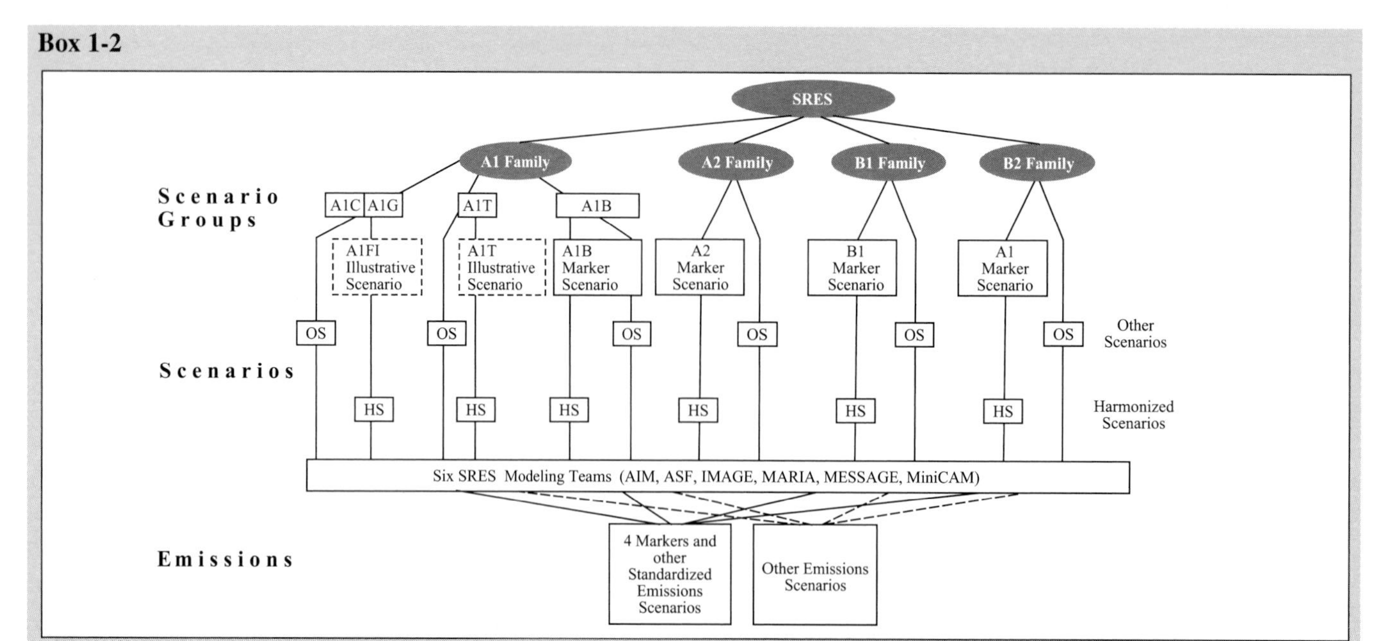

Figure 1-5: Schematic illustration of SRES scenarios. The set of scenarios consists of four scenario families: A1, A2, B1 and B2. Scenario family A1 is further subdivided into four scenario groups: A1C, A1G, A1B and A1T, (see also note below), resulting in seven scenario groups together with the other three scenario families. Each family and group consists of a number of scenarios. Some of them have "harmonized" driving forces and share the same prespecified population and gross world product (a few that also share common final energy trajectories are called "fully harmonized"). These are marked as "HS" for harmonized scenarios. One of the harmonized scenarios, originally posted on the open-process web site, is called a "marker scenario." All other scenarios of the same family based on the quantification of the storyline chosen by the modeling team are marked as "OS." Six modeling groups developed the set of 40 emissions scenarios. The GHG and SO_2 emissions of the scenarios were standardized to share the same data for 1990 and 2000 on request of the user communities. The time-dependent standardized emissions were also translated into geographic distributions.

SRES Terminology

Model: a formal representation of a system that allows quantification of relevant system variables.
Storyline: a narrative description of a scenario (or a family of scenarios) highlighting the main scenario characteristics, relationships between key driving forces, and the dynamics of the scenarios.
Scenario: a description of a potential future, based on a clear logic and a quantified storyline.
Family: scenarios that have a similar demographic, societal, economic and technical-change storyline. Four scenario families comprise the SRES: A1, A2, B1 and B2.
Group: scenarios within a family that reflect a variation of the storyline. The A1 scenario family includes four groups designated by A1T, A1C, A1G and A1B (see also note below) that explore alternative structures of future energy systems. In the Summary for Policymakers, the A1C and A1G groups have been combined into one "fossil intensive" A1FI scenario group, thus reducing the number of groups constituting the A1 scenario family to three. The other three scenario families consist of one group each.
Category: scenarios are grouped into four categories of cumulative CO_2 emissions between 1990 and 2100: low, medium-low, medium-high, and high emissions. Each category contains scenarios with a range of different driving forces yet similar cumulative emissions.
Marker: a scenario that was originally posted on the SRES web site to represent a given scenario family. A marker is not necessarily the median or mean scenario.
Illustrative: a scenario that is illustrative for each of the six scenario groups reflected in the Summary for Policymakers of this report (after combining A1G and A1C into a single A1FI group). They include four revised "scenario markers" for the scenario groups A1B, A2, B1 and B2, and two additional illustrative scenarios for the A1FI and AIT groups. See also "(Scenario) Groups" and "(Scenario) Markers."
Harmonized: harmonized scenarios within a family share common assumptions for global population and GDP while fully harmonized scenarios are within 5% of the population projections specified for the respective marker scenario, within 10% of the GDP and within 10% of the marker scenario's final energy consumption.
Standardized: emissions for 1990 and 2000 are indexed to have the same values.
Other scenarios: scenarios that are not harmonized.

Note: During the approval process of the Summary for Policymakers at the 5th Session of WGIII of the IPCC from 8-11 March 2000 in Katmandu, Nepal, it was decided to combine the A1C and A1G groups into one "fossil intensive" group A1FI in contrast to the non-fossil group A1T, and to select two illustrative scenarios from these two A1 groups to facilitate use by modelers and policy makers. This leads to six scenario groups that constitute the four scenario families, three of which are in the A1 family. These six groups all have "illustrative scenarios," four of which are marker scenarios. All scenarios are equally sound. See also Figure SPM-1.

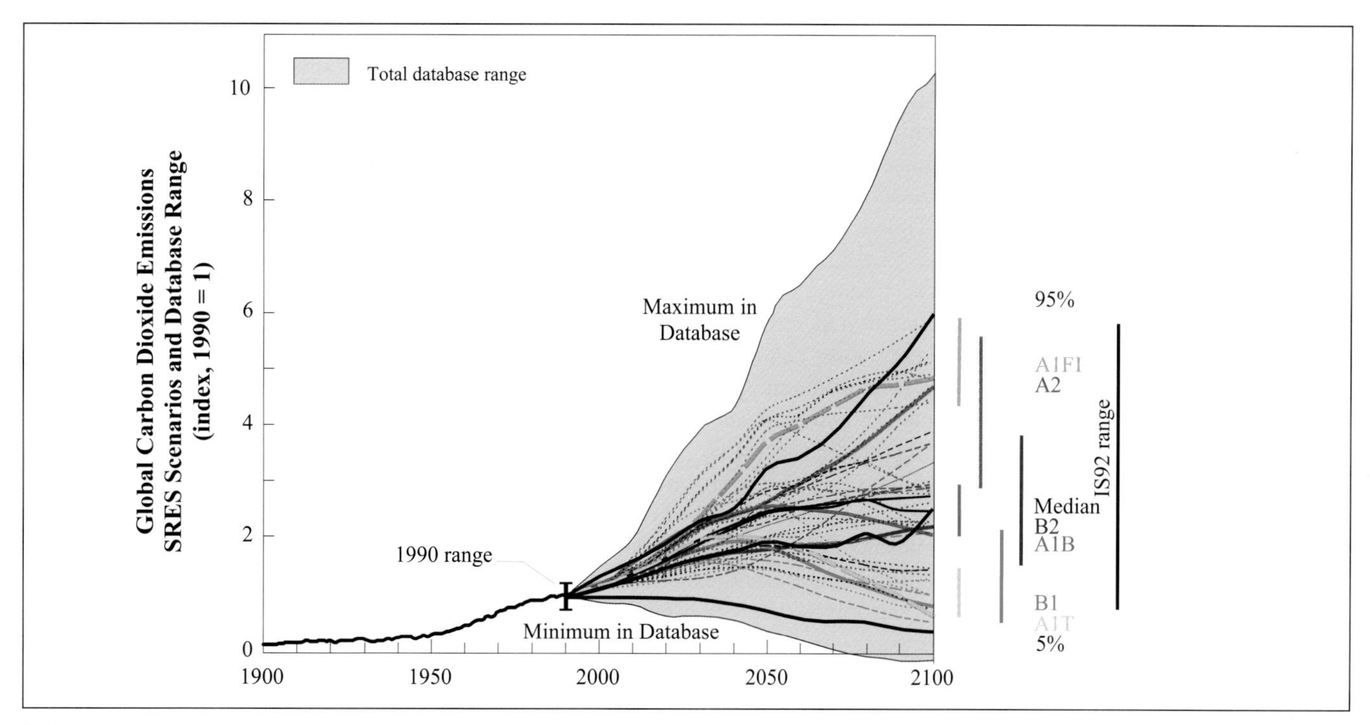

Figure 1-6: Range of global energy-related and industrial CO_2 emissions for the 40 SRES scenarios. The dashed time-paths depict individual SRES scenarios and the shaded area the range of scenarios from the SRES database. The median (50th), 5th, and 95th percentiles of the frequency distribution are shown. The statistics associated with scenarios from the literature do not imply probability of occurrence (e.g., the frequency distribution of the scenarios may be influenced by the use of IS92a as a reference for many subsequent studies). The 40 SRES scenarios are classified into groups that constitute four scenario families. Jointly the scenarios span most of the range of the scenarios in the literature. The emissions profiles are dynamic, ranging from continuous increases to those that curve through a maximum and then decline. The colored vertical bars indicate the range of the four SRES scenario families in 2100. The black vertical bar shows the range of the IS92 scenarios. See also the note in Box 1.2.

that each of the four scenario families covers a sizable part of this distribution, which implies that a similar quantification of driving forces can lead to a wide range of future emissions. More specifically, a given combination of the main driving forces is not sufficient to uniquely determine a future emissions path. There are too many uncertainties. The fact that each of the scenario families covers a substantial part of the literature range also leads to an overlap in the emissions ranges of the four families. This implies that a given level of future emissions can arise from very different combinations of driving forces. This result is of fundamental importance for the assessment of climate-change impacts and possible mitigation and adaptation strategies. Thus, it warrants some further discussion. The emissions paths of the A1 and B2 scenario families perhaps best illustrate these implications.

The A1 scenario family has explored variations in energy systems most explicitly and hence covers the largest part of the scenario distribution shown in Figure 1-6, from the 95th to just above the 10th percentile. The A1 marker (A1B) scenario represents a structure of the future energy mix, balanced in the sense that it does not rely too heavily on one particular energy source. The A1 scenario family includes different groups of scenarios that explore different specific structures of future energy systems, from carbon-intensive development paths to high rates of decarbonization as captured by the two illustrative scenarios that span most of the emissions range for the A1 family. All groups otherwise share the same assumptions about the main driving forces. This indicates that different structures of the energy system can lead to basically the same variation in future emissions as can be generated by different combinations of the other main driving forces – population, economic activities and energy consumption levels. The implication is that decarbonization of energy systems – the shift from carbon-intensive to less carbon-intensive and carbon-free sources of energy – is of similar importance in determining the future emissions paths as other driving forces. Sustained decarbonization requires the development and successful diffusion of new technologies. Thus investments in new technologies during the coming decades might have the same order of influence on future emissions as population growth, economic development and levels of energy consumption taken together.

For example, the comparison of the A1B and B2 marker scenarios indicates that they have similar emissions of about 13.5 and 13.7 GtC by 2100, respectively. The dynamics of the paths are different so that they have different cumulative CO_2 emissions. To facilitate such comparisons, the scenarios were grouped into four categories of cumulative emissions between

1990 and 2100. This categorization can guide comparisons using either scenarios with different driving forces yet similar emissions, or scenarios with similar driving forces but different emissions. This characteristic of SRES scenarios also has very important implications for the assessment of climate-change impacts, mitigation and adaptation strategies. Two future worlds with fundamentally different characteristic features, such as A1B and B2 marker scenarios, also have different cumulative CO_2 emissions and radiative forcing, but very similar CO_2 emissions in 2100. In contrast, scenarios that are in the same category of cumulative emissions can have fundamentally different driving forces and different CO_2 emissions in 2100, but very similar cumulative emissions and radiative forcing. Presumably, adverse impacts and effective adaptation measures would vary among the scenarios from different families that share similar cumulative emissions but have different demographic, socio-economic and technological driving forces. This is another reason for considering the entire range of future emissions in future assessments of climate change. There is no single "best guess" or central scenario.

The SRES emissions scenarios also have different emissions for other GHGs and chemically active species such as carbon monoxide, nitrogen oxides, and volatile organic hydrocarbons. The emissions of other gases follow dynamic patterns much like those shown in Figure 1-6 for CO_2 emissions. Further details about GHG emissions are given in Chapter 5. Emissions of sulfur aerosol precursors portray even more dynamic patterns in time and space than the CO_2 emissions shown in Figure 1-6. Factors other than climate change, namely regional and local air quality, and transformations in the structure of the energy system and end use intervene to limit future emissions. In view of the significant adverse impacts, SO_2 emissions in the scenarios are increasingly controlled outside countries of the OECD. As such the SRES scenarios reflect both recent legislation in North America and in Europe and recent policy initiatives in a number of developing countries aimed at reducing SO_2 emissions (reviewed in more detail in Chapters 3 and 5). As a result, in the second half of the 21st century both the trends and regional patterns of SO_2 emissions evolve differently from those of CO_2 emissions in the SRES scenarios. Emissions outside OECD90 rise initially, most notably in ASIA, and compensate for declining OECD90 emissions. Over the long term, however, SO_2 emissions decline throughout the world, but the timing and magnitude vary across the scenarios. One important implication of this varying pattern of SO_2 emissions is that the historically important, but uncertain negative radiative forcing of sulfate aerosols may decline in the very long run.

An important feature of the SRES scenarios is their implications for radiative forcing. A vigorous increase of global SO_2 emissions during the next few decades across most of the scenarios followed by a decline thereafter will lead to a cooling effect that will differ from the effect that results from the continuously increasing SO_2 emissions in the IS92 scenarios. On one hand, the reduction in global SO_2 emissions reduces the role of sulfate aerosols in determining future climate toward the end of the 21st century and therefore reduces one aspect of uncertainty about future climate change (because the precise forcing effect of sulfate aerosols is highly uncertain). On the other hand, uncertainty increases because of the diversity in spatial patterns of SO_2 emissions in the scenarios. Future assessments of possible climate change need to account for these different spatial and temporal dynamics of GHG and SO_2 emissions, and they need to cover the whole range of radiative forcing associated with the scenarios.

1.8. Structure of the Report

The report consists of six chapters and 11 appendices. After this introductory chapter, Chapters 2 and 3 present the scenario literature review and analysis. Chapters 4 and 5 describe the new SRES scenarios, Chapter 6 summarizes the main findings, and the appendices present the methodologic approach and statistical background material.

Chapter 2 presents the assessment of anthropogenic GHG emissions scenarios and their main driving forces based on an extensive literature review. It describes the unique scenario database developed for this study, which contains over 400 global and regional scenarios. The chapter presents the range of emissions from the scenarios in the literature with associated statistics such, as medians, percentiles and histograms. The main scenario driving forces are analyzed in the same way, from population and economic development to energy.

Chapter 3 reviews the main driving forces of past and possible future anthropogenic GHG emissions. These include demographic, economic and social development, changes in resources and technology, agriculture and land-use change, and policy issues other than those related to climate. The relationships and possible interactions among the driving forces are highly complex and heterogeneous. The focus of the chapter is to provide an overview of the main driving forces and their possible relationships that are particularly relevant for the SRES scenarios.

Chapter 4 presents the narrative scenario storylines and the quantification of the main scenario driving forces with the six SRES IA models. First, an overview of the four storylines is given which describes their main characteristics, relationships and implications. Then, the 40 scenario quantifications of the four storylines with the six models are presented. For each storyline one scenario is designated as a representative marker scenario. Together the 40 scenarios span the range of scenario driving forces and their relationships presented in the previous two chapters.

Chapter 5 documents anthropogenic GHG and SO_2 emissions for the 40 SRES scenarios highlighting the four marker scenarios. First, CO_2 emissions are presented, followed by other GHGs, and by the assessment of indirect effects and aerosols. Together the 40 scenarios span the emissions ranges from the literature and the four marker scenarios jointly characterize both the dynamics of emissions patterns and their ranges.

Chapter 6 summarizes the main characteristics of the SRES scenarios and findings and compares the new scenarios with the IS92 set as well as with other scenarios from the literature. The chapter addresses possible implications of the new scenarios for future assessments of climate change and concludes with recommendations from the writing team for user communities.

Finally, 11 appendices conclude the report. They include for example the SRES Terms of Reference, a technical appendix that describes the six modeling approaches used to formulate the 40 scenarios, the scenario database and tables with further statistics that describe the new scenarios.

References:

Alcamo, J., A. Bouwman, J. Edmonds, A. Grübler, T. Morita, and A. Sugandhy, 1995: An evaluation of the IPCC IS92 emission scenarios. In *Climate Change 1994, Radiative Forcing of Climate Change* and *An Evaluation of the IPCC IS92 Emission Scenarios,* J.T. Houghton, L.G. Meira Filho, J. Bruce, Hoesung Lee, B.A. Callander, E. Haites, N. Harris and K. Maskell (eds.), Cambridge University Press, Cambridge, pp. 233-304.

Alcamo, J., R. Leemans, and E. Kreileman (eds.) 1998: *Global Change Scenarios of the 21st Century. Results from the IMAGE 2.1 model.* Elsevier Science, London.

Alcamo, J., and R. Swart, 1998: Future trends of land-use emissions of major greenhouse gases. *Mitigation and Adaptation Strategies for Global Change,* **3**(2-4), 343-381.

Anderson, D., 1998: On the effects of social and economic policies on future carbon emissions. *Mitigation and Adaptation Strategies for Global Change*, **3**(2-4), 419-453.

Davis, G.R., 1999: Foreseeing a refracted future. *Scenario & Strategy Planning*, **1**(1), 13-15.

De Jong, A., and G. Zalm, 1991: Scanning the future: A long-term scenario study of the world economy 1990-2015. In *Long-term Prospects of the World Economy.* OECD, Paris, pp. 27-74.

De Vries, H.J.M., J.G.J. Olivier, R.A. van den Wijngaart, G.J.J. Kreileman, and A.M.C. Toet, 1994: Model for calculating regional energy use, industrial production and greenhouse gas emissions for evaluating global climate scenarios. *Water, Air Soil Pollution*, **76**, 79-131.

De Vries, B., M. Janssen, and A. Beusen, 1999: Perspectives on global energy futures – simulations with the TIME model. *Energy Policy,* **27**, 477-494.

De Vries, B., J. Bollen, L. Bouwman, M. den Elzen, M. Janssen, and E. Kreileman, 2000: Greenhouse gas emissions in an equity-, environment- and service-oriented world: An IMAGE-based scenario for the next century. *Technological Forecasting & Social Change*, **63**(2-3). (In press).

Dowlatabadi, H., and M. Kandlikar, 1995: *Key Uncertainties in Climate Change Policy: Results from ICAM-2*. Proceedings of the 6th Global Warming Conference, San Francisco, CA.

Eatwell, J., M. Murray, and P. Newman (eds.), 1998: *The New Palgrave, A Dictionary of Economics,* **3**.

Edmonds, J., M. Wise, and C. MacCracken, 1994: *Advanced energy echnologies and climate change.* An Analysis Using the Global Change Assessment Model (GCAM). PNL-9798, UC-402, Pacific Northwest Laboratory, Richland, WA, USA.

Edmonds, J., M. Wise, H. Pitcher, R. Richels, T. Wigley, and C. MacCracken, 1996a: An integrated assessment of climate change and the accelerated introduction of advanced energy technologies: An application of MiniCAM 1.0. *Mitigation and Adaptation Strategies for Global Change*, **1**(4), 311-339.

Edmonds, J., M. Wise, R. Sands, R. Brown, and H. Kheshgi, 1996b: *Agriculture, land-use, and commercial biomass energy.* A Preliminary integrated analysis of the potential role of Biomass Energy for Reducing Future Greenhouse Related Emissions. PNNL-11155, Pacific Northwest National Laboratories, Washington, DC.

Fenhann, J., 2000: Industrial non-energy, non-CO_2 greenhouse gas emissions. *Technological Forecasting & Social Change*, **63**(2-3). (In press).

Funtowicz, S.O., and J.R. Ravetz, 1990: *Uncertainty and Quality in Science for Policy*. Kluwer Academic Publishers, Dordrecht, the Netherlands.

Gaffin, S.R., 1998: World population projections for greenhouse gas emissions scenarios. *Mitigation and Adaptation Strategies for Global Change*, **3**(2-4), 133-170.

Gregory, K., 1998: Factors affecting future emissions of methane from non-land-use sources. *Mitigation and Adaptation Strategies for Global Change*, **3**(2-4), 321-341.

Gregory, K., and H.-H. Rogner, 1998: Energy resources and conversion technologies for the 21st century. *Mitigation and Adaptation Strategies for Global Change*, **3**(2-4), 171-229.

Grübler, A., 1998: A review of global and regional sulfur emission scenarios. *Mitigation and Adaptation Strategies for Global Change*, **3**(2-4), 383-418.

Houghton, J. T., G.J. Jenkins, and J.J. Ephraums (eds.), 1990: *Climate Change: The IPCC Scientific Assessment*. Cambridge University Press, Cambridge, 365 pp.

Houghton, J.T., L.G. Meira Filho, B.A. Callander, N. Harris, A. Kattenberg, and K. Maskell (eds.), 1996: *Climate Change 1995. The Science of Climate Change*. Contribution of Working Group I to the Second Assessment Report of the Intergovernmental Panel on Climate Change, Cambridge University Press, Cambridge.

Jefferson, M., 1983: Economic uncertainty and business decision-making. In *Beyond Positive Economics?* J. Wiseman (ed.), Macmillan Press, London, pp. 122-159.

Jiang, K., T. Masui, T. Morita, and Y. Matsuoka, 2000: Long-term GHG emission scenarios of Asia-Pacific and the world. *Technological Forecasting & Social Change*, **63**(2-3). (In press).

Jung, T.-Y., E.L. La Rovere, H. Gaj, P.R. Shukla, and D. Zhou, 2000: Structural changes in developing countries and their implication to energy-related CO_2 emissions. *Technological Forecasting & Social Change*, **63**(2-3). (In press).

Kram, T., K. Riahi, R.A. Roehrl, S. van Rooijen, T. Morita, and B. de Vries, 2000: Global and regional greenhouse gas emissions scenarios. *Technological Forecasting & Social Change*, **63**(2-3). (In press).

Lashof, D., and Tirpak, D.A., 1990: *Policy Options for Stabilizing Global Climate*. 21P-2003. U.S. Environmental Protection Agency, Washington, DC.

Leggett, J., W.J. Pepper, and R.J. Swart, 1992: Emissions Scenarios for IPCC: An Update. In: *Climate Change 1992. The Supplementary Report to the IPCC Scientific Assessment*. J.T. Houghton, B.A. Callander and S.K. Varney (eds.), Cambridge University Press, Cambridge, pp. 69-95.

Marland, G., R.J. Andres, and T.A. Boden, 1994: Global, regional, and national CO_2 emissions. pp.505-584. In *Trends '93: A Compendium of Data on Global Change,* ORNL/CDIAC-65. T.A. Boden, D.P. Kaiser, R.J. Sepanski, and F.W. Stoss (eds).,Carbon Dioxide Information Analysis Center, Oak Ridge National Laboratory, Oak Ridge, TN, pp. 505-584.

Messner, S., and M. Strubegger, 1995: *User's Guide for MESSAGE III*. WP-95-69, International Institute for Applied Systems Analysis, Laxenburg, Austria, 155 pp.

Michaelis, L., 1998: Economic and technological development in climate scenarios. *Mitigation and Adaptation Strategies for Global Change*, **3**(2-4), 231-261.

Morgan, M.G., and M. Henrion, 1990: *Uncertainty: A Guide to Dealing with Uncertainty in Quantitative Risk and Policy Analysis*. Cambridge University Press, Cambridge.

Mori, S., and M. Takahashi, 1999: An integrated assessment model for the evaluation of new energy technologies and food productivity. *International Journal of Global Energy Issues*, **11**(1-4), 1-18.

Mori, S., 2000: The development of greenhouse gas emissions scenarios using an extension of the MARIA model for the assessment of resource and energy technologies. *Technological Forecasting & Social Change*, **63**(2-3). (In press).

Morita, T., Y. Matsuoka, I. Penna, and M. Kainuma, 1994: *Global Carbon Dioxide Emission Scenarios and Their Basic Assumptions: 1994 Survey*, CGER-1011-94. Center for Global Environmental Research, National Institute for Environmental Studies, Tsukuba, Japan.

Morita, T., and H.-C. Lee, 1998a: Appendix to Emissions Scenarios Database and Review of Scenarios. *Mitigation and Adaptation Strategies for Global Change*, **3**(2-4), 121-131.

Morita, T., and H.-C. Lee, 1998b: *IPCC SRES Database*, Version 0.1, Emission Scenario Database prepared for IPCC Special Report on Emissions Scenarios (http:www-cger.nies.go.jp/cger-e/db/ipcc.html).

Nakićenović, N., A. Grübler, H. Ishitani, T. Johansson, G. Marland, J.R. Moreira, and H.-H. Rogner, 1996: Energy primer. In *Climate Change 1995. Impacts, Adaptations and Mitigation of Climate Change: Scientific Analyses*. R. Watson, M.C. Zinyowera, R. Moss (eds.), Cambridge University Press, Cambridge, 75-92.

Nakićenović, N., N. Victor, and T. Morita, 1998: Emissions Scenarios Database and Review of Scenarios. *Mitigation and Adaptation Strategies for Global Change*, **3**(2-4), 95-120.

Nakićenović, N., 2000: Greenhouse gas emissions scenarios. *Technological Forecasting & Social Change*, **65**(3). (In press).

Parikh, J.K., 1992: IPCC strategies unfair to the south. *Nature*, **360**, 507-508.

Peck, S.C., and T.J. Teisberg, 1995: International CO_2 emissions control - An analysis using CETA. *Energy Policy*, **23**(4), 297-308.

Pepper, W.J., J. Leggett, R. Swart, J. Wasson, J. Edmonds, and I. Mintzer, 1992: Emissions scenarios for the IPCC. An update: Assumptions, methodology, and results. Support document for Chapter A3. In *Climate Change 1992: Supplementary Report to the IPCC Scientific Assessment*. J.T. Houghton, B.A. Callandar, S.K. Varney (eds.), Cambridge University Press, Cambridge.

Pepper, W.J., Barbour, W., Sankovski, A., and Braaz, B., 1998: No-policy greenhouse gas emission scenarios: revisiting IPCC 1992. *Environmental Science & Policy*, **1**, 289-312.

Price, L., L. Michaelis, E. Worell and M. Khrushch, 1998: Sectoral trends and driving forces of global energy use and greenhouse gas emissions. *Mitigation and Adaptation Strategies for Global Change*, **3**(2-4), 263-319.

Riahi, K., and R.A. Roehrl, 2000: Greenhouse gas emissions in a dynamics-as-usual scenario of economic and energy development. *Technological Forecasting & Social Change*, **63**(2-3). (In press).

Roehrl, R.A., and K. Riahi, 2000: Technology dynamics and greenhouse gas emissions mitigation - a cost assessment. *Technological Forecasting & Social Change*, **63**(2-3). (In press).

Sankovski, A., W. Barbour, and W. Pepper, 2000: Quantification of the IS99 emission scenario storylines using the atmospheric stabilization framework (ASF). *Technological Forecasting & Social Change*, **63**(2-3). (In press).

UNFCCC (United Nations Framework Convention on Climate Change), 1992: Convention Text, IUCC, Geneva.

UNFCCC (United Nations Framework Convention on Climate Change), 1997: *Kyoto Protocol to the United Nations Framework Convention on Climate Change*. FCCC/CP/L7/Add.1, 10 December 1997, UN, New York, NY.

Wigley, T.M.L., and S.C.B. Raper, 1992: Implications for climate and sea level of revised IPCC emissions scenarios. *Nature*, **357**, 293-300.

Wigley, T.M.L., *et al.*, 1994: *Model for the Assessment of Greenhouse-gas Induced Climate Change Version 1.2*. Climate Research Unit, University of East Anglia, UK.

2

An Overview of the Scenario Literature

CONTENTS

2.1. Introduction

Presented in this chapter is the assessment of more than 400 global and regional greenhouse gas (GHG) emissions scenarios based on an extensive literature review. Emissions scenarios provide an important input for the assessment of future climate change. Future anthropogenic GHG emissions depend on numerous driving forces, including population growth, economic development, energy supply and use, land-use patterns, and a host of other human activities. These main driving forces that determine the emissions trajectories in the scenarios often also provide input to assess possible emissions mitigation strategies and possible impacts of unabated emissions. In view of the many different uses, it is not surprising that numerous emissions scenarios are presented in the literature and that the number of regional and global emissions scenarios is growing.

An important characteristic of the scenarios in this Special Report on Emissions Scenarios (SRES) is that they reflect the underlying uncertainty, part of which derives from the range of emissions in the literature. The objective was to encompass the variation within the most important scenario driving forces and emissions, the complexity of possible relationships between driving forces and emissions, and the associated uncertainties that characterize alternative future developments. The SRES scenarios cover most of the range of the GHG emissions scenarios found in the literature, including the International Panel on Climate Change (IPCC) 1992 Scenarios (IS92) series (Leggett *et al.*, 1992). The writing team considered the literature in creating a new set of scenarios. Importantly, however, the literature on existing scenarios provides only a general framework to aid analysis; it is informative, but not determinative.

The literature review consists of four parts:

- documentation of as many as possible of the quantitative global and regional emissions scenarios available both in the open literature and from international activities that involve documentation of submitted scenarios;
- development of a scenario database to document the more than 400 emissions scenarios collected during the literature review;
- evaluation of the ranges and relationships of the main scenario driving forces and the resultant emissions for the scenarios documented in the database; and
- assessment of the scenario submissions received through the SRES "open process."

These four components of the literature review are well suited to document and assess the (quantitative) scenarios that assign numeric values to describe the future evolution of driving forces and emissions.

Central to this assessment of emissions scenarios and their main driving forces is a unique scenario database developed by the National Institute for Environmental Studies (NIES) in Japan for SRES (Morita and Lee, 1998). The database version of 3 April 1998, which is assessed in this chapter, includes 416 different scenarios. The current database version can be accessed through an ftp-site (www-cger.nies.go.jp/cger-e/db/ipcc.html). It is the most comprehensive collection of emissions scenarios in the publicly available literature. It includes most of the recent global and regional scenarios and all of the scenarios used in the latest IPCC evaluation of emissions scenarios (Alcamo *et al.*, 1995). Therefore, the emissions scenarios documented in the database are representative of the literature in general. However, there are a number of ways in which the coverage of the scenarios in the database could be extended in the future. For example, inclusion of long-term emissions scenarios for individual countries, when available, would improve the regional coverage (e.g., Parikh, 1996; Murthy *et al.*, 1997). Also, a large majority of the scenarios report only energy-related carbon dioxide (CO_2) emissions (230 scenarios), while only some report non-energy CO_2 and other GHG emissions. This shortcoming of the emissions scenarios in the literature was identified in the last IPCC evaluation (Alcamo *et al.*, 1995).

The scenarios in the database were collected from 171 different literature sources and other scenario-evaluation activities, such as the Energy Modeling Forum (EMF; see Weyant, 1993) and the International Energy Workshop (IEW; see Manne and Schrattenholzer, 1996, 1997). The scenarios span a wide range of assumptions about demographic trends, levels of economic development, energy consumption and efficiency patterns, and other factors. The aim of this chapter is to show how the database can be utilized for the analysis of GHG emissions ranges and their main driving forces. Part of this assessment of the emissions scenarios is based on an earlier publication on the analysis of scenarios documented in the SRES database (Nakićenović *et al.*, 1998a).

The scenarios in the database display a large range of future GHG emissions. Part of the range can be attributed to the different methods and models used to formulate the scenarios, which include simple spreadsheet models, economic models, and systems-engineering models. However, most of the range results from differences in the input assumptions of the scenarios, in particular those of the main scenario driving forces. In addition, simply to compare alternative emissions levels across different scenarios is not sufficient to shed light on internal consistency, plausibility, and comparability of the assumptions behind the scenarios. Analysis of the underlying driving forces is thus also an important part of the evaluation. This chapter provides an analysis of the main driving forces, such as population growth, economic growth, energy consumption, and energy and carbon intensities. Some of these driving forces are specified as model inputs, and some are derived from model outputs, so it is necessary to determine the assumed relationships among the main driving forces.

Although the scenario database is well suited for the documentation of quantitative scenarios, there is also a significant literature on narrative scenarios. Both scenario types have in common that they are generally carefully constructed descriptions of possible future developments within the bounds of explicit assumptions and circumstances (see Chapter 1 for a more detailed discussion about scenarios). The difference is that the quantitative scenarios are usually developed with the help of formal models so as to assign internally consistent values to the various scenario characteristics.

The SRES scenarios employ both approaches – a storyline that gives a broad, narrative, and qualitative scenario description plus a number of quantifications of each storyline with six different models. Thus, even though both narrative stories and quantitative scenarios are an integral part of the SRES emissions scenarios, the literature review focused on the documentation and the assessment of quantitative scenarios, for two reasons. First, it was not possible to devise a classification system that would allow the documentation of many different forms of narrative scenarios. Second, the SRES objective was to develop a set of numeric emissions scenarios for use in the IPCC and other assessments of climate change. Therefore, in this chapter the focus is only on the literature review of quantitative scenarios. A more detailed discussion of narrative scenarios is given in Chapter 4; it deals with the four SRES storylines and how they are related to recent work in the area of qualitative scenarios.

The literature on quantitative scenarios is large indeed. This assessment is focused on the scenarios that extend at least to 2020, but about 10 scenarios with a shorter time horizon of 2010 are also included in the database. In addition, most of the scenarios have a global coverage, although a few regional scenarios are included to enhance the coverage of some parts of the world. These criteria narrowed considerably the number of global and regional GHG emissions scenarios with sufficient information to be included in the scenario literature review.

This scenario literature review and evaluation is the second undertaken by the IPCC. The first was conducted to evaluate the IS92 set of scenarios in comparison to other GHG emissions scenarios found in the literature (Alcamo *et al.*, 1995)[1]. It was completed in 1994 and included a comprehensive evaluation of GHG emissions and their main driving forces. This second review and evaluation builds upon and extends the earlier IPCC assessment. Consequently, an effort was made in the present review to include especially the GHG emissions scenarios published since the presentation of the IS92 scenarios.

2.2. General Overview of Scenarios

The construction of scenarios to investigate alternative future developments under a set of assumed conditions dates far back. Scenarios are one of the main tools used to address the complexity and uncertainty of future challenges. The first scenarios were probably designed to help plan military operations, often called "war games." Today, scenarios are used regularly by military organizations around the world for training and planning purposes. Military strategists and teachers often use very sophisticated computer models to develop scenarios for a multitude of different purposes.

Scenarios are also increasingly used by enterprises around the world for many commercial purposes. Perhaps the most famous example is that of the Shell Group in the wake of the so-called oil crisis, which used scenarios to plan the corporate response strategies (Jefferson, 1983; Schwartz, 1991). Today, the use of scenarios is quite widespread.. Many scenarios, particularly those developed for enterprises in the energy sector, are quantitative and include GHG emissions. Recently, the World Business Council for Sustainable Development (WBCSD) presented a set of scenarios developed in collaboration with 35 major corporations (WBCSD, 1998). The SRES scenario database documents a number of such scenarios that are in the public domain and have been published.

During the past three decades many global studies have used scenarios as a tool to assess future CO_2 (and in a few cases also other GHG) emissions. One of the first such global studies was *Energy in a Finite World*, conducted by the International Institute of Applied Systems Analysis (IIASA) during the late 1970s (Häfele *et al.*, 1981). Another influential series of scenarios that included the assessment of CO_2 emissions was developed by the World Energy Council (WEC, 1993). Recently, IIASA and WEC jointly presented a set of global and regional scenarios that were developed with a set of integrated assessment models and then reviewed and revised through 11 regional expert groups (Nakićenović *et al.*, 1998b). Another recent set of three scenarios, based on elaborate narrative stories that described alternative futures, was developed by the Global Scenario Group (Raskin *et al.*, 1998) and received considerable attention.

Scenarios of future emissions played an important role from the beginning of the IPCC work. In 1990, the IPCC initiated the development of its first set of GHG emissions scenarios designed to serve as inputs to general circulation models (GCMs) and facilitate the assessments of climate-change impacts (Houghton *et al.*, 1990). Two years later, in 1992, the IPCC approved six new emissions scenarios (IS92) that provided alternative emissions trajectories for the years 1990 through 2100 for such radiatively active gases as CO_2, carbon monoxide (CO), methane (CH_4), nitrous oxide (N_2O), nitrogen oxides (NO_x), and sulfur dioxide (SO_2) (Leggett *et al.*, 1992). They were widely used by atmospheric and climate scientists in the preparation of scenarios of atmospheric composition and climate change (Alcamo *et al.*, 1995). In many ways, the IS92 scenarios were pathbreaking. They were the first global scenarios to provide estimates of the full suite of GHGs and at the time were the only scenarios to provide emissions

[1] The IS92 scenarios have also been analyzed with regard to short-term adequacy (see Gray, 1998).

trajectories for SO_2. The IS92 scenarios are marked for reference in many of the illustrations herein that show the variation of emissions and their driving forces across the scenarios in the SRES database.

An important group of emissions scenarios included in this literature review was compiled from two international scenario and model comparison activities. This first group is from the IEW and involves structured comparisons of energy and emissions scenarios since 1981 (Manne and Schrattenholzer, 1996, 1997). The participating groups provide information for a standardized scenario poll from which the ranges and other sample statistics are reported for the main driving forces and emissions. The other group is the EMF (Weyant, 1993) and also involves regular scenario comparisons, in addition to standardized input assumptions, such as the international oil price or carbon emissions taxes. Both of these international scenario comparison activities provide a large share of the data for this scenario review and comparison. They include most of the global and regional emissions scenarios developed by formal modeling approaches. A large part of these activities is based on the use of scenarios for the purpose of climate-change research. A third scientific effort that involves scenario comparisons is the Energy Technology Systems Analysis Programme (ETSAP; Kram, 1993) supported by the International Energy Agency (IEA). The ETSAP work involves scenario analysis by more than 40 scientific groups from about 20 countries using the same modeling approach.

In addition to the many scientific, governmental, and private organizations throughout the world engaged in scenario-building, some international governmental organizations regularly develop global and regional scenarios that include GHG emissions. For example, the IEA regularly publishes global energy scenarios that include CO_2 emissions (IEA, 1998). Most of these scenarios are of shorter term and so are not suitable for the requirements of IPCC (see Chapter 1 for further details). Nevertheless, they are included in this assessment to facilitate a more comprehensive evaluation of emissions and their driving forces during the next few decades.

Some studies consider scenarios that involve explicit policies and measures to reduce emissions of GHGs or adapt to climate change. Such climate change intervention, control, or mitigation scenarios are an important tool for the assessment of policies and measures that would be required to reduce future GHG emissions. In this report, we use the terminology from the most recent IPCC evaluation of emissions scenarios (Alcamo *et al.*, 1995). Those scenarios that include some form of policy intervention are referred to as *intervention scenarios*, while those that do not assume any climate policy measures, such as the 40 SRES scenarios, are referred to as *non-intervention scenarios*. In some cases, intervention scenarios go even further and investigate more radical emissions reductions required to stabilize atmospheric concentrations of these gases (in accordance with Article 2 of the United Nations Framework Convention on Climate Change (UNFCCC, 1992)). In contrast, the SRES scenarios do not include any explicit additional climate policy initiatives in accordance with the Terms of Reference (see Appendix I).

The SRES writing team used a general approach to identify intervention scenarios. According to this approach, a scenario is identified as an intervention scenario if it meets one of the following two conditions:

- it incorporates specific climate change targets, which may include absolute or relative GHG limits, GHG concentration levels (e.g., CO_2 stabilization scenarios), or maximum allowable changes in temperature or sea level; and
- it includes explicit or implicit policies and/or measures of which the primary goal is to reduce GHG emissions (e.g., a carbon tax or a policy encouraging the use of renewable energy).

Note that this classification system is only a first step, and further work is needed to refine this taxonomy.

Some scenarios in the literature are difficult to classify as intervention or non-intervention, such as those developed to assess sustainable development. These studies consider futures that require radical policy and behavioral changes to achieve a transition to a sustainable development path; Greenpeace formulated one of the first (Lazarus *et al.*, 1993). This class of scenarios describes low emissions futures that sometimes, but not always, result from specific climate policy measures. Such sustainable development scenarios are also included in this assessment of the scenario literature. Where they do not include the explicit policies of the SRES criteria, they can be classified as non-intervention scenarios. However, there is a great deal of ambiguity as to what constitutes policies directed at climate change, as opposed to those directed at achieving sustainable development in general. Thus, some of these sustainable development scenarios are "non-classified" (i.e., the information available is insufficient to determine whether or not the scenarios included any additional climate policy initiatives).

2.3. Emissions Scenario Database

The SRES Emissions Scenario Database (ESD) was designed to fulfill several objectives:

- To facilitate a thorough review and analysis of the literature.
- To enable a statistical analysis of all scenarios in the database – to generate distribution functions of the main scenario driving forces, calculate mean and median values, percentiles, and other sample statistics (the use of such statistical analyses of the scenarios in the database ensures that the new SRES emissions scenarios generally reflect the range of emissions and input assumptions currently found in the open literature).

Table 2-1: *Number of regional and global GHG emissions scenarios in the SRES ESD. The database (from 3 April 1998) included a total of 416 regional and global scenarios from 171 sources. The individual number of scenarios per region or country exceeds the global total because some scenarios include both global and regional data. There are also more scenarios at the regional level than at the global level. In addition to the original sources of individual emissions scenarios, the database utilized the large number of scenarios compiled in the following assessments: International Energy Workshop Poll (Manne and Schrattenholzer, 1995, 1996, 1997); Energy Modeling Forum (EMF-14; see, e.g., Weyant, 1993) data; and the previous database compiled for the IPCC (Alcamo et al., 1995).*

Region ID	Number of Matching Scenarios	Region ID	Number of Matching Scenarios
World	340	Europe	12
OECD	164	OECD West[f]	13
Non-OECD	158	Middle East/North Africa	12
China	153	East Asia	12
USA	136	Extra[g]	12
FSU	121	West Europe	11
EEC	85	DC	7
Japan	69	OSEAsia[h]	7
FSU+EEU[a]	61	SubSAfrica	6
Annex 1[2]	46	Annex 2[2]	6
Non-Annex 1[2]	46	Opacific	6
Latin America	42	Poland	5
India	36	OPEC	4
Africa	34	United Kingdom	4
CPAsia[b]	32	LDC[i]	4
East Europe[c]	31	Non-OPEC DC[j]	3
ALM[d]	30	Hungary	3
CANZ	29	Switzerland	3
Mexico and OPEC	29	INDUS[k]	3
Non-OECD Annex 1	29	Asia Pacific	2
Middle East	27	Austria	2
Oceania	25	Brazil	2
Canada	24	Germany	2
OECD Pacific[e]	23	Korea	2
SouthAsia (incl. India)	23	Netherlands	2
OECD Europe	22	Sweden	2
SEAsia (South and East Asia)	16	Nigeria	2
North America	15	Other regions	26

Abbreviations: ALM, Africa, Latin America, and Middle East; CANZ, Canada, Australia, New Zealand; DC, Developing Countries; EEC, European Economic Community; EEU, Eastern Europe; FSU, Former Soviet Union; OECD, Organization for Economic Development and Cooperation; Opacific, Other Pacific Asia; OPEC, Organization of Petroleum Exporting Countries; SubSAfrica , Sub Saharan Africa; USA, United States of America

a. Economies under transition, Former Soviet Union and Eastern Europe.
b. Central Planning Asia including China.
c. Eastern and Central Europe.
d. Africa, Latin America and the Middle East.
e. Japan, Australia, New Zealand.
f. OECD Europe and Canada.
g. South Pole and other regions with very small populations.
h. Asia including Japan and China
i. Developing World (less than US$(1985)1700/capita).
j. Non-OPEC Developing Countries.
k. Developed World (more than US$(1985)1700/capita).

[2] As defined in UNFCCC, 1992.

- The third objective was to make the database accessible through a website (www-cger.nies.go.jp/cger-e/db/ipcc.html) so that data queries, browsing, data retrieval, and entry of new scenarios would be possible by remote users, and necessitated designing a database that could manage flexibly large amounts of data as well as diverse data types.

The database serves primarily to document the GHG emissions, including CO_2, CH_4, N_2O, CFCs (chlorofluorocarbons), HFCs (hydrofluorocarbons), and other radiatively active gases such as SO_2, CO, and NO_x. In addition, it includes information about the main driving forces of GHG emissions, such as population growth and economic development, usually expressed in terms of gross domestic product (GDP), energy consumption, and land use. Each of these scenario characteristics has subcategories and different values in time and space. The temporal dimension is often in steps of 10 years, but this is not standardized across the scenarios in the database. The spatial dimension refers to the regional disaggregation of the scenarios. Priority was given to covering all accessible quantitative scenarios with global and regional coverage. The main scenario characteristics are documented by the name and aggregation given in the original study. In some cases, regional and national scenarios are also included to improve the coverage of some parts of the world. (Table 2-1 lists the number of scenarios in the database that include a given region, from the global level through to some individual countries.) There is great diversity with respect to regional aggregation of scenarios in the database. Inclusion of long-term emissions scenarios for individual countries, when available, would improve the regional coverage of the database. Sectoral studies in developing countries, such as power system emissions (e.g., Chattopadhyay and Parikh, 1993) or transport system emissions (e.g., Ramanathan and Parikh, 1999), were also considered in this assessment to develop SRES emissions scenarios.

A list of scenario characteristics and their frequency of occurrence across the 416 scenarios is given in Appendix V. Most of these scenarios were created after 1994. Of the 416 scenarios 340 provide data on the global level, and 256 scenarios of these 340 report information on CO_2 emissions.

A large majority (230) of the scenarios report only energy-related CO_2 emissions, while only some report non-energy CO_2 and other GHG emissions. For example, only three models estimate land use-related emissions: the Atmospheric Stabilization Framework (ASF) (Lashof and Tirpak, 1990) model that was used to formulate the IS92 scenarios; the Integrated Model to Assess the Greenhouse Effect 2 (IMAGE 2) model (Alcamo, 1994); and the Asian-Pacific integrated model (AIM; Morita *et al.*, 1994). Only a few scenarios report regional and global SO_2 and sulfur aerosol emissions that are also climatically important because of their cooling effect (negative radiative forcing of climate change). Box 2-1 in Section 2.4.1 summarizes the set of scenarios that report non-energy-related CO_2 emissions.

The information documented in the database about emissions scenarios illustrates both areas that are well covered in the scenario literature and areas with substantial gaps in knowledge. For example, the information in the database strongly confirms the findings of the latest IPCC scenario assessment and evaluation (Alcamo *et al.*, 1995). One of the key findings is that of all GHG emissions, CO_2 emissions are by far the most frequently studied, and that of all the CO_2 emissions sources, fossil fuel is the source most extensively analyzed in the literature. In part, this is because energy-related sources of CO_2 emissions contribute more to the current and potential future climate forcing than any other single GHG released by any other human activity. In part, this is also because of improved data, assessment methods, and models for energy-related activities than for other emissions sources. Another information gap example is the rather diverse regional disaggregation chosen for different scenarios. Even when the regions are similar or equivalent in terms of this assessment, the names are sometimes different, which hampers comparisons. Such gaps in knowledge limit the range and effectiveness of the various policy options that logically follow from the discussion. This creates a level of uncertainty that can only be addressed by concentrated research efforts.

2.4. Analysis of Literature

Individual scenarios are considered independent entities in the database. Clearly, in practice, individual scenarios are often related to each other and are not always developed independently. Some are simply variants of others generated for a particular purpose. Many "new" scenarios are designed to track existing benchmark scenarios. A good example is the set of IS92 scenarios, especially the "central" IS92a scenario, which was often used as a reference from which to develop other scenarios. A further consideration is that not all scenarios are created in an equal fashion. Some are the result of elaborate effort, which includes extensive reviews and revisions; others are simply the outcome of input assumptions without much significant reflection. Some are based on extensive formal models, while others are generated using simple spreadsheets or even without any formal tools at all.

Numerous factors influence future emissions paths in the scenarios. Clearly, demographic and economic developments play a crucial role in determining emissions. However, many other factors are involved also, from human resources, such as education, institutional frameworks, and lifestyles, to natural resource endowments, technologic change, and international trade. Many of these important factors are not documented in the database, and sometimes not even in the respective scenario reports and publications. Some are neither quantified in the scenarios nor explicitly assumed in a narrative form.

For this analysis, a simple scheme is used to decompose the main driving forces of GHG emissions. This scheme is based on the Kaya identity (Kaya, 1990; Yamaji *et al.*, 1991), which gives the main emissions driving forces as multiplicative factors on one side of the identity and total CO_2 (or GHG) emissions on the other. It multiplies population growth, per capita value added (i.e., per capita gross world product), energy consumption per unit value added, and emissions per unit energy on one side of the identity, and total CO_2 emissions on the other side (Yamaji *et al.*, 1991);[3] it is a specific application of a frequently used approach to organize discussion of the drivers of emissions through the so-called IPAT identity that relates impacts (I) to population (P) multiplied by affluence (A) and technology (T), (see Chapter 3 for a more detailed discussion). The same approach can be used for other emissions such as SO_2. However, the driving forces might be different for some species of anthropogenic emissions.

Apart from its simplicity, an advantage of analysis that uses the Kaya identity to decompose emissions into four main driving forces is that it facilitates at least some standardization in the comparison and analysis of many diverse emissions scenarios. This decomposition is very useful because it indicates where to seek differences in scenario assumptions that may account for differences in the resultant GHG emissions (Alcamo *et al.*, 1995). However, the identity is not used here to suggest causality. An important caveat is that these driving forces are not independent of each other; in many scenarios they explicitly depend on each other. For example, scenario builders often assume that high rates of economic growth lead to high capital turnover. This favors more advanced and more efficient technologies, which result in lower energy intensities. Sometimes a weak inverse relationship is assumed between population and economic growth. Thus, the scenario ranges for these main driving forces are not (necessarily) independent of each other. (See also the discussion of the relationships between the main scenario driving forces in Chapter 3.)

In the following sections, scenario ranges are presented for each of the four factors in the Kaya identity that represent the main (energy-related) emissions driving forces: population, gross world product, energy consumption, energy intensity (energy per unit of gross world product) and carbon intensity (CO_2 emissions per unit of energy). The ranges for CO_2 and SO_2 emissions are presented first because they represent the "dependent variable" in the Kaya identity. These are followed by scenario ranges for the other factors in the decomposition that represent the "independent variables" (main emissions scenario driving forces) in the identity. This sequence was chosen to present the main scenario driving forces because it corresponds to their representation in the Kaya identity; it does not imply *a priori* any causal relationships among the driving forces themselves or between the driving forces and CO_2 emissions.

Four complementary methods of analysis are used:

- charts that show the distributions of scenarios in terms of their main characteristics and driving forces, including CO_2 emissions, population growth, global GDP, energy consumption, energy intensity, and carbon intensity;
- histograms that show the range of values of main scenario driving forces, together with associated statistics such as the mean, minimum, and maximum values;
- "snowflake" diagrams, in which each of the axes represents the range of one of the key driving forces; and
- analysis of the relationships among the main driving forces of energy-related CO_2 emissions.

The main findings of this scenario analysis are reported in Nakićenović *et al.* (1998a).

2.4.1. Carbon Dioxide Emissions Ranges

The span of CO_2 emissions across all scenarios in the database is indeed large, with a range from more than seven times the current emissions levels to below current levels in 2100 (see Figures 2-1a and 2-1b). The possible interpretations of this large range are many. The most important is the great uncertainty as to how the main driving forces, such as population growth, economic development, and energy production, conversion and end use, might unfold during the 21st century. A large majority of the scenarios in the database report only energy-related CO_2 emissions (230 of the 256 that report CO_2 emissions), while only some report non-energy CO_2 and other GHG emissions. Therefore, these comparisons of emissions scenarios focus mostly on energy-related emissions. Box 2-1 summarizes the set of 26 scenarios that report land-use CO_2 emissions.

Figure 2-1a shows the global energy-related and industrial CO_2 emissions pathways from the database in the form of spaghetti curves for the period 1990 to 2100 against the background of historical emissions from 1900 to 1990. These curves are plotted against an index on the vertical axis rather than as absolute values, because of large differences and discrepancies for the values assumed for the base year 1990 in various scenarios. The discrepancies may result from genuine differences among the scenarios (e.g., different base years, data sources, and definitions[4]) or from simple errors in calibration.

[3] $CO_2 = (CO2/E) \times (E/GDP) \times (GDP/P) \times P$, where E represents energy consumption, GDP the global domestic product (or global value added) and P population. Changes in CO_2 emissions can be described by changes in these four factors or driving forces.

[4] The 1990 emissions from energy production and use are estimated by Marland *et al.* (1999) at 5.9 GtC excluding cement production. It appears as if four scenarios also include deforestation, which might explain relatively large differences in the base-year emissions compared to the other scenarios. Excluding these four scenarios, the 1990 base-year values in the scenarios reviewed range from 4.8 to 6.4 GtC. With the four scenarios the range is from 4.8 to 7.4 GtC.

Box 2-1: Range of Land Use CO_2 Emissions in the Database

About 23% of the current total anthropogenic CO_2 emissions arise from land-use change (Pepper *et al.*, 1992), which makes it an important driving force. Direct comparison of absolute levels of land-use CO_2 emissions between scenarios in the database is difficult because of variations in how models depict deforestation and in how modelers classify anthropogenic and natural land-use fluxes. In addition, models are based on different base-year data, which further complicates comparison. For example, the 1990 base-year emissions estimates from land-use change (e.g., deforestation) range from 0.6 to 1.4 gigatons of elemental carbon (GtC). However, by indexing emissions to 1990, it is possible to make more meaningful comparisons of trends among the 26 scenarios in the SRES database that report land-use CO_2 emissions. It is important to note that only three modeling groups produced the 26 scenarios described here. Clearly, emissions from land-use change have not been as well explored by the modeling community as energy-related emissions.

All 26 scenarios show a decrease in CO_2 emissions from land-use change over time and are below current levels by 2100; some models even report emissions reductions below zero, which suggests CO_2 sequestration (e.g., through afforestation). The emissions range is very wide during the next few decades, but narrows considerably around mid-century. For example, there is more than a factor ten difference (after normalizing for base-year differences) between the highest and lowest scenarios in 2020. (For reference, the IS92a scenario falls slightly below the median of the range.) By 2050, however, the gap between the extremes narrows, and by 2100 the range is very small indeed: the highest scenario shows CO_2 emissions from land use only 2.4 times as great as those found in the lowest scenario. All 26 scenarios in the database report CO_2 emissions of less than 1 GtC originating from land-use change in 2100, and 23% of the scenarios indicate CO_2 sequestration by the end of the 21st century.

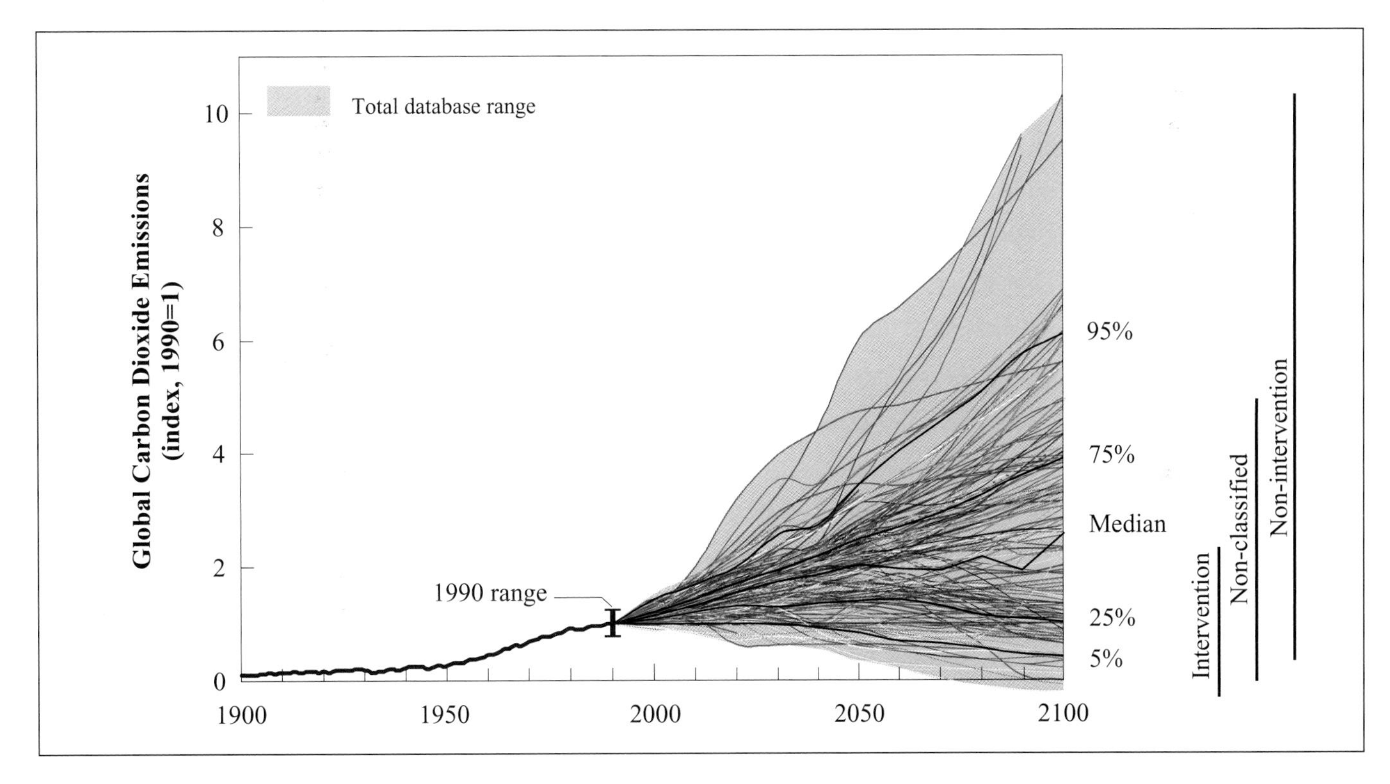

Figure 2-1a: Global energy-related and industrial CO_2 emissions[5] – historical development and future scenarios, shown as an index (1990 = 1). Median (50th), 5th, 25th, 75th, and 95th percentiles of the frequency distribution are shown. The statistics associated with scenarios from the literature do not imply probability of occurrence (e.g., the frequency distribution of the scenarios may be influenced by the use of IS92a as a reference for many subsequent studies). The vertical bars on the right side of the figure indicate the ranges for intervention, non-intervention, and non-classified scenario samples, respectively. The emissions paths indicate a wide range of future emissions. The range is also large in the base year 1990 and is indicated by an "error" bar (see also Figure 2-1b). To separate the variation due to base-year specification from different future paths, emissions are indexed for the year 1990, when actual global energy-related and industrial CO_2 emissions were about 6 GtC. The actual coverage of CO_2 emissions sources may vary across the 256 different scenarios from the database included in the figure. The scenario samples used vary across the time steps (for 1990 256 scenarios, for 2020 and 2030 247, for 2050 211, and for 2100 190 scenarios). As a result of software limitations, only 250 scenarios can be plotted on the graph. However, the scenarios not shown are included in the assessment and have almost identical trajectories to the sample of 250 scenarios shown. Data sources: Morita and Lee, 1998; Nakićenović *et al.*, 1998a.

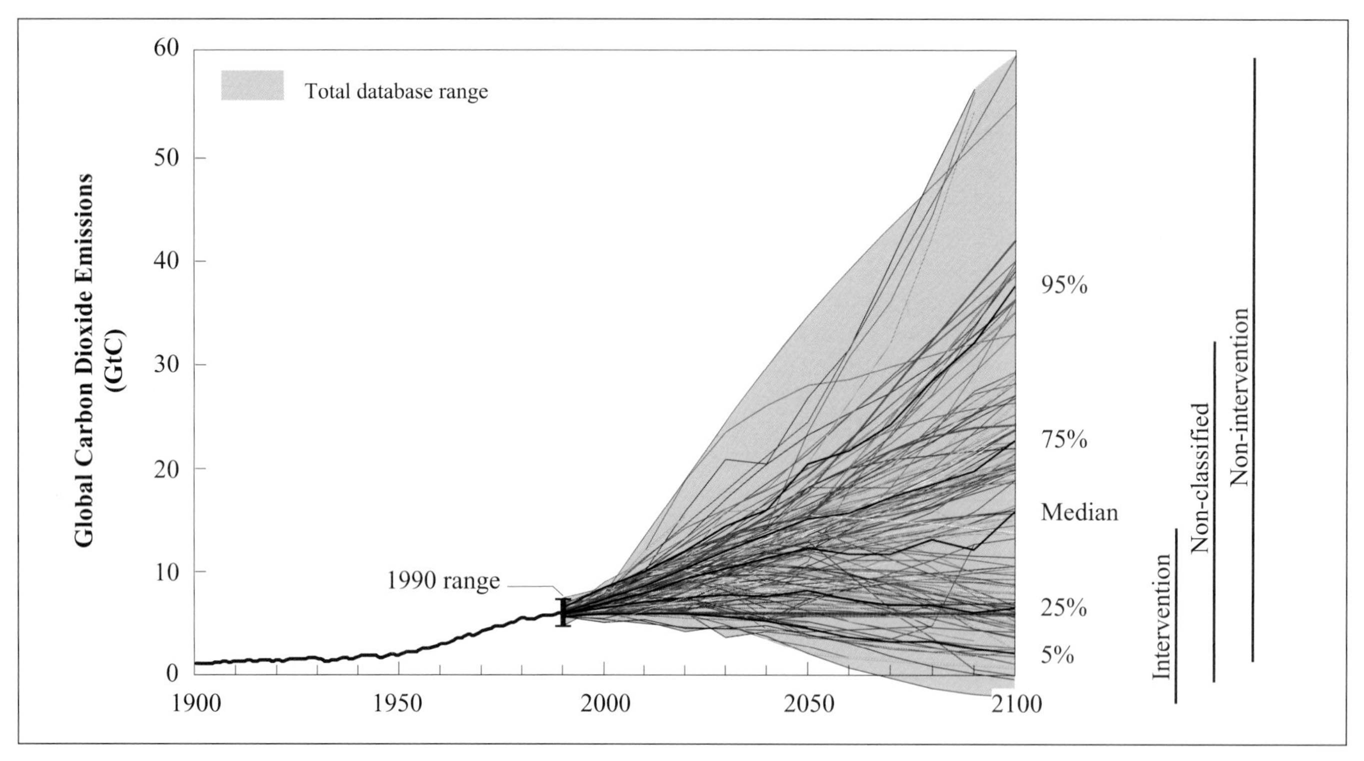

Figure 2-1b: Global energy-related and industrial CO_2 emissions[5] – historical development and future scenarios (used to derive indexed values in Figure 2-1a), shown as absolute values in GtC. Median (50th), 5th, 25th, 75th, and 95th percentiles of the frequency distribution are shown. The statistics associated with scenarios from the literature do not imply probability of occurrence (e.g., the frequency distribution of the scenarios may be influenced by the use of IS92a as a reference for many subsequent studies). The vertical bars on the right side of the figure indicate the ranges for the intervention, non-intervention, and non-classified scenario samples, respectively. The emissions paths indicate a wide range of future emissions. The range is also large in the base year 1990 and is clearly discernable. The actual coverage of CO_2 emissions sources may vary across the 256 scenarios from the database included in the figure. The scenario samples used vary across the time steps (for 1990 256 scenarios; for 2020 and 2030 247 scenarios; for 2050 211 scenarios, and for 2100 190 scenarios were analyzed). As a result of software limitations, only 250 scenarios could be plotted on the graph. However, the scenarios not shown are included in the assessment and have almost identical trajectories to the sample of 250 scenarios shown in the graphic. Data sources: Morita and Lee, 1998; Nakićenović *et al.*, 1998a.

Figure 2-1b gives absolute values of CO_2 emissions from various scenarios in the database. It shows the magnitude of differences between the scenarios in the base year 1990 and gives the resultant range of emissions in 2100 directly in GtC.[6] Between 1900 and 1990, global CO_2 emissions have increased at an average rate of about 1.7% per year (Nakićenović *et al.*, 1998b). Global emissions would double during the next three to four decades if this historical trend were to continue. Many scenarios in the database describe such a development. However, even by 2030 the range is very large around this value of possible doubling of global emissions. The highest scenarios have emissions four times the 1990 level by 2030, while the lowest are barely above half that level. The divergence increases with time so that the highest scenarios envisage a tenfold increase of global emissions by 2100. The lowest scenarios continue to decrease and some of them are consistent with emissions trajectories that lead to an eventual stabilization of atmospheric GHG concentrations.

Figure 2-1a indicates a large range between the highest and the lowest scenarios. Some of the scenarios on both extremes can be characterized as clear outliers in the far tails of the scenario distribution. Often, such scenarios are normative in nature, having been formulated for a particular purpose. However, even if these extreme tails are discarded the range is very wide. The 95th percentile corresponds to a sixfold increase (about 37 GtC) by 2100, while the 5th percentile leads to a decrease to

[5] Some of the scenarios may also include CO_2 emissions from industrial sources. Since non-energy-related industrial emissions are very low compared to the energy-related CO_2 emissions, their impact on the results of the statistical analyses is negligible. It also appears as if four scenarios also include deforestation. These scenarios tend to cluster around the median and none occur in the tails of the scenario frequency distribution. Therefore, they have very little influence on the range.

[6] The issue of large differences in the base year quantifications across the range of scenarios is discussed in Chapter 4, in which are presented the new SRES scenarios as developed by six different modeling approaches utilizing different base-year specifications.

about a third (2 GtC) compared to the 1990 level. This range of 2–37 GtC in 2100 indicates the high degree of uncertainty with respect to the level of future GHG emissions. The emissions range is somewhat smaller, from 6.5 to 22 GtC, for the 25th and 75th percentiles, respectively. However, the statistics associated with scenarios from the literature do not imply probability of their occurrence.

Some additional information about the range of future emissions can be obtained by examining the results from detailed analyses (see, e.g., Alcamo *et al.*, 1996, 1998) conducted by Nordhaus and Yohe (1983), Edmonds *et al.* (1986), de Vries *et al.* (1994), and Manne and Richels (1994). These analyses confirm that the range is very large. The analysis of Edmonds *et al.* (1986) shows a range of 87 to 2 GtC by the year 2070 for the 95th and 5th percentiles, respectively; the range is from 27 to 4 GtC for the 75th and 25th percentiles. Nordhaus and Yohe (1983) estimated the range to be from 55 to 7 GtC for the 95th and 5th percentiles, and from 27 to 12 GtC for the 75th and 25th percentiles, respectively. In contrast, de Vries *et al.* (1994) estimated a fundamentally smaller range of 24 to 11 GtC by 2050 for the 95th and 5th percentiles, respectively, but they standardized some of the scenario assumptions, such as population and economic growth, relative to a base case.

The SRES team also applied the criteria presented in Section 2.2 to identify intervention scenarios from among the CO_2 emissions scenarios in the database. Of the 190 scenarios that reported CO_2 emissions through the year 2100, 62 were classified as intervention scenarios and 88 as non-intervention scenarios. For 40 scenarios the information available was insufficient to determine whether or not they included any climate policies. These are referred to as non-classified scenarios. The statistics for these three scenario samples are given in Table 2-2. The medians for these samples are also shown in Figures 2-1c to 2-1f. The analysis indicates that many of the 88 non-intervention scenarios are emulations of IS92 scenarios, and many of these are emulations of IS92a. For example, 42 of the 88 non-intervention scenarios were produced by EMF-14, of which 25 are based on IS92a.

Figure 2-1c shows spaghetti curves for the non-intervention scenarios found in Table 2-2. In the year 2100 these scenarios cover almost the same range (1.2 GtC to 59.4 GtC) as does the entire sample of 190 scenarios (–2.1 GtC to 59.4 GtC). The sample of non-intervention scenarios includes some that have very low emissions, which suggests that emissions can be low even in the absence of explicit climate intervention policies. However, the non-intervention sample also has a higher median

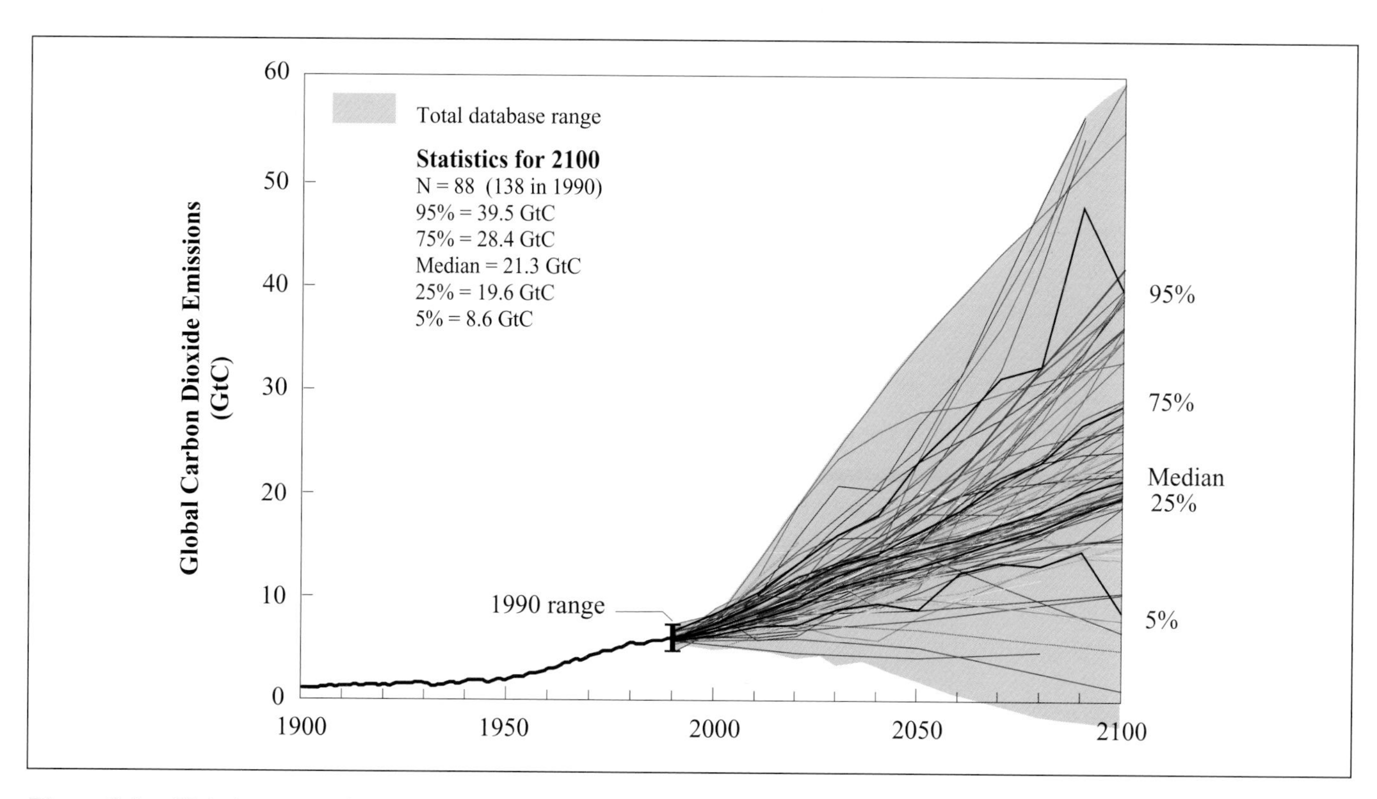

Figure 2-1c: Global energy-related and industrial CO_2 emissions for 138 non-intervention scenarios identified from the SRES database. Historical development and future scenarios are shown as absolute values in GtC. Median (50th), 5th, 25th, 75th, and 95th percentiles of the frequency distribution are shown. The statistics associated with scenarios from the literature do not imply probability of occurrence (e.g., the frequency distribution of the scenarios may be influenced by the use of IS92a as a reference for many subsequent studies). Again, the emissions paths indicate a wide range of future emissions. The actual coverage of CO_2 emissions sources may vary across the 138 scenarios from the database included in the figure. The scenario samples used vary across the time steps (for 1990 138 scenarios; for 2100 88 scenarios were analyzed). Data sources: Morita and Lee, 1998; Nakićenović *et al.*, 1998a.

Table 2-2: *Database minimum, maximum, and median CO_2 emissions levels in the year 2100 (in GtC). Data source: Morita and Lee, 1998.*

	Number of scenarios	Minimum	Maximum	Median
Intervention scenarios	62	–2.1	14.4	6.0
Non-intervention scenarios	88	1.2	59.4	21.3
Non-classified scenarios	40	–0.4	32.4	15.2
Total sample	**190**	**–2.1**	**59.4**	**15.7**

than the total sample – 21 GtC in 2100 compared with the 15.7 GtC median found in the total sample. The lower median of the total sample may result from downward pressure exerted by the inclusion of some intervention scenarios in the total sample. However, it could also result from the influence of IS92a-like scenarios (upward pressure) in the non-intervention sample (see also Figure 2-1f).

Figure 2-1d likewise depicts spaghetti curves for the intervention scenarios. The entire range for the intervention scenarios is small compared to those of the no policy scenarios and the total set of scenarios. In 2100 the maximum (14.4 GtC) and the median (6 GtC) of this sample are significantly lower than maximum and median values of the full and non-intervention sets.

Figure 2-1e shows the range for scenarios that could not be classified into these two groups because of insufficient information in the database. The range of scenarios in this category is similar to that of the total sample in that it is very broad. The median of this set in 2100 is also similar to that of the total sample (15.7 GtC) and follows a similar trajectory.

Finally, the IS92a scenario (Leggett *et al.*, 1992; Pepper *et al.*, 1992) appears to have influenced subsequent emissions scenarios in the literature. The median population, gross world product, and primary energy consumption trends in many scenarios in the literature track very closely the developments in the IS92a scenario (Morita and Lee, 1998). This is because IS92a constituted a reference baseline that was emulated by different

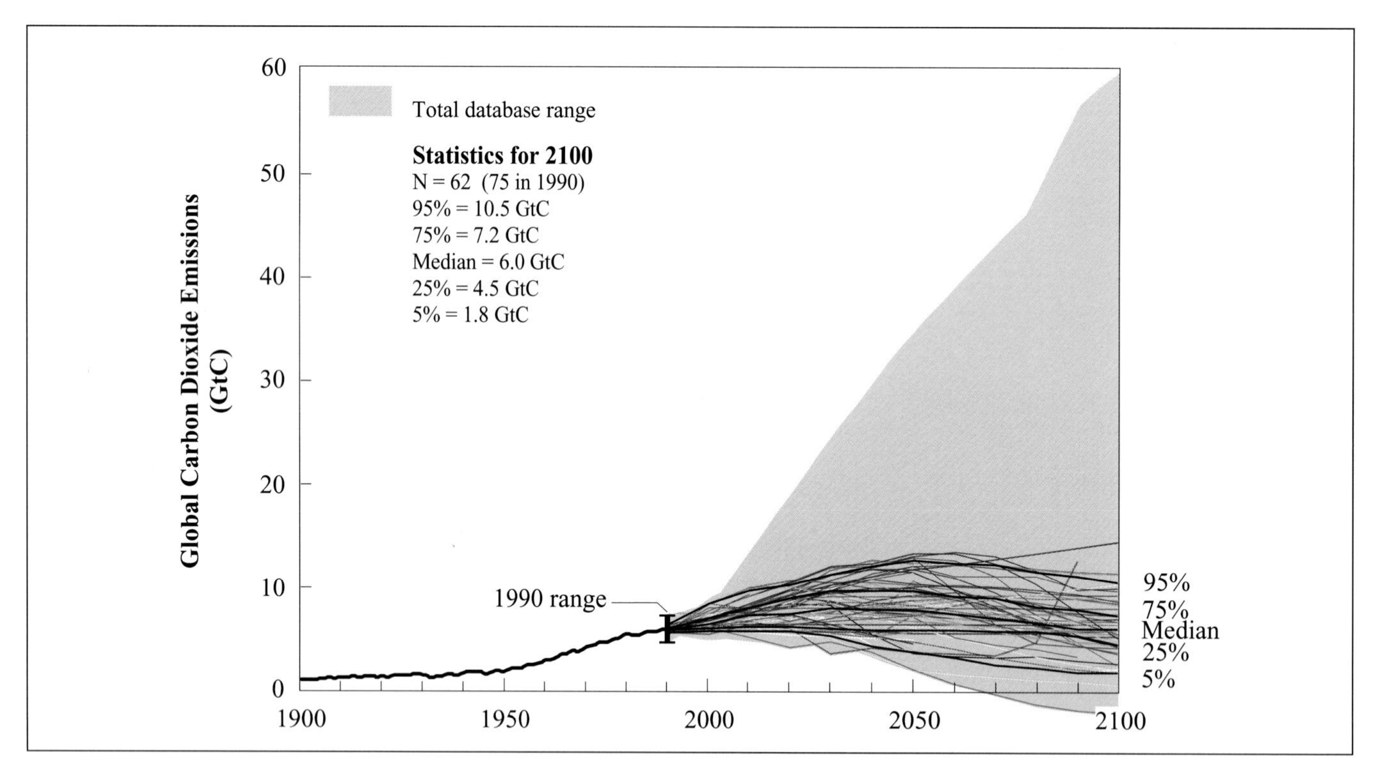

Figure 2-1d: Global energy-related and industrial CO_2 emissions for 75 intervention scenarios identified from the SRES database. Historical development and future scenarios are shown as absolute values in GtC. Median (50th), 5th, 25th, 75th, and 95th percentiles of the frequency distribution are shown. The statistics associated with scenarios from the literature do not imply probability of occurrence. The emissions paths for intervention scenarios show a more limited range of future emissions than do those for the non-intervention scenarios (see Figure 2-1c). The actual coverage of CO_2 emissions sources may vary across the 75 scenarios from the database included in the figure. The scenario samples used vary across the time steps (for 1990 75 scenarios; for 2100 62 scenarios were analyzed). Data sources: Morita and Lee, 1998; Nakićenović *et al.*,1998a.

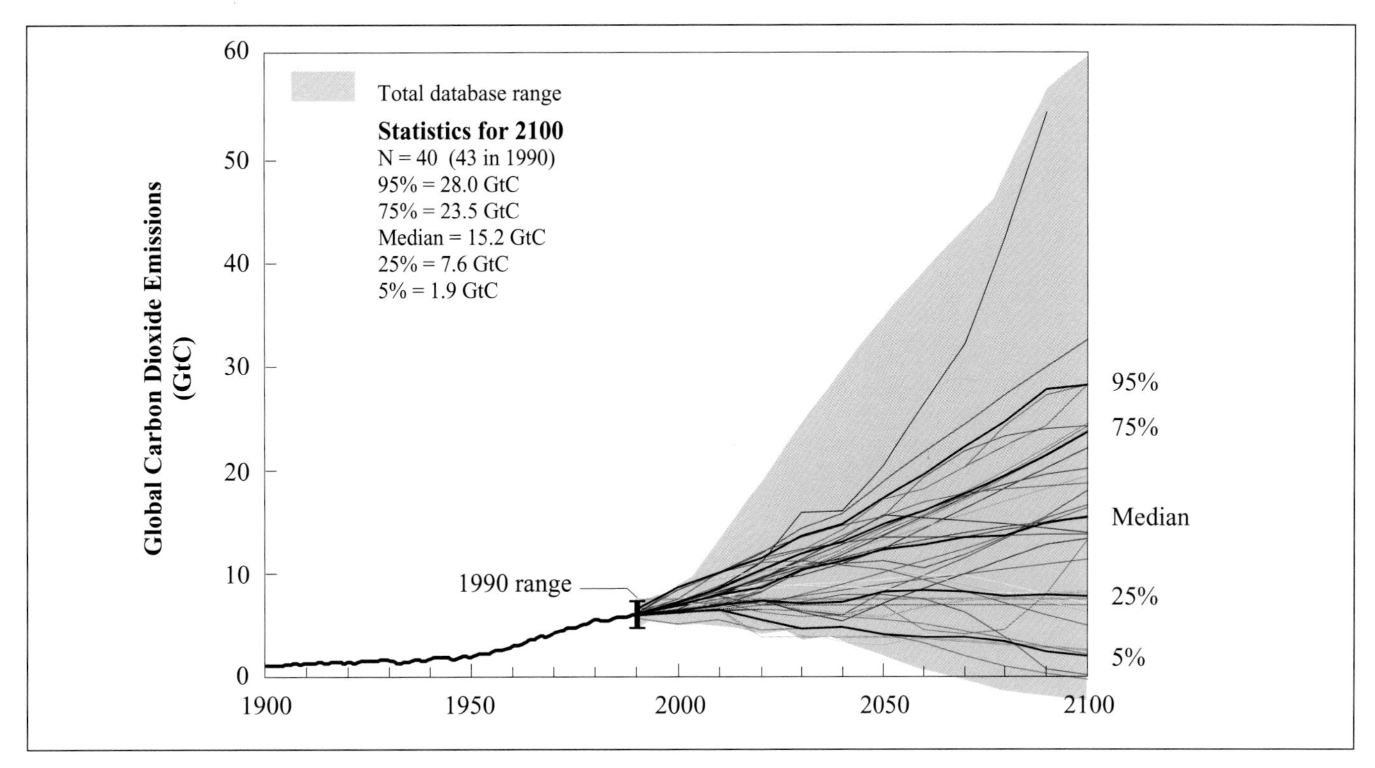

Figure 2-1e: Global energy-related and industrial CO_2 emissions for 43 scenarios that could not be classified as intervention or non-intervention scenarios from the SRES database. Historical development and future scenarios are shown as absolute values in GtC. Median (50th), 5th, 25th, 75th, and 95th percentiles of the frequency distribution are shown. The statistics associated with scenarios from the literature do not imply probability of occurrence. The emissions paths for the "non-classified" scenarios indicate a wide range of future emissions. The actual coverage of CO_2 emissions sources may vary across the 43 scenarios from the database included in the figure. The scenario samples used vary across the time steps (for 1990 43 scenarios; for 2100 40 scenarios were analyzed). Data sources: Morita and Lee, 1998; Nakićenović *et al.*,1998a.

modeling groups in a number of scenario evaluation and comparison activities. Figure 2-1f shows the set of 35 IS92a-like scenarios that could be classified from the set of non-intervention scenarios. As these scenarios appear to emulate IS92a, they show little variation around the median of 20.3 GtC, which is about the emission level in that scenario (20.4 GtC) in 2100.

The analyses in the following sections focus on the total set of scenarios in the database only. The distinction between intervention and non-intervention scenarios applies only to analyses of CO_2 emissions.

2.4.2. *Carbon Dioxide Emissions Histograms*

The first two histograms (Figures 2-2a and 2-2b) give the global CO_2 emissions ranges for 2050 and 2100.[7] The total range from the highest to the lowest scenario in 2100 is between 59 and 2 GtC,[8] from about seven times the current emissions to below zero. For about 10% of the scenarios, emissions in the year 2100 are half the current emissions or below this level. Presumably, some of the scenarios that have low future emissions include some policy interventions to reduce GHG emissions.

The distribution of emissions in 2050 is asymmetric; most of the scenarios cluster in the range between 20 to 6 GtC. The thin tail that extends above this emissions level includes only 46 of a total of 211 scenarios. Altogether, the distribution implies a substantial increase in global CO_2 emissions during the next 50 years.

The distribution of emissions in 2100 is even more asymmetric than that in 2050. The emissions portray a structure that resembles a trimodal distribution: those that show emissions of more than 30 GtC (20 scenarios), those with emissions between 12 and 30 GtC (88 scenarios), and those that show emissions of less than 12 GtC (82 scenarios). That this is quite similar to the structure of primary energy consumption distribution for 2100 is not by chance. The lowest cluster may have been influenced by many analyses of stabilization of atmospheric concentrations at levels of 450–550 parts per million (10^6) by volume (ppmv). The middle cluster appears to echo the many analyses that took IS92a as a reference scenario; it indicates the possible influence of IS92 scenarios on other scenarios in the literature. It is very likely that the majority of

[7] Not all 256 scenarios that report global CO_2 emissions cover the whole period to the year 2100.

[8] Carbon sequestration exceeds carbon emissions in the negative emission scenarios.

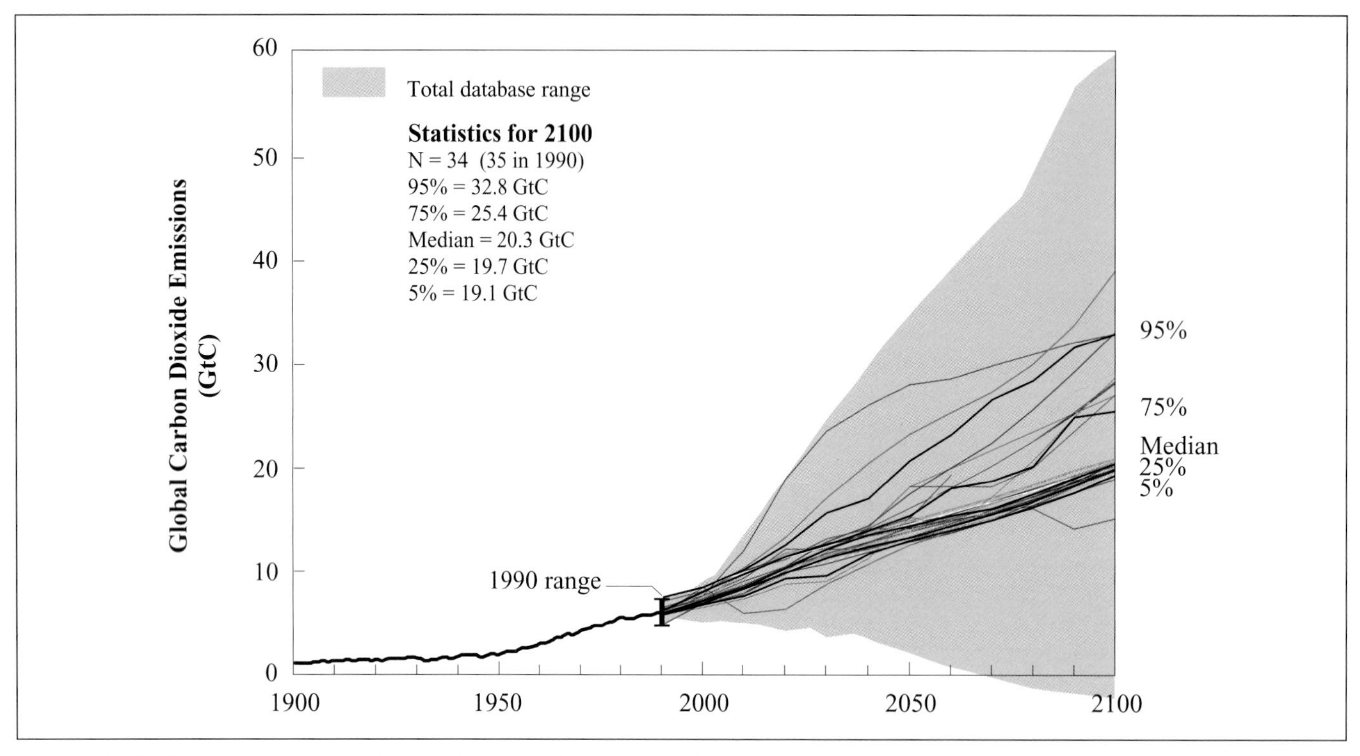

Figure 2-1f: Global energy-related and industrial CO_2 emissions for 35 "IS92a-like" scenarios identified from the set of non-intervention scenarios included in the SRES database. Historical development and future scenarios are shown as absolute values in GtC. Median (50th), 5th, 25th, 75th, and 95th percentiles of the frequency distribution are shown. The statistics associated with scenarios from the literature do not imply probability of occurrence. The actual coverage of CO_2 emissions sources may vary across the 35 scenarios from the database included in the figure. The scenario samples used vary across the time steps (for 1990 35 scenarios; for 2100 34 scenarios were analyzed). Data sources: Morita and Lee, 1998; Nakićenović *et al.*, 1998a.

the scenarios in the middle cluster foresee a substantial contribution of fossil energy sources to total energy consumption in the year 2100; thus, in the first approximation, CO_2 emissions can be expected to be proportional to energy consumption. The median is 15.7 GtC, surrounded by the centers of the other two modes, the first at about 21 GtC and the second at about 9 GtC.

Published energy-related emissions vary by a factor of 17 between the highest and lowest scenarios for 2050 and from –2 to 60 GtC between the highest and lowest scenarios for 2100.

(a)
Non-intervention
Non-classified
Intervention
IS92f
IS92a,b
IS92d
IS92c
IS92e
2050
N=211
Mean=11.9
Median=12.0
Standard Deviation=4.8
Relative Frequency (%)
0 5 10 15 20 25 30
0.5 1 1.5 2 2.5 3 3.5 4 4.5 5 5.5 6 6.5 7
Range (index, 1990=1)
3 6 9 12 15 18 21 24 27 30 33 36 39
Range (GtC)

Figure 2-2a: Histogram showing the frequency distribution of global CO_2 emissions in 2050 for 211 scenarios. The first horizontal axis shows indexed emissions (1990 = 1); the second axis indicates approximately absolute values by multiplying the index by the 1990 value (5.9 GtC, see footnote 4). For reference, the emissions of the IS92 scenarios are indicated. The horizontal bars indicate the ranges for the intervention, non-intervention, and non-classified scenario samples, respectively. The frequency distribution associated with scenarios from the literature does not imply probability of occurrence.

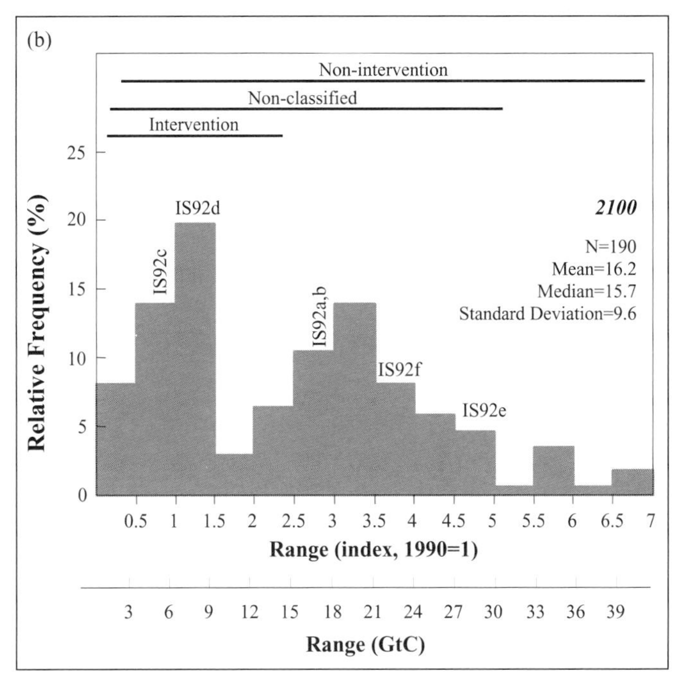

Figure 2-2b: Histogram showing the frequency distribution of global CO_2 emissions in 2100 for 190 scenarios. The first horizontal axis shows indexed emissions (1990 = 1); the second axis indicates approximate absolute values by multiplying the index by the 1990 value (5.9 GtC, see footnote 4). For reference, the emissions of the IS92 scenarios are indicated. The horizontal bars indicate the ranges for the intervention, non-intervention and non-classified scenario samples respectively. The frequency distribution associated with scenarios from the literature does not imply probability of occurrence.

2.4.3. *Sulfur Dioxide Emissions*

An overview of global long-term scenarios of SO_2 emissions is shown in Figure 2-3. Altogether 81 scenarios in the scenario database report SO_2 emissions. Most scenarios were published after 1995, which indicates the importance of the influential and innovative SO_2 emissions included in the previous IPCC scenario series IS92 (Pepper *et al.*, 1992). Apparently, they stimulated research on long-term trends and impacts of SO_2 emissions.

The 1990 base-year emissions estimates in the database range from 55 to 91 megatons of sulfur (MtS), a seemingly large difference that can be explained partially by the different extent of coverage of SO_2 emissions in different models and scenario studies, in addition to uncertainties in emissions factors. Typically, lower range emissions derive from models that report only (the dominant) energy sector emissions, higher ranges also include other anthropogenic sources such as SO_2 emissions from metallurgic processes. Differences in 1990 base-year values across scenario studies and a review of available SO_2 emissions inventories are discussed in more detail in Chapter 3. Indexed to a common 1990 basis, future SO_2 emissions trends reveal a number of remarkable characteristics. First, contrary to other trends discussed in this chapter, increases are generally modest; numerous scenarios even depict a long-term decline in emissions. Thus, SO_2 emissions are invariably projected to be decoupled progressively from their underlying driving forces of increases in population and economic activity, and hence energy demand. The median across all scenarios indicates a gradual increase of some 22% over 1990 levels by 2050, and a return to 1990 levels by 2100. Only two scenarios exceed the range of increases in long-term SO_2 emissions spanned by the IS92 scenario series.

A detailed review of long-term global and regional SO_2 emissions scenarios is given in Grübler (1998) and is summarized in Chapter 3. The most important new finding from the scenario literature is recognition of the significant impacts of continued unabated high SO_2 emissions on human health, food production, and ecosystems. As a result, scenarios published since 1995 all assume various degrees of SO_2 emissions control and interventions to be implemented in the future, and are thus substantially lower than previous projections, including the IS92 series. In most of these scenarios, such low levels of SO_2 emissions are not simply the result of direct SO_2 emissions control measures, such as flue gas desulfurization. They also result from other interventions in which SO_2 emissions reduction is more a secondary benefit than a primary goal (e.g., structural changes for various reasons other than SO_2 control).

2.4.4. *Population Projection Ranges*

Population is one of the fundamental driving forces of future emissions. Most models used to formulate population projections for the emissions scenarios are taken from the literature and are exogenous inputs. Today three main research groups project global population – United Nations (UN, 1998, 1999), World Bank (Bos and Vu, 1994), and IIASA (Lutz *et al.*, 1997). (For more details see the discussion in Chapter 3.) Most of the "central" population projections lead to a doubling of global population by 2100 (to about 10 billion people compared to 5.3 billion in 1990). In recent years the central population projections for the year 2100 have declined somewhat, but are still in line with a doubling by 2100. For example, the latest UN (1998) medium–low and medium–high projections indicate a range of between 7.2 and 14.6 billion people by 2100, with the medium scenario at 10.4 billion. The IIASA central estimate for 2100 is also 10.4 billion, with a 95% probability that world population would exceed six and be lower than 17 billion (Lutz *et al.*, 1997).

While all scenarios require some kind of population assumptions, relatively few are reported explicitly in the SRES database. Figure 2-4 illustrates global population projections in the database. Of the 416 scenarios currently documented, only 46 report their underlying population projections. This limited number indicates that, even though population is an extremely important driving force for emissions, it is typically either not reported or not well explored in most models. For the small

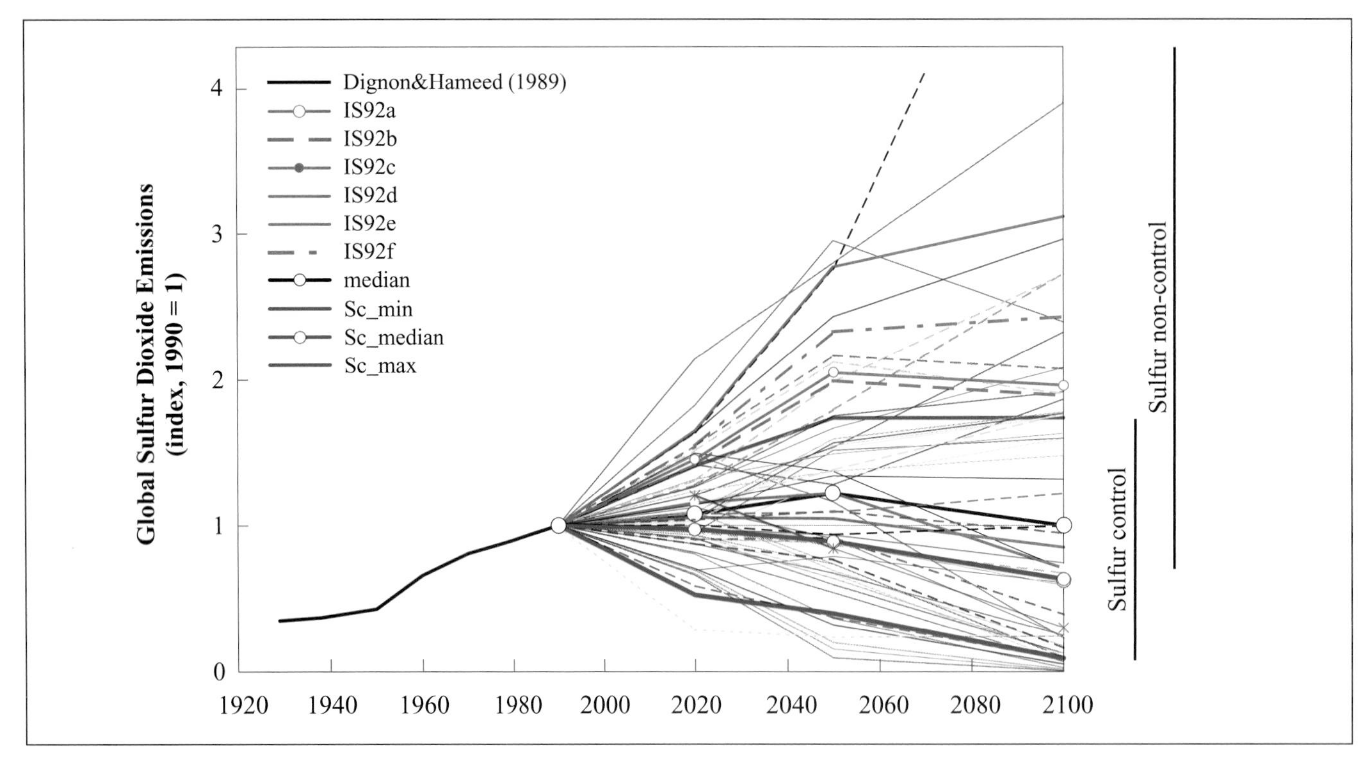

Figure 2-3: Global SO_2 emissions – historical development and 81 scenarios from the database, shown as an index (1990 = 1). For comparison, the IS92 scenarios are shown, as is the range and median of SO_2 emissions control (Sc_median, Sc_min, Sc_max) scenarios (see also Chapter 3). The vertical bars on the right side of the figure indicate the ranges for the SO_2 emissions control (intervention) scenarios and for SO_2 non-control scenarios for 2100, respectively. Data sources: Dignon and Hameed, 1989; Grübler, 1998; Morita and Lee, 1998.

sample of population projections, the range is from about 20 to 6 billion people in 2100, with the central or median estimates at about 10 billion. Thus, population assumptions in the emissions scenarios appear to be broadly consistent with the recent population projections, with the caveat that only a few underlying projections are reported in the database.

Figure 2-4 contrasts the alternative population projections from the SRES database with historical developments. The long-term historical population growth rate has been on average about 1% per year during the past two centuries. Between 1800 and 1900 the global population growth rate was about 0.5% per year. The average annual growth rate since 1900 has been 1.3%. Between 1990 and 1995 the rate was 1.46%; and since 1995 world population has been growing at a rate of 1.3% annually (UN, 1998). All scenarios reviewed here envisage that population growth will slow in the future. The most recent doubling of the world's population took approximately 40 years. Even the highest population projections in Figure 2-4 require 70 years or more for the next doubling, while in roughly half of the scenarios the global population does not double during the 21st century. The highest average population growth across all projections is 1.2% per year, the lowest is 0.1% per year, and the median is about 0.7% per year.

Interestingly, the population projections in Figure 2-4 are not evenly distributed across the full range. Instead, they are grouped into three clusters. The middle cluster is representative of the central projections, with a range of about 10 to less than 12 billion people by 2100. The other two clusters mark the highest and the lowest population projections available in the literature, with between 15 and 20 billion at the high end and about 6 billion at the low end.

Despite these large ranges among alternative global population projections, the variation in this factor compared to the base year is the smallest of all the scenario driving forces considered in this comparison. Compared with 1990 values, the factor increase varies from 3.3 to 1.2.

2.4.5. *Gross World Product*

Economic development and growth are fundamental prerequisites to achieve an increase in living standards. It is thus not surprising that assumptions about economic development constitute among the most important determinants of emissions levels in the scenarios. However, economic growth prospects are among the most uncertain determinants of future emissions. Figure 2-5 shows the future increase in gross world product compared with the historical experience since 1950. As the differences in base-year data are relatively large, the gross world product paths are plotted as an index and spliced to historical data in 1990.

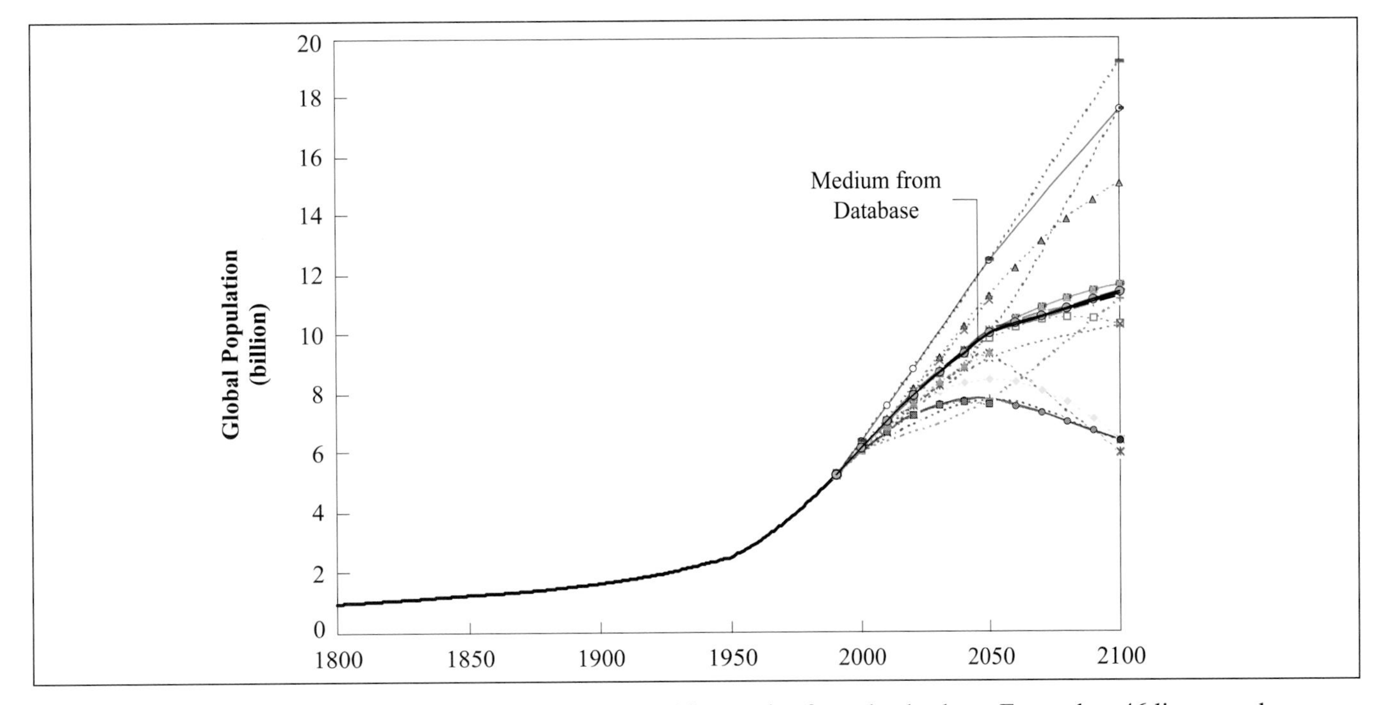

Figure 2-4: Global population – historical development and 46 scenarios from the database. Fewer than 46 lines are shown because many of the 46 scenarios use identical population projections. Only 46 of the scenarios in the database identify their population projections. Data source: Durand, 1967; Demeny, 1990; UN, 1996; Morita and Lee, 1998.

The historical gross world product growth rate has been about 4% per year since 1950; in the scenarios the average annual growth rates to 2100 range from 3.2% per year to 1.1% per year, with the median value at 2.3% per year. Table 2-3 summarizes the future economic growth rates of the 148 scenarios in the database that report gross world product for 2100. This translates into a gross world product level in 2100 that varies from 3.2 (IS92c, Pepper *et al.*, 1992) to more than 32 (FUND/EMF Modeler's choice, Tol, 1995) times the 1990 gross world product. The 1990 gross world product was about US$20 trillion (all values in 1990 US dollars), which translates into a range from more than US$700 to about US$65 trillion by 2100.

Figure 2-5 also indicates that this full range includes a few noticeable outliers toward the high and low of future gross world product development. The rest of the scenarios are grouped much more closely together, which compresses the range to a factor increase of about 17 to 7 (129 out of 148 scenarios) times compared to 1990. The degree of clustering is discussed in greater detail in the histograms that follow.

2.4.6. Gross World Product Histograms

Figures 2-6a and 2-6b depict the range of gross world product in 2050 and 2100 across all scenarios in the database.[9] The lower horizontal axis shows the factor increase of gross world product compared with the 1990 value (about US$20 trillion). The second horizontal axis multiplies the index by the 1990 gross world product to indicate absolute values for each histogram.

[9] Values are shown also as indexed to 1990 values, since model base years and base-year values differ.

Table 2-3: Future economic growth rates for the maximum, minimum, median, and percentiles of the 148 scenarios in the database that report gross world product for 2100 (percent per year).

	1990–2020	1990–2050	1990–2100
Maximum	3.8	3.6	3.2
95%	3.2	2.9	2.4
75%	2.8	2.6	2.3
Median	2.7	2.5	2.3
25%	2.6	2.4	2.1
5%	2.0	1.7	1.3
Minimum	1.5	1.4	1.1

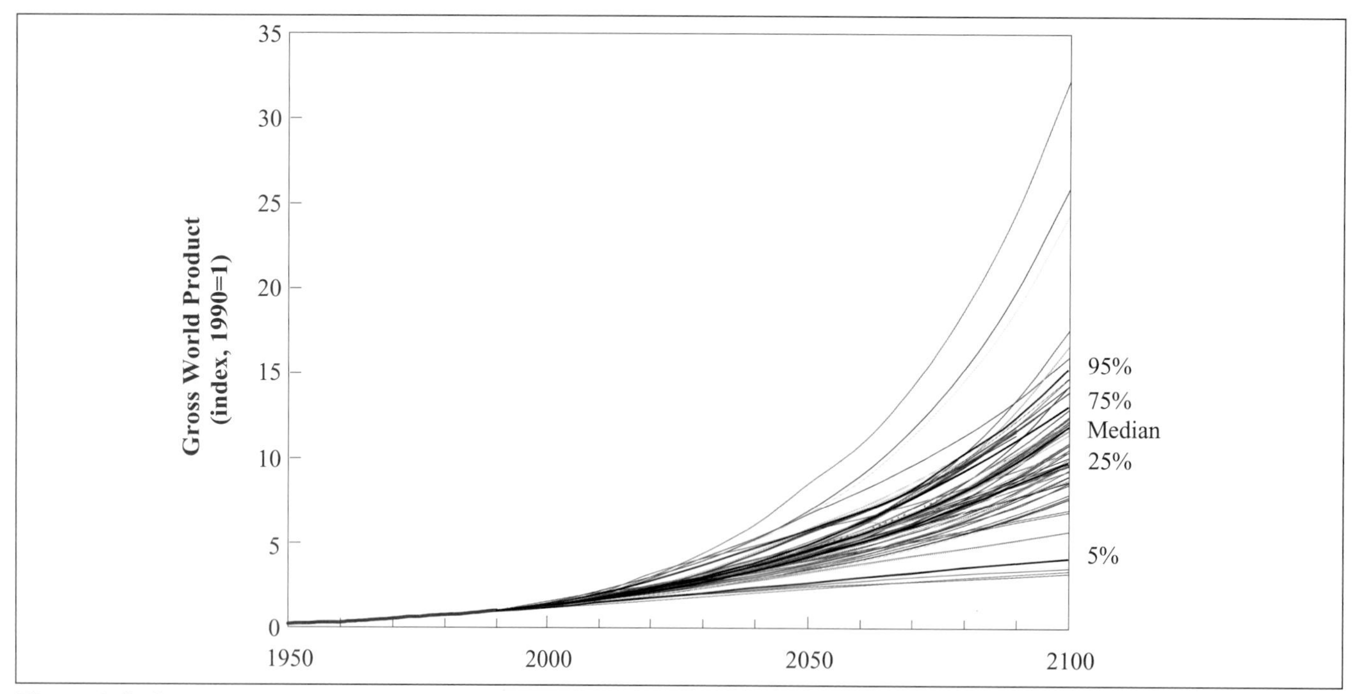

Figure 2-5: Gross world product – historical development and 193 scenarios, shown as an index (1990 = 1). Data source: UN, 1993a, 1993b; Morita and Lee, 1998.

For 2050 most of the scenarios cluster in a rather narrow range around a value of about US$100 trillion. In total, 166 scenarios were used to derive the histogram for 2050. This picture changes radically for the year 2100, with a very wide variation of gross world product values, from about US$700 to US$70 trillion, and a median of US$250 trillion. As expected, the distribution of emissions becomes significantly wider as the scenarios extend further into the future. Most of the distribution is concentrated between about US$320 and US$160 trillion, with very thin and asymmetric tails. A very strong peak of values lies at around US$250 trillion, which apparently represents an apparent consensus among modelers based on an average economic growth rate of about 2.3% per year. The frequency of the mode is smaller in 2100 than in

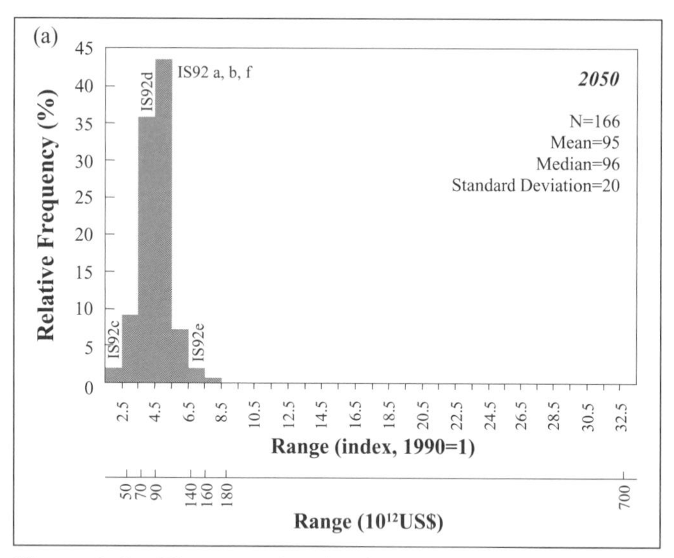

Figure 2-6a: Histogram showing frequency distribution of gross world product in 2050 for 166 scenarios. The upper horizontal axis shows indexed gross world product (1990 = 1); the lower axis indicates approximate absolute values by multiplying the index by the 1990 value (US$20 trillion). For reference, the gross world products of the IS92 scenarios are indicated. The frequency distribution associated with scenarios from the literature does not imply probability of occurrence.

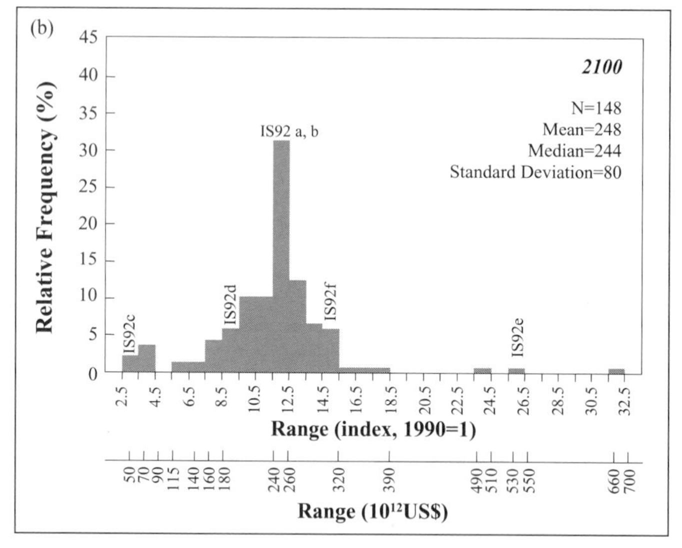

Figure 2-6b: Histogram showing frequency distribution of gross world product in 2100 for 148 scenarios. The upper horizontal axis shows indexed gross world product (1990 = 1); the lower axis indicates approximate absolute values by multiplying the index by the 1990 value (US$20 trillion). For reference, the gross world products of the IS92 scenarios are indicated. The frequency distribution associated with scenarios from the literature does not imply probability of occurrence.

2050, which indicates that the scenarios agree less about the central estimated gross world product (Tol, 1995; Yohe, 1995). For 2050 and 2100 the gross world products for the IS92a and b scenarios are the same as the median for all scenarios reviewed (Pepper *et al.*, 1992). 148 different scenarios were used to derive the histogram for the year 2100.

2.4.7. *Population and Gross World Product Relationships*

The scenarios in the database portray a weak relationship between population and economic growth; the correlation is slightly negative. Scenarios that lead to a very high gross world product are generally associated with central to low population projections, while high population projections do not lead to the highest gross world product scenarios. At extremely high levels of average global income the correlation is strongly negative. The highest per capita incomes in 2100 – in the range between US$30,000 and US$45,000 – are achieved with a low-to-medium population growth.

Figure 2-7 illustrates some of the relationships between population and gross world product in the scenarios. It compares only 39 scenarios as information about population and gross world product assumptions is available for only a few scenarios. In most of these, global population transition is achieved during the 21st century and stabilization occurs at a population between 10 to 12 billion people in the year 2100. Generally, this is associated with relatively high levels of economic development, in the range US$200–500 trillion in the year 2100. Scenarios at the lower end of this scale are labeled collectively as the "mid-range cluster," which includes all IIASA–WEC scenarios (IIASA–WEC, 1995; Nakićenović *et al.*, 1998b), IS92a and b (Pepper *et al.*, 1992), and AIM96 (Matsuoka *et al.*, 1994). The two highest scenarios are labeled as the "extra high growth" cases, namely IS92e (Pepper *et al.*, 1992) and IMAGE 2.1, Baseline-C (Alcamo and Kreileman, 1996).

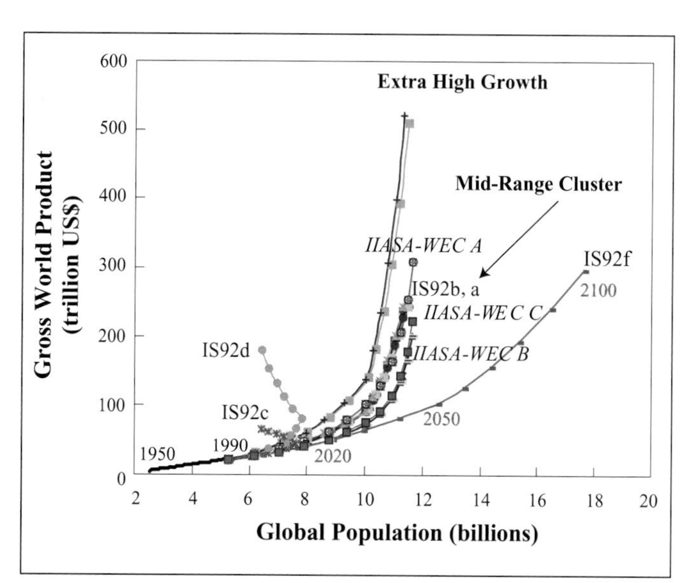

Figure 2-7: Gross world product and population growth, historical development, 1950 to 1990, and scenarios in the database to 2100, in trillion US dollars and billion people. All endpoints of the curves correspond to 2100.

One scenario, IS92f, shows high population growth (over 18 billion people by 2100) with comparatively low economic growth (about the same level as the mid-range cluster of scenarios, approximately US$300 trillion). At the other side of the scale are the two IS92 variants (c and d (Pepper *et al.*, 1992)) with low population projections (about 6 billion people by 2100).

2.4.8. *Primary Energy Consumption Ranges*

Primary energy consumption is another fundamental determinant of GHG emissions. Clearly, high energy consumption leads to high emissions. However, what is more important for emissions is the structure of future energy systems. High carbon intensities of energy – namely high shares of fossil energy sources, especially coal, in total energy consumption – lead to scenarios with the highest CO_2 emissions. The primary energy paths of different scenarios are compared here, and the issue of energy carbon intensity is considered in the next section.

Figure 2-8 shows the primary energy consumption paths in the scenarios and its historical development since 1900. It gives the whole distribution of the 153 scenarios in the SRES database that report primary energy consumption, the median, and the 95th, 75th, 25th, and 5th percentiles. As a result of the relatively large differences in the base-year values, the primary energy consumption paths are plotted as an index and spliced to the historical data in 1990. In 1990, primary energy was about 370 EJ, including non-commercial energy (Nakićenović *et al.*, 1996).

On average the global primary energy consumption has increased at more than 2% per year (fossil energy alone has risen at almost 3% per year) since 1900. Also, the short-term trend from 1975 to 1995 shows a similar increase. In the scenarios the average growth rates to 2100 range from 2.4% per year to –0.1% per year, with a median value of 1.3% per year.

For the full range of scenarios, the factor increase above the 1990 level is 0.9 to 10 by 2100.[10] However, Figure 2-8 indicates that this full range includes a few noticeable outliers, especially toward the high end of energy consumption levels. The rest of the scenarios are grouped more closely together, which compresses the range to a factor increase of about 1.5 to 7.5 times the 1990 level. The degree of clustering is discussed in greater detail below.

[10] Note that the highest scenario in the database reports 3400 EJ for primary energy consumption by 2100. Relative to the base year of this scenario (340 EJ), this level corresponds to a 10-fold increase. However, relative to the base-year value including non-commercial biomass (370 EJ), this level corresponds to a nine-fold increase only.

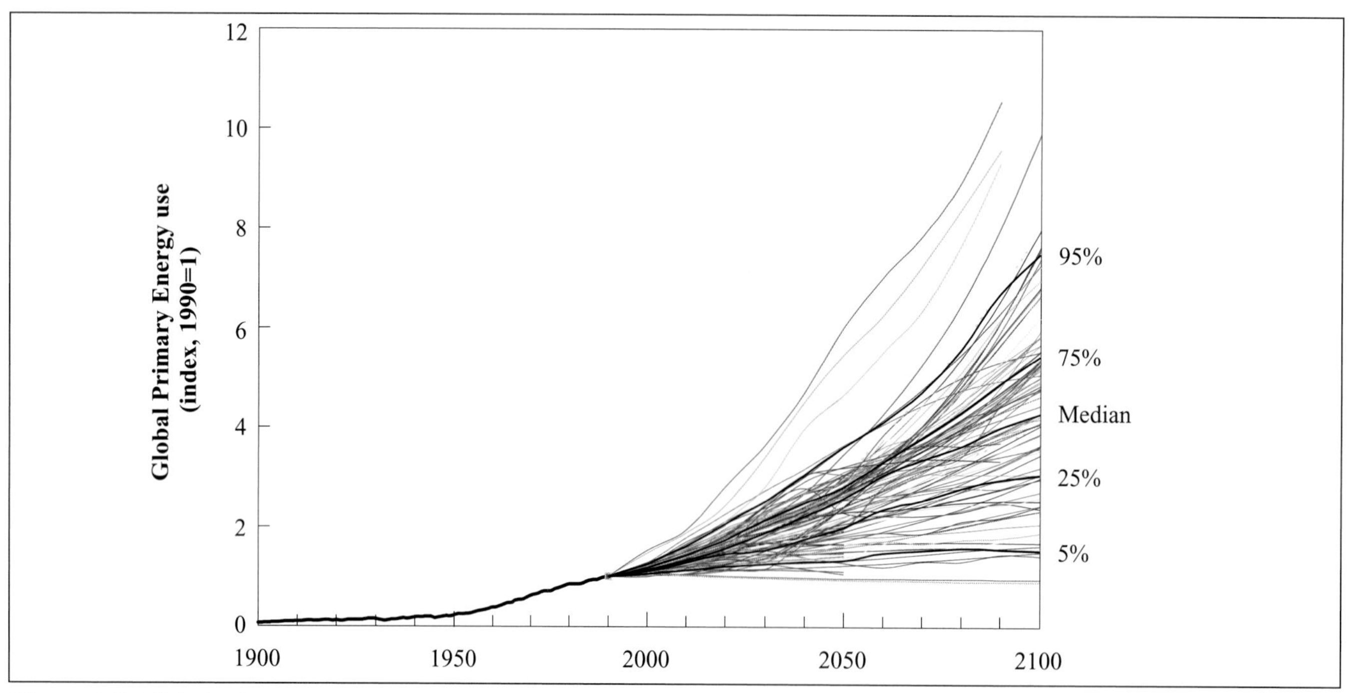

Figure 2-8: Global primary energy consumption – historical development and in future scenarios, shown as an index (1990 = 1). Data sources: Morita and Lee, 1998; Nakićenović *et al.*, 1998a.

2.4.9. *Primary Energy Consumption Histograms*

The final two histograms, Figures 2-9a and 2-9b, give the global primary energy consumption ranges for 2050 and 2100, indexed to 1990. The total range in 2100 is between 330 EJ and 3400 EJ, which corresponds to difference of more than a factor of ten. This factor range is about the same as the factor range for gross world product (10.7) and much larger than that for population.

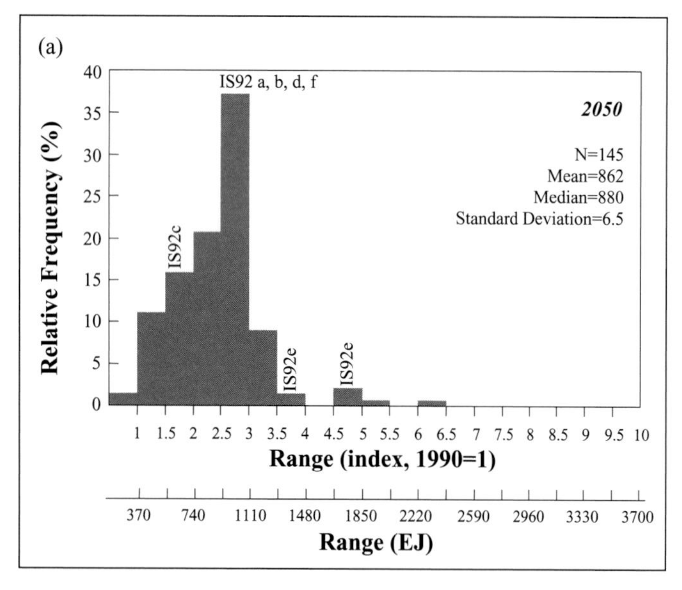

Figure 2-9a: Histogram showing frequency distribution of global primary energy consumption in 2050 for 145 scenarios. The upper horizontal axis shows indexed primary energy consumption (1990 = 1); the lower axis indicates approximate absolute values by multiplying the index by the 1990 value (370 EJ). For reference, primary energy consumptions of the IS92 scenarios are indicated. The frequency distribution associated with scenarios from the literature does not imply probability of occurrence.

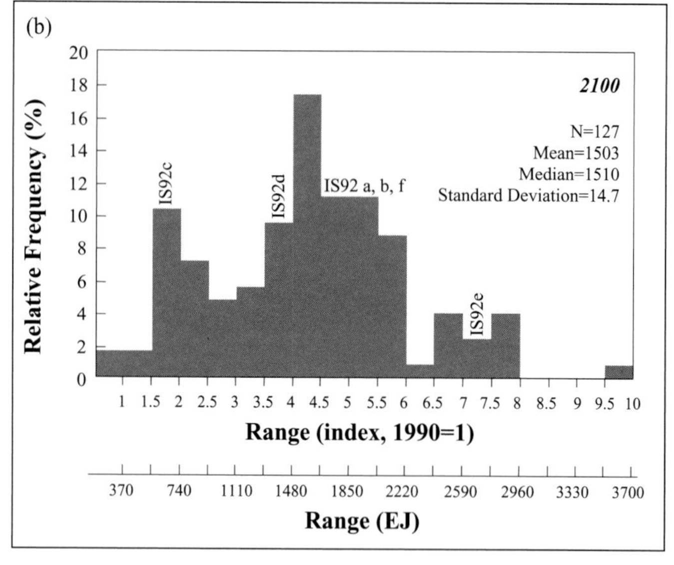

Figure 2-9b: Histogram showing frequency distribution of global primary energy consumption in 2100 for 127 scenarios. The upper horizontal axis shows indexed primary energy consumption (1990 = 1); the lower axis indicates approximate absolute values by multiplying the index by the 1990 value (370 EJ). For reference, primary energy consumptions of the IS92 scenarios are indicated. The frequency distribution associated with scenarios from the literature does not imply probability of occurrence.

The distribution of primary energy consumption in 2050 is asymmetric, with a long but thin tail toward high levels. Most of the distribution is between 600 and 1300 EJ, or about 1.5–3.5 times the 1990 level of consumption. The higher value corresponds to the continuation of a historical growth rate of global primary energy consumption of about 2.2% per year. Most of the scenarios cluster around 1000 EJ, which is about three times that of 1990. A total of 145 different scenarios were used to derive this histogram.

In 2100, the distribution of primary energy consumption is much less concentrated and covers the full range, from about 10 times the 1990 level (to 3400 EJ) to a slight decrease by a factor of 0.9 (340 EJ). However, only a few scenarios occur toward these extreme values of the two distribution tails. The rest of the observations portray an interesting structure that resembles a trimodal distribution. The first mode is around 2600 EJ, the second around 1700 EJ, and the third at about 700 EJ. The median of the whole distribution is at 1500 EJ. Interestingly, the continuation of the historical growth rate of about 2.2% per year corresponds to about 11 times the 1990 level (about 4050 EJ in 2100), well above any values observed in the database. Thus, in contrast to data for 2050, all the scenarios for 2100 foresee a level of primary energy consumption that is lower than that of trend extrapolation. Altogether, 127 different scenarios were used to derive the histogram for the year 2100.

2.4.10. Relationships between Primary Energy and Gross World Product

In all scenarios, economic growth outpaces the increase in energy consumption, which leads to substantial reductions in the ratio of primary energy consumption to gross world product, also known as "energy intensity." Individual technologies progress, inefficient technologies are retired in favor of more efficient ones, the structure of the energy-supply system and patterns of energy services change; these factors reduce the amount of primary energy needed per unit of gross world product. With all other factors being equal, the faster the economic growth, the higher the turnover of capital, and the greater the decline in energy intensity.

These long-term relationships between energy and economic development are reflected in the majority of scenarios and are consistent with historical experience across a range of alternative development paths observed in different countries.

Figure 2-10 shows the historical relationship since 1970 between energy intensity and GDP per capita for various world regions (see IIASA–WEC, 1995). This shorter record of development is contrasted with the experience of the USA since 1800. Although there are consistent differences among regions and energy paths, the level of energy intensities in developing countries today is generally comparable with the

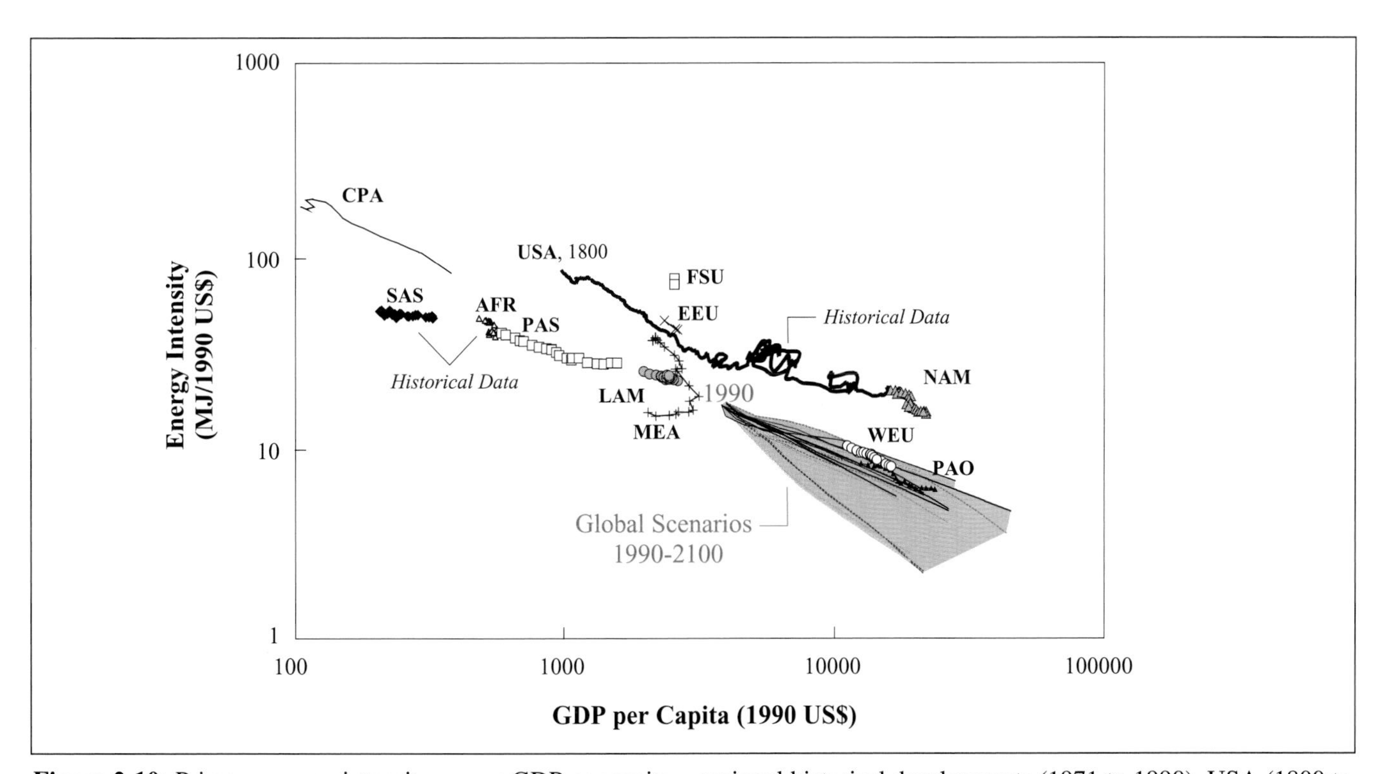

Figure 2-10: Primary energy intensity versus GDP per capita – regional historical developments (1971 to 1990), USA (1800 to 1990), and in global scenarios; energy intensity in MJ per US dollar at 1990 prices and per capita in US dollars. Abbreviations: AFR, Sub-Saharan Africa; CPA, Centrally Planned Asia and China; EEU, Eastern Europe; FSU, Former Soviet Union; LAM, Latin America and the Caribbean; MEA, Middle East and North Africa; NAM, North America; PAO, Pacific OECD; PAS, Other Pacific Asia; SAS, South Asia; WEU, Western Europe. Data sources: IEA, 1993; World Bank, 1993; Morita and Lee, 1998; Nakićenović *et al.*, 1998b.

range of the now-industrialized countries when they had the same level of per capita GDP (see Figure 2-10). The historical experiences illustrate that different countries and regions can follow different development paths; moreover, there are some persistent differences in energy intensities even at similar levels of per capita GDP.

Global energy intensities diverge across different scenarios, as shown in the shaded wedge in Figure 2-10. The wedge clearly illustrates a persistent inverse relationship between economic development and energy intensity across the wide range of scenarios in the database, despite numerous differences among them.

2.4.11. Carbon Intensity and Decarbonization

Decarbonization denotes the declining average carbon intensity of primary energy over time (see Kanoh, 1992). Although the decarbonization of the world's energy system shown in Figure 2-11 is comparatively slow, at the rate of 0.3% per year, the trend has persisted throughout the past two centuries (Nakićenović, 1996). The overall tendency toward lower carbon intensities results from the continuous replacement of fuels with high carbon content by those with low carbon content.

The carbon intensities of the scenarios are shown in Figure 2-11 as an index spliced in the base year 1990 to the historical development. The median of all the scenarios indicates a continuation of the historical trend, with a decarbonization rate of about 0.4% per year, which is similar to the trend in the IS92a scenario (Pepper *et al.*, 1992).

The scenarios that are most intensive in the use of fossil fuels lead to practically no reduction in carbon intensity. The highest rates of decarbonization (up to 3.3% per year) are from scenarios that envision a complete transition to non-fossil sources of energy.

Figure 2-12 illustrates the relationships between energy intensities of gross world product and carbon intensities of energy across the scenarios in the database. Both intensities are shown on logarithmic scales. The starting point is the base year 1990 normalized to an index (1990 = 100) for both intensities. Scenarios that unfold horizontally are pure decarbonization cases with little structural change in the economy; scenarios that unfold vertically indicate reduction in the energy intensity of economic activities with little change in the energy system. Most scenarios stay away from these extremes and develop a fan-shaped pattern – marked by both decarbonization and declining energy intensity – across the graph in Figure 2-12.

The fan-shaped graph illustrates the notable differences in policies and structures of the global energy system among the scenarios. For example, in a number of scenarios decarbonization is achieved largely through energy efficiency improvements, while in others it is mainly the result of lower carbon intensity because of the vigorous substitution of fuels. A few scenarios follow a path opposite to the other scenarios: decarbonization of primary energy with decreasing energy efficiency until 2040. After 2040 the ratio of CO_2 per unit of primary energy increases – in other words, recarbonization.

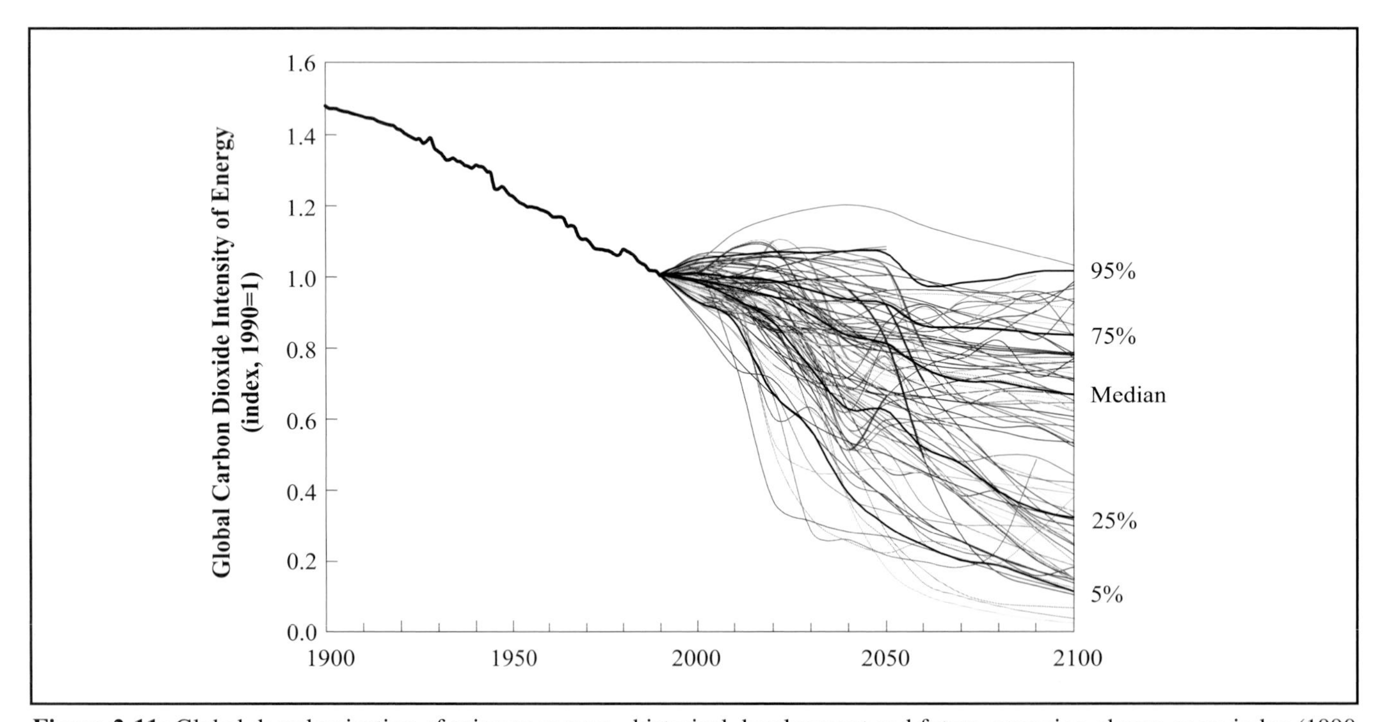

Figure 2-11: Global decarbonization of primary energy – historical development and future scenarios, shown as an index (1990 = 1). The median (50th), 5th, 25th, 75th, and 95th percentiles of the frequency distribution are shown. Statistics associated with scenarios from the literature do not imply probability of occurrence. Data source: Nakićenović, 1996; Morita and Lee, 1998.

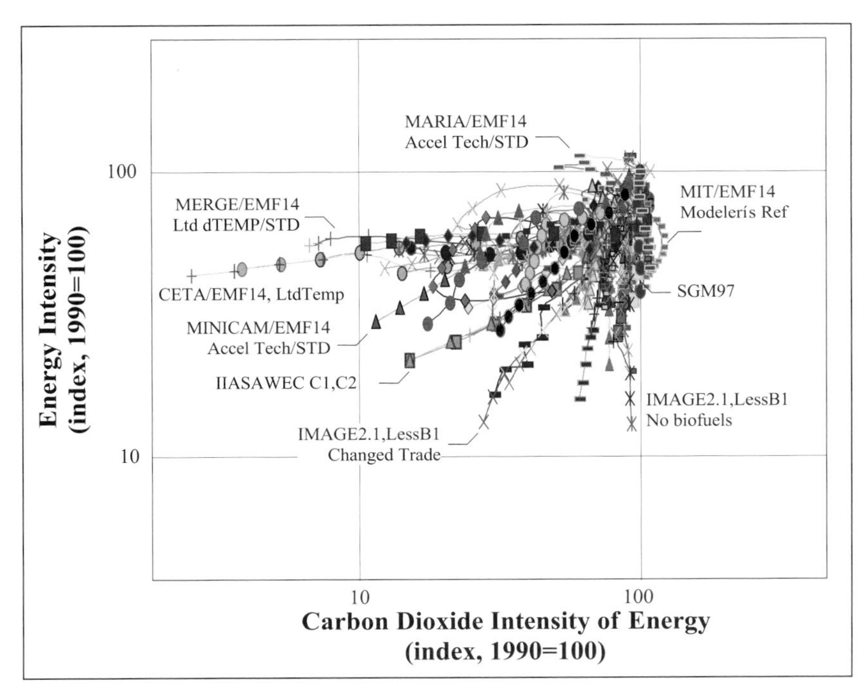

Figure 2-12: Global decarbonization and deintensification of energy in the scenarios, 1990 to 2100; energy and carbon intensities shown as an index (1990 = 1). Some of the scenarios are identified in the figure. Data source: Morita and Lee, 1998.

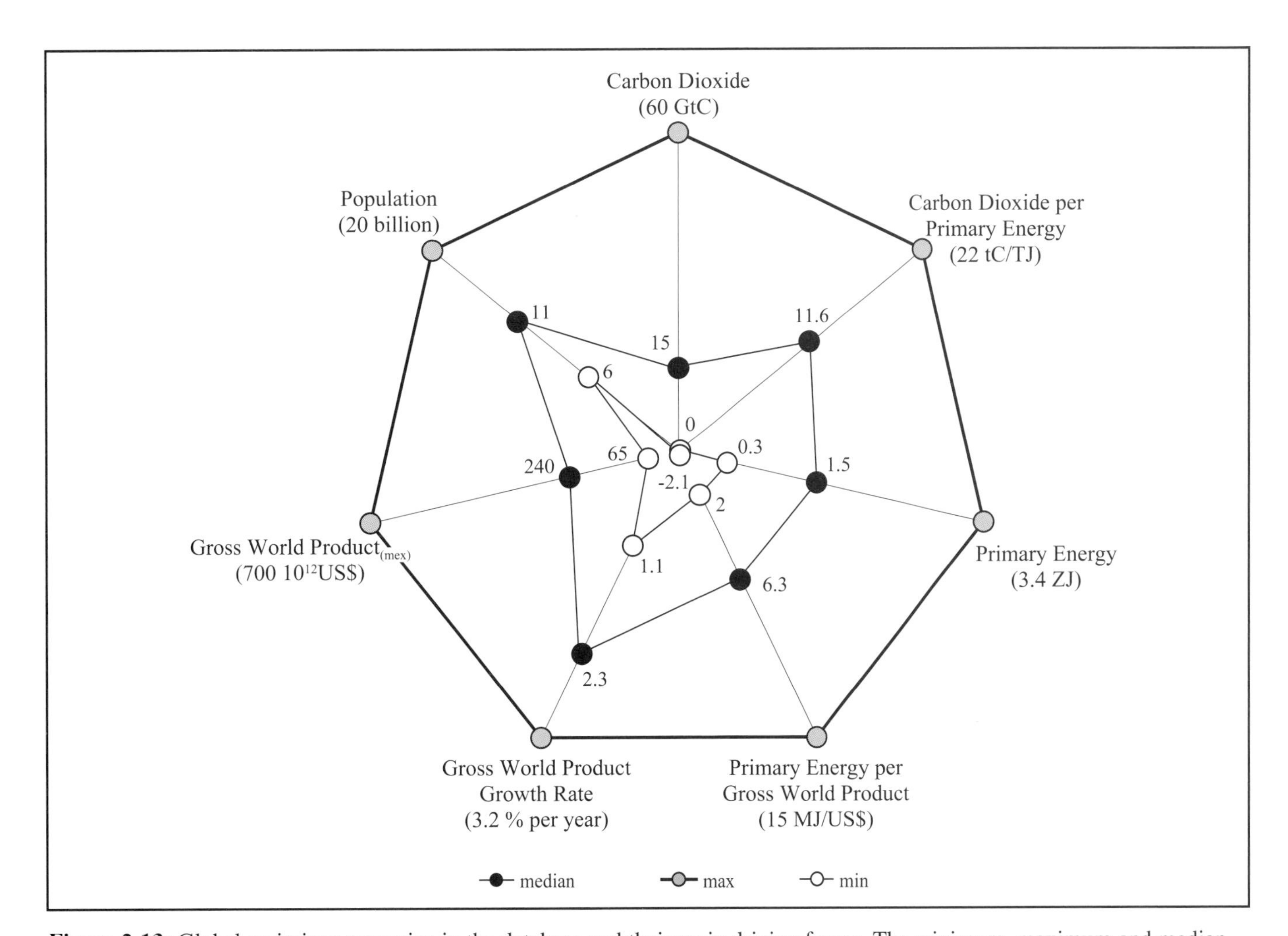

Figure 2-13: Global emissions scenarios in the database and their main driving forces. The minimum, maximum and median (50th percentile) values for 2100 are shown on seven axes of the heptagon based on the scenarios in the database. The seven axes show the ranges across the scenarios in 2100 of CO_2 emissions in GtC, population in billions, gross world product in trillion US dollars at 1990 prices, gross world product growth rates in percent per year, energy intensity in MJ per US dollar at 1990 prices, primary energy in ZJ (1000 EJ) and carbon intensity in tC per TJ.

2.4.12. *Comparison of Indicators*

Figure 2-13 illustrates the database distributions of CO_2, population, gross world product, primary energy, and carbon intensity of energy. The circles closest to the center denote the minimum value of the distribution; the solid circles denote the median value; and the shaded circles represent the maximum database value for each variable. While the values are connected in the form of a snowflake, it is important to note that those of a given range (e.g., minimum, median, and maximum), taken together, do not necessarily yield a consistent or logically possible scenario. As is shown in Chapter 4 (Figure 4-4), actual scenarios may fall into a median range on some axes and into a higher or lower range on others. Snowflake diagrams are useful because they allow the reader to see at a glance the full range of values encompassed by the database. Subsequent snowflake diagrams plot SRES scenario values on the various axes to illustrate where the scenarios fall relative to the database minimum, median, and maximum values. Snowflake diagrams should be used only for purposes of scenario classification and interpretation and not for scenario design, since the latter could lead to logical inconsistencies.

2.5. Conclusions

In this chapter a brief overview and evaluation of the emissions scenarios is presented. Much of the quantitative analysis of the emissions scenarios is based on the unique scenario database developed for SRES. This database, its structure, and the process of assembling the data that it now contains are described. The database provides a valuable overview of the various emissions modeling approaches and scenarios in the literature. It is the most comprehensive ESD available, and it can be accessed through a website (www-cger.nies.go.jp /cger-e/db/ipcc.html).

The SRES database represents the basis for evaluations of emissions scenarios, their main driving forces, and their uncertainty ranges. The CO_2 emissions trajectories of the scenarios in the SRES database are presented and their distribution and the associated sample statistics assessed. In the same way, the main driving forces of future emissions – population, economic development and energy consumption – are analyzed. Finally, the possible relationships among these driving forces for the collection of emissions scenarios in the database are considered.

Future levels of CO_2 emissions from the energy sector are a function of population, gross world product, the structure and efficiency of the economy, and the costs and availability of different sources of energy.

The factor range for population increases by 2100 across the scenarios is between 1.2 and 3.3 times 1990 levels (see Figure 2-4). This is the smallest factor increase of all the emissions driving forces. This probably reflects a relatively high consensus among demographers as to future population growth. However, this observation is based on a relatively small number of reported population projections in the database (46). The range in projected gross world product values is between 3.2 and 35 times the 1990 levels (see Figure 2-5). The range of the factor increase of primary energy consumption is from 0.9 to 10 times 1990 levels (see Figure 2-8). For energy intensity the range in 2100 is from 9.3 MJ/US$ down to 2.3 MJ/US$.

The range in carbon intensity of primary energy is the widest range of all the driving forces considered here. It varies from 0.025 to 1.1 times 1990 levels (about a factor of 45) in the year 2100 (see Figure 2-11). Emissions trajectories are extremely sensitive to a number of driving forces, which include including population growth, economic growth, and energy intensity improvement. Variation in carbon intensity is the main indicator of the wide variation in energy-related CO_2. However, it is important to recognize that this is a result of the inputs, assumptions, methods, and types of models used to calculate the scenarios.

The findings of the analysis of the scenarios in the literature suggest the following general conclusions:

- the ranges over which the driving forces vary are large, which contributes to the wide range of future emissions;
- of these driving forces, gross world product and population are often exogenous, while carbon and energy intensity are not;
- the frequency distributions of driving forces and emissions are asymmetric, with long tails and often more than one peak;
- for many driving forces the median and mean are closer to the minimum than to the maximum;
- lack of information as to whether climate intervention policies and measures were included in the scenarios, for about 40 scenarios, means that it is not possible to distinguish between intervention and non-intervention scenarios;
- it is hoped that IPCC or another international institution will, in the future, maintain this or a similar database so as to assure a continuity of knowledge and scientific progress in GHG emissions scenarios (an equivalent database to document narrative and other qualitative scenarios would also be very useful for future climate-change assessments); and
- it would be useful in the future to evaluate the consistency of the driving forces and results of all scenarios included in the scenario database. In addition, it would be helpful to extend the database to include land-use change data where currently only land-use emissions are presented. As noted above, only a small percentage of scenarios contain information on land use.

References:

Alcamo, J., A. Bouwman, J. Edmonds, A. Grübler, T. Morita, and A. Sugandhy, 1995: An evaluation of the IPCC IS92 emission scenarios. In *Climate Change 1994, Radiative Forcing of Climate Change* and *An Evaluation of the IPCC IS92 Emission Scenarios,* J.T. Houghton, L.G. Meira Filho, J. Bruce, Hoesung Lee, B.A. Callander, E. Haites, N. Harris and K. Maskell (eds.), Cambridge University Press, Cambridge, pp. 233-304.

Alcamo, J., A. Bouwman, J. Edmonds, A. Grübler, T. Morita, and A. Sugandhy, 1995: An Evaluation of the IPCC IS92 Emission Scenarios. In *Climate Change 1994, Radiative Forcing of Climate Change and An Evaluation of the IPCC IS92 Emission Scenarios,* Cambridge University Press, Cambridge, UK, pp. 233–304.

Alcamo, J., and E. Kreileman, 1996: Emission scenarios and global climate protection. *Global Environmental Change*, **6**(4), 305-334.

Alcamo, J., G.J.J. Kreileman, J.C. Bollen, G.J. van den Born, R. Gerlagh, M.S. Krol, A.M.C. Toet, and H.J.M. De Vries, 1996: Baseline scenarios of global environmental change. *Global Environmental Change*, **6**(4), 261–303.

Alcamo, J., G.J.J. Kreileman, J.C. Bollen, G.J. van den Born, R. Gerlagh, M.S. Krol, A.M.C. Toet, and H.J.M. De Vries, 1998: Baseline scenarios of global environmental change. In *Global Change Scenarios of the 21st Century, Results from the IMAGE 2.1 Model.* J. Alcamo, R. Leemans, E. Kreileman (eds.), Elsevier Science, Kidlington, Oxford, pp. 97-139.

Bos, E., and M.T. Vu, 1994*: World Population Projections: Estimates and Projections with Related Demographic Statistics,* 1994-1995 Edition, World Bank, Washington, DC.

Chattopadhyay, D. and J.K. Parikh, 1993: CO_2 emissions reduction from power system in India, *Natural Resources Forum*, **17**(4), 251-261.

Demeny, P., 1990: Population. In *The Earth As Transformed by Human Action.* B.L. Turner II *et al.* (ed.), Cambridge University Press, Cambridge, pp. 41-54.

De Vries, H.J.M., J.G.J. Olivier, R.A. van den Wijngaart, G.J.J. Kreileman, and A.M.C. Toet, 1994: Model for calculating regional energy use, industrial production and greenhouse gas emissions for evaluating global climate scenarios. *Water, Air Soil Pollution*, **76**, 79-131.

Dignon, J., and S. Hameed, 1989: Global emissions of nitrogen and sulfur oxides from 1860 to 1989. *Journal of the Air and Waste Management Association*, **39**(2), 180-186.

Dowlatabadi, H., and M. Kandlikar, 1995: *Key Uncertainties in Climate Change Policy: Results from ICAM-2*. Proceedings of the 6th Global Warming Conference, San Francisco, CA.

Durand, J.D., 1967: The modern expansion of world population. *Proceedings of the American Philosophical Society*, **111**(3), 136–159.

Edmonds, J.A., J. Reilly, J.R. Trabalka, D.E. Reichle, D. Rind, S. Lebedeff, J.P. Palutikof, T.M.L. Wigley, J.M. Lough, T.J. Blasing, A.M. Salomon, S. Seidel, D. Keyes and M. Steinberg, 1986: *Future Atmospheric Carbon Dioxide Scenarios and Limitation Strategies.* Noyes Publications, Park Ridge, NJ.

EMF (Energy Modeling Forum), 1995: *Second Round Study Design for EMF 14: Integrated Assessment of Climate Change*. Report prepared for the Energy Modeling Forum, Terman Engineering Center, Stanford University, CA 94305.

Gray, V., 1998: The IPCC future projections: are they plausible? *Climate Research,* **10,** 155-162.

Grübler, A., 1998: A review of global and regional sulfur emission scenarios. *Mitigation and Adaptation Strategies for Global Change,* **3**(2-4), 383-418.

Häfele, W., J. Anderer, A. McDonald, and N. Nakićenović, 1981: *Energy in a Finite World: Paths to a Sustainable Future*. Ballinger, Cambridge, MA,

Häfele, W., 1981: *Energy in a Finite World: A Global Systems Analysis*. Ballinger, Cambridge, MA, USA, 834 pp. (ISBN 0-88410-642-X).

Houghton, J.T., G.J. Jenkins, and J.J. Ephraums (eds.), 1990: *Climate Change: The IPCC Scientific Assessment*. Cambridge University Press, Cambridge, 365 pp.

IEA (International Energy Agency), 1993: *Energy Statistics and Balances for OECD and non-OECD countries 1971–1991*, OECD, Paris.

IEA (International Energy Agency), 1998: *Energy Statistics and Balances for OECD and non-OECD countries*. OECD, Paris.

IIASA–WEC (International Institute for Applied Systems Analysis – World Energy Council) 1995: *Global Energy Perspectives to 2050 and Beyond*, International Institute for Applied Systems Analysis, Laxenburg, Austria, 106 pp.

Jefferson, M., 1983: Economic Uncertainty and Business Decision-Making. In *Beyond Positive Economics?* Wiseman, Jack (ed.). Macmillan Press, London, pp. 122-159.

Kanoh, T., 1992: Toward dematerialization and decarbonization. *Science and Sustainability, Selected Papers on IIASA's 20th anniversary*. International Institute for Applied Systems Analysis, Laxenburg, Austria.

Kaya, Y., 1990: *Impact of Carbon Dioxide Emission Control on GNP Growth: Interpretation of Proposed Scenarios*. Paper presented to the IPCC Energy and Industry Subgroup, Response Strategies Working Group, Paris, (mimeo).

Kram, T., 1993: *National Energy Options for Reducing CO_2 Emissions*. Report of the Energy Technology Systems Analysis Programme (ETSAP), Annex IV (1990–1993). ECN-C-93-101, Netherlands Energy Research Foundation, Petten, the Netherlands.

Lazarus, M.L., L. Greber, J. Hall, C. Bartels, S. Bernow, E. Hansen, P. Raskin, and D. von Hippel, 1993: *Towards a Fossil Free Energy Future: The Next Energy Transition*. A Technical Analysis for Greenpeace International. Stockholm Environmental Institute Boston Center, Boston.

Leggett, J., W.J. Pepper and R.J. Swart, 1992: Emissions Scenarios for IPCC: An Update. In *Climate Change 1992. The Supplementary Report to the IPCC Scientific Assessment*. J.T. Houghton, B.A. Callander and S.K. Varney (eds.), Cambridge University Press, Cambridge.

Lutz, W., W. Sanderson, and S. Scherbov, 1997: Doubling of world population unlikely. *Nature*, **387**(6635), 803-805.

Manne, A. and R. Richels, 1994: *The Costs of Stabilizing Global CO_2 Emissions – a Probabilistic Analysis Based on Expert Judgement*. Electric Power Research Institute, Palo Alto, CA.

Manne, A., and L. Schrattenholzer, 1995: *International Energy Workshop January 1995 Poll Edition*. International Institute for Applied Systems Analysis, Laxenburg, Austria.

Manne, A., and L. Schrattenholzer, 1996, *International Energy Workshop January 1996 Poll Edition*. International Institute for Applied Systems Analysis, Laxenburg, Austria.

Manne, A., and L. Schrattenholzer, 1997: *International Energy Workshop, Part I: Overview of Poll Responses, Part II: Frequency Distributions, Part III: Individual Poll Responses,* February, 1997, International Institute for Applied Systems Analysis, Laxenburg, Austria.

Marland, G., T.A. Boden, R.J. Andres, A.L. Brenkert, and C. Johnston, 1999: *Global, Regional, and National CO_2 Emissions*. In *Trends Online: A Compendium of Data on Global Change*. Carbon Dioxide Information Analysis Center, Oak Ridge National Laboratory, U.S. Department of Energy, Oak Ridge, TN. (http://cdiac.esd.ornl.gov/trends/abstract.htm).

Matsuoka, Y., M. Kainuma, and T. Morita, 1994: Scenario analysis of global warming using the Asian-Pacific integrated model (AIM). *Energy Policy*, **23**(4/5), 357-371.

Morita, T., Y. Matsuoka, I. Penna, and M. Kainuma, 1994: *Global Carbon Dioxide Emission Scenarios and Their Basic Assumptions: 1994 Survey*. CGER-1011-94. Center for Global Environmental Research, National Institute for Environmental Studies, Tsukuba, Japan.

Morita, T., and H.-C. Lee, 1998: *IPCC SRES Database*, Version 0.1, Emission Scenario Database prepared for IPCC Special Report on Emissions Scenarios. (http:www-cger.nies.go.jp/cger-e/db/ipcc.html).

Murthy, N.S., M. Panda, J. Parikh, 1997: Economic development, poverty reduction and carbon emissions in India, *Energy Economics*, **19**, 327-354.

Nakićenović, N., A. Grübler, H. Ishitani, T. Johansson, G. Marland, J.R. Moreira, H.-H. Rogner, 1996: Energy primer. In: *Climate Change 1995. Impacts, Adaptations and Mitigation of Climate Change: Scientific Analyses*. R. Watson, M.C. Zinyowera and R. Moss (eds.), Cambridge University Press, Cambridge, UK, pp. 75-92.

Nakićenović, N., 1996: Freeing Energy from Carbon. *Daedalus*, **125**(3), 95–112.

Nakićenović, N., N. Victor, and T. Morita, 1998a: Emissions scenarios database and review of scenarios. *Mitigation and Adaptation Strategies for Global Change,* **3**(2-4), 95-120.

Nakićenović, N., A. Grübler, and A. McDonald (eds.), 1998b: *Global Energy Perspectives*. Cambridge University Press, Cambridge, 299 pp.

Nordhaus, W.D., and G.W. Yohe, 1983: Future paths of energy and carbon dioxide emissions. In: Changing Climate: Report of the Carbon Dioxide Assessment Committee. National Academy Press, Washington, DC.

Parikh J. (ed.) 1996: *Energy Models: Beyond 2000*. Tata McGraw Hills Publishing Co. Ltd., New Delhi, India.

Peck, S.C., and T.J. Teisberg, 1995: International CO_2 emissions control - An analysis using CETA. *Energy Policy*, **23**(4), 297-308.

Pepper, W.J., J. Leggett, R. Swart, J. Wasson, J. Edmonds, and I. Mintzer, 1992: Emissions Scenarios for the IPCC. An Update: Assumptions, Methodology, and Results. Support Document for Chapter A3. In *Climate Change 1992: Supplementary Report to the IPCC Scientific Assessment*. J.T. Houghton, B.A. Callandar and S.K. Varney (eds.), Cambridge University Press, Cambridge.

Ramanathan, R. and J.K. Parikh, 1999: Transport sector in India: An analysis in the context of sustainable development. *Transport Policy*, **6**(1), 35-45.

Raskin, P., G. Gallopin, P. Gutman, A. Hammond, and R. Swart, 1998: *Bending the Curve: Toward Global Sustainability*. A report of the Global Scenario Group, PoleStar Series Report 8. Stockholm Environment Institute, Stockholm, 128 pp.

Schwartz, P., 1991: *The Art of the Longview: Three global scenarios to 2005*. Doubleday Publications.

Tol, R.S.J., 1995: *The climate fund sensitivity, uncertainty, and robustness analyses*. W-95/02, Institute for Environmental Studies, Climate Framework, Vrije Universiteit, Amsterdam, the Netherlands.

UN (United Nations), 1993a, *Macroeconomic Data Systems, MSPA Data Bank of World Development Statistics, MSPA Handbook of World Development Statistics*, MEDS/DTA/1 and 2 June, UN, New York, NY.

UN (United Nations), 1993b, *UN MEDS Macroeconomic Data Systems, MSPA Data Bank of World Development Statistics*, MEDS/DTA/1 MSPA-BK.93, Long-Term Socioeconomic Perspectives Branch, Department of Economic and Social Information and Policy Analysis, UN, New York, NY.

UN (United Nations), 1996: *Annual Populations 1950–2050: The 1996 Revision*. UN Population Division, United Nations, New York, NY. (Data on diskettes.)

UN (United Nations), 1998: *World Population Projections to 2150*. UN Department of Economics and Social Affairs, Population Division, United Nations, New York, NY.

UN (United Nations), 1999: *World Population Prospects: 1998 Revision*, Vol. 1. UN Department of Economics and Social Affairs, Population Division, United Nations, New York, NY.

UNFCCC (United Nations Framework Convention on Climate Change), 1992: *Convention Text*, IUCC, Geneva.

US EPA (United States Environmental Protection Agency), 1990: *Policy Options for Stabilizing Global Climate*. Report to Congress. Washington, DC.

WBCSD (World Business Council for Sustainable Development), 1998: *Exploring Sustainable Development*, 29 pp.

WEC (World Energy Council), 1993: *Energy for Tomorrow's World*. Kogan Page Ltd., London, 320 pp.

Weyant, J.P. 1993: Costs of reducing global carbon emissions. *Journal of Economic Perspectives,* **7**(4), 7–46.

Weyant, J.P. (ed), 1999: The Costs of the Kyoto Protocol: A Multi-Model Evaluation. *Energy Journal* Special Issue, **June**, 448 pp.

World Bank, 1993: *World Development Report*. Oxford University Press, New York, NY.

Yamaji, K., R. Matsuhashi, Y. Nagata, and Y. Kaya, 1991: *An Integrated Systems for CO_2 /Energy/GNP Analysis: Case Studies on Economic Measures for CO_2 Reduction in Japan*. Workshop on CO_2 Reduction and Removal: Measures for the Next Century, 19–21 March 1991. International Institute for Applied Systems Analysis, Laxenburg, Austria.

Yohe, G., 1995: *Exercises in hedging against extreme consequences of global change and the expected value of information*, Department of Economics, Wesleyan University, CT.

3

Scenario Driving Forces

CONTENTS

3.1. Introduction

Some of the major driving forces of past and future anthropogenic greenhouse gas (GHG) emissions, which include demographics, economics, resources, technology, and (non-climate) policies, are reviewed in this chapter. Economic, social, and technical systems and their interactions are highly complex and only a limited overview is provided in this chapter. The discussion of major scenario driving forces herein is structured by considering the links from demography and the economy to resource use and emissions. A frequently used approach to organize discussion of the drivers of emissions is through the so-called IPAT identity, equation (3.1).

$$Impact = Population \times Affluence \times Technology \quad (3.1)$$

The IPAT identity states that environmental impacts (e.g., emissions) are the product of the level of population times affluence (income per capita, i.e. gross domestic product (GDP) divided by population) times the level of technology deployed (emissions per unit of income). The IPAT identity has been widely discussed in analyses of energy-related carbon dioxide (CO_2) emissions (e.g., Ogawa, 1991; Parikh *et al.*, 1991; Nakićenović *et al.*, 1993; Parikh, 1994; Alcamo *et al.*, 1995; Gaffin and O'Neill, 1997; Gürer and Ban, 1997; O'Neill *et al.*, 2000), in which it is often referred to as the Kaya identity (Kaya, 1990), equation (3.2).

$$CO_2\ Emissions = Population \times (GDP/Population) \times \times (Energy/GDP) \times (CO_2/Energy) \quad (3.2)$$

The Kaya multiplicative identity also underlies the analysis of the emissions scenario literature (Chapter 2). It can be broken down into further subcomponents. For instance, the energy component can be decomposed into fossil and non-fossil shares, and emissions can be expressed as carbon emissions per unit of fossil energy, as shown in Figure 3-1 (Gürer and Ban, 1997). A property of the multiplicative identity is that component growth rates are additive. For instance, global energy-related CO_2 emissions since the middle of the 19th century are estimated to have increased by approximately 1.7% per year (Watson *et al.*, 1996). This growth rate can be decomposed roughly into a 3% growth in gross world product (the sum of a 1% growth in population and a 2% growth in per capita income) minus a 1% per year decline in the energy intensity of world GDP (the third term in equation (3.2)) and a decline in the carbon intensity of primary energy (the fourth term) of 0.3% per year (Nakićenović *et al.*, 1993; Watson *et al.*, 1996).

While the Kaya identity above can be used to organize discussion of the primary driving forces of CO_2 emissions and, by extension, emissions of other GHGs, there are important caveats. Most important, the four terms on the right-hand side of equation (3.2) should be considered neither as fundamental driving forces in themselves, nor as generally independent from each other.

Global analysis is often not instructive and even misleading, because of the great heterogeneity among populations with respect to GHG emissions. The ratios of per capita emissions of the world's richest countries to those of its poorest countries approach several hundred (Parikh *et al.*, 1991; Engelman, 1994). Of course, some level of aggregation is necessary. In practice, the models used to produce emissions scenarios in this report, for example, operate on the basis of 9–15 regions (see Appendix IV, Table IV-1). This level of detail isolates the most important differences, particularly with respect to industrial versus developing countries (Lutz, 1993).

The spatial and temporal heterogeneity of emission growth that becomes masked in the global aggregates is shown in Figure 3-1, in which the growth in energy-related CO_2 emissions since 1970 is broken down into a number of subcomponents. For industrial countries the population growth has been modest and their emissions have evolved roughly in line with increases (or declines) in economic activity. For developing countries both population and income growth appear as important drivers of emissions. However, even in developing countries the regional heterogeneity becomes masked in the aggregate analysis (Grübler *et al.*, 1993a).

Although, at face value, the IPAT and Kaya identities suggest that CO_2 emissions grow linearly with population increases, this depends on the real (or modeled) interactions between demographics and economic growth (see Section 3.2) as well as on those between technology, economic structure, and affluence (Section 3.3). In principle, such interactions preclude a simple linear interpretation of the role of population growth in emissions.

Demographic development interacts in many ways with social and economic development. Fertility and mortality trends depend, among other things, on education, income, social norms, and health provisions. In turn, these determine the size and age composition of the population. Many of these factors combined are recognized as necessary to explain long-run productivity, economic growth, economic structure, and technological change (Barro, 1997). In turn, long-run per capita economic growth and structural change are closely linked with advances in knowledge and technological change. In fact, long-run growth accounts (e.g., Solow, 1956; Denison, 1962, 1985; Maddison, 1989, 1995; Barro and Sala-I-Martin, 1995) confirm that advances in knowledge and technology may be the most important reason for long-run economic growth; more important even than growth in other factors of production such as capital and labor. Abramovitz (1993) demonstrates that capital and labor productivity cannot be treated as independent from technological change. Therefore, it is not possible to treat the affluence and technology variables in IPAT as independent of each other.

Pollution abatement efforts appear to increase with income, growing willingness to pay for a clean environment, and progress in the development of clean technology. Thus, as incomes rise, pollution should increase initially and later

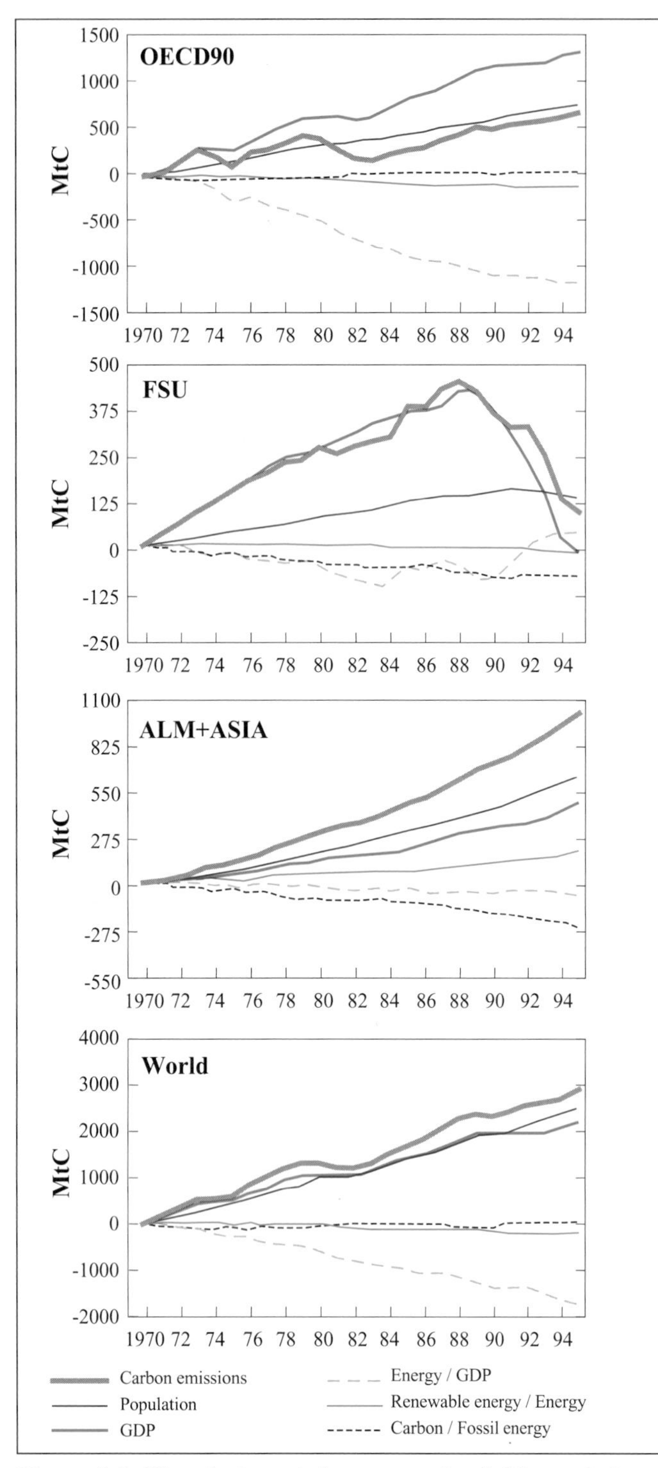

Figure 3-1: Historical trends in energy-related CO_2 emissions ("carbon emissions" shown as bold gray line) and broken down into the components of emission growth: growth or declines of population, gross domestic product (GDP) at purchasing power parities (PPPs), energy use per unit of GDP (Energy/GDP), share of renewables in energy use (Renewable energy/Energy), and carbon intensity per fossil energy (Carbon/Fossil energy) since 1970, in million tons elemental carbon (MtC). From top to bottom: Organization for Economic Cooperation and Development (OECD90, countries that belong to the OECD as of 1990), former USSR (FSU), Developing Countries (ASIA and Africa, Latin America and the Middle East (ALM)), and World. Source: Gürer and Ban, 1997.

decline, a relationship often referred to as the "environmental Kuznets curve." This process seems well established for traditional pollutants, such as particulates and sulfur (e.g., World Bank, 1992; Kato, 1996; Viguier, 1999), and there have been some claims that it might apply to GHG emissions. Schmalensee *et al.* (1998) found that CO_2 emissions have flattened and may have reversed for highly developed economies such as the US and Japan. Other researchers argue that the Kuznets curve does not apply to GHG emissions (Pearce, 1995; Galeotti and Lanza, 1999, Viguier, 1999). The flattening in emissions can be explained by normal market processes and does not appear to result from a willingness to pay to protect the global environment. Urbanization, infrastructure, poverty, and income distribution are other factors in the complex interplay between population, economy, and environment (see, e.g., Rotmans and de Vries, 1997; de Vries *et al.*, 1999; O'Neill *et al.*, 2000).

Technological, economic, and social innovation have long been means by which a greater number of people can live from the same environmental resources. The best known historical examples of major periods of innovation include the Neolithic revolution (beginnings of organized agriculture from around 10,000 years ago); and the industrial revolution that began two centuries ago (Rosenberg and Birdzell, 1986). In each case, changes in patterns of primary production (food, energy, materials) are linked to changes in social organization, institutions, economy, and technology (e.g., Mumford, 1934; Campbell, 1959; Landes, 1969; Hill, 1975; Wilber, 1981; Buchanan, 1992; Reynolds and Cutcliffe, 1997). The most remarkable change in recent decades is the so-called demographic transition, which has led to a stabilization of population in many parts of the world. No single one of these changes can be considered as the primary driver, and they cannot be considered as independent from each other: each play a role in an interconnected system.

Most innovative efforts in the past two centuries were devoted to improving labor productivity and the human ability to harness resources for economic purposes. While material and energy efficiency improved slowly, economic growth was faster and thus aggregate resource use increased.

Finally, and importantly, the high uncertainty with regard to the nature and extent of the relationships between driving forces of GHG emissions means that, with current knowledge, it is not possible to develop probabilistic future emission scenarios. Even if it were possible to derive (subjective) probability distributions of the future evolution of individual scenario driving-force variables (like population, economic growth, or technological change), the nature of their relationships is known only qualitatively at best or remains uncertain (and controversial) in many instances.

The next five sections review the major driving forces of GHG emissions within the IPAT identity. Section 3.2 discusses the role of population, Section 3.3 addresses economic and social development processes (including technological change),

Section 3.4 examines energy resources and technology in more detail, and Section 3.5 addresses agriculture, forestry, and land-use change. Section 3.6 considers other sources of non-CO_2 GHGs. The chapter concludes with a discussion of non-climate policies and their potential impact on the principal driving forces of future emissions. Each section briefly reviews past trends, available scenarios, and important new methodological and empirical advances since the publication of previous International Panel on Climate Change (IPCC) emissions scenarios in 1992 (IS92). This chapter provides the background to establish recommendations for the range of driving-force variables to be explored in the new set of scenarios. The available literature and current understanding of the inherent uncertainties in developing very long-term scenarios are reflected. Each section elucidates in detail the important relationships between scenario driving forces, as the question of relationships is a new and important mandate for SRES. Nonetheless, most attention is paid to the possible relationships between population and economic growth, because this is the area most intensively discussed in the literature.

3.2. Population

3.2.1. *Introduction*

Population projections are arguably the backbone of GHG emissions scenarios, and are comparable in some ways with them. Population projections cover timeframes of a century or more, and they involve social and economic considerations and uncertainties similar to those in GHG emissions scenarios. Population projections are among the most commonly cited indicators of the future state of the world. Compared to the multitude of projection efforts they have a relatively high accuracy in the near-to-medium term. Even so the future is always unknowable and surprises are in store, as confirmed by a cursory review of the past history of population projections in which fundamental events were largely unforeseen (post-World War II baby boom, acquired immunodeficiency syndrome (AIDS) or the recent rapidity of fertility decline in developing countries)).

To be useful for the development of emissions scenarios, population projections need a timeframe of a century or more, global coverage and regional disaggregations, and an appropriate treatment of uncertainty reflected in the variants of the projections. Although other "demographic units" more immediately linked to GHG emissions than people, such as automobiles or households, can be considered, the integrated assessment models used in this report are all based on regional population and, in some cases, labor-force projections.

3.2.2. *Past Population Trends*

World annual population growth rates probably averaged less than 0.6% during the 18th and 19th centuries, passed the 1% rate around 1920, and peaked at 2.04% in the late 1960s (UN, 1998). This peak coincided with growing international concern about population growth in general. World population reached 1 billion in 1804, 2 billion in 1927, 3 billion in 1960, 4 billion in 1974, and 5 billion in 1987, reaching the 6 billion level shortly before the millenium (UN, 1998).

The population of the developing regions increased from 1.71 billion in 1950 to 4.59 billion in 1996, with annual growth rates dropping from a peak of 2.5% in 1965 to 1.7% presently. The population of the more-developed regions increased from 813 million to 1.18 billion over the same period, with annual growth rates dropping from 1.2% in 1950 to 0.4% presently (UN, 1998). Population distribution and growth thus differ markedly among major geographic regions. Latin America and the Caribbean was the fastest growing region between 1950 and 1970, followed by Africa, and this is projected to remain the case until 2050 (UN, 1998). Table 3-1 shows the population levels of the major geographic areas between 1800 and the present.

3.2.3. *Population Scenarios*

3.2.3.1. Population Projections Used in Emission Scenarios

Since the IPCC was first convened in 1988, its Working Group III has generated two distinct series of emissions scenarios: the 1990 Scientific Assessment (SA90) series of four scenarios (Houghton *et al.*, 1990), and the IS92 series of six scenarios (Houghton *et al.*, 1992, 1995; Pepper *et al.*, 1992). The four

Table 3-1: *Population of the world and by major areas between 1800 and 1996 in millions. Data source: UN, 1998.*

	1800	1850	1900	1950	1996
World	978	1262	1650	2524	5768
Africa	107	111	133	224	739
Asia	635	809	947	1402	3488
Europe	203	276	408	547	729
Latin America and Caribbean	24	38	74	166	484
Northern America	7	26	82	172	299
Oceania	2	2	6	13	29

SA90 scenarios all used the same median population projection – the World Bank 1987 projection (Zachariah and Vu, 1988). The IS92a–f series made use of three different projection variants, the World Bank (World Bank, 1991) 1991 projection and the United Nations (UN, 1992) 1992 medium–high and medium–low projections.

Wexler (1996) surveyed the world population projections used in GHG emissions scenarios since 1990. Of the models surveyed, all but one employed the World Bank central projections. The sole exception – the DICE model of Nordhaus (1993; Nordhaus and Yohe, 1983) – used algorithmic projections based on assumed declining population growth rates. As noted in Chapter 2, many long-term emission scenarios available in the literature do not even report their underlying population projections. In general, World Bank projections have been more heavily employed than the UN projections, apparently because of the shorter time horizon and longer cycle time of the UN Long Range projections. UN Revisions until 1994 extended only to 2025 (now to 2050), which is too short for emissions scenarios. The UN Long Range series is revised less frequently, with a six-year interval between the previous two UN Long Range projections. The World Bank, in contrast, updates its published projection every two years, and it has always been a long-range projection, out to year 2150. In addition, the World Bank maintains a country-level disaggregation throughout its projection, unlike the UN Long Range series, which switches to a nine-region summary.

In 1994, however, the World Bank discontinued the publication of population projections, even if these continue to be generated for internal uses. In contrast to the past dominance of World Bank projections, the SRES scenario series instead employs published projections from the International Institute for Applied Systems Analysis (IIASA) along with the UN's Medium Long Range projection.

3.2.3.2. Currently Available Population Projections

World population projections are currently generated by the following institutions:

- United States Census Bureau (USCB);
- World Bank;
- UN; and
- IIASA.

The main defining characteristics of the individual projections and a detailed description of each projection is given in Gaffin (1998).

Figure 3-2 displays these latest world population numbers from all four demographic organizations. Two of the four projections, from the World Bank and USCB, contain only one central estimate and are unpublished. These are generated currently for internal organizational purposes and are not further considered here.

The only population projections that currently incorporate variation of the long-range fertility rates are those produced by UN and IIASA (Gaffin, 1998). For the UN, these variants result in four additional projections to the UN medium projection and are referred to as "low," "medium low," "medium high" and "high." The designations reflect the average world fertility rate relative to the medium projection. For IIASA, the primary variants also refer to world average fertility (and mortality) rate relative to the medium projection and are referred to in Figure 3-2 as "high" and "low."

The central projections in Figure 3-3 show excellent agreement over the next 100 years, with the exception of the "overshoot" in the IIASA projection, discussed below. Although significant, such an agreement does not imply certainty or accuracy. Rather, it reflects the use of similar methodologies and

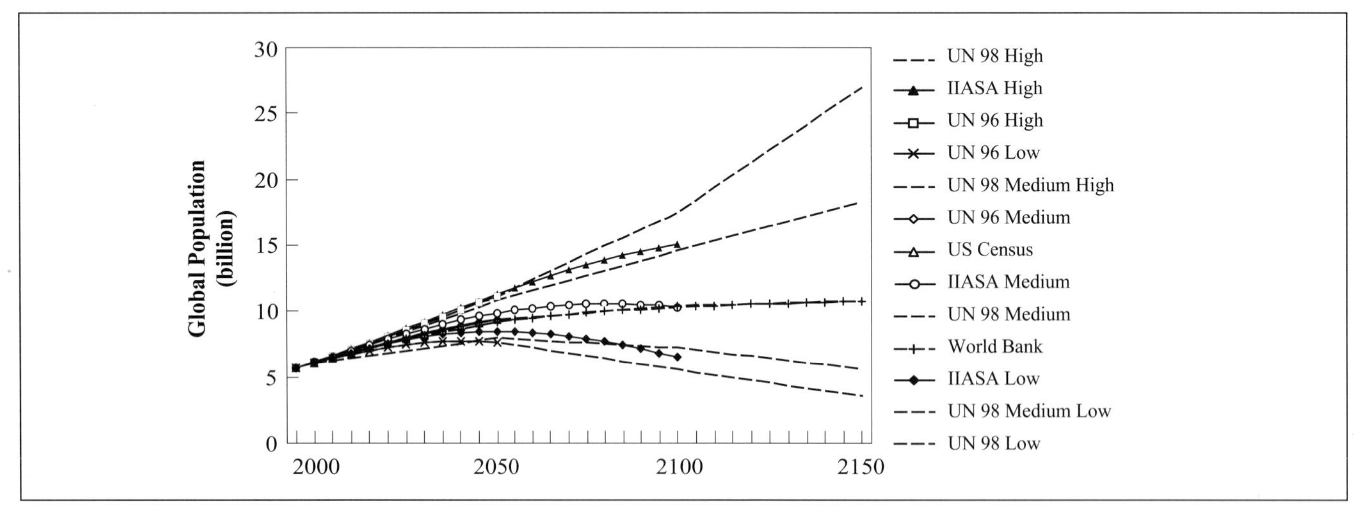

Figure 3-2: Most recent world population projections from the four main demographic organizations, including high and low variants. Two of the projections (USCB and World Bank) will not be published. Variations in future world populations are largely determined by different assumptions concerning the demographic transition in developing countries as well as long-range fertility rates worldwide.

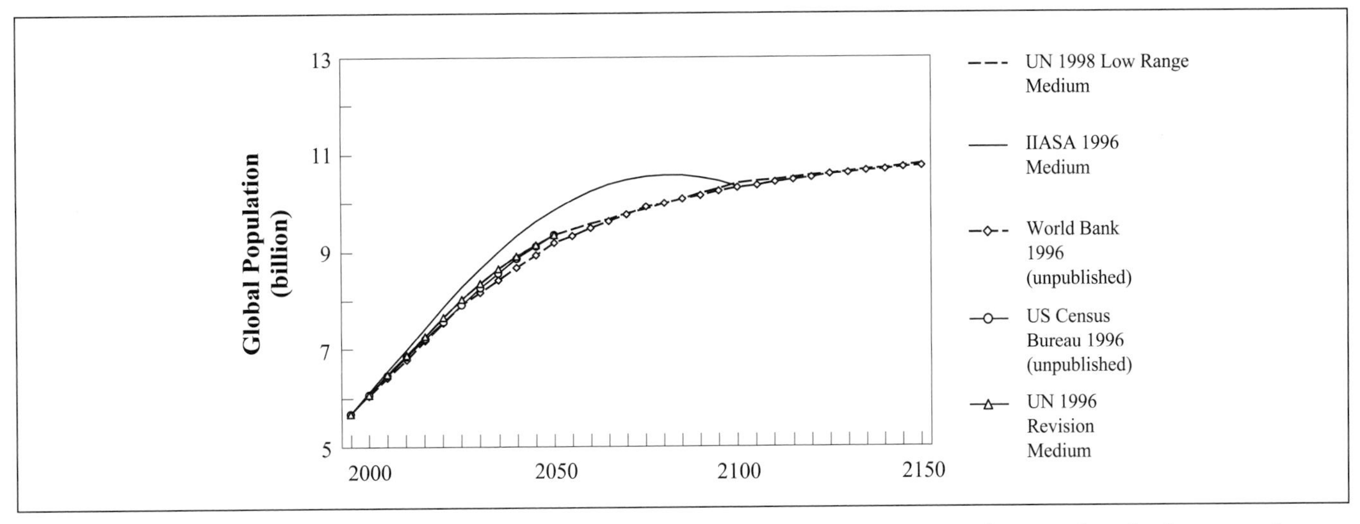

Figure 3-3: Central projections extracted from Figure 3-2 show excellent agreement among the central projections over the next 100 years (with the possible exception of the IIASA projection).

databases, the dominance of population momentum associated with a young population age structure, and the assumption of replacement-level fertility in the long term.

Figure 3-4 shows the breakdown of population growth projections in industrial countries (IND) and developing countries (DEV) as defined by the UN. The key conclusion to draw from Figure 3-4 is the dominance of the DEV population on future world population growth.

The most important variable to determine future population levels is fertility. To a lesser extent, future population also depends on mortality and migration rates. Figure 3-5 shows the world-average total fertility rate (TFR), the average number of births per woman, assumed for the various projections. Overall, a fairly broad range of assumptions about future world average fertility is encompassed by the projections. Future population size is very sensitive (see Figure 3-2) to comparatively small changes in the long-range fertility rate. For example, within the UN projections, a decrease of the asymptotic fertility rate of less than half-a-birth per woman (from 2.1 to 1.7) decreases population in 2100 by 46% (UN, 1998). Only the replacement-level fertility of about 2.1 results in a stable population in the long run.

Such sensitivity to small asymptotic fertility-rate changes indicates the high and low projections from UN and IIASA are all feasible scenarios of the future population (Gaffin and O'Neill, 1998).

Up to 2050, the IIASA central TFR is high compared with the UN and World Bank central projection. Later the IIASA central TFR declines to below replacement-level fertility rate, as depicted in Figure 3-5. As a result, the IIASA central forecast lies considerably above all the others throughout the 21st century. After that, IIASA's below-replacement, long-term

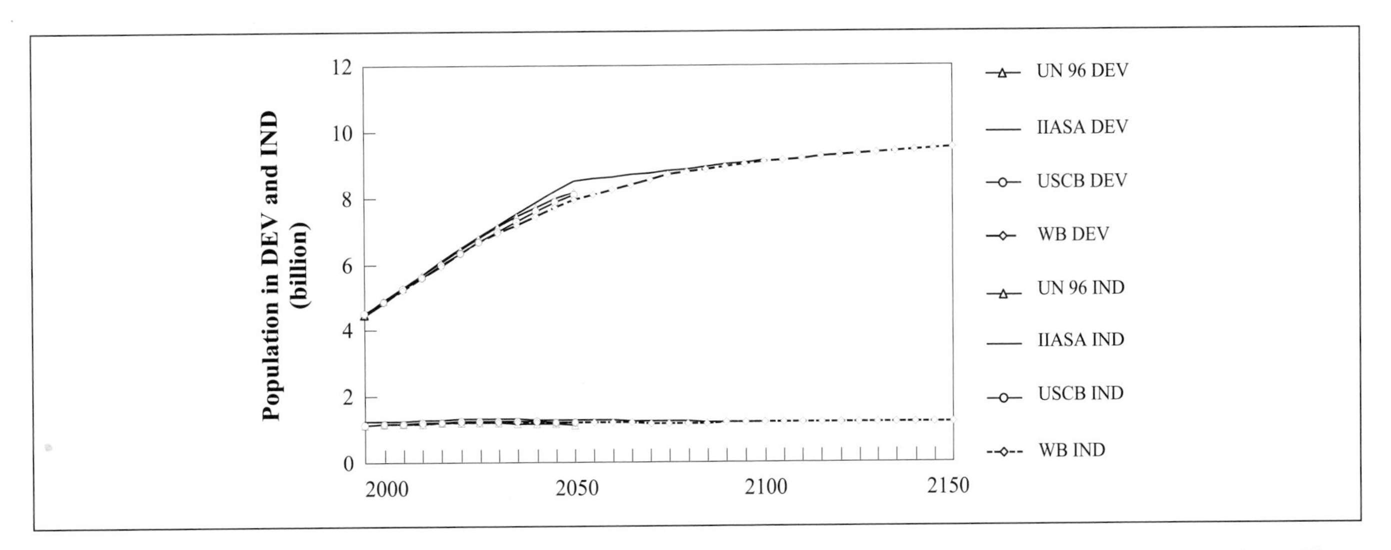

Figure 3-4: Industrial (IND) and developing (DEV) countries' population projections from the medium (central) variants. Note that regional definitions differ in IIASA compared to UN projections.

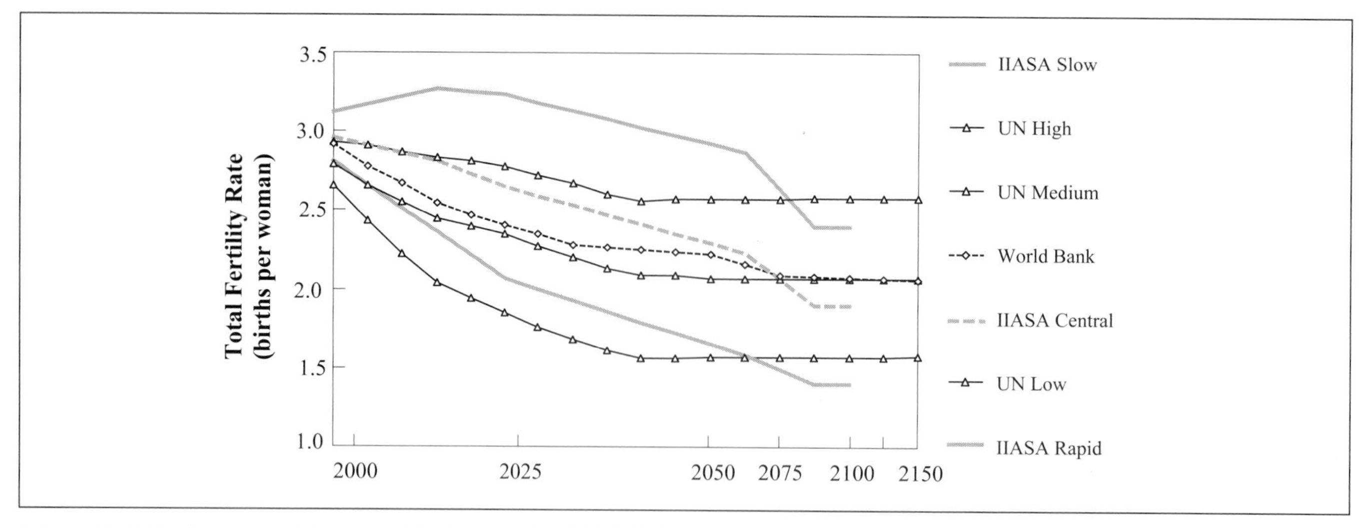

Figure 3-5: Projected world average TFRs from the UN, IIASA, and World Bank. The older 1994 World Bank projections are shown, because the latest (1996) projection methodology is not published. USCB fertility rates are not readily available and so are omitted. UN fertility assumptions are shown only for the 1996 Revision as full detail of the 1998 data were unavailable at the time of drafting this report.

fertility assumption causes the projection to decline and to "hit" the World Bank and UN curves in 2100. This "overshoot" occurs because the IIASA central scenario assumptions rely on an expert poll conducted in 1993 (Lutz, 1994), whereas other projections incorporate more recent information into their future assumptions (more rapid recent fertility declines in many developing countries; see Courbage, 1998). However, among demographers disagreement persists on the timing and rates of demographic transition in the developing countries, in particular between the demographers at IIASA and those at the UN and World Bank (see the discussion in Lutz, 1994) on the differences among alternative population projections). This disagreement reflects important uncertainties in the projection of future demographic developments. The below-replacement, long-term fertility level in the IIASA central estimate is a hallmark of the institute's work. Demographers at IIASA strongly adhere to the view that there is little reason to expect developed nations to return to replacement-level fertility, while much evidence suggests that low long-term rates will persist (Lutz *et al.*, 1996). One issue here, however, is that the TFR measure is currently depressed in Europe, in part because women are delaying their childbearing until later ages (Bongaarts, 1998). Thus, the TFR may rise if these women do have children, and such a correction has to be incorporated into the long-term TFR assumptions (e.g., see Courbage, 1998). Important uncertainties are by how fast and how much the TFR will rise.

3.2.3.3. Recent Developments in Demographic Projections

3.2.3.3.1. Downward revisions in population projections

The UN 1996 Revision generated substantial press attention in 1997 because it forecast nearly 500 million fewer people in 2050 than it had in 1994. The base year of data for the UN 1996 Revision is 1995, whereas in the 1994 Revision fertility, mortality, and migration rates for 1995 are forecast. The reduction in the population projection largely results from more accurate data available for 1995 (UN, 1997b, 1997c). The major change was the lower-than-anticipated world average fertility of 2.96 children per woman during the period 1990–1995, as compared with 3.10 children per woman assumed in the 1994 Revision. The main reason for this decrease is a faster-than-anticipated decline in fertility in a number of countries in south central Asia, Bangladesh and India, and Sub-Saharan Africa, Kenya, and Rwanda (UN, 1997c). Other important regional declines in fertility also took place in Brazil, the former Soviet republics, and the newly independent states in eastern and southern Europe (Haub, 1997). Higher-than-anticipated mortality rates in a number of countries afflicted by wars and the spread of AIDS also contributed to the downward population revision.

3.2.3.3.2. Demographic impact of the HIV and AIDS epidemic

Both the UN and the World Health Organization (WHO) conduct ongoing surveys of the global HIV (human immunodeficiency virus) and AIDS epidemic. The most recent surveys indicate that at the beginning of 1998 30.6 million people were infected with HIV, the virus that causes AIDS, and 11.7 million people have already lost their lives to the disease (UNAIDS and WHO, 1998).

Since the impact of HIV and AIDS on mortality rates is greatest in the sub-Saharan region, the impact of the epidemic on population growth will be greatest there. One study (Bongaarts, 1996) suggests that by 2005 the annual population growth rates (expressed as "persons increase per thousand population") in sub-Saharan Africa will be about 1.4 persons per thousand lower than would have occurred in the absence of the disease. Other regions, however, will experience a much smaller impact in population growth so that the AIDS-related deficit in world population growth rate will be only –0.4 persons per thousand in 2005. The reason for the modest impact is that birth rates in many developing countries are much higher than the death rates, so the AIDS increase in mortality only partially offsets this larger difference.

3.2.3.3.3. The correlation between fertility and mortality

In the recently published IIASA population projections (Lutz, 1996), the correlations of fertility and mortality rates are also different to those of the UN projections. Within IIASA the main scenarios are labeled "central," "rapid transition," "slow transition," "high," and "low." In the two "transition" scenarios, mortality rates and fertility rates are correlated so that low or high mortality accompanies low or high fertility, respectively, in line with conventional wisdom among demographers that fertility declines are associated with mortality declines. This correlation narrows the range of projected population size as compared with an anticorrelation assumption. The "high" and "low" scenarios, by contrast, anticorrelate mortality and fertility and are considered quite unlikely.

3.2.3.3.4. Probabilistic population projections

Another recent development in demographic projections is that of probabilistic scenarios. Lutz *et al.* (1997) consider, based on their probabilistic population projections, a doubling of world population unlikely. Their scenarios use fertility, mortality, and migration rate assumptions based on a Gaussian fit to a survey of demographic experts who were asked to give a range of rates for each region that they considered to cover the 90th percentile probability range. Given the Gaussian curve fits to the expert data, a Monte Carlo simulation was run to generate 4000 scenarios, with five-year timesteps, which have a probability distribution attached to them. The branch points in the fertility, mortality, and net migration rate curves, based on the expert data, were set in 1995, 2000, 2030, and 2080. The probabilistic projections extend to 2100. The 5th and 95th percentile intervals are between 6.7 to 15.6 billion people by 2100, a range that usefully covers current knowledge on the uncertainty of future world population levels to be considered in SRES. It should be noted that such probability assignments derived from expert opinion are inherently subjective and do not necessarily suggest a corresponding likelihood of future occurrence.

3.2.4. *Other Aspects of Population: Aging and Urbanization*

3.2.4.1. Aging

Population aging has widely discussed implications for social planning, health care, labor force structural changes, and entitlement programs. As shown in Figure 3-6, percentage growth in the elderly age cohorts is predicted strongly by all projections. The figure shows the percentage of elderly age cohorts using the medium projection data. Importantly, it will be a continuous process over the entire 21st century, even though total population size is forecast in these cases to stabilize during the latter half of the century.

The detailed economic effects of such a profound and rapid change in social structure are not well understood (Eberstadt, 1997). The problem is considered below in the discussion of the impact of population dynamics on economic development. In short, conventional wisdom takes a more or less neutral view of the effect of population growth, including the impacts of aging, on the rate of economic growth (Hammer, 1985; Kelley, 1988; National Research Council, 1986). Lowered population growth rates (and concomitant aging) might have a beneficial effect on the economy through reduced youth dependency ratios, which result in higher savings rates (Higgins and Williamson, 1997). Also, population aging could reduce labor supply and thus reduce potential economic growth. Against this argument, labor scarcity induces higher wages that in turn are a powerful incentive to increase labor productivity. Increasing

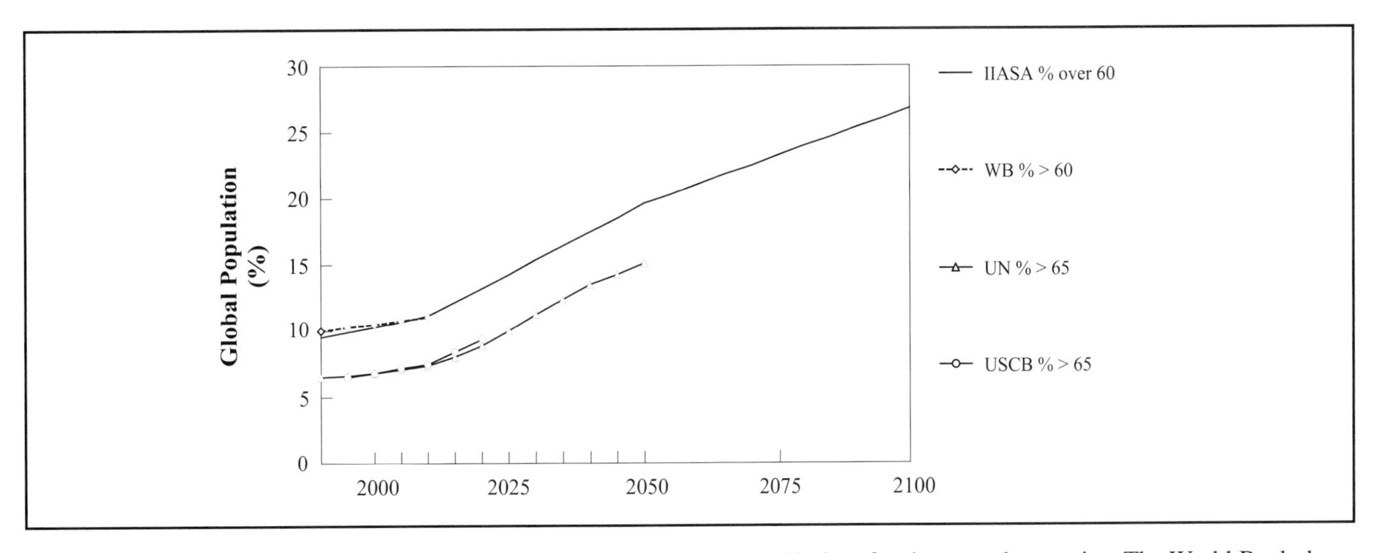

Figure 3-6: Percentage of world population over 60 or 65 years of age with time for the central scenarios. The World Bank data are from 1994 (Bos *et al.*, 1994) because the results from 1996 are not published. The USCB data are from McDevitt (1996).

productivity (per capita economic output) would balance shortfalls from possible reductions in the labor force (Disney, 1996). However, lowered fertility and mortality rates will also cause the elderly dependency ratio to increase, which might lead to a trend of lower savings, characteristic of elderly cohorts, which have a reduced incentive to save. In terms of public savings, without institutional reforms aging could eventually lead to severe strains on social security and health care programs supported by governments and thus to reduced governmental savings (US Council of Economic Advisors, 1997).

Interestingly, a case has been made that aging may have significant impacts on future CO_2 emissions. The suggested mechanism for this relates to household formation rate (MacKellar *et al.*, 1995). An aging population has a greater proportion of people in older age groups. Assuming age-specific household formation rates remain constant over time, as more people enter the older age cohorts the overall household formation rates will increase. This increase will be accompanied by a decline in the number of people per household (a process already observed in industrialized countries) and is related to reduced fertility rates. As small households consume significantly more energy per person than large households (Ironmonger *et al.*, 1995), the various effects suggest CO_2 emissions will increase with increased aging (MacKellar *et al.*, 1995). Important uncertainties of this effect remain, not least because household formation rates of aging populations are not well understood.

3.2.4.2. Urbanization

Urbanization is also a strongly anticipated demographic trend. Since 1970, most urban growth has taken place in developing countries. It is caused by both internal increases of the existing urban population and rural-to-urban migration (UN, 1997b).

Urbanization, though, is not a rigorously modeled phenomenon within the projections. Essentially, future urban and rural growth and decline rates are simply assumed and applied to the projected population levels. Thus, the projections contain no explicit feedback mechanism from urbanization to population growth, even though urbanization is an important factor in fertility rate changes (urban populations generally have lower fertility rates than rural populations). Instead, urbanization rates are considered implicitly within the projections of future fertility. It is estimated that by 2010 more than half of the world's population will live in urban areas (UN, 1996).

Urbanization will lead to a rapid expansion of infrastructure and especially transportation uses (Wexler, 1996). In addition, urban households in developing countries use significantly more fossil fuels, as opposed to biofuels, than do rural households. However, the choice of fuel is predominantly an income effect rather than a function of locale (see Murthy *et al.*, 1997), even within urban settings. Hence, urbanization exerts its influence on emissions primarily via higher urban incomes compared to rural ones. Generally, opportunities for higher income are considered an important driver of rural-to-urban migration, and so contribute to rising urbanization rates (HABITAT, 1996). Urbanization is obviously an important factor for future GHG emissions.

3.2.5. Relationships

3.2.5.1. Introduction

Within the caveats in Section 3.1, a number of demographic studies show that population change does exert a strong first-order scaling effect on CO_2 emissions models (O'Neill, 1996; Gaffin and O'Neill, 1997; O'Neill *et al.*, 2000; Wexler, 1996). These studies support the notion that population growth and the policies that affect it are key factors for future emissions. Balanced against this, however, are other studies that take a more skeptical view (Kolsrud and Torrey, 1992; Birdsall, 1994; Preston, 1996). A full review of model results that address this question is given in O'Neill (1996) and Gaffin and O'Neill (1997) and is not be reproduced here because of space limitations (for a review see Gaffin, 1998). In essence, the controversy is one of the relationships between population growth and economic development, as well as other salient factors that influence emissions. These relationships were first discussed within the context of IPCC emissions scenarios by Alcamo *et al.* (1995) and are discussed in more detail in the following sections.

3.2.5.2. The Effect of Economic Growth on Population Growth

Figure 3-7 shows the long-established negative correlation between fertility rates and per capita income. Clearly, richer countries uniformly have a relatively low fertility rate. Poorer countries, *on average*, have a higher fertility rate. Lower fertility, however, does exist in some poor countries or regions, which illustrates the importance of social and institutional structures.

Barro (1997) reports a statistically significant correlation between per capita GDP growth and the variables life expectancy and fertility in his analysis of post-1960 growth performance of 100 countries. Other things being equal, growth rates correlate positively (higher) with increasing life expectancy and negatively (lower) with high fertility, which confirms the view that the affluent live longer and have fewer children.

Figure 3-7, a snapshot of many countries passing through the "demographic fertility transition" can be explained from both economic and socio-demographic points of view (Easterlin, 1978). Economically, Figure 3-7 can be interpreted as a reflection of the *substitution* that families make – away from having children and toward consuming more goods and services. With greater wealth, both goods and services become increasingly available as part of the families' "basket of choices" for consumption and, accordingly, they shift away

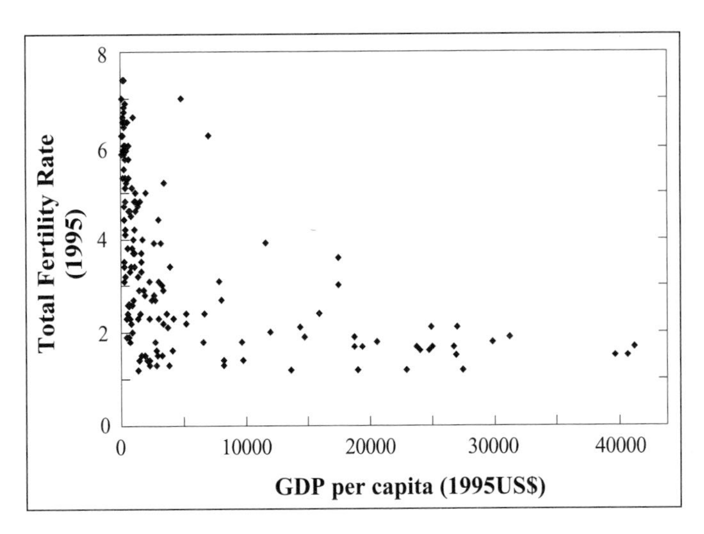

Figure 3-7: TFRs in 1995 versus GDP per capita in US dollars at 1995 prices for most of the world's countries. Data source: World Bank, 1997b.

from higher fertility to lower fertility. This move is further prompted by the rising relative costs of childcare, which include preferences that increase a child's quality of life, such as better schooling and extracurricular activities.

This income effect is primarily interpreted with respect to fertility changes in currently developing countries. In currently industrial countries, any change of fertility in response to increasing wealth is likely to be different from and probably even opposite to that of developing countries' fertility. Indeed, there is evidence of this in Eastern Europe, Sweden, Russia, and (recently) the United States (UN, 1997a), and it is linked to the question of long-term fertility rates in industrial countries (see Section 3.2.3.2.). However, it is evident that future world population levels will be dominated by growth in developing countries. Thus, if it is accepted that fertility is lower with greater affluence, then in emissions scenarios lower populations will still tend to correlate with higher per capita incomes.

Another countervailing factor is that mortality rates should also decline with wealth; as an isolated effect, this obviously results in higher population levels. However, the combined impact of both fertility and mortality reduction on population size is a net reduction in population levels (Lutz, 1996).

From a demographic point of view, the primary effect seen in Figure 3-7 is interpreted as infant and child mortality decline with increasing affluence. Families, in a sense, have to use birth control to achieve their desired number, which has always been lower than fertility rates. Increased affluence results in increased knowledge of, access to, and use of birth control, and accordingly families shift their reproductive behavior to lower fertility rates. Accompanying this basic premise is a host of complex social changes, including increased opportunities for education, employment, and non-maternal roles for women.
Incorporation of the inverse relationship between economic and population growth in long-term emission scenarios is recent and mostly carried out in a qualitative way. Alcamo *et al.* (1995) reviews the literature available up to 1994. Nakićenović *et al.* (1998a) report a long-term scenario study in which timing and extent of economic catch-up in developing countries were found to be tied to timing and pace of their demographic transition. Patterns of the range of per capita GDP growth rates for developing countries available in the literature (and given in Figure 3-10) also appear to reflect this relationship. Growth trends for the period 2020–2050 are generally higher than those for earlier or later periods; it is in this period that, according to demographic projections, the fastest change in demographic variables (especially fertility) will take place.

3.2.6. *Conclusions*

From the available population projections, only those from the UN and IIASA fulfill the characteristics needed for use in long-term emission scenarios. First, the UN and IIASA data are published and available in the public domain, and second (more importantly) the scenarios consider uncertainty by developing more than just one, central demographic projection.

We use the medium UN projections in the SRES emissions scenarios because they have greater recognition internationally, and garner considerable attention as evident from the press focus devoted to the 1996 Revision (mentioned above). In addition, the UN assumption of replacement-level fertility in the long term, in contrast to the IIASA below-replacement assumption, is an important normative approach widely used heretofore in projections.

The rapid and slow demographic transition variants from IIASA projections remain attractive as the "high" and "low" population variants to be considered for the new IPCC emissions scenarios. The incorporation of a correlation between mortality rates and fertility (Lutz, 1996) is a logical first-order relationship not used in previous population variants and, in particular, not a feature of the UN variants. The two IIASA variants also represent well the uncertainty range as spanned by the probabilistic projections of Lutz *et al.* (1997), which represent an important methodological advance in the field. As shown in Figure 3-2, the resultant IIASA population range falls within the range of the UN projections.

Based on the above recommendations, Figure 3-8 compares the older IS92 population range with the population range described in this section. The population projections in IS92 scenarios comprise the UN 1992 medium–high and medium–low variants for the high and low ranges with the World Bank 1991 projection as the central case. As seen in Figure 3-8, the new range for SRES is somewhat narrower and lower than the IS92 range. The cause is partly the positive correlation between mortality rates and fertility rates within the IIASA variants, which mildly offset each other in terms of future population size. Another reason is the recent downward

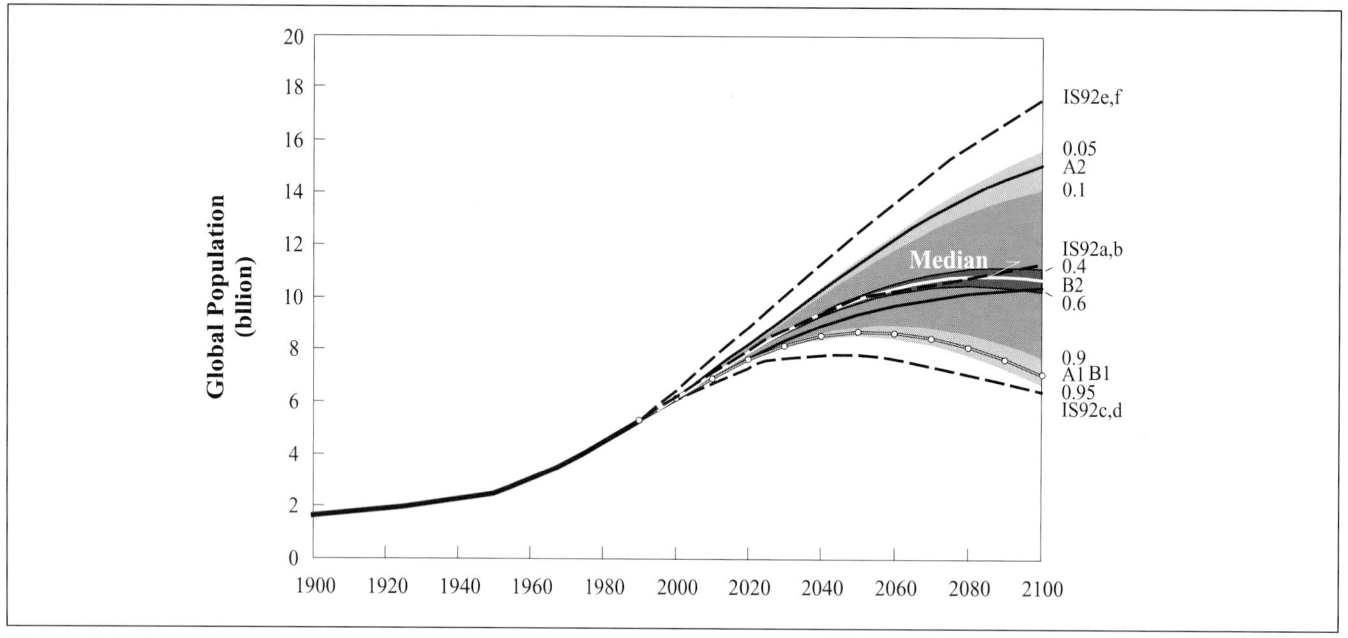

Figure 3-8: Comparison of the IS92 population range (dashed lines) with the population range adopted for SRES (solid lines), which uses the two IIASA variants (low, high) and the UN 1998 Long Range medium projection. The correspondence to the four SRES scenario families (A1, A2, B1 and B2) is also shown. Generally, the new range is narrower and has significantly lower medium and high variant population levels, reflecting recent advances in demographic projections. Also for comparison, the probabilistic range of world population projections given by Lutz *et al.* (1997) is shown.

revision of the UN medium projection compared to earlier UN scenarios (as outlined above).

The literature suggests a general inverse correlation between economic growth rates and population growth rates. Higher economic growth rates in developing countries should correlate with lower population growth rates in long-term scenarios and vice versa, because of the importance of economic development in bringing about the demographic fertility transition. This represents a distinctive change from the IS92 trajectories. Importantly, there is still no quantitative basis for associating any *particular* economic growth curve with a particular population curve; this is a qualitative negative correlation only. Even more important is that income is not necessarily the best predictor of future fertility rates and many countries are currently moving through the demographic transition without a clear economic cause. Alternatively, some countries have failed to begin a fertility decline even though economic and social conditions have improved (e.g., Sathar and Casterline, 1998).

The inclusion of a household demographic unit, in addition to population, should be encouraged in future studies. The effect is linked to a strongly predicted demographic trend – aging. Also important are that smaller households are more energy intensive, per person, and that aging may continue to increase more rapidly than population in the future. These factors may increase CO_2 emissions (MacKellar *et al.*, 1995), although senior citizens group-living is a tendency in some industrialized countries. Urbanization might also have a strong effect on emissions because of its effect on income distribution and thus energy consumption patterns around the world, although many of these effects are included implicitly in the models and parameters used in this report.

3.3. Economic and Social Development

3.3.1. Introduction

Economic and social development comprises many dimensions and a number of indicators have been devised to assess progress and setbacks in human development (see Box 3-1). The UN defines development as the furthering of human choices. Such choices are neither finite nor static. Yet, regardless of the level of development, the three essential choices are to have access to the resources needed for a decent standard of living, to lead a long and healthy life, and to acquire knowledge (UNDP, 1997). Other valued choices range from political, economic, and social freedom to opportunities for being creative and productive, and to enjoy human rights (UNDP, 1997).

Arguably, choices are only possible once basic human needs for food, shelter, health care, safety, and education have been met. Poverty is therefore an important indicator of the absence of satisfactory economic development. Alleviation of poverty is an essential prerequisite for human development. Beyond the satisfaction of basic needs, the issue of what constitutes "development" involves many cultural, social, and economic dimensions that cannot be resolved by scientific methods, but are inherently a question of values, preferences, and policies.

Box 3-1: On Measures of Human and Economic Development

Writing 220 years ago in *The Wealth of Nations*, Adam Smith noted that: "whatever the soil, climate, or extent of territory of any particular nation, the abundance or scantiness of its annual output fundamentally depends on its human resources – the skill, dexterity, and judgement of its labour" (Smith, 1970). Although economists recognized the importance of land, labor, and capital in explaining economic growth and national wealth, in the post-World War II period national well-being has usually been measured by GDP or gross national product (GNP). GDP is defined as the monetary equivalent of all products and services generated in a given economy in a given year. GNP equals GDP plus the net balance of international payments to and from that economy. Few questions were asked about the underlying resource base for GDP growth and whether or not it was sustainable. Further, since GDP does not reflect all economic transactions it does not provide a full measure of human well-being. Nevertheless, GDP is very widely used because it is universally accepted as the monetary indicator of all products and services generated in a given economy within a given year.

Environmental and Social Modifications to GDP

More recently, several new approaches have been developed to address the inherent shortcomings of GDP measures. These include "green" national accounts that incorporate the role of the stocks and flows of renewable and non-renewable resources, and the related concept of genuine savings (UN, 1993). "Green" GDP is the informal name given to national income measures that are adjusted for the depletion of natural resources and degradation of the environment. The types of adjustment made to standard GDP include a measure of the user costs of exploiting natural resources and a value for the social costs of pollution emissions. In terms of measuring the sustainability of development, the green accounting aggregate with the most policy relevance is "genuine saving." This represents the value of the net change in assets that are important for development – produced assets, natural resources, environmental quality, foreign assets, and human resources, which include returns to education and raw labor and the strength and scope of social institutions. Human resources turn out, not unexpectedly, to be the dominant form of wealth in the majority of countries (World Bank, 1997a).

Purchasing Power Parities

A further problem arises in international comparisons, in which economic indicators are converted from local currencies into a common currency, such as dollars. Traditionally, market exchange rates are used to make these conversions. In theory, exchange rates adjust so that the local currency prices of a group of identical goods and services represent equivalent value in every nation. In practice, such adjustments can lag far behind changing economic circumstances. Policies, such as currency controls, may further distort the accuracy of market-based rates. Moreover, many goods and services are not traded internationally so market-based exchange rates may not reflect the relative values of such goods and services, even in theory. An alternative approach is based on estimates of the purchasing power of different currencies. The International Comparison Project compared prices for several hundred goods and services in a large number of countries. On the basis of this comparison, the relative values of local currencies are adjusted to reflect PPP (see UNDP, 1993). In effect, the PPP currency values reflect the number of units of a country's currency required to buy the same quantity of comparable goods and services in the local market as one US dollar would buy in an "average" country. The average country is based on a composite of all participating countries. In 1996 the World Bank initiated the ranking of countries by GDP converted at PPP rates; the effect was to reduce the income spread between the poorest and richest countries (WRI, 1997a).

UN Human Development Index

The UN has tried to address the shortcomings of GDP by developing *The Human Development Index* (UNDP, 1997). This index, produced since 1990, combines three factors to measure overall development:

- Income as measured by real GDP per capita at PPP to represent command over resources to enjoy a decent standard of living.
- Longevity as measured by life expectancy at birth.
- Educational attainment as measured by adult literacy and school enrolment.

The UN has also developed other measures, such as the Gender-related Development Index (GDI) and the Gender Empowerment Measure (GEM) to assess conditions such as gender equality.

The difficulty of incorporating these alternative measures into long-term scenarios is that, with the exception of life expectancy, underlying data or base projections (e.g. on future PPPs, levels of educational attainment, etc.) needed to develop these alternative indicators for future projections are not available. Therefore, this report largely focuses on traditional measures of economic development like GDP. Projections of PPPs are calculated by one of the six SRES models and are presented in Chapter 4.

Given the inherent ambiguities of such a complex, multidimensional issue, it is easiest to define and develop indicators of no development. Estimates indicate, for instance, that 1.3 billion people in developing countries live on incomes of less than US$1 (PPP based) per day, a level used to define the absolute poverty cut-off in international comparisons (UNDP, 1997). An equal number of people are estimated to have no access to safe drinking water (UNDP, 1997) and 2 billion people are estimated to have no access to services provided by the use of modern energy forms (WEC, 1993).

Income is not an end in itself, but a way to enable human choices, or to foreclose them in the case of poverty. Therefore, levels of per capita income (GDP or GNP) have been widely used as a measure of the degree of economic development, as in many instances such levels correlate closely (as lead or lag indicator) with other indicators and dimensions of social development, such as mortality, nutrition, and access to basic services, etc. Average income values also do not indicate the distribution of income, which is an important quantity. Composite measures, such as the UN Human Development Index, are also used in historical analyses (see Box 3-1). Note, however, that the overall nature of scenario results may not vary much even if some other measure could be used, because often-used components, such as literacy rates, are generally correlated with income levels.

In fact, per capita income is *the* (and often only) development indicator used in the literature for long-term energy and GHG emissions scenarios. This explains why this review chapter, while recognizing the importance of alternative dimensions and indicators to describe long-term human development, almost exclusively embraces an economic perspective.

The widespread use of GDP or GNP per capita (however measured) should not distract from the fact that, while a powerful indicator, it does not describe all aspects of economic development (see Box 3-1). GDP and GNP are indicators of financial *flows* (see Box 3-1), and are not designed to measure stock variables such as the size of the capital stock in an economy. GDP and GNP relate only to goods and services that are subject to market transactions, *that is* only those activities that are part of the formal economy. Subsistence and other "gray" economic activities and socially important obligatory activities, such as childcare or household work, are not included; also, depletion allowances for natural capital and resources are not considered.

Although PPP comparison (see Box 3-1) is considered a valid indicator of relative wealth, it is sometimes quite uncertain and dependent on detailed comparison exercises. As a result, even if studies and scenarios consistently use the same indicator of economic development, numeric values are often not directly comparable because of large differences in base-year values. Comparison of growth rates are more robust, but even here many difficulties exist (see, e.g., Alcamo *et al.*, 1995, and Chapter 2).

3.3.2. *Historical Trends*

Rostow (1990) described several stages in the economic development process:

- First, the pre-industrial economy, in which most resources must be devoted to agriculture because of the low level of productivity.
- Second, the phase of capacity-building that leads to an economic acceleration.
- Third, the acceleration itself, which requires about two decades.
- Fourth, about six decades of industrialization and catch-up to the "productivity frontiers" prevailing in the industrialized countries.
- Fifth, the period of mass-consumerism and the welfare state.

It is important not to conceptualize economic development as a quasi-autonomous, linear development path. Numerous socio-institutional preconditions have to be met before any "take off" into accelerated rates of productivity and economic growth can materialize (see Section 3.3.4). "Leading sectors" (Fogel, 1970) that drive productivity and output growth change over time (Freeman and Perez, 1988; Freeman, 1990), and different "industrialization paths" (Chenery *et al.*, 1986) have been identified in historical analyses. Still, historical evidence, consistent with neoclassic growth theory, allows a number of generalizations as to the patterns of advances in productivity and economic growth.

By and large, growth rates are lower for economies at the technology and productivity frontier, compared to those approaching it. For instance, in the 19th century productivity and per capita GDP growth in the rapidly industrializing US far exceeded those of England, then at the technology and productivity frontier. Likewise, in the post-World War II period growth rates in Japan and most of Western Europe exceeded those of the US (by then at the technology and productivity frontier) (Maddison, 1991, 1995). High human capital (education), a favorable institutional environment, free trade, and access to technology are acknowledged as key factors for rapid economic catch-up (see Section 3.3.4). Likewise, entrenchment in progressively outdated capital and technology vintages acts as a retarding force against growth (Frankel, 1955), and rapid capital turnover and possibilities to "leapfrog" (Goldemberg, 1991) outdated technologies and infrastructures provide the potential for faster economic catch-up.

Perhaps the most comprehensive compilation of data on historical economic development is that of Maddison (1995). Table 3-2 shows Maddison's per capita GDP growth rate estimates for selected regions and time periods. Since 1820 global GDP has increased by a factor of 40, or at a rate of about 2.2% per year. Per capita GDP growth was 1.2% per year faster than population growth. In the past 110 years (a time frame comparable to that addressed in this report) global GDP increased by a factor of 20, or at a rate of 2.7% per year, and

Table 3-2: Per capita GDP growth rates for selected regions and time periods, in percent per year. Data source: Maddison, 1995.

	1870–1913	1913–1950	1950–1980	1980–1992
Western Europe	1.3	0.9	3.5	1.7
Australia, Canada, New Zealand, USA	1.8	1.6	2.2	1.3
Eastern Europe	1.0	1.2	2.9	-2.4
Latin America	1.5	1.5	2.5	-0.6
Asia	0.6	0.1	3.5	3.6
Africa	0.5	1.0	1.8	-0.8
World (sample of 199 countries)	1.3	0.9	2.5	1.1

global per capita GDP grew by a factor of more than five, or at a rate of 1.5% per year. There is substantial variation in the rates of economic growth over time and across countries. Even for the present OECD countries modern economic growth dates only from approximately 1870 onward, and for developing economies comparable conditions for economic growth and catch-up in productivity levels existed only in the second half of the 20th century. These approximate dates also set the time frame for drawing useful comparisons between historical experiences and future projections (see Section 3.3.3).

Historically, economic growth has been concentrated in Europe, the Americas, and Australasia. Sustained high-productivity growth (per capita GDP growth) resulted in the current high levels of per capita income in the OECD countries. Latecomers (such as Austria, Japan, Scandinavia) rapidly caught up to the productivity frontier of the other OECD economies (most notably that of the US) in the post-World War II period. Per capita GDP growth rates of 3.5% per annum were, for instance, achieved in Western Europe between 1950 and 1980. Similarly, high per capita GDP growth rates were achieved in the developing economies of Asia. Per capita GDP growth rates of individual countries have even been higher – 8% per annum in Japan over the period 1950–1973, 7% in Korea between 1965 and 1992, and 6.5% per year in China since 1980 (Maddison, 1995). Progress in increasing per capita income levels has been significant in many regions and countries over the long term, although income gaps in both absolute and relative terms have not been reduced in the aggregate. For instance, per capita GDP in Africa is estimated to have been about 20% of the level of the most affluent OECD region in 1870; by 1990 this ratio had decreased to 6% (Maddison, 1995).

Other key indicators of human development have also improved strongly in recent decades. For instance, since 1960 (in little more than a generation) infant mortality rates in developing countries have more than halved, malnutrition rates have declined by one-third, primary school attendance rates have increased from about half to three-quarters, and the share of rural families with access to safe drinking water has increased from 25% to 65% for low-income families and to more than 95% for high-income families (UNDP, 1997; World Bank, 1999).

Equally noticeable in Table 3-2 is the slowdown of per capita GDP growth in the OECD countries since the end of the 1970s and the serious setbacks in Eastern Europe and developing countries outside Asia over the same time period. After a decade of decline in economic output, a trend reversal to positive growth rates is expected to occur only after the year 2000 (World Bank, 1998b). Even optimistic scenarios indicate that the pre-crises (1989) levels of per capita income cannot be achieved again until 2010 (Nakićenović *et al.*, 1998a).

The effects of the recent Asian financial crises are estimated to reduce significantly short-term economic growth in the region, but longer-term growth prospects remain solid. After a sluggish growth to the year 2000, economic growth in the developing countries of Asia is anticipated to resume at 6.6% per year for the period 2001–2007 (World Bank, 1998b).

The past two centuries have seen major structural shifts, with inter-related changes in demography, economic structure, and technology. The world's largest economies have seen a continuous shift in economic structure, from agricultural production to industry and, to a greater extent, services (see Figure 3-9). During the past half-century, the industry share declined to leave these economies dominated by service sectors. A similar pattern has occurred in most other economies in Europe, North America, and Australasia. Another major feature of the past two centuries, not shown in Figure 3-9, is the growing role of government in the economy. The literature on structural change is reviewed in more detail in Jung *et al.* (2000).

Differences in statistical definitions make it difficult to compare structural change in the former centrally planned economies with that in mature market economies. Also, in the less-developed regions a marked decline in the contribution of agriculture to GDP has occurred in recent years, but the contribution of services is larger than it was historically in industrial countries at the same level of income. This indicates the dangers of using past history too literally as a guide to future behavior.

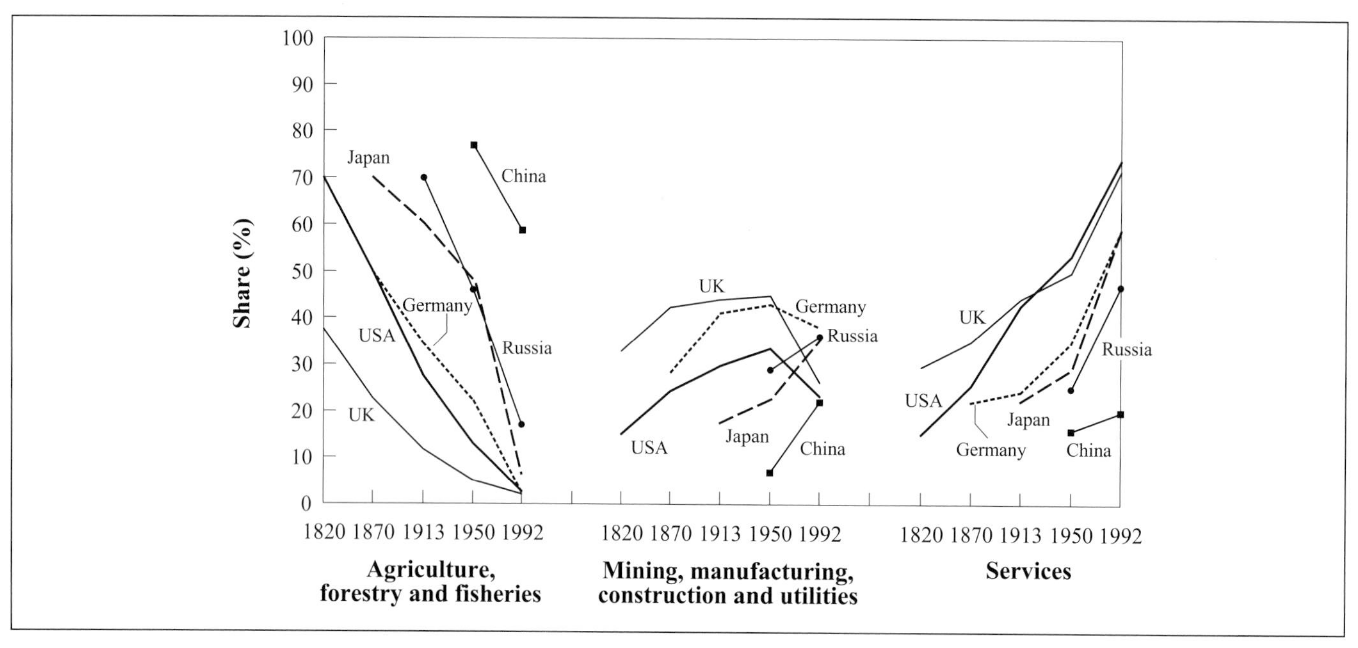

Figure 3-9: Changes in economic structure for selected countries. Data source: Maddison, 1995.

These economic developments have accompanied the processes of urbanization, increased access to education, improved health care, and longer life expectancies. They have been linked to increasing complexity in economic, legal, and social institutions (Tainter, 1996). The pattern seen in the six countries shown in Figure 3-9 is reproduced in most parts of the world. In general, increasing income per capita is associated with a shift in production patterns, first from agriculture to industry, and then more gradually away from industry into services. Succession processes beneath these macro-level changes are important. Industrial sectors and technologies have risen and fallen in importance over the past few centuries. These processes and analysts' various attempts to find patterns to describe them are further discussed in Section 3.3.4.

3.3.3. *Scenarios of Economic Development*

Unlike population projections, no long-term economic-development scenarios are available in the literature (for an earlier review, see Jefferson, 1983). In fact, for economic projections "long term" means time horizons of up to a decade (e.g., World Bank, 1997b), 1998b, far too short for the time frame addressed in this report. The longest time frames for economic growth projections available in the literature extend to 2015 (e.g., Maddison, 1998) and 2020 (World Bank, 1997b). The need for long-term economic growth scenarios has arisen primarily in connection with long-term energy and environmental impact analyses. Earlier reviews on the related economic growth assumptions are contained in Nordhaus and Yohe (1983), Keepin (1986), Grübler (1994), and Alcamo *et al.* (1995). An expert poll on uncertainty in future GDP growth projections is reported in Manne and Richels (1994). Recent scenario assumptions are reviewed in Chapter 2 above.

The current state of modeling long-term economic growth is not well developed, not least because the dominant forces of long-run productivity growth, such as the role of institutions and technological change (see Section 3.3.4), remain exogenous to models. As a result, productivity growth assumptions enter scenario calculations as exogenous input assumptions. The structural changes in the economy discussed in the previous paragraph result in additional difficulties; notably that service sector productivity growth is difficult to evaluate and project.

Figure 3-10 summarizes the results of the analysis of available literature data on per capita economic growth, disaggregated into global as well as industrial and developing countries.

Overall, uncertainty concerning productivity and hence per capita GDP growth is considerable. Uncertainties in productivity growth rates become amplified because even small differences in productivity growth rates in all scenarios, when compounded over a time frame of a century or more into the future, translate into enormous differences in absolute levels of per capita GDP. For instance, in the scenarios reviewed in Alcamo *et al.* (1995) and Grübler (1994) per capita GDP growth rates range typically between 0.8 and 2.8% per year over the period 1990–2100. On the basis of an average global per capita income of US$4000 in 1990, global per capita GDP could range anywhere between about US$10,000 to about US$83,000 by 2100. Such uncertainties are amplified even more when regional disaggregations are considered, in particular future productivity growth in developing countries. The range of views spans all the extremes between developing countries that lag perennially behind current income levels in the OECD, to scenarios in which they catch up.

These ranges are reflected in the SRES scenarios shown in Figure 3-10. Exogenously assumed productivity growth rates

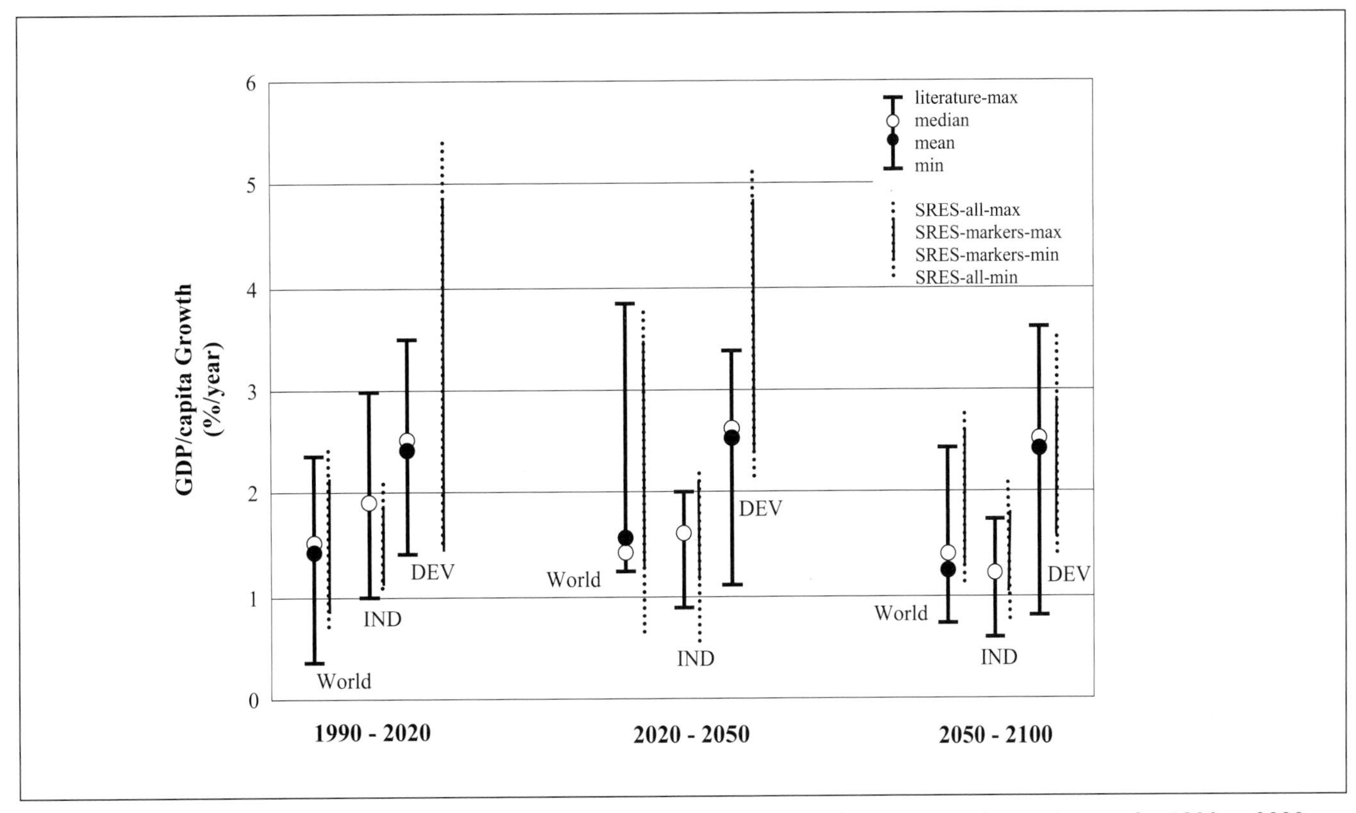

Figure 3-10: Per capita GDP or GNP growth rates, a review of the literature. Average annual growth rates for 1990 to 2020, 2020 to 2050, and 2050 to 2100 for world, industrial (IND), and developing (DEV) regions. Literature mean, median and ranges compared to SRES ranges(see Chapter 4).

correspond to alternative qualitative interpretations as to how the future could unfold, ranging from SRES low (all-min) to SRES high (all-max) rates. Extreme scenarios of productivity growth or lack of growth have not been explored because the SRES terms of reference cover a qualified range from the literature; methodological (and model) pluralism is mandatory (extreme scenarios can be reflected across a wide range of modeling approaches only to a limited degree). Furthermore, it is not possible to treat uncertainties of future demographic, economic, and technological developments as independent. This is shown by the conclusions of recent scenario evaluation exercises (Alcamo *et al.*, 1995) as well as by theoretical and empirical evidence (e.g. Abramovitz, 1993; Barro, 1997). Thus, contrary to the previous IPCC IS92 scenario series (that varied salient scenario driving forces independently of each other), the SRES scenarios attempt to incorporate advances in the understanding of the relationships between important scenario drivers. From this perspective, uncertainties about future productivity and hence economic growth are not parametric, but rather are related to the uncertainties in current understanding and modeling of the interactions between demographics, productivity growth, and socio-institutional and technological change. These are addressed in Section 3.3.4.

3.3.4. Relationships

3.3.4.1. Introduction

Economic growth can either be achieved by increasing the factor inputs to production, such as capital and labor, or by increasing productivity (i.e., the efficiency by which factors of production are used to generate economic output). Without productivity growth, long-run output growth cannot be maintained with limited or depletable resource inputs; as a result, complex societies become increasingly vulnerable (Tainter, 1988). Changes between inputs and outputs are usually analyzed by drawing upon the production function approach pioneered by Tinbergen (1942) and Solow (1957). Yet, empirical analyses (e.g., Denison, 1962, 1985) quickly identified that quality and composition of factor inputs are more important in explaining long-run output growth than merely the quantitative growth in available factor inputs. For instance, at first sight population growth might be considered as central for economic growth, because it increases the labor force. Upon closer examination, however, institutional and social factors that govern working-time regulation, female workforce participation, and above all the qualification of the workforce (education) have been more important determinants of long-run economic growth (Denison, 1962, 1985) than simple growth in the numbers of the potential workforce (usually calculated as the population in the age bracket 15 to 65 years). Another puzzling finding of Solow (1957) is that, even when changes in quality and composition of factors of production are accounted for, increases in per capita economic output (productivity) remain largely unexplained, a "residual" in the analysis remains unclear (for a review, see Griliches, 1996). The "residual," is usually ascribed to "advances in knowledge and technology" which, unlike capital and labor, cannot be measured directly. However, it might also be the result of other influences, which potentially include growing contributions to the economy by non-market or under-priced natural resources. Thus, considerable measurement and interpretative uncertainties remain in the explanation of productivity growth.

New approaches and models extended the neoclassic growth model (e.g., Romer, 1986; Lucas, 1988; Grossmann and Helpman, 1991, 1993). In these, increases in human capital through education and the importance of technological innovation via directed activity (research and development (R&D)) complement more traditional approaches, which represents a return to the earlier work of Schumpeter (1943), Kuznets (1958), Nelson *et al.* (1967), and Landes (1969).

3.3.4.2. *Influence of Demographics*

Neoclassic economic growth theory embraces as a general principle the notion that long-term per capita income *growth rate* is independent of population *growth rate*. Thus, a rapidly growing population should not necessarily slow down a countries' economic development. Blanchet (1991) summarizes the country-level data. Prior to 1980, the overwhelming majority of studies showed no significant correlation between population growth and economic growth (National Research Council, 1986). Recent correlation studies, however, suggest a statistically significant, but weak, inverse relationship for the 1970s and 1980s, despite no correlation being established previously (Blanchet, 1991). As noted in Section 3.2, the reverse effect of income growth on demographics is much clearer.

Population aging is another consideration advanced as having significant influence on economic growth rates. Reductions in workforce availability and excessive social security and pension expenditures are cited as possible drivers. Section 3.2 above concluded that evidence for a strong negative impact is rather elusive. Two additional points deserve consideration. First, population aging is not necessarily the best indicator for workforce availability, because while the percentage of the elderly, in particular those of retirement age, increases, the proportion of younger people (of pre-work or -career age) decreases. As a result, the percentage of the working age population (age 15 to 65 years) in the total population changes less dramatically, even in scenarios of pronounced aging. For instance, in the IIASA low population scenario (7 billion world population by 2100) discussed in Section 3.2, the percentage of age categories 15 to 65 years changes from 62% in 1995 to 54% by 2100. This percentage falls to 48% in the regions with the highest population aging (Lutz *et al.*, 1996).

A second point is that these demographic variables only indicate potential workforce numbers. Actual gainfully employed workforce numbers are influenced by additional important variables – unemployment levels, female workforce participation rates, and finally working time. The importance of these variables can be illustrated by a few statistics. Currently, about 40 million people are unemployed in the OECD countries (UNDP, 1997). The female workforce participation ratios vary enormously, from about 10% to 48% of the workforce (as in Saudi Arabia and Sweden, respectively; UNDP, 1997), and have been changing dramatically over time. For the US, for instance, female workforce participation rates increased from 17% in 1890 (US DOC, 1975) to 45% in 1990 (UNDP, 1997). Similar dramatic long-term changes have occurred in the number of working hours in all industrial countries. Compared to the mid-19th century, the number of average working hours has declined from about 3000 to about 1500 (Maddison, 1995; Ausubel and Grübler, 1995). However, in most OECD countries the trend in working time reductions has slowed to a halt since the early 1980s (Marchand, 1992).

Thus, unless the rather implausible assumption is made that with population aging all these other important determinants of labor input remain unchanged, the impacts of aging are likely to be compensated by corresponding changes in these variables (e.g. greater female workforce participation, earlier retirement, etc.). Finally, it must be reiterated that qualitative labor force characteristics, most notably education, are a more important determinant for long-run productivity and hence economic growth than mere workforce numbers.

3.3.4.3. *Influence of Social and Institutional Changes*

The importance of social and institutional changes to provide conditions that enabled the acceleration of the Industrial Revolution is widely acknowledged (Rosenberg and Birdzell, 1986, 1990). Rostow (1990) and Landes (1969) identify many social and cultural factors in the "preconditions for economic acceleration" and in the process of economic development.

The importance of institutions and stable social environments is also increasingly discussed in the literature concerned with current economic growth (World Bank, 1991, 1998a). Barro (1997), and Barro and Sala-I-Martin (1995) report a statistically significant relationship between rule-of-law and democracy indices with per capita GDP growth. Law enforcement and legal rights are important indicators for human development in their own right, but enforceable legal contracts are equally important for markets to function. Other socio-institutional factors have been identified that are important to productivity and economic growth: education is mentioned above. Income inequality (and resultant social tensions) also appears to correlate negatively with economic development (World Bank, 1998a; Maddison, 1995).

Strong parallels run between social, institutional, and technological changes (Grübler, 1998a; OECD, 1998a). In particular, many features common to the processes of evolution in biologic organisms have been found (e.g., Teilhard de Chardin, 1959; Hayek, 1967; Matthews, 1984; Dawkins, 1986; Michaelis, 1997c). Thus, to understand these processes would involve:

- A search for new behaviors, institutions, and social or cultural patterns.
- Experimentation with those that are found.
- Various methods of selecting the "fit" or "desirable" changes.
- Various methods of perpetuating and diffusing those changes that are selected.

Many aspects of the processes of technical change (e.g., its unpredictability and the importance of mechanisms such as path-dependence and "lock-in") also apply to social change.

It is obviously difficult to evaluate the role of social, cultural, and institutional changes in economic and technical development. Whereas the monetary and technological aspects of change are often measurable and can be observed on a relatively "objective" basis, social, cultural, and institutional processes are hard to measure and often subjective. They tend to involve personal interactions among people, sometimes large numbers of people, over long periods.

Nonetheless, these factors must be taken into account in the scenarios. The SRES approach to develop qualitative scenario "storylines" that provide an overall framework and background for quantitative scenario assumptions and model runs can be considered a particularly valuable strength. Storylines allow these issues to be addressed explicitly, even if current knowledge does not allow social, cultural, and institutional factors to be treated in a rigid, quantitative (not to mention deterministic) way.

3.3.4.4. Influence of International Trade and Investment

International trade is recognized as an important source of economic gains, as it enables comparative advantages to be exploited and the diffusion of new technologies and practices. Equally important are domestic and international investments. From an economic perspective, trade reflects gains from increasing division of labor, and the historical evidence indicates an associated phenomenal growth. Updated (Grübler, 1998a) estimates from Rostow (1978) indicate an increase of more than a thousand-fold in international trade since the beginning of the 19th century. In 1990, world trade accounted for close to US$3.4 trillion ($10^{12}$), or 13% of world GDP. Chenery *et al.* (1986) and Barro (1997) indicate strong empirical evidence of a positive relationship between trade ("openness" of economies) and terms of trade (of the economies) on productivity, industrialization, and economic growth. Dosi *et al.* (1990) highlight also the critical roles of policies and institutions in the relative success or failure of realizing economic gains from the international division of labor. Finally, openness to trade could have negative economic impacts on countries that experience a deterioration of terms of trade.

Globalization is an increasingly popular term for an ill-defined collection of processes. The most important of these processes appear to be the liberalization of markets for goods, services, and capital, and the increasing flow of information and capital around the world. According to the neoclassic model, globalization should benefit everyone; it helps industry in all countries to move closer to the productivity frontier and gives consumers access to a wider choice of goods and services at lower prices. In practice, of course, the world is more complicated than is assumed in the neoclassic model. Economists such as Dosi *et al.* (1990) and Grossman and Helpman (1991) have experimented with alternative models, and shown that some countries may not benefit from free trade or from freely available information. Huntington (1996) observes that by "globalization," Americans and Europeans often mean "westernization" in the sense of global adoption of western social and cultural norms. This process of cultural convergence is a source of great concern in many parts of the world. However, Huntington suggests that economic globalization may be possible without cultural globalization.

3.3.4.5. Influence of Innovation and Technological Change

The importance of "advances in knowledge" and technology in explaining the historical record of productivity growth is mentioned above. In the original study by Solow (1957) this was estimated to account for 87% of per capita productivity growth (the remainder was attributed to increases in capital inputs). Since then further methodological and statistical refinements have reduced the unexplained "residual" of productivity growth that is equated to advances in knowledge and technology, but it remains the largest single source of long-run productivity and economic growth. It is estimated to account for more than one-third of total GDP growth in the USA since 1929 (Denison, 1985), and for between 34% and 63% of GDP growth in the OECD countries over the period 1947 to 1973 (Barro and Sala-I-Martin, 1995).

The observed slowdown in productivity growth rates since the early 1970s is generally interpreted as a weakening of the technological frontier in the OECD countries (Maddison, 1995; Barro, 1997), although quantitative statistics (and even everyday experience) do not corroborate the perception of a slowdown in technological innovation and change. An alternative interpretation of slower recent productivity growth is that the OECD countries have moved out of a long period of industrialization and into post-industrial development as service economies. In such economies, productivity is extremely hard to measure, partly because services comprise a mixture of government, non-market, and market activities, partly because economic accounts measure services primarily via inputs (e.g. cost of labor) rather than outputs, and partly because it is difficult to define service *quality*. Nevertheless, labor productivity in the service sector appears to grow more slowly than that in the agricultural and industrial sectors (Millward, 1990; Baumol, 1993). The traditional concept of labor productivity may need revision to be applied usefully to the service economy.

Finally, another interpretation is that productivity growth lags behind technological change, because appropriate institutional and social adjustment processes take considerable time to be implemented (Freeman and Perez, 1988; David, 1990). Once an appropriate "match" (Freeman and Perez, 1988) between institutional and technological change is achieved, productivity growth could accelerate. Maddison (1995) observes that the 19th century productivity surge in the USA was preceded by a long period of investment in infrastructure. Landes (1969) notes that both the German and Japanese economic acceleration was preceded by a long period of investment in education. Maddison (1995) further suggests that recent developments in information technology involve considerable investment, both in hardware and in human learning, which may result in higher efficiency improvements and productivity gains in the future only. This perspective is consistent with the possible emergence of a new "techno-economic paradigm" or "Kondratiev wave." This complementary theory of long-term economic development focuses on the interplay and interrelationships between institutional and technological change. Following Kondratiev's (1926) observation of "long waves" in the American economy, Schumpeter (1935) developed a theory to explain these waves on the basis of discontinuities in entrepreneurial innovations. Freeman (1990) emphasizes Schumpeter's view that long waves are not so much evident in economic statistics as in qualitative features of the economy. Freeman and Perez (1988) describe five historical waves of technical and economic change and identify associated economic booms and recessions (successive "techno-economic paradigms").

Work at IIASA (e.g., Häfele *et al.*, 1982; Grübler and Nakićenović, 1991; Nakićenović, 1996; Grübler *et al.*, 1993b; Grübler, 1998a) sought to identify regularities in the market succession of technology in energy supply, transport, and the iron and steel industry. In fact, there may be a close parallel between the waves of technology they observe and Kondratiev's economic long waves. However, both empirical and theoretical implications of "long waves" remain unclear (see the review in Freeman, 1996; and especially Rosenberg and Frischtak, 1984). Equally, no approach can hope to foresee reliably the form of the next "wave." It is possible that solar energy and nuclear power might play a strong role, perhaps combined with hydrogen as an energy carrier, as suggested by Häfele *et al.* (1982) and others. The next wave in transport technology might well be an increasing share for aviation and high speed rail (Grübler and Nakićenović, 1991), a transition to 0.5 litre/100 km "hypercars" (Lovins *et al.*, 1993), or a radical change in urban planning to minimize transport needs (Newman and Kenworthy, 1990). These possibilities highlight in particular the need for technology diffusion to avoid technological "lock-in" to older technological vintages (Parikh *et al.*, 1997).

3.3.4.6. Do the Poor Get Richer and the Rich Slow Down?

Neoclassic growth theory suggests that different capital and labor productivities across countries lead to differential productivity growth rates and hence to conditional convergence across different economies. Rostow (1990) coined the term that the "poor get richer and the rich slow down." The convergence theorem of neoclassic theory arises from diminishing returns on capital. Economies that tend to have less capital per worker tend to have higher rates of return and hence higher growth rates (see Abramovitz, 1986). Conversely, economies with high capital intensity (which, because of the relationship between capital intensity and productivity, are closer to or at the productivity frontier) tend to have lower growth rates. Evidently, economies differ in more respects than their capital intensities, and hence even neoclassic theory only postulates conditional convergence (after accounting for all other factors).

The neoclassic concept of capital is usefully extended to include also human capital in the form of education, experience, and health (see, e.g., Lucas, 1988; Barro and Sala-I-Martin, 1995). Thus, additional convergence potentials accrue for economies with a proportionally higher ratio of human-to-physical capital. Equally, the generation and adoption of new technologies is facilitated by high human capital. Yet, even with the inclusion of human capital, long-term per capita growth must eventually cease in the absence of continuous improvements in technology. This cessation, however, mostly affects economies at the productivity frontier and not those that lag behind. For the latter, both theory and empirical evidence seem, all else being equal, to indicate conditional convergence (i.e., "the poor can get richer."). The *ceteris paribus* condition is an important qualifier for the convergence theorem. Evidently, the potential for conditional convergence and economic catch-up cannot be realized in an economy struck by civil war, poor institutions, or even low savings rates (related to the demographic transition discussed in Section 3.2). The recent Asian financial crisis also demonstrated that differential capital productivity indeed can lead to vast influxes of capital into developing economies, and appropriate assimilative capacities (banking systems, functioning legal system, institutions, etc.) need to be in place to use such capital flows productively (World Bank, 1998b).

In terms of a functional relationship, therefore, per capita GDP growth rates are expected to be higher for economies with low per capita GDP levels. Notwithstanding many frustrating setbacks, such as the recent "lost decade" for economic catch-up in Africa and Latin America, empirical data indicate that the convergence theorem holds. Figure 3-11 illustrates some empirical evidence put forward by Barro (1997) based on the experiences of some 100 countries in the period 1960 to 1985.

Similar convergence trends have also been identified *within* economies. For instance, Barro and Sala-I-Martin (1995) find significant convergence trends across individual states in the USA, between prefectures in Japan, and between different regions in Europe. Barro (1997) concludes in his analysis that the conditional convergence rates across these countries is statistically highly significant, and proceeds rather slowly at

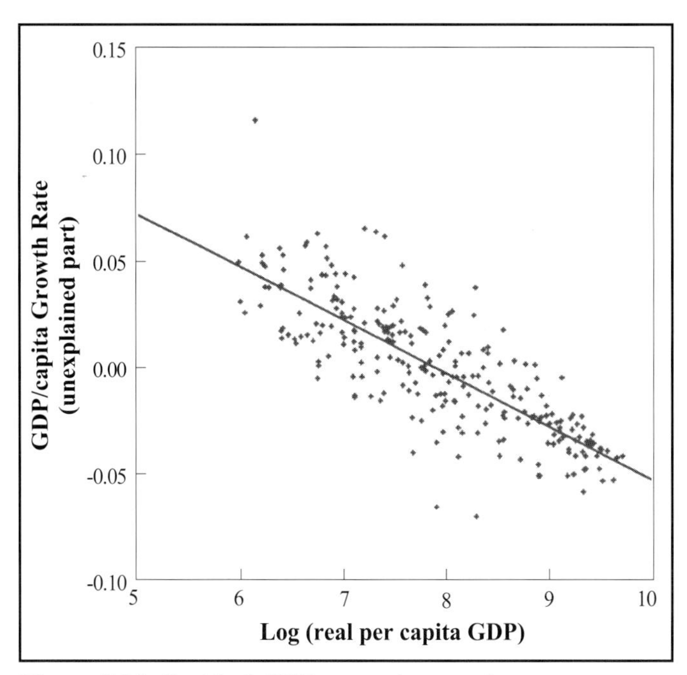

Figure 3-11: Residual GDP per capita growth rates as a function of GDP per capita (log scale). The residual growth rate is that per capita GDP growth not explained by other factors such as education, terms-of-trade, institutional factors, etc., in Barro's multi-factor analysis of per capita GDP growth. Data source: Barro, 1997.

2–3% per year. It may take an economy 27 years to reach 50% of steady-state levels (the productivity frontier) and some 90 years to achieve 90% of that level. Based on this convergence criterion alone, it may well take a century (given all other factors set favorably) for a poor economy to catch-up to levels that prevail in the industrial countries today, never mind the levels that might prevail in affluent countries 100 years in the future. Barro's analysis indicates a threshold GDP per capita level at approximately US$3000 per year. Below that level, additional productivity growth potentials result from catch-up; beyond that level, higher per capita GDP levels make further productivity growth ever more difficult to achieve (as indicated by the negative values of the residual GDP per capita growth rates in Figure 3-11).

Given the wide range in historical experiences and the slow rates of convergence suggested by neoclassic growth theory, it is not surprising that the available scenario literature takes a cautious view on economic catch-up. Whereas convergence tendencies are generally evident in scenario assumptions (see the significantly higher GDP per capita growth rates for currently developing countries compared to industrial countries in Figure 3-10), long-term convergence rates are low. For instance, from all six IS92 scenarios only one (IS92e) assumes that developing countries outside China may eventually reach present OECD income levels, and even in this most optimistic scenario it is assumed to occur only after 2080 (Pepper *et al.*, 1992). Even in this convergence scenario per capita income differences remain large – a factor of five by the end of the simulation horizon (US$31,000 per capita GDP per year in developing countries outside China versus US$150,000 OECD average). In an influential critique Parikh (1992) referred to the IS92 scenario series as being "unfair to the South," a point also taken up in the evaluation of the IS92 scenarios. Alcamo *et al.* (1995) concluded that new IPCC scenarios "will be needed for exploring a wide variety of economic development pathways, for example, a closing of the income gap between industrial and developing countries." With a few notable exceptions (e.g., the scenario developed by Lazarus *et al.* (1993) and the Case C scenarios presented in IIASA–World Energy Conference (WEC) (IIASA–WEC, 1995) and Nakićenović *et al.* (1998a)), the challenge to explore conditions and pathways that close the income gap between developing and industrial regions appears to have been insufficiently taken up in the scenario literature, a gap this report aims to begin to fill. Chapter 4 describes two scenarios in which the ratios between regions of GDP/capita decline and the absolute differences increase.

3.3.4.7. Economic Productivity and Energy and Materials Intensity

Evidence suggests that the physical input of energy or materials per unit of monetary output (materials or energy intensity) follows an inverted U-curve (IU hypothesis) as a function of income. For some materials the IU-hypothesis (Moll, 1989; Tilton, 1990) holds quite well. The underlying explanatory factors are a mixture of structural change in the economy along with technology and resource substitution and innovation processes. Recent literature illustrates material consumption that rises faster than GDP in well-developed countries in a relationship better described as N-shaped (de Bruyn and Opschoor, 1994; de Bruyn, *et al.*, 1995; Suri and Chapman, 1996; Ansuategi *et al.*, 1997). A similar IU curve is observed for modern, commercial energy forms (Darmstadter *et al.*, 1977; Goldemberg *et al.*, 1988; Martin, 1988; IIASA–WEC, 1995; Watson *et al.*, 1996; Judson *et al.*, 1999), although the initially rising part of commercial energy intensity stems from the substitution of traditional (inefficient) energy forms and technologies by modern commercial energy forms (see also the discussion of the "environmental Kuznets curve" for traditional air pollutants in Section 3.1, also an inverted U-shaped curve). The resultant aggregate total (commercial plus non-commercial) energy intensity shows a persistent declining trend over time, especially with rising incomes (Watson *et al.*, 1996; Nakićenović *et al.*, 1998a). Empirical evidence thus suggests that, all else being equal, energy and materials intensities are closely related to overall macroeconomic productivity. In other words, higher productivity (GDP per capita) is associated with lower energy and materials intensity (lower use of energy and materials per unit of GDP).

Figure 3-12 shows material intensity versus per capita income data for 13 world regions for some metals (Van Vuuren *et al.*, 2000; see also the discussion in de Vries *et al.*, 1994). Figure 3-13 shows a similar curve for total energy intensity (including traditional non-commercial energy forms) for 11 world

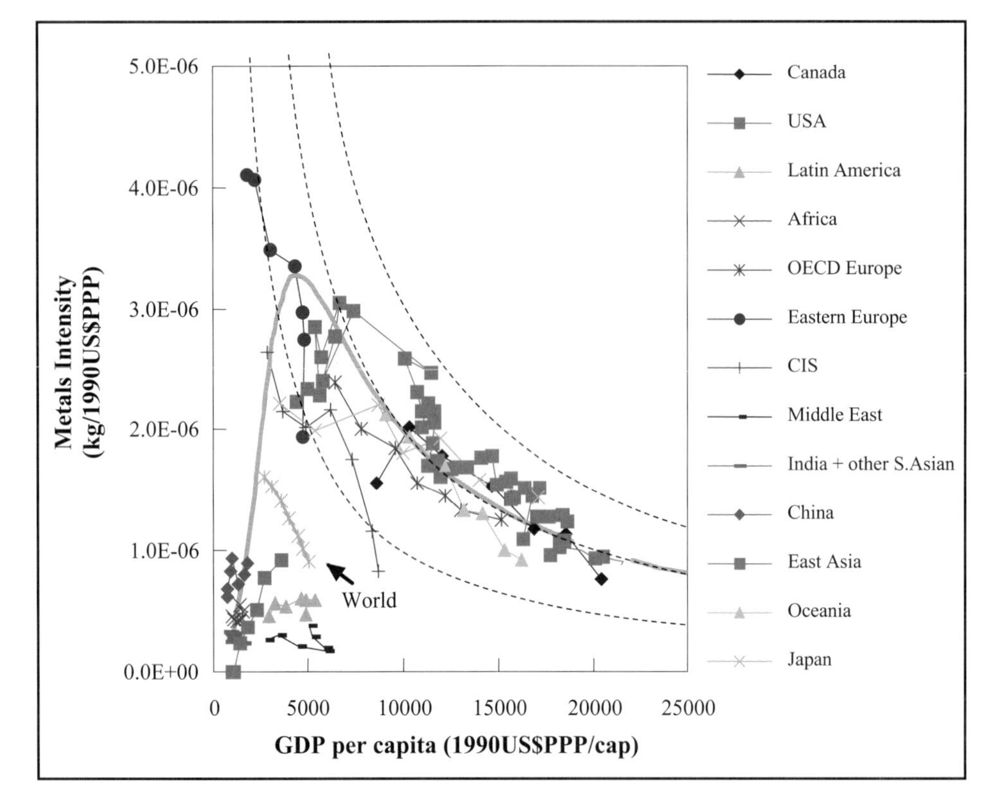

Figure 3-12: Metals intensity of use per unit of GDP as a function of GDP per capita for 13 world regions. Metals include refined steel and MedAlloy (the sum of copper, lead, zinc, tin, and nickel). GDP here is measured in terms of purchasing power parities (PPP). The dashed curves are isolines that represent a constant per capita consumption of metals. The thick pink line indicates the inverse U-shaped curve that best describes the trends in the different regions as part of a global metal model. Data source: Van Vuuren *et al.*, 2000.

regions[1], again as a function of per capita income (Nakićenović *et al.*, 1998a).

The most important conclusion to retain from Figures 3-12 and 3-13 is that energy and materials intensity (i.e. energy use per unit of economic output) tend to decline with rising levels of GDP per capita. Thus, energy and material productivity – the inverse of energy intensity – improve in line with overall macroeconomic productivity (as represented, e.g., by GDP per capita levels). Also, they tend to deteriorate when productivity levels fall, as happened during the start of the deep economic recession in Central and Eastern Europe (Government of Russian Federation, 1995). A corollary of this relationship is that material and energy intensities decline the faster per capita income grows. Thus, overall economic productivity growth (GDP per capita growth) and reductions in materials and energy use per unit GDP (materials and energy productivity growth) are closely related. The fundamental reason is that high macroeconomic growth presupposes accelerated rates of technical change and corresponds with a fast turnover of capital stock and, hence, a faster incorporation in the economy. This represents an important new finding for long-term energy and emissions scenarios that have, to date, largely treated economic and resource productivity growth as independent of each other. According to recent empirical and theoretical findings, they no longer can.

Two final caveats are important. First, growth in productivity and intensity improvement growth have historically been outpaced by economic output growth. Hence, materials and energy use has risen in absolute terms (see Nriagu, 1996; Watson *et al.*, 1996; Grübler, 1998a). The second caveat is that energy and material intensity are affected by many factors other than macroeconomic productivity growth and resultant income. OECD (1998b) notes that high rates of productivity increase have been associated in the past with new competitive pressures, strong price or regulatory incentives, catching up or recovery, and a good "climate for innovation." Also, the emergence of new technologies and resource–strategic considerations has led to rapid productivity growth.

Table 3-3 summarizes selected macroeconomic, labor, energy, and material productivity increases that have been achieved in a range of economies and sectors at different times. Principally, it demonstrates that the least likely assumption for future scenarios from historical evidence is absence of productivity growth. Human ingenuity (as reflected in new technologies and new practices) historically has responded to purposeful action (R&D and inventive activities) and to a wide range of policies to improve vastly productivity in the use of *all* factors of production. Of course, it remains uncertain how future productivity growth rates will diffuse nationally, regionally, and internationally. The essential lesson provided by history is that change is continuous and pervasive. From that perspective, static or "business-as-usual" scenarios need to be replaced by "dynamics-as-usual" scenarios, which divide into "fast" and

[1] The regional definition used in Nakićenović *et al.*, 1998a, includes: AFR: Sub-saharan Africa; CPA: China and Centrally Planned Asia; EEU: Central and Eastern Europe; FSU: Newly Independent States of the former Soviet Union; LAM: Latin America and the Caribbean; MEA: Middle East and North Africa; PAO: Pacific OECD; PAS: Other Pacific Asia; SAS: South Asia; WEU: Western Europe.

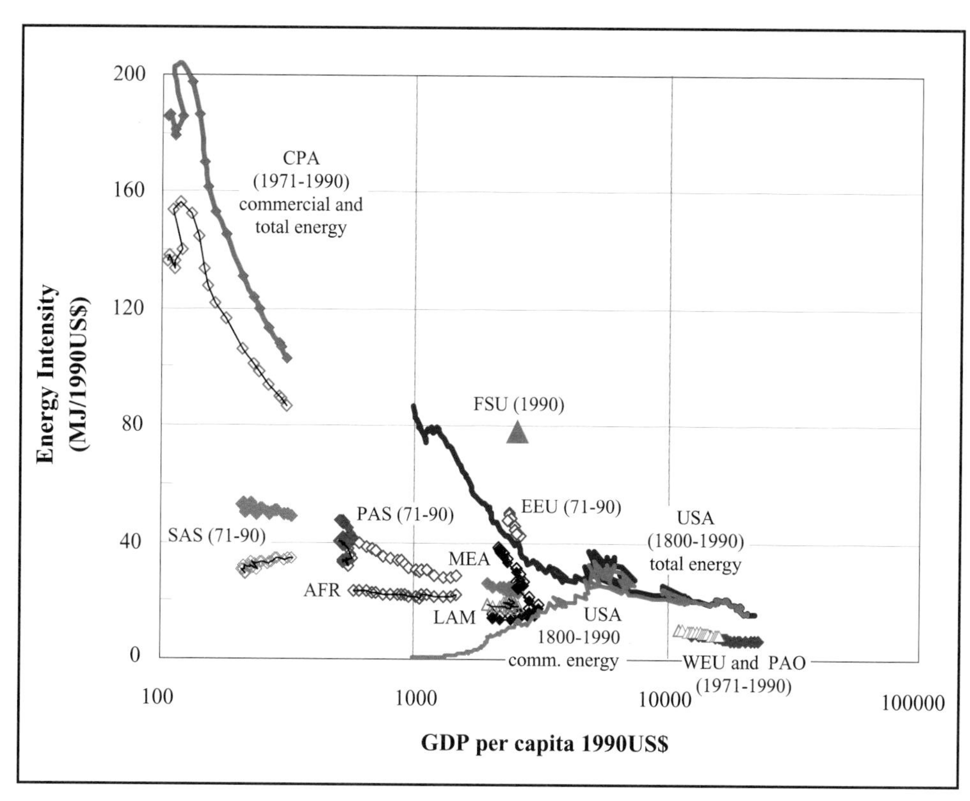

Figure 3-13: Energy intensity (all energy forms in the higher curves, and commercial energy only in the lower curves) as a function of GDP per capita for 11 world regions since 1970. For an explanation of regional abbreviations see text. Historical data for the USA since 1800 are equally shown. Source: adapted from Nakićenović *et al*., 1998a.

"slow" alternatives as a function of perceived opportunities, new institutional configurations, policies, and environmental constraints, even if these do not include climate policies (as mandated by the SRES Terms of Reference, see Appendix I).

For instance, low historical rates of energy intensity improvement reflect the low priority placed on energy efficiency by most producers and users of technology. On average energy costs account for only about 5% of GDP. Energy intensity reductions average about 1% per year, in contrast to improvements in labor productivity above 2% per year over the period 1870 to1992. Over shorter time periods, and given appropriate incentives, energy intensity improvement rates can be substantially higher, as in the OECD countries after 1973 or in China since 1977, where energy intensity improvement rates of 5% have been observed. Rapid productivity growth can also occur during periods of successful economic catch-up; for instance, Japanese labor productivity grew at 7.7% annually during 1950–1973 (Maddison, 1995). Similar high-productivity growths were also achieved in industrial oil usage in the OECD or US car fuel economies after 1973. Of the examples given in Table 3-3, productivity increases are the highest for communication. Not surprisingly, many observers consider that given a continuation of historical trends communication may become a similarly important driver of economic growth in the future as traditional, resource- and energy-intensive industries have been in the past.

3.3.4.8. *Development Patterns*

The key questions about how future development patterns determine GHG emissions thus include the following.

Material and energy content of development in industrial countries:

- Will structural change toward services and increasing importance of information as a "raw material" reduce the energy and matter content of economic activity?
- Will telecommunications substitute significantly for travel or encourage more of it?
- Will growth in transportation and other energy-using activities, stimulated by trade liberalization, be offset by less material intensive development patterns?
- Will the tendency to saturation in some energy end-use requirements be offset by new energy or GHG intensive goods and services (e.g., in leisure activities)?

Development patterns in the developing countries:

- Will developing countries reproduce the development paths of industrial countries with respect to energy use and GHG emissions?
- Is there a potential for technological "leapfrogging" whereby developing countries bypass dirty intermediate technologies and jump straight to cleaner technologies?

Links among energy, transport and urban planning:

- Will modal choices and urban-form decisions tend toward less or greater energy intensity?
- How are significant differences within and between industrial and developing countries going to evolve in the future?

Land use and human settlements:

- What are the links between agriculture, forestry, rural-to-urban migration, energy use, GHG, and sulfur emissions, particularly in developing countries?

Table 3-3: Examples of productivity growth for the entire economy and for selected sectors and countries. Data sources: see table footnote.

Sector/Technology	Region	Productivity Indicator	Period	Annual Productivity Change (%)
Whole economy[a]	12 countries Europe	GDP/capita	1870–1992	1.7
Whole economy[a]	12 countries Europe	GDP/hour worked	1870–1992	2.2
Whole economy[a]	USA	GDP/hour worked	1870–1973	2.3
Whole economy[a]	USA	GDP/hour worked	1973–1992	1.1
Whole economy[a]	Japan	GDP/hour worked	1950–1973	7.7
Whole economy[a]	South Korea	GDP/hour worked	1950–1992	4.6
Whole economy[b]	World	GDP/primary energy	1971–1995	1.0
Whole economy[b]	OECD	GDP/primary energy	1971–1995	1.3
Whole economy[b]	USA	GDP/primary energy	1800–1995	0.9
Whole economy[b]	United Kingdom	GDP/primary energy	1890–1995	0.9
Whole economy[b]	China	GDP/primary energy	1977–1995	4.9
Whole economy[c]	Japan	GDP/material use	1975–1994	2.0
Whole economy[c]	USA	GDP/material use	1975–1994	2.5
Agriculture[g, f]	Ireland	Tons wheat/hectare	1950–1990	5.3
Agriculture[g]	Japan	Tons rice/hectare	1950–1996	2.2
Agriculture[g]	India	Tons rice/hectare	1950–1996	2.0
Industry[a]	OECD (6 countries)	Value added/hour worked	1950-1984	5.3
Industry[a]	Japan	Value added/hour worked	1950-1973	7.3
Industry[b]	OECD	Industrial production/energy	1971–1995	2.5
Industry[b]	OECD	Industrial production/oil use	1974–1986	8.0
New cars[d]	USA	Vehicle fuel economy	1972–1982	7.0
New cars[d]	USA	Vehicle fuel economy	1982–1992	0.0
Commercial aviation[e]	World	Ton-km/energy	1974–1988	3.8
Commercial aviation[e]	World	Ton-km/energy	1988–1995	0.3
Commercial aviation[e]	World	Ton-km/labor	1974–1995	5.6
Telephone call costs[f]	Transatlantic	London–NY, costs for 3 min	1925–1995	8.5
Telephone cables[f]	Transatlantic	Telephone calls/unit cable mass	1914–1994	25.0

Data sources:
a: Maddison, 1995.
b: OECD and IEA statistics.
c: WRI, 1997a.
d: Including light trucks, Schipper, 1996.
e: International Civil Aviation Organization (ICAO) statistics.
f: OECD, 1998b; Waggoner, 1996; Hayami and Ruttan, 1985.
g: FAO (various years 1963–1996) Production Statistics.

Distribution issues:

- What are the links between development patterns, distribution of income, energy use and GHG emissions?

The informal economy:

- What is the link between informal economic activities and GHG emissions?

The effect of development pattern assumptions in the assessment of future GHG emissions is greater for developing countries. As a major part of the needed infrastructure to meet development needs is still to be built in the developing countries, the spectrum of future options is considerably wider than that in industrial countries. The traditional approach to assume "business-as-usual" as the baseline is particularly meaningless in such cases; instead there is a need for multiple baselines for different scenarios built to cover the range of possible futures. It cannot be assumed that developing countries will automatically follow the past development paths of industrial countries. The significant transformations that recently intervened in the international economy and energy markets highlight the important dangers of such a double analogy, both in space and time. It can also be argued that many developing countries may have passed already any developmental bifurcation point, in which case developments could follow the patterns of industrialized countries.

Both the GDP structure and the physical basis on which it is to be achieved in developing countries have to be considered. A crucial question regards their share in the world production of highly energy- and pollution-intensive goods, such as steel and aluminum. As the recent shift of heavy industries from the industrial toward the developing countries ends, long-term economic output could come from services and other less energy-intensive activities.

Moreover, technological choices, both in production and consumption apparatus, can substantially decrease the energy demand per GDP. For instance, Chinese households are not bound to adopt the same model of energy-intensive refrigerators that have equipped American families. Similarly, future cement factories in developing countries should not fail to include up-to-date technological improvements, such as the dry process.

The spatial distribution of the population and its economic activities is still unclear, which raises the possibility of adopting urban and/or regional planning. Industrial policies directed at rural development and strengthening the role of small and medium cities would reduce the extent of rural exodus and the degree of demographic concentration in large cities.

These examples show that developing countries could adopt anticipative strategies to avoid, in the long-term, the problems faced today by industrial societies ("leapfrogging"). Such alternative development patterns highlight the technical feasibility of emission futures in the developing regions that can be compatible with national objectives. However, the barriers to a more sustainable development in the developing regions can hardly be underestimated, and range from financial constraints to cultural behaviors (in both industrial and developing countries), which include the lack of appropriate institutional structures.

3.3.5. Conclusion

The process of economic and social development depends both on the ability of the current lead countries in productivity to maintain their technological and institutional creativity, and on the ability of other countries to adopt leading-edge technologies and institutions or to develop their own. The crucial issues of "how much" and "what kind" of productivity growth can be addressed only by describing alternative scenarios of future development. To develop alternative scenarios it is necessary to recall the important qualitative relationship between demographic transition and social and economic development. Causality links could be in either direction, but the importance of the relationship is recognized in both theoretical and empirical studies. Hence, as summarized in Section 3.2, this relationship should be incorporated into the SRES scenarios. Scenarios of accelerated rates of economic and social development should therefore be the scenarios with an accelerated demographic transition. This corresponds to a linking of high per capita development with comparatively low population levels.

There is a need to explore in particular pathways that close the development gap (see Parikh, 1992; Alcamo *et al.*, 1995). As the likelihood of zero or even negative productivity growth in the developed countries is low, closure of the development gap requires accelerated rates of productivity growth and the need to overcome or avoid setbacks in per capita income growth in many developing countries. Scenarios that explore this possibility will necessarily extend beyond the range of futures spanned by the IS92 scenarios, as well as beyond the range of the majority of the "conventional wisdom" scenario literature on the future of developing countries.

As a major part of the needed infrastructure to meet development needs still has to be built, the spectrum of future options is considerably wider in developing than in industrial countries. For instance, the technical possibilities for low emission futures in the developing countries are many. The extent of the spectrum of future options depends on the changes discussed in Sections 3.2 and 3.3, but also on the outcome of crucial issues. These include political power structure, national governance and institutional structure; income distribution; cultural attitudes and consumption patterns (diets, housing, etc.), development of and access to modern technologies (energy, production, distribution, etc.), and the geographic distribution of activities (land use, urban settlement, transportation needs, etc.).

Particular sets of technological and behavioral options can be clustered into alternative, internally consistent packages to represent different choices over time and so define different development paths for any economy. Such clusters can give rise to self-reinforcing loops between technical choice, consumer demand, and geographic distribution, which create "lock-in" effects and foreclosures of options in technology and socio-institutional innovations. The time-dependent nature of these choices gives rise to bifurcations and irreversibilities in which the shift from one development path to another entails important economic and political costs.

Globalization of markets, technologies, and information networks may help accelerate productivity growth in the future. However, both economic and social losers could result from the globalization process. The financial instability during 1998 has cast further doubt on the inevitability of global convergence as a standard model of political economy. Hence, a further important dimension of uncertainty to be explored in the scenarios is the degree of globalization or regionalization in economic, social, and technological development.

The various perspectives on economic history discussed in Section 3.3.4 reveal several possible options for the future:

- Perhaps the most extreme view may be that the development process is nearing completion in Europe, North America, and Australasia, so that the main prospect for growth is through the diffusion of existing best-practice technologies to the rest of the world.
- The nature of economic development may have changed significantly in high-income countries, with a new emphasis on services, quality, and information.

Such development is hard to measure in material terms. Many writers refer to "dematerialization" and the emergence of the "knowledge-based" economy.
- The start of a new "Kondratiev wave" may be underway, to be revealed in the early 21st century in a surge of economic growth, with the massive development of high-technology industries leading to new products of increasing value and renewed opportunities for fast developmental catch-up.

As a result, scenarios can span from low dematerialization to high dematerialization futures associated with a wide range of income levels. In the former, the shift toward more value-added products in industry would be compensated by rising labor productivity and hence lower product costs. Economic production remains material oriented. It may be a world with huge underground cities, air-conditioned tourist resort areas with indoor beaches, a significant fraction of people in low-density regions may fly their own airplanes, and robots may do housework in most homes. In the latter, much of the money flow would be associated with exchange of information and services. Industrial value added would be, to a large extent, generated from R&D and know-how, and less from increasing productivity in traditional industries. Educational, childcare, and medical services would make up a large part of personal expenditures. Already, all kinds of artistic and handicraft work have become part of the formal monetary economy, partly because of the booming world tourist industry. Much "economic growth" may revolve around the (re)distribution of scarce, positional goods such as space and valuable artworks.

Therefore, the task of future scenario development entails more than just the adoption of alternative quantitative assumptions. The overall context within which alternative assumptions on productivity growth or energy and materials intensity take place needs to be made explicit. This is simply because many key influencing factors (e.g., institutions) cannot be assessed quantitatively, or the relationship between factors is known only qualitatively. The development of alternative qualitative scenario "storylines" (see Chapter 4) is therefore an important advance over previous IPCC scenario methodologies.

3.4. Energy and Technology

3.4.1. *Introduction*

In this section, energy end-uses, resources, and technologies are reviewed. Their future evolution is of critical importance to future emissions levels. First, major patterns of energy end-use and emissions by sector are considered, followed by a discussion of energy resources; then energy supply technologies that might become of greater importance in the future are reviewed briefly before the current understanding and modeling of technological change are discussed.

3.4.2. *Energy Use and Emissions by Major Sectors*

3.4.2.1. *Overview*

Sectoral energy use and GHG emissions changes are often discussed in terms of trends in the major end-use sectors (e.g., Sathaye *et al.*, 1989; IEA, 1997c; Schipper *et al.*, 1997a; Price *et al.*, 1998). Trends reveal striking differences between sectors and regions of the world. The key sectors of the economy that use energy are industry (including agriculture), commercial, residential, and institutional buildings, and transportation. Key drivers of energy use and carbon emissions include activity drivers (total population growth, urbanization, building, and vehicle stock, commodity production), economic drivers (total GDP, income, and price elasticities), energy intensity trends (energy intensity of energy-using equipment, appliances, vehicles), and carbon intensity trends. These factors are in turn driven by changes in consumer preferences, energy and technology costs, settlement and infrastructure patterns, technical progress, and overall economic conditions.

Table 3-4 shows that global primary energy use grew from 191 EJ in 1971 to 307 EJ in 1990 at an average annual growth rate of 2.5% per year. This growth tapered off in all sectors after 1990, and total global primary energy increased to only 319 EJ by 1995, mainly because of the large declines experienced in the REF region (see Chapter 1 for definition of SRES world regions) as a result of the political and economic restructuring of the countries within it. Table 3-4 shows that the industrial sector clearly dominates total primary energy use, followed by the buildings sector (commercial, residential, and institutional buildings combined), transport sector, and agriculture sector.

Energy intensity is the amount of energy used to perform a particular service, such as to produce a ton of steel, power a refrigerator, or propel a vehicle. Technical progress generally leads to improved energy efficiency in technologies such as lights, vehicles, refrigerators, and manufacturing processes. Many studies show that considerable energy efficiency improvement can be realized (technically and economically) in the short term (10–15 years) with available technologies (Szargut and Morris, 1987; Ayres, 1989; Jochem, 1989; Lovins and Lovins, 1991; Nakićenović *et al.*, 1993; WEC, 1995b; Watson *et al.*, 1996; Worrell *et al.*, 1997).

In 1990, industry accounted for two-fifths of global primary energy use, residential and commercial buildings for a slightly smaller amount, and transportation for one-fifth of the total. These shares vary according to economic structures in each region (see below). Carbon emissions that result from energy use depend on the carbon intensity of the energy source. Changes in carbon intensity mainly result from fuel substitution, but can also arise from changes in technology or process. The largest shifts in carbon intensity over the long term are associated with changes in the energy sources used for power generation since 1850 (Nakićenović and Grübler, 1996). Smaller but still significant shifts resulted from fuel switching

***Table 3-4**: Primary energy (EJ per year) use by sector and region, 1971 to 1995, and average annual growth rates (AAGR) for 1971 to 1990 and 1990 to 1995. Source: Price et al., 1998, based on IEA, 1997a; IEA, 1997b; BP, 1997 (see Chapter 1 for definitions of SRES world regions).*

	1971	1975	1980	1985	1990	1995	AAGR 1971–1990	AAGR 1990–1995
Industrial Sector:								
OECD90	48.6	49.3	55.0	52.3	54.3	56.8	0.6%	0.9%
REF	26.0	31.6	34.0	36.9	38.0	26.0	2.0%	-7.3%
ASIA	8.8	11.5	15.5	20.0	26.1	34.8	5.9%	5.9%
ALM	4.6	6.2	8.9	10.5	11.0	13.0	4.7%	3.5%
World	88.0	98.5	113.5	119.8	129.4	130.8	2.1%	0.2%
Buildings Sector:								
OECD90	44.4	48.9	52.3	56.8	62.3	68.5	1.8%	1.9%
REF	10.7	13.0	18.2	21.0	23.0	16.2	4.1%	-6.8%
ASIA	3.6	4.6	5.6	7.9	10.2	12.9	5.7%	4.8%
ALM	2.7	3.7	5.1	6.9	10.1	12.1	7.1%	3.8%
World	61.5	70.3	81.3	92.6	105.6	109.8	2.9%	0.8%
Transport Sector:								
OECD90	26.2	29.4	32.5	33.8	39.4	43.3	2.2%	1.9%
REF	6.0	7.3	8.0	9.2	10.0	7.3	2.7%	-6.0%
ASIA	2.0	2.4	3.3	4.3	6.0	8.7	5.9%	7.6%
ALM	3.3	4.6	6.3	7.2	7.8	9.6	4.6%	4.2%
World	37.5	43.6	50.1	54.4	63.3	69.0	2.8%	1.7%
Agriculture Sector								
OECD90	1.8	1.8	2.1	2.6	2.7	3.0	2.2%	1.6%
REF	1.3	1.6	1.8	2.4	3.0	1.7	4.5%	-10.6%
ASIA	0.9	1.3	1.6	1.7	2.3	3.0	4.8%	5.6%
ALM	0.4	0.5	0.7	0.8	0.9	1.6	4.7%	12.6%
World	4.4	5.1	6.1	7.5	8.9	9.3	3.8%	0.8%
All Sectors:								
OECD90	121.0	129.3	141.8	145.5	158.8	171.7	1.4%	1.6%
REF	44.0	53.5	62.0	69.5	74.0	51.3	2.8%	-7.1%
ASIA	15.4	19.7	26.0	33.9	44.7	59.5	5.8%	5.9%
ALM	11.0	14.9	21.1	25.4	29.8	36.4	5.4%	4.1%
WORLD	191.4	217.5	251.0	274.2	307.2	318.8	2.5%	0.7%

in industrial, commercial, and residential energy consumption. The relationship between total sector energy use and economic drivers such as GDP per capita varies across countries depends upon the sector. In 1995, the relationship in the transport and buildings sectors was relatively strong and that in the industrial sector was moderate (Price *et al.*, 1998). Income elasticities vary widely among the different types of energy services and the country or region under consideration. For example, the income elasticity of refrigerator ownership in most countries in the IND region (see Chapter 1 for definition of SRES world regions) is extremely low, as most households already own a refrigerator. The elasticity is much higher in medium-income countries in which refrigerator ownership is low. Other economic indicators, such as level of economic development in the industrial sector and personal consumption expenditures in residential buildings, are more closely correlated with energy use in these sectors.

3.4.2.2. Industry and Agriculture

Driving forces behind energy use and carbon emissions in the industrial sector include the state of economic development, consumption and trade patterns, relative costs of labor, capital, and energy, and availability of resources. In 1990, industry accounted for 42% (129 EJ) of global primary energy use. Between 1971 and 1990, industrial energy use grew at a rate of 2.1% per year, slightly less than the world total energy demand growth of 2.5% per year. This growth rate has slowed in recent years, and was virtually flat between 1990 and 1995, primarily because of declines in industrial output in the REF region.

Energy use in the industrial sector is dominated by the industrialized countries, which accounted for 42% of world industrial energy use in 1990. Countries in the REF, ASIA, and ALM regions used 29%, 20%, and 9% of world industrial energy, respectively, that year. The share of industrial sector energy consumption within the industrialized countries declined from 40% in 1971 to 33% in 1995, which partly reflects the transition toward a less energy-intensive manufacturing base. The industrial sector dominates in the REF region, accounting for more than 50% of total primary energy demand, a result of the long-term policy that emphasized materials production and was promoted under years of central planning. Average annual growth in industrial energy use in this region was 2% between 1971 and 1990, but dropped by an average of 7.3% per year between 1990 and 1995 (IEA, 1997a; IEA, 1997b; BP, 1997).

The agriculture sector used only 3% of global primary commercial energy in 1990. Unlike the other sectors, the REF region dominated agricultural energy use in 1990, using 34% of the total, followed by the IND (30%), ASIA (26%), and ALM (10%). Between 1971 and 1990, the average annual growth in primary energy used for agriculture was slower in the industrialized countries (2.2% per year) than in the three other regions, for which growth ranged between 4.5% and 4.8% per year. Trends in agricultural primary energy use changed significantly in the REF and ALM regions after 1990, with REF consumption dropping to an average of 10.6% per year and ALM consumption increasing to an average of 12.6% per year by 1995.

Energy use in the industrial sector is dominated by the production of a few major energy-intensive commodities, such as steel, paper, cement, and chemicals. Rapidly industrializing countries have higher demands for these infrastructure materials and more mature markets have declining or stable levels of consumption. Studies of material consumption in industrialized countries show increases in the initial development of society to a maximum consumption level, which then remains constant or even declines as infrastructure needs are met and material recycling increases. Absolute and per capita consumptions of some materials appear to have reached levels of stabilization in many industrialized countries, although this is not true of all materials (e.g. paper). Expressed as a function of unit GDP, material intensity generally declines after reaching a maximum (Williams *et al*., 1987; Wernick, 1996; WRI, 1997b; see also Section 3.3). Although the use of all materials in developing countries will certainly grow, per capita consumption may not reach that in the industrialized countries, because more efficient processes and substitutes are available.

Carbon intensities with respect to GDP (CO_2 emissions as a function of GDP) in the industrial sector have been relatively stable in most countries except for those that are rapidly industrializing (Houghton *et al*., 1995). This trend results from the changing economic structure, reduced energy intensity, and reduced carbon intensity of the fuel mix. A shift toward less carbon-intensive fuels took place between 1971 and 1992 in most industrialized countries, as well as in South Korea (Ang and Pandiyan, 1997; Schipper *et al*., 1997a). The industrial sector fuel mix has become more carbon-intensive in some developing countries, such as China and Mexico (Ang and Pandiyan, 1997; Sheinbaum and Rodriguez, 1997), although a trend away from coal to other fuels has also occurred in some developing countries (Han and Chatterjee, 1997). The contribution of fuel-mix changes to CO_2 emissions reduction has been small in most industrialized countries (Golove and Schipper, 1997; Schipper *et al*., 1997a).

Technical energy-intensity reductions of 1 to 2% per year are possible in the industrial sector and have occurred in the past (Ross and Steinmeyer, 1990). The annual change in energy intensity in the industrial sector varied between –0.1% and –6.6% per year for a variety of countries from the early 1970s to the early 1990s. Generally, electricity intensity remained constant and fuel intensity declined, which reflects the increasing importance of electricity (IEA, 1997c).

3.4.2.3. Residential, Commercial, and Institutional Buildings

In the buildings sector, household expenditure levels, appliance and equipment penetration levels, and the share of population that lives in urban areas all affect energy use. In 1990, residential, commercial, and institutional buildings consumed almost 100 EJ of primary energy, about one-third of the total global primary energy. Uncertainties persist with respect to quantities and structure of non-commercial fuel use in developing countries. Primary energy use in the buildings sector worldwide grew at an average annual rate of 2.9% between 1971 and 1990. Growth in buildings energy use varied widely by region, ranging from 1.8% per year in the IND region to 7.1% per year in the ALM region. Growth in commercial buildings was higher than growth in residential buildings in all regions of the world, averaging 3.5% per year globally. In 1990, the IND region used about 60% of global building energy, followed by REF (22%), ASIA (10%), and ALM (9%) countries, respectively. Between 1990 and 1995, growth in the use of primary energy in buildings slowed in all regions except the industrialized countries, where buildings primary energy use climbed at an average of 1.9% per year. The greatest decline occurred in the REF region, where buildings energy use declined by an average of 6.8% annually between 1990 and 1995, dominated by a 7.2% per year average drop in residential primary energy use. Growth in buildings energy use in the other two regions – ASIA and ALM – slowed during this period, but growth rates were still high, averaging 4.8% and 3.8%, respectively (BP, 1997; IEA, 1997a; IEA, 1997b).

Along with population size, key activity drivers of energy demand in buildings are the rate of urbanization, number of dwellings, per capita living area, persons per residence, and commercial floor space. As populations become more urbanized and areas develop electrification, the demand for energy services such as refrigeration, lighting, heating, and cooling increases. In the residential buildings sector, the level of energy demand is further influenced by population age distribution, household income, number of households, size of dwellings, and number of people per household. In the commercial buildings sector, factors that influence energy demand include the overall population level (i.e., the number of people who desire commercial services), the size of the labor force, and commercial sector income.

The number of people living in urban areas increased from 1.35 billion, or 37% of the total, in 1970 to 2.27 billion, or 43% of the total, in 1990 (see also Section 3.2). Growth in urbanization was strongest in the ASIA and ALM regions, where the average annual increase in urban population was nearly 4.0% per year. This increase and the resultant income effects led to increased usage of commercial fuels, such as kerosene and liquefied petroleum gas (LPG), for cooking instead of traditional biomass fuels. In general, higher levels of urbanization are associated with higher incomes and increased household energy use (Sathaye *et al.*, 1989; Nadel *et al.*, 1997).

Energy consumption in residential buildings is strongly correlated with household income levels. Between 1973 and 1993, increases in total private consumption translated into larger homes, more appliances, and an increased use of energy services (water heating, space heating) in most industrialized countries (IEA, 1997c). Dwelling size is a key determinant of residential energy use, and has grown with personal consumption expenditures in most industrialized countries. In the DEV region, urban areas are generally associated with higher average incomes (Sathaye and Ketoff, 1991). Wealthier populaces in developing countries exhibit consumption patterns similar to those in industrialized countries – purchases of appliances and other energy-using equipment increase with gains in disposable income (WEC, 1995b).

In the commercial sector, the ratio of primary energy use to commercial sector GDP fell in a number of industrialized countries between 1970 and the early 1990s, despite a large growth in energy-using equipment in commercial buildings. Almost certainly this is an effect of improved equipment efficiencies combined with economic growth in the commercial sector unrelated to energy consumption. Electricity use in the commercial sector shows a relatively strong correlation with commercial sector GDP, although there is a wide range of electricity use at any given level of commercial sector GDP (IEA, 1997c).

Overall energy intensity in the buildings sector can be measured using energy consumption per capita values. Between 1971 and 1990, global primary energy use in the buildings sector grew from 16.5 GJ per capita to 20 GJ per capita. Buildings per capita energy use varied widely by region, with the IND and REF regions dominating globally. Energy use per capita was higher in the residential sector than in the commercial sector in all regions, although average annual growth in commercial energy use per capita was higher during the period, averaging 1.7% per year globally compared to 0.6% per year for the residential sector (Price *et al.*, 1998).

Space heating is an important end-use in the IND and REF regions and in some developing countries; it accounts for half of China's residential and commercial building energy demand (Nadel *et al.*, 1997). The penetration of central heating doubled, from about 40% of dwellings to almost 80% of dwellings, in many industrialized countries between 1970 and 1992 (IEA, 1997c). District heating systems are common in some areas of Europe and in the REF region. Space heating is not common in most developing countries, with the exception of China, South Korea, South Africa, Argentina, and a few other South American countries (Sathaye *et al.*, 1989). Residential space-heating energy intensities declined in most industrial countries (except Japan) between 1970 and 1992 because of reduced heat losses in buildings, lowered indoor temperatures, more careful heating practices, and improvements in efficiency of heating equipment (Schipper *et al.*, 1996; IEA, 1997c). Water heating, refrigeration, space cooling, and lighting are the next largest residential energy uses, respectively, in most industrialized countries (IEA, 1997c). In developing countries, cooking and water heating dominate, followed by lighting, small appliances, and refrigerators (Sathaye and Ketoff, 1991). Appliance penetration rates increased in all regions between 1970 and 1990. The energy intensity of new appliances has declined over the past two decades – new refrigerators in the US were 65% less energy-intensive in 1993 than in 1972 (Schipper *et al.*, 1996).

Primary energy use per square meter of commercial sector floor area has gradually declined in most industrial countries, despite countervailing trends such as growth in the share of electricity, increases in electricity intensity, and reduction in fuel use and fuel intensity. Electricity use and intensity per unit area increased rapidly in the commercial buildings sector as the penetration of computers, other office equipment, air conditioning, and lighting grew. Fuel intensity per unit area declined rapidly in industrialized countries as the share of energy used for space heating in commercial buildings dropped because of thermal improvements in buildings (Krackeler *et al.*, 1998).

Carbon intensity of the residential sector declined in most industrialized countries between 1970 and the early 1990s (LBL, 1998). In the commercial sector, CO_2 emissions per square meter of commercial floor area also dropped in most industrialized countries during this period, even though carbon intensity per unit area for electric end-uses increased in most industrialized countries, dropping only in France, Sweden, and Norway – countries that moved away from fossil fuels (Krackeler *et al.*, 1998). In developing countries, (non-commercial) biomass is often used in the residential sector, especially in rural areas. Increased urbanization, as well as increases in incomes and rural electrification, can lead to rising carbon intensities in buildings when sustainable biomass use is replaced with carbon-intensive fuels such as LPG, coal, and electricity generated by fossil fuels. Conversely, if biomass use was previously on an unsustainable basis, the shift toward commercial fuels can lower carbon intensities. The trend toward replacement of biomass fuels by commercial fuels is expected to continue in developing countries (IEA, 1995).

In addition to the energy use in buildings, Tiwari and Parikh (1995) drew attention to energy use for buildings construction, which accounts for 17% of India's carbon emissions in terms of embodied energy in steel, cement, glass, bricks, etc. Typically, this embodied buildings energy is accounted for as industrial energy use in energy statistics. Tiwari and Parikh (1995) found that in India alternative construction methods could save 23% of energy use at 0.03% increase in costs.

3.4.2.4. Transport

The transport sector consumed slightly over 63 EJ, or about 20% of global primary energy, in 1990. Transport sector primary energy use grew at a relatively rapid average annual rate of 2.8% between 1971 and 1990, slowing to 1.7% per year between 1990 and 1995. Industrialized countries clearly dominate energy consumption in this sector, using 62% of the world's transport energy in 1990, followed by REF (16%), ALM (12%), and ASIA (10%) regions. The most rapid growth was seen in the ASIA countries (5.9% per year) and the ALM region (4.6% per year). Transport energy use dropped dramatically in the REF region after 1990; by 1995 this region only consumed 11% of global transport energy use. Growth in transport primary energy use also declined slightly in the IND region, dropping from an average of 2.2% per year between 1971 and 1990 to 1.9% per year between 1990 and 1995. High growth continued in the ASIA and ALM regions, with the ASIA countries increasing to an average of 7.6% per year between 1990 and 1995 (BP, 1997; IEA, 1997a; IEA, 1997b).

Influences on GHG emissions from the transport sector are often divided into those that affect activity levels (travel and freight movements) and those that affect technology (energy efficiency, carbon intensity of fuel, emission factors for nitrous oxide (N_2O), etc.). The various driving forces and their effects are reviewed in detail in the IPCC Working Group II (WGII) Second Assessment Report (SAR) (Michaelis *et al.*, 1996).

In aggregate, transport patterns are closely related to economic activity, infrastructure, settlement patterns, and prices of fuels and vehicles. They are also related to communication links. At the household level, travel is affected by transport costs, income, household size, local settlement patterns, the occupation of the head of the household, household make-up, and location (Jansson, 1989; Hensher *et al.*, 1990; Walls *et al.*, 1993). People in higher-skilled occupations that require higher levels of education are more price- and income-responsive in their transport energy demand than people in lower-skilled occupations (Greening and Jeng, 1994; Greening *et al.*, 1994).

Urban layout both affects and is affected by the predominant transport systems. It is also strongly influenced by other factors such as people's preference for living in low-density areas, close to parks or other green spaces, away from industry, and close to schools and other services. Travel patterns may be influenced by many factors, including the size of the settlements, proximity to other settlements, location of workplaces, provision of local facilities, and car ownership. A survey of cities around the world (Newman and Kenworthy, 1990) found that population density strongly and inversely correlates with transport energy use.

Many studies have examined the response of car travel and gasoline demand to gasoline price, and are reviewed, for example, in Michaelis (1996) and Michaelis *et al.* (1996). Such studies typically find a measurable reduction in fuel demand, distance traveled, car sales, and energy intensity in response to fuel price increases. Studies of freight transport found relatively small short-term impacts of diesel price increases, and often produced results that were inconclusive or statistically insignificant. Over the longer term, price responsiveness is generally assumed to be larger because of possible technology responses.

An important influence on future travel may be the development of telecommunication technologies. In some instances, improved communication can substitute for travel as people can work at home or shop via the internet. In others, communication can help to increase travel by enabling friendships and working relationships to develop over long distances, and by permitting people to stay in touch with their homes and offices while traveling. To the extent that improvements in telecommunication technology stimulate the economy, they are likely to result in increased freight transport.

Energy intensity in the transport sector is measured as energy used per passenger-km for passenger transport and per ton-km for freight transport. Transport energy projections typically incorporate a reduction in fleet energy intensity in the range 0.5 to 2% per year (Grübler *et al.*, 1993b; IEA, 1993; Walsh, 1993). On-road energy intensity (fuel consumption per kilometer driven) of light-duty passenger vehicles in North America fell by nearly 2% per year between 1970 and 1990, to about 13 to 14 liters per 100 kilometers, but it is now stationary or rising. In other industrialized countries, changes in on-road fuel consumption from 1970 to the present were quite small. The average on-road energy intensity in North America was 85% higher than that in Europe in 1970, but only 25 to 30% higher by the mid-1990s (Schipper, 1996).

In some countries, such as Italy and France, where fleet average energy intensity has fallen during the past 20 years, the energy intensity of car travel (MJ/passenger-km) has increased as a result of declining car occupancy and the increasing use of more efficient diesel vehicles (Schipper *et al.*, 1993). However, conversion to diesel has been encouraged by low duties on diesel fuel relative to those on gasoline. The lower costs of driving diesel vehicles may have acted as a significant stimulus to travel by diesel car owners, and so offset much of the energy saved from their high-energy efficiency. A more recent trend, though, is toward higher energy intensity in new cars in countries such as the US, Germany, and Japan (IEA, 1993). Factors in the recent increases in energy intensity include the trend toward larger cars, increasing engine size, and the use of increasingly power-hungry accessories (Martin and Shock, 1989; Difiglio *et al.*, 1990; Greene and Duleep, 1993; IEA, 1993).

Average truck energy use per ton-km of freight moved has shown little sign of reduction during the past 20 years in countries for which data are available (Schipper *et al.*, 1993). Energy use is typically in the region 0.7 to 1.4 MJ/ton-km for the heaviest trucks but can be in excess of 5 MJ/ton-km for smaller trucks. In countries where services and light industry are growing faster than heavy industry, the share of small trucks or vans in road freight is increasing. Along with the increasing power-to-weight ratios of goods vehicles, these trends offset, and in some cases outweigh, the benefits of improved engine and vehicle technology (Delsey, 1991). Energy intensity tends to be lower in countries with large heavy-industry sectors, because a high proportion of goods traffic is made up of bulk materials or primary commodities.

Air traffic grew about three times as fast as GDP in the early 1970s, but only about twice as fast since the early 1980s. After allowing for the effects of continually falling prices, the elasticity of the 10-year average growth rate with respect to the 10-year average GDP growth is not much more than 1.0 (Michaelis, 1997a). Over the 30 years to 1990, the average energy intensity of the civil aircraft fleet fell by about 2.7% per year. The fastest reduction, of about 4% per year, was in the period 1974 to 1988. The large reductions in energy intensity during the 1970s and 1980s resulted partly from developments in the technology used for new aircraft in the rapidly expanding civil aircraft fleet and partly from increases in aircraft load factor (passengers per seat or percentage of cargo capacity filled). The aircraft weight load factor increased from 49% in 1972 to 59% in 1990, but nearly all of this rise occurred during the 1970s (ICAO, 1995a, 1995b).

Transport sector carbon intensities for personal travel, measured as the ratio of emissions to passenger-km traveled, increased in most European countries and Japan between 1972 and 1994. This increase resulted from falling load factors (persons per vehicle), which were greater than improvements in vehicle energy intensity. The only exception among industrialized countries was the US, where carbon intensities dropped from 55 kgC/passenger-km in 1972 to 46 kgC/passenger-km in 1994 (IEA, 1997c; Schipper *et al.*, 1997a). Carbon intensity of freight travel, measured as the ratio of emissions to ton-km transported, rose slightly in a number of industrialized countries between 1972 and 1994, mostly because of modal shifts to more carbon-intensive trucks (Schipper *et al.*, 1997b). As mobility increases in developing countries, transport emissions could rise dramatically. Ramanathan and Parikh (1999) indicate passenger traffic growth at 8% per year and train traffic growth at 5% per year for India. They found that efficiency improvements could reduce future energy demand by 26%. If, in addition, the modal split changes in favor of public transport modes, these authors estimate a 45% reduction in energy demand (Ramanathan and Parikh, 1999).

Fuels used to power transport are typically oil-based, except for rail, for which shifts toward electrified systems can lower carbon intensities depending upon the source of fuel for electricity generation in the country. In France, for example, the move toward electrified rail based on electricity generated by nuclear power led to lower carbon intensities (IEA, 1997c). Increased use of diesel engines can reduce CO_2 emissions, but leads to greater emissions of other gaseous pollutants, such as N_2O and carbon monoxide (CO). Use of alternative fuels, such as compressed natural gas, LPG, and ethanol, can significantly reduce CO_2 emissions from transport (IEA, 1995).

3.4.3. Energy Resources

3.4.3.1. *Fossil and Fissile Resources*

The term energy resource can be defined as "the occurrence of material in recognizable form" (WEC, 1995a) – it is essentially the amount of oil, gas, coal, etc., in the ground. In the IPCC WGII SAR (energy primer, see Nakićenović *et al.*, 1996) a further definitional distinction was made. Resources were defined as those occurrences considered "potentially recoverable with foreseeable technological and economic developments" and any additional amounts not considered as potentially recoverable were referred to as "occurrences." An energy reserve is a portion of the total, and depends on exploration to locate and evaluate a resource and on the availability of a technology to extract some of the resource at acceptable cost. Proved oil reserves, for example, are defined as "those quantities which geological and engineering information indicates with reasonable certainty can be recovered in the future from known reservoirs under existing economic and operating conditions" (BP, 1996). Thus, reserves can increase with exploration (new or better information), engineering advances (better economic and operating conditions), and higher prices (better economic conditions). In essence, reserves are "replenished" by shifting volumes from the resource into the reserve category. Reserves can also be depleted through production and can decrease with lower prices. Throughout this section the size of reserve and resource figures are expressed in EJ or ZJ (i.e. 10^{21} J, or 1000 EJ).

For SRES the fossil resource categorization used is reserves, resources, and additional occurrences. The definition of BP (1996) was adopted for reserves. Resources are those hydrocarbon occurrences with uncertain geologic assurance or that lack economic attractiveness. Finally, all other hydrocarbons that do not fall within the reserve and resource categories are aggregated in the category "additional occurrences" (i.e., occurrences that have a high degree of

Table 3-5: *Global fossil and fissile energy reserves, resources, and occurrences (in ZJ (10^{21}J)). Global and regional estimates are discussed in detail in Rogner (1997) and Gregory and Rogner (1998).*

	Consumption 1860–1990	Consumption 1990	Reserves Identified	Conventional Resources Remaining to be Discovered Low	Conventional Resources Remaining to be Discovered High	Recoverable with Technological Progress	Additional Occurrences
Oil							
Conventional	3.35	0.13	6.3	1.6	5.9		
Unconventional	—	—	7.1			9	>15
Gas							
Conventional	1.70	0.07	5.4	9.4	22.6		>10
Unconventional	—	—	6.9			20	>22
Hydrates	—	—					>800
Coal	5.20	0.09	22.9			80	>150
Total	10.25	0.29	48.6	>11.0	>28.5	>109	>987
Nuclear	0.21	0.02	2.0			>11	>1,000

geologic uncertainty, are not recoverable with current or foreseeable technology, or are economically unwarranted at present).

The assessment is summarized in Table 3-5. This account of fossil resources needs to be put in context with the long-run demand for these fuels and their relative production economics. It is the specific demand for these fuels that "converts" resources into reserves (Odell, 1997, 1998, 1999). Obviously, this is a dynamic process that, in addition to future demand trajectories, depends on advances in knowledge and technological progress. The discussion of oil reserves below applies to all hydrocarbon and nuclear resources.

In terms of exploration, the oil industry is relatively mature and the quantity of additional reserves that remain to be discovered is unclear . One group argues that few new oil fields are being discovered, despite the surge in drilling activity from 1978 to 1986, and that most of the increases in reserves results from revisions of underestimated existing reserves (Ivanhoe and Leckie, 1993; Laherrere, 1994; Campbell, 1997; Hatfield, 1997). Laherrere (1994) puts ultimately recoverable oil resources at about 10 ZJ (1800 billion barrels), including production to date. Adelman and Lynch (1997), while accepting some aspects in the propositions behind the pessimistic view of reserves, point to previous pessimistic estimates that have been wrong. They argue that "there are huge amounts of hydrocarbons in the earth's crust" and that "estimates of declining reserves and production are incurably wrong because they treat as a quantity what is really a dynamic process driven by growing knowledge." Smith and Robinson (1997) note improvements in technology, such as 3D seismic surveys and extended reach (e.g. horizontal) drilling, that have improved recovery rates from existing reservoirs and made profitable the development of fields previously regarded as uneconomic. Both of these increase reserves and lower costs. The various arguments and assessments are reviewed in greater detail in Gregory and Rogner (1998). To include all these views and to reflect uncertainty, future reserves availability cannot be represented by single numbers. Instead, a range of values that reflect the optimistic and pessimistic assumptions on extent and success rates of exploration activities, as well as the future evolution of prices and technology, needs to be considered for a scenario approach. To this end, the estimates of Masters *et al.* (1994) reflect the current state of knowledge as to the uncertainties in future potentials for conventional oil resources. These estimates assess conventional oil reserves at slightly above 6 ZJ, and a corresponding range of additionally recoverable resources between 1.6 and 5.9 ZJ. The figures include estimates of oil that is yet to be discovered.

In addition to conventional oil reserves and resources, oil shales, natural bitumen, and heavy crude oil, together called unconventional oil resources, have previously been defined as occurrences that cannot be tapped by conventional production methods for technical or economic reasons, or both (Rogner, 1996, 1997). In part these resources represent some of the huge amounts of hydrocarbons in the earth's crust that Adelman and Lynch (1997) refer to. Technologies to extract some of these resources competitively at current market conditions are now developed and production has started in countries such as Canada and Venezuela. Masters *et al.* (1987) put total recoverable resources of heavy and extra heavy crude oil at 3 ZJ, recoverable resources of bitumen at 2 ZJ, and ultimate resources of shale oil in place at 79 ZJ (they do not estimate the proportion of shale oil that might be recovered and hence give resources in place). The extent to which these unconventional resources might be defined as reserves in the future depends on the continued development of technologies to extract them at acceptable costs. Nakićenović *et al.* (1996) in IPCC WGII SAR assess all unconventional oil *reserves* at 7.1 ZJ, with an additional 20 ZJ of unconventional oil resources estimated to be recoverable with foreseeable technological progress.

Estimates of ultimately recoverable reserves of gas are less controversial than those for oil. Proved reserves are high, both in relation to current production (BP, 1996) and to cumulative production to date (Masters *et al.*, 1994). Masters *et al.* (1994) and Ivanhoe and Leckie (1993) note that gas discoveries need to be matched to an infrastructure for gas consumption, which is currently lacking in many parts of the world. Hence, exploration has been limited and the potential for discoveries of major quantities of gas in the 21st century is high. Estimates of gas reserves and resources are being revised continuously. The most up-to-date information is represented by the figures of the International Gas Union (IGU, 1997a), which give conventional gas reserves of 5.4 ZJ plus 9.4 ZJ additional reserves, including gas yet to be discovered. On the basis of IGU comments that some of their regional estimates of reserves are extremely conservative, Gregory and Rogner (1998) suggest an optimistic estimate for ultimately recoverable reserves of 28 ZJ (5.4 ZJ reserves plus 22.6 ZJ additional reserves, including quantities to be discovered), using the same ratio of optimistic to pessimistic reserves as Masters *et al.* (1994).

In addition to conventional reserves, reviews of the literature indicate very substantial amounts of unconventional gas occurrences. Rogner (1996, 1997) estimated resources in place for coal-bed methane (CH_4) of 10 ZJ, gas from fractured shale of 17 ZJ, tight formation gas of 7 ZJ, gas remaining *in situ* after production of 5 ZJ, and clathrates at some 980 ZJ. The magnitude of these estimates is also confirmed in IPCC WGII SAR (Nakićenović *et al.*, 1996), which gives 6.9 ZJ unconventional gas as current reserves, and an additional 20 ZJ as recoverable with current or foreseeable improvements in technologies. The largest resource occurrence of all fossil fuels (even exceeding coal) is estimated to be methane clathrates. Also called hydrates, methane clathrates represent gas locked in frozen ice-like crystals that probably cover a significant proportion of the ocean floor and have been found in numerous locations in continental permafrost areas. Technologies to recover these resources economically could be developed in the future, if demand for natural gas continues to grow in the longer run, in which case gas resource availability would increase enormously. The implications of such developments are considered in some of the SRES scenarios.

Coal reserves are different in character to oil and gas – coal occurs in seams, often covers large areas, and relatively limited exploration is required to provide a reasonable estimate of coal in place. Total coal in place is estimated at about 220–280 ZJ (WEC, 1995a; Rogner, 1996; 1997; Gregory and Rogner, 1998). Of this total, about 22.9 ZJ are classified as recoverable reserves (WEC, 1995a; 1998), over 200 times current production levels. The question is the extent to which additional resources can be upgraded to reserves. WEC (1995a; 1998) estimates additional recoverable reserves at about 80 ZJ, although it is not clear under what conditions these reserves would become economically attractive. Over 90% of their estimate of total reserves occur in just six countries, with 70% in the Russian Federation alone. Further coal resources are known to exist in various countries, some of which might be exploitable in the future, perhaps at high cost. However, in some countries the environmental damage from coal mining will prevent possible additional reserves being developed. In the IPCC WGII SAR, Nakićenović *et al* (1996) estimate that, in addition to today's reserves, a further 89 ZJ could, at least in principle, be mined with technological advances, a figure in agreement with the WEC (1998) estimates.

The picture for uranium and thorium reserves is different again. Current proved uranium reserves recoverable at less than US$130/kg amount to some 3.38 million tons (WEC, 1998) or 2 ZJ in once-through fuel cycles. This extractable thermal energy would be some 60 times larger if reprocessing and fast breeder reactors are used (Ishitani and Johansson, 1996). These reserves are sufficient to meet the needs of an expanded nuclear program well into the 21st century, even without reprocessing and fast breeder reactors. The ultimately recoverable global natural uranium resource base is currently estimated at around 29 million tons, which corresponds to 17 ZJ without reprocessing and about 1000 ZJ with reprocessing and fast breeder reactors (Nakićenović *et al.*, 1996). Additionally, very limited exploration for new reserves has occurred in recent years because of the relative abundance of existing known reserves and the drop of real uranium prices from US$150/kg in 1980 to about US$30/kg in 1996. The exploration and development of uranium deposits today is probably on a par with that of the oil industry 100 years ago, while thorium occurrences have hardly been assessed. Uranium and thorium are minerals contained in deposits in the Earth's crust and their long-term availability will be determined by the same process dynamics, in terms of knowledge and technology advances, as for their hydrocarbon counterparts. Once new reserves are required, given the comparison with the oil industry over the past 100 years, the potential for exploration to yield major discoveries at acceptable cost is enormous. From the perspective of occurrence alone, uranium resources are already known to be immense, especially if low-concentration sources such as seawater or granite rock are considered. In summary, the development of nuclear power throughout the 21st century, even based only on once-through reactors, is unlikely to be constrained by uranium (or thorium) resource limitations.

3.4.3.2. Renewable Resources

A review of medium-term (to 2025) and long-term (up to 2100) potentials of renewable energy is given in IPCC WGII SAR

Table 3-6: *Global renewable energy potentials for 2020 to 2025, maximum technical potentials, and annual flows, in EJ. Data sources: Watson et al., 1996; Enquete-Kommission, 1990.*[2]

	Consumption		Potentials by	Long-term Technical	Annual
	1860–1990	1990	2020–2025	Potentials	Flows
Hydro	560	21	35–55	>130	>400
Geothermal	–	<1	4	>20	>800
Wind	–	–	7–10	>130	>200,000
Ocean	–	–	2	>20	>300
Solar	–	–	16–22	>2,600	>3,000,000
Biomass	1,150	55	72–137	>1,300	>3,000
Total	1,710	76	130–230	>4,200	>3,000,000

(Nakićenović *et al.*, 1996) and shown in Table 3-6. A summary of the literature of renewable resource development potentials consistent with IPCC WGII SAR, including a detailed regional breakdown, is given in Christiansson (1995) and Neij (1997).

Hydropower currently provides some of the cheapest electricity available in the world, although the potential for new capacity is limited in some regions. WEC (1994, 1995a) estimates the gross world potential for hydroelectric schemes at about 144 EJ per year, of which about 47 EJ per year is technically feasible for development, about 32 EJ per year is economically feasible at present, and about 8 EJ per year is currently in operation. IPCC WGII SAR (Nakićenović *et al.*, 1996) gives a comparable medium-term potential of between 13 and 55 EJ, and a maximum technical potential above 130 EJ.

Other important renewable energy resources are wind and solar, as well as modern forms of biomass use. Biomass resources are potentially the largest renewable global energy source, with an annual primary production of 220 billion oven dry tons (ODT) or 4500 EJ (Hall and Rosillo-Calle, 1998). The annual bioenergy potential is estimated to be in the order of 2900 EJ, of which 270 EJ could currently be considered available on a sustainable basis (Hall and Rosillo-Calle, 1998). Hall and Rao (1994) conclude that the biomass challenge is not one of availability but of the sustainable management, conversion, and delivery to the market place in the form of modern and affordable energy services. It is also important to distinguish between harvesting and deforestation; the former results in afforestation, and the latter in conversion of forest land for other uses, such as agriculture or urban development.

The use of biomass as an energy source necessitates the use of land. Based on estimates by IIASA–WEC (1995), by 2100 about 690–1350 million hectares of additional land would be needed to support future biomass energy requirements for a high-growth scenario. However, the additional land requirement for agriculture is estimated to reach 1700 million hectares during the same period. These land requirements can be fulfilled if the potential additional arable land is taken into account (at present this is mostly covered by forest). Hence, land-use conflicts could arise, and particularly for Asia which is projected to require its entire potential of arable land by 2100. Africa and Latin America may have sufficient land to support an expanded biomass program. One estimate (WEC, 1994) shows that Africa can support the production of biomass energy equivalent to 115% of its current energy consumption (8.6 EJ).

Some authors stress that increased demand for bioenergy could compete with food production (Azar and Berndes, 1999). They note that the competitiveness between food and bioenergy production is not realistic in most energy–economy models; rather it is treated in an *ad hoc* fashion with the assumption that enough land is secured for food production. In reality an increasing competitiveness of bioenergy plantations may cause food prices to jump. Some developing regions, in particular Africa, are often assumed in scenarios to become major importers of food (Azar and Berndes, 1999).

Unlike hydropower, most of the technologies that could harness these renewable energy forms are in their infancy and are generally still high cost (although wind power is becoming increasingly competitive in some areas). Conversely, the potential for improvement in technical performance and costs is substantial. Thus, the future resource potential of these renewables is largely determined by advances in technologies and economics (discussed in Section 3.4.4).

Advances in renewable energy technologies could materialize to a significant extent even in the absence of climate policies, *albeit* conventional wisdom holds that such policies could accelerate their diffusion considerably. According to IPCC WGII SAR (Nakićenović *et al.*, 1996), in the medium-term (to 2025) the largest renewable energy potentials lie in the development of modern biomass (70 to 140 EJ), solar (16 to 22 EJ), and wind energy (7 to 10 EJ) as indicated in Table 3-6. In the long term the maximum technical energy supply potential for renewable energy is evidently solar (>2,600 EJ), followed by biomass (>1,300 EJ).

[2] All estimates, excluding biomass, have been converted into thermal equivalents with an average factor of 38.5%.

3.4.3.3. Conclusion

Comparatively few scenarios in the literature explicitly consider the interplay between resource availability and technological change, and hence the possibilities of wide-ranging alternative futures of fossil and renewable resource use. For fossil fuels, alternative resource development scenarios are described in Ausubel *et al.* (1988), Edmonds and Barns (1992), IIASA–WEC (1995), Nakićenović *et al.* (1998a), and Schollenberger (1998). For renewable energy resources, scenarios of enhanced resource development are described in Goldemberg *et al.* (1988), Johansson *et al.* (1993), Lazarus *et al.* (1993), Watson *et al.* (1996), and Nakićenović *et al.* (1998a).

A critical issue in the context of this report is how to capture alternative future interplays between energy technology and resource development, in contrast to the more traditional approach of assuming fixed resource quantities across all scenarios. This is important because the literature reviewed above indicates that resource availability can vary widely. For instance, generally oil and gas are considered the most constrained fossil fuel resources. Yet, a representative range from the literature gives cumulative production levels between 1990 and 2100 of between 21 and 65 ZJ, with typical intermediate scenarios of 30–35 ZJ (Nakićenović *et al.*, 1998a; Schollenberger, 1998). The extreme values of cumulative resource use are evidently inversely related between different energy sources across the range of alternative scenarios. For instance, in scenarios with high availability of oil and gas, typically the use of coal or renewable resources is more limited, whereas in scenarios of rapid development of renewable alternatives, the use of fossil resources is more limited. In other words, future resource availability of fossil fuels as well as renewables is *constructed* in scenarios based on current understanding and the available literature. Resource availability results from alternative policies and strategies in exploration, R&D, investments, and the resultant resource development efforts. The long lead-times and enormous investments involved result in such strategies yielding a cumulative effect, referred to in the technological literature as "lock-in" (see Section 3.4.4). Development of alternatives can be furthered, but it can also be blocked when policies and investments favor existing resources and technologies. The most important long-term issue is how the transition away from easily accessible conventional oil (and to a lesser extent conventional gas) reserves will unfold. Will it lead to a massive development of coal in the absence of alternatives or, conversely, to a massive development of unconventional oil and gas? Alternatively, could the development of post-fossil alternatives make the recourse to coal and unconventional oil and gas (such as methane clathrates) obsolete?

3.4.4. Energy Supply Technologies

3.4.4.1. Introduction

The recent literature on long-term energy and emission scenarios increasingly emphasizes that both resource availability and technology are interrelated and inherently dynamic (see, e.g., IIASA–WEC, 1995; Watson *et al.* 1996; Nakićenović *et al.*, 1998a). The state of the art of theories and models of technological change is reviewed in Section 3.4.5. The literature suggests that models of endogenous technological change are still in their infancy, and that no methodologies are established that reduce the substantial uncertainties with respect to direction and rates of change of future technology developments. Differences in opinions as to the likelihood and dynamics of change in future technologies will therefore persist. Such future uncertainties are best captured by adopting a scenario approach. The following discussion on changes in energy supply and end-use technologies therefore reviews the literature with emphasis on empirically observed historical and conjectured future changes. The principal message is that while the future is uncertain, the certainty is that future technologies will be different from those used today. Hence the most unlikely scenario of future development is that of stagnation, or absence of change.

3.4.4.2. Fossil and Fissile Energy Supply Technologies

Fossil-fueled power stations traditionally have been designed around steam turbines to convert heat into electricity. Conversion efficiencies of new power stations can exceed 40% (on a lower heating value basis – when the latent heat of steam from water in the fuel or the steam arising from the hydrogen content of the fuel has been excluded). New designs, such as supercritical designs that involve new materials to allow higher steam temperatures and pressures, enable efficiencies of close to 50%. In the long run, further improvements might be expected. However, the past decade or so has seen the dramatic breakthrough of combined cycle gas turbines (CCGTs). The technology involves expanding very hot combustion gases through a gas turbine with the waste heat in the exhaust gases used to generate steam for a steam turbine. The gas turbine can withstand much higher inlet temperatures than a steam turbine, which produces considerable increases in overall efficiency. The latest designs currently under construction can achieve efficiencies of over 60%, a figure that has been rising by over 1% per year for a decade. The low capital costs and high availability of CCGTs also make them highly desirable to power station operators. Gregory and Rogner (1998) estimate that maximum efficiencies of 71 to 73% are achievable within a reasonable period (on a lower heating basis; around 65 to 68% on a higher heating basis).

CCGTs can also be used with more difficult fuels, such as coal and biomass, by adding a gasifier to the front end to form an integrated gasification combined cycle (IGCC) power station. The gases need to be hot cleaned prior to combustion to avoid energy losses and this is one of the key areas of development. The added benefit is that coal flue gas desulfurization (FGD) becomes unnecessary as sulfur is removed before the combustion stage. In addition to FGD and IGCC, fluidized bed combustion (FBC) technology facilitates sulfur abatement (adding limestone during combustion to retain sulfur) and allows the utilization of low quality fossil fuels because of the high sulfur retention capacity. However, FGD reduces overall conversion efficiencies (i.e., increases CO_2 emissions). Also, both FGD and FBC sulfur abatement technologies use calcium carbonate to reduce emissions, which increases CO_2 emissions because of CaO liberation from $CaCO_3$ capture. The use of high sulfur coals in FBC requires high limestone consumption, which results in increased CO_2 emissions. Only IGCC reduces SO_2 and CO_2 emissions, because of the higher power generation efficiency.

Biomass is particularly suited to gasification. Stoll and Todd (1996) and Willerboer (1997) estimate that current designs for coal are around 9% less efficient than a standard CCGT burning gas. Developments to reduce heat losses through better heat recovery and by hot gas cleaning could potentially increase efficiency significantly in the next 10 to 15 years, and the technology can yield efficiencies of 51% now and perhaps 65% in the longer run.

The major potential competitor to CCGT technology is the fuel cell, which may be able to offer similar efficiencies at much lower plant sizes and so may be an ideal candidate for distributed combined heat and power generation. Another promising fuel cell application is vehicle propulsion. In contrast to other fossil-sourced electricity generation, fuel cells (similar to batteries) convert the chemical energy of fuels electrochemically (i.e., without combustion) into electricity and heat, and thus offer considerably higher conversion efficiencies than internal combustion engines. In the past, high costs and durability problems have restricted their use to some highly specialized applications, such as electricity generation in space. Recent advances in fuel cell technology, however, have led to their commercial production and application in niche markets for distributed combined heat and power production (Penner *et al.*, 1995).

Although operating internally on hydrogen, fuel cells can be fueled with a hydrocarbon fuel, such as natural gas, methanol, gasoline, or even coal. Before entering the fuel cells, these fuels would be converted on-site or on-board into hydrogen via steam reforming, partial oxidation, or gasification and hydrogen separation. In the longer run and to make fuel cells truly zero-emission devices, non-fossil derived pure hydrogen, supplied and stored as compressed gas, cryogenic liquid, metal hydrate, or other storage form, could replace hydrocarbon fuels. Current fuel cell conversion efficiencies (45 to 50%) have yet to approach their potentials; some designs report efficiencies as high as 60% electrical efficiency in simple systems and 74% total efficiency in hybrid fuel cell and gas combined cycle systems (FETC, 1997). Hydrogen production efficiencies range from 65 to 85% for fossil-based systems, 55 to 73% for biomass-based systems, and 80% to close to 90% for electrolysis (Ouellette *et al.*, 1995; Williams, 1998).

Nuclear power is a proven technology that provides 17% of global electricity supply. There is currently no consensus concerning the future role of nuclear power. While it stagnates in Europe and North America, it continues as a strong option in a number of Asian countries and countries undergoing economic reform. Economics and security of supply are considerations in the choice of nuclear power, along with its environmental trade-offs – on a full energy chain basis, from mining to waste disposal and decommissioning, nuclear power emits little GHG emissions.

Public opinion is opposed to the use of nuclear power in many countries because of concerns regarding operating safety, final disposal of high-level radioactive waste, and proliferation of weapon-grade fissile materials, as well as uranium mining and its environmental implications. These concerns, perceived or real, cannot be ignored if nuclear power is to regain the position of an accepted technology. In addition, changed market conditions call for new reactor technologies – smaller in scale, reduced construction periods, and improved economics with no compromise in safety. The industry is striving continuously to develop advanced reactor designs of much lower cost and with inherent safety concepts (i.e., designs that make safety less dependent on specifically activated technology components and human performance). The latest reactor technology already prevents the release of fission products or health-damaging radiation to the environment even under highly unlikely severe accident conditions.

The 1998 update of the OECD cost comparison for power plants expected to be commercially available around 2005 reports nuclear generating cost in the range 2.5 to 9.0 c/kWh (US cents per kilowatt-hour), depending on location, type of reactor, plant factor, plant life time, discount rate, and underlying fuel price escalation (OECD, 1998c). Reconciliation of the assumptions of the OECD report with those used in the IPCC WGII SAR leads to an almost identical cost range of 2.9 to 5.4 c/kWh, at 5% discount rate, and 7.7 c/kWh, at 10% discount rate (Ishitani and Johansson, 1996).

3.4.4.3. Renewable Energy Technologies

Since the birth of the modern wind power industry in the mid-1970s, it has seen a continuous chain of innovations and cost reductions. Christiansson (1995) discusses a learning curve that relates cost reductions to installed capacity. She notes that the experience curve for the USA indicates a progress ratio of 0.84 (i.e., for each doubling of installed capacity, costs of new installations are reduced to 0.84 of the previous level). In the UK, developers have the opportunity to bid to supply renewable electricity to the grid under the Non-Fossil Fuel Obligation. In 1998, the average bid price for large wind power schemes was about 5 c/kWh, nearly 20% lower than in 1996 (ENDS, 1998). The bids are made on a full commercial cost basis and they indicate that wind energy is cheaper than most of its competitors for new schemes in the UK for modest increments of capacity. (The exception is gas-powered CCGT generation.) However, wind, by nature, is intermittent and back-up capacity will be required as the proportion of electricity provided by wind increases, which reduces its economic attractiveness. Large integrated electricity systems and systems that contain significant amounts of hydropower (especially with storage) are able to cope with the fluctuations in wind power best (Grubb and Meyer, 1993; Johansson *et al.*, 1993).

Solar voltaic power is on a similar learning curve to wind, with progress ratios of 0.82 in the USA and 0.81 in Japan (Christiansson, 1995; Watanabe, 1997). Various alternative technologies are being developed and solar voltaics can be envisaged as providing electricity through large arrays of cells in central power generation, through arrays built into cladding or roofing on buildings, or through single arrays to meet specific purposes. In this last form, solar cells have already established themselves as economic and reliable power sources in the provision of light, clean water, and improved health services to isolated rural communities (WEC, 1994). Based on the principle of a learning curve, solar cell costs are expected to fall as capacity builds. For example, the US Department of Energy projects costs to fall from 38–55 c/kWh in 1995 to 3.5–5 c/kWh in 2030 (US DOE, 1994). For another system, WEC estimates costs to fall from 13–23 c/kWh in 1990 to 5–10 c/kWh in 2020 (WEC, 1994).

Biomass, particularly wood, has been the main initial source of energy as countries develop and remains a substantial energy source in many developing countries (WEC, 1994). In addition to this, byproducts of agriculture and forestry can be useful sources of energy. The extent to which biomass can contribute beyond this to provide energy crops for use in, say, power stations or for conversion into liquid fuels depends firstly on the competition for land with agriculture for food production. This, in turn, depends on improved productivity in food production, the amount and type of meat and other animal (e.g., dairy) products in the diet, and the growth in human population. The future contribution of biomass will also depend on increased productivity in biomass production coupled with limitations in energy input to grow, harvest, and use energy crops (Ishitani and Johansson, 1996; Leemans, 1996). Current costs of biomass (for a eucalyptus plantation in Uruguay) are typically put at US$1.8/GJ with projected costs of US$1.4/GJ and productivity at 360GJ/hectare per year (Shell International Ltd., 1996). Agricultural productivity is also growing at 2% per year with the prospects of a similar productivity gain for energy crops (Shell International Ltd., 1996). WEC (1994) notes that Brazil has had a program to grow sugar cane to produce transport fuel since the mid 1970s and that production has reached 62% of the country's needs; this may be one option for wider use as the availability of conventional oil starts to fall.

Finally, other forms of renewable energy, such as geothermal energy, tidal energy, wave energy, ocean thermal energy conversion, and solar thermal power plants, could make significant contributions at some stage in the future, as geothermal energy already does in specific markets. Also, technologies and processes that lead to carbon sequestration as a by-product include enhanced oil recovery using CO_2 to improve the viscosity of crude oil or reforestation for reasons of soil preservation (Ishitani and Johansson, 1996).

3.4.4.4. Combined Heat and Power Production (Cogeneration)

The fuel effectiveness of all energy conversion processes that involve combustion, but also of fuel cells, can be raised substantially by combined heat and power generation. Utilities located in the vicinity of urban areas may divert the waste heat from combustion for residential or commercial heating purposes. Industrial producers of high-temperature process heat may consider the generation of electricity when process temperature requirements are lower than the temperatures supplied. Combined heat and power production can accomplish fuel utilization rates of 90% or more (Ishitani and Johansson, 1996).

3.4.5. Understanding and Modeling Technological Change

The future direction and rates of technological change are uncertain and therefore need to be explored when developing a *range* of alternative futures (i.e., scenarios). However, it would be misleading to resort to simplistic parametric variations of scenario assumptions without considering some basic elements of the nature of technological change, briefly reviewed here.

Technological change has often been pictured in linear terms that involve several sequential steps:

- Scientific discovery – an addition to knowledge.
- Invention – a tested combination of already existing knowledge to a useful end.
- Innovation – an initial and significant application of an invention.
- Improvement of technology characteristics and reduction of costs.
- Spread of an innovation, usually accompanied by improvement.

However, this model places undue emphasis on the role of basic R&D and scientific knowledge as precursors and determinants of innovation. It also understates the role of interactions among different actors and between the five functions listed above. The emphasis in recent innovation literature is placed more on a "chain-link" model of innovation, exploiting interactions between firms' R&D departments, and various stages of production and marketing (Dosi, 1988; Freeman, 1994). Lane and Maxfield (1995) emphasize the role of "generative" relationships in creativity.

Technological change is linked to the economic and cultural environment beyond the innovating firm in many ways, as described by Landes (1969), Mokyr (1990), Rosenberg (1982, 1994, 1997), Rostow (1990), and Grübler (1998a). Innovations are highly context-specific; they emerge from local capabilities and needs, evolve from existing designs, and conform to standards imposed by complementary technologies and infrastructure. Successful innovations may spread geographically and also fulfill much broader functions. The classic example is the steam engine, developed as a means of pumping water out of deep mines in Cornwall, England, but to become the main source of industrial motive power and the key technology in the rail revolution worldwide.

Numerous examples can be used to demonstrate the messiness, or complexity, of innovation processes (e.g., Grübler, 1998a; Rosenberg, 1994). But even if the innovation process is messy, at least some general features or "stylized facts" can be identified (Dosi, 1988; Grübler, 1998a):

- The process is fundamentally uncertain: outcomes cannot be predicted.
- Innovation draws on underlying scientific or other knowledge.
- Some kind of search or experimentation process is usually involved.
- Many innovations depend on the exploitation of "tacit knowledge" obtained through "learning by doing" or experience.
- Technological change is a cumulative process and depends on the history of the individual or organization involved.

These five features render some individuals, firms, or countries better at innovation than others. Innovators must be willing and able to take risks; have some level of underlying knowledge; have the means and resources to undertake a search process; may need relevant experience; and may need access to an existing body of technology. Many of these features introduce positive feedback into the innovation process, so that countries or firms that take the technological lead in a market or field can often retain that lead for a considerable time.

Technological change may be supply driven, demand driven, or both (Grübler, 1998a). Some of the most radical innovations are designed to respond to the most pressing perceived needs. Many technologies have been developed during wartime to address resource constraints or military objectives. Alternatively, some innovation (e.g., television) is generated largely through curiosity or the desire of the innovator to meet a technical and intellectual challenge. Market forces (including those anticipated in the future) can act as a strong stimulus for innovation by firms and entrepreneurs aiming either to reduce costs or to gain market share. For example, Michaelis (1997a) shows the strong relationship between fuel prices and the rate of energy efficiency improvement in the aviation industry; Michaelis (1997b) also discusses the effects of the introduction of competition on the organizational efficiency of the British nuclear industry.

All innovations require some social or behavioral change (OECD, 1998a). At a minimum, changes in production processes require some change in working practices. Product innovations, if they are noticeable by the user, demand a change in consumer behavior and sometimes in consumer preferences. Some product innovations – such as those that result in faster computers or more powerful cars – provide consumers with more of what they already want. Nevertheless, successful marketing may depend on consumer acceptance of the new technology. Other innovations – such as alternative fuel vehicles or compact fluorescent lights – depend on consumers accepting different performance characteristics or even redefining their preferences. An important perspective on technical change is that of the end-user or consumer of products and services. Technology can be seen as a means of satisfying human needs. Several conceptual models have been developed to describe needs and motivation, although their empirical foundations are weak (Douglas *et al.*, 1998; Maslow, 1954; Allardt, 1993). In many cases, a given technology helps to satisfy several different types of need, particularly evident in two of the most significant areas of energy use: cars and houses. This tendency of successful technologies to serve multiple needs contributes to lock-in by making it harder for competing innovations to replace them fully. Hence, many attempts to introduce new energy efficient or alternative fuel technologies, especially in the case of the car, have failed because of a failure to meet *all* the needs satisfied by the incumbent technology. Different individuals may interpret the same fundamental needs in different ways, in terms of the technology attributes they desire (OECD, 1996). Deep-seated cultural values or "metarules" for behavior can be considered to be filtered through a variety of influences at the societal, community, household, and individual level (Douglas *et al.*, 1998; Strang, 1997). Commercial marketing of products usually aims to adjust the filters, and encourages people to associate their deep-seated values with specific product attributes (Wilhite, 1997). These associations are likely to be more flexible than the values themselves, and provide a potential source of future changes in technology choice.

Technology diffusion is an integral part of technical change. Uptake of a technology that is locally "new" can be viewed as an innovation. Often, when technology is adopted it is also adapted in some way, or used in an original way. Just as technology development is much more complicated than the simple exploitation of scientific knowledge, Landes (1969), Wallace (1995), Rosenberg (1997), and others emphasize that technology diffusion is highly complex. Wallace emphasizes the importance of an active and creative absorption process in the country that takes up the new technology. The implication of this complexity is that no general rules define "what works." The process of technology adoption is as context dependent as that of the original innovation. Rosenberg (1997) also emphasizes the role of movements of skilled people in the diffusion of technology. Transnational firms often play a strong role in such movements. Other factors that influence the technology transfer process include differences in economic developmental, social and cultural processes, and national policies, such as protectionist measures.

Grossman and Helpman (1991), Dosi *et al.* (1990), and others have attempted to capture some of the complexities in "new growth" and "evolutionary" economic models. They have been able to demonstrate the flaws in some of the simpler solutions to technology diffusion often advocated – for example, they show how free trade might sometimes exacerbate existing gaps in institutions, skills, and technology.

The complex interactions that underpin technology diffusion may give rise to regularities at an aggregate level. The geographical and spatial distribution of successive technologies displays patterns similar to those found in the succession of biological species in ecosystems, and also in the succession of social institutions, cultures, myths, and languages. These processes have been analyzed, for example, in Campbell (1959), Marchetti (1980), Grübler and Nakićenović (1991), and Grübler (1998a). An extensive review of the process of international technology diffusion is available in the IPCC Special Report on *Methodological and Technological Issues in Technology Transfer* (IPCC, 2000). That report provides a synthesis of the available knowledge and experience of the economic, social and institutional processes involved.

Many attempts to endogenize technical change in economic models rely on a linear approach in which technical change is linked to the level of investment in R&D (e.g., Grossman and Helpman, 1991, 1993). More importantly, this linear model has been the basis of many governments' strategies for technological innovation. As mentioned above, important additional features of technological change include uncertainty, the reliance on sources of knowledge other than R&D, "learning by doing" and other phenomena of "increasing returns" that often lead to technological "lock in" and hence great difficulties in introducing new alternatives.

These features can be captured to some degree in models and a great deal of experimentation has taken place with different model specifications. However, the first feature, uncertainty, means that models cannot be used to predict the process of technical change. This uncertainty stems partly from lack of knowledge – the outcomes of cutting-edge empirical research simply cannot be predicted. It also stems from the complexity of the influences on technological change, and in particular the social and cultural influences that are extremely difficult to describe in formal models. Recent attempts to endogenize technical change in energy and economic models are reviewed by Azar (1996). Optimization models usually treat technology *development* as exogenous, but technology *deployment* as endogenous and driven by relative technology life-cycle costs. A few GHG emission projection models (e.g., Messner, 1997) were developed to incorporate "learning by doing" – the reduction in technology costs and improvement in performance that can result from experience (Arrow, 1962). Models have also been developed that explicitly include technological uncertainty to analyze robust technology policy options (e.g., Grübler and Messner, 1996; Messner *et al.*, 1996). Other models developed more recently incorporate the effects of investment in knowledge and R&D (Goulder and Mathai, 1998). Economists and others who study technological change have developed models that take a variety of dynamics into account (Silverberg, 1988). Some models focus on technologies themselves, for example examining the various sources of "increasing returns to scale" and "lock-in" (Arthur, 1989, 1994). Other models focus on firms and other decision-makers, and their processes of information assimilation, imitation, and learning (Nelson and Winter, 1982; Silverberg, 1988; Andersen, 1994). Few of these dynamics, apart from "increasing returns to scale," have been applied to the projection of GHG emissions from the energy sector.

3.5. Agriculture and Land-Use Emissions

3.5.1. Introduction

The most important categories of land-use emissions are CO_2 from net deforestation, CH_4 from rice cultivation, CH_4 from enteric fermentation of cattle, and N_2O from fertilizer application. These sources account for nearly all the land-use emissions of CO_2 (Schimel *et al.*, 1995), about 53% of the land-use emissions of CH_4 (Prather, *et al.*, 1995), and about 80% of land-use emissions of N_2O (Prather, *et al.*, 1995). These estimates, however, have a high uncertainty. Measurements and analyses of other sources of CH_4 and N_2O (notably biomass burning, landfills, animal waste, and sewage) are relatively rare, but increasing (Bogner *et al.*, 1997, in the literature. Of the scenarios reviewed for this report (see Table 3.7), about 20 address emissions from agriculture and land-use change (Lashof and Tirpak, 1990; Houghton, 1991; Leggett *et al.*, 1992; Matsuoka and Morita, 1994; Alcamo *et al.*, 1998; Alcamo and Kreileman, 1996; Leemans *et al.*, 1996).

Current assessments of GHG emissions indicate that land use or land cover activities make an important contribution to the concentration of GHGs in the atmosphere;[3] these are referred to as "land-use emissions" in this report.[4] Of the three most important GHGs, the contribution of land-use emissions to total global CO_2 is relatively small (23%), but it is very large for CH_4 (74%) and N_2O. Furthermore, although land-use emissions make up only a small percentage of global CO_2 emissions, they comprise a large part (45%) of CO_2 emissions from developing countries, and an even larger percentage of their total CH_4 (78%) and N_2O (76%) emissions (Pepper *et al.*,1992). Hence, from a variety of perspectives, the contribution of land-use emissions to total emissions of GHGs is important, and consequently their future trends are relevant to the estimation of climate change and its mitigation.

[3] These activities include deforestation, afforestation, changes in agricultural management, and other anthropogenic land-use changes that result in a net flow of GHGs to or from the atmosphere. They exclude natural biogenic emissions and emissions that are not related to anthropogenic activity such as CO_2 from volcanoes or volatile organic compounds from forests.

[4] We include deforestation in this category of emissions even though this is a process of land-cover change rather than a land-use activity.

Table 3-7: *Overview of scenarios presented in Section 3.5.*

Scenario Number	Scenario Identification	Type(*)	Reference
1-1	IS92a IPCC 1992	G R	Leggett *et al.* (1992)
1-2	IS92b IPCC 1992	G R	Leggett *et al.* (1992)
1-3	IS92c IPCC 1992	G R	Leggett *et al.* (1992)
1-4	IS92d IPCC 1992	G R	Leggett *et al.* (1992)
1-5	IS92e IPCC 1992	G R	Leggett *et al.* (1992)
1-6	IS92f IPCC 1992	G R	Leggett *et al.* (1992)
1-7	IS92 S1 : IPCC 1992 Sensitivity 1 (High Deforestation, High Biomass)	G R	Leggett *et al.* (1992)
1-8	IS92 S4 : IPCC 1992 Sensitivity 4 (Halt Deforestation, High Plantation)	G R	Leggett *et al.* (1992)
2-1	Baseline A IMAGE 2.1	G R	Alcamo, *et al.* (1996)
2-2	Baseline B IMAGE 2.1	G R	Alcamo, *et al.* (1996)
2-3	Baseline C IMAGE 2.1	G R	Alcamo, *et al.* (1996)
2-4	Less B1 Changed Trade	G R	Leemans, *et al.* (1996)
2-5	Less B1 No Biofuels	G R	Alcamo and Kreileman (1996)
2-6	Stab 350 All	G R	Alcamo and Kreileman (1996)
3-1	AIM, Asian Pacific Integrated Model Land use emission scenario	G R	Matsuoka and Morita (1994)
7-1	EPA-SCW EPA (Slowly Changing World)	G R	Lashof and Tirpak (1990)
7-2	EPA-RCW EPA (Rapidly Changing World)	G R	Lashof and Tirpak (1990)
7-3	EPA-High Reforestation EPA (Halt Deforestation, High Reforestation)	G R	Lashof and Tirpak (1990)
8-1	H1 Houghton-Population	G	Houghton (1991)
8-2	H2 Houghton-Exponential Extrapolation	G	Houghton (1991)

* G = global, R = regional.

3.5.2. *Carbon Dioxide Emissions from Anthropogenic Land-Use Change*

A variety of changes in land use can result in anthropogenic CO_2 emission or absorption. These changes most obviously include permanent deforestation or afforestation. However, many changes in land-management practices also contribute to CO_2 fluxes because of changes in standing biomass densities or in soil carbon. Empirical studies of such CO_2 fluxes are rare, so that information on current emissions is very poor. Whereas comprehensive information exists for forests globally, only a few countries have detailed information on forest and agricultural land-management practices. Hence, global estimates of CO_2 emissions from anthropogenic land-use change, including those in the SRES, tend to be based entirely on net deforestation–afforestation and on average figures for carbon storage per hectare in forests.

Emissions of CO_2 from deforestation arise mostly from the burning of trees and other vegetation in tropical forests cleared for agricultural use. These emissions also stem from the decomposition of trees harvested for lumber, the burning of wood for fuel, and soil respiration. If harvested wood is replaced by new seedlings, it is normally assumed that the amount of CO_2 released by decomposition or burning is compensated by the CO_2 taken up during growth of the seedlings and, therefore, that the net emissions of harvested trees is zero. Where net afforestation occurs, net emissions are taken to be negative (i.e. afforestation acts as a sink).

As a consequence of inconsistencies in base-year estimates for net changes in forest biomass, emission estimates are normalized relative to their 1990 value before comparison with each other (Figure 3-14). Figure 3-14 also clearly depicts the relative change of emissions with time (Alcamo, *et al.*, 1995; Nakićenović *et al.*, 1998b).

The scenarios of CO_2 emissions from land-use change have quite different temporal paths, and show their widest range before the middle of the 21st century (Figure 3-14). Nearly all the scenarios then converge to very low emissions by the end of the century. At their widest point, the scenarios span about a factor of 14.

The different sets of scenarios can be grouped into roughly two typical paths: One set declines smoothly after 1990, while the other sharply increases for a few decades after 1990. After the middle of the 21st century most scenarios stabilize or continue to decline because either the driving forces of deforestation equilibrate or because forests are depleted. These processes are discussed further below. By 2100, CO_2 scenarios of

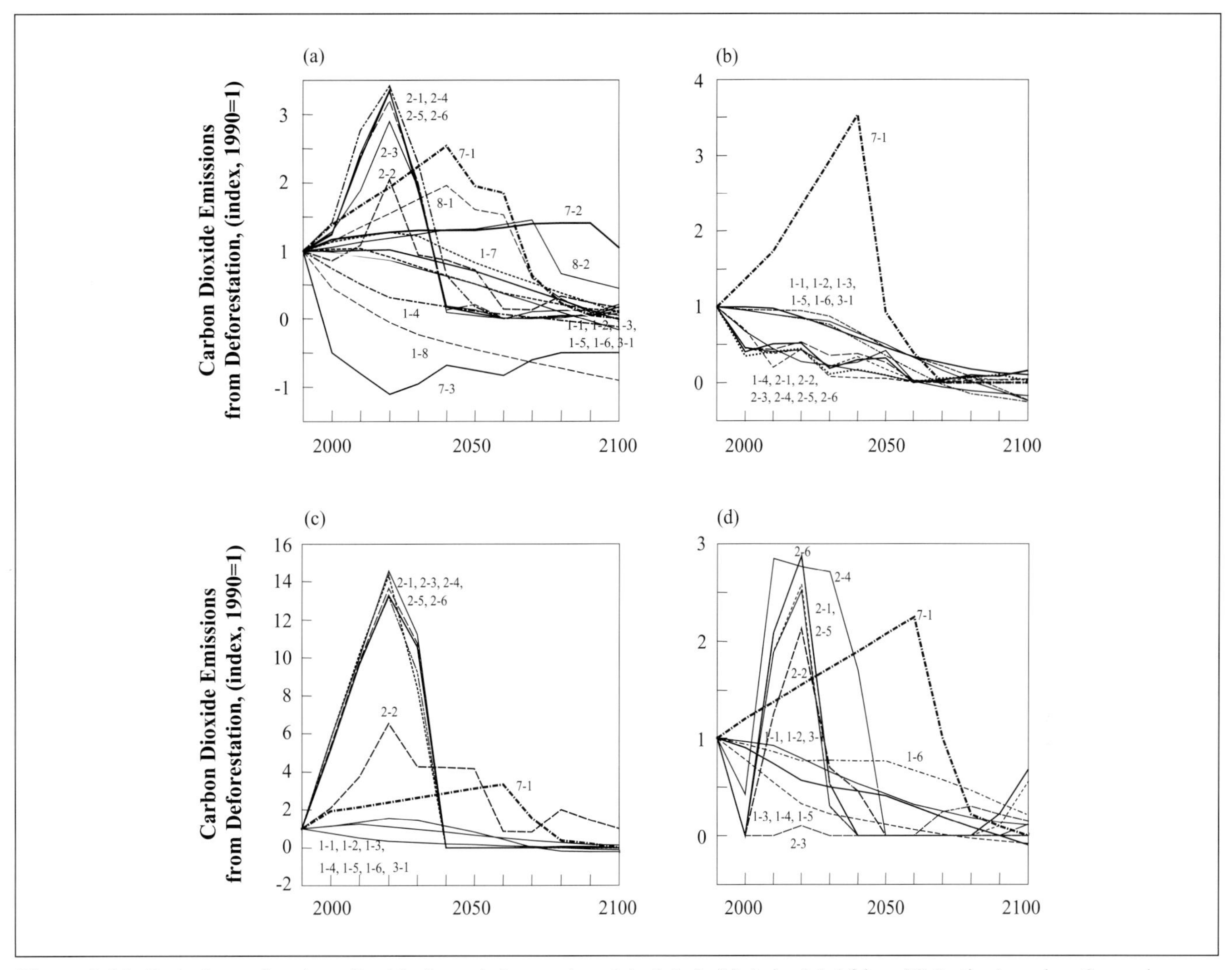

Figure 3-14: Emissions of carbon dioxide from deforestation: (a) global, (b) Asia, (c) Africa, (d) Latin America. Scenarios numbers are given in Table 3-7.

deforestation range from zero to 30% of their base-year estimates.

As emissions from deforestation are more significant in developing regions than industrial regions, few scenarios of CO_2 from deforestation are available for industrial regions, so the focus here is on developing regions only. For Asia, emissions in all but one scenario (EPA-SCW, 7-1 in Table 3-7) steadily decline after the base year, and reach 25% or less of their 1990 value in 2100 (Figure 3-14b). For Africa, emissions first increase before eventually decreasing in 2100 to a small fraction of their 1990 value (Figure 3-14c). In this scenario, deforestation rates decline in Asia and Africa because of the depletion of their forests. For Latin America, the wide range of scenarios reflects the wide range of views about its future rates of deforestation (Figure 3-14d).

One of the main factors that affects estimates of CO_2 from deforestation is the assumed deforestation rate, which is estimated by a wide variety of methods. For example, the IS92a–IS92f scenarios assume that deforested area is proportional to population, with a time lag of 25 years, and that deforestation continues until 25 years after the population stabilizes or until forests are exhausted (Leggett *et al.*, 1992). The IMAGE 2.1 emission estimates are based on computed changes in global land cover, which take into account changing demand for agricultural commodities. Trexler and Haugen (1995) compute the rate of tropical deforestation on a country-by-country basis, and include information from questionnaires. Jepma (1995) uses a combination of three models (a socio-economic model, a wood demand–supply model, and a land-use model), while Palo *et al.* (1997) correlate deforestation rates with income levels. To further confound this situation, the factors that affect deforestation vary greatly from place to place, and therefore need to be defined as locally as possible.

Figure 3-15 presents population assumptions of various deforestation scenarios, together with the assumed or implied deforestation rates on a per capita basis. The shape of the emission curves (Fig. 3-14) follows the shape of the deforestation rate curves, rather than that of the population assumptions. However, the range of emissions is much larger

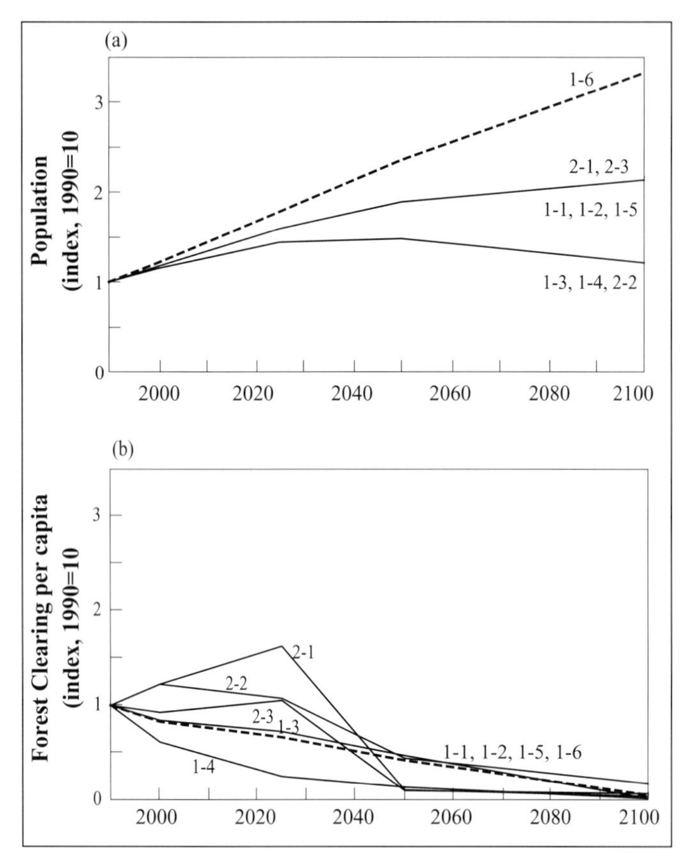

Figure 3-15: Selected driving forces of deforestation: (a) Population assumptions, and (b) Forest clearing rate per unit population. Both are normalized. Scenarios numbers are given in Table 3-7.

than the range of deforestation rates. Some conclusions can be drawn from these data:

- When given on a per capita basis, the range of deforestation rates is perhaps not as great as expected (about a factor of eight) considering the wide range of assumptions and methods used to estimate these rates.
- The temporal trend of emissions arises from the temporal trend of deforestation estimates.
- The range of emissions is strongly, but not only, influenced by the deforestation rates, and at its widest (a factor 14) can be better explained by the combined range of per capita deforestation rates, population, and carbon density.

The role that assumed carbon density of vegetation plays in the estimation of CO_2 emissions must be considered. To estimate CO_2 emissions, it is necessary to estimate the carbon density of vegetation that is burned and decayed when forests are cleared. However, there is wide disagreement in the literature as to typical values of carbon density because of both the many methods used to measure density and the wide variation in the mix of tropical forest vegetation.

3.5.3. Methane Emissions from Rice Production

CH_4 emissions from rice paddies are an important emission category, and are reviewed in more detail in Wassmann *et al.* (1993, 1997, 1998), Houghton *et al.* (1995), and Olivier *et al.* (1996). CH_4 emissions are primarily a function of emission factors and assumed rice cropland area. In turn, emission factors depend on cultivation method (wet versus dry cultivation), water management practices, type of rice variety planted, and cropping patterns. Most long-term scenarios assume that the emission factor of CH_4 per unit area of rice cropland remains constant with time, although estimates vary greatly from one scenario reference to another.

The many approaches used to estimate the future extent of rice fields result in increases in the global area of rice fields from a factor of 0.8 to one of 1.8 by 2100 (see Alcamo and Swart, 1998, for a brief review). One of the main factors to affect the future area of rice cropland is the assumed long-term improvement in rice productivity. The typical range of estimates for this variable is between 1.0 and 1.6% per year, depending on the region, time horizon, and reference.

3.5.4. Methane Emissions from Enteric Fermentation

Estimated emissions of CH_4 from enteric fermentation depend on assumptions about emission factors per animal and the number of livestock. As summarized by IPCC (1995), emission factors vary greatly depending on the type of cow, their feed regime, and their productivity. Assumptions for the change in meat production from 1990 to 2100 in existing scenarios vary greatly, by a factor of 1.2 to 4.2. Despite the wide range of assumptions about meat production, emissions of the various scenarios do not vary by more than a factor of two, which indicates that other assumptions (e.g., animal productivity) must compensate for the differences in assumed meat production.

As noted above, global estimates can mask significant differences in assumptions about industrial and developing regions. For industrial regions, nearly all scenarios assume a decline in beef production per capita, which is consistent with the current shift away from the consumption of beef to poultry and other protein sources. Meanwhile, the scenarios for developing countries assume a continuing increase in beef consumption, which grew by 3.1% per year between 1982 and 1994, leading to an overall growth of 1.1% per year globally (Rosegrant *et al.*, 1997).

Another factor that influences the future number of livestock is the change in animal productivity, that is, the weight of meat or dairy product per animal. The rate of increase in beef productivity dropped in industrial countries from 1.25% per year in 1967–1982 to 0.69% in 1982–1994, but increased from 0.11% per year to 0.61% per year in the developing countries. Similar to emissions from rice fields, emissions from livestock are influenced not only by number of livestock (equivalent to the extent of rice area), but also by changes in the productivity of animals as these alter the CH_4 emission factor.

Some authors doubt that assumed increases in meat production and animal productivity can be sustained indefinitely. For example, Brown and Kane (1995) argue that livestock production cannot be increased greatly because nearly all of the world's suitable rangelands are intensively exploited already. They claim that the rapidly growing demand for meat and dairy products can only be met by livestock production in feedlots, which would result in a rising demand for feed that requires further development of agricultural land and further GHG emissions.

3.5.5. Nitrous Oxide Emissions from Agriculture

N_2O budgets are associated with considerable uncertainties. Agricultural activities and animal production systems are the largest anthropogenic sources of these emissions. Recent calculations using IPCC 1996 revised guidelines indicate that N_2O emission from agriculture is 6.2 MtN as N_2O per year (IPCC, 1996; Mosier *et al.*, 1998). About one-third is related to direct emissions from the soil, another third is related to N_2O emission from animal waste management, and the final third originates from indirect N_2O emissions through ammonia (NH_3), nitrogen oxides (NO_x), and nitrate losses. This compares to earlier estimates of total anthropogenic emissions that range between 3.7 and 7.7 MtN (Houghton *et al.*, 1995). Industrial sources contribute between 0.7 to 1.8 MtN (Houghton *et al.*, 1995; see also Chapter 5, Table 5-3 and Section 3.6.2).

Total natural emissions amount to 9.0 ± 3.0 MtN as N_2O, so oceans, tropical, and temperate soils are together the most important source of N_2O today. Atmospheric concentrations of N_2O in 1992 were 311 parts per billion (10^9) by volume (ppbv) (Houghton *et al.*, 1995); the 1993 rate of increase was 0.5 ppbv, somewhat lower than that in the previous decade of approximately 0.8 ppbv per year (Houghton *et al.*, 1996).

Among the anthropogenic sources, cultivated soils are the most important, contributing 50 to 70% of the anthropogenic total (see Chapter 5, Table 5-3). This source of N_2O is particularly uncertain as the emission level is a complex function of soil type, soil humidity, species grown, amount and type of fertilizer applied, etc. The second largest anthropogenic source of N_2O is industry; two processes account for the bulk of industrial emissions – nitric acid (HNO_3) and adipic acid production. In both cases N_2O is released with the off-gases from the production facilities. Recently, N_2O release from animal manure was identified as another significant source of N_2O emissions.

N_2O emissions from agricultural soils occur through the nitrification and denitrification of nitrogen in soils, particularly that from mineral or organic fertilizers. Emissions are very dependent on local management practices, fertilizer types, and climatic and soil conditions, and are calculated by multiplying an emission factor by the sum of mineral and organic nitrogen applied as fertilizer. The emission factor depends on the fertilizer type and local environmental circumstances, and those used in IPCC (1996) result in an assumed loss of 1.25% (range 0.25 to 2.25%) of nitrogen as N_2O per year.

To estimate the trend in fertilizer use, different references employ different approaches. For example, Leggett *et al.* (1992) directly estimate the amount of fertilizer used, whereas Alcamo *et al.* (1996) back-calculate fertilizer use from the future amount of agricultural land. Despite these different approaches, estimates of future fertilizer use are quite consistently given as an increase by about a factor 1.4 to 2.8 between 1990 and 2100.

Although the different references are consistent in their findings about future global fertilizer use, the question arises whether these are at all reasonable guesses. Some researchers assume that fertilizer use will increase even more. For example, Kendall and Pimentel (1994) in their "business-as-usual" scenario assume a 300% increase in the use of nitrogen and other fertilizers by 2050. Moreover, most studies of future world food production assume improvements in crop yield. These yield improvements may imply higher overall rates of fertilizer use because many high-yielding crop varieties depend on large amounts of fertilizer.

However, some authors question whether global average fertilizer use will grow. For example, Brown and Kane (1995) note that world fertilizer use has actually fallen in recent years and Kroeze (1993) assumes that per capita N_2O emissions from fertilizer consumption decrease by 50% in 2100 relative to 1990 through policies that promote the more efficient use of synthetic fertilizers. Future fertilizer use may also be lower than in the "business-as-usual" scenarios because farmers have other incentives to reduce nitrogen fertilizer use, such as to reduce farming costs and avoid nitrate contamination of groundwater.

This brief review of the literature on prognoses of fertilizer use indicates that the N_2O emission scenarios depicted in Figure 3-16 do not take into account the full range of views about future trends in fertilizer use. Additional uncertainty in future emissions occurs because changes in the number of livestock, as discussed above for CH_4 emissions, and animal husbandry practices will also affect N_2O emissions.

3.5.6. Findings Regarding Driving Forces

Herein, some of the many specific factors that affect scenarios of land-use emissions have been discussed. From a correlation analysis that compared the influence of changing population, economic activity, and technological change on land-use emission scenarios, Alcamo and Swart (1998) concluded that population was the most influential driving force. The reason is the relationship between population and increasing food demand, which leads to more cows that produce CH_4 and more extensive fertilized croplands that release N_2O. Although most

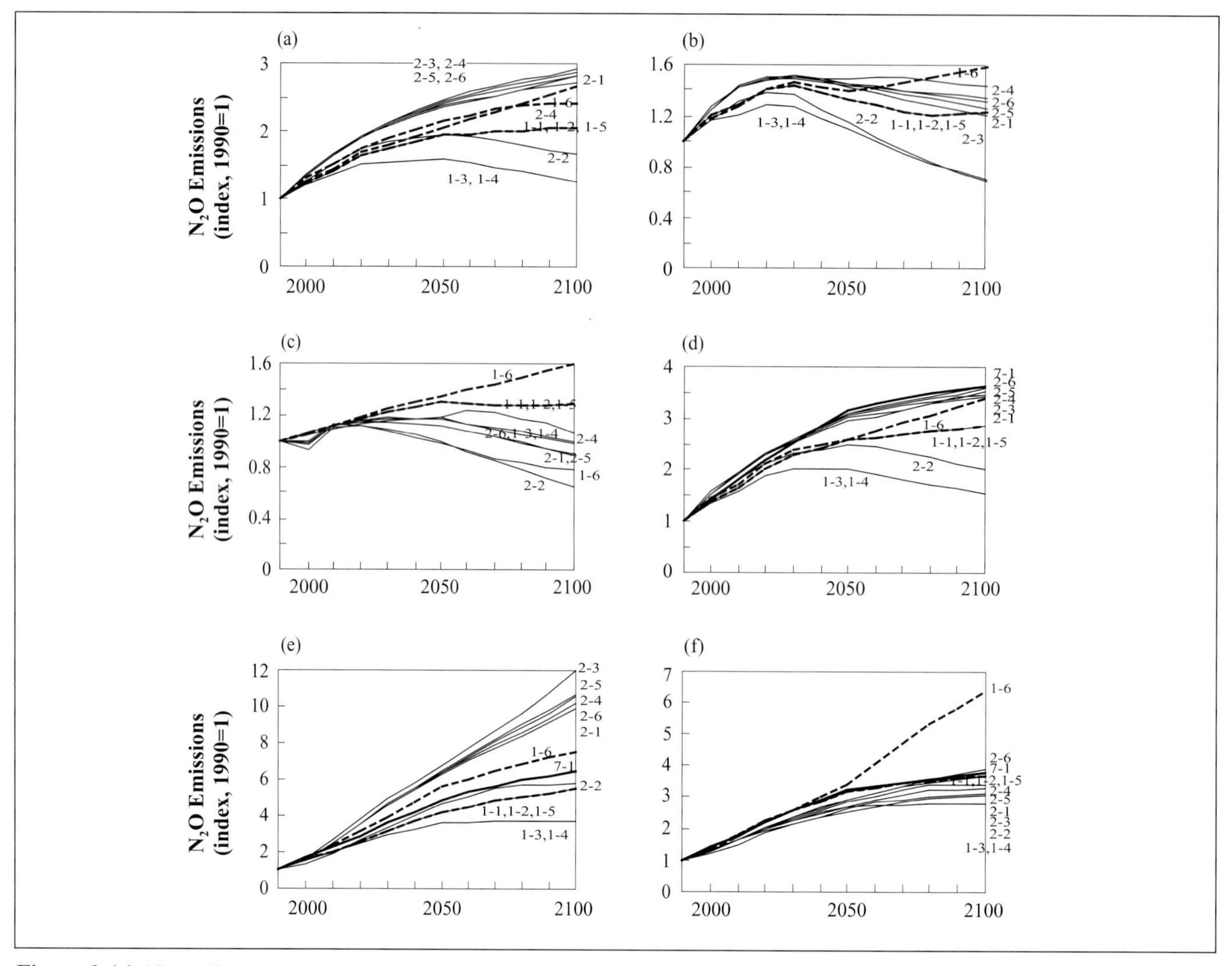

Figure 3-16: Normalized N_2O emissions from fertilized soils for (a) global, (b) OECD, (c) former Soviet Union, (d) Asia, (e) Africa, (f) Latin America, based on scenarios in the literature. Scenarios numbers are given in Table 3-7.

scenarios also assume that improvements in crop and animal productivity will partly compensate for increasing food demand, some authors do not believe that productivity increases can be sustained. For example, Kendall and Pimentel (1994) show a decrease, rather than increase, in per capita grain production because of less optimistic assumptions about the increase in crop productivity (0.7% per year). Brown and Kane (1995) point out some developments that may slow down productivity:

- The backlog of unused agricultural technology (and hence the potential for further agricultural productivity increases) is shrinking.
- Demands for water are reducing the ability of the hydrologic cycle to supply irrigation water.
- In many countries, the use of additional fertilizer on current crop varieties does not significantly increase their yields.
- Industrial countries are losing fertile land to urbanization.

Brown and Kane (1995) argue that in recent years rice yields have either stabilized or fallen in many key rice-producing countries, and suggest that dramatically boosting rice yields above a level of 4 tons per hectare may require new technological advances. If the skeptics are right, and the assumed productivity increases of the "business-as-usual" scenarios are not realized, then a greater expanse of cropland would be needed to satisfy the same agricultural demand. This expansion would lead to still higher emissions of CH_4 from rice fields and livestock, and more rapid deforestation and earlier peaks in the emissions of CO_2. However, the lower increase in productivity probably also implies lower fertilizer use per hectare, which may lower emissions of N_2O from fertilized soils.

3.5.7. Conclusions

Some of the main findings regarding different categories of land-use emissions are outline below.

Global scenarios of *CO_2 emissions from deforestation* have their widest range around the middle of the 21st century and converge on zero toward the end of the century. The eventual decrease in emissions computed by the scenarios results in part from the assumed slowing of agricultural land expansion in tropical regions. Another reason is some scenarios assume that forests will nearly disappear in Asia and Africa before or around the middle of the 21st century.

Most global scenarios of *CH_4 emissions from rice cultivation* show an upward trend until the middle of the 21st century and then stabilize. The global trend is chiefly influenced by estimates for Asia, where more than 80% of these emissions currently originate. Normalized global emissions range by a factor of three in the year 2100. The wide range has mostly to do with different estimates of the future rice cropland area, which is influenced largely by different assumptions about future rice productivity.

All global scenarios of *CH_4 emissions from enteric fermentation* show an upward trend until the end of the 21st century. The maximum range of normalized emissions is by a factor of 2.0 (which occurs in the year 2100), the smallest range of the four categories of emissions examined. Also, these emissions have the smallest range of current estimates in the literature, and the smallest range of base-year estimates in the scenarios. Most scenarios of emissions in industrial regions show a stabilizing or decreasing trend, because of the assumption that the number of livestock will continue to decline with decreasing demand for beef and increasing animal productivity. Meanwhile, the assumed economic development in the developing regions will stimulate demand for beef, which leads to an increase in livestock (despite improvements in animal productivity) and higher emissions.

Most global scenarios show that N_2O emissions from fertilized soils continue to increase up to the end of the 21st century, and the range of estimates of normalized emissions in 2100 exceeds a factor of two.

Three of the four categories of emissions show increasing global trends up to the end of the 21st century. The exception is CO_2 from deforestation (see above). Hence it is likely that land-use emissions will continue to contribute significantly to the build-up of GHGs in the atmosphere, especially to levels of CH_4 and N_2O. Studies of mitigation of climate change should take this into account and scenarios of land-use emissions should be included in these studies.

Regarding regional scenarios, land-use emissions stabilize or decrease in industrial regions, and increase substantially in Africa, but less so in Asia. Emission trends in Latin America are between those of industrial and developing regions. These regional trends reflect the stabilizing demand for agricultural products and agricultural land in industrial countries, and the assumed continuation of agricultural development elsewhere.

3.6 Other Gas Emissions

3.6.1. Introduction

Driving forces of emissions other than CO_2 or those of agriculture or land-use changes are discussed here. The direct GHGs N_2O and CH_4 are discussed first, followed by the indirect GHGs, which include sulfur and the ozone precursors NO_x, CO, and volatile organic compounds (VOCs). Finally, the many various powerful GHGs, including ozone-depleting substances (ODS), are discussed.

The sources and sinks for these gases continue to be highly uncertain. Little research has been carried out to evaluate the influences of socio-economic and technological driving forces on long-term emission trends of these gases. As a rule, future emissions of these gases are included in long-term emission models on the basis of simple relationships to aggregate economic or sector-specific activity drivers, not least because individual source strengths continue to be highly uncertain. Notable exceptions are emissions of sulfur and ODS, which have been more intensively studied in connection with non-climate policy analysis in the domains of regional acidification and stratospheric ozone depletion.

3.6.2. Nitrous Oxide

Natural and agricultural soils are the dominant sources of N_2O emissions, so future emission levels are governed by the land-use changes and changes in agricultural output and practices discussed in Section 3.5.2. Nevertheless, other sources are also important and are discussed here.

The dominant industrial sources are the production of HNO_3 and adipic acid. The key driver for the production of HNO_3 is the demand for fertilizer. Hence this emission source is closely related to the agricultural production driving forces discussed in Section 3.5, as well as to improvements in production technologies. Adipic acid, $(CH_2)_4(COOH)_2$, is a feedstock for nylon production and one of the largest-volume synthetic chemicals produced in the world each year – current annual global production is 1.8 million metric tons (Stevens III, 1993). Production has an associated by-product of 0.3 kg N_2O/kg adipic acid for unabated emission, which at present results in a global emission of about 0.4 MtN as N_2O annually. Emissions mostly arise in the OECD countries, which accounted for some 95% of global adipic acid production in 1990 (Davis and Kemp, 1991). Fenhann (2000) reviews the (sparse) scenario literature and concludes that future emissions will be determined mostly by two variables – demand growth as a result of growth in economic activity and progressively phased-in emission controls.

By the early 1990s, it was estimated that about one-third of OECD emissions had been abated (Stevens III, 1993). This abatement is an accidental result of the treatment of flue-gases in a reductive furnace (thermal destruction) to reduce NO_x

emissions, which coincidentally also converts about 99% of the N_2O into nitrogen gas (N_2). In other regions only about 20% of emissions had been abated by the early 1990s.

Major adipic acid producers worldwide have agreed to substantially reduce N_2O emissions by 1996 to 1998. In July 1991 they formed an inter-industry group to share information on old and new technologies developed for N_2O abatement, such as improved thermal destruction, conversion into nitric oxide for recycling, and the promising low-temperature N_2O catalytic decomposition into N_2 currently being developed by DuPont. The introduction of all three technologies could result in a 99% reduction of N_2O emissions from adipic acid production (Storey, 1996). They are expected to be introduced at plants owned by Asahi (Japan), BASF and Bayer (Germany), DuPont (US), and Rhône-Poulenc (France) (*Chemical Week*, 1994). After the planned changes, US producers will have abated over 90% of the N_2O emissions from adipic acid production. In recent years nylon-6.6 production dropped in the US, Western Europe, and Japan, largely in response to capacity and production in other Asian countries. By 2000 production is expected to recover in these countries (Storey, 1996).

Another major source of N_2O is the transport sector. Gasoline vehicles without catalytic converters have very low, sometimes immeasurably small, emissions of N_2O. However, vehicles equipped with three-way catalytic converters have N_2O emissions that range from 0.01 to 0.1 g/km in new catalysts, and from 0.16 to 0.22 g/km in aging catalysts (IPCC, 1996). Emission levels also depend on precise engine running conditions. At the upper end of the emission range from aging catalysts, N_2O emissions contribute around 25% of the in-use global warming impact of driving (Michaelis *et al.*, 1996).

The introduction of catalytic converters as a pollution control measure in the majority of industrialized countries is resulting in a substantial increase in N_2O emissions from gasoline vehicles. Several Annex I countries include projections of N_2O from this source in their national communications to the UNFCCC, using a variety of projection methods (for example, Environment Canada, 1997; UNFCCC, 1997; VROM, 1997). The projections from these counties differ substantially in the contribution that transport is expected to make to their national N_2O emissions in 2020, ranging from about 10% in France to over 25% in Canada. They anticipate that mitigation measures will be much more effective in reducing industrial and agricultural emissions of N_2O than mobile source emissions. Indeed, little research has been carried out to identify catalytic converter technologies that result in lower N_2O emissions. However, emissions are likely to be lower in countries that require regular emission inspections and replacement of faulty pollution control equipment.

3.6.3. *Methane*

Agricultural and land-use change emission drivers are discussed in Section 3.5.2. The other major sources are from the use of fossil fuels and the disposal of waste, for which the driving forces are briefly reviewed here. The earlier literature is reviewed in Barnes and Edmonds (1990). A more detailed recent literature review is given in Gregory (1998).

Emissions from the extraction, processing, and use of fossil fuels will be driven by future fossil fuel use. CH_4 emissions from venting during oil and gas production may decrease because of efforts to reduce them (IGU, 1997b). Flaring and venting volumes from oil and gas operations peaked in 1976 to 1978, but a gradual reduction in volumes of gas flared and vented has occurred over the past 20 years (Boden *et al.*, 1994, Marland *et al.*, 1998; Stern and Kaufmann, 1998). Shell International Ltd. (1998) estimated a reduction in its own emissions from venting by 1 $MtCH_4$ per year to 0.367 $MtCH_4$ in the five years to 1997. The IEA Greenhouse Gases R&D Programme (1997) notes that emission reductions from the oil and gas sector would yield a high economic return. Additionally, new natural gas developments generally use the latest technology and are almost leak free compared to older systems. Taking all these factors into account, it seems plausible that CH_4 emissions from the oil and gas sector should fall as the 21st century progresses. Nonetheless, the primary driver (oil and gas production) is likely to expand significantly in the future, depending on resource availability and technological change. A representative range from the literature, for example the scenarios described in Nakićenović *et al.* (1998a), indicates substantial uncertainty in which future levels of oil and gas production could range between 130 and some 900 EJ. Assuming a constant emission factor, future CH_4 emissions from oil and gas could range from a decline compared to current levels to a fourfold increase. With the more likely assumption of declining emission factors, future emission levels would be somewhat lower than suggested by this range.

The concentrations of CH_4 in coal seams are low close to the surface, and hence emissions from surface mining are also low (IEA CIAB, 1992). Concentrations at a few hundred meters or deeper can be more significant; releases from these depths are normally associated with underground mining. Emissions per ton of coal mined can vary widely both from country to country and at adjacent mines within a country (IEA Greenhouse Gases R&D Programme, 1996a). CH_4 mixed with air in the right proportions is an explosive mixture and a danger to miners. Measures to capture and drain the CH_4 are common in many countries – the captured CH_4, if of adequate concentration, can be a valuable energy source. The techniques currently used reduce total emissions by about 10%. Many older, deeper coal mines in Europe are being closed, which will reduce emissions. Replacement coal mines tend to be in exporting countries with low cost reserves near the surface, so the emissions will be low. For the future, emissions will depend principally on the proportion of coal production from deep mines and on total coal production.

A representative range of future coal production scenarios given in Nakićenović *et al.* (1998a) indicates a very wide range of uncertainty. Future coal production levels could range anywhere from 14 to well over 700 EJ, between a sevenfold decrease to an eightfold increase compared to 1990 levels. Conversely, CH_4 capture, either during mining or prior to mining, not only reduces risk to miners but also provides a valuable energy source. Thus, rising levels of CH_4 capture for non-climate reasons are likely to characterize the 21st century. This would in particular apply to high coal production scenarios, in which most of the coal will need to come from deep mining once the easily accessible surface mine deposits have become exhausted. Growth in future emissions from coal mining is therefore likely to be substantially lower than growth in coal production.

Domestic and some industrial wastes contain organic matter that emits a combination of CO_2 and CH_4 on decomposition (IEA Greenhouse Gases R&D Programme, 1996b). If oxygen is present, most of the waste degrades by aerobic micro-organisms and the main product is CO_2. If no oxygen is present, different micro-organisms become active and a mixture of CO_2 and CH_4 is produced. Decay by this mechanism can take months or even years (US EPA, 1994). Traditionally, waste has been dumped in open pits and this is still the main practice in most developing countries. Thus, oxygen is present and the main decay product is CO_2. In recent decades, health and local environmental concerns in developed countries have resulted in better waste management, with lined pits and a cap of clay, for example, added regularly over newer dumps. This prevents fresh supplies of oxygen becoming available so the subsequent decay process is anaerobic and CH_4 is produced. Williams (1993) notes that landfill sites are complex and highly variable biologic systems and many factors can lead to a wide variability in CH_4 production. For the future, increasing wealth and urbanization in developing countries may lead to more managed landfill sites and to more CH_4 production. However, the CH_4 produced can be captured and utilized as a valuable energy source, or at least flared for pollution and safety reasons; indeed, this is a legal requirement in the USA for large landfills. Future emissions are therefore unlikely to evolve linearly with population growth and waste generation, but the scenario literature is extremely sparse on this subject – the major source remains the previous IS92 scenario series (Pepper *et al.*, 1992).

Different methods are used to treat domestic sewage, some of which involve anaerobic decomposition and the production of CH_4. Again, capture and use of some of the CH_4 produced limits emissions. For the future, emissions will depend on the extension of sewage treatment in developing countries, the extent to which the techniques used enhance or limit CH_4 production, and the extent to which the CH_4 produced is captured and used.

Several authors, including Rudd *et al.* (1993) and Fearnside (1995), note that some hydroelectric schemes result in emissions of CH_4 from decaying vegetation trapped by water as the dams fill; these emissions climatically exceed those of a thermopower plant delivering the same electricity. Rosa *et al.* (1996), Rosa and Schäffer (1994), and Gagnon and van de Vate (1997) point out that the two schemes discussed by Rudd *et al.* (1993) and Fearnside (1995) may be exceptional, with very large reservoir surface areas, a high density of organic matter, and low power output. Gagnon and van de Vate (1997) estimate the combined CH_4 and N_2O emissions from hydroelectric schemes at 5.5 gC equivalent per kWh compared to a range of 80 to 200 gC equivalent per kWh for a modern fossil power station (Rogner and Khan, 1998); that is, hydroelectric power emits less than 3% and 7%, respectively. While some GHG emissions from new hydroelectric schemes are expected in the future, especially in tropical settings (Galy-Lacaux *et al.*, 1999), in the absence of more comprehensive field data, such schemes are regarded as a lower source of CH_4 emissions compared to those of other energy sector or agricultural activities. Hydroelectric power is therefore not treated as a separate emission category in SRES.

In summary, numerous factors could lead to increases in emissions of CH_4 in the future, primarily related to the expansion of agricultural production and greater fossil fuel use. Recent studies also identify a number of processes and trends that could reduce CH_4 emission factors and hence may lead to reduced emissions in the future. These trends are not yet sufficiently accounted for in the literature, in which CH_4 emission factors typically are held constant. The overall consequence is to introduce additional uncertainty into projections, as the future evolution of such emission factors is unclear. However, from the above discussion, the least likely future is one of constant emission factors and the range of future emissions is likely to be lower than those projected in previous scenarios with comparable growth in primary activity drivers.

3.6.4. *Sulfur Dioxide*

Two major sets of driving forces influence future SO_2 emissions:

- Level and structure of energy supply and end-use, and (to a lesser extent) levels of industrial output and process mix.
- The degree of SO_2-control policy intervention assumed (i.e., level of environmental policies implemented to limit SO_2 emissions).

Grübler (1998c) reviewed the literature and empirical evidence, and showed that both clusters of driving forces are linked to the level of economic development. With increasing affluence, energy use per capita rises and its structure changes away from traditional solid fuels (coal, lignite, peat, fuelwood) toward cleaner fuels (gas or electricity) at the point of end-use. This structural shift combined with the greater emphasis on urban air quality that accompanies rising incomes results in a roughly inverted U (IU) pattern of SO_2 emissions and/or concentrations. Emissions rise initially (with growing per capita energy use), pass through a maximum, and decline at higher income levels due to structural change in the end-use fuel mix and also control measures for large point sources. This pattern emerges also from the literature on environmental Kuznets curves (e.g., World Bank, 1992; IIASA–WEC, 1995) and is corroborated by both longitudinal and cross-sectional empirical data reviewed in detail in Grübler (1998c). Historically, the decline in sulfur pollution levels was achieved simply by dispersion of pollutants (tall stacks policy). Subsequently, the actual emissions also started to decline, as a result of both structural change (substitution of solids by gas and electricity as end-use fuels) and sulfur reduction measures (oil product desulfurization and scrubbing of large point sources).

Emissions for 1990 reported in the scenarios reviewed in Chapter 2 and in Grübler (1998c) indicate a range from 55 to 91 MtS. The upper range is explained largely by a lack of complete coverage of SO_2 emission sources in long-term scenario studies and models. Lower values correspond to studies that include only the dominant energy sector emissions (range of 59.7 to 65.4 MtS), and higher estimates also include other sources, most notably metallurgical and from biomass burning. None of the long-term scenario studies appears to include SO_2 emissions from international bunker (shipping) fuels, estimated at 3 ± 1 MtS in 1990 (Olivier *et al.*, 1996; Corbett *et al.,* 1999; Smith *et al.*, 2000). Historical global sulfur emissions estimates are given in Dignon and Hameed (1989).

Grübler (1998c) also argues that SO_2 control and intervention policies in many rapidly industrializing countries (particularly those with high population densities) are highly likely to be phased in more quickly than the historical experience of Europe, North America, Japan, or Korea. This analysis is supported by existing policies and trends in Brazil, China, and India (Shukla *et al.*, 1999; Rosa and Schechtman, 1996; Qian and Zhang, 1998). Most recent SO_2 emission inventory data suggest that since 1990 SO_2 emission growth has significantly slowed in East Asia compared to earlier forecasts, in response to the first SO_2 control measures implemented in China, South Korea and Thailand (Streets and Waldhoff, 2000). Dadi *et al.* (1998) estimate that in 1995 about 11% (1.5 MtS of a total of 13.5 MtS gross emissions) of China's SO_2 emissions were removed through various control measures.

The evaluation of the IS92 scenarios (Alcamo *et al.*, 1995) concluded that the projected SO_2 emissions in the IS92 scenarios do not reflect recent changes in sulfur-related environmental legislation, in particular the amendments to the Clean Air Act in the USA, and the Second European Sulfur Protocol. Increasingly, many developing countries are adopting sulfur control legislation that ranges from reduction of sulfur contents in oil products (e.g. China, Thailand, and India; see Streets *et al.*, 2000), through a maximum sulfur content in coal (e.g. in China; see Streets and Waldhoff, 2000), to SO_2 controls at coal-fired power plants (e.g. China, South Korea, Thailand; for a review see IEA, 1999). For instance, an estimated 3575 MW of coal-fired electricity China is generated by plants already equipped with sulfur control devices (IEA, 1999).

Since publication of the IS92 scenarios a number of important new sulfur impact studies have become available, and analyzed in particular:

- Implications of acidic deposition levels of high SO_2 emissions scenarios such as IS92a (Amann *et al.*, 1995; Posch *et al.*, 1996).
- Aggregate ecosystems impacts, especially whether critical loads for acidification are exceeded given deposition levels and different buffering capacities of soils (Amann *et al.*, 1995; Posch *et al.*, 1996).
- Direct vegetation damage, particularly on food crops (Fischer and Rosenzweig, 1996).

These studies provide further information on the impacts of high concentrations and deposition of SO_2 emissions, beyond the well-documented impacts on human health, ecosystems productivity, and material damages (for reviews see Crutzen and Graedel, 1986; WHO and UNEP, 1993; WMO, 1997). These studies are particularly important because they document environmental changes of high-emission scenarios by using detailed representations of the numerous non-linear dose–response relationships between emissions, atmospheric concentrations, deposition, ecosystems sensitivity thresholds, and impacts. All recent studies agree that unabated high SO_2 emissions along the lines of IS92a or even above would yield high impacts not only for natural ecosystems and forests, but also for economically important food crops and human health, especially in Asia where emissions growth is projected to be particularly high.

A representative result (based on Amann *et al.*, 1995) is shown in Figure 3-17, which contrasts 1990 European sulfur deposition levels with those of Asia by 2050 in a high SO_2 emission scenario (very close to IS92a). Typically, in such scenarios, SO_2 emissions in Asia alone could surpass current global levels as early as 2020 (Amann *et al.*, 1995; Posch *et al.*, 1996). Sulfur deposition above 5 g/m^2 per year occurred in Europe in 1990 in the area of the borders of the Czech Republic, Poland, and Germany (the former GDR), often referred to as the "black triangle." In view of its ecological impacts it was officially designated by UNEP as an "ecological disaster zone." In a scenario such as IS92a (or even higher emissions), similar high sulfur deposition would occur by around 2020 over more than half of Eastern China, large parts of southern Korea, and some smaller parts of Thailand and southern Japan.

Fischer and Rosenzweig (1996) assessed the combined impacts of climate change and acidification of agricultural crops in Asia for such a scenario. Their overall conclusion was that the projected likely regional climate change would largely benefit agricultural output in China, whereas it would lower agricultural productivity on the Indian subcontinent (the combined effect of projected temperature and precipitation changes would have differential impacts across various crops and subregions). However, projected high levels of acidic deposition in China would reduce agricultural output to an

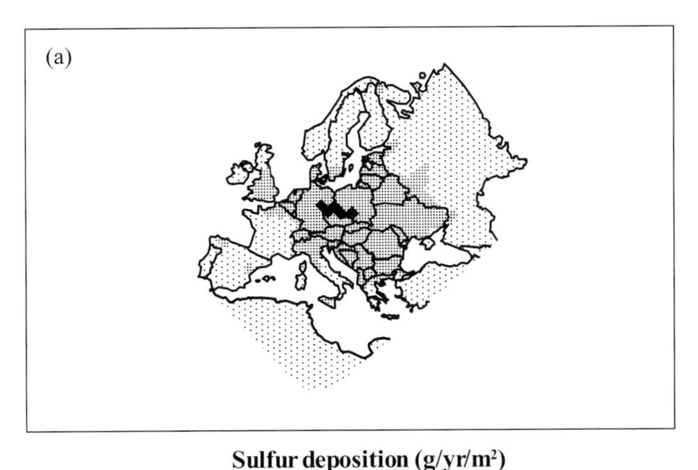

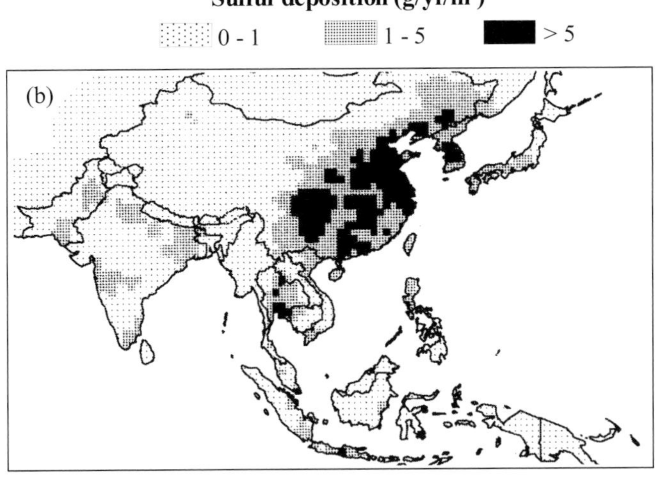

Figure 3-17: Current sulfur deposition in Europe (a) and projections for a high growth, coal-intensive scenario similar to IS92a for Asia in 2020 (b), in gS/m^2. Source: Grübler, 1998c, based on Amann *et al.*, 1995.

extent that would more than offset any possible beneficial impacts of regional climate change. This is primarily because sulfur (and nitrogen) deposition, while acting as fertilizer for plant growth at lower deposition levels, negatively affects plant growth at higher deposition levels. Projections in a scenario such as IS92a are that the threshold levels will be surpassed between 2020 and 2050 for all major Asian food crops.

The review of recent literature on acidification impact studies given in Grübler (1998c) concludes that the impacts on human health, on economically important food crops, and on ecosystems are so substantial as to render any scenario with SO_2 emissions as high as, or higher than, IS92a very unlikely. Grübler (1998c) carried out a detailed comparison of SO_2 emissions scenarios at the global and regional level. He concluded that the range of future SO_2 emissions spanned by the previous high-demand IPCC scenarios (all IS92 scenarios except IS92c and IS92d) corresponded well with scenarios available in the literature that do not include any direct sulfur-emissions control or indirect intervention measures and policies (Figure 3-18). Typically, in such scenarios, global SO_2 emissions could rise to between 130 and 250 MtS by 2100, and in some older scenarios (Matsuoka *et al.*, 1994; Morita *et al.*, 1994) are projected to rise above that level. Interestingly, all long-term sulfur scenarios published since 1995 do not judge this to be a likely (not to mention environmentally desirable) possibility. Representative sulfur-emissions control and intervention scenarios (Amann *et al.*, 1995; Posch *et al.*, 1996; Nakićenović *et al.*, 1998a) suggest instead an upper range of global emissions below 100 MtS by 2050 and below 120 MtS by 2100, a range covered in the two low variants IS92c and IS92d only. These patterns were also confirmed by Pepper *et al.* (1998) in a recent re-analysis of the previous IS92 scenarios. Using the same methodology as deployed in developing the IS92 scenarios, the revised scenarios have maximum global SO_2 emissions below 142 MtS by 2020, and 56 MtS by 2100 (EPA3 and EPA5 scenarios, respectively). The median from the more recent scenario literature analyzed in Grübler (1998c) indicates near-constant global SO_2 emissions – 77 MtS by 2020, 68 MtS by 2060, and 57 MtS by 2100. This global stability, however, masks decisive regional differences (discussed above). Emissions in the OECD countries will continue their declining trends in line with their sulfur reduction policies. Emissions outside OECD will rise initially with increasing energy demand, but sulfur controls will be progressively phased in to mitigate against impacts of high unabated SO_2 emissions on health, agriculture, ecosystems, and tourism.

The need to abate local air pollution, including SO_2 emissions, is not only environmental, but also economic. For example, according to the World Bank (1997c, 1997d) the current damage by environmental pollution is about 8% of GDP in China (and up to 20% of production in urban areas), while abatement costs would be between 1 and 2.5% of GDP. According to one World Bank (1997c) report, the costs are "so high under the business-as-usual scenario that it is hardly necessary to consider the amenity and ecosystem benefits of cleaner air to justify action." Therefore, it is no surprise that in several developing regions, policies are already being developed and implemented to abate SO_2 emissions. In China, by 1995 coal-cleaning technology had been developed, and de-sulfurizing technology introduced and applied (Government of People's Republic of China, 1996). The SO_2 emissions target set by the Chinese government is 12.3 MtS by 2000 as compared to 11.9 MtS in 1995. Economic instruments, such as pollution charges, pricing policy, favorable terms of investment for environmental technology, market creation, and ecological compensation fees, are being introduced in China now (UNEP, 1999). By June 1997, some 64,000 enterprises with heavy pollutant emissions had been closed for refurbishment or had ceased production. As a consequence, ambient concentrations of sulfur have been relatively stable in medium-size and small cities, and they have actually decreased in large cities. This change is occurring at significantly lower levels of income as compared to income levels in the USA and Europe at the time when their sulfur abatement started.

In India, several sulfur policies are being introduced currently, including mandatory washing of coal used 500 km away from the mine mouth, a policy that is expected to significantly reduce SO_2 emissions from coal use. Some refineries have already advertised their investments (and efforts) to meet this standard. In a landmark case on the Taj trapezium (a 10,400 km^2 area surrounding the Taj Mahal), the Supreme Court of India has ordered a limit on the sulfur content of diesel sold in this area to within 0.25% (Shukla, personal communication). The Indian government has spent US$1.34 billion to reduce the sulfur content in diesel from 1% to 0.25% by weight (Mr. K.P. Shahi, Advisor to the Ministry of Petroleum and Natural Gas, as quoted in *Down to Earth*, February 15, 1999, page 15).

These policies already have had an effect on emissions. Streets *et al.* (2000) analyzed the impact since 1990 of these policies and new energy and emissions factor data on emissions in Asia.[5] They found that emissions may have increased in Asia from 16.9 MtS in 1990 to only 19.3 MtS in 1995, rather than to the 26.7 MtS projected in the earlier studies with the RAINS-ASIA model. The authors conclude that SO_2 emissions in Asia have not grown nearly as fast as was thought likely in the early 1990s, with major implications for projections beyond the year 2000. It is probable that the emissions trajectory will be even lower as a result of increasing environmental awareness in many countries of Southeast Asia and East Asia, the implementation of China's "two-control-zone" policy, and the downturn of Southeast Asia economies in the late 1990s (Streets *et al.*, 2000).

In Latin America, the contribution of coal-fired power plants to total power generation is relatively low and consequently SO_2 emissions are lower than those in other regions. This contribution is not expected to increase significantly in the future. It is expected that the rate of increase of SO_2 emissions in Latin America will be reduced because environmental agencies in several Latin American countries are already enforcing strict SO_2 emissions standards. Also, increases in power generation are expected to be mainly from combined cycle natural gas plants (La Rovere and Americano, 1998).

Different methodologies have been developed to assess the sulfur control scenarios in integrated assessment models. Grübler (1998c) has summarized the literature, and classifies three main modeling approaches:

- Ecological targets, and analysis of events when critical acidification loads are exceeded (e.g., Amann *et al.*, 1995; Foell *et al.*, 1995; Hettelingh *et al.*, 1995; Posch *et al.*, 1996; Nakićenović *et al.*, 1997).
- The pollutant burden approach (e.g., Alcamo *et al.*, 1997).
- Income driven approaches (e.g., Smith *et al.*, 2000).

Although models differ in their analytical representation of the driving forces of sulfur reduction policies and also provide a range of possible futures, invariably all scenarios yield comparatively low future SO_2 emissions. Alcamo *et al.* (1997) estimates a 95% probability that global SO_2 emissions will be below 120 MtS by 2050 and decline thereafter. Their 50% probability level suggests emissions of 90 MtS by 2050 and 57 MtS by 2100, the latter being identical to the median from the scenario literature analyzed in Chapter 2.

Increasingly, energy sector and integrated assessment models are able to link regional acidification models with simplified climate models, which enables joint analysis of sulfur and climate policies and impacts. Examples include the IMAGE model (Posch *et al.*, 1996) and the IIASA model (Rogner and

[5] East Asia including China, Southeast Asia, and the Indian subcontinent

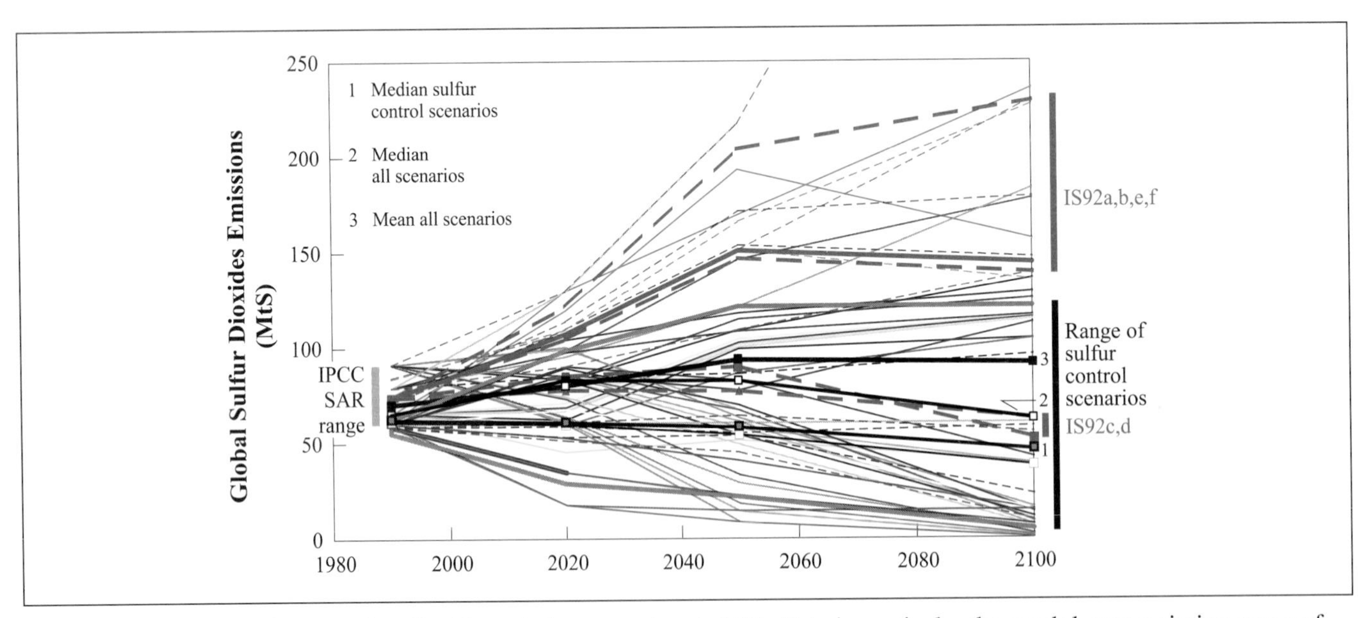

Figure 3-18: Range of future sulfur dioxide emission scenarios, in MtS. Note in particular the much lower emission range of post-1995 sulfur dioxide emissions control and intervention scenarios compared with the earlier high-growth IS92a, b, e, and f scenarios. Source: Grübler, 1998c.

Nakićenović, 1996; McDonald, 1999; Riahi and Roehrl, 2000), which are linked with the acidification model RAINS for Europe and Asia, or the AIM (Morita *et al.*, 1994) model for Asia. These models extend earlier energy sector models that dealt with a comparative costs assessment of isolated sulfur and carbon reductions and joint mitigation, such as the OECD GREEN model (Complainville and Martins, 1994) or the IIASA MESSAGE model (Grübler, 1998b; Nakićenović *et al.*, 1998a). The state of knowledge and availability of models to study the joint benefits of sulfur and carbon emission reductions was reviewed in the 1995 IPCC SAR WGIII Report (Bruce *et al.*, 1996) and is expanding rapidly (CIRED *et al.*, 1997; Nakićenović *et al.*, 1997; Grübler, 1998c).

3.6.5. *Ozone Precursors*

IPCC Working Group I (WGI) SAR (Houghton *et al.*, 1996) confirmed the importance of tropospheric ozone as a greenhouse gas. Ozone is produced in the troposphere in a complex chain of reactions that involve the ozone precursors nitrogen oxides (NO_x), non-CH_4 hydrocarbons or volatile organic compounds (NMVOCs), and CO. Therefore, it is important to explore possible future developments of emissions of these substances to analyze the evolution of tropospheric ozone levels.

3.6.5.1. Nitrogen Oxides

NO_x are released through fossil fuel combustion (24 MtN per year around 1990), natural and anthropogenic soil release (12 MtN per year), biomass burning (8 MtN per year), lightning (5 MtN per year), NH_3 oxidation (3 MtN per year), aircraft (0.4 MtN/year), and transport from the stratosphere (0.1 MtN per year). These figures are mean estimates within a range; for fossil fuel combustion, aircraft emissions, and stratospheric input the ranges may be as narrow as 30%, but for natural sources the ranges may be up to a factor of 2 (Prather *et al.*, 1995). The uncertainties in the estimates are illustrated by comparison of the detailed emissions inventory of Olivier *et al.* (1996) with the 1994 IPCC estimates – while global total emissions estimates by source are very similar, at the regional level emissions estimates show pronounced differences, particularly in Asia.

Fossil fuel combustion in the electric power and transport sectors is the largest source. Emissions from fossil fuel use in North America and Europe have barely increased since 1979 because fossil fuel consumption leveled off and air quality abatement was enacted, but in Asia emissions are believed to increase by 4% annually (Prather *et al.*, 1995). As a result of the first NO_x Protocol in Europe, NO_x emissions in Europe had decreased from 1987 levels by 13% in 1994, but the European Union is unlikely to meet its target of the 5th European Action Plan of a 30% reduction (EEA, 1999). An important reason is that it is difficult to abate NO_x emissions in the growing transport sector. Perhaps critically, there are significant differences between the characteristics of abatement of SO_2 and NO_x emissions. While both substances have regional acidification effects, a priority for SO_2 abatement is induced by its important local health effects. Also, whereas SO_2 emissions relate closely to the type of fuel, NO_x emissions are more dependent on the combustion technology and conditions.

Few scenarios for NO_x emissions exist beyond the studies for Europe, North America, and Japan (IS92 scenarios are a notable exception). New scenarios, such as those by Bouwman and van Vuuren (1999) and Collins *et al.* (1999) often still use IS92a as a "loose" baseline, with new abatement policies added as they were introduced in the OECD countries after 1992, according to current reduction plans (CRP). Collins *et al.* (1999) also explore a maximum feasible reduction scenario, in which European NO_x emissions decrease by 60% by 2015 and North American emissions by 5%. In the related CRP scenario of Bouwman and van Vuuren (1999), NO_x emissions in the developing countries are assumed to decrease also (by more than 10%) by 2015. These studies, however, should be used with care as the authors developed their somewhat arbitrary scenarios primarily for atmospheric chemistry analysis; they are not based on an in-depth analysis of the characteristics of the emissions sources and potential policies in the various regions outside the OECD.

3.6.5.2. Carbon Monoxide and Non-Methane Hydrocarbons

Prather *et al.* (1995) estimates the total global emissions of CO at 1800 to 2700 MtC per year in the decade before 1994. The most important of the approximately 1000 TgC anthropogenic sources are technological (300 to 550 MtC per year) and biomass burning (300 to 700 MtC per year). Technological sources dominate in the northern hemisphere, and include transport, combustion, industrial processes, and refuse incineration. Biomass burning dominates in the southern hemisphere, and includes burning of agricultural waste, savanna burning, and deforestation. The detailed, geographically explicit EDGAR database (Olivier *et al.*, 1996) has similar emissions estimates for CO. Other sources are biogenics (60 to 160 MtC per year), oceans (20 to 200 MtC per year), and oxidation of CH_4 (400 to 1000 MtC per year) and of NMVOCs (200 to 600 MtC per year). To a large extent, this oxidation may be considered anthropogenic in origin because many emissions sources of CH_4 and NMVOCs are of an anthropogenic nature.

Global emissions estimates of NMVOCs are also very uncertain. Prather *et al.* (1995) indicate a global total for anthropogenic NMVOCs of about 140 MtC per year, from road transport (25%), solvent use (14%), fuel production and distribution (13%), fuel consumption (34%), and the rest from uncontrolled burning and other sources. The EDGAR inventory by Olivier *et al.* (1996) suggests that global emissions may be higher (178 MtC per year) because of higher estimates of emissions from energy production and use. As with CO, emissions in the northern hemisphere are dominated by transport and industry, while in the southern hemisphere biomass and biofuel burning is often the dominant source. In Europe, emissions of NMVOCs are controlled under the 5th Environmental Action Programme of the European Union and the VOC Protocol of the UN Convention on Long-Range Transboundary Air Pollution. However, for reasons similar to those for NO_x, the current reduction of 11% with respect to 1990 levels and 15% with respect to 1987 levels suggests that the planned reduction to 30% in 1999 may not be reached (EEA, 1999). As a consequence, threshold values for ozone continue to be exceeded in Europe.

No long-term global scenarios for emissions of NMVOCs and CO were identified beyond IS92, which assumes increasing emissions. The important role of biomass combustion in these emissions means that a scenario with low carbon emissions because of an increased used of biomass energy does not automatically lead to low emissions of NMVOCs and CO. Also, emissions trends are influenced significantly by assumptions as to the type of combustion or other conversion technology (e.g. gasification) deployed in the future. If biomass fuel is used in modern large power plants or boilers, or is to be converted into modern energy carriers, CO emissions will be almost negligible compared to those of traditional uses. As with sulfur, however, it seems plausible that with rising incomes, abatement of the ozone precursors may be initiated in non-OECD regions to address local and particularly regional air pollution (photochemical smog). Since control of these substances is more difficult than that of sulfur, it may not be implemented until later.

3.6.6. Halocarbons and Other Industrial Gases

This category of GHG emissions comprises a wide basket of different gas species that originate from a multitude of processes. Generally, their common characteristic is that they are released into the atmosphere in comparatively small amounts, but on a molecular basis most of the gases are long-lived, with atmospheric lifetimes up to 50,000 years. Generally they have a strong greenhouse forcing per molecule (see Chapter 5, Table 5-7).

Anthropogenic emissions of gases that cause stratospheric ozone depletion (chlorofluorocarbons (CFCs), hydrochlorofluorocarbons (HCFCs), halons, methylchloroform, carbon tetrachloride, and methylbromide) are controlled by consumption restrictions (production plus imports minus exports) in the Montreal Protocol. No special SRES scenarios were developed for these gases because their future emission levels (phase out) are primarily policy driven and hence unrelated to scenario variations of important driving-force variables such as population, economic growth, or industrial output. Instead, the Montreal Protocol scenario (A3, maximum allowed production scenario) from the 1998 WMO/UNEP *Scientific Assessment of Ozone Depletion* is used (WMO/UNEP, 1998).

The procedures for constructing scenarios for hydrofluorocarbon (HFC), polyfluorocarbon (PFC), and sulfur hexafluoride (SF_6) emissions – for which there is an extreme paucity of scenario literature – are based on Fenhann (2000) and are described in greater detail in Chapter 5, Section 5.3.3. In this approach, future total demand for CFCs, HFCs, and other CFC substitutes is estimated on the basis of historical trends. HFC emissions are calculated using an assumed future replacement of CFCs by HFCs and other substitutes. The main drivers for the emissions are population and GDP growth. The sparse literature available (reviewed in Fenhann, 2000) indicates that emissions are related non-linearly to these driving forces, with important possibilities for saturation effects and long-term decoupling between growth in driving force variables and emissions. The emissions have been tuned to agree with emissions scenarios presented at the joint IPCC–TEAP expert meeting (WMO/UNEP, 1999). Material from the March Consulting Group (1999) has also been used.

For PFCs (CF_4 and C_2F_6) the emissions driver is primary aluminum production, which is generally modeled using GDP and a consumption elasticity. Recycling rates are increasingly important, as reflected in the SRES scenarios (see Chapter 5). Aluminum production by the Soederberg process resulted, on average, in the emission of 0.45 kgCF_4 per tAl and 0.02 kgC_2F_6 per tAl in 1998 in Norway. The effect of future technological change on the emissions factor can be assumed to be large, since the costs of modifications in process technology can be offset by the costs of saved energy. A considerable reduction in the emission factors has already taken place and the present emission factor of 0.5 kgCF_4 per tAl is expected to fall to 0.15 kgCF_4 per tAl at various rates (see Chapter 5). An emission factor for C_2F_6, 10 times lower than that of CF_4 was used in the calculations. The present trend of not replacing CFCs and HCFCs with high global warming compounds like PFCs (or SF_6) is also assumed to continue, which might underestimate the effect of future emissions. The only other source included for PFC emissions is semiconductor manufacturing, for which the industry has globally adopted a voluntary agreement to reduce its PFC emissions by 10% in 2010 relative to 1995 levels.

SF_6 emissions originate from two main activities – the use of SF_6 as a gas insulator in high-voltage electricity equipment, and its use in magnesium foundries, in which SF_6 prevents the oxidation of molten magnesium. The driver for the former is electricity demand and for the latter it is future magnesium production, which will depend on GDP and a consumption elasticity. Emission factor reductions over time that result from more careful handling, recovery, recycling, and substitution of SF_6 are assumed for both sources. Fenhann (2000) assumes that in low future scenarios SF_6 emissions factors decline to one-tenth their present values between 2020 and 2090. In high future scenarios, Fenhann (2000) assumes reduction levels are somewhat lower, ranging from 55% to 90% depending on the region. In the absence of scenario literature, these assumptions are retained here (see Chapter 5). Other applications of SF_6 include as a tracer gas in medical surgery and the production of semiconductors, and as an insulator in some windows. However, these sources are assumed to be cause less than 1% of the global emissions.

3.7. Policies

3.7.1. *Introduction*

"Policies" in this report are government policies. They are formulated against the larger background of national and international events and trends, and result from millions of decisions within the existing and only slowly changing cultural, economic, and military balances. Their implementation often poses considerable problems if they represent longer-term interests and insights.

Government policies are among the dynamics that influence population growth, economic and social development, technological change, resource exploitation, and pollution management. While the role of policy has been touched upon occasionally in earlier sections, government policy development can be thought of as a process in itself. The role of policies in SRES needs to be considered, partly because governments are one of the primary audiences for the scenarios and partly because the scenarios are intended to form a reference against which mitigation strategies can be assessed (although, as stated earlier, the SRES terms of reference require the SRES scenarios to *not* consider any explicit climate policies).

GHG emissions are affected by policies designed for a wide variety of purposes. Perhaps the most obvious are energy policies, but other important policy areas are those of economic development, technology development, education, health, social welfare, transport, industry, agriculture, and forestry. Policies in each of these areas also affect other areas. In each policy area various instruments are used. The choice of instrument may influence both the policy's success in achieving its primary objective and its effect on GHG emissions. Taxes, subsidies, regulations, information-based instruments, and R&D all bring different mechanisms into play and so have different affects.

The remainder of this section is organized around specific policy areas or objectives. It considers major policy issues in each area, and discusses the possible implications for GHG emissions in reference (non-mitigation) scenarios.

3.7.2. *Policy Areas*

3.7.2.1. Population and Social Welfare Policies

Given the interactions between demographics and social and economic development discussed in Sections 3.2 and 3.3, population and social welfare policies that currently exist or are options for various countries can also be viewed as "non-climate" policies (in the sense that they are not motivated by climate concerns, but will affect future GHG emissions). Studies support the notion that reduced population growth significantly abates GHG emissions. Indeed, some integrated assessment models suggest that emissions scenarios may be more sensitive to population changes, with respect to normalized uncertainty analysis, than to other factors that affect emissions (Nordhaus, 1993). Therefore, social policies that affect fertility rates (and mortality and migration rates) also could have a significant impact on future emissions. By the same token, demographic policies for health and education may also affect productivity growth in a positive manner. Thus, the desirable objective to further development may result in higher economic growth, consumption, and emissions per capita. The overall effects are likely to vary from country to country.

For instance, efforts can be made to help women avoid unwanted pregnancies or to reduce infant mortality. Demographic health surveys suggest that more than 100 million women in less developed countries do not want to become pregnant, but they do not practice contraception (Bongaarts, 1994). The Cairo Program of Action (UN, 1995) estimates that US$17 billion annually would successfully deliver family planning and reproductive health services to the majority of people in developing countries who desire them. Family planning assistance today contributes to the observed recent declines in fertility rates in many developing countries. In one study, it was estimated that such programs over the past two decades reduced the present population by about 40 million persons, which in itself may reduce future population levels by some 400 million people in the year 2100 (Bongaarts *et al.*, 1990).

Other policy measures are less direct, but also exert important influences on fertility rates. These include improvements in health care and female education, especially primary school education, which is a factor that correlates highly with fertility rates in young women (Bongaarts, 1994). Similarly, measures that improve gender equality reduce fertility rates as they encourage non-maternal roles and increase employment and empowerment opportunities for women. Their implementation is currently unrelated to concerns about global warming, yet their effect on this environmental issue may be significant.

3.7.2.2. Policies that Target Economic Development and Technological Innovation

A wide range of policies and circumstances may contribute toward the desirable objective of furthering development and economic growth (see Section 3.3). In the short term, fiscal, monetary, and interest rate management policies are among the main instruments used by governments. In the longer term, economic growth may be affected more by measures that influence fundamental capabilities, such as policies in education, and in the development of physical infrastructure, social and economic institutions, and national systems for innovation.

As emphasized in Sections 3.3 and 3.4, the effects of economic growth on GHG emissions depend on economic structure and technology. Governments generally aim to encourage the development of particular sectors that are perceived to contribute to national goals for security, food and energy supply, high employment, and long-term economic growth (Maddison, 1995). The encouragement may take many forms, such as direct subsidies and protection from foreign competition, public investment in infrastructure, training, or R&D, and support for collaborative development programs and information networks (OECD, 1997a). If governments support sectors that are fossil-fuel intensive, the tendency to increase GHG emissions is clear. However, protectionist policies may also reduce national economic efficiency, which dampens income growth and tends to restrict growth in GHG emissions. Conversely, if governments support the development of rapid-growth sectors, the tendency may be to promote long-term economic growth, increase household income and consumption, and hence increase GHG emissions.

Over a period of 100 years the policies that most influence the development of GHG emissions are probably those that contribute to the processes of technical and social innovation, which themselves contribute to economic development. Innovation policies mostly emphasize the development of technologies that improve international competitiveness with new products and improved performance or reduced costs of existing products. The policies are not usually designed to achieve these and other (e.g., environmental and social) objectives in an integrated way (OECD, 1998b). Hence, their impact on GHG emissions is hard to predict, but as currently constituted many national systems for innovation could tend to increase emissions by stimulating economic growth.

3.7.2.3. *Energy, Agriculture and Other Resource Management Policies*

Government policies on energy and agriculture have, on the whole, paralleled global trends during the 20th century. Early in the century there was a move toward protectionism, which aimed to secure national self-sufficiency, especially in food and energy. Governments established import quotas and tariffs, subsidies for domestic production, and research and investment programs to improve agricultural productivity and develop new energy sources. During the 1980s policy emphasis shifted in many countries, and has continued into the 1990s, toward open borders and reduced subsidies and R&D. Nonetheless, numerous energy and agricultural policies persist that influence production and trade patterns and hence also GHG emissions.

3.7.2.3.1. *Energy policies*

Various policies exist to promote energy efficiency and the adoption of energy-efficient technologies and practices. Government standards, such as appliance efficiency standards, motors standards, and the automobile fuel economy standards in the US, prescribe the energy consumption levels of particular commodities. Residential and commercial building standards require the use of energy-efficient construction practices and components. Information dissemination programs, such as the Green Lights program in the USA or similar programs in other countries, provide consumers with the information required to make purchase decisions as well as to install and operate energy-efficient equipment. Subsidy or investment credit programs are often used to promote the adoption of a particular technology; combined heat and power was promoted in The Netherlands through such a program in the 1980s (Farla and Blok, 1995). Other energy efficiency policies or programs include audits and assessments, rebate programs, government procurement programs, benchmarking programs, labeling programs, and technology demonstration programs (Worrell *et al.*, 1997).

Many reports point to government subsidies as a major impediment to cleaner production of energy (Burniaux *et al.*, 1992; Larsen and Shah, 1992; de Moor and Calamai, 1996; Roodman, 1996; Greenpeace, 1997). In addition to direct subsidies, governments use a wide variety of measures to support domestic or regional industries, or to protect legal monopolies. These policies inhibit innovation and can lead to higher levels of pollution or resource intensity than would occur in a less constrained market. A recent OECD study found that reform of supports to coal, electricity, and transport could substantially reduce CO_2 and acid rain emissions in some countries (OECD, 1997a). In other countries, subsidy reform would have minimal direct environmental benefits, but would increase the effectiveness or reduce the cost of environmental policies such as eco-taxes and emission limits. Where subsidies support nuclear power or other non-fossil energy sources, their reform could conversely lead to increased GHG emissions. Energy taxes also have an important influence on energy demand and hence GHG emissions. The majority of energy taxes are intended as a pure fiscal instrument or, in the case of road fuel taxes in some countries, to raise funds for road provision and maintenance. Many countries are raising these taxes, or considering doing so.

3.7.2.3.2. *Agriculture policies*

Agricultural policy reform has received more attention than energy policy reform in recent years. Most OECD countries support domestic agriculture, whether through direct subsidies, import tariffs, or price controls. The general trend is toward a reduction in these supports, in part as a result of trade negotiations, but also as part of the broader trend toward policies that reduce budget deficits and improve market efficiency. Supports are also being reformed to reduce their linkage to production volumes. Where subsidies are linked to the volume of production, they provide an incentive to increase output beyond the level of demand, which leads to surpluses. This incentive may tend to increase GHG emissions as a result of soil carbon depletion and oxidation, excessive use of nitrogen fertilizer leading to N_2O emissions, and over-intensive animal farming that results in excess CH_4 emissions from manure and from the animals themselves (OECD, 1997b; Storey, 1997). The overall impact of agriculture subsidy reforms on GHG emissions will depend on associated fiscal changes in other parts of the economy.

Overproduction in one country may be compensated to some extent by lower production elsewhere. However, in general, incentives for higher agricultural output are likely to lead to more production globally, with a shift from consumption of plant products to animal products, which are land-, resource-, and GHG-intensive. In a few industrialized countries a small trend has developed to support organic farming and regional marketing of foods. Future policies may thus lead to agricultural subsidies that are linked more to ecological and social factors than to the volume of production.

3.7.2.3.3. Dematerialization policies

GHG emissions are likely to be reduced by other policies for the sustainable use of resources, such as land, forest ecosystems, mineral resources, water, and soil. Instruments may include direct planning, regulations, establishing property rights and obligations, information, education, and persuasion, and a broad range of policies to support or influence the innovation process to encourage dematerialization (OECD, 1998b).

3.7.2.4. Environmental Policies

While environmental objectives often form part of the rationale for agriculture and energy policy reforms, many instruments are focused entirely on environmental objectives. The most obvious of these are pollution regulations and standards, eco-taxes, and voluntary and other measures.

In the context of non-mitigation GHG emission scenarios, probably the most important environmental policies are those related to sulfur emissions (see Section 3.4.3). Sulfur emissions are controlled for local and regional environmental reasons, but sulfur oxides do have a radiative impact, and sulfur controls can lead to the switching of fuel away from coal and oil. Thus, almost paradoxically, environmental policies to combat urban air pollution and acid rain may (via reduced sulfate aerosol "cooling") exacerbate climate change. Most sulfur control policies to date have involved either regulations that limit the concentration of sulfur oxides in flue gas from large combustion plants, or give standards for the sulfur content of fuel. Recently, sulfur control policies have become more sophisticated, and aim to limit aggregate emissions on a national or regional basis to minimize acidic deposition in a trans-national context. New policy instruments have also been introduced. The USA has pioneered a "cap and trade" system with tradable emission permits (for a review see Joskow *et al.*, 1998).

Other environmental policies with a greenhouse impact include controls on ODSs; urban air pollution precursor compounds (CO, NO_x, CH_4, and NMVOCs), especially from transport and domestic solid fuels; and controls on agricultural practice to reduce water pollution and soil erosion. Policies in all of these areas are likely to contribute to GHG mitigation. However, some options, such as an accelerated shift to electric vehicles to reduce local air pollution, could result in higher GHG emissions in the short term in certain circumstances (Michaelis *et al.*, 1996).

3.7.2.5. Transportation and Infrastructure Policies

Policies on infrastructure may have a very long-term influence on GHG emissions. although in many cases the causal relationships are complex and not understood well enough to justify quantitative analyses of the policy options. These include urban planning guidance, construction regulations, policies on ownership and financing of infrastructure, and user pricing for roads and parking. The most significant impacts on GHG emissions are likely to derive from policies that influence demand for travel by car and for freight transport by truck (Newman and Kenworthy, 1990; Michaelis *et al.*, 1996; Watson *et al.*, 1996), those that influence energy use in buildings (Levine *et al.*, 1996; Watson *et al.*, 1996), and those that influence the conversion of forest for agriculture, or agricultural land for urban development.

3.7.3. *Quantification of Impacts and Implementation of Policies in SRES*

Few of the policies and instruments identified above can be represented directly in the models typically used to produce GHG emission scenarios. In general, the impacts of policies are highly uncertain (Houghton *et al.*, 1996). Price-based instruments have been analyzed in greater detail than other types of measure, and many empirical studies have been carried out to determine the response to price changes of demand for various commodities, especially energy. However, such research and analysis usually yields very large ranges of uncertainty in the magnitude of the price response, and often reveals a strong dependence on specific circumstances. Even for price-based policies, national and global effects over 20–100 years are very uncertain. For the SRES, it is not possible to make a precise link between governments' application of specific policies and the outcome in the various scenarios.

Instead, the qualitative SRES scenario storylines give a broad characterization of the areas of policy emphasis thought to be associated with particular economic, technological, and environmental outcomes, as reflected in alternative scenario assumptions in the models used to generate long-term GHG emission scenarios. In some selected areas, such as sulfur control policies, a wide body of literature can be drawn upon to derive specific pollution control levels or maximum emission trajectories consistent with a particular interpretation of a scenario storyline. In other areas, such as GHG gases controlled by the Montreal Protocol, existing scenarios that reflect the most up-to-date information are used as direct input to SRES.

References:

Abramovitz, M., 1986: *Catching Up and Falling Behind*. Economic Research Report No. 1, Trade Union Institute for Economic Research, Stockholm, Sweden.

Abramovitz, M., 1993: The search for the sources of growth: Areas of ignorance, old and new. *Journal of Economic History*, **52**(2), 217-243.

Adelman M.A., and M.C. Lynch, 1997: Fixed view of resource limits creates undue pessimism. *Oil and Gas Journal*, **April**, 56-60.

Alcamo, J., A. Bouwman, J. Edmonds, A. Grübler, T. Morita, and A. Sugandhy, 1995: An evaluation of the IPCC IS92 emission scenarios. In *Climate Change 1994, Radiative Forcing of Climate Change* and *An Evaluation of the IPCC IS92 Emission Scenarios*, J.T. Houghton, L.G. Meira Filho, J. Bruce, Hoesung Lee, B.A. Callander, E. Haites, N. Harris and K. Maskell (eds.), Cambridge University Press, Cambridge, pp. 233-304.

Alcamo, J., and G.J.J. Kreileman, 1996: Emission scenarios and global climate protection. *Global Environmental Change*, **6**(4), 305-334.

Alcamo, J., G.J.J. Kreileman, J.C. Bollen, G.J. van den Born, R. Gerlagh, M.S. Krol, A.M.C. Toet, and H.J.M. de Vries, 1996: Baseline scenarios of global environmental change. *Global Environmental Change*, **6**(4), 261–303.

Alcamo, J., Onigkeit, J., and Kaspar, F., 1997: *The Pollutant Burden Approach for Computing Global and Regional Emissions of Sulfur Dioxide*. Center for Environmental Systems Research, University of Kassel, Germany.

Alcamo, J., and R. Swart, 1998: Future trends of land-use emissions of major greenhouse gases. *Mitigation and Adaptation Strategies for Global Change*, **3**(2-4), 343-381.

Alcamo, J., E. Kreileman, M. Krol, R. Leemans, J. Bollen, J. van Minnen, M. Schäfer, S. Toet, and B. de Vries, 1998: Global modelling of environmental change: an overview of IMAGE 2.1. In *Global change scenarios of the 21st century. Results from the IMAGE 2.1 Model*. J. Alcamo, R. Leemans, E. Kreileman (eds.), Elsevier Science, Kidlington/Oxford, pp. 3-94.

Allardt, E., 1993: Having, loving, being: an alternative to the Swedish model of welfare research. In *The Quality of Life*. M. Nussbaum, A.K. Sen, (eds.), Clarendon Press, Oxford, pp. 88-94.

Amann, M., J. Cofala, P. Dörfner, F. Gyarfas, and W. Schöpp, 1995: Impacts of energy scenarios on regional acidifications. In *WEC Project 4 on Environment, Working Group C, Local* and *Regional Energy Related Environmental Issues*. World Energy Council, London, pp. 291-317.

Andersen, E.S., 1994: *Evolutionary Economics: Post-Schumpeterian Contributions*. Pinter, London.

Ang, B.W., and G. Pandiyan, 1997: Decomposition of energy-induced CO_2 emissions in manufacturing. *Energy Economics*, **19**, 363-374.

Ansuategi, A., E. Barbier, and C. Perrings, 1997: *The Environmental Kuznets Curve*. USF Workshop on Economic Modelling of Sustainable Development: Between Theory and Practice, Tinbergen Institute, Amsterdam.

Arrow, K., 1962: The economic implications of learning by doing. *Review of Economic Studies*, **29**, 155-173.

Arthur, W.B., 1989: Competing technologies, increasing returns, and lock-in by historical events. *The Economic Journal*, **99**, 116-131.

Arthur, W.B., 1994: *Increasing Returns and Path Dependence in the Economy*. Michigan University Press, Ann Arbor, MI.

Ausubel, J.H., A. Grübler, and N. Nakićenović, 1988: Carbon dioxide emissions in a methane economy. *Climatic Change*, **12**, 245-263.

Ausubel, J.H., and A. Grübler, 1995: Working less and living longer: Long-term trends in working time and time budgets. *Technological Forecasting and Social Change*, **50**(3), 195-213.

Ayres, R.U., 1989: *Energy Efficiency in the U.S. Economy: A New Case for Conservation*, RR-89-12, International Institute for Applied Systems Analysis, Laxenburg, Austria.

Azar, C., 1996: *Technological Change and the Long-Run Cost of Reducing CO_2 Emissions*. Working Paper 96/84/EPS, INSEAD Centre for the Management of Environmental Resources, Fontainebleau, France.

Azar, C., and G. Berndes, 1999: The implication of CO_2-abatement policies on food prices. In *Sustainable Agriculture and Environment: Globalization and Trade Liberalisation Impacts*. A. Dragun, C. Tisdell (eds.), Edward Elgar Publishing Ltd., Cheltenham, UK.

Barnes, D.W., and J.A. Edmonds, 1990: *An Evaluation of the Relationship between the Production and Use of Energy and Atmospheric Methane Emissions*. Report DOE/NBB-0088P, US Department of Energy, Washington, DC.

Barro, R.J., and X. Sala-I-Martin, 1995: *Economic Growth*. McGraw-Hill, New York, NY.

Barro, R.J., 1997: *Determinants of Economic Growth*. The MIT Press, Cambridge, MA.

Baumol, W.J., 1993: *Social Wants and Dismal Science: the Curious Case of the Climbing Costs of Health and Teaching*. WP-64-93, Fondazione Enrico Mattei, Milan, Italy.

Birdsall, N., 1994: Another look at population and global warming. In *Population, Environment and Development*. Proceedings of the United Nations Expert Group Meeting on Population, Environment and Development, United Nations Headquarters, New York NY, 20–24 January 1992, pp. 39-54.

Blanchet, D., 1991: Estimating the relationship between population growth and aggregate economic growth in developing countries: Methodological problems. In *Consequences of Rapid Population Growth in Developing Countries*. United Nations (ed.), Taylor & Francis, New York, NY, pp. 67-98.

Boden, T.A., D.P. Kaiser, R.J. Sepanski, and F.W. Stoss (eds), 1994: *Carbon Dioxide Emissions, Trends '93*. A compendium of data on global change, Carbon Dioxide Information Analysis Center, World Data Center A for Atmospheric Trace Gases, Center for Global Environmental Studies, Oak Ridge National Laboratory, Oak Ridge, TN.

Bogner, J., M. Meadows, and P. Czepiel, 1997: Fluxes of methane between landfills and the atmosphere: natural and engineering controls. *Soil Use and Management*, **13**, 268-277.

Bongaarts, J., W.P. Maudlin, and J.R. Phillips, 1990: The demographic impact of family planning programs. *Studies in Family Planning*, **21**(29), 299-310.

Bongaarts, J., 1994: Population policy options in the developing world. *Science*, **263**, 771–776.

Bongaarts, J., 1996: Global trends in AIDS mortality. *Population and Development Review*, **22**(1), 21-45.

Bongaarts, J., 1998: Global population growth: Demographic consequences of declining fertility. *Science*, **282**, 419-420.

Bos, E., M.T. Vu, E. Massiah, and R. Bulatao, 1994: *World Population Projections 1994–95 Edition*. Johns Hopkins University Press, Baltimore, MD.

Bouwman, A.F., and D.P. van Vuuren, 1999: *Global Assessment of Acidification and Eutrophication of Natural Ecosystems*. RIVM Report No. 802001012, RIVM, Bilthoven, the Netherlands.

BP (British Petroleum), 1996: *BP Statistical Review of World Energy 1996*. British Petroleum, London. (http://www.bp.com).

BP (British Petroleum), 1997: *BP Statistical Review of World Energy 1997*. British Petroleum, London. (http://www.bp.com).

Brown, L.R., and H. Kane. 1995: *Full House: Reassessing the Earth's Population Carrying Capacity*. Earthscan, London.

Bruce, J.P., H. Lee, and E.F. Haites (eds.), 1996: *Climate Change 1995. Economic and Social Dimensions of Climate Change*. Contribution of Working Group III to the Second Assessment Report of the Intergovernmental Panel on Climate Change, Cambridge University Press, Cambridge.

Buchanan, R.A., 1992: *The Power of the Machine: The Impact of Technology from 1700 to the Present*. Penguin Books, London.

Burniaux, J.-M., J. Martin, and J. Oliveira-Martins, 1992: The effects of existing distortions in energy markets on the cost of policies to reduce CO_2 emissions: Evidence from GREEN. *OECD Economic Studies*, **Winter**, pp. 141-165.

Campbell, J., 1959: *The Masks of God: Primitive Mythology, 1-4*. Arkana, Penguin Group, New York, NY.

Campbell, C.J., 1997: Better understanding urged for rapidly depleting reserves. *Oil and Gas Journal*, **7 April**, 51-54.

Chemical Week, 1994: Sixth international workshop on N_2O emissions N_2O abatement by adipic acid producers. *Chemical Week*, **13**, 95-96.

Chenery, H., S. Robinson, and M. Syrquin (eds.), 1986: *Industrialization and Growth: A Comparative Study*. Oxford University Press, Oxford.

Christiansson L., 1995: *Diffusion and Learning Curves of Renewable Energy Technologies*. WP-95-126, International Institute for Applied Systems Analysis, Laxenburg, Austria.

CIRED (Centre International de Recherche sur l'Environnement et le Developpement), IVM, IIASA, SMASH, RIIA, 1997: *Integrated Assessment Modelling of Global Environmental Policies and Decision Patterns*. Interim Report ENV4-CT96-0197, CIRED, Paris.

Collins, W.J., D.E. Stevenson, C.E. Johnson, and R.G. Derwent, 1999: *The European Regional Ozone Distribution and its Links with the Global Scale for the Years 1992 and 2015*. Climate Research Division, Meteorological Office, Bracknell, UK.

Complainville, C., and J.O. Martins, 1994: *NO_x and SO_x Emissions and carbon Abatement*. Economic Department Working Paper 151, OECD, Paris.

Corbett, J.J., P. Fischbeck, and S.N. Pandis, 1999: Global nitrogen and sulfur inventories for oceangoing ships. *Journal of Geophysical Research*, **104**(D3), 3457-3470.

Courbage, Y., 1998: *Nuovi Scenari Demographici Mediterranei*. Fondazione Giovanni Agnelli, Turin, Italy.

Crutzen, P.J., and T.E. Graedel, 1986: The role of atmospheric chemistry in environment-development interactions. In *Sustainable Development of the Biosphere*. W.C. Clark, T. Munn (eds.), Cambridge University Press, Cambridge, pp. 213-250.

Dadi, Z., L. Xueyi, and X. Huaqing, 1998: *Estimation of Sulfur Dioxide Emissions in China in 1990 and 1995*. Energy Research Institute, Beijing, China.

Darmstadter, J., J. Dunkerley, and J. Alterman, 1977: *How Industrial Societies Use Energy - A comparative analysis. Resources for the Future*. John Hopkins University Press, Baltimore, MD.

David, P.A., 1990: The dynamo and the computer: A historical perspective on the modern productivity paradox. *American Economic Review*, **80**(2), 355-361.

Davis, D.D., and D.R. Kemp, 1991: Adipic Acid. In *Encyclopedia of Chemical Technology*, 4th Edition.

Dawkins, R., 1986: *The Blind Watchmaker*. Longman, London.

Delsey, J., 1991: How to Reduce Fuel Consumption of Road Vehicles. In *Low Consumption/Low Emission Automobile*. Proceedings of an expert panel Rome 14th-15th February 1990, OECD/IEA, Paris.

Denison, E.F., 1962: *The Sources of Economic Growth in the United States and the Alternatives Before Us*. Supplementary Paper No. 13, Committee for Economic Development, New York, NY.

Denison, E.F., 1985: *Trends in American Economic Growth, 1929-1982*. The Brookings Institution, Washington, DC.

De Bruyn, S.M., and J.B. Opschoor, 1994: *Is the Economy Ecologizing? De- or Re-linking Economic Development with Environmental Pressure*. TRACE Discussion Paper TI 94-65, Tinbergen Institute, Amsterdam.

De Bruyn, S.M., J. van den Bergh, and J.B. Opschoor, 1995: *Empirical Investigations in Environmental-Economic Relationships: Reconsidering the Empirical Basis of Environmental Kuznets Curves and the De-linking of Pollution from Economic Growth*. TRACE Discussion Paper TI-95-140, Tinbergen Institute, Amsterdam.

De Moor, A., and P. Calamai, 1996: *Subsidising Unsustainable Development: Undermining the Earth with Public Funds*. Institute for Research on Public Expenditure, The Hague and Earth Council, San José, Costa Rica.

De Vries, H.J.M., J.G.J. Olivier, R.A. van den Wijngaart, G.J.J. Kreileman, and A.M.C. Toet, 1994: Model for calculating regional energy use, industrial production and greenhouse gas emissions for evaluating global climate scenarios. *Water, Air and Soil Pollution*, **76**, 79-131.

De Vries, B., M. Janssen, and A. Beusen, 1999: Perspectives on energy. *Energy Policy*, **27**(8), 477-494.

Difiglio, C., K.G. Duleep, and D.L. Greene, 1990: Cost effectiveness of future fuel economy improvements. *The Energy Journal*, **11**(1), 65-86.

Dignon, J., and S. Hameed, 1989: Global emissions of nitrogen and sulfur oxides from 1860 to 1989. *Journal of the Air and Waste Management Association*, **39**(2), 180-186.

Disney, R., 1996: *Can We Afford to Grow Older?* MIT Press, Cambridge, MA.

Dosi, G., 1988: The nature of the innovation process (Chapter 10). In *Technical Change and Economic Theory*. G. Dosi, C. Freeman, R. Nelson, G. Silverberg, and L. Soete (eds.). Pinter, London/New York, NY.

Dosi, G., K. Pavitt, and L. Soete, 1990: *The Economics of Technical Change and International Trade*. Harvester Wheatsheaf, London.

Douglas, M., D. Gasper, S. Ney, and M. Thompson, 1998: Human needs and wants. In *Human Choice and Climate Change*, **1**(3), S. Rayner, E.L. Malone (eds.), The Societal Framework, Battelle Press, Columbus, OH, pp. 195-263.

Easterlin, R.A., 1978: The economics and sociology of fertility: A synthesis. In *Historical Studies of Changing Fertility*. C. Tilly, (ed.), Princeton University Press, Princeton, NJ, pp. 57-113.

Eberstadt, N., 1997: World Population Implosion? *Public Interest*, **129**, 3-22.

Edmonds, J.A., and D.W. Barns, 1992: Factors affecting the long-term cost of fossil fuel CO_2 emissions reductions. *International Journal of Global Energy Issues*, **4**(3), 140-166.

EEA (European Environmental Agency), 1999: *Environment in the European Union at the Turn of the Century*. EEA, Copenhagen.

ENDS, 1998: *Renewables Win Major Boost under Largest NFFO Order*. The ENDS Report 284, Environmental Data Services Ltd, London.

Engelman, R., 1994: *Stabilizing the Atmosphere: Population, Consumption and Greenhouse Gases*. Population Action International, Washington, DC.

Enquete-Kommission, 1990: *Protection of the Atmosphere*. Report of the "Enquete Kommission: Schutz der Erdatmosphäre" of the German Parliament, Economica Verlag, Bonn.

Environment Canada, 1997: *Canada's Second Report on Climate Change*. Environment Canada, Ottawa and Secretariat of the UN Framework Convention on Climate Change, Bonn.

FAO (UN Food and Agriculture Organization), 1963-1996: *Production Yearbook* (yearly volumes). FAO, Rome.

Farla, J.C.M., and K. Blok, 1995: Energy conservation investment of firms: Analysis of investments in energy efficiency in the Netherlands in the 1980s. In *Proceedings of the American Council for an Energy-Efficient Economy 1995 Summer Study on Energy Efficiency in Industry*, ACEEE, Washington, DC.

Fenhann, J., 2000: Industrial non-energy, non-CO_2 greenhouse gas emissions. *Technological Forecasting and Social Change*, **63**(2-3). (In press).

Fearnside, P.M., 1995: Hydroelectric dams in the Brazilian Amazon as sources of greenhouse gases. *Environmental Conservation*, **22**(1), 7-19.

FETC (Federal Energy Technology Center), 1997: *Fuel Cell Overview*. USDOE/EIA, Washington, DC.

Fischer, G., and C. Rosenzweig, 1996: *The Impacts of Climate Change, CO_2, and SO_2 on Agricultural Supply and Trade*. WP-96-5, International Institute for Applied Systems Analysis, Laxenburg, Austria.

Foell W., M. Amann, G. Carmichael, M. Chadwick, J.-P. Hettelingh, L. Hordijk, and Z. Dianwu, 1995: *Rains Asia: An assessment model for air pollution in Asia*. Report on the World Bank sponsored project "Acid Rain and Emission Reductions in Asia", World Bank, Washington, DC.

Fogel, R.W., 1970: *Railroads and American Economic Growth: Essays in Economics History*. The Johns Hopkins University Press, Baltimore, MD.

Frankel, M., 1955: Obsolescence and technological change in a maturing economy. *American Economic Review*, **45**, 296-319.

Freeman, C., and C. Perez, 1988: Structural crises of adjustment: business cycles and investment behaviour (Chapter 3). In *Technical Change and Economic Theory*. G. Dosi *et al*. (eds.), Pinter, London, pp. 38-66.

Freeman, C., 1990: Schumpeter's business cycles revisited. In *Evolving Technology and Market Structure — Studies in Schumpeterian Economics*. A. Heertje, M. Perlman (eds.), University of Michigan, Ann Arbor, MI, pp. 17-38.

Freeman, C., 1994: The economics of technical change. *Cambridge Journal of Economics*, **18**, 463-514.

Freeman, C. (ed.), 1996: *Long Wave Theory*. Elgar Reference, Cheltenham, UK.

Gaffin, S. R., and B.C. O'Neill, 1997: Population and global warming with and without CO_2 targets. *Population and Environment*, **18**(4), 389-413.

Gaffin, S.R., 1998: World population projections for greenhouse gas emissions scenarios. *Mitigation and Adaptation Strategies for Global Change*, **3**(2-4), 133-170.

Gaffin, S. R., and B.C. O'Neill, 1998: Combat climate change by reducing fertility, *Nature*, **396**(6709), 307.

Gagnon, L., and J. F. van de Vate, 1997: Greenhouse gas emissions from hydropower. *Energy Policy*, **25**(1), 7-13.

Galeotti, M., and A. Lanza 1999: *Desperately Seeking (Environmental) Kuznets*. WP 2.99, Fondazione Eni Enrico Mattei, Milan, Italy.

Galy-Lacaux, C., R. Delmas, G. Kouadio, S. Richard, and P. Gosse, 1999: Long-term greenhouse gas emissions from hydroelectric reservoirs in tropical forest regions. *Global Biogeochemical Cycles*, **13**(2), 503-517.

Goldemberg, J., T.B. Johansson, A.K. Reddy, and R.H. Williams, 1988: *Energy for a Sustainable World*. Wiley Eastern, New Delhi.

Goldemberg, J., 1991: Leap-frogging: A new energy policy for developing countries. *WEC Journal*, **December**, 27-30.

Golove, W., and L. Schipper, 1997: Restraining carbon emissions: Measuring energy use and efficiency in the USA. *Energy Policy*, **25**(7-9), 803-812.

Goulder, L.H., and K. Mathai, 1998: *Optimal CO_2 Abatement in the Presence of Induced Technological Change*. WP 6494, National Bureau of Economic Research (NBER), Cambridge, MA.

Government of Russian Federation, *1995:* Russian Economic Trends 1995, **4**(1-2). Whurr Publishers, London.

Government of the People's Republic of China, 1996: *National Report on Sustainable Development*. http://www.acca21.edu.cn/nrport.html.

Greene, D.L., and K.G. Duleep, 1993: Costs and benefits of automotive fuel economy improvement: A partial analysis. *Transportation Research-A*, **27A**(3), 217-235.

Greening, L.A., and H.T. Jeng, 1994: Life-cycle analysis of gasoline expenditure patterns. *Energy Economics*, **16**(3), 217-228.

Greening, L.A., H.T. Jeng, J. Formby, and D.C. Cheng, 1994: Use of region, life-cycle and role variables in the short-run estimation of the demand for gasoline and miles travelled. *Applied Economics*, **27**(7), 643-655.

Greenpeace, 1997: *Energy Subsidies in Europe: How Governments Use Taxpayers Money to Promote Climate Change and Nuclear Risk*. Greenpeace International Climate Campaign, Amsterdam.

Gregory, K., 1998: Factors affecting future emissions of methane from non-land use sources. *Mitigation and Adaptation Strategies for Global Change*, **3**(2-4), 321-341.

Gregory, K., and H.-H. Rogner, 1998: Energy resources and conversion technologies for the 21st century. *Mitigation and Adaptation Strategies for Global Change*, **3**(2-4), 171-229.

Griliches, Z., 1996: The discovery of the residual: A historical note. *Journal of Economic Literature*, **34** (September), 1324-1330.

Grossman, G.M., and E. Helpman, 1991: *Innovation and Growth in the Global Economy*. MIT Press, Cambridge, MA.

Grossman, G.M., and E. Helpman, 1993: *Endogenous Innovation in the Theory of Growth*. WP 4527, National Bureau of Economic Research (NBER), Cambridge, MA.

Grubb M.J., and N.I. Meyer, 1993: Wind energy: resources, systems and regional strategies. In *Renewable Energy: Sources for Fuels and Electricity*. T.B. Johansson, H. Kelly, A.K.N. Reddy, R.H. Williams (eds.), Island Press, Washington, DC.

Grübler, A., and N. Nakićenović, 1991: *Evolution of Transport Systems: Past and Future*. RR-91-8, International Institute for Applied Systems Analysis, Laxenburg, Austria.

Grübler, A., S. Messner, L. Schrattenholzer, and A. Schäfer, 1993a: Emission reduction at the global level. *Energy*, **18**(5), 539-581.

Grübler, A, N. Nakićenović, and A. Schäfer, 1993b: *Dynamics of Transport and Energy Systems: History of development and a scenario for the future*. RR-93-19, International Institute for Applied Systems Analysis, Laxenburg, Austria.

Grübler, A., 1994: *A Comparison of Global and Regional Energy Emission Scenarios*. WP-94-132, International Institute for Applied Systems Analysis, Laxenburg, Austria.

Grübler, A., and S. Messner, 1996: Technological uncertainty. In *Climate Change: Integrating Science, Economics, and Policy*. CP-96-1, N. Nakićenović, W.D. Nordhaus, R. Richels, F.L. Toth (eds.), International Institute for Applied Systems Analysis, Laxenburg, Austria, pp. 295-314.

Grübler, A., 1998a: *Technology and Global Change*. Cambridge University Press, Cambridge.

Grübler, A., 1998b: *Integrated Assessment Modeling of Global Environmental Policies and Decision Patterns*. Report to EC DG XII/D-5 No. ENV4-CT-96-0197, International Institute for Applied Systems Analysis, Laxenburg, Austria.

Grübler, A., 1998c: A review of global and regional sulfur emission scenarios. *Mitigation and Adaptation Strategies for Global Change*, **3**(2-4), 383-418.

Gürer, N., and J. Ban, 1997: Factors affecting energy-related CO_2 emissions: past levels and present trends. *OPEC Review*, **XXI**(4), 309-350.

HABITAT (United Nations Centre for Human Settlements), 1996: *An Urbanizing World: Global Report on Human Settlements 1996*. Oxford University Press, Oxford.

Häfele, W., J. Anderer, A. McDonald, and N. Nakićenović, 1982: *Energy in a Finite World: Paths to a Sustainable Future*. Ballinger, Cambridge, MA.

Hall, D.O., and K.K. Rao, 1994: *Photosynthesis, 5th Edition*. Cambridge University Press, Cambridge.

Hall, D.O., and F. Rosillo-Calle, 1998: *Biomass Resources Other than Wood*. World Energy Council, London.

Hammer, J.S., 1985: *Population Growth and Savings in Developing Countries*. WP 687, World Bank Staff, Population and Development Series 12, World Bank, Washington, DC.

Han, X., and L. Chatterjee, 1997: Impacts of growth and structural change on CO_2 emissions of developing countries. *World Development*, **5**(2), 395-407.

Hatfield, C.B., 1997: Oil back on the global agenda. *Nature*, **387**(8 May), 121.

Haub, C., 1997: New UN projections depict a variety of demographic futures. *Population Today: News Numbers and Analysis*, **25**(4), 1-3.

Hayami, Y., and V.W. Ruttan, 1985: *Agricultural Development: An International Perspective, Revised 2nd Edition*. John Hopkins University Press, Baltimore MD.

Hayek, F., 1967: Notes on the Evolution of Systems of Rules of Conduct. In *Studies in Philosophy, Politics and Economics*, Routledge and Kegan Paul, London, pp. 66-81.

Hensher, D.A., F.W. Milthorpe, and N.C. Smith, 1990: The demand for vehicle use in the urban household sector: Theory and empirical evidence. *Journal of Transport Economics and Policy*, **24**(2), 119-137.

Hettelingh, J.P., M.J. Chadwick, H. Sverdrup, and D. Zhao, 1995: Assessment of environmental effects of acidic deposition. In *Rains Asia: An assessment model for airpollution in Asia*. W. Foell, M. Amann, G. Carmichael, M. Chadwick, J.-P. Hettelingh, L. Hordijk, and Z. Dianwu (eds.), Report on the World Bank sponsored project "Acid Rain and Emission Reductions in Asia", World Bank, Washington, DC, pp. VI-1-64.

Higgins, M., and J.G. Williamson, 1997: Age structure dynamics in Asia and dependence on foreign capital. *Population and Development Review*, **23**(2), 261-293.

Hill, C. 1975: *The World Turned Upside Down: Radical Ideas During the English Revolution*. Penguin Books, London.

Houghton, J.T., G.J. Jenkins, and J.J. Ephraums (eds.), 1990: *Climate Change: The IPCC Scientific Assessment*. Cambridge University Press, Cambridge.

Houghton, R.A., 1991: Tropical deforestation and climate change. *Climatic Change*, **19**, 99-118.

Houghton, J.T., B.A. Callander, and S.K. Varney (eds.), 1992: *Climate Change 1992*. The Supplementary Report to the IPCC Scientific Assessment, Cambridge University Press, Cambridge.

Houghton, J.T., L.G. Meira Filho, J. Bruce, Hoesung Lee, B. A. Callander, E. Haites, N. Harris, and K. Maskell (eds.), 1995: *Climate Change 1994: Radiative Forcing of Climate Change and an Evaluation of the IPCC IS92 Emissions Scenarios*. Cambridge University Press, Cambridge.

Houghton, J.T., L.G. Meira Filho, B.A. Callander, N. Harris, A. Kattenberg, and K. Maskell (eds.), 1996: *Climate Change 1995. The Science of Climate Change*. Contribution of Working Group I to the Second Assessment Report of the Intergovernmental Panel on Climate Change, Cambridge University Press, Cambridge.

Huntington, S.P, 1996: *The Clash of Civilizations and the Remaking of World Order*. Simon & Schuster, New York, NY.

ICAO (International Civil Aviation Organization), 1995a: *Outlook for Air Transport to the Year 2003*. Circular 252-AT/103, Montreal, Canada.

ICAO (International Civil Aviation Organization), 1995b: *Civil Aviation Statistics of the World, 1994*. Doc. 9180/20, Montreal, Canada.

IEA (International Energy Agency), 1993: *Cars and Climate Change*. OECD, Paris.

IEA (International Energy Agency), 1995: *World Energy Outlook, 1995 Edition*. IEA/OECD, Paris.

IEA (International Energy Agency), 1997a: *Energy Balances of OECD Countries, 1960-1995*. IEA/OECD, Paris.

IEA (International Energy Agency), 1997b: *Energy Balances of Non-OECD Countries, 1960-1995*. IEA/OECD, Paris.

IEA (International Energy Agency), 1997c: *Indicators of Energy Use and Efficiency: Understanding the Link Between Energy and Human Activity*. IEA/OECD, Paris.

IEA (International Energy Agency), 1999. *Non-OECD coal-fired power generation – trends in the 1990s*. IEA Coal Research, London.

IEA CIAB (Coal Industry Advisory Board), 1992: *Global Methane Emissions from the Coal Industry*. International Energy Agency , Paris.

IEA Greenhouse Gases R&D Programme, 1996a: *Methane Emissions from Coal Mining*. IEA Greenhouse Gas R&D Programme, Stoke Orchard, Cheltenham, UK.

IEA Greenhouse Gases R&D Programme, 1996b: *Methane Emissions from Land Disposal of Solid Waste*. IEA Greenhouse Gas R&D Programme, Stoke Orchard, Cheltenham, UK.

IEA Greenhouse Gases R&D Programme, 1997: *Methane Emissions from the Oil and Gas Industry*. IEA Greenhouse Gas R&D Programme, Stoke Orchard, Cheltenham, UK.

IGU (International Gas Union), 1997a: *World Gas Prospects, Strategies and Economics*. Proceedings of the 20th World Gas Conference, Copenhagen. (http://www.wgc.org/proceedings).

IGU (International Gas Union), 1997b: *Gas and the Environment - Methane Emissions*. Report of IGU Task Force I, 20th World Gas Conference Proceedings, Copenhagen. (http://www.wgc.org/proceedings/rep/tf1r.html).

IIASA–WEC (International Institute for Applied Systems Analysis – World Energy Council), 1995: *Global Energy Perspectives to 2050 and Beyond*. WEC, London.

IPCC (Intergovernmental Panel on Climate Change), 1995*: Greenhouse Gas Inventory Reporting Instructions, Workbook and Reference Manual*. IPCC Guidelines for National Greenhouse Gas Inventories, Vols. 1-3. *UNEP, WHO, OECD, IEA, Geneva*.

IPCC (Intergovernmental Panel on Climate Change)**,** 1996: Greenhouse Gas Inventory Reference Manual. *Revised 1996 IPCC Guidelines for National Greenhouse Gas Inventories, Vol. 3*. IPCC/OECD/IEA, Geneva.

IPCC (Intergovernmental Panel on Climate Change), 2000: *IPCC Special Report on Methodological and Technological Issues in Technology Transfer*. Cambridge University Press, Cambridge. (In press).

Ironmonger, D., C., Aitkane, and B. Erbas, 1995: Economies of scale in energy use in adult-only households. *Energy Economics*, **17**(4), 301-310.

Ishitani H., and T.B. Johansson, 1996: Energy supply mitigation options. In *Climate Change 1995 - Impacts, Adaptation and Mitigation of Climate Change: Scientific Analysis*. R.T. Watson, M.C. Zinyowera, R.H. Moss (eds.), IPCC, Cambridge University Press, Cambridge, pp. 589-647.

Ivanhoe L.F., and G.G. Leckie, 1993: Global oil, gas fields, sizes tallied, analyzed. *Oil and Gas Journal*, **91**(7), 87-91.

Jansson, J.O., 1989: Car demand modelling and forecasting. *Journal of Transport Economics and Policy*, **23**(2), 125-140.

Jefferson, M., 1983: Economic uncertainty and business decision-making. In *Beyond Positive Economics?* J. Wiseman (ed.), Macmillan Press, London, pp. 122-159.

Jepma, C.J., 1995: *Tropical Deforestation: A Socio-Economic Approach*. Earthscan, London.

Jochem, E., 1989: *Rationelle Energienutzung in den Industrieländern – Praxis und zukünftige Chancen, Vol. 53*. Physikertagung der Deutschen Physikalischen Gesellschaft, Bonn.

Johansson, T.B., H. Kelly, A.K.N. Reddy, and R.H. Williams, 1993: Renewable fuels and electricity for a growing world economy: Defining and achieving the potential. In *Renewable Energy: Sources for Fuels and Electricity*. T.B. Johansson, H. Kelly, A.K.N. Reddy, R.H. Williams, Island Press, Washington, DC.

Joskow, P.L., R. Schmalensee, and E.M. Bailey, 1998: The market for sulfur dioxide emissions. *American Economic Review*, **88**(4), 669-685.

Judson, R.A., R. Schmalensee, and T.M. Stoker, 1999: Economic development and the structure of the demand for commercial energy. *The Energy Journal*, **20**(2), 29-57.

Jung, T.Y., E. Lebre La Rovere, H. Gaj, P.R. Shukla, and D. Zhou, 2000: Structural changes in developing countries and their implication to energy-related CO_2 emissions. *Technological Forecasting and Social Change*, **63**(2-3). (In press).

Kato, N., 1996: Analysis of structure of energy consumption and dynamics of emission of atmospheric species related to global environmental change (SO_x, NO_x, CO_2) in Asia. *Atmospheric Environment*, **30**(5), 757-785.

Kaya, Y., 1990: *Impact of Carbon Dioxide Emission Control on GNP Growth: Interpretation of Proposed Scenarios*. Paper presented to the IPCC Energy and Industry Subgroup, Response Strategies Working Group, Paris (mimeo).

Keepin, B. 1986: A review of global energy and carbon dioxide projections. *Annual Review of Energy*, **11**, 357-392.

Kelley, A., 1988: Economic Consequences of Population Change in the Third World. *Journal of Economic Literature*, **26**(4), 685-728.

Kendall, H.W., and D. Pimentel, 1994: Constraints on the expansion of the global food supply. *Ambio*, **23**(3), 198-205.

Kolsrud, G., and B.B. Torrey, 1992: The importance of population growth in future commercial energy consumption. In *Global Climate Change: Linking Energy, Environment, Economy and Equity*. J.C. White (ed.), Plenum, New York, NY.

Kondratiev, N.D., 1926: Die langen Wellen in der Konjunktur. *Archiv für Sozialwissenschaft und Sozialpolitik*, **56**, 573-609.

Krackeler, T., L. Schipper, and O. Sezgen, 1998: Carbon dioxide emissions in OECD service sectors: the critical role of electricity. *Energy Policy*, **26**(15), 1137-1152. See also **Krackeler**, T., L. Schipper, and O. Sezgen, 1998: *The Dynamics of Service Sector Carbon Dioxide Emissions and the Critical Role of Electricity Use: A Comparative Analysis of 13 OECD Countries from 1973-1995*. LBNL-41882, Lawrence Berkeley National Laboratory, Berkeley, CA.

Kroeze, C., 1993: Global Warming and Nitrous Oxide. Dissertation. University of Amsterdam, the Netherlands.

Kuznets, S.S., 1958: *Six Lectures on Economic Growth*. Free Press, New York, NY.

Laherrere, J., 1994: Published figures and political reserves. *World Oil*, **January**, 33.

Landes, D.S., 1969: *The Unbound Prometheus: Technological Change and Industrial Development in Western Europe from 1750 to the Present*. Cambridge University Press, Cambridge.

Lane, D., and R. Maxfield, 1995: *Foresight, Complexity and Strategy*. Santa Fe Institute, New Mexico, USA.

La Rovere, E.L., and B. Americano, 1998: *Environmental Impacts of Privatizing the Brazilian Power Sector*. Proceedings of the International Association of Impact Assessment Annual Meeting, Christchurch, New Zealand, April 1998.

Larson B., and A. Shah, 1992: *World Fossil Fuel Subsidies and Global Carbon Emissionss*, Policy Research Working Paper Serres No. 1002, World Bank, Washington DC, USA.

Lashof, D., and Tirpak, D.A., 1990: *Policy Options for Stabilizing Global Climate*. 21P-2003. US Environmental Protection Agency, Washington DC.

Lazarus, M.L., L. Greber, J. Hall, C. Bartels, S. Bernow, E. Hansen, P. Raskin, and D. von Hippel, 1993: *Towards a Fossil Free Energy Future: The Next Energy Transition*. A Technical Analysis for Greenpeace International, Stockholm Environmental Institute Boston Center, Boston, MA.

LBL (Lawrence Berkeley National Laboratory), 1998: *OECD Database*. Lawrence Berkeley National Laboratory, International Energy Studies, Berkeley, CA.

Leemans, R., 1996: Mitigation: Cross-sectional and other issues. In *Climate Change 1995, Impacts Adaptation and Mitigation of Climate Change: Scientific-Technical Analysis*. R.T. Watson, M.C. Zinyowera, R.H. Moss (eds.), Contribution of Working Group II to the Second Assessment Report of the Intergovernmental Panel on Climate Change, Cambridge University Press, Cambridge, pp. 801-819.

Leemans, R., van Amstel., A., Battjes, C., Kreileman, E., and S. Toet, 1996: The land cover and carbon cycle consequences of large scale utilizations of biomass as an energy source. *Global Environmental Change*, **6**(4), 335-357.

Leggett, J., W.J. Pepper, and R.J. Swart, 1992: Emissions scenarios for IPCC: An update. In *Climate Change 1992. The Supplementary Report to the IPCC Scientific Assessment*. J.T. Houghton, B.A. Callander, S.K. Varney (eds.), Cambridge University Press, Cambridge.

Levine, M.D., H. Akbari, J. Busch, G. Dutt, K. Hogan, P. Komor, S. Meyeres, and H. Tsuchiya, 1996: *Mitigation Options for Human Settlements, Climate Change 1995: Impacts, Adaptations and Mitigation of Climate Change: Scientific-Technical Analyses*. Contribution of Working Group II to the Second Assessment Report of the Intergovernmental Panel on Climate Change, R.T. Watson, M.C. Zinyowera, R.H. Moss (eds.), Cambridge and New York, Cambridge University Press, Cambridge, UK, pp. 399-426.

Lovins, A.B., and L.H. Lovins, 1991: Least-cost climatic stabilization. *Annual Review of Energy*, **16**, 433-531.

Lovins, A.B., J.W. Barnett, and L.H. Lovins, 1993: *Supercars, The Coming Light-Vehicle Revolution*. Rocky Mountain Institute, Snowmass, CO.

Lucas, R., 1988: On the mechanics of economic development. *Journal of Monetary Economics*, **22**(1), 3-42.

Lutz, W., 1993: Population and environment – What do we need more urgently: Better data, better models, or better questions? In *Environment and Population Change*. B. Zaba, J. Clarke (eds.), Derouaux Ordina Editions, Liege, Belgium.

Lutz, W. (ed.), 1994: *The Future Population of the World: What can we assume today*? 1st Edition, Earthscan, London.

Lutz, W. (ed.), 1996: *The Future Population of the World: What can we assume today?* 2nd Edition, Earthscan, London.

Lutz, W., W. Sanderson, S. Scherbov, and A. Goujon, 1996: World population scenarios in the 21st century. In *The Future Population of the World: What Can We Assume Today?* 2nd Rev. ed. W. Lutz (ed.), Earthscan, London, pp. 361-396.

Lutz, W., W. Sanderson, and S. Scherbov, 1997: Doubling of world population unlikely. *Nature*, **387**(6635), 803-805.

MacKellar, F.L., W. Lutz, C. Prinz, and A. Goujon, 1995: Population, households and CO_2 emissions. *Population and Development Review*, **21**(4), 849-865.

Maddison, A., 1989: *The World Economy in the 20th Century*. OECD Development Centre Studies, Organisation for Economic Co-Operation and Development, Paris.

Maddison, A., 1991: *Dynamic Forces in Capitalist Development - A Long-run Comparative View*. Oxford University Press, Oxford.

Maddison, A., 1995: *Monitoring the World Economy 1820-1992*. OECD Development Centre Studies, Organisation for Economic Co-operation and Development, Paris.

Maddison, A., 1998: *Chinese Economic Performance in the Long Run*. OECD Development Centre Studies, Organisation for Economic Co-operation and Development, Paris.

Manne, A., and R. Richels, 1994: The costs of stabilizing global CO_2 emissions: A probabilistic analysis based on expert judgements. *The Energy Journal*, **15**(1), 31-56.

March Consulting Group, 1999: Opportunities to Minimize Emissions of HFCs from EU. European Communities DG III, Brussels.

Marchand, O, 1992: Une comparison internationale de temps de travail. *Futuribles*, **165-166**(5-6), 29-39.

Marchetti, C., 1980: Society as a learning system: discovery, invention, and innovation cycles revisited. *Technological Forecasting and Social Change*, **18**, 267-282.

Marland, G., R.J. Andres, T.A. Boden, C. Johnston, and A. Brenkert, 1998: *Global, Regional and National CO_2 Emissions Estimates from Fossil Fuel Burning, Cement Production, and Gas Flaring: 1951-1995 (revised January 1998)*. Carbon Dioxide Information Analysis Center, Oak Ridge National Laboratory/U.S. Department of Energy, Oak Ridge, TN.

Martin, J.M., 1988: L'intensité énergetique de l'activité economique dans les pays industrialisés: Les evolutions de très longue periode liverent-elles des enseignements utiles? *Economies et Societés*, **4**, 9-27.

Martin, D.J., and R.A.W. Shock, 1989: *Energy Use and Energy Efficiency in UK Transport up to the Year 2010*. Energy Efficiency Series No. 10, Energy Efficiency Office, Department of Energy, HMSO, London.

Martinerie, P., G. Brasseur, and C. Granier, 1995: The chemical composition of ancient atmospheres: A model study constrained by ice core data. *Journal of Geophysical Research*, **100**, 14,291-14,304.

Maslow, A., 1954: *Motivation and Personality*. Harper and Row, New York, NY.

Masters, C.D., E.D. Attanasi, W.D. Dietzman, R.F. Meyer, R.W. Mitchell, and D.H. Root, 1987: *World Resources of Crude Oil, Natural Gas, Natural Bitumen and Shale Oil*, Proceedings of the 12th World Petroleum Congress, Houston, TX.

Masters, C.D., E.D. Attanasi, and D.H. Root, 1994: *World petroleum assessment and analysis*. Proceedings of the 14th World Petroleum Congress, Stavanger, Norway, John Wiley, Chichester, UK, pp. 1-13.

Matsuoka, Y., and T. Morita, 1994: *Estimation of Carbon Dioxide Flux from Tropical Deforestation*. CGER-I013-94. Center for Global Environmental Research, National Institute for Environmental Studies, Tskuba, Japan.

Matsuoka, Y., M. Kainuma, and T. Morita, 1994: Scenario analysis of global warming using the Asian-Pacific integrated model (AIM). *Energy Policy*, **23**(4/5), 357-371.

Matthews, R.C.O., 1984: Darwinism and economic change. In *Economic Theory and Hicksian Themes*. D.A. Collard, N.H. Dimsdale, C.L. Gilbert, D.R. Helm, M.F.G. Scott, A.K. Sen (eds.), Oxford University Press, Oxford.

McDevitt, T.M., 1996: *World Population Profile*. Report WP/96, U.S. Bureau of Census, Government Printing Office, Washington, DC.

McDonald, A., 1999: Combating acid deposition and climate change. *Environment*, **41**(3), 4-11, 43-41.

Messner, S., A. Golodnikov, and A. Gritsevskii, 1996: *A Stochastic Version of the Dynamic Linear Programming Model MESSAGE III*. RR-97-002, International Institute for Applied Systems Analysis, Laxenburg, Austria.

Messner, S., 1997: Endogenized technological learning in an energy systems model. *Journal of Evolutionary Economics*, **7**(3), 291-313.

Michaelis, L., 1996: *CO_2 Emissions from Road Vehicles: Policies and Measures for Common Action under the UNFCCC*. Working Paper 1 in the Series on Policies and Measures for Common Action Under the UNFCCC, OECD, Paris.

Michaelis, L, D.L. Bleviss, J.-P. Orfeuil, and R. Pischinger, 1996: Mitigation Options in the Transportation Sector. In *Climate Change 1995: Impacts, Adaptations and Mitigation of Climate Change: Scientific-Technical Analyses*. Contribution of Working Group II to the Second Assessment Report of the Intergovernmental Panel on Climate Change, R.T.Watson, M.C. Zinyowera, R.H. Moss (eds.), Cambridge University Press, Cambridge, pp. 679-712.

Michaelis, L., 1997a: *Special Issues in Carbon/Energy Taxation: Carbon Charges on Aviation Fuels*. Working Paper 12 in the Series on Policies and Measures for Common Action Under the UNFCCC, OECD, Paris.

Michaelis, L., 1997b: Case study on electricity in the United Kingdom. In *Supports to the Coal Industry and the Electricity Sector: Environmental implications of energy and transport subsidies, Vol. 2*. OCDE/GD(97)155, OECD, Paris.

Michaelis, L., 1997c: Technical and behavioral change: Implications for energy end-use. In *Energy Modelling: Beyond Economics and Technology*. B. Giovannini, A. Baranzini (eds.), International Academy of the Environment, Centre for Energy Studies, University of Geneva, Geneva.

Millward, R., 1990: Productivity in the UK services sector: Historical trends 1865-1985 and comparisons with the USA 1950-85. *Oxford Bulletin Economic Stat.*, **52**(4), 423-436.

Mokyr, J., 1990: *The Lever of Riches: Technological Creativity and Economic Progress*. Oxford University Press, Oxford.

Moll, H.C., 1989: *Aanbod van en vraag naar metalen; ontwikkelingen, implicaties en relaties; een methodische assesment omtrent subsitutie van materialen*. Working report no. 9, Institute for Energy and Environment (IVEM), University of Groningen, Groningen, the Netherlands.

Morita, Y., Y. Matsuoka, M. Kainuma, and H. Harasawa, 1994: AIM - Asian Pacific integrated model for evaluating policy options to reduce GHG emissions and global warming impacts. In *Global Warming Issues in Asia*. S. Bhattacharya *et al.* (eds.), AIT, Bangkok, pp. 254-273.

Mosier, A., C. Kroeze, C. Nevison, O. Oenema, S. Seitzinger, and O. van Cleemput, 1998: Closing the global N_2O budget: nitrous oxide emissions through the agricultural nitrogen cycle: OECD/IPCC/IEA phase II development of IPCC guidelines for national greenhouse gas inventory methodology. *Nutrient Cycling in Agrosystems*, **52**, 225-248.

Mumford, L., 1934: *Technics and Civilation*. Harcourt, New York, NY.

Murthy, N.S., M. Panda, and J.K. Parikh, 1997: Economic development, poverty reduction and carbon emissions in India. *Energy Economics*, **19**, 327-354.

Nadel, S.M., D. Fridley, J. Sinton, Y. Zhirong, and L. Hong, 1997: *Energy Efficiency Opportunities in the Chinese Building Sector.* American Council for an Energy-Efficient Economy, Washington, DC.

Nakićenović, N., A. Grübler, A. Inaba, S. Messner, S. Nilson, *et al.*, 1993: Long-term strategies for mitigating global warming. *Energy,* **18**(5), 401-609.

Nakićenović, N., 1996: Technological change and learning. In *Climate Change: Integrating Science, Economics, and Policy.* N. Nakićenović, W.D. Nordhaus, R. Richels, F.L. Tol (eds.), International Institute for Applied Systems Analysis, Laxenburg, Austria, pp. 271-294.

Nakićenović, N., and A. Grübler, 1996: *Energy and the Protection of the Atmosphere.* United Nations Department for Policy Coordination and Sustainable Development, New York, NY.

Nakićenović, N., A. Grübler, H. Ishitani, T. Johansson, G. Marland, J.R. Moreira, and H.-H. Rogner, 1996: Energy primer. In *Climate Change 1995. Impacts, Adaptations and Mitigation of Climate Change: Scientific Analyses.* R. Watson, M.C. Zinyowera, R. Moss (eds.), Cambridge University Press, Cambridge, pp. 75-92.

Nakićenović, N., M. Amann, and G. Fischer, 1997: *Global Energy Supply and Demand and their Environmental Effects.* International Institute for Applied Systems Analysis, Laxenburg, Austria (see also: McDonald, 1999).

Nakićenović, N., A. Grübler, and A. McDonald (eds.), 1998a: *Global Energy Perspectives.* Cambridge University Press, Cambridge.

Nakićenović, N., N. Victor, and T. Morita, 1998b: Emissions scenarios database and review of scenarios. *Mitigation and Adaptation Strategies for Global Change,* **3**(2-4), 95-131.

National Research Council, 1986: *Population Growth and Economic Development: Policy Questions.* National Academy Press, Washington, DC.

Neij, L., 1997: Use of experience curves to analyse the prospects for diffusion and adoption of renewable energy technology. *Energy Policy,* **23**(13), 1099-1107.

Nelson, R.R., M.J. Peck, and E.D. Kalachek, 1967: *Technology, Economic Growth and Public Policy.* The Brookings Institute, Washington, DC.

Nelson, R.R., and S.G. Winter, 1982: *An Evolutionary Theory of Economic Change.* Harvard University Press, Cambridge, MA.

Newman, P., and M. Kenworthy, 1990: *Cities and Automobile Dependence.* Gower, London.

Nordhaus, W.D., 1993: Rolling the DICE: An optimal transition path for controlling greenhouse gases. *Resource and Energy Economics* **15**, 27-50.

Nordhaus, W.D., and G.W. Yohe, 1983: Future paths of energy and carbon dioxide emissions. In *Changing Climate: Report of the Carbon Dioxide Assessment Committee.* National Academy Press, Washington, DC.

Nriagu, J.O., 1996: A history of global metal pollution. *Science,* **272**, 223-224.

Odell, P.R., 1997: Oil reserves: Much more than meets the eye. *Petroleum Economist,* **64**, 29-31.

Odell, P.R., 1998: *Fossil Fuel Resources in the 21st Century.* Report submitted to the International Atomic Energy Agency, Vienna, Austria (see also Odell, 1999).

Odell, P.R., 1999: Dynamics of energy technologies and global change. *Energy Policy,* **27**, 737-742.

OECD (Organisation for Economic Co-operation and Development), 1996: *Values, Welfare and Quality of Life.* Final Report of the First OECD Workshop on Individual Travel Behaviour , OCDE/GD(96)199, OECD, Paris.

OECD (Organisation for Economic Co-operation and Development), 1997a: *Environmental Implications of Energy and Transport Subsidies.* OECD, Paris.

OECD (Organisation for Economic Co-operation and Development), 1997b: *Agriculture and Forestry: Identification of Options for Net Greenhouse Gas Reduction.* Annex I, Expert Group on the United Nations Framework Convention on Climate Change, WP 7, OCDE/GD(97)74, OECD, Paris.

OECD (Organisation for Economic Co-operation and Development), 1998a: *Twentyfirst Century Technologies: Promises and Perils of a Dynamic Future.* OECD, Paris.

OECD (Organisation for Economic Co-operation and Development), 1998b: *Eco-Efficiency.* OECD, Paris.

OECD (Organisation for Economic Co-operation and Development), 1998c: *Projected Costs of Generating Electricity – Update 1998.* Nuclear Energy Agency (NEA), International Energy Agency (IEA) and Organisation for Economic Co-operation and Development (OECD), Paris.

Ogawa, Y. 1991: Economic activity and greenhouse effect. *The Energy Journal,* **12**(1), 23-34.

Olivier, J.G.J, A.F. Bouwman, C.W.M. van der Maas, J.J.M. Berdowski, C. Veldt, J.P.J. Bloos, A.J.H. Visschedijk, P.Y.J. Zanfeld, and J.L. Haverlag, 1996*: Description of EDGAR Version 2.0: "A Set of Global Emission Inventories of Greenhouse Gas Gases and Oxon-depleting Substances for all Anthropogenic and most Natural Scources on a per Country Basis and on a 1 x 1 grid".* RIVM Report 771060 002, RIVM, Bilthoven, the Netherlands.

O'Neill, B.C., 1996: *Greenhouse Gases: Time scales, Response Functions and the Role of Population Growth in Future Emissions.* PhD dissertation, New York University, New York, NY.

O'Neill, B.C., F.L. MacKellar, and W. Lutz, 2000: *Population and Climate Change.* Cambridge University Press, Cambridge. (In press).

Ouellette N., H-H. Rogner, and D.S. Scott, 1995: Hydrogen from remote excess hydro-electricity. Part I: Production plant capacity and production costs. *International Journal of Hydrogen Energy*, **20**(11), 865-872.

Palo, M., E. Lehto, and E.E. Enroth, 1997: *Scenarios on Tropical Deforestation and Carbon Fluxes.* Paper presented at the Global Modelling Forum, Tokyo, Japan.

Parikh, J.K., K.S. Parikh, S. Gokarn, J.P. Painuly, B. Saha, and V. Shukla, 1991: *Consumption Patterns: The Driving Force of Environmental Stress.* Report prepared for the United Nations Conference on Environment and Development (UNCED), IGIDR-PP-014, Indira Ghandi Institute for Development Research, Mumbai, India.

Parikh, J.K., 1992: IPCC strategies unfair to the south. *Nature*, **360** (10 December), 507-508.

Parikh, J.K., 1994: North-south issues for climate change. *Economic and Political Weekly*, **November 5-12**, 2940-2943.

Parikh, J.K., R. Culpeper, D. Runnalls, and J.P. Painuly (eds.), 1997: *Climate Change and North-South Cooperation: Indo-Canadian Cooperation in Joint Implementation.* Tata McGraw Hill Publishing, New Dehli.

Pearce, D., 1995: *Blueprint 4 – Capturing Global Environmental Value.* Earthscan, London.

Penner, S.S. *et al.*, 1995: Commercialization of fuel cells. *Energy,* **20**(5), 331-470.

Pepper, W.J., J. Leggett, R. Swart, J. Wasson, J. Edmonds, and I. Mintzer, 1992: Emissions Scenarios for the IPCC. An update: Assumptions, methodology, and results, Support document for Chapter A3. In *Climate Change 1992: Supplementary Report to the IPCC Scientific Assessment.* J.T. Houghton, B.A. Callandar, S.K. Varney (eds.), Cambridge University Press, Cambridge.

Pepper, W.J., W. Barbour, A. Sankovski, and B. Braaz, 1998: No-policy greenhouse gas emission scenarios: revisiting IPCC 1992. *Environmental Science & Policy,* **1**, 289-312.

Posch, M., J-P. Hettelingh, J. Alcamo, and M. Krol, 1996: Integrated scenarios of acidification and climate change in Asia and Europe. *Global Environmental Change*, **6**(4), 375-394.

Prather, M., R. Derwent, D. Erhalt, P. Fraser, E. Sanhueza, and X. Zhou, 1995: Other trace gases and atmospheric chemistry. In *Climate Change 1994 - Radiative Forcing of Climate Change*, IPCC (Intergovernmental Panel on Climate Change) and Cambridge University Press, Cambridge, pp. 73-126.

Preston, S.H., 1996: The effect of population growth on environmental quality. *Population Research and Policy Review,* **15**, 95-108.

Price, L., L. Michaelis, E. Worrell, and M. Khrushch, 1998: Sectoral trends and driving forces of global energy use and greenhouse gas emissions. *Mitigation and Adaptation Strategies for Global Change,* **3**(2-4), 263-319.

Qian, J., and Zhang, K., 1998: China's desulfurization potential. *Energy Policy,* **26**(4), 345-351.

Ramanathan, R., and J.K. Parikh, 1999: Transport sector in India: An analysis in the context of sustainable development. *Transport Policy,* **6**(1), 35-45.

Reynolds, T.S., and S.H. Cutcliffe, 1997: *Technology and the West*. Chicago University Press, Chicago, USA.

Riahi, K., and R.A. Roehrl, 2000: Greenhouse gas emissions in a dynamics-as-usual scenario of economic and energy development. *Technological Forecasting and Social Change*, **63**(2-3). (In press).

Rogner, H.-H., 1996: *An Assessment of World Hydrocarbon Resources*. WP-96-56, International Institute for Applied Systems Analysis, Laxenburg, Austria.

Rogner, H.-H., and N. Nakićenović, 1996: Zur Rolle des Schwefels in der Klimadebatte. *Energiewirtschaftliche Tagesfragen*, **46**(11), 731-736.

Rogner, H.-H., 1997: An assessment of world hydrocarbon resources. *Annual Review of Energy and the Environment*, **22**, 217-262.

Rogner, H.-H., and Khan, A.M., 1998: Comparing energy options. *IAEA Bulletin*, **40**(1), 2-6.

Romer, P.M., 1986: Increasing returns and long-run growth. *Journal of Political Economy*, **94**(5), 1002-1037.

Roodman, D., 1996: *Paying the Piper: Subsidies, Politics, and the Environment*. Worldwatch Paper 133, The Worldwatch Institute, Washington, DC.

Rosa, L.P., and R. Schäffer, 1994: Greenhouse gas emissions from hydroelectric reservoirs. *Ambio*, **23**(2), 164-165.

Rosa, L.P., and R. Schechtman, 1996: Avaliação de Custos Ambientais da Geração Termelétrica: inserção de variáveis ambientais no planejamento da expansão do setor elétrico. *Cadernos de Energia*, **2**(9), Edição Especial, PPE/COPPE/UFRJ, Rio de Janeiro, Brazil

Rosa, L.P., R. Schäffer, and M.A. dos Santios, 1996: Comment: Are hydroelectric dams in the Brazilian Amazon significant sources of greenhouse gases? *Environmental Conservation*, **23**(1), 2-6.

Rosegrant M.W., M.A. Sombilla, R.V. Gerpacio, and C. Ringler, 1997: *Global Food Markets and US Exports in the 21st Century*. Revisions of a paper presented at the Illinois World Food and Sustainable Agriculture Program Conference "Meeting the Demand for Food in the 21st Century: Challenges and Opportunities for Illinois Agriculture (IFPRI)". Washington, DC.

Rosenberg, N., 1982: *Inside the Black Box: Technology and Economics*. Cambridge University Press, Cambridge.

Rosenberg, N., and C. Frischtak, 1984: Technological innovation and long waves. *Cambridge Journal of Economics*, **8**(1), 7-24.

Rosenberg, N., and L.E. Birdzell, 1986: *How the West Grew Rich: The Economic Transformation of the Industrial World*. I.B. Tauris & Co., London.

Rosenberg, N., and L.E. Birdzell, 1990: Science, Technology, and the Western Miracle. *Scientific American*, **263**, 18-25.

Rosenberg, N., 1994: *Exploring the Black Box: Technology, Economics and History*. University Press, Cambridge.

Rosenberg, N., 1997: Economic development and the transfer of technology: some historical perspectives. In *Technology and the West*. T.S. Reynolds, S.H. Cutcliffe (eds.), University Press, Chicago, IL, pp. 251-276.

Ross, M.H., and D. Steinmeyer, 1990: Energy for industry. *Scientific American*, **263**, 89-98.

Rostow, W.W., 1978: *The World Economy: History and Prospect*. University of Texas Press, Austin, TX.

Rostow, W.W., 1990: *The Stages of Economic Growth, Third Edition*. Cambridge University Press, Cambridge.

Rotmans, J., and H.J.M. de Vries (eds.), 1997: *Perspectives on Global Futures: The TARGETS approach*. Cambridge University Press, Cambridge.

Rudd, J.W.M., R. Harris, C.A. Kelly, and R.E. Hecky, 1993: Are hydroelectric reservoirs significant sources of greenhouse gases? *Ambio*, **22**, (4 June), 246-248.

Sathar, Z.A., and J.B. Casterline, 1998: The onset of fertility transition in Pakistan. *Population and Development Review*, **24**(4), 773-796.

Sathaye, J., A. Ketoff, L. Schipper, and S. Lele, 1989: *An End-Use Approach to Development of Long-Term Energy Demand Scenarios for Developing Countries*, LBL-25611. Lawrence Berkeley National Laboratory, Berkeley, CA.

Sathaye, J., and A. Ketoff, 1991: CO_2 emissions from major developing countries: Better understanding the role of energy in the long term. *The Energy Journal*, **12**(1), 161-196.

Schimel, D., D. Alves, I. Enting, M. Heimann, F. Joos, D. Raynaud, T. Wigley, M. Prather, R. Derwent, D. Ehhalt, P. Fraser, E. Sanhueza, X. Zhou, P. Jonas, R. Charlson, H. Rodhe, S. Sadasivan, K.P. Shine, Y. Fouquart, V. Ramaswamy, S. Solomon, J. Srinivasan, D. Albritton, I. Isaksen, M. Lal, and D. Wuebbles, 1995: Radiative Forcing of Climate Change. In *Climate Change 1995 – The Science of Climate Change*, IPCC, Cambridge University Press, Cambridge, pp. 65-131.

Schipper, L., M.J. Figueroa, L. Price, and M. Epsey, 1993: Mind the gap: the vicious circle of measuring automobile fuel use. *Energy Policy*, **21**, 1173-1190.

Schipper, L.J., 1996: Excel spreadsheets containing transport energy data, PASSUM.XLS and FRTSUM.XLS. Versions of 9 February 1996, International Energy Agency, Paris/ France/Lawrence Berkeley National Laboratory, Berkeley, CA (Personal communication).

Schipper, L., R. Hass, and C. Sheinbaum, 1996: Recent trends in residential energy use in OECD countries and their impact on carbon dioxide emissions: A comparative analysis of the period 1973-1992. *Journal of Mitigation and Adaptation Strategies for Global Change*, **1**, 167-196.

Schipper, L., M. Ting, M. Khrushch, and W. Golove, 1997a: The evolution of carbon dioxide emissions from energy use in industrial countries: An end-use analysis. *Energy Policy*, **25**(7-9), 651-672.

Schipper, L., L. Scholl, and L. Price, 1997b: Energy use and carbon emissions from freight in 10 industrial countries: An analysis of trends from 1973 to 1992. *Transportation Research-D*, **2D**(1), 57-76.

Schmalensee, R., T. Stoker, and R. Judson, 1998: World carbon dioxide emissions: 1950-2050. *The Review of Economics and Statistics*, **LXXX**(1), 15-27.

Schumpeter, J., 1935: The analysis of economic change. *Review of Economic Statistics*, **May**, 2-10.

Schumpeter, J., 1943: *Capitalism, Socialism and Democracy*. Harper & Brothers, New York, NY.

Schollenberger, W.E., 1998: Gedanken über die Kohlenwasserstoffereserven der Erde: Wie lange können sie vorhalten? In *Energievorräte und mineralische Rohstoffe: Wie lange noch?* J. Zeemann (ed.), Austrian Academy of Sciences, Vienna, pp. 75-126.

Sheinbaum, C., and Rodriguez, L., 1997: Recent trends in Mexican industrial energy use and their impact on carbon dioxide emissions. *Energy Policy*, **25**(7-9), 825-831.

Shell International Ltd., 1996: *The Evolution of the World's Energy Systems*. Shell International Ltd., The Hague.

Shell International Ltd., 1998: *People and the Environment*. The 1997 Shell International Exploration and Production Health, Safety and Environment Report, Shell International Ltd., The Hague.

Shukla, P.R., W. Chandler, D. Ghosh, and J. Logan, 1999: *Developing Countries and Global Climate Change – Electric Power Choices for India*. Report for the Pew Center on Global Climate Change by the Indian Institute of Management and Battelle Advanced International Studies Unit, (Photocopy, publication forthcoming in 2000).

Silverberg, G., 1988: Modelling economic dynamics and technical change: Mathematical approaches to self-organisation and evolution (Chapter 24). In *Technical Change and Economic Theory*. G.C. Dosi, C. Freeman, R. Nelson, G. Silverberg, L. Soete (eds.), Pinter, London/ New York, NY, pp. 531-559.

Smith, A., 1970: *The Wealth of Nations*. Penguin, Harmondsworth, UK, 538 pp.

Smith N.J., and G.H. Robinson, 1997: Technology pushes reserves "crunch" date back in time. *Oil and Gas Journal*, **April 7**, 43-50.

Smith, S., H. Pitcher, and T.M.L. Wigley, 2000: Global and regional anthropogenic sulfur dioxide emissions. *Global Biogeochemical Cycles*. (In press).

Solow, R., 1956: A contribution to the theory of economic growth, *Quarterly Journal of Economics*, **70**, 56-94.

Solow, R., 1957: Technical change and the aggregate production function. *Review of Economics and Statistics*, **39**, 312-320.

Stern, D.I., and R.K. Kaufmann, 1998: Estimates of global anthropogenic methane emissions:1860 – 1994. In *Trends Online: A Compendium of Data on Global Change*. Carbon Dioxide Information Analysis Center, Oak Ridge National Laboratory, U.S. Department of Energy, Oak Ridge, TN.

Stevens III, W.R., 1993: Abatement of nitrous oxide emissions produced in the adipic acid industry. In *White House Conference on Global Climate Change, Nitrous Oxide Workshop June 11 1993*. The DuPont Company, Wilmington DE (mimeo).

Stoll H., and D.M. Todd, 1996: *Current IGCC Market Competitiveness*. General Electric Co, Schenactady, NY.

Storey, M., 1996: *Policies and Measures for Common Action. Demand Side Efficiency: Voluntary Agreements with Industry*. Annex I Expert Group on the UNFCCC, WP 8, OECD, Paris.

Storey, M., 1997: *The Climate Implications of Agricultural Policy Reform*. Annex I. Expert Group on the UNFCCC, WP16, OECD, Paris.

Strang, V., 1997: Cultural theory and modelling practice. In *Energy Modelling: Beyond Economics and Technology*. B. Giovannini, A. Baranzini (eds.), International Academy of the Environment, Geneva, and the Centre for Energy Studies of the University of Geneva, Geneva, pp. 53-70.

Streets, D.G., and S.T. Waldhoff, 2000: Present and future emissions of air pollutants in China: SO_2, NO_x, and CO. *Atmospheric Environment*, **34**(3), 363-374.

Streets, D.G, N. Tsai, S. Waldhoff, H. Akimoto, and K. Oka, 2000: *Sulfur Dioxide Emission Trends for Asian Countries 1985-1995*, Proceedings of the Workshop on the Transport of Air Pollutants in Asia, July 22-23 1999, IIASA, Laxenburg, Austria. R

Suri, V., and D. Chapman, 1996: *Economic Growth, Trade and Environment: An Econometric Evaluation of the Environmental Kuznets Curve*. Working Paper WP-96-05, Cornell University, Ithaca, NY.

Szargut, J., and D.R. Morris, 1987: Cumulative energy consumption and cumulative degree of perfection in chemical processes. *Energy Research*, **11**, 245-261.

Tainter, J., 1988: *The Collapse of Complex Societies*. Cambridge University Press, Cambridge.

Tainter, J., 1996: Complexity, problem solving and sustainable societies. In *Getting Down to Earth: Practical Applications of Ecological Economics*. R. Costanza, O. Segura, J. Martinez-Alier (eds.), Island Press, Washington DC, pp. 61-76.

Teilhard de Chardin, P., 1959: *The Phenomenon of Man*. English translation published by William Collins Sons & Co., London.

Thompson, A.M., 1992: The oxidising capacity of the earth's atmosphere: Probable past and future changes. *Science*, **256**, 1157-1165.

Tilton, J.E., 1990: *World Metal Demand: Trends and Prospects*. Resources for the Future, Washington, DC.

Tinbergen, J., 1942: Zur Theorie des langfristigen Wirtschaftsentwicklung. *Weltwirtschaftliches Archiv*, **1**, 511-549.

Tiwari, P., and J.K. Parikh, 1995: Cost of carbon dioxide reduction in building construction. *Energy – The International Journal*, **20**(6), 531-547.

Trexler, M.C., and C. Haugen, 1995: *Keeping it Green: Tropical Forestry Opportunities for Mitigating Climate Change*. WRI/EPA, Washington D.C.

UN (United Nations), 1992: *Long-Range World Population Projections: Two Centuries of Population Growth: 1950-2150*. Department of International Economic and Social Affairs, UN, New York, NY.

UN (United Nations), 1993: *Integrated Environmental and Economic Accounting, Studies in Methods*. Series F, No. 61, Handbook of National Accounting. UN, New York, NY.

UN (United Nations), 1995: *Population and Development*. Program of Action adopted at the International Conference on Population and Development, Cairo, 5–13 September 1994, **1**, United Nations Publication no. E.95.XIII.7, New York, NY.

UN (United Nations), 1996: *World Population Prospects: 1996 Revision*. United Nations, New York, NY.

UN (United Nations), 1997a: *Proceedings of Expert Group Meeting on Below Replacement Fertility*. United Nations Population Division, United Nations Publication no. ESA/P/WP.140, New York, NY.

UN (United Nations), 1997b: *Critical Trends: Global Change and Sustainable Development*, United Nations Publication no. ST/ESA/255, New York, NY.

UN (United Nations), 1997c: *Report of the Secretary General on Progress on Work in the Field of Population in 1996*. United Nations Commission on Population and Development, Thirteenth Session. United Nations Publication, New York, NY, pp. 8–10.

UN (United Nations), 1998: *World Population Projections to 2150*. United Nations Department of Economic and Social Affairs Population Division, New York, NY.

UNAIDS and **WHO**, 1998: *Report on the Global HIV/AIDS Epidemic*. UNAIDS and WHO, Geneva, Switzerland.

UNDP (United Nations Development Programme), 1993: *Human Development Report 1993*, Oxford University Press, New York, NY.

UNDP (United Nations Development Programme), 1997: *Human Development Report 1997*, Oxford University Press, New York, NY.

UNEP (United Nations Environmental Programme), 1993: *Environmental Data Report 1993-1994*. Blackwell Publishers, Oxford.

UNEP, 1999: *Global Environmental Outlook 2000*, UNEP, Nairobi, Kenya.

UNFCCC (United Nations Framework Convention on Climate Change), 1997: *Second National Communication of France under the Climate Convention*. Secretariat of the UN Framework Convention on Climate Change, Bonn.

US Council of Economic Advisors, 1997: Excerpts on the challenge of an aging population. *Population and Development Review*, **23**(2), 443-451.

US DOC (US Department of Commerce), 1975: *Historical Statistics of the United States: Colonial Times to 1970, Vols. I and II*. USDOC, Washington, DC.

US DOE (US Department of Energy), 1994: *Technology Characterizations*. US DOE Office of Utility Technologies, Washington, DC.

US EPA (Environmental Protection Agency), 1994: *International Anthropogenic Methane Emissions: Estimates for 1990*. EPA 230-R-93-010, US EPA Office of Policy Planning and Evaluation, Washington, DC.

Van Vuuren, D.P., B. Strengers, and H.J.M. de Vries, 2000: Long-term perspectives on world metal use - a systems dynamics model. *Resources Policy* (accepted for publication).

Viguier, L., 1999. Emissions of SO_2, NO_x, and CO_2 in transition economies: Emission inventories and Divisa index analysis. *Energy Economics*, **20**(2), 59-75.

VROM (Netherlands Ministry of Housing, Spatial Planning and the Environment), 1997: *Second Netherlands' National Communication on Climate Change Policies*. Distribution Centre of Government Policies, Zoetermeer, the Netherlands and Secretariat of the UN Framework Convention on Climate Change, Bonn.

Waggoner, P.E., 1996: How much land can ten billion people spare for Nature? *Daedalus* **125**(3), 73-93.

Wallace, D., 1995: *Environmental Policy and Industrial Innovation: Strategies in Europe, the USA, and Japan*. Earthscan, London.

Walls, M.A., A.J. Krupnick, and H.C. Hood, 1993: *Estimating the Demand for Vehicle-Miles Travelled Using Household Survey Data: Results from the 1990 Nationwide Personal Transportation Survey*. Resources for the Future discussion paper ENR 93-25, Resources for the Future (RFF), Washington, DC.

Walsh, M.P. 1993: Highway vehicle activity trends and their implications for global warming: the United States in an international context. In *Transportation and Global Climate Change*. D.L. Greene, D.J. Santini (eds.). American Council for an Energy-Efficient Economy, Washington, DC.

Wassmann, R., P. Papen, and H. Rennberg, 1993: Methane emission from rice paddies and possible mitigation strategies. *Chemosphere*, **26**(1-4), 2010-2117.

Wassmann, R., H.K. Kludze, W. Bujun, and R.S. Lantin, 1997: Factors and processes controlling methane emissions from rice fields. *Nutrient Cycling in Agrosystems*, **49**(1-3), 111-117.

Wassmann, R., H.U. Neue, C. Bueno, R.S. Lantin, M.C.R. Alberto, L.V. Buendia, K. Bronson, H. Papen, and H. Rennenberg, 1998: Methane production capacities in different rice soils derived from inherent and exogenous substrates. *Plant and Soil*, **203**(2), 227-237.

Watanabe, C. (Tokyo Institute for Technology), 1997: Personal communication.

Watson, R., M.C. Zinyowera, and R. Moss (eds.), 1996: *Climate Change 1995. Impacts, Adaptations and Mitigation of Climate Change: Scientific Analyses*. Contribution of Working Group II to the Second Assessment Report of the Intergovernmental Panel on Climate Change, Cambridge University Press, Cambridge, 861 pp.

WEC (World Energy Council), 1993: *Energy for Tomorrow's World: The Realities, the Real Options and the Agenda for Achievements*. Kogan Page, London.

WEC (World Energy Council), 1994: *New Renewable Energy Resources*. World Energy Council (WEC), London.

WEC (World Energy Council), 1995a: *Survey of Energy Resources*. World Energy Council (WEC), London.

WEC (World Energy Council), 1995b: *Energy Efficiency Utilizing High Technology: An Assessment of Energy Use in Industry and Buildings*. Prepared by M.D. Levine, E. Worrell, N. Martin, and L. Price, World Energy Council (WEC), London.

WEC (World Energy Council), 1998: *Survey of Energy Resources*. World Energy Council (WEC), London.

Wernick, I., 1996: Consuming materials: The American way. *Technological Forecasting and Social Change*, **53**(1), 111-122.

Wexler, L., 1996: *Improving Population Assumptions in Greenhouse Emissions Models*, WP-96-099, International Institute for Applied Systems Analysis, Laxenburg, Austria.

WHO (World Health Organization) and UNEP (United Nations Environment Programme), 1993: *Urban Air Pollution in Megacities of the World*. 2nd Edition, Blackwell Publishers, Oxford.

Wilber, K., 1981: *Up From Eden: A Transpersonal View of Human Evolution*. Quest Books, Wheaton, IL.

Wilhite, H., 1997: Framing the socio-cultural context for analyzing energy consumption. In *Energy Modelling: Beyond Economics and Technology*. B. Giovannini, A. Baranzini (eds.), International Academy of the Environment, Geneva, and the Centre for Energy Studies of the University of Geneva, Geneva, pp. 35-52.

Willerboer, W., 1997: *Future IGCC Concepts*. Demkolec BV, Moerdijk, the Netherlands.

Williams, A. (ed.), 1993: *Methane Emissions*. The Watt Committee on Energy, London.

Williams, R.H., 1998: Fuel decarbonization for fuel cell applications and sequestration of the separated CO_2. In *Ecorestructuring: Implications for Sustainable Development*. R.U. Ayres, P.M. Weaver (eds.), United Nations University Press, Tokyo, pp. 180-222.

Williams, R.H., E.D. Larson, and M.H. Ross, 1987: Materials, affluence, and industrial energy use. *Ann. Rev. Energy*, **12**, 99-144.

WMO (World Meteorological Organisation) (D.M. Whelpdale and M.S. Kaiser, eds), 1997: *Global Acid Deposition Assessment*. WMO Global Atmospheric Watch No. 106, WMO, Geneva.

WMO/UNEP, (World Meteorological Organisation and United Nations Environment Programme), 1998: *Scientific Assessment of Ozone Depletion: 1998*. WMO Global Ozone Research & Monitoring, World Meteorological Organisation (WMO), Geneva.

WMO/UNEP, 1999: *Options for the Limitation of Emissions of HFCs and PFCs*. Proceedings of the Joint IPCC/TEAP Expert Meeting, Petten, the Netherlands, 26-28 May, 1999.

World Bank, 1991: *World Development Report 1991: The Challenge of Development*. Oxford University Press, Oxford.

World Bank, 1992: *World Development Report 1992: Development and the Environment*. Oxford University Press, Oxford.

World Bank, 1997a: *Monitoring Environmental Progress, Expanding the Measure of Wealth*. World Bank, Washington, DC.

World Bank, 1997b: *World Development Indicators 1997*. World Bank, Washington, DC.

World Bank,1997c: *Can the Environment Wait: Priorities for East Asia*. World Bank, Washington, DC.

World Bank, 1997d: *Clear Water, Blue Skies: China 2020, China's Environment in the New Century*. World Bank, Washington, DC.

World Bank, 1998a: *World Development Report 1998-1999: Knowledge for Development*. Oxford University Press, Oxford.

World Bank, 1998b: *Global Economic Prospects and the Developing Countries: Beyond Financial Crisis*. World Bank, Washington, DC.

World Bank, 1999: *1999 World Development Indicators*. World Bank, Washington, DC.

Worrell, E., M.D. Levine, L. Price, N. Martin, R. van den Broek, and K. Blok, 1997: *Potentials and Policy Implications of Energy and Material Efficiency Improvement*. United Nations Department for Policy Coordination and Sustainable Development. New York, NY.

WRI (World Resources Institute), 1997a: *World Resources 1996-97*. WRI, Washington, DC.

WRI (World Resources Institute), 1997b: *Resource Flows: The Material Basis of Industrial Economies*. WRI, Washington, DC.

Zachariah, K.C., and M.T. Vu, 1988: *World Population Projections 1987-88 Edition*. Johns Hopkins University Press, Baltimore, MD.

4

An Overview of Scenarios

CONTENTS

4.1. Introduction

In Chapter 4 the main characteristics of the scenarios developed for the Special Report on Emissions Scenarios (SRES scenarios) are presented. These scenarios cover a wide range of driving forces from demographic to social and economic developments, and they encompass a wide range of future greenhouse gas (GHG) emissions (see Chapter 5). Chapters 2 and 3 provide an overview and assessment of the scenario literature, the main driving forces of future GHG emissions, and their relationships. How the driving forces are combined to produce a set of scenarios that cover the ranges of GHG emissions from the literature is described in this chapter.

Chapters 2 and 3 demonstrate the large uncertainty in the literature that surrounds both future emissions and the possible developments of their underlying driving forces. The uncertainties range from inadequate scientific understanding of the problems, through data gaps or lack of data, to the inherent uncertainties of future events in general. Hence alternative scenarios are used to describe the range of possible future emissions.

The SRES approach involved the development of a set of four alternative scenario "families" (see Chapter 1, Section 1.7.2). Each family of SRES scenarios includes a descriptive part (called a "storyline") and a number of alternative interpretations and quantifications of each storyline developed by six different modeling approaches (see also Box 1-1 on terminology). Each storyline describes a demographic, social, economic, technological, and policy future for each of the scenario families. Within each family different *scenarios* explore variations of global and regional developments and their implications for GHG, ozone precursors, and sulfur emissions. Each of these scenarios is consistent with the broad framework specified by the storyline of the scenario family.

Each storyline is basically a short "history" of a possible future development expressed as a combination of key scenario characteristics. These descriptions are stylized and designed to facilitate specification and further interpretation of scenario quantifications. The storylines identify particular dynamics, visible in the world today, that might have important influences on future GHG emissions. They deliberately explore what might happen if social, economic, technical, and policy developments take a particular direction at the global level; they also pay attention to regional differences and interactions, especially between developing and industrialized countries.

The broad consensus among the SRES writing team is that the current literature analysis suggests the future is inherently unpredictable and so views will differ as to which of the storylines and representative scenarios could be more or less likely. Therefore, the development of a single "best guess" or "business-as-usual" scenario is neither desirable nor possible. Nor should the storylines and scenarios be taken as policy recommendations. The storylines represent the playing out of certain social, economic, technological, and environmental paradigms, which will be viewed positively by some people and negatively by others. The SRES writing team decided on four storylines – an even number helps to avoid the impression of a "central" or "most likely" case. The team wanted more than two storylines to help illustrate that the future depends on many different underlying dynamics; the team wanted no more than four to avoid complicating the process with too many alternatives. The scenarios cover a wide range of, but not all possible, futures. In particular, it was decided that possible "surprises" would not be considered and that there would be no "disaster" scenarios. The team decided to carry out sensitivity tests within some of the storylines by considering alternative scenarios with different fossil-fuel reserves, rates of economic growth, or rates of technical change. These sensitivity analyses resulted in groups of scenarios within a given scenario family and alternative scenario interpretations within a scenario group or family (see Section 4.2 below for a description of scenario terminology and taxonomy).

The titles of the storylines are deliberately simple – A1, A2, B1, and B2. There is no particular order among the storylines (they are listed alphabetically). Figure 4-1 shows that the SRES scenarios build on the main driving forces of GHG emissions. Each scenario family is based on a common specification of the main driving forces.

All four storylines and scenario families describe future worlds that are generally more affluent compared to the current situation. They range from very rapid economic growth and technologic change to high levels of environmental protection, from low-to-high global populations, and from high-to-low GHG emissions. Perhaps more importantly, all the storylines describe dynamic changes and transitions in generally different directions. The storylines do not include specific climate-change policies, but they do include numerous other socio-economic developments and non-climate environmental policies. As time progresses, the storylines diverge from each other in many of their characteristic features. In this way they span the relevant range of GHG emissions and different combinations of their main sources.

After the basic features and driving forces for each of the four storylines had been determined, the team quantified the storylines into individual scenarios with the help of formal (computer) models. While the writing team and the modeling groups included experts from around the world, all six modeling groups are based in Europe, North America, and Japan. As indicated above, each model quantification of a storyline constitutes a scenario, and all scenarios of one storyline constitute a "scenario family." The six models are representative of different approaches to emissions-scenario modeling and different integrated assessment frameworks in the literature, and include so-called top-down and bottom-up models. The use of different models reflects the SRES Terms of Reference call for methodologic pluralism and for an open process (see Appendix I). The number and type of models chosen in the open process was on a voluntary basis. In January 1997 the Intergovernmental Panel on Climate Change Working

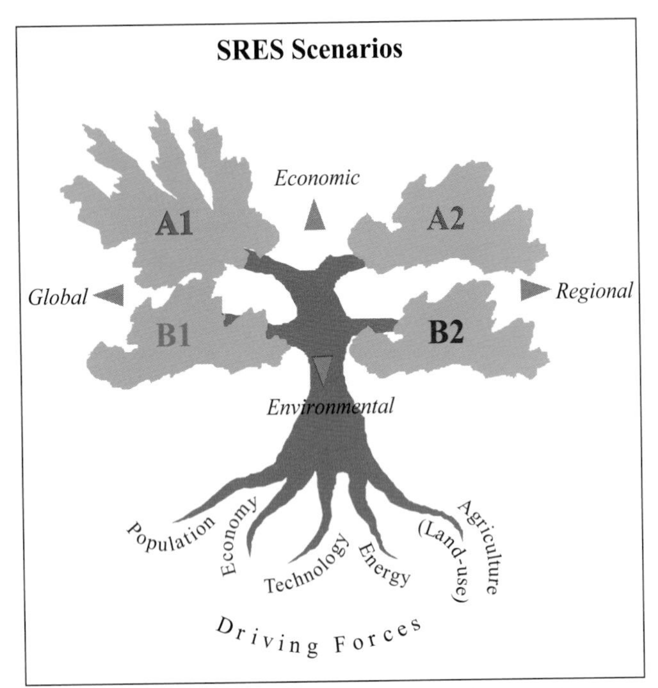

Figure 4-1: The four SRES scenario families that share common storylines are illustrated as branches of a two-dimensional tree. The two dimensions indicate the relative orientation of the different scenario storylines toward economic or environmental concerns and global and regional scenario development patterns, respectively. There is no implication that these two are mutually exclusive or incompatible. In reality, the four scenarios share a space of a much higher dimensionality given the numerous driving forces and other assumptions needed to define any given scenario in a particular modeling approach. The A1 storyline branches out into different groups of scenarios to illustrate that alternative development paths are possible within one scenario family.

Group II Technical Support Unit (IPCC WGII TSU) sent letters to Governments asking for nominations of modeling teams to contribute to SRES and advertised in a number of scientific journals for modelers to participate in SRES. Six different modeling groups from Europe, North America, and Japan volunteered to participate in the formulation and development of the scenarios in response to the call. It is fortunate that they are from three different continents and also include different methodological approaches used in the literature to develop quantitative emissions scenarios.

The six models have different regional aggregations. The writing team decided to group the various global regions into four "macro-regions" common to all the regional aggregations across the six models (Box 4-1).

In response to a number of requests from potential user groups within IPCC and in accordance with a decision of the IPCC Bureau in 1998 to release draft scenarios to climate modellers for their input in the Third Assessment Report, the writing team chose one model run to characterize each scenario family. Scenarios resulting from these runs are called "marker" scenarios or simply "markers." There are four marker scenarios, each considered characteristic for one of the four scenario families. The rationale and process for designating marker scenarios is discussed in more detail in Section 4.4.1.

The SRES scenario quantifications of the main indicators (such as population and economic growth, characteristics of the energy system, and the associated GHG emissions) all fall within the range of studies published in the literature and scenarios documented in the SRES database (see Chapter 2). Quantitative indicators form an important part of each scenario description. These indicators include gross world product, population, supply and demand for principal energy forms, energy resource characteristics, the breakdown of land use, and emissions of various GHGs. The scenarios are designed so that the evolution of their indicators over the 21st century falls well within the range represented by scenarios from the literature and included in the SRES database (see Chapter 2 and Morita and Lee, 1998; Nakićenović, *et al.*, 1998). More importantly, they correspond to the qualitative characteristics of the respective storylines. Also, they were revised iteratively within the six modeling approaches to achieve internal consistency on the basis of inputs from the entire SRES writing team and the SRES open process.

Each storyline was characterized initially by two quantitative "targets," namely global population (15, 10, and 7 billion by 2100 in scenarios A2, B2, and both A1 and B1, respectively) and global gross domestic product (GDP) by 2100 (in 1990 US dollars, US$550 trillion for A1, US$250 trillion for A2, US$350 trillion for B1, and US$250 trillion for B2). These quantitative targets guided the subsequent quantification of the SRES scenarios with different model approaches. Generally, the orders of magnitude of these original quantitative scenario "guideposts" are reflected in the final SRES scenarios (see Table 4-2) and have been adopted in a majority of SRES scenarios. Evidently, the quantitative characteristics of the four SRES scenario families comprise many more dimensions than this, in particular regional patterns, differences in resource and technology availability, land-use changes, non-carbon dioxide (CO_2) GHGs, etc. These are discussed in the subsequent Sections.

4.2. SRES Scenario Taxonomy

4.2.1. Storylines

The primary purpose of developing multiple scenario families was to explore the uncertainties behind potential trends in global developments and GHG emissions, as well as the key drivers that influence these (see also Chapter 1, Section 1.7.2). The writing team decided that narrative storylines, based on the futures and scenario literature, would be the most coherent way to describe their scenarios, for the following reasons.

- To help the team to think more coherently about the complex interplay between scenario driving forces

within and across alternative scenarios and to enhance the consistency in assumptions for different parameters.
- To make it easier to explain the scenarios to the various user communities by providing a narrative description of alternative futures that goes beyond quantitative scenario features.
- To make the scenarios more useful, in particular, to analysts contributing to IPCC WGs II and III. The demographic, social, political, and technological contexts described in the scenario storylines are all-important in the analysis of the effects of policies to either adapt to climate change or to reduce GHG emissions.
- To provide a guide for additional assumptions to be made in detailed climate-impact and mitigation analyses, because at present no model or scenario can possibly respond to the wide variety of informational and data needs of the different user communities of long-term emissions scenarios.

The four scenario families presented in this report are representative of a broad range of scenarios found in the literature, but they are not directly based on any particular published scenario taxonomy or set of scenarios. Rather, the storylines of each scenario family were developed on the basis

Box 4-1: Four SRES World Regions

The six modeling frameworks used to develop the SRES scenarios have different regional aggregations. The writing team decided to group the various global regions into four "macro-regions" common to all the different regional aggregations across the six models (Figure 4.2; see Appendix IV, Table IV-1). The individual scenarios were formulated with the respective regional aggregation of each model. Afterward, the input assumptions and results were summed to correspond to the four macro-regions:

- **OECD90** region groups together all member countries of the Organization for Economic Cooperation and Development as of 1990, the base year of the participating models, and corresponds to the Annex II countries originally defined in UNFCCC (1992).
- **REF** region consists of countries undergoing economic reform and groups together the East and Central European countries and the Newly Independent States of the former Soviet Union; it roughly corresponds to Annex I outside the Annex II countries as defined in UNFCCC (1992).
- **ASIA** region stands for all developing (non-Annex I) countries in Asia (excluding the Middle East).
- **ALM** region stands for the rest of the world and corresponds to developing (non-Annex I) countries in Africa, Latin America, and Middle East.

In other words, the OECD90 and REF regions together roughly correspond to Annex I or industrialized (developed) countries (IND), while the ASIA and ALM regions together roughly correspond to the non-Annex I, or developing countries (DEV). Developing, or non-Annex I countries (i.e., ASIA and ALM), are sometimes referred to in the text as the "South" to distinguish them from the industrialized, or Annex I countries, of the "North" (i.e., OECD90 and REF). A detailed description of each region is provided in Appendix III.

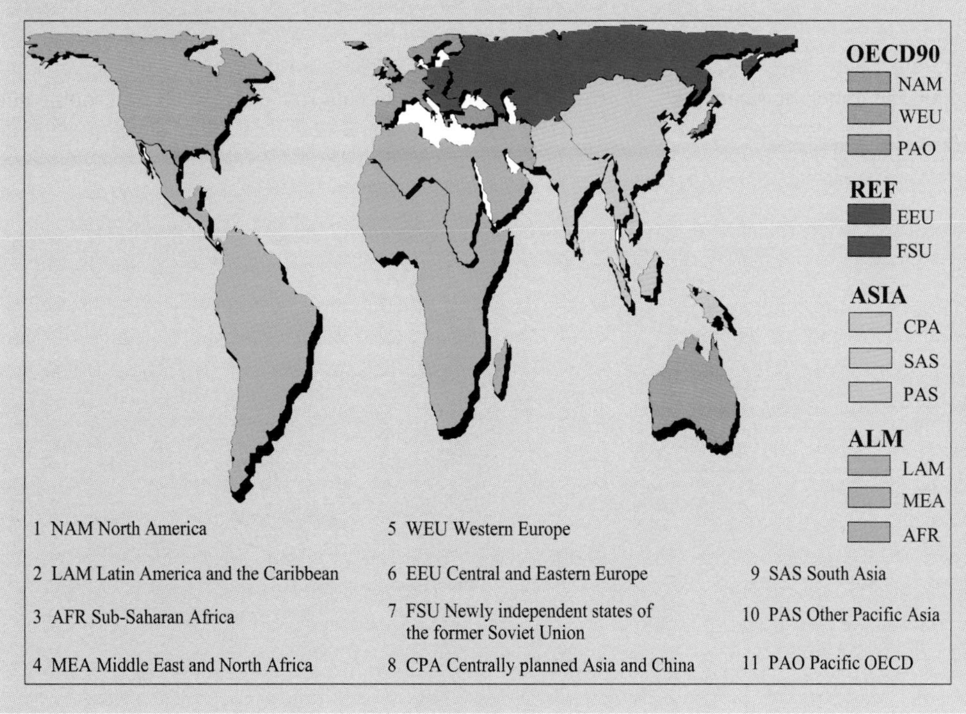

Figure 4-2: SRES world regions ALM, ASIA, OECD90, and REF. The developing (DEV) countries, comprising the ALM and ASIA regions, roughly correspond to non-Annex I countries of the UNFCCC (1992). The industrialized (IND) countries, comprising the OECD90 and REF regions, roughly correspond to Annex I countries of the UNFCCC.

of the general knowledge of this literature, and on the personal experience and creativity within the writing team. The writing team spent the better part of the first year (1997) formulating the storylines, which during the following two years were revised iteratively with the scenario development until the completion of the report.

Four brief "future histories" captured by the SRES storylines differ in how global regions interrelate, how new technologies diffuse, how regional economic activities evolve, how protection of local and regional environments is implemented, and how demographic structure changes. The "qualitative" storyline characteristics include various political, social, cultural, and educational conditions (e.g., type of governance, social structure, and educational level) that often cannot be defined in strictly quantitative terms and do not directly "drive" GHG emissions. These qualitative variables, however, participate in complex "cause–effect" relationships with quantitative emission drivers (e.g., economic activities, population levels, energy consumption). Their explicit inclusion in the scenario development process not only makes scenarios more "plausible" and "believable," but also ensures they do not become an arbitrary numeric combination of quantitative parameters.

The SRES storylines do *not* include explicit policies to limit GHG emissions or to adapt to the expected global climate change, reflecting the SRES Terms of Reference (see Appendix I). However, the storyline demographic, social, economic and technological profiles can be used in other studies to develop and evaluate climate-change mitigation and adaptation measures and policies. Such evaluation would require additional (prescriptive) assumptions about policies and measures to affect future climates and human responses to climate change now absent from the storylines.

All four SRES "futures" represented by the distinct storylines are treated as equally possible and there are no "central," "business-as-usual," "surprise," or "disaster" futures (examples of which are given in Box 4.2). All of the storylines have features that can be interpreted as "positive" or "negative" and they play out different tendencies and changes in part visible in the world today. To avoid the tendency to overemphasize "positive" or "negative" features of individual storylines, their titles were kept simple. Many attempts were made to capture the spirit of each storyline with a short and snappy title, but no single title was found to reflect adequately the complex mix of characteristics of any storyline.

By 2100 the world will have changed in ways that are difficult to imagine, as difficult as it was at the end of the 19th century to imagine the changes of the 20th century. However, each storyline takes a different direction of future developments so that they differ in an increasingly irreversible way. They describe divergent futures that reflect a significant portion of the underlying uncertainties in the main driving forces. The differences among the storylines cover a wide range of the key "future" characteristics, such as technology, governance, and behavioral patterns. Hence the plausibility or feasibility of the storyline assumptions should be viewed with an "open mind," not from a narrow interpretation of current situations and trends in economic conditions, technology developments, and social and governing structures.

The main characteristics of future developments that take distinct development paths in the four storylines include (see also Table 4-2 for an overview):

- Nature of the global and regional demographic developments in relation to other characteristics of the storyline.
- Extent to which economic globalization and increased social and cultural interactions continue over the 21st century.
- Rates of global and regional economic developments and trade patterns in relation to the other characteristics of the storyline.
- Rates and direction of global and regional technological change, especially in relation to the economic development prospects.
- Extent to which local and regional environmental concerns shape the direction of future development and environmental controls.

Box 4-2: "Neutrality" of the SRES Scenarios

The SRES scenarios are intended to exclude catastrophic futures. Such catastrophic futures feature prominently in the literature. They typically involve large-scale environmental or economic collapses, and extrapolate current unfavorable conditions and trends in many regions. Prominent examples of such scenarios include "Retrenchment" (Kinsman, 1990), "Dark Side of the Market World" or "Change without Progress" (Schwartz, 1991), "Black and Grey" (Godet *et al.*, 1994), "Global Incoherence Scenario" (Peterson, 1994), "New World Disorder" (Schwartz, 1996), "A Visit to Belindia" (Pohl, 1994), the future evoked by the description of the current situation in parts of West-Africa and Central Asia (Kaplan, 1996), "Barbarization" (Gallopin *et al.*, 1997), "Dark Space" (Glenn and Gordon, 1999), "Global Fragmentation" (Lawrence *et al.*, 1997), and "A Passive Mean World" (Glenn and Gordon, 1997, 1999). In this last scenario the world is carved up into three rigid and distinct trading blocs, with fragmented political boundaries and out-of-control ethnic conflicts. In "Global Crisis" (de Jong and Zalm, 1991; CPB, 1992) protectionism leads to a vicious circle of slowing economic growth and eventually breakdown. Many of these scenarios suggest that catastrophic developments may draw the world into a state of chaos within one or two decades. In such scenarios GHG emissions might be low because of low or negative economic growth, but it seems unlikely they would receive much attention in the light of more immediate problems. Hence, this report does not analyze such futures.

- Degree to which human and natural resources are mobilized globally and regionally to achieve multiple development objectives of each storyline.
- Balance of economic, social, technological, or environmental objectives in the choices made by consumers, governments, enterprises, and other stakeholders.

Thus, the storylines describe developments in many different economic, technical, environmental, and social dimensions. Consequently, they occupy a multidimensional space and no simple metric can be used to classify them. Even though they occupy such a multidimensional space along many driving forces relevant for GHG emissions, it is useful here to highlight just two dimensions. The first refers to the extent of economic convergence and social and cultural interactions across the regions and the second to the balance between economic objectives and environmental and equity objectives. Possible names for these two dimensions could be "globalization" (Box 4-3) and "sustainability," respectively (Box 4-4). As these two expressions are not necessarily viewed by everyone as being value-free, the two dimensions could alternatively be designated simply as a more *global* or more *regional* orientation and as a more *economic* or a more *environmental* orientation (see Figure 4-1). These dimensions are important in the SRES scenarios. Nevertheless, there was considerable resistance in the SRES writing team against such a simplistic classification of storylines, so it is presented here for illustrative purposes only. These distinctions are, in a sense, artificial. For example, both economic and environmental objectives are pursued in all scenarios, albeit with different levels of relative emphasis.

The extent to which the currently observed global and regional orientations will prevail in the 21st century is pertinent to the distinction between the A1 and B1 scenario families on one side and A2 and B2 families on the other side. While the A1 and B1 storylines, to different degrees, emphasize successful economic global convergence and social and cultural interactions, A2 and B2 focus on a blossoming of diverse regional development pathways (see Box 4-3).

Box 4-3: Globalization Issues

With the convergence in governments' economic policies in the 1990s, combined with the rapid development of communication networks, it is perhaps not surprising that an extensive poll of scenarios by the Millennium Institute suggested "globalization" as the main driving force that will shape the future (Glenn and Gordon, 1997, 1999). However, some scenarios in the literature explore the possibility that unfettered markets, usually seen as an integral element of "globalization," might destabilize society in ways that endanger the process (Mohan Rao, 1998). In UNESCO's 1998 World Culture Report, it is noted that communities are increasingly emphasizing their cultural individuality; meanwhile, communication and travel are resulting in interactions between communities that result in the evolution of new "local" cultures (UNESCO, 1998). Huntington (1996) asserts that continental regional cultures may determine the shape of future geopolitical developments rather than globalization.

Box 4-4: Sustainability Issues

Recent decades have seen considerable growth in discourse of environmental and social issues, represented at the global level by several high-level United Nations (UN) meetings on social and economic development and environmental sustainability (UNCED, 1992; UN, 1994, 1995; Leach, 1998; Munasinghe and Swart,2000). The range of participants has expanded from the most closely involved government ministries, businesses, and environmental NGOs to include a broad range of representation by different ministries, local government, businesses, professions, and community groups. Increased interest in sustainability issues can lead to all kinds of socio-economic and technological changes that may not be aimed explicitly at reducing GHG emissions, but which may in effect contribute significantly to such reductions.

The extent to which the currently observed economic and environmental orientations will prevail in the 21st century is pertinent to the distinction between A1 and A2 scenario families on one side and B1 and B2 scenario families on the other side. In the B1 and B2 storylines this transition is pursued, to different degrees, through a successful translation of global concerns into local actions to promote environmental sustainability. Alternatively, in the A1 and A2 storylines the emphasis remains, again to different degrees, on sustained economic development and achievement of high levels of affluence throughout the world, where environmental priorities are perceived as less important than those of economic development (see Box 4-4).

In short, each of the storylines can be summarized as follows:

- The A1 storyline and scenario family describes a future world of very rapid economic growth, low population growth, and the rapid introduction of new and more efficient technologies. Major underlying themes are convergence among regions, capacity building, and increased cultural and social interactions, with a substantial reduction in regional differences in per capita income. The A1 scenario family develops into four groups that describe alternative directions of technological change in the energy system.[1]

[1] During the approval process of the Summary for Policymakers at the 5th Session of WGIII of the IPCC from 8-11 March 2000 in Katmandu, Nepal, it was decided to combine two of thiese groups (A1C and A1G) into one "fossil intensive" group A1FI, in contrast to the non-fossil group A1T, and to select two illustrative scenarios from these two A1 groups to facilitate use by modelers and policy makers. This leads to six scenario groups that constitute the four scenario families, three of which are in the A1 family. All scenarios are equally sound.

- The A2 storyline and scenario family describes a very heterogeneous world. The underlying theme is self-reliance and preservation of local identities. Fertility patterns across regions converge very slowly, which results in high population growth. Economic development is primarily regionally oriented and per capita economic growth and technological change are more fragmented and slower than in other storylines.
- The B1 storyline and scenario family describes a convergent world with the same low population growth as in the A1 storyline, but with rapid changes in economic structures toward a service and information economy, with reductions in material intensity, and the introduction of clean and resource-efficient technologies. The emphasis is on global solutions to economic, social, and environmental sustainability, including improved equity, but without additional climate initiatives.
- The B2 storyline and scenario family describes a world in which the emphasis is on local solutions to economic, social, and environmental sustainability. It is a world with moderate population growth, intermediate levels of economic development, and less rapid and more diverse technological change than in the B1 and A1 storylines. While the scenario is also oriented toward environmental protection and social equity, it focuses on local and regional levels.

These storylines are presented in more detail in Section 4.3, which includes their original quantitative indicators that served as input to the scenario quantification process.

4.2.2. Scenarios

All SRES scenarios were designed as quantitative "interpretations" (quantifications) of the SRES qualitative storylines. Each scenario is a particular quantification of one of the four storylines. The quantitative inputs for each scenario involved, for instance, regionalized measures of population, economic development, and energy efficiency, the availability of various forms of energy, agricultural productivity, and local pollution controls. Each participating modeling group (see above) used computer models and their experience in the assessment of long-range development of economic, technological, and environmental systems to generate quantifications of the storylines. The models used to develop the scenarios are:

- Asian Pacific Integrated Model (AIM) from the National Institute of Environmental Studies (NIES) in Japan (Morita *et al.*, 1994).
- Atmospheric Stabilization Framework Model (ASF) from ICF Consulting in the US (Lashof and Tirpak, 1990; Pepper *et al.*, 1998; Sankovski *et al.*, 2000).
- Integrated Model to Assess the Greenhouse Effect (IMAGE) from the National Institute for Public Health and Hygiene (RIVM) in the Netherlands (Alcamo *et al.*, 1998; de Vries *et al.*, 1994, 1999, 2000), used in connection with the Central Planning Bureau (CPB) WorldScan model (de Jong and Zalm, 1991), the Netherlands.
- Multiregional Approach for Resource and Industry Allocation (MARIA) from the Science University of Tokyo in Japan (Mori and Takahashi, 1999; Mori, 2000).
- Model for Energy Supply Strategy Alternatives and their General Environmental Impact (MESSAGE) from the International Institute of Applied Systems Analysis (IIASA) in Austria (Messner and Strubegger, 1995; Riahi and Roehrl, 2000).
- The Mini Climate Assessment Model (MiniCAM) from the Pacific Northwest National Laboratory (PNNL) in the USA (Edmonds *et al.*, 1994, 1996a, 1996b).

A more detailed description of the modeling approaches is given in Appendix IV. Some modeling teams developed scenarios that reflected all four storylines, while some presented scenarios for fewer storylines. Some scenarios share harmonized[2] input assumptions of main scenario drivers, such as population, economic growth, and final energy use, with their respective designated marker scenarios of the four scenario families and underlying storylines (see Section 4.4.1). Others explore scenario sensitivities in these driving forces through alternative interpretations of the four scenario storylines. Table 4-1 lists all SRES scenarios, by modeling group and by scenario family, and indicates which scenarios share harmonized input assumptions of important driving forces of emissions at the global level and at the level of the four SRES regions. Altogether, the six modeling teams formulated 40 alternative SRES scenarios

All the qualitative and quantitative features of scenarios that belong to the same family were set to conform to the corresponding features of the underlying storyline. Quantitative storyline targets recommended for use in all scenarios within a given family included, in particular, population and GDP growth assumptions. Most scenarios developed within a given family follow these storyline recommendations, but some scenarios offer alternative interpretations. Scenarios within each family vary quite substantially in such characteristics as the assumptions about availability of fossil-fuel resources, the rate of energy-efficiency improvements, the extent of renewable-energy development, and, hence, resultant GHG emissions. This variation reflects the modeling teams' alternative views on the plausible global and regional developments and also stems from differences in the underlying modeling approaches. After the modeling teams had quantified the key driving forces and made an effort to harmonize them with the storylines by

[1] The harmonization criteria agreed by the writing team are indicated in Table 4-1. The classification of scenarios is quite robust against varying the percentage deviation harmonization criteria (see Section 4.4.1).

***Table 4-1:** Characteristics of SRES scenario quantifications. Shown for each scenario is the name of the storyline and scenario family, full scenario name (ID), descriptive scenario name, and which of the driving forces are harmonized at the global and regional level, and on the global level only, respectively. The listed harmonized driving forces are population (POP), gross domestic product (GDP), and final energy (FE), see also Section 4.4.1. and Table 4-4. Marker scenarios are indicated in bold and are harmonized by definition, and additional illustrative scenarios, that are also harmonized are given in italics. The lower table indicates the harmonization criteria in terms of the maximum deviation (%) from the specified common population, gross world product, and final energy development at the global and regional levels.*

Storyline	Scenario ID	Scenario Name	Harmonized Drivers (on World *and* SRES Regional Level)	Harmonized Drivers (on World Level)
A1	**A1B-AIM**	A1	FE, GDP, POP by definition	FE, GDP, POP by definition
	A1B-ASF	A1	POP	GDP, POP
	A1B-IMAGE	A1	POP	GDP, POP
	A1B-MARIA	A1	–	POP, GDP[d]
	A1B-MESSAGE	A1	FE, GDP, POP	FE, GDP, POP
	A1B- MiniCAM	A1	POP	POP, GDP[d]
	A1C-AIM	A1 coal	FE, GDP, POP	FE, GDP, POP
	A1C-MESSAGE	A1 coal	POP	FE, GDP, POP
	A1C-MiniCAM	A1 coal	POP	POP
	A1G-AIM	A1 oil and gas	FE, GDP, POP	FE, GDP, POP
	A1G-MESSAGE	A1 oil and gas	POP	FE, GDP, POP
	A1G-MiniCAM	A1 oil and gas	POP	POP, GDP[d]
	A1T-AIM[a]	A1 technology	GDP, POP	GDP, POP
	A1T-MESSAGE[a]	A1 technology	POP	GDP, POP
	A1T-MARIA[a]	A1 technology	–	POP
	A1v1-MiniCAM[b]	A1v1	POP	POP
	A1v2-MiniCAM[b]	A1v2	–	–
A2	A2-AIM	A2	POP	FE, POP
	A2-ASF	**A2**	FE, GDP, POP by definition	FE, GDP, POP by definition
	A2G-IMAGE[c]	A2 gas	–	POP
	A2-MESSAGE	A2	FE, GDP, POP	FE, GDP, POP
	A2-MiniCAM	A2	POP	POP
	A2-A1-MiniCAM[b]	A2-A1	–	–
B1	B1-AIM	B1	POP	GDP, POP
	B1-ASF	B1	POP	GDP, POP
	B1-IMAGE	B1	FE, GDP, POP by definition	FE, GDP, POP by definition
	B1-MARIA	B1	–	POP
	B1-MESSAGE	B1	FE, GDP, POP	FE, GDP, POP
	B1-MiniCAM	B1	POP	GDP, POP
	B1T-MESSAGE	B1 technology	FE, GDP, POP	FE, GDP, POP
	B1High-MESSAGE	B1 high	POP	GDP, POP
	B1High-MiniCAM	B1 high	POP	POP
B2	B2-AIM	B2	FE, GDP, POP	FE, GDP, POP
	B2-ASF	B2	POP	POP
	B2-IMAGE[c]	B2	—	—
	B2-MARIA	B2	—	FE, GDP, POP
	B2-MESSAGE	B2	FE, GDP, POP by definition	FE, GDP, POP by definition
	B2-MiniCAM	B2	—	GDP
	B2C-MARIA	B2 coal	—	FE, GDP, POP
	B2High-MiniCAM	B2 high	—	GDP

continued on next page

Table 4.1 continued

Harmonization criteria:

		1990–2020	*2020–2050*	*2050–2100*
Population	World	5%	5%	5%
	4 SRES regions	10%	10%	10%
GDP	World	10%	10%	10%
	4 SRES regions	25%	25%	25%
Final Energy	World	15%	15%	15%
	4 SRES regions	25%	20%	15%

[a] The A1T scenarios explored cases of increased energy end-use efficiency and therefore share similar levels of energy services, but not final energy, with the A1 marker scenario. As this was an agreed upon (different) feature of this particular scenario group compared to that of the A1 marker, the final energy harmonization criteria does not apply by design. If final energy use is excluded as harmonization criteria for the scenarios of the A1T scenario group the number of harmonized scenarios increases to 13 (four SRES regions and world level) and 17 (world level only), respectively.

[b] A1v1-MiniCAM, A1v2-MiniCAM, and A2-A1-MiniCAM became available only late in the process (after the 15 July 1999 deadline). Intentionally, they describe futures that are quite different in character from the other scenarios in their respective families and are therefore only to a limited degree comparable to other scenarios of the A1 and A2 scenario families.

[c] The IMAGE-results for the A2 and B2 scenarios are based on preliminary model experiments done in March 1998. Due to limited resources it has not been possible to redo these experiments. Hence, the IMAGE-team is not able to provide background data and details for these scenario calculations and the population and economic growth assumptions are not fully harmonized, as is the case for the IMAGE A1 and B1 scenarios.

[d] Deviations from harmonization criteria in one time period are not considered in this classification.

adjusting control parameters, possible diversity still remained (see Section 4.4.1).

In addition, the A1 scenario family developed into different distinct scenario groups, each based on the A1 storyline that describes alternative developments in future energy systems, from carbon-intensive development to decarbonization (see footnote 1). (Similar storyline variations were considered for other scenario families, but were pursued only to a limited degree in scenario sensitivity analysis in order to limit the number of scenarios.) This further increased the richness in different GHG emissions paths, because this variation in the structure of the future energy system in itself resulted in a range of emissions almost as large as that generated through the variation of other main driving forces, such as population and economic development. The differentiation into various scenario groups was introduced into the A1 storyline because of its "high growth with high technology" nature, in which differences in alternative technology developments translate into large differences in future GHG emission levels.

As for the storylines, no single scenario was treated as more or less "probable" than others belonging to the same family. However, after requests from various user communities to reduce the number of scenarios to a manageable size, a single scenario within a family was selected as a representative case to illustrate a particular storyline on the basis of the modeling teams' consensus. These scenarios were named "marker scenarios" or simply "markers" and were put on the SRES open process webpage for review. The marker scenario for the A1 scenario storyline was developed using the AIM model; for the A2 storyline using the ASF model; for the B1 storyline using the IMAGE model; and finally for the B2 storyline using the MESSAGE model (see Table 4-1).

The choice of the markers was based on extensive discussion within the SRES team:

- Which of the initial quantifications (by the models) reflected the story best.
- Preference of some of the modeling teams and features of specific models.
- Range of emissions across all the markers.
- Use of different models for the four markers.

In 1998, the preliminary descriptions and quantifications of the marker scenarios were posted on the SRES website for the open process and, in accordance with a decision of the IPCC Bureau, were in this way made available to climate modelers for their input in the Third Assessment Report. As a result of the inputs and comments received through the open process and by the entire writing team, the marker scenarios have been successively refined and improved without changing their fundamental characteristics in terms of important scenario driving forces (population, GDP) and order of magnitude of GHG emissions. Subsequently, additional scenarios within each scenario family were developed to explore the sensitivity of adopting alternative quantitative scenario input assumptions on future GHG emissions. As a result the markers are not necessarily the median or mean of a scenario family (nor would it be possible to construct such a median or mean scenario by taking all salient scenario characteristics and regional results into account). The markers are simply those scenarios considered by the SRES writing team as illustrative of a particular storyline. They are not singled out as more likely than alternative quantitative interpretations of a particular scenario family and its underlying storyline. Perhaps they may be best described as "first among equals." However, as a result of time and resource limitations the marker scenarios have received the closest scrutiny from the entire writing team and through the SRES open process compared to other scenario quantifications. The marker scenarios are also the SRES scenarios most intensively tested in terms of reproducibility. As a rule, at least four different models were used in attempts to replicate the model quantification of a particular marker scenario. Available time and resources did not allow a similar exercise to be conducted for all SRES scenarios, albeit a more limited effort was devoted to reproduce the A1 scenario groups (next to the A1 marker) with different models.

To enable a comparison of the resultant GHG emissions, the writing team decided to define a subset of harmonized scenarios within each family that share common main scenario driving-force assumptions, such as population or GDP growth. Two harmonization criteria were developed (see also Section 4.4.1). This procedure and the harmonization criteria were adopted in a joint agreement among the six SRES modeling teams.

"Fully harmonized" scenarios are those that share important driving force variables, including population, GDP, and final energy use for each of the four SRES regions and the world (according to the quantitative criteria listed in Table 4-1). Fully harmonized scenarios by definition include the respective marker scenario. From 40 scenarios 11 are classified as scenarios with "full harmonization." For each scenario family at least two scenarios are harmonized using the most restrictive criteria. This also applies to the scenario groups within the A1 scenario family, which correspondingly has the highest number of fully harmonized scenarios. This subset of "fully harmonized" scenarios serves to provide a better correspondence between the development of the three main driving forces and the resultant GHG emissions. The "fully harmonized" scenarios thus demonstrate the degree by which a particular marker scenario is reproducible by alternative modeling approaches. Therefore, "fully harmonized" scenarios are not independent from each other within a particular scenario family (or scenario group in case of A1).

"Globally harmonized" scenarios are those that share global population and GDP profiles within the agreed upon bounds of 5% and 10%, respectively, over the period 1990–2100[3] (see Table 4-1). Altogether 26[4] scenarios are categorized into this category and can be considered to capture the main global development characteristics over time for each respective scenario family and storyline. Again these 26 scenarios are not independent from each other, constituting seven distinct scenario groups (see also footnote 1).

Thus, there are three different types of scenarios within each family:

- One marker and a set of "fully harmonized" scenarios that attempt to replicate the marker scenario quantification.
- A set of "globally harmonized" scenarios.
- A set of non-harmonized scenarios.

In addition, two illustrative scenarios have been selected in the Summary for Policymakers (SPM) from the additional A1 scenario groups (see also footnote 1).

For the sake of simplicity, the term "harmonized" is used herein to describe global harmonization of population and GDP growth. "Fully harmonized" scenarios, for which the main objective is to assess the reproducibility of particular marker scenario quantifications and any remaining uncertainty in GHG emissions from internal model parametrizations, are referred to in the text where appropriate. Figure 4-3 shows the SRES scenarios as a set consisting of subsets that correspond to the four families. The A1 family is in addition divided into different groups of scenarios. The detailed descriptions of inputs and outputs (other than GHG emissions) of the SRES marker scenarios, other harmonized scenarios, and other scenarios are presented in Section 4.4 (see Appendix VII for further numeric detail). The emissions of GHGs and other radiatively important species of gases are described in Chapter 5 (more detail is again presented in Appendix VII).

Table 4-2 summarizes the main characteristics of the four SRES scenario families and their scenario groups, and gives an overview about the number of scenarios that were developed for each scenario group (see also Table 4-1 and Section 4.4.1).

4.3. Scenario Storylines

Although each of the four SRES storylines can unfold only if certain values are emphasized more than others, no explicit

[3] Deviations within each 10-year time period are not considered.

[4] Additionally, three scenarios (A2-AIM, A2-MiniCAM, and B2-MiniCAM) deviate only slightly from the global harmonization criteria for between two to three time steps. Hence these scenarios can be considered as 'almost' harmonized and comparable with the other harmonized scenarios.

Table 4-2: *Overview of SRES scenario quantifications. Shown for each scenario is the name of the storyline and scenario family, the name of the scenario group, number of harmonized and total scenarios in the respective group, by how many different modeling approaches they were developed, and the main (qualitative) characteristics of each of the scenario groups. Please note that A1C and A12G were combined into one fossil-intensive A1FI group in the SPM (see also footnote 1).*

Set	SRES							Total
Family	A1				A2	B1	B2	
Scenario Group	**A1C**	**A1G**	**A1B**	**A1T**	**A2**	**B1**	**B2**	
Globally Harmonized Scenarios[a]	2	3	6	2	2	7	4	26
Other Scenarios[b]	1	0	2	1	4	2	4	14
Total Scenarios *(Different Models Used)*	3 *(3)*	3 *(3)*	8 *(6)*	3 *(3)*	6 *(5)*	9 *(6)*	8 *(6)*	40 *(6)*
Scenario characteristics:[c]								
Population growth	low	low	low	low	high	low	medium	
GDP growth	very high	very high	very high	very high	medium	high	medium	
Energy use	very high	very high	very high	high	high	low	medium	
Land-use changes	low–medium	low–medium	low	low	medium/high	high	medium	
Resource availability[d]	high	high	medium	medium	low	low	medium	
Pace and direction of technological change favoring	rapid coal	rapid oil & gas	rapid balanced	rapid non-fossils	slow regional	medium efficiency & dematerialization	medium "dynamics as usual"	

[a]Globally Harmonized Scenarios share common major input assumptions that describe a particular scenario family at the global level (i.e., global population and GDP within agreed bounds of 5% and 10%, respectively) compared to the marker scenarios over the entire time horizon 1990 to 2100 (deviation in one time period being tolerated).
To further scenario comparability more stringent harmonization criteria were applied where population, GDP, and final energy trajectories were harmonized at the level of the four SRES regions.

[b]Other Scenarios offer alternative interpretations of a scenario storyline for global population and GDP either in its time path or in their levels (or both). Scenarios A2-AIM, A2-MiniCAM and B2-MiniCAM deviate only slightly from the global harmonization criterion for between two to three time steps. Hence these scenarios can be considered as "almost" harmonized and comparable with the other harmonized scenarios.

[c]Scenario characteristics as applied to harmonized scenarios. Other scenarios explore sensitivities of adopting alternative input assumptions than captured in this classification.

[d]Resource availability of conventional and unconventional oil and gas.

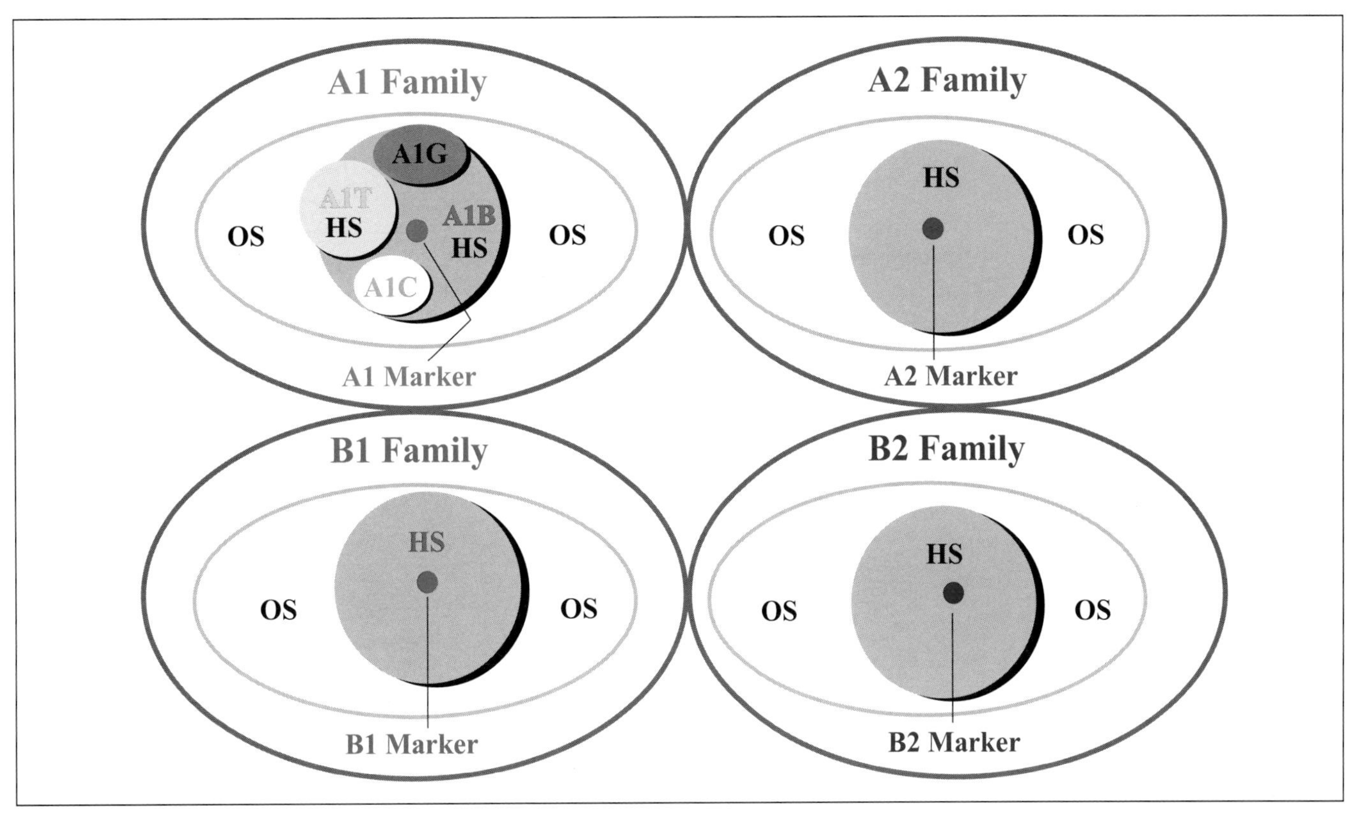

Figure 4-3: Schematic illustration of the multidimensional classification space of SRES scenarios. The set of scenarios consists of four scenario families (A1, A2, B1, and B2), each of which consists of a number of scenarios. Some of these have "harmonized" inputs – they share similar pre-specified global population and GDP trajectories. They are marked as "HS" for (globally) harmonized scenarios. All other scenarios of the same family based on the quantification of the storyline chosen by the modeling team are marked as "OS." The A1 family is divided into four scenario groups that explore alternative developments in the future energy sector. These were merged into three groups in the SPM (see also footnote 1). Finally, one of the harmonized scenarios is designated as the characteristic representative of each family and is the marker scenario.

judgments have been made by the SRES team as to their desirability or probability. Storylines in the literature, however, are often explicitly intended by their authors to have a positive or negative connotation, and sometimes explicitly include assumed dominant or preferred types of governance. Thus, the four SRES storylines are fundamentally different in this respect compared to many narrative scenarios in the underlying literature.

Two arguments are pertinent to the linkage between the scenario storylines and the underlying literature. First, a future regarded as negative by some people may be perceived as positive by others. Second, the storylines represent families of scenarios that can include both success and failure, depending on the perspective of the beholder. To quote Wilkerson (1995): "Like the real life from which they are drawn, the scenarios are mixed bags, at once wonderfully dreadful and dreadfully wonderful." Importantly, the "neutral" or "agnostic" character of SRES scenarios is an explicit departure from most of the underlying literature about storylines and narrative scenarios.

Another important departure and an innovation unique to the SRES approach is the use of the storylines in conjunction with multiple (six) modeling approaches to develop and formulate a set of quantifications or scenarios that are overall consistent with the underlying storylines. This approach provides a rigorous modeling test of the underlying logic and structure of the storyline, while the narrative aspect of the storyline provides a broader, descriptive context for better understanding and interpretation of the scenarios.

4.3.1. A1 Storyline and Scenario Family

The A1 storyline is a case of rapid and successful economic development, in which regional average income per capita converge – current distinctions between "poor" and "rich" countries eventually dissolve. The primary dynamics are:

- Strong commitment to market-based solutions.
- High savings and commitment to education at the household level.
- High rates of investment and innovation in education, technology, and institutions at the national and international levels.
- International mobility of people, ideas, and technology.

The transition to economic convergence results from advances in transport and communication technology, shifts in national policies on immigration and education, and international cooperation in the development of national and international institutions that enhance productivity growth and technology diffusion.

This may be the type of scenario best represented in recent literature (e.g., Shinn, 1985; UN, 1990; Schwartz, 1991; Peterson, 1994; Gallopin *et al.*, 1997; Glenn and Gordon, 1997, 1999; Lawrence *et al.*, 1997; Hammond, 1998; Raskin *et al.*, 1998). Such scenarios are dominated by an American or European entrepreneurial, progress-oriented perspective in which technology, especially communication technology, plays a central role. Wilkerson (1995) designed various scenarios that share features with A1. They emphasize market-oriented solutions, high consumption of both tangible and intangible commodities, advanced technology, and intensive mobility and communication. In some examples of this type of scenario, high economic growth leads to shifts of economic power from traditional core countries to the current economic "periphery," as in the "Global Shift" scenario by CPB (1992) and de Jong and Zalm (1991). The Shell scenario "New Frontiers" (Shell, 1993) is also representative of this family. The IPCC Scenarios IS92a and IS92e are well-known examples of futures with high levels of economic growth (Leggett *et al.*, 1992). IIASA and World Energy Council (WEC) jointly developed three High Growth Scenarios that share assumptions on rapid technological progress, liberalized trade markets, and rising income levels (Nakićenović *et al.*, 1998).

In the A1 scenario family, demographic and economic trends are closely linked, as affluence is correlated with long life and small families (low mortality and low fertility). Global population grows to some nine billion by 2050 and declines to about seven billion by 2100. Average age increases, with the needs of retired people met mainly through their accumulated savings in private pension systems.

The global economy expands at an average annual rate of about 3% to 2100, reaching around US$550 trillion (all dollar amounts herein are expressed in 1990 dollars, unless stated otherwise). This is approximately the same as average global growth since 1850, although the conditions that lead to this global growth in productivity and per capita incomes in the scenario are unparalleled in history. Global average income per capita reaches about US$21,000 by 2050. While the high average level of income per capita contributes to a great improvement in the overall health and social conditions of the majority of people, this world is not necessarily devoid of problems. In particular, many communities could face some of the problems of social exclusion encountered in the wealthiest countries during the 20th century, and in many places income growth could produce increased pressure on the global commons.

Energy and mineral resources are abundant in this scenario family because of rapid technical progress, which both reduces the resources needed to produce a given level of output and increases the economically recoverable reserves. Final energy intensity (energy use per unit of GDP) decreases at an average annual rate of 1.3%. Environmental amenities are valued and rapid technological progress "frees" natural resources currently devoted to provision of human needs for other purposes. The concept of environmental quality changes in this storyline from the current emphasis on "conservation" of nature to active "management" of natural and environmental services, which increases ecologic resilience.

With the rapid increase in income, dietary patterns shift initially toward increased consumption of meat and dairy products, but may decrease subsequently with increasing emphasis on the health of an aging society. High incomes also translate into high car ownership, sprawling suburbia, and dense transport networks, nationally and internationally.

Several scenario groups considered in the A1 scenario family reflect uncertainty in the development of energy sources and conversion technologies in this rapidly changing world. Some scenario groups evolve along the carbon-intensive energy path consistent with the current development strategy of countries with abundant domestic coal resources. Other scenario groups intensify the dependence on (unconventional) oil and (in the longer-run) natural-gas resources[5]. A third group envisages a stronger shift toward renewable energy sources and conceivably also toward nuclear energy. A fourth group (which includes the A1B marker scenario) assumes a balanced mix of technologies and supply sources, with technology improvements and resource assumptions such that no single source of energy is overly dominant. The implications of these alternative development paths for future GHG emissions are challenging: the emissions vary from the carbon-intensive to decarbonization paths by at least as much as the variation of all the other driving forces across the other SRES scenarios.

4.3.2. A2 Storyline and Scenario Family

The A2 scenario family represents a differentiated world. Compared to the A1 storyline it is characterized by lower trade flows, relatively slow capital stock turnover, and slower technological change. The A2 world "consolidates" into a series of economic regions. Self-reliance in terms of resources and less emphasis on economic, social, and cultural interactions between regions are characteristic for this future. Economic growth is uneven and the income gap between now-industrialized and developing parts of the world does not narrow, unlike in the A1 and B1 scenario families.

[5]The coal and gas/oil intensive groups were merged into one fossil-intensive group in the Summary for Policymakers. More detailed information on these two groups is presented here, in Chapter 5 and in Appendix VII.

The A2 world has less international cooperation than the A1 or B1 worlds. People, ideas, and capital are less mobile so that technology diffuses more slowly than in the other scenario families. International disparities in productivity, and hence income per capita, are largely maintained or increased in absolute terms. With the emphasis on family and community life, fertility rates decline relatively slowly, which makes the A2 population the largest among the storylines (15 billion by 2100). Global average per capita income in A2 is low relative to other storylines (especially A1 and B1), reaching about US$7200 per capita by 2050 and US$16,000 in 2100. By 2100 the global GDP reaches about US$250 trillion. Technological change in the A2 scenario world is also more heterogeneous than that in A1. It is more rapid than average in some regions and slower in others, as industry adjusts to local resource endowments, culture, and education levels. Regions with abundant energy and mineral resources evolve more resource-intensive economies, while those poor in resources place a very high priority on minimizing import dependence through technological innovation to improve resource efficiency and make use of substitute inputs. The fuel mix in different regions is determined primarily by resource availability. High-income but resource-poor regions shift toward advanced post-fossil technologies (renewables or nuclear), while low-income resource-rich regions generally rely on older fossil technologies. Final energy intensities in A2 decline with a pace of 0.5 to 0.7% per year.

In the A2 world, social and political structures diversify; some regions move toward stronger welfare systems and reduced income inequality, while others move toward "leaner" government and more heterogeneous income distributions. With substantial food requirements, agricultural productivity in the A2 world is one of the main focus areas for innovation and research, development, and deployment (RD&D) efforts, and environmental concerns. Initial high levels of soil erosion and water pollution are eventually eased through the local development of more sustainable high-yield agriculture. Although attention is given to potential local and regional environmental damage, it is not uniform across regions. Global environmental concerns are relatively weak, although attempts are made to bring regional and local pollution under control and to maintain environmental amenities.

As in other SRES storylines, the intention in this storyline is not to imply that the underlying dynamics of A2 are either good or bad. The literature suggests that such a world could have many positive aspects from the current perspective, such as the increasing tendency toward cultural pluralism with mutual acceptance of diversity and fundamental differences. Various scenarios from the literature may be grouped under this scenario family. For example, "New Empires" by Schwartz (1991) is an example of a society in which most nations protect their threatened cultural identities. Some regions might achieve relative stability while others suffer under civil disorders (Schwartz, 1996). In "European Renaissance" (de Jong and Zalm, 1991; CPB, 1992), economic growth slows down because of a strengthening of protectionist trade blocks. In "Imperial Harmonization" (Lawrence *et al.*, 1997), major economic blocs impose standards and regulations on smaller countries. The Shell scenario "Global Mercantilism" (1989, see Schwartz, 1991) explores the possibility of regional spheres of influence, whereas "Barricades" (Shell, 1993) reflects resistance to globalization and liberalization of markets. Noting the tensions that arise as societies adopt western technology without western culture, Huntington (1996) suggests that conflicts between civilizations rather than globalizing economies may determine the geo-political future of the world.

4.3.3. B1 Storyline and Scenario Family

The central elements of the B1 future are a high level of environmental and social consciousness combined with a globally coherent approach to a more sustainable development. Heightened environmental consciousness might be brought about by clear evidence that impacts of natural resource use, such as deforestation, soil depletion, over-fishing, and global and regional pollution, pose a serious threat to the continuation of human life on Earth. In the B1 storyline, governments, businesses, the media, and the public pay increased attention to the environmental and social aspects of development. Technological change plays an important role. At the same time, however, the storyline does not include any climate policies, to reflect the SRES terms of reference. Nevertheless, such a possible future cannot be ruled out.

The "Conventional Worlds-Policy Reform" scenario by Gallopin *et al.* (1997) is a good example of such a future, although it includes climate policies. In "Ecotopia," Wilkerson (1995) describes a reaction to early decades of crime and chaos, in which community values triumph over individualist ones and lead to resource-friendly lifestyles based on clean and light technologies. This scenario includes a voluntary embrace of cohesion, cooperation, and reduced consumption, backed by legislation and even corporate policies (Wilkerson, 1995). In the normative "Human Development Success" (Glenn and Gordon, 1997), the world achieves an environmentally sustainable economy by 2050, primarily through education to develop human potential. In "Balanced Future" (de Jong and Zalm, 1991; CPB, 1992), economic equilibrium and innovation lead to sustainable development. The "Ecologically Driven" scenarios by WEC (1993) and IIASA–WEC (Nakićenović *et al.*, 1998) – with accelerated efficiency improvements in resource use – share several of the characteristics of the B1 type of future, as does the egalitarian utopia scenario in the TARGETS approach (Rotmans and de Vries, 1997).

Many additional scenarios in the literature could be seen as examples of this family, but may describe the changes as more fundamental than those of B1. The "Transformed World" of Hammond (1998), based on the "Great Transitions" scenario of Gallopin *et al.* (1997), stresses the role of global technological innovation in addition to enlightened corporate actions, government policies, and empowerment of local groups. In

"Shared Space" by the Millennium Institute (Glenn and Gordon, 1997), resources are shared more equitably to the benefit of all and the greater safety of humanity. The Shell scenario "Sustainable World" (1989, see Schwartz, 1991) and the World Business Council for Sustainable Development scenarios (WBCSD, 1998), "Geopolity" and "Jazz," examine sustainable futures.

Economic development in B1 is balanced, and efforts to achieve equitable income distribution are effective. As in A1, the B1 storyline describes a fast-changing and convergent world, but the priorities differ. Whereas the A1 world invests its gains from increased productivity and know-how primarily in further economic growth, the B1 world invests a large part of its gains in improved efficiency of resource use ("dematerialization"), equity, social institutions, and environmental protection.

A strong welfare net prevents social exclusion on the basis of poverty. However, counter-currents may develop and in some places people may not conform to the main social and environmental intentions of the mainstream in this scenario family. Massive income redistribution and presumably high taxation levels may adversely affect the economic efficiency and functioning of world markets.

Particular effort is devoted to increases in resource efficiency to achieve the goals stated above. Incentive systems, combined with advances in international institutions, permit the rapid diffusion of cleaner technology. To this end, R&D is also enhanced, together with education and the capacity building for clean and equitable development. Organizational measures are adopted to reduce material wastage by maximizing reuse and recycling. The combination of technical and organizational change yields high levels of material and energy saving, as well as reductions in pollution. Labor productivity also improves as a by-product of these efforts. Alternative scenarios considered within the B1 family include different rates of GDP growth and dematerialization (e.g., decline in energy and material intensities).

The demographic transition to low mortality and fertility occurs at the same rate as in A1, but for different reasons as it is motivated partly by social and environmental concerns. Global population reaches nine billion by 2050 and declines to about seven billion by 2100. This is a world with high levels of economic activity (a global GDP of around US$350 trillion by 2100) and significant and deliberate progress toward international and national income equality. Global income per capita in 2050 averages US$13,000, one-third lower than in A1. A higher proportion of this income is spent on services rather than on material goods, and on quality rather than quantity, because the emphasis on material goods is less and also resource prices are increased by environmental taxation.

The B1 storyline sees a relatively smooth transition to alternative energy systems as conventional oil and gas resources decline. There is extensive use of conventional and unconventional gas as the cleanest fossil resource during the transition, but the major push is toward post-fossil technologies, driven in large part by environmental concerns.

Given the high environmental consciousness and institutional effectiveness in the B1 storyline, environmental quality is high, as most potentially negative environmental aspects of rapid development are anticipated and effectively dealt with locally, nationally, and internationally. For example, transboundary air pollution (acid rain) is basically eliminated in the long term. Land use is managed carefully to counteract the impacts of activities potentially damaging to the environment. Cities are compact and designed for public and non-motorized transport, with suburban developments tightly controlled. Strong incentives for low-input, low-impact agriculture, along with maintenance of large areas of wilderness, contribute to high food prices with much lower levels of meat consumption than those in A1. These proactive local and regional environmental measures and policies also lead to relatively low GHG emissions, even in the absence of explicit interventions to mitigate climate change.

4.3.4. B2 Storyline and Scenario Family

The B2 world is one of increased concern for environmental and social sustainability compared to the A2 storyline. Increasingly, government policies and business strategies at the national and local levels are influenced by environmentally aware citizens, with a trend toward local self-reliance and stronger communities. International institutions decline in importance, with a shift toward local and regional decision-making structures and institutions. Human welfare, equality, and environmental protection all have high priority, and they are addressed through community-based social solutions in addition to technical solutions, although implementation rates vary across regions.

Like the other scenario families, the B2 scenario family includes futures that can be seen as positive or negative. While the B2 storyline is basically neutral, Kinsman (1990) in his "Caring Autonomy" scenario clearly paints a positive world with emphasis on decentralized governments and strong interpersonal relationships. In the "New Civics" scenario by Wilkerson (1995), values are only shared within small competing groups, which results in a decentralized world of tribes, clans, families, networks, and gangs. The IIASA–WEC "Middle Course" scenario (Nakićenović *et al.*, 1998), with slow removal of trade barriers, may also be grouped in this family. On the positive side, this storyline appears to be consistent with current institutional frameworks in the world and with the current technology dynamics. On the negative side is the relatively slow rate of development in general, but particularly in the currently developing parts of the world.

Education and welfare programs are pursued widely, which reduces mortality and, to a lesser extent, fertility. The population reaches about 10 billion people by 2100, consistent with both the UN and IIASA median projections. Income per capita grows at an intermediate rate to reach about US$12,000 by 2050. By 2100 the global economy might expand to reach some US$250 trillion. International income differences decrease, although not as rapidly as in storylines of higher global convergence. Local inequity is reduced considerably through the development of stronger community-support networks.

Generally, high educational levels promote both development and environmental protection. Indeed, environmental protection is one of the few truly international common priorities that remain in B2. However, strategies to address global environmental challenges are not of a central priority and are thus less successful compared to local and regional environmental response strategies. The governments have difficulty designing and implementing agreements that combine global environmental protection, even when this could be associated with mutual economic benefits.

The B2 storyline presents a particularly favorable climate for community initiative and social innovation, especially in view of the high educational levels. Technological frontiers are pushed less than they are in A1 and B1, and innovations are also regionally more heterogeneous. Globally, investment in energy R&D continues its current declining trend (EIA, 1997, 1999), and mechanisms for international diffusion of technology and know-how remain weaker than in scenarios A1 and B1 (but higher than in A2). Some regions with rapid economic development and limited natural resources place particular emphasis on technology development and bilateral cooperation. Technical change is therefore uneven. The energy intensity of GDP declines at about 1% per year, in line with the average historical experience since 1800.

Land-use management becomes better integrated at the local level in the B2 world. Urban and transport infrastructure is a particular focus of community innovation, and contributes to a low level of car dependence and less urban sprawl. An emphasis on food self-reliance contributes to a shift in dietary patterns toward local products, with relatively low meat consumption in countries with high population densities.

Energy systems differ from region to region, depending on the availability of natural resources. The need to use energy and other resources more efficiently spurs the development of less carbon-intensive technology in some regions. Environment policy cooperation at the regional level leads to success in the management of some transboundary environmental problems, such as acidification caused by sulfur dioxide (SO_2), especially to sustain regional self-reliance in agricultural production. Regional cooperation also results in lower emissions of nitrogen oxides (NO_x) and volatile organic compounds (VOCs), which reduce the incidence of elevated tropospheric ozone levels. Although globally the energy system remains predominantly hydrocarbon-based to 2100, a gradual transition occurs away from the current share of fossil resources in world energy supply, with a corresponding reduction in carbon intensity.

4.4. Scenario Quantification and Overview

4.4.1. Scenario Terminology

In this section representative quantifications of the four scenario storylines described in Section 4.3 are summarized, and the evolution of the main scenario driving forces and associated quantitative scenario characteristics are described. Their resultant GHG and other emissions are discussed in more detail in Chapter 5.

To elucidate differences in uncertainties that stem both from adopting alternative (exogenous) scenario driving-force assumptions and from the uncertainties that arise from different model representations, alternative scenario quantifications are differentiated into harmonized and unharmonized scenarios (see Section 4.2, Tables 4-1 and 4-2, and Box 1-1 for terminology description).

To achieve harmonization across six different modeling approaches is not a trivial task. For example, most of the models have different regional disaggregations, so that harmonization at the level of the four SRES regions required some "inverse" solutions, often achieved through iterative model runs and adjustments of input assumptions. Also, in some modeling frameworks the harmonized "input" parameters are actually outputs of components of the modeling framework (e.g., GDP as an output of economic general equilibrium models, or final energy as an output variable after considering endogenous energy prices and exogenously pre-specified energy-intensity improvement rates). Therefore, harmonization of important scenario driving-force inputs was neither possible for all scenarios and for all participating modeling teams, and nor was it judged desirable, as the adoption of any harmonization criterion somewhat artificially compresses uncertainty. This is also why simpler harmonization criteria were adopted (see Section 4.2. above) that focused on global population and GDP growth profiles. These are referred to as "globally harmonized" scenarios in the subsequent Subsections.

From the 40 SRES scenarios, 26 are classified as "globally harmonized" scenarios and 14 are classified as "other" scenarios. (The latter category includes three scenarios that only deviate slightly from the harmonization criteria.) Harmonized scenarios are thus comparable in that they describe similar global development patterns with respect to demographics and economic growth. In the subsequent discussion of scenario driving forces a three-tiered structure is adopted. First, for each scenario family (and where applicable for each scenario group in the A1 scenario family), the discussion starts with a presentation of the respective marker and "fully harmonized" scenarios. Subsequently, "globally harmonized" scenarios and "other" scenarios are discussed. "Globally harmonized" scenarios shed additional light into uncertainties that stem from adopting different regional assumptions (see above). Finally, "other" scenarios are presented that offer a different quantitative interpretation of a particular scenario storyline compared to the previous scenario categories. In some cases, differences in interpretation relate to uncertainties in rates of change – "other" scenarios yield similar global demographic and economic outcomes by 2100 (e.g. the B2-ASF scenario compared to the B2 marker), but illustrate different dynamics of how these could unfold. In other cases, the "other" scenario category comprises scenario quantifications that deliberately explore alternative interpretations of a scenario storyline in terms of global population and GDP growth altogether (e.g. in the A2-A1-MiniCAM scenario). The reason is to indicate that quantitative scenario descriptions entail a high degree of uncertainty (and subjectivity from different modeling teams) when it comes to interpret the four different qualitative SRES scenario storylines and to translate them into the quantitative assumptions that drive emission models. When comparing GHG emissions results for the four SRES marker scenarios (see Chapter 5) with those of the other SRES scenarios, it is illustrative to distinguish the effects of different model methodologies and parametrizations from variations of important scenario drivers that often serve as exogenous input to models.

Of the total of 40 SRES scenarios, 29 (including the marker scenarios) satisfy the harmonization criteria for population on the world level *and* for all four SRES regions, 12 scenarios are harmonized for population and GDP, and 11 (13 including the A1T scenario group) scenarios are harmonized for population, GDP *and* final energy (see Table 4-1). Also, 35 scenarios are harmonized for population on the world level and 26 scenarios are harmonized for global population and GDP (see Table 4-1). The status of harmonization is also relatively stable to changes in the harmonization criteria. For example, if the above harmonization criteria were increased by 50% (i.e. GDP for the four SRES regions may differ by up to ±38% from the respective GDP of the marker scenario), the sample of 11 *harmonized* scenarios does not change; however, the number of scenarios harmonized on the *global* level increases from 15 to 20.

Thus, as mentioned above not all scenario quantifications comply with the adopted harmonization criteria differences in regional coverage and definition among models. In some instances modeling teams also deliberately chose not to follow harmonized input assumptions, but instead explored scenario sensitivities by emphasizing alternative developments than suggested in the marker scenario quantification. The writing team recognizes that this increases the number of scenarios as well as complexity in the interpretation of results. These additional scenarios are the result of the SRES terms of reference of proceeding via an open process soliciting as wide participation and viewpoints as possible and also serve the purpose of highlighting important uncertainties of the future that are necessarily compressed by limiting scenario quantification to four illustrative marker scenarios. Thus, while unharmonized scenarios illustrate the impact on GHG emissions of expanding the uncertainty range of main scenario drivers within any particular scenario family, the "globally harmonized" scenarios indicate the range of GHG emissions uncertainty that remains *after* most important global driving force assumptions (population and GDP) have been harmonized. (Finally, the range of GHG emissions resulting from comparing "fully harmonized" scenarios is indicative of the uncertainty of internal model parametrizations such as energy technology change, dietary patterns, and agricultural productivity changes that influence structural changes in energy supply and end-use and land-use changes, see Table 4-1.)

Harmonization of input assumptions increases the comparability across scenarios and can serve as an additional guide for choosing a particular SRES scenario subset, and to illustrate different degrees of scenario uncertainty. The latter is an important aspect, considering the different user communities of SRES scenarios. Given the comparatively narrow variation as defined by the harmonization criteria, differences in population, GDP, and final energy use between harmonized scenarios of the same scenario family need not to be considered in subsequent analyses and are also not discussed separately below.

In the A1 scenario family, the scenarios within one group were also harmonized. In one A1 scenario group the transition away from conventional oil and gas either leads to a massive development of unconventional oil and gas resources (A1G) or to a large-scale synfuel economy based on coal (A1C). Please note that A1C and A1G were combined into one fossil intensive group A1FI in the Summary for Policymakers during its approval process (see also footnote 1). GHG emissions in these scenarios approach emissions characteristic of the A2 scenario family (i.e. are much higher than in the case of the A1 marker scenario). In another A1 scenario group, dwindling conventional oil and gas resources lead to fast development of post-fossil alternatives and enhanced energy conservation. In this technology-intensive scenario group (A1T), energy demands are lower than in the other A1 scenario groups and, because of radical technological change in energy systems, GHG emissions are much lower than in the other A1 scenario groups (including the A1B marker scenario), approaching those of the B1 scenario family.

The six modeling teams also produced other scenarios as part of the SRES open process. These modeling runs were generally not harmonized and are presented as appropriate later in the report.

Table 4-3 gives an overview of the 40 SRES scenario quantifications as they were developed to describe the four scenario families and the seven different scenario groups.

4.4.2. *Translation of Storylines into Scenario Drivers*

Table 4-4 gives a summary overview of the main scenario assumptions and characteristics (see also Table 4-2 above). To facilitate comparability, the summary format adopted is similar to the previous IS92 scenario series (Pepper *et al.*, 1992). Specific assumptions about the quantification of particular

Table 4-3: *Overview of SRES scenarios subdivided into the four scenario families and seven scenario groups (four for the A1 family, one for each of the other scenario families) (see also footnote 1). Each scenario represents a quantitative interpretation of a particular qualitative scenario storyline with the help of one model. Scenarios are named after their respective scenario family (A1, A2, B1, and B2) or scenario groups in case of the A1 scenario family (A1C, A1G, A1B, and A1T) followed by the name of the model that was used for the scenario quantification. Additional scenarios are labeled according to the specifications provided by the modeling teams contributing to the SRES open process. The scenarios are additionally classified as "harmonized" and "other" scenarios with respect to whether they share harmonized input assumptions on global population and GDP growth within their respective scenario family or whether they offer an alternative scenario interpretation. Scenarios denoted by an asterisk share harmonized input assumptions for population, GDP, and final energy use at both the global level and the level of the four SRES regions (i.e. are classified as "fully harmonized").*

Family	**A1**				**A2**	**B1**	**B2**
Scenario Group *(Different Models Used)*	**A1C** *(3)*	**A1G** *(3)*	**A1B** *(6)*	**A1T**[c] *(3)*	**A2** *(5)*	**B1** *(6)*	**B2** *(6)*
Total Scenarios	3	3	8	3	6	9	8
Globally Harmonized Scenarios[a]	2	3	6	2	2	7	4
Other Scenarios[b]	1	0	2	1	4	2	4
Marker and Globally Harmonized Scenarios	A1C-AIM* A1C-MESSAGE*	A1G-AIM* A1G-MESSAGE* *A1G-MiniCAM*	**A1B-AIM*** A1B-ASF A1B-IMAGE A1B-MARIA A1B-MESSAGE* A1B-MiniCAM	A1T-AIM* *A1T-MESSAGE**	**A2-ASF*** A2-MESSAGE*	**B1-IMAGE*** B1-AIM B1-ASF B1-MESSAGE* B1-MiniCAM B1T-MESSAGE* B1High-MESSAGE	**B2-MESSAGE*** B2-AIM* B2-MARIA* B2C-MARIA*
Other Scenarios	A1C-MiniCAM		A1v1-MiniCAM A1v2-MiniCAM	A1T-MARIA	A2-AIM A2G-IMAGE A2-MiniCAM A2-A1-MiniCAM	B1-MARIA B1High-MiniCAM	B2-ASF B2-IMAGE B2-MiniCAM B2High-MiniCAM

[a]Globally Harmonized Scenarios share common major input assumptions that describe a particular scenario family at the global level (i.e., global population and GDP within agreed bounds of 5% and 10%, respectively) compared to the marker scenarios over the entire time horizon 1990–2100 (deviation in one time period of ten years being tolerated).
To further scenario comparability more stringent harmonization criteria were applied where population, GDP, and final energy trajectories were harmonized at the level of the four SRES regions ("fully (global + regional) harmonized" scenarios are indicated with an asterisk).
[b]Other Scenarios offer alternative interpretations of a scenario storyline for global population and GDP either in its time path or in their levels (or both). Scenarios A2-AIM, A2-MiniCAM and B2-MiniCAM deviate only slightly from the global harmonization criterion for between two to three time steps. Hence these scenarios can be considered as "almost" harmonized and comparable with the other harmonized scenarios.
[c]Harmonization criteria for final energy does not apply by design as scenario explores sensitivity of technological change in improving end-use efficiency compared to other A1 scenario groups.

Table 4-4a: *Overview of main driving forces for the four SRES marker scenarios for 2100 if not indicated otherwise. Numbers in brackets show the range across all other scenarios from the same scenario family as the marker. Units are given in the table. (IND regions includes industrialized countries consisting of OECD90 and REF regions; and DEV region includes developing countries consisting of ASIA and ALM regions, see Appendix IV).*

	Population In Billion	Economic Growth, GDP_{mex}[a]	Per Capita Income, GDP_{mex}/capita	Primary Energy Use	Hydrocarbon Resource Use[b]	Land-Use Change[c]
A1	Lutz (1996) Low ~7 billion 1.4 IND 5.6 DEV	Very high 1990-2020: 3.3 (2.8-3.6) 1990-2050: 3.6 (2.9-3.7) 1990-2100: 2.9 (2.5-3.0)	Very high in IND: US$107,300 (60,300-113,500) in DEV: US$ 66,500 (41,400-69,800)	Very high 2.226 (1,002-2,683) EJ Low energy intensity of 4.2 MJ/US$ (1.9-5.1)	Varied in four scenario groups: Oil: Low to very high 20.8 (11.5-50.8) ZJ Gas: High to very high 42.2 (19.7-54.9) ZJ Coal: Medium to very high 15.9 (4.4-68.3) ZJ	Low. 1990-2100: 3% cropland, 6% grasslands 2% forests
A2	Lutz (1996) High ~15 billion 2.2 IND 12.9 DEV	Medium 1990-2020: 2.2 (2.0-2.6) 1990-2050: 2.3 (1.7-2.8) 1990-2100: 2.3 (2.0-2.3)	Low in DEV Medium in IND in IND: US$46,200 (37,100-64,500) in DEV: US$11,000 (10,300-13,700)	High 1,717 (1,304-2.040) EJ High energy intensity of 7.1 MJ/US$ (5.2-8.9)	Scenario dependent: Oil: Very low to medium 17.3 (11.0-22.5) ZJ Gas: Low to high 24.6 (18.4-35.5) ZJ Coal: Medium to Very high 46.8 (20.1-47.7) ZJ	Medium n.a. from ASF
B1	Lutz (1996) Low ~7 billion 1.4 IND 5.7 DEV	High 1990-2020: 3.1 (2.9-3.3) 1990-2050: 3.1 (2.9-3.5) 1990-2100: 2.5 (2.5-2.6)	High in IND: US$72,800 (65,300-77,700) In DEV: US$40,200 (40,200-45,200)	Low. 514 (514-1,157) EJ Very low energy intensity of 1.6 EJ/US$ (1.6-3.4)	Scenario dependent: Oil: Very low to high 19.6 (15.7-19.6) ZJ Gas: Medium to high 14.7 (14.7-31.8) ZJ Coal: Very low to high 13.2 (3.3-27.2) ZJ	High 1990-2100: -28% cropland -45% grassland +30% forests
B2	UN (1998) Median ~10 billion 1.3 IND 9.1 DEV	Medium 1990-2020: 3.0 (2.2-3.1) 1990-2050: 2.8 (2.1-2.9) 1990-2100: 2.2 (2.0-2.3)	Medium in IND: US$54,400 (42,400-61,100) In DEV: US$18,000 (14,200-21,500)	Medium 1,357 (846-1,625) EJ Medium energy intensity of 5.8 MJ/US$ (4.3-6.5)	Oil: Low to medium 19.5 (11.2-22.7) ZJ by 2100 Gas: Low to medium 26.9 (17.9-26.9) ZJ by 2100 Coal: Low to very high 12.6 (12.6-44.4) ZJ by 2100	Medium 1990-2100: +22% cropland +9% grasslands +5% forests[d]

[a] Exponential growth rates after World Bank (1999) method (given on pages 371 to 372) are calculated using the *different base years* from the models.
[b] Resource availability is generally combined with scenario specific rates of technological change.
[c] Residual and other land-use categories are not shown in the Table.
[d] Land-use data for B2 marker taken from AIM land-use B2 scenario run.

scenario drivers, such as population and economic growth, technological change, resource availability, land-use changes, and local and regional environmental policies, are summarized in this Section (GHG emissions are reported in detail in Chapter 5). The assumptions are based on the range of driving forces identified in Chapter 2 and their relationships as summarized in Chapter 3. For simplicity these drivers are presented separately, but it is important to keep in mind that the evolution of these scenario drivers is to a large extent interrelated, as reflected in the SRES scenarios.

As discussed above, the SRES scenarios were designed to reflect inherent uncertainties of future developments by adopting a range of salient input assumptions, but without attempting to cover the extremes from the scenario literature. Given the nature of the SRES open process and its multi-model approach, as well as the need for documented input assumptions, published scenario extremes are difficult to reproduce using alternative model approaches or insufficiently documented input data. (For instance, many long-term emission scenarios do not report their underlying population assumptions (see Chapters 2 and 3), which is especially true for extreme scenarios that are usually performed within the context of model sensitivity analysis.)

Compared to the previous IS92 scenario series there are important similarities, but also important differences. For instance, three different future population scenarios were adopted, albeit that the future population levels are somewhat lower and the range more compressed than those in IS92 this reflects advances in demographic modeling and population projections. Conversely, the range of assumptions that concern resource availability and future technological change is much wider compared to earlier scenarios, reflecting in particular the results of the IPCC WGII Second Assessment Report (SAR; Watson *et al.*, 1996). Another distinguishing characteristic of the SRES scenarios is an attempt to reflect the most recent understanding on the *relationships* between important scenario driving-force variables. For instance, no scenario combines low fertility with high mortality assumptions, which reflects the consensus view from demographers (see Chapter 3). Equally, all SRES scenarios assume a qualitative relationship between demographics and social and economic development trends, which reflects both the literature (see Chapter 3) and the results of the evaluation of the IS92 scenario series (Alcamo *et al.*, 1995). All else being equal, fertility and mortality trends are thus lower in scenarios with high-income growth assumptions, but the multidimensionality of the causal linkages must be recognized and so no particular cause–effect model is postulated here. Finally, the scenarios also attempt to reflect recent advances (as reviewed in Chapter 3) in understanding of the evolution of macro-economic and material productivity (e.g., their coupling via capital turnover rates), uncertainties in future levels of "dematerialization" (reflected in the difference between the B1 and A1 scenarios), and the likely evolution of local and regional environmental policies (e.g., all scenarios assume various degrees of sulfur-control policies).

The main aspects of translating the storylines into scenario drivers are summarized below. For each scenario family an overview of all scenario quantifications is given. Scenarios that share harmonized input assumptions with the respective scenario marker in terms of global population and GDP profiles (see Tables 4-1 and 4-3) are indicated in *italics* in the subsequent discussion. Altogether, 26 scenarios in the four scenario families share similar assumptions about population and GDP at the global level. The other 14 scenarios either do not fully comply with the agreed common input assumptions concerning global population and GDP or explore important sensitivities of future demographic and economic developments beyond that described in the 24 scenarios. These sensitivities include resource availability, technology development, or land-use changes and describe similar demographic and economic development patterns as other scenarios within a family, even if they do not fall within the range suggested by the harmonization criteria (see Table 4-1). Combined, the SRES scenario set comprises 40 scenarios grouped into four scenario families and different scenario groups (see Table 4-3).

Each scenario family is illustrated by a designated marker scenario. A marker is not necessarily the mean or mode of comparable scenario quantifications, nor would it be possible to construct an internally consistent scenario reflecting medians/modes of all salient scenario characteristics (both in terms of scenario input assumptions as well as scenario outcomes, i.e. emissions). Marker scenarios should also not be interpreted as being the more likely alternative scenario quantifications. However, only the four marker scenarios were subjected to the SRES open process through the SRES website and they have also received closest scrutiny by the entire writing team.

4.4.2.1. A1 Scenarios

The A1 marker scenario (Jiang *et al.*, 2000) was created with the AIM model, an integrated assessment model developed by NIES, Japan (see Appendix IV). The A1 scenario family is characterized by:

- An affluent world, with a rapid demographic transition (declining mortality and fertility rates) and an increasing degree of international development equity.
- Very high productivity and economic growth in all regions, with a considerable catch-up of developing countries.
- Comparatively high energy and materials demands, moderated however by continuous structural change and the diffusion of more efficient technologies, consistent with the high productivity growth and capital turnover rates of the scenario.

The first group of A1 scenarios, which includes the A1B marker, assumes "balanced"[6] progress across all resources and technologies from energy supply to end use, as well as "balanced" land-use changes. Three other groups of A1 scenarios were identified which describe three alternative pathways according to different resource and technology development assumptions:

- A1C: "clean coal" technologies that are generally environmentally friendly with the exception of GHG emissions.
- A1G: an "oil- and gas–rich" future, with a swift transition from conventional resources to abundant unconventional resources including methane clathrates.
- A1T: a "non-fossil" future, with rapid development of solar and nuclear technologies on the supply side and mini-turbines and fuel cells used in energy end-use applications.

The divergence between the various scenario groups (in terms of resource availability and the direction of technological change) results in a wide range of GHG emissions. The two fossil-fuel dominated alternatives, A1C and A1G (combined into the fossil-intensive A1FI scenario group in the SPM, see Footnote 1), have higher, and the A1T alternatives have lower, GHG emissions than the A1 marker scenario (see Chapter 5).

"Balanced" A1 scenarios quantifications were also calculated by the models[7] *A1B-ASF, A1B-IMAGE, A1B-MARIA, A1B-MESSAGE*, and *A1B-MiniCAM*. Additional scenarios representing A1 scenario groups were developed using the AIM (*A1C-AIM, A1G-AIM, A1T-AIM*[8]), MARIA (A1T-MARIA), MESSAGE (*A1C-MESSAGE, A1G-MESSAGE, A1T-MESSAGE*), and MiniCAM (A1C-MiniCAM, *A1G-MiniCAM*) models. The MiniCAM modeling team also evaluated alternative interpretations of the A1 scenario storyline with different demographic, economic, and energy development patterns (A1v1-MiniCAM and A1v2-MiniCAM) on top of the alternative technology–resource developments examined in the other A1 scenarios.

[6] Different modeling teams provided different interpretations of what such a "balanced" resource–technology portfolio could be in the 21st century. The assumed rapid technology dynamics that underlie the A1 scenario storyline necessitate that technologies and resource exploitation profiles change significantly. Hence, the concept of "balanced" development does not necessarily apply to any particular future date, but rather to the entirety of the scenario's development path throughout the 21st century. As a consequence of the slow turnover rates in the capital stock of the energy sector, all scenarios necessarily rely more heavily on currently dominant resources and technologies in the near-term and project more radical departures only in the long-term.

[7] Scenarios denoted by *italics* indicate scenario quantifications that share harmonized global population and GDP trajectories with the respective marker scenario (i.e. "globally harmonized" scenarios).

[8] By design, the A1T scenario group explores the possibility of technological change in energy end-use technologies and hence lower energy demand compared to the A1 marker scenario. Considering that these scenarios provide similar levels of energy service, albeit at lower levels of final energy use, they are classified as "harmonized" based on population and GDP profiles only.

4.4.2.2. A2 Scenarios

The A2 marker scenario (A2-ASF) was developed using ASF (see Appendix IV), an integrated set of modeling tools that was also used to generate the first and the second sets of IPCC emission scenarios (SA90 and IS92). Overall, the A2-ASF quantification is based on the following assumptions (Sankovski *et al.*, 2000):

- Relatively slow demographic transition and relatively slow convergence in regional fertility patterns.
- Relatively slow convergence in inter-regional GDP per capita differences.
- Relatively slow end-use and supply-side energy efficiency improvements (compared to other storylines).
- Delayed development of renewable energy.
- No barriers to the use of nuclear energy.

Additional scenario quantifications of A2 were developed using the AIM (A2-AIM)[9], IMAGE (A2-IMAGE),[10] MESSAGE (*A2-MESSAGE*), and MiniCAM (A2-MiniCAM)[11] models. An alternative interpretation of the A2 scenario storyline in the form of a "delayed development" or "transitional" scenario between the A2 and A1 scenario families was developed by the MiniCAM modeling team (A2-A1-MiniCAM).

4.4.2.3. B1 Scenarios

The B1 marker scenario (de Vries *et al.*, 2000) was developed using the IMAGE 2.1 model (see Appendix IV). Earlier versions of the model were used in the first IPCC scenario development effort (SA90). B1 illustrates the possible emissions implications of a scenario in which the world chooses consistently and effectively a development path that favors efficiency of resource use and "dematerialization" of economic activities. The scenario entails in particular:

- Rapid demographic transition driven by rapid social development, including education.

[9] The scenario deviates only slightly from the global population and GDP assumptions of other "harmonized" scenarios within this scenario family.

[10] The IMAGE results for the A2 and B2 scenarios are based on preliminary model experiments carried out in March 1998. As a result of limited resources it has not been possible to redo these experiments. Hence, the IMAGE team is not able to provide background data and details for these scenario calculations and the population and economic growth assumptions are not fully harmonized, as is the case for the IMAGE A1 and B1 scenarios.

[11] The scenario deviates only slightly from the global population and GDP assumptions of other "harmonized" scenarios within this scenario family.

- High economic growth in all regions, with significant catch-up in the presently less-developed regions that leads to a substantial reduction in present income disparities.
- Comparatively small increase in energy demand because of dematerialization of economic activities, saturation of material- and energy-intensive activities (e.g., car ownership), and effective innovation and implementation of measures to improve energy efficiency.
- Timely and effective development of non-fossil energy supply options in response to the desire for a clean local and regional environment and to the gradual depletion of conventional oil and gas supplies.

Additional scenarios of B1 were developed using the AIM (*B1-AIM*), ASF (*B1-ASF*), MARIA (B1-MARIA), MESSAGE (*B1-MESSAGE*), and MiniCAM (*B1-MiniCAM*) models. Some of these scenarios explore alternative technological developments (akin to the A1 scenario, e.g. *B1T-MESSAGE*) or alternative interpretations on rates and potentials of future dematerialization and energy-intensity improvements (e.g., *B1High-MESSAGE* and B1High-MiniCAM explore scenario sensitivities of higher energy demand compared to the B1 marker).

4.4.2.4. B2 Scenarios

The B2 marker scenario (Riahi and Roehrl, 2000) was developed using the MESSAGE model (see Appendix IV), an integrated set of energy-sector simulation and optimization models used to generate the IIASA–WEC long-term energy and emission scenarios (IIASA–WEC, 1995; Nakićenović *et al.*, 1998). Compared to the other storylines (A1 and B1), the B2 future unfolds with more gradual changes and less extreme developments in all respects, including geopolitics, demographics, productivity growth, technological dynamics, and other salient scenario characteristics. A more fragmented pattern of future development (not that different from present trends) precludes any particularly strong convergence tendencies in the scenario quantification:

- Model parameter values for projections to 2100 were derived typically from long-term historical data series where applicable (Marchetti and Nakićenović, 1979; Nakićenović, 1987; Grübler, 1990; Nakićenović *et al.*, 1996; Grübler, 1998a; Nakićenović *et al.*, 1998), or adopted from the medians of the analysis of the scenario literature (see Chapter 2).
- The scenario quantification assumes effective policies in solving local and regional problems such as traffic congestion, local air pollution, and acid rain impacts.

Additional B2 scenarios were developed using the AIM (*B2-AIM*), ASF (B2-ASF)[12], IMAGE (B2-IMAGE),[10] MARIA (*B2-MARIA*; Mori, 2000), and MiniCAM (B2-MiniCAM)[13] models. Again, more than one B2 scenario interpretation was generated. Some models (e.g., *B2-MARIA* or B2High-MiniCAM) offered additional perspectives of both inter- and intra-model variability in the interpretation of the B2 storyline, particularly with respect to resource availability and technology development assumptions (see Section 4.4.7) and their resultant impact on GHG emissions (see Chapter 5).

Figure 4-4 summarizes the main global scenario indicators of the four SRES marker scenarios by 2100, including population and global GDP levels, final energy intensities, final energy use, corresponding carbon intensities, land-use changes,[14] and energy-related CO_2 emissions. It illustrates that the range of the most important scenario characteristics spanned by the four SRES marker scenarios and the entire SRES scenario set covers the uncertainty range well, as reflected in the scenario literature. The scenario space defined by the lines "SRES-max" and "SRES-min" lies well within the range spanned by the scenario literature contained in the SRES scenario database and analyzed in Chapter 2. The two exceptions are:

- The low range of future CO_2 emissions from the literature is not reflected in the SRES scenarios, consistent with the SRES Terms of Reference to consider only scenarios that assume no "additional climate policy initiatives" (see Appendix I).
- The low end of the range of global GDP and energy use from the literature is equally not reflected in the SRES scenarios. Very low global GDP values arise from a combination of rapid demographic transition with low per capita productivity growth, a combination for which there is little theoretic or empiric support in the available literature on demographic and economic growth reviewed in Chapter 3. Low GDP scenarios can also reflect a combination of average population growth and low economic growth; this type of future usually depicts low-income, inequitable, and possibly unstable worlds that are not analyzed in this report (see Box 4-2).

Equally, while the SRES scenarios cover the range from the literature, the four marker scenarios cannot and do not replicate the frequency distributions of individual scenario variables as discussed in Chapter 2. Nor can their quantitative characteristics segment the relevant distributions in

[12] The B2-ASF shares global population and GDP assumptions with the B2 marker by 2100, but explores different dynamics of growth in the intervening time period.

[13] The scenario deviates only slightly from the global population and GDP assumptions of other "harmonized" scenarios within this scenario family.

[14] The dynamic profiles of land-use changes mean that scenario comparisons for any given year, such as 2100, are somewhat misleading. Hence, cumulative carbon emissions that result from land-use changes over the 1990 to 2100 period are used as a proxy indicator in Figure 4-4 (see Chapter 5 for a more detailed discussion).

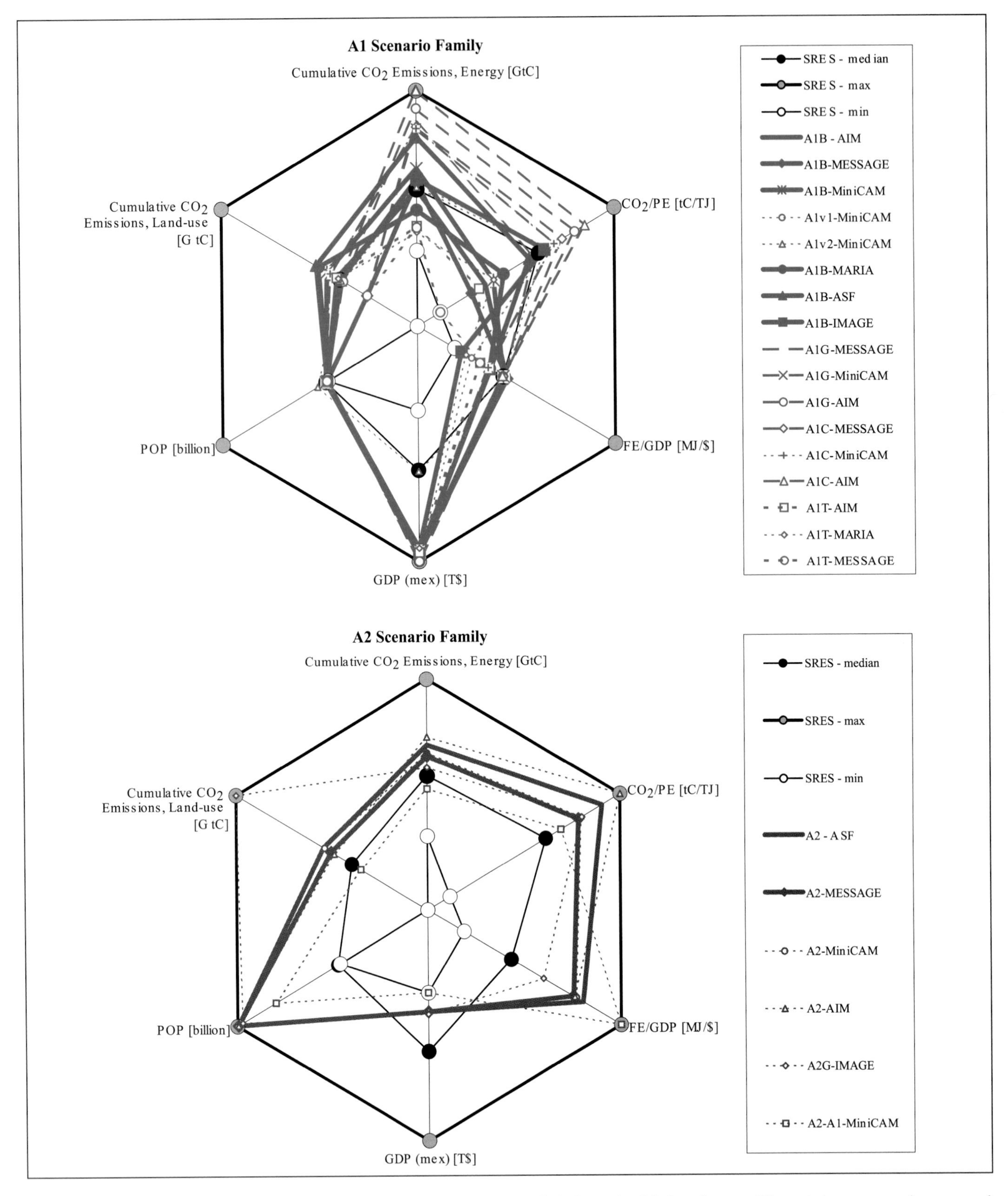

Figure 4-4: Global cumulative CO_2 emissions in the scenarios and their main driving forces. The minimum, maximum, and median (50th percentile) values shown on the six axes of each hexagon, for the cumulative energy and land-use CO_2 emissions from 1990 to 2100 and 2100 values for the four driving forces, are based on the distribution of scenarios in the literature (see Chapter 2). The four hexagons show the ranges across the four scenario families (A1, A2, B1, and B2), cumulative CO_2 emissions in GtC, population (POP) in billions, gross world product (GDP) in trillion US dollars (T$) at 1990 prices, final energy intensity of the gross world product (FE/GDP) in MJ per US dollar at 1990 prices (MJ/$), and CO_2 emissions intensity of primary energy (PE) (tC/TJ).

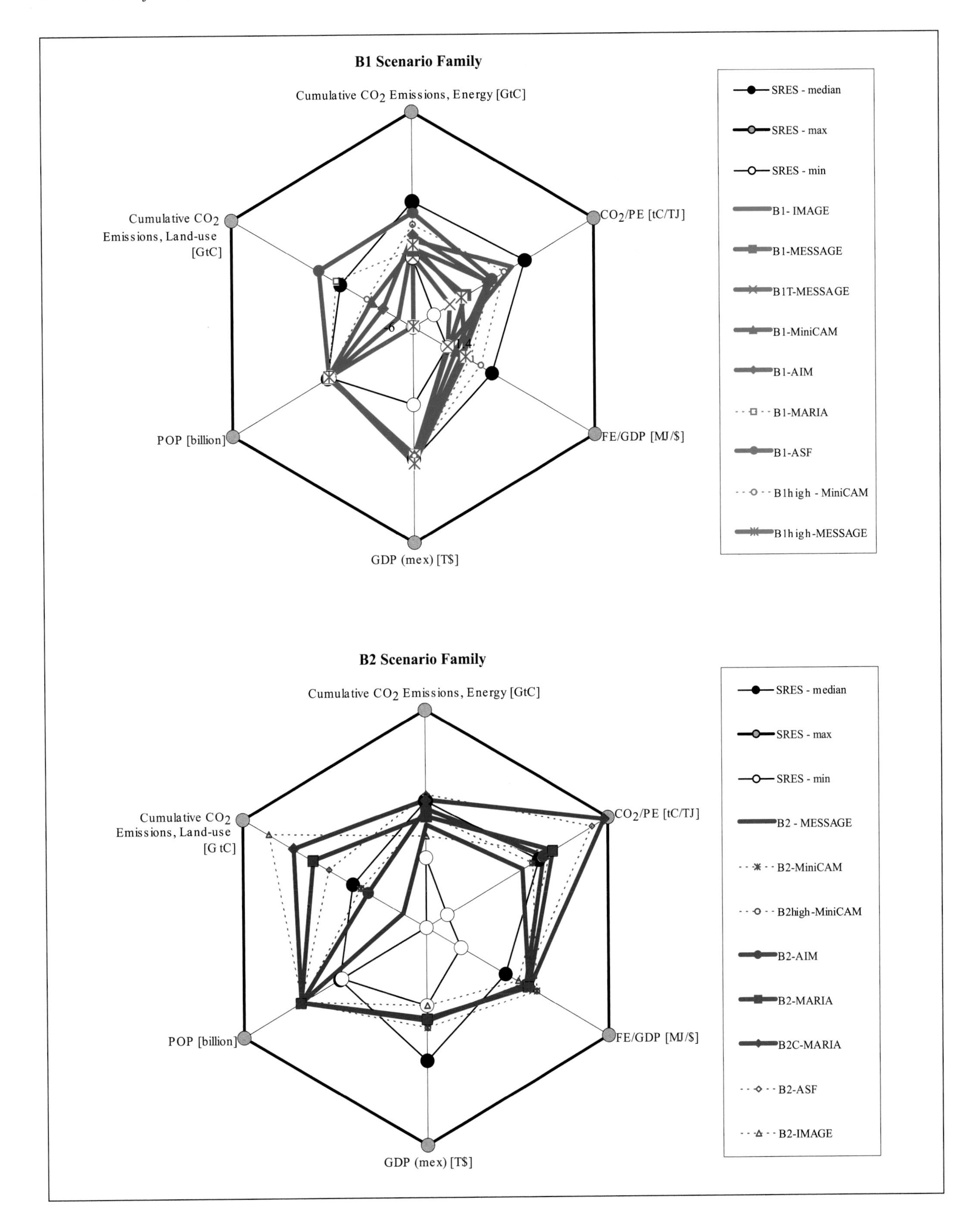
B1 Scenario Family
Cumulative CO_2 Emissions, Energy [GtC]
CO_2/PE [tC/TJ]
FE/GDP [MJ/$]
GDP (mex) [T$]
POP [billion]
Cumulative CO_2 Emissions, Land-use [GtC]
SRES - median
SRES - max
SRES - min
B1- IMAGE
B1-MESSAGE
B1T-MESSAGE
B1-MiniCAM
B1-AIM
B1-MARIA
B1-ASF
B1high - MiniCAM
B1high-MESSAGE
B2 Scenario Family
Cumulative CO_2 Emissions, Energy [GtC]
CO_2/PE [tC/TJ]
FE/GDP [MJ/$]
GDP (mex) [T$]
POP [billion]
Cumulative CO_2 Emissions, Land-use [G tC]
SRES - median
SRES - max
SRES - min
B2 - MESSAGE
B2-MiniCAM
B2high-MiniCAM
B2-AIM
B2-MARIA
B2C-MARIA
B2-ASF
B2-IMAGE

approximately equal intervals. Two distinguishing features characterize the SRES scenarios. First, probabilities or likelihood are not assigned to any quantitative scenario characteristics (inputs or outputs). Thus, that two of the SRES marker scenarios deploy the same (low) demographic projection does not imply that such a scenario is considered more likely. It only indicates that such a demographic scenario was judged by the SRES writing team to be consistent with two of the four SRES storylines, as opposed to arbitrarily assigning different population projections to other "high" or "low" scenario characteristics. Second, the SRES scenarios incorporate current understanding of important interrelations between various scenario-driving forces (see Chapter 3). Thus, a "free," or "modeler's choice," numeric combination of scenario indicators is simply not possible. For instance, intermediary levels of global GDP or energy use could result both from a medium population projection combined with intermediate per capita GDP or energy use growth, or alternatively from low or high population projections combined with high or low GDP and energy per capita values, respectively. The fact that for some quantitative scenario characteristics a number of SRES marker scenarios cluster more toward the upper or lower range spanned by the scenario literature merely indicates the existence of important relationships between scenario characteristics. It also indicates that a limited number of scenarios (four markers) cannot replicate the distribution of individual scenario values arising out of an analysis of more than 400 scenarios published in the literature[15]. Hence, it is important to consider always the entire range across all 40 SRES scenarios when analyzing uncertainties in all driving-force variables and the resultant emission categories.

4.4.3. Population Prospects

For the SRES scenario quantification three different population trajectories were chosen to reflect future demographic uncertainties based on published population projections (Lutz, 1996; UN, 1998; see Chapter 3). Global population ranges between 7 and 15 billion people by 2100 (Figure 4-5) depending on the speed and extent of the demographic transition. Fertility rates were assumed to converge to replacement levels (UN, 1998 medium scenario) in the B2 scenario or to below replacement levels in the A1 and B1 scenario families that adopt a variant of the low population scenario of Lutz (1996). The A2 scenario family is based on the high population projection described in Lutz (1996), which is characterized by heterogeneous fertility patterns that remain above replacement levels in many regions but nonetheless decline compared to current levels.

Across all scenarios, the concentration of future population growth and its uncertainty lies primarily in the developing countries. An equally pervasive trend across all scenarios is urbanization (see Chapter 3). Since the population trajectories are exogenous input to all the models used for this report, and are subject to a stringent harmonization criterion across different models, no further demographic variants to these are reported here, with exception of the A2 scenario family.[16]

4.4.3.1. A1 and B1 Scenarios

The population trajectory assumed for the A1 and B1 scenario families is based on a variant of the low population projection reported in Lutz (1996), which combines low fertility with low mortality and central migration rate assumptions. As is the case for other population scenarios used in this report, it is well within the uncertainty ranges as discussed in the demographic literature and the UN (1998) long-range population projections. After peaking at 8.7 billion in the middle of the 21st century, world population declines to 7.1 billion in the year 2100 (for comparison, the lowest UN Long Range projection indicates 5.6 billion by 2100; UN, 1998). As discussed in Chapter 3, the scenario population is somewhat higher than the previous low population scenario used in the IS92 scenario series, as only a combination of low fertility with *high* mortality rates could result in a global population as low as six billion people by 2100. Such a development is judged to be inconsistent with demographic theory (see Chapter 3) and to be inconsistent with the scenario storylines (see Section 4.2).

The pace of demographic transition in developing countries is fastest among all the SRES scenarios, reflecting the emphasis on social and educational development (scenario B1) and economic development (A1). The use of the same population projection for two SRES scenario families thus reflects different views of the driving forces of the demographic transition (with causality links in both directions), but it does not imply that such a demographic scenario is considered more likely compared to that of other projections. Better education and unproved social development, in particular concerning the role of women in society, lead the demographic transition (and consequently also economic development) in scenario B1. Accelerated rates of economic development and its required favorable social environment (education, reduction of income disparities, etc.) in turn lead the demographic transition in scenario A1. In both scenarios, low (infant) mortality rates are a necessary precondition to lower fertility rates (consistent with the Cairo targets of the UN Conference on Population and Development, discussed in Chapter 2). A distinguishing feature of the IIASA low population projection is the assumption of below replacement fertility levels, on the basis of the actual experience of industrial countries.[17] The implications of this trend are visible in both the absolute decline of global population (from its peak at 8.7 billion people in the middle of

[15] To address this issue, during the approval process of the Summary for Policymakers at the 5th Session of WGIII of the IPCC from 8-11 March 2000 in Katmandu, Nepal, it was decided to select two illustrative scenarios in the SPM from two additional A1 groups, in addition to the marker scenarios.

[16] The different population projection that underlies the "delayed development" scenario A2–A1-MiniCAM is described in Box 4-6.

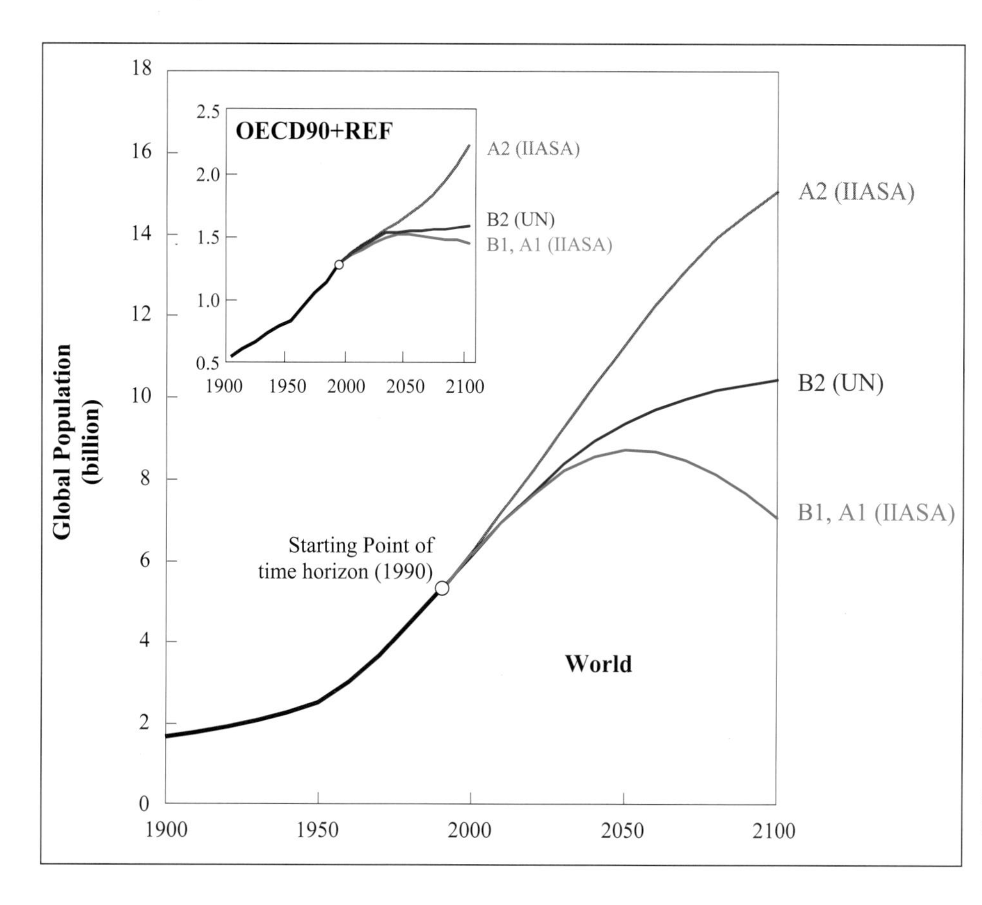

Figure 4-5: Population projections – historical data from 1900 to 1990 (based on Durand, 1967; Demeny, 1990; UN, 1998) and in SRES scenarios for 1990 to 2100 (based on Lutz, 1996 high and low; UN, 1998 medium).

the 21st century) and the significant population aging. In the long-term this trend affects not only the industrialized countries but also the currently developing countries (see the discussion in Chapter 3).

For A1 and B1 scenarios, regional population trajectories are (almost for all years) within the proposed 5% interval of their respective marker scenarios, except for two of the scenarios.[18]

4.4.3.2. A2 Scenarios

For the A2 scenario family, future population levels are based on the high scenario (15 billion) reported in Lutz (1996),[19] which is somewhat lower than the high population projection used in the previous IS92 scenarios (17.6 billion by 2100) for the reasons outlined in Chapter 3. The A2 population trajectory is the highest among the SRES scenarios, but well within the range of the UN long-range population projections[20] and

[17] The SRES writing team gratefully acknowledges the assistance of Anne Goujon of IIASA's Population Project in providing a numeric interpretation of the IIASA low population scenario that is consistent with the assumption of convergence of social and economic development underlying the B1 and A1 scenario storylines. Numeric scenario values and documentation is given in www.iiasa.ac.at/Research/POP/IPCC-special-report/.

[18] Scenarios A1v1-MiniCAM and A1v2-MiniCAM adopt a slightly different population projection, generated endogenously in the MiniCAM model. Its main differences are an asymptotic fertility rate of 1.75 compared to the 1.5 of Lutz *et al.* (1996) and a slightly different temporal pattern of the demographic transition. The resultant demographic projection is about 5% lower in the first half of the 21st century and about 10% higher toward 2100 compared to those of the other A1 family scenarios. As such, the demographic scenario is well within the uncertainty range characteristic of any long-term demographic projection.

[19] For a detailed description and the scenario's underlying assumptions and numeric values see www.iiasa.ac.at/Research/POP/IPCC-special-report.

[20] It was noted in the government review process that A2's population projection by 2050 is 11.3 billion people, higher than the highest UN projection (10.7 billion) published in the report *The State of World Population 1999* (UNFPA, 1999). As this UN projection extends only to 2050, it is necessary to consider the corresponding UN long-range population projections that extend to 2150 (UN, 1998). In these, the corresponding high/medium scenario is projected to have 10.8 billion people by 2050, which increases to 14.6 billion by 2100. The UN presents two additional scenarios that result in yet higher population levels: the UN high scenario projects 11.2 billion people by 2050 and 17.5 billion by 2100, and the UN constant fertility scenario projects 14.9 and 57.2 billion people by 2050 and 2100, respectively. For comparison, the A2 population scenario adopted from Lutz *et al.* (1996) indicates population levels of 11.3 billion by 2050 and 15.1 billion by 2100. Thus, A2 population levels by 2050 are comparable to those of the UN high scenario but remain significantly below the UN constant fertility scenario. By 2100, A2's population assumption is comparable to that of the UN high/medium scenario (14.6 billion) and significantly below that of the other two UN scenarios (high and constant fertility, with 17.5 and 57.2 billion, respectively). Hence, the adopted values for the A2 scenario are well within the range of the UN long-range population projections. As discussed in Chapter 3, the reason for the comparatively higher 2050 population in the Lutz *et al.* (1996) scenario is lower assumed mortality rates compared to the UN projections in the medium/high scenario. The lower mortality rates characteristic of A2's demographic scenario are judged to be more consistent with the A2 scenario storyline than alternative UN projections, such as the UN high/medium variant, that results in lower population by 2050, albeit at the expense of higher mortality. Yet by 2100 the differences in global population between the two scenarios is rather small (15.1 versus 14.6 billion, or 3%).

corresponding uncertainties estimated by demographers (see Chapter 3). For instance, Lutz *et al.* (1997) attach a probability of about 90% that actual world population will be lower (and 10% that it will be higher) than the value adopted for the A2 scenario family. Thus, the scenario represents well an upper bound of population growth scenarios found in the current scenario literature, although higher population scenarios exist in the demographic literature (see Chapter 3). As mentioned above, the SRES writing team is not in a position to attach any judgment concerning probability or likelihood to this or to any other demographic scenario. Population growth in the A2 world remains uninterrupted across all the SRES regions (Figure 4-5). The average global population growth rate over the 21st century is 0.96% per year, half that observed during the period between 1950 and 1990 (1.86%; UN, 1998). In the A2 marker, fertility rates vary considerably from one region to another; this reflects the regional orientation toward specific values, lifestyles, etc. described in the A2 scenario storyline (see Section 4.3). In the A2 world, in the year 2100 less than one-tenth of the world population lives in OECD90 countries, and toward the end of the 21st century a pronounced shift occurs in the population distribution, from ASIA to the ALM region (specifically Africa).

4.4.3.3. Harmonized and Other A2 Scenarios

The A2 scenarios share, with one exception, the same global population trajectory, but some of them show variation in population across the four SRES regions because of differences in the regional breakdown of the underlying models. For example, the A2-IMAGE scenario has a smaller population in the REF region as compared with the marker, 405 million versus 706 million in the marker. An alternative demographic interpretation at the global level was attempted in the "transitional" A2-A1-MiniCAM scenario, in which the implications of delayed development patterns are explored. In this scenario global population is assumed to reach 10 billion by 2050, and 12 billion by 2100 (see Box 4-6).

4.4.3.4. B2 Scenarios

The B2 marker scenario adopted the UN median 1998 population projection (UN, 1998), wherein global population increases to about 9.4 billion people by 2050 and to about 10.4 billion by 2100[21]. The scenario is characteristic of recent median global population projections (see discussion in Chapter 3), and describes a continuation of historical trends toward a completion of the demographic transition in the 21st century. The projection is consistent with recent demographic data and scenarios; it reflects faster declines in world fertility together with declining mortality rates. Hence, the scenario is somewhat lower than previous UN median projections, as used in the previous IS92 scenario series. A distinguishing feature of the UN population projections is the assumption that, in the long-term, fertility levels converge toward replacement levels globally (see Chapter 3). Future population growth is assumed to be slow in today's industrialized countries. In Asia, population size stabilizes in the second half of the 21st century, and in the rest of the world population growth slows down toward the end of that century.

The UN median population projection is shared across all B2 scenario quantifications, although differences remain at the regional level. The different regional aggregations used across various models did not coincide with the regional aggregation of the original UN projection, which suggests that a more detailed regional breakdown of demographic projections is highly desirable for long-term global scenario studies.

4.4.4. Economic Development

The SRES scenarios span a wide range of future economic growth rates (Table 4-5) and resultant levels of economic output. The A1 scenario family, with a global GDP of US$520 to 550 trillion in 2100, delineates the SRES upper bound, whereas the A2 and B2 scenarios, with a range of US$230 to 250 trillion in 2100, represent its lower bound. The B1 scenario family is intermediary. Although the SRES scenarios span a wide range, both lower and higher global GDP levels can be found in the literature (see Chapter 2).

Uncertainties in future GDP levels are governed by the rates of future productivity growth and population growth, especially those in developing countries. Different assumptions on conditions and possibilities for development "catch-up" and for narrower per capita income gaps in particular explain the wide range in projected future economic growth rates. Given the weak inverse relationship between population growth and per capita income growth discussed in Chapter 2, uncertainties in future population growth rates tend to restrict the range of associated GDP projections. High population growth, all else being equal, lowers per capita income growth, whereas low population growth tends to increase this growth. This relationship is evident in empiric data – high per capita income countries are generally also those that have completed their demographic transition. The affluent live long and generally have few children. Notable exceptions are countries with small populations and significant income from commodity exports. Yet even assuming this relationship holds for an extended time into the future, its quantification is subject to considerable theoretic and empiric uncertainties (Alcamo *et al.*, 1995).

As outlined above, two of the SRES scenarios explicitly explore alternative pathways of the gradually closure of existing income gaps. As a reflection of uncertainty, the development "catch-up" diverges in terms of geographically distinct economic growth patterns across the four SRES scenario families, as summarized in Tables 4-5, 4-6, and 4-7. The scenarios of rapid development and "catch-up" remain in

[21] The SRES writing team gratefully acknowledges the assistance of Thomas Büttner of the UN Population Division, New York, in developing more detailed regional population projections based on the UN 1998 medium projection, and in making these data available to the SRES writing team in electronic form.

Table 4-5: *Historical economic growth rates (% per annum) from 1950 (Maddison, 1989, 1995; UN, 1993a, 1993b), and SRES scenarios for 1990 to 2100. Growth rates were calculated on the basis of GDP at 1990 prices and market exchange rates[a]. Long-term growth rates are lower than those from 1950 to 1990 (e.g., the average annual growth rate for OECD90 countries from 1850 to 1990 was about 2.8%, and for the reforming economies in Eastern Europe and the Former Soviet Union it was about 1%; Maddison, 1989). Numbers in brackets give the minimum and maximum values of all SRES scenarios.*

Economic Growth Rates (% per annum)									
		1990–2050				1990–2100			
Region	**1950–1990**	**A1**	**A2**	**B1**	**B2**	**A1**	**A2**	**B1**	**B2**
OECD90	3.9	2.0	1.6	1.8	1.4	1.8	1.6	1.5	1.1
		(1.2-2.2)	(1.0-2.1)	(1.7-2.0)	(1.3-1.6)	(0.9-1.9)	(0.9-1.7)	(1.4-1.5)	(1.0-1.3)
REF	4.8	4.1	2.3	3.1	3.0	3.1	2.5	2.7	2.3
		(2.8-4.6)	(0.6-2.3)	(2.7-3.7)	(1.9-3.3)	(2.2-3.5)	(1.6-2.5)	(2.4-2.7)	(1.6-2.5)
IND	3.9	2.2	1.6	1.9	1.6	2.0	1.7	1.6	1.3
		(1.4-2.4)	(1.0-2.1)	(1.8-2.0)	(1.4-1.8)	(1.1-2.1)	(1.0-1.7)	(1.5-1.6)	(1.1-1.4)
ASIA	6.4	6.2	3.9	5.5	5.5	4.5	3.3	3.9	3.8
		(5.8-6.6)	(3.8-4.8)	(5.3-6.2)	(4.2-5.7)	(4.2-4.7)	(3.3-3.7)	(3.8-4.2)	(3.6-3.9)
ALM	4.0	5.5	3.8	5.0	4.1	4.1	3.2	3.7	3.2
		(4.8-5.8)	(3.3-4.1)	(4.5-5.3)	(3.3-4.4)	(3.9-4.2)	(3.1-3.4)	(3.5-3.9)	(3.0-3.6)
DEV	4.8	5.9	3.8	5.2	4.9	4.3	3.3	3.8	3.5
		(5.3-6.2)	(3.5-4.4)	(4.9-5.7)	(3.7-5.0)	(4.1-4.4)	(3.3-3.6)	(3.7-4.1)	(3.3-3.7)
WORLD	4.0	3.6	2.3	3.1	2.8	2.9	2.3	2.5	2.2
		(2.9-3.7)	(1.7-2.8)	(2.9-3.5)	(2.1-2.9)	(2.5-3.0)	(2.0-2.3)	(2.5-2.6)	(2.0-2.3)

Note: independent rounding.
[a] In the calculations the concept of logarithmic growth rates is used.

Table 4-6: *Income per capita (1000 US dollars at 1990 prices and exchange rates) in the world and by SRES region. Numbers in brackets give minimum and maximum values of the SRES scenarios. The range for 1990 illustrates differences in base-year calibration across models.*

Income per Capita by World and Regions (10^3 1990US$ per capita)									
		2050				2100			
Region	**1990**	**A1**	**A2**	**B1**	**B2**	**A1**	**A2**	**B1**	**B2**
OECD90	17.8-20.6	50.1	34.6	49.8	39.2	109.2	58.5	79.7	61.0
		(39.4-62.3)	(32.3-54.0)	(40.3 -52.0)	(35.1-42.2)	(69.8-115.7)	(48.0-78.7)	(70.6-84.7)	(50.1-73.2)
REF	2.2-2.7	29.3	7.1	14.3	16.3	100.9	20.2	52.2	38.3
		(13.5-32.5)	(3.3-9.0)	(12.4-23.4)	(7.8-16.8)	(39.9-119.3)	(13.5-20.2)	(41.2-56.4)	(14.0-38.3)
IND									
	12.8-14.4	44.2	26.1	39.1	32.5	107.3	46.2	72.8	54.4
		(30.7-50.0)	(22.4-41.9)	(32.5-40.8)	(27.0-34.7)	(60.3-113.5)	(37.1-64.5)	(65.3-77.7)	(42.4-61.1)
ASIA	0.4-0.6	14.9	2.6	9.0	8.9	71.9	7.8	35.7	19.5
		(10.8-15.7)	(2.5-4.5)	(7.2-14.3)	(3.6-9.5)	(38.8-76.8)	(7.4-12.9)	(35.7-46.1)	(14.8-20.6)
ALM	1.3-2.1	17.5	6.0	13.6	6.9	60.9	15.2	44.9	16.1
		(12.2-18.0)	(4.2-6.0)	(8.0-15.3)	(4.4-7.7)	(44.2-69.5)	(11.3-15.2)	(41.3-45.8)	(13.6-22.6)
DEV	0.7-1.1	15.9	3.9	10.9	8.1	66.5	11.0	40.2	18.0
		(11.4-16.7)	(3.3-5.1)	(7.5-14.8)	(3.9-8.4)	(41.4-69.8)	(10.3-13.7)	(40.2-45.2)	(14.2-21.5)
WORLD	3.7-4.0	20.8	7.2	15.6	11.7	74.9	16.1	46.6	22.6
		(14.3-21.5)	(6.0-9.9)	(12.7-19.1)	(7.7-11.9)	(43.7-77.9)	(15.9-16.9)	(46.3-49.6)	(19.2-24.5)

Table 4-7: *Growth rates (% per year) of income per capita (using GDP at 1990 prices and exchange rates) in the world and by region. Historical data from 1950 to 1990 from Maddison (1989, 1995), UN (1993a, 1993b), and Klein Goldewijk and Battjes (1995). Numbers in brackets give minimum and maximum values of all SRES scenarios.*

Growth Rates of Income Per Capita (%)

		1990–2050				1990–2100			
Region	**1950–1990**	**A1**	**A2**	**B1**	**B2**	**A1**	**A2**	**B1**	**B2**
OECD90	2.8	1.6	1.1	1.5	1.2	1.6	1.1	1.2	1.1
		(1.2-1.8)	(0.8-1.6)	(1.2-1.6)	(1.0-1.4)	(1.2-1.7)	(0.8-1.2)	(1.2-1.3)	(0.9-1.3)
REF	3.7	4.0	1.9	3.0	3.0	3.3	2.0	2.8	2.4
		(2.8-4.5)	(0.5-2.2)	(2.7-3.6)	(1.9-3.3)	(2.5-3.4)	(1.5-2.0)	(2.6-2.8)	(1.6-2.6)
IND	2.9	2.0	1.2	1.7	1.4	1.9	1.2	1.5	1.2
		(1.3-2.1)	(0.8-1.8)	(1.5-1.8)	(1.1-1.6)	(1.3-2.0)	(0.9-1.4)	(1.4-1.5)	(1.0-1.4)
ASIA	4.4	5.5	2.7	4.8	4.7	4.4	2.5	3.9	3.3
		(5.1-5.9)	(2.7-3.6)	(4.6-5.5)	(3.3-4.8)	(3.9-4.7)	(2.4-2.9)	(3.8-4.2)	(3.1-3.4)
ALM	1.6	4.0	1.9	3.5	2.4	3.3	1.9	3.0	2.1
		(3.5-4.4)	(1.7-2.2)	(3.1-3.9)	(1.7-2.7)	(3.1-3.5)	(1.8-2.1)	(2.8-3.2)	(1.9-2.5)
DEV	2.7	4.9	2.4	4.2	3.8	4.0	2.2	3.5	2.8
		(4.4-5.2)	(2.3-3.0)	(3.9-4.8)	(2.5-3.9)	(3.6-4.1)	(2.2-2.6)	(3.4-3.7)	(2.6-3.0)
WORLD	2.2	2.8	1.1	2.3	1.8	2.7	1.3	2.2	1.6
		(2.2-2.9)	(0.7-1.5)	(2.1-2.6)	(1.1-1.9)	(2.2-2.8)	(1.3-1.5)	(2.2-2.4)	(1.4-1.7)

dispute within the SRES writing team because they imply high productivity growth (see Box 4-5 and, for a contrasting viewpoint and scenario interpretation, Box 4-6). However, it is agreed that such scenarios of high productivity growth and smaller income-per-capita disparities cannot be ruled out, even if they certainly are very challenging from the perspective of recent growth experiences in a number of regions, most notably Africa. There is also agreement that the assumptions deployed for the SRES scenarios are within the range suggested by the literature (see Chapter 2). In this the highest GDP growth is up to US$700 trillion by 2100 compared to US$550 trillion in the highest SRES scenario. For scenarios developed within the context of sustainability analyses, reductions in per capita income gaps also occur faster than for any of the scenarios presented here.

Important differences remain between models in terms of 1990 base-year data on economic activity levels. Even after differences in regional definitions are accounted for, 1990 regional GDP differences between models range up to ±32% in a few cases. Such differences are particularly pronounced for developing countries, where in many cases national currencies are not freely convertible and thus important uncertainties on the applicable conversion rates remain (World Bank, 1999). Differences for OECD countries are much smaller (±3% across the models) and because of their current dominance in global economic activity (and counterbalancing effects), 1990 global GDP numbers agree well across the models (±5%). Scenario comparisons, especially at the regional level, are therefore best based on a comparison of growth rates (see Chapter 2), and the SRES scenarios are no exception.

Historical data indicate that, even though the process of economic growth is heterogeneous across countries and over time, the patterns of growth show certain similarities. Economic "catch-up" follows a general dynamic pattern, characterized by initially accelerating economic growth rates that pass through a maximum, and decline once the industrial base of an economy becomes established. This overall feature of growth dynamics is reflected in all the SRES scenarios, albeit timing and magnitude vary across the four scenario families. This variation reflects the scenario-specific storylines, as well as particular relationships to other driving-force variables, such as demographics, described in the scenario.

4.4.4.1. A1 Scenarios

By design (see Section 4.3) the "High Growth" scenario family A1 explores a world in which future economic development follows the patterns of the most successful historical examples of economic development catch-up. Free trade, continued innovation, and a stable political and social climate enable developing regions to access knowledge, technology, and capital. Combined with a rapid demographic transition, this is assumed to lead to acceleration in time and space of economic growth compared to the historical OECD experience since the 19th century. The global economy is projected to expand at an average annual rate of 2.9% to 2100 (see Table 4-5), roughly in line with historical experience over the past 100 years (of 2.7% per year, see Chapter 3). Such growth rates are considered high by the current scenario literature (see Chapter 2). Compared to historical experience, however, the broad-based nature of economic development catch-up (i.e., no region "is left behind") is without precedent. The 2.9% per year economic growth rate translates into a 25-fold expansion of global GDP that would reach US$529 trillion by 2100.

As a byproduct of rapid economic development and fast demographic transition, income inequities between industrial and developing countries are virtually eradicated. Per capita income ratios are 1:1.6 in 2100, compared to a ratio of 1:16 in 1990 in terms of the GDP/capita difference between current developed (IND) and developing (DEV) regions across the four SRES regions. However, even if relative income differences are reduced drastically, absolute differences remain large, not least because of the high incomes characteristic of the A1 scenario family (per capita income differences are also larger when considered at a more disaggregated level). When measured across the four SRES regions in 1990, income per capita differences are nearly 1:40 (between ASIA and OECD90). Per capita income differences are yet higher for differences across countries or between different social strata. The poorest 20% of Bangladesh's population, for instance, earn per capita incomes that are a factor of 700 lower than that of the 20% richest Swiss population (UNDP, 1993). A distinguishing feature of the A1 scenarios is to explore pathways of reductions in present disparities. In A1, per capita income in industrial countries (IND) increases to about US$107,300 and in now developing countries (DEV) to US$66,500. Non-OECD GDP growth rates rise to a peak of about 8% between 2010 to 2030 in scenario A1, and decline once the industrial and infrastructural bases of their economies are established. By and large, the A1 scenario implies a replication across all developing regions of the post-World War II experience of Japan and Korea or the recent economic development of China.

4.4.4.2. Harmonized and Other A1 Scenarios

The high economic growth characteristics of the A1B marker scenario reproduce well in the scenarios calculated with different models. The A1B-MESSAGE scenario tracks the A1B marker scenario closely at the global and regional levels ("fully harmonized" input assumptions). At the global GDP level, most growth trajectories agree within a range up to 15%, except the A1v2-MiniCAM scenario (see Box 4-5).[22] Differences at the regional level are larger. All models with comparable regional aggregation levels agree well for ASIA (except A1v2-MiniCAM). Differences in economic growth prospects also agree well for OECD90 (except A1v2-MiniCAM). For REF the A1B scenarios group into two clusters – one group reproduces the GDP growth scenario of the A1B marker, while the other group suggests a GDP level by 2100 about one-third lower than that of the A1B marker. Mostly this reflects different assumptions used in the models on future labor productivity growth that have not been harmonized with the values adopted for the A1B marker scenario. For ALM, again one group of scenarios tracks closely the A1B marker, whereas other groups indicate either higher (A1B-ASF, A1B-MiniCAM) or lower (A1v2-MiniCAM, A1T-MARIA) GDP growth. The reasons are similar to those discussed above. For the SRES region REF, the different regional aggregations across the models required complicated "inverse" calculations on regional growth rates for harmonization with the respective marker scenario at the level of the aggregated SRES region. For ALM, such calculations were neither possible for all models nor considered desirable by various modeling teams, which preferred to emphasize the inherent uncertainties in regional economic growth perspectives even for an otherwise shared vision of rapid global economic growth and development catch-up.

[22] The A1v2-MinCAM scenario reaches only about 64% of the A1B marker global GDP in 2100.

4.4.4.3. A2 Scenarios

As compared to the other SRES scenario families, the A2 world is characterized by relatively slower productivity growth rates and resultant lower per capita incomes (see Table 4-6). Yet, the global average (1990–2100) growth rate in per capita income of 1.3% is still somewhat higher than that observed from 1970 to 1995 (1.2%; World Bank, 1998). The comparatively conservative assumptions on per-capita-income growth reflect both the more fragmented economic outlook of the A2 storyline (see Section 4.3) and the slow pace of the demographic transition that underlies A2's high population growth trajectory. The fastest growth in per capita incomes (on average over 2.3% per year) occurs in the ASIA region, while the slowest growth is observed in the OECD90 region (on average 1.0% per year). In a reversal of current short-term trends, the REF and ALM regions experience a stable increase of their per capita income levels over the 21st century at a rate that is almost twice as high as in the OECD90 region (see Table 4-7). The A2 world is also characterized by a slow convergence of incomes among regions. Nonetheless, present income disparities become narrower, from a factor of 40 difference in 1990 per capita income levels between the richest and the poorest of the four SRES regions, to a factor of seven or eight by 2100. The increase of global population from 6 to 15 billion by 2100 translates into an increase of global GDP by a factor of 12 over a century. The average (1990 to 2100) annual growth rate of total GDP is 2.2%, which is lower than the 2.9% average annual growth rate observed between 1970 and 1995 (World Bank, 1998) and the 4% rate observed from 1950 to 1990 (see Table 4-5).

4.4.4.4. Harmonized and Other A2 Scenarios

Four non-marker A2 scenarios (A2-AIM, A2-IMAGE with the exception of 2050, A2-MiniCAM, and A2-MESSAGE) have global GDPs within 5% of the A2 marker. The A2-A1-MiniCAM scenario has much lower global GDP than the A2 marker, reflecting a different viewpoint on future labor productivity growth (see Box 4-5) and a different interpretation of the A2 scenario storyline altogether (see Box 4-6). While this particular scenario illustrates important uncertainties with respect to economic growth and development catch-up for developing countries, it remains controversial within the writing team, especially as to whether it reflects the overall

Box 4-5: Labor Productivity Growth Rates in the SRES Scenarios

The high income growths assumed in the scenarios, especially in the A1 and B1 scenario families, imply large labor productivity increases. According to one member of the writing team such increases might not be plausible, but the other members find the assumptions plausible, especially in view of historical precedents. In line with IPCC practice, the dissenting view is elaborated here. Its corresponding implications on scenario quantifications are discussed with one example in Box 4-6.

Long-term economic growth rates can be expressed as the sum of the labor force growth rate and the growth rate of labor productivity. This framework can be used to understand more clearly the kind of technological and demographic assumptions present in the SRES scenarios. A simplified measure of labor productivity, the average economic output per member of the labor force, is used to examine the SRES scenarios.

Productivity assumptions are not comparable across the different models used to quantify the SRES emissions scenarios, so these issues were examined by running quantifications of the SRES scenarios with the MiniCAM model. Population projections were taken as exogenous inputs and regional GDP growth paths were taken to be similar to those in the SRES marker scenarios. From these, the implied labor force productivity growth rates are determined using assumptions about labor force participation rates. In these calculations, total labor force participation rates were assumed to be asymptotic to a participation rate of 80% for a working age population of all persons aged 15 to 65.

The result of this exercise is that increases in labor force productivity range between 0.79% and 5.85% per year (calculated for the periods 1990 to 2020, 2020 to 2050, and 2050 to 2100) for the four scenario families and four SRES macro-regions considered separately. This compares with historical experience between 1970 and 1995 of growth in regional labor force productivity of 0.69 to 4.13% per year and the longer-term productivity growth rates of between 1.1 and 7.7% per year. The implied future growth *in global* labor productivity for the MiniCAM scenario calculations ranges between 1.12% and 3.49% per year for the four scenarios (again from 1990 to 2020, 2020 to 2050, and 2050 to 2100). The historical growth in global labor force productivity between 1970 and 1995 was 1.04% per year. Historical rates of GDP per capita growth (a macro-economic proxy for labor productivity; see Table 3-2 and Table 4-7) were 1.1% per year between 1980 and 1992 and 2.5% per year between 1950 and 1980.

At the upper end of the these ranges, the SRES scenarios exhibit a growth in global labor force productivity that is higher than recent historical global experience, particularly for the SRES regions REF and ALM. This indicates that none of the four SRES scenario families envisions a recurrence of the current economic crisis in Eastern Europe and Russia or a recurrence of the "lost decade" of negative GDP per capita growth in Africa or Latin America. In addition, the period over which some developing regions exhibit high growth rates in these scenarios is longer than any historical record of high growth rates. However, there is limited analogous historical experience, as Japan is the only country that can be said to have completed such an economic "catch-up."

The assumptions on labor force participation used in the calculations reported in this box result in a decline in the growth of the labor force through the 21st century. The total labor force actually declines in all regions for the last simulation period. This reflects the demographic pattern of an overall decline in the population and the assumed stability in labor force participation rates. However, the scenarios describe generally affluent worlds in which people live longer. Thus, the labor participation patterns are likely to change with respect to current practices. People may work much longer over their lifetime, but this trend is countered by the probable need for an increase in education levels, which act to delay entry into the labor force. Pushing the asymptotic labor participation rates upward from the value assumed here results in only a small increase in total productivity, on the order of 0.2% per year.

The high economic growth branches (A1 and B1 families) of the SRES scenarios may represent upper bounds for future increases in labor productivity. Different assumptions about future rates of labor force participation do not appear to change this conclusion substantially. These scenarios essentially assume that, within the next few decades, most developing regions will experience an extended period of successful economic development analogous to the historical experience of Japan and the "Asian Tigers." An alternative view on future labor productivity growth is provided in a number of scenarios developed with the MiniCAM model, most notably A1v2-MiniCAM and the "transitional" scenario A2-A1-MiniCAM (see Box 4-6). In these scenarios global economic output is between 20% (A2-A1-MiniCAM) and 36% (A1v2-MiniCAM) lower than in the other scenarios of their respective scenario families.

Box 4-6: Possible Transitions between Scenario Families: The A2-A1-MiniCAM Scenario

The four scenario storylines have been stylized as global socio-economic developments that evolve in different directions, but globally and continuously. In reality, different regions may follow the developments pictured in the scenarios in different time periods. For example, the world may in reality develop according to one of the storylines and after some time move toward another. As an illustration of this, one scenario (A2-A1-MiniCAM) was elaborated by a member of the writing team. This scenario is described here even though some members of the writing team considered its inclusion undesirable and possibly confusing as it was submitted too late to have the team thoroughly discuss its consistency[24] and to clarify its relationship to the four storylines. However, the point that in reality transitions between scenarios are possible is a valid one and the contrasting viewpoint is presented here for consideration of the reader, following IPCC practice.

The A2-A1-MiniCAM transition scenario explores a world in which the prerequisites for development, such as education, effective institutions, and high saving rates, take some time to develop, so that rapid development does not begin to occur until between 2020 and 2050 depending on the region. In this scenario, total GDP by 2100 is below the median of the historical scenario data base and, with the population 20% higher than the current median UN forecast, average per capita income is at the lower end of the historical data base range. In such a relatively poor world the economic structure is more sensitive to environmental change than in the marker scenario, and the institutional structure is less capable. Thus, the impacts of climate change are larger and the ability to adapt less than those in the A1 world. The primary driving forces for the A2-A1-MiniCAM transition scenario follow the logic of this story line, as detailed below.

Population is lower in the A2-A1-MiniCAM scenario variant than in the A2 marker, since its total completed fertility early in the 21st century is lower. This reflects the continued rapid historical declines and leads to slower population growth rates than in the population scenario adopted for the A2 scenario family. Total completed fertility declines slowly in the forecast period with a long-term asymptotic level of 2.25 for all regions, which results in a global population that is still growing by over 100 million per year in 2100. The values for migration and death rates used to generate the population trajectory follow those of the UN median forecast. The population scenario of A2-A1-MiniCAM is quite close to the UN medium population projection (see the discussion of the B2 scenario family below) until about 2060. Thereafter, however, A2-A1-MiniCAM's population scenario continues to grow linearly to about 12 billion, whereas the UN median projection stabilizes at about 10 billion by 2080 and slightly declines thereafter.

The regionally heterogeneous ("delayed") pattern of development of the precursors to rapid economic growth (e.g. education) means that some developing regions experience stagnant or very slowly growth in per capita incomes well into the 21st century. As regions begin rapid development they approach and follow the average OECD labor productivity pattern, so GDP growth rates accelerate post-2050 and average nearly 2.5% per year over the second half of the 21st century. GDP thus rises more slowly early in the 21st century in this scenario, reaching just over US$50 trillion by 2050. After 2050, GDP rises more rapidly to reach just under US$200 trillion in 2100.

Per capita final energy demands are limited by per capita income, and rise only as economic growth occurs. After growing slowly until 2050, per capita energy demands grow by more than 1.5% annually to reach 120 GJ per capita by 2100. With increases in the efficiency with which services are provided, this results in a global average level of energy services similar to that currently seen in Western Europe. Global per capita income and energy use by 2100 approach that of Western Europe in 1990. Therefore global energy intensities in 2100 approach those of Western Europe in 1990, a value lower than in other scenarios but in line with current observations.

Natural gas and oil dominate the primary energy system, and contribute slightly more than half of the primary energy. Non-fossil sources contribute about 30% of total primary energy, with coal providing the remainder. Total fossil energy carbon emissions reach 22 GtC by 2100. Sulfur controls are delayed in this implementation until economic growth takes off after 2050. With high levels of fossil fuel use, and relatively low rates of control, sulfur emissions are about ten million tons higher in 2100 than they are today.

The relatively slow growth in output and productivity in the economy in general is mirrored in the agricultural sector, with lower growth in agricultural productivity until post 2050. The large population complicates this problem, leading to large-scale expansion of agricultural lands and a resultant decrease in forested and unmanaged lands, especially in developing regions. Land-use changes do not offset any of these emissions, since the relatively high population, the rapid growth in income, and the growth in modern biomass result in essentially zero carbon emissions from land use and agriculture, rather than the substantial uptake seen in many other scenarios.

[24] In particular, the consistency of a continued fast demographic transition to 2050 combined with a scenario of stagnating per capita income growth for as much as five decades is questioned by a number of members from the writing team.

development tendencies captured in the A2 scenario storyline. Nonetheless, in view of the spirit of the SRES open process this contrasting scenario is presented in Box 4-6. Regional GDP growth differs more than global values across the different model interpretations of the A2 scenario storyline. A2-MESSAGE and A2-AIM scenarios are harmonized, based on the marker scenario, at the regional level also, while the A2-IMAGE and A2-MiniCAM scenarios have significant regional deviations from the A2 (ASF) marker. In particular, these scenarios assume stronger GDP/capita growth in the ASIA region and slower growth in the ALM and REF regions.

4.4.4.5. B1 Scenarios

The B1 scenario storyline assumes high levels of social consciousness and successful governance that result in strong reductions in income inequalities and social inequity. Growth in GDP, while being substantial, is qualitatively different compared to that of other scenarios, as social activities and environmental conservation are emphasized. Concepts of "green" GDP, including socially desirable activities such as childcare, apply in particular in the B1 scenario and qualify its similarity to other scenarios in terms of monetary value of GDP. In contrast to the world of scenario A1, the reduction of income inequalities is not a byproduct, but rather the result of constant domestic and international efforts. Global GDP reaches US$328 trillion, which corresponds to an average annual growth rate of 2.5%, slightly less than the long-term historical average. Per capita income differences between the IND and DEV regions are reduced from 1:16 in 1990 to 1:1.8 by 2100; income disparities within particular regions are assumed to be even further reduced, consistent with the thread of the B1 storyline described in Section 4.3. For the IMAGE model simulation, convergence assumptions were applied in the following domains:

- Technology convergence was toward the level of the productivity frontier region (either Japan or the US).
- Economic structure convergence was toward long-term sectoral shares of OECD economies (e.g., the US), which over the long-term implies a decline of the share of manufacturing sectors, and hence convergence to a service-oriented economy.
- Education convergence was to the OECD ratio of highly skilled workers within the total workforce.

4.4.4.6. Harmonized and Other B1 Scenarios

The global GDP trajectories are all within the proposed harmonization intervals, except for the B1High-MiniCAM (see Box 4-5) and the B1-MARIA scenarios in the middle period of the 21st century. On the regional level, differences in GDP trajectories from different models are larger because of the differences in regional aggregations outlined above in the discussion of the A1 scenario family.

4.4.4.7. B2 Scenarios

Global GDP in B2 is assumed to increase by more than a factor of 10 during the 21st century, or at an average annual growth rate of 2.2%. This growth rate of GDP is similar to the median GDP growth in the scenario database reviewed in Chapter 2. Stabilization of global population at less than double current levels, as projected in the UN median scenario adopted for B2, combined with a sustained pace of development implies that a B2 world generally achieves high levels of affluence. Average per capita income reaches about US$18,000 by 2100 in the developing countries, which exceeds the current OECD averages. In comparison, average per capita GDP reaches US$54,400 by 2100 in the developed regions, which corresponds to an income ratio of about 3:1 between industrial (IND) and now developing (DEV) regions, a considerable improvement in interregional equity by 2100 (Riahi and Roehrl, 2000). Nonetheless, given the nature of the B2 scenario storyline (Section 4.3), per capita income differences among the world regions are higher than those in the A1 and B1 scenarios, but much smaller than those in A2.

4.4.4.8. Harmonized and Other B2 Scenarios

The economic growth paths described by the B2-MESSAGE marker scenario are closely tracked by the B2 quantifications derived from alternative models at the global level, with the exception of the B2-IMAGE scenario, which is slightly below.[23] Differences at the regional level are larger. For B2-ASF and B2-MiniCAM, GDP is higher in the ALM and lower in the ASIA regions, respectively, compared to the B2 marker scenario and its "fully harmonized" companion B2-AIM. The latter – like all AIM scenarios – was developed by an interdisciplinary group of researchers from different countries in the ASIA region and therefore helps to guide readers as to which scenario better reflects regional perspectives.

4.4.5. ***Energy Intensities, Energy Demand, and Structure of Energy Use***

Population and GDP assumptions, along with structural change and technological change that affect energy efficiency and energy costs (and prices), drive the demand for energy services. Given the different model representations of energy service demands, in this section final energy use is discussed as a common measurement point across all SRES models and scenarios. Final energy use per unit economic activity, that is, energy intensity, is a frequently used measure of comparative efficiency of resource use, and reflects a whole range of structural, technological, and lifestyle factors (Schipper and Meyers, 1992).

[23] In B2-IMAGE, economic output is higher in OECD90 and lower in the other regions compared to the B2 marker.

Figure 4-6 illustrates the evolution of final energy intensities for the four SRES marker scenarios. Instead of time, per capita income is shown on the horizontal axis, to illustrate a conditional convergence of regional final energy intensities. Invariably, intensities are projected to decline with increasing income levels. As discussed in Chapter 3, the main reason for this trend stems from the common source of economic growth and energy intensity improvements – technological change. All else being equal, the faster intensity improvements are, the faster aggregate productivity (per capita income) grows. An important methodologic improvement over previous studies is the explicit inclusion of non-commercial energy forms in some SRES models, drawing on estimates as reported in IPCC WGII SAR (Watson *et al.*, 1996) and in Nakićenović *et al.* (1998).

In the A1 and B1 scenarios, per capita income differences are substantially narrowed and convergent because of increased economic integration and rapid technological change. Therefore, differences in energy intensities are also narrowed significantly and are convergent, as shown in Figure 4.6. The B1 storyline describes a development path to a less material-intensive economy. Hence, the final energy intensities in the B1 marker are lowest among the four SRES marker scenarios for a given per capita income level. The A2 storyline reflects a world with less rapid technological change, as shown by the smallest rate of energy intensity improvement among the four marker scenarios.[25] Different interpretations of the four scenario storylines, as well as alternative rates of energy intensity improvement to the four marker scenarios, are discussed below.

Owing to methodologic differences across the six models (see Box 4-7) it is not possible to disaggregate energy intensity improvements into various components, such as structural change, price effects, technological change, etc., in a consistent way. In some models (macro-economic) price effects are differentiated from "everything else" (frequently labeled AEEI, or autonomous energy intensity improvements). As a rule, the importance of non-price factors is an inverse function of the time horizon considered. Over the short-term, the impacts of economic structural change and technology diffusion are necessarily low. Hence, prices assume a paramount importance in driving alternative energy demand patterns in short-term (to 2010–2020) scenario studies (e.g., IEA, 1998; EIA, 1997, 1999). Over the longer term (i.e., the time horizon considered by the SRES scenarios), economic structural and technological changes become more pronounced, as does their influence on energy intensity improvements and energy demand. This does not imply that prices do not matter over the long term, but simply that "everything else" (e.g., AEEI) is likely to outweigh the impacts of prices, as indeed suggested by quantitative scenario analyses performed within the Energy Modeling Forum EMF-14 (Weyant, 1995).

Important feedback mechanisms between technological change and costs (and thus also prices) exist over the long term. These are as a rule treated endogenously in the models, for instance when modeling long-run resource extraction costs or structural changes in energy supply options (see Sections 4.4.6 and 4.4.7). Energy prices are also strongly affected by policies (e.g., taxation), but to project these far into the future is both outside the capability of currently available methodologies and outside the general "policy neutral" stance of the SRES scenarios. Therefore, most models treat dynamic changes in (average and marginal) *costs* as the driving force for energy intensity improvements and for technology choice (see Sections 4.4.6 and 4.4.7).

4.4.5.1. A1 Scenarios

Improvements in energy efficiency on the demand side are assumed to be relatively low in the A1B marker scenario, because of low energy prices caused by rapid technological progress in resource availability and energy supply technologies (see Sections 4.4.6 and 4.4.7). These low energy prices provide little incentive to improve end-use-energy efficiencies and high income levels encourage comfortable and convenient(and often energy intensive) lifestyles (especially in the household, service, and transport sectors). Efficient technologies are not fully introduced into the end-use side, dematerialization processes in the industrial sector are not well promoted, lifestyles become energy intensive, and private motor vehicles are used more in developing countries as per capita GDP increases. Conversely, fast rates of economic growth and capital turnover and rising incomes also enable the diffusion of more efficient technologies

[25] Note that this statement only indicates the relative position of the A2 scenario compared to other SRES scenario families. In absolute terms the scenario's decline in energy intensity is very substantial – on average, energy use per unit of GDP declines by a factor of more than two as a result of the compounding effect of an improvement rate of final energy intensity of 0.8% per year. Comparison of this improvement rate with the SRES scenario range calculated by the ASF model indicates that A2's energy intensity improvement rates are one-third lower compared to the B2 scenario and less than half compared to the B1 scenario. By 2100, A2's final energy intensity is calculated by the ASF model at 5.9 MJ/$, which compares to the literature range of up to 7 MJ/$, and a value of 7.3 MJ/$ in the A2-A1-MiniCAM scenario, which contains the highest energy-intensity trajectory within the 40 SRES scenarios. Thus, the A2 scenario's energy-intensity improvement rates are well within the uncertainty range as indicated by the scenario literature and are not considered overly pessimistic by the writing team. During the government review process, comment was made on the fact that energy intensities in A2 are one-third higher than those in B2. This figure is classified as "reasonable" for an inter-family scenario variation by the writing team because it is consistent both with the underlying differences in per capita GDP (i.e. productivity) growth between the two scenario families and with the relationship between energy intensity improvements and macro-economic productivity growth identified in the literature assessment in Chapters 2 and 3.

Box 4-7: The Role of Prices in SRES Scenarios

The price of energy comprises many components:

- Costs to establish and maintain the production, conversion, transport, and distribution infrastructure of energy supply.
- Profit margins.
- A whole host of levies such as royalties and taxes raised at the points of energy production or use.
- Consumers' willingness to pay for quality and convenience of energy services.

Furthermore, given the importance of energy and the vast volumes traded, prices are influenced by a whole range of additional factors, from inevitable elements of speculation to geopolitical considerations, all of which can decouple energy price trends from any underlying physical balance between supply and demand. Taxes are especially significant. In a number of OECD countries, up to 80% of the consumer price of gasoline is taxes (OECD, 1998), and the differences between countries are enormous. In 1997, 27% of the price of gasoline in the USA was taxes, compared with 78% in France. Taxes vary substantially even between large oil producers (and exporters). In Mexico taxes are 13% of gasoline prices, but in Norway they are 75% (OECD, 1998).

Currently, no methodologies exist to project future energy prices taking all of above mentioned factors into account, nor were the SRES scenarios intended to make explicit assumptions on such factors such as future energy taxation. Price information enters long-term emission models either in the form of exogenous scenario assumptions, or it is derived internally in models based on simplified representations of price formation mechanisms usually based on (marginal) *cost* information.

The six models used for SRES range from detailed "bottom-up" models (e.g., AIM, IMAGE), through macro-economic (partial equilibrium) models (e.g., MARIA, MiniCAM), to hybrid approaches (successive iteration between the engineering model MESSAGE with a macro-economic model, or using the Worldscan model with IMAGE). Each has different representations of price formation mechanisms and their relationship to macro-economic or sectoral energy demand. These are summarized in Appendix IV. As a rule, "bottom-up" (optimization) models calculate only (average and marginal) *costs* endogenously. As a result of their sectoral perspective (energy, agriculture, etc.), these models cannot determine macro-economic feedbacks on other sectors or the entire economy and thus are unable to represent a consistent picture of *price* formation. Conversely, price formation is endogenized in "top-down" models; however, these rely on the stringent assumption that demand and supply must be in equilibrium and in addition provide little sectoral detail. Over recent years this simplified modeling dichotomy has progressively weakened because of further advances in methodology and the development of "hybrid" modeling approaches. To illustrate the methodologies deployed in the six SRES models, two (MARIA and MESSAGE) are discussed here, but (for space limitations) only in terms of one scenario (B2). (Table 4-9 gives additional details of an inter-scenario comparison of energy prices for the MiniCAM and ASF models. Owing to methodologic differences, a comparison of prices across scenarios is only possible within a consistent approach (i.e. be comparing scenarios quantified with the same model).)

The energy prices represented in MARIA (see also Mori, 2000) consist of energy production and energy utilization costs. Market prices are determined endogenously by model-calculated shadow prices (for further model details see Appendix IV and Mori and Takahashi, 1999). Among various parameters, the extraction costs of fossil fuel resources and the coefficients of utilization costs and their evolution over time are the most important determinants. For the MARIA runs, the resource estimates of Rogner (1997) were used as input. For the sake of simplicity, all fossil resource categories of Rogner (1997) were aggregated into two classes and a quadratic production function was used to interpolate the extraction costs of reserves and all other occurrences. For coal, long-term extraction costs range up to US$6.3 per GJ in 1990US$ prices, for gas up to US$25 per GJ, and for oil up to US$28 per GJ (see Appendix IV for further details). The energy cost coefficients (representing 16 different energy conversion technologies) are based on Manne and Richels (1992). For the B2 scenario quantification, the Manne and Richels (1992) estimates were largely retained. For instance, electricity generation costs range between 14 mills[26]/kWh for gas to 51 mills/kWh for coal. (For the other scenario quantifications these cost values were modified to conform to the different interpretations of a particular scenario storyline.) Together these assumptions determined long-run costs and shadow prices that were set equal to energy prices in the macro-economic production function of MARIA. The energy prices were combined with assumed (low) AEEI values and potential GDP growth rates (the latter from the B2 marker) to calculate the resultant aggregate energy demand in the model. The resultant primary energy demand was (with exception of the REF region) within 15% of the respective B2 marker quantification at the regional level and within 5% of global energy demand. As a result of different model structures, comparable price data for the MESSAGE model are only available for internationally traded primary energy forms (these are given in Table 4-8).

[26] 1 mill is 0.1 USCents (US$0.001).

The bottom-up, systems engineering (optimization) model MESSAGE does not compute energy prices. Instead, the model is entirely based on cost information, but such costs are treated as dynamic. Their overall treatment follows the lines outlined above for the MARIA model, except that greater technology-specific detail is contained in the model. Altogether 19 different fossil resource grades are differentiated, based on the estimates of Rogner (1997). The resultant (levelized) extraction costs for the B2 marker are in the range US$1.1 to US$5.4 per GJ for coal, US$1.2 to US$5.3 per GJ for oil, and US$1.2 to US$5.7 per GJ for gas (range indicates costs variations between lowest and highest costs of the four SRES regions for 2020, 2050, and 2100 respectively, see Appendix IV). Technology-specific cost assumptions cannot be summarized here as MESSAGE contains literally hundreds of energy supply and end-use technologies. Examples of cost assumptions are given in Section 4.4.7 and more detail is reported in Riahi and Roehrl (2000). However, as in MARIA, MESSAGE also calculates shadow prices for internationally traded primary energy forms and therefore these two indicators can be compared (Table 4-8).

Table 4-8: *International price (MARIA) and calculated shadow price (MESSAGE) of internationally traded energy (1990US$/GJ) by 2020, 2050, 2100 for the SRES B2 scenario.*

	Coal		Oil		Gas		Biofuels	Synfuels
	MARIA	MESSAGE[a]	MARIA	MESSAGE[b]	MARIA	MESSAGE[c]	MARIA	MESSAGE[d]
2020	0.5	3.4	3.5	3.9-4.4	2.9	2.8-4.4	4.8	n.a.
2050	0.8	2.5	4.9	7.5-8.2	4.3	5.1-6.4	6.5	10.4-16.2
2100	1.4	8.1	6.3	17.3-18.2	5.4	5.2-11.4	6.3	17.1-20.7

[a] Costs include export and/or import infrastructure and transport.
[b] Range between crude oil and light and heavy oil products.
[c] Range between liquid natural gas and direct pipeline imports to North America, Europe, Japan, and North Africa.
[d] Range between methanol, ethanol, and liquid hydrogen.

To achieve consistency between model-calculated energy cost dynamics and energy demand assumptions an iterative modeling procedure between MESSAGE and MACRO (a macro-economic production function model based on Manne and Richels, 1992) was used, on the basis of model-calculated shadow prices as indicators of future price dynamics. The methodology is described in more detail in Wene (1996). This approach requires time-intensive model iterations, but has the advantage that the impact of price increases can be separated from efficiency improvements through fuel substitution (e.g., a gas-fired cook stove energy end-use efficiency is up to 10 times higher than a traditional cook stove fired with fuelwood) as well as from "everything else," i.e., the AEEI in the traditional sense). Aggregated, the impact of (shadow) price increases in MESSAGE's B2 scenario accounts for 8% of global primary energy demand by 2020, 23% by 2050, and 30% by 2100. This impact is calculated as a reduction in energy demand compared to a hypothetical scenario with constant 1990 prices (and correspondingly higher energy demand). The impact of price increases on future energy demand in the B2 scenario is thus relatively small compared to that of other factors, although far from negligible. This also explains why the two B2 scenario quantifications by MARIA and MESSAGE have quite similar energy demand figures, even if international trade prices may differ. First, trade prices are only one component of the cost–price mechanism treated in the models (which also includes domestic energy production, conversion, and end-use costs). Second, models differ in their parametrizations of the "everything else" (AEEI) model parameters, for which a wide range of views on applicable ranges exists. Therefore it is one of the model calibration parameters frequently used to replicate existing scenarios or to standardize inter-model comparison projects such as EMF-14 (Weyant, 1995).

(Box 4.7 continues)

Box 4.7 (continued)

Table 4-9: *Energy prices (1990US$/GJ) across SRES scenarios as calculated in the ASF (top) and MiniCAM models for their respective A1, A2, B1, and B2 (cf. Table 4-8) scenarios. Note in particular significant base-year differences in fuel prices because of different cost accounting definitions used in models (c.i.f. versus f.o.b.[27]), in particular with respect to transportation costs (included in the price figures given for ASF, but excluded in the numbers given for MiniCAM).*

		A1	A2	B1	B2
ASF[a]					
Coal	2000	1.5	1.5	1.5	1.5
	2020	1.6	1.5	1.6	1.5
	2050	1.9	1.7	1.7	1.6
	2100	2.0	1.8	1.6	1.7
Oil	2000	4.4	4.4	4.4	4.4
	2020	5.3	4.7	5.1	4.7
	2050	7.1	6.2	6.3	6.1
	2100	7.7	7.5	6.1	7.1
Gas	2000	5.0	5.0	5.0	5.0
	2020	5.0	5.0	4.9	5.0
	2050	5.3	5.0	4.8	4.9
	2100	7.9	6.1	4.9	5.8
MiniCAM[b]					
Coal	1990	1.0	1.0	1.0	1.0
	2020	1.6	1.7	1.6	1.6
	2050	1.9	2.0	1.7	1.7
	2100	2.5	2.5	1.9	2.0
Oil	1990	3.9	3.9	3.9	3.9
	2020	8.6	10.2	6.4	7.3
	2050	10.4	13.3	9.9	10.4
	2100	9.6	15.2	8.5	10.2
Gas	1990	1.6	1.6	1.6	1.6
	2020	2.8	3.2	2.0	2.4
	2050	3.8	5.7	2.5	3.0
	2100	6.8	8.7	1.9	2.3
Biofuels	1990	n.a.	n.a.	n.a.	n.a.
	2020	2.1	2.2	2.0	2
	2050	2.4	2.6	2.0	2.1
	2100	2.3	3.2	1.5	2.0

[a]ASF: global average supply price, including transportation.
[b]MiniCAM: as determined by solution to a partial equilibrium supply and demand model.

and economic structural changes, with consequent improvements in energy intensity. As a result, the rate of energy intensity improvement in Annex I countries is around 1.16% per year, and in non-Annex I countries 1.44% over the 100 years to 2100. Thus, final energy use for A1 is much higher than those in the A2, B1, and B2 scenarios, with a substantial long-term convergence in final energy use per capita between Annex 1 countries and non-Annex 1 countries.

[27] c.i.f., cost, insurance, freight (included in price); f.o.b., free on board (i.e. insurance and transport costs not included in fuel price delivered "free on board" transport vessel only). These different cost-accounting methods for international energy trade are particular important for transport and infrastructure intensive fuels such as natural gas.

4.4.5.2. Harmonized and Other A1 Scenarios

The various A1 scenarios indicate a wide range in energy intensity improvements and resultant energy demand. A1B-MESSAGE tracks closely the global and regional energy demand patterns of the A1B marker and so satisfies the criteria of a "fully harmonized" scenario, albeit that ASIA deviations from the A1B marker amount to about 20% during the period 2030 to 2060. Other scenarios indicate, for example, higher energy demand in 2050 (e.g., A1-ASF) and/or lower energy demand in 2100 (e.g., A1B-IMAGE, A1B-ASF, and A1B-MARIA). Differences are regionally heterogeneous and result from:

- Higher GDP growth rates assumptions compared to those of the A1B marker (e.g., A1-MiniCAM for OECD90).

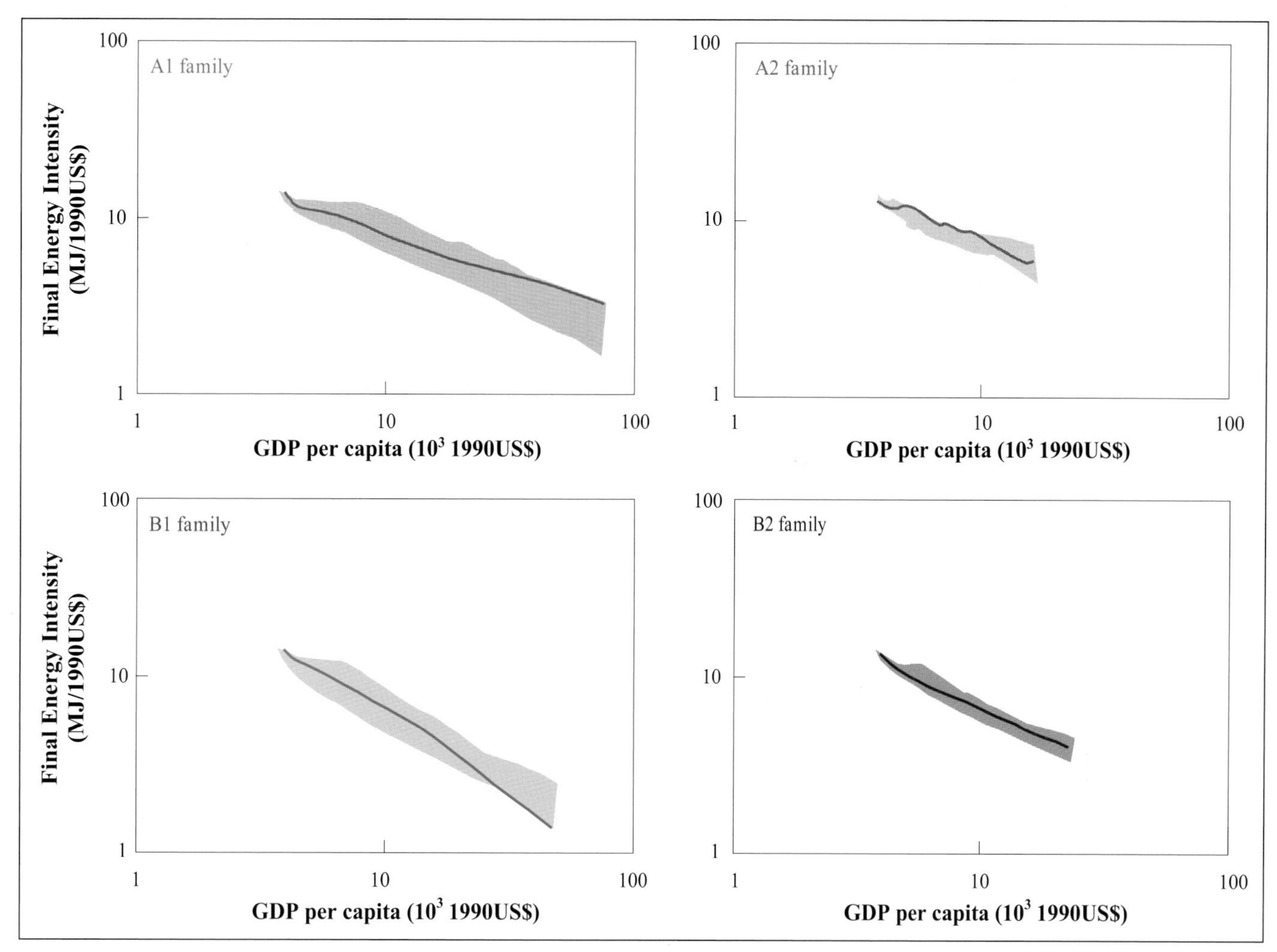

Figure 4-6: Relationship between final energy intensity and per capita income in the four marker scenarios. The data points represent values in 1990, 2020, 2050, and 2100. The 1990 value is at the top of each curve and the 2100 value at the bottom.

- Higher or lower (non-price induced) assumptions in energy intensity improvement rates in the various scenarios.
- Differences in technology assumptions that lead to differences in projected energy prices
- Combination of all three above factors.

The differences in model representation of these factors along with the available time and resources have not allowed a detailed analysis of the numerous underlying differences in energy intensities across the scenarios. For instance, final energy use in the A1v1-MiniCAM and A1v2-MiniCAM scenarios reaches about 60% only of the final energy use described in the A1 marker. However, both scenarios describe very different worlds with respect to economic growth but share the assumption of saturating per capita energy use at levels of 150 GJ/capita. All else being equal, scenarios of lower GDP per capita growth (e.g. A1v2-MiniCAM) generally also assume lower improvement rates in energy intensity, consistent with the literature (as discussed in Chapter 3). Generally, the range between energy intensity across scenarios is larger in the 1990 base year than toward the end of the simulation horizon. This results from alternative estimates and model specifications of non-commercial energy use (excluded in some models), which yield large differences in 1990 energy intensities. Over time – and consistent with the high-income characteristics of an A1 world – the use of non-commercial energy declines and is ultimately phased out altogether. Thus, by 2100 differences in energy intensities across models become much smaller. An interesting set of scenarios was explored in the A1T group. In these the diffusion of a whole host of new energy end-use technologies (e.g., microturbines, fuel cells) results in substantial additional gains in efficiency and hence higher energy intensity improvements and lower final energy demand at equal or lower energy costs compared to the other A1 scenarios.

4.4.5.3. A2 Scenarios

Final energy intensities in the A2 (ASF) marker scenario improve steadily, with the exception of the ALM-region between 1990 and 2020 because of material and energy intensive infrastructure build-up. The fastest reduction occurs in the REF region as its energy-intensive economy progressively restructures. The slowest improvements are projected for the OECD90 region, because of slow capital turnover rates (GDP growth of 1.6% per year between 1990 and 2100). The final energy demand across regions is determined by the product of regional GDP growth and energy intensity improvements. For example, the higher energy intensities in ASIA compared to the ALM region lead to a higher absolute final energy demand in the former in spite of a lower GDP. This is explained by differences in initial conditions and by delayed diffusion of more efficient energy end-use technologies, because of lower GDP per capita growth.

4.4.5.4. Harmonized and Other A2 Scenarios

The global primary energy use and final energy use in the A2 scenarios created with the AIM, MESSAGE, and MiniCAM models are quite close to those of the marker scenario, while A2-IMAGE scenario projects lower primary and final energy use compared to the A2 marker. As mentioned above, the A2-A1-MiniCAM scenario explores a very different unfolding of driving forces in terms of population and GDP growth, combined with an assumption about saturating energy demand at current Western European levels. Combined, these assumptions translate into the lowest energy intensity improvement rates across the SRES scenario set. Global primary energy use per unit of GDP (intensities) improves from 14.7 MJ/US$90 in 1990 to 8.9 MJ/US$90 in 2100. This reflects mainly the low per capita GDP (productivity) growth of this scenario which (other factors being equal) translates into low rates of energy intensity improvement. Nonetheless, the resultant energy intensities are comparable to current Western European levels, as are income and energy use per capita. In other words, the scenario describes a global picture by 2100 quite similar to that of Western Europe of today, 100 years earlier.

4.4.5.5. B1 Scenarios

Energy intensity improvements in the B1 marker result from energy efficiency investments brought about by increases in fuel and electricity prices and technological innovations (including assumptions on taxes and perceived premium values for clean fuels). The rather high rates in energy intensity reduction in B1 stem also from the explicit assumption that less industrialized regions catch-up. Another factor is the assumption that monetary economic growth in less developed regions initially largely replaces activities in the informal economy, which leads to a replacement of traditional non-commercial energy forms by high-efficiency modern applications and fuels – and hence substantial energy intensity improvements. In the developed regions the high economic growth in the B1 scenario may, for instance, be in the form of increasing monetization of human activities previously not included in GDP accounts (e.g., childcare, household work). Such monetary GDP growth does not result in additional demands for energy services, and hence again results in significant energy intensity improvements. The demand for electricity is assumed to rise faster than that for non-electricity energy, and may pose one of the capital availability constraints in this scenario.

4.4.5.6. Harmonized and Other B1 Scenarios

Various alternative scenario quantifications were developed for B1 by the modeling teams. For the first four to five decades most model runs show a global final energy use within the proposed bounds of the B1 marker, except B1-ASF which is higher. By 2100 most scenarios assume higher final energy use than the marker run, except B1-MESSAGE which reproduces closely the final energy use of the B1 marker (and is correspondingly classified as a "fully harmonized" scenario). In particular, B1-MARIA, B1High-MiniCAM, and B1High-MESSAGE show a global final energy use in 2100 nearly twice that of the marker. These scenarios explored the implications for energy demand of less rapid "dematerialization" tendencies of the economy, especially for developing countries, with trends in line with historical energy intensity experiences in the OECD countries. Regional trends differ most dramatically for the MiniCAM and MARIA runs for ASIA and ALM, with the MiniCAM simulations assuming a saturating (converging) energy use on a per capita basis at 125 GJ/capita. However, current knowledge about rates and direction of dematerialization of economic activities is limited. Therefore, both historical OECD trends and their applicability to the future economies of currently developing countries may not necessarily reflect future developments. The use of alternative modeling approaches in the quantification of the B1 scenario storyline has helped to shed light on this important area of uncertainty of the future.

4.4.5.7. B2 Scenarios

Final energy demand for the B2 marker was derived by applying efficiencies of end-use technologies to the demand of electric and non-electric energy services. These in turn depend on the economic development rates, income levels, and sectoral economic structure of each region. The evolution of the final energy demand levels and structure in the developing regions follows patterns that are similar to the historical development in the now-industrialized regions of the world. This again is consistent with the "dynamics-as-usual" interpretation of the B2 storyline. By successive iterations with a macro-economic model, marginal cost increases are taken into account in the energy demand projections of the MESSAGE model for the B2 marker (see Appendix IV). The result is an aggregate energy-intensity improvement rate at the global level of about 1% per year until 2100, about the same as has prevailed over the past 100 years in countries for which such long-term time series data are available (see Chapter 3). This aggregate global improvement rate masks important differences in the temporal and spatial evolution of energy intensities. Improvements are generally higher in regions far away from the energy-intensity frontier and also faster in those for which the capital turnover rate (i.e., GDP growth) is higher. Consistent with the more imperfect realization of future trends characteristic of the B2 scenario, energy intensity improvements are slower than in the A1 or B1 scenario families, but higher than in the A2 scenario family.

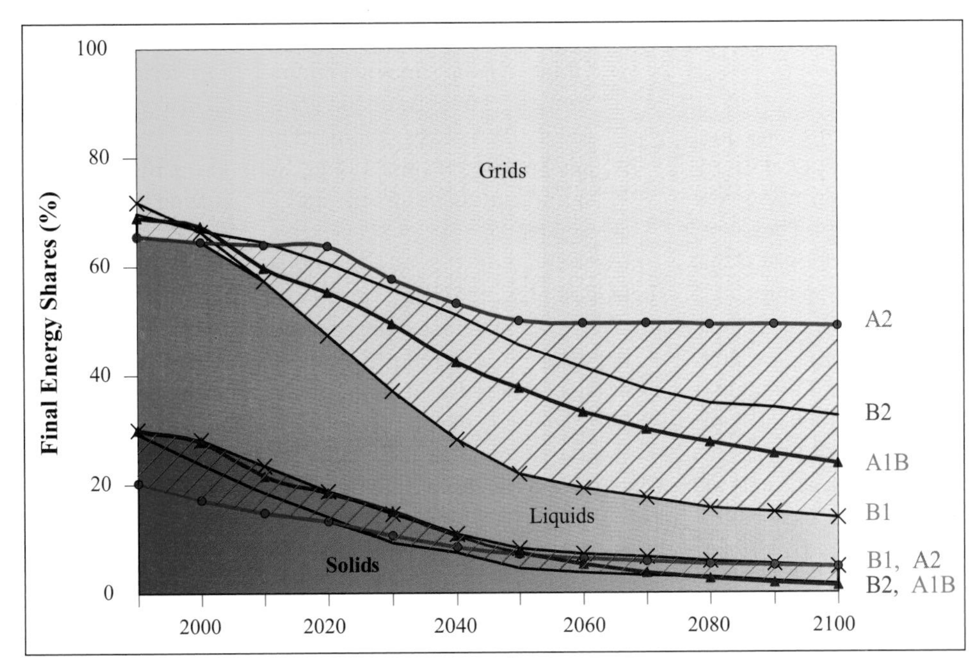

Figure 4-7: World final energy (%) by form of delivery. Direct use of solids, direct use of liquids, and delivery of grids (gas, district heat, electricity, and hydrogen) for the four SRES marker scenarios. Overlapping shaded areas indicate variation across the four marker scenarios. Liquids include oil products, methanol, and ethanol. Solids include coal and biomass.

4.4.5.8. Harmonized and Other B2 Scenarios

The energy intensity and resultant global growth in final energy demand of the B2 marker is matched reasonably well by other scenario quantifications until about 2050; the B2-AIM scenario shares "fully harmonized" input assumptions at the global and regional levels alike over the entire time horizon. By 2100, however, differences become larger, and in particular the two MiniCAM scenarios show lower final energy intensity improvements compared to the B2 marker and other B2 scenarios. The MiniCAM scenarios were developed to explore high-end sensitivities for GHG emissions in the B2 scenario family by assuming higher energy demand, combined with a high reliance on coal in the B2High-MiniCAM scenario. Different scenarios at the regional level illustrate alternative interpretations, with particularly large variations in the REF and ALM regions. Alternative scenarios indicate up to 50% lower energy use in REF (e.g., B2-IMAGE) and up to 50% higher energy use in ALM (B2-MiniCAM, B2High-MiniCAM) compared to the B2 marker. The alternative scenarios illustrate that uncertainties in future energy intensities and resultant final energy demand are generally larger for developing and transition economies compared to the OECD90 region. Among other reasons, this is a function of the much higher uncertainties with respect to future economic growth rates in these regions.

An interesting observation (considering the SRES multi-model approach) is that the changes in structure of final energy are similar in the four markers of the SRES scenario families, even though these are derived from four different modeling approaches and describe very different futures in terms of demographic, socio-economic, and energy development (Figure 4-7).[28] The trend is toward energy reaching the consumer in more flexible, more convenient, and cleaner forms. This reflects that people with higher income are willing to pay more for more convenient energy forms (e.g., even if coal were cheaper than gas, everybody would rather heat with gas than coal). Therefore, the final energy mix is characterized by growing importance and dominance of grid-dependent fuels, such as electricity, district heat, and gas. Consistent with the storylines and the higher income levels of A1 and B1, this change in final energy structure is faster in these scenario families than in the other two scenario families. The structural shift is slowest in scenario A2, with scenario B2 taking an intermediate position. These scenario differences mainly reflect differences in per capita income levels.

4.4.6. Resource Availability

Section 3.4 in Chapter 3 reviews energy resources and technologies. Here existing *reserves* (identified quantities recoverable at today's prices and with today's technologies), *resources* that have yet to be discovered or that need foreseeable techno-economic progress to become available in the future, and other *occurrences* of hydrocarbons in the Earth's crust are considered. Oil, gas, and uranium occur in deposits that need to be located, and the exploration for new resources is related to the needs for production over the next few decades rather than to a need to define what might ultimately be available for exploitation. Thus the ultimate resource base is uncertain. Coal, on the other hand, occurs in seams over wide areas and very little exploration is needed to give an estimate of potentially available resources. Whether or not they could be mined with given technologies and economics remains the most important uncertainty. Finally,

[28] Corresponding data were not available for all scenarios developed with other models, and hence a detailed comparison across the entire SRES scenario set was not possible.

new renewable sources of energy are dependent on ongoing technological development and cost reductions.

The conventional oil industry is relatively mature and the question is at what point in the 21st century will the current reserves start to run out. However, unconventional resources are also available – shale oil, bitumen, and heavy oil. These are starting to be exploited and they will extend current conventional oil reserves. The gas industry is less mature and much more remains to be discovered, particularly in areas that do not currently have the infrastructure to utilize gas and consequently exploration has been unattractive. Additionally, large amounts of unconventional gas have been identified, some of which are already in commercial production (e.g., in the US). Also, huge quantities of natural gas are believed to exist as methane hydrates on the ocean floor (see Chapter 3) and it is possible that technology to exploit these will be developed at some stage. For uranium and thorium, the amount of exploration to date has been very limited, and hence the possibilities of discovering new deposits are enormous. It is likely that even a major expansion of the nuclear industry will not be limited by the amount of available uranium or thorium. With coal, the question is not one of discovery but one of economics, accessibility, and environmental acceptability.

To consider future resource availability as a dynamic process, however, does not resolve the inherent uncertainties in terms of future success rates of hydrocarbon exploration, technology development for either non-conventional fossil resources or non-fossil alternatives, or future energy prices. Therefore, these uncertainties are explored by adopting different scenario assumptions that range from low to (very) high resource availability (see Table 4-4), consistent with the interpretation of the various scenario storylines presented in Section 4.3. This scenario approach is especially important given that hydrocarbon occurrences are the largest storage of carbon. IPCC WGII SAR (Watson *et al.*, 1996) estimates the size of the total carbon "pool" in the form of hydrocarbon occurrences to be up to 25,000 GtC. How much of this eventually could become atmospheric emissions is at present unknown, and depends on the future evolution of technology, prices, and other incentives for future hydrocarbon use and their alternatives.

Given that long-term emission scenarios invariably rely on quantification by formal models, an important distinction needs to be made between assumptions concerning the ultimate resource base and projected actual resource use. Typically, assumptions on the ultimate resource base enter models as exogenously specified *constraints* – cumulative future production simply cannot exceed values specified as the resource base. Actual resource use, or what is frequently termed the "call on resources" conversely depends on numerous other factors represented in models, such as:

- Future price levels (either assumed as exogenous inputs or determined endogenously in the model).
- Assumptions on future technology improvements that either enable unconventional hydrocarbons to be "mined" economically or, conversely, that draw on non-fossil alternatives and/or non-climate environmental and social constraints (e.g., limits on particulates and sulfur emissions or on land degradation and mining accidents).

Their complex interplay results in scenarios of future cumulative resource *use* being the most appropriate indicator, as opposed to exogenously pre-specified resource-base constraints, especially in view of the multi-model approach adopted to develop the SRES scenarios. Table 4-10 and Figures 4-8 to 4-10 summarize the results for the four SRES marker scenarios and of the ensemble of SRES scenarios for their respective scenario families and scenario groups (in the case of the A1 scenario family). It is evident that, in the absence of climate policies, none of the SRES scenarios depicts a premature end to the fossil-fuel age. Invariably, cumulative fossil-fuel use to 2050 (not to mention 2100) exceeds the quantities of fossil fuels extracted since the onset of the Industrial Revolution, even though the "call on" fossil resources differs significantly across the four marker scenarios. This increase is higher in the scenarios that explore a wider domain of uncertainty on future fossil-resource availability.

For *non-fossil resources*, like uranium and renewable energies, future resource potentials are primarily a function of the assumed rates of technological change, energy prices, and other factors such as safety and risk considerations for nuclear power generation. Generally, absolute resource constraints do not become binding in the marker scenarios or other scenarios. The contribution of these resources is substantially below the physical flows identified in Section 3.4, and therefore results mainly from scenario-specific assumptions concerning technology availability, performance, and costs. These are summarized in Section 4.4.7.

4.4.6.1. A1 Scenarios

Energy resources are taken to be plentiful by assuming a large future availability of coal, unconventional oil, and gas as well as high levels of improvement in the efficiency of energy exploitation technologies, energy conversion technologies, and transport technologies. The grades of energy resources used in the model differ on the basis of extraction costs. When combined with the level of improvement in efficiency of exploitation technology (expressed as the rate of improvement in marginal production costs), the graded costs of energy-resource exploitation determine the energy production costs (prices) and hence the ultimate resource extraction quantities. For A1, large amounts of unconventional oil and natural gas availability were assumed. Cumulative (1990 to 2100) extraction of oil ranges between 15 and 30 ZJ in the A1 scenarios (A1B marker, 17 ZJ); for gas the range is between 23 and 48 ZJ (A1B marker, 36 ZJ) and for coal the range is between 8 and 50 ZJ (A1B marker, 12 ZJ). Resource availability and reliance uncertainties are also explored through

Table 4-10: Cumulative hydrocarbon use, historical data from 1800 to 1994 (Nakićenović et al., 1993, 1996; Rogner, 1997) and range for SRES scenarios (markers and range across all scenarios) for the four scenario families and their scenario groups. The numbers in brackets give minimum and maximum values of scenario variants. Note in particular the large variation within the A1 scenario family as a result of its branching out into four scenario groups, each with a different reliance on particular resource categories and technologies that range from carbon-intensive developments to decarbonization. A1C and A1G have been combined into one fossil-intensive group A1FI in the SPM (see also footnote 1).

World Cumulative Hydrocarbon Use, in ZJ (1,000 EJ)								
	1800–1994	1990–2100						
		A1B	A1C	A1G	A1T	A2	B1	B2
Oil	4.6	20.8				17.2	19.6	19.5
		(17.0-29.9)	(11.5-20.4)	(29.6-50.8)	(16.6-20.8)	(11.0-22.5)	(15.7-19.6)	(11.2-22.7)
Gas	2.0	42.2				24.6	14.7	26.9
		(22.8-45.2)	(19.7-22.4)	(40.9-54.9)	(23.9-29.9)	(18.4-35.5)	(14.7-31.8)	(17.9-26.9)
Coal	5.6	15.9				46.8	13.2	12.6
		(8.5-51.5)	(48.4-68.3)	(18.8-37.9)	(4.4-12.4)	(20.1-47.7)	(3.3-27.2)	(12.6-44.4)

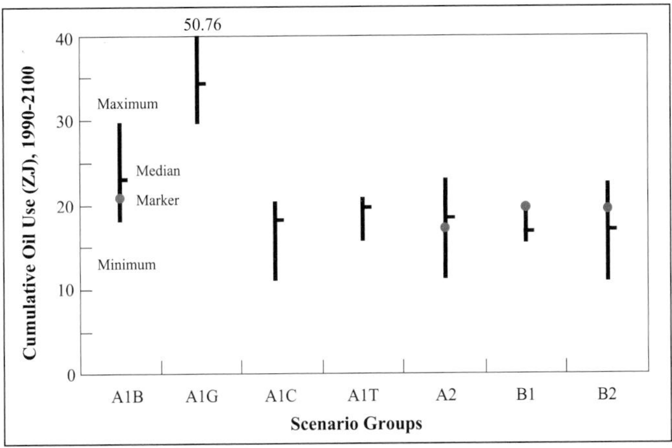

Figure 4-8: Cumulative oil resource use 1990 to 2100 in the SRES scenario families, including the four scenario groups within the A1 scenario family. The bars show the spread of total oil extraction over all scenarios in the respective scenario family; the resultant medians and the values of the respective marker scenarios are also shown. A1C and A1G have been combined into one fossil-intensive group A1FI in the SPM (see also footnote 1).

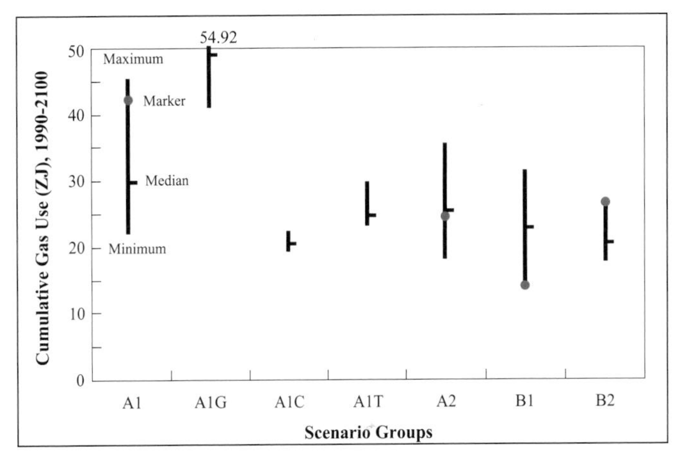

Figure 4-9: Cumulative gas resource use 1990 to 2100 in the SRES scenario families, including the four scenario groups within the A1 scenario family. The bars show the spread of total gas extraction over all scenarios in the respective scenario family; the resultant medians and the values of the respective marker scenarios are also shown. A1C and A1G have been combined into one fossil-intensive group A1FI in the SPM (see also footnote 1).

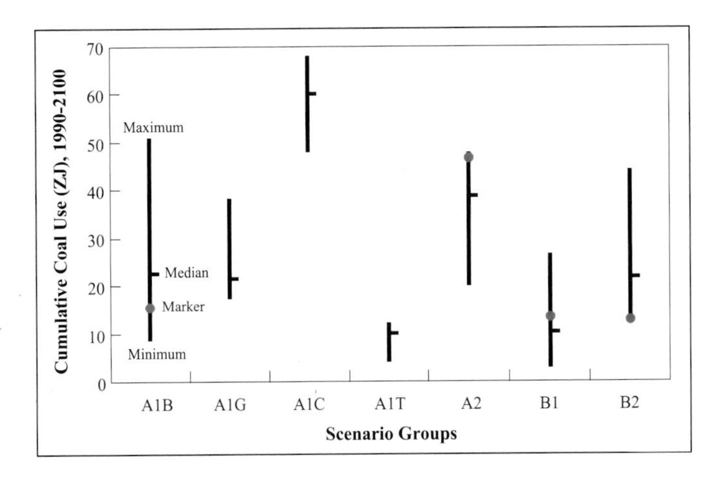

Figure 4-10: Cumulative coal resource use 1990 to 2100 in the SRES scenario families, including the four scenario groups within the A1 scenario family. The bars show the spread of total coal extraction over all scenarios in the respective scenario family; the resultant medians and the values of the respective marker scenarios are also shown. A1C and A1G have been combined into one fossil-intensive group A1FI in the SPM (see also footnote 1).

additional scenario groups. Three of these (A1C, A1G, and A1T) explore more extreme patterns of reliance on particular resources and technologies compared to the more "balanced" tendencies described in the A1B scenarios, including the A1B marker. As discussed in Chapter 3 and Section 4.3.1, this characteristic of the A1 scenario family stems from the interpretation of technological change and resource availability as being cumulative and path dependent.

4.4.6.2. A1 Scenario Groups

Besides the A1B marker scenario group, alternative pathways unfold within the A1 family, according to diverging technology and resource assumptions (Figures 4-8 to 4-10). Two of these groups (A1C and A1G) were merged into one fossil-intensive group (A1FI) in the SPM. The more detailed information on these two groups is presented here, in Chapter 5 and Appendix VII (see also footnote 1).

The coal-intensive scenario group A1C is restricted mainly to conventional oil and gas, which results in the lowest cumulative oil and gas use (15 to 19 ZJ) of all scenarios; it is even slightly lower than in the B2 scenario, which has much lower energy demand. As such, the scenario illustrates the long-term GHG emission implications of quickly "running out of conventional oil and gas" combined with rapid technological progress in developing coal resources and clean coal winning and conversion technologies. As a result, cumulative coal use is very high – between 48 and 62 ZJ (median, 60 ZJ) between 1990 and 2100.

Conversely, oil and gas resources are assumed to be plentiful in the world of scenario group A1G because of the assumed development of economic extraction methods for unconventional oil and gas, including methane clathrates. Cumulative oil and gas extraction amounts to 76 to 88 ZJ, about twice as high as in the A1C scenario group. Mainly this reflects current perceptions that radical technological change needs to occur to translate a more significant portion of the resource base of unconventional oil and gas into potentially recoverable reserves, a development evidently also cross-checked by possible developments in non-fossil alternatives. Cumulative coal extraction in A1G is relatively low at 15 to 38 ZJ (median, 19 ZJ) across the scenarios of this scenario group.

As a result of fast technological progress in post-fossil alternatives in the technology-dynamic A1T scenario group, the call on oil and gas resources is comparatively modest – cumulative extraction to 2100 ranges between 36 and 46 ZJ, quite similar to the A1C scenario group. The main difference is that because of the improvements in non-fossil alternatives the call on coal resources remains modest – cumulative coal use of 4 to 12 ZJ (median: 10 ZJ) in A1T is the lowest of all the scenarios.

4.4.6.3. A2 Scenarios

Resource availability assumptions for the A2-ASF world are generally rather conservative, essentially that current conventional estimates of petroleum resource availability are not expanded.[29] Unconventional hydrocarbons, such as methane clathrates and heavy oils, do not come into large-scale use. As a result, coal resource use is the highest among the SRES marker scenarios. The ASF marker scenario quantification of oil, natural gas, and coal resource availability reflects the Rogner (1997) estimates for conventional oil and coal resource availability and the recent IGU (1997) estimates for conventional gas reserves (optimistic scenario, see Chapter 3). Resource extraction costs in the ASF depend on the resource "grade" and vary from US$2.6 to 5.2 per GJ for oil (in 1990 dollars), from US$1.2 to 4.6 per GJ for gas, and US$0.7 to 6.0 per GJ for coal.

4.4.6.4. Harmonized and Other A2 Scenarios

The primary energy structure of the A2 family scenarios is also reflected in the cumulative fossil fuel resource use, characterized by an increasing reliance on coal resources (see Figures 4-8 to 4-10). The cumulative oil use varies by a factor of two across the A2-family, between 11 and 24 ZJ (median, 18 ZJ; A2 marker, 17 ZJ). Cumulative gas use ranges between 20 and ZJ 36 (median, 23 ZJ; A2 marker, 25 ZJ). The higher end of the range of gas resource use occurs in the A2G-IMAGE scenarios, which explored the scenario sensitivity to assuming that a significant fraction of methane hydrate occurrences become technically and economically recoverable in an A2 world. Given the regional orientation of the A2 scenario storyline and the resultant quest for energy independence, the possibility of tapping even currently "exotic" fossil resources certainly merits such a scenario sensitivity analysis. The opposite end of the resource availability spectrum is explored in the MiniCAM scenarios of the A2 scenario family. First, methane clathrates are assumed not to become available. As a result, the call on resources focuses on coal (A2-MiniCAM) or, in a scenario sensitivity analysis, more on unconventional oil and gas (A2-A1-MiniCAM). The range of reliance on coal resources is thus an inverse image of the range of oil and gas resource availability. Cumulative coal extraction varies between 22 and 53 ZJ (median, 35 ZJ; A2 marker, 47 ZJ) across the scenarios of the A2 scenario family. This picture mainly represents what used to be termed "conventional wisdom" in much of the scenario literature (including the previous IS92 scenario series). Importantly, while the probabilities of alternative developments of fossil and non-fossil resource availability cannot be assessed at present, the multi-model, multi-scenario approach described here demonstrates that the uncertainties in fossil resource

[29] Even with this "conservative" assumption cumulative oil extraction in the A2 marker scenario totals 16 ZJ, or 2.7 times currently identified, recoverable oil reserves (6 ZJ or 143.3 billion tons; BP, 1999). Thus, the A2 scenario also assumes that in future it will be possible to continue the historical trend in which large quantities of (undiscovered or presently uneconomic) oil *resources* are transferred into recoverable *reserves*. Some analysts consider such a future trend as definitely optimistic (see the literature review in Chapter 3).

availability might be much larger than assumed a decade ago. This finding also reflects the results of IPCC WGII SAR (Watson *et al.,* 1996).

4.4.6.5. B1 Scenarios

Assumptions on the fossil fuel resource-base used in the B1 marker scenario quantification are based on the estimates of ultimately recoverable conventional and unconventional fossil resources described in Rogner (1997). The capital output ratio of resource exploitation is assumed to rise with progressive resource depletion, but this is counteracted by learning curve effects in the marker scenario quantification provided by the IMAGE model. Regional estimates of the exploitation costs of conventional and unconventional resources of Rogner (1997) were used to construct long-term supply cost curves as of 1971. These values, rather than absolute upper bounds on resource base availability, define future resource availability in the IMAGE model. The supposed availability of huge non-conventional occurrences of oil and natural gas, with a geographic distribution markedly different from the distribution of conventional oil and gas, has significant implications for fuel supply and trade patterns in the long term. For coal resources, Rogner's (1997) estimates were also adopted; of the total of 262 ZJ, 58 ZJ belong to the categories of proved recoverable, additional recoverable, and additional identified coal resources. The production costs of coal were assumed to rise with increasing depth and rising labor wages, but these costs are largely offset by mechanization (in underground mining) and economies of scale (in surface mining).

4.4.6.6. Harmonized and Other B1 Scenarios

The call on oil resources in the scenarios that comprise the B1 scenario family ranges between 11 and 20 ZJ, with a median of 17 ZJ (B1 marker, 20 ZJ). For gas the range is 15 to 33 ZJ (median, 20 ZJ; B1 marker, 15 ZJ), and for coal the corresponding range is between 3 and 27 ZJ (median, 11 ZJ; B1 marker, 13 ZJ). An overview is given in Figures 4-8 to 4-10.

4.4.6.7. B2 Scenarios

The availability of fossil energy resources in the B2 marker scenario is assumed to be conservative, in line with the gradual, incremental change philosophy of the B2 scenario storyline. Consequently, oil and gas availability expands only gradually while coal continues to be abundant. Assumed oil and gas resource availability does not extend much beyond current conventional and unconventional *reserves*. Through gradual improvements in technology, a larger share of unconventional *reserves* and some additional resource categories are assumed to become available at improved costs over the 21st century. The availability of oil and gas, in particular, is limited compared to the estimated magnitude of global fossil resources and occurrences (Watson *et al.*, 1996). This translates into relatively limited energy options in general and extends also to non-fossil energy options.

4.4.6.8. Harmonized and Other B2 Scenarios

Alternative B2 scenario implementations assumed similar order of magnitudes of resource availability as the B2-marker scenario, except for B2High-MiniCAM. The resultant cumulative resource use (1990–2100) ranged between 9 and 23 ZJ (median, 17 ZJ; B2 marker, 19 ZJ) for oil, between 18 and 27 ZJ (median, 21 ZJ; B2 marker, 27 ZJ) for gas, and between 12 and 55 ZJ (median, 21 ZJ; B2 marker, 13 ZJ) for coal (see Figures 4-8 to 4-10). The largest uncertainties relate to different interpretations of the more gradual changes under a "dynamics-as-usual" philosophy that characterizes the B2 scenario storyline. One group of scenarios (including the B2 marker) assumed a gradual expansion in the availability of conventional and unconventional oil and gas, whereas another group of scenarios adopted more conservative assumptions (akin to the A2 and B1 scenario families).[30] All else being equal, lower resource-availability assumptions for oil and natural gas lead to a higher reliance on coal and non-fossil alternatives and explain, together with technology assumptions, the differences in emissions between alternative B2 scenario quantifications discussed in Chapter 5.

4.4.7. Technological Change

Chapter 3 highlights the importance of technological change in long-run productivity growth, but also for the historical transformations of energy end-use and supply systems. The importance of technological change in explaining wide-ranging outcomes in future emissions has been highlighted by Alcamo *et al.*, (1995) and Grübler and Nakićenović (1996), among others. The latter reference also provides a critical assessment of the previous IS92 scenario series and its comparison to the literature. Prominent scenario studies of possible technological change in future energy systems in the absence of climate policies include Ausubel *et al.* (1988), Edmonds *et al.* (1994, 1996a), IIASA–WEC (1995), and Nakićenović *et al.* (1998). Future technology characteristics must therefore be treated as dynamic, with future improvement rates subject to considerable uncertainty. This is reflected in the SRES scenarios that adopt a wide range of improvement rates for energy extraction, conversion, and end-use technologies (Table 4-11). The actual representation of technological change in the six SRES models ranges from exogenously prescribed availability, through cost and performance profiles (which in some cases also include consumer or end-use costs for technology use), to stylized representation of learning processes.[31] Yet, as summarized in Chapter 3, model representations of technological change are poorly developed, although evolving rapidly.

---

[30] Resource availability assumptions also appear to be rather model specific in this scenario family. For instance, in many scenarios patterns of resource availability resemble the hypotheses retained by a particular model used for quantification of a marker scenario in one of the other three scenario families.

[31] Roehrl and Riahi (2000) provide a description of the methodology of representing technological change in MESSAGE as used here in the SRES scenarios.

***Table 4-11:** Summary of technology improvements for extraction, distribution, and conversion technologies assumed for the SRES scenarios. The classification reviews technology dynamics across the four marker scenarios and the four A1 scenario groups relative to each other. Illustrative, scenario-specific technology assumptions are discussed in the text. A1C and A1G have been combined into one fossil-intensive group A1FI in the SPM (see also footnote 1).*

	Technology Improvement Rates			
Scenario	**Coal**	**Oil**	**Gas**	**Non-fossil**
A1B	High	High	High	High
A2[a]	Medium	Low	Low	Low
B1[b]	Medium	Medium	Medium	Moderate–high
B2[c]	Low	Low–medium	Moderate–high	Medium
A1G	Low	Very high	Very high	Medium
A1C	High	Low	Low	Low
A1T	Low	High	High	Very high

[a] Technology improvement rates in the A2 scenario are heterogeneous among the world regions.

[b] B1: The assumed time-dependent learning coefficients range from 0.9 (i.e. a 10% reduction in the capital:output ratio on a doubling of cumulated production) for oil, 0.9–0.95 for gas, and 0.9–0.95 for surface coal mining to about 0.94–0.96 for non-fossil electric power generation options and 0.9–0.95 for commercial biofuels.

[c] In the specific model implementations, "inconvenience costs" of energy-end use, including social externalities costs, are expected to be particularly important for traditional coal technologies (e.g., underground mining, cooking with coal stoves).

4.4.7.1. A1 Scenarios

The A1B marker scenario represents the "balanced" technology development group of A1 scenarios; it assumes significant innovations in energy technologies, which improve energy efficiency and reduce the cost of energy supply. Consistent with the A1 scenario storyline, such improvements occur across the board and neither favor nor penalize particular groups of technologies. A1 assumes, in particular, drastic reductions in power-generation costs, through the use of solar, wind, and other modern renewable energies, and significant progress in gas exploration, production, and transport. For a different view, alternative scenario groups embedded within the overall A1 scenario family explore pathways of cumulative technological change; that is, path-dependent scenarios in which technologies evolve on mutually largely exclusive development paths. In general this has been the historical experience, in which the success of particular energy technologies (the steam engine in the 19th century, or internal combustion in the 20th) have "locked out" other technological alternatives. These scenario groups explore alternative spectra of technology dynamics in the domains of unconventional oil and gas, coal, as well as post-fossil technologies. Salient technology assumptions are described below.

Keeping in mind the very different degrees of technological detail and the mechanisms for technology improvements represented in the different models, a consistent inter-scenario comparison of technology assumptions is best achieved within the framework of one particular model. An overview of different technology developments for the scenario groups of the A1 scenario is given in Box 4-8 for the AIM model, which was also used to develop the A1B marker scenario. (A comparison with the MARIA model indicated that technology cost assumptions and their dynamics are quite congruent.) To illustrate differences in technology characteristics that drive the four different SRES scenario families, corresponding scenario-specific data based on MESSAGE data are presented at the end of this Section.

4.4.7.2. A1 Scenario Groups

As outlined above, besides the marker, three different groups of A1 scenarios were developed by the different modeling groups (combined into two in the SPM, see also footnote 1). In total, nine alternative runs are clustered in three scenario groups based on the AIM, MARIA, MESSAGE, and MiniCAM models.

In the A1G scenario group, technological change enables a larger fraction of the large occurrences of unconventional oil and gas, including oil shales, tar sands, and especially methane hydrates (clathrates) to be tapped. High technological learning and cost reduction effects could lower unconventional oil and gas extraction costs by approximately 1% per year and conversion technology costs by about factor of two (A1G-MESSAGE, see Roehrl and Riahi, 2000). As mentioned in Section 4.4.6, although these assumptions yield higher extractions of unconventional oil and gas resources, they are not sufficient to tap significant fractions of unconventional resources such as gas clathrates. Future scenario studies might reassess the current state of knowledge on possible technology development of these "exotic" fossil-fuel occurrences and the conditions under which they could become a major future source of unconventional hydrocarbon supply (and a massive source of carbon emissions). For the A1G scenario group, substantial improvements and extensions of the present pipeline grids and entirely new natural gas pipelines systems

Box 4-8: Technological Change in the AIM-based Quantifications for the A1 Scenario Family

The A1 storyline describes a world with rapid economic development. High economic growth results in pressures on resource availability, counterbalanced by technological progress, which is assumed to be highest among the four scenario families. In the AIM quantifications of the A1 storyline, rates of technological change are high both with respect to "supply push" factors (most notably RD&D) as well as with respect to "demand pull" factors (most notably high capital turnover rates). Since large resource availability and high incomes stimulate demand growth, technological change in energy supply receives a higher emphasis compared to changes in energy end-use technologies. Common technology assumptions in the A1 scenarios can be summarized as follows.

The supply of oil, gas, and biomass in the A1 scenario family is assumed to be very high and results from high rates of technological progress for fossil fuel and biomass exploitation technologies. Unconventional oil and gas, such as deep-sea methane hydrates, oil shale, etc., become available at relatively low cost. Also, large amounts of biomass are utilized through well-developed biomass farm plantations and harvest technologies, and biomass utilization technologies, such as biomass power generation and biofuel conversion technologies, become available at low costs through RD&D and other mechanisms of technology improvements (learning by doing and learning by using). High levels in the use of other renewable energy are reached when technologies for solar photovoltaics and thermal utilization, wind farms, geothermal energy utilization, and ocean energy are introduced at low cost. Energy end-use technologies are assumed to progress at medium rates compared with the fast rates of technological change in energy supply technologies.

The A1B marker and A1T scenarios assume drastic reductions in cost for solar, wind, and other renewable energies. A1C assumes lower coal costs and emphasizes coal exploitation technology progress and the introduction of advanced coal-fired power generation technology, such as integrated gasification combined cycle (IGCC). A1G assumes lower oil and gas costs than other A1 scenarios. The cost of nuclear power is assumed to be the lowest in A1G and A1T, and highest in A1C. The different cost assumptions that drive and result from technological change in the A1 scenario family are summarized in Table 4-12.

Table 4-12: *Technology costs (1990US$/GJ and 1990USCents/kWh) in AIM-based A1 scenarios.*

	Scenarios	2020	2050	2100
Coal (1990US$/GJ)	A1B	2.6	3.2	3.1
	A1C	1.5	1.5	1.1
	A1G	3.5	3.8	3.7
	A1T	2.6	2.8	2.8
Natural Gas, ConventionalA1B	2.0	1.9	1.4	
& Unconventional (1990US$/GJ)	A1C	3.5	5.0	4.6
	A1G	1.8	1.6	1.6
	A1T	1.5	1.5	1.4
Crude Oil, Conventional	A1B	7.3	10.1	14.9
& Unconventional (1990US$/GJ)	A1C	9.4	13.1	14.0
	A1G	7.3	8.2	8.4
	A1T	7.9	8.4	15.7
Nuclear (1990UScent/kWh)	A1	5.4	3.9	2.3
	A1C	5.7	4.4	2.8
	A1G	5.9	4.7	3.1
	A1T	5.6	4.1	2.5
Solar, wind, geothermal	A1B	12.2	5.9	2.0
electricity (1990UScent/kWh)	A1C	13.1	6.9	3.3
	A1G	15.2	9.3	5.2
	A1T	12.4	6.2	2.7

As mentioned above, improvements in energy efficiency on the demand side are assumed to be comparatively lower in the A1 scenario family, except for the A1T scenario, because the low energy prices give very little incentive to improve end-use energy efficiencies. Efficient technologies are not fully introduced into the end-use side, dematerialization processes in the industrial

(Box 4.8 continues)

Box 4.8 (continued)

sector are not promoted, lifestyles become energy intensive, and private motor vehicles are used more in developing countries as per capita incomes increase. As a result, energy efficiency improvement in the industrialized countries (IND) is around 0.8% per year, and in developing countries (DEV) it is 1.0% per year over the next 100 years to 2100. Only A1T assumes greater efficiency improvements (1.1% per year for IND and 1.5% per year for DEV), as a result of the diffusion of new highly efficient energy end-use devices such as fuel cell vehicles.

Technology progress is also assumed for land-use changes and sulfur emissions. Higher productivity increases in biomass and crop land (1.5% per year) in comparison to 0.5–1.0%) are assumed for the A1 world in the AIM quantification compared to those in the A2 and B2 scenario families. Desulfurization technologies could be introduced because of concerns of economic damage caused by acid rain and there would be strong financial support to install these technologies with the rapid income growth associated with the A1 world.

from Siberia and the Caspian to South East Asia, China, Korea, and Japan after 2010/2020 would be needed. Since unconventional oil and gas resources are distributed unevenly geographically, the scenario implies both capital-intensive infrastructure investments and unprecedented large-scale gas and oil trade flows. There is also little pressure to develop non-fossil alternatives in such scenarios, so costs of non-fossil alternatives remain comparatively high, even after significant technological improvements. For instance, solar electricity costs could drop to US$0.05 per kWh (A1G-AIM).

The high-growth coal-intensive scenario group A1C assumes relatively large cost improvements in new and clean coal technologies, such as coal high-temperature fuel cells, IGCC power plants, and coal liquefaction. More modest assumptions are made for all the other technologies, except for nuclear technologies in A1C-MESSAGE, as this requires zero-carbon options to ease resource and environmental constraints. The relative costs between coal and oil- or gas-related technologies also shift in A1C-AIM. Progress in renewables is also assumed to be substantial. For instance, solar photovoltaic costs would decline to USCents3/kWh (A1C-AIM).

In the dynamic technology scenario group A1T, technological change, driven by market mechanisms and policies to promote innovation, favors non-fossil technologies and synfuels, especially hydrogen from non-fossil sources. Liquid fuels from coal, unconventional oil and gas sources, and renewables become available at less than US$30 per barrel, with costs that fall further, by about 1% per year, through exploitation of learning-curve effects (A1T-MESSAGE). A1T-MARIA also projects declining costs for biofuels, from about US$30 to US$20, after the 2020 period (and in comparison to the A1-MARIA scenario biofuels substitute coal-derived synfuels). Non-fossil electricity (e.g., photovoltaics) begin massive market penetration at costs of about USCents1 to 3 per kWh (A1T-MARIA, A1T-MESSAGE, A1T-AIM), and could continue to improve further (perhaps as low as USCents0.1/kWh in A1T-MESSAGE) as a result of learning-curve effects. An important difference between the marker scenario A1B and the A1T group is that in A1T additional end-use efficiency improvements are assumed to take place with the diffusion of new end-use devices for decentralized production of electricity (fuel cells, microturbines). As a result, final energy demand in the A1T scenario group is between 30% (A1T-AIM, A1T-MESSAGE) and 40% (A1T-MARIA) lower compared to the A1B marker scenario.

4.4.7.3. A2 Scenarios

The A2 scenario family includes slow improvements in the energy supply efficiency and a relatively slow convergence of end-use energy efficiency in the industrial, commercial, residential, and transportation sectors between regions. A combination of slow technological progress, more limited environmental concerns, and low land availability because of high population growth means that the energy needs of the A2 world are satisfied primarily by fossil (mostly coal) and nuclear energy. However, in some cases regional energy shortages force investments into renewable alternatives, such as solar and biomass. For instance, intermittent renewable electricity supply options, such as solar and wind, are assumed to decline in costs to about USCents4/kWh and (because of storage requirements) to about twice that value when these intermittent sources are used for medium load applications (50% of electricity supply).

4.4.7.4. B1 Scenarios

Consistent with the general environmentally conscious and resource-conservation thrust of the B1 scenario storyline, technological change is largely directed at improving conversion efficiency rather than costs for fossil technologies. Within the SRES Terms of Reference, no additional climate initiatives are assumed that could bar the application of certain technologies or yield forced diffusion of others. The thermal efficiency of centrally generated electricity is assumed to rise to 45% (conventional coal) or to 65% (gas combined cycles) by 2100, while specific investment costs decline slightly from 1990 levels. It is assumed that subsidies on coal for electricity generation are removed entirely. A specific feature of the IMAGE model used to generate the B1 marker scenario is that it treats non-fossil electricity generation technologies as highly generic; for instance, it does not distinguish between nuclear, solar, or wind-power generation technologies. The specific investment costs of generation options for non-fossil electricity and of the production and conversion of commercial biofuels are assumed to fall by 5–10% for every doubling of cumulated production. Cost decreases down to USCents2.5/kWh are anticipated once non-fossil options penetrate on a large scale. The costs of gaseous biofuels in the major producing regions (Latin America, Africa, NIS) are assumed to be in the order of US$3 to 5 per GJ from 2020 to 2030 onward. Liquid biofuels are produced in small amounts in almost all regions at costs in the order of US$3 to 6 per GJ. In all regions a gradual transition occurs from fossil fuels to non-fossil options in electric-power generation, because of rising fuel prices and declining specific investment costs for fossil alternatives. Learning rates were assumed, conservatively, to yield 2 to 6% cost reductions for every doubling of cumulative production. The shift would start in resource-poor industrialized regions such as Japan and Western Europe, but is somewhat tempered by rising conversion efficiencies of fossil-fueled power plants. One of the factors that constrains the use of natural gas in the scenario is the assumption that only a limited part of the transport market is open to competition from non-liquid fuels (between 50% around 2050 to 80% around 2100). Also, the market share of coal in industry is fixed exogenously at 10 to 15% in some regions, to reflect the decreasing environmental and social attractiveness of the more "dirty" coal.

4.4.7.5. B2 Scenarios

The approach that underlies the B2 scenario storyline translates into important future improvements of technologies, albeit at more conservative rates than in scenarios A1 or B1, but with higher rates than in scenario A2. Compared to A1 and B1, cost improvements are more modest, because of the regionally fragmented technology policies assumed to characterize a B2 world. Hence, technology-spillover effects and benefits from shared development expenditures are more limited in the scenario. The high emphasis of environmental protection at the local and regional levels is reflected in faster development and diffusion of energy technologies with lower emissions, including advanced coal technologies, nuclear, and renewables. For instance, solar and wind electricity-generating costs are assumed to decline to USCents3/kWh, that is, a similar level as assumed for the long-term costs of advanced, clean coal technologies (such as IGCCs). As conventional oil supplies dwindle, initially high-cost synfuels from coal and also biofuels are introduced as substitutes. With increasing production volume, costs are assumed to decline from initial levels of some US$7/GJ to US$2.6/GJ. Conventional coal technologies undergo the lowest aggregate rates of improvement in the scenario and are also subject to increasing controls of social and environmental externalities (mining safety, particulates, and sulfur emissions). Increasingly, therefore, only advanced coal technologies are deployed. Nonetheless, extraction and conversion costs increase, especially in regions with a large share of deep-mined coal and in high population density agglomerations. In regions with abundant surface minable coal reserves (e.g., North America and Australia), coal extraction costs remain relatively low.

4.4.7.6. Harmonized and Other Scenarios

As a consequence of the "multi-model approach" used in SRES, detailed improvement assumptions and scenario implementations for individual technologies vary greatly from one model to another, although the same storyline characteristics were used as guiding principles and many scenarios share similar assumptions on improvement potentials for different technologies. Detailed quantitative comparisons are difficult because of different time profiles of technology improvements assumed in the different models, different representations of regional technology, and the modeling of the international diffusion of technology. For instance, many models assume aggregate regional rates of technological change (e.g., MARIA, MiniCAM, ASF), whereas others attempt to represent spatial and temporal diffusion patterns more explicitly (e.g., MESSAGE, AIM).

It is difficult to quantify the influence of varying technology-specific scenario assumptions on scenario outcomes, because in most model simulations the technology assumptions were varied in conjunction with other salient scenario characteristics, such as economic growth and resource availability (e.g. in the MiniCAM simulations). Therefore, the impact of alternative assumptions with respect to technological change can be best quantified within a particular scenario family and with "fully harmonized" scenario quantifications (i.e. with comparable energy demand), as discussed for the A1 scenario groups above. In some scenarios within other scenario families, technology-specific sensitivity analyses were performed, such as in the B2C-MARIA scenario variant of the B2-MARIA quantification. The main differences between the two scenarios are the respective costs of coal and nuclear power. In B2C-MARIA, the price of coal was assumed to be US$1.4/GJ, while that in B2-MARIA is US$1.8/GJ. In contrast, the capital costs of nuclear power stations are US$1400/kW in B2-MARIA, while those in B2C-MARIA are assumed to remain at US$1800/kW. Thus, even comparatively small variations in relative technology characteristics such as costs and efficiencies can lead to wide differences in scenario outcomes. As discussed in Chapter 5, for instance, changing the relative economics between coal and nuclear in the two MARIA scenarios results in a difference of more than 200 GtC cumulative emissions[32] over the 21st century

An illustration of inter-scenario variability in technology costs and diffusion is given in Box 4-9 for the MESSAGE model simulations for one representative scenario of each scenario family and scenario group. As stated above, differences in technology diffusion across scenarios are influenced by many more factors than just alternative technology characteristics and cost assumptions. Growth of energy demand, resource availability and costs, and local circumstances (local air-quality regulations that require desulfurization of fuels or stack

[32] Cumulative carbon emissions (all sources) are 1359 GtC for B2-MARIA and 1573 GtC for B2C-MARIA (see Chapter 5).

gases, or land availability and prices that influence biomass costs) are also important determinants of speed and potentials for the diffusion of new energy technologies.

4.4.8. *Prospects for Future Energy Systems*

In the energy systems models used to generate the scenarios reported here, the entire energy systems structure is represented from primary energy extraction, through conversion, transport, and distribution, all the way to the provision of energy services. Primary energy harnessed from nature (e.g., coal from a mine, hydropower, biomass, solar radiation, produced crude oil, or natural gas) is converted in refineries, power plants, and other conversion facilities to give secondary energy in the form of fuels and electricity. This secondary energy is transported and distributed (including trade between regions) to the point of final energy use. Final energy is transformed into useful energy (i.e., work or heat) in appliances, machines, and vehicles. Finally, application of useful energy results in delivered energy services (e.g., the light from a light bulb, mobility).

Important differences exist in accounting conventions on how to calculate the primary energy equivalent of particularly renewable and nuclear energy (see Watson *et al.*, 1996). To assure comparability of model results, the SRES writing team agreed to adopt as a common accounting methodology the

Box 4-9: Dynamics of Technological Change in the MESSAGE-Based Quantifications for the Four SRES Marker Scenarios.

Technological change in energy supply and end-use technologies has historically been a main driver of structural changes in energy systems, efficiency improvements, and improved environmental compatibility. Yet, despite its crucial role, the mechanisms that underlie technological innovation and diffusion of new technologies remain poorly understood, so modeling technological change as an endogenous process to the economy and society is still in its infancy. Historically, the track record of technology forecasts has at best been mixed, with a number of notable failures particularly in the energy sector. In the 1960s, for instance, R&D in the US attempted to develop nuclear-propelled aircraft, and nuclear electricity was anticipated to become "too cheap to meter." Conversely, the dynamic technological changes in microprocessors, information technologies, and aeroderivative turbines (and their combination with the steam cycle in the form of combined cycle gas turbines) were largely underestimated. This is similar to the pessimistic market outlook for gasoline-powered cars at the end of the 19th and start of the 20th centuries.

In recognition of the considerable uncertainty in describing future technological trends, a scenario approach was adopted to vary technology-specific assumptions in the MESSAGE model runs of the SRES scenarios. Depending on the specific interpretation of the four SRES scenario storylines, alternative technologies and alternative ranges of their future characteristics were assumed as model inputs.

Two guiding principles determined the choice of particular technology assumptions in MESSAGE.

First, technologies not yet demonstrated to function on a prototype scale were excluded. Therefore, for instance, nuclear fusion is excluded from the technology portfolio of all SRES scenarios calculated with the MESSAGE model. However, production of hydrogen- or biomass-based synfuels (e.g. ethanol) or advanced nuclear and solar electricity generation technologies are included, as they have demonstrated their physical feasibility at least on a laboratory or prototype scale, or in some specific niche markets (even if they are uneconomic at currently prevailing energy prices). Second, the range of technology-specific assumptions is empirically derived. Statistical distributions of technology characteristics based on a large technology inventory (consisting of 1600 technologies) and developed at IIASA (Messner and Strubegger, 1991; Strubegger and Reitgruber, 1995) were used. Means, maxima, and minima from these distributions (e.g. of estimated future technology costs) guided which particular values to adopt across scenarios on the basis of the scenario taxonomy suggested by the scenario storylines (ranging from conservative to optimistic).

Tables 4-13a to 4-13e summarize the technology characteristics and resultant diffusion rates across the four SRES scenario families and their scenario groups. Table 4-13a presents a brief overview of a selection of major energy technologies represented in the MESSAGE model. (Being a detailed "bottom-up" model, MESSAGE literally contains hundreds of individual technologies, too many to summarize here; instead, only the most important technology groups, aggregated across many individual technologies, are presented.) Table 4-13b summarizes salient technology characteristics in terms of levelized costs (investment and operating costs levelized per unit energy output, excluding fuel costs) and Table 4-13c summarizes the resultant marker deployment (diffusion) of these technologies by 2050 and 2100 for the B2-MESSAGE marker scenario. This scenario is characterized by intermediate levels of growth in energy demand and conservative assumptions as to future technological change. The latter were adopted based on a literature survey (Strubegger and Reitgruber, 1995) as well as an expert opinion poll.

In particular, the B2-MESSAGE scenario adopted technology characteristics of the equally conservative IIASA–WEC Scenario B (Nakicenovic *et al.*, 1998), which was based on the Strubegger and Reitgruber (1995) analysis, complemented by a review of some 100 energy experts assembled by WEC. Table 4-13d indicates how technology costs in the other MESSAGE scenarios differ from those of the B2 scenario. (The prevalence of negative values in Table 4-13d indicates that most scenarios are more optimistic concerning cost improvements of future technology than the MESSAGE B2 scenario.) Finally, Table 4-13e indicates the difference in market deployment (diffusion) of the other MESSAGE SRES scenarios compared to that of the B2 scenario. Positive values indicate higher market deployment, and negative ones show lower diffusion. However, differences across scenarios in terms of technology diffusion are not governed by technology costs alone. Other technology characteristics (such as efficiency and infrastructure availability) and market (demand) growth are also important in determining market deployment rates and diffusion potentials of energy technologies.

Simplifying the complex dynamic patterns of technological change across the scenario groups, one conclusion is that perhaps the single most important dichotomy of energy technologies of relevance for future GHG emissions is between advancements in "clean coal" and other fossil (e.g. methane hydrate) technologies (delivering electricity, gas, or synliquids) and those of decentralized hydrogen-powered fuel cells, combined with nuclear and renewable energy (for hydrogen production). The first, GHG-intensive technology cluster largely follows traditional centralized technological configurations of the energy sector. The latter represents both radical organizational and technological changes. The revolutionary change may well be less the hydrogen-powered fuel cell car itself, but rather that it could generate electricity when parked, dispensing entirely the need for centralized power plants and utilities.

Table 4-13a: *Overview of selected energy technologies represented in MESSAGE.*

Technology Aggregates	Including:
Centralized Electricity Generation:	
Coal conventional	Conventional coal power plants with DESOX (flue-gas desulfurization, FGD) and DENOX (flue-gas denitrification)
IGCC	Integrated coal Gasification Combined Cycle
Coal fuel cell	Coal-based high-temperature fuel cell (internal reforming)
Oil	New standard oil power plant (Rankine cycle, low NO_x and with FGD); existing crude oil and light oil engine-plants; light oil combined cycle power plants
Gas standard	Standard gas power plant (Rankine cycle, potential for cogeneration)
NGCC	Natural Gas-fired Combined Cycle power plant with DENOX
NGFC	Natural Gas-powered high-temperature Fuel Cell, cogeneration possibilities
Bio	New biomass-fired power plant (Rankine cycle, cogeneration possibilities); advanced biomass power plants (gasified biomass is burned in combined cycle gas turbines)
Nuclear	Conventional, existing nuclear power plants
Advanced nuclear/other	Nuclear high-temperature reactors for electricity and hydrogen coproduction, future inherently safe nuclear reactor designs, and other future zero-carbon electricity-generating technologies for base load
Hydro	Hydropower plants (low and high cost)
Wind	Wind power plant
Other renewables	Geothermal power plant (cogeneration potential); grid-connected solar photo-voltaic power plant (no storage); solar thermal power plants with storage, and solar thermal power plant for hydrogen production
Decentralized Electricity Generation:	
Hydrogen fuel cell	Decentralized stationary and mobile hydrogen fuel cells (cogeneration systems or off-hours electricity generation)
Photo-voltaics	On-site solar photo-voltaic power plant in the residential and/or commercial sectors, and in the industrial sector
Synfuels:	
Coal synliquids	Light oil and methanol production from coal
Biomass synliquids	Ethanol production from biomass
Gas synliquids	Methanol production from natural gas
Syngases	Syngases from various sources, including biomass and coal gasification
Hydrogen, H2(1)	Hydrogen production from fossil fuels (coal or gas)
Hydrogen, H2(2),(3),	Non-fossil hydrogen production: H2(2): from biomass and electricity, H2(3): from nuclear and solar

(Box 4.9 continues)

Box 4.9 (continued)

Table 4-13b: *Levelized costs (1990US$/GJ) of selected energy technologies (excluding fuel costs) in B2-MESSAGE (minima and maxima for eleven world regions).*

	1990		2050		2100	
	min	**max**	**min**	**max**	**min**	**max**
Coal conversion	3.6	7.5	4.4	7.8	4.4	7.8
IGCC	9.4	9.4	8.3	8.6	6.9	8.6
Coal fuel cell	11.9	11.9	11.9	11.9	11.9	11.9
Oil	3.9	28.9	3.3	5.3	3.3	5.3
Gas standard	3.6	8.3	3.9	4.7	3.9	4.7
NGCC	4.9	5.0	3.3	3.3	2.8	2.8
NGFC	8.4	8.4	6.7	6.7	6.7	6.7
Biofuel	5.8	9.2	5.8	8.3	5.8	8.3
Nuclear	6.7	9.7	7.2	9.7	7.2	9.7
Advanced nuclear/other	10.8	10.8	10.6	10.6	10.6	10.6
Hydro	2.5	15.8	2.5	22.2	2.5	22.2
Wind	15.8	15.8	9.4	9.4	9.4	9.4
Other renewables	6.4	29.8	7.2	10.8	7.2	10.8
Hydrogen fuel cell	8.4	8.4	6.7	6.7	6.3	6.3
Photo-voltaic	20.4	29.8	8.1	11.7	8.1	11.7
Coal synliquids	6.9	6.9	6.4	7.0	6.4	7.0
Biomass synliquids	7.1	7.1	4.8	4.8	4.8	4.8
Gas synliquids	3.7	3.7	2.6	2.6	2.6	2.6
Syngases	4.6	4.6	3.4	4.1	3.4	4.1
Hydrogen H2(1)	5.6	5.6	1.7	3.9	1.7	3.9
Hydrogen H2(2)	4.9	4.9	1.5	3.2	1.5	3.2
Hydrogen H2(3)	11.9	11.9	8.4	12.6	8.4	12.6

Table 4-13c: *Energy output (EJ) of selected energy technologies in B2-MESSAGE.*

	1990	2050	2100
Coal conversion	16.2	9.7	0.0
IGCC	0.0	15.9	65.1
Coal fuel cell	0.0	0.0	0.0
Oil	4.8	0.1	0.0
Gas standard	5.7	0.4	0.0
NGCC	0.6	45.0	72.7
NGFC	0.0	5.1	0.0
Biofuel	0.5	2.8	22.3
Nuclear	5.7	18.9	50.6
Advanced nuclear/other	1.3	27.9	88.7
Hydro	7.9	19.7	28.4
Wind	0.011	11.5	17.2
Other renewables	0.11	18.4	50.4
Hydrogen fuel cell	0.000	10.5	11.4
Photo-voltaic	0.001	25.2	57.3
Coal synliquids	0.0	4.2	71.8
Biomass synliquids	1.5	31.8	34.9
Gas synliquids	0.0	13.0	39.5
Syngases	0.0	0.1	0.0
Hydrogen H2(1)	0.0	36.4	0.0
Hydrogen H2(2)	0.0	10.4	0.0
Hydrogen H2(3)	0.0	0.0	0.0

Table 4-13d: *Levelized costs (1990US$/GJ) of selected energy technologies (excluding fuel costs) in MESSAGE scenarios relative to the costs in the B2-MESSAGE marker scenario (minima and maxima for eleven world regions). The A1C and A1G scenario groups have been combined into the fossil-intensive A1FI group in the SPM (see also footnote 1).*

	2050												2100											
	B1		A1B		A1C		A1G		A1T		A2[a]		B1		A1B		A1C		A1G		A1T		A2[a]	
	min	max	min	max	min	max	min	max	min	max	min	max	min	max	min	max	min	max	min	max	min	max	min	max
Coal conversion	0	0	0	0	0	0	0	0	0	0	-0.6	0	0	0	0	0	0	0	0	0	0	0	-0.6	0
IGCC	-0.8	-1.1	-0.6	-0.8	-0.6	-0.8	1.1	0.8	-0.6	-0.8	-0.8	-0.3	0.0	-1.7	0.3	-1.1	0.8	-0.8	2.5	0.8	0.3	-1.1	0	-0.6
Coal fuel cell	-2.2	-2.2	-2.2	-2.2	-2.2	-2.2	-0.3	-0.3	-2.2	-2.2	0	0	-2.5	-2.5	-2.5	-2.5	-2.5	-2.5	-0.3	-0.3	-2.5	-2.5	0	0
Oil	-1.1	0	-1.1	0	0	0	-1.1	0	-1.1	0	0.3	-0.6	-1.1	0	-1.1	0	0	0	-1.1	0	-1.1	0	0.3	-0.6
Gas Standard	0	0	0	0	0	0	0	0	0	0	0	0	0	0	0	0	0	0	0	0	0	0	0	0
NGCC	-1.4	-0.6	-1.4	-0.6	-0.3	0.6	-1.1	-0.6	-1.1	-0.6	-0.3	0.6	-0.8	0	-0.8	0	0.3	1.1	-0.6	0.0	-0.6	0	0	1.1
NGFC	-1.4	-1.4	-1.4	-1.4	0	0	-1.1	-1.1	-1.4	-1.4	0	0	-2.2	-2.2	-2.2	-2.2	0	0	-1.1	-1.1	-2.2	-2.2	0	0
Biofuel	-0.6	-1.7	-0.6	-1.7	0.0	0.0	-0.3	-1.1	-0.6	-1.7	0	0	-1.4	-2.8	-1.4	-2.8	0	0	-0.3	-1.1	-1.4	-2.8	0	0
Nuclear	0	0	0	0	1.4	0	0	0	0	0	0	0	0	0	0	0	1.4	0	0	0	0	0.0	0	0
Advanced																								
Nuclear/other	-2.2	0.6	-3.9	0.6	-1.1	3.3	-1.1	1.9	-3.6	0.6	0	0	-5.3	2.2	-6.4	-2.5	-2.2	3.3	-2.2	-0.3	-5.8	-1.4	0	0
Hydro	0	0	0	0	0	0	0	0	0	0	0.3	-5.6	0	0	0	0	0	0	0	0	0	0	0.3	-5.6
Wind	-2.8	-2.8	-4.7	-4.7	3.3	3.3	-2.8	-2.8	-4.7	-4.7	0.0	0.6	-4.2	-4.2	-6.4	-6.4	3.3	3.3	-2.8	-2.8	-6.4	-6.4	0.0	0.6
Other renewables	-3.3	-2.8	-4.4	-2.8	0.8	12.8	-3.9	-2.8	-4.4	-2.8	0.8	13.3	-5.3	-2.8	-6.1	-2.8	0.8	12.8	-4.2	-2.8	-6.1	-2.8	0.8	10.8
Hydrogen fuel cell	-1.3	-1.3	-1.2	-1.2	0	0	-0.9	-0.9	-1.5	-1.5	0	0	-2.1	-2.1	-1.8	-1.8	0	0	-0.8	-0.8	-2.3	-2.3	0	0
Photo-voltaic	-4.2	-5.8	-5.3	-7.5	8.1	11.9	0.0	0.0	-5.3	-7.5	0.6	11.1	-6.1	-5.8	-6.7	-9.4	8.1	11.9	0.0	0.0	-6.7	-9.4	0.6	11.1
Coal synliquids	-1.3	-1.8	-1.7	-0.9	-2.1	-0.9	-1.7	-0.9	-1.7	-0.9	-1.3	-0.9	-1.3	-1.8	-1.7	-0.9	-2.4	-0.9	-1.7	-0.9	-1.7	-0.9	-1.3	-0.9
Bio synliquids	-1.7	-1.7	-1.7	-1.7	0.0	0.0	-0.8	-0.8	-1.7	-1.7	0.0	0.0	-2.4	-1.7	-2.3	-1.7	0.0	0.0	-0.8	-0.8	-2.4	-1.7	0.0	0.0
Gas synliquids	-0.5	-0.5	1.1	1.1	0.0	0.0	1.0	1.0	1.1	1.1	0.0	0.0	-0.8	-0.5	1.1	1.1	0.0	0.0	1.0	1.0	1.1	1.1	0.0	0.0
Syngases	-0.5	-1.0	-0.5	-1.0	0.0	-0.5	-0.5	0.6	-0.5	-1.0	0.0	0.0	-0.6	-1.0	-0.5	-1.0	0.0	-0.5	-0.5	0.6	-0.6	-1.0	0.0	0.0
Hydrogen H2(1)	-0.3	-0.7	-0.3	-0.7	0.8	1.6	-0.3	-0.6	-0.3	-0.7	0.0	0.0	-0.7	-0.7	-0.7	-0.7	0.8	1.6	-0.3	-0.6	-0.7	-0.7	0.0	0.0
Hydrogen H2(2)	0.0	-0.4	-0.2	-0.4	0.3	1.4	0.0	-0.3	-0.2	-0.4	0.0	0.0	0.0	-0.4	-0.5	-0.4	0.1	1.4	0.0	-0.3	-0.5	-0.4	0.0	0.0
Hydrogen H2(3)	-3.0	-1.6	-5.0	-7.4	5.5	1.3	-8.4	-12.6	-5.0	-2.3	7.1	8.6	-3.0	-1.6	-5.5	-7.4	5.5	1.3	-8.4	-12.6	-5.5	-2.3	7.1	8.6

[a] Cost variations refer to a four-region model only. The spread across regions is therefore somewhat smaller than in the other scenarios.

Box 4.9 continues

Box 4.9 (continued)

***Table 4-13e**: Energy output (EJ) of selected energy technologies in MESSAGE scenarios relative to the B2-MESSAGE marker scenario. The A1C and A1G scenario groups have been combined into the fossil-intensive A1FI group in the SPM (see also footnote 1).*

	2050						2100					
	B1	**A1B**	**A1C**	**A1G**	**A1T**	**A2[a]**	**B1**	**A1B**	**A1C**	**A1G**	**A1T**	**A2[a]**
Coal conversion	-8.6	6.5	36.0	26.9	4.7	16.9	0.0	1.8	6.5	7.5	0.1	7.2
IGCC	-15.6	8.8	-4.7	-2.9	-12.3	25.2	-65.1	-26.3	-17.6	-65.1	-65.1	62.7
Coal fuel cell	0.0	2.2	12.9	2.4	0.0	0.0	0.0	0.1	204.9	0.0	0.0	0.0
Oil	0.0	0.0	0.0	7.8	0.0	0.0	0.0	0.0	0.0	1.0	0.0	0.0
Gas standard	-0.3	-0.3	-0.3	3.8	0.7	-0.3	0.0	0.0	0.0	0.0	0.0	0.0
NGCC	-15.1	34.4	-7.7	13.0	5.5	-18.1	-64.7	102.8	-67.8	143.4	-43.5	-2.0
NGFC	-5.1	10.0	10.3	26.2	-1.6	-3.6	0.0	0.0	0.0	60.5	0.0	0.0
Biofuel	-2.2	5.0	6.7	12.1	0.4	-0.5	-19.1	55.3	-19.3	-13.4	-15.9	3.9
Nuclear	-13.1	-9.2	-11.1	-5.3	-11.3	13.4	-50.0	-41.2	-50.6	-22.5	-50.6	31.2
Advanced nuclear/other	-4.0	44.8	83.6	47.2	44.8	-14.7	-52.0	255.0	253.5	211.1	16.2	-38.4
Hydro	0.4	8.7	7.2	7.3	3.5	1.4	-3.6	11.3	12.5	8.2	-2.6	3.8
Wind	-0.2	5.5	-0.6	2.5	4.4	1.7	-7.9	14.6	1.7	1.9	0.3	1.0
Other renewables	1.9	9.2	-12.8	3.0	8.5	-10.0	3.8	21.2	-10.4	2.0	18.8	7.2
Hydrogen fuel cell	57.0	5.2	-10.5	9.8	41.3	-8.7	84.2	92.7	-11.4	28.5	309.7	-9.4
Photo-voltaic	-1.8	12.9	8.5	14.2	9.7	-2.9	-29.3	58.3	42.2	45.9	34.3	-0.4
Coal synliquids	1.2	34.7	177.0	29.3	24.8	48.4	-69.4	-57.0	421.3	-48.9	-57.7	258.0
Biomass synliquids	-5.6	-30.4	1.3	-30.7	-22.2	-17.1	-9.0	-34.6	-6.5	7.9	-20.4	-34.9
Gas synliquids	-3.3	-3.2	24.7	-6.5	-5.3	-5.1	-27.9	-33.1	137.3	-5.1	-29.1	64.0
Syngases	0.5	6.4	2.0	0.5	-0.1	0.5	60.6	854.7	84.3	0.0	990.2	35.6
Hydrogen H2(1)	75.2	-8.4	-24.3	45.9	-6.5	-15.2	0.0	0.0	0.0	0.0	0.0	0.0
Hydrogen H2(2)	17.3	-5.6	-9.6	-10.1	-9.5	-5.8	0.0	0.0	0.0	0.0	0.0	0.0
Hydrogen H2(3)	62.7	97.9	5.1	0.0	125.0	0.0	0.0	0.0	0.0	0.0	0.0	0.0

[a] Calculated with a four-region model. The spread across regions is therefore somewhat smaller than in the other scenarios.

Table 4-14: *Primary energy use (EJ) for the four SRES marker scenarios and all SRES scenarios in 1990, 2020, 2050, and 2100. The range for 1990 illustrates the differences in base-year calibration across the models and uncertainties that stem from the inclusion or exclusion of non-commercial energy use, which is particularly important for developing countries.*

Primary Energy (EJ) by World and Regions

		2050				2100			
Region	**1990**	**A1**	**A2**	**B1**	**B2**	**A1**	**A2**	**B1**	**B2**
OECD90	151-182	267	266	166	236	397	418	126	274
		(184-315)	(207-300)	(134-233)	(189-236)	(181-607)	(267-496)	(126-274)	(197-274)
REF	69-95	103	93	64	97	139	155	39	125
		(83-267)	(57-116)	(50-79)	(53-117)	(70-290)	(61-457)	(25-80)	(40-328)
IND	227-252	370	359	230	334	536	573	164	399
		(303-532)	(264-406)	(203-303)	(255-339)	(275-896)	(385-847)	(164-345)	(237-593)
ASIA	49-79	440	335	272	319	838	581	154	521
		(293-789)	(249-449)	(204-537)	(284-411)	(308-965)	(477-753)	(154-434)	(309-562)
ALM	35-49	538	278	312	217	852	563	196	437
		(235-634)	(166-354)	(176-312)	(137-254)	(391-1109)	(437-662)	(196-446)	(300-538)
DEV	84-123	977	612	583	536	1639	1144	350	959
		(606-1278)	(415-740)	(406-837)	(421-660)	(700-2074)	(914-1375)	(350-880)	(609-1096)
WORLD	326-368	1347	971	813	869	2226	1717	514	1357
		(913-1611)	(679-1059)	(642-1090)	(679-966)	(1002-2737)	(1304-2040)	(515-1157)	(846-1625)

direct equivalent method for all non-thermal uses of renewables and nuclear. The primary energy equivalence of these energy forms is accounted for at the level of secondary energy, that is, the first usable energy form or "currency" available to the energy system. For instance, the primary energy equivalence of electricity generated from solar photovoltaics or nuclear power plants is set equal to their respective gross electricity output, not to the heat equivalent of radiation energy from fissile reaction, the solar radiance that falls onto a photo-voltaic panel and is converted into electricity with efficiencies that range from 10% to 15%, or the heat that would have to be generated by burning fossil fuels to produce the same amount of electricity as generated in a photo-voltaic cell or a nuclear reactor (as used in the so-called "substitution" accounting method). This common[33] SRES accounting convention must be borne in mind when comparing the primary energy-use figures of this report with those of other studies, which invariably use different accounting methods depending on the organization that produces the scenario. An illustration of the sensitivity of different accounting methods on estimates of primary energy use in long-term energy scenarios is given in Nakićenović *et al.* (1998). (See also the discussion in Chapter 2, in which scenario comparisons are based on index numbers rather than absolute figures to account for these definitional differences.)

Table 4-14 gives an overview of primary energy use in the four SRES marker scenarios and the range of all SRES scenarios.

Figure 4-11 illustrates both the historical change of world primary energy structure over time and future changes as given in the SRES scenarios. Each corner of the triangle corresponds to a hypothetical situation in which all primary energy is supplied by a single source – oil and gas at the top, coal at the left, and non-fossil sources, renewables (including wood), and nuclear at the right. The historical change reflects major technology shifts from the traditional use of renewable energy flows to the coal and steam age of the 19th century, and subsequently to the dominance of oil and internal combustion engines in the 20th century. In around 1850 (lower right of Figure 4-11), only about 20% of world primary energy was provided by coal; the other 80% was provided by traditional renewable energies (biomass, hydropower, and animal energy). With the rise of industrialization, coal substituted for traditional renewable energy forms, and by 1910 (lower left of Figure 4-11) around three-quarters of world primary energy use relied on coal. The second major transition was the replacement of coal by oil and later by gas. By the early 1970s (see 1970 point labeled on Figure 4-11), 56% of global primary energy use was based on oil and gas. From the early 1970s to 1990, the global primary energy structure has changed little, although efforts to substitute for oil imports have led to an increase in the absolute amount of coal used and to the introduction of non-fossil alternatives in the OECD countries (e.g., nuclear energy in France). Rapid growths in energy demand and coal use, particularly in Asia, have outweighed structural changes in the OECD countries.

[33] Adopting a common accounting convention avoids misrepresentation of the contribution of renewable and other new energy forms, which can be both under- or over-represented by inconsistent accounting conventions, as continues to be the case in energy statistics and scenario studies.

Figure 4-11 also gives an overview of the evolution of the global energy system between 1990 and 2100 as reflected in the SRES scenarios. The four marker scenarios are shown as thick lines. In addition, for each scenario family the area spanned by all the SRES scenarios in that family is marked in the same color as the trajectory for the respective marker. The SRES scenarios cover a wider range of energy structures than the previous IS92 scenario series, reflecting advances in knowledge on the uncertainty ranges of future fossil resource availability and technological change. Scenarios B1, B2, A1T, and to some extent A1B follow a trend toward increasing shares of zero-carbon options in the long term. A1G more or less follows an oil–gas isoshare line that perpetuates the current dominance of oil and gas in the global energy balance far into the 21st century. Scenarios in group A1C indicate a near doubling of coal's share in primary energy use. Also of interest is the trajectory of the A2 marker scenario, which returns in its energy structure by 2100 (over 50% coal share) to the situation that prevailed almost 200 years before (i.e., around 1900). However, even with similar fuel shares, the technologies, end-use fuels, and applications projected in the A2 scenario are radically different from those of the past.

4.4.8.1. A1 Scenarios

The most significant change in the long-term primary energy mix in the A1B-AIM marker scenario is the fast market penetration of (new) renewable energy. Its share increases from

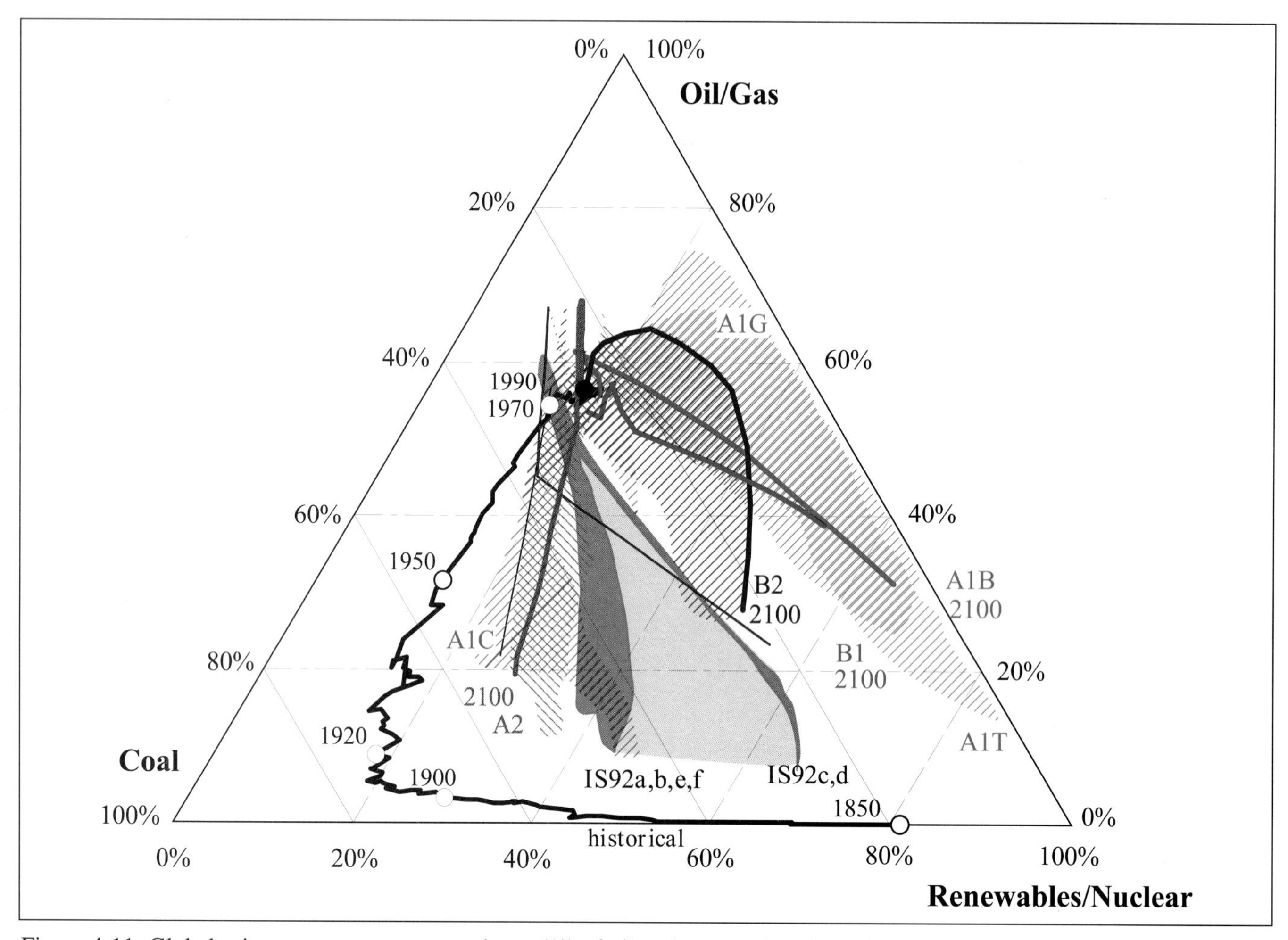

Figure 4-11: Global primary energy structure, shares (%) of oil and gas, coal, and non-fossil (zero-carbon) energy sources – historical development from 1850 to 1990 and in SRES scenarios. Each corner of the triangle corresponds to a hypothetical situation in which all primary energy is supplied by a single source – oil and gas, coal at the left, and non-fossil sources (renewables and nuclear) to the right. Constant market shares of these energies are denoted by their respective isoshare lines. Historical data from 1850 to 1990 are based on Nakićenović *et al.* (1998). For 1990 to 2100, alternative trajectories show the changes in the energy systems structures across SRES scenarios. They are grouped by shaded areas for the scenario families A1, A2, B1, and B2 with respective markers shown as lines. In addition, the four scenario groups within the A1 family (A1, A1C, A1G, and A1T) that explore different technological developments in the energy systems are shaded individually. The A1C and A1G scenario groups have been merged into one fossil-intensive A1FI scenario group in the SPM (see footnote 1). For comparison the IS92 scenario series are also shown, clustering along two trajectories (IS92c,d and IS92a,b,e,f). For model results that do not include non-commercial energies, the corresponding estimates from the emulations of the various marker scenarios by the MESSAGE model were added to the original model outputs.

a current 3% (excluding traditional non-commercial biomass use) to some 66% by 2100. Given the assumption of rapid technology progress, the costs of modern renewable energy technologies (solar, wind, commercial biomass, etc.) decline significantly in the long-term (see Section 4.4.7). Such low costs could make solar energy the largest primary energy source by 2100. Commercial, "high-tech" biomass also increases substantially, and contributes 18% of primary energy supply by 2100 globally. In the meantime, the shares of coal and oil decrease from 25% and 36% in 1990 to 12% and 15%, respectively, by 2050, and decline thereafter either because of depletion of conventional oil resources or because of fast market penetration of post-fossil technologies (in the case of coal). Gas increases its market share initially (from 20% in 1990 to 33% by 2050) and declines thereafter, but still maintains an important market share (24%) by the end of the 21st century. Nuclear is mainly a transient "backstop" technology – its share increases from 2% in 1990 to some 10% by 2050, and declines to 4% by 2100 because its economics increasingly fall behind those of new renewables. Overall, consistent with the high-income characteristics of the A1 scenario family, the share of traditional, non-commercial biomass use declines. By 2100 its use has virtually disappeared.

4.4.8.2. A1 Scenario Groups

Embedded in the overall storyline of the A1 scenario family are the possible widely ranging technological bifurcations, each of which spans a "corridor" of the future evolution of primary energy shares (see Figure 4-11). The A1C scenario group spans a range of structural change in future energy supply between the extremes of the A2 marker scenario and of the previous coal-intensive IS92 scenarios (IS92a,b,e,f). Conversely, the A1T scenario group spans a range of structural change in future energy systems delineated by the A1B and B1 marker scenarios, respectively. In all of these the global energy system completes a structural shift initiated with the onset of the Industrial Revolution. This could draw to a close more than 100 years from now, around 2100, with an energy system that predominantly relies on non-fossil energy sources, but evidently with a radically different technology portfolio of "high-tech" non-fossil energy compared to the "low-tech" non-fossil energy of 1800. The second major difference is that even with somewhat similar primary energy structures, *absolute* levels of demand would have increased by a factor 200 – 2000 EJ by 2100 in the A1T scenario, compared to a mere 10 EJ in 1800. Compared to all other scenarios, the A1G scenario group represents a distinct cluster in which the current dominance of oil and gas is perpetuated throughout much of the 21st century. This scenario cluster is somewhat difficult to discern in Figure 4-11, because of the absence of structural change as a result of which the scenario cluster in terms of primary energy shares moves horizontally along the top part of the energy triangle. The only long-term fuel substitution that takes place is between coal and non-fossil alternatives.

4.4.8.3. A2 Scenarios

Major global trends in the A2 (ASF) marker scenario include an increase in the coal share (from 29% in 1990, through 30% by 2050, to 53% in 2100) and a reduction in the share of conventional oil (from 43% in 1990 to 23% by 2050, from where it declines asymptotically toward zero at the end of the 21st century). The progressive depletion of oil resources in the scenario reflects the prevailing view as to the finiteness of conventional oil resources,[34] a picture also confirmed by other quantifications of the storyline (see discussion below). Nonetheless, the decline in oil market share should not be interpreted as a physical "running out," but rather as a gradual replacement process via price competition with synfuels and other alternatives as oil prices rise, along with the need to access ever more remote and expensive petroleum deposits. Once oil becomes increasingly expensive (starting from 2030) a substantial proportion of coal is converted into synthetic liquid fuels. Nuclear and renewable energy sources gradually increase in importance from 1990 to 2100, while the share of natural gas remains almost constant. Substantial changes in the primary energy mix also occur at the regional level. Coal shifts from a currently dominant primary energy source only in ASIA and becomes the most important fuel in all the SRES regions. Natural gas remains the second or the third major fuel. Nuclear energy becomes important in OECD90 and ASIA regions, and biomass and other renewables in the ALM region.

4.4.8.4. Harmonized and Other A2 Scenarios

The shares of different fuel types in all the A2 scenarios, except A2-IMAGE, are close to that of the marker scenario. As in the A2 marker, and with the exception of A2-IMAGE, coal becomes a major fuel by 2100 (45 to 50% of primary energy) in all the scenarios, followed by renewables (19 to 31%) and natural gas (9 to 18%). A higher share of renewables (biomass plus other) and a lower share of nuclear is the major difference between the scenarios (e.g., A2-MESSAGE shows the lowest nuclear share in primary energy of 7% in 2100). The A2G-IMAGE scenario has a quite different energy supply structure in which natural gas is the most important source, followed by renewables and coal, as it was designed to explore the implications of larger gas availability in an A2 world. Table 4-15 presents the primary energy structure of different A2 scenarios.

4.4.8.5. B1 Scenarios

The B1 marker scenario in the IMAGE model describes a structural transition in energy systems toward increasing shares of non-fossil energy. This long-term transition features an interim reliance on fossil fuels, in particular with natural gas as the preferred transitional fuel. Structural changes in energy

[34] Nonetheless, cumulative oil extraction to 2100 equals 2.7 times the currently identified reserves of conventional oil in the scenario.

Table 4-15: *Global primary energy supply structure (%) in the A2 scenarios for the year 2100. (See also Figures 4-8 to 4-10 for cumulative fossil resource use.)*

	ASF	AIM	MiniCAM	IMAGE [a]	MESSAGE
Coal	53	50	50	18	45
Oil	<1	3	3	14	2
Gas	19	18	9	35	15
Nuclear	14	11	11		7
Biomass	9	11	8	6	16
Other renewables	5	8	18	26	15
Total	100	100	100	100	100

Note that columns may not add due to independent rounding.
[a] A2-IMAGE "Other renewables" category includes nuclear and renewable sources except for biomass.

supply are comparatively fast because of both the dynamic outlook on energy-efficiency improvements and structural change and the "dematerialization" of economic activities, characteristic of the B1 world. With lower energy demand than in the other scenarios, technological innovation and diffusion initially translate into slower rates of structural change in primary energy supply (lower demand growth leads to lower investment in capacity expansion and hence fewer opportunities for technological learning). However, once underway, structural change translates into radical systems restructuring in the long term (i.e. post-2050). A persistent long-term trend of B1 is also the continuously declining share of coal in the global primary energy mix, caused by local and regional environmental considerations (airborne emissions, social and environmental impacts from large-scale mining activities, etc.). Until 2050, the global energy system remains fossil-fuel dominated (with an important shift away from coal to gas use within fossil fuels). By 2020, fossils still account for some 79% of global primary energy (23% coal, and 56% oil and gas combined). By 2050, fossils account for 69% (20% coal and 49% oil and gas). By 2100, however, the transition away from fossil fuels is well underway; they only account for some 47% of primary energy use, mostly natural gas. The non-fossil share mirrors in its growth the declining trajectory of fossil fuels. Generally, even this scenario of significant structural change illustrates the long lead times needed for an "orderly" transition away from the current dominance of fossil fuels. Energy efficiency and "dematerialization" of the economy are integral parts of this transition in the B1 scenario.

A distinguishing feature of the IMAGE model used to develop the B1 marker scenario is the generic treatment of non-fossil fuel alternatives, which recognizes that the technology portfolio is particularly diverse and that numerous combinations are possible because of regional resource endowments, economics, technology policies, etc. A second important characteristic of the IMAGE modeling approach is the link between energy sector investments with future improvements in the form of a learning-curve approach. The model structure and methodology is described in more detail in Alcamo *et al.* (1998) and de Vries *et al.* (1994, 1999, 2000), (see also Appendix IV).

4.4.8.6. Harmonized and Other B1 Scenarios

Other models have (in most instances[35] with harmonized input assumptions) examined the uncertainties associated with the structural shift patterns described by the B1 scenario family. Whereas long-term trends all point in similar directions, considerable scenario variability remains, a function of differences in energy demand, resource availability, and technology assumptions (Table 4-16 and see Sections 4.4.6 and 4.4.7). Relatively robust patterns across all B1 scenarios include the continued importance of oil until about 2050 (15 to 28%) and a subsequent decline thereafter. Three scenarios (B1-ASF, B1-AIM, and B1-MESSAGE) indicate a more rapid decline to less than 10% market share by 2100, and three scenarios (the B1-IMAGE marker, B1-MARIA, and B1-MiniCAM) depict higher oil shares between 16% and 20% by 2100, largely as a function of higher oil resource availability assumptions (see Section 4.4.6). For gas, all scenarios suggest a relatively robust market share range of between 21% and 39% by 2050 and between 17% and 39% by 2100. Coal, biomass, and other non-fossil sources are largely substitutes for each other, depending on the specific cost assumptions used in the models. Whenever nuclear, renewable, and biomass costs are low, the share of coal declines (e.g., B1-AIM, B1-MARIA, B1-MESSAGE) and that of other sources increases, the degree of interfuel substitution being scenario (i.e. model) specific. Drastic shifts, however, are not anticipated before 2050. In most scenarios (except B1-MARIA and B1-MESSAGE) coal's share is around 20% by 2050, and declines to below 10% by 2100 as biomass and other non-fossil sources gain respective market shares. Their diffusion is described more conservatively in the B1-ASF scenario, in which coal maintains a market share of about 22% until 2100.

[35] Except B1-MARIA and B1High-MiniCAM.

4.4.8.7. B2 Scenarios

The B2 (MESSAGE) marker scenario first follows a trend toward increasing shares of gas, followed by renewables, and finally – as oil and gas start to become scarce – increasingly returns to coal. By 2100, the B2 scenario ends up somewhere in the middle of the triangle in Figure 4-11 (i.e., it relies on a broad, diversified mixture of different primary energy sources). This global diversification results from heterogeneous trends at the regional level and is largely a function of the more modest assumptions concerning technology improvements and oil and gas resource availability (compared to other scenario families, in particular A1 and B1) that are characteristic of the B2 scenario family. By 2100, the main primary energy carriers are biomass (23%), coal (22%), oil and gas (29%), and other renewables and nuclear (26%). Countries with low income and high resource availability continue to rely on fossil fuels up to the end of the 21st century, such as China (mainly coal), Former Soviet Union (mainly gas), and Middle East (first oil and later gas). Regions with low resource availability, such as Africa and South America, rely on renewables and nuclear. The decreasing share of coal and oil in the primary energy structure of OECD countries is substituted by the growing share of renewables, gas, and nuclear. A major characteristic of the B2 scenario is the increasing importance of synthetic liquid fuels in the second half of the 21st century, because of a continuous phase out of conventional oil in all regions.

4.4.8.8. Harmonized and Other B2 Scenarios

Alternative B2 scenarios show a great diversity in changes in energy systems structures compared to the B2 marker. Common to all scenarios is their gradual transition away from conventional oil and gas, which are assumed to be comparatively scarce in the B2 scenario storyline. However alternative scenarios depict very different trends for this structural change, ranging from increased reliance on coal and coal-derived synfuels (B2-ASF, B2High-MiniCAM) to more biomass- and nuclear-intensive scenarios (B2-AIM, B2-MARIA, and B2-IMAGE). Generally, this reflects the considerable uncertainty as to direction and pace of technological change in the technologically more fragmented world described in the B2 scenario storyline. B2-MiniCAM anticipates a strong reliance on oil and natural gas as transitional fuels, with a share in primary energy of about 50% over the next 100 years to 2100 (i.e., gas shares in B2-MiniCAM are as high as in the A1G scenario group). In turn, B2-ASF suggests an increasing reliance on coal and coal-derived synfuels. A third group of scenarios tends to follow similar directions of structural change as those of the B2-marker – a gradual introduction of post-fossil alternatives (with different weights for nuclear and renewables as a function of technological progress), along with gas (or in some scenarios coal-derived synfuels) as transitional technology options. Structural changes in energy systems of the various B2 scenarios largely follow the main directions of the marker scenario developed with a particular model. That is, differences in alternative B2 scenarios appear to relate strongly to differences in model parametrizations derived from the respective marker scenario runs, most notably in the domains of resource availability and technology (see Sections 4.4.6 and 4.4.7). Thus, B2-ASF depicts structural changes in energy technologies and systems akin to the trends of the A2 marker scenario, whereas B2-AIM and B2-IMAGE largely follow the patterns of change of their marker scenarios (A1 and B1, respectively). Alternative patterns of change are illustrated by the B2-MiniCAM and B2-MARIA scenarios, which have also explored scenario sensitivities by developing alternative B2 scenario quantifications (B2C-MARIA, B2High-MiniCAM) that show a higher reliance on coal (and hence higher GHG emissions).

4.4.9. *Land-Use Changes*

The main driving forces for land-use changes are related to increasing demands for food because of a growing population and changing diets. In addition, numerous other social, economic, and institutional factors govern land-use changes, such as deforestation, expansion of cropland areas, or their re-conversion back to forest cover (see Chapter 3). Global food production can be increased, either through intensification (e.g., using multi-cropping, raising cropping intensity, applying fertilizers, new seeds, improved farming technology) or through land expansion (e.g., cultivating land, converting forests). Especially in less developed countries, many examples show the potentials for intensification of food production in a more or less ecological way that may not lead to higher GHG emissions (e.g., multi-cropping; agro-forestry).

Different assumptions on these processes translate into alternative scenarios of future land-use changes and GHG emissions, most notably CO_2, methane (CH_4), and nitrous oxide (N_2O). A distinguishing characteristic of several models (e.g., AIM, IMAGE, MARIA, and MiniCAM) used in SRES is the explicit modeling of land-use changes from expanding biomass uses, and hence exploration of possible land-use conflicts between the energy and agricultural sectors. The corresponding scenarios of land-use changes are summarized in Table 4-17 and Figure 4-12 for the four SRES marker scenarios and all SRES scenarios. The distinction in scenario groups is related to the energy system and is thus not relevant in this section.

As discussed further in Chapter 5, model treatment of land-use change and base-year parameterization differ substantially. Therefore, comparisons between different models can yield substantial differences. Land-use change assumptions for each of the marker scenarios are described below. More detailed inter-model comparisons of land-use change and emissions models, as well as a deeper analysis of potentials and rates of change of main driving force variables, such as agricultural productivity growth and dietary changes, remain an important area for future research.

Table 4-16: *Global primary energy supply structure (%) in the B1 scenarios for 2050 and 2100. Ranges indicate more than one B1 scenario calculated with a particular model, typically to explore uncertainties in resource availability and technology assumptions. (See also Figures 4-8 to 4-10 for cumulative fossil resource use.)*

	ASF		AIM		IMAGE		MARIA		MESSAGE [b]		MiniCAM [b]	
	2050	2100	2050	2100	2050	2100	2050	2100	2050	2100	2050	2100
Coal	37.0	21.7	22.9	8.4	20.5	8.5	2.0	0.5	4.4-7.0	0.3-5.2	18.1-19.8	6.9-7.3
Oil	14.7	0.2	15.4	11.6	28.0	19.3	27.1	16.5	20.9-23.2	4.5-6.7	15.5-22.4	19.9-27.4
Gas	30.7	23.2	34.5	19.4	21.3	20.0	36.6	21.3	31.5-35.4	21.1-28.4	25.8-39.0	22.0-39.4
Biomass	9.4	25.0	13.1	19.1	11.6	13.0	19.8	28.2	13.5-15.0	18.0-31.1	3.6-8.6	3.2-6.7
Other non-fossil [a]	8.2	29.9	14.1	41.5	18.6	39.2	14.5	33.5	22.9-25.5	31.6-51.8	18.8-28.5	26.8-40.6
Total	100	100	100	100	100	100	100	100	100	100	100	100

Note that columns may not add due to independent rounding.

[a] Following the B1-IMAGE marker disaggregation that treats nuclear and other renewable electricity sources as a generic technology.

[b] Min/max percentages do not add up to 100% as they describe different scenarios.

Table 4-17: *Global land cover in 1990, and land-use changes between 1990 and 2050, and 2050 and 2100 (in million ha) for the four SRES marker scenarios, and ranges across all four scenario families for all SRES scenarios (minimum and maximum in brackets). In particular, the different model representations of related processes among the various models are shown. MESSAGE (B2 marker) does not include a land-use change and related GHG emissions module, ASF (A2 marker) models changes in carbon fluxes only, whereas AIM and IMAGE (A1 and B1 markers, respectively) model both land-use changes and related emissions. Appropriate land-use change and emission scenarios calculated with alternative models with consistent socio-economic driving-force assumptions have been adopted for the corresponding scenarios based on the judgment of the individual modeling teams that developed the respective marker scenarios. In these cases (A2 and B2), land-use change scenarios represent first-order approximations only.*

	Land-Use (million ha)	Land-Use Change (million ha)							
		1990–2050				1990–2100			
Type	1990	A1	A2	B1	B2	A1	A2	B1	B2
Cropland	1434–1472	–17	n.a.	–7	167	–39	n.a.	325	–394
		(–113, +904)	=(–187, +267)	(–305, +461)	(–49, 628)	(–826, –39)	(–422, +420)	(–979, –30)	(–582, 325)
Grassland	3209–3435	109	n.a.	–650	155	188	n.a.	–1537	307
		(–794, +1714)	(+194, +1218)	(–650, +1335)	(–491, +1331)	(–1087, +622)	(+313, +1262)	(–1537, +320)	(–491, +823)
Energy Biomass	0–8	418	n.a.	263	288	495	n.a.	196	307
		(+12, +745)	(+18, +311)	(0, +260)	(0, +288)	(+3, +1932)	(+67, +396)	(0, +1095)	(+4, +597)
Forests	4138–4296	–106	n.a.	274	57	–92	n.a.	1260	227
		(–1146, +175)	(–778, +302)	(–667, +274)	(–732, +57)	(–464, +480)	(–673, –19)	(274, +1266)	(–116, +227)
Others	3805–4310	–405	n.a.	122	–667	–552	n.a.	482	–1166
		(–1072, +15)	(–833, –431)	(–579, +122)	(–667, -98)	(–873, +566)	(–1085, –278)	(–983, -482)	(–1166, –137)

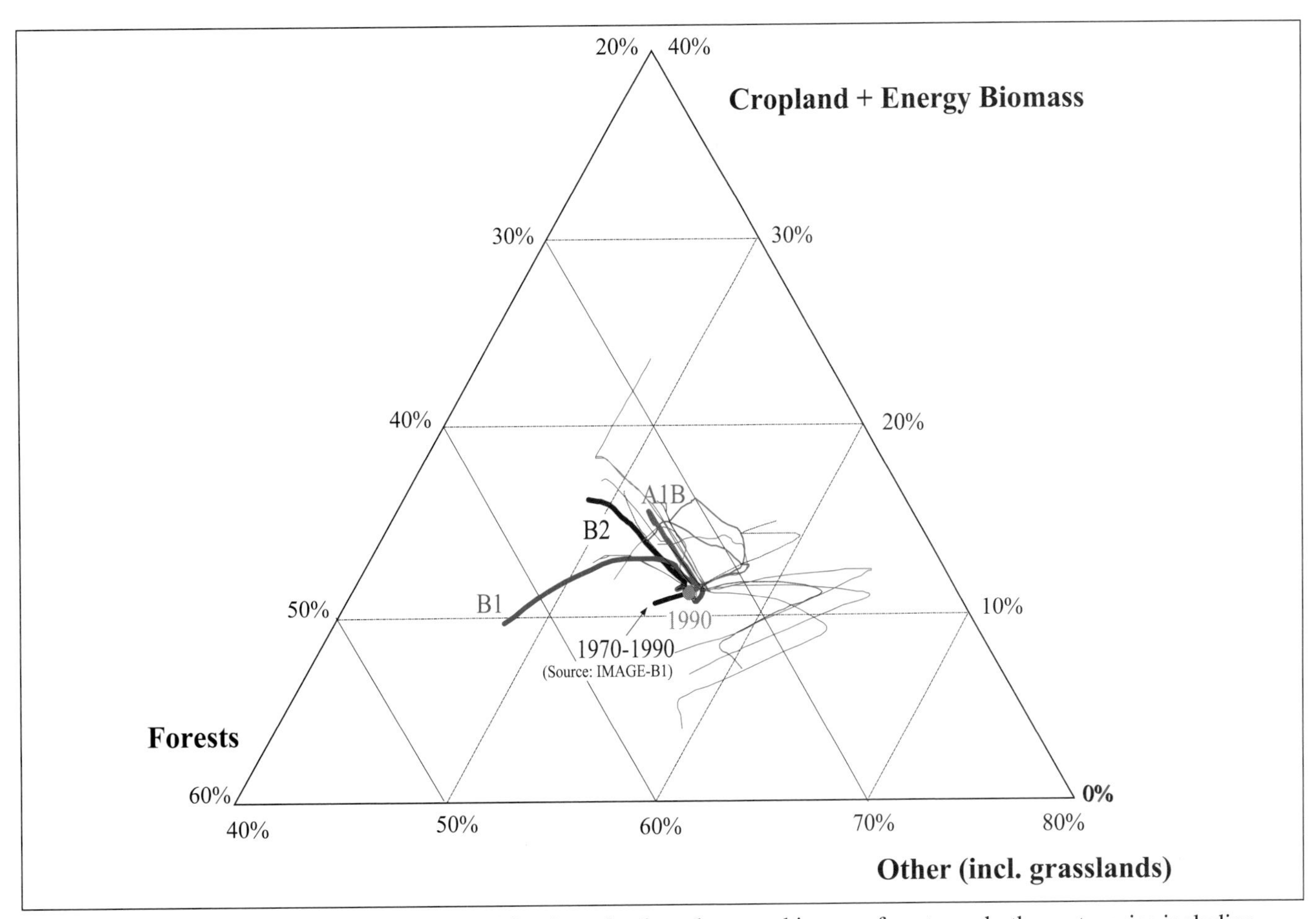

Figure 4-12: Global land-use patterns, shares (%) of croplands and energy biomass, forests, and other categories including grasslands – historical development from 1970 to 1990 (based on B1-IMAGE) and in SRES scenarios. As for the energy triangle in Figure 6-3, each corner corresponds to a hypothetical situation in which land use is dedicated to a much greater extent than today to one category – 60% to cropland and energy biomass at the top, 80% to forests to the left, and 80% to other categories (including grasslands) to the right. Constant shares in total land area of cropland and energy biomass, forests, and other categories are denoted by their respective isoshare lines. For 1990 to 2100, alternative trajectories are shown for the SRES scenarios. The three marker scenarios A1B, B1, and B2 are shown as thick colored lines, and other SRES scenarios as thin colored lines. The ASF model used to develop the A2 marker scenario projects only land-use change related GHG emissions. Comparable data on land cover changes are therefore not available.The trajectories appear to be largely model specific and illustrate the different views and interpretations of future land-use patterns across the scenarios (e.g. the scenario trajectories on the right that illustrate larger increases in grasslands and decreases in cropland are MiniCAM results).

4.4.9.1. A1 Scenarios

In the AIM model, land-use changes at the beginning of the 21st century follow largely historical trends. Over longer time horizons, the assumption is that land-use changes are driven primarily by economic forces, consistent with the A1 scenario storyline. Expected land rents and agricultural prices determine long-run land-use changes, based on an equilibrium approach of international agricultural markets. The AIM land-use model is linked to the AIM energy module via biomass energy demand. In the A1 scenario, the rapid increase in the demand for biomass energy raises the expected rent of biomass farmland. Reduction in forest area occurs until 2020 because of population growth and rapid increases of meat demand in the developing countries. Rising meat demands also result in a substantial expansion of grasslands and pasture. However, high incomes in scenario A1 also increase the demand for environmental amenities. Hence, "demand" for forests also increases with economic growth, and the expected rent of forestland is assumed to increase after 2020. These rising rents reduce the rate of deforestation and increase the area of managed tree-covered land in the latter half of the 21st century. Rising food productivity also counterbalances the pressure on cropland and pastureland in the latter half of the 21st century. For instance, crop productivity is assumed to grow on average by about 1.5% per year in the A1 scenario family.[36] However, despite these counterbalances the demand for pastureland continues to increase throughout the 21st century because of the high income growth and associated changes in diets. The resultant land-use changes for the A1 marker scenario between 1990 and 2100 are:

- Largely stationary trend of global cropland areas (–39 million ha between 1990 and 2100, i.e., 3% of 1990 cropland areas).
- Decline in global forest cover by some 92 million ha.
- Increase of grasslands and biomass land-use of 188 and 552 million ha, respectively.

Land-use change patterns are more dynamic in the intermediate time periods, and also display a wide variation across different regions.

4.4.9.2. Harmonized and Other A1 Scenarios

By and large, land-use changes in other A1 scenarios show a very wide range, being mostly model specific. As a rule, the A1 land-use change scenarios take an intermediate position between the more extreme tendencies described by the respective storylines of the A2 and B1 scenario families (see Figure 4-12), which both result in large land-use changes (albeit of an entirely different qualitative nature). Corresponding scenario (model) differences are thus discussed below for these two scenario families where they are largest.

4.4.9.3. A2 Scenarios

The ASF model used to produce the A2 marker scenario does not generate estimates of the area covered by different ecosystem types (e.g., forests, grasslands, etc.). However, the ASF deforestation module estimates the area of different tropical-forest types cleared annually for agricultural and other purposes (Lashof and Tirpak, 1990). This information is subsequently used to estimate GHG emissions from deforestation. The A2-ASF scenario also includes estimates of natural carbon sinks and additional sinks attributable to re-forestation and afforestation activities. These estimates are based on an extensive survey of available literature (Pepper *et. al.*, 1998).

By and large the land-use changes in the A2 marker scenario reflect conventional wisdom – in a high-population and low-income world, natural land cover becomes progressively depleted, which is only partially counterbalanced by re- and afforestation activities. The size of the natural terrestrial carbon sink (including forests, grasslands, wetlands, and other ecosystem types) in 1990 is estimated in ASF-A2 to amount to 1.8 GtC (1800 MtC). By 2050 this sink reduces to 0.8 GtC and by the end of the 21st century terrestrial ecosystems become a net source of carbon. On the other side of the natural carbon balance is carbon sequestered by re- and afforestation. In A2-ASF the respective (negative) carbon fluxes are estimated to increase from 0.003 GtC in 1990, to 0.171 GtC by 2050, and 0.205 GtC by 2100.

4.4.9.4. Harmonized and Other A2 Scenarios

Out of the five A2 non-marker scenarios, A2-AIM and A2-MiniCAM (and A2-A1-MINICAM) include explicit land-use patterns. Table 4-18 illustrates that the major difference between these quantifications is a higher energy biomass area assumption in A2-AIM. Among other differences is a higher percentage of cropland and grasslands in A2-MiniCAM and of forestland in A2-AIM.

4.4.9.5. B1 Scenarios

The most important indicators and assumptions made in the IMAGE land-use model relate to agricultural yields and diets that influence meat production, cattle population, and, in turn, grasslands land cover. Cereal yields (to 2100) for REF, ASIA, ALM, and the world average are assumed to increase by about a factor of four, while cereal yields for OECD90 start from a higher initial value and increase by only a factor of two. The total cattle population in the IMAGE model includes dairy and non-dairy cattle. Dairy cattle populations change as a result of milk production per animal and the demand for milk. Non-dairy cattle populations change as a result of meat demand, slaughter weight, and off-take rate. The number of slaughtered animals (beef) increases in the period 1995 to 2060, the net result of increasing animal productivity and increasing human consumption of meat. The total number of cattle (dairy and beef) shows a decreasing trend beyond the year 2000, with the near-term increase in beef cattle more than offset by a decrease in the number of dairy cattle. The forest area reflects the result of increasing agricultural production and increasing productivity. The major increase in demand for food in the initial period 1995 to 2030 is nearly compensated by increasing productivity, with a resultant slow decline in forest area, while beyond 2030 forest areas start expand over abandoned agricultural land. Overall, the net balance of land-use changes in scenario B1 between 1990 and 2100 suggests a considerable "greening" of the planet – a net increase in forest cover by some 1260 million ha, a decrease in cropland of 390 million ha, and an increase of about 200 million ha devoted to biomass production. Grasslands decline by 1540 million ha, partly because of the afforested areas mentioned above, but also because they are converted to other land uses. Thus, while the B1 scenario family is characterized by "high" rates of land-use changes, the quality of these changes ("greening") is entirely different compared to those of other scenario families (e.g. A2).

[36] For comparison, the corresponding productivity growth rates in the other scenarios range from 1% per year (B2-AIM) to 2% per year (B1-AIM), consistent with their respective storylines. This emphasizes productivity and efficiency (B1) and more fragmented technology and productivity growth (B2). For A2-AIM, crop productivity growth rates range between 1% and 1.5 % per year in the DEV and IND regions, respectively; the difference is explained by only slowly closing productivity gaps (approximated by GDP/capita), characteristic of this scenario storyline. Similar differences also characterize other salient scenario assumptions of importance to land-use changes (like biomass yields, animal productivity, or the distribution of grain- versus range-fed cattle). For instance, feed and protein yields from pasture land are assumed to grow at 1.5 % per year and biomass yields at 0.5 % per year in the A1 scenario.

Table 4-18: Global land use patterns (% global land cover) in A2-AIM and A2-MiniCAM scenarios. The columns may not total to 100% because of independent rounding errors.

	1990		2020		2050		2100	
Land Use Type	**AIM**	**MiniCAM**	**AIM**	**MiniCAM**	**AIM**	**MiniCAM**	**AIM**	**MiniCAM**
Cropland	11	11	12	12	13	13	15	12
Grasslands	26	24	27	29	28	34	29	34
Energy Biomass	0	0	1	0	2	0	2	3
Forest	33	32	33	30	33	26	33	27
Other	29	33	28	29	24	26	22	24

4.4.9.6. B2 Scenarios

Since the MESSAGE model does not include a land-use-change module, an external scenario for land-use change consistent with the underlying socio-economic assumptions of the B2 scenario was adopted. This scenario was developed by the AIM model in an iterative process between the two respective modeling teams to ensure consistency and harmonization in assumptions of driving-force variables (see AIM land-use model description above). In B2, population growth adds pressure on cropland expansion. Changing diets (higher meat demand) with rising incomes cause an expansion of pastureland. Land productivity increases were assumed to be more modest (e.g. compared to the A1 scenarios), consistent with the more cautious expectations of the B2 scenario storyline and the intermediate values adopted for productivity increases of energy use. As a result, land-use changes are substantial. Stabilization of population growth combined with continued gradual increases in agricultural productivity relieves some of the land-use-change pressure, particularly for forests. As a result, post-2050 forest cover expands through reforestation and afforestation activities, consistent with the local environmental orientation of the B2 scenario storyline. Emission patterns related to land-use change closely follow these trends in land-cover changes. By 2100, global forest cover in B2 shows a net increase of 230 million ha, croplands by 330 million ha, and grasslands and biomass plantations by 300 million ha each. This "greening" of long-term land-use patterns is at the expense of "other" land uses, and primarily involve reclamation of degraded lands, consistent with the emphasis on improving local and regional environmental quality and food and energy security outlined in the B2 scenario storyline.

4.4.9.7. Harmonized and Other B2 Scenarios

Compared to the B2 marker scenario, alternative scenarios project a very wide range of possible future patterns of land-use change. In fact, the range of resultant emissions for the B2 scenarios is as large as the entire range spanned by the four SRES marker scenarios altogether. Similarly, as observed above for structural changes in energy systems, different B2 scenarios are rather model specific and follow the general trends indicated by the respective marker scenario developed with a particular model. Overall, land-use related emissions decline, reflecting changes in driving forces such as increases in agricultural productivity. B2-MARIA and, of course, B2-AIM are structurally similar to the B2 marker – declining deforestation and long-term shifts to maintain (B2-MARIA) or increase forest cover (B2-AIM) – whereas B2-MiniCAM projects a stabilization of deforestation only after 2050. Generally, B2-MiniCAM shows a substantial decrease in cropland area and an increase of grassland area.

4.4.10. Environmental Policies

The SRES scenarios quantified emissions of CO_2, CH_4, N_2O, NO_x, carbon monoxide, VOCs, SO_x, chlorofluorocarbons (CFCs), hydrofluorochlorocarbons, hydrofluorocarbons, perfluorocarbons, and sulfur hexafluoride. A detailed discussion of the trajectories and the most relevant driving forces is presented in Chapter 5. In all cases herein, future emissions of these gases are affected by environmental policies **not** related to climate-change concerns, as in the SRES Terms of Reference (see Appendix I). Efforts to protect the ozone layer will result in a significant decline in CFCs and some of their substitutes. Urban air quality concerns will have a significant influence on future emissions of indirect GHGs and of sulfate aerosols. As discussed in Chapter 3, for sulfur emissions the SRES scenarios reflect the evaluation of the IS92 scenario series (Alcamo *et al.*, 1995), the recent sulfur scenario literature (for a review see Grübler, 1998b), and sulfur control policies in the OECD countries and the beginning of similar developments in many developing countries (see Chapter 3). Thus, a range of environmental measures is assumed in all four SRES scenario families, although the magnitude and timing differ as a function of the four different scenario storylines.

4.4.10.1. A1 Scenarios

In the A1 scenario family, environmental policies are implemented as incomes rise and societies choose to protect environmental amenities. This first of all would concern water and air quality (health concerns), but also traffic congestion and noise, as well as land-use policies (preservation of recreational spaces). Economic instruments are assumed to be the preferred policy instrument in an A1 world. In some of the more "technology intensive" A1 scenario groups, cleaner energy systems and lower GHG emissions are achieved as a by-product of high economic growth coupled with rapid technological change.

4.4.10.2. A2 Scenarios

Environmental concerns in their own right are perhaps the least important in the A2 world and they are mostly local in nature. For instance, no environmentally related barriers to nuclear energy development or environmental costs of fossil energy use are assumed. However, A2 scenarios include indirect control options for several GHGs that adversely effect local air quality. They also require vigorous environmental controls on pollutants that affect the availability of water, quality of soils, and productivity of agricultural crops to ensure food security is not jeopardized in a 15 billion person world. Hence, in terms of a number of traditional pollutants, A2 is far from an environmental "worst case" scenario, even if it generally has the highest GHG emissions.

4.4.10.3. B1 Scenarios

The B1 scenario family assumes the most comprehensive environmental measures (see Section 4.3). Sulfur and nitrogen oxide emissions decline with vigorous efficiency improvements and "dematerialization" of the economy. Economic structural change (toward services) and cleaner energy systems (electricity, gas, district heat) to improve urban air quality also result in lower emissions levels compared to other scenarios. Finally, additional "add-on" technologies, such as flue-gas desulfurization and de-nitrification equipment, catalytic converters, etc., bridge the transition gap before inherently clean energy systems (e.g. IGCC based on coal or biomass, hydrogen-powered fuel cell vehicles) come into use on a global scale. Environmental consciousness would also lead less-developed regions to take such measures earlier and at lower income levels than occurred in the history of presently industrialized regions (as indeed already actually occurs; see Chapter 5) .

4.4.10.4. B2 Scenarios

In B2 scenarios, environmental concerns are prominent, but in contrast to the B1 world it is assumed that measures are implemented effectively only at local and regional scales. For instance, it is assumed that sulfur emissions would be progressively controlled and other local and regional problems concerning, for example, air quality, human health, and food supply would also be the focus of policies.

4.5. Regional Scenario Patterns

The scenarios imply different regional patterns of socio-economic driving forces and resultant emissions. These are summarized in Table 4-19.

Following the definition of the four world regions adopted in this report, the industrial region (IND) corresponds to the SRES regions OECD90 and REF, and the developing region (DEV) corresponds to the SRES regions ASIA and ALM. In 1990, the base year of the SRES scenarios, driving-force variables as well as emissions are distributed unevenly. According to the statistics reported in Nakićenović *et al.* (1998), DEV countries account for 76% of global population. However, they only account for 16% to 36% of global economic activity (considering GDP at market exchange rates and purchasing power parities, respectively), 34% of primary energy use, and 42% of global CO_2 emissions (including all sources, from energy plus land-use changes).

Over time, the regional distribution of socio-economic activities and emission shifts in the scenarios, albeit because of different driving forces. These shifts range from rapid population growth (scenario family A2) and rapid economic development and catch-up (scenario family A1) in developing countries, to rapid "dematerialization" of economic activities in industrial countries (scenario family B1). Generally, scenario family B2 depicts the most gradual changes, and hence also the slowest rates of change in the global distribution of activities and emissions. Of all the SRES scenarios, A2-A1-MiniCAM exhibits the slowest dynamics of development catch-up.

Table 4-19 summarizes the main scenario indicators differentiated between industrial and developing countries. Two measures are given (following La Rovere, 1998):

- Date when activities and/or emissions in developing countries reach those that prevailed in industrial countries in 1990.
- Date when developing country values equalize those of industrial countries.

In this comparison, developing countries in 1990 already accounted for 76% of world population and, because of the inevitable demographic momentum, their share is likely to increase further over the next few decades before alternative demographic projections branch out into different regional distributions. However, the basic pattern of dominance of developing countries in global population remains unchanged across all demographic projections. The share of developing countries in global population ranges between 80% (A1, B1) and 86% (A2, B2) by 2100 for demographic projections between 7 and 15 billion people. From the distribution of global population, the scenarios indicate an increasing importance of developing countries in the future, albeit at different rates.

In consequence, as shown in Table 4-19, total GDP expressed at purchasing power parities in developing countries could reach 1990 levels in industrial regions between 2000 and 2010 and equalize future industrial country levels between 2010 and 2030. For GDP expressed at market exchange rates, the 1990 "parity date" is reached between 2015 and 2030 and equalization in absolute terms even later, between 2030 and

Table 4-19: *Date (rounded to nearest 5 years) when DEV countries reach 1990 levels of IND countries (top panel), and date when they reach parity (and overtake) projected IND country levels (bottom panel). Dates are given for the four SRES marker scenarios.*

Reaching 1990 IND levels	**A2**	**B2**	**A1B**	**B1**
GDP (mex)	~ 2030	~ 2020	~ 2015	~ 2020
GDP (ppp) (IIASA runs)	~ 2010	~ 2005	~ 2000	~ 2000
GDP (mex) per capita	> 2100	~ 2080	~ 2050	~ 2060
Primary energy	~ 2010	~ 2010	~ 2005	~ 2005
Primary energy per capita	–	–	~ 2070	–
Annual CO_2	~ 2000	> 2000	~ 2000	~2005
Cumulative CO_2 since 1800	~ 2020	~ 2030	~ 2015	~ 2020
CO_2 per capita	–	–	–	–
Overtaking IND	**A2**	**B2**	**A1B**	**B1**
GDP (mex)	~ 2060	~ 2035	~ 2030	~ 2035
GDP (ppp) (IIASA runs)	~ 2030	~ 2020	~ 2015	~ 2010
GDP (mex) per capita	–	–	–	–
Primary energy	~ 2015	~ 2020	~ 2010	~ 2005
Primary energy per capita	–	–	> 2100	–
Annual CO_2	~ 2000	~ 2005	~ 2000	~2005
Cumulative CO_2 since 1800	~ 2050	~ 2110	~ 2040	~ 2050
CO_2 per capita	–	–	–	–

mex, market exchange rate; ppp, purchasing power parities.
– Denotes that no date can be given within the time horizon of the SRES scenarios (to 2100) or short-term trend extrapolations after that date.

2060. Total primary energy use in developing countries could reach 1990 industrial levels between 2005 and 2010, and parity in absolute terms between 2005 and 2020. Possible spatial distributions of economic activities based on satellite night imagery data that correlate highly with, for example, GDP and electricity use are discussed in Box 4-10.

Conversely, scenario per capita indicators converge only slowly, sometimes well after 2100. Per capita GDP (expressed at market exchange rates) of developing countries reaches the 1990 level of industrial countries at the earliest around 2050 (scenario family A1) and well after 2100 in scenario family A2. None of the four marker scenarios projects a situation in which per capita income in developing countries surpasses future levels of per capita incomes of Annex I countries. Energy use per capita shows a similar pattern. Only scenario family A1 depicts a development in which per capita energy use in developing countries could approach that prevailing in industrial countries in 1990 (by 2070) and in which it could reach parity with industrial countries in the very long-term (after 2100). In all other scenarios energy use per capita remains below 1990 or future per capita energy use levels of industrial countries. By and large, these scenario features reflect the very large differences in present per capita levels of economic activity and energy use, which require many decades, even a century, to narrow. The scenarios thus portray a feature known as "slow conditional convergence only" in the development literature (see, e.g., Shin, 1996).

Concerning GHG emissions, discussed in more detail in Chapter 5, trends reflect the evolution of scenario-driving forces discussed above. CO_2 emissions of developing countries reach levels in industrial countries around or shortly after the year 2000. This reflects the continued growth in energy use (and emissions), slow recovery of economies in transition (and thus modest growth in aggregate emission from industrial countries), and a continuation of current trends in land-use changes (deforestation); these only diverge in the medium- to long-term across the SRES scenarios.

Yet, even with equalizing total emission levels, regional differences in cumulative and per capita emissions remain pronounced. Based on the estimates of cumulative CO_2 (all sources) emissions since 1800 given in Grübler and Nakićenović (1994), developing countries reach 1990 levels of industrial countries only between 2015 and 2030. They reach parity (historical data from 1800 to 1990, and scenario values from 1990 onward) at earliest by 2040 (scenario family A1), and between 2050 (A2 and B1) and post-2100 (scenario family B2) in the other scenarios. None of the SRES scenarios reaches the 1990 per capita CO_2 emissions levels of industrial countries in the developing countries.

In scenarios with vigorous climate policies (not in the SRES terms of reference) per capita emissions levels in industrial countries may approach levels as projected for developing countries. However, no scenario without climate policies could be found in the literature in which per capita emissions in industrialized and developing countries reached similar levels. Thus the absence of such convergence in the SRES scenarios reflects the current literature, and results from the nature of the SRES scenarios as "no climate policy" scenarios.

SRES scenarios do follow the recommendation to explore possible pathways of closing the income gap between the industrial and now developing regions (Alcamo *et al.*, 1995). For reasons of plausibility and foundation in the reviewed no-climate-policy literature, the SRES scenarios do not achieve full income convergence in the scenario period analyzed. However, income levels in developing countries do reach the 1990 levels of the industrial countries in the second half of the next century in three out of four scenario families.

4.6. A Roadmap to the SRES Scenarios

In the preceding sections the characteristics of the SRES scenarios are summarized in terms of scenario driving forces such as population, economic development, resources, technology, land-use changes, and other factors. The scenarios were designed in such a way as to deliberately span a wide range, reflecting uncertainties of the future, but not cover the very extremes from the scenario literature concerning driving forces. A distinguishing feature of the SRES scenarios is that various driving-force variables are not combined numerically (or arbitrarily), but instead try to reflect current understanding of the *interrelationships* between important scenario driving forces. For instance, according to the literature review of Chapter 3 it would be rather inconsistent to develop scenarios of rapid technological change in a macro-economic and social context of low labor productivity and stagnant income per capita. Scenario *storylines* were the method developed within SRES to help guide the scenario quantifications and to assure scenario consistency in terms of the main relationships between scenario driving forces.

The different quantifications discussed in the previous sections demonstrate that even if scenarios share important main input assumptions in terms of population and GDP growth,

Box 4-10: Spatial Distributions of Economic Activities Based on Nighttime Satellite Imagery Data.

Spatially explicit data on socio-economic activity is sparse. The reason arises mainly in that Systems of National Accounts and similar socio-economic statistics are available only at high levels of spatial aggregations defined by administrative boundaries (countries, provinces, or regions). As a result, gridded emission inventories largely rely on estimations of current population density distributions (e.g. Olivier *et al.*, 1996) and modeling approaches to date have also relied exclusively on rescaling *future* socio-economic activities (economic output, energy use, etc.) based on current or future population density distribution patterns (see e.g., Sørensen and Meibom, 1998; Sørensen *et al.*, 1999). Population density is, however, not necessarily a good indicator for spatial patterns of socio-economic activities (Sutton *et al.*, 1997). At higher levels of spatial aggregation, it is estimated that about two billion people remain outside the formal economy, most of them in rural areas of developing countries (UNDP, 1997). At lower levels of spatial aggregations, locations such as airports, industrial zones, and commercial centers have low resident population densities, but high levels of economic activity. Also, with increasing urbanization future population distribution will be markedly different from present ones (HABITAT, 1996).

Night satellite imagery from the US Air Force Defense Meteorological Satellite Program (DMSP) Operational Linescan System (OLS) offers an interesting alternative based on direct observations. Early nighttime lights data were analyzed from analog film strips (Croft, 1978, 1979; Foster, 1983; Sullivan, 1989). Digital DMSP-OLS data have recently become available with global coverage (Elvidge *et al.*, 1997a, 1997b, 1999). Nocturnal lighting can be regarded as one of the defining features of concentrated human activity, such as flaring of natural gas in oil fields (Croft, 1973), fishing fleets, or urban settlements (Tobler, 1969; Lo and Welch, 1977; Foster, 1983; Gallo *et al.*, 1995; Elvidge *et al.*, 1997c). Consequently, extent and brightness of nocturnal lighting correlate highly with indicators of city size and socio-economic activities such as GDP, and energy and electricity use (Welch, 1980; Gallo *et al.*; 1995; Elvidge *et al.*, 1997a).

Figure 4-13 (bottom panel) shows a 1995/1996 night-luminosity map of the world developed by National Oceanic and Atmospheric Administration's National Geophysical Data Center. The map was derived from composites of cloud-free visible band observations made by the DMSP-OLS (see Elvidge *et al.*, 1997b; Imhoff *et al.*, 1997). The DMSP-OLS is an oscillating scan radiometer that generates images with a swath width of 3000 km. The DMSP-OLS is unique in its capability to perform low-light imaging of the entire earth on a nightly basis. With 14 orbits per day, the polar orbiting DMSP-OLS is able to generate global daytime and nighttime coverage of the Earth every 24 hours. The "visible" bandpass straddles the visible and near-infrared (VNIR) portion of the spectrum. The thermal infrared channel has a bandpass that covers 10^{-13} μm of the spectrum. Satellite altitude is stabilized using four gyroscopes (three-axis stabilization), a starmapper, Earth limb sensor, and a solar detector. Image time series analysis is used to distinguish lights produced by cities, towns, and industrial facilities from sensor noise and ephemeral lights that arise from fires and lightning. The time series approach is required to ensure that each land area is covered with sufficient cloud-free observations to determine the presence or absence of VNIR emission sources.

Dietz *et al.* (2000) performed illustrative simulations of possible future light patterns. Considering the high correlation between observed radiance-calibrated night satellite imagery data and economic activity variables like GDP, these simulations indicate future spatial patterns of economic activity. As a basis of the simulation, Dietz *et al.* (2000) rescaled present night luminosity patterns using global and regional GDP growth patterns of the preliminary A1B marker scenario reported in the SRES open-process web site combined with a simple stochastic model of spatial evolution and interaction. An illustrative simulation for the year 2070 is given in Figure 4-13 (top panel). The resultant changes in spatial light patterns indicate socio-economic activities and provide useful information for infrastructure planning, such as expansion of gas and electricity networks. When combined with topographical information, like latitude, the data can also be used as input to climate impact and vulnerability assessments (e.g., extent of socio-economic activities that may be affected by sea-level rise).

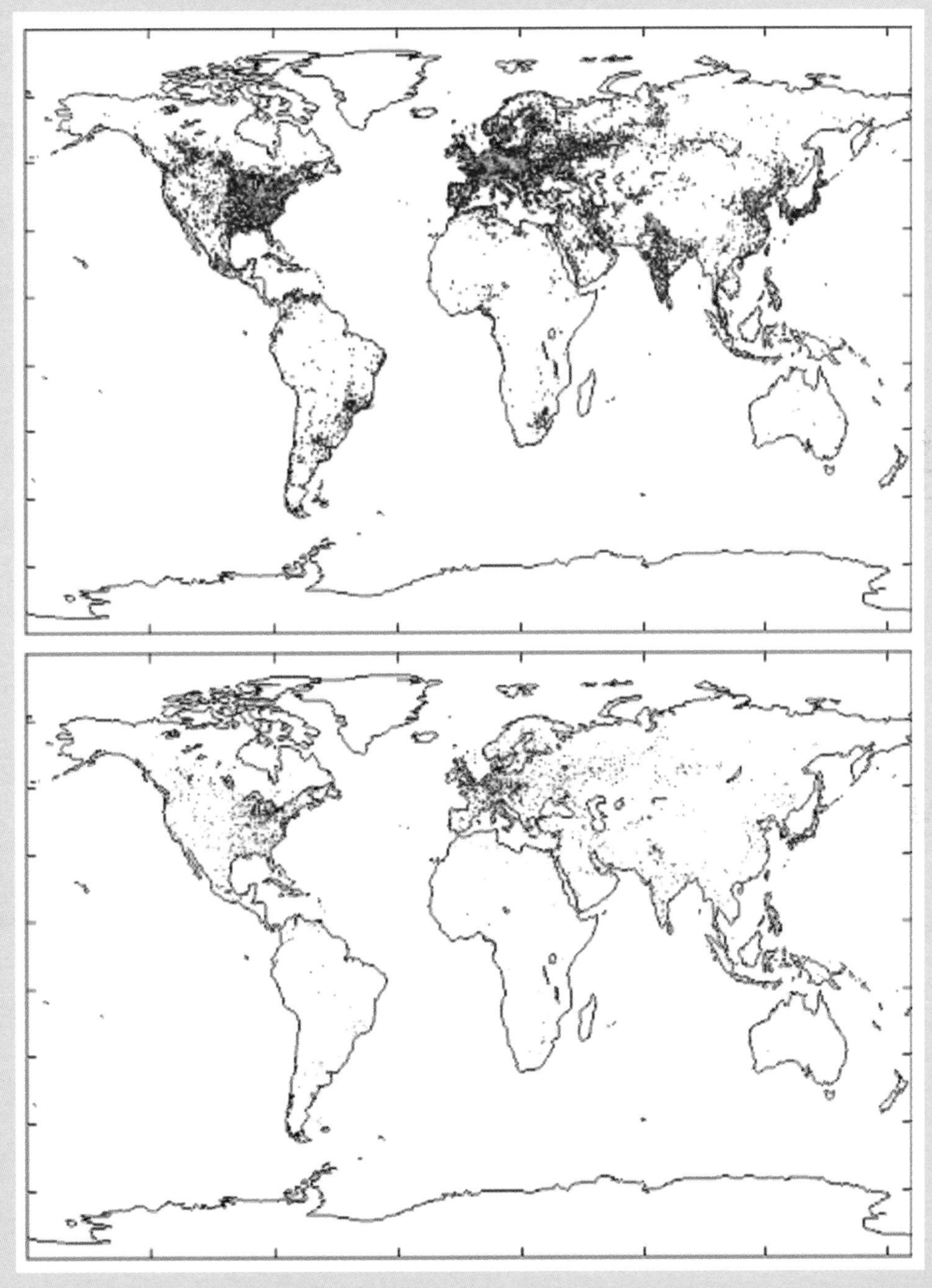

Figure 4-13: Radiance calibrated lights obtained from night satellite imagery. Situation in 1995/1996 (bottom panel) and illustrative simulation for the SRES A1 scenario's implied GDP growth for 2070 (top panel). Color codes refer to radiance units (DN), where radiance = DN3/2 $\times 10^{-10}$ W/cm^2 per sr/µm (Watts per square centimeter per steradian per micrometer, the brightness units to which the US Air Force Defense Meteorological Satellite Program (DMSP) Operational Linescan System (OLS) is calibrated; it normalizes for the bandpass (µm) and solid angle of the optics (cm^2/sr)).

Table 4-20: *Overview of SRES scenarios categorized into the four scenario families and associated scenario groups (four for the A1 family, combined into three in the SPM one for each of the other scenario families). The scenarios are classified as "harmonized" and "other" scenarios with respect to whether they share harmonized input assumptions on global population and GDP growth (see also Tables 4-1 to 4-4). A second layer of classification relates to scenario outcomes in terms of cumulative emissions (see Chapter 5). Four categories are distinguished: Low (<1100 GtC), Medium-Low (1100-1450 GtC), Medium-High (1450-1800 GtC), and High (>1800 GtC).*

Family			**A1**		**A2**	**B1**	**B2**
Scenario Group	**A1C**	**A1G**	**A1B**	**A1T**	**A2**	**B1**	**B2**
Marker			**A1B-AIM**		A2-ASF	*B1-IMAGE*	***B2-MESSAGE***
Globally	A1C-AIM	A1G-AIM	A1B-ASF	***A1T-AIM***	**A2-MESSAGE**	*B1-AIM*	***B2-AIM***
Harmonized	A1C-MESSAGE	A1G-MESSAGE	**A1B-IMAGE**	*A1T-MESSAGE*		***B1-ASF***	***B2-MARIA***
Scenarios[a]		A1G-MiniCAM	***A1B-MARIA***			*B1-MESSAGE*	**B2C-MARIA**
			A1B-MESSAGE			*B1-MiniCAM*	
			A1B-MiniCAM			*B1T-MESSAGE*	
						B1High-MESSAGE	
Other Scenarios	A1C-MiniCAM		**A1v1-MiniCAM**	*A1T-MARIA*	A2-AIM	*B1-MARIA*	**B2-ASF**
			A1v2-MiniCAM		**A2G-IMAGE**	***B1High-MiniCAM***	***B2-IMAGE***
					A2-MiniCAM		***B2-MiniCAM***
					A2-A1-MiniCAM		**B2High-MiniCAM**
Scenario outcomes: Cumulative emissions, GtC 1990–2100							
Marker			**1499**		1862	*983*	***1164***
Harmonized	2127-2538	2178-2345	***1301***-2073	*1068*-***1114***	**1732**	*773*-***1390***	***1359***-**1573**
Other scenarios	2148		**1519-1547**	*1049*	***1352***-1938	*947*-***1201***	***1186***-**1599**

Classification according to cumulative CO_2 emissions (Chapter 5):

High (>1800 GtC)

Medium–High (1450-1800 GtC)

Medium–Low (1100-1450 GtC)

Low (<1100 GtC)

considerable uncertainty and scenario variability remains. For instance, features of technological and land-use changes can be interpreted quite differently within the framework of different models, even if they conform to the overall conceptual description and "scenario logic" described in a particular scenario family. In some instances the broad outlines of scenario driving forces were not followed entirely in particular scenario quantifications, but alternative scenario interpretations were submitted. These highlight important scenario uncertainties or express scientific disagreements within the writing team, as for future labor productivity growth and economic "catch-up" possibilities of currently developing countries. These alternative scenario interpretations and different model quantifications are presented here to reflect the SRES Terms of Reference for an open process and the use of multiple modeling approaches, even if this necessarily increases complexity and reduces simplicity and transparency in discussion of a large number of scenario quantifications.

To guide readers through the different driving-force assumptions that characterize the various scenarios, Tables 4-2 and 4-3 give an overview of the SRES scenario set. They classify scenarios that share important input assumptions (harmonized scenarios share global population and GDP assumptions) from scenarios that offer alternative quantifications. Table 4-4 summarizes the main quantitative scenario descriptors for each of the four SRES scenario families of the "harmonized" scenario category. Here an attempt is made to link this information with the resultant scenario outcomes (emissions) that are discussed in more detail in Chapter 5.

In Chapter 5, an additional, complementary scenario classification scheme to that used in this chapter is presented and focuses on driving forces. Scenarios are classified according to their cumulative carbon emissions (1990 to 2100, all sources), the best single quantitative indicator available to compare emission scenarios that portray widely different dynamics and different combinations and magnitude of a variety of emission categories. Four categories of cumulative emissions, Low (<1000 GtC), Medium–Low (1100 to 1450 GtC), Medium–High (1450 to 1800 GtC), and High (>1800 GtC) are presented. Table 4-20 links the scenario overview from Tables 4-2 and 4-3 with this information to guide readers through the differences in scenarios.

Table 4-20 indicates that in most cases there is an easily discernable direct connection between main scenario characteristics of a particular scenario family or scenario group and the resultant outcomes in terms of cumulative emissions. For instance, in the high GDP, high energy demand scenario family A1, all scenarios within the two scenario groups that are fossil fuel and technologies intensive (A1C and A1G combined into A1FI in the SPM) result in high cumulative carbon emissions. Conversely, cumulative emissions of the "efficiency and dematerialization" (without additional climate initiatives) scenario family B1 are generally in the "low" emissions category, but two model quantifications indicate medium–low emissions. For the scenario family B2, outcomes in terms of cumulative carbon emissions can also be related clearly to scenario characteristics. One group of scenarios (which includes the B2 marker) adopts an incrementalist perspective of technological change ("dynamics as usual") applied to medium levels of population and GDP (and resultant energy demand) and results in medium–low cumulative carbon emissions. Another group of scenarios explored the sensitivity of a gradual return to coal-based technologies (B2C-MARIA, B2-ASF), in one case combined with higher energy demand than in the other scenarios (B2High-MiniCAM); and results in higher cumulative emissions (Medium–High category in Table 4-20).

Equally discernable in Table 4-20 is the wide range in cumulative carbon emissions that characterize the various scenario groups within the A1 scenario family. By design, the different scenario groups within this family explored the implications of different directions of technological change, ranging from carbon-intensive developments (A1C and A1G, combined into A1FI in the SPM) to decarbonization (A1T), with the "balanced" technology development scenario group taking an intermediary position. Different developments concerning fossil or non-fossil resource and technology availability in a less populated but affluent and thus high energy demand world (such as A1) can lead to widely different outcomes in terms of cumulative emissions, with a range as wide as that spanned by all four scenario families together. Technology can thus be as important a driving force as population and GDP growth combined. In other words, very different emissions outcomes are possible for future worlds that otherwise share similar developments of main driving forces such as population and economic growth and high rates of technological change.

However, areas of overlap and uncertainties of scenario outcomes (cumulative emissions) occur even for scenario quantifications that share otherwise similar assumptions for the main scenario drivers. Not surprisingly, differences in quantifications are largest within the A1 "balanced" technological progress scenario group, which includes the A1B marker scenario. Most model interpretations result in cumulative carbon emissions within the Medium–High category (1450–1800 GtC). However, there are also scenario quantifications in which technological change tilts more in the direction of the A1C (A1-ASF) or A1T (A1-MARIA) scenario groups that favor fossil (coal) or post-fossil (nuclear, renewables, and biomass) technologies, respectively. This leads to very wide differences in cumulative emissions, from the Medium–Low through to the High categories. A similar range of scenario outcomes between Medium–High to High categories also characterizes the A2 scenario family that otherwise describes an entirely different world (high population and comparatively low per capita income compared to low population with high per capita income for the A1 scenario family; see Table 4-4). Departing from the main scenario characteristics of the A2 scenario family in terms of population and income in direction of lower values (such as in the A2-A1-MiniCAM scenario) could even yield emissions in the Medium–Low category. Thus, the A2 scenario family also indicates that a wide range of emissions outcomes is possible for any given development path of main scenario driving forces, such as population and income per capita.

Finally, the categorization of scenarios in terms of their (cumulative carbon) emission outcomes illustrates that similar emission outcomes could arise from very different developments of main scenario drivers. For instance, High category cumulative emissions could arise from scenarios of low population growth, combined with high incomes (and energy use) and globalized technological developments that favor accessibility and economics of fossil fuels (coal, unconventional oil and gas; e.g., A1C and A1G scenario groups). Alternatively, similar High category cumulative emissions could also arise from scenarios of high population growth combined with slower per capita income growth and more regionally oriented technology development trends (scenario family A2). A comparison of the B1 and A1T scenario groups (see Table 4-20) also confirms this conclusion. Both scenarios explore pathways that reduce current income disparities between regions. They indicate that such a tendency does not necessarily lead to high emissions, but could be achieved with Low to Medium–Low category cumulative emissions (as scenario groups A1C and A1G also indicate that High category emission pathways are possible).

Perhaps the most important conclusion from the SRES multi-model, open process, and the large number of scenarios it has generated is the recognition that there is no simple, linear relationship between scenario driving forces and outcomes or between emission outcomes and scenario driving forces. High or low population scenarios need not automatically lead to high or low emissions; similar statements also hold for economic growth and for closing regional income gaps.

References:

Alcamo, J., A. Bouwman, J. Edmonds, A. Grübler, T. Morita, and A. Sugandhy, 1995: An evaluation of the IPCC IS92 emission scenarios. In *Climate Change 1994, Radiative Forcing of Climate Change* and *An Evaluation of the IPCC IS92 Emission Scenarios,* J.T. Houghton, L.G. Meira Filho, J. Bruce, Hoesung Lee, B.A. Callander, E. Haites, N. Harris and K. Maskell (eds.), Cambridge University Press, Cambridge, pp. 233-304.

Alcamo, J., E. Kreileman, M. Krol, R. Leemans, J. Bollen, J. van Minnen, M. Schaefer, S. Toet, and B. de Vries, 1998: Global modelling of environmental change: an overview of IMAGE 2.1. In *Global change scenarios of the 21st century. Results from the IMAGE 2.1 Model.* J. Alcamo, R. Leemans, E. Kreileman (eds.), Elsevier Science, Kidlington, Oxford, pp. 3-94.

Ausubel, J.H., A. Grübler, and N. Nakićenović, 1988: Carbon dioxide emissions in a methane economy. *Climatic Change*, **12**, 245-263.

BP (British Petroleum), various years, 1972-1999: *BP Statistical Review of World Energy.* BP, London. (http://www.bpamoco.com/worldenergy/oil/index.htm).

CPB (Bureau for Economic Policy Analysis), 1992: *Scanning the Future: A Long-term study of the World Economy 1990-2015.* Sdu Publishers, The Hague.

Croft, T.A., 1973: Burning waste gas in oil fields. *Nature,* **245** (16 July), 375-376.

Croft, T.A., 1978: Nighttime images of the earth from space. *Scientific American,* **239**, 68-79.

Croft, T.A., 1979: *The brightness of lights on earth at night, digitally recorded by DMSP satellite.* Stanford Research Institute final report prepared for the U.S. Geological Survey, SRI, Stanford, CA.

Demeny, P., 1990: Population. In *The Earth As Transformed by Human Action.* B.L. Turner II *et al.* (eds.), Cambridge University Press, Cambridge, pp. 41-54.

De Jong, A., and G. Zalm, 1991: Scanning the Future: A long-term scenario study of the world economy 1990-2015. In *Long-term Prospects of the World Economy.* OECD, Paris, pp. 27-74.

De Vries, H.J.M., J.G.J. Olivier, R.A. van den Wijngaart, G.J.J. Kreileman, and A.M.C. Toet, 1994: Model for calculating regional energy use, industrial production and greenhouse gas emissions for evaluating global climate scenarios. *Water, Air Soil Pollution*, **76**, 79-131.

De Vries, B., M. Janssen, and A. Beusen, 1999: Perspectives on global energy futures – simulations with the TIME model. *Energy Policy,* **27**, 477-494.

De Vries, B., J. Bollen, L. Bouwman, M. den Elzen, M. Janssen, and E. Kreileman, 2000: Greenhouse gas emissions in an equity-, environment- and service-oriented world: An IMAGE-based scenario for the next century. *Technological Forecasting & Social Change*, **63**(2-3). (In press).

Dietz, J., C.D. Elvidge, A. Gritsevskyi, and A. Grübler, 2000: *Night satellite imagery data and socio-economic activity variables: Current patterns and long-term simulations.* IR-00-XX, International Institute for Applied Systems Analysis, Laxenburg, Austria. (http://www.iiasa.ac.at/Publications/Catalogue).

Durand, J.D., 1967: The modern expansion of world population. *Proceedings of the American Philosophical Society*, **111**(3), 136–159.

Edmonds, J., M. Wise, and C. MacCracken, 1994: *Advanced energy echnologies and climate change. An Analysis Using the Global Change Assessment Model (GCAM).* PNL-9798, UC-402, Pacific Northwest Laboratory, Richland, WA.

Edmonds, J., M. Wise, H. Pitcher, R. Richels, T. Wigley, and C. MacCracken, 1996a: An integrated assessment of climate change and the accelerated introduction of advanced energy technologies: An application of MiniCAM 1.0. *Mitigation and Adaptation Strategies for Global Change*, **1**(4), 311-339.

Edmonds, J., M. Wise, R. Sands, R. Brown, and H. Kheshgi, 1996b: *Agriculture, land-use, and commercial biomass energy. A Preliminary integrated analysis of the potential role of Biomass Energy for Reducing Future Greenhouse Related Emissions.* PNNL-11155, Pacific Northwest National Laboratories, Washington, DC.

EIA (Energy Information Administration), 1997: *International Energy Outlook.* EIA, Washington, DC.

EIA (Energy Information Administration), 1999: *International Energy Outlook.* EIA, Washington, DC.

Elvidge, C.D., K.E. Baugh, E.A. Kihn, H.W. Kroehl, E.R. Davis, and C. Davis, 1997a: Relation between satellite observed visible to near infrared emissions, population, and energy consumption. *International Journal of Remote Sensing,* **18**, 1373-1379.

Elvidge, C.D., K.E. Baugh, K.E. Hobson, E.A. Kihn, H.W. Kroehl, H.W, Davis, E.R, Davis, and D. Cocero, 1997b: Satellite inventory of human settlements using nocturnal radiation emissions: A contribution for the global toolchest. *Global Change Biology,* **3**, 387-395.

Elvidge, C.D., K.E. Baugh, E.A. Kihn, H.W. Kroehl, and E.R. Davis, 1997c: Mapping of city lights using DMSP Operational Linescan System data. *Photogrammetric Engineering and Remote Sensing,* **63**, 727-734.

Elvidge, C.D., K.E. Baugh, J.B. Dietz, T. Bland, P.C. Sutton, and H.W. Kroehl, 1999: Radiance calibration of DMSP-OLS low-light imaging data of human settlements. *Remote Sensing of Environment,* **68**, 77-88.

Foster, J.L., 1983: Observations of the earth using nighttime visible imagery. *International Journal of Remote Sensing,* **4**, 785-791.

Gallo, K.P., J.D. Tarpley, A.L. McNab, and T.R. Karl. 1995: Assessment of urban heat islands: a satellite perspective. *Atmospheric Research,* **37**, 37-43.

Gallopin, G., A. Hammond, P. Raskin, and R. Swart, 1997: *Branch Points*. PoleStar Series Report 7, Stockholm Environment Institute, Boston, MA.

Glenn, J.C., and T.J. Gordon (eds.), 1997: *State of the Future: Implications for Actions Today*. The Millennium project, American Council for the United Nations University, Washington, DC.

Glenn, J.C., and T.J. Gordon, 1999: The Millennium project: issues and opportunities for the future. *Technological Forecasting and Social Change*, **61**(2), 97-208.

Godet, M., P. Chapay, and G. Comyn, 1994: Global scenarios: Geopolitical and economic context to the year 2000. *Futures*, **26**(3), 275-288.

Grübler, A., 1990: *The Rise and Fall of Infrastructures*. Physica Verlag, Heidelberg, Germany.

Grübler, A., and N. Nakićenović, 1994: *International burden sharing in greenhouse gas reduction. Environment*. Environment Working Paper 55, World Bank, Washington, DC.

Grübler, A., and N. Nakićenović, 1996: Decarbonizing the global energy system. *Technological Forecasting and Social Change*, **53**(1), 97-110.

Grübler, A., 1998a: *Technology and Global Change*. Cambridge University Press, Cambridge, 452 pp.

Grübler, A., 1998b: A review of global and regional sulfur emission scenarios. *Mitigation and Adaptation Strategies for Global Change*, **3**(2-4), 383-418.

HABITAT (United Nations Centre for Human Settlements), 1996: *An Urbanizing World: Global Report on Human Settlements 1996*. Oxford University Press, Oxford.

Hammond, A., 1998: *Which World? Scenarios for the 21st Century, Global Destinies, Regional Choices*. Earthscan Publications Ltd., London.

Huntington, S.P., 1996: *The Clash of Civilizations and the Remaking of World Order*. Simon and Schuster, New York, NY.

IEA (International Energy Agency), 1997: *IEA Energy Technology R&D Statistics*. IEA/OECD, Paris.

IEA (International Energy Agency), 1998: *World Energy Outlook*. IEA/OECD, Paris.

IGU (International Gas Union), 1997: *World Gas Prospects, Strategies and Economics*. Proceedings of the 20th World Gas Conference, Copenhagen. (http://www.wgc.org/proceedings).

IIASA–WEC (International Institute for Applied Systems Analysis – World Energy Council) 1995: *Global Energy Perspectives to 2050 and Beyond*. WEC, London.

Imhoff, M.L., W.T. Lawrence, D.C. Stutzer, and C.D. Elvidge, 1997: A technique for using composite DMSP/OLS "city lights" satellite data to accurately map urban areas. *Remote Sensing of Environment*, **61**, 361-370.

Jiang, K., T. Masui, T. Morita, and Y. Matsuoka, 2000: Long-term GHG emission scenarios of Asia-Pacific and the world. *Technological Forecasting & Social Change*, **63**(2-3). (In press).

Kaplan, R., 1996: *The Ends of the Earth: A Journey at the End of the 21st Century*. Random House, New York, NY.

Kinsman, F., 1990: *Millenium: Towards Tomorrow's Society*. W.H. Allen, London.

Klein Goldewijk, C.G.M., and J.J. Battjes, 1995: *The Image 2 Hundred Year (1890-1990) Data Base of the Global Environment (HYDE)*. Report No. 481507008, RIVM, Bilthoven, the Netherlands.

La Rovere, E.L., 1998: Climate change convention: a tool of sustainable development? *Economies et Sociétés*, Série F, **36**(1), 247-259.

Lashof, D., and Tirpak, D.A., 1990: *Policy Options for Stabilizing Global Climate*. 21P-2003. U.S. Environmental Protection Agency, Washington, D.C.

Lawrence, R.Z., A. Bressand, and T. Ito, 1997: *A Vision for the World Economy - Openness, Diversity, and Cohesion*. The Brookings Institution, Washington, DC.

Leach, M.A., 1998: Culture and sustainability. In *UNESCO: 1998 World Culture Report: Culture, Creativity and Markets*. UNESCO Publishing, Paris.

Leggett, J., W.J. Pepper, and R.J. Swart, 1992: Emissions Scenarios for IPCC: An Update. In *Climate Change 1992. The Supplementary Report to the IPCC Scientific Assessment*. J.T. Houghton, B.A. Callander, S.K. Varney (eds.). Cambridge University Press, Cambridge.

Lo, C.P., and R. Welch, 1977: Chinese urban population estimates. *Annals Association of American Geographers*, **67**, 246-253.

Lutz, W. (ed), 1996: *The Future Population of the World: What can we assume today? 2nd Edition*. Earthscan, London.

Lutz, W., W. Sanderson, and S. Scherbov, 1997: Doubling of world population unlikely. *Nature*, **387**(6635), 803-805.

Maddison, A., 1989: *The World Economy in the 20th Century*, OECD Development Centre Studies, Organisation for Economic Cooperation and Development, Paris.

Maddison, A., 1995: *Monitoring the World Economy 1820-1992*. OECD Development Centre Studies, Organisation for Economic Cooperation and Development, Paris.

Manne, A.S., and R. Richels, 1992: *Buying Greenhouse Insurance: The Economic Costs of CO_2 Emission Limits*. The MIT Press, Cambridge.

Marchetti, C., and N. Nakićenović, 1979: *The Dynamics of Energy Systems and the Logistic Substitution Model*. RR-79-13, International Institute for Applied Systems Analysis, Laxenburg, Austria.

Messner, S., and M. Strubegger, 1991: *User's Guide to CO2DB: The IIASA CO_2 Technology Data Base Version 1.0*. WP-91-31a, International Institute for Applied Systems Analysis, Laxenburg, Austria.

Messner, S., and M. Strubegger, 1995: *User's Guide for MESSAGE III*. WP-95-69, International Institute for Applied Systems Analysis, Laxenburg, Austria.

Mohan Rao, J., 1998: Culture and economic development. In *1998 World Culture Report: Culture, Creativity and Markets*, UNESCO Publishing, Paris.

Mori, S., and M. Takahashi, 1999: An integrated assessment model for the evaluation of new energy technologies and food productivity. *International Journal of Global Energy Issues*, **11**(1-4), 1-18.

Mori, S., 2000: The development of greenhouse gas emissions scenarios using an extension of the MARIA model for the assessment of resource and energy technologies. *Technological Forecasting & Social Change*, **63**(2-3). (In press).

Morita, Y., Y. Matsuoka, M. Kainuma, and H. Harasawa, 1994: AIM - Asian Pacific integrated model for evaluating policy options to reduce GHG emissions and global warming impacts. In *Global Warming Issues in Asia*. S. Bhattacharya *et al.* (eds.), AIT, Bangkok, pp. 254-273.

Morita, T., and H.-C. Lee, 1998: *IPCC SRES Database, Version 0.1*. Emission Scenario Database prepared for IPCC Special Report on Emissions Scenarios. (http:www-cger.nies.go.jp/cger-e/db/ipcc.html).

Munasinghe, M and R. Swart, 2000: Climate change and its linkages with development, equity and sustainability, Proceedings of the IPCC expert meeting, Colombo, Sri Lanka, April 1999.

Nakićenović, N., 1987: Technological substitution and long waves. In *The Long Wave Debate*. T. Vasko (ed), Springer Verlag, Berlin, pp. 76-103.

Nakićenović, N., A. Grübler, A., Inaba, S. Messner, S. Nilsson, *et al.*, 1993: Long-term strategies for mitigating global warming. *Energy*, **18**(5), 401-609.

Nakićenović, N., 1996: Freeing energy from carbon. *Daedalus*, **125**(3), 95–112.

Nakićenović, N., A. Grübler, H. Ishitani, T. Johansson, G. Marland, J.R. Moreira, and H-H. Rogner, 1996: Energy primer. In *Climate Change 1995. Impacts, Adaptations and Mitigation of Climate Change: Scientific Analyses*. R. Watson, M.C. Zinyowera, R. Moss (eds.), Cambridge University Press, Cambridge, pp. 75-92.

Nakićenović, N., A. Grübler, and A. McDonald (eds.), 1998: *Global Energy Perspectives*. Cambridge University Press, Cambridge.

OECD (Organization for Economic Cooperation and Development), 1998: *Energy Prices and Taxes*. OECD, Paris.

Olivier, J.G.J., A.F. Bouwman, C.W.M. van der Maas, J.J.M. Berdowski, C. Veldt, J.P.J. Bloos, A.J.H. Visschedijk, P.Y.J. Zandveld, and J.L. Haverlag, 1996: *Description of EDGAR Version 2.0: A set of global emission inventories of greenhouse gases and ozone-depleting substances for all anthropogenic and most natural sources on a per country basis and on $1^o \times 1^o$ grid*. Report 771060002, National Institute of Public Health and the Environment, Bilthoven, the Netherlands.

Pepper, W.J., J. Leggett, R. Swart, J. Wasson, J. Edmonds, and I. Mintzer, 1992: Emissions scenarios for the IPCC. An update: Assumptions, methodology, and results. Support document for Chapter A3. In *Climate Change 1992: Supplementary Report to the IPCC Scientific Assessment*. J.T. Houghton, B.A. Callandar, S.K. Varney (eds.), Cambridge University Press, Cambridge.

Pepper, W.J., Barbour, W., Sankovski, A., and Braaz, B., 1998: No-policy greenhouse gas emission scenarios: revisiting IPCC 1992. *Environmental Science & Policy,* **1**, 289-312.

Peterson, J.L., 1994: *The Road to 2015: Profiles of the Future*. Waite Group Press, Publishers Group West, Mill Valley, CA.

Pohl, F., 1994: A Visit to Belindia. In *The World of 2044: Technological Development and the Future of Society*. C. Sheffield, M. Alonso, M.A. Kaplan (eds.), Paragon House, St. Paul, MN, pp.189-197.

Raskin, P., G. Gallopin, P. Gutman, A. Hammond, and R. Swart, 1998: *Bending the Curve: Toward Global Sustainability*. A report of the Global Scenario Group, PoleStar Series Report 8, Stockholm Environment Institute, Stockholm, 90 + A-38 pp.

Riahi, K., and R.A. Roehrl, 2000: Greenhouse gas emissions in a dynamics-as-usual scenario of economic and energy development. *Technological Forecasting & Social Change*, **63**(2-3). (In press).

Roehrl, R.A., and K. Riahi, 2000: Technology dynamics and greenhouse gas emissions mitigation - a cost assessment. *Technological Forecasting & Social Change*, **63**(2-3). (In press).

Rogner, H.-H., 1997: An assessment of w hydrocarbon resources. *Annual Review of Energy and the Environment*, **22**, 217-262.

Rotmans, J., and H.J.M. de Vries (eds.), 1997: *Perspectives on Global Futures: the TARGETS approach*. Cambridge University Press, Cambridge.

Sankovski, A., W. Barbour, and W. Pepper, 2000: Quantification of the IS99 emission scenario storylines using the atmospheric stabilization framework (ASF). *Technological Forecasting & Social Change*, **63**(2-3). (In press).

Schipper, L., and S. Meyers, 1992: *Energy Efficiency and Human Activity: Past Trends, Future Prospects*. Cambridge University Press, Cambridge.

Schwartz, P., 1991: *The Art of the Longview: Three global scenarios to 2005*. Doubleday Publications, New York, NY.

Schwartz, P., 1996: The New World Disorder. *WIRED Special Edition 1.10*. (See also: Global Business Network. (http://www.gbn.org/scenarios).

Shell, 1993: *Global Scenarios 1992-2020*. PL-93-S-04, Group Planning, Shell International, London.

Shin, J.S., 1996: *The Economics of the Latecomers*. Routledge Studies in the Growth Economics of Asia, Routledge, London.

Shinn, R.L., 1985: *Forced Options: Social Decisions for the 21st century*, 2nd edition. The Pilgrim Press, New York, NY.

Sørensen, B., and P. Meibom, 1998: GIS Tools for Renewable Energy Modelling. Paper presented at the World Renewable Energy Congress V, Florence, 19-25 September 1998, Italy. Published in: *Renewable Energy*, **16**(1-4) (1999), 1262-1267.

Sørensen, B., R. Kümmel, and P. Meibom, 1999: *Long-term Scenarios for Global Energy Demand and Supply: Four Global Greenhouse Mitigation Scenarios*. IMFUFA 359, Roskilde University, Denmark.

Strubegger, M., and I. Reitgruber, 1995: *Statistical Analysis of Investment Costs for Power Generation Technologies*. WP-95-109, International Institute for Applied Systems Analysis, Laxenburg, Austria, 21 pp.

Sullivan, W.T. III, 1989: A 10 km resolution image of the entire night-time earth based on cloud-free satellite photographs in the 400-1100 µm band. *International Journal of Remote Sensing*, **10**, 1-5.

Sutton, P., D. Roberts, C. Elvidge, and H. Meij, 1997: A comparison of nighttime satellite imagery and population density for the continental United States. *Photogrammetric Engineering and Remote Sensing* **63**, 1303-1313.

Tobler, W.R., 1969: Satellite confirmation of settlement size coefficients. *Area*, **1**, 30-34.

UN (United Nations), 1990: *Global Outlook 2000: an Economic, Social and Environmental Perspective*. United Nations Publications, New York, NY.

UN (United Nations), 1993a: *Human Development Report 1993*, United Nations, New York, NY.

UN (United Nations), 1993b: *UNMEDS Macroeconomic Data System, MSPA Data Bank of World Development Statistics*. MEDS/DTA/1 MSPA-BK.93, Long-Term Socio-Economic Perspectives Branch, Department of Economic and Social Information & Policy Analysis, United Nations, New York, NY.

UN (United Nations), 1994: *Report on the International Conference on Population and Development*. Cairo, 5-13 September 1994, New York, NY.

UN (United Nations), 1995: *Report on the World Summit for Social Development*. Copenhagen, 6-12 March 1995, New York, NY.

UN (United Nations), 1998: *World Population Projections to 2150*. United Nations Department of Economic and Social Affairs Population Division, New York, NY.

UNCED (United Nations Conference of Environment and Development), 1992: *Report of the United Nations Conference of Environment and Development*, New York, NY.

UNDP (United Nations Development Programme), 1993: *Human Development Report 1993*. Oxford University Press, New York, NY.

UNDP (United Nations Development Programme), 1997: *Human Development Report 1997*. Oxford University Press, New York, NY.

UNESCO, 1998: *World Culture Report: Culture, Creativity and Markets*. UNESCO Publishing, Paris.

UNFCCC (United Nations Framework Convention on Climate Change), 1992: *United Nations Framework Convention on Climate Change*. Convention text, UNEP/WMO Information Unit of Climate Change (IUCC) on behalf of the Interim Secretariat of the Convention. IUCC, Geneva.

UNFPA (United Nations Population Fund), 1999: *6 Billion: A Time for Choices. The State of World Population 1999*. UNFPA, New York. (http://www.unfpa.org/swp/1999/pdf/swp99.pdf).

Watson, R., M.C. Zinyowera, and R. Moss (eds.), 1996: *Climate Change 1995. Impacts, Adaptations and Mitigation of Climate Change: Scientific Analyses*. Contribution of Working Group II to the Second Assessment Report of the Intergovernmental Panel on Climate Change, Cambridge University Press, Cambridge, 861 pp.

WBCSD (World Business Council for Sustainable Development), 1998: *Exploring Sustainable Development*, WBCSD, Geneva.

WEC (World Energy Council), 1993: *Energy for Tomorrow's World: The Realities, the Real Options and the Agenda for Achievements*. Kogan Page, London.

Welch, R., 1980: Monitoring urban population and energy utilization patterns from satellite data. *Remote Sensing of Environment*, **9**, 1-9.

Wene, C-O., 1996: Energy-economy analysis: linking the macroeconomic and systems engineering approaches. *Energy*, **21**(9), 809-824.

Weyant, J., 1995: *Second Round Study Design for EMF 14: Integrated Assessment of Climate Change*. Energy Modeling Forum, Terman Engineering Center, Stanford University, Stanford, CA (mimeo, deposited with TSU).

Wilkerson, L., 1995: How to build scenarios. *WIRED* (Special Edition: The Future of the Future, October 13 1995), Condé Nast Publications, San Francisco, CA. (http://wired.lycos.com/wired/scenarios/build.htm).

World Bank, 1998: *Global Economic Prospects and the Developing Countries: Beyond Financial Crisis*. World Bank, Washington, DC.

World Bank, 1999: *World Development Indicators*. World Bank, Washington, DC.

5

Emission Scenarios

CONTENTS

5.1 Introduction

In this chapter emission estimates for radiatively important gases generated in 40 Special Report on Emission Scenarios (SRES) scenarios are present. These gases are carbon dioxide (CO_2), methane (CH_4), nitrous oxide (N_2O), nitrogen oxides (NO_x), carbon monoxide (CO), non-methane volatile organic compounds (NMVOCs), sulfur dioxide (SO_2), chlorofluorocarbons (CFCs) and hydrochlorofluorocarbons (HCFCs),[1] hydrofluorocarbons (HFCs), perfluorocarbons (PFCs), and sulfur hexafluoride (SF_6) (see Table 5-1). Emission estimates presented here span the interval from 1990 to 2100 at the global level and at the level of four SRES macro-regions (OECD90, REF, ASIA, and ALM; see Appendix IV). In addition, sulfur emission estimates are presented in the regional gridded format to assist in quantifying the effects at the local level. Links between emissions and the underlying driving forces presented in Chapter 4 are illustrated also.

As a result of differences in modeling and estimation approaches, base-year (1990) emission values in SRES scenarios developed using different models show substantial variation. To facilitate both use of the scenarios and comparisons across scenarios and families, base-year emissions in all 40 SRES scenarios were standardized using one common set of values (see Box 5-1). With a few clearly indicated exceptions, only standardized emission values are discussed in this chapter. A complete set of standardized emissions, along with other quantitative scenario information, is provided in Appendix VII.

The first section in this chapter presents a "roadmap" that serves as an orientation to the 40 SRES scenarios. The roadmap gives a simple taxonomy that compares input parameters, as represented by the four scenario families, and emission outputs, as represented by the 1990 to 2100 cumulative CO_2 emissions. The subsequent sections discuss in detail emissions of each gas over the next 100 years to 2100.

5.2. Roadmap to the Scenarios

A classification scheme is presented here to assist the reader in understanding the links between driving forces and scenario outputs. This scheme can also be used to help select appropriate scenarios for further analysis (see Chapter 6).

The SRES scenarios were developed as quantitative interpretations of the four alternative storylines that represent possible futures with different combinations of driving forces. These broad scenario families are broken down further into seven scenario groups,[2] used here to classify the input driving forces (see also Table 4-20 in Chapter 4).

The scenario outputs of most interest are emissions of GHGs, SO_2, and other radiatively important gases. However, the categorization of scenarios based on emissions of multiple gases is quite difficult. All gases that contribute to radiative forcing should be considered, but methods of combining gases such as the use of global warming potentials (GWP) are appropriate only for near-term GHG inventories.[3] In addition, emission trajectories may display different dynamics, from monotonic increases to non-linear trajectories in which a subsequent decline from a maximum occurs. This particularly diminishes the significance of a focus on any given year, such as 2100. In light of these difficulties, the classification approach presented here uses cumulative CO_2 emissions between 1990 and 2100. CO_2 is the dominant GHG and cumulative CO_2 emissions are expected to be roughly proportional to CO_2 radiative forcing over the time scale considered (Houghton *et al.* 1996).

Total cumulative CO_2 emissions from the 40 SRES scenarios fall into the range from 773 to 2538 gigatonnes of carbon (GtC) with a median of 1509 GtC. To represent this range, the scenario classification uses four intervals:

- Less than 1100 GtC (low).
- Between 1100 and 1450 GtC (medium–low).
- Between 1450 and 1800 GtC (medium–high).
- Greater than 1800 GtC (high).

Each CO_2 interval contains multiple scenarios and scenarios from more than one family. Each category also includes one of the four marker scenarios. Figure 5-1 shows how cumulative CO_2 emissions from the 40 SRES scenarios fit within the selected emission intervals.

Table 5-2 provides an overview of this scenario classification and links the scenario outcomes with factors that drive them, organized by family and scenario group (see also Table 4-20). The rows in Table 5-2 represent the emission categories, while the columns represent the scenario families. The analysis of Table 5-2 reveals two key results:

[1] Emission trajectories of these ozone-depleting substances (ODSs) are not developed by the SRES team, but adopted from WMO/UNEP (1998).

[2] During the approval process of the Summary for Policymakers at the 5th Session of Working Group III (WGIII) of the IPCC from 8 to 11 March 2000 in Katmandu, Nepal, it was decided to combine the A1C and A1G groups into one "fossil intensive" group A1FI in contrast to the non-fossil group A1T, and to select two illustrative scenarios from these two A1 groups to facilitate use by modelers and policy makers. This leads to six scenario groups that constitute the four scenario families, three of which are in the A1 family. All scenarios are equally sound.

[3] In particular, the IPCC Working Group I (WGI) Second Assessment Report (SAR) GWPs are calculated for constant concentrations. In long-term scenarios, concentrations may change significantly, as do GWP values. It is unclear how to apply GWPs to long-term scenarios in a meaningful manner. In addition, the GWP approach is not applicable to gases such as SO_2 and ozone precursors.

Table 5-1: *Overview of greenhouse gases (GHGs), ozone precursors, and sulfur emissions for the SRES scenario groups. Numbers are for the four markers and (in brackets) for the range across all scenarios from the same scenario group (standardized emissions, see also footnote 2).*

	CO_2 (GtC)	CH_4 ($MtCH_4$)	N_2O (MtN)	HFC, PFC, SF_6 (MtC equiv.)	CO (Mt CO)	NMVOCs (Mt)	NO_x (MtN)	SO_x (MtS)
A1B	13.5 (13.5-17.9)	289 (289-640)	7.0 (5.8-17.2)	824 combined	1663 (1080-2532)	193 (133-552)	40.2 (40.2-77.0)	27.6 (27.6-71.2)
A1C	(25.9-36.7)	(392-693)	(6.1-16.2)	same as in A1	(2298-3766)	(167-373)	(63.3 -151.4)	(26.9-83.3)
A1G	(28.2-30.8)	(289-735)	(5.9-16.6)	same as in A1	(3260-3666)	(192-484)	39.9 -132.7)	(27.4-40.5)
A1T	(4.3-9.1)	(274-291)	(4.8-5.4)	same as in A1	(1520-2077)	(114 -128)	(28.1-39.9)	(20.2-27.4)
A2	29.1 (19.6-34.5)	889 (549-1069)	16.5 (8.1-19.3)	1096 combined	2325 (776-2646)	342 (169-342)	109.2 (70.9-110.0)	60.3 (60.3-92.9)
B1	4.2 (2.7-10.4)	236 (236-579)	5.7 (5.3-20.2)	386 combined	363 (363-1871)	87 (58-349)	18.7 (16.0-35.0)	24.9 (11.4-24.9)
B2	13.3 (10.8-21.8)	597 (465-613)	6.9 (6.9-18.1)	839 combined	2002 (661-2002)	170 (130-304)	61.2 (34.5-76.5)	ß47.9 (33.3-47.9)

Box 5-1: Scenario Standardization

One of the primary reasons for developing emissions scenarios is to enable coordinated studies of climate change, climate impacts, and mitigation options and strategies. With the multi-model approach used in the SRES process, 1990 and 2000 emissions do not agree in scenarios developed using different models. In addition, even with agreed reference values, it is time consuming and often impractical to fine-tune most integrated assessment models to reproduce a particular desired result.

Nevertheless, differences in the base year and 2000 emissions may lead to confusion among the scenario users. Therefore, the 1990 and 2000 emission estimates were standardized in all the SRES scenarios, with emissions diverging after the year 2000. The procedure for selecting 1990 and 2000 emission values and the subsequent adjustments to scenario emissions are described in this box.

The standardized scenarios share the same values for emissions in both 1990 and 2000. Emissions for the year 2000 are, of course, not yet known and 1990 emissions are also uncertain. The 1990 and 2000 emission estimates for all gases, except SO_2, were set to be equal to the averages of initial values in the unadjusted four marker scenarios. This was carried out at the four-region level, and summed to obtain the standardized global totals. The resultant estimates are within relevant uncertainty ranges for each substance, but should not be interpreted as "endorsed" by the Intergovernmental Panel on Climate Change (IPCC) to represent values for either global or regional emissions. Rather, they are the standardized base-year estimates used for the emissions scenarios.

From 2000 to 2100, emissions in all the scenarios (except CO_2 emissions from land-use and SO_2 emissions) were adjusted by applying a constant offset equal to the difference between the standardized 2000 value and the scenario-specific 2000 value. The purpose was to smooth scenario trajectories between 2000 and 2010.

This procedure results in small distortions for those emissions that rise with time, or at least that do not ultimately decrease by a large amount as compared to the base year. However, for emissions that fall significantly over time, such as those from deforestation or of SO_2, this procedure can cause more significant distortions and can even change the sign of the emission estimates at later times (i.e., change positive emission estimates into negative ones and vice versa). To avoid these distortions, for the aforementioned emissions, the year 2000 offset was reduced by 10% per decade, cumulatively, to make the offset zero by the year 2100. This allows preservation of the shape of emission trajectories and still ensures the 2000 standardization.

The non-standardized scenario values are available from the modeling teams upon request, although the standardized values should be used for most purposes.

- Similar future cumulative CO_2 emissions may emerge from different sets of driving forces;
- Conversely, similar future states of the world with respect to socio-economic development may yield different outcomes in terms of cumulative CO_2 emissions.

Similar cumulative CO_2 emissions can be attained in very different social, economic, and technological circumstances. *High* emission levels of the A2 marker scenario (A2-ASF) are also attained in all the A1 family scenarios with high fossil-fuel use (A1C and A1G groups), for example in A1G-MiniCAM. *Medium–high* emissions are attained in most of the A1 group scenarios (including the A1B marker, A1B-AIM), but also in scenarios from the B2 scenario family with high fossil-fuel use (e.g., B2-ASF). *Medium–low* emissions, which are characteristic of the B2 family, including the B2 marker (B2-MESSAGE), are also attained in the A2-A1-MiniCAM scenario, which illustrates the transition between the A2 and A1 families. Finally, *low* emission levels result from almost all the B1 family scenarios (including the B1 marker, B1-IMAGE) as well as from scenarios that belong to the A1T high-technology scenario group (e.g., A1T-MARIA). In Table 5-2, italics are used for examples of scenarios within each emission category that illustrate alternative ways to achieve cumulative CO_2 emissions similar to those of the marker scenarios.

5.3. Carbon Dioxide

CO_2 is the largest contributor to anthropogenic radiative forcing of the atmosphere. As described in more detail in Chapter 3, the main sources of anthropogenic CO_2 emissions are fossil fuel combustion and the net release of carbon from changes in terrestrial ecosystems, commonly referred to as land-use changes. To a lesser extent, CO_2 is emitted by industrial activities, in particular by cement production (Table 5-3).

5.3.1. Carbon Dioxide Emissions from Fossil Fuels and Industry

As shown in Table 5-3, fossil fuels were the main source of CO_2 emissions in 1990. Therefore, it is expected that future CO_2 emission levels will depend primarily on the total energy consumption and the structure of energy supply. The total

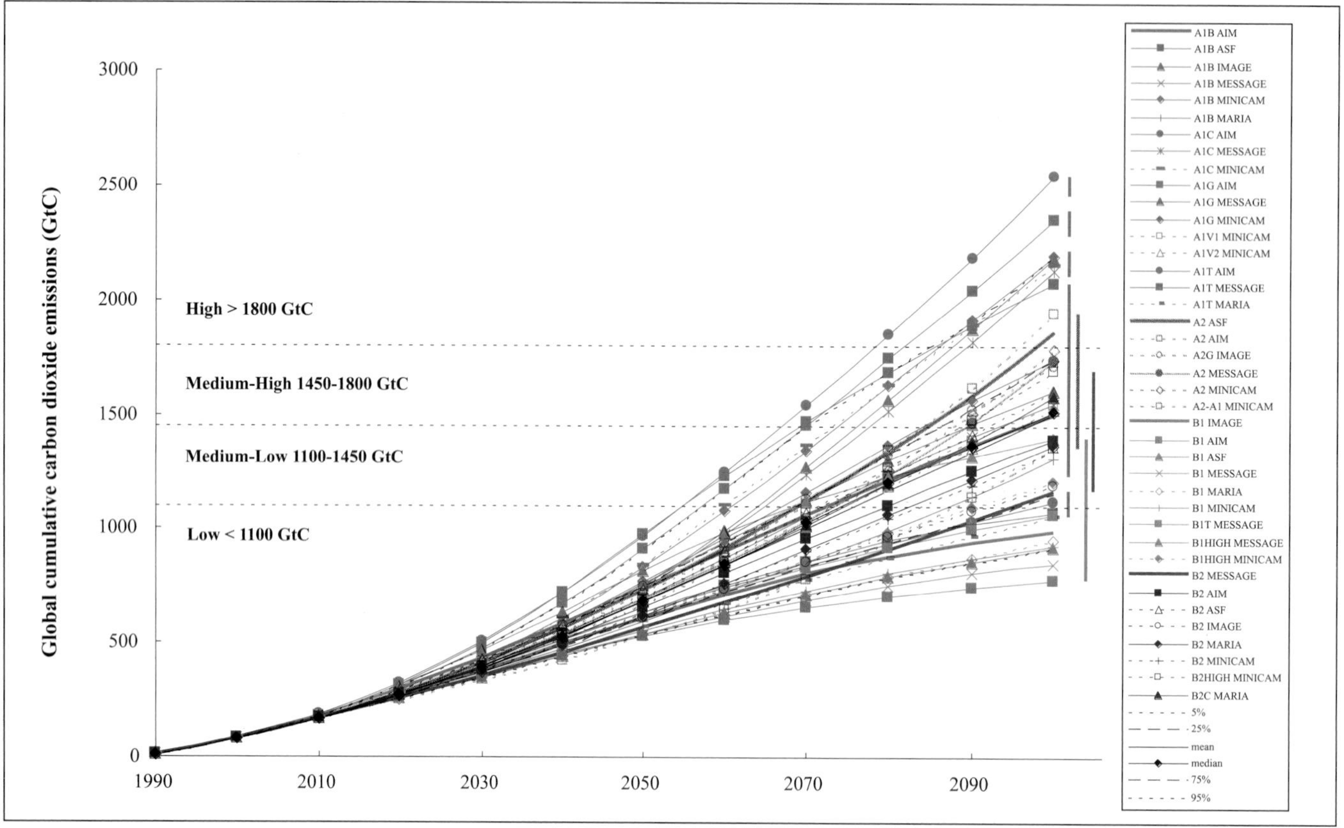

Figure 5-1: Global cumulative CO_2 emissions in the 40 SRES scenarios, classified into four scenario families (each denoted by a different color code: A1, red; A2, brown; B1, green; B2, blue). Marker scenarios are shown with thick lines without ticks, globally harmonized scenarios with thin lines, and non-harmonized scenarios with thin, dotted lines. Black lines show percentiles, means, and medians for the 40 SRES scenarios. For numbers on the two additional illustrative scenarios A1FI and A1T see Appendix VII.

energy consumption is driven by population size, level of affluence, technological development, environmental concerns, and other factors. The composition of energy supply is determined by estimated reserves of fossil fuel and the availability, relative efficiency, and cost of supply technologies.

Emissions from gas flaring and industrial emissions are much lower in comparison with energy-related emissions; for simplification, in this discussion they are added to the latter. In 1990, the global emissions from cement made up about 2.5% of the total global CO_2 emissions (Houghton *et al.*, 1995).

Figure 5-2 shows standardized carbon emissions from fossil energy and industry for the 40 SRES scenarios. Sample statistics (in terms of percentiles, means, and medians) are indicated against the background of 40 individual scenarios that make up the SRES scenario set. The figure also presents emissions ranges spanned by each of the four scenario families in 2100.

SRES scenarios cover a wide range of annual emissions, and the uncertainties in future emission levels increase with time. Up to about the 2040s and the 2050s, emissions tend to rise in all scenarios, albeit at different rates. Across scenarios this reflects changes in the underlying driving forces, such as population, economic output, energy demand, and the share of fossil fuels in energy supply. By 2050, the emissions range covered by the 40 SRES scenarios is from about 9 to 27 GtC, with the mean and median values equal to about 15 GtC. The range between the 25th and 75th percentiles of emissions (the "central tendencies") extends from 12 to 18 GtC (i.e. from twice to thrice that in 1990). Within this interval lie three of the four marker scenarios. However, a fair number of scenarios (eight out of 40) also indicate the possibility of much higher emissions (in the 18 to 27 GtC range) that reflect an increase by a factor of up to 4.5 over 60 years (1990 to 2050). Another eight SRES scenarios have 2050 emissions below the 25th percentile (Figure 5-2).

Beyond 2050, the uncertainties in energy and industrial CO_2 emissions continue to increase. By 2100, the range of emissions across the 40 SRES scenarios is between 3 and 37 GtC, which reflects either a decrease to half the 1990 levels or an increase by a factor of six. Emissions between the 25th and 75th percentiles range from 9 to 24 GtC, while the range of the four marker scenarios is even wider, 5 to 29 GtC. The 2100 median and mean of all 40 scenarios are 15.5 and 17 GtC, respectively.

As time passes in the scenarios, uncertainties not only increase with respect to absolute levels of CO_2 emissions, but also with

Table 5-2: *Scenario classification according to scenario family and cumulative total carbon dioxide emissions (fossil, industrial, and net deforestation) from 1990 to 2100. Markers are in bold print, the two additional illustrative scenarios of the A1 scenario group (see footnote 2 of text) in italics; all harmonized scenarios have shaded background; and examples of scenarios with similar CO_2 emissions (see text) are underlined.*

Cumulative Carbon Emission 1990-2100

Family	**A1**				**A2**	**B1**	**B2**
Scenario group	**A1C**	**A1G**	**A1B**	**A1T**	**A2**	**B1**	**B2**
High **>1800GtC**	A1C-AIM A1C-MESSAGE A1C-MiniCAM	A1G-AIM A1G-MESSAGE *A1G-MiniCAM*	A1B-ASF		**A2-ASF** A2-AIM		
Medium-High **1450-1800GtC**			**A1B-AIM** A1B-IMAGE A1B-MESSAGE A1B-MiniCAM A1v1-MiniCAM A1v2-MiniCAM		A2-MESSAGE A2G-IMAGE A2-MiniCAM		B2C-MARIA B2-ASF B2High-MiniCAM
Medium-Low **1100-1450 GtC**			A1B-MARIA	A1T-AIM	A2-A1-MiniCAM	B1-ASF B1High-MiniCAM	**B2-MESSAGE** B2-AIM B2-MARIA B2-IMAGE B2-MiniCAM
Low **<1100 GtC**				*A1T-MESSAGE* A1T-MARIA		**B1-IMAGE** B1-AIM B1-MESSAGE B1-MiniCAM B1High-MESSAGE B1T-MESSAGE B1-MARIA	
Marker scenario			**1499**		**1862**	**983**	**1164**
Illustr. scenario	*2189*			*1068*			
Harmonized scenarios	2127-2538	2178-2345	1301-2073	1068-1114	1732	773-1390	1359-1573
Other scenarios	2148	2189	1519-1731	1049	1352-1938	947-1201	1186-1686

***Table 5-3**: Global CO_2, CH_4, N_2O, and CO emissions in the base year (1990) by source.*

CO_2 [GtC]	SAR	SRES Range	SRES Standardized
Total anthropogenic	*6.0-8.2*	*7.0-7.5*	*7.1*
Fossil fuel and cement production*	5.0-6.0	5.8-6.5	6.0
Land-use change	0.6-2.6	1.0-1.4	1.1
CH_4 (Mt CH_4)			
Total anthropogenic	*300-450*	*298-337*	*310*
Fossil fuel related	70-120	68-94	
Total biospheric	*200-350*	*204-250*	
Enteric fermentation	65-100	}80-97	
Animal waste	20-30		
Rice paddies	20-100	29-61	
Biomass burning	20-80	27-46	
Landfills	20-70	}51-62	
Domestic sewage	15-80		
Total natural	*110-210*		
Total identified	*410-660*		
N_2O (MtN)			
Total anthropogenic	*3.7-7.7*	*6.0-6.9*	*6.7*
Cultivated soils	1.8-5.3	}4.2-4.8	
Cattle and feed lots	0.2-0.5		
Biomass burning	0.2-1.0	0.4-1.3	
Industrial sources	0.7-1.8	0.9-1.2	
Total natural	*6-12*		
Total identified	*10-17*		
CO (Mt CO)			
Technological	300-500	*}752-1000*	*}879*
Biomass burning	300-700		
Biogenics	60-160		
Oceans	20-200		
Methane oxidation	400-1000		
NMVOC oxidation	200-600		
Total	*1800-2700*		

*Of which around 0.16 GtC are from cement production.

respect to their trajectories. Scenarios portray different emission patterns that range from continuous increases up to 2100, through emissions that gradually level off by 2100, to trend reversals in which emissions begin to decline in the second half of the 21st century. These dynamic emission patterns diminish the significance of emission levels in any particular year, such as 2100.

Over a long time horizon, it also becomes increasingly difficult to perceive the future in terms of "central tendencies." For instance, between 2050 to 2100, up to eight different scenarios from all four SRES scenario families have CO_2 emission levels within 10% of the median of all 40 SRES scenarios. Thus, there is no single scenario family or individual scenario that has "median" emissions with respect to the entire uncertainty space described by the 40 SRES scenarios. Similar emission levels can arise from very different combinations of driving-force variables that are embedded in the SRES scenario families and groups.

The wide ranges of energy and industry-related CO_2 emissions in the SRES scenarios reflect the fact that the "best" or the "most likely" quantifications are nearly impossible to identify. The following discussion suggests that even scenarios with very similar input parameters (e.g., population and GDP) may produce a large variation in resultant CO_2 emissions.

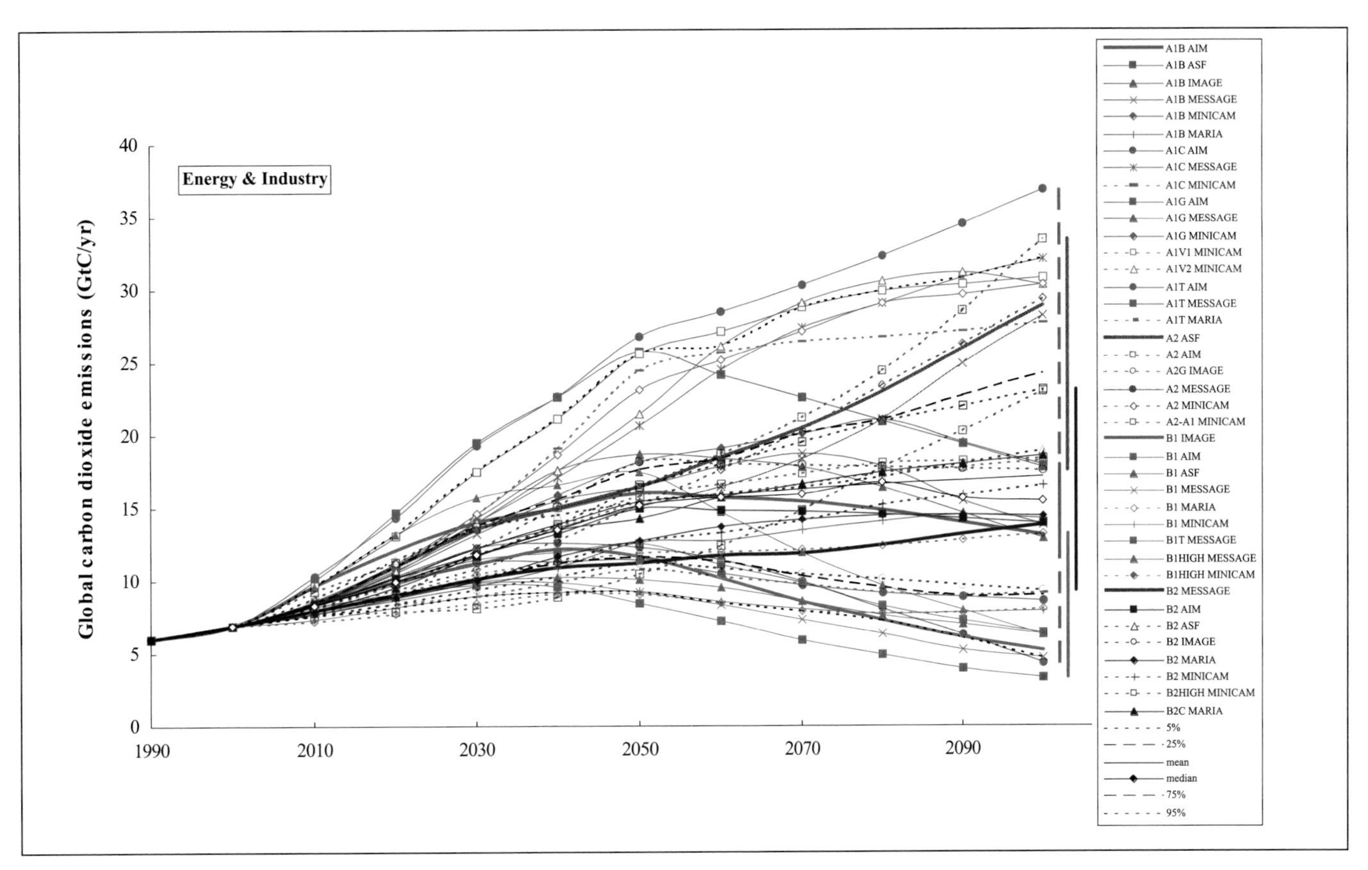

Figure 5-2: Standardized global energy-related and industrial CO_2 emissions for 40 SRES scenarios, classified into four scenario families (each denoted by a different color code: A1, red; A2, brown; B1, green; B2, blue). Marker scenarios are shown with thick lines without ticks, globally harmonized scenarios with thin lines, and non-harmonized scenarios with thin, dotted lines (see Table 4-3). Black lines show percentiles, means, and medians for the 40 SRES scenarios. For numbers on the two additional illustrative scenarios A1FI and A1T see Appendix VII.

5.3.1.1. A1 Scenario Family

Rapid economic growth in the A1 family scenario leads to high energy demand and hence to a steep increase in CO_2 emissions in the first decades of the 21st century. Structural changes in the energy supply become effective only in the longer term because of the inertia caused by long periods of capital turnover. With respect to alternative energy supply technologies, the A1B scenario group assumes a "balanced" approach, in which none of these technologies gain an overwhelming advantage. This scenario group includes the A1B marker scenario developed using the AIM model (Jiang *et al.*, 2000). In addition to the A1 scenario group, variations that assumed fast advancements in specific energy supply sectors were explored (renewables, A1T; oil and gas, A1G; coal, A1C) (see Chapter 4).

In the A1B-AIM marker scenario, the global average per capita final energy demand grows from 54 GJ in 1990 to 115 GJ in 2050 and to 247 GJ in 2100. Meanwhile, the final energy carbon intensity declines relatively slowly until 2050 (from the current 21 tC per TJ of final energy to 16 tC per TJ), which results in a steep increase in CO_2 emissions in the first decades of the 21st century. After 2050, when structural changes in the energy sector take effect, carbon intensity declines rapidly to reach 7.5 tC per TJ. This more than offsets growing energy demand from a contracting but increasingly prosperous population, so that carbon emissions decline between the years 2050 and 2100. Emissions peak around 2050 at a level 2.7 times (16 GtC) that of 1990 and fall to around 13 GtC by 2100, which is about twice the current level (Figure 5-3a). The total, cumulative 1990 to 2100 carbon emissions[4] in the A1B-AIM scenario equal 1499 GtC (cumulative emissions from energy and industry only amount to 1437 GtC).

Energy-related carbon emissions vary widely across alternative A1 group scenarios. The smallest differences with respect to the A1B marker scenario (A1B-AIM) are in the A1B-MESSAGE (Roehrl and Riahi, 2000) scenario, which used harmonized input assumptions for population, GDP, and final energy use at the level of the four regions (see Chapter 4). Resultant energy-related carbon emissions in A1B-MESSAGE closely match those of the A1B-AIM scenario, reaching 13.8 GtC by 2100, which indicates that energy supply technologies agree well between the AIM and MESSAGE models.

[4] Note that unless stated otherwise cumulative emissions reported in this chapter always refer to total carbon, while annual emissions in this section refer to energy and industry emissions only.

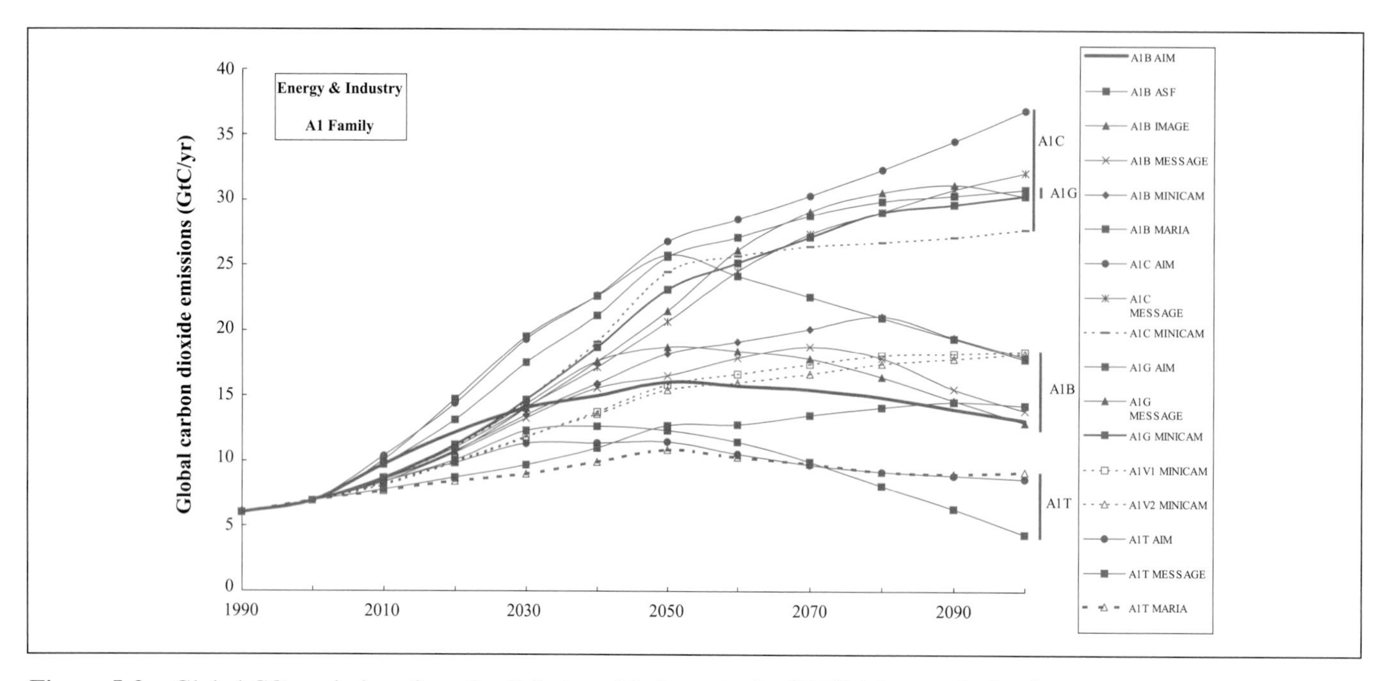

Figure 5-3a: Global CO_2 emissions from fossil fuels and industry in the SRES A1 scenario family (standardized). The marker scenario is shown with a thick line without ticks, the globally harmonized scenarios with thin lines, and the non-harmonized scenarios with thin, dotted lines (see Table 4-3). In the SPM, A1C and A1G scenarios are merged into one fossil-intensive A1FI scenario group (see also footnote 2).

While still adhering to the "balanced" supply-mix assumption (see Section 4.3.1), other scenarios within the A1 balanced group span a larger range of future emissions around the A1B-AIM scenario. In part this results from different assumptions on global and regional GDP growth and energy-intensity improvements, but also from alternative interpretations of how to translate the concept of a "balanced" pathway of technological change (as described in the A1 scenario storyline) into model-specific technology assumptions.

The A1B-ASF and A1B-IMAGE scenarios depict futures with a stronger reliance on fossil fuels (especially coal) than that in A1B-AIM. In 2050 this results in higher emissions (19 to 26 GtC versus 16 GtC for the A1B-AIM; see statistical tables in Appendix VII for more details). The emission pattern in the A1B-MARIA scenario (Mori, 2000) is quite different. This scenario does not reproduce an initial rapid growth in emissions, which reach a maximum around 2050 and subsequently decline. Instead, emissions grow more slowly and then level off. By 2050, the emissions in A1B-MARIA are well below those in A1B-AIM (13 versus 16 GtC). Two out of three A1 scenarios developed using the MiniCAM model (A1V1-MiniCAM and A1V2-MiniCAM) were based on somewhat different assumptions with respect to key scenario driving forces, which included lower energy intensity, less rapid decline in population after 2050, and lower incomes per capita (see Chapter 4 for more details). Emissions in these scenarios increase continuously during the 21st century to reach about 18 MtC by 2100 (see statistical tables in Appendix VII for more details).

The emissions range spanned by all the A1 balanced scenario group in 2100 is from 13 to 18 GtC. Total, cumulative carbon emissions range from 1301 to 2073 GtC by 2100. This illustrates the importance of different assumptions on rates and direction of technological change in the energy sector for long-term emissions scenarios, which is further explored in the three additional scenario groups discussed below.

Two alternative A1 scenario groups, A1G (oil and gas) and A1C (coal), explore high-fossil futures and one group, A1T, explores advanced technology futures. A1G and A1C have been combined into one fossil intensive scenario group A1FI in the SPM (see also footnote 2). Energy-related carbon emissions vary widely across these scenario groups in conjunction with the scenario assumptions as to roles of different energy sources (Figure 5-3a). The high fossil scenarios explore worlds that consume vast amounts of coal, or oil and gas, while with the advanced technology assumptions the role of fossil fuel declines strongly as nuclear and/or renewable sources are favored to supply the reduced final energy demand.

The A1 fossil scenarios have essentially the same population, GDP, and final energy demand as the A1B marker, but have a primary energy system that relies mainly on fossil fuels. Based on cumulative CO_2 emissions, scenarios from the A1 fossil groups (A1G and A1C) fall within the high emissions category, with A1G-MiniCAM falling roughly in the middle of the range spanned by the six A1 fossil scenarios (Table 5-2). The primary energy system in the A1G-MiniCAM scenario emerges as a result of rapid increases in the efficiency with which fossil resources can be recovered, especially in the most expensive grades of these resources; unconventional resources play an increasingly important role toward the end of the 21st century.

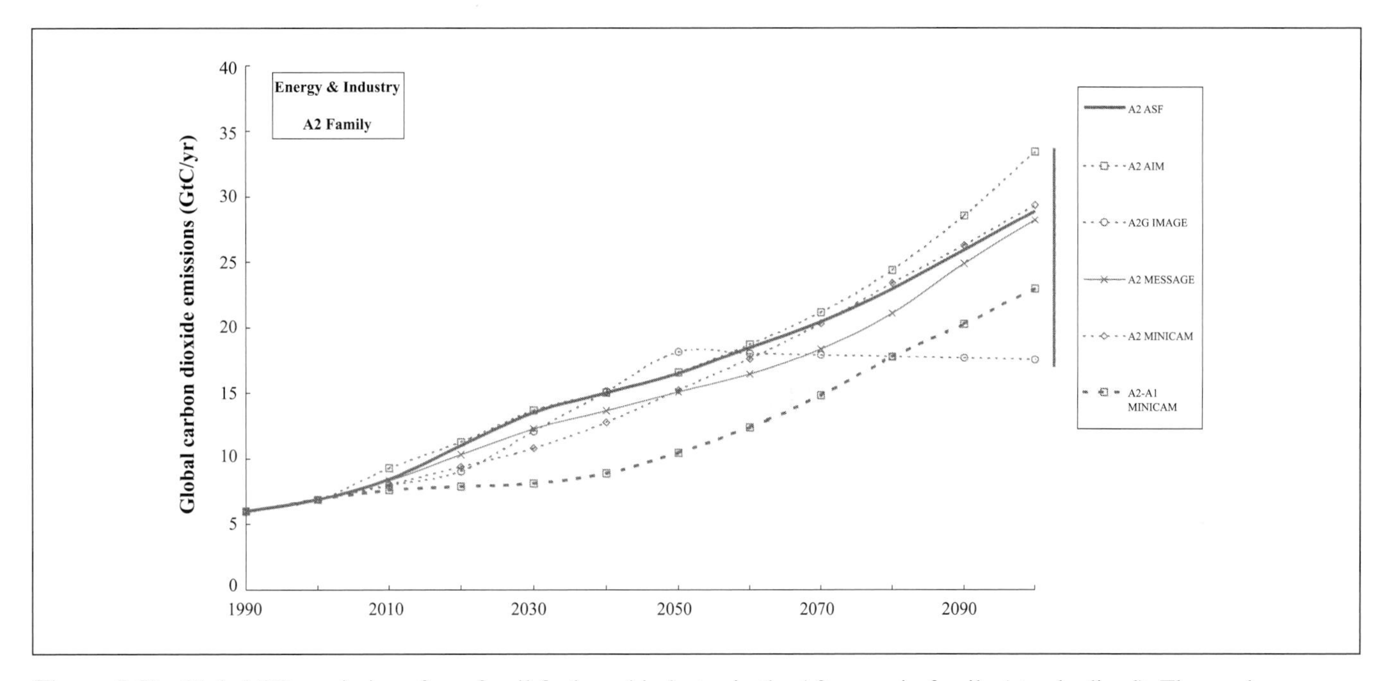

Figure 5-3b: Global CO_2 emissions from fossil fuels and industry in the A2 scenario family (standardized). The marker scenario is shown with a thick line without ticks, the globally harmonized scenarios with thin lines, and the non-harmonized scenarios with thin, dotted lines (see Table 4-3).

These rapid technical changes, as well as relatively low rates of growth in the productivity of biomass and other non-fossil sources, imply a world that is heavily reliant on oil and gas. The three coal-based scenarios in this group (A1C-MESSAGE, A1C-MiniCAM, and A1C-AIM) assume rapid increases in the efficiency of synthetic fuel production, as well as relatively limited oil and gas resources.

Three advanced technology or A1T scenarios fall in either the "low" or the "medium–low" cumulative emissions categories (Table 5-2). The A1T-MARIA scenario (Mori, 2000) achieves very low carbon emissions through the combination of a low final energy use (54% of that in A1B-AIM) and a rapid increase in the importance of non-fossil energy technologies in the primary energy system (81% by 2100).

The resultant CO_2 emissions range of all the A1 family scenarios is so wide that most of the remaining SRES scenarios fall within its bandwidth, from 4.3 to 37 GtC in 2100. The total cumulative carbon emissions of the A1 family scenarios also span a very wide range, from around 1000 GtC to more than 2500 GtC (Figure 5-1, Table 5-2).

5.3.1.2. A2 Scenario Family

In the A2 scenario family, alternative energy technologies develop relatively slowly and fossil fuels maintain their dominant position in the energy supply mix. As oil and gas resources become scarcer and non-fossil alternatives remain underdeveloped, coal gains the leading role. Its share in the energy mix ranges from 45% to 52% in the harmonized scenarios (A2-ASF, A2-MESSAGE) and in other scenarios with similar input assumptions (A2-MiniCAM and A2-AIM).

Global CO_2 emissions in the A2 marker scenario implemented using the ASF model (Sankovski *et al.*, 2000) increase by more than fourfold over their 1990 level, to reach 29 GtC by 2100 (Figure 5.3b). Total cumulative carbon emissions from the A2-ASF scenario amount to 1860 GtC by 2100.

In other A2 scenarios, CO_2 emissions range from 17 GtC (A2G-IMAGE[5]) to 33 GtC (A2-AIM) by 2100. The harmonized A2-MESSAGE scenario and the A2-MiniCAM scenario have very similar global CO_2 emissions as compared to the marker. Slightly higher emissions, primarily caused by higher primary energy use, are generated in the A2-AIM scenario. Unlike the rest of the A2 family scenarios, the A2G-IMAGE scenario yields constant emissions after 2050. This emission trajectory is explained by the combination of a lower energy demand and a larger share of natural gas in the energy supply mix (see Chapter 4 for more details). Total cumulative carbon emissions in the A2 scenario group range between 1710 and 1860 GtC by 2100.

A variant of the A2 storyline, developed with the MiniCAM model (A2-A1-MiniCAM), explores a world in which economic growth and population follow a "hybrid" trajectory, which have traits of both the A2 and A1 storylines. As

[5] The IMAGE results for the A2 and B2 scenarios are based on preliminary model experiments carried out in March 1998. As a result of limited resources it has not been possible to re-run these experiments. Hence, unlike for the IMAGE A1 and B1 scenarios, the IMAGE team has not been able to provide background data and details for these scenario calculations and the population and economic growth assumptions are not harmonized fully.

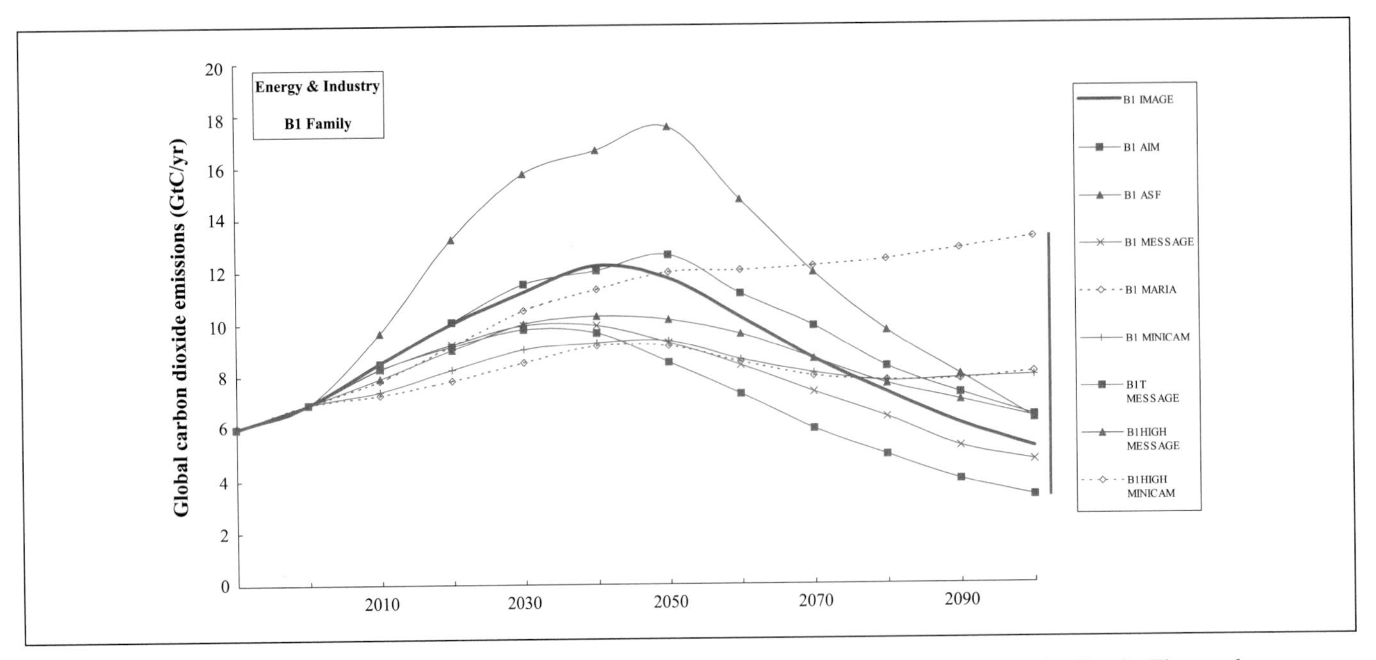

Figure 5-3c: Global CO_2 emissions from fossil fuels and industry in the B1 scenario family (standardized). The marker scenario is shown with a thick line without ticks, the globally harmonized scenarios with thin lines, and the non-harmonized scenarios with thin, dotted lines (see Table 4-3).

described in Box 4-6, A2-A1-MiniCAM has a lower population than other scenarios of the A2 family, and a slower rate of economic growth in the first half of the 21st century. These in turn limit the growth in final energy demand until after the onset of rapid economic growth around 2050. Emissions also rise rapidly during the post-2050 period, as the large growth in energy consumption results mostly from coal, oil, and gas.

Two out of the six A2-family scenarios, A2-ASF and A2-AIM, fall into the "high" cumulative CO_2 emissions category. Three scenarios, A2-MESSAGE, A2-IMAGE, and A2-MiniCAM, belong to the "medium–high" category, and only the A2-A1-MiniCAM scenario with its drastically different assumptions about key driving forces falls in the "medium–low" category.

5.3.1.3. B1 Scenario Family

The strong trend toward ecologically more compatible consumption and production patterns in the B1 family is reflected by structural changes toward less energy- and material-intensive activities, which lead to a partial decoupling of welfare and energy demands. In the B1 marker scenario (B1-IMAGE; de Vries *et al.*, 2000) the rapid technological change toward resource saving and ecologically sound solutions is assumed to spread very quickly, facilitated by high capital stock turnover rates in currently less developed regions. As a result, energy requirements in B1-IMAGE increase slowly and a shift away from fossil fuels eventually breaks the already slow upward trend in carbon emissions (Figure 5-3c). Emissions peak around 2040 at 12 GtC, twice the 1990 level, and by 2100 the emissions fall below the base-year level to 5 GtC. Total cumulative carbon emissions in the B1-IMAGE scenario amount to 983 GtC by 2100. As for A1, the population projection adopted for this scenario family declines after 2050.

Rates of energy-intensity improvement in the first half of the 21st century range quite widely in B1 family scenarios, and lead to emission levels from 8.5 (B1T-MESSAGE) to 17.5 (B1-ASF) GtC in 2050. By 2100 the gap in annual emissions narrows again, with final emissions between 3 and 8 GtC in all the B1 scenarios (except B1-High-MiniCAM).

The B1 family also includes one scenario in which energy-related emissions continue to increase throughout the modeling period, B1-High-MiniCAM. In this scenario the final energy demand is assumed to rise more rapidly with increasing income than in the rest of the B1 scenarios. (The total carbon emissions in B1High-MiniCAM decline slightly later in the modeling period because of reduced land under management and associated carbon sequestration.)

Total cumulative carbon emissions in the B1 scenario group ranges between 770 and 1390 GtC by 2100. All but two B1 scenarios fall in the low cumulative emissions category (Table 5-2).

5.3.1.4. B2 Scenario Family

In the B2 world, dynamics of technological change continue along historical trends ("dynamics as usual"). The exploitation of comparative regional advantages in energy resources and technologies leads to regionally different mixes of clean fossil and non-fossil supply. With the continued growth of population and of income per capita, a steady increase of CO_2 emissions

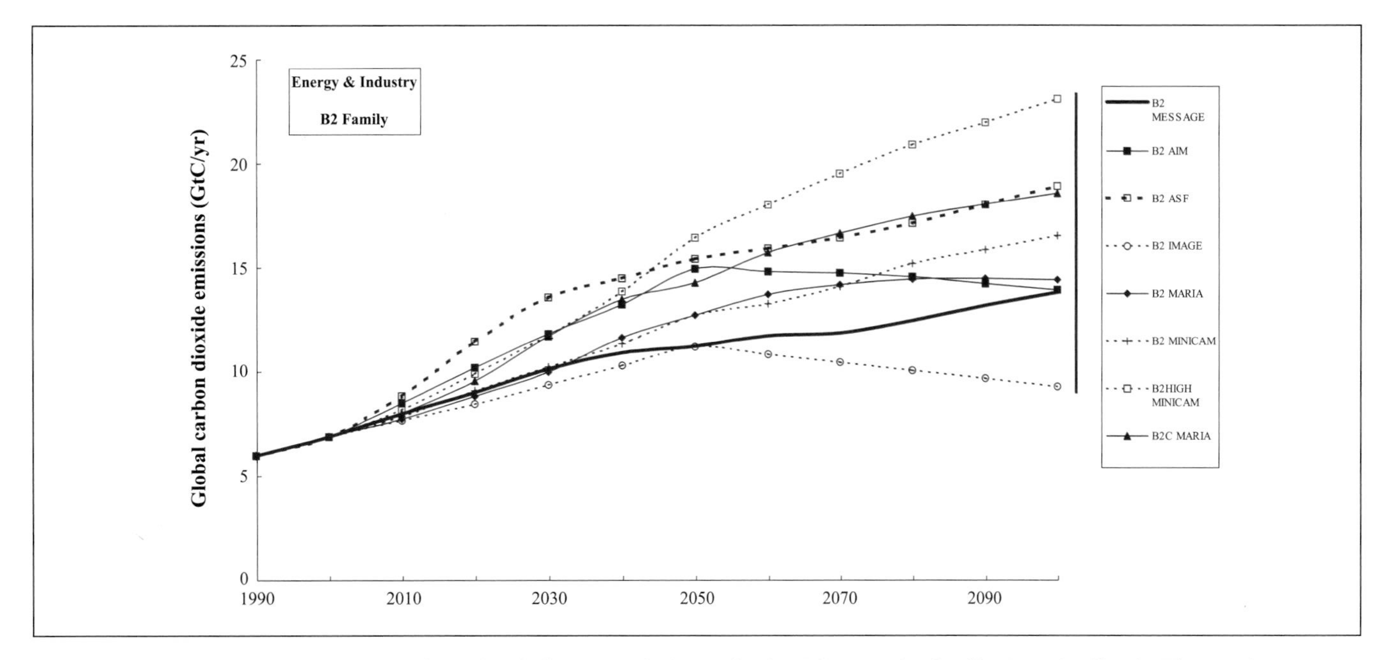

Figure 5-3d: Global CO_2 emissions from fossil fuels and industry in the B2 scenario family (standardized). The marker scenario is shown with a thick line without ticks, the globally harmonized scenarios with thin lines, and the non-harmonized scenarios with thin, dotted lines (see Table 4-3).

emerges in the B2 marker (Riahi and Roehrl, 2000), developed using the MESSAGE model (B2-MESSAGE). By 2050 emissions reach 11 GtC and by 2100 they reach 14 GtC (Figure 5-3d). Total cumulative CO_2 emissions in the B2 marker scenario amount to 1160 GtC by 2100.

Emissions in the B2 scenarios with harmonized global input assumptions (population, GDP, final energy; B2-MESSAGE, B2-AIM, B2-MARIA) are very close in 2100. Differences in emissions are largest around 2050, which reflects the different patterns of structural change in the energy systems in anticipation of depletion of conventional oil and gas. The emissions range for non-harmonized scenarios in the B2 scenario group is larger – between 11.2 (B2-IMAGE[6]) and 15.4 (B2-ASF) GtC by 2050. Relatively high emissions in the B2-ASF scenario are explained by a large share of coal in the fuel mix, because of high oil and gas prices. In this scenario, coal is also widely used for synthetic liquid and gaseous fuel production. In 2100, B2-IMAGE emissions (9 GtC) drop below the B2 marker level, while in B2-ASF emissions continue to grow and reach 19 GtC by 2100. The B2 high scenario variant developed using MiniCAM assumes far less efficiency gains, smaller available resources of oil and gas, and less favorable development of solar power costs than in the B2 marker. As in B2-ASF, the near exhaustion of oil supplies in B2High-MiniCAM leads to a heavy reliance on synthetic fuels to supply the needs of the transportation sector. Emissions in this scenario increase to 23 GtC by 2100. An additional alternative with more coal use (B2C-MARIA) was explored using the MARIA model. Emissions in this scenario are considerably higher than in the original B2-MARIA case, but are very close to those of B2-ASF (19 GtC).

Total cumulative carbon emissions across the B2 scenario group range between 1164 and 1686 GtC by 2100. Three of the eight B2 family scenarios fall into the "medium–high" cumulative emissions category, and the five others fall into the "medium–low" category (Table 5-2).

5.3.1.5. Inter-family Comparison

Table 5-2 suggests fairly strong contrasts in the level of cumulative emissions across the four scenario families (note that estimates in Table 5-2 also include emissions from land use). A variety of available energy supply and demand options means that cumulative emissions in the A1 family span the entire set of emission categories. The A2 family scenarios, characterized by a large population and a relatively carbon-intensive energy system, fall into either the "high" scenario category or in the upper part of the "medium–high" category. Only the A2-A1-MiniCAM scenario falls in the "medium–low" category, because of low cumulative emissions prior to 2050. As a consequence of the low energy consumption and non-fossil energy systems associated with its sustainable development theme, the B1 family scenarios are concentrated in the "low" emissions category. Finally, representatives of the B2 family are present in the two middle categories.

[6] The IMAGE results for the A2 and B2 scenarios are based on preliminary model experiments carried out in March 1998. As a result of limited resources it has not been possible to re-run these experiments. Hence, unlike for the IMAGE A1 and B1 scenarios, the IMAGE team has not been able to provide background data and details for these scenario calculations and the population and economic growth assumptions are not harmonized fully.

In addition to their cumulative emissions, the scenario families are characterized by very different emissions trajectories. The A1 fossil fuel scenarios have continuously increasing emissions, with rapid growth before 2050 and slower growth thereafter. The A1 scenarios with the "balanced" energy mix (e.g. the marker A1B-AIM) typically have emissions that decline after 2050, while the A1T technology scenarios have slower growth prior to 2050, and a steeper decline after 2050. As for the A1 fossil fuel scenarios, A2 family scenarios are characterized by high rates of growth in emissions prior to 2050 and subsequently continued growth but at lower rates. Unlike the rest of the A2 scenarios, the A2-A1-MiniCAM has more rapid growth in emissions after 2050 than before 2050, because per capita incomes in a number of developing regions do not reach a level at which per capita energy demands rise rapidly until the middle of the 21st century. The B1 family is characterized by lower growth in emissions prior to 2050, mostly because of lower rates of growth in energy demand, followed by declining emissions after 2050. Finally, the B2 family is characterized by relatively stable emissions post-2050, after roughly doubling emissions between 1990 and 2050.

5.3.2. *Land-Use Carbon Dioxide Emissions*

Most changes in land use are induced by the demand for cropland and grassland, which is driven by the demand for food products, the extent of biomass energy use, and policies and practices associated with forest management. The 1990 land-use CO_2 emissions remain fairly uncertain, estimated at 1.6 ± 1.0 GtC (Watson *et al.*, 1996a); a similar level of uncertainty is attached to current land-use emissions. This uncertainty is reflected by the quantification of the SRES storylines – the 1990 emission estimates from different models range between 1.0 and 1.6 GtC, while the spread of estimates at the four-region level is even larger. For the sake of comparability, common, standardized (see Box 5-1 on standardization) emissions are established at 1.1 GtC in 1990 and 1.0 GtC in 2000, to reflect the net carbon flux from contemporary changes in forest cover.

Generally, the SRES models use different approaches to estimate land-use change emissions – in some cases the only source of emissions is tropical deforestation, while in other cases more sources and sinks (including natural) are included.[7] Moreover, methodologic differences and uncertainties in carbon content, carbon cycling, and land classification result in seemingly inconsistent results between models that cover the same land-use sources. These features complicate a straightforward comparison between land-use emissions in scenarios generated by different SRES models.

Future trajectories of land-use CO_2 emissions are shown in Figure 5-4. Emissions in the 40 SRES scenarios range widely in the same year and change significantly over time. In scenarios with continued deforestation, emissions rise initially, then reach a maximum, and finally decline with depletion of forestland that can be cleared. At the other end of the scenario spectrum, emissions turn negative and land-use changes become an increasing CO_2 sink through afforestation. By 2020, the resultant uncertainty ranges between 0 and 3 GtC across all 40 SRES scenarios with a median of 1.1 GtC. By 2050, the scenario range shifts to between –0.7 and 1.2 GtC (median, 0.5 GtC). By 2100 the scenario range lies between –2.8 and 2.2 GtC, with a median of 0.0 GtC. Interestingly, in specific years (e.g., in 2050) scenarios from all four SRES families fall within a relatively narrow emissions corridor (i.e. at least one scenario from each of the four SRES scenario families falls within the 25th and 75th percentiles of the emission range). This indicates that similar levels of carbon fluxes related to land use could arise from widely different socio-economic driving forces, depending on future trends in food demand and dietary patterns, agricultural productivity growth, forest practices, etc.

In general, the SRES CO_2 land-use emissions follow the same pattern as found in the literature (see Chapter 3) – initially emissions increase because of continuing deforestation in developing regions and subsequently they decrease following a drop in population growth and increases in agricultural productivity. The main difference between the SRES scenarios and the literature reviewed in Chapter 3 lies in the maximum emission values, which are significantly lower in the SRES scenarios. Possibly this arises because the SRES models explicitly simulate land-use change as a function of pressure on the land (itself a function of agricultural productivity), while in the literature land-use CO_2 emission scenarios are often based on trend extrapolation or statistical relationships (e.g. with population growth). The rapid economic development and technological advances assumed in the SRES storylines thus tend to mitigate CO_2 emissions from deforestation, and in some cases lead to their reversal (e.g., turning deforested lands into carbon sinks).

As suggested by Figure 5-4, no simple relation exists between the scenario families and land-use emissions. As in the case of energy-related emissions, the A1 family scenarios cover the widest range of emissions and trajectories. In most cases, emissions in these scenarios decline after 2030 to zero or negative (carbon sinks) values. However, in two scenarios (A1B-IMAGE and A1B-MARIA), emissions increase later in the 21st century, reflecting assumptions on additional land pressures. As a rule, scenarios of the A2 family follow the same convex trajectory as A1 scenarios, also with two exceptions (A2-IMAGE and A2-AIM). In accordance with the sustainability emphasis of the B1 family storyline, scenarios from this family have the lowest initial emissions, which by 2080 or earlier drop to zero or negative levels. The B1 family marker scenario (B1-IMAGE) has negative land-use emissions for most of the modeling period, a property directly related to B1's "environmental conservation" emphasis.

[7] The most comprehensive treatment is embedded in the IMAGE model. For the sake of comparison the IMAGE team has made a tentative estimate of those (net) emissions that are reasonably consistent with what other models reported. These derived values are used in this report.

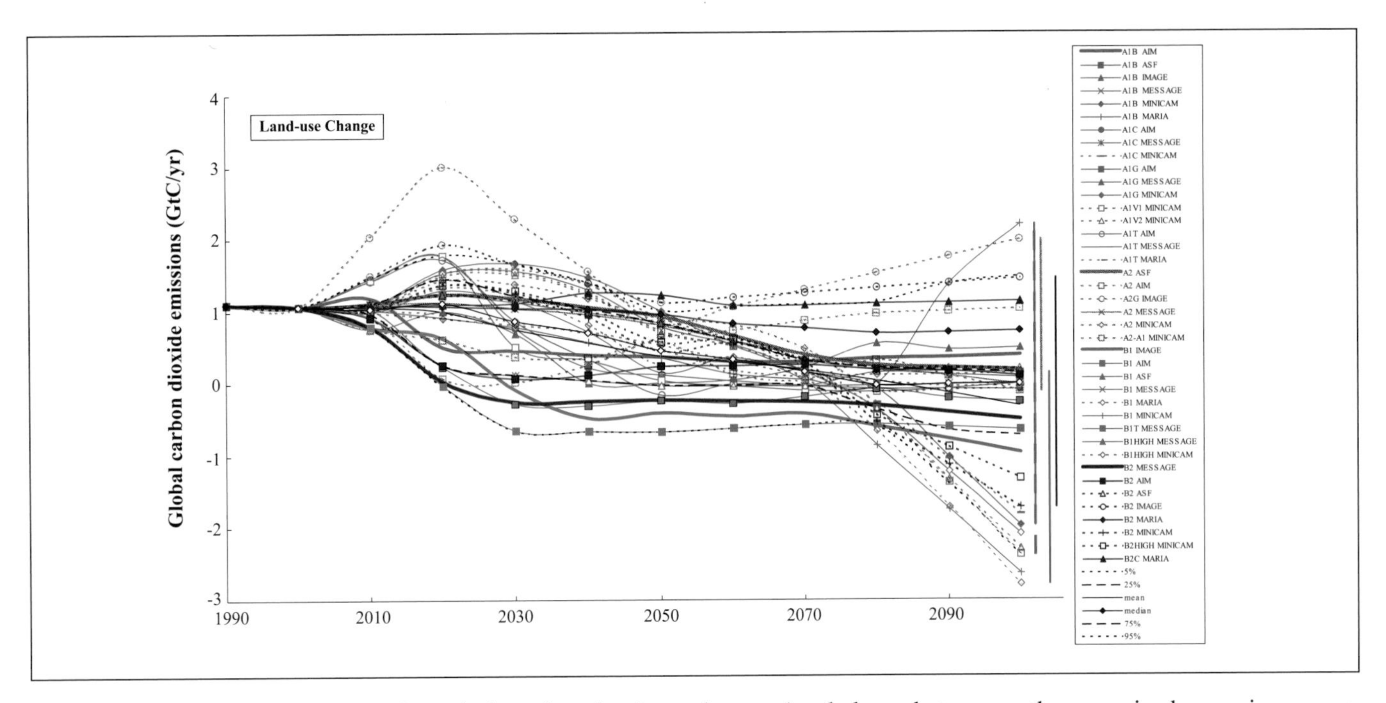

Figure 5-4: Standardized global CO_2 emissions from land-use changes (net balance between anthropogenic changes in sources and sinks) for 40 SRES scenarios, classified into the four scenario families (each denoted by a different color code – A1, red; A2, brown; B1, green; B2, blue). Marker scenarios are shown with thick lines without ticks, globally harmonized scenarios with thin lines, and non-harmonized scenarios with thin, dotted lines (see Table 4-3). Black lines show percentiles, means, and medians for the 40 SRES scenarios. For numbers on the two additional illustrative scenarios A1FI and A1T see Appendix VII.

5.3.3. *Overview of Sectoral Carbon Dioxide Emissions*

As a result of different model specifications and detail, it is not possible to draw up consistent comparisons between sectoral emissions across different models. An overview of sectoral CO_2 emissions by sector and source category is summarized in Box 5-2 on the basis of the results of the MESSAGE (for energy and industrial sources) and AIM (for land-use change sources) models.

5.4. Other Greenhouse Gases

Non-CO_2 GHGs (CH_4, N_2O, and halocarbons) account for about 40% of the total induced additional radiative forcing compared to pre-industrial times (Houghton *et al.*, 1996). Whereas CO_2 emissions are largely attributable to two major sources (energy and land use), other gases arise from many different sectors and applications as shown in Table 5-3. Consequently, their emission levels are more uncertain. Also

Box 5-2: CO_2 Emissions by Sector and Source for MESSAGE Scenarios

Table 5-4 gives an overview of CO_2 emissions by sector and source category according to the IPCC reporting format given in Watson *et al.* (1996b). The differences in sectoral detail across models mean a consistent comparison and sectoral CO_2 emission balances are only possible within one particular modeling framework. Table 5-4 presents the scenario results as calculated with the MESSAGE model for 1990, 2050, and 2100 and for the four scenario families and their scenario groups. Emissions related to land-use change were derived from consistent model runs with the AIM model.

As in Watson *et al.* (1996b), emissions are presented by sector, and emission categories adopt both supply and demand perspectives for energy-related CO_2 emissions. The supply side CO_2 balance accounts emissions at the point of energy combustion, that is at a coal-fired power plant (electric generation) or by burning coal in industrial boilers (direct fuel use by industry). Conversely, the demand side CO_2 balance accounts emissions per end-use category, irrespective of whether emissions originate directly at the point of end-use or upstream in the energy conversion sector. For example, for residential and commercial energy uses, CO_2 emissions include those from direct fuel combustion as well as those emissions that originate from the generation of electricity consumed by the residential and commercial sectors. Finally, an emissions balance by source category is given, in which emissions are accounted for at the level of primary energy (solids, liquids, and gases), again after Watson *et al.* (1996b). Non-energy emissions are included in a separate "others" category. Combined, these different emission balances can serve as data input for subsequent mitigation analyses at the sectoral level or at that of the entire economy.

Box 5.2 continues

Box 5.2 continued

Table 5-4: *Global CO_2 (MtC) emissions by sector and source category for seven scenarios calculated with the MESSAGE model for 1990, 2050, and 2100. In the SPM, A1C and A1G scenarios are merged into one fossil-intensive A1FI scenario group (see also footnote 2).*

	1990	2050							2100						
Supply Side		A1B	A2	B1	B2	A1T	A1G	A1C	A1B	A2	B1	B2	A1T	A1G	A1C
Energy Supply/ Transformation															
Electric generation	1773	4783	4875	763	2844	2192	6519	5924	7541	9283	207	5323	293	8854	11166
Synfuels production	0	1162	1613	2207	929	1278	2294	3958	170	7245	2071	2848	441	3619	9788
Other conversion*	680	2277	1312	999	1170	1506	3012	719	2901	2393	735	1394	807	7682	622
Direct Use of Fuels by Sector															
Residential/commercial	880	2782	1618	1361	1494	2362	3606	2458	1402	925	367	973	656	3085	1255
Industry	1289	1540	1515	843	1401	845	987	1165	273	1601	176	1166	503	984	631
Transportation	1310	3823	2952	2669	2587	3445	4513	5587	1881	4524	1135	1845	1577	5924	8640
Feedstocks	303	912	1311	490	891	972	766	768	91	1963	66	398	448	535	422
Non-Energy Emissions															
Cement prod./gas flaring	68	172	192	65	104	141	206	249	136	479	36	187	64	224	462
Land-use change	1010	–139	–104	–902	–436	–139	–139	–139	2	81	–646	–501	2	2	2
TOTAL	7312	16789	15044	8367	10983	12601	21802	21086	14397	28493	4147	13634	4789	30909	32988
Demand Side															
Residential/commercial	1995	7212	4631	3126	4059	4846	10290	7032	9407	7827	1698	5830	1209	14971	10939
Industry**	2784	3916	5734	1975	4042	2996	4388	4434	1471	9165	601	4909	1244	4341	3493
Transportation	1523	5800	4783	4168	3318	4897	7262	9759	3516	11421	2494	3396	2334	11596	18554
Land-use change	1010	–139	–104	–902	–436	–139	–139	–139	2	81	–646	–501	2	2	2
TOTAL	7312	16789	15044	8367	10983	12601	21802	21086	14397	28493	4147	13634	4789	30909	32988
By Source															
Solids	2346	5356	7106	841	2230	2643	7553	12471	6166	22586	565	7765	188	3904	29596
Liquids	2787	5618	4100	3832	4542	5007	7310	4185	2751	933	910	1039	1542	7813	1118
Gases	1102	5782	3750	4532	4544	4949	6872	4320	5342	4415	3283	5144	2993	18967	1810
Others***	1078	33	88	–837	–332	3	67	110	138	559	–611	314	66	226	464
TOTAL	7312	16789	15044	8367	10983	12601	21802	21086	14397	28493	4147	13634	4789	30909	32988

*Includes emissions from district-heat production, energy transmission/distribution, oil refining, fuel extraction, and other conversion losses.

**Includes emissions from feedstocks, cement production, and gas flaring.

***Emissions from land-use change, cement production, and gas flaring.

the base-year emissions of non-CO_2 GHGs are subject to considerable uncertainty, in particular when it comes to regional and sectoral breakdowns.

Emissions of non-CO_2 radiatively important gases are subject to considerable and unresolved uncertainties and are driven by a more complex set of forces than CO_2 emissions. Therefore, the types of models employed for the SRES analyses are not expected to produce unambiguous and widely approved estimates of emissions of these gases for a period of over a century. Despite the limited knowledge, at some point in time causal relationships between driving forces and non-CO_2 emissions need to be crafted into the models for the sake of completeness. Even if new insights are generated by specialist researchers in certain fields of environmental science, and these become accepted as the mainstream view, their adoption in the models is often far from straightforward, as appropriate links to drivers may not be readily available in the underlying structure. Limited manpower and resources imply that priorities must be assigned when deciding on further model development, and as a consequence the models lag behind "common wisdom" in certain areas. Of course, this does not necessarily limit their abilities to capture major trends at a more aggregate level, the main purpose of these models.

In the following sections emission trajectories generated in the SRES scenarios are presented and discussed. However, model structures and properties, and exogenous assumptions made by the modelers involved, may give rise to systematic deviations within scenario families that may prove very significant compared to average inter-family differences. Further investigation and analysis is required to understand these issues more fully.

5.4.1. Methane

In the IPCC WGI SAR, anthropogenic CH_4 emissions in the year 1990 were estimated at 375 ± 75 $MtCH_4$ (Houghton *et al.*, 1996) and are shown in Table 5-3. These emissions arise from a variety of activities, dominated by biogenic processes that are often subject to considerable uncertainties (see Chapter 3). CH_4 emissions across the six models used to generate the SRES scenarios for 1990 range between 298 and 337 $MtCH_4$. After standardization (see Box 5-1), the base-year emissions in the SRES scenarios were set to 310 $MtCH_4$, within the range mentioned above. About one-quarter of the total emissions are related to fossil fuel extraction (CH_4 emissions from coal mines, CH_4 venting from oil extraction), transport and distribution (e.g., leakage from pipelines), and consumption (incomplete combustion). The biogenic sources include agriculture (enteric fermentation, rice paddies, and animal waste), biomass burning, and waste management (landfills, sewage). Based on this source list, future CH_4 emission trajectories depend upon such variables as volumes of fossil fuels used in the scenarios, regional demographic and affluence developments, and assumptions on preferred diets and agricultural practices.

Results from the 40 SRES scenarios indicate that uncertainties surrounding future CH_4 emission levels are likely to increase over time (Figure 5-5). By 2050 the range across all scenarios

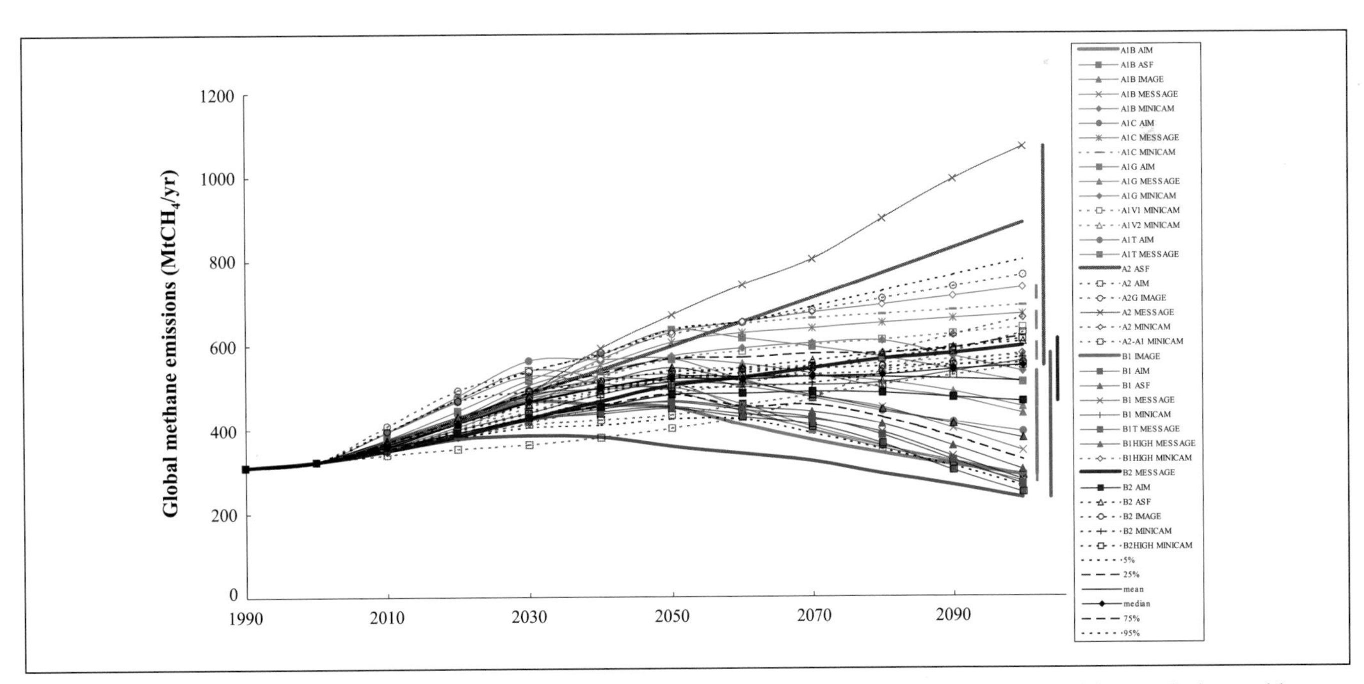

Figure 5-5: Standardized global CH_4 emissions for SRES scenarios, classified into four scenario families (each denoted by a different color code – A1, red; A2, brown; B1, green; B2, blue). Marker scenarios are shown with thick lines without ticks, globally harmonized scenarios with thin lines, and non-harmonized scenarios with thin, dotted lines (see Table 4-3). Black lines show percentiles, means, and medians for the SRES scenarios. For numbers on the two additional illustrative scenarios A1FI and A1T see Appendix VII.

***Table 5-5**: CH_4 emissions ($MtCH_4$) by region and emissions sector for the four marker scenarios (A1B-AIM, A2-ASF, B1-IMAGE, and B2-MESSAGE). Categories are agricultural animals (enteric fermentation and animal waste), rice production, biomass burning, landfills and sewage treatment, and fossil fuel use (extraction, distribution, and consumption). Emissions are provided in non-standardized or "raw" format and are not comparable with standardized emission estimates used in the figures and other tables.*

		1990				2050				2100			
Source	***Region***	**A1B**	**A2**	**B1**	**B2**	**A1B**	**A2**	**B1**	**B2**	**A1B**	**A2**	**B1**	**B2**
Agricultural Animals													
	OECD90	18	27	21	18	19	38	18	19	20	42	14	20
	REF	12	14	11	12	13	18	9	13	13	22	6	13
	ASIA	23	25	23	23	23	75	34	23	23	102	26	24
	ALM	27	30	30	27	28	92	41	28	28	151	33	29
	Sub-total	**80**	**97**	**84**	**82**	**83**	**223**	**104**	**84**	**84**	**317**	**81**	**87**
Rice													
	OECD90	2	2	2	2	2	2	1	2	1	1	1	2
	REF	1	1	0	1	1	1	0	1	1	1	0	1
	ASIA	52	53	25	54	51	80	21	58	51	80	15	64
	ALM	5	5	2	5	5	8	3	6	5	11	3	6
	Sub-total	**60**	**60**	**29**	**61**	**59**	**91**	**25**	**66**	**58**	**93**	**19**	**73**
Biomass Burning													
	OECD90	7	1	2	7	7	2	1	8	7	2	1	9
	REF	2	0	1	2	2	0	0	2	1	0	0	2
	ASIA	16	7	7	16	8	10	1	9	8	11	1	10
	ALM	20	18	19	20	4	23	7	5	5	23	7	5
	Sub-total	**45**	**27**	**30**	**46**	**21**	**35**	**10**	**24**	**21**	**36**	**9**	**26**
Landfill + Sewage													
	OECD90	23	22	23	23	9	32	22	10	5	44	18	7
	REF	7	5	6	7	6	6	6	14	1	8	4	4
	ASIA	19	16	20	19	110	32	49	131	17	41	32	111
	ALM	12	10	13	12	66	30	42	67	18	44	33	97
	Sub-total	**61**	**51**	**61**	**62**	**191**	**99**	**119**	**222**	**42**	**137**	**85**	**218**
Fossil Fuels													
	OECD90	31	22	28	19	23	33	21	20	17	77	13	31
	REF	37	21	35	24	34	51	17	27	30	110	9	30
	ASIA	14	13	14	14	25	36	28	27	22	81	8	78
	ALM	10	12	17	11	27	42	41	29	26	51	16	47
	Sub-total	**91**	**69**	**94**	**68**	**110**	**162**	**104**	**104**	**95**	**319**	**46**	**187**
Total													
	OECD90	81	74	76	70	60	106	63	60	51	166	47	69
	REF	57	42	53	47	55	76	32	56	47	141	19	50
	ASIA	123	113	89	127	218	233	132	249	121	314	80	287
	ALM	75	75	81	76	131	195	134	135	82	280	92	185
	Total	**337**	**304**	**298**	**318**	**464**	**611**	**361**	**500**	**300**	**902**	**239**	**592**

Per-capita Methane Emissions (kg/person)

Total ($MtCH_4$)	1990				2050				2100			
Region	**A1B**	**A2**	**B1**	**B2**	**A1B**	**A2**	**B1**	**B2**	**A1B**	**A2**	**B1**	**B2**
OECD90	95	87	95	83	56	92	63	61	46	111	46	74
REF	139	100	128	110	131	147	74	139	139	200	55	133
ASIA	44	41	32	46	52	40	31	53	42	43	28	58
ALM	63	61	63	61	44	51	44	41	30	51	33	4
Total	**64**	**58**	**56**	**60**	**53**	**54**	**41**	**53**	**43**	**60**	**34**	**57**

is between 359 and 671 $MtCH_4$, and the range increases further to levels between 236 and 1069 $MtCH_4$ by 2100. This wide range in the SRES emissions reflects new information and additional uncertainties concerning certain source categories such as sewage systems. As for CO_2 emissions related to land use (see Section 5.3.2), at least one scenario from all four SRES scenario families falls within the 25th and 75th percentiles of the emission range. Thus, very different future developments in energy and agricultural systems could lead to similar outcomes in terms of global CH_4 emissions, even if the source categories and regional patterns of these emissions are very different. At the same time, the uncertainty range for any given scenario family is also substantial, as indicated by the range of 2100 emissions for the A1, A2, and B1 scenario families in Figure 5-5.

The subsequent sections discuss CH_4 emission trajectories for individual scenario families, with sectoral and regional patterns described on the basis of the output of marker scenarios.

5.4.1.1. A1 Scenario Family

The A1 family of scenarios covers close to the full range of CH_4 emissions in all the SRES scenarios (Figure 5-5). A1C-AIM shows the highest emissions before 2050 and A1G-MiniCam scenario shows them after 2050 (Figure 5-6a). In three A1 fossil scenarios (A1G-MiniCAM, A1C-MiniCAM, and A1C-MESSAGE) and in both alternative MiniCAM scenarios (A1V1-MiniCAM and A1V2-MiniCAM) emissions increase continuously through the 21st century, while in the rest of the A1 scenarios emissions peak between 2030 and 2050 and decline thereafter.

In the A1B marker scenario (A1B-AIM) emissions increase through 2030 and subsequently decline to levels similar to those in 1990 (Figure 5-6a). Almost all of the emissions dynamics in A1B-AIM are explained by a rise in emissions from landfills, sewage, and fossil fuel production. At the same time emissions from agriculture are relatively flat because better management of animal wastes and high productivity are assumed to offset the effect of increased food requirements (Table 5-5). Growing population and per capita income combined with increased use of landfills generates increasing emissions from landfills and sewage in developing countries through 2030. After 2030, declining population levels, the introduction of modern management techniques, and increased recycling reduces waste sent to landfills and thus emissions from these wastes. Emissions from biomass burning in A1B-AIM are assumed to decline steadily through the adoption of bio-recycling and other "no-waste" agricultural practices. Similarly, CH_4 emissions from fossil fuel production and use grow through 2030 and subsequently decline as fossil fuel production falls.

5.4.1.2. A2 Scenario Family

The A2 family of scenarios contributes the upper half of the full range of CH_4 emissions in the SRES scenarios (Figure 5-5). The global CH_4 emissions in the A2 family scenarios grow continuously throughout the 21st century and range from 550 to 1070 $MtCH_4$ in 2100 (Figure 5-6b). The rate of growth depends on the scenario-specific dynamics of major CH_4 emission drivers. Emissions in the A2-ASF scenario, which are close to the upper end of the range, are driven mainly by increases in coal production, livestock population, and waste management capacity to satisfy the needs of an expanding population (Table 5-5). At the lower end of the emission range are the two MiniCAM scenarios and A2-AIM. Relatively slow emission growth in the MiniCAM scenarios is attributed primarily to an increase in rice productivity (which offsets an increase in the area of rice fields) and a shift in livestock production from cattle to animal groups that have notably lower emission factors per animal (e.g. poultry and swine). In the A2-AIM scenario low CH_4 emissions are caused by relatively low CH_4 emission factors for coal production and a relatively slow increase in livestock production.

5.4.1.3. B1 Scenario Family

The B1 family of scenarios covers the lower half of the full range of CH_4 emissions in the SRES scenarios, with the B1-IMAGE marker scenario positioned at the low end of this range (Figure 5-5). Global emissions in B1-IMAGE grow through to 2030, driven primarily by increased emissions from landfills and sewage (Figure 5-6c). This growth is partially offset by declines in emissions from biomass burning (Table 5-5). After 2030, emissions level off and subsequently decline from 2050, a reflection of reductions in fossil fuel production and use. Emissions from other sources, which include enteric fermentation and rice production, also decline, primarily from the combination of stabilizing and declining populations with continued improvements in slaughter weight and off-take rate. The B1-AIM scenario lies in the middle of the range for the B1 family. CH_4 emissions in this scenario increase through to 2050, with most of the increase associated with landfills and sewage systems. After 2050, emissions from landfills decline because of a combination of factors that include recycling and a declining population. Emissions from sewage decrease after 2050 through the decline in population. Emissions from rice production stay relatively flat because of increases in rice productivity. Emissions from fossil fuel production increase through to 2030 and subsequently decline, which mirrors the increase in production of fossil fuels through to 2030 and the increased use of renewable energy after 2030. Unlike other B1 scenarios, both MiniCAM cases have emissions that continuously increase (Figure 5-6c). The reason is a slow rise in non-energy system emissions for the B1-MiniCAM scenario, which offsets the decrease in energy system emissions and the nearly constant emissions in the agriculture sector. In the B1High-MiniCAM scenario, the additional energy demand designed into this scenario is met largely by an increased use of natural gas, which thus explains the faster rise in CH_4 emissions in this scenario as compared to the base B1 MiniCAM scenario.

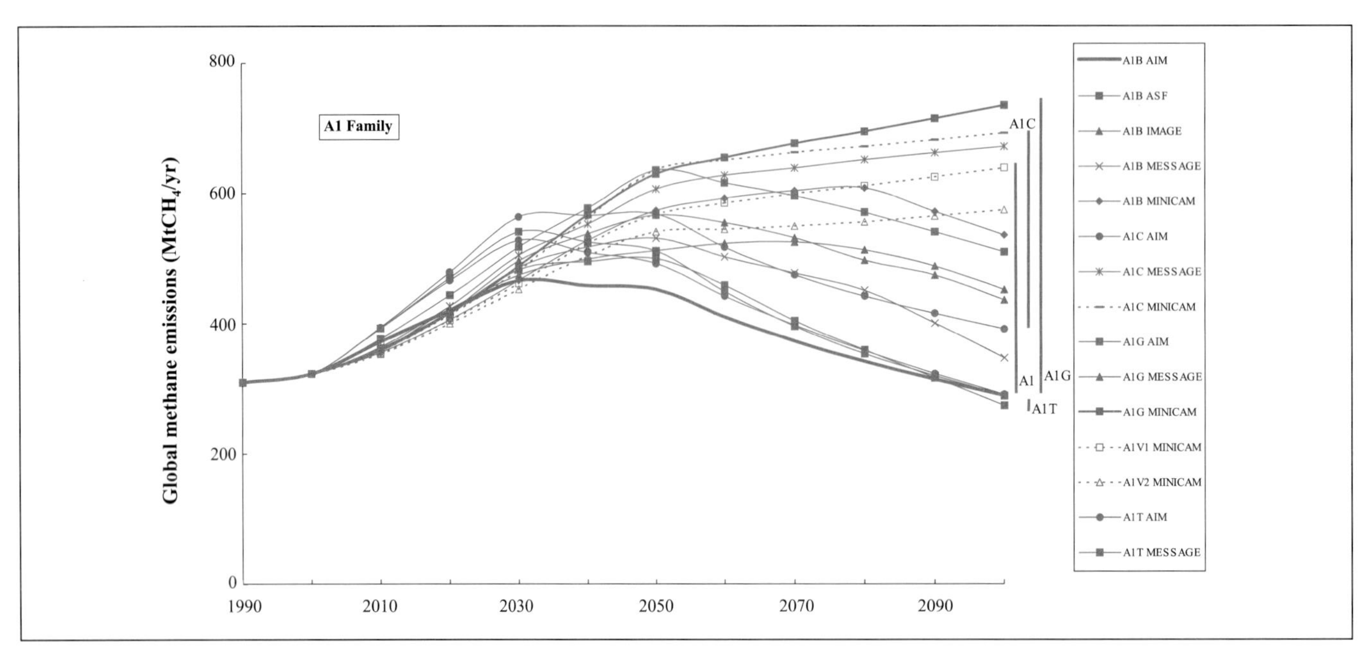

Figure 5-6a: Standardized global CH_4 emissions in the A1 family scenarios. The marker scenario is shown with a thick line without ticks, the globally harmonized scenarios with thin lines, and the non-harmonized scenarios with thin, dotted lines (see Table 4-3). In the SPM, A1C and A1G scenarios are merged into one fossil-intensive A1FI scenario group (see also footnote 2).

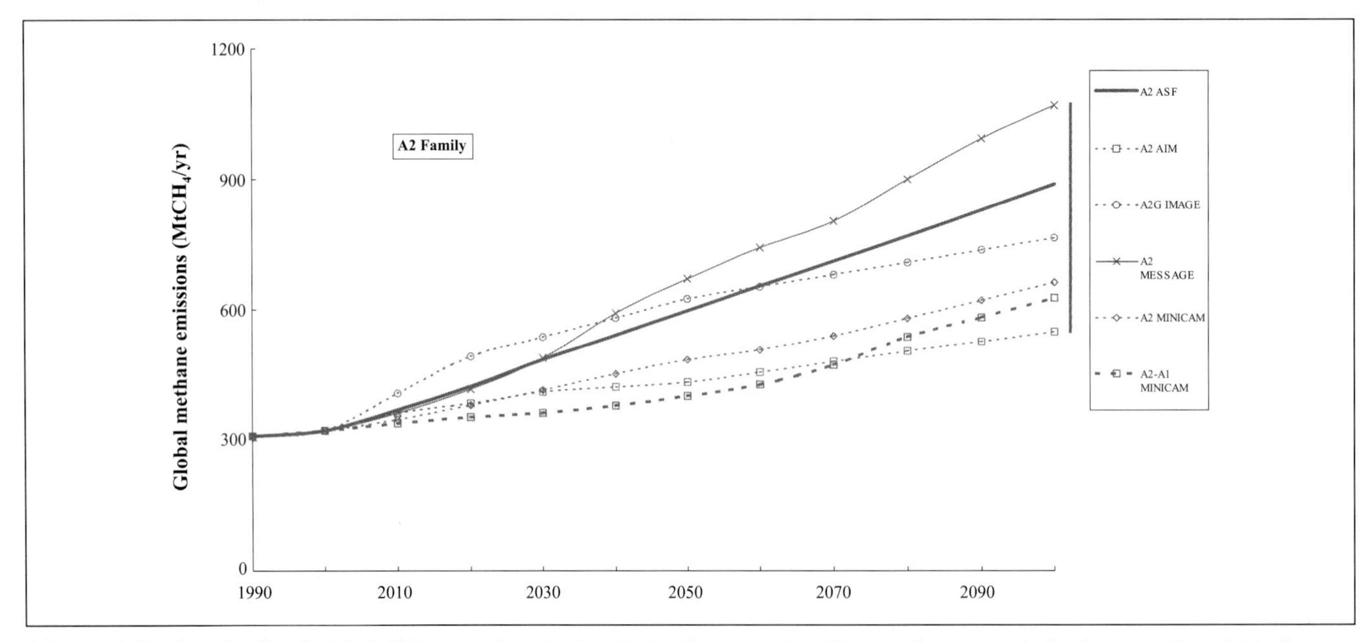

Figure 5-6b: Standardized global CH_4 emissions in the A2 family scenarios. The marker scenario is shown with a thick line without ticks, the globally harmonized scenarios with thin lines, and the non-harmonized scenarios with thin, dotted lines (see Table 4-3).

5.4.1.4. B2 Scenario Family

CH_4 emission trajectories of the B2 family of scenarios are located in the middle of the SRES range and (except for the B2-AIM scenario) follow very similar paths (Figure 5-6d). Similar emission profiles in B2 scenarios may have different explanations. The increase in emissions from 1990 to 2100 in the B2-MESSAGE scenario is predominantly from landfills and sewage, with fossil-fuel energy production as the second major source. However, in the B2-ASF scenario increasing volumes of CH_4 are generated mostly in the energy and agricultural sectors. Unlike the rest of the B2 family, the B2-AIM scenario emissions after 2070 decline because of an increase in the CH_4 recovery rate from energy and waste management systems.

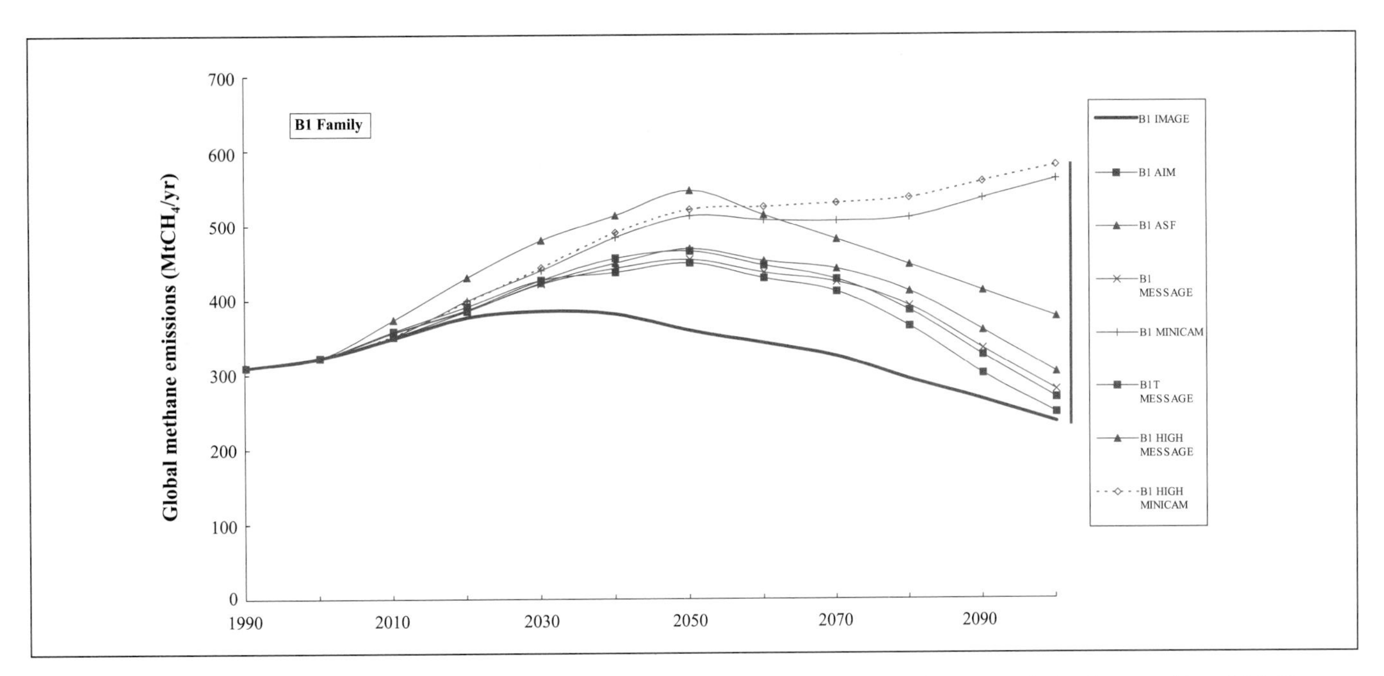

Figure 5-6c: Standardized global CH_4 emissions in the B1 family scenarios. The marker scenario is shown with a thick line without ticks, the globally harmonized scenarios with thin lines, and the non-harmonized scenarios with thin, dotted lines (see Table 4-3).

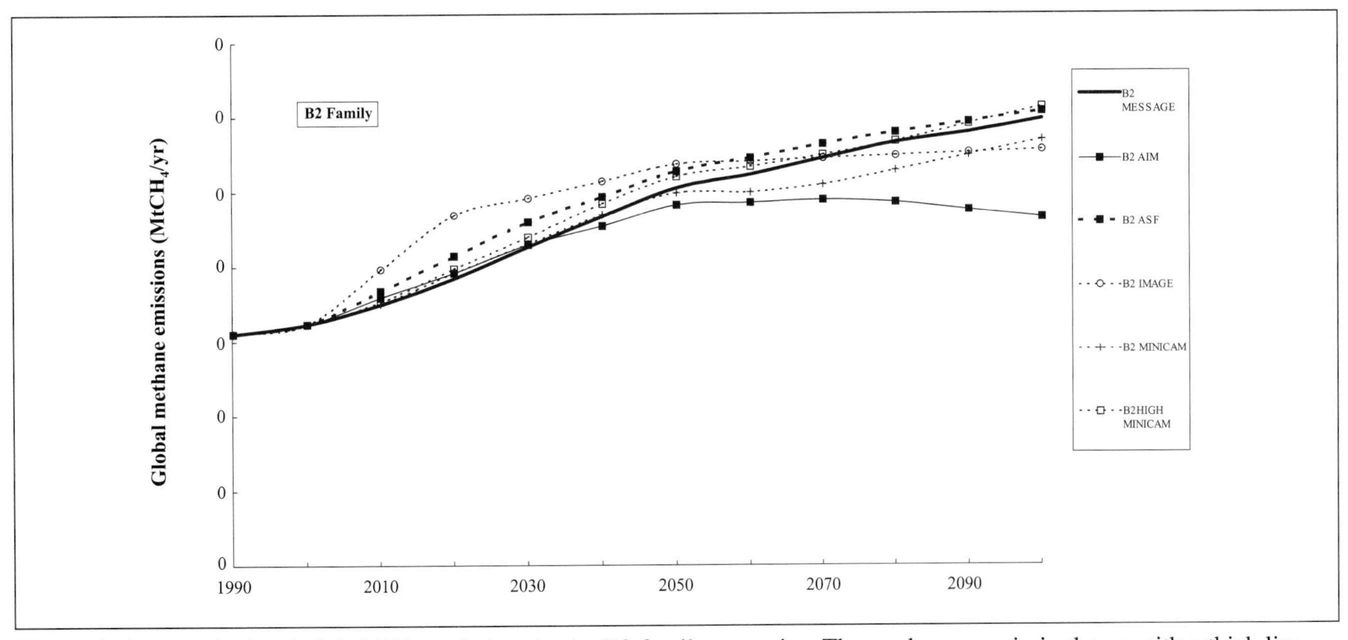

Figure 5-6d: Standardized global CH_4 emissions in the B2 family scenarios. The marker scenario is shown with a thick line without ticks, the globally harmonized scenarios with thin lines, and the non-harmonized scenarios with thin, dotted lines (see Table 4-3).

5.4.1.5. Inter-Family Comparison

Global CH_4 emissions in all the four markers increase at about the same rate up to 2020 (Figure 5-5). After the early and rapid increase in agricultural productivity and a shift away from fossil fuels, emissions in the B1 marker (B1-IMAGE) level off and subsequently decline. Emissions in the other three markers continue to increase. Emissions in the A1B marker (A1B-AIM) grow to 2030 because of an increasing and much more affluent population in developing regions. The emissions subsequently decline until 2100 as the population growth slows and eventually reverses. Increases in efficiency and productivity, through rapid technological change, offset increases in consumption (A1 global GDP is the largest among the SRES storylines), so that by 2100 global emissions per person are lower than in 1990 even though average incomes grow very significantly (Table 5-5). In the A2 (A2-ASF) and B2 (B2-MESSAGE) markers, emissions increase throughout the whole time horizon to the year 2100. This increase is most pronounced in the A2 marker scenario, in which emissions reach about 900 $MtCH_4$ by 2100 (about a three-fold increase of 1990 levels). The emission level by 2100 for the B2 marker, 600 $MtCH_4$, is about twice as high as in 1990.

The four SRES marker scenarios illustrate a complexity of relative impacts of technology, GDP, and population on CH_4 emissions. In 2050 and 2100 the A2 and B2 markers have very similar per capita emissions, which indicates a certain similarity between the cumulative impact of economic, technological, and structural changes on emissions in both of these scenarios (in spite of significant differences in the global average per capita incomes). Hence, differences in the absolute emission levels in these scenarios can be explained primarily by differences in population trajectories, with the A2 population in 2100 being 1.5 times larger than the B2 population. At the same time, differences in the absolute emissions between the A1B and B1 markers cannot be explained by population size (both markers have the same population trajectory), but instead result from the greater emphasis on "dematerialization" and "sustainability" in the B1 storyline. The two illustrative scenarios A1G-MiniCAM and A1T-MESSAGE display similar patterns of methane emissions as the A2 and B1 marker scenarios respectively and are therefore not discussed separately here.

At the sectoral level, CH_4 emissions in the A1B marker originate primarily from landfills and sewage followed by fossil fuel production and agricultural animals (Table 5-5). Agricultural animals are the major source of emissions in the A2 marker, with fossil fuels being second, and landfills and sewage third in importance. In the B1 marker, fossil fuels produce the largest volumes of CH_4, followed by agricultural sources (Table 5-5). Finally, the B2 marker emissions originate primarily from landfills and sewage, followed by fossil fuels

5.4.2. Nitrous Oxide

Uncertainties in the estimates of current N_2O emissions (Table 5-3) are also reflected in base-year 1990 differences, between 4.8 and 6.9 MtN across the six models used to develop the SRES scenarios. The range across the four markers (Table 5-3) is not as wide, between 6.0 and 6.9 MtN, and leans more toward the higher end of the uncertainty range reported in Houghton *et al.* (1995). After standardization, 1990 N_2O emissions in the SRES scenarios were set to 6.7 MtN, well within the IPCC WGI SAR range (Table 5-3). Even more so than for CH_4, food supply is assumed to be a key determinant of future N_2O emissions. Size, age structure, and regional spread of the global population are likely to affect future emission trajectories, together with diets and rates of improvement in agricultural productivity. The representation of how these driving forces translate into N_2O emissions varies across the models because of differences in both sectoral coverage and emission factors. As a result, differences in emission trajectories are not only scenario dependent, but also model dependent, which illustrates additional uncertainties in our understanding and representation of driving forces and their influence on N_2O emissions.

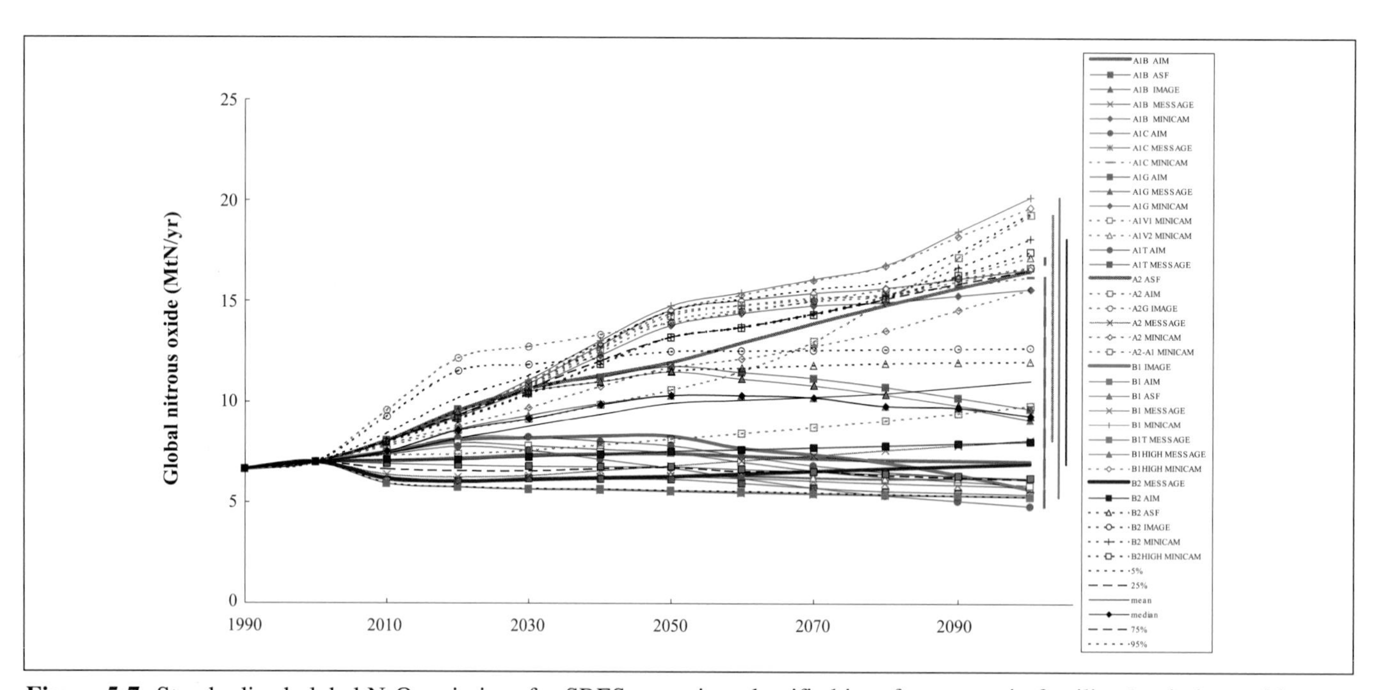

Figure 5-7: Standardized global N_2O emissions for SRES scenarios, classified into four scenario families (each denoted by a different color code – A1, red; A2, brown; B1, green; B2, blue). Marker scenarios are shown with thick lines without ticks, globally harmonized scenarios with thin lines, and non-harmonized scenarios with thin, dotted lines (see Table 4-3). Black lines show percentiles, means, and medians for SRES scenarios. For numbers on the two additional illustrative scenarios A1FI and A1T see Appendix VII.

The range of emissions across all 40 SRES scenarios increases continuously over the 21st century (Figure 5-7). By 2100, N_2O emissions range between 5 and 20 MtN. Emission ranges also tend to be comparable across the four scenario storylines, an indication of the decisive impact of uncertainty in the modeling (mentioned above). Future N_2O emissions in the SRES scenarios tend to cluster into two broad groups – those that project relatively flat, even slightly declining, emissions and those that indicate continuously rising trends toward high emission levels. Some other scenarios (e.g. A1B-ASF, B1-ASF, and B1-IMAGE) indicate the possibility of transitional emission patterns in which N_2O emissions peak around 2050 and decline thereafter, more or less in step with the population size. As a result of differences in modeling approaches, no individual scenario tracks the mean or median across all 40 scenarios of the SRES set.

5.4.2.1. A1 Scenario Family

The A1 family of scenarios covers close to the full range of N_2O emissions from the SRES scenarios (Figure 5-7). The A1G-MiniCAM scenario is at the high end of the A1 range (Figure 5-8a). Most of the emissions increase in this scenario is associated with the agriculture sector, primarily with animal manure management. Emissions are driven by a rapid increase in income that induces steep increases in meat and dairy consumption. The A1B-AIM marker scenario has emissions near the low end of the range. This scenario assumes that fertilizer use in developing countries is nearly saturated and that increased productivity comes from better management. A portion of nitrogen from fertilizers is also assumed to be stored indefinitely in underground water. However, the emissions from energy use in A1B-AIM increase until the third quarter of the 21st century, and reach a level about three times the 1990 volume. In the middle of the A1 family range lie emissions from the A1B-ASF and A1B-IMAGE scenarios (Figure 5-8a). In these scenarios, emissions growth to the 2050s is driven by growth in the agricultural and transportation sectors. Thereafter, emissions decline, primarily because of declining population levels and increases in agricultural productivity (lower production factor input per unit of output).

As suggested by Figure 5-8a, differences in the energy supply system reflected by the four A1 scenario groups (A1G, A1C, A1T, and A1 or "balance") are only partially translated into differences in N_2O emissions. For example, scenarios that rely on coal as the primary energy source span the range from 6 MtN (A1C-MESSAGE) to 16 MtN (A1C-MiniCAM). At the same time, trajectories of A1T "technology" scenarios are located in the lower end of the emissions range.

5.4.2.2. A2 Scenario Family

N_2O emission trajectories of the A2 scenario family fall into two major groups. The first group includes the A2-AIM and A2-MESSAGE scenarios, and the rest of scenarios form the second group. The A2-AIM N_2O emissions increase at a relatively low rate and do not exceed 10 MtN in 2100 (Figure 5-8b), because of a relatively slow increase in nitrogen fertilizer input and omission of N_2O sources associated with animal wastes. N_2O emissions in the second group of scenarios (which includes the family marker A2-ASF) grow steadily throughout the 21st century, driven by an increased demand for food and associated increases in the animal waste and nitrogen fertilizer use (Table 5-6). In 2100 the emissions from this group of scenarios range from 15 to 20 MtN.

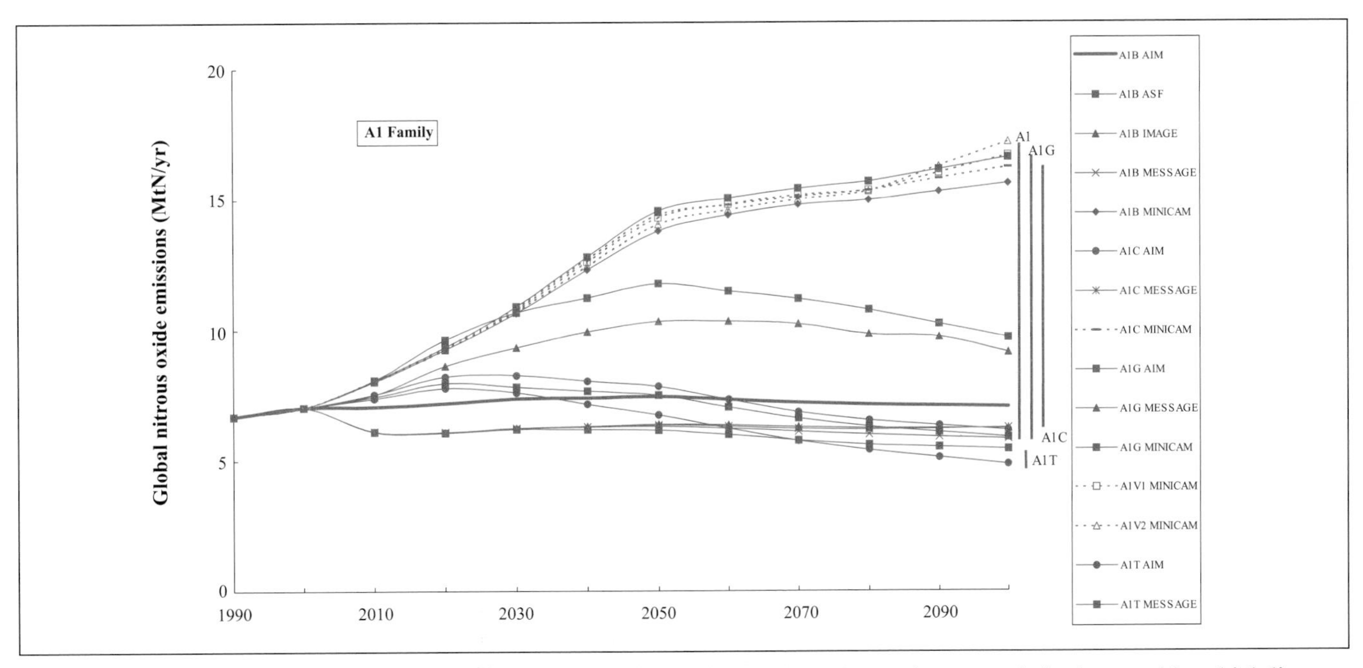

Figure 5-8a: Standardized global N_2O emissions in the A1 family scenarios. The marker scenario is shown with a thick line without ticks, the globally harmonized scenarios with thin lines, and the non-harmonized scenarios with thin, dotted lines (see Table 4-3). In the SPM, A1C and A1G scenarios are merged into one fossil-intensive A1FI scenario group (see also footnote 2).

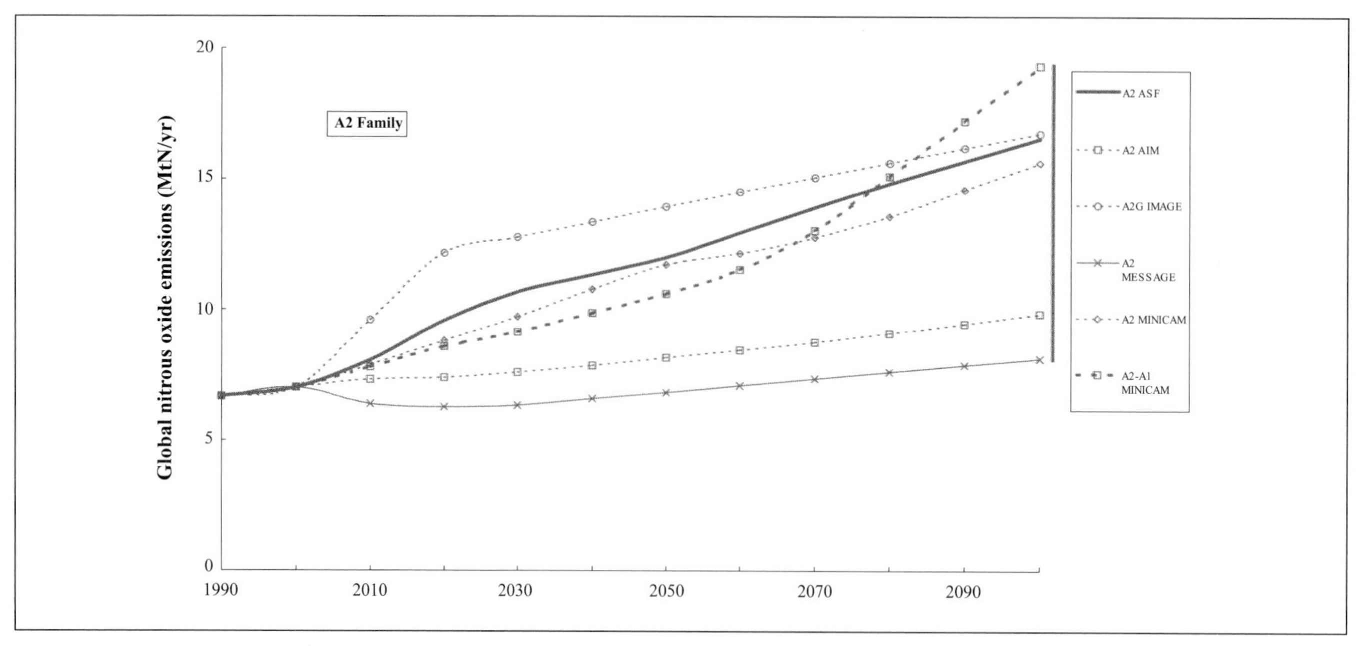

Figure 5-8b: Standardized global N_2O emissions in the A2 family scenarios. The marker scenario is shown with a thick line without ticks, the globally harmonized scenarios with thin lines, and the non-harmonized scenarios with thin, dotted lines (see Table 4-3).

Table 5-6: *N_2O emissions (MtN) by source for the marker scenarios (A1B-AIM, A2-ASF, B1-IMAGE, and B2-MESSAGE). Categories are fertilized soils and manure, fossil fuel use and industrial processes (adipic and nitric acid production), and other emissions. The "other" category includes biomass burning, land-use change, and sewage treatment (in some models). Emissions are provided in a non-standardized or "raw" format and are not comparable with standardized emission estimates used in the figures and other tables.*

		1990	2000	2010	2020	2030	2040	2050	2060	2070	2080	2090	2100
Fertilized Soils													
+ Manure	A1B	4.41	4.32	4.30	4.29	4.28	4.26	4.23	4.20	4.17	4.15	4.15	4.14
	A2	4.83	6.20	7.19	8.44	9.33	10.20	10.48	11.37	12.26	13.04	13.61	14.39
	B1	4.70	5.10	5.60	6.10	6.30	6.30	6.10	5.90	5.60	5.20	4.90	4.50
	B2	4.15	5.03	4.45	4.54	4.63	4.70	4.78	4.86	4.95	5.03	5.12	5.21
Industry													
+ Fossil Fuels	A1B	1.22	1.29	1.46	1.80	2.05	2.13	2.21	2.11	2.02	1.95	1.92	1.89
	A2	0.99	0.73	0.70	0.85	0.95	1.02	1.04	1.07	1.10	1.13	1.19	1.22
	B1	0.90	0.90	1.00	1.20	1.30	1.30	1.40	1.30	1.20	1.10	1.00	0.90
	B2	0.95	0.94	0.90	0.85	0.90	0.90	0.89	0.94	0.98	1.03	1.06	1.08
Other													
	A1B	1.32	0.64	0.52	0.33	0.26	0.24	0.23	0.23	0.24	0.25	0.25	0.25
	A2	0.75	0.86	0.95	1.42	1.16	1.22	1.24	1.28	1.33	1.42	1.60	1.69
	B1	0.40	0.60	0.70	0.50	0.50	0.50	0.40	0.40	0.30	0.30	0.30	0.30
	B2	1.21	0.68	0.48	0.30	0.24	0.25	0.26	0.26	0.26	0.27	0.27	0.28
Total													
	A1B	6.95	6.25	6.28	6.42	6.59	6.63	6.67	6.54	6.43	6.35	6.32	6.28
	A2	6.57	7.79	8.84	10.71	11.44	12.44	12.76	13.72	14.69	15.59	16.40	17.30
	B1	6.00	6.60	7.30	7.80	8.10	8.10	7.90	7.60	7.10	6.60	6.20	5.70
	B2	6.31	6.65	5.83	5.69	5.77	5.85	5.93	6.06	6.19	6.33	6.45	6.57

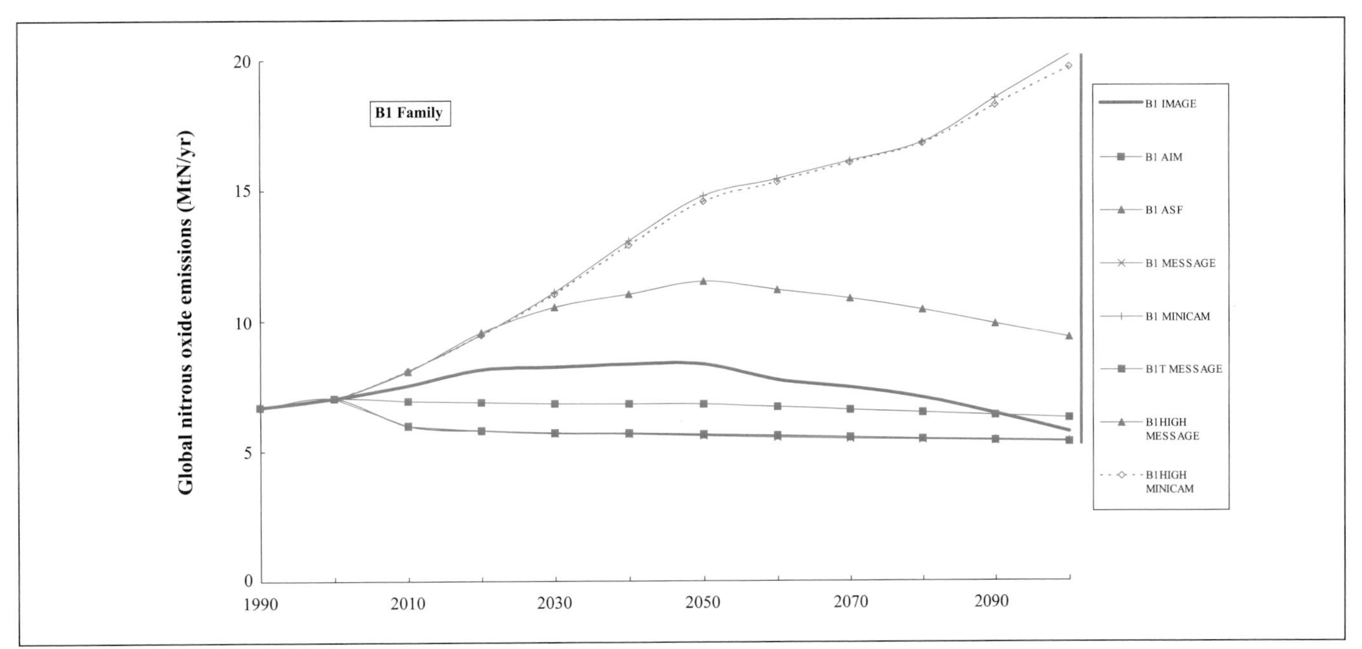

Figure 5-8c: Standardized global N_2O emissions in the B1 family scenarios. The marker scenario is shown with a thick line without ticks, the globally harmonized scenarios with thin lines, and the non-harmonized scenarios with thin, dotted lines (see Table 4-3).

5.4.2.3. B1 Scenario Family

The spread of N_2O emissions in the B1 family is quite large. The high-end reflects the agricultural assumptions used in the MiniCAM model, in which continued growth in meat production is driven by increases in per capita consumption that more than offset the reduction in population size (Figure 5-8c). The B1-AIM scenario has emissions near the low end of the range, because fertilizer use in developing countries is assumed to level off and better management increases productivity. Emissions from animal waste, mobile sources, and other sources in B1-AIM are nearly constant. The B1-IMAGE scenario (B1 family marker) has slowly increasing emissions through to 2050, which predominately originate from increases in livestock and the use of synthetic fertilizer (Table 5-6). After 2050, these emissions decline as the population size and associated demand for animal products declines (Figure 5-8c). The B1-ASF scenario has an intermediate emission trajectory with a maximum of 11.5 MtN in 2050 and a subsequent decline to 9.3 MtN by 2100 (Figure 5-8c). The two illustrative scenarios A1G-MiniCAM and A1T-MESSAGE display similar patterns of methane emissions as the A2 and B1 marker scenarios, respectively, and are therefore not discussed separately here.

5.4.2.4. B2 Scenario Family

The B2 family of scenarios covers a large range of N_2O emissions, similar to that found for the A1 family (see Figure 5-7). In the B2-MESSAGE scenario (B2 family marker), which has the lowest emissions in the family, improvements in agricultural productivity exceed increases in the demand for agricultural products and significantly slow emission growth in the sector (Table 5-6). Slow growth in agriculture and energy-related emissions and declines of emissions from other sectors result in a relatively flat emissions profile (Figure 5-8d). Emissions in the B2-ASF scenario are between the middle of the family range and its high end. The continuous increase in emissions in the B2-ASF scenario (which lie between the middle and high end of the family range) occurs through growth in nitrogen fertilizer use, in animal wastes, and in mobile sources. It is driven by increases in population and, to a smaller extent, by per capita increases in consumption of meat and dairy products.

5.4.2.5. Inter-Family Comparison

The allocation of N_2O emissions into source categories for the SRES marker scenarios is shown in Table 5-6. Separate emissions estimates for fertilized soils and manure were not available for all scenarios. As the emissions in Table 5-6 have not been standardized (see Box 5-1), base-year 1990 (and 2000) emissions are not the same for the different scenarios. In general, base-year emissions are likely to have greater differences at finer levels of detail. These differences result from different model calibrations, different model methodologies, and different classification schemes.

Emissions from fertilized soils and manure from agricultural animals dominate N_2O emissions in all the SRES markers. The fraction of total emissions from these two sources increases to about 80% in the A2, B1, and B2 marker scenarios, while the agricultural fraction remains roughly the same over the 21st century in the A1B marker. As discussed in Section 5.4.2, the range of future emissions is similar across each scenario family when all storyline interpretations are considered. Therefore,

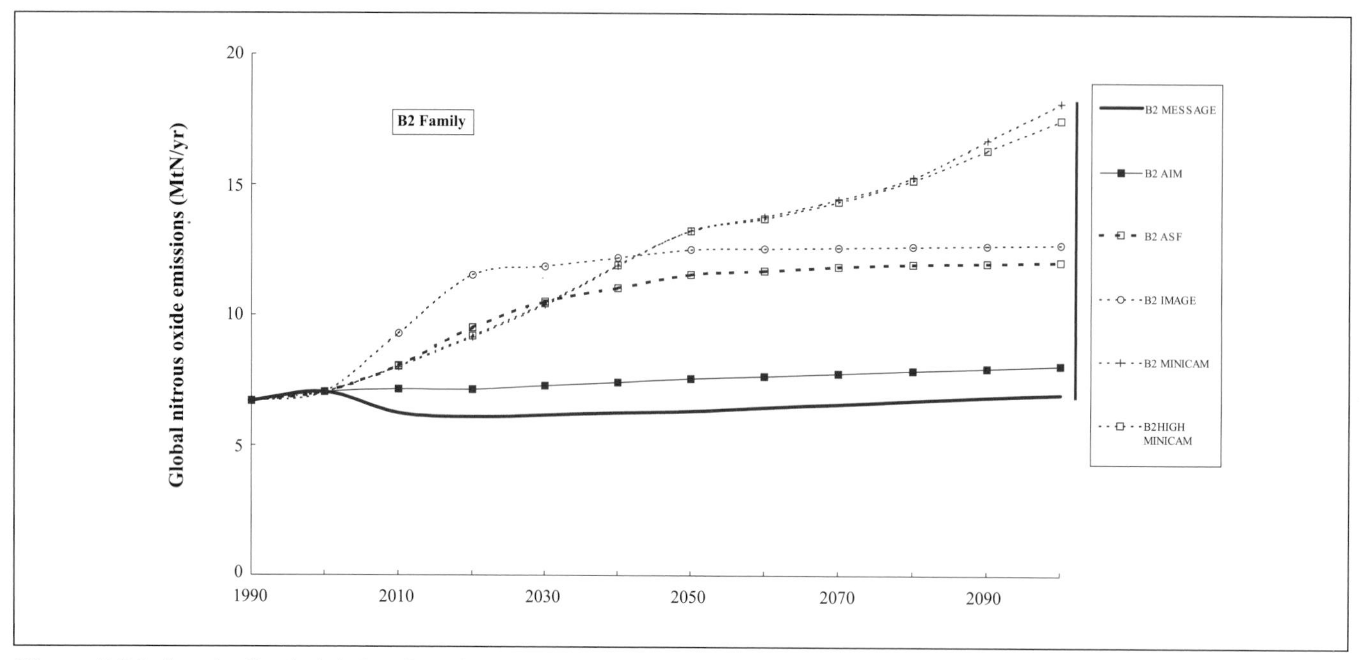

Figure 5-8d: Standardized global N_2O emissions in the B2 family scenarios. The marker scenario is shown with a thick line without ticks, the globally harmonized scenarios with thin lines, and the non-harmonized scenarios with thin, dotted lines (see Table 4-3).

while it is useful to discuss emissions from the marker scenarios individually, clearly differences in modeling approaches and model assumptions have a particularly strong effect on future N_2O emissions.

Agricultural emissions in the A1B-AIM marker scenario decrease over the 21st century, with increased productivity offsetting additional food demand. Agricultural emissions in the A2-ASF marker scenario increase substantially because of the food demands of a large population, coupled with less technological change. Emissions in the B1-IMAGE marker scenario increase and subsequently fall. In the B2-MESSAGE marker, agricultural emissions of N_2O increase steadily throughout most of the 21st century, with a decrease between 2000 and 2010. These patterns differ because of the different driving forces (population and demand), model assumptions, and modeling approaches.

Emissions from the categories "Industry/Fossil Fuels" and "Other" show a monotonic rise in the A2-ASF marker, while emissions in the B1-IMAGE marker rise and subsequently fall. Industry and fossil fuel emissions fall and then increase slightly in the B2-MESSAGE marker scenario, while "Other" emissions, which largely arise from land-use changes, fall. Industry and fossil fuel emissions in the A1B-AIM marker scenario increase through the 21st century, but are countered by a decrease in emissions from the "Other" category.

The combined dynamics of N_2O emissions in the A1, B1, and B2 markers leads to nearly stable or declining emissions during most of the 21st century (Figure 5-7). The B2 marker shows the lowest emission level (from the year 2010 to 2080), despite a larger population than in A1 and B1. One possible reason is inter-model differences in treatment of land-use changes. Unlike the other markers, the A2-ASF scenario yields a continuous growth in emissions, which corresponds to its assumptions of high population and slow technological change.

5.4.3. *Halocarbons and Other Halogenated Compounds*

Emissions of halocarbons (CFCs, HCFCs, halons, PFCs, and HFCs) and other halogenated compounds (SF_6) on a substance-by-substance basis are described in detail in Fenhann (2000). A list of the substances covered, together with their GWPs and lifetimes (as in IPCC SAR; Houghton, *et al.* 1996), is given in Table 5-7.

Importantly, future emissions of halocarbons and other halogenated compounds strongly depend on the technologies involved in their production and use. New uses for these substances may arise or new products or technologies may replace current uses. It is assumed here that the current mix of products continues to exist for the next 100 years to 2100, with some generic technological improvements as described below. This assumption, however, means that emissions projections for industrial gases discussed in this section carry a substantial uncertainty.

Halocarbons are carbon compounds that contain fluorine, chlorine, bromine, and iodine. Halocarbons that contain chlorine (CFCs and HCFCs) and bromine (halons) cause ozone depletion, and their emissions are controlled under the Montreal Protocol and its Adjustment and Amendments. According to the 1987 Montreal Protocol and its subsequent

Table 5-7: *GWPs and atmospheric lifetimes of halocarbons and other halogenated compounds.*

Species	Chemical Formula	100 Years GWP	Atm. Lifetime Years
CFC-11	CCl_3F	4000	50
CFC-12	CCl_2F_2	8500	102
CFC-113	CCl_2FCClF_2	5000	85
CFC-114	$CClF_2CClF_2$	9300	300
CFC-115	CF_3CClF_2	9300	1700
Carbon tetrachloride	CCl_4	1400	42
Methyl chloroform	CH_3CCl_3	110	4.9
Halon-1211	$CBrClF_2$	No data	20
Halon-1301	$CBrF_3$	5600	65
Halon-2402	$CBrF_2CBrF_2$	No data	20
HCFC-22	$CHClF_2$	1700	12.1
HCFC-141b	CH_3CFCl_2	630	9.4
HCFC-142b	CH_3CF_2Cl	2000	18.4
HCFC-123	CF_3CHCl_2	93	1.4
HFC-23	CHF_3	11700	264
HFC-32	CH_2F_2	650	5.6
HFC-43-10	$C_5H_2F_{10}$	1300	17.1
HFC-125	C_2HF_5	2800	32.6
HFC-134a	CH_2FCF_3	1300	14.6
HFC-143a	$C_2H_3F_3$	3800	48.3
HFC-152a	$C_2H_4F_2$	140	1.5
HFC-227ea	C_3HF_7	2900	36.5
HFC-236fa	$C_3H_2F_6$	6300	209
HFC-245ca	$C_3H_3F_5$	560	6.6
Perfluoromethane	CF_4	6500	50000
Perfluoroethane	$C2F_6$	9200	10000
Perfluorobutane	C_4F_{10}	7000	2600
Sulfur hexafluoride	SF_6	23900	3200

amendments, consumption (the balance of production plus imports minus exports) of CFCs is largely banned in developed countries after January 1996 (and developing countries after 2010), although some countries have failed to meet the deadline. Furthermore, HCFC consumption will be subjected to a gradual phase-out, with cuts from the 1986 base-year values of 35%, 65%, and 90% in 2004, 2010, and 2015, respectively. Final HCFC consumption phase-out will occur in 2020 (2040 for developing countries).

The six modeling teams participating in the SRES process did not develop their own projections for emissions of ODS and their substitutes. Hence, a different approach for the development of long-range estimates for halocarbons and other halogenated compounds was adopted. First, for ODSs the external Montreal Protocol A3 maximum production scenario was used as a direct input to all SRES scenarios (WMO/UNEP, 1998), since most measures in this A3 scenario have been implemented already or are well established and under way (and so no large scenario variation is expected). For other gas species, a simple methodology to develop different emission trajectories consistent with aggregate SRES scenario driving-force assumptions (population, GDP, etc.) was developed. Also, the assumed future control rates have been adopted to conform to the SRES storylines presented in Chapter 4. The underlying literature, scenario methodology, and data are documented in more detail in Fenhann (2000) and are summarized in this section.

The resultant emissions of Montreal gases, HFCs, PFCs, and SF_6 are summarized in Table 5-8. The effect on climate of each of the substances listed in Table 5-9 varies greatly because of differences in both the atmospheric lifetime and the radiative effect per molecule of each gas. A good measure of the net climate effect of halocarbons and other halogenated compounds is provided by their radiative forcing. Radiative forcing will be addressed in IPCC's Third Assessment Report, but is not discussed in this report.

Emissions of individual groups of halocarbons and other halogenated compounds in the four families of SRES scenarios are presented below.

***Table 5-8**: Global anthropogenic emissions (kt) projections for ODS, HFC, PFC, and SF_6 emissions in the four marker scenarios.*

	1990	2020				2050				2100			
Marker Scenario		A1	A2	B1	B2	A1	A2	B1	B2	A1	A2	B1	B2
ODS	1864	253				21				1			
HFC-23	6.4	4.9	4.9	4.9	4.9	1.0	1.0	1.0	1.0	1.0	1.0	1.0	1.0
HFC-32	0.0	8.3	6.4	6.0	6.2	24.3	14.0	13.9	14.1	30.3	32.8	12.9	25.9
HFC-43-10	0.0	8.8	7.6	6.9	7.2	18.1	10.7	10.7	11.1	30.3	21.8	10.4	17.9
HFC-125	0.0	27.1	20.7	20.6	21.5	80.4	45.6	47.9	48.7	100.8	106.5	44.3	89.1
HFC-134a	0.0	325.5	252.2	248.8	261.9	931.0	506.4	547.4	561.2	980.3	1259.8	486.0	1079.3
HFC-143a	0.0	20.6	16.0	15.0	15.6	60.9	35.1	34.8	35.4	75.7	82.1	32.2	64.7
HFC-227ea	0.0	22.2	16.6	18.5	19.7	62.1	31.5	39.4	40.7	60.6	80.4	34.4	80.0
HFC-245ca	0.0	100.5	78.7	80.3	85.4	292.3	149.2	172.6	178.5	288.5	388.0	150.2	352.7
HFCs-total	6.4	517.9	403.0	401.0	422.4	1470.2	793.5	867.8	890.8	1567.3	1972.4	771.4	1710.5
CF_4	15.8	21.1	25.2	15.7	27.1	43.8	45.6	20.9	52.7	57.0	88.2	22.2	59.9
C_2F_6	1.6	2.1	2.5	1.6	2.7	4.4	4.6	2.1	5.5	5.7	8.8	2.2	6.0
PFCs, total	17.4	23.2	27.7	17.3	29.8	48.2	50.2	23.0	58.2	62.7	97.0	24.4	65.9
SF_6	5.8	7.3	9.7	5.7	8.4	18.3	16.0	10.4	12.1	14.5	25.2	6.5	10.6

Notes: ODS emissions are from scenario A3 in the *UNEP/WMO Scientific Assessment of Ozone Depletion* (UNEP/WMO, 1998).

5.4.3.1. Hydrofluorocarbons

HFCs are beginning to be produced as replacements for CFCs and HCFCs. Unlike the CFCs and the HCFCs, HFCs do not convey chlorine to the stratosphere and thus do not contribute to ozone depletion.

For the development of future HFC emissions, Fenhann (2000) used a procedure based on the work by Kroeze (1995) that includes two steps:

- "Virtual" future CFC emissions are first calculated assuming a situation without the Montreal Protocol.
- CFCs are substituted with HFCs according to substitution percentages adopted from the literature (Table 5-9) and also the various degrees of emission reduction potentials from better housekeeping measures and technological change.

Concerning the first step of the methodology used in Fenhann (2000), 1990 CFC emissions were taken from the *Scientific Assessment of the Ozone Depletion* (WMO/UNEP, 1998). Pre-Montreal 1986 emissions were obtained from McCulloch *et al.* (1994). Future "virtual" (assuming no Montreal protocol) emissions of CFCs were assumed to be proportional to their consumption, for which GDP numbers in the four marker scenarios were used as a driver (see Chapter 4). The saturation level of per capita demands was assumed to be the same in all four SRES scenario families.

The projection of CFC emissions in the absence of the Montreal Protocol shows how emissions would change under conditions of unrestricted production. However, with the Montreal Protocol in place, other chemical compounds will be used to replace the Montreal gases. To compute the amount of CFCs replaced with these other compounds, future CFC emissions with the Montreal Protocol in place (according to the WMO/UNEP A3 ODS scenario) were first subtracted from the "virtual" CFC emissions.

Different assumptions about CFC applications as well as substitute candidates were developed (Fenhann, 2000). These were initially based on Kroeze and Reijnders (1992) and Midgley and McCulloch (1999), and subsequently updated using the latest information from the *Joint IPCC/TEAP Expert Meeting on Options for the Limitation of Emissions of HFCs and PFCs* (WMO/UNEP, 1999).

An important assumption (based on the latest information from the industry) used in the current analysis is that relatively few Montreal gases will be replaced completely by HFCs. Currently, HFC-134a is favored, and it is the only HFC with sufficiently large sales to be included in the current production and sales statistics (AFEAS, 1998). The global emissions of this gas are estimated to be 0.1 kt HFC-134a in 1990 and 42.7 kt HFC-134a in 1997. Current data indicate that substitution rates of CFCs by HFCs will be less than 50%. It was shown recently that in the European Union the substitution rate of CFCs by HCFCs was 26%, and the HFC share was 6% or a total of 32% (McCulloch and Midgley, 1998). Time series data for the global sales from AFEAS (1998) confirm a 763 kt per year reduction in CFC production and use from the peak production year of 1987 through 1996. An increase in the total HFC and HCFC production and use was 340 kt per year, or a

***Table 5-9**: Substitution of CFCs by HFCs and PFCs.*

Application	From	HFC-23	HFC-32	HFC-43-10	HFC-125	HFC-134a	HFC-143a	HFC-152a	HFC-227ea	HFC-236fa	HFC-245fa	C_4F_{10}	Total
Aerosols	CFC					4.0%			4.0%				8.0%
Cleaning/drying	CFC			0.5%									0.5%
Open cell foams	CFC												0.0%
Closed cell foams	CFC					25.0%					25.0%		50.0%
Stationary cooling	CFC		2.0%		5.0%	25.0%	5.0%						37.0%
Stationary cooling	HCFC -22		2.0%		5.0%	25.0%	5.0%						37.0%
Mobile cooling	CFC					25.0%							25.0%
Fire extinguisher (portable)	Halon-1211								1.0%				1.0%
Fire extinguisher (fixed)	Halon-1301								25.0%				25.0%
Other uses	CFC				5.0%	5.0%							10.0%

44% substitution up to 1996. In Fenhann (2000) future technological developments are assumed to result in about 25% of the CFCs ultimately being substituted by HFCs (Table 5-9). This low percentage not only reflects the introduction of non-HFC substitutes, but also the notion that smaller amounts of halocarbons are used in many applications when changing to HFCs and that emissions are reduced by increased containment and recycling. A general assumption is that the present trend to not substitute CFCs with high GWP substances, including PFCs and SF_6, will continue. The substitution rates shown in Table 5-9 were used in all four scenarios; the technological options adopted are those known at present. Further substitution away from HFCs is assumed to require a climate policy.

Hydrocarbons are expected to be the substitutes used in the *aerosols/propellant* sector, except for situations in which the flammability of hydrocarbons would be a problem and also in metered dose inhalers (to avoid possible adverse clinical effects). HFC-227ea and HFC-134a, and possibly HFC-152a are expected to replace hydrocarbons (Table 5-9; WMO/UNEP, 1999).

CFC-113 was used extensively as a *cleaning* solvent for metal, electronics, and textiles. The general trend in this area now is toward water-based systems. However, as suggested by Table 5-9 a small fraction (0.5%) of the CFC in this sector is substituted by HFC-43-10 (Kroeze, 1995).

The WMO/UNEP (1998) report states that no fluorocarbons are now used for *open cell foams*, an assumption also adopted in the scenarios.

It is expected that closed *cell foams* and *refrigeration* will be the largest demand sectors for HFCs in the future. For *closed cell* foams, the substitution is expected to be 50%, one-half as HFC-134a and the other half as the liquid HFC-245fa (expected to be commercially available by 2002; Table 5-9) (Ashford, 1999). In some cases, HFC-365mfc will be used instead of HFC-245fa. However, all the calculations in Table 5-9 were carried out for HFC245fa, since these two substances have almost the same climate effect.

Prior to 1986, the main *refrigerants* in use were CFCs, HCFCs, and ammonia. In response to the Montreal Protocol, HFC and hydrocarbon refrigerants have been promoted as the primary alternatives (WMO/UNEP, 1999). The main HFC assumed to be used for *stationary cooling* is HFC-134a, with 5% of the demand substituted by HFC-125 and another 5% by HFC-143a (Kroeze, 1995). This would agree with the reported measurements of these two substances in the atmosphere. According to Kroeze (1995) about 2% might be substituted by HFC-32.

Before 1993, all air-conditioned cars were equipped with systems using CFC-12 as a refrigerant. Over the lifetime of a car, 0.4 kg of this halocarbon was emitted every year. In 1994, two years after the new refrigerant HFC-134a had become available globally in sufficient quantities, almost all major vehicle manufacturers began to use HFC-134a. This conversion was accompanied by a significant reduction in annual losses of refrigerants per car, down to 0.096 kg of halocarbon (Preisegger, 1999). Therefore the substitution rate in Table 5-9 for *mobile cooling* is assumed to be 25%.

In the *fixed fire extinguishers* sector, only about 25% of the systems that formerly used halons now use HFCs, mainly HFC-227ea. The rest use CO_2, inert gas mixtures, water-based systems, foam, dry powder, etc. (WMO/UNEP, 1999). Increased environmental awareness in the industry is assumed to have resulted in the reduction of HFC emissions by a factor of three, compared to former practice.

For *portable fire extinguishers* the substitution rate is assumed to be only 1%, even less than the 2% assumed by Kroeze (1995).

CFCs have also been used for other purposes, such as sterilants, tobacco expansion, and others. Kroeze (1995) assumes a 30% substitution by HFCs. However, in the SRES scenarios this value is reduced to 10% to remain consistent with the above assumption that HFCs ultimately will substitute for about 25% of the CFCs.

As well as using non-halocarbon substitutes, HFC emissions can be avoided by better housekeeping, for instance by reduced spilling of cooling agents. Leakage control equipment can also serve this purpose. Finally, halocarbons can be recovered for recycling or destruction when equipment is discarded. Some of this emission reduction potential is likely to be implemented as a result of technological changes introduced to control ODSs. In the SRES scenarios, reduction rates were varied over time and between industrialized and developing countries to reflect the definitive features of the underlying storylines (Chapter 4). Generally, the reduction rates are assumed highest in scenarios that emphasize sustainability and environmental policies (B1 family). These reductions, however, were not associated with any explicit GHG reduction policies, as required by the SRES Terms of Reference (see Appendix I). In one scenario family, A2, no reductions were assumed, whereas in the A1 and B2 families reduction rates were set at intermediate levels.

In addition to consumption-related emissions of HFCs, HFC-23 is emitted as an undesired by-product from the HCFC-22 production process. As a result of the Montreal Protocol, the direct use of HCFC-22, and hence the related HFC-23 emissions, will come to a halt in 2050. To calculate the HFC-23 emissions, information from Oram *et al.* (1998) was used (estimated emissions of HFC-23 at 6.4 kt in 1990). By relating this value to 178.1 kt HCFC-22 emitted in 1990 (WMO/UNEP, 1999), an emissions factor of 0.036 tons of HFC-23 per ton of HCFC-22 was calculated and applied to estimate future emissions. Since this estimation procedure does not take into account any pollution control regulations (that are not driven by climate considerations), it may result in an overestimation of HFC-23 during the early decades of the 21st century, until HCFC production is phased out under the Montreal Protocol. After the phase-out of HCFC-22 consumption, some HFC-23 emissions will still occur because of the continued HCFC-22 feedstock production allowed under the Montreal Protocol. The resultant projections are shown by individual HFC in Table 5-8.

In general, the SRES scenarios might underestimate HFC emissions if the substitution of CFCs with alternatives that have no radiative forcing effect and with more efficient HFCs-based technologies does not penetrate as quickly as is assumed, especially in developing countries. However, more effective technologies and/or suitable non-HFC alternatives may be developed, which would lead to even lower emissions.

5.4.3.2. Perfluorocarbons

PFCs, fully fluorinated hydrocarbons, have extremely long atmospheric lifetimes (2600 to 50,000 years) and particularly high radiative forcing (Table 5-7). The production of aluminum is thought to be the largest source of PFCs (CF_4, and C_2F_6) emissions. These emissions are generated, primarily, by the anode effect, which occurs during the reduction of alumina (aluminum oxide) in the primary smelting process as alumina concentrations become too low in the smelter. Under these conditions, the electrolysis cell voltage increases sharply to a level sufficient for bath electrolysis to replace alumina electrolysis. This causes substantial energy loss and the release of fluorine, which reacts with carbon to form CF_4 and C_2F_6.

In 1990, the total annual global primary aluminum production was 19.4 Mt. Secondary aluminum production from recycling accounted for 21.5% of the total consumption in 1990. The production statistics from the World Bureau of Metal Statistics (1997) show that the total aluminum production was 27.5 Mt, and recycling has increased to 25.6%, or by about 3.5 percentage points, in 10 years.

The scenarios developed by Fenhann (2000) adopt a methodology of projecting future aluminum demand based on:

- Aluminum consumption elasticity with respect to GDP.
- Use of alternative assumptions concerning recycling rates.
- Varying emission factors to reflect future technological change.

These assumptions are altered to be in consistent with the four SRES scenario storylines described in Chapter 4.

For instance, in Fenhann (2000) the aluminum consumption elasticity varies between 0.8 and 0.96, and the range of increases in aluminum recycling rates varies between 1.5 and 3.5 percentage points per decade. The PFC emission factor varies according to the aluminum production technology used. The default emission factor from the Revised IPCC Guidelines (IPCC, 1997) is 1.4 kgCF_4/t aluminum. However, Harnisch (1999) gives evidence that the average specific emissions of CF_4 per ton of aluminum has decreased from about 1.0 kg to 0.5 kg between 1985 and 1995. Accordingly, an emissions factor of 0.8 kgCF_4/t was used for 1990 and this was assumed to decrease to 0.5 kg CF_4/t in the future. This is also in agreement with the value of 0.51 kgCF_4/t recommended by the IPCC Expert Meeting on Good Practices in Inventory Preparation for Industrial Processes and the New Gases (January 1999, Washington, DC). The same sources also agree on an emission factor for C_2F_6 that is 10 times lower than that for CF_4. This assumption was also used in the calculations presented here (Table 5-8).

Aluminum production is being upgraded from highly inefficient smelters and practices to reduce the frequency and duration of the anode effect. Since aluminum smelters are large consumers of energy, the costs of these modifications are offset by savings in energy costs and are therefore assumed to occur in all scenarios. The ultimate reduction of the anode effect frequency and duration was assumed to reach the same level in all the SRES scenarios. However, scenarios vary with respect to the rate of introducing the underlying modifications. It is technically possible to reduce the anode emissions by a factor of 10 (EU, 1997). This technically feasible reduction can be achieved by changing from the Söderberg cells currently in use to more modern pre-bake cells. It is assumed that this will happen in the A1 and B1 family scenarios, in which specific emissions of 0.15 $kgCF_4/t$ are achieved by 2040 in the OECD90 region and by 2090 in the other regions. In the A2 and B2 family scenarios the same specific emissions are achieved later in the century in the OECD90 region and not until after 2100 in the other regions.

PFCs are consumed in small amounts in such sectors as electronics (tracers), cosmetics, and medical applications. However, the only emissions included in Fenhann (2000) beyond aluminum production were PFCs (as CF_4) from semiconductor production. In all SRES scenarios the emission estimates used are those given by Harnisch *et al.* (1999) of 0.3 kt CF_4 per year in 1990, 1.1 $ktCF_4$ in 2000, 1.0 $ktCF_4$ in 2010, and constant thereafter. The use of these estimates reflects the voluntary agreement, in April 1999, of the World Semiconductor Council, which represent manufacturers from Europe, Japan, Korea , and the US, among others. According to this agreement, manufacturers have adopted the emission reduction target for PFCs of 10% absolute reduction from 1995 emission levels by 2010. This target encompasses over 90% of the total semiconductor production (WMO/UNEP, 1999). The total PFC emissions in the four SRES scenario families cover a range from 24 to 97 kt PFC in 2100 (Table 5-8; Fenhann, 2000).

5.4.3.3. *Sulfur hexafluoride*

SF_6 is an extremely stable atmospheric trace gas. All studies concur that this gas is entirely anthropogenic. Its unique physico-chemical properties make SF_6 ideally suited for many specialized industrial applications. Its 100-year GWP of 23,900 is the highest of any atmospheric trace gas. In 1994, atmospheric concentrations of SF_6 were reported to rise by 6.9% per year, which is equivalent to annual emissions of 5,800 tSF_6 (Maiss *et al.*, 1996).

According to several sources (Kroeze, 1995; Maiss *et al.*, 1996; Victor and MacDonald, 1998), about 80% of SF_6 emissions originate from its use as an insulator in high-voltage electrical equipment. The remaining 20% of the present global SF_6 emissions (1200 tons per year) are emitted from magnesium foundries, in which SF_6 is used to prevent oxidation of molten magnesium. The global annual production of magnesium is about 350,000 tons (US Geological Survey, 1998), and developing countries account for about 15% of the total. SF_6 is also used to de-gas aluminum, but since SF_6 reacts with aluminum, little or no atmospheric emissions result from this process.

Major manufacturers of SF_6 agreed voluntarily to co-operate on the compilation of worldwide SF_6 sales data by end-use markets. Six companies from the US (three), Japan, Italy, and Germany participated in the data survey. The companies do not expect the total sales for magnesium foundries to increase before 2000 (Science & Policy Services Inc., 1997). Based on this information, the 1996 statistical production values were used for the year 2000 in the formulation of the scenarios reported in Fenhann (2000). Future production was projected assuming the same consumption elasticity to GDP as for aluminum (see discussion above). In 1996, about 41% of the world magnesium was produced in the US; of this, only 16% was processed in foundries for casting that resulted in emissions of SF_6 (Victor and MacDonald, 1998). Since the distribution of world foundry capacity appears to be roughly similar to that of world magnesium production, Fenhann (2000) assumes that, presently, 16% of the produced magnesium is processed in foundries across all regions. Relating this amount of the processed magnesium to the aforementioned emission of 1200 tSF_6 per year yields an emission factor of 21 $kgSF_6$ per ton of magnesium processed in foundries. The demand for magnesium in automotive applications as a strong lightweight replacement for steel is growing quickly. Hence, it is expected that the fraction of total magnesium production processed in foundries by 2050 will grow to between two to three times the present level.

As mentioned above, no less than 80% of SF_6 emissions (or 4600 tons of SF_6 per year at present) originate from the use of SF_6 as a gaseous insulator in high-voltage electrical equipment. The unique ability of SF_6 to quench electric arcs has enabled the development of safe, reliable gas-insulated high-voltage breakers, substations, transformers, and transmission lines. The demand for such electrical equipment is assumed to grow proportionally to electricity demand (Victor and MacDonald, 1998; Fenhann, 2000) with an emission factor of 132.6 tSF_6/EJ electricity. Fenhann (2000) used preliminary electricity generation projections from the four SRES marker scenarios and assumed additional various other potentials for emission reductions that result from more careful handling, recovery, recycling, and substitution of SF_6. Reduction rates vary in the different SRES scenario storylines; the detailed assumptions are reported in Fenhann (2000). The SF_6 emissions for the four scenarios given in Fenhann (2000) range from 7 to 25 $ktSF_6$ in 2100. The main driver is electricity consumption, since the bulk of emissions originate from electric power transmission (for transformers).

SF_6 is also emitted from other minor sources, but for the purposes of this report it is assumed that uncertainty ranges factored into the alternative scenario formulations cover the emissions from these sources.

5.5. Aerosols and Ozone Precursors

In addition to the GHGs discussed above, aerosol particles and tropospheric ozone also change the radiative balance of the atmosphere, albeit in a spatially heterogeneous manner. Sulfate aerosol particles, which form as a consequence of SO_2 emissions, act as a cooling agent. Their net effect is quite uncertain, but is thought to offset the forcing from all non-CO_2

GHGs to date (Houghton *et al.*, 1996). As interest in the role of sulfur has increased the since the previous IPCC assessment, and to encapsulate recent trends and expectations, sulfur emissions are discussed here in substantial detail. Nitrates, ammonia, organic compounds, and black carbon also contribute to the formation of atmospheric aerosols. Carbonaceous aerosols exert a small positive forcing effect, while the effects of other compounds and aerosols are less clear. Tropospheric ozone is a GHG also, with a small net positive forcing effect. Future tropospheric ozone levels will be determined by emissions of CH_4, CO, NO_x, and NMVOCs. The last three groups are reported and discussed here in a more aggregated and stylized form only, because these gases are short lived, their potential to form ozone is highly non-linear; NMVOCs are not distinguished by their reactivity, and data problems associated with including key sources in aggregated long-term models are large.

5.5.1. *Ozone Precursors: Nitrogen Oxides, Non-Methane Volatile Organic Compounds, and Carbon Monoxide*

5.5.1.1. *Nitrogen Oxides*

Emissions of NO_x primarily result from the combustion of fossil fuels. The NO_x concentration in exhaust gases depends on combustion conditions (temperature, residence time, air-to-fuel ratio, mixing) and varies widely across different applications. In particular, internal combustion engines used in road vehicles and ships have very high emissions, although new designs and exhaust-gas treatment offer much lower specific emission levels. Recent research (Davidson and Kingerlee, 1997; Delmas *et al.*, 1997; Mosier *et al.*, 1998) indicates that soil may be a significant source of NO_x emissions also. This source, however, is not included in the models used in the current report.

The 1990 NO_x emissions in the six SRES models range between 26.5 and 34.2 MtN, but not all the models provide a comprehensive description of NO_x emissions. Some models do not estimate NO_x emissions at all (MARIA, MiniCAM[8]), whereas others only include energy-related sources (MESSAGE) and have adopted other source categories from corresponding model runs derived from other models (i.e. AIM). Standardized (see Box 5-1 on Standardization) 1990 NO_x emissions in the SRES scenarios, measured as nitrogen, amount to 31 MtN (Figure 5-9).

As mentioned in Chapter 4, the volume of fossil fuels used for various energy purposes varies widely in the SRES scenario families. In addition, the level and timing of emission controls, inspired by local air quality concerns, is assumed to differ. As a result the spread is largest within the A1 scenario family, in which it is almost as large as the range across all 40 SRES scenarios. Up to the 2020s, all scenarios project rising NO_x emissions (Figure 5-9). The 25th and 75th percentile emissions corridor spans between 40 and 60 MtN by the 2020s, which can be interpreted as a "central tendency" among the entire spectrum of the 40 SRES scenarios. Beyond 2030, uncertainties in emission levels increase significantly. By 2100, the SRES range is between 16 and 150 MtN (i.e. emissions decrease by a factor of two or increase by a factor of five compared with 1990 levels). The median and mean emissions are tracked by a number of scenarios, most notably by B2-MESSAGE (B2 family marker) and A1B-IMAGE. In these scenarios, NO_x emissions tend to increase up to 2050 and stabilize thereafter, the result of a gradual substitution of fossil fuels by alternatives as well as by the increasing diffusion of NO_x control technologies. Low emission futures are described by various B1 family scenarios, whereas the upper bound for future NO_x emissions is represented by scenarios of the fossil fuel intensive A1 scenario groups (e.g. A1C- and A1G-MESSAGE) and the high population, high fossil energy A2 scenario family (A2-ASF, A2-MESSAGE, or A2G-IMAGE) (Figure 5-9).

The fossil fuel dominated A2-ASF (A2 family marker) with limited environmental concern has a rapidly increasing NO_x trajectory (Figure 5-9). Emissions in other A2 scenarios also continue to grow, except in A2-AIM for which emissions level off by the last decades of the 21st century. In the A1B marker (A1B-AIM), the emissions growth is initially about as strong as in A2-ASF, but emissions peak in 2030, and decline as the fossil fuel share of total primary energy falls and the remaining fossil fuel technologies become more advanced (Figure 5-9). Scenarios from other A1 family groups that assume a much larger and continued role of fossil fuels yield much higher NO_x emissions, which reach 150 MtN by 2100 in the coal-based A1C-MESSAGE scenario. Emission growth in the B2 family scenarios is less steep than in the A1 family, but persists throughout the entire period, albeit at a declining rate. By 2100, emissions in the B2-MESSAGE scenario (B2 family marker) are about twice as large as in 1990 (Figure 5-9). B2-ASF has a similar trajectory, while the B2-AIM scenario has essentially constant NO_x emissions over the entire period. Emissions in the B1 marker (B1-IMAGE) are among the lowest of all the 40 scenarios (Figure 5-9). In this scenario, emissions increase stops around 2050 and subsequently declines toward the end of the 21st century to 60% of the current level. Other scenarios within the B1 group coincide well with the B1 marker in 2100, although the maximum emission levels in these scenarios are much higher than in B1-IMAGE (Figure 5-9).

5.5.1.2. *Non-Methane Volatile Organic Compounds*

NMVOCs arise from fossil fuel combustion (as with NO_x, wide ranges of emission factors are typical for internal combustion engines), and also from industrial processes, fuel storage (fugitive emissions), use of solvents (e.g., in paint and cleaners), and a variety of other activities. As the chemical reactivities of the various substances grouped under the

[8] For the A1G-MiniCAM scenario emissions from congruent model runs derived from other models have been estimated.

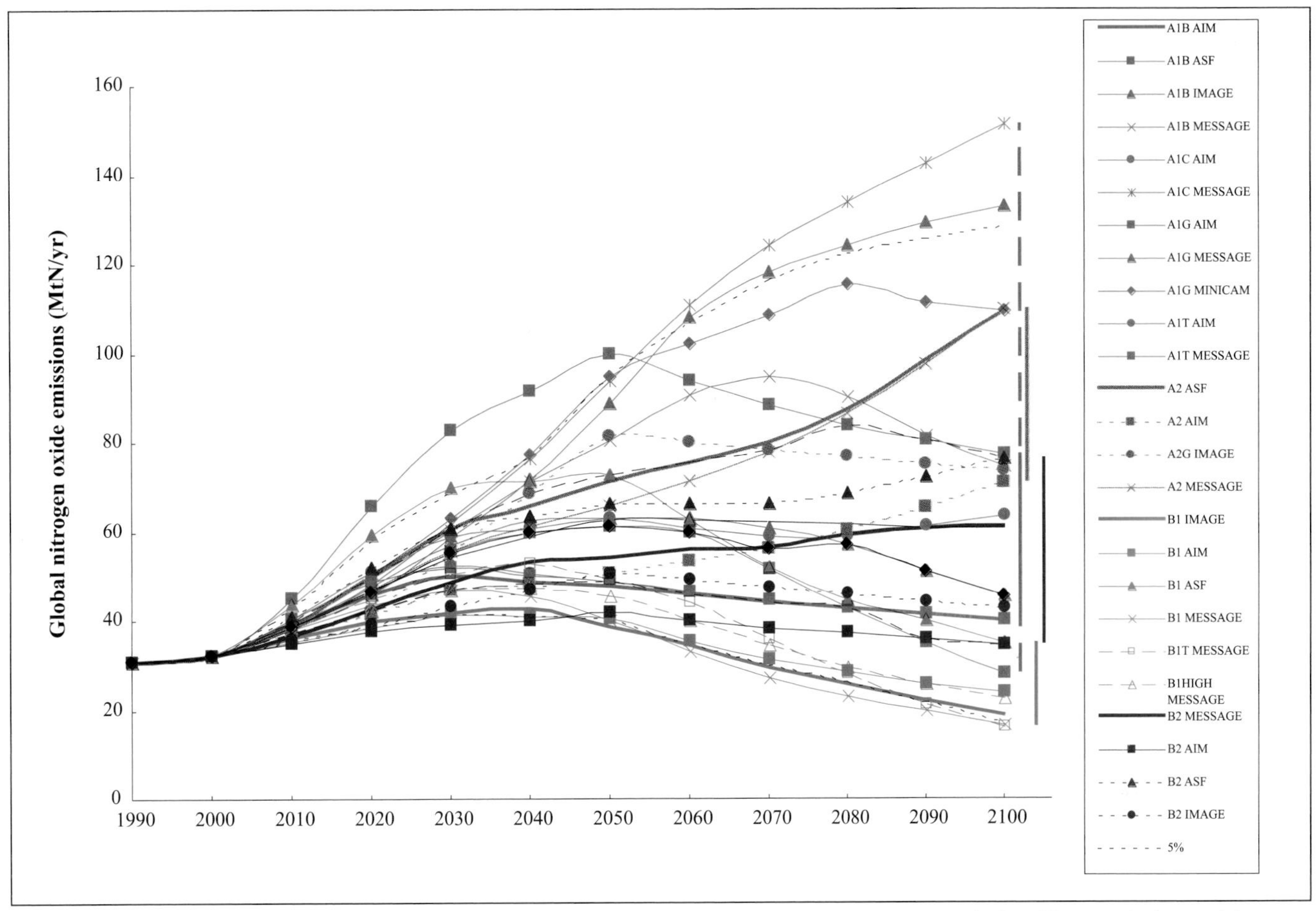

Figure 5-9: Standardized global NO_x emissions in SRES scenarios, classified into four scenario families (each denoted by a different color code – A1, red; A2, brown; B1, green; B2, blue). Marker scenarios are shown with thick lines without ticks, globally harmonized scenarios with thin lines, and non-harmonized scenarios with thin, dotted lines (see Table 4-3). Black lines show percentiles, means, and medians for SRES scenarios. For numbers on the two additional illustrative scenarios A1FI and A1T see Appendix VII.

NMVOCs category are very different, so are their roles in ozone formation and the (potential) health hazards associated with NMVOCs. In this report NMVOCs are reported as one group. In 1990, the estimated NMVOC emissions range between 83 and 178 Mt, which after standardization (see Box 5-1) translates into 140 Mt (Figure 5-10). As discussed above for NO_x emissions, not all models include this emissions category or all of its sources; the most detailed treatment of NMVOC emissions is given in the ASF model.

A relatively robust trend across all 40 scenarios (see Figure 5-10) is a gradual increase in NMVOC emissions up to about 2050, as indicated by the 25th and 75th percentile corridor, with the range between 190 and 260 Mt by that year. Beyond 2050, uncertainties increase with respect to both emission levels and trends. As for NO_x emissions discussed above, the upper bounds of NMVOC emissions are formed by fossil fuel intensive scenarios within the A1 scenario family (e.g. A1B-ASF), and the lower bounds by the scenarios within the B1 family (with an important alternative higher scenario B1-ASF). Characteristic ranges are between 60 and 90 Mt by 2100 in the low emissions cluster and between 370 and 550 Mt in the high emissions cluster. All other scenario families and individual scenarios fall between these two emissions clusters, with the B2 marker scenario (B2-MESSAGE) closely tracking the median of global NMVOC emissions from all the SRES scenarios.

In the B1 family marker (B1-IMAGE) emissions gradually decline to 60% of the 1990 level by 2100 (Figure 5-10). The B1-AIM and B1-MESSAGE trajectories are similar, but differ somewhat from the B1 marker. They increase until the 2020s (AIM) or even the 2050s (MESSAGE) at modest rates, but subsequently decline to around or even below the B1 marker in 2100. The B1-ASF profile, however, is radically different, growing faster and continuously throughout the 21st century. By 2100, the NMVOC emissions in B1-ASF increase to 350 Mt, or 2.5 times the 1990 level. This indicates the adoption in this scenario of fairly different assumptions with respect to technological development and emission controls as compared to the rest of B1 scenarios. In the A2 family marker (A2-ASF) and B2 family marker (B2-MESSAGE), emissions grow steadily until 2050, by which time they are around 50% higher than today. Thereafter, emissions in B2–MESSAGE decline gradually, while in the A2 marker the growth continues to reach

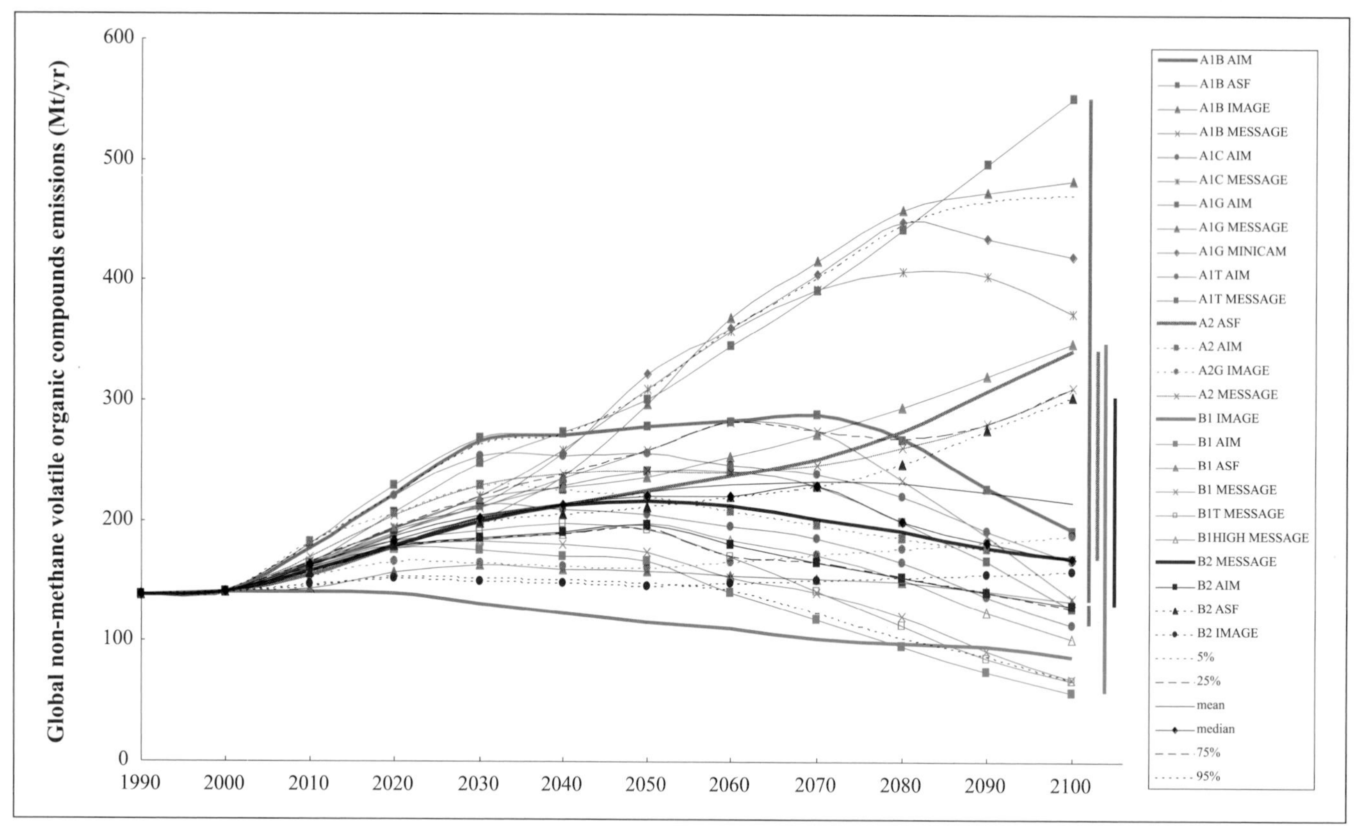

Figure 5-10: Standardized global emissions of NMVOCs for SRES scenarios, classified into four scenario families (each denoted by a different color code – A1, red; A2, brown; B1, green; B2, blue). Marker scenarios are shown with thick lines without ticks, globally harmonized scenarios with thin lines, and non-harmonized scenarios with thin, dotted lines (see Table 4-3). Black lines show percentiles, means, and medians for SRES scenarios. For numbers on the two additional illustrative scenarios A1FI and A1T see Appendix VII.

2.5 times the current level by 2100 (similar to B1-ASF). Emissions in A2-MESSAGE are fairly similar to those of the A2 marker, while emissions in A2-AIM decline to 170 Mt by 2100, only half of the A2 marker level. The trajectory in the A1B marker (A1B-AIM) is very distinct – up to 2060 a fast growth is observed to more than twice the 1990 level, after which emissions decline to the B2 family marker level (Figure 5-10). A decrease of NMVOC emissions in A1B-AIM after 2060 is explained mainly by the substitution of fossil fuels with renewables, especially in the transport sector. A very similar trajectory emerges in the A1B-MESSAGE scenario, while the A1B-ASF scenario emissions grow continuously up to 2100, by when they are the highest of the set at 550 Mt, almost four times the 1990 level.

5.5.1.3. Carbon Monoxide

CO emissions in 1990 are estimated to range between 752 and 984 MtCO across the models used to derive the SRES scenarios. The same caveats as for NO_x and NMVOC emissions (see above) also apply to CO emissions – the number of models that represent all the emission source categories is limited and modeling and data uncertainties, such as emission factors, are considerable. As a result, CO emission estimates across the scenarios are highly model specific, as indicated by the overlapping ranges of the four scenario families (Figure 5-11). From a standardized (see Box 5-1) 1990 level of 880 MtCO the range of future emissions is rather wide for both medium-term and long-term time horizons. By 2020, emissions range from 630 to 1550 MtCO, by 2050 they range between 470 and 2300 MtCO, and by 2100 the range is between 360 and 3760 MtCO (i.e. one order of magnitude difference between the highest and the lowest projections). Focusing on the 25th and 75th percentile intervals reduces uncertainty ranges somewhat, but nonetheless they remain substantial – between 1260 and 2300 MtCO by 2100. The median of all 40 scenarios is tracked quite closely by the B2-ASF scenario at the global level.

Emissions of CO follow rather different trajectories in the SRES markers than emissions of NMVOCs, except the A2 marker (A2-ASF). Starting from the standardized (see Box 5-1) level of 880 MtCO in 1990, emissions in the A1B marker (A1B-AIM) grow continuously and almost double between 1990 and 2100 (Figure 5-11). Emissions in A1B-MESSAGE (Roehrl and Riahi, 2000) increase at a higher rate than in the marker and reach thrice the current level by 2100. Emissions in A1B-ASF grow as fast as in A1B-MESSAGE until 2040 (1700 MtCO versus 1650 MtCO for A1B-MESSAGE), when the growth stops and emissions more or less stabilize. Emissions in the B1-IMAGE

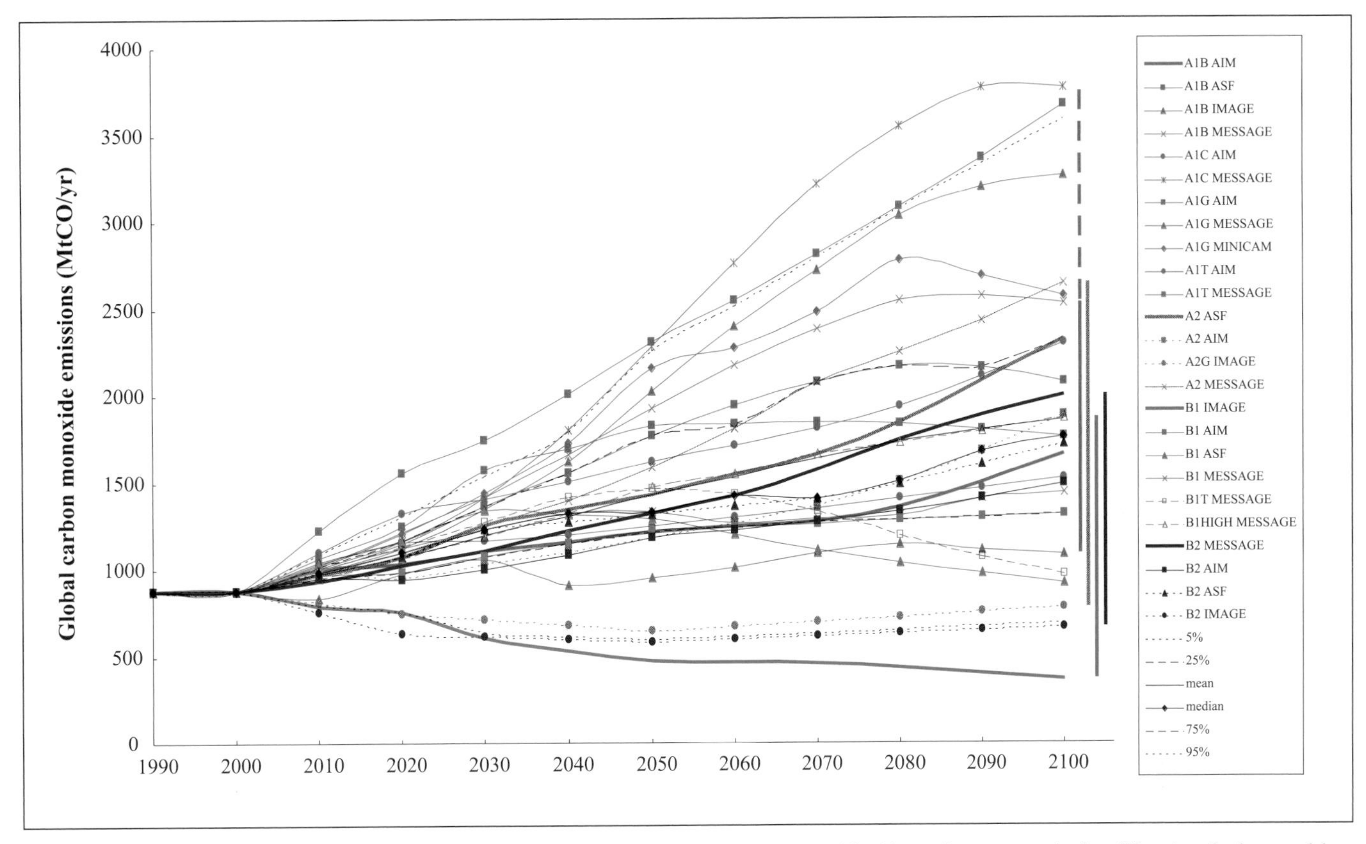

Figure 5-11: Standardized global emissions of CO for SRES scenarios, classified into four scenario families (each denoted by a different color code – A1, red; A2, brown; B1, green; B2, blue). Marker scenarios are shown with thick lines without ticks, globally harmonized scenarios with thin lines, and non-harmonized scenarios with thin, dotted lines (see Table 4-3). Black lines show percentiles, means, and medians for SRES scenarios. For numbers on the two additional illustrative scenarios A1FI and A1T see Appendix VII.

(B1 family marker) gradually decline to 40% of the current level by 2100 (Figure 5-11). Again, other B1 family scenarios have quite different trajectories and the patterns generated by AIM, ASF, and MESSAGE models for the B1 scenario family are similar to those derived by the same models for other storylines. The shape of the A2 family marker (A2-ASF) trajectory is very similar to that for NMVOCs, which suggests that these two substances are governed mainly by the same drivers and that similar assumptions with regard to improved emissions factors are adopted. The emission trajectory in the B2 marker (B2-MESSAGE) is just below the A2 marker trajectory and is about 2000 MtCO by 2100, more than twice the 1990 level. An increase of biomass energy use after 2050 becomes an important factor that affects emissions in this scenario. Other B2 family scenarios produced using the AIM and ASF models closely track the B2 marker, although emissions in these scenarios are somewhat lower, with those of B2-ASF at 1700 MtCO by 2100 and of B2-AIM falling just below 1500 MtCO.

5.5.2. *Sulfur*

Aerosols result from complex atmospheric processes in which sulfur emissions play an important role. Besides sulfur, other substances, like NO_x, ammonia, and small particles from the burning of fossil fuels and biomass, are involved in these processes.

Global anthropogenic sulfur emissions are estimated to range between 65 and 90 MtS in 1990 (Houghton *et al.*, 1995; Benkovitz *et al.*, 1996; Olivier *et al.*, 1996; WMO, 1997). Reviews of most recent inventories, given in Smith *et al.* (2000) and Grübler (1998), indicate a most likely value of 75 ± 10 MtS. These reviews draw on a large body of literature sources and sulfur inventories (in particular the EMEP and CORINAIR inventories for Europe, NAPAP for North America, and the most recent inventories available for Asia, including Akimoto and Narita (1994), Foell *et al.* (1995), and Kato (1996)). Anthropogenic emissions add to natural sulfur flows, which are estimated to range between 4 and 45 MtS (Houghton *et al.*, 1995). Pepper *et al.* (1992) adopted an intermediary, constant natural sulfur flux of 22 MtS for the IS92 scenario series.

Even with a comparatively good agreement on global sulfur emission levels, important uncertainties remain at the sector and regional levels (discussed in more detail in Alcamo *et al.* (1997) and Grübler (1998)). The main sources of uncertainties are a lack of detailed inventory data (especially for developing countries outside Asia, but also for the non-European part of

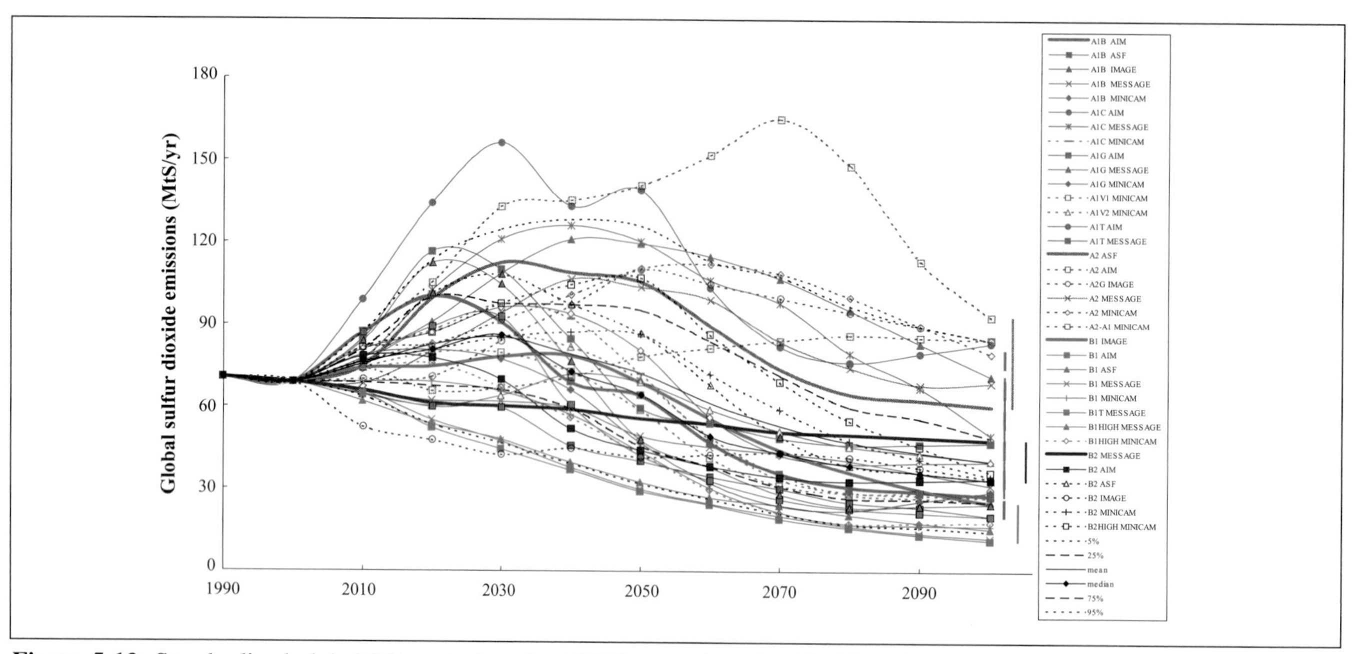

Figure 5-12: Standardized global SO_2 emissions for SRES scenarios, classified into four scenario families (each denoted by a different color code – A1, red; A2, brown; B1, green; B2, blue). Marker scenarios are shown with thick lines without ticks, globally harmonized scenarios with thin lines, and non-harmonized scenarios with thin, dotted lines (see Table 4-3). Black lines show percentiles, means, and medians for SRES scenarios. For numbers on the two additional illustrative scenarios A1FI and A1T see Appendix VII.

Russia), uncertainties in sulfur contents of fuels (especially coal) in many regions, and the use of different base years for development of sulfur inventories. For instance, inventories and scenario studies for China and Centrally Planned Asia give a range of sulfur emissions that differ by more than a factor of two (8.4 to 18 MtS) for the year 1990 (Grübler, 1998).

Base-year differences in the available data sources are especially important because regional sulfur emissions trends have changed drastically in the past decade. Although they decreased strongly in Europe and North America as a result of sulfur control policies, they increased rapidly in Asia with growing energy demand and coal use. For instance, between 1980 and 1995 sulfur emissions declined by 59% in Western Europe and Russia (albeit for entirely different reasons – environmental policy limiting sulfur emissions in Western Europe versus a massive economic depression in Russia), by 37% in Eastern Europe, and by 36% in North America (ECE, 1997). Conversely, emissions in China rose rapidly, from an estimated 6.6 MtS in 1985 to 9.1 MtS in 1994, or by 38% (Sinton, 1996; Dadi *et al.*, 1998). These diverging emission trends and their rapid changes also require a continuous updating of available gridded sulfur emission inventories (e.g., Dignon and Hameed, 1989; Spiro *et al.*, 1992; Benkovitz *et al.*, 1996; Olivier *et al.*, 1996) that in, some instances, still rely on outdated 1980 emissions data.

Global base-year (1990) sulfur emission values from the SRES models range from 63 to 77 MtS, with the addition of 3 MtS from international shipping.[9] This difference reflects the existing uncertainty in sulfur emission estimates, particularly at the regional level. The range is within the range of values given by global inventories.

Model differences at the regional level are even larger, which reflects the greater uncertainty of emission inventories at this level, particularly outside the OECD countries. To standardize sulfur emissions, the number of SRES reporting regions was increased to six regions, by splitting Latin America from the ALM region and Centrally Planned Asia and China from ASIA. Important differences in economic development status and resource endowments lead to different patterns of sulfur emissions across all SRES scenarios. Regional emissions were standardized (see Box 5-1) and then aggregated to the global level. Global standardized base-year emissions for 1990 for the SRES scenarios are equal to 70.9 MtS, in line with the literature range of global emission inventories given above. The regional sulfur emission profiles were also used to generate spatially gridded emission patterns (see Section 5.6.2 below).

Concerning future emissions of sulfur, the SRES scenarios reflect recent literature and trends of sulfur control scenarios, as well as the conclusions from the 1994 evaluation of the IS92 scenarios (Alcamo *et al.*, 1995). Despite considerable scenario variability, all scenarios portray similar emission dynamics – at various future dates (between 2020–2030 and 2070, depending

[9] Following UN energy statistics and IPCC inventory practices, international bunker fuels are included in global totals, but not in national/regional subtotals (and their aggregates to global totals). Hence, bunker fuels are reported separately here.

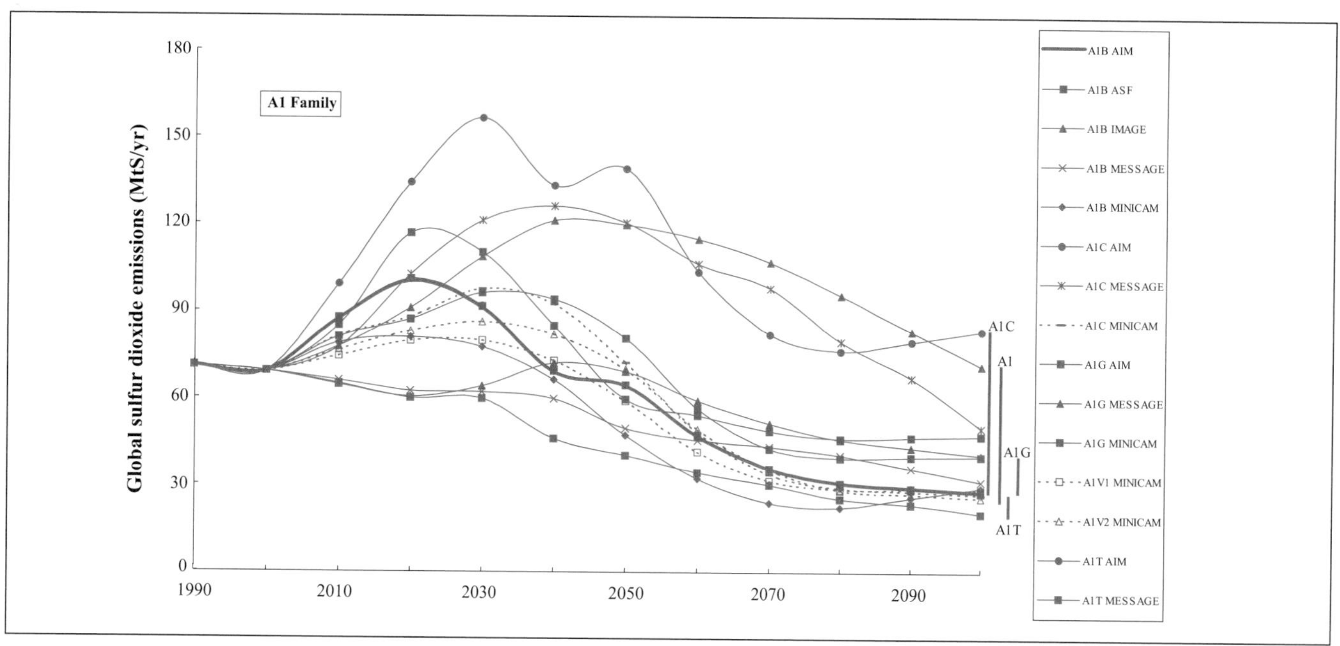

Figure 5-13a: Standardized global SO_2 emissions in the A1 scenario family. The marker scenario is shown with a thick line without ticks, the globally harmonized scenarios with thin lines, and the non-harmonized scenarios with thin, dotted lines (see Table 4-3). In the SPM, A1C and A1G scenarios are merged into one fossil-intensive A1FI scenario group (see also footnote 2).

on the scenario and its underlying storyline), global SO_2 emissions reach a maximum level and decline thereafter (Figure 5-12). By 2030 sulfur emissions range between 40 and 160 MtS,[10] by 2070 between 20 and 165 MtS, and by 2100 between 10 and 95 MtS. Emission trajectories of the SRES scenarios reflect a combined impact of different scenario driving forces (local air quality concerns, structural change in energy supply and end-use, etc.), which lead to a gradual decline in sulfur emissions in the second half of the 21st century.

Importantly, all SRES scenarios are sulfur-control scenarios *only* and do not assume any additional climate policy measures. There is, however, an indirect effect of GHG emission reduction from sulfur-control policies that result in energy conservation and inter-fuel substitution from high sulfur to low sulfur fuels (e.g., from coal to gas).

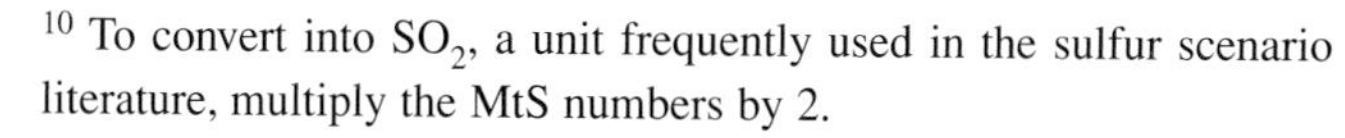

[10] To convert into SO_2, a unit frequently used in the sulfur scenario literature, multiply the MtS numbers by 2.

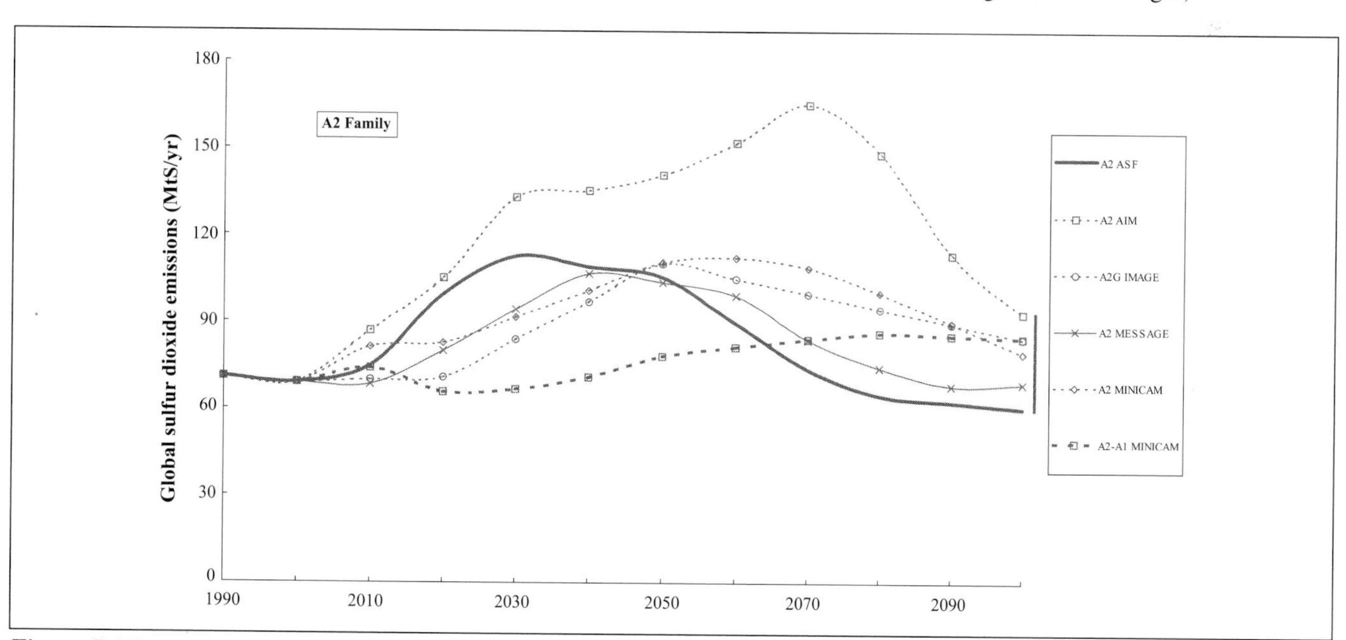

Figure 5-13b: Standardized global SO_2 emissions in the A2 scenario family. The marker scenario is shown with a thick line without ticks, the globally harmonized scenarios with thin lines, and the non-harmonized scenarios with thin, dotted lines (see Table 4-3).

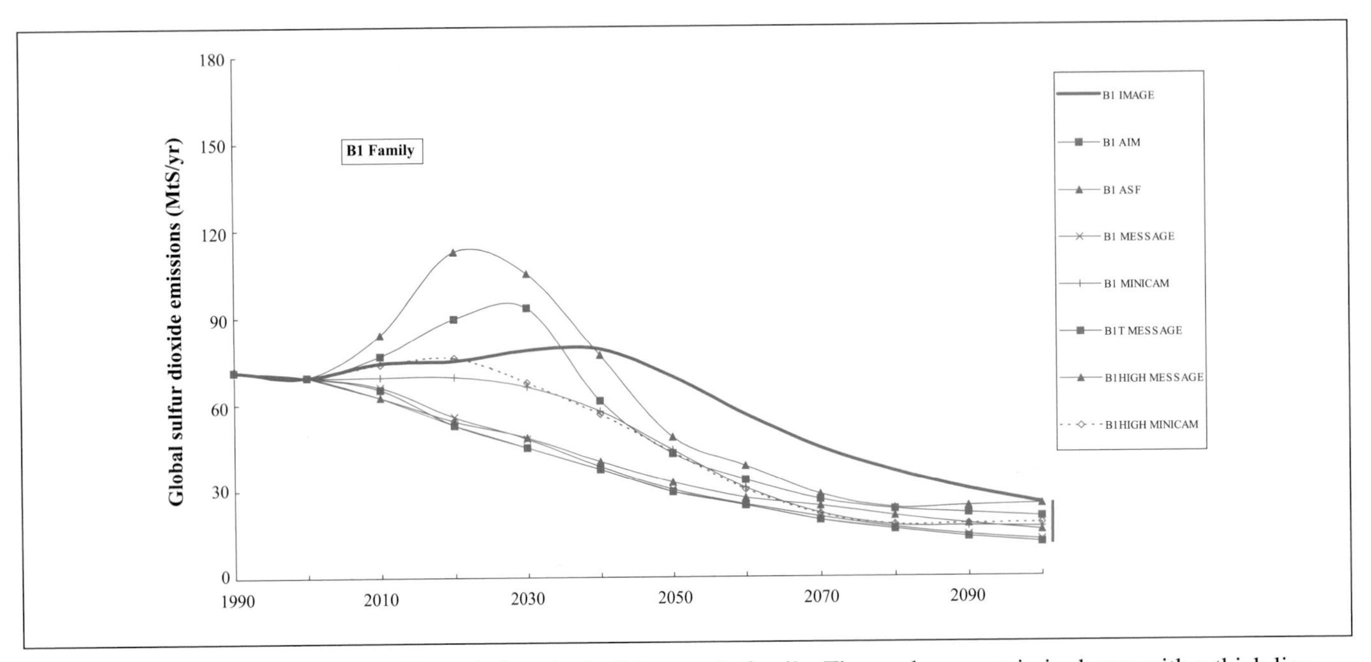

Figure 5-13c: Standardized global SO_2 emissions in the B1 scenario family. The marker scenario is shown with a thick line without ticks, the globally harmonized scenarios with thin lines, and the non-harmonized scenarios with thin, dotted lines (see Table 4-3).

5.5.2.1. A1 Scenario Family

The A1 family of scenarios covers most of the range of the 40 SRES scenarios (Figure 5-13a). The A1B-AIM marker is in the middle of the range and its trajectory is similar to those of many other scenarios – a rapid increase in the near term followed by a decline. Increasing fossil fuel use in developing countries combined with low levels of SO_2 controls produces the near-term increase in emissions. After 2025, per capita incomes reach levels at which countries place more emphasis on the environment, resulting in emission controls on SO_2 (see Section 4.4.10). These controls, combined with a transition from fossil fuels to non-fossil energy, result in declining emissions. More detailed information as to how SO_2 emissions are treated in the A1B-AIM scenario is provided in Box 5-3 .

Another A1 family scenario, A1G-MiniCAM, shows similar behavior for the same reasons, although the increase and

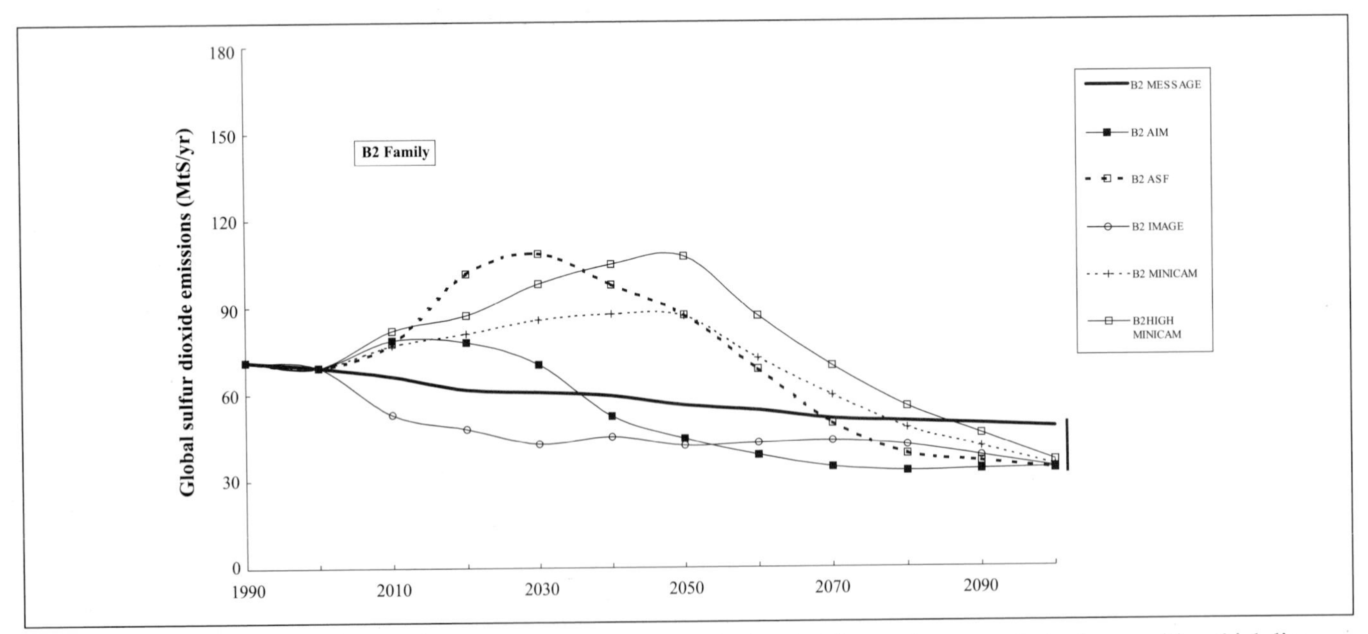

Figure 5-13d: Standardized global SO_2 emissions in the B2 scenario family. The marker scenario is shown with a thick line without ticks, the globally harmonized scenarios with thin lines, and the non-harmonized scenarios with thin, dotted lines (see Table 4-3).

Box 5-3: Sulfur Emissions in the A1B-AIM Marker Scenario

In the A1B-AIM marker scenario, global SO_2 emissions increase rapidly from 70.6 MtS in 1990 to reach a peak around 2020 at 101 MtS, and subsequently decline to around 30 MtS after 2050. The global SO_2 emission trajectory follows that of the developing countries, for which emissions increase because of the rapid growth of fossil fuel use driven by rapid economic development, combined with a lack of sulfur controls that reflect investment and infrastructure limitations. With an increase in personal incomes in developing countries and associated concerns about local air pollution, SO_2 becomes controlled in a similar way to that in developed countries, for which SO_2 emissions have declined in the past two decades.

The sources of SO_2 emissions tracked by the AIM model include energy use, industrial production processes (steel and cement production), and biomass burning.

The major factors of SO_2 emission reduction are changes in the fuel mix, use of advanced energy technology, and implementation of desulfurization technology. In the A1B-AIM scenario, the volume and share in total primary energy of coal and crude oil changes drastically over time (Table 5-10).

***Table 5-10**: Coal and crude oil in the A1-AIM scenario in terms of primary energy requirements (EJ) and as shares of total primary energy (%).*

	1990		**2050**		**2100**	
Coal	85 EJ	*24.6%*	140 EJ	*11.7%*	41 EJ	*2.0%*
Crude oil	126 EJ	*36.5%*	181 EJ	*15.0%*	107 EJ	*5.0%*

Before 2050, both advanced energy technology and desulfurization play a key role in the SO_2 emissions reduction. In the power generation sector, integrated gasification combined cycle (IGCC), fluidized bed combustion (FBC), and flue-gas desulfurization (FGD) technologies are adopted. In the industry and commercial sectors, FBC boilers, fuel desulfurization and FGD are adopted. SO_2 emission reductions are also achieved in the transport sector. Table 5-11 summarizes the SO_2 abatement effect of these technologies.

***Table 5-11**: SO_2 emission reductions (%) by technology and sector in the A1-AIM scenario.*

Technologies	**SO_2 Emission reduction rate**	**Note**
Power generation		
IGCC	96%	Introduced after 2015
FBC	95%	
FGD	98%	
Industry and commercial sector		
FBC	92%	
FGD	95%	
Fuel desulfurization	70%	
Transport sector		
SO_2 emission control	75%	

Income levels determine the time at which these technologies are introduced in the A1B-AIM scenario. According to the experience of SO_2 emission controls in developed countries, low-income developing countries will start to introduce SO_2 emission control technologies when GDP reaches around US$3500 per capita in 1990 dollars. In A1B-AIM this threshold level is reached in a period of about 30 years.

subsequent decline in emissions is less pronounced. The MiniCAM model has a higher level of SO_2 controls, but also a greater fossil fuel use, resulting in slightly higher SO_2 emissions in 2100.

The A1T, A1G, and A1C scenario groups assume various roles for the development of energy resources and technologies (the coal-intensive A1C scenario group, the oil- and gas-intensive A1G group, and the A1T scenario group with accelerated non-fossil technology diffusion). Consequently, they span a very wide range of future sulfur emissions, ranging from 20 MtS by 2100 in the A1T-MESSAGE scenario to 83 MtS in the A1C-AIM scenario.

5.5.2.2. *A2 Scenario Family*

Sulfur emission trajectories in all the A2 family scenarios, except A2-A1-MiniCAM with its much lower energy consumption, have the same general convex shape (Figure 5-13b). However, the time when maximum emissions are attained varies quite widely across the scenarios. Emissions peak in the A2-ASF marker scenario in 2030, in A2-MESSAGE in 2040, in A2G-IMAGE in 2050, in A2-MiniCAM in 2060, and in A2-AIM not until 2070. These differences are explained primarily by different assumptions about the mechanisms of SO_2 reduction adopted in different models. For example, the SO_2 reduction in A2-MESSAGE is explained by fundamental structural changes in the electricity generation technologies, while the A2-ASF scenario assumes a rapid introduction of inexpensive end-of-the-pipe sulfur scrubbers and shifts to low-sulfur fuel qualities.

The A2-A1-MiniCAM scenario yields a nearly flat sulfur emission profile, explained by a relatively slow growth in population and GDP in the first half of the 21st century (compared to other scenarios of the A2 family) and by expedited technological progress in the second half.

Numeric estimates of SO_2 emissions developed by different models vary substantially, especially in the middle of the modeling period (Figure 5-13b). Emissions are largest in the A2-AIM scenario, exceeding 160 MtS in 2070. From 2020 to 2060 the lowest emissions are produced by the A2-A1-MiniCAM scenario, while after 2070 the lowest emissions are achieved in the A2-ASF scenario.

5.5.2.3. *B1 Scenario Family*

Sulfur emissions in the B1 family scenarios fall within the lower third of the full range of emissions from the SRES scenarios (Figure 5-12). The B1-IMAGE trajectory is generally positioned in the middle of the B1 family range in the first decades of the 21st century, which corresponds to a 75% reduction of emissions in industrialized countries between 2000 and 2050. The near-term increase in emissions in B1-IMAGE occurs predominantly in developing countries and is associated with increases in fossil fuel use in those regions, combined with relatively low levels of emission controls. After 2040, increased emission controls and declines in fossil fuel use result in decreasing SO_2 emissions.

Emissions in other B1 family scenarios decline even faster than in B1-IMAGE. In B1-AIM, a decline in emissions is directly related to per capita incomes. As developing countries reach levels of US$3500 per capita in 1990 dollars, they start to apply more stringent SO_2 controls that lead to quickly dropping emission levels (Figure 5-13c).

5.5.2.4. *B2 Scenario Family*

Emissions in the B2 family scenarios cover close to the full range of SO_2 emissions (Figure 5-12). The B2-MESSAGE scenario yields steadily declining global emissions from 1990. This overall trajectory includes a tripling of emissions from non-energy sources by 2050, with subsequent stabilization and eventual decline after that (Figure 5-13d). The increase in emissions from non-energy sources is more than offset by very rapid reductions in emissions from energy sources between 2015 and 2050. Developing countries have rising emissions through to 2025, which stabilize by 2050, and decline thereafter. This increase in emissions in developing countries is offset by reductions in developed countries. These results are explained largely by regional measures and technological changes to minimize critical loads of acidic deposition (see Box 5-4). The B2-ASF scenario projects very rapid growth in emissions from energy use through 2025, primarily from increases in fossil fuel use in developing countries. After 2025, the growth in fossil fuel declines and developing countries become wealthier and are more aggressive on SO_2 emission controls, which in the ASF model are directly linked to GDP/capita levels. Emissions from other industrial sources increase steadily throughout the period as economic activity increases.

5.5.2.5. *Inter-Family Comparison*

The relatively rapid desulfurization in the A1B-AIM marker compared to the other SRES markers mainly results from high capital turnover rates and, therefore, rapid diffusion of new and clean technologies combined with high income levels in the developing world by the middle of the 21st century (Figure 5-12). The structure and patterns of sulfur emissions for the two illustrative scenarios in the two scenario groups A1FI and A1T are similar to those of the A1B and B1 marker scenarios, respectively, and are therefore not discussed separately here.

As technological progress and income growth are the slowest in A2 among all the SRES scenario families, the primary energy mix in the A2-ASF marker by 2100 is still dominated by fossil fuels, with about 50% of the primary energy supplied by coal. Although measures are adopted to limit local and regional environmental damages, sulfur-mitigation measures in the A2 world are less pronounced than in the other SRES scenarios. Therefore, global sulfur emissions in the A2 marker are highest as compared to the other markers (Figure 5-12).

Box 5-4: Future Sulfur Dioxide Emissions in the B2 Marker Scenario

Future global emissions of SO_2 are generally lower across the SRES emissions scenarios compared to most earlier projections, because of three factors:

- Switch to cleaner fuels, such as natural gas and renewable sources.
- Transition to cleaner, more efficient coal technologies, such as IGCC generation and pressurized FBC.
- Utilization of direct emissions-reduction technologies, such as FGD.

The scope of reductions from fuel switching is illustrated by the following. At present, 60% of anthropogenic SO_2 emissions in developing regions are from the direct use of fossil fuels in buildings and industry (Smith *et al.*, 2000). Reductions in these emissions will be realized by shifting from an energy structure that relies on the direct use of solid fuels to one that increasingly relies on distributed energy grids (electricity, natural gas, etc.). This shift, and a resultant decrease in SO_2 emissions, has already occurred in Europe and North America. Further reductions are expected in the future as more efficient, and inherently low emission, fossil fuel technologies (such as IGCC) become commercialized. Explicit policies for sulfur emissions reductions, such as the use of FGD devices, will be needed to meet emissions targets in the near term, but are likely to be less necessary in the longer term.

To illustrate these trends and the resultant decrease in SO_2 emissions over the 21st century, emissions in the B2-MESSAGE scenario (Riahi and Roehrl, 2000) are analyzed. Global energy-related SO_2 emissions in this B2 scenario decline from 59 MtS in 1990 to about 12 MtS in 2100. The primary causes for this reduction are the transition to more advanced coal technologies and desulfurization. Consistent with the characteristics of gradual change in the B2 storyline, the aggregated emissions coefficient[11] for power production from coal declines from about 5.3 kgS/MWh in 1990 to 0.04 kgS/MWh in the year 2100. The latter emissions intensity is similar to that of the most advanced current technologies, for example, the Siemens IGCC98 power plant featuring 0.032 kgS/MWh (Baumann *et al.*, 1998). In the second half of the 21st century, desulfurization of the energy system also takes place because of the production of methanol from coal. Sulfur removal is a process-inherent feature. Methanol is mainly used in the transport sector as a substitute for oil products that become scarce after 2050.

The percentage reductions in energy-related SO_2 emissions in the B2 scenario are shown in Table 5-12 as a function of time for five technologies. These percentages were obtained by re-calculating the B2-MESSAGE emissions scenario with the energy structure and emissions coefficients held constant in time at those of 1990. In this hypothetical case, SO_2 emissions in 2100 would be 227 MtS. Accordingly, energy-related SO_2 emissions in 2100 are reduced by about 215 MtS in the B2 scenario, 68% of which results from technological change and 32% from fuel switching. By the year 2100 the contribution to these emissions reductions by flue-gas scrubbing (FGD) is negligible by the year 2100, because more-advanced coal technologies eliminate any need for it. Furthermore, a shift in refining technology to lighter oil products, currently underway worldwide, contributes to a reduction in SO_2 emissions, particularly in the early part of the 21st century.

***Table 5-12**: Sources of energy-related SO_2 emissions reductions in the B2 marker scenario.*

Year	Scrubbing (%)	IGCC (%)	Synfuel (%)	Light oil shift (%)	Fuel switching (%)	Total reduction (MtS)
1990	100	0	0	0	0	5.1
2020	26	16	0	24	33	58.6
2050	11	15	2	17	54	128.9

Following the B1 storyline, in the B1-IMAGE marker the emphasis is on global solutions to environmental sustainability and improved welfare and development equity. High technological development rates in the renewable energy sector result in a continued structural shift away from fossil fuels. Combined with dematerialization of the economy and with the most pronounced sulfur mitigation measures assumed among the SRES scenarios, this results in emissions that peak around 2020 and subsequently decline continuously to 2100 (Figure 5-12).

In the B2 scenario family, and thus in the B2-MESSAGE marker, strong emphasis is placed on regional environmental protection. Dynamics of technological change continue along historical trends ("dynamics as usual"), which are slower than

[11] All emissions coefficients were calculated with the MESSAGE model.

in the A1 or B1 families, but faster than in the A2 family. Sulfur emission projections for the B2 marker scenario were generated on the basis of minimization of critical loads of acidic deposition using the methodology described in Amann *et al.* (1996). No explicit link between income levels and sulfur control regime was made in this scenario. The resultant sulfur emissions are 61 MtS in 2020, 56 MtS in 2050, and 48 MtS in 2100 (see Figure 5-12).

All the SRES marker scenarios anticipate increasing levels of sulfur control, with rates and timing ranging from rapid introduction of stringent controls in B1-IMAGE to more gradual, later, and less stringent controls in A2-ASF. To illustrate the impacts of sulfur controls in the scenarios, an "uncontrolled sulfur" variant of the B2 marker scenario was calculated at the International Association for Applied Systems Analysis (IIASA; for details see Box 5-4). In this hypothetical scenario, sulfur emissions amount to 182 MtS in 2050 and 227 MtS in 2100 (compared to 57 and 47 MtS, respectively, in the B2 marker scenario). These emission levels are much higher even than those in the A2 marker scenario (105 MtS in 2050 and 60 MtS in 2100).

5.6. Regional Distribution and Gridding

Regional information on emissions serves at least two major purposes – to identify the contribution of world regions[12] to the global total and to track shifts in the relative weight of different regions. This information is especially relevant for the development of mitigation scenarios. For climate modeling, the regional distribution of emissions for well-mixed GHGs (CO_2, CH_4, N_2O, and halocarbons) may not be that important. However, short-lived gases such as SO_2 are radiatively important close to the point of origin only; their local and regional concentrations may significantly change the future climate outlook. The same is true for the group of ozone precursors (CO, NO_x, and NMVOCs). To be able to estimate tropospheric ozone concentration levels, regionalized information is indispensable.

The initial evaluation showed that the 40 SRES scenarios have a very substantial regional variability in emissions of all radiatively important substances. The detailed and rigorous analysis of this variability falls outside the scope of the current report. Therefore, this section merely illustrates possible regional patterns based on standardized regional emissions in the four SRES marker scenarios (see also Kram *et al.*, 2000). Standardized regional outputs from the 40 SRES scenarios are provided in Appendix VII.

Subsection 5.6.1 describes emissions of GHGs and SO_2 in the four SRES macro-regions, followed by the description of "gridded" SO_2 emissions (distributed over a $1^{\circ}x1^{\circ}$ grid) in 5.6.2.

[12] In this report represented by four macro-regions – OECD90, REF, ASIA, and ALM.

5.6.1. Regional Distribution

As Tables 5-13a to 5-13d clearly illustrate, the distribution of emissions over the four regions in the base year (1990) is very uneven. For example, while in industrialized regions (OECD90 and REF) fossil and industrial CO_2 emissions are dominant, in the developing regions (ASIA and ALM) the contribution of land-use emissions (deforestation) is also very important. In 1990, developing regions produced much lower volumes of CO_2 and high-GWP gases than the industrialized world, while their relative share of N_2O, CH_4, and NO_x emissions was much more substantial (see Figures 5-13a to 5-13d).

5.6.1.1. Carbon Dioxide Emissions from Fossil Fuels and Industry

As suggested by Figure 5-14, in all the SRES scenario families the share of industrialized regions (OECD90 and REF) in global total becomes progressively smaller and by 2100 these regions emit from 23% to 32% of the total (Table 5-14, Figure 5-14).

In the OECD90 region, standardized fossil fuel and industrial CO_2 emissions in the A1B marker scenario (A1B-AIM) increase from 2.8 GtC in 1990 to 3.4 GtC in 2050, and subsequently decline to 2.2 GtC in 2100 (Figure 5-14). Compared to other scenarios, the growth in primary energy use in this region is relatively high, spurred by rapid economic development (see also Chapter 4). However, after 2050 the increases in the use of primary energy are accompanied by declining emissions through the combination of a lower use of fossil fuels and a switch from coal to gas. The share of non-fossil fuels in the OECD90 region of the A1B marker scenario also increases drastically. In 2100, the contribution of non-fossil energy amounts to 68% of the total primary energy use of the OECD90 countries, the largest non-fossil fuel share for this region of all the SRES marker scenarios.

The fossil fuel and industrial CO_2 emission trajectory of the REF region is even less linear than in the OECD90 region. Initially, emissions decline from the base year level of 1.3 GtC to 1.1 GtC in 2020 because of economic restructuring. After 2020, emissions increase, driven by an increased energy demand to support renewed economic growth (Figure 5-14). However, after 2050 emissions decline again primarily through a decrease in population and improved energy efficiency. By 2100, non-fossil fuels in REF contribute 58% of the total primary energy use and the share of natural gas reaches almost 40%.

The energy and industry CO_2 emission growth in the ASIA region of the A1B marker scenario is very high, reflecting rapid economic growth and high energy demand. By 2100 the total primary energy use in this region exceeds the 1990 level more than 10 fold. Standardized CO_2 emissions increase from 1.15 GtC in 1990 to 5.73 GtC in 2050 and then drop to 5.27 GtC in 2100 (Figure 5-14, Table 5-13c). By 2100 contributions from the two major energy sources, non-fossil fuels and natural gas, are 69% and 25%, respectively.

Table 5-13a: *Standardized anthropogenic emissions (CO_2, CH_4, N_2O, NO_x, CO, NMVOCs, SO_2, HFCs, PFCs and SF_6) for the four SRES marker scenarios, OECD90 region.*[13]

		1990	2020				2050				2100			
Marker scenarios			**A1B**	**A2**	**B1**	**B2**	**A1B**	**A2**	**B1**	**B2**	**A1B**	**A2**	**B1**	**B2**
Region								**OECD90**						
Fossil CO_2	GtC	2.83	3.51	3.96	3.20	3.71	3.36	4.74	2.00	3.26	2.24	6.91	1.10	3.10
Land-use CO_2	GtC	0.00	0.03	0.00	0.06	–0.06	0.00	0.00	–0.09	–0.05	0.01	0.00	–0.11	–0.19
Total CO_2	GtC	2.83	3.54	3.96	3.26	3.64	3.36	4.74	2.01	3.22	2.25	6.91	0.99	2.81
CH_4 total	Mt CH_4	73.0	68.9	83.8	71.5	71.3	51.5	105.4	55.5	68.8	42.4	165.7	39.5	77.8
N_2O total	Mt N_2O-N	2.6	2.6	3.0	2.6	2.4	2.4	3.0	2.4	2.5	2.2	3.9	2.0	2.6
SO_x total	MtS	22.7	6.9	8.7	7.9	6.7	6.3	9.8	2.5	4.1	4.6	11.8	2.6	3.5
CFC/HCFC*	MtC equiv.													
HFC	MtC equiv.	19	108	103	103	99	122	125	116	102	125	160	120	97
PFC	MtC equiv.	18	11	14	10	13	10	14	7	10	16	17	6	7
SF_6	MtC equiv.	23	5	28	5	23	9	28	7	13	20	16	8	10
CO	Mt CO	179.4	204.7	175.3	137.5	172.5	240.8	140.6	85.5	183.9	262.0	243.1	56.5	197.4
NMVOCs	Mt	42.4	39.1	44.4	33.0	39.2	28.2	42.3	21.0	43.4	14.6	66.6	13.0	30.2
NO_x	MtN	12.8	11.5	15.6	10.4	14.7	6.2	16.2	4.7	15.3	4.9	21.5	2.2	11.4

*Montreal gases are not distributed over regions.

[13] The numbers of the two additional illustrative scenarios for the A1FI and A1T scenario groups can be found in Appendix VII.

Table 5-13b: *Standardized anthropogenic emissions (CO_2, CH_4, N_2O, NO_x, CO, NMVOCs, SO_2, HFCs, PFCs and SF_6) for the four SRES marker scenarios, REF region.*[13]

		1990	2020				2050				2100			
Marker scenarios Region			**A1B**	**A2**	**B1**	**B2**	**A1B**	**A2 REF**	**B1**	**B2**	**A1B**	**A2**	**B1**	**B2**
Fossil CO_2	GtC	1.30	1.11	1.22	0.91	0.81	1.18	1.52	0.91	1.24	0.78	2.41	0.41	1.18
Land-use CO_2	GtC	0.00	0.03	0.00	–0.10	–0.18	–0.13	0.00	–0.36	–0.04	–0.03	0.00	–0.29	-0.04
Total CO_2	GtC	1.30	1.14	1.22	0.81	0.63	1.05	1.52	0.55	1.20	0.75	2.41	0.12	1.14
CH_4 total	Mt CH_4	47.1	61.3	45.8	41.9	39.8	42.4	78.0	33.9	52.9	34.2	143.2	20.9	47.0
N_2O total	Mt N_2O-N	0.6	0.6	0.7	0.6	0.6	0.6	0.8	0.5	0.6	0.5	1.0	0.3	0.7
SO_x total	MtS	17.0	10.8	12.0	7.7	3.5	2.4	10.2	6.5	2.9	1.6	3.2	2.5	3.6
CFC/HCFC*	MtC equiv.													
HFC	MtC equiv.	0	19	13	15	15	32	31	26	25	31	52	26	27
PFC	MtC equiv.	7	8	10	6	10	21	20	9	25	24	42	8	27
SF_6	MtC equiv.	8	10	10	7	7	21	19	9	15	11	38	4	14
CO	Mt CO	68.9	40.7	39.9	23.3	47.9	43.0	55.6	19.3	74.2	43.8	119.0	9.3	78.6
NMVOCs	Mt	15.7	14.3	19.0	11.9	16.5	16.4	30.9	10.9	32.1	17.8	37.4	8.9	26.0
NO_x	MtN	4.7	3.2	4.0	2.6	3.2	2.2	5.3	2.6	5.2	1.2	7.6	1.4	3.7

*Montreal gases are not distributed over regions.

Table 5-13c: *Standardized anthropogenic emissions (CO_2, CH_4, N_2O, NO_x, CO, NMVOCs, SO_2, HFCs, PFCs, and SF_6) for the four SRES marker scenarios, ASIA region.*[13]

		1990	2020				2050				2100			
Marker scenarios			**A1B**	**A2**	**B1**	**B2**	**A1B**	**A2**	**B1**	**B2**	**A1B**	**A2**	**B1**	**B2**
Region							**ASIA**							
Fossil CO_2	GtC	1.15	4.10	3.52	3.18	3.02	5.73	6.26	3.68	4.12	5.27	10.71	1.28	5.69
Land-use CO_2	GtC	0.37	0.05	0.39	0.22	–0.15	0.25	0.22	0.18	–0.03	0.19	0.02	–0.35	–0.06
Total CO_2	GtC	1.52	4.15	3.92	3.40	2.87	5.98	6.48	3.86	4.10	5.46	10.73	0.93	5.63
CH_4 total	Mt CH_4	112.9	170.7	162.9	148.4	171.1	214.2	226.5	157.4	234.0	117.3	307.5	105.4	271.9
N_2O total	Mt N_2O-N	2.3	2.7	4.0	3.2	2.4	3.0	5.3	3.3	2.6	2.9	7.2	1.9	2.9
SO_x total	MtS	17.7	54.2	51.5	29.1	32.9	8.4	48.9	21.4	26.4	6.4	20.5	4.2	20.6
CFC/HCFC*	MtC equiv.													
HFC	MtC equiv.	0	45	18	20	40	224	54	93	130	262	204	64	302
PFC	MtC equiv.	3	15	15	9	22	35	32	17	46	46	67	18	51
SF_6	MtC equiv.	4	19	16	17	18	50	34	30	36	37	67	16	30
CO	Mt CO	234.8	359.7	361.2	288.9	375.4	491.8	522.5	244.9	517.9	678.3	905.6	213.9	650.9
NMVOCs	Mt	32.7	69.9	45.5	40.3	53.0	105.5	55.8	37.3	58.8	73.1	82.1	29.3	39.1
NO_x	MtN	6.9	16.1	16.3	13.5	14.9	18.8	25.9	13.3	19.4	13.1	40.0	5.5	25.7

*Montreal gases are not distributed over regions.

***Table 5-13d**: Standardized anthropogenic emissions (CO_2, CH_4, N_2O, NO_x, CO, NMVOCs, SO_2, HFCs, PFCs, and SF_6) for the four SRES marker scenarios, ALM region.*[13]

		1990	2020				2050				2100			
Marker scenarios Region			**A1B**	**A2**	**B1**	**B2**	**A1B ALM**	**A2**	**B1**	**B2**	**A1B**	**A2**	**B1**	**B2**
Fossil CO_2	GtC	0.72	3.40	2.31	2.71	1.48	5.73	3.96	5.11	2.60	4.81	8.87	2.41	3.84
Land-use CO_2	GtC	0.73	0.40	0.85	0.45	0.42	0.26	0.71	–0.13	–0.10	0.16	0.16	0.22	-0.20
Total CO_2	GtC	1.45	3.80	3.16	3.16	1.90	5.99	4.68	4.98	2.50	4.96	9.03	2.63	3.64
CH_4 total	Mt CH_4	76.7	119.8	131.9	115.2	101.7	144.2	187.6	112.2	148.7	95.3	272.4	70.2	199.9
N_2O total	Mt N_2O-N	1.2	1.3	1.9	1.8	0.6	1.4	2.8	2.2	0.6	1.4	4.4	1.6	0.8
SO_x total	MtS	10.5	25.3	24.4	26.8	15.2	44.1	33.5	35.6	19.4	12.1	21.8	12.6	17.2
CFC/HCFC*	MtC equiv.													
HFC	MtC equiv.	0	39	32	28	20	184	98	98	84	196	336	89	223
PFC	MtC equiv.	4	9	12	6	10	23	26	9	27	30	52	12	37
SF_6	MtC equiv.	3	14	10	8	7	40	23	22	16	26	43	14	14
CO	Mt CO	395.9	426.8	498.6	301.5	426.7	438.8	709.0	121.5	541.5	678.8	1057.9	83.5	1075.1
NMVOCs	Mt	48.3	98.7	69.7	55.2	71.6	129.0	96.2	47.2	82.6	88.1	156.3	36.2	75.0
NO_x	MtN	6.6	15.3	14.3	13.4	10.0	20.8	23.8	18.2	14.5	21.0	40.1	9.6	20.4

*Montreal gases are not distributed over regions.

Table 5-14: *Regional allocation of CO_2 emissions in the SRES marker scenarios (IND region includes OECD90 and REF regions; and DEV includes region ASIA and ALM, see Appendix IV).*

		World emissions (GtC)	IND (%)	DEV (%)
1990	Fossil fuel & industry	6.0	69	31
	Total	7.1	58	42
2020	Fossil fuel & industry	9.0–12.1	38–50	50–62
	Total	9.1–12.6	37–47	53–63
2050	Fossil fuel & industry	11.2–16.5	25–40	60–75
	Total	11.0–17.4	22–40	60–78
2100	Fossil fuel & industry	5.2–28.9	23–32	68–77
	Total	4.2–29.1	23–32	68–77

In the A1B marker scenario, the increase in energy demand in the ALM region is even higher than in the ASIA region. The primary energy use of 47 EJ in the base year increases to a level of 802 EJ in 2100, with 72% of energy from non-fossil sources. The emission path in this region is in line with trends observed in ASIA. Emissions grow from 0.72 GtC in 1990 to 5.72 GtC in 2050. After this peak they decline to 4.81 GtC in 2100 (Figure 5-14, Tables 5-13d).

In the A2 marker scenario (A2-ASF), technological development is relatively slow and fossil fuels maintain their dominant position to supply the rapidly expanding population. By 2100, the contributions of coal to the total primary energy mix in the OECD90, REF, ASIA, and ALM regions are 52%, 38%, 61%, and 48%, respectively, the largest shares across all the SRES marker scenarios. Relatively slow rates of technological improvements in the A2 scenario family result in the lowest contribution of non-fossil fuels compared to the other scenarios. In the A2 marker, CO_2 emissions grow continuously in all SRES regions (except REF from 1990 to 2020; Figure 5-14, Tables 5-13a–d). The fastest growth occurs in the ASIA and ALM regions as a result of the fast population growth in these regions. The contribution of CO_2 emissions by ASIA increases from 19% to 38% of the global total, and that by ALM from 12% to 31%.

The strong trend toward more ecologically compatible consumption and production in the B1 storyline is reflected by structural changes that lead to fewer energy- and material-intensive activities and result in a relatively limited growth of energy requirements in the B1 marker scenario (B1-IMAGE). In all the regions the shift is away from fossil fuels. In 2100, non-fossil sources supply more than 50% of the global energy requirements, with regional shares ranging from 41% (REF) to 64% (ASIA). Drastic changes in energy systems lead to an eventual decline in OECD90 emissions starting from 2020; this

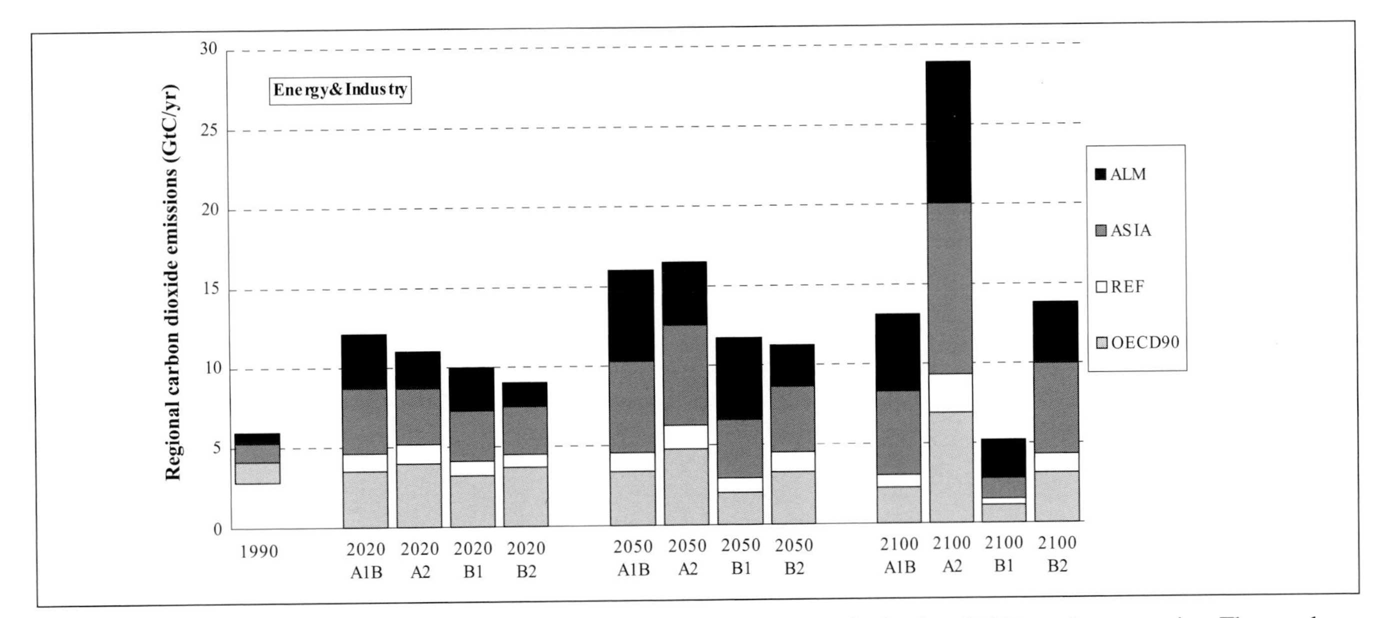

Figure 5-14: Regional CO_2 emissions from fossil fuels and industrial sources in the four SRES marker scenarios. The numbers for the additional two illustrative scenarios for the A1FI and A1T scenario groups noted in the Summary for Policymakers can be found in Appendix VII.

starts from 2050 in other regions (Figure 5-14, Tables 5-13a-d). By 2100 emissions in all regions but ALM are smaller than they were in 1990. A decline in emissions is less pronounced in the developing regions – ASIA and ALM combine to produce around 70% of CO_2 emissions by in 2100.

In the B2 world (illustrated by the B2-MESSAGE marker), the regions exploit comparative resource and technology advantages to structure their energy systems. Combined emissions in the OECD90 and REF regions remain more or less stable, changing from 4.1 GtC in 1990 to 4.3 GtC in 2100. The relative share of these two regions decreases from 69% in the base year to 31% in 2100 (Figure 5-14, Tables 5-13a–d). In the B2 marker scenario, fossil fuel and industrial CO_2 emissions in the OECD90 region increase to 3.71 GtC by 2020. Thereafter, emissions decline to 3.3 GtC in 2050 and to 3.1 GtC in 2100. This dynamic is caused by a decline in the use of fossil fuels and by the replacement of oil with natural gas, as pressure on the oil resource base increases considerably after 2050. In the REF region, standardized fossil CO_2 emissions decline to 0.8 GtC in 2020, after which they return to the 1990 level by 2100 (1.2 GtC). Toward the end of the 21st century, primary energy use in this region decreases while emissions increase because of a switch to coal (mainly to produce liquid substitutes for oil). In ASIA, both primary energy use and carbon emissions increase during the 21st century. Although the use of non-fossil fuels becomes more important, the contribution of fossil fuels to emissions remains high. The use of coal, oil, and gas increases until 2050, after which the use of oil and gas decreases, while the use of coal grows rapidly. Population, energy use, and emissions in the ALM region constantly increase during the 21st century. Again, the fossil fuels retain a dominant role and supply 47% of the energy requirements in 2100. Gas use increases until 2100, while the use of coal is rather stable until 2050 and shows a rapid increase afterward. Oil use drops sharply after 2050 as resources become depleted.

5.6.1.2. Land-Use Carbon Dioxide Emissions

Changes in land use are influenced primarily by the demand for cropland and grassland (to supply plant and animal food to the world population) and by the role of biomass energy. The uncertainty of emission estimates is reflected in the models used to quantify the SRES scenarios – in 1990 they range between 1.0 and 1.6 GtC and the spread at the regional level is even larger. In all the SRES marker scenarios, most emissions related to land use originate from the ASIA and ALM regions (Tables 5-13a–d). In the industrialized regions, the land-use change emissions in 2100 vary from –0.40 to +0.04 GtC. In the developing regions emissions from land-use change span a larger range (from –0.56 to +0.35 GtC).

5.6.1.3. Total Carbon Dioxide Emissions

Adding land-use CO_2 emissions to the energy- and industry-related emissions does not make significant changes to the distribution of emissions across regions (Figures 5-14 and 5-15). Table 5-14 provides an overview of the relative shares of the industrialized and developing regions within global CO_2 emissions. On average, the SRES marker scenarios project a shift in relative contribution in both energy- and industry-related and total CO_2 emissions from the industrialized to developing regions. In general, the relative contribution of industrialized regions is the lowest in A1 and the highest in B2.

Shifts in the regional emission shares (Table 5-13a–d) result from different developments in regional emission trajectories. To illustrate this, the trajectories were normalized to the base year (1990 = 100 for each region) and are presented in Figure 5-16.

Figure 5-16 confirms that CO_2 emissions in the ASIA and ALM regions in all the SRES markers grow much faster than

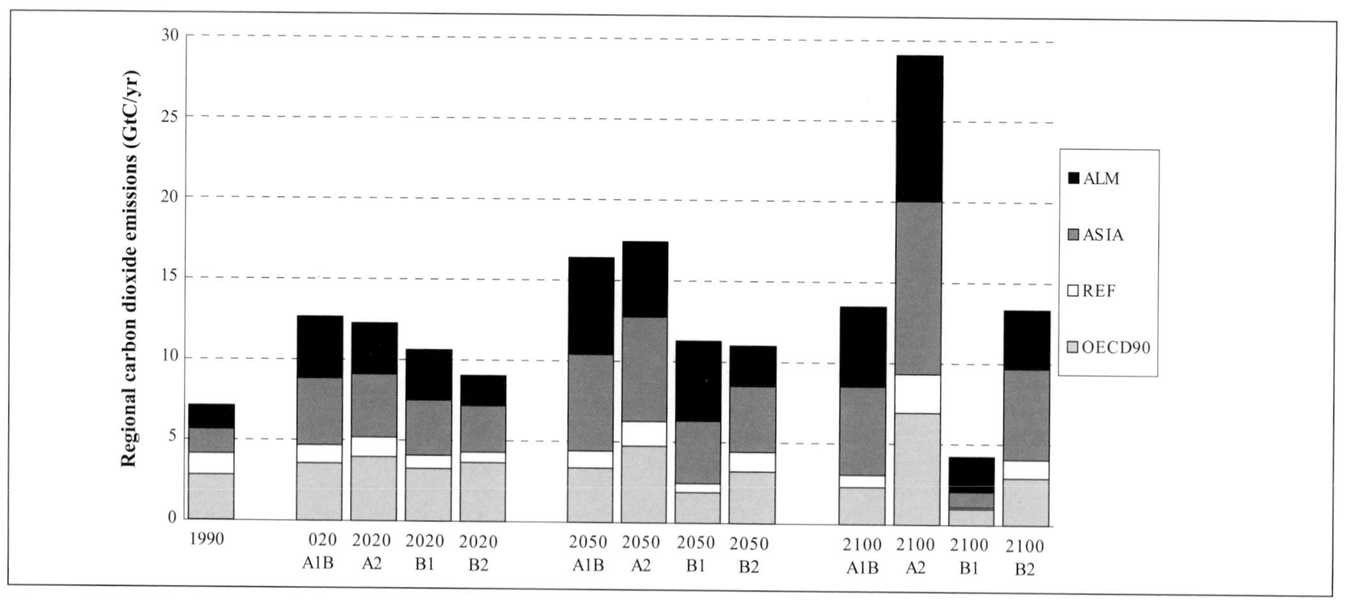

Figure 5-15: Total CO_2 emissions in the SRES marker scenarios by region. The numbers for the additional two illustrative scenarios for the A1FI and A1T scenario groups noted in the Summary for Policymakers can be found in Appendix VII.

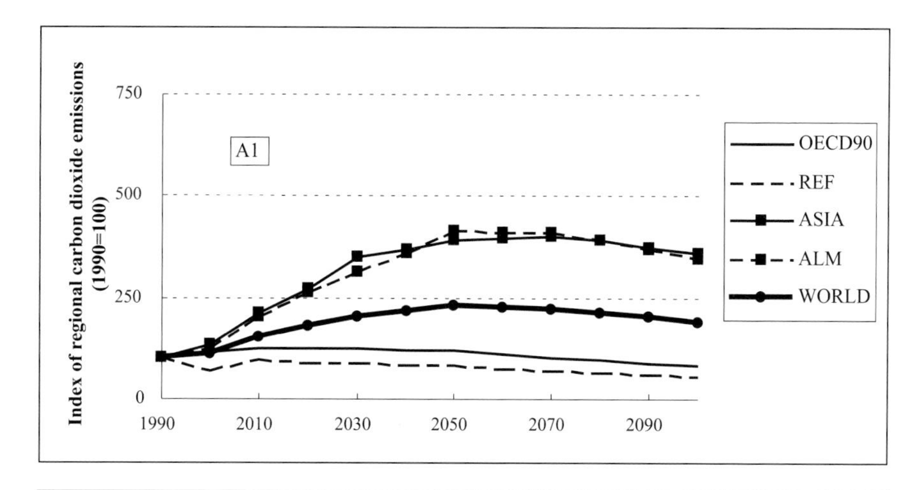

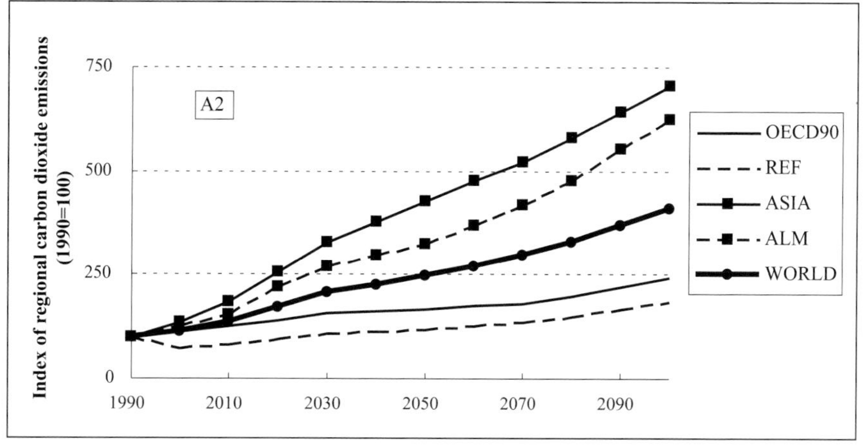

Figure 5-16: Regional and global CO_2 emissions in the four SRES markers scenarios A1B, A2, B1, and B2, shown as an index (1990 = 100). The numbers for the additional two illustrative scenarios for the A1FI and A1T scenario groups noted in the Summary for Policymakers can be found in Appendix VII.

in the industrialized regions. It also illustrates that the global pattern is strongly influenced by the developing region trajectories. Furthermore, reflecting different development perspectives in the four SRES families, CO_2 emissions grow differently in ASIA and ALM. In the A1B marker, emission trajectories in ALM and ASIA are roughly parallel over the entire time horizon. In the B1 marker, this is only true in the earlier years. As the emission of ALM peaks and then declines later than that of ASIA, emission trajectories diverge strongly in the second half of the 21st century. In the A2 marker, emissions in ALM start to grow at a lower rate than in ASIA, but subsequently catch up and later the two are again fairly close. Finally, in the B2 marker, ALM emissions initially grow at a modest rate, close to those for the OECD90 region and the world average. In later years, the growth in ALM exceeds the global rate, but total carbon emissions remain far below those in the ASIA region (Figures 5-15, 5-16).

5.6.1.4. Methane

The resultant CH_4 emission trajectories in the four SRES markers are displayed in Figure 5-17. By 2020, regional differences between the four markers are minimal. In 2050, the largest difference is the relative share of the REF region in the A2 marker, attributable primarily to an increased coal and gas production in this region. By 2100, the A2 marker has the largest CH_4 emissions in all the regions as compared to the other markers (Tables 13a–d, Figure 5-17). This arises from the "heterogeneous" nature of the A2 storyline, in which each region has to rely primarily on its own resources and progress in the renewable energy sector is quite limited. The second highest methane emissions are attained in the B2 marker, which also has a "regional" orientation, but with a more environmentally sustainable emphasis as compared to the A2 marker. Starting from 2100, both A1B and B1 markers have notably lower CH_4 emissions in all the regions in comparison with the A2 and B2 markers (Figure 5-17). The regional emission allocation changes considerably from 1990 to 2100; all four markers project much greater percentages of emissions in the developing regions (ASIA and ALM).

5.6.1.5. Nitrous Oxide

The relative shares of the OECD90, REF, ASIA, and ALM regions in the base year N_2O emissions are 39%, 9%, 34%, and 18%, respectively (Figure 5-18). The OECD90 emissions remain quite stable over the 21st century in all the markers, except A2 in which emissions increase from 2.6 MtN in 1990 to almost 4 MtN in 2100. Emissions in the ASIA and REF regions increase in the A2 marker, decline in the B1 marker (after an initial increase in ASIA), and do not change significantly in the A1 and B2 markers. Finally, the ALM N_2O emissions grow quickly in the A2 marker and remain relatively flat in the other markers. The relatively small changes in the

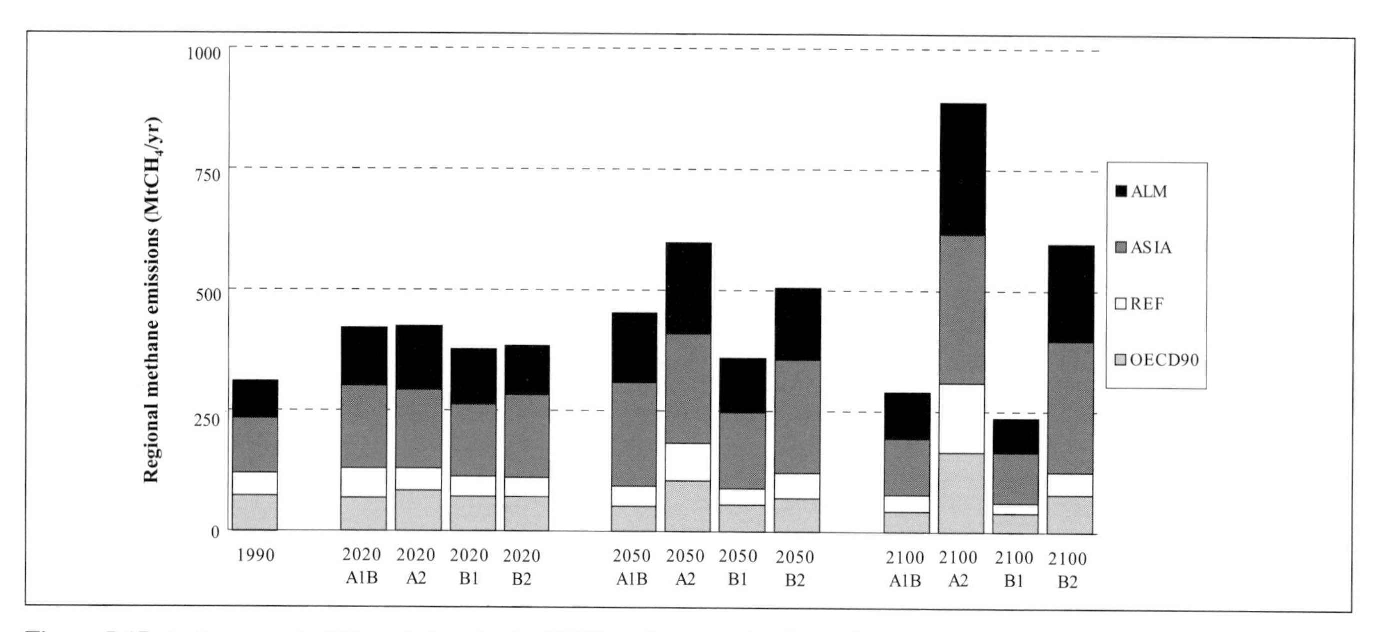

Figure 5-17: Anthropogenic CH_4 emissions in the SRES marker scenarios by region. The numbers for the additional two illustrative scenarios for the A1FI and A1T scenario groups noted in the Summary for Policymakers can be found in Appendix VII.

N_2O emissions across regions and scenarios are explained, in part, by a limited capacity of the SRES models to capture drastic shifts in technologies and practices (e.g., new catalytic converters or new manure management systems) that directly impact emission levels.

5.6.1.6. Halocarbons and Other Halogenated Compounds

In 1990, emissions of halocarbons and other halogenated compounds occurred almost exclusively in the OECD90 region, which contributed 95% to the world total (Figure 5-19). By 2020, OECD90 still remains a major emitter, but emissions in ASIA and ALM are increasing at much higher rates. The continued growth of the production and use of halocarbons and other halogenated compounds in the developing regions after 2020 makes them primary emitters of these substances in all the markers, except the B1 marker, in 2100 (Figure 5-19).

The A1B marker has the largest emissions in all the regions in 2050, while in 2100 the largest emissions across all the regions are produced in the A2 marker. Emissions in all the regions are smallest in the B1 marker, which reflects its sustainability features (e.g., increased recycling and "dematerialization").

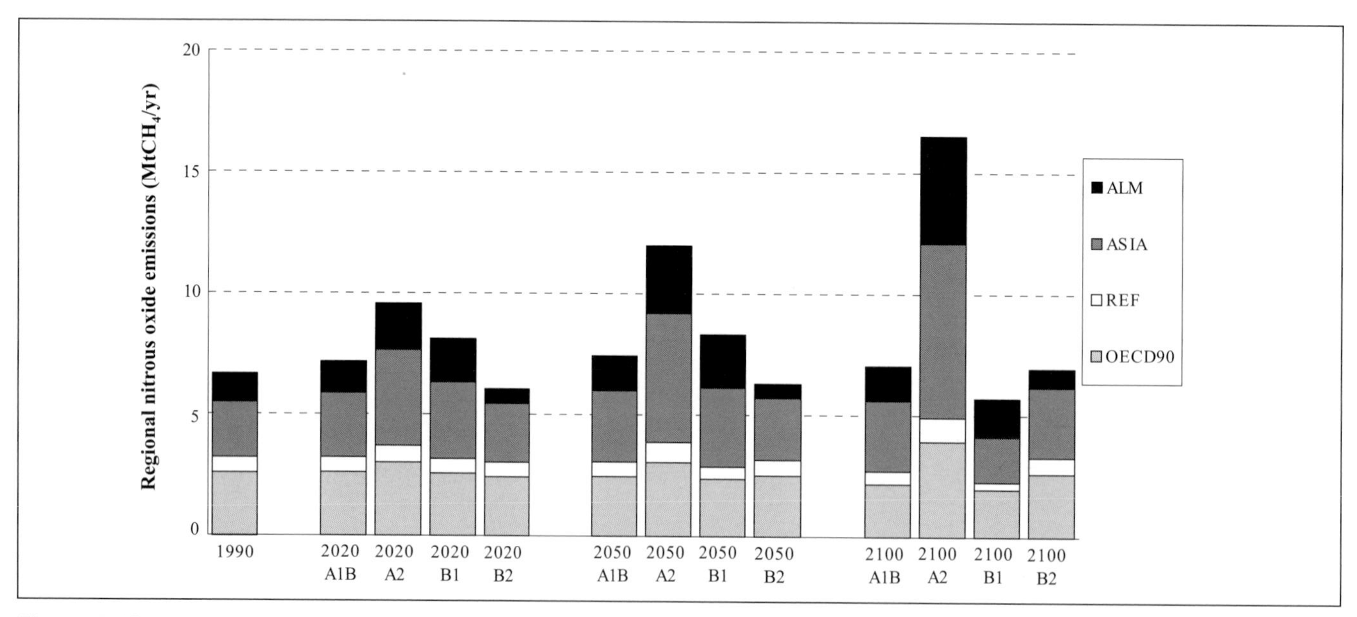

Figure 5-18: Anthropogenic N_2O emissions in the SRES marker scenarios by region. The numbers for the additional two illustrative scenarios for the A1FI and A1T scenario groups noted in the Summary for Policymakers can be found in Appendix VII.

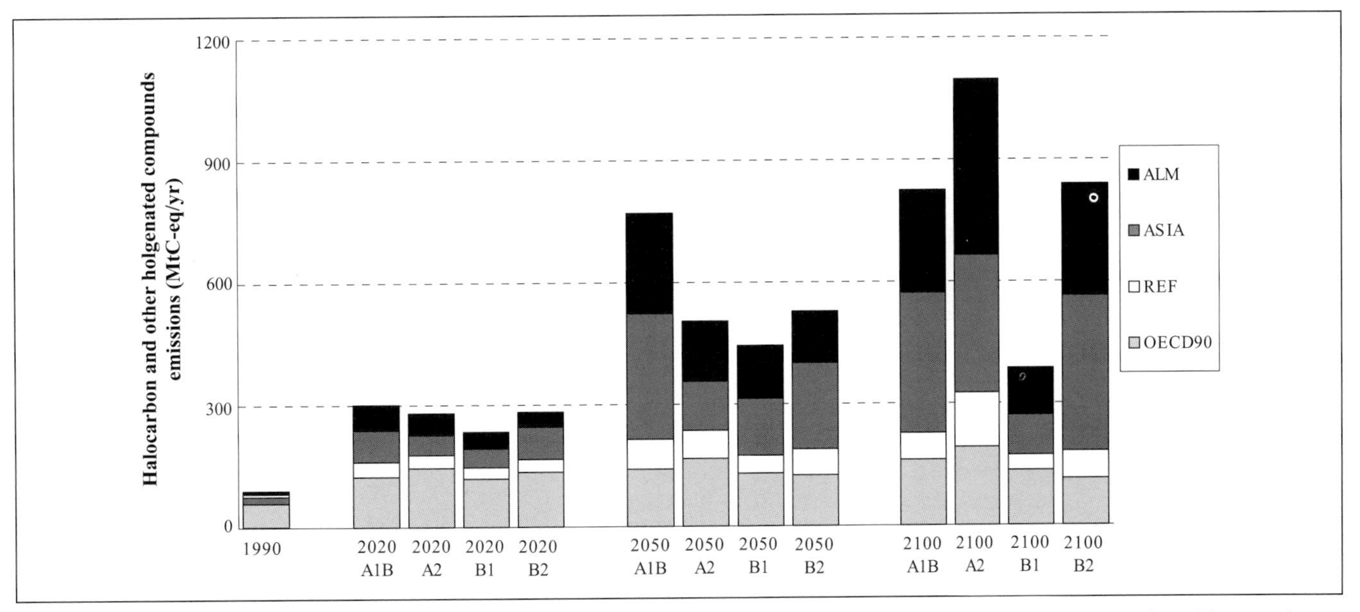

Figure 5-19: Halocarbons and other halogenated compounds emissions in the SRES marker scenarios by region. The numbers for the additional two illustrative scenarios for the A1FI and A1T scenario groups noted in the Summary for Policymakers can be found in Appendix VII.

5.6.1.7. Sulfur

As noted in Section 5.5.2.1, even with a comparatively good agreement on global sulfur emission levels, important uncertainties remain at the sectoral and regional levels. The base-year uncertainties are especially important because regional sulfur emissions trends have changed drastically during the past decade. While declining strongly in the industrialized regions as a result of sulfur control policies in Europe and North America, and because of economic reforms in Russia and Eastern Europe, emissions increase rapidly in Asia with an increase in the energy demand and coal use.

As a general rule, in the SRES scenarios an increasing affluence causes energy use per capita to rise and leads to the substitution of solid fuels, such as coal and fuelwood, with energy forms of higher quality. This relationship determines the sulfur emission dynamics across the SRES markers and regions (Figure 5-20).

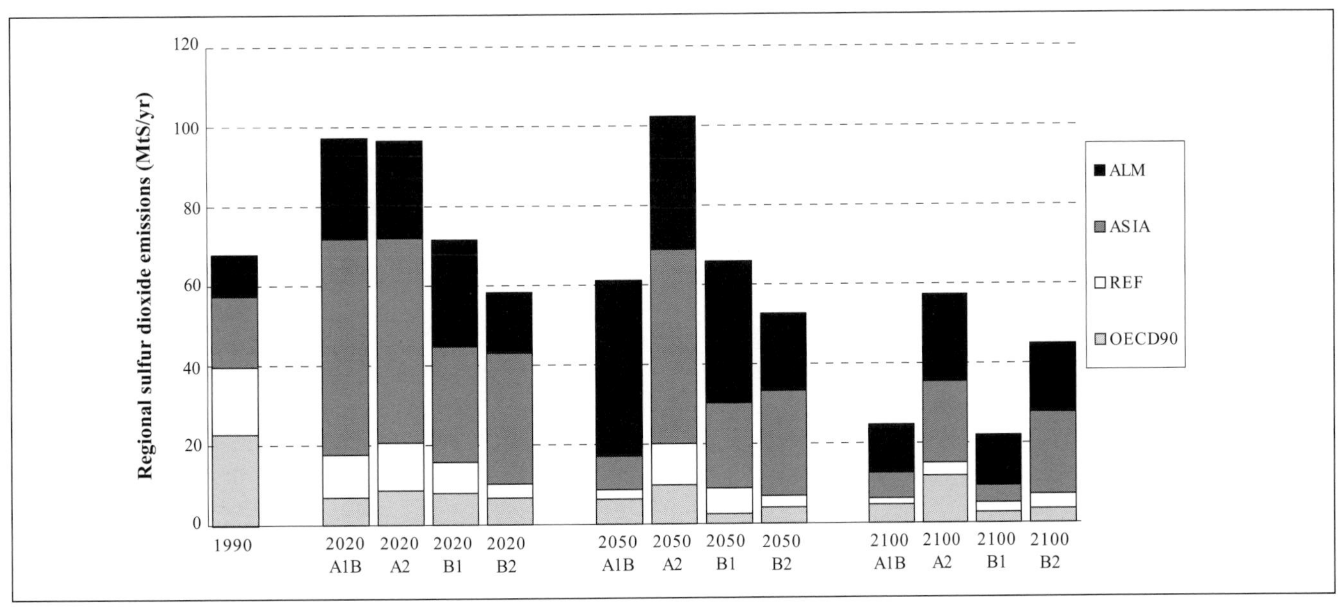

Figure 5-20: Anthropogenic SO_2 emissions in the SRES marker scenarios by region. The numbers for the additional two illustrative scenarios for the A1FI and A1T scenario groups noted in the Summary for Policymakers can be found in Appendix VII.

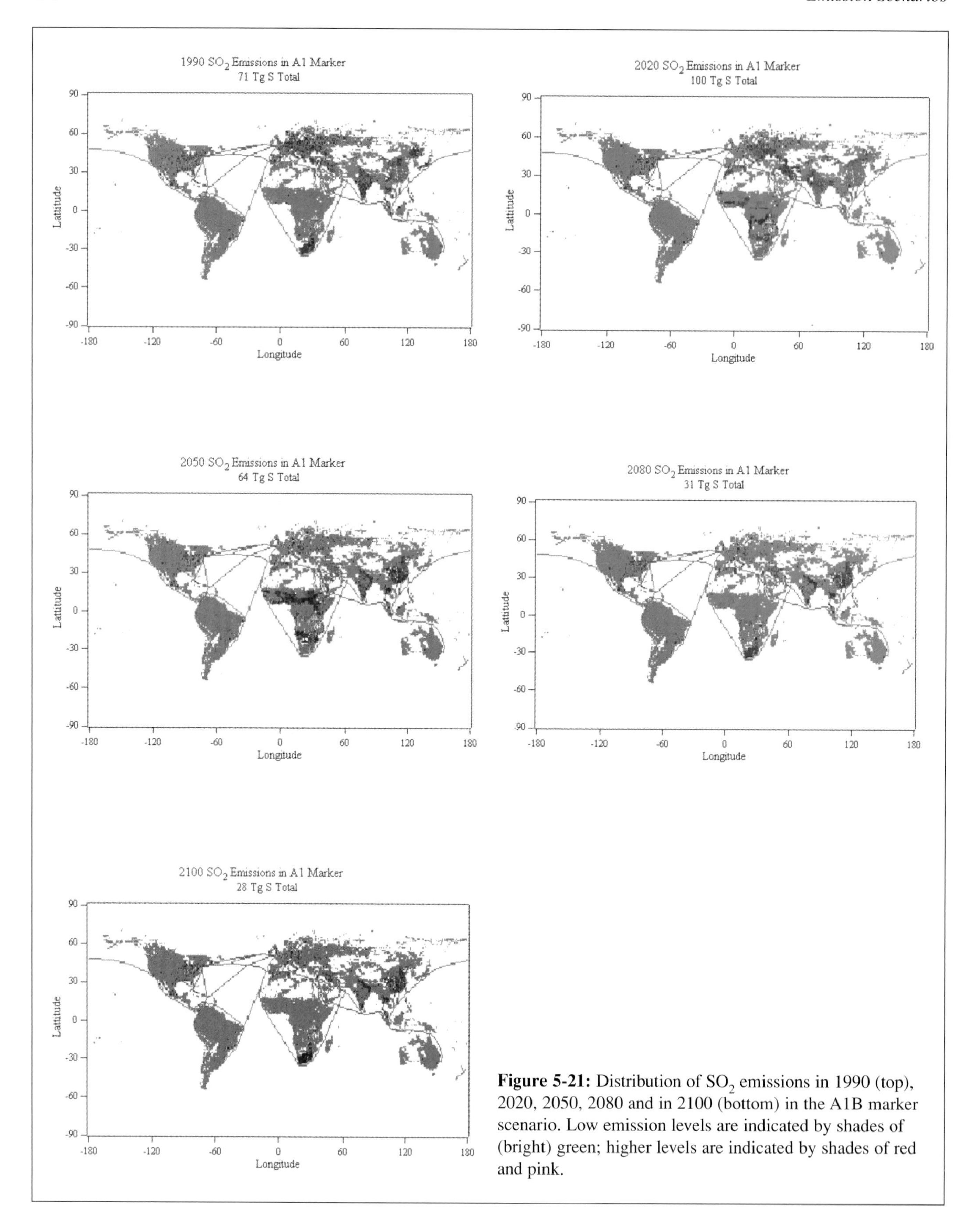

Figure 5-21: Distribution of SO_2 emissions in 1990 (top), 2020, 2050, 2080 and in 2100 (bottom) in the A1B marker scenario. Low emission levels are indicated by shades of (bright) green; higher levels are indicated by shades of red and pink.

Box 5-5: Gridding of Emission Data

The climate effects of SO_2 are intrinsically regional and emissions on a latitude–longitude grid are required as input to climate models. Emissions of SO_2 were first standardized for four world regions as described above. Then, emissions from the marker scenarios for six regions (OECD90, REF, Centrally Planned Asia, Rest of Asia, Latin America, and Africa/Middle East, scaled to match the standardized emissions) were used for gridding purposes. For the Annex II countries, a value of 23 MtS was taken for 1990 emissions, a figure derived from a compilation of country-level emissions inventories (Smith *et al.*, 2000).

These emissions were mapped to a global 1° x 1° emissions grid. For each region, the pattern of total SO_2 emissions from the EDGAR database (Olivier *et al.*, 1996) was scaled by the total emissions for that region and a time period. Emissions for OECD90 countries were first scaled individually to their country-specific values. The value of 3 MtS was added to reflect international shipping, with the pattern and magnitude of these emissions held constant.

Emissions of other short-lived gases (CO, NO_x, NMVOCs, and CH_4) also needed to be mapped to a global grid for use in atmospheric chemistry models. The approach taken was essentially the same, with the EDGAR database used to establish the spatial pattern. Standardization and subsequent gridding were carried out at the level of the original four world regions, and no specific adjustments were made for international shipping.

In "high income regions" (OECD90, REF) sulfur emissions have already passed their peaks and are actually declining at present. This trend is expected to continue in all the markers, except A2 in which an increased use of coal "counters" a decline in specific emissions in OECD90 (Figure 5-20). Emissions in ASIA grow in all the markers by 2020, and then decline by 2050, and further decline by 2100. The most dramatic decline is registered in the A1B marker; this is related to its aggressive assumptions on the introduction of low-sulfur technologies and fuel switching in the ASIA region (see Box 5-3 for more details). Unlike ASIA, the ALM region sees increases in emissions in all four markers from 2020 to 2050, because of the somewhat "mixed" nature of this region, which combines countries with substantially different affluence levels and development trends. However, by 2100, when low-sulfur technology becomes widely available everywhere, emissions in the ALM decline in all markers (Figure 5-20).

5.6.2. *Gridded Sulfur Emissions*

As discussed above, global sulfur emissions eventually decline in all SRES scenario families and associated groups. In addition, the regional distribution of emissions changes drastically over time. While in previous decades major sulfur emitters were located primarily in industrialized regions of the world, presently emissions for these sources are declining because of the introduction of cleaner fuels and the conversion to low-sulfur technologies to comply with environmental regulations. In the majority of SRES scenarios, this trend is expected to continue. Meanwhile, less-developed regions are anticipated to experience strong economic growth associated with an increased demand for energy. Especially in the short term, fossil fuels are likely to satisfy the major share of this new demand, which may lead to a steep initial growth in sulfur emissions. As mentioned earlier (see Section 5.5.2), at some point in time sulfur emissions will be controlled in all the scenarios and, together with shifts to essentially sulfur-free energy resources, they will decrease in the developing regions as they are decreasing now in the industrialized world. As a consequence of these complex dynamics, different countries and regions are bound to experience very different levels of sulfur emissions over the 21st century. To illustrate this, Figure 5-21 shows gridded sulfur emissions in 1990 and 2050 in the A1B marker (see Box 5-5).

References:

AFEAS (Alternative Fluorocarbons Environmental Acceptability Study), 1998: *Production, Sales and Atmospheric Release of Fluorocarbons Through 1997*. AFEAS, Washington, DC.

Akimoto, H., and H. Narita, 1994: Distribution of SO_2, NO_X, and CO_2 emissions from fuel combustion and industrial activities in Asia with 1°x1° resolution. *Atmospheric Environment*, **28**(2), 213-225.

Alcamo, J., A. Bouwman, J. Edmonds, A. Grübler, T. Morita, and A. Sugandhy, 1995: An evaluation of the IPCC IS92 emission scenarios. In *Climate Change 1994, Radiative Forcing of Climate Change* and *An Evaluation of the IPCC IS92 Emission Scenarios,* J.T. Houghton, L.G. Meira Filho, J. Bruce, Hoesung Lee, B.A. Callander, E. Haites, N. Harris and K. Maskell (eds.), Cambridge University Press, Cambridge, pp. 233-304.

Alcamo, J., J. Onigkeit, and F. Kaspar, 1997: *The Pollutant Burden Approach for Computing Global and Regional Emissions of Sulfur Dioxide*. Center for Environmental Systems Research, University of Kassel, Kassel, Germany.

Amann, M., I. Bertok, J. Cofala, F. Gyarfas, C. Heyes, Z. Klimont, M. Makowski, W. Schoepp, and S. Shibayev, 1996: *Cost-Effective Control of Acidification and Ground-Level Ozone*.Second interim report to the European Commission (DGXI), International Institute of Applied Systems Analysis (IIASA), Laxenburg, Austria.

Ashford, P., 1999: *Consideration for the responsible use of HFCs in Foams*. Joint IPCC/TEAP Expert Meeting on Options for the Limitation of Emissions of HFCs and PCFs, Petten, the Netherlands.

Baumann H., N. Ulrich, G. Haupt, G. Zimmermann, R. Pruschek, and G. Oeljeklaus, 1998: *Development of a Cost Effective IGCC 98 Power Plant*. Power Gen Europe '98, Milan, June 9-11.

Benkovitz, C.M., M.T. Scholtz, J. Pacyna, L. Tarrason, J. Dignon, E.C. Voldner, P.A. Spiro, J.A. Logan, and T.E. Graedel, 1996: Global gridded inventories of anthropogenic emissions of sulfur and nitrogen. *Journal of Geophysical Research*, **101**(D22), 29,239-29,253.

Dadi, Z., L. Xueyi, and X. Huaqing, 1998: *Estimation of Sulfur Dioxide Emissions in China in 1990 and 1995*. Energy Research Institute, Beijing, China.

Davidson, EA., and W. Kingerlee, 1997: A global inventory of nitric oxide emissions from soils. *Nutrient Cycling in Agroecosystems*, **48**(1/2), 37-50.

Delmas, R, D. Serca, and C. Jambert, 1997: Global inventory of NO_x sources. *Nutrient Cycling in Agroecosystems*, **48**(1/2), 51-60.

De Vries, B., J. Bollen, L. Bouwman, M. den Elzen, M. Janssen, and E. Kreileman 2000: Greenhouse gas emissions in an equity-, environment- and service-oriented world: An IMAGE-based scenario for the next century. *Technological Forecasting & Social Change*, **63**(2-3). (In press).

Dignon, J., and S. Hameed, 1989: Global emissions of nitrogen and sulfur oxides from 1860 to 1989. *Journal of the Air and Waste Management Association*, **39**(2), 180-186.

ECE (UN Economic Commission for Europe), 1997: *Anthropogenic Emissions of Sulfur 1980 to 2010 in the ECE Region*. EB.AIR/GE.1/1997/3 and Addendum 1, ECE, Geneva.

EU (European Commission), 1997: Estimated European Union (EU) fluorocarbon emissions now and in the future & EU reduction potential of fluorocarbon emissions. Draft, January 28, 1997, EC, Brussels.

Fenhann, J., 2000: Industrial non-energy, non-CO_2 greenhouse gas emissions. *Technological Forecasting & Social Change*, **63**(2-3). (In press).

Foell W., M. Amann, G. Carmichael, M. Chadwick, J.-P. Hettelingh, L. Hordijk, and Z. Dianwu, 1995: *Rains Asia: An assessment model for air pollution in Asia*. Report on the World Bank sponsored project "Acid Rain and Emission Reductions in Asia," World Bank, Washington, DC.

Grübler, A., 1998: A review of global and regional sulfur emission scenarios. *Mitigation and Adaptation Strategies for Global Change*, **3**(2-4), 383-418.

Harnisch, J., R. Borchers, P. Fabian, and M. Maiss, 1999: CF_4 and the age of mesopheric polar vortex air. *Geophysical Research Letters*, **26**(3), 295-298.

Houghton, J.T., L.G. Meira Filho, J. Bruce, Hoesung Lee, B. A. Callander, E. Haites, N. Harris, and K. Maskell (eds.), 1995: *Climate Change 1994: Radiative Forcing of Climate Change and an Evaluation of the IPCC IS92 Emissions Scenarios*. Cambridge University Press, Cambridge.

Houghton, J.T., L.G. Meira Filho, B.A. Callander, N. Harris, A. Kattenberg, and K. Maskell (eds.), 1996: *Climate Change 1995. The Science of Climate Change*, Contribution of Working Group I to the Second Assessment Report of the Intergovernmental Panel on Climate Change, Cambridge University Press, Cambridge.

IPCC (Intergovernmental Panel on Climate Change), 1997: *Revised Guidelines for National Greenhouse Gas Inventories*. The Intergovernmental Panel on Climate Change (IPCC), the Organisation for Economic Cooperation (OECD), and the International Energy Agency (IEA), Paris.

Jiang, K., T. Masui, T. Morita, and Y. Matsuoka, 2000: Long-term GHG emission scenarios of Asia-Pacific and the world. *Technological Forecasting & Social Change*, **63**(2-3). (In press).

Kato, N., 1996: Analysis of the structure of energy consumption and dynamics of emission of atmospheric species related to global environmental change (SO_x, NO_x, CO_2) in Asia. *Atmospheric Environment* **30**(5), 757-785.

Kram, T., K. Riahi, R.A. Roehrl, S. van Rooijen, T. Morita, and B. de Vries, 2000: Global and regional greenhouse gas emissions scenarios. *Technological Forecasting & Social Change*, **63**(2-3). (In press).

Kroeze, C., and Reijnders, L., 1992: Halocarbons and global warming III. *The Science of Total Environment*, **112**, 291-314.

Kroeze, C., 1995: *Fluorocarbons and SF_6, Global Emission Inventory and Options for Control*. RIVM report 773001007, Bilthoven, the Netherlands.

Maiss, M., L.P. Steele, R.J. Francey, P.J. Fraser, R.L. Langenfelds, N.B.A Trivett, and I. Levin, 1996: Sulfur hexafluoride - a powerful new atmospheric tracer. *Atmospheric Environment*, **30**, 1621-1629.

McCulloch, A., P.M. Midgley, and D.A. Fisher, 1994: Distribution of emissions of chlorofluorocarbons (CFCs) 11, 12, 113, 114 and 115 among reporting and non-reporting countries in 1986. *Atmospheric Environment*, **28**(16), 2567-2582.

McCulloch, A., and P.M. Midgley, 1998: Estimated historic emissions of fluorocarbons from the European Union. *Atmospheric Environment*, **32**(9), 1571-1580.

Midgley, P.M., and A. McCulloch, 1999: Properties and applications of industrial halocarbons. In *Reactive Halogen Compounds in the Atmosphere*, **4**, *Part E*. P. Fabian, and O.N. Singh (eds.), *The Handbook of Environmental Chemistry*, Springer-Verlag, Berlin/Heidelberg.

Mori, S., 2000: The development of greenhouse gas emissions scenarios using an extension of the MARIA model for the assessment of resource and energy technologies. *Technological Forecasting & Social Change*, **63**(2-3). (In press).

Mosier, A., C. Kroeze, C. Nevison, O. Oenema, S. Seitzinger, and O. van Cleemput, 1998: Closing the global N_2O budget: nitrous oxide emissions through the agricultural nitrogen cycle: OECD/IPCC/IEA phase II development of IPCC guidelines for national greenhouse gas inventory methodology. *Nutrient Cycling in Agrosystems*, **52**, 225-248.

Olivier, J.G.J., A.F. Bouwman, C.W.M. van der Maas, J.J.M. Berdowski, C. Veldt, J.P.J. Bloos, A.J.H. Visschedijk, P.Y.J. Zandveld, and J.L. Haverlag, 1996: *Description of EDGAR Version 2.0: A set of global emission inventories of GHGs and ozone-depleting substances for all anthropogenic and most natural sources on a per country basis and on $1^o x 1^o$ grid*. Report 771060002, National Institute of Public Health and the Environment, Bilthoven, the Netherlands.

Oram, D.E., W.T. Sturges, S.A. Penkett, A. McCulloch, and P.J. Fraser, 1998: Growth of fluoroform (CHF_3, HFC-23) in the background atmosphere. *Geophysical Research Letters*, **25**(1), 35-38.

Pepper, W.J., J. Leggett, R. Swart, J. Wasson, J. Edmonds, and I. Mintzer, 1992: Emissions Scenarios for the IPCC. An update: Assumptions, methodology, and results. Support document for Chapter A3. In *Climate Change 1992. Supplementary Report to the IPCC Scientific Assessment*. J.T. Houghton, B.A. Callandar, and S.K. Varney (eds.), Cambridge University Press, Cambridge.

Preisegger, E., 1999: *Automotive Air Conditioning Impact of Refrigerant on Global Warming*. Joint IPCC/TEAP Expert Meeting on Options for the Limitation of Emissions of HFCs and PFCs, Petten, the Netherlands.

Riahi, K., and R.A. Roehrl, 2000: Greenhouse gas emissions in a dynamics-as-usual scenario of economic and energy development. *Technological Forecasting & Social Change*, **63**(2-3). (In press).

Roehrl, R.A., and K. Riahi, 2000: Technology dynamics and greenhouse gas emissions mitigation - a cost assessment. *Technological Forecasting & Social Change*, **63**(2-3). (In press).

Sankovski, A., W. Barbour, and W. Pepper, 2000: Quantification of the IS99 emission scenario storylines using the atmospheric stabilization framework (ASF). *Technological Forecasting & Social Change*, **63**(2-3). (In press).

Science & Policy Services Inc., 1997: Sales of sulphur hexafluoride (SF_6) by end-use applications; Annual sales for 1961 through 1996; Sales projections for 1997 through 2000, Washington, DC.

Sinton, J.E., (ed.), 1996: China Energy Databook. LBL-32822 Rev.4 UC-900, Lawrence Berkeley National Laboratory, University of California, Berkeley, CA.

Smith, S.J., H. Pitcher, and T.M.L. Wigley, 2000: Global and regional anthropogenic sulfur dioxide emissions. *Global Biogeochemical Cycles*. (In press).

Spiro, P.A., D.J. Jacob, and J.A. Logan, 1992: Global inventory of sulfur emissions with $1^o x 1^o$ resolution. *Journal of Geophysical Research*, **97**, 6023-6036.

US Geological Survey, 1998: *Mineral Commodities Information* (http://minerals.er.gov/minerals/pubs/commodity/).

Victor, D.G., and G.J. MacDonald, 1998: *A Model for Estimating Future Emissions of Sulfur Hexafluoride and Perfluorocarbons*. IR-98-053, International Institute for Applied Analysis, Laxenburg, Austria, July 1998, pp. 33.

Watson, R., M.C. Zinyowera, and R. Moss (eds.), 1996a: *Climate Change 1995. Impacts, Adaptations and Mitigation of Climate Change: Scientific Analyses*. Contribution of Working Group II to the Second Assessment Report of the Intergovernmental Panel on Climate Change, Cambridge University Press, Cambridge, 861 pp.

Watson, R., M.C. Zinyowera, and R. Moss (eds.), 1996b: *Technologies, Policies and Measures for Mitigating Climate Change*. IPCC Technical Paper I, IPCC Working Group II, 85pp.

WMO (World Meteorological Organisation), 1997: Global acid deposition assessment. In *WMO Global Atmospheric Watch, No. 16*. D.M. Whelpdate, and M.S. Kaiser (eds.), WMO, Geneva.

WMO/UNEP (World Meteorological Organisation and United Nations Environment Programme), 1998: *Scientific Assessment of Ozone Depletion: 1998*. WMO Global Ozone Research & Monitoring Project, December 1998, WMO, Geneva.

WMO/UNEP, 1999: *Conference Report*. Joint IPCC/TEAP Expert Meeting on Options for the Limitation of Emissions of HFCs and PFCs, Petten, the Netherlands, 26-28 May 1999.

World Bureau of Metal Statistics, 1997: *World Metal Statistics Yearbook*. Hertfordshire, UK.

6

Summary Discussions and Recommendations

CONTENTS

6.1. Introduction and Background

The set of 40 emissions scenarios in this Special Report on Emissions Scenarios (SRES) is based on an extensive assessment of the literature, six alternative modeling approaches, and an "open process" that solicited wide participation and feedback from many groups and individuals. The set of scenarios includes all relevant species of greenhouse gases (GHGs)[1]. This chapter provides a summary of the SRES emissions scenarios and compares them with the previous set of Intergovernmental Panel on Climate Change (IPCC) IS92 scenarios and the underlying literature.

The first step in the formulation of the scenarios was the review and analysis of the published literature and the development of the database with more than 400 emissions scenarios (accessible on the web site, www-cger.nies.go.jp/ cger-e/db/ipcc.html). One of the recommendations of the writing team is that IPCC or a similar international institution should maintain such a database to ensure continuity of knowledge and scientific progress in any future assessments of GHG scenarios. An equivalent database to document narrative and other qualitative scenarios would also be very useful for future climate-change assessments. One difficulty encountered in the analysis of the emissions scenarios is that the distinction between climate policy scenarios, non-climate policy scenarios, and other scenarios appeared to be to a degree arbitrary and was often impossible to make. Therefore, the writing team recommends that an effort should be made in the future to develop an appropriate emissions scenario classification scheme. Chapters 2 and 3 give a more detailed description of the very wide range of future emissions paths, their driving forces, and their relationships as reflected in the literature; the wide rage indicates that their possible developments are highly uncertain. The sources of inherent uncertainties range from data and modeling uncertainties through to inadequate scientific understanding of the underlying problems.

Scenarios are appropriate tools for dealing with such uncertainty. Scenarios are images of the future, or alternative futures. As an integration tool in the assessment of climate change they allow a role for intuition, analysis, and synthesis; thus we turn to scenarios in this report to take advantage of these features to aid the assessment of future climate change, impacts, vulnerabilities, adoption, and mitigation. Scenarios are not predictions. A set of scenarios can assist in the understanding of possible future developments, and hence the development of a set of alternative scenarios (see Chapters 1 and 4 for more detail).

[1] Included are anthropogenic emissions of carbon dioxide (CO_2), methane (CH_4), nitrous oxide (N_2O), hydrofluorcarbons (HFCs), perfluorocarbons (PFCs), sulfur hexafluoride (SF_6), hydrochlorofluorocarbons (HCFCs), chlorofluorocarbons (CFCs), sulfur dioxide (SO_2), carbon monoxide (CO), nitrogen oxides (NO_x), and non-methane volatile organic compounds (NMVOCs).

The SRES approach involved the development of a set of four alternative scenario "families" that encompass the 40 scenarios. Each family of SRES scenarios includes a descriptive part called a "storyline," and a number of alternative interpretations and quantifications of each storyline developed by six different modeling approaches. All the interpretations and quantifications of one storyline together are called a scenario family (see Chapter 1 for terminology). Each storyline describes a demographic, social, economic, technological, and policy future for one of these scenario families. Within each family different scenarios explore variations of global and regional developments and their implications for GHG and sulfur emissions. Each of these scenarios is consistent with the broad framework of that scenario family as specified by the storyline. Chapters 4 and 5 give a more detailed description of the storylines, their quantifications, and the resultant 40 emissions scenarios.

The SRES writing team reached a broad consensus that there could be no "best guess" scenarios; that the future is inherently unpredictable and that views will differ as to which storylines could be more likely. There is no "business-as-usual" scenario. The storylines represent the playing out of certain social, economic, technological and environmental paradigms that will be viewed positively by some people and negatively by others. The writing team decided on four storylines – an even number helps to avoid the impression that there is a "central" or "most likely" case. The team wanted more than two storylines to help illustrate that the future depends on many different underlying dynamics, but no more than four, as they wanted to avoid complicating the process with too many alternatives. The scenarios cover a wide range, but not all possible futures. In particular, it was decided that possible "surprises" would not be considered and that there would be no "disaster" scenarios.

The storylines describe developments in many different social, economic, technological, environmental, and policy dimensions. The titles of the storylines have been kept simple – A1, A2, B1, and B2. There is no particular order among the storylines, which are listed in Box 6-1 in alphabetic order. The team decided to carry out sensitivity tests within some of the storylines by considering alternative scenarios with different fossil-fuel reserves, rates of economic growth, or rates of technological change within a given scenario family.

All four storylines and scenario families describe future worlds that are generally more affluent compared to the current situation. They range from very rapid economic growth and technological change to high levels of environmental protection, from low to high global populations, and from high to low GHG emissions. What is perhaps even more important is that all the storylines describe dynamic changes and transitions in generally different directions. Although they do not include additional climate initiatives, none of them are policy free. As time progresses, the storylines diverge from each other in many of their characteristic features. In this way they span the relevant range of GHG emissions and different combinations of their main sources.

Box 6-1: The Main Characteristics of the Four SRES Storylines and Scenario Families.

By 2100 the world will have changed in ways that are hard to imagine – as hard as it would have been at the end of the 19th century to imagine the changes of the 100 years since. Each storyline assumes a distinctly different direction for future developments, such that the four storylines differ in increasingly irreversible ways. Together they describe divergent futures that encompass a significant portion of the underlying uncertainties in the main driving forces. They cover a wide range of key "future" characteristics such as population growth, economic development, and technological change. For this reason, their plausibility or feasibility should not be considered solely on the basis of an extrapolation of *current* economic, technological, and social trends.

- The A1 storyline and scenario family describes a future world of very rapid economic growth, low population growth, and the rapid introduction of new and more efficient technologies. Major underlying themes are convergence among regions, capacity building and increased cultural and social interactions, with a substantial reduction in regional differences in per capita income. The A1 scenario family develops into four groups that describe alternative directions of technological change in the energy system. Please note that in the Summary for Policymakers, two of these groups were merged into one.[2]
- The A2 storyline and scenario family describes a very heterogeneous world. The underlying theme is self-reliance and preservation of local identities. Fertility patterns across regions converge very slowly, which results in high population growth. Economic development is primarily regionally oriented and per capita economic growth and technological change are more fragmented and slower than in other storylines.
- The B1 storyline and scenario family describes a convergent world with the same low population growth as in the A1 storyline, but with rapid changes in economic structures toward a service and information economy, with reductions in material intensity, and the introduction of clean and resource-efficient technologies. The emphasis is on global solutions to economic, social, and environmental sustainability, including improved equity, but without additional climate initiatives.
- The B2 storyline and scenario family describes a world in which the emphasis is on local solutions to economic, social, and environmental sustainability. It is a world with moderate population growth, intermediate levels of economic development, and less rapid and more diverse technological change than in the B1 and A1 storylines. While the scenario is also oriented toward environmental protection and social equity, it focuses on local and regional levels.

After determining the basic features and driving forces for each of the four storylines, the team began modeling and quantifying the storylines. This resulted in 40 scenarios, each of which constitutes an alternative interpretation and quantification of a storyline. All the interpretations and quantifications associated with a single storyline are called a scenario "family" (see Chapter 1 for terminology and Chapter 4 for further details).

After determining the basic features and driving forces for each of the four storylines, the team quantified the storylines into individual scenarios with the help of formal (computer) models. The six modeling groups that quantified the storylines are listed in Box 6-2. The six models are representative of different approaches to modeling emissions scenarios and different integrated assessment (IA) frameworks in the literature and include so-called top-down and bottom-up models. The writing team recommends that IPCC or a similar international institution should ensure participation of modeling groups around the world, and especially those from developing countries, in future scenario development and assessment efforts. Clearly, this would also require resources specifically to assist modeling groups from developing countries. Indeed, a concerted effort was made to engage modeling groups and experts from developing countries in SRES as a direct response to recommendations of the IPCC scenario evaluation (Alcamo *et al.*, 1995).

The six models have different regional aggregations. The writing team decided to group the various global regions into four "macro-regions" common to all different regional aggregations across the six models. The four macro-regions (see Appendix III) are broadly consistent with the allocation of countries in the United Nations Framework Convention on Climate Change (UNFCCC, 1997) although the correspondence is not exact because of changes in the countries listed in Annex I of UNFCCC.

All the qualitative and quantitative features of scenarios that belong to the same family were set to conform to the corresponding features of the underlying storyline. Together,

[2] During the approval process of the Summary for Policymakers at the 5th Session of Working Group III of the IPCC from 8-11 March 2000 in Katmandu, Nepal, it was decided to combine the A1C and A1G groups into one "fossil intensive" group A1FI in contrast to the non-fossil group A1T, and select two illustrative scenarios from these two A1 groups to facilitate use by modelers and policy makers. This leads to six scenario groups that constitute the four scenario families, three of which are in the A1 family. The six groups all have " illustrative scenarios", four of which are marker scenarios. All scenarios are equally sound. See also Figure SPM-1.

Box 6-2: SRES Modeling Teams

In all, six models were used to generate the 40 scenarios:

- Asian Pacific Integrated Model (AIM) from the National Institute of Environmental Studies in Japan (Morita *et al.*, 1994);
- Atmospheric Stabilization Framework Model (ASF) from ICF Consulting in the USA (Lashof and Tirpak, 1990; Pepper *et al.*, 1992, 1998; Sankovski *et al.*, 2000);
- Integrated Model to Assess the Greenhouse Effect (IMAGE) from the National Institute for Public Health and Environmental Hygiene (RIVM) (Alcamo *et al.*, 1998; de Vries *et al.*, 1994, 1999, 2000), used in connection with the Dutch Bureau for Economic Policy Analysis (CPB) WorldScan model (de Jong and Zalm, 1991), the Netherlands;
- Multiregional Approach for Resource and Industry Allocation (MARIA) from the Science University of Tokyo in Japan (Mori and Takahashi, 1999; Mori, 2000);
- Model for Energy Supply Strategy Alternatives and their General Environmental Impact (MESSAGE) from the International Institute of Applied Systems Analysis (IIASA) in Austria (Messner and Strubegger, 1995; Riahi and Roehrl, 2000); and the
- Mini Climate Assessment Model (MiniCAM) from the Pacific Northwest National Laboratory (PNNL) in the USA (Edmonds *et al.*, 1994, 1996a, 1996b).

For a more detailed description of the modeling approaches see Appendix IV.

26 scenarios were "harmonized" to share agreed common assumptions about population and gross domestic product (GDP) developments (a few that also share common final energy trajectories are called "fully harmonized," see Section 4.1. in Chapter 4). Thus, the harmonized scenarios are not independent of each other within each family, but they are independent across the four families. However, scenarios within each family vary quite substantially in characteristics such as the assumptions about availability of fossil-fuel resources, the rate of energy-efficiency improvements, the extent of renewable-energy development, and, hence, the resultant GHG emissions. Thus, after the modeling teams had quantified the key driving forces and made an effort to harmonize them with the storylines by adjusting control parameters, there still remained diversity in the assumptions about the driving forces and in the resultant emissions (see Chapter 4).

The remaining 14 scenarios adopted alternative interpretations of the four scenario storylines to explore additional scenario uncertainties beyond differences in methodologic approaches, such as different rates of economic growth and variations in population projections. These variations reflect the "modeling teams' choice" of alternative but plausible global and regional developments compared to those of the "harmonized" scenarios; they also stem from the differences in the underlying modeling approaches. This approach generated a large variation and richness in different scenario quantifications, often with overlapping ranges of main driving forces and GHG emissions across the four families.

In addition, the A1 scenario family branched out into different distinct scenario groups, based on alternative technological developments in future energy systems, from carbon-intensive development to decarbonization. Similar storyline variations were considered for other scenario families, but they did not result in genuine scenario groupings within the respective families. However, if future energy systems variations were applied fully to other storylines, they may evolve differently from those in A1. They have been introduced into the A1 storyline because of its "high growth with high technology" nature, for which differences in alternative technology developments translate into large differences in future GHG emission levels. The A1 groups further increased the richness in different GHG and SO_2 emissions paths. Indeed, this variation in the structure of future energy systems in itself resulted in a range of emissions almost as large as that generated through the variation of other main driving forces, such as population and economic development. Altogether the 40 SRES scenarios fall into seven groups: the three scenario families A2, B1, and B2, plus four groups within the A1 scenario (see footnote 2).

As in the case of the storylines, no single scenario – whether it represents a modeler's choice or harmonized assumptions – was treated as being more or less "probable" than others belonging to the same family. However, one preliminary harmonized scenario from each family, referred to as a "marker," was used in 1998 to solicit comments during the "open process" and as input for climate modelers in accordance with a decision of the IPCC Bureau. The four marker scenarios were posted on the IPCC web site (sres.ciesin.org) in June 1998, and the open scenario review process through the IPCC web site lasted until January 1999. The choice of markers was based on extensive discussion of:

- Range of emissions across all of marker scenarios.
- Which of the initial quantifications (by the modelers) reflected the storyline.
- Preference of some of the modeling teams and features of specific models.
- Use of different models for the four markers.

Markers were not intended to be the median or mean scenarios from their respective families. Indeed, in general it proved impossible to develop scenarios in which all relevant characteristics matched mean or median values. Thus, marker scenarios are no more or less likely than any other scenarios, but are those scenarios considered by the SRES writing team as illustrative of a particular storyline. These scenarios have received much closer scrutiny, not only from the entire writing team, but also via the SRES open process, than other scenario quantifications. The marker scenarios are also the SRES scenarios that have been most intensively tested in terms of reproducibility. As a rule, different modeling teams have attempted to replicate the model quantification of marker scenarios. Available time and resources have not allowed a similar exercise to be conducted for all SRES scenarios, although some effort was devoted to reproduce the scenario groups that constitute different interpretations of the A1 storyline with different models.

Additional scenarios using the same harmonized assumptions as the marker scenarios developed by different modeling teams and other scenarios that give alternative quantitative interpretations of the four storylines constitute the final set of 40 SRES scenarios. However, differences in modeling approaches mean that not all the scenarios provide estimates for all the direct and indirect GHG emissions for all the sources and sectors. The four SRES marker scenarios cover all the relevant gas species and emission categories comprehensively and thus constitute the smallest set of independent and fully documented SRES scenarios.

The scenario groups and cumulative emissions categories were developed as the smallest subsets of SRES scenarios that capture the range of uncertainties associated with driving forces and emissions. Together, the scenario groups constitute the set of SRES scenarios that reflects the uncertainty ranges in the emissions and their driving forces. Furthermore, the writing team recommends that, to the extent possible, these scenarios, but at least the four markers, be used to capture the range of uncertainties of driving forces and in addition, the two additional illustrative scenarios in A1 be used to capture the range of GHG emissions, and these should always be used together, and that no individual scenario should be singled out for any purpose. Multiple baselines and overlapping emissions ranges have important implications for making policy analysis (e.g., similar policies might have different impacts in different scenarios). Combinations of policies might shape the future development in the direction of certain scenarios. Box 6-4 (see later) summarizes the recommendations of the writing team for consideration by the user communities within and outside the IPCC.

Thus, there are three different types of scenarios within each family – one marker (and two illustrative scenarios in the A1 family), a set of harmonized scenarios, and a set of other (non-harmonized) scenarios. In addition, the A1 family of scenarios is subdivided into groups that describe alternative technological developments in the energy system. Together with the other three scenario families the SRES scenarios build seven distinct scenario groups (see footnote 2). Figure 6-1 illustrates this scenario terminology schematically. The

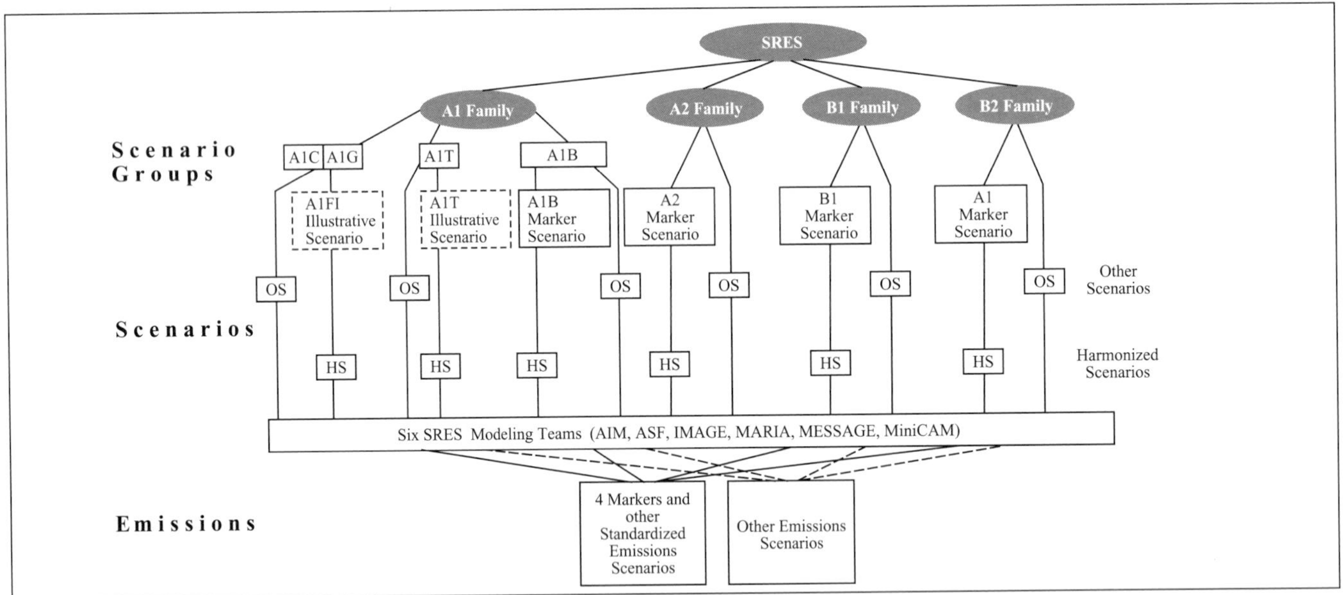

Figure 6-1: Schematic illustration of SRES scenarios. The set of scenarios consists of the four scenario families A1, A2, B1, and B2. Each family consists of a number of scenarios, some of which have "harmonized" driving forces and share the same prespecified population and gross world product (a few that also share common final energy trajectories are called "fully harmonized"). These are marked as "HS" for harmonized scenarios. One of the harmonized scenarios, originally posted on the open-process web site, is called a "marker scenario." All other scenarios of the same family based on the quantification of the storyline chosen by the modeling team are marked as "OS." Six modeling groups developed the set of 40 emissions scenarios. The GHG and SO_2 emissions of the scenarios were standardized to share the same data for 1990 and 2000 on request of the user communities. The time-dependent standardized emissions were also translated into geographic distributions. See also footnote 2.

detailed descriptions of inputs and outputs (other than GHG emissions) of the SRES marker scenarios, other harmonized scenarios, and all other scenarios are presented in Chapter 4 and the Appendices, while the emissions of GHGs and other radiatively important species of gases are described in Chapter 5 and Appendices.

The writing team considers that the SRES scenario set (in all the richness of scenario families, groups, markers, and illustrative and harmonized scenarios) is based on a "neutral" choice of scenario drivers; no driver is unduly emphasized as being more important than others. The scenarios do not suggest that future population growth alone is *the* driver of future emissions, nor do they suggest that technological change *alone* in any one sector could drive future emissions in one way or the other. While recognizing the importance of any of these driving forces *per se,* this report illustrates the critical role of relationships and interdependencies between scenario driving forces. To an extent it is the nature of these relationships that drives the future more than the possible evolution of any individual driving forces by itself. In other words, the uncertainty of the future is not simply parametric, but deeply functional; uncertainties and incomplete understanding exist for both. Qualitative scenario storylines add transparency and consistency to the relationships assumed in any particular scenario. The storylines also allow for additional interpretation of scenario results by different user communities.

6.2. Scenario Driving Forces

The scenarios cover a wide range of driving forces, from demographic to social and economic developments. This section summarizes the assumptions on important scenario drivers. For simplicity, only three important driving forces are presented separately following the exposition in Chapters 2, 3, and 4. Nonetheless, it is important to keep in mind that the future evolution of these and other main driving forces is interrelated in the SRES scenarios (see Table 6.2a for a summary of the ranges of the main driving forces across the scenario groups in 2020, 2050, and 2100).

The SRES scenarios span a wide range of assumptions for the most salient scenario drivers, and thus reflect the uncertainty of the future. Evidently, views of the future are a time-specific phenomenon, and this report and its scenarios are no exception. However, it is important to emphasize that this is an explicit part of the Terms of Reference for the SRES writing team – to reflect a range of views, based on current knowledge and the most recently available literature (see Appendix I). The scenario quantification results reflect well the literature range, except for extreme scenarios.

6.2.1. Population Projections

Three different population trajectories were chosen for SRES scenarios to reflect future demographic uncertainties based on published population projections (Lutz, 1996; UN, 1998; see Chapter 3). The population projections are exogenous input to all the models used to develop the SRES scenarios. The models used do not develop population from other assumptions within the model. Figure 6-2 shows the three population projections in comparison with the three population projections used in the IS92 scenarios. Global population ranges between 7 and 15 billion people by 2100 across the scenarios, depending on the rate and extent of the demographic transition. The insert in Figure 6.2 shows population development in the industrialized (i.e., developed) regions. The range of future populations is smaller than in the IS92 scenarios, particularly in the industrialized regions, for which the lowest scenario indicates a very modest population decline compared to IS92 scenarios. The greatest uncertainty about future growth lies in the developing regions across all scenarios in the literature. An equally pervasive trend across all scenarios is urbanization (see Chapter 3). Altogether three different population projections were used in the 26 harmonized scenarios. Other scenarios explored alternative population projections consistent with the storylines.

The lowest population trajectory is assumed for the A1 and B1 scenario families and is based on the low population projection in Lutz (1996), which combines low fertility with low mortality and central migration rate assumptions. After peaking at 8.7 billion in the middle of the 21st century, world population declines to 7.1 billion by the year 2100. As discussed in Chapters 3 and 4, this population development is somewhat higher than the previous low population used in the IS92 scenarios. The B2 scenario family is based on the UN median 1998 population projection (UN, 1998). The global population increases to about 9.4 billion people by 2050 and to about 10.4 billion by 2100. This population scenario is characteristic of recent median global population projections, which describe a continuation of historical trends towards a completion of the demographic transition that would lead to a level global population, and is consistent with recent faster fertility declines in the world together with declining mortality rates. Hence, the population is somewhat lower than previous UN median projections, as used in the IS92 scenarios. This median scenario projects very low population growth in today's industrialized countries, with stabilization of growth in Asia in the second half of the 21st century and in the rest of the world towards the end of the 21st century. The A2 scenario family is based on the high population growth of 15 billion by 2100 reported in Lutz (1996), which assumes a significant decline in fertility for most regions and a stabilization at above replacement levels. It falls below the long-term 1998 UN high projection of 18 billion. It is also lower than in the highest IS92 scenario (17.6 billion by 2100). Nevertheless, this scenario represents very high population growth compared with that in current demographic literature. Demographers attach a probability of more than 90% that actual population will be lower than the trajectory adopted in the A2 scenario family (Lutz *et al.*, 1997). A more detailed discussion of the population projections used to quantify the four scenario families is given in Chapters 3 and 4.

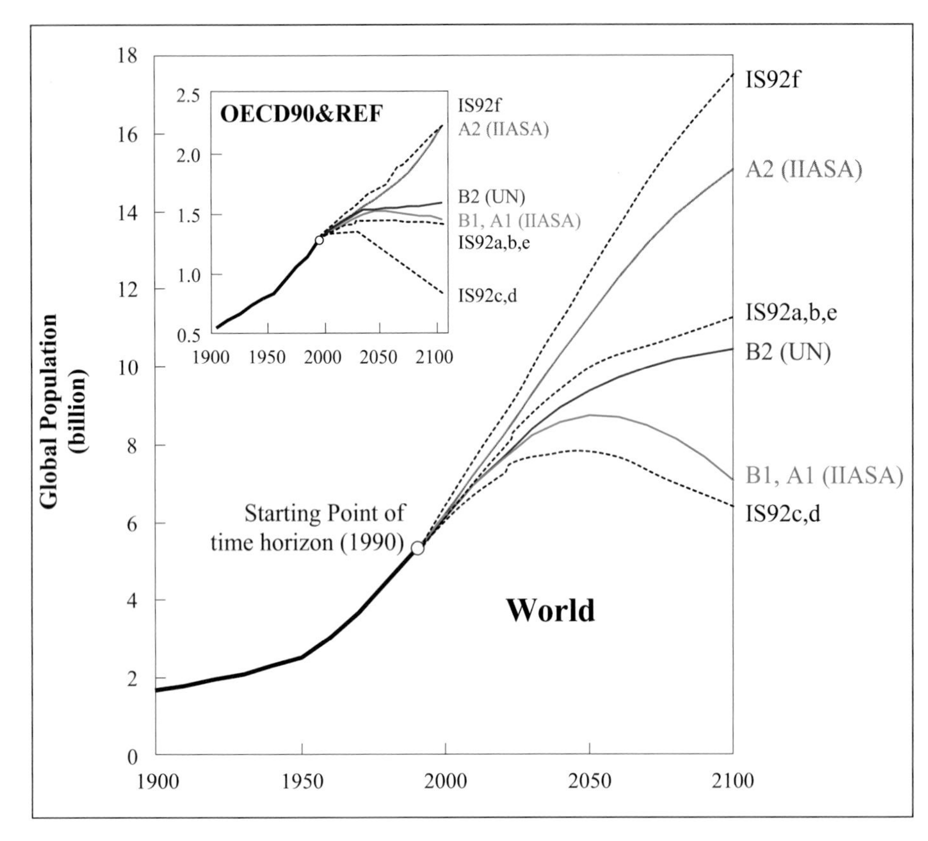

Figure 6-2: Population projections – historical data from 1900 to 1990 (based on Durand, 1967; Demeny; 1990; UN, 1998), SRES scenarios (based on Lutz, 1996, for high and low, and UN, 1998, for medium), and IPCC IS92 scenarios (Leggett *et al.*, 1992; Pepper *et al.*, 1992) from 1990 to 2100.

6.2.2. *Economic Development*

The SRES scenarios span a wide range of future levels of economic activity (expressed in gross world product). The A1 scenario family with a ("harmonized") gross world product of US$529 trillion (all values in 1990 US dollars unless otherwise indicated) in 2100 delineates the SRES upper bound, whereas B2 with ("harmonized") US$235 trillion in 2100 represents its lower bound. The range of gross world product across all scenarios is even higher, from US$197 to US$550 by 2100.

Although the SRES scenarios span a wide range, still lower and higher gross world product levels can be found in the literature (see Chapters 2, 3, and 4). Uncertainties in future gross world product levels are governed by the pace of future productivity growth and population growth, especially in developing regions. Different assumptions on conditions and possibilities for development "catch-up" and for narrowing per capita income gaps in particular explain the wide range in projected future gross world product levels. Given a qualitatively negative relationship between population growth and per capita income growth discussed in Chapters 2 and 3, uncertainties in future population growth rates tend to narrow the range of associated gross world product projections. High population growth would, *ceteris paribus*, lower per capita income growth, whereas low population growth would tend to increase it. This relationship is evident in empiric data – high per capita income countries are generally also those that have completed their demographic transition. The affluent live long and generally have few children. (Exceptions are some countries with small populations, high birth rates, and significant income from commodity exports.) This relationship between affluence and longevity again identifies development as one of the most important indicators of human well being. Yet even assuming this relationship holds for an extended time into the future, its quantification is subject to considerable theoretic and empiric uncertainties (Alcamo *et al.*, 1995).

Two of the SRES scenario families, A1 and B1, explicitly explore alternative pathways to gradually close existing income gaps. As a reflection of uncertainty, development "catch-up" diverges in terms of geographically distinct economic growth patterns across the four SRES scenario families. Table 6-1 summarizes per capita income for SRES and IS92 scenarios for the four SRES world regions. SRES scenarios indicate a smaller difference between the now industrialized and developing countries compared with the IS92 scenarios. This tendency toward a substantially narrower income "gap" compared with the IS92 scenarios overcomes one of the major shortcomings of the previous IPCC scenarios cited in the literature (Parikh, 1992).

6.2.3. *Structural and Technological Change*

In this brief summary of the SRES scenarios, structural and technological changes are illustrated by using energy and land use as examples. These examples are characteristic for the driving forces of emissions because the energy system and land use are the major sources of GHG and sulfur emission. Chapter 4 gives a more detailed treatment of the full range of emissions driving forces across the SRES scenarios.

***Table 6-1:** Income per capita in the world and by SRES region for the IS92 (Leggett et al., 1992) and four marker scenarios by 2050 and 2100, measured by GDP per capita in 1000 US dollars (at 1990 prices and exchange rates). The additional illustrative scenarios A1FI and A1T have GDP assumptions similar to the A1B marker, shared with all harmonized scenarios in the A1 family.*

Income per Capita by World and Regions (103 1990US$ per capita)

		Regions						
Year	**Scenario**	**OECD90**	**REF**	**IND**	**ASIA**	**ALM**	**DEV**	**WORLD**
1990	SRES MESSAGE	19.1	2.7	13.7	0.5	1.6	0.9	4.0
2050	IS92a,b	49.0	23.2	39.7	3.7	4.8	4.1	9.2
	IS92c	35.2	14.6	27.4	2.2	2.9	2.5	6.3
	IS92d	54.4	25.5	43.4	4.1	5.4	4.6	10.5
	IS92e	67.4	38.3	56.9	5.9	7.7	6.6	13.8
	IS92f	43.9	21.5	35.8	3.3	4.1	3.6	8.1
	A1B	50.1	29.3	44.2	14.9	17.5	15.9	20.8
	A2	34.6	7.1	26.1	2.6	6.0	3.9	7.2
	B1	49.8	14.3	39.1	9.0	13.6	10.9	15.6
	B2	39.2	16.3	32.5	8.9	6.9	8.1	11.7
2100	IS92a,b	85.9	40.6	69.5	15.0	14.2	14.6	21.5
	IS92c	49.2	17.6	36.5	6.4	5.8	6.1	10.1
	IS92d	113.9	51.3	88.8	20.3	17.7	19.1	28.2
	IS92e	150.6	96.6	131.0	34.6	33.0	33.8	46.0
	IS92f	69.7	31.3	54.9	11.9	10.7	11.4	16.8
	A1B	109.2	100.9	107.3	71.9	60.9	66.5	74.9
	A2	58.5	20.2	46.6	7.8	15.2	11.0	16.1
	B1	79.7	52.2	72.8	35.7	44.9	40.2	46.6
	B2	61.0	38.3	54.4	19.5	16.1	18.0	22.6

6.2.3.1. Energy Systems

Figure 6-3 illustrates that the change of world primary energy structure diverges over time. It shows the contributions of individual primary energy sources – the percentage supplied by coal, that by oil and gas, and that by all non-fossil sources taken together (for simplicity of presentation and because not all models distinguish between renewables and nuclear energy). Each corner of the triangle corresponds to a hypothetical situation in which all primary energy is supplied by a single source – oil and gas, coal at the left, and non-fossil sources (renewables and nuclear) to the right. Historically, the primary energy structure has evolved clockwise according to the two "grand transitions" (discussed in Chapter 3) that are shown by the two segments of the "thick black" curve. From 1850 to 1920 the first transition can be characterized as the substitution of traditional (non-fossil) energy sources by coal. The share of coal increased from 20% to about 70%, while the share of non-fossils declined from 80% to about 20%. The second transition, from 1920 to 1990, can be characterized as the replacement of coal by oil and gas (while the share of non-fossils remained essentially constant). The share of oil and gas increased to about 50% and the share of coal declined to about 30%.

Figure 6-3 gives an overview of the divergent evolution of global primary energy structures between 1990 and 2100, regrouped into their respective scenario families and four A1 scenarios groups that explore different technological developments in the energy systems. The SRES scenarios cover a wider range of energy structures than the previous IS92 scenario series, which reflects advances in knowledge on the uncertainty ranges of future fossil resource availability and technological change.

In a clockwise direction, A1 and B1 scenario groups map the structural transitions toward higher shares of non-fossil energy in the future, which almost closes the historical "loop" that started in 1850. The B2 scenarios indicate a more "moderate" direction of change with about half of the energy coming from non-fossil sources and the other half shared by coal on one side and oil and gas on the other. Finally, the A2 scenario group marks a stark transition back to coal. Shares of oil and gas decline while non-fossils increase moderately. What is perhaps more significant than the diverging developments in these three marker scenarios is that the whole set of 40 scenarios covers virtually all possible directions of change, from high shares of oil and gas to high shares of coal and non-

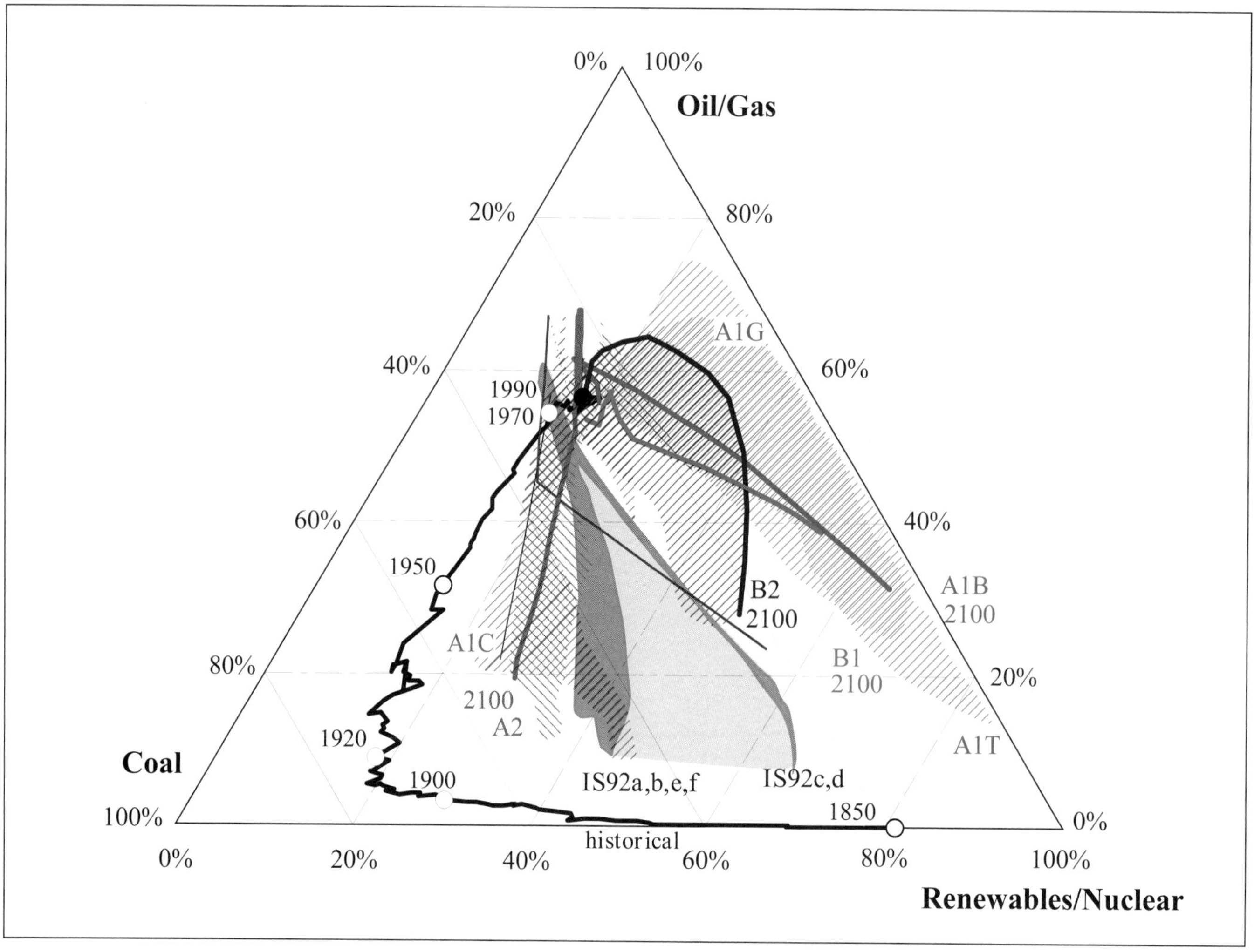

Figure 6-3: Global primary energy structure, shares (%) of oil and gas, coal, and non-fossil (zero-carbon) energy sources – historical development from 1850 to 1990 and in SRES scenarios. Each corner of the triangle corresponds to a hypothetical situation in which all primary energy is supplied by a single source – oil and gas on the top, coal to the left, and non-fossil sources (renewables and nuclear) to the right. Constant market shares of these energies are denoted by their respective isoshare lines. Historical data from 1850 to 1990 are based on Nakićenović *et al.* (1998). For 1990 to 2100, alternative trajectories show the changes in the energy systems structures across SRES scenarios. They are grouped by shaded areas for the scenario families A1, A2, B1, and B2 with respective markers shown as lines. In addition, the four scenario groups within the A1 family, A1, A1C, A1G, and A1T, that explore different technological developments in the energy systems, are shaded individually. In the SPM, the A1C and A1G scenario groups are combined into the fossil-intensive A1FI scenario group. For comparison the IS92 scenario series are also shown, clustering along two trajectories (IS92c,d and IS92a,b,e,f). For model results that do not include non-commercial energies, the corresponding estimates from the emulations of the various marker scenarios by the MESSAGE model were added to the original model outputs.

fossils. In particular, the A1 scenario family covers basically the same range of structural change as all other scenarios together. In contrast, the IS92 scenarios cluster into two groups; one contains IS92c and IS92d and the other the four others. In all of these the share of oil and gas declines, and the main structural change occurs between coal on the one hand and non-fossils on the other. This divergent nature in the structural change of the energy system and in the underlying technological base of the SRES results in a wide span of future GHG and sulfur emissions.

6.2.3.2. *Land-use Patterns*

Figure 6-4 illustrates that the land-use patterns diverge over time. It shows the main land-use categories – the percentages of total land area use that constitute the forests, the joint shares of cropland and energy biomass, and all other categories including grasslands. As for the energy triangle in Figure 6-3, in Figure 6-4 each corner corresponds to a hypothetical situation in which land use is dedicated to a much greater extent than today to two of the three land-use categories: 40%

to cropland and energy biomass and 20% to forests at the top, 60% to forests and 40% to other categories (including grasslands) to the left, and 80% to other categories (including grasslands) to the right.

In most scenarios, the current trend of shrinking forests is eventually reversed because of slower population growth and increased agricultural productivity. Reversals of deforestation trends are strongest in the B1 and A1 families. In the B1 family pasture lands decrease significantly because of increased productivity in livestock management and dietary shifts away from meat, thus illustrating the importance of both technological and social developments.

The main driving forces for land-use changes are related to increasing demands for food because of a growing population and changing diets. In addition, numerous other social, economic, and institutional factors govern land-use changes such as deforestation, expansion of cropland areas, or their reconversion back to forest cover (see Chapter 3). Global food production can be increased, either through intensification (by multi-cropping, raising cropping intensity, applying fertilizers, new seeds, improved farming technology) or through land expansion (cultivating land, converting forests). Especially in developing countries, there are many examples of the potential to intensify food production in a more or less ecologic way (e.g. multi-cropping; agroforestry) that may not lead to higher GHG emissions.

Different assumptions on these processes translate into alternative scenarios of future land-use changes and GHG emissions, most notably CO_2, CH_4, and N_2O. A distinguishing

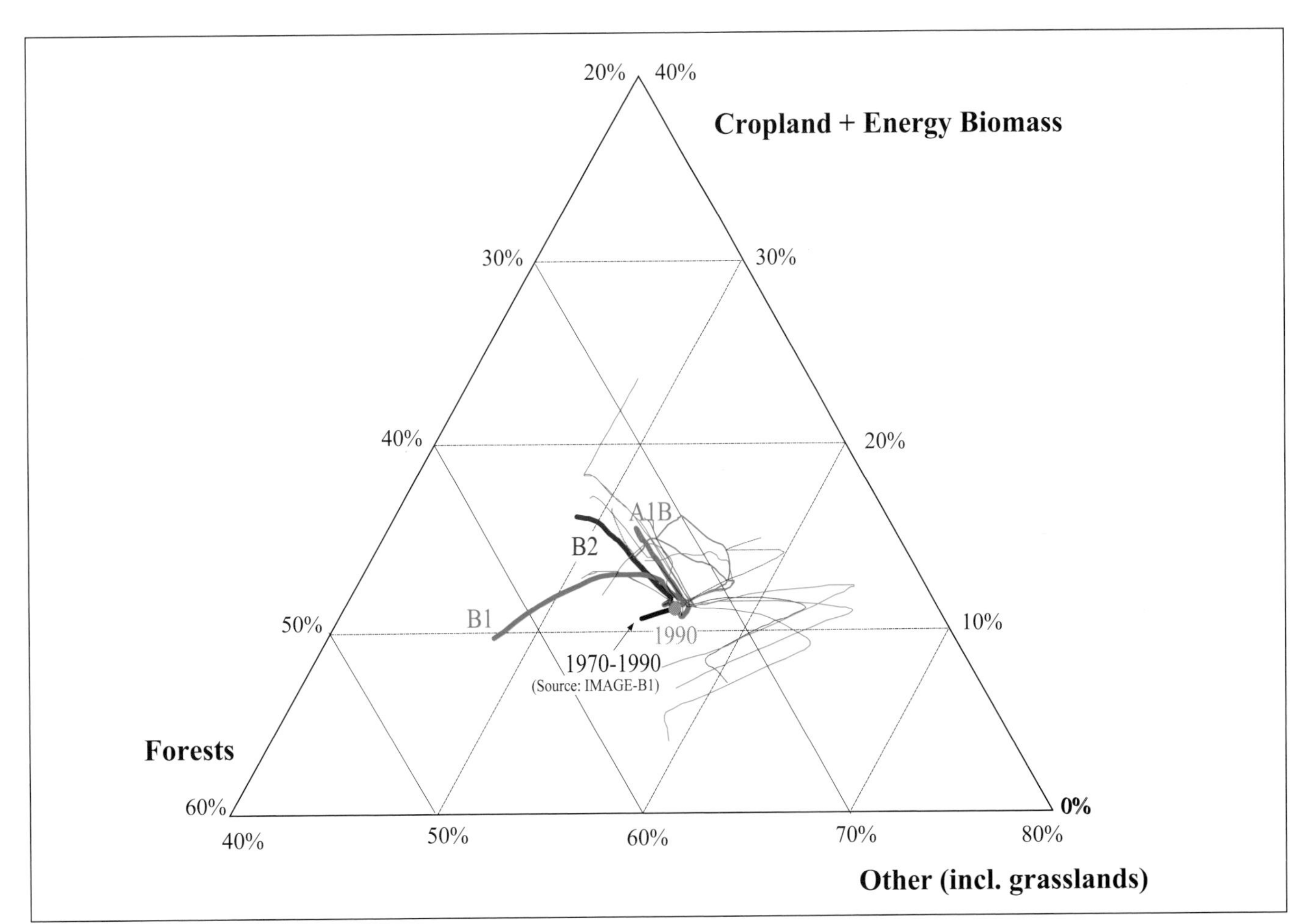

Figure 6-4: Global land-use patterns, shares (%) of croplands and energy biomass, forests, and other categories including grasslands – historical development from 1970 to 1990 (based on B1-IMAGE) and in SRES scenarios. As for the energy triangle in Figure 6-3, each corner corresponds to a hypothetical situation in which land use is dedicated to a much greater extent than today to one category – 60% to cropland and energy biomass at the top, 80% to forests to the left, and 80% to other categories (including grasslands) to the right. Constant shares in total land area of cropland and energy biomass, forests, and other categories are denoted by their respective isoshare lines. For 1990 to 2100, alternative trajectories are shown for the SRES scenarios. The three marker scenarios A1B, B1, and B2 are shown as thick colored lines, and other SRES scenarios as thin colored lines. The ASF model used to develop the A2 marker scenario projects only land-use change related GHG emissions. Comparable data on land cover changes are therefore not available.The trajectories appear to be largely model specific and illustrate the different views and interpretations of future land-use patterns across the scenarios (e.g. the scenario trajectories on the right that illustrate larger increases in grasslands and decreases in cropland are MiniCAM results).

characteristic of several models (e.g., AIM, IMAGE, MARIA, and MiniCAM) used in SRES is the explicit modeling of land-use changes caused by expanding biomass uses and hence exploration of possible land-use conflicts between energy and agricultural sectors. The corresponding scenarios of land-use changes are illustrated in Figure 6-4 for all SRES scenarios. In some contrast to the structural changes in energy systems shown in Figure 6-3, different land-use scenarios in Figure 6-4 appear to be rather model specific, following the general trends as indicated by the respective marker scenario developed with a particular model.

6.3. Greenhouse Gases and Sulfur Emissions

The SRES scenarios generally cover the full range of GHG and sulfur emissions consistent with the storylines and the underlying range of driving forces from studies in the literature, as documented in the SRES database. This section summarizes the emissions of CO_2, CH_4, and SO_2. For simplicity, only these three important gases are presented separately, following the more detailed exposition in Chapter 5 (see Table 6.2b for a summary of the ranges of emissions across the scenario groups in 2020, 2050, and 2100).

6.3.1. Carbon Dioxide Emissions

6.3.1.1. Emissions from Energy, Industry, and Land Use

Figure 6-5 illustrates the range of CO_2 emissions for the 40 SRES scenarios against the background of all the emissions scenarios in the SRES scenario database shown in Figure 1-3. For simplicity, only energy-related and industrial sources of CO_2 emissions are shown.

Figure 6-5 shows that the marker scenarios by themselves cover a large portion of the overall scenario distribution. This is one reason why the SRES writing team recommends the use of at least the four marker scenarios. Together, they cover a large range of future emissions, both with respect to the scenarios in the literature and the full SRES scenario set.

The SRES scenarios cover rather evenly the range of future emissions found in the literature, from high to low levels over the whole time horizon. In contrast, the distribution for emissions by 2100 of scenarios in the literature is very asymmetric. It has a structure that resembles a tri-modal frequency distribution – those showing emissions of more than 30 gigatons of carbon (GtC; 20 scenarios), those with emissions between 12 and 30 GtC (88 scenarios), and those

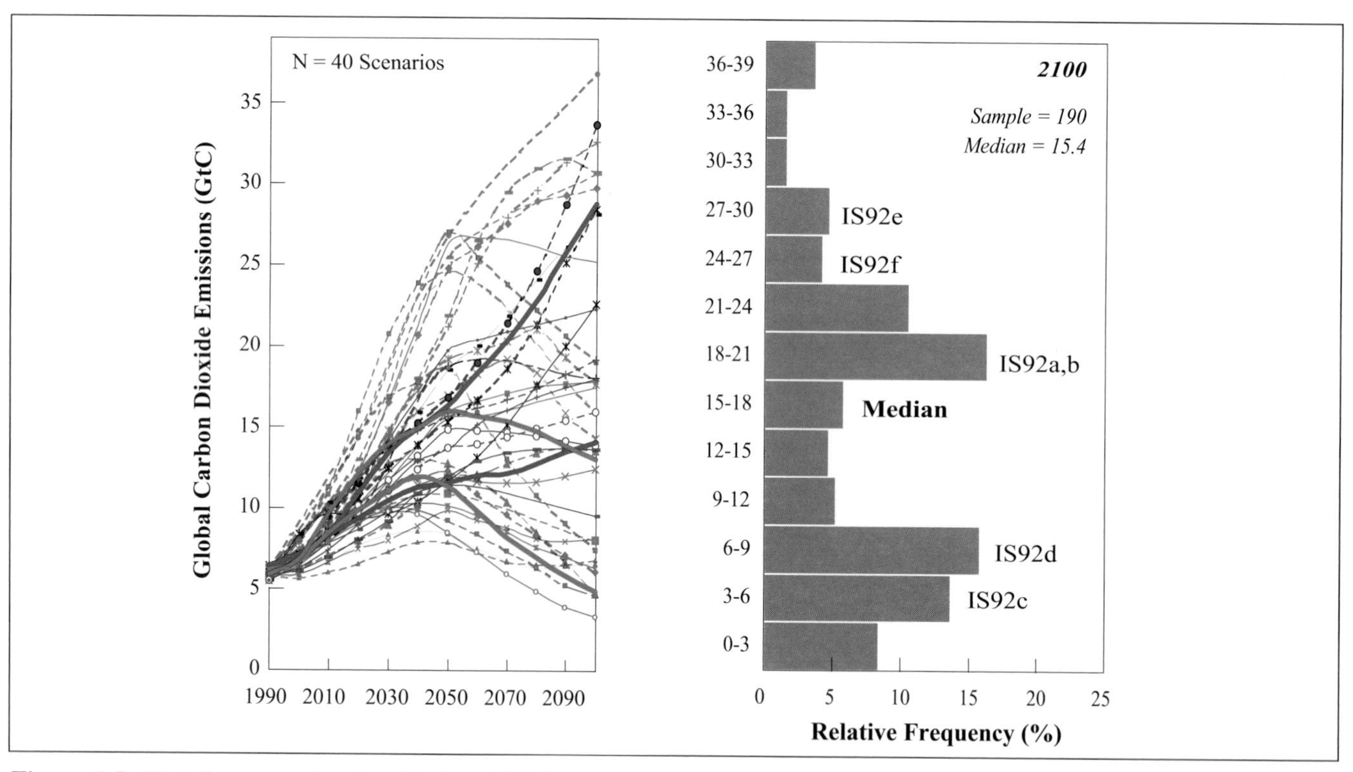

Figure 6-5: Global energy-related and industrial CO_2 emissions for the 40 SRES scenarios. The individual scenarios are shown grouped into four scenario families. Marker scenarios are shown as bold continuous lines and the other 36 scenarios as dashed lines. The emissions profiles are dynamic, ranging from continuous increases to those that curve through a maximum and then decline. The relative positions of the scenarios change in time, with numerous cross-overs among the individual emissions trajectories. The histogram on the right shows, for comparison, the frequency distribution of energy-related and industrial CO_2 emissions based on the scenario database. The histogram indicates the relative position of the four marker scenarios and the six IS92 scenarios compared to the emissions in the literature. Jointly, the SRES scenarios span most of the range of scenarios in the literature.

showing emissions of less than 12 GtC (82 scenarios). As discussed in Chapter 2, the lowest cluster appears to include many of the intervention scenarios; the second and third clusters are most likely the non-intervention cases. The lowest cluster may have been influenced by many analyses of stabilizing atmospheric concentrations. The middle cluster echoes the many analyses that took IS92a as a reference and is testament to the enormous influence of the IS92 series on emissions assessments in general.

The range of CO_2 and other GHG emissions for the four marker scenarios is generally somewhat lower than that of the six IS92 scenarios.[3] However, the IS92 scenarios do not cover the "middle" range of emissions where the median and the average of all scenarios in the literature are situated. Adding the other 36 scenarios to the four SRES markers increases the covered emissions range beyond the IS92 series at the high end of the distribution but not at the low end. SRES scenarios stop short of the lower literature emissions because they are scenarios without additional climate initiatives (as per the Terms of References, see Appendix I).

Figure 6-6 illustrates the range of CO_2 emissions of the SRES scenarios against the background of all the IS92 scenarios and other emissions scenarios from the literature documented in the SRES scenario database. The shaded areas depict the range of the scenarios in the database that exceeds the SRES emissions range. The range of future emissions is very large so that the highest scenarios envisage more than a sevenfold increase of global emissions by 2100, while the lowest have emissions lower than today.

The literature includes scenarios with additional climate initiatives and policies, which are also referred to as mitigation or intervention scenarios. As shown in Chapter 2, many ambiguities are associated with the classification of emissions scenarios into those that include additional climate initiatives and those that do not. Many cannot be classified in this way on basis of the information available from the SRES scenario database and the published literature.

Figure 6-6a indicates the ranges of emissions from energy and industry in 2100 from scenarios that apparently include additional climate initiatives (designated as intervention emissions range), those that do not (non-intervention), and those that cannot be assigned to either of these two categories (non-classified). This classification is based on the subjective evaluation of the scenarios in the database by the members of the writing team and is explained in Chapter 2. The range of the whole sample of scenarios has significant overlap with the range of those that cannot be classified and they share virtually the same median (15.7 and 15.2 GtC in 2100, respectively), but the non-classified scenarios do not cover the high part of the range. Also, the range of the scenarios that apparently do not include climate polices (non-intervention) has considerable overlap with the other two ranges (lower bound is slightly higher), but with a significantly higher median (of 21.3 GtC in 2100).

The median of all energy and industry emissions scenarios from the literature is 15.7 GtC by 2100. This is lower than the median of the IS92 set and is lower than the IS92a scenario often (inappropriately) considered as the "central" scenario. Again, the distribution of emissions is asymmetric (see the emissions histogram in Figure 6-5) and the thin tail that extends above 30 GtC includes only a few scenarios.

Figure 6-6 shows the range of emissions of the four families (vertical bars next to each of the four marker scenarios), which illustrate that the scenarios groups by themselves cover a large portion of the overall scenario distribution. Together, they cover much of the range of future emissions, both with respect to the scenarios in the literature and all SRES scenarios. Adding all other scenarios increases the covered range. For example, the SRES scenarios span jointly from the 95th percentile to just above the 5th percentile of the distribution of energy and industry emissions scenarios from the literature. This illustrates again that they only exclude the most extreme emissions scenarios found in the literature, which are situated out in the tails of the distribution. What is perhaps more important is that each of the four scenario families covers a substantial part of this distribution. This leads to a substantial overlap in the emissions ranges of the four scenario families. In other words, a similar quantification of driving forces can lead to a wide range of future emissions and a given level of future emissions can result from different combinations of driving forces. This result is of fundamental importance for the assessments of climate change impacts and possible mitigation and adaptation strategies. Thus, it warrants some further discussion.

Another interpretation is that a given combination of the main driving forces, such as population and economic growth, is not sufficient to determine the future emissions paths. Different modeling approaches and different specifications of other scenario assumptions overshadow the influence of the main driving forces. A particular combination of driving forces, such as specified in the A1 scenario family, is associated with a whole range of possible emission paths for energy and industry. The nature of climate change impacts and adaptation and mitigation strategies would be fundamentally different depending on whether emissions are high or low, given a particular combination of scenario driving forces. Thus, the implication is that the whole range needs to be considered in the assessments of climate change, from high emissions and driving forces to low ones.

The A1 scenario family explored variations in energy systems most explicitly and hence covers the largest part of the scenario distribution shown in Figures 6-5 and 6-6a, from the 95th to just above the 10th percentile. The A1 scenario family includes

[3] This is still true when the two illustrative cases in the A1 family - as selected for the Summary for Policymakers, see footnote 2 - are added.

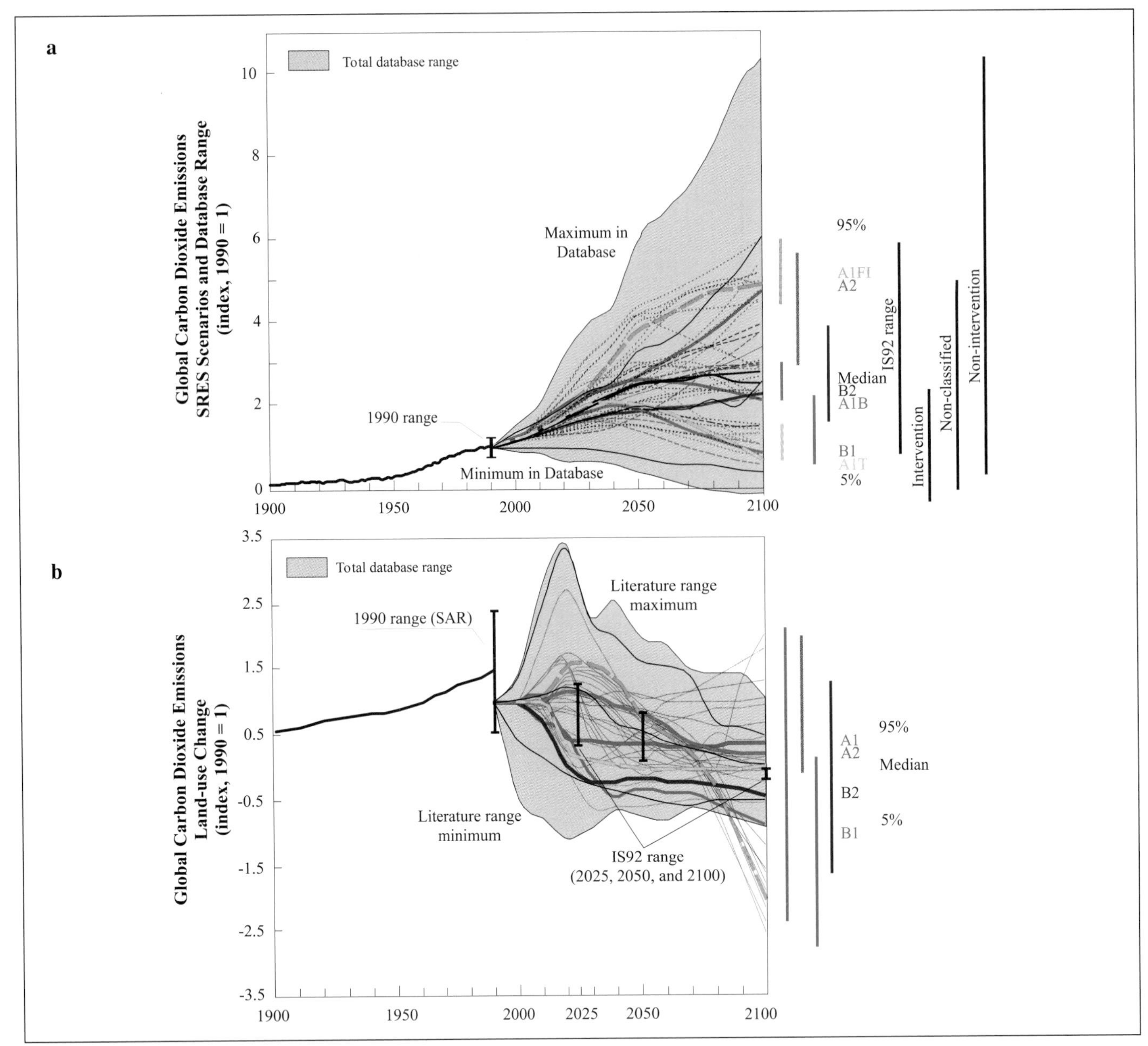

Figure 6-6: Global CO_2 emissions from energy and industry in Figure 6-6a and from land-use change in Figure 6-6b – historical development from 1900 to 1990 and in 40 SRES scenarios from 1990 to 2100, shown as an index (1990 = 1). The range is large in the base year 1990, as indicated by an "error" bar, but is excluded from the indexed future emissions paths. The dashed time-paths depict individual SRES scenarios and the shaded area the range of scenarios from the literature (as documented in the SRES database). The median (50th), 5th, and 95th percentiles of the frequency distribution are shown. The statistics associated with the distribution of scenarios do not imply probability of occurrence (e.g., the frequency distribution of the scenarios in the literature may be influenced by the use of IS92a as a reference for many subsequent studies). The 40 SRES scenarios are classified into seven groups that constitute four scenario families. Jointly the scenarios span most of the range of the scenarios in the literature. The emissions profiles are dynamic, ranging from continuous increases to those that curve through a maximum and then decline. The colored vertical bars indicate the range of the four SRES scenario families in 2100. Also shown as vertical bars on the right of Figure 6-6a are the ranges of emissions in 2100 of IS92 scenarios and of scenarios from the literature that apparently include additional climate initiatives (designated as "intervention" scenarios emissions range), those that do not ("non-intervention"), and those that cannot be assigned to either of these two categories ("non-classified"). This classification is based on a subjective evaluation of the scenarios in the database by the members of the writing team and is explained in Chapter 2. It was not possible to develop an equivalent classification for land-use emissions scenarios. Three vertical bars in Figure 6-6b indicate the range of IS92 land-use emissions in 2025, 2050 and 2100. Classification of land-use change emission scenarios similar to that for energy and industry emissions was not possible.

different groups of scenarios that explore different structures of future energy systems, from carbon-intensive development paths to high rates of decarbonization. All groups otherwise share the same assumptions about the main driving forces (see Section 6.2.3 and, for further detail, Chapters 4 and 5). This indicates that different structures of the energy system can lead to basically the same variation in future emissions as generated by different combinations of the other main driving forces – population, economic activities, and energy consumption levels. The implication is that decarbonization of energy systems – the shift from carbon-intensive to less carbon-intensive and carbon-free sources of energy – is of similar importance in determining the future emissions paths as other driving forces. Sustained decarbonization requires the development and successful diffusion of new technologies. Thus investments in new technologies during the coming decades might have the same order of influence on future emissions as population growth, economic development, and levels of energy consumption taken together.

Figure 6-6b shows that CO_2 emissions from deforestation peak in many SRES scenarios after several decades and subsequently gradually decline. This pattern is consistent with many scenarios in the literature and can be associated with slowing population growth and increasing agricultural productivity. These allow a reversal of current deforestation trends, leading to eventual CO_2 sequestration. Emissions decline fastest in the B1 family. Only in the A2 family do net anthropogenic CO_2 emissions from land use remain positive through to 2100. As was the case for energy-related emissions, CO_2 emissions related to land-use in the A1 family cover the widest range. The range of land-use emissions across the IS92 scenarios is narrower in comparison.

6.3.1.2. *Four Categories of Cumulative Emissions*

This comparison of some of the SRES scenario characteristics implies that similar future emissions can result from very different socio-economic developments, and similar developments of driving forces can result in different future emissions. Uncertainties in the future development of key emission driving forces create large uncertainties in future emissions, even within the same socio-economic development paths. Therefore, emissions from each scenario family overlap substantially with emissions from other scenario families. Figure 6-6 shows this for CO_2 emissions. For example, comparison of the A1B and B2 marker scenarios indicates that they have similar emissions of about 13.5 and 13.7 GtC by 2100, respectively. The dynamics of the paths, however, are different so that they have different cumulative CO_2 emissions and different emissions of other GHG gases and SO_2.

To facilitate comparisons of emissions and their driving forces across the scenarios, the writing team grouped them into four categories of cumulative emissions between 1990 and 2100. However, any categorization of scenarios based on emissions of multiple gases is quite difficult. Figure 6-7 shows total CO_2 emissions from all sources (from Figures 6-6a and b). Most of the scenarios are shown aggregated into seven groups, the four A1 groups and the other three families. The scenarios that remain outside the seven groups adopted alternative interpretations of the four scenario storylines. The emission trajectories ("bands") of the seven groups display different dynamics, from monotonic increases to non-linear trajectories in which there is a subsequent decline from a maximum. The dynamics of the individual scenarios are also different across gasses, sectors, or world regions. This particularly diminishes the significance of focusing scenario categorization on any given year, such as 2100. In addition, all gases that contribute to radiative forcing should be considered, but methods of combining gases such as the use of global warming potentials (GWP) are appropriate only for near-term GHG inventories.[4] In light of these difficulties, the classification approach presented here uses cumulative CO_2 emissions between 1990 and 2100. CO_2 is the dominant GHG and cumulative CO_2 emissions are expected to be roughly proportional to CO_2 radiative forcing over the time scale of a century. According to the IPCC SAR, "any eventual stabilised concentration is governed more by the accumulated anthropogenic CO_2 emissions from now until the time of stabilisation than by the way emissions change over the period" (Houghton *et al.*, 1996). Therefore, the writing team also grouped the scenarios according to their cumulative emissions (see Figure 6.8).

This categorization can guide comparisons using either scenarios with different driving forces yet similar emissions, or scenarios with similar driving forces but different emissions. This characteristic of SRES scenarios also has very important implications for the assessment of climate-change impacts, mitigation, and adaptation strategies. Two future worlds with fundamentally different characteristic features, such as the A1 and B2 marker scenarios, also have different cumulative CO_2 emissions, but very similar CO_2 emissions in 2100. In contrast, scenarios that are in the same category of cumulative emissions can have fundamentally different driving forces and different CO_2 emissions in 2100, but very similar cumulative emissions. Presumably, adverse impacts and effective adaptation measures would vary among the scenarios from different families that share similar cumulative emissions, but have different demographic, socio-economic, and technological driving forces. This is another reason for considering the entire range of emissions in future assessments of climate change.

Figure 6-9 shows the histogram of cumulative CO_2 emissions from 1990 to 2100 for the SRES scenarios subdivided into the four emissions categories. Relative positions of the four marker scenarios, and the ranges of the four families and the six IS92 scenarios, are marked. The SRES scenarios have a bimodal

[4] In particular, the IPCC WGI Second Assessment Report (SAR) GWPs are calculated for constant concentrations (Houghton *et al.*, 1996). In long-term scenarios, concentrations may change significantly, as do GWP values. It is unclear how to apply GWPs to long-term scenarios in a meaningful manner. In addition, the GWP approach is not applicable to gases such as SO_2 and ozone precursors.

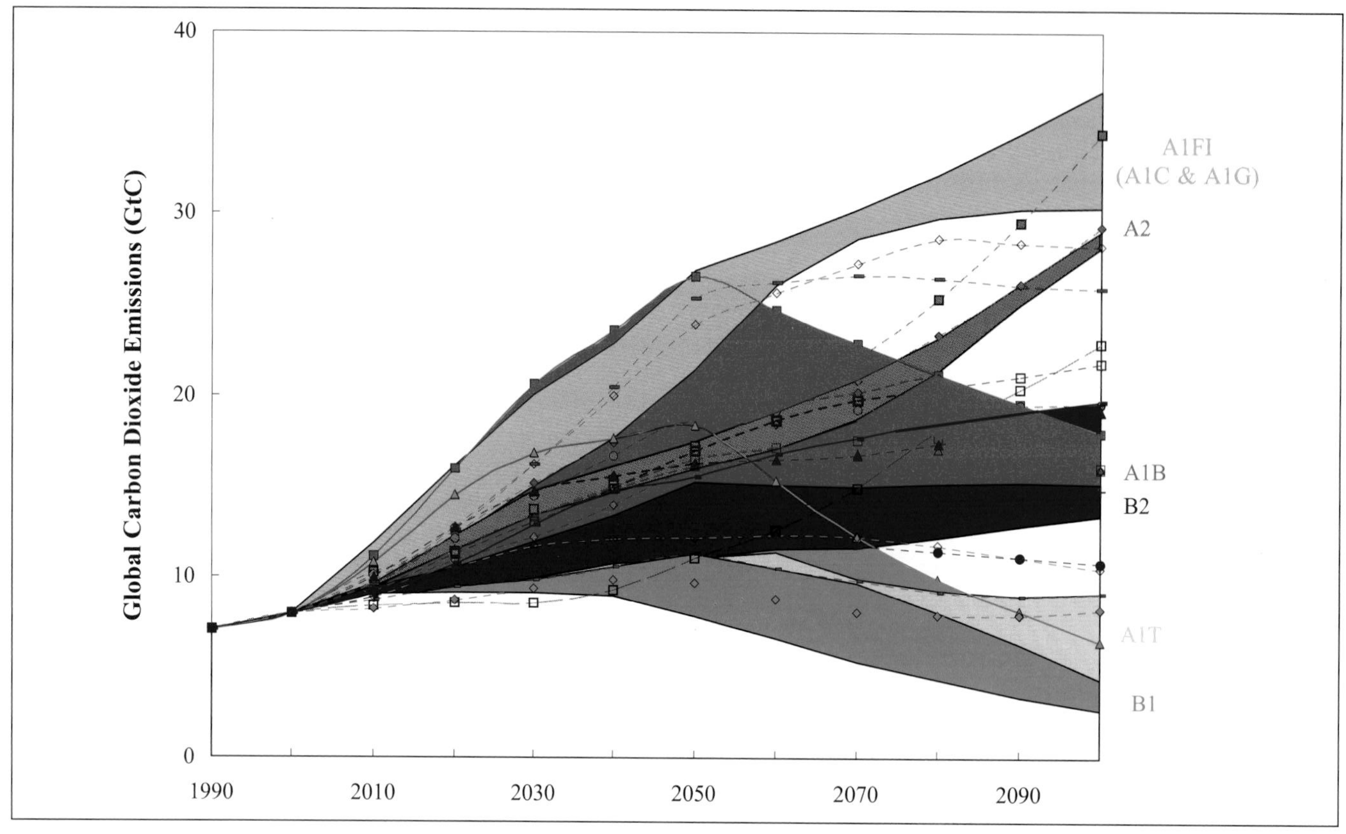

Figure 6-7: Global CO_2 emissions (GtC, standardized) from all sources for the four scenario families from 1990 to 2100. Scenarios are also presented for the four constituent groups of the A1 family (high-coal A1C, high oil and gas A1G, high non-fossil fuel A1T, and the balanced A1B) and for the other three families (A2, B1, and B2), forming seven scenario groups altogether. The emissions of A1C and A1G scenario groups are combined into A1FI (see footnote 2). Each colored emission band shows the range of the harmonized scenarios within one group that share common global input assumptions for population and GDP. The scenarios remaining outside the seven groups adopted alternative interpretations of the four scenario storylines.

structure similar to the distribution of cumulative emissions from the scenarios in the literature. The groups of SRES scenarios span the whole range of cumulative emissions from the scenarios in the literature. The range of the four markers is from 1000 to 1900 GtC. In comparison, the IS92 cumulative emissions range from 700 GtC for IS92c to 2140 GtC for IS92e.

The SRES emissions scenarios encompass emissions of other GHGs and chemically active species such as carbon monoxide, nitrogen oxides, and non-methane volatile organic compounds. The emissions of other gases follow dynamic patterns much like those shown in Figures 6-5 and 6-6 for CO_2 emissions. Further details of GHG emissions are given in Chapter 5.

6.3.2. Other Greenhouse Gases

Of the GHGs, CO_2 is the main contributor to anthropogenic radiative forcing because of changes in concentrations from pre-industrial times. According to Houghton *et al.* (1996) well-mixed GHGs (CO_2, CH_4, N_2O, and the halocarbons) induced additional radiative forcing of around 2.5 W/m^2 on a global and annually averaged basis. CO_2 accounted for 60% of the total, which indicates that the other GHGs are significant as well. Whereas CO_2 emissions are by-and-large attributable to two major sources, energy consumption and land-use change, other emissions arise from many different sources and a large number of sectors and applications (e.g. see Table 5-3 in Chapter 5).

The SRES emissions scenarios also have different emissions for other GHGs and chemically active species such as carbon monoxide, nitrogen oxides, and non-methane volatile organic compounds. The uncertainties that surround the emissions sources of these gases, and the more complex set of driving forces behind them are considerable and unresolved. Hence, model projections of these gases are particularly uncertain and the scenarios presented here are no exception. Improved inventories and studies linking driving forces to changing emissions in order to improve the representation of these gases in global and regional emission models remain an important future research task. Therefore, the models and approaches

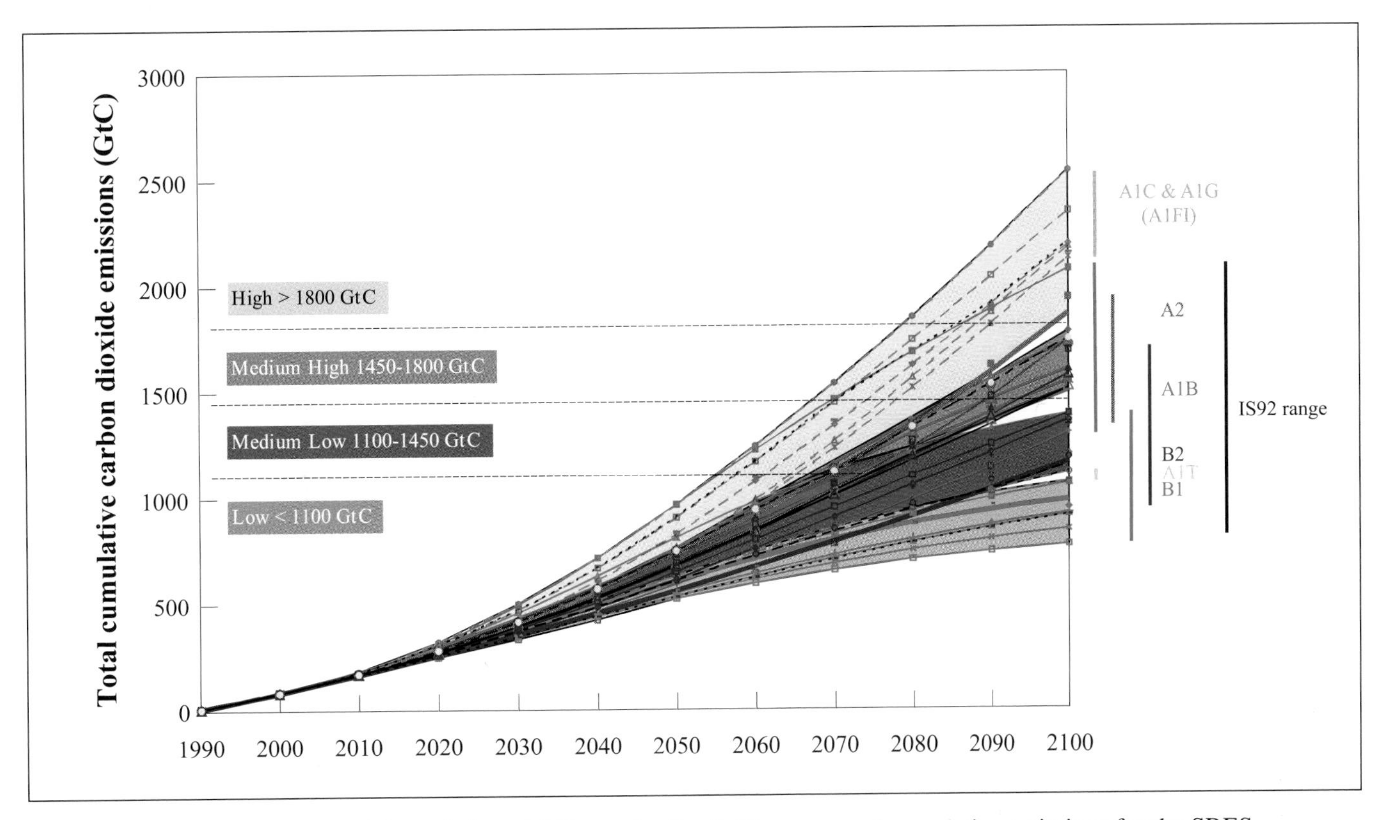

Figure 6-8: Global cumulative CO_2 emissions (GtC, standardized). The ranges of cumulative emissions for the SRES scenarios are shown. Scenarios are grouped into four categories: low, medium–low, medium–high, and high emissions. Each category contains one marker scenario plus alternatives that lead to comparable cumulative emissions, although often through different driving forces. The ranges of cumulative emissions of the SRES scenario groups are shown as colored vertical bars and the range of the IS92 scenarios as a black vertical bar. (See also footnote 2).

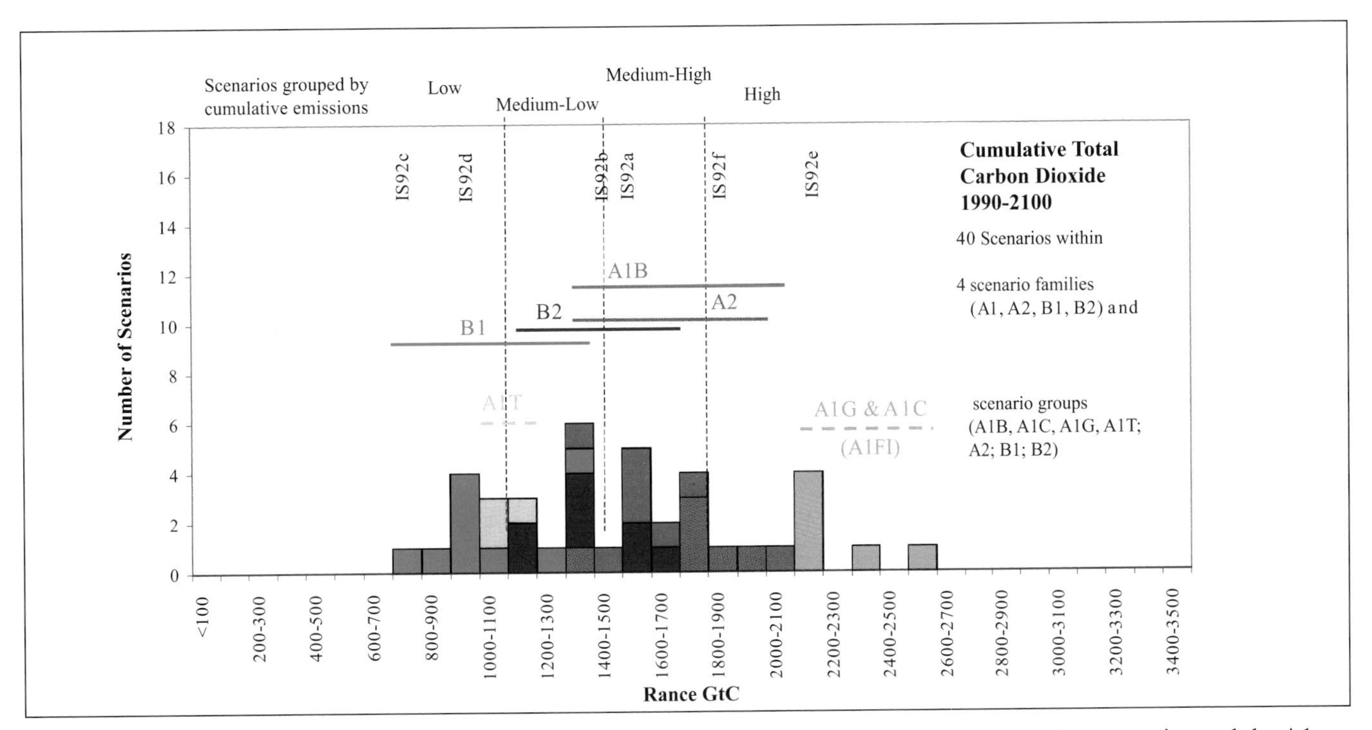

Figure 6-9: Global cumulative CO_2 emissions of the SRES scenarios. Relative positions of the marker scenarios and the A1 scenario groups are shown on the histogram. The 40 SRES scenarios have median cumulative emissions of about 1500 GtC. (See also footnote 2).

employed for the SRES analyses cannot produce unambiguous and generally approved estimates for different sources and world regions over a century. Despite the limited knowledge, at some point in time causal relationships between driving forces and emissions need to be crafted into the models for the sake of completeness. Even if new insights are generated by research specialists in certain fields of environmental science, and these become accepted as mainstream view, adopting them in the models is often far from straightforward as appropriate links to drivers may not be readily available in the underlying model structures. Limited personnel and resources imply that priorities must be assigned when deciding on further model development, and as a consequence the models lag behind "common wisdom" in certain areas. Of course, this does not necessarily limit their capabilities to capture major trends at a more aggregate level, which is the main purpose of these models.

Keeping the caveats above in mind, Table 6-2b (see later) shows the emissions in 2100 of all relevant direct and indirect GHGs for the four marker scenarios and, in brackets, the range of the other scenarios in the same family (or scenario groups for the A1 family). Chapter 5 gives further detail about the full range of GHG emissions across the SRES scenarios. Table 6-2b also compares the SRES scenarios emissions range to that of the IS92 scenario series (Pepper *et al.*, 1992).

6.3.2.1. Methane Emissions

Anthropogenic CH_4 emissions in the year 1990 are estimated at 375 ± 75 Mt CH_4 in the second IPCC assessment (Prather *et al.*, 1995). They arise from a variety of activities, dominated by biologic processes, each associated with considerable uncertainty. Future CH_4 emissions in the scenarios depend in part on the consumption of fossil fuels, adjusted for assumed changes in technology and operational practices, but more strongly on scenario-specific, regional demographic and affluence developments, together with assumptions on preferred diets and agricultural practices. For example, it is noted in Chapter 5 that the observed slowing of the rate of increase of CH_4 concentrations in recent years might indicate that the emission factors that link emissions to changes in their drivers could be changing. The writing team recommends further research into the sources and modeling approaches to capture large uncertainties surrounding future CH_4 emissions.

The resultant CH_4 emissions trajectories for the four SRES markers and other scenarios in the four families portray complex patterns (as displayed in Figure 5-5 in Chapter 5). For example, the emissions in A2 and B2 marker scenarios increase throughout the whole time horizon to the year 2100. This increase is most pronounced in the A2 marker scenario, in which emissions reach about 900 Mt CH_4 by 2100 (about a three-fold increase since 1990). The range for other scenarios in the A2 scenario family is between 549 and 1069 Mt CH_4 by 2100. The emissions level by 2100 for the B2 marker (600 Mt CH_4) is about twice as high as in 1990 (310 Mt CH_4) and ranges between 465 and 613 Mt CH_4 for the other scenarios of the B2 family. In the A1B and B1 marker scenarios, the CH_4 emissions level off and subsequently decline sooner or later in the 21st century. This phenomenon is most pronounced in the A1B marker, in which the fastest growth in the first few decades is followed by the steepest decline; the 2100 level ends up slightly below the current emission of 310 Mt CH_4. The range of emissions in Table 6-2b indicate that alternative developments in energy technologies and resources could yield a higher range in CH_4 emissions compared to the "balanced" technology A1 scenario group. In the two fossil fuel intensive scenario groups (A1C and A1G, combined into the non-fossil A1FI group in the Summary for Policymakers of this report), CH_4 emissions could reach some 735 Mt CH_4 by 2100, whereas in the post-fossil A1T scenario group emissions are correspondingly lower (some 300 Mt CH_4 by 2100). Interestingly, the A1 scenarios generally have comparatively low CH_4 emissions from non-energy sources because of a combination of low population growth and rapid advances in agricultural productivity. Hence the SRES scenarios extend the uncertainty range of the IS92 scenario series somewhat toward lower emissions. However, both scenario sets indicate an upper bound of emissions of some 1000 Mt CH_4 by 2100.

6.3.2.2. Nitrous Oxide Emissions

Even more than for CH_4, the assumed future food supply will be a key determinant of future N_2O emissions. Size, age structure, and regional spread of the global population will be reflected in the emissions trajectories, together with assumptions on diets and improvements in agricultural practices. Again, as for CH_4 in the SRES scenarios (see Section 5.4.1 in Chapter 5), continued growth of N_2O emissions emerges only in the A2 scenario, largely because of high population growth. In the other three marker scenarios, emissions peak and then decline sooner or later in the course of the 21st century. Importantly, as the largest anthropogenic source of N_2O (cultivated soils) is already very uncertain in the base year, all future emissions trajectories are affected by large uncertainties, especially if calculated with different models, as is the case in this SRES report. Therefore, the writing team recommends further research into the sources and modeling of long-term N_2O emissions. Uncertainty ranges are correspondingly large, and are sometimes asymmetric. For example, while the range in 2100 reported in all A1 scenarios is between 5 and 10 MtN (7 MtN in the A1B marker), the A2 marker reports 17 MtN in 2100. Other A2 scenarios report emissions that fall within the range reported for A1 (from 8 to 19 MtN in 2100). Thus, different model representations of processes that lead to N_2O emissions and uncertainties in source strength can outweigh easily any underlying differences between individual scenarios in terms of population growth, economic development, etc. Different assumptions with respect to future crop productivity, agricultural practices, and associated emission factors, especially in the very populous regions of the world, explain the very different global emission levels even for otherwise shared main scenario drivers. Hence, the SRES scenarios extend the uncertainty range of future emissions significantly toward higher emissions (4.8 to 20.2

MtN by 2100 in SRES compared to 5.4 to 10.8 MtN in the IS92 scenarios. (Note that natural sources are excluded in this comparison.)

6.3.2.3. Halocarbons and Halogenated Compounds

The emissions of halocarbons (chlorofluorocarbons (CFCs), hydrochlorofluorocarbons (HCFCs), halons, methylbromide, and hydrofluorocarbons (HFCs)) and other halogenated compounds (polyfluorocarbons (PFCs) and sulfur hexafluoride (SF_6)) across the SRES scenarios are described in detail on a substance-by-substance basis in Chapter 5 and Fenhann (2000). However, none of the six SRES models has its own projections for emissions of ozone depleting substances (ODSs), their detailed driving forces, and their substitutes. Hence, a different approach for scenario generation was adopted.

First, for ODSs, an external scenario, the Montreal Protocol scenario (A3, maximum allowed production) from WMO/UNEP (1998) is used as direct input to SRES. In this scenario corresponding emissions decline to zero by 2100 as a result of international environmental agreements, a development not yet anticipated in some of the IS92 scenarios (Pepper *et al.*, 1992). For the other gas species, most notably for CFC and HCFC substitutes, a simple methodology of developing different emissions trajectories consistent with the aggregate SRES scenario driving force assumptions (population, GDP, etc.) was developed. Scenarios are equally further differentiated as to assumed future technological change and control rates for these gases, varied across the scenarios consistently within the interpretation of the SRES storylines presented in Chapter 4. The literature, as well as the scenario methodology and data, are documented in more detail in Fenhann (2000) and are summarized in Chapter 5.

Second, different assumptions about CFC applications as well as substitute candidates were developed. These were initially based on Kroeze and Reijnders (1992) and information given in Midgley and McCulloch (1999), but updated with the most recent information from the Joint IPCC/TEAP Expert Meeting on Options for the Limitation of Emissions of HFCs and PFCs (WMO/UNEP, 1999) as described below. An important assumption, on the basis of the latest information from the industry, is that relatively few Montreal gases will be replaced fully by HFCs. Current indications are that substitution rates of CFCs by HFCs will be less than 50% (McCulloch and Midgley, 1998). In Fenhann (2000) a further technological development is assumed that would result in about 25% of the CFCs ultimately being substituted by HFCs (see Table 5-9 in Chapter 5). This low percentage not only reflects the introduction of non-HFC substitutes, but also the notion that smaller amounts of halocarbons will be used in many applications when changing to HFCs (efficiency gains with technological change). A general assumption is that the present trend, not to substitute with high GWP substances (including PFCs and SF_6), will continue. As a result of this assumption, the emissions reported here may be underestimates. This substitution approach is used in all four scenarios, and the technological options adopted are those known at present. Further substitution away from HFCs is assumed to require a climate policy and is therefore not considered in SRES scenarios. Policy measures that may indirectly induce lower halocarbon emissions in the scenarios are adopted for reasons other than climate change. For one scenario (A2) no reductions were assumed, whereas in the other scenarios intermediary reduction rates and levels were assumed. Expressed in HFC-134a equivalents (based on SAR equivalents), HFCs in the SRES scenarios range between 843 and 2123 kt HFC-134a equivalent by 2100, compared to 1188 to 2375 kt HFC-134a equivalent in IS92. The range of emissions of HFCs in the SRES scenario is initially generally lower than in earlier IPCC scenarios because of new insights about the availability of alternatives to HFCs as replacements for substances controlled by the Montreal Protocol. In two of the four scenarios in the report, HFC emissions increase rapidly in the second half of the 21st century, while in two others the growth of emissions is significantly slowed down or reversed in that period.

Aggregating all the different halocarbons (CFCs, HCFCs, HFCs) as well as halogenated compounds (PFCs and SF_6) into MtC-equivalents (using SAR GWPs) indicates a range between 386 and 1096 MtC-equivalent by 2100 for the SRES scenarios. This compares (see Table 6-2b) with a range of 746 to 875 MtC-equivalent for IS92 (which, however, does not include PFCs and SF_6). (The comparable SRES range, excluding PFCs and SF_6, is between 299 and 753 MtC-equivalent by 2100.) The scenarios presented here indicate a wider range of uncertainty compared to IS92, particularly toward lower emissions (because of the technological and substitution reasons discussed above).

The effect on climate of each of the substances aggregated to MtC-equivalents given in Table 6-2b varies greatly, because of differences in both atmospheric lifetime and the radiative effect per molecule of each gas. The net effect on climate of these substances is best determined by a calculation of their radiative forcing – which is the amount by which these gases enhance the anthropogenic greenhouse effect. The net radiative effect of all halocarbons, PFCs, and SF_6 from 1990 to 2100, including a current estimate of the radiative effect of stratospheric ozone depletion and subsequent recovery, ranges from 6% to 9% of the total radiative forcing from all GHGs and SO_2. Preliminary calculations indicate that the net radiative effect of PFCs and SF_6 in SRES scenarios will be no greater, relative to total anthropogenic forcing, by 2100 than it is at present.

6.3.3. Sulfur Dioxide Emissions

Emissions of sulfur portray even more dynamic patterns in time and space than the CO_2 emissions shown in Figures 6-5 and 6-6. Factors other than climate change (namely regional and local air quality, and transformations in the structure of the energy system and end use) intervene to limit future emissions. Figure 6-10 shows the range of global sulfur emissions for all

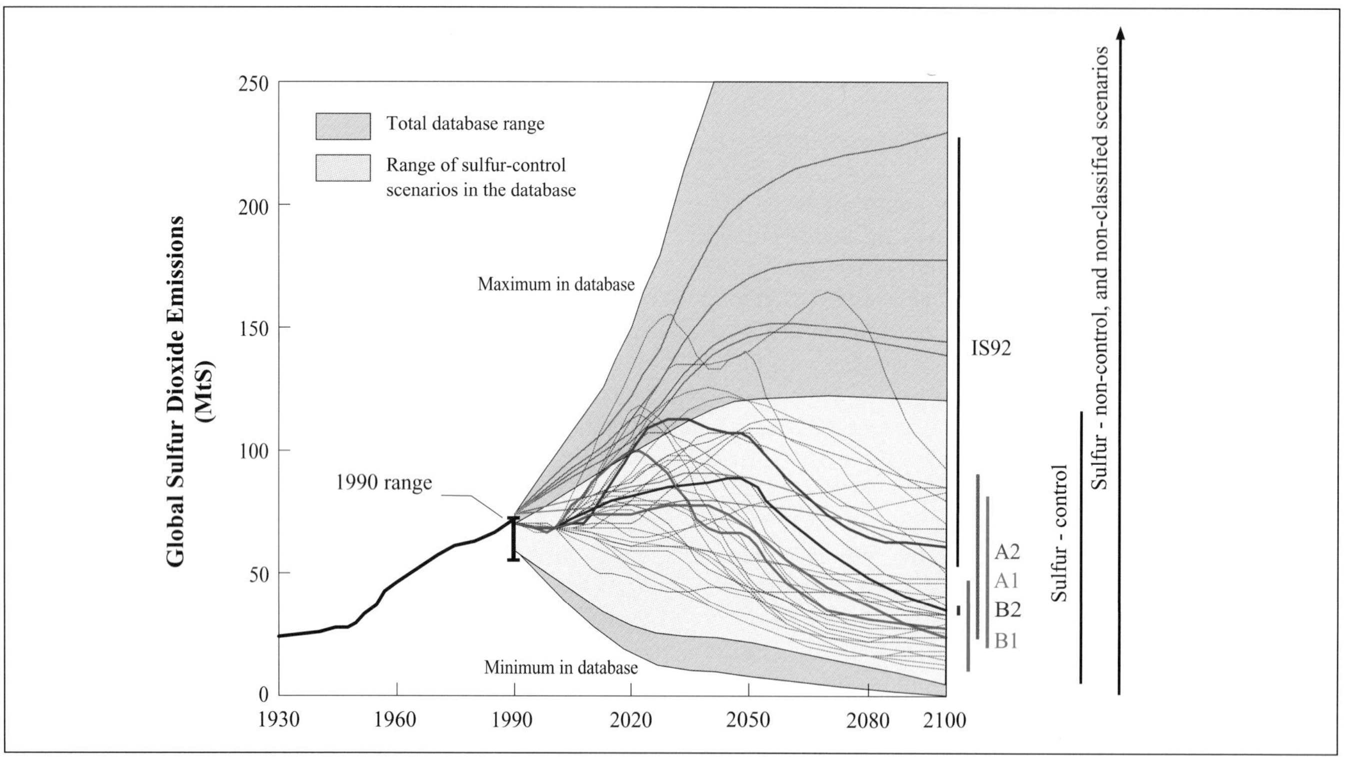

Figure 6-10: Global anthropogenic SO_2 emissions (MtS) – historical development from 1930 to 1990 and (standardized) in the SRES scenarios. The dashed colored time-paths depict individual SRES scenarios, the solid colored lines the four marker scenarios, the solid thin curves the six IS92 scenarios, the shaded areas the range of 81 scenarios from the literature, the gray shaded area the sulfur-control and the blue shaded area the range of sulfur-non-control scenarios or "non-classified" scenarios from the literature that exceeds the range of sulfur control scenarios. The colored vertical bars indicate the range of the SRES scenario families in 2100. Database source: Grübler (1998).

SRES scenarios and the four markers against the emissions range of the IS92 scenarios, more than 80 scenarios from the literature, and the historical development.

A detailed review of long-term global and regional sulfur emission scenarios is given in Grübler (1998) and summarized in Chapter 3. The most important new finding from the scenario literature is recognition of the significant adverse impacts of sulfur emissions on human health, food production, and ecosystems. As a result, scenarios published since 1995 generally assume various degrees of sulfur controls to be implemented in the future, and thus have projections substantially lower than previous ones, including the IS92 scenario series. Of these, only the two low-demand scenarios IS92c and IS92d fall within the range of more recent long-term sulfur emission scenarios. A related reason for lower sulfur emission projections is the recent tightening of sulfur-control policies in the Organization for Economic Cooperation and Development (OECD) countries, such as the Amendments of the Clean Air Act in the USA and the implementation of the Second European Sulfur Protocol. Such legislative changes were not reflected in previous long-term emission scenarios, as noted in Alcamo *et al.* (1995) and Houghton *et al.* (1995). Similar sulfur control initiatives due to local air quality concerns are beginning to impact sulfur emissions also in a number of developing countries in Asia and Latin America (see IEA, 1999; La Rovere and Americano, 1998; Streets and Waldhoff, 2000; for a more detailed discussion see Chapter 3). As a result, the median from recent sulfur scenarios (see Chapter 3) is consequently significantly lower compared to IS92, indicating a continual decline in global sulfur emissions in the long-term. The median and mean of sulfur control scenarios are almost identical. As mentioned above, even the highest range of recent sulfur-control scenarios is significantly below that of comparable, high-demand IS92 scenarios (IS92a, IS92b, IS92e, and IS92f). The scenarios with the lowest ranges project stringent sulfur-control levels that lead to a substantial decline in long-term emissions and a return to emission levels that prevailed at the beginning of the 20th century.

Reflecting recent developments and the literature (reviewed in Chapter 3), it is assumed that sulfur emissions in the SRES scenarios will also be controlled increasingly outside the OECD. As a result, both long-term trends and regional patterns of sulfur emissions evolve differently from carbon emissions in the SRES scenarios. As a general pattern, global sulfur emissions do not rise substantially, and eventually decline, even in absolute terms, during the second half of the 21st century (see also Chapters 2 and 3). The spatial distribution of emissions changes markedly. Emissions in the OECD countries continue their recent declining trend (reflecting the tightening of control measures). Emissions outside the OECD rise initially, most notably in Asia, which compensates for the declining OECD emissions. Over the long term, however, sulfur emissions decline throughout the world, but the timing and magnitude vary across the scenarios

The SRES scenario set brackets global anthropogenic sulfur emissions between 27 and 169 MtS by 2050 and between 11 and 93 MtS by 2100 (see Table 6-2b). The range of emissions for the four markers is smaller. In contrast, the range of the IS92 scenarios (Pepper *et al.*, 1992; Alcamo *et al.*, 1995) is substantially higher, starting at 80 MtS and extending all the way to 200 MtS by 2050 and from 55 to 230 MtS by 2100. The two lowest scenarios, IS92c and IS92d, approach the higher end estimates of the SRES scenarios in 2100, while others are above the SRES range. As mentioned, this difference reflects the expected future consequences of recent policies that aim to achieve a drastic reduction in sulfur emissions in OECD countries, as well as an anticipated gradual introduction of sulfur controls in developing regions in the long-term, as reported in the underlying literature (see Chapter 3). In other words, all SRES scenarios assume sulfur control measures, although the uncertainty in timing and magnitude of implementation is reflected in the variation across different scenarios. Importantly, SRES scenarios assume sulfur controls *only* and do not assume any additional climate policy measures. Nevertheless, one important implication of this varying pattern of sulfur emissions is that the historically important, but uncertain, negative radiative forcing of sulfate aerosols may decline in the very long run. This view is also confirmed by model calculations reported in Subak *et al.* (1997) and Nakićenović *et al.* (1998), on the basis of recent long-term GHG and sulfur emission scenarios.

6.3.4. Nitrogen Oxides and Volatile Organic Compounds

6.3.4.1. Nitrogen Oxides Emissions

The 1990 NO_x emissions in the six SRES models range between 26.5 and 34.2 MtN, but not all the models provide a comprehensive description of NO_x emissions. Some models do not estimate NO_x emissions at all (MARIA, MiniCAM), whereas others only include energy-related sources (MESSAGE) and have adopted other sources of emissions from corresponding scenarios derived from other models (i.e. MESSAGE uses corresponding AIM scenarios). Standardized (see Box 5-1 on Standardization) 1990 NO_x emissions in the SRES scenarios, measured as nitrogen, amount to 31 MtN (Figure 5-9 in Chapter 5).

As mentioned in Chapter 4, the volume of fossil fuels used for various energy purposes varies widely in the SRES scenario families. In addition, the level and timing of emission controls, inspired by local air quality concerns, is assumed to differ. As a result the spread of NO_x emissions is largest within the A1 scenario family (28–151 MtN by 2100), almost as large as the range across all 40 SRES scenarios (see Table 6-2b). Only in the highest emission scenarios (the fossil fuel intensive A1C and A1G scenario groups within the A1 scenario family and the high population, coal intensive A2 scenario family) do emissions rise continuously throughout the 21st century. In the A1 ("balanced") scenario group and in the B2 scenario family, NO_x emission levels rise less. NO_x emissions tend to increase up to 2050 and stabilize thereafter, the result of a gradual substitution of fossil fuels by alternatives as well as of the increasing diffusion of NO_x control technologies. Low emission futures are described by various B1 family scenarios, as well as in the A1T scenario group that describe futures in which NO_x emissions are controlled because of either local air quality concerns or rapid technological change away from conventional fossil technologies. Overall, the SRES scenarios describe a similar upper range of NO_x emissions as the previous IS92 scenarios (151 MtN versus 134 MtN, respectively, by 2100), but extend the IS92 uncertainty range toward lower emission levels (16 versus 54 MtN by 2100 in the SRES and IS92 scenarios, respectively).

6.3.4.2. Volatile Organic Compounds, Excluding Methane

Non-methane volatile organic compounds (NMVOCs) arise from fossil fuel combustion (as with NO_x, wide ranges of emission factors are typical for internal combustion engines), and also from industrial processes, fuel storage (fugitive emissions), use of solvents (e.g., in paint and cleaners), and a variety of other activities. As chemical reactivities of the various substances grouped under the NMVOCs category are very different, so are their roles in ozone formation and the (potential) health hazards associated with them. In this report, NMVOCs are discussed as one group. In 1990, the estimated emissions range was between 83 and 178 Mt NMVOC, which after standardization (see Box 5-1) translates into 140 Mt NMVOC. As for NO_x emissions, not all models include this emissions category or all of its sources; the most detailed treatment of NMVOC emissions is given in the ASF model.

A relatively robust trend across all 40 scenarios (see Figure 5-10 in Chapter 5) is a gradual increase in NMVOC emissions up to about 2050, with a range of between 190 and 260 Mt. Beyond 2050, uncertainties increase with respect to both emission levels and trends. By 2100, the range is between 58 and 552 Mt, which extends the IS92 scenario range of 136 to 403 Mt by 2100 toward both higher and lower emissions (see Table 6-2b). As for NO_x emissions, the upper bounds of NMVOC emissions are formed by fossil fuel intensive scenario groups within the A1 scenario family (A1C, A1G, combined into one fossil intensive scenario group A1FI in the Summary for Policymakers, see also footnote 2), and the lower bounds by the scenarios within the B1 scenario family. Characteristic ranges are between 60 and 90 Mt NMVOC by 2100 in the low emissions cluster and between 370 and 550 Mt NMVOC in the high emissions cluster. All other scenario families and individual scenarios fall between these two emissions clusters; the B2 marker scenario (B2-MESSAGE) closely tracks the median of global NMVOC emissions from all the SRES scenarios (see Figure 5-10 in Chapter 5).

6.3.4.3. Carbon Monoxide

CO emissions in 1990 are estimated to range between 752 and 984 Mt CO (880 Mt CO after standardization) across the models used to derive the SRES scenarios. The same caveats as for NO_x and NMVOC emissions also apply to CO emissions – the number of models that represent all the emission source categories is limited and modeling and data uncertainties, such as emission factors, are considerable. As a result, CO emission estimates across scenarios are highly model specific and future emission levels overlap considerably between the four SRES scenario families (see Table 6-2b). By 2100, emissions range between 363 and 3766 Mt CO, a considerably larger uncertainty range, particularly toward higher emissions, than in IS92, for which the 2100 emission range was between 450 and 929 Mt CO (see Table 6-2b).

As for the NO_x and NMVOC emissions discussed above, the highest CO emission levels are associated with the high-growth fossil fuel intensive scenarios (A1C and A1G scenario groups, combined into one fossil intensive scenario group A1FI in the Summary for Policymakers; see also footnote 2) within the A1 scenario family, and the lowest emission levels are generally associated with the B1 and B2 scenario families. However, inter-model variability is considerable, which indicates that uncertainties are equally large with respect to scenario driving forces (such as energy demand and supply growth) and other factors that influence CO emissions (such as local air quality concerns or technological change).

6.4. Summary and Conclusions

In summary, the SRES scenarios lead to the following findings:

- Alternative combinations of driving forces can lead to similar levels and structure of energy and land-use patterns, as illustrated by different scenarios and groups. Hence, even for a given scenario outcome (e.g. in terms of GHG emissions) there are alternative combinations of driving forces and pathways that could lead to that outcome. For instance, significant global changes could result from a scenario of high population growth, even if per capita incomes rise only modestly, as well as from a scenario in which a rapid demographic transition (to low population levels) coincides with high rates of income growth and affluence.
- Important possibilities for further bifurcations in future development trends exist within one scenario family, even when particular values are adopted for the important scenario driving force variables to illustrate a particular development path. The four technology scenario groups in A1 family (combined into three in the Summary for Policymakers) illustrate such alternative development paths with similar quantifications of the main driving forces.
- Emissions profiles are dynamic across the range of SRES scenarios. They portray trend reversals and indicate possible emissions crossover among different scenarios. They do not represent mere extensions of continuous increase of GHGs and SO_2 emissions into the future. This more complex pattern of future emissions across the range of SRES scenarios, time periods, world regions, and sectors reflects recent scenario literature.
- Describing potential future developments involves inherent ambiguities and uncertainties. One and only one possible development path (as alluded to, for instance, in concepts such as "business-as-usual scenario") simply does not exist alone. And even for each alternative development path described by any given scenario, there are numerous combinations of driving forces and numeric values that can be consistent with a particular scenario description. The numeric precision of any model result should not distract from the basic fact that uncertainty abounds. However, the multi-model approach increases the value of the SRES scenario set, since uncertainties in the choice of model input assumptions can be separated more explicitly from the specific model behavior and related modeling uncertainties.
- Any scenario has subjective elements and is open to various interpretations. While the writing team as a whole has no preference for any of the scenarios, and has no judgment as to the probability or desirability of different scenarios, the open process and initial reactions to draft versions of this report show that individuals and interest groups do have such judgments. The writing team hopes that this will stimulate an open discussion in the policy making arena about potential futures and choices that can be made in the context of climate change response. For the scientific community, the SRES scenario exercise has led to the identification of a number of recommendations for future research that can further increase the understanding of potential developments of socio-economic driving forces and their interactions, and the associated GHG emissions. A summary of main findings and recommendations for potential users of the SRES scenarios is given in Box 6-3 and Box 6-4. The writing teams' suggestions for consideration by the IPCC are summarized in Box 6-5.
- Finally, the writing team believes that the SRES scenarios largely fulfill all the specifications set out in Chapter 1. To support reproducibility, more detailed information than can be included in this report will be made available by individual modeling groups and members of the writing team through other means, such as web sites, peer-reviewed literature, or background documentation if additional resources can be made available.

In conclusion, Tables 6-2a and 6-2b summarize the main characteristics of the seven scenario groups that constitute the four families (combined into six groups in the Summary for Policymakers). The tables give the global ranges of driving

Box 6-3 Main Findings and Implications of SRES Scenarios

- The four scenario families each have a narrative storyline and consist of 40 scenarios developed by six modeling groups.
- The 40 scenarios cover the full range of GHGs and SO_2 emissions consistent with the underlying range of driving forces from scenario literature.
- The 40 SRES scenarios fall into different groups – the three scenario families A2, B1, and B2, plus four groups within the A1 scenario family, two of which (A1C and A1G) have been combined into one fossil-intensive group A1FI in the Summary for Policymakers; see also footnote 2. The four A1 groups are distinguished by their technological emphasis – on coal (A1C), oil and gas (A1G), non-fossil energy sources (A1T), or a balance across all sources (A1).
- The scenarios are grouped into four categories of cumulative CO_2 emissions, which indicate that scenarios with different driving forces can lead to similar cumulative emissions and those with similar driving forces can branch out into different categories of cumulative emissions.
- Four from 40 scenarios are designated as marker scenarios that are characteristic of the four scenarios families. Together with the two additional illustrative scenarios selected from the scenario groups in the A1 family, they capture most of the emissions and driving forces spanned by the full set of the scenarios.
- There is no single central or "best guess" scenario, and probabilities or likelihood are not assigned to individual scenarios. Instead, the writing team recommends that the smallest set of scenarios used should include the four designated marker scenarios and the two additional illustrative scenarios selected from the scenario groups in the A1 family.
- Distinction between scenarios that envisage stringent environmental policies and those that include direct climate policies was very difficult to make, a difficulty associated with many definitional and other ambiguities.
- All scenarios describe futures that are generally more affluent than today. Many of the scenarios envisage a more rapid convergence in per capita income ratios in the world compared to the IS92 scenarios while, at the same time, they jointly cover a wide range of GHG and SO_2 emissions.
- Emissions profiles are more dynamic than the IS92 scenarios, which reflects changes in future emissions trends for some scenarios and GHG species.
- The levels of GHG emissions are generally lower than the IS92 levels, especially toward the end of the 21st century, while emissions of SO_2, which have a cooling effect on the atmosphere, are significantly lower than in IS92.
- Alternative combinations of main scenario driving forces can lead to similar levels of GHG emissions by the end of the 21st century. Scenarios with different underlying assumptions can result in very similar climate changes.
- Technology is at least as important a driving force of GHG emissions as population and economic development across the set of 40 SRES scenarios.

Box 6-4: Recommendations for Consideration by the User Communities

The writing team recommends that the SRES scenarios be the main basis for the assessment of future emissions and their driving forces in the Third Assessment Report (TAR). Accordingly, the SRES writing team makes the following recommendations regarding the emissions scenarios to be used in the atmosphere/ocean general circulation models (A/O GCMs) simulations for Working Group I (WGI), for the models that will be used in the assessment of climate change impacts by Working Group II (WGII), and for the mitigation and stabilization assessments by WGIII:

- *It is recommended that a range of SRES scenarios from more than one family be used in any analysis.* The scenario groups – the scenario families A2, B1, and B2, plus the groups within the A1 scenario family, and four cumulative emissions categories were developed as the smallest subsets of SRES scenarios that capture the range of uncertainties associated with driving forces and emissions.
- *The important uncertainties may be different in different applications – for example climate modeling; assessment of impacts, vulnerability, mitigation, and adaptation options; and policy analysis.* Climate modelers may want to cover the range reflected by the cumulative emissions categories. To assess the robustness of options in terms of impacts, vulnerability, and adaptation may require scenarios with similar emissions but different socio-economic characteristics, as reflected by the scenario groups. For mitigation analysis, variation in both emissions and socio-economic characteristics may be necessary. For analysis at the national or regional scale, the most appropriate scenarios may be those that best reflect specific circumstances and perspectives.
- *There is no single most likely, "central", or "best-guess" scenario, either with respect to other SRES scenarios or to the underlying scenario literature.* Probabilities or likelihoods are not assigned to individual SRES scenarios. None of the SRES scenarios represents an estimate of a central tendency for all driving forces and emissions, such as the mean or

(Box 6.4 continues)

(Box 6.4 continued)

median, and none should be interpreted as such. The statistics associated with the frequency distributions of SRES scenarios do not represent the likelihood of their occurrence. The writing team cautions against constructing a central, "best-estimate" scenario from the SRES scenarios; instead it recommends use of the SRES scenarios as they are.

- *Concerning large-scale climate models, the writing team recommends that the minimum set of SRES scenarios should include the four designated marker scenarios and the two additional illustrative scenarios selected from the scenario groups in the A1 family.* At the minimum (a) a simulation for one and the same SRES marker or illustrative scenario should be performed by every TAR climate model for a given stabilization ceiling, and (b) the set of simulations performed by the TAR climate models and stabilization runs for a given ceiling should include all four of the SRES marker scenarios.
- *The driving forces and emissions of each SRES scenario should be used together.* To avoid internal inconsistencies, components of SRES scenarios should not be mixed. For example, the GHG emissions from one scenario and the SO_2 emissions from another scenario, or the population from one and economic development path from another, should not be combined.
- *The SRES scenarios can provide policy makers with a long-term context for near-term decisions.* This implies that they are not necessarily well suited for the analysis of near-term developments. When analyzing mitigation and adaptation options, the user should be aware that although no additional climate initiatives are included in the SRES scenarios, various changes have been assumed to occur that would require other policy interventions.
- *More detailed information on assumptions, inputs, and the results of the 40 SRES scenarios should be made available at a web site and on a CD-ROM.* Regular maintenance of the SRES web site is

Box 6-5: Recommendations for Consideration by the IPCC

- Extend the SRES web site and CD-ROM to provide, if appropriate, time-dependent geographic distributions of driving forces and emissions, and concentrations of GHGs and sulfate aerosols.
- Development of a classification scheme for classifying scenarios as intervention or non-intervention scenarios.
- Establish a programme for on-going evaluations and comparisons of long-term emissions scenarios, including a regularly updated scenario database.
- An effort should be made in the future to develop an appropriate emissions scenario classification scheme.
- Identify resources for capacity building in the area of emissions scenarios for future IPCC assessments, with a particular emphasis to involve strong participation from developing countries.
- Promote activities within and outside the IPCC to extend the SRES multi-baseline and multi-model approach in future assessments of climate change impacts, adaptation, and mitigation.
- Initiate new programs to assess GHG emissions from land use and sources of emissions other than energy-related CO_2 emissions, to go beyond the effort of SRES, which was limited by time and resources.
- Initiate new programs to assess future developments of driving forces and GHG emissions for different regions and for different sectors (taking the set of SRES scenarios as reference for overall global and regional developments) to provide more regional and sectorial detail than time and resources allowed SRES to achieve.

forces and emissions in 2020, 2050, and 2100. Table 6-2a summarizes the ranges of the main scenario driving forces: global population, economic development, per capita income levels and income ratios, energy intensity, primary energy use, and structure of energy supply. Table 6-2b summarizes the emissions of GHGs, SO_2, and ozone precursors emissions for the years 2020, 2050, and 2100 as well as cumulative 1990-2100 CO_2 emissions broken down into energy- and land-use related sources. Together, the two tables provide a concise summary of the new SRES scenarios.

Table 6-2a: *Overview of main driving forces for the four SRES marker scenarios for 2100 if not indicated otherwise. Numbers in brackets show the range across all other scenarios from the same scenario family as the marker. Units are given in the table. (IND regions includes industrialized countries consisting of OECD90 and REF regions; and DEV region includes developing countries consisting of ASIA and ALM regions, see Appendix IV).*

	Population In Billion	Economic Growth, GDP_{mex} [a]	Per Capita Income, GDP_{mex}/capita	Primary Energy Use	Hydrocarbon Resource Use[b]	Land-Use Change[c]
A1	Lutz (1996) Low ~7 billion 1.4 IND 5.6 DEV	Very high 1990-2020: 3.3 (2.8-3.6) 1990-2050: 3.6 (2.9-3.7) 1990-2100: 2.9 (2.5-3.0)	Very high in IND: US$107,300 (60,300-113,500) in DEV: US$ 66,500 (41,400-69,800)	Very high 2.226 (1,002-2,683) EJ Low energy intensity of 4.2 MJ/US$ (1.9-5.1)	Varied in four scenario groups: Oil: Low to very high 20.8 (11.5-50.8) ZJ Gas: High to very high 42.2 (19.7-54.9) ZJ Coal: Medium to very high 15.9 (4.4-68.3) ZJ	Low. 1990-2100: 3% cropland, 6% grasslands 2% forests
A2	Lutz (1996) High ~15 billion 2.2 IND 12.9 DEV	Medium 1990-2020: 2.2 (2.0-2.6) 1990-2050: 2.3 (1.7-2.8) 1990-2100: 2.3 (2.0-2.3)	Low in DEV Medium in IND in IND: US$46,200 (37,100-64,500) in DEV: US$11,000 (10,300-13,700)	High 1,717 (1,304-2.040) EJ High energy intensity of 7.1 MJ/US$ (5.2-8.9)	Scenario dependent: Oil: Very low to medium 17.3 (11.0-22.5) ZJ Gas: Low to high 24.6 (18.4-35.5) ZJ Coal: Medium to Very high 46.8 (20.1-47.7) ZJ	Medium n.a. from ASF
B1	Lutz (1996) Low ~7 billion 1.4 IND 5.7 DEV	High 1990-2020: 3.1 (2.9-3.3) 1990-2050: 3.1 (2.9-3.5) 1990-2100: 2.5 (2.5-2.6)	High in IND: US$72,800 (65,300-77,700) In DEV: US$40,200 (40,200-45,200)	Low. 514 (514-1,157) EJ Very low energy intensity of 1.6 EJ/US$ (1.6-3.4)	Scenario dependent: Oil: Very low to high 19.6 (15.7-19.6) ZJ Gas: Medium to high 14.7 (14.7-31.8) ZJ Coal: Very low to high 13.2 (3.3-27.2) ZJ	High 1990-2100: -28% cropland -45% grassland +30% forests
B2	UN (1998) Median ~10 billion 1.3 IND 9.1 DEV	Medium 1990-2020: 3.0 (2.2-3.1) 1990-2050: 2.8 (2.1-2.9) 1990-2100: 2.2 (2.0-2.3)	Medium in IND: US$54,400 (42,400-61,100) In DEV: US$18,000 (14,200-21,500)	Medium 1,357 (846-1,625) EJ Medium energy intensity of 5.8 MJ/US$ (4.3-6.5)	Oil: Low to medium 19.5 (11.2-22.7) ZJ by 2100 Gas: Low to medium 26.9 (17.9-26.9) ZJ by 2100 Coal: Low to very high 12.6 (12.6-44.4) ZJ by 2100	Medium 1990-2100: +22% cropland +9% grasslands +5% forests[d]

[a] Exponential growth rates after World Bank (1999) method (given on pages 371 to 372) are calculated using the *different base years* from the models.
[b] Resource availability is generally combined with scenario specific rates of technological change.
[c] Residual and other land-use categories are not shown in the Table.
[d] Land-use data for B2 marker taken from AIM land-use B2 scenario run.

Table 6-2b: *Overview of GHG, ozone precursors and sulfur emissions (standardized) for the four SRES scenario families and seven scenario groups for 2100. Numbers are for the four markers, the two additional illustrative A1FI and A1T scenarios (A1G and A1C are combined into one fossil-intensive scenario group A1FI in the SPM, A1FI is from the A1G group), and (in brackets) for the range across all scenarios within the same scenario group[1]. Range of SRES scenarios are also compared to the range from the IS92 scenarios. Units are given in the table.*

	CO_2 (GtC)	CH_4 (Mt CH4)	N_2O (MtN)	CFC, HCFC, HFC, PFC, SF_6 (MtC equiv.)[a]	CO (Mt CO)	NMVOCs (Mt)	NO_x (MtN)	SO_x (MtS)
A1B	Medium 13.5 (13.5-17.9)	Low 289 (289-640)	Medium 7.0 (5.8-17.2)	Medium Total of 824	Medium 1663 (1080-2532)	Medium 194 (133-552)	Medium 40.2 (40.2-77.0)	Low 28 (26-71)
A1C	High (25.9-36.7)	Medium (392-693)	Medium (6.1-16.2)	as A1	High (2298-3766)	Medium (167-373)	High (63.3 -151.4)	High (27-83)
A1G	High 28.2 (28.2-30.8)	Medium 735 (289-735)	Medium 16.6 (5.9-16.6)	as A1	High 2570 (3260-3666)	Medium 420 (192-484)	High 110 (39.9 -132.7)	Low 40 (27-41)
A1T	Low 4.3 (4.3-9.1)	Low 274 (274-291)	Low 5.4 (4.8-5.4)	as A1	Medium 2077 (1520-2077)	Low 128 (114 -128)	Low 28 (28.1-39.9)	Very low (20.2-27.4)
A2	High 29.1 (19.6-34.5)	High 889 (549-1069)	High 16.5 (8.1-19.3)	High Total of 1096	High 2325 (776-2646)	High 342 (169-342)	Very high 109.2 (70.9-110.0)	High 60 (60.3-93)
B1	Low 4.2 (2.7-10.4)	Low 236 (236-579)	Low 5.7 (5.3-20.2)	Low Total of 386	Low 363 (363-1871)	Low 87 (58-349)	Low 18.7 (16.0-35.0)	Low 25 (11-25)
B2	Medium 13.3 (10.8-21.8)	Medium 597 (465-613)	Medium 6.9 (6.9-18.1)	Medium Total of 839	Medium 2002 (661-2002)	Medium 170 (130-304)	High 61.2 (34.5-76.5)	Low-Medium 48 (33-48)
SRES	2.7-36.7	236-1069	4.8-20.2	386-1096	363-3766	58-552	16-151	11-93
IS92[b]	4.6-35.8	546-1168	13.7-19.1	746-875	450-929	136-403	54-134	55-232

[a] Based on SAR GWPs. Emissions of ozone depleting substances (CFCs, HCFCs) decline to zero by 2100, reflecting the Montreal Protocol (WMO/UNEP, 1998).
[b] Anthropogenic emissions only (i.e. excluding natural sources).

[1] Please note that in the Summary for Policymakers and the Technical Summary additional information has been included on the ranges for the harmonised scenarios only, as derived from the information in the various tables in this report and its Appendices

References:

Alcamo, J. (ed.), 1994: *IMAGE 2.0: Integrated Modelling of Global Change*. Kluwer Academic Publishers, Dordrecht, the Netherlands.

Alcamo, J., A. Bouwman, J. Edmonds, A. Grübler, T. Morita, and A. Sugandhy, 1995: An evaluation of the IPCC IS92 emission scenarios. In *Climate Change 1994, Radiative Forcing of Climate Change* and *An Evaluation of the IPCC IS92 Emission Scenarios,* J.T. Houghton, L.G. Meira Filho, J. Bruce, Hoesung Lee, B.A. Callander, E. Haites, N. Harris and K. Maskell (eds.), Cambridge University Press, Cambridge, pp. 233-304.

Alcamo, J., E. Kreileman and R. Leemans (eds.), 1998: *Global Change Scenarios of the 21st Century. Results from the IMAGE 2.1 model*. Elsevier Science, London.

De Jong, A. and G. Zalm. 1991: Scanning the Future: A long-term Scenario Study of the World Economy 1990-2015. In: *Long-term Prospects of the World Economy.* OECD, Paris, France, pp. 27-74.

Demeny, P., 1990: Population. In *The Earth As Transformed by Human Action*. B.L. Turner II *et al*., (ed.), Cambridge University Press, Cambridge.

De Vries, H.J.M., J.G.J. Olivier, R.A. van den Wijngaart, G.J.J. Kreileman, and A.M.C. Toet, 1994: Model for calculating regional energy use, industrial production and greenhouse gas emissions for evaluating global climate scenarios. *Water, Air Soil Pollution*, **76**, 79-131.

De Vries, B., M. Janssen, and A. Beusen, 1999: Perspectives on global energy futures – simulations with the TIME model. *Energy Policy,* **27**, 477-494.

De Vries, B., J. Bollen, L. Bouwman, M. den Elzen, M. Janssen, and E. Kreileman, 2000: Greenhouse gas emissions in an equity-, environment- and service-oriented world: An IMAGE-based scenario for the next century. *Technological Forecasting & Social Change*, **63**(2-3). (In press).

Durand, J.D., 1967: The modern expansion of world population. *Proceedings of the American Philosophical Society*, **111**(3), 136-159.

Edmonds, J., M. Wise, and C. MacCracken, 1994: *Advanced Energy Technologies and Climate Change. An Analysis Using the Global Change Assessment Model (GCAM)*. PNL-9798, UC-402, Pacific Northwest Laboratory, Richland, WA, USA.

Edmonds, J., M. Wise, H. Pitcher, R. Richels, T. Wigley, and C. MacCracken, 1996a: An integrated assessment of climate change and the accelerated introduction of advanced energy technologies: An application of MiniCAM 1.0. *Mitigation and Adaptation Strategies for Global Change*, **1**(4), 311-339.

Edmonds, J., M. Wise, R. Sands, R. Brown, and H. Kheshgi, 1996b: *Agriculture, Land-Use, and Commercial Biomass Energy. A Preliminary integrated analysis of the potential role of biomass energy for reducing future greenhouse related emissions*. PNNL-11155, Pacific Northwest National Laboratories, Washington, DC.

Fenhann, J., 2000: Industrial non-energy, non-CO_2 greenhouse gas emissions. *Technological Forecasting & Social Change,* **63**(2-3). (In press).

Grübler, A., 1998: A review of global and regional sulfur emission scenarios. *Mitigation and Adaptation Strategies for Global Change,* **3**(2-4), 383-418.

Houghton, J.T., L.G. Meira Filho, J. Bruce, Hoesung Lee, B.A. Callander, E. Haites, N. Harris, and K. Maskell (eds.), 1995: *Climate Change 1994: Radiative Forcing of Climate Change and an Evaluation of the IPCC IS92 Emissions Scenarios*. Cambridge University Press, Cambridge, 339 pp.

Houghton, J.T., L.G. Meira Filho, B.A. Callander, N. Harris, A. Kattenberg, and K. Maskell (eds.), 1996: *Climate Change 1995. The Science of Climate Change*. Contribution of Working Group I to the Second Assessment Report of the Intergovernmental Panel on Climate Change, Cambridge University Press, Cambridge.

IEA (International Energy Agency), 1999. *Non-OECD Coal-Fired Power Generation –trends in the 1990s*. IEA Coal Research, London.

Kroeze, C. and Reijnders, L., 1992: Halocarbons and global warming III, *The Science of Total Environment*, **112**, 291-314.

La Rovere, E.L., and B. Americano, 1998: *Environmental Impacts of Privatizing the Brazilian Power Sector.* Proceedings of the International Association of Impact Assessment Annual Meeting, Christchurch, New Zealand, April 1998.

Lashof, D., and Tirpak, D.A., 1990: *Policy Options for Stabilizing Global Climate*. 21P-2003. U.S. Environmental Protection Agency, Washington, DC.

Leggett, J., W.J. Pepper, and R.J. Swart, 1992: Emissions scenarios for IPCC: An update. In *Climate Change 1992*. Supplementary Report to the IPCC Scientific Assessment, J.T. Houghton, B.A. Callander, S.K.Varney (eds.), Cambridge University Press, Cambridge, pp. 69-95.

Lutz, W. (ed.), 1996: *The Future Population of the World: What can we assume today?* 2nd Edition. Earthscan, London.

Lutz, W., W. Sanderson, and S. Scherbov, 1997: Doubling of world population unlikely. *Nature*, **387**(6635), 803-805.

McCulloch, A., P.M. Midgley, 1998: Estimated historic emissions of fluorocarbons from the European Union, *Atmospheric Environment,* **32**(9), 1571-1580.

Messner, S., and M. Strubegger, 1995: *User's Guide for MESSAGE III*. WP-95-69, International Institute for Applied Systems Analysis, Laxenburg, Austria, 155 pp.

Midgley, P.M., and A. McCulloch, 1999: Properties and applications of industrial halocarbons. In *Reactive Halogen Compounds in the Atmosphere,* **4**, *Part E*. P. Fabian, and O.N. Singh (eds.), *The Handbook of Environmental Chemistry*, Springer-Verlag, Berlin/Heidelberg.

Mori, S., and M. Takahashi, 1999: An integrated assessment model for the evaluation of new energy technologies and food productivity. *International Journal of Global Energy Issues*, **11**(1-4), 1-18.

Mori, S., 2000: The development of greenhouse gas emissions scenarios using an extension of the MARIA model for the assessment of resource and energy technologies. *Technological Forecasting & Social Change*, **63**(2-3). (In press).

Morita, T., Y. Matsuoka, I. Penna, and M. Kainuma, 1994: *Global Carbon Dioxide Emission Scenarios and their Basic Assumptions: 1994 Survey*. CGER-1011-94, Center for Global Environmental Research, National Institute for Environmental Studies, Tsukuba, Japan.

Nakićenović, N., N. Victor, and T. Morita, 1998: Emissions scenarios database and review of scenarios. *Mitigation and Adaptation Strategies for Global Change*, **3**(2-4), 95-120.

Parikh, J.K., 1992: IPCC strategies unfair to the south. *Nature*, **360**, 507-508.

Pepper, W.J., J. Leggett, R. Swart, J. Wasson, J. Edmonds, and I. Mintzer, 1992: Emissions scenarios for the IPCC. An update: Assumptions, methodology, and results: Support Document for Chapter A3. In *Climate Change 1992: Supplementary Report to the IPCC Scientific Assessment,* J.T. Houghton, B.A. Callandar, S.K. Varney (eds.), Cambridge University Press, Cambridge.

Pepper, W.J., Barbour, W., Sankovski, A., and Braaz, B., 1998: No-policy greenhouse gas emission scenarios: revisiting IPCC 1992. *Environmental Science & Policy,* **1**, 289-312.

Prather, M., R. Derwent, D. Erhalt, P. Fraser, E. Sanhueza, and X. Zhou, 1995: Other trace gases and atmospheric chemistry. In *Climate Change 1994, Radiative Forcing of Climate Change and An Evaluation of the IPCC IS92 Emission Scenarios,* Cambridge University Press, Cambridge, pp. 77-119.

Riahi, K., and R.A. Roehrl, 2000: Greenhouse gas emissions in a dynamics-as-usual scenario of economic and energy development. *Technological Forecasting & Social Change,* **63**(2-3). (In press).

Sankovski, A., W. Barbour, and W. Pepper, 2000: Quantification of the IS99 emission scenario storylines using the atmospheric stabilization framework (ASF). *Technological Forecasting & Social Change*, **63**(2-3). (In press).

Streets, D.G., and S.T. Waldhoff, 2000: Present and future emissions of air pollutants in China: SO_2, NO_x, and CO. *Atmospheric Environment,* **34**(3), 363-374.

Subak, S., M. Hulme and L. Bohn, 1997: *The Implications of FCCC Protocol Proposals for Future Global Temperature: Results Considering Alternative Sulfur Forcing*. CSERGE Working Paper GEC-97-19, CSERGE University of East Anglia, Norwich, UK.

UN (United Nations), 1998: *World Population Projections to 2150*. UN Department of Economics and Social Affairs (Population Division), United Nations, New York, NY.

UNFCCC (United Nations Framework Convention on Climate Change), 1997: *Kyoto Protocol to the United Nations Framework Convention on Climate Change*, FCCC/CP/L7/Add.1, 10 December 1997. UN, New York , NY.

WMO/UNEP (World Meteorological Organisation and United Nations Environment Programme), 1998: *Scientific Assessment of Ozone Depletion: 1998.* WMO Global Ozone Research & Monitoring Project, December 1998, WMO, Geneva.

WMO/UNEP (World Meteorological Organisation and United Nations Environment Programme). 1999: *Conference Report,* Joint IPCC/TEAP Expert Meeting on Options for the Limitation of Emissions of HFCs and PFCs, 26-28 May 1999, Petten, the Netherlands.

World Bank, 1999: *1999 World Development Indicators.* World Bank,

SPECIAL REPORT ON EMISSIONS SCENARIOS: APPENDICES

I

SRES Terms of Reference: New IPCC Emission Scenarios

SRES Terms of Reference
New IPCC Emissions Scenarios

It is proposed that Working Group III coordinate the development of new emissions scenarios that assume no additional climate policy initiatives.

I.1. Background

In 1992 the Intergovernmental Panel on Climate Change (IPCC) released six emissions scenarios (Leggett *et al.*, 1992) providing alternative emissions trajectories spanning the years 1990 through 2100 for greenhouse-related gases, Carbon dioxide (CO_2), carbon monoxide (CO), methane (CH_4), nitrous oxide (N_2O), nitrogen oxyde (NO_x), and sulfur dioxide (SO_2). These scenarios were intended for use by atmospheric and climate scientists in the preparation of scenarios of atmospheric composition and climate change. The work updated and extended earlier work prepared for the IPCC first assessment report. These six scenarios are referred to as the IS92 scenarios.

In many ways the IS92 scenarios were pathbreaking. They were the first global scenarios to provide estimates of the full suite of greenhouse gases. At the time, they were the only scenarios to provide emission trajectories for SO_2. Alcamo *et al.* (1995) reviewed the scenarios and found that the fossil fuel carbon emissions trajectories spanned more than half of the open literature emissions scenarios reviewed. Other emissions trajectories had received less scrutiny in the open literature and, while the IS92 cases were not dissimilar to those in the open literature, the open literature was extremely sparse in many instances.

Much has changed in the period following the creation of the IS92 scenarios. Sulfur emissions have been recognized as a more important radiative forcing factor than other non-CO_2 greenhouse-related gases, and some regional control policies have been adopted. Restructuring in the states of Eastern Europe and the Former Soviet Union has had far more powerful effects on economic activity and emissions than were foreseen in the IS92 scenarios. For some regions these scenarios are not representative of those found in the literature. The advent of integrated assessment (IA) models has made it possible to construct self-consistent emissions scenarios that jointly consider the interactions between energy, economy, and land-use changes.

Alcamo *et al.* (1995) found that for the purposes of driving atmospheric climate models, the CO_2 emissions trajectories of the IS92 scenarios provided a reasonable reflection of variations found in the open literature. However, scenarios are also required for other purposes, and the IS92 scenarios are not suitable for purposes for which they were not developed. It was concluded that, if the scenarios were intended to have broader uses than simply a set of emissions trajectories to drive climate models, new scenarios should be developed. Further, a new approach should be adopted. The new approach should open the process to the broader research community.

I.2. Approach

It is proposed that new scenarios should be developed through a coordinated effort that draws upon the expertise of all researchers in the relevant community. A three-step process is envisaged. First, key input assumptions would be reviewed and provided to modelers. Second, modelers would be asked to construct emissions scenarios based on the input assumptions provided. Finally, the model results will be used to develop new emissions scenarios in the form of average results for participating models or results from a representative model.

A writing team would be established to consider key input assumptions (such as population projections and technologic change) and emissions from specific sources (such as SO_2 emissions and CO_2 emissions due to land-use change), possibly with the assistance of specialized task groups. The writing team will also stipulate a set of geographical reporting regions, reporting years, units of measure, etc., designed to provide climate modelers, impact assessment analysts, and other users with the detail they need for their work. Finally, the writing team would ensure that the range of results reflects the underlying uncertainty and, to the extent possible, that the assumptions for specific scenarios are internally consistent.

Scenario development will be an open process. There will be no "official" model. There will be no "expert teams." Any research group with the capability of preparing scenarios for any region can participate. This means that, while modeling teams which employ global coverage will be able to participate, so too will regional modelers. By opening the process in this way, developing and developed region researchers with local expertise can participate even if they do not have global coverage. Modeling teams will be provided with information on the input assumptions and other necessary information such as, for example, the world oil price, to regional modeling teams.

Once the modeling teams have completed their work, a set of scenarios will be chosen. This will likely be the inputs and outputs of a "representative" model, but it could also be the average of the participating models or some other representation of the model results.

I.3. Reporting and Distribution of Results

To maximize the usefulness of the new scenarios, two steps should be taken. First, arrangements should be made with an organization whose mission it is to disseminate information to provide a means by which users can access scenario results. All results from research institutions will be included in the database along with associated assumptions. In addition, for research teams willing to participate, the associated models will also be made available so that users can not only have access to scenario assumptions and outputs, but have the capability of independently creating derivative scenarios.

I.4. Timing and Coordination

It is proposed that the writing team begin work before the end of 1996. The team should establish the parameters – geographic reporting regions, reporting years, time horizon, units, etc. – by the end of the first quarter of 1997. Reports of the expert groups on the range of values for each of the input assumptions should be available by the end of the third quarter of 1997. The scenario results corresponding to these input assumptions should be available from participating modeling groups during the first quarter of 1998. Peer and government review should be complete by the end of 1998.

References:

Alcamo, J., A. Bouwman, J. Edmonds, A. Grübler, T. Morita, and A. Sugandhy, 1995: An Evaluation of the IPCC IS92 Emission Scenarios. In: *Climate Change 1994, Radiative Forcing of Climate Change and An Evaluation of the IPCC IS92 Emission Scenarios,* Cambridge University Press, Cambridge, UK, pp. 233–304.

Leggett, J., W.J. Pepper and R.J. Swart, 1992: *Emissions Scenarios for IPCC: An Update*. In: *Climate Change 1992*. The Supplementary Report to the IPCC Scientific Assessment [Houghton, J.T., B.A. Callander and S.K. Varney (eds.)], Cambridge University Press, Cambridge, UK, pp. 69-95.

II

SRES Writing Team and SRES Reviewers

SRES Writing Team and SRES Reviewers

II.1 SRES Writing Team

Joseph M. Alcamo	University of Kassel Kassel, Germany
Dennis Anderson	Oxford University Oxford, UK
Johannes Bollen	National Institute for Public Health and Environmental Hygiene (RIVM) Bilthoven, The Netherlands
Lex Bouwman	National Institute for Public Health and Environmental Hygiene (RIVM) Bilthoven, The Netherlands
Ogunlade R. Davidson	Co-chair of IPCC-WGIII University of Sierra Leone Fourah Bay College Freetown, Sierra Leone
Gerald R. Davis	Shell International Petroleum London, UK
Bert de Vries	National Institute for Public Health and Environmental Hygiene (RIVM) Bilthoven, The Netherlands
Michel den Elzen	National Institute for Public Health and Environmental Hygiene (RIVM) Bilthoven, The Netherlands
Jae Edmonds	Pacific Northwest National Laboratory (PNNL) Washington, DC, USA
Christopher Elvidge	NOAA National Geophysical Data Center Boulder, CO, USA
Jørgen Fenhann	Risø National Laboratory Roskilde, Denmark
Stuart R. Gaffin	Environmental Defense Fund New York, NY, USA
Henryk Gaj	ENERGSYS Ltd. Warsaw, Poland
Kenneth Gregory	Centre for Business and the Environment Middlesex, UK
Arnulf Grübler	International Institute for Applied Systems Analysis (IIASA) Laxenburg, Austria
William Hare	Greenpeace International Amsterdam, The Netherlands
Marco Janssen	National Institute for Public Health and Environmental Hygiene (RIVM) Bilthoven, The Netherlands
Kejun Jiang	National Institute for Environmental Studies (NIES) Tsukuba, Japan
Anne Johnson	US Environmental Protection Agency Washington, DC, USA
Tae-Yong Jung	Institute for Global Environmental Strategies (IGES) Kanagawa, Japan
Tom Kram	Netherlands Energy Research Foundation (ECN) Petten, The Netherlands
Eric Kreileman	National Institute for Public Health and Environmental Hygiene (RIVM) Bilthoven, The Netherlands
Emilio Lebre La Rovere	Universidade Federal do Rio de Janeiro Rio de Janeiro, Brazil
Mathew Luhanga	University of Dar es Salaam Dar es Salaam, United Rep. of Tanzania
Nicolette Manson	Inform, Inc. New York, NY, USA
Toshihiko Masui	National Institute for Environmental Studies (NIES) Tsukuba, Japan
Alan McDonald	International Institute for Applied Systems Analysis (IIASA) Laxenburg, Austria
Douglas McKay	Shell International Petroleum London, UK
Bert Metz	Co-chair of IPCC-WGIII National Institute for Public Health and Environmental Hygiene (RIVM) Bilthoven, The Netherlands
Laurie Michaelis	Oxford Centre for the Environment, Ethics and Society (OCEES) Oxford, UK
Shunsuke Mori	Science University of Tokyo Chiba, Japan
Tsuneyuki Morita	National Institute for Environmental Studies (NIES) Tsukuba, Japan
Nebojsa Nakićenović	International Institute for Applied Systems Analysis (IIASA) Laxenburg, Austria
William Pepper	ICF Consulting Fairfax, VA, USA
Hugh Martin Pitcher	Pacific Northwest National Laboratory (PNNL) Washington, DC, USA
Lynn Price	Lawrence Berkeley National Laboratory (LBNL) Berkeley, CA, USA
Keywan Riahi	International Institute for Applied Systems Analysis (IIASA) Laxenburg, Austria
R Alexander Roehrl	International Institute for Applied Systems Analysis (IIASA) Laxenburg, Austria
Hans-Holger Rogner	International Atomic Energy Agency (IAEA) Vienna, Austria
Alexei Sankovski	ICF Consulting Washington, DC, USA

Michael Schlesinger	University of Illinois Urbana, IL, USA
Priyadarshi Shukla	Indian Institute of Management Ahmedabad, India
Steven Smith	National Center for Atmospheric Research (NCAR) Boulder, CO, USA
Leena Srivastava	Tata Energy Research Institute (TERI) New Delhi, India
Robert Swart	National Institute for Public Health and Environmental Hygiene (RIVM) Bilthoven, The Netherlands
Sascha van Rooijen	Netherlands Energy Research Foundation (ECN) Petten, The Netherlands
Nadejda Victor	Rockefeller University New York, NY, USA
Cees Volkers	Netherlands Energy Research Foundation (ECN) Petten, The Netherlands
Robert Watson	IPCC Chairman World Bank Washington, DC, USA
John P. Weyant	Stanford University Stanford, CA, USA
Ernst Worrell	Lawrence Berkeley National Laboratory (LBNL) Berkeley, CA, USA
Xioashi Xing	Center for International Earth Science Information Network (CIESIN) New York, NY, USA
Zhou Dadi	State Planning Commission Chinese Academy of Sciences Beijing, China

II.2 SRES Reviewers

M. Abduli	Faculty of Environment Iran
H. Ahlgrimm	Institut für Technologie Germany
M. Aho	VTT Energy Finland
C. Albrecht	State Secretariat for Economic Affairs Switzerland
K. Alfsen	University of Oslo Norway
M. Anderson	US Department of Agriculture USA
P. Ashford	Arran Cottage United Kingdom
H. Audus	CRE Group Ltd. United Kingdom
C. Azar	Chalmers University of Technology Sweden
A. Baker	IEA Coal Industry Advisory Board United Kingdom
S. Baldwin	Office of Science and Technology Policy USA
W. Barbour	US Environmental protection Agency USA
W. Bjerke	IPAI - International. Primary Aluminiaar. Institute United Kingdom
P. Boeck	University of Gent Belgium
E. Bruci	Hydrometeorological Institute Albania
B. Burdick	Wuppertal Institut für Klima, Umwelt, Energie Germany
P. Burschel	Universität München Germany
T. Bye	Statistics Norway Norway
C.C. Change	US Department of Energy USA
R.C. Dahlman	US Department of Energy USA
R. Darwin	US Department of Agriculture USA
B. DeAngelo	US Environmental Protection Agency USA
U. Dethlefsen	Vattenfall Utveckling AB Sweden
Y. Ding	China Meteorological Administration China
C. Doyle	The Scottish Agricultural College United Kingdom
E.L. Fletzor	Department of State USA
R. Forte	US Environmental Protection Agency USA
L. Fresco	FAO Italy
T.R. Gerholm	Physics, prof.em. Sweden
M. Giroux	International Atomic Energy Agency Austria
R. Gommes	FAO Italy
A. Grambsch	US Environmental Protection Agency USA
V.R. Gray	Consultant New Zealand
K. Green	Volpe Center USA
H. Gruenspecht	Department of Energy USA
G. Guilpain	ELF ATOCHEM France
P. Hall	Canadian Forest Service, Natural Resources Canada

K. Heinloth	Universität Bonn Germany
W. Hohenstein	US Environmental Protection Agency USA
R. House	US Department of Agriculture USA
G. Hovsenius	ELFORSK - Swedish Electrical Utilities R&D Company Sweden
J. Hrubovcak	US Department of Agriculture USA
M. Hulme	University of East Anglia United Kingdom
H. Jacoby	Sloan School of Management USA
M. J. Jawson	US Department of Agriculture USA
A. Jhaveri	Seattle Regional Support Office USA
M. Jefferson	World Energy Council United Kingdom
V. Kagramanian	International Atomic Energy Agency Austria
S. Kane	National Oceanic and Atmospheric Administration USA
A. Khan	International Atomic Energy Agency Austria
F. Krambeck	Mobil USA
S. Kononov	International Atomic Energy Agency Austria
C. Kroeze	Wageningen Agricultural University The Netherlands
D. Kruger	US Environmental Protection Agency Washington DC, USA
L.L. Langlois	International Atomic Energy Agency Austria
D. Lashof	Natural Resources Defense Council USA
M. MacCracken	US Global Change Research Program USA
P. MacLaren	New Zealand Forest Research Institute New Zealand
J. Maues	Petrobras/Conpet Brazil
M. Mazur	US Department of Energy USA
A. McCulloch	ICI United Kingdom
P. Menna	ENEA Italy
S. Montzka	National Oceanic and Atmospheric Administration USA
L. Nadler	Mobil USA
P. Nagelhout	US Environmental Protection Agency USA
A.A. Niederberger	State Secretariat for Economic Affairs Switzerland
A. Olecka	Polish UNFCCC Executive Bureau Poland
Z. Ozdogan	Tubitak-Marmara Research Center Turkey
J. Pacyna	Norwegian Institute for Air Research (NILU) Norway
D. Pearlman	Climate Council USA
L. Perez	Ministerio del Ambiente y de los Recursos Naturales Venezuela
T. Pieper	US Global Change Research Program USA
R. Piltz	US Global Change Research Program USA
J. Pretel	Czech Hydrometeorological Institute Czech Republic
X. Querol	Spanish Council for Scientific Research Spain
B. Rhodes	US Environmental Protection Agency USA
D. Rothman	Colombia University USA
E. Scheehle	US Environmental Protection Agency USA
J. Shrouds	Federal Highway Administration USA
B. Sorensen	Roskilde University Denmark
T. Tajima	Ministry of International Trade and Industry Japan
T. Terry	US Department of Energy USA
D. Trilling	US Department of Transportation USA
P. Tseng	US Environmental Protection Agency USA
P. Tsui	Mobil USA
O. van Cleemput	Univeresity of Gent Belgium
E. Wilson	US Environmental Protection Agency USA
I. Yesserkepova	KazHydromet Kazakhstan
H. Yokota	Environment Agency of Japan Japan

III

Definition of SRES World Region

Definition of SRES World Regions

OECD90 REGION

North America (NAM)
Canada
Guam
Puerto Rico
United States of America
Virgin Islands

Western Europe (WEU)
Andorra
Austria
Azores
Belgium
Canary Islands
Channel Islands
Cyprus
Denmark
Faeroe Islands
Finland
France
Germany
Gibraltar
Greece
Greenland
Iceland
Ireland
Isle of Man
Italy
Liechtenstein
Luxembourg
Madeira
Malta
Monaco
Netherlands
Norway
Portugal
Spain
Sweden
Switzerland
Turkey

Pacific OECD (PAO)
Australia
Japan
New Zealand

REF REGION (countries undergoing economic reform)

Central and Eastern Europe (EEU)
Albania
Bosnia and Herzegovina
Bulgaria
Croatia
Czech Republic
The former Yugoslav Republic of Macedonia
Hungary
Poland
Romania
Slovak Republic
Slovenia
Yugoslavia

Newly independent states (NIS) of the former Soviet Union (FSU)
Armenia
Azerbaijan
Belarus
Estonia
Georgia
Kazakhstan
Kyrgyzstan
Latvia
Lithuania
Republic of Moldova
Russian Federation
Tajikistan
Turkmenistan
Ukraine
Uzbekistan

ASIA REGION

Centrally planned Asia and China (CPA)
Cambodia
China
Hong Kong
Korea (DPR)
Laos (PDR)
Mongolia
Viet Nam

South Asia (SAS)
Afghanistan
Bangladesh
Bhutan
India
Maldives
Nepal
Pakistan
Sri Lanka

Other Pacific Asia (PAS)
American Samoa
Brunei Darussalam
Fiji
French Polynesia
Gilbert-Kiribati
Indonesia
Malaysia
Myanmar
New Caledonia
Papua New Guinea
Philippines
Republic of Korea
Singapore
Solomon Islands
Taiwan, province of China
Thailand
Tonga
Vanuatu
Western Samoa

ALM REGION (Africa and Latin America)

Middle East and North Africa (MEA)
Algeria
Bahrain
Egypt (Arab Republic)
Iraq
Iran (Islamic Republic)
Israel
Jordan
Kuwait
Lebanon
Libya/SPLAJ
Morocco
Oman
Qatar
Saudi Arabia
Sudan
Syria (Arab Republic)
Tunisia
United Arab Emirates
Yemen

Latin America and the Caribbean (LAM)
Antigua and Barbuda
Argentina
Bahamas
Barbados
Belize
Bermuda
Bolivia
Brazil
Chile
Colombia
Costa Rica
Cuba
Dominica
Dominica
Dominican Republic
Ecuador
El Salvador
French Guyana
Grenada
Guadeloupe
Guatemala
Guyana
Haiti
Honduras
Jamaica
Martinique

Mexico
Netherlands Antilles
Nicaragua
Panama
Paraguay
Peru
Saint Kitts and Nevis
Santa Lucia
Saint Vincent and the Grenadines
Suriname
Trinidad and Tobago
Uruguay
Venezuela

Sub-Saharan Africa (AFR)

Angola
Benin
Botswana
British Indian Ocean Territory
Burkina Faso
Burundi
Cameroon
Cape Verde
Central African Republic
Chad
Comoros
Cote d'Ivoire Congo
Djibouti
Equatorial Guinea
Eritrea
Ethiopia
Gabon
Gambia
Ghana
Guinea
Guinea-Bissau
Kenya
Lesotho
Liberia
Madagascar
Malawi
Mali
Mauritania
Mauritius
Mozambique
Namibia
Niger
Nigeria
Reunion
Rwanda
Sao Tome and Principe
Senegal
Seychelles
Sierra Leone
Somalia
South Africa
Saint Helena
Swaziland
Tanzania
Togo
Uganda
Zaire
Zambia
Zimbabwe

IV

Six Modeling Approaches

Six Modeling Approaches

The SRES Terms of Reference call for a multi-model approach for developing emissions scenarios (see Appendix I). In all, six different modeling approaches were used to generate the 40 SRES scenarios. These six models are representative of the approaches to emissions scenario modeling and the different integrated assessment frameworks used in the scenario literature and include both macro-economic (so-called top-down) and systems-engineering (so-called bottom-up) models. Some modeling teams developed scenarios to reflect all four storylines, while some presented scenarios for fewer storylines. Chapter 4 lists all the SRES scenarios, by modeling group and by scenario family. The six modeling approaches include:

- Asian Pacific Integrated Model (AIM) from the National Institute of Environmental Studies in Japan (Morita *et al.*, 1994);
- Atmospheric Stabilization Framework Model (ASF) from ICF Consulting in the USA (Lashof and Tirpak, 1990; Pepper *et al.*, 1992, 1998; Sankovski *et al.*, 2000);
- Integrated Model to Assess the Greenhouse Effect (IMAGE) from the National Institute for Public Health and Environmental Hygiene (RIVM) (Alcamo *et al.*, 1998; de Vries *et al.*, 1994, 1999, 2000), used in connection with the Dutch Bureau for Economic Policy Analysis (CPB) WorldScan model (de Jong and Zalm, 1991), the Netherlands;
- Multiregional Approach for Resource and Industry Allocation (MARIA) from the Science University of Tokyo in Japan (Mori and Takahashi, 1999; Mori, 2000);
- Model for Energy Supply Strategy Alternatives and their General Environmental Impact (MESSAGE) from the International Institute of Applied Systems Analysis (IIASA) in Austria (Messner and Strubegger, 1995; Riahi and Roehrl, 2000); and the
- Mini Climate Assessment Model (MiniCAM) from the Pacific Northwest National Laboratory (PNNL) in the USA (Edmonds *et al.*, 1994, 1996a, 1996b).

IV.1. Asian Pacific Integrated Model

The Asian Pacific Integrated Model (AIM) is a large-scale computer simulation model for scenario analyses of greenhouse gas (GHG) emissions and the impacts of global warming in the Asian–Pacific region. This model is being developed mainly to examine global warming response measures in the region, but it is linked to a world model so that it is possible to make global estimates. AIM comprises three main models – the GHG emission model (AIM/emission), the global climate change model (AIM/climate), and the climate change impact model (AIM/impact).

The AIM-based quantification was conducted as an Asian collaborative project using a new linked version of the AIM/emission model, which covers the world but has a more detailed structure for the Asian–Pacific region than for other regions. The new linked version couples bottom-up models and top-down models (Figure IV-1).

The bottom-up models were prepared using the original AIM bottom-up components, which can reproduce detailed processes of energy consumption, industrial production, land-use changes, and waste management as well as technology development and social demand changes. However, two kinds of top-down models were prepared for this quantification:

- An energy–economic model based on the revised Edmonds–Reilly–Barns (ERB) Model, which can estimate interactions between energy sectors and economic sectors.
- An original land equilibrium model that can reproduce interactions between land-use changes and economic sectors.

The original AIM bottom-up components were integrated with these two top-down models through a newly developed linkage module. This new structure maximizes the ability to simulate a variety of inputs at a variety of levels, and to calculate future GHG emissions in a relatively full-range analysis.

The AIM model has nine regions for the energy–economic model and 17 regions for the bottom-up and land equilibrium models (see Table IV-1). Its time horizon is from 1990 to 2100. Before 2030, it uses 5-year time steps, but then jumps to 2050, 2075, and 2100. The GHGs and related gases include:

- Carbon dioxide (CO_2), methane (CH_4), nitrous oxide (N_2O), carbon monoxide (CO), non-methane volatile organic compounds (NMVOCs), nitrogen oxides (NO_x), and sulfur dioxide (SO_2) emissions from energy combustion–production processes.
- CO_2 from deforestation.
- CH_4 and N_2O from agricultural production.
- NMVOCs and SO_2 from biomass combustion.
- CO_2, CH_4, N_2O, NO_x, CO, NMVOCs, and SO_2 emissions from industrial processes, waste management, and land-use changes.

More detailed information can be obtained by referring to the web site: www-cger.nies.go.jp/ipcc/aim/

The ASF energy model consists of four end-use sectors (residential, commercial, industrial, and transportation). These sectors consume liquid fuels, solid fuels, gaseous fuels, and electricity. An electricity generation sector converts liquid fuels, solid fuels, gaseous fuels, nuclear energy, hydro energy, and solar energy into electricity. A synfuels sector converts coal and/or biomass into either a liquid or gaseous fuel. There is no direct consumption of solar energy or biomass by the end-use sectors.

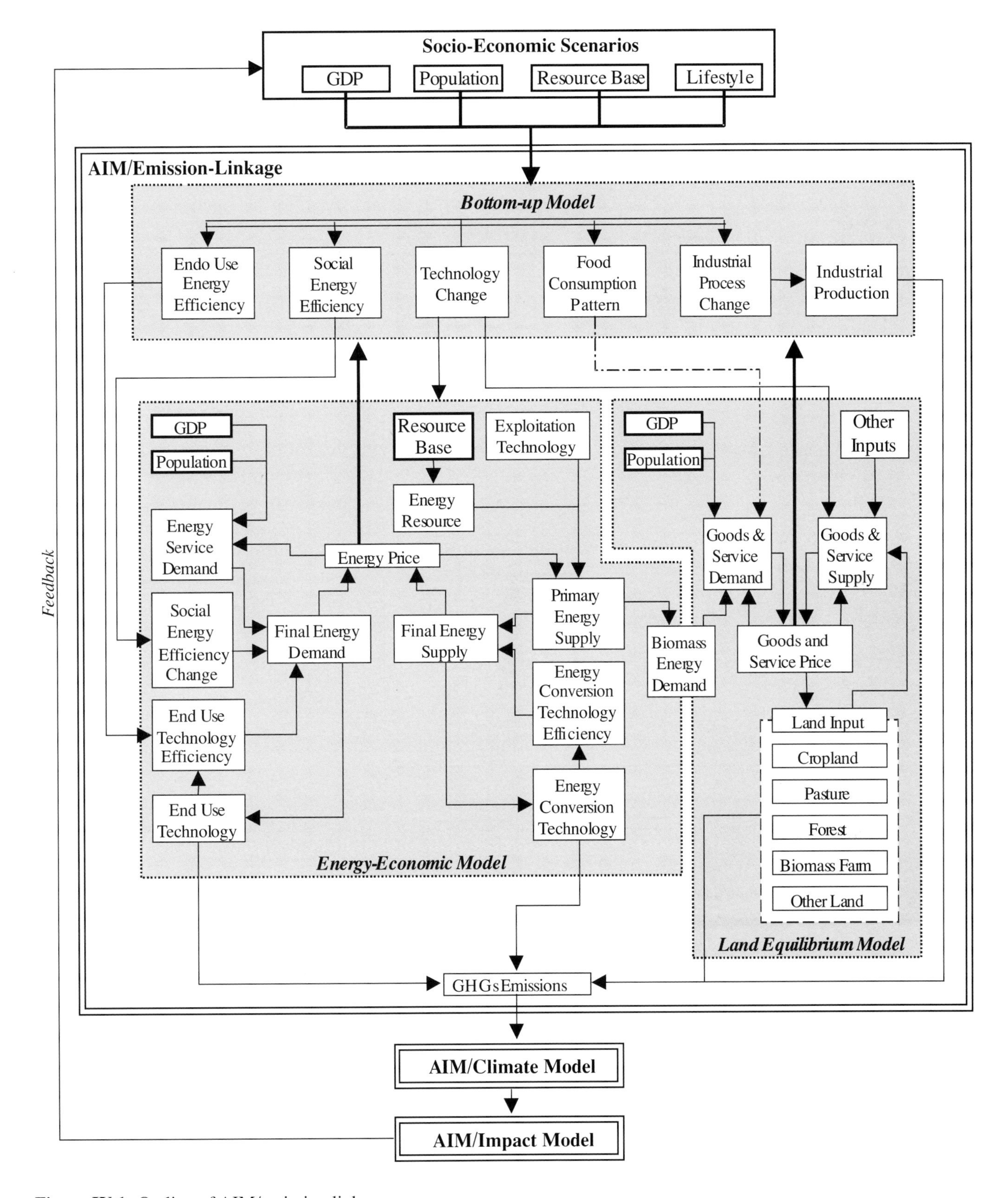

Figure IV-1: Outline of AIM/emission linkages.

Table IV-1: *Regional disaggregation of the six modeling approaches.*

<table>
<tr><th>SRES</th><th colspan="2">REF</th><th colspan="5">OECD90</th><th colspan="9">ASIA</th><th colspan="3">ALM</th></tr>
<tr><td>AIM</td><td colspan="2">Economies in Transition</td><td colspan="2">OECD-West</td><td>USA</td><td>Oceania</td><td>Japan</td><td>Korea</td><td>Indo-nesia</td><td>Thai-land</td><td>Mal-aysia</td><td>Other East Asia</td><td>India</td><td>Other South-Asia</td><td>China</td><td>Other Centrally Planned Asia</td><td>Middle East</td><td>Africa</td><td>Latin America</td></tr>
<tr><td>ASF</td><td colspan="2">Centrally Planned Europe</td><td colspan="2">OECD-West</td><td>USA</td><td colspan="2">OECD Asia-Pacific</td><td colspan="7">South East Asia</td><td colspan="2">Centrally Planned Asia</td><td>Middle East</td><td>Africa</td><td>Latin America</td></tr>
<tr><td>IMAGE</td><td>Former Soviet Union</td><td>Eastern Europe</td><td>OECD-Europe</td><td>Canada</td><td>USA</td><td>Oceania</td><td>Japan</td><td colspan="5">East Asia</td><td colspan="2">South Asia</td><td colspan="2">Centrally Planned Asia</td><td>Middle East</td><td>Africa</td><td>Latin America</td></tr>
<tr><td>MESSAGE</td><td>Former Soviet Union</td><td>Eastern Europe</td><td>Western Europe</td><td colspan="2">North America</td><td colspan="2">Pacific OECD</td><td colspan="5">Pacific Asia</td><td colspan="2">South Asia</td><td colspan="2">Centrally Planned Asia</td><td>Middle East & North Africa</td><td>Sub-Saharan Africa</td><td>Latin America</td></tr>
<tr><td>MARIA</td><td colspan="2">Eastern Europe and Former Soviet Union</td><td colspan="2">Other OECD</td><td>North Ame-rica</td><td>(Other OECD)</td><td>Japan</td><td colspan="5">ASEAN and Other Asia</td><td colspan="2">South Asia</td><td colspan="2">China</td><td colspan="3">ALM and others</td></tr>
<tr><td>MiniCAM</td><td colspan="2">Centrally Planned Europe</td><td>OECD-Europe</td><td>Canada</td><td>USA</td><td>Oceania</td><td>Japan</td><td colspan="7">South East Asia</td><td colspan="2">Centrally Planned Asia</td><td>Middle East</td><td>Africa</td><td>Latin America</td></tr>
</table>

Table IV-2: *ASF regions.*

Region	Countries
Africa (AFRICA)	All African countries
Centrally Planned Asia (CPASIA)	China, Laos, Mongolia, Korea (DPR), Vietnam
Eastern Europe and newly independent states (EENIS)	Albania, Bulgaria, Czech Republic, Hungary, Poland, Romania, former USSR, former Yugoslavia
Latin America (LAMER)	All Latin American countries (including Mexico, Central and South America)
Middle East (MEAST)	All Middle Eastern countries including Iran, Iraq, Kuwait, Qatar, Saudi Arabia, and UAE
OECD-East (OECDA)	Australia, Japan, New Zealand
OECD-West (OECDW)	Austria, Belgium, Canada, Denmark, Finland, France, Germany, Greece, Iceland, Ireland, Italy, Luxembourg, Netherlands, Norway, Portugal, Spain, Sweden, Switzerland, Turkey, United Kingdom
South East Asia and Oceania (SEASIA)	Afghanistan, Bangladesh, Bhutan, India, Indonesia, Malaysia, Republic of Korea, Burma, Pakistan, Philippines, Singapore, Thailand, and other countries of the region
USA (USA)	USA, Puerto Rico, and other US territories

IV.2. The Atmospheric Stabilization Framework (ASF) Model

The current version of ASF includes energy, agricultural, and deforestation GHG emissions and atmospheric models and provides emission estimates for nine world regions (Tables IV-1 and IV-2).

In the ASF model balancing the supply and demand for energy is achieved ultimately by adjusting energy prices. Energy prices differ by region to reflect regional market conditions, and by type of energy to reflect supply constraints, conversion costs, and the value of the energy to end users. ASF estimates the supply–demand balance by an iterative search technique to determine supply prices. These supply prices, which energy producers charge for the fuel at the wellhead or at the mine, are used to estimate the secondary energy prices in each region. These secondary prices are based on the supply price for the marginal export region, the interregional transportation cost, refining and distribution costs, and regional tax policies. For electricity, the secondary prices reflect the relative proportions of each fuel used to produce the electricity, the secondary prices of those fuels, the non-fossil costs of converting the fuels into electricity, and the conversion efficiency.

The agricultural ASF model estimates the production of major agricultural products, such as meat, milk, and grain, which is driven by population and gross national product (GNP) growth. This model is linked with the ASF deforestation model, which estimates the area of land deforested annually as a function of population growth and demand for agricultural products.

The ASF GHG emissions model uses outputs of the energy, agricultural, and deforestation models to estimate GHG emissions in each ASF region. These emissions are estimated by mapping GHG emission sources to the corresponding emission drivers and changing them according to changes in these drivers. For example, CH_4 emissions from landfills are mapped to population, while CO_2 emissions from cement production are mapped to GNP.

Finally, the ASF atmospheric model uses GHG emission estimates to calculate GHG concentrations, and corresponding radiative forcing and temperature effects. A detailed description of the ASF is provided in the ASF 1990 Report to Congress (Lashof and Tirpak, 1990), and recent applications of the model are reported in Pepper *et al.* (1998) and Sankovski *et al.* (2000).

IV.3. Integrated Model to Assess the Greenhouse Effect

The Integrated Model to Assess the Greenhouse Effect (IMAGE 2) consists of three fully linked systems of models:

- The Energy–Industry System (EIS).
- The Terrestrial Environment System (TES).
- The Atmosphere–Ocean System (AOS).

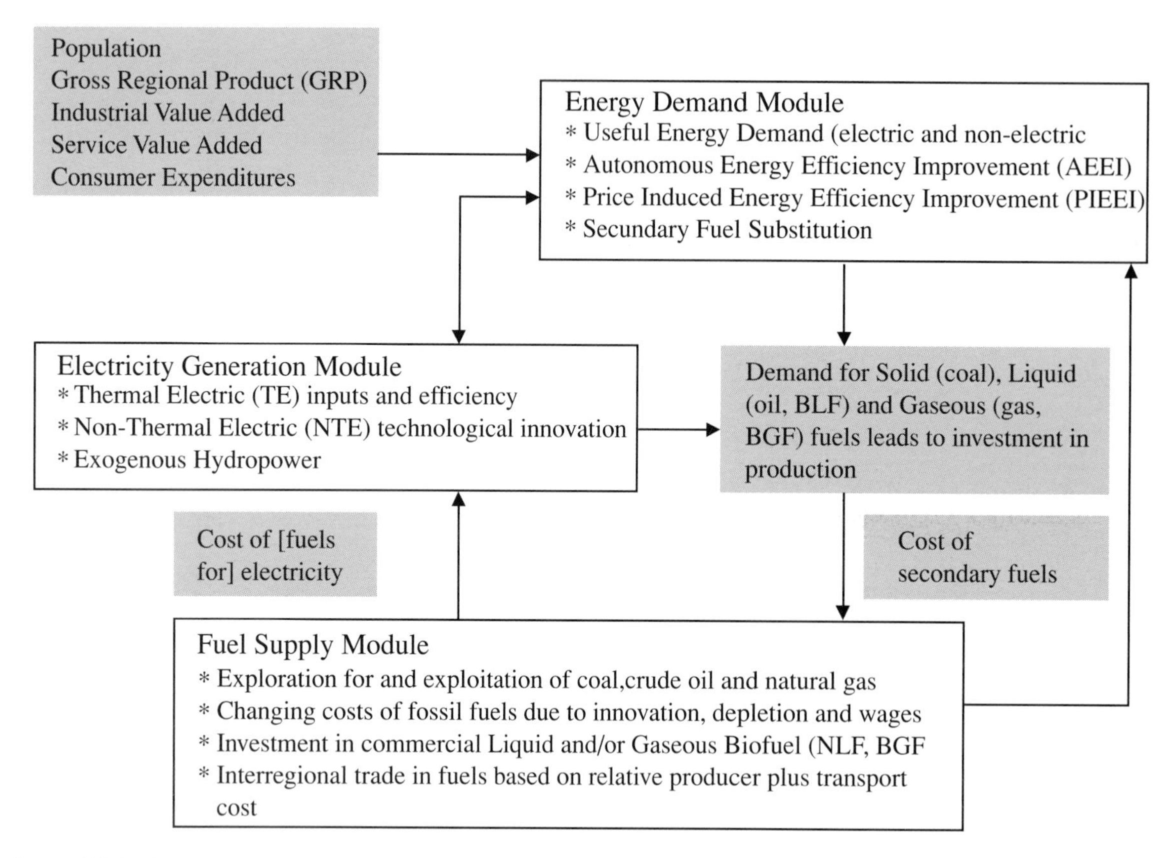

Figure IV-2: Overview of the IMAGE EIS/TIMER model.

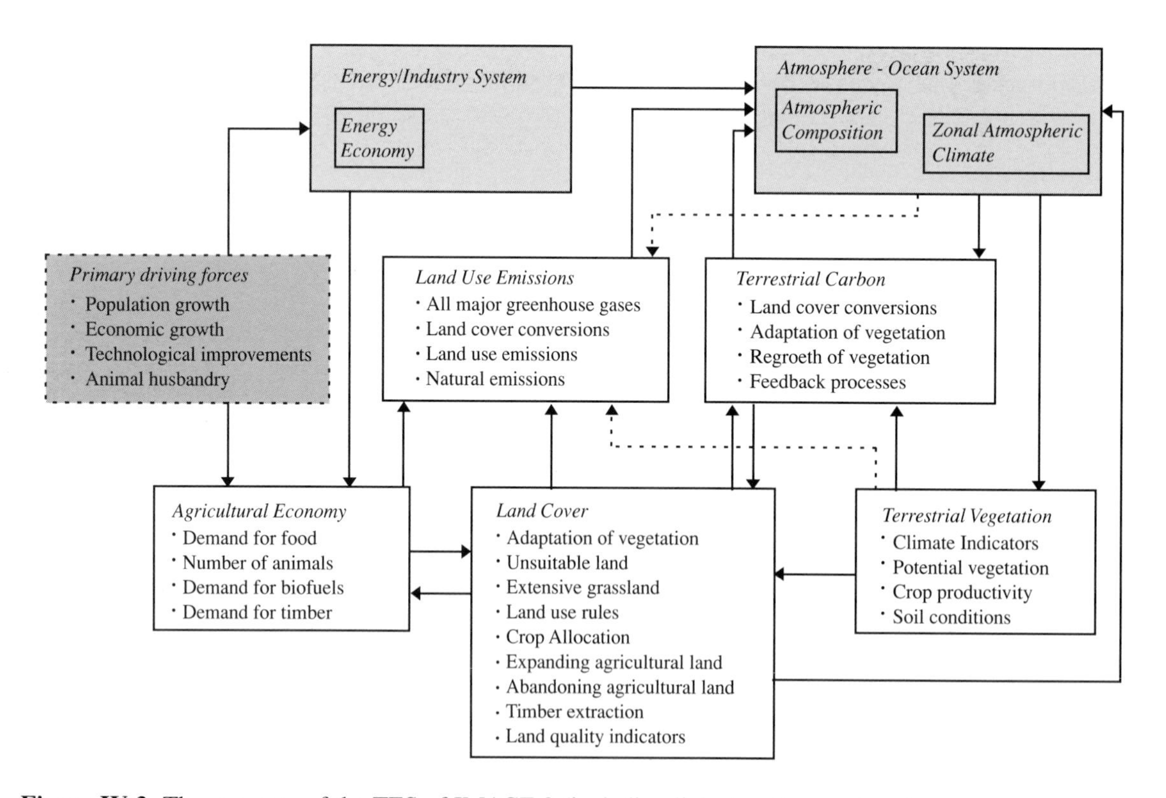

Figure IV-3: The structure of the TES of IMAGE 2 (including links to other modules).

***Table IV-3:** IMAGE 2 regions (see also Table IV-1).*

Canada
USA
Latin America (Central and South)
Africa
OECD Europe
Eastern Europe
CIS (former Soviet Union)
Middle East
India (including Bangladesh, Bhutan, India, Myanmar, Nepal, Pakistan, Sri Lanka)
China (including China, Korea (DPR), Kampuchea, Laos, Mongolia, Vietnam)
East Asia South (Indonesia, Republic of Korea, Malaysia, Philippines, Thailand)
Oceania
Japan

EIS computes the emissions of GHGs in 13 world regions (Tables IV-1 and IV-3). The energy-related emissions are based on the Targets Image Energy Regional (TIMER) simulation model (Figure IV-2). TIMER is a systems dynamics model with investment decisions in energy efficiency, electricity generation, and energy supply based on anticipated demand, relative costs or prices, and institutional and informational delays. The model uses five economic sectors. Technological change and fuel price dynamics influence energy intensity, fuel substitution, and the penetration of non-fossil options such as solar electricity and biomass-based fuels.

The objective of TES is to simulate global land-use and land-cover changes and their effect on emissions of GHGs and ozone precursors, and on carbon fluxes between the biosphere and the atmosphere (Figure IV-3). This subsystem can be used to:

- Evaluate the effectiveness of land-use policies to control the build-up of GHGs.
- Assess the land consequences of large-scale use of biofuels.
- Evaluate the impact of climate change on global ecosystems and agriculture.
- Investigate the effects of population, economic, and technological trends on changing global land cover.

More detailed information can be obtained by referring to the following web site: http://sedac.ciesin.org/mva/.

IV.4. Model for Energy Supply Strategy Alternatives and their General Environmental Impact (MESSAGE)

A set of integrated models was used to formulate the SRES scenarios at IIASA (Nakićenović, *et al.*, 1998). Model for Energy Supply Strategy Alternatives and their General Environmental Impact (MESSAGE) is one of the six models that constitute IIASA's integrated modeling framework (Messner and Strubegger, 1995; Riahi and Roehrl, 2000; Roehrl and Riahi, 2000).

The scenario formulation process starts with exogenous assumptions about population and per capita economic growth by region. Energy demand (defined at the useful energy level) is derived using the Scenario Generator (SG) model, a dynamic model of future economic and energy development. It combines extensive historical data about economic development and energy systems with empirically estimated equations of trends to determine future structural change. For each scenario, SG generates future paths of energy use consistent with historical dynamics and with the specific scenario features (e.g., high or moderate economic growth, rapid or more gradual energy intensity improvements).

The economic and energy development profiles serve as inputs for the energy systems engineering model MESSAGE (Messner and Strubegger, 1995; Riahi and Roehrl, 2000; Roehrl and Riahi, 2000) and the macro-economic model MACRO (Manne and Richels, 1992). MESSAGE is a dynamic linear programming model that calculates cost-minimal supply structures under the constraints of resource availability, the menu of given technologies, and the demand for useful energy. It estimates detailed energy system structures, including energy demand, supply, and emissions patterns, consistent with the evolution of the energy demand produced by SG. MACRO is a modified version of the Global 2100 model, originally published in 1992 (Manne and Richels, 1992) and subsequently used widely in many energy studies around the world. MACRO maximizes the inter-temporal utility function of a single representative producer–consumer in each world region and estimates the relationships between macro-economic development and energy use. MESSAGE and MACRO are linked and used in tandem to test scenario consistency because they correspond to the two different perspectives from which energy modeling is usually carried out – top-down (MACRO) and bottom-up (MESSAGE).

The impacts of energy price changes on energy demand and gross domestic product (GDP) growth are estimated by iterating shadow prices from MESSAGE and energy demands from the MACRO model. The iteration is repeated until energy intensities and GDP growth rates are consistent with the output of the SG model adopted as exogenous input assumptions at the beginning of the scenario formulation process. The demand reductions caused by increasing energy prices in the B2 marker compared to a hypothetical case with constant energy prices were calculated with MACRO. Compared to this hypothetical case the price-induced energy demand savings in the B2 marker are 8% by 2020, 23% by 2050, and 30% by 2100. Table IV-4 gives the shadow prices for international trade for gas, oil, and coal in the B2 marker. Table IV-5 summarizes the regional ranges for extraction costs of gas, oil, and coal in the B2 marker.

The atmospheric concentrations of GHGs and the resultant warming potentials can be estimated by the Model for the

Table IV-4: Shadow prices for international trade in the B2 marker (1990US$/GJ).

Year	Gas	Coal	Oil
2020	0.4	0.3	0.5
2050	0.7	0.4	1.1
2100	0.7	1.1	2.3

Table IV-5: Ranges of extraction costs for the four SRES regions in the B2 marker (1990US$/GJ).

Year	Gas	Coal	Oil
2020	(0.2-0.3)	(0.2-0.3)	(0.1-0.4)
2050	(0.3-0.6)	(0.2-0.3)	(0.4-0.6)
2100	(0.5-0.8)	(0.4-0.7)	(0.5-0.7)

Assessment of Greenhouse Gas-Induced Climate Change (MAGICC), a carbon cycle and climate change model developed by Wigley *et al.* (1994).

Figure IV-4 illustrates the IIASA integrated modeling framework and shows how the models are linked (Nakićenović, *et al.*, 1998). Of the six models shown in Figure IV-4, four (SG, MESSAGE, MACRO, and MAGICC) were used for the formulation and analysis of SRES scenarios, including the B2 marker scenario. In addition the MESSAGE model was used to quantify all four scenario groups of the A1 storyline and scenario family and a number of scenarios of the B1 storyline and scenario family. Altogether, the IIASA team formulated nine SRES scenarios, including the B2 marker.

The other two models shown in Figure IV-4, RAINS and BLS, were not used to model the SRES scenarios. RAINS (Alcamo *et al.*, 1990) is a simulation model of sulfur and NO_x emissions, their subsequent atmospheric transport, chemical transformations of those emissions, deposition, and ecological impacts. BLS (Fischer *et al.*, 1988, 1994) is a sectorial macro-economic model that accounts for all major inputs (such as land, fertilizer, capital, and labor) required for the production of 11 agricultural commodities.

The IIASA model set covers energy sector and industrial emission sources only. Agricultural and land-use related emissions for the B2 marker scenario and other SRES

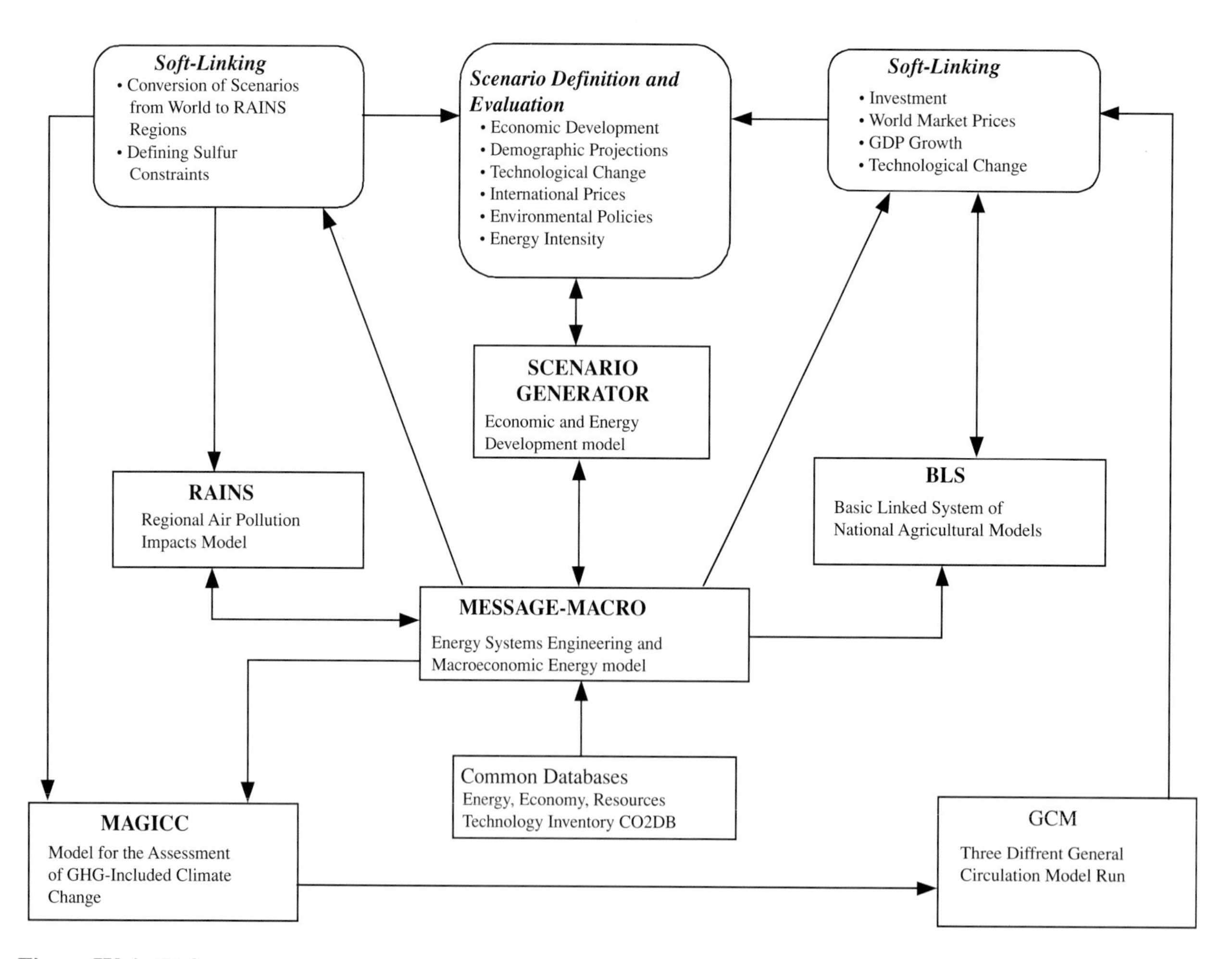

Figure IV-4: IIASA integrated modeling framework (Nakićenović, *et al.*, 1998).

Table IV-6: *Assumptions on cumulative resources and extraction costs as used in MARIA (source: based on Rogner, 1997).*

	Coal		Oil		Natural Gas	
	Grade A-C	Grade D-E	Grade I-III	Grade IV-VIII	Grade I-III	Grade IV-VIII
World occurrences	53	205	12	98	16	820
Cost	0.2-2.8	2.8-6.3	< 4.4	4.4-28.0	< 4.4	4.4-25.4

(1) Resources are in ZJ and extraction costs are in 1990US$/GJ (in the model itself costs are given in 1990US$/barrel oil equivalent).

(2) Coal resources include brown coal.

(3) Grade I–III and Grade A–C, conventional resources; Grade I and A, proved recoverable reserves; Grade II and B, additional recoverable resources; Grade III and C, additional speculative (identified) reserves.

(4) Grade IV, enhanced recovery, Grade V–VIII, unconventional resources and reserves; Grade VII–VIII, additional occurrences; Grade D-E, additional resources.

scenarios were derived from corresponding quantifications by the AIM model. They are consistent with the energy-related emissions because they are based on assumptions about the main driving forces that are in line with those in the quantifications with the MESSAGE model.

More detailed information can be obtained by referring to the web site: http://www.iiasa.ac.at/Research/ECS/.

IV.5. The Multiregional Approach for Resource and Industry Allocation Model (MARIA)

The Multiregional Approach for Resource and Industry Allocation Model (MARIA) is a compact integrated assessment model to assess the interrelationships among the economy, energy, resources, land use, and global climate change (Mori and Takahashi, 1999; Mori 2000). The origin of the model is the Dynamic Integrated Model of Climate and the Economy (DICE) model, developed by Nordhaus (1994). Involving energy flows and dividing the world into regions, MARIA has been developed to assess technology and policy options to address global warming. Like Global 2100 developed by Manne and Richels (1992), MARIA is currently an intertemporal non-linear optimization model that deals with international trading among eight regions – NAM (USA and Canada), Japan, Other OECD countries, China, ASEAN countries (Indonesia, Malaysia, Philippines, Singapore, Republic of Korea, Thailand), SAS (India, Bangladesh, Pakistan, Sri-Lanka), EEFSU (Eastern Europe and the Former Soviet Union), and ALM (Africa and Latin America). It also encompasses energy flows and simplified food production and land use changes to show the potential contribution of biomass.

Economic activities are represented by a constant elasticity of substitution (CES) production function with capital stock, labor, electricity, and non-electric energy set for the above eight world regions. Future GDP growth is projected by considering potential GDP growth rates (the product of two exogenous assumptions – population and potential per capita GDP growth) as well as endogenously determined energy costs and prices. The energy module in MARIA involves three fossil primary energy resources (i.e., coal, natural gas, and oil), biomass, nuclear power, and renewable energy technologies (e.g., hydropower, solar, wind, and geothermal). Energy demand consists of industry, transportation, and other public uses. Nuclear fuel recycling technologies are simply but explicitly formulated. Carbon sequestration technologies are also taken into account. Typically, MARIA basically generates resource extraction profiles in which gas is mainly used in the first half of the 21st century, and subsequently carbon-free sources (e.g., solar, nuclear, and biomass) and coal assume the main roles in the second half of the 21st century.

Energy costs in the model consist of energy production and utilization costs. Market prices are determined endogenously on the basis of model-calculated shadow prices. Among various parameters, the extraction costs of fossil fuel resources and energy conversion cost coefficients contribute substantially to determining the model's energy mix and emissions. The latest model version, MARIA-8, applied Rogner's estimates on fossil resource availability (Rogner, 1997). For the sake of simplicity, the fossil resource and reserve categories are aggregated into two classes, assuming a quadratic production function to interpolate the relationships between resource occurrences and extraction costs. Corresponding model parameters are summarized in Table IV-6. The cost coefficients of energy conversion technologies are basically extracted from the GLOBAL 2100 model (Manne and Richels, 1992). The basic values used in the case of the B2 scenario are illustrated in Table IV-7, and important model parameters deployed for MARIA's other SRES scenario quantifications are shown in Table IV-8. Other energy-related cost parameters correspond to renewable energy sources, methanol and ethanol processes, nuclear fuel recycle, carbon sequestration, etc. They are described in more detail in Mori and Takahashi (1999).

International trade prices are generated by the Lagrange multipliers of the corresponding constraints as a feature of optimization models. The Negishi weight technique was

Table IV-7: *Illustration of basic energy conversion cost coefficients in MARIA used for calculating the SRES B2 Scenario.*

	COAL	OIL	GAS	BIO
IND	6.00	2.50	3.25	4.15
TRN	8.58	3.43	4.56	5.02
PUB	6.00	2.50	3.25	4.15
ELC	51.00	12.20	13.70	15.76

IND, TRN, PUB, and ELC denote industry, transportation, public and other services, and electric power generation sectors. The values in the first three rows (non-electricity) are millions per MJ. Those in the last row are millions per kWh.

employed to assess the international equilibrium prices of tradable goods under the budget constraints (Negishi, 1972). Illustrative international energy trade prices for scenario B2 are summarized in Chapter 4 and are not repeated here.

The Global Warming Subsystem in MARIA is based on Wigley's five-time constant model for the emission-concentration mechanism. A two-level thermal reservoir model is also employed following the DICE model (Wigley, 1994; Nordhaus, 1994). Only global carbon emissions are currently treated in this model component.

MARIA's Food and Land Use module serves to assess the potential contributions of biomass. A simplified food demand and land-use subsystem was included. Nutrition, calorie, and protein demand is a function of per capita income. Either directly or via meat, crop and pasture supply these demands. Forests are a source of biomass and wood products, but also their function as a carbon sink is evaluated. The relationships among the above-mentioned subsystems are shown in Figure IV-5.

Since MARIA is designed for macro-level evaluation of various options consistently, detailed information, such as gridded SO_2 emissions, industrial structure change, and urbanization issues, is not generated. However, MARIA can provide long-term profiles of fuel mix changes and possible trade premiums under various scenarios.

More detailed information can be obtained by referring to the following web site: http://shun-sea.ia.noda.sut.ac.jp/indexj.html.

IV.6. The Mini Climate Assessment Model

The Mini Climate Assessment Model (MiniCAM) is a small rapidly running Integrated Assessment Model that estimates global GHG emissions with the ERB model (Edmonds *et al.*, 1994, 1996a) and the agriculture, forestry and land-use model (Edmonds *et al.*, 1996b). MiniCAM uses the Wigley and Raper MAGICC (Wigley and Raper, 1993) model to estimate climate changes, the Hulme *et al.* (1995) SCENGEN tool to estimate regional climate changes, and the Manne *et al.* (1995) damage functions to examine the impacts of climate change. MiniCAM, developed by the Global Change Group at Pacific Northwest Laboratory, undergoes regular enhancements. Recent changes include the addition of an agriculture land-use module and the capability to estimate emissions of all the Kyoto gases.

At present the model consists of 11 regions (USA, Canada, Western Europe, Japan, Australia, Eastern Europe and the Former Soviet Union, Centrally Planned Asia, the Mid-East, Africa, Latin America, and South and East Asia) that provide complete world coverage (see Table IV-1). A 14-region version is nearing completion.

MiniCAM uses a straightforward population times labor productivity process to estimate aggregate labor productivity levels. The resultant estimate of GNP is corrected for the impact of changes in energy prices using GNP/energy elasticity. For the scenario exercise, an extended economic activity level process was developed to allow a clearer understanding of the potential impacts of the new population scenarios. First, a detailed age breakdown was included so working age populations could be computed. Second, a labor force participation rate was added to estimate the labor force, and third an external process was created to estimate the long-term evolution of the rate of labor productivity increase.

ERB is a partial equilibrium model that uses prices to balance energy supply and demand for the seven major primary energy categories (coal, oil, gas, nuclear, hydro, solar, and biomass) in the eleven regions in the model.

The energy demand module initially estimates demands for three categories of energy services (residential/commercial, industrial, and transportation) as a function of price and income. Energy services are provided by four secondary fuels (solids, liquids, gases, and electricity). Demand for the secondary fuels depends upon their relative costs and the

Table IV-8: *Parameter adjustments to meet the key driving forces interpretation of the SRES scenario storylines.*

Storylines	Potential economic growth rates	Autonomous energy efficiency	Potential cropland	Energy cost coefficients of coal
A1	High	Middle	High	260% of gas
B1	Middle	High	High	250% of gas
B2	Low	Low	Low	185% of gas

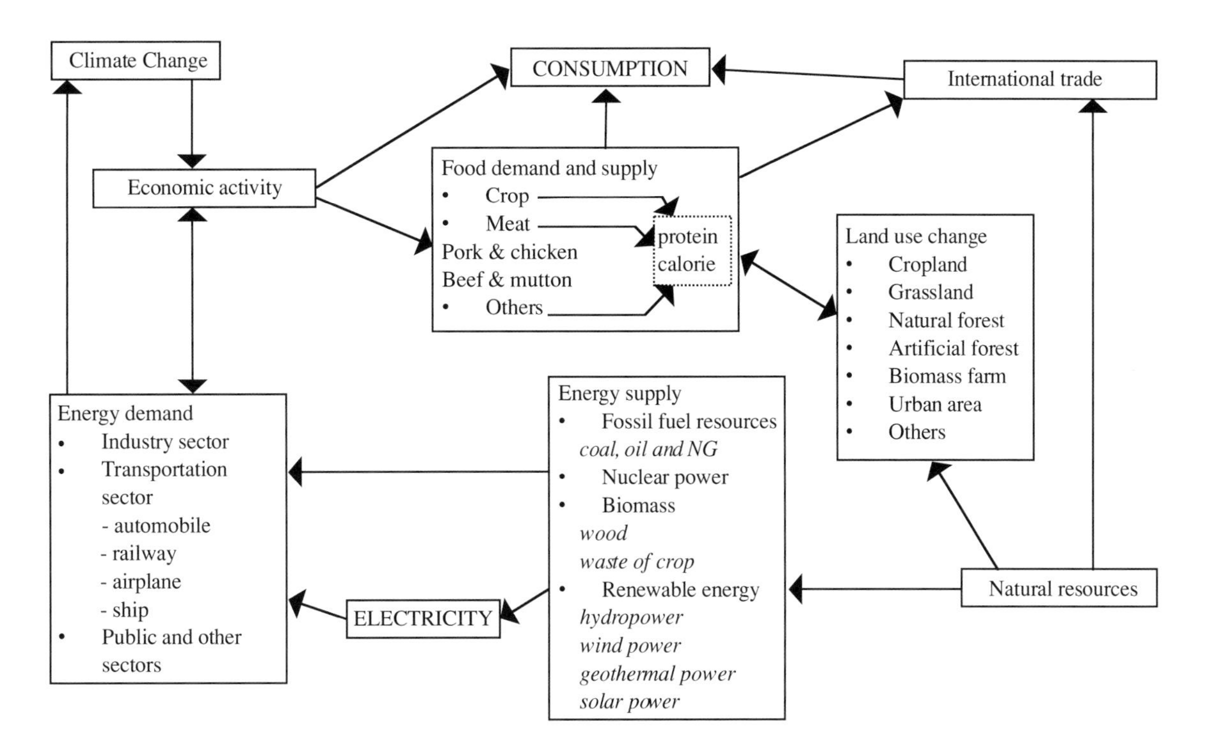

Figure IV-5: Structure of the MARIA model of one region.

evolution of the end-use technologies, represented by the improvement in end-use energy efficiency. Demand for primary fuels is determined by the relative costs of transforming them into the secondary fuels. Nuclear, solar, and hydro are directly consumed by the electricity sector, while coal and biomass can be transformed into gas and liquids if the fossil oil and gas become too expensive or run out. Hydrogen has recently been added to the model, and it, like refined gas and oil, can be used to generate electricity or as a secondary fuel for the three final demand sectors.

The energy supply sector provides both renewable (hydro, solar, and biomass) and non-renewable (coal, oil, gas, and nuclear) resources. The cost of the fossil resources relates to the resource base by grade, the cost of production (both technical and environmental), and to historical production capacity. The introduction of a graded resource base for fossil fuel allows the model to test explicitly the importance of fossil fuel resource constraints as well as to represent unconventional fuels such as shale oil and methane hydrates. For unconventional fuels only small amounts are available at low costs, but large amounts are potentially available at high cost, or after extensive technology development. Fuel-specific rates of technical change are available for primary fuel production and conversion, as are technical change coefficients for each category of electricity production.

Biomass is supplied by the agriculture sector, and provides the link between the agriculture, forestry, and land-use module and the energy module. The former module estimates the allocation of land to one of five activities (crops, pasture, forestry, modern biomass, and other) in each region. This allocation reflects the relative profitability of each of these uses. Profitability is determined by the prices for crops, livestock, forest products, and biomass, which reflect regional demand and supply functions for each product. There are separate technical change coefficients for crops, livestock/pasture, forestry, and modern biomass production.

Once the model has reached equilibrium for a period, emissions of GHGs are computed. For energy, emissions of CO_2, CH_4, and N_2O reflect fossil fuel use by type of fuel, while agriculture emissions of these gases reflect land-use change, the use of fertilizer, and the amount and type of livestock produced. The high global warming gases (chlorofluorocarbons, hydrochlorofluorocarbons, hydrofluorocarbons, and perfluorocarbons) are estimated only for each category and not by their individual components. Sulfur emissions are estimated as a function of fossil fuel use and reflect sulfur controls, the effectiveness of which is determined by a Kuznets curve that relates control levels to per capita income.

The emissions estimates are aggregated to a global level and used as inputs to MAGICC to produce estimates of GHG concentrations, changes in radiative forcing, and consequent changes in global mean temperature. The global mean temperature change is used to drive SCENGEN-derived changes in climate patterns and to produce estimates of regional change in temperature, precipitation, and cloud cover. Finally, the regional changes in temperature are used to estimate market and non-market based damages. Developing-region damage functions produce higher damages than those for developed regions, reflecting the higher vulnerability of regions with low per capita income.

More detailed information can be obtained by referring to the following web site: http://sedac.ciesin.org/mva/.

References:

Alcamo, J., R. Shaw, and L. Hordijk, 1990: *The RAINS model of acidification: Science and Strategies in Europe*, Kluwer Academic Publishers, Dordrecht, The Netherlands.

Alcamo, J., R. Leemans, and E. Kreileman (eds.), 1998. *Global Change Scenarios of the 21st Century. Results from the IMAGE 2.1 model.* Elsevier Science, London, UK.

De Jong, A., and G. Zalm, 1991: Scanning the Future: A long-term scenario study of the world economy 1990-2015. In *Long-term Prospects of the World Economy.* OECD, Paris, pp. 27-74.

De Vries, H.J.M., J.G.J. Olivier, R.A. van den Wijngaart, G.J.J. Kreileman, and A.M.C. Toet, 1994: Model for calculating regional energy use, industrial production and greenhouse gas emissions for evaluating global climate scenarios. *Water, Air Soil Pollution*, **76**, 79-131.

De Vries, B., M. Janssen, and A. Beusen, 1999: Perspectives on global energy futures – simulations with the TIME model. *Energy Policy* **27**, 477-494.

De Vries, B., J. Bollen, L. Bouwman, M. den Elzen, M. Janssen, and E. Kreileman, 2000: Greenhouse gas emissions in an equity-, environment- and service-oriented world: An IMAGE-based scenario for the next century. *Technological Forecasting & Social Change*, **63**, (2-3) (in press).

Edmonds, J., M. Wise, and C. MacCracken, 1994: *Advanced Energy Technologies and Climate Change: An Analysis Using the Global Change Assessment Model (GCAM)*, PNL-9798, UC-402. Pacific Northwest National Laboratory, Richland, WA, USA.

Edmonds, J., M. Wise, H. Pitcher, R. Richels, T. Wigley, and C. MacCracken, 1996a: An integrated assessment of climate change and the accelerated introduction of advanced energy technologies: An application of MiniCAM 1.0. *Mitigation and Adaptation Strategies for Global Change*, **1(4)**, 311-339.

Edmonds, J., M. Wise, R. Sands, R. Brown, and H. Kheshgi, 1996b: *Agriculture, land-use, and commercial biomass energy.* A Preliminary integrated analysis of the potential role of Biomass Energy for Reducing Future Greenhouse Related Emissions. PNNL-11155, Pacific Northwest National Laboratories, Washington, DC.

Fischer, G., K. Frohberg, M.A. Keyzer, and K.S. Parikh, 1988*: Linked National Models: A Tool for International Policy Analysis,* Kluwer Academic Publishers, Dordrecht, Netherlands.

Fischer G., K. Frohberg, M.L. Parry, and C. Rosenzweig, 1994: Climate Change and World Food Supply, Demand and Trade: Who Benefits, Who Looses? *Global Environmental Change*, **4/1**, 7-23.

Hulme, M., T. Jiang, and T. Wigley, 1995: *SCENGEN: A Climate Change SCENario GENerator: Software User Manual, Version 1.0.* Climate Change Research Unit, School of Environmental Sciences, University of East Anglia, Norwich, United Kingdom.

Lashof, D., and Tirpak, D.A., 1990: *Policy Options for Stabilizing Global Climate.* 21P-2003. U.S. Environmental Protection Agency, Washington D.C.

Manne, A.S., and R.G. Richels, R, 1992: *Buying Greenhouse Insurance, The Economic Costs of CO_2 Emissions Limits*, MIT Press, Cambridge, MA, USA.

Manne, A.S., R. Mendelsohn, and R. Richels, 1995: MERGE — A Model for Evaluating Regional and Global Effects of GHG Reduction Policies. *Energy Policy*, **23(1),** 17-34.

Messner, S., and M. Strubegger, 1995: *User's Guide for MESSAGE III*, WP-95-69, International Institute for Applied Systems Analysis, Laxenburg, Austria.

Mori, S., and M. Takahashi, 1999: An integrated assessment model for the evaluation of new energy technologies and food productivity. *International Journal of Global Energy Issues*, **11**(1-4), 1-18.

Mori, S., 2000: The development of greenhouse gas emissions scenarios using an extension of the MARIA model for the assessment of resource and energy technologies. *Technological Forecasting & Social Change*, **63,** (2-3) (in press).

Morita, T., Y. Matsuoka, I. Penna, and M. Kainuma, 1994: *Global Carbon Dioxide Emission Scenarios and Their Basic Assumptions: 1994 Survey.* CGER-1011-94. Center for Global Environmental Research, National Institute for Environmental Studies, Tsukuba, Japan.

Nakićenović, N., A. Grübler, and A. McDonald (eds.), 1998: *Global Energy Perspectives.* Cambridge University Press, Cambridge. (see also http://www.iiasa.ac.at/cgi-bin/ecs/book_dyn/bookcnt.py)

Negishi, T., 1972: *General Equilibrium Theory and International Trade*, American Elsevier, New York.

Nordhaus, W., 1994: *Managing the Global Commons*, MIT Press, Cambridge, MA, USA.

Pepper, W.J., J. Leggett, R. Swart, J. Wasson, J. Edmonds, and I. Mintzer, 1992: Emissions Scenarios for the IPCC. An update: Assumptions, methodology, and results, Support document for Chapter A3. In *Climate Change 1992: Supplementary Report to the IPCC Scientific Assessment.* J.T. Houghton, B.A. Callandar, S.K. Varney (eds.), Cambridge University Press, Cambridge.

Pepper, W.J., Barbour, W., Sankovski, A., and Braaz, B., 1998: No-policy greenhouse gas emission scenarios: revisiting IPCC 1992. *Environmental Science & Policy* **1**:289-312.

Riahi, K., and R.A. Roehrl, 2000: Greenhouse gas emissions in a dynamics as usual scenario of economic and energy development. *Technological Forecasting & Social Change*, **63**, (2-3) (in press).

Roehrl, R.A., and K. Riahi, Greenhouse gas emissions mitigation and the role of technology dynamics and path dependency – a cost assessment. *Technological Forecasting & Social Change*, **63**, (2-3) (in press).

Rogner, H-H., 1997: An assessment of world hydrocarbon resources. *Ann. Rev. Energy Environ.*, **22**, 217-262.

Sankovski, A., W. Barbour, and W. Pepper, 2000: Quantification of the IS99 emission scenario storylines using the atmospheric stabilization framework (ASF). *Technological Forecasting & Social Change*, **63,** (2-3) (in press).

Timmer, H., 1998: WorldScan – A world model. *Quarterly Review Netherlands Bureau for Economic Policy Analysis* **3,** 37-40.

Wigley, T.M.L., and S.C.B. Raper, 1993: Future changes in global mean temperature and sea level. In *Climate and Sea Level Change: Observations, Projections and Implementation*, R.A. Warwick, E.M. Barrow, and T.M.L. Wigley (eds). Cambridge University Press, Cambridge, UK, pp. 111-113.

Wigley, T.M.L., 1994: Reservoir timescales for anthropogenic CO_2 in the atmosphere. *TELLUS*, **46B**, 378-389

Wigley, T.M.L., M. Solomon, and S.C.B. Raper, 1994: *Model for the Assessment of Greenhouse-gas Induced Climate Change* Version 1.2, Climate Research Unit, University of East Anglia, UK.

V

Database Description

Database Description

The SRES Emission Scenario Database (ESD) was developed to manage and access a large number of data sets and emissions scenarios documented in the literature. The SRES Terms of Reference call for the assessment of emissions scenarios in the literature (see Appendix I). The database was developed for SRES by the National Institute for Environmental Studies (NIES) of Japan and can be accessed via the ftp site www-cger.nies.go.jp/cger-e/db/ipcc.html. This section summarizes the database structure and the data collection for the database. Chapter 2 gives further detail about the quantitative assessment of the scenarios in the database. At the time of writing the database included 416 scenarios from 171 sources.

V.1. Database Structure

The main purpose behind the development of the new database is to make it easier to manage and utilize the vast amounts of data related to emission scenarios of greenhouse gases (GHGs), which include carbon dioxide (CO_2), nitrous oxide (N_2O), methane (CH_4), sulfur oxides (SO_x), and related gases, (such as carbon monoxide (CO), nitrogen oxides (NO_x), and hydrofluorocarbons (HFCs). The need for such a database is a function of both the increasing number of emission scenarios (because of increasing political and research interests in this topic) and the necessity to identify the strengths and weaknesses of current scenarios (to allow research to be focused on the most crucial or under-investigated areas).

These emission scenarios have been quantified mainly using computer simulation models, which in turn utilize many assumptions on factors such as population growth, gross domestic product (GDP) growth, technology efficiency improvements, land-use changes, and the energy resource base. The assumptions used in incorporating these factors often differ between simulations, as do the actual factors represented in the simulations. As a result, the database was designed to organize and store the input assumptions behind the scenarios as well as GHG emissions and other output.

Given the diversity of data types that must be accommodated, the database was designed with a relational database structure (using MS Access '97). The data represent large samples, and it is important that they be stored according to a structure that also allows the relationships between different data types to be represented and stored. A detailed description of the database structure is given in Morita and Lee (1998).

Each individual data entry is stored in the DATAMOM. Using the relational structure, it is possible to call data from within any of four main fields (Source ID, Scenario ID, Region ID, or Variable) using a number of subcategories specific to the individual fields. For example, the Source ID data entry field has the following subcategories:

- Source ID (an abbreviated model or organization name with multiple data sets distinguished by the year of publication).
- Authors (individual name or organization name).
- Reference (publication in which the data are found).
- Model (main simulation models).
- Category (of simulation model, such as bottom-up or top-down, dynamic optimization of general equilibrium, etc.).
- Update date (of the most recent publication).
- Notes (if any).

Table V-1 briefly summarizes the subdivisions in the other key fields.

The database has the primary function of acting as a data storage tool, and as an interface that will allow the user easy access to the data sets contained therein. Thus, it only provides data, and analyses are conducted using other tools such as spreadsheets. However, the relational structure of the database makes it possible to call up comparable data sets across the key fields, giving maximum flexibility in manipulation, extraction, and presentation of all the data in the database. Similarly, there is great flexibility in importing new data, or making a data set from the database using combinations of specific sources, specific scenarios (or categories), specific regions, and specific variables. The extraction screen in Figure V-1, for example, shows the settings used to extract all information on all scenarios that are generated with the AIM Japan source model and to examine global sea level rise.

The writing team recommends that this database or a new revised one should also be maintained by some institution in the future to facilitate comparisons and assessments of emissions scenarios. However, this would require additional resources.

V.2. Data Collection

The main sources of data used in ESD were International Energy Workshop Polls (Manne and Schrattenholzer, 1995, 1996, 1997), Energy Modeling Forum (EMF-14 comparison studies) data, the previous database compiled for the IPCC Supplement Report, "Climate Change 1994" (Alcamo *et al.*, 1995), which examined emission scenarios produced prior to 1994, and individual emission scenarios collected by the SRES writing team. The current database used in this report includes the results of a total of 416 scenarios from 171 sources. Most of these scenarios date after 1994.

Most of the total of 416 scenarios focus on energy-related CO_2 emissions (256). Only three models estimated land-use related emissions – the ASF model, the IMAGE 2 model, and the AIM model. Very few scenarios considered global SO_2 emissions.

Table V-1: *Overview of Subdivisions in Key Fields in the Scenario Database. The main data fields for the SRES ESD are shown for the field names that identify the scenario by reference and name of world region and variable type.*

Field Name	Subdivision (and brief description)		
Scenario	Scenario ID (For reference scenario, or specific name of scenario, if one exists)	Category (Of scenario: non-intervention, intervention, or uncertain(ty). Blank if scenario not identifiable. Non-intervention means a scenario with no reduction policies for carbon emission, but which might include policies on other GHGs)	Description: (Of scenarios storyline and main assumptions)
Region	Region ID (Many scenarios use different names for the same region. These are converted into one unique Region ID. The full country name is used for national studies)		Definition (Of each Region ID)
Variable	Description (Of each variable for each source)	Variable (A common variable name is used for the same data item when names vary among sources)	Unit

Table V-2. *List of Data Categories in the 416 Scenario*

Variable	No. of Scenarios	Variable	No. of Scenarios
CO_2 emissions	372	Reduction in macro-economic consumption	29
Total primary energy consumption	243	Oil primary energy consumption	26
GDP or GNP	228	CH_4 emissions	25
Electricity generation	164	Coal	29
CO_2 concentration	161	Nuclear energy primary energy consumption	24
Temperature change	140	Biomass energy production	21
Coal consumption	107	Natural gas electricity generation	20
Oil consumption	101	Global mean temperature increase	20
Control costs	100	CO_2 emissions from deforestation	19
Natural gas consumption	99	CO_2 emissions in industrial sector	19
CH_4 concentration	97	N_2O emissions	19
Climate change costs	97	CH_4 emissions from animal wastes	18
Carbon tax	96	CH_4 emissions from biomass burning	18
Nuclear energy	93	CH_4 emissions from domestic sewage	18
Coal production	93	CH_4 emissions from enteric fermentation	18
Oil production	92	CH_4 emissions from landfills	18
Renewables, electric	90	Coal primary energy consumption	18
Oil exports–imports	88	N_2O emissions from biomass burning	18
Renewables, nonelectric	87	N_2O emissions from land clearing	18
Natural gas production	86	Total electricity generation	16
Natural gas exports–imports	79	Other electricity generation	14
Coal exports–imports	78	Final energy consumption in industry	13
Crude oil price, international	77	Final energy consumption in residential and commercial	13
Coal/shale consumption	70	Final energy consumption in transport	13
Total primary energy production	66	NO_x emissions	13
Sea level rise	63	Autonomous Energy Efficiency Index	13
Sulfur emissions	61	Biomass commercial production	13
Total fossil fuel consumption	61	Electricity generation primary energy consumption	13
Population	52	N_2O concentration	13
Carbon intensity	33	CH_4 emissions from agricultural waste burning	12
Natural gas primary energy consumption	32	CH_4 emissions from deforestation	12
Energy intensity	31	CH_4 emissions from energy production	12
Carbon permits	29	CH_4 emissions from energy/industry	12
CO_2 emissions fossil fuel	29	CH_4 emissions from industry	12
Incremental value of carbon permit/carbon tax	29	CH_4 emissions from nature	12
Natural gas	29	CH_4 emissions from savanna burning	12
Oil	29	CH_4 emissions from wet rice field	12

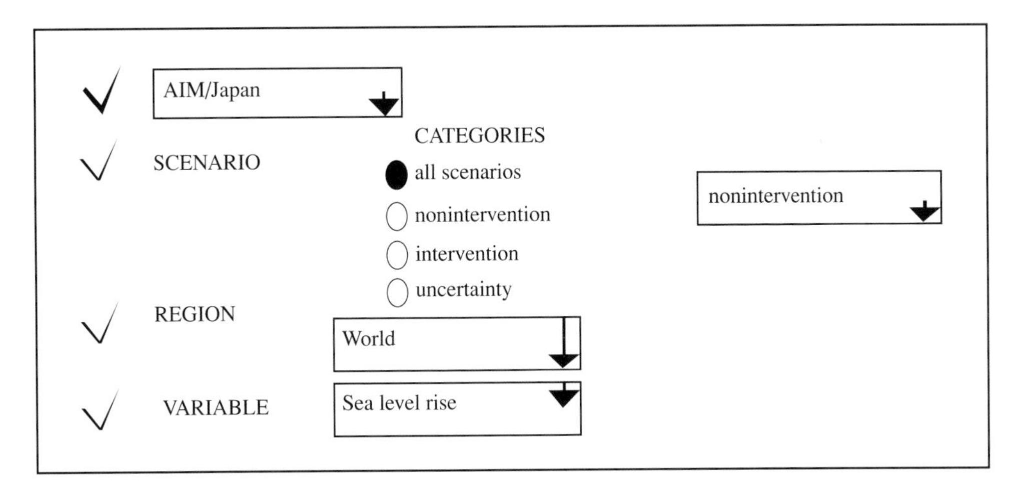

Figure V-1: Example of an extraction screen, showing the settings used to extract all information on all scenarios that are generated with the AIM Japan source model and to examine global sea level rise.

The variables considered while collating scenario data, and the frequency with which such they are found in the 416 scenarios (and thus stored in the SRES database), are listed in Table V-2.

References:

Alcamo, J., A. Bouwman, J. Edmonds, A. Grübler, T. Morita, and A. Sugandhy, 1995: An Evaluation of the IPCC IS92 Emission Scenarios. In: *Climate Change 1994, Radiative Forcing of Climate Change and An Evaluation of the IPCC IS92 Emission Scenarios,* Cambridge University Press, Cambridge, UK, pp. 233–304.

Manne, A., and L. Schrattenholzer, 1995: *International Energy Workshop January 1995 Poll Edition.* International Institute for Applied Systems Analysis, Laxenburg, Austria.

Manne, A., and L. Schrattenholzer, 1996, *International Energy Workshop January 1996 Poll Edition.* International Institute for Applied Systems Analysis, Laxenburg, Austria.

Manne, A., and L. Schrattenholzer, 1997: *International Energy Workshop, Part I: Overview of Poll Responses, Part II: Frequency Distributions, Part III: Individual Poll Responses,* February, 1997, International Institute for Applied Systems Analysis, Laxenburg, Austria.

Morita, T., and H.-C. Lee, 1998: Appendix to Emissions Scenarios Database and Review of Scenarios. *Mitigation and Adaptation Strategies for Global Change*, **3**(2-4), 121-131.

VI

Open Process

Open Process

The Terms of Reference of this Special Report on Emissions Scenarios (SRES) include a so-called "open process" to stimulate input from a community of experts much broader than the writing team (see Appendices I and II). IPCC documents should take into account as many scientific perspectives as possible. This is particularly important in the area of scenarios, for which views on the plausibility of various aspects of the described futures and their interactions can differ between regions, between different sectors of society, and between individual experts. The SRES web site (sres.ciesin.org) was created to facilitate the open process and to help gain input from a community of experts much broader than the writing team.

The web site:

- Includes a description of SRES activities and the scenario development process.
- Provides detailed information on the four marker scenarios and their storylines.
- Offers facilities to view and plot scenario driving forces and emissions.
- Offers facilities to receive feedback from the open process.

The open process lasted from June 1998 to January 1999. As a result of the interest in SRES scenarios, the web site is accessible to acquire updated information about SRES marker scenarios. For other reasons and the input received so far, the information on the web site has been improved and updated considerably. The writing team recommends that the web site also be maintained in the future so that it is available to access updated information on SRES scenarios. However, this will require additional resources.

The four marker scenarios were posted on the IPCC web site (sres.ciesin.org) in June 1998. The submissions invited through the open process and web site fell into three categories:

- Additional scenarios published in the reviewed literature that had not been included in the scenario database (see Appendix V).
- New scenarios based on the SRES marker scenarios.
- General suggestions to improve the work of the SRES writing team as posted on the web site (preferably based on referenced literature).

The submissions were used to revise the marker scenarios and to develop additional alternatives within each of the four scenario families. The result is a more complete, refined set of new scenarios that reflects the broad spectrum of modeling approaches and regional perspectives. The preliminary scenarios posted on the web site were provided to climate modelers also, with the approval of the IPCC Bureau.

***Table VI-1**: SRES web site access summary by month, from July 1998 to March 1999.*

Month	Unique non-CIESIN hosts connected	Total non-CIESIN www accesses
July 1998 (20-31)	17	65
August 1998	143	2,214
September 1998	610	6,217
October 1998	425	4,083
November 1998	313	3,696
December 1998	455	5,170
January 1999	497	5,946
February 1999	468	5,764
March 1999(1-5)	103	1,064
Total	3,031	34,219

Most of the submissions received fall into the first two categories above. Altogether, more than 34,000 accesses to the SRES web site were registered by April 1999 from some 3,000 unique hosts that were connected. Tables VI-1 and VI-2 and Figure VI-1 give more detail about the number of monthly accesses between July 1998 and March 1999 and about accesses from different countries and territories during the same period. Tables VI-3 and VI-4 give details of the preliminary marker scenarios.

The web site is managed by the Center for International Earth Science Information Network (CIESIN) in the US, in collaboration with the Energy Research Foundation (ECN) in the Netherlands, the Technical Support Unit (TSU) of Working Group III on Mitigation of IPCC at the National Institute of Public Health and Environment (RIVM) in the Netherlands, and the International Institute for Applied Systems Analysis (IIASA) in Austria.

***Table VI-2**: SRES web site access summary by countries and territories, from July 1998 to March 1999.*

Accesses	Internet domain	Country/Territory	Accesses	Internet domain	Country/Territory
36	ar	Argentina	6	lb	Lebanon
676	at	Austria	8	lu	Luxembourg
373	au	Australia	3	my	Malaysia
26	ba	Bosnia and Herzegovina	17	mx	Mexico
333	be	Belgium	22	lk	Sri Lanka
81	br	Brazil	3424	nl	Netherlands
715	ca	Canada	803	no	Norway
301	ch	Switzerland	154	nz	New Zealand
14	cn	China	206	pl	Poland
10	cl	Chile	94	pt	Portugal
16	cr	Costa Rica	47	ru	Russian Federation
11	cz	Czech Republic	3	sa	Saudi Arabia
1085	de	Germany	155	se	Sweden
203	dk	Denmark	13	sg	Singapore
3	ec	Ecuador	10	si	Slovenia
9	ee	Estonia	49	za	South Africa
204	es	Spain	23	th	Thailand
297	fi	Finland	15	tw	Taiwan, province of China
3	fj	Fiji	4	tt	Trinidad and Tobago
262	fr	France	12	tz	Tanzania
13	gb	Great Britain (UK)	3	ua	Ukraine
12	gr	Greece	1921	uk	United Kingdom
7	hr	Croatia	74	us	United States
23	hk	Hong Kong	3	uy	Uruguay
36	hu	Hungary	2894	com	US Commercial
20	id	Indonesia	2948	edu	US Educational
11	ie	Ireland	1736	gov	US Government
34	il	Israel	9	int	International
35	in	India	78	mil	US Military
6	is	Iceland	1723	net	Network
320	it	Italy	806	org	Non-Profit Organization
1991	jp	Japan	8613	unresolved	IP addresses only
20	kr	Republic of Korea (South)			

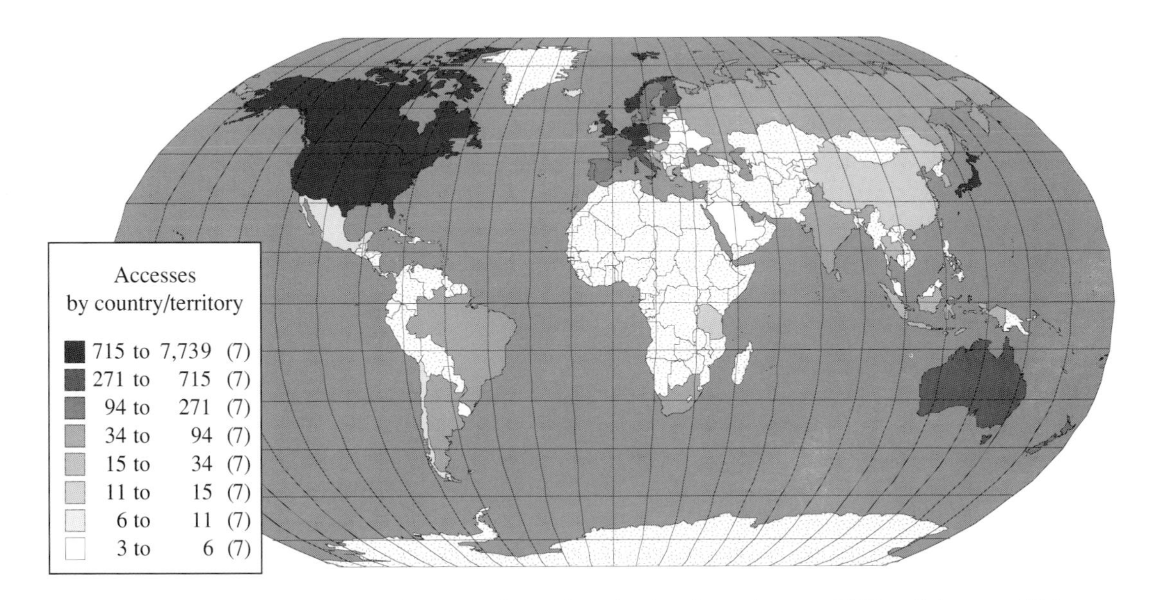

Figure VI-1: SRES web site access summary by countries and territories, from August 1998 to February 1999.

Table VI-3: *Standardized GHG emissions for the four preliminary markers of the SRES open process, as posted on the CIESIN web page (sres.ciesin.org). These are **not** necessarily identical with the final marker scenarios.*

Preliminary A1 Marker Scenario – Standardized Global GHG Emissions (Open Process Version, date of submission: 12/10/98)

Greenhouse Gas	Units	1990	2000	2010	2020	2030	2040	2050	2060	2070	2080	2090	2100
CO_2 Fossil & Industry	GtC	6.2	6.8	9.7	12.2	14.2	15.2	16.2	15.9	15.6	15.0	14.1	13.2
CO_2 Deforestation	GtC	1.1	1.6	1.5	1.6	0.7	0.3	-0.2	-0.3	-0.3	-0.4	-0.5	-0.6
CH_4	Mt CH_4	322.2	346.8	416.8	483.8	547.2	530.6	514.1	463.6	413.0	370.3	335.5	300.6
N_2O	Mt N_2O-N	6.3	6.9	7.3	7.7	7.5	7.1	6.8	6.3	5.9	5.5	5.2	4.9
SO_x	MtS	72.6	69.0	87.4	100.8	91.4	77.9	64.3	51.2	44.9	30.7	29.1	27.4
NO_x	MtN	29.9	32.5	41.0	48.9	52.5	50.9	49.3	47.2	45.1	43.3	41.8	40.3
NMVOC	Mt	137.6	150.8	177.8	207.3	229.2	255.0	285.0	324.4	301.1	263.3	223.2	174.0
CO	Mt CO	892.5	1035.9	1272.6	1530.5	1641.2	1815.4	1989.7	2174.4	2359.2	2455.4	2463.0	2470.7
CFCs	GtC equiv.	1.563	0.614	0.413	0.050	0.014	0.006	0.002	0.001	0.001	0.000	0.000	0.000
HCFCs	GtC equiv.	0.087	0.201	0.226	0.075	0.038	0.035	0.002	0.000	0.000	0.000	0.000	0.000
HFCs	GtC equiv.	0.034	0.120	0.164	0.245	0.380	0.540	0.651	0.684	0.694	0.687	0.671	0.645
PFCs	GtC equiv.	0.057	0.068	0.075	0.109	0.157	0.196	0.217	0.221	0.200	0.155	0.170	0.184
SF_6	GtC equiv.	0.038	0.037	0.043	0.048	0.066	0.099	0.119	0.127	0.113	0.075	0.084	0.095

Preliminary A2 Marker Scenario – Standardized Global GHG Emissions (Open Process Version, date of submission: 12/14//98)

Greenhouse Gas	Units	1990	2000	2010	2020	2030	2040	2050	2060	2070	2080	2090	2100
CO_2 Fossil & Industry	GtC	6.2	6.8	8.4	10.9	13.3	14.7	16.4	18.2	20.2	22.7	25.6	28.8
CO_2 Deforestation	GtC	1.1	1.6	1.6	1.7	1.5	1.3	1.2	0.7	0.4	0.3	0.2	0.2
CH_4	Mt CH_4	322.2	346.8	394.1	448.3	506.2	560.4	621.4	674.0	732.3	790.4	848.2	912.7
N_2O	Mt N_2O-N	6.3	6.9	7.9	9.4	10.5	11.1	11.8	12.7	13.7	14.6	15.5	16.4
SO_x	MtS	72.6	69.0	74.7	99.5	111.9	108.1	105.4	86.3	71.7	64.2	61.9	60.3
NO_x	MtN	29.9	32.5	39.6	50.7	60.8	65.8	71.5	75.6	80.1	87.3	97.9	109.7
NMVOC	Mt	137.6	150.8	164.1	188.0	210.3	221.3	234.6	245.8	259.5	282.2	314.7	351.9
CO	Mt CO	892.5	1035.9	1135.5	1233.6	1413.0	1494.3	1586.3	1696.1	1815.8	1985.4	2218.3	2484.3
CFCs	GtC equiv.	1.563	0.614	0.413	0.050	0.014	0.006	0.002	0.001	0.001	0.000	0.000	0.000
HCFCs	GtC equiv.	0.087	0.201	0.226	0.075	0.038	0.035	0.002	0.000	0.000	0.000	0.000	0.000
HFCs	GtC equiv.	0.035	0.116	0.152	0.182	0.228	0.278	0.341	0.427	0.512	0.621	0.753	0.883
PFCs	GtC equiv.	0.057	0.070	0.078	0.094	0.114	0.134	0.158	0.191	0.223	0.259	0.297	0.323
SF_6	GtC equiv.	0.038	0.041	0.053	0.067	0.079	0.092	0.110	0.122	0.129	0.135	0.153	0.165

Preliminary B1 Marker Scenario – Standardized Global GHG Emissions (Open Process Version, date of submission: 10/7/98)

Greenhouse Gas	Units	1990	2000	2010	2020	2030	2040	2050	2060	2070	2080	2090	2100
CO_2 Fossil & Industry	GtC	6.2	6.8	7.7	8.3	8.4	9.1	9.8	10.4	10.1	8.7	7.5	6.5
CO_2 Deforestation	GtC	1.1	1.6	0.8	1.3	0.7	0.6	0.5	0.7	0.8	1.0	1.2	1.4
CH_4	Mt CH_4	322.2	346.8	366.8	395.8	402.8	423.3	443.8	444.8	445.8	446.8	412.8	378.8
N_2O	Mt N_2O -N	6.3	6.9	7.4	8.1	8.3	8.6	8.9	8.8	8.7	8.6	8.3	8.0
SO_x	MtS	72.6	69.0	59.8	56.2	53.5	53.3	51.4	51.2	49.2	42.2	33.9	28.6
NO_x	MtN	29.9	32.5	34.8	39.3	40.7	44.8	48.9	48.9	48.9	48.9	41.2	33.6
NMVOC	Mt	137.6	150.8	142.8	150.8	143.8	146.8	149.8	154.8	159.8	164.8	159.3	153.8
CO	Mt CO	892.5	1035.9	848.9	984.9	863.9	902.9	941.9	983.9	1025.9	1067.9	1008.9	949.9
CFCs	GtC equiv.	1.563	0.614	0.413	0.050	0.014	0.006	0.002	0.001	0.001	0.000	0.000	0.000
HCFCs	GtC equiv.	0.087	0.201	0.226	0.075	0.038	0.035	0.002	0.000	0.000	0.000	0.000	0.000
HFCs	GtC equiv.	0.033	0.113	0.151	0.182	0.232	0.296	0.391	0.390	0.383	0.373	0.357	0.337
PFCs	GtC equiv.	0.056	0.066	0.069	0.078	0.086	0.094	0.107	0.108	0.095	0.074	0.075	0.073
SF_6	GtC equiv.	0.038	0.036	0.037	0.038	0.048	0.059	0.069	0.071	0.062	0.040	0.043	0.043

Preliminary B2 Marker Scenario – Standardized Global GHG Emissions (Open Process Version, date of submission: 11/2/98)

Greenhouse Gas	Units	1990	2000	2010	2020	2030	2040	2050	2060	2070	2080	2090	2100
CO_2 Fossil & Industry	GtC	6.2	6.8	7.9	8.9	10.0	10.8	11.1	11.6	11.8	12.4	13.1	13.7
CO_2 Deforestation	GtC	1.1	1.6	1.8	1.6	0.3	0.0	-0.3	-0.2	-0.2	-0.2	-0.2	-0.2
CH_4	Mt CH_4	322.2	346.8	388.7	447.5	500.9	528.3	537.9	544.4	541.5	529.4	508.5	508.4
N_2O	Mt N_2O -N	6.3	6.9	7.1	7.1	6.7	6.4	6.0	5.8	5.5	5.4	5.2	5.1
SO_x	MtS	72.6	69.0	68.2	65.0	59.9	58.8	57.2	53.7	51.9	49.1	48.0	47.3
NO_x	MtN	29.9	32.5	37.6	43.4	48.4	52.8	53.7	55.4	55.6	58.5	60.1	60.4
NMVOC	Mt	137.6	150.8	171.8	191.7	201.5	215.0	216.9	214.2	202.4	191.6	178.1	170.3
CO	Mt CO	892.5	1035.9	1138.3	1211.2	1175.1	1268.3	1350.5	1466.3	1624.7	1802.7	1948.3	2066.6
CFCs	GtC equiv.	1.563	0.614	0.413	0.050	0.014	0.006	0.002	0.001	0.001	0.000	0.000	0.000
HCFCs	GtC equiv.	0.087	0.201	0.226	0.075	0.038	0.035	0.002	0.000	0.000	0.000	0.000	0.000
HFCs	GtC equiv.	0.033	0.122	0.165	0.212	0.279	0.353	0.441	0.533	0.607	0.645	0.677	0.707
PFCs	GtC equiv.	0.057	0.070	0.080	0.101	0.126	0.155	0.186	0.210	0.228	0.235	0.234	0.229
SF_6	GtC equiv.	0.038	0.047	0.057	0.071	0.088	0.110	0.128	0.140	0.150	0.156	0.163	0.166

Table VI-4a: *Standardized GHG emissions for the preliminary A1 marker (AIM) as posted on the CIESIN web page (sres.ciesin.org) during the open process. These data are* ***not*** *necessarily identical with the final marker scenarios.*

Preliminary A1 Marker Scenario – WORLD (Open process version, date of submission: 12/10/98)

	1990	2020	2050	2100
Population in Million	5262	7493	8704	7056
GNP/GDP in Trillion US$90*	20.9	56.5	181.3	528.5
Final Energy by Fuel in EJ				
Non-commercial	49.8	35.6	0	0
Solids	36	62	74.5	24.6
Liquids	110.7	196.9	301.8	386.5
Gas	50.5	134	295.1	431.7
Electricity	38.3	103.7	330.7	898.1
Others				
Total	285.3	532.2	1002	1741
Primary Energy by Fuel				
Coal	85	144.5	140.3	41.1
Oil	125.8	200.5	181	107
Gas	67.6	186.5	400.3	490.4
Nuclear	6.5	30.4	122.7	77.9
Biomass	49.8	62.8	192.8	375.9
Other Renewables	10.3	23	167.1	986.9
Total	345	647.8	1204	2079
Cumulative Resource Use in ZJ				
Coal	0.1	3.4	7.9	12.2
Oil	0.1	5	10.8	17.7
Gas	0.1	3.5	12.2	36.1
Cumulative CO_2 Emissions in GtC	7.1	309.5	768.7	1517
Land Use in Million ha				
Cropland	1437	1553	1325	858
Grasslands	3290	3750	4065	3551
Energy Biomass	0	200	793	1208
Forest	4249	3811	3874	4326
Others	3966	3628	2885	2999
Total	12942	12942	12942	12942
Anthropogenic Emissions (not standardized)				
CO_2 (as C) in GtC	6	12.1	16	13.1
Other CO_2 (as C) in GtC	1.2	1.9	0.2	-0.2
Total CO_2 (as C) in GtC	7.1	14	16.2	12.9
CH_4 total (as CH_4) in Mt CH_4	341.7	499.3	529.7	316.2
N_2O total (as N) in Mt N_2O-N	6.6	7.7	6.8	4.8
SO_x total (as S) in MtS	69.1	101.8	63.8	24.4
CO (as CO) in Mt CO	751.7	1328	1787	2268
VOCs in Mt				
NO_x (as N) in MtN	27.6	46.9	47.2	38.3

* at market exchange rate

Preliminary A1 Marker Scenario – OECD90 (Open process version, date of submission: 12/10/98)

	1990	**2020**	**2050**	**2100**
Population in Million	858.5	1002	1081	1110
GNP/GDP in Trillion US$90*	16.4	31	54.1	121
Final Energy by Fuel in EJ				
Non-commercial	6.1	0	0	0
Solids	10	8.5	6.1	1.4
Liquids	64.1	81.8	74.5	66.5
Gas	24.8	50.4	68.1	78.1
Electricity	21.8	39.5	69.7	175
Others				
Total	126.8	180	218	321
Primary Energy by Fuel				
Coal	37.7	32.5	26	6.2
Oil	70	80.8	50.6	29.3
Gas	31.4	65.2	88.4	89.4
Nuclear	5.2	16.5	24.3	15
Biomass	6.1	4.1	29.1	56.8
Other Renewables	5.5	10.8	35.3	192
Total	155.9	210	254	389
Cumulative Resource Use in ZJ				
Coal	0	1.2	2	2.7
Oil	0.1	2.5	4.5	6.3
Gas	0	1.5	3.9	8.5
Cumulative CO_2 Emissions in GtC	2.9	103	207	344
Land Use in Million ha				
Cropland	410	381	300	198
Grasslands	787	788	815	730
Energy Biomass	0	15	95	179
Forest	1056	1105	1243	1370
Others	886	851	686	663
Total	3140	3140	3140	3140
Anthropogenic Emissions (not standardized)				
CO_2 (as C) in GtC	2.9	3.6	3.5	2.3
Other CO_2 (as C) in GtC	0	0	-0.2	0
Total CO_2 (as C) in GtC	2.9	3.6	3.2	2.3
CH_4 total (as CH_4) in Mt CH_4	81.1	78.9	61.7	47.5
N_2O total (as N) in Mt N_2O-N	2.7	2.8	2.3	1.5
SO_x total (as S) in MtS	22.0	6.1	5.8	4.6
CO (as CO) in Mt CO	173.5	225	247	261
VOCs in Mt				
NO_x (as N) in MtN	12.9	12.8	7.3	5.9

* at market exchange rate

Preliminary A1 Marker Scenario – REF (Open process version, date of submission: 12/10/98)

	1990	2020	2050	2100
Population in Million	413	430	423	339
GNP/GDP in Trillion US$90*	1.1	2.9	12.4	34.2
Final Energy by Fuel in EJ				
Non-commercial	2	0	0	0
Solids	9.4	5.5	4	1.2
Liquids	18.9	8.9	7	6.2
Gas	19.3	22.9	45.5	49.9
Electricity	8.1	12.7	27.7	55.4
Others				
Total	57.7	50	84.2	113
Primary Energy by Fuel				
Coal	18.4	14.1	11	2.7
Oil	22.2	11.6	6.7	2.2
Gas	26.3	32.7	55.3	53.5
Nuclear	1.1	3.9	8.4	4.8
Biomass	2	1.1	7.6	14.8
Other Renewables	1.3	1.7	15	60.7
Total	71.3	65	104	139
Cumulative Resource Use in ZJ				
Coal	0	0.4	0.8	1.1
Oil	0	0.4	0.7	0.9
Gas	0	0.8	2.1	4.9
Cumulative CO_2 Emissions in GtC	1.2	30.3	62.6	102
Land Use in Million ha				
Cropland	279	299	287	180
Grasslands	346	454	566	478
Energy Biomass	0	0	31	49
Forest	960	970	973	1114
Others	720	582	447	485
Total	2305	2305	2305	2305
Anthropogenic Emissions (not standardized)				
CO_2 (as C) in GtC	1.2	1	1.1	0.7
Other CO_2 (as C) in GtC	0	0	0	-0.1
Total CO_2 (as C) in GtC	1.2	1	1.1	0.6
CH_4 total (as CH_4) in Mt CH_4	58.7	79.1	64.5	51.9
N_2O total (as N) in Mt N_2O-N	0.7	0.7	0.6	0.4
SO_x total (as S) in MtS	16.5	11.7	2.9	1.6
CO (as CO) in Mt CO	69.8	51.1	54.5	47.4
VOCs in Mt				
NO_x (as N) in MtN	4	3.1	2.1	1

* at market exchange rate

Preliminary A1 Marker Scenario – ASIA (Open process version, date of submission: 12/10/98)

	1990	2020	2050	2100
Population in Million	2798.2	3851	4220	2882
GNP/GDP in Trillion US$90*	1.5	12.3	62.7	207
Final Energy by Fuel in EJ				
Non-commercial	27.8	14.1	0	0
Solids	15.4	44	57.3	20.1
Liquids	10.8	48.3	86.9	112
Gas	2	17.3	70	160
Electricity	4.6	27.1	117	363
Others				
Total	60.6	151	331	655
Primary Energy by Fuel				
Coal	25.8	80.4	77.5	25.1
Oil	13.1	48.3	40.4	20.9
Gas	3	26.9	108	184
Nuclear	0.1	5.7	45.3	31.8
Biomass	27.8	20.5	45.7	89
Other Renewables	1.3	4.2	57.6	400
Total	71.3	186	374	750
Cumulative Resource Use in ZJ				
Coal	0	1.5	4.1	6.6
Oil	0	0.9	2.3	3.8
Gas	0	0.3	2.3	10.3
Cumulative CO_2 Emissions in GtC	1.5	87.8	253	546
Land Use in Million ha				
Cropland	390	437	367	246
Grasslands	521	586	621	584
Energy Biomass	0	44	197	257
Forest	527	411	405	472
Others	576	536	424	456
Total	2014	2014	2014	2014
Anthropogenic Emissions (not standardized)				
CO_2 (as C) in GtC	1.1	4	5.6	5.2
Other CO_2 (as C) in GtC	0.3	0.6	0.2	0.1
Total CO_2 (as C) in GtC	1.5	4.6	5.9	5.2
CH_4 total (as CH_4) in Mt CH_4	126.6	211	249	127
N_2O total (as N) in Mt N_2O-N	2.3	2.8	2.7	2
SO_x total (as S) in MtS	19.2	56.2	9.5	6.3
CO (as CO) in Mt CO	265.5	583	847	979
VOCs in Mt				
NO_x (as N) in MtN	6	16.4	18.6	12.3

* at market exchange rate

Preliminary A1 Marker Scenario – ALM (Open process version, date of submission: 12/10/98)

	1990	2020	2050	2100
Population in Million	1192.1	2211	2980	2727
GNP/GDP in Trillion US$90*	1.9	10.3	52	166
Final Energy by Fuel in EJ				
Non-commercial	13.9	21.5	0	0
Solids	1.2	4	7.1	1.9
Liquids	16.9	57.9	133	202
Gas	4.4	43.4	112	144
Electricity	3.8	24.4	117	304
Others				
Total	40.2	151	369	652
Primary Energy by Fuel				
Coal	3.1	17.5	25.8	7.1
Oil	20.5	59.9	83.4	54.5
Gas	6.9	61.7	149	164
Nuclear	0	4.3	44.7	26.3
Biomass	13.9	37.1	110	215
Other Renewables	2.1	6.4	59.2	335
Total	46.5	187	472	802
Cumulative Resource Use in ZJ				
Coal	0	0.3	0.9	1.7
Oil	0	1.1	3.3	6.7
Gas	0	0.9	4	12.5
Cumulative CO_2 Emissions in GtC	1.6	88.7	246	525
Land Use in Million ha				
Cropland	357	436	371	234
Grasslands	1636	1921	2062	1759
Energy Biomass	0	141	470	724
Forest	1706	1326	1253	1371
Others	1784	1659	1328	1396
Total	5483	5483	5483	5483
Anthropogenic Emissions (not standardized)				
CO_2 (as C) in GtC	0.8	3.5	5.8	4.9
Other CO_2 (as C) in GtC	0.8	1.3	0.2	-0.2
Total CO_2 (as C) in GtC	1.6	4.8	6	4.8
CH_4 total (as CH_4) in Mt CH_4	75.3	130	154	90.1
N_2O total (as N) in Mt N_2O-N	1	1.5	1.2	0.9
SO_x total (as S) in MtS	11.4	27.7	45.5	11.9
CO (as CO) in Mt CO	242.9	469	639	981
VOCs in Mt				
NO_x (as N) in MtN	4.6	14.6	19.3	19

* at market exchange rate

Table VI-4b: *Standardized GHG emissions for the preliminary A2 marker (ASF) as posted on the CIESIN web page (sres.ciesin.org) during the open process. These data are* ***not*** *necessarily identical with the final marker scenario.*

Preliminary A2 Marker Scenario – WORLD (Open process version, date of submission: 12/14//98)

	1990	**2020**	**2050**	**2100**
Population in Million	5263	8191	11296	15068
GNP/GDP in Trillion US$90*	20.9	40.5	81.6	242.8
Final Energy by Fuel in EJ				
Non-commercial				
Solids	55.6	64.7	54.9	66.5
Liquids	123.1	245.9	333.5	635.0
Gas	51.1	92.4	186.7	261.8
Electricity	43.0	85.4	203.6	468.0
Other (e.g. H_2)				
Total	272.8	488.4	778.7	1431.3
Primary Energy by Fuel				
Coal	96.6	129.3	293.8	903.7
Oil	140.8	291.0	227.7	0.5
Gas	74.0	125.8	274.9	331.2
Nuclear	8.3	16.8	61.8	234.2
Biomass	0.0	12.2	71.5	161.6
Other Renewables	9.6	19.9	41.6	86.0
Total	329.3	594.9	971.4	1717.1
Cumulative Resource Use in ZJ				
Coal	0	3.2	9.2	38.6
Oil	0	6.2	13.8	16.1
Gas	0	2.7	8.4	24.2
Cumulative CO_2 Emissions in GtC	0	291	738	1862
Land Use in Million ha				
Forests				
Grasslands				
Cropland				
Energy Biomass				
Other				
Total				
Anthropogenic Emissions (not standardized)				
CO_2 (as C) in MtC	6137	10498	15731	27823
Other CO_2 (as C) in MtC	1584	2010	1802	1128
Total CO_2 (as C) in MtC	7721	12508	17533	28950
CH_4 total (as CH_4) in Mt CH_4	307	437	611	902
N_2O total (as N) in Mt N_2O-N	7	10	13	17
SO_x total (as S) in MtS	77	102	106	57
CO (as CO) in Mt CO	428	565	716	1101
VOCs in Mt	118	166	212	330
NO_x (as N) in MtN	35	56	77	115
Total CO_2 – Sinks, in MtC	7719	12438	17361	28745

* at market exchange rate

Preliminary A2 Marker Scenario – OECD90 (Open process version, date of submission: 12/14//98)

	1990	2020	2050	2100
Population in Million	848	1030	1151	1496
GNP/GDP in Trillion US$90*	15.7	26.0	39.9	87.6
Final Energy by Fuel in EJ				
Non-commercial				
Solids	12.7	12.7	14.2	16.3
Liquids	68.6	99.0	92.1	146.0
Gas	27.7	32.9	45.6	59.6
Electricity	23.9	39.0	58.8	130.4
Other (e.g. H_2)				
Total	132.9	183.6	210.7	352.3
Primary Energy by Fuel				
Coal	32.2	37.0	92.0	217.7
Oil	77.4	116.4	49.4	0.0
Gas	34.9	43.3	69.3	77.0
Nuclear	6.9	10.8	20.2	72.0
Biomass	0.0	3.6	22.8	28.5
Other Renewables	5.5	8.4	12.0	22.6
Total	156.9	219.5	265.7	417.8
Cumulative Resource Use in ZJ				
Coal	0	0.99	2.8	10.4
Oil	0	3.06	5.4	5.7
Gas	0	1.17	2.8	6.5
Cumulative CO_2 Emissions in GtC	0	102	229	500
Land Use in Million ha				
Forests				
Grasslands				
Cropland				
Energy Biomass				
Other				
Total				
Anthropogenic Emissions (not standardized)				
CO_2 (as C) in MtC	2758	3721	4272	6659
Other CO_2 (as C) in MtC	78.6	139	373	156
Total CO_2 (as C) in MtC	2837	3860	4644	6815
CH_4 total (as CH_4) in Mt CH_4	73.9	84.4	106	166
N_2O total (as N) in Mt N_2O-N	2.67	3.2	3.24	4.13
SO_x total (as S) in MtS	29.2	14.5	13.4	11.8
CO (as CO) in Mt CO	78	98	83	127
VOCs in Mt	43	58	56	80
NO_x (as N) in MtN	12	17	18	23

* at market exchange rate

Preliminary A2 Marker Scenario – REF (Open process version, date of submission: 12/14//98)

	1990	2020	2050	2100
Population in Million	420	455	519	706
GNP/GDP in Trillion US$90*	1.0	1.4	3.7	14.2
Final Energy by Fuel in EJ				
Non-commercial				
Solids	17.9	10.7	10.8	11.2
Liquids	16.3	12.8	17.1	36.5
Gas	15.8	17.1	32.0	46.3
Electricity	8.3	11.2	18.6	42.5
Other (e.g. H_2)				
Total	58.3	51.8	78.5	136.5
Primary Energy by Fuel				
Coal	23.4	16.5	22.7	59.1
Oil	18.3	15.3	20.7	0.2
Gas	26.5	29.4	40.4	52.3
Nuclear	1.0	1.6	5.9	21.6
Biomass	0.0	-0.1	0.4	15.0
Other Renewables	1.1	1.7	3.2	7.0
Total	70.3	64.4	93.4	155.1
Cumulative Resource Use in ZJ				
Coal	0	0.46	1.0	2.8
Oil	0	0.42	1.0	1.3
Gas	0	0.71	1.8	4.1
Cumulative CO_2 Emissions in GtC	0	31	69	160
Land Use in Million ha				
Forests				
Grasslands				
Cropland				
Energy Biomass				
Other				
Total				
Anthropogenic Emissions (not standardized)				
CO_2 (as C) in MtC	1325	1139	1560	2281
Other CO_2 (as C) in MtC	32	-3.41	-120	45.6
Total CO_2 (as C) in MtC	1357	1136	1440	2327
CH_4 total (as CH_4) in Mt CH_4	43.2	43.92	76.07	141
N_2O total (as N) in Mt N_2O-N	0.64	0.852	0.982	1.15
SO_x total (as S) in MtS	15.3	9.637	8.8	3.24
CO (as CO) in Mt CO	25	17	24	51
VOCs in Mt	11	15	27	34
NO_x (as N) in MtN	6	5	6	8

* at market exchange rate

Preliminary A2 Marker Scenario – ASIA (Open process version, date of submission: 12/14//98)				
	1990	**2020**	**2050**	**2100**
Population in Million	2779	4308	5764	7340
GNP/GDP in Trillion US$90*	1.7	5.3	15.0	57.1
Final Energy by Fuel in EJ				
Non-commercial				
Solids	23.7	38.1	21.5	23.2
Liquids	15.3	63.8	103.0	214.7
Gas	2.6	18.8	45.2	60.2
Electricity	6.2	20.2	86.3	180.6
Other (e.g. H_2)				
Total	47.8	140.9	256.0	478.7
Primary Energy by Fuel				
Coal	36.3	62.2	134.0	355.0
Oil	19.1	76.9	57.1	0.0
Gas	4.0	23.8	78.6	84.2
Nuclear	0.4	3.1	26.5	90.3
Biomass	0.0	3.4	22.9	19.8
Other Renewables	1.3	4.9	15.6	31.9
Total	61.0	174.2	334.8	581.2
Cumulative Resource Use in ZJ				
Coal	0	1.46	4.3	16.5
Oil	0	1.28	3.2	3.6
Gas	0	0.33	1.7	6.1
Cumulative CO_2 Emissions in GtC	0	82	239	666
Land Use in Million ha				
Forests				
Grasslands				
Cropland				
Energy Biomass				
Other				
Total				
Anthropogenic Emissions (not standardized)				
CO_2 (as C) in MtC	1328	3340	5615	10238
Other CO_2 (as C) in MtC	400	706	956	522.6
Total CO_2 (as C) in MtC	1728	4045	6572	10761
CH_4 total (as CH_4) in Mt CH_4	115	170	233	314.3
N_2O total (as N) in Mt N_2O-N	2.31	4.44	5.81	7.67
SO_x total (as S) in MtS	19.4	51.6	49	20.51
CO (as CO) in Mt CO	97	155	224	389
VOCs in Mt	21	32	42	69
NO_x (as N) in MtN	9	18	28	42

* at market exchange rate

Preliminary A2 Marker Scenario – ALM (Open process version, date of submission: 12/14//98)

	1990	2020	2050	2100
Population in Million	1217	2398	3862	5526
GNP/GDP in Trillion US$90*	2.6	7.8	23.0	83.8
Final Energy by Fuel in EJ				
Non-commercial				
Solids	1.3	3.2	8.4	15.8
Liquids	22.9	70.3	121.3	237.8
Gas	5.0	23.6	63.9	95.7
Electricity	4.6	15.0	39.9	114.5
Other (e.g. H_2)				
Total	33.8	112.1	233.5	463.8
Primary Energy by Fuel				
Coal	4.7	13.7	45.1	271.9
Oil	26.0	82.5	100.5	0.3
Gas	8.6	29.3	86.6	117.7
Nuclear	0.1	1.3	9.2	50.4
Biomass	0.0	5.3	25.4	98.3
Other Renewables	1.7	4.8	10.8	24.6
Total	41.1	136.8	277.6	563.0
Cumulative Resource Use in ZJ				
Coal	0	0.25	1.1	8.8
Oil	0	1.45	4.2	5.4
Gas	0	0.48	2.1	7.5
Cumulative CO_2 Emissions in GtC	0	76	200	536
Land Use in Million ha				
Forests				
Grasslands				
Cropland				
Energy Biomass				
Other				
Total				
Anthropogenic Emissions (not standardized)				
CO_2 (as C) in MtC	725	2299	4284	8644
Other CO_2 (as C) in MtC	1074	1169	593	403
Total CO_2 (as C) in MtC	1799	3467	4877	9048
CH_4 total (as CH_4) in Mt CH_4	74.9	139	195	280
N_2O total (as N) in Mt N_2O-N	0.97	1.85	2.73	4.37
SO_x total (as S) in MtS	13.2	25.8	34.4	21.8
CO (as CO) in Mt CO	229	294	384	534
VOCs in Mt	43	61	87	147
NO_x (as N) in MtN	8	16	26	42

* at market exchange rate

Table VI-4c: *Standardized GHG emissions for the preliminary B1 marker (IMAGE) as posted on the CIESIN web page (sres.ciesin.org) during the open process. These data are **not** necessarily identical with the final marker scenarios.*

Preliminary B1 Marker Scenario – WORLD (Open process version, date of submission: 10/7/98)

	1990	2020	2050	2100
Population in Million	5297	7767	8933	7239
GNP/GDP in Trillion US$90*	21	48.21	113.94	338.29
GNP/GDP (US$/cap)	3965	6208	12755	46729
Final Energy by Fuel in EJ				
Non-commercial	49.34	34.04	15.74	9.65
Solids	35.76	24.07	24.7	35.2
Liquids	96.8	124.14	149.2	106.1
Gas	45.06	76.35	123.7	174.5
Electricity	39.76	106.28	233.2	425.6
Others				
Total	266.7	364.9	546.5	751.1
Primary Energy by Fuel				
Coal	93.71	68.26	87	62.1
Oil (excluding feedstocks)	115.9	157.1	187.1	62.2
Gas (excluding feedstocks)	70.27	145.06	157.4	127
Non-Fossil Electricity (Nuclear/Solar)	8.56	43.85	144	413.4
Biomass	1.98	13.62	76.3	118.2
Other Renewables (Hydro/Fuelwood)	57.26	46.75	37.3	44.3
Total	347.7	474.6	689.1	827.2
Cumulative Resource Use in ZJ				
Coal	5.55	7.83	10.06	14.94
Oil	2.8	6.75	11.85	18.27
Gas	1.65	4.73	9.25	16.45
Cumulative CO_2 Emissions in GtC	100	296	537	947
Carbon Sequestration in GtC	2.0	3.7	3.3	2.9
Land Use in Million ha				
Cropland	1436.3	1268.5	1362.4	1119.1
Grasslands	3435.6	3934.8	3428.1	1914.8
Energy Biomass	6.4	32.8	201.9	373.4
Forest	4277.0	4095.0	4207.7	5075.5
Others	3915.7	3739.9	3870.8	4588.2
Total	13071.0	13071.0	13071.0	13071.0
Anthropogenic Emissions (not standardized)				
CO_2 (as C) in GtC	6.1	7.5	9	5.7
Other CO_2 (as C) in GtC	0.9	1.9	0.9	1.4
Total CO_2 (as C) in GtC	7.0	9.4	9.9	7.1
CH_4 total (as CH_4) in Mt CH_4	429	495	543	478
N_2O total (as N) in Mt N_2O-N	10.5	12.4	13.2	12.3
SO_x total (as S) in MtS	70.6	42.7	41.8	25.6
CO (as CO) in Mt CO	865	1098	1055	1063
VOCs in Mt	77	94	93	97
NO_x (as N) in MtN	27.5	32.9	42.5	27.2

* at market exchange rate

Preliminary B1 Marker Scenario – OECD90 (Open process version, date of submission: 10/7/98)

	1990	2020	2050	2100
Population in Million	801	950	1023	1055
GNP/GDP in Trillion US$90*	16.51	32.22	52.32	78.19
GNP/GDP (US$/cap)	20613	33917	51144	74114
Final Energy by Fuel in EJ				
Non-commercial	5.3	3	1.2	0.8
Solids	8.2	5.7	5.5	8.4
Liquids	49.7	57.9	53.5	39
Gas	28.6	38.3	47.3	69.4
Electricity	24.1	53.1	63.6	72.3
Others				
Total	115.9	158	171.1	189.9
Primary Energy by Fuel				
Coal	36.9	15	9.8	11.6
Oil (excluding feedstocks)	53.2	64.2	52.9	23.4
Gas (excluding feedstocks)	39.9	68.4	55.7	65.1
Non-Fossil Electricity (Nuclear/Solar)	6.7	30.4	51.9	63.9
Biomass	1.5	4.4	13.5	22.4
Other Renewables (Hydro/Fuelwood)	9.4	8.8	7.9	8.1
Total	147.6	191.2	183.8	186.4
Cumulative Resource Use in ZJ				
Coal	2.42	3.17	3.66	4.69
Oil	0.95	1.8	2.56	3.33
Gas	0.94	2.38	3.94	6.78
Cumulative CO_2 Emissions in GtC	49	139	215	325
Carbon Sequestration in GtC	0.4	0.6	0.8	0.8
Land Use in Million ha				
Cropland	379.3	415.4	501.8	437.7
Grasslands	785.7	687.0	513.2	439.1
Energy Biomass	2.0	3.1	22.9	73.0
Forest	1114.8	1160.2	1198.7	1199.8
Others	956.4	972.5	1001.7	1088.6
Total	3238	3238	3238	3238
Anthropogenic Emissions (not standardized)				
CO_2 (as C) in GtC	2.7	2.9	2.4	2.1
Other CO_2 (as C) in GtC	0.2	0.3	0.4	0.5
Total CO_2 (as C) in GtC	2.9	3.1	2.7	2.5
CH_4 total (as CH_4) in Mt CH_4	102	114	115	117
N_2O total (as N) in Mt N_2O-N	1.9	2.2	2.2	2.2
SO_x total (as S) in MtS	25.0	6.1	4.5	4.4
CO (as CO) in Mt CO	158	137	146	141
VOCs in Mt	27	30	29	28
NO_x (as N) in MtN	9.8	7.4	6.7	6.5

* at market exchange rate

Preliminary B1 Marker Scenario – REF (Open process version, date of submission: 10/7/98)

	1990	2020	2050	2100
Population in Million	413	442	437	352
GNP/GDP in Trillion US$90*	0.97	1.78	5.12	15.4
GNP/GDP (US$/cap)	2353	4029	11727	43750
Final Energy by Fuel in EJ				
Non-commercial	2.1	0.7	0.2	0.1
Solids	9.5	2.3	1.3	1.5
Liquids	19.1	10.6	9.7	5.6
Gas	11.8	5.8	6.3	8.6
Electricity	8.5	9.6	17.3	21.2
Others				
Total	50.9	29	34.8	37
Primary Energy by Fuel				
Coal	27.1	7.2	5.6	4
Oil (excluding feedstocks)	28.4	16.6	15.3	5.8
Gas (excluding feedstocks)	23.2	13.4	11.1	8.3
Non-Fossil Electricity (Nuclear/Solar)	1.4	3.1	10.6	17.1
Biomass	0.2	0.7	2.7	6.8
Other Renewables (Hydro/Fuelwood)	3.1	2.1	2.3	2.7
Total	83.4	43.1	47.6	44.7
Cumulative Resource Use in ZJ				
Coal	1.83	2.2	2.46	2.97
Oil	0.55	0.99	1.33	2.25
Gas	0.45	0.87	1.3	2.17
Cumulative CO_2 Emissions in GtC	28	54	74	107
Carbon Sequestration in GtC	0.4	0.7	1.2	1.2
Land Use in Million ha				
Cropland	278.3	147.1	98.7	62.4
Grasslands	391.2	306.1	198.7	122.6
Energy Biomass	0.1	1.0	15.6	65.2
Forest	1147.0	1326.2	1414.9	1450.9
Others	461.1	497.3	549.8	576.6
Total	2278	2278	2278	2278
Anthropogenic Emissions (not standardized)				
CO_2 (as C) in GtC	1.7	0.8	0.7	0.5
Other CO_2 (as C) in GtC	0.1	0.1	0.2	0.4
Total CO_2 (as C) in GtC	1.8	0.9	0.9	0.8
CH_4 total (as CH_4) in Mt CH_4	99	68	64	59
N_2O total (as N) in Mt N_2O-N	0.8	0.8	0.8	0.7
SO_x total (as S) in MtS	22.8	1.9	1.6	0.8
CO (as CO) in Mt CO	109	69	73	65
VOCs in Mt	11	6	5	4
NO_x (as N) in MtN	8.2	4.1	3.9	2.5

* at market exchange rate

Preliminary B1 Marker Scenario – ASIA (Open process version, date of submission: 10/7/98)

	1990	2020	2050	2100
Population in Million	2790	3924	4209	2875
GNP/GDP in Trillion US$90*	1.42	6.57	29.92	119.6
GNP/GDP (US$/cap)	508	1675	7108	41615
Final Energy by Fuel in EJ				
Non-commercial	28.4	21.3	6.1	2.7
Solids	16.4	12.9	12.9	11.9
Liquids	13.9	28.9	41.3	23.3
Gas	2	20.5	42.9	45
Electricity	4.6	29.5	82.1	157.4
Others				
Total	65.3	113.1	185.3	240.3
Primary Energy by Fuel				
Coal	26.1	36.7	46.6	18.3
Oil (excluding feedstocks)	17	39.3	48.8	13
Gas (excluding feedstocks)	3.9	45.4	56.8	26
Non-Fossil Electricity (Nuclear/Solar)	0.4	5.6	40.3	157.4
Biomass	0.1	6	41.1	37.5
Other Renewables (Hydro/Fuelwood)	29.4	23.6	10	7.5
Total	76.9	156.5	243.6	259.7
Cumulative Resource Use in ZJ				
Coal	1.09	2.01	3.01	4.56
Oil	0.2	0.84	1.58	2.03
Gas	0.07	0.86	2.38	3.82
Cumulative CO_2 Emissions in GtC	13	66	151	273
Carbon Sequestration in GtC	0.2	0.2	0.3	0.2
Land Use in Million ha				
Cropland	382.3	316.8	336.7	237.0
Grasslands	560.5	777.7	768.3	366.2
Energy Biomass	1.0	11.9	104.7	134.8
Forest	487.7	375.6	318.5	571.6
Others	544.5	493.8	447.7	666.3
Total	1976	1976	1976	1976
Anthropogenic Emissions (not standardized)				
CO_2 (as C) in GtC	1.1	2.5	3.1	1.3
Other CO_2 (as C) in GtC	0.3	0.2	0.3	0.3
Total CO_2 (as C) in GtC	1.4	2.6	3.4	1.6
CH_4 total (as CH_4) in Mt CH_4	122	149	179	137
N_2O total (as N) in Mt N_2O-N	2.6	3.2	3.7	2.9
SO_x total (as S) in MtS	15.3	20.9	16.5	5.2
CO (as CO) in Mt CO	204	260	365	264
VOCs in Mt	14	18	24	18
NO_x (as N) in MtN	5.4	12.3	16.6	7.0

* at market exchange rate

Preliminary B1 Marker Scenario – ALM (Open process version, date of submission: 10/7/98)

	1990	2020	2050	2100
Population in Million	1293	2450	3265	2958
GNP/GDP in Trillion US$90*	2.1	7.64	26.58	125.1
GNP/GDP (US$/cap)	1625	3118	8142	42286
Final Energy by Fuel in EJ				
Non-commercial	13.6	9	8.2	6.1
Solids	1.7	3.2	5	13.4
Liquids	14.1	26.8	44.7	38.2
Gas	2.7	11.7	27.2	51.5
Electricity	2.6	14.1	70.2	174.7
Others				
Total	34.7	64.7	155.3	283.9
Primary Energy by Fuel				
Coal	3.6	9.3	25	28.2
Oil (excluding feedstocks)	17.3	37	70.1	20
Gas (excluding feedstocks)	3.2	17.9	33.8	27.6
Non-Fossil Electricity (Nuclear/Solar)	0.1	4.8	41.2	175
Biomass	0.2	2.6	19	51.5
Other Renewables (Hydro/Fuelwood)	15.4	12.3	11.5	11.7
Total	39.8	83.9	200.6	314
Cumulative Resource Use in ZJ				
Coal	0.21	0.45	0.93	2.72
Oil	1.1	3.12	6.38	10.66
Gas	0.19	0.62	1.63	3.68
Cumulative CO_2 Emissions in GtC	10	37	97	242
Carbon Sequestration in GtC	1.1	2.2	1.0	0.7
Land Use in Million ha				
Cropland	396.4	389.3	425.3	382.0
Grasslands	1698.2	2163.9	1947.9	987.0
Energy Biomass	3.2	16.8	58.7	100.4
Forest	1527.6	1232.9	1275.6	1853.2
Others	1953.7	1776.3	1871.6	2256.7
Total	5579	5579	5579	5579
Anthropogenic Emissions (not standardized)				
CO_2 (as C) in GtC	0.5	1.4	2.8	1.8
Other CO_2 (as C) in GtC	0.3	1.4	0.1	0.2
Total CO_2 (as C) in GtC	0.9	2.8	2.9	2.1
CH_4 total (as CH_4) in Mt CH_4	105	164	184	166
N_2O total (as N) in Mt N_2O-N	5.2	6.3	6.5	6.5
SO_x total (as S) in MtS	7.5	13.8	19.2	15.2
CO (as CO) in Mt CO	394	631	470	593
VOCs in Mt	25	40	35	47
NO_x (as N) in MtN	4.1	9.1	15.3	11.1

* at market exchange rate

***Table VI-4d:** Standardized GHG emissions for the preliminary B2 marker (MESSAGE) as posted on the CIESIN web page (sres.ciesin.org) during the open process. These data are **not** necessarily identical with the final marker scenarios.*

Preliminary B2 Marker Scenario – WORLD (Open process version, date of submission: 11/2/98)

	1990	2020	2050	2100
Population in Million	5262	7672	9367	10414
GNP/GDP (mex) in Trillion US$90 [a]	20.9	50.7	109.5	234.9
GNP/GDP (ppp) in Trillion (1990 prices) [a]	25.7	60.2	113.9	231.8
Final Energy by Fuel in EJ				
Non-commercial	38.4	23.9	10.7	6.8
Solids	42.2	35.8	19.4	7.0
Liquids	110.9	200.1	267.9	294.0
Gas	40.9	58.7	104.9	111.2
Electricity	34.7	85.1	188.2	409.0
Other (e.g. H_2)	7.8	25.4	63.2	123.2
Total	274.9	429.0	654.3	951.2
Primary Energy by Fuel				
Coal	91.1	98.3	85.6	300.2
Oil	128.3	214.3	227.1	51.9
Gas	70.5	150.3	297.4	336.6
Nuclear	7.3	15.6	47.6	142.0
Biomass	46.0	52.6	104.7	316.0
Other Renewables	8.1	34.4	107.2	211.8
Total	351.3	565.5	869.6	1358.5
Cumulative Resource Use in ZJ				
Coal	0.0	2.8	5.7	12.6
Oil	0.0	5.1	12.0	19.5
Gas	0.0	2.7	8.6	26.9
Cumulative CO_2 Emissions in GtC [b]	0.0	215.1	519.6	1129.2
Land Use in Million ha [c]				
Forests	4249.5	3775.9	3906.7	4121.7
Grasslands	3289.8	3766.0	4014.4	3988.9
Cropland	1436.6	1583.4	1403.5	1113.2
Energy Biomass	0.0	15.5	497.9	717.7
Other	3966.4	3801.4	3119.7	3000.7
Total	12942.2	12942.2	12942.2	12942.2
Anthropogenic Emissions (not standardized)[d]				
CO_2 (as C)[b] in GtC	6.5	9.4	11.6	14.2
Other CO_2 (as C)[e] in GtC	1.0	1.4	-0.4	-0.2
Total CO_2 (as C) in GtC	7.5	10.7	11.2	14.0
CH_4 total (as CH_4) in Mt CH_4	318.3	447.6	538.1	508.6
N_2O total (as N) in Mt N_2O-N	6.3	6.8	5.6	4.7
SO_x total (as S) in MtS	69.0	59.0	51.9	43.0
CO (as CO) in Mt CO	974.4	1216.8	1356.0	2072.2
VOCs in Mt	177.5	233.9	259.1	212.4
NO_x (as N) in MtN	31.0	46.3	56.6	63.4

[a] mex = market exchange rate; ppp= purchasing power parity
[b] CO_2 emissions from fossil fuel and industrial processes (MESSAGE data).
[c] Land-use data taken from AIM B2 emulation run.
[d] Other non-energy emissions categories were calculated based on AIM B2 land-use change estimates and/or AIM B2 non-energy emissions.
[e] CO_2 emissions from land-use changes (AIM B2 emulation run)
Note: Subtotals may not add due to independent rounding.

Preliminary B2 Marker Scenario – OECD90 (Open process version, date of submission: 11/2/98)

	1990	2020	2050	2100
Population in Million	859	982	976	928
GNP/GDP (mex) in Trillion US$90 [a]	16.4	30.3	38.3	56.6
GNP/GDP (ppp) in Trillion (1990 prices) [a]	14.1	26.3	33.5	50.4
Final Energy by Fuel in EJ				
Non-commercial	0.0	0.0	0.0	0.0
Solids	13.1	3.2	0.2	0.0
Liquids	65.9	86.4	76.0	57.5
Gas	22.4	24.8	35.4	17.3
Electricity	21.5	43.7	60.5	89.1
Other (e.g. H_2)	0.7	3.7	10.3	17.9
Total	123.6	161.8	182.4	181.8
Primary Energy by Fuel				
Coal	38.0	38.9	18.9	46.8
Oil	72.1	90.4	65.0	16.7
Gas	32.9	60.6	92.7	127.9
Nuclear	5.9	10.8	16.5	29.3
Biomass	5.6	7.2	10.3	43.3
Other Renewables	4.2	11.2	24.7	40.8
Synfuel Trade	0	0	8.2	-30.7
Total	158.7	219.1	236.3	274.1
Cumulative Resource Use in ZJ				
Coal	0.0	1.1	2.0	3.3
Oil	0.0	2.6	5.2	7.3
Gas	0.0	1.3	3.4	9.2
Cumulative CO_2 Emissions in GtC C) [b]	0.0	98.6	209.4	371.2
Land Use in Million ha [c]				
Forests	1056.3	1107.3	1181.0	1290.3
Grasslands	787.2	775.9	793.0	753.4
Cropland	410.2	377.3	324.0	238.3
Energy Biomass	0.0	0.0	100.0	150.3
Other	886.2	879.5	742.0	707.6
Total	3140.0	3140.0	3140.0	3140.0
Anthropogenic Emissions (not standardized)[d]				
CO_2 (as C)[b] in GtC	3.1	3.8	3.3	3.2
Other CO_2 (as C)[e] in GtC	0.0	-0.1	-0.1	-0.1
Total CO_2 (as C) in GtC	3.0	3.7	3.3	3.1
CH_4 total (as CH_4) in Mt CH_4	68.9	62.7	59.0	62.0
N_2O total (as N) in Mt N_2O-N	2.5	2.2	1.9	1.3
SO_x total (as S) in MtS	26.2	9.3	5.7	3.5
CO (as CO) in Mt CO	192.1	211.6	221.4	231.7
VOCs in Mt	46.6	49.6	52.5	39.5
NO_x (as N) in MtN	13.2	16.5	16.1	12.0

[a] mex = market exchange rate; ppp= purchasing power parity

[b] CO_2 emissions from fossil fuel and industrial processes (MESSAGE data).

[c] Land-use data taken from AIM B2 emulation run.

[d] Other non-energy emissions categories were calculated based on AIM B2 land-use change estimates and/or AIM B2 non-energy emissions.

[e] CO_2 emissions from land-use changes (AIM B2 emulation run)

Note: Subtotals may not add due to independent rounding.

Preliminary B2 Marker Scenario – REF (Open process version, date of submission: 11/2/98)

	1990	2020	2050	2100
Population in Million	413	418	406	379
GNP/GDP (mex) in Trillion US$90 [a]	1.1	1.8	6.6	14.5
GNP/GDP (ppp) in Trillion (1990 prices) [a]	2.6	3.3	7.2	16.2
Final Energy by Fuel in EJ				
Non-commercial	0.0	0.0	0.0	0.0
Solids	9.3	2.2	0.1	0.0
Liquids	15.4	17.2	19.2	21.6
Gas	13.5	15.4	19.2	9.5
Electricity	5.7	6.8	17.9	32.1
Other (e.g. H_2)	6.5	7.2	11.5	15.4
Total	50.4	48.8	67.9	78.6
Primary Energy by Fuel				
Coal	18.6	7.3	12.4	32.4
Oil	20.4	18.9	19.6	0.1
Gas	26.7	31.4	54.8	44.8
Nuclear	1.0	0.7	2.4	9.3
Biomass	1.8	0.8	4.8	36.2
Other Renewables	1.1	2.6	8.2	21.2
Synfuel Trade	0	0	-5	-19.4
Total	69.6	61.7	97.2	124.6
Cumulative Resource Use in ZJ				
Coal	0.0	0.4	0.7	1.5
Oil	0.0	0.6	1.1	1.6
Gas	0.0	0.8	2.0	5.2
Cumulative CO_2 Emissions in GtC [b]	0.0	34.6	70.7	144.4
Land Use in Million ha [c]				
Forests	960.0	940.3	967.2	1004.4
Grasslands	345.7	453.4	531.4	561.3
Cropland	279.1	300.2	283.4	238.1
Energy Biomass	0.0	0.0	43.4	35.0
Other	719.9	610.7	479.3	465.9
Total	2304.7	2304.7	2304.7	2304.7
Anthropogenic Emissions (not standardized)[d]				
CO_2 (as C)[b] in GtC	1.4	1.1	1.5	1.4
Other CO_2 (as C)[e] in GtC	0.0	0.0	0.0	0.0
Total CO_2 (as C) in GtC	1.4	1.0	1.5	1.4
CH_4 total (as CH_4) in Mt CH_4	46.5	48.3	59.8	58.2
N_2O total (as N) in Mt N_2O-N	0.6	0.6	0.6	0.5
SO_x total (as S) in MtS	16.8	6.1	4.5	3.6
CO (as CO) in Mt CO	82.0	86.9	111.3	114.8
VOCs in Mt	20.2	23.3	38.4	32.2
NO_x (as N) in MtN	4.1	3.8	5.5	3.9

[a] mex = market exchange rate; ppp= purchasing power parity

[b] CO_2 emissions from fossil fuel and industrial processes (MESSAGE data).

[c] Land-use data taken from AIM B2 emulation run.

[d] Other non-energy emissions categories were calculated based on AIM B2 land-use change estimates and/or AIM B2 non-energy emissions.

[e] CO_2 emissions from land-use changes (AIM B2 emulation run)

Note: Subtotals may not add due to independent rounding.

Preliminary B2 Marker Scenario – ASIA (Open process version, date of submission: 11/2/98)

	1990	2020	2050	2100
Population in Million	2798	4008	4696	4968
GNP/GDP (mex) in Trillion US$90 [a]	1.5	13.2	41.8	97.1
GNP/GDP (ppp) in Trillion (1990 prices) [a]	5.3	22.4	49.3	100.4
Final Energy by Fuel in EJ				
Non-commercial	24.2	13.2	6.7	4.1
Solids	18.5	27.8	16.7	2.4
Liquids	12.6	59.8	112.0	121.6
Gas	1.5	8.7	22.0	32.9
Electricity	4.1	24.1	69.8	168.6
Other (e.g. H_2)	0.6	10.2	24.8	55.4
Total	61.5	143.8	252.0	385.0
Primary Energy by Fuel				
Coal	29.8	47.4	47.7	178.1
Oil	15.3	62.6	92.8	21.4
Gas	2.8	30.9	55.4	38.1
Nuclear	0.3	3.7	20.9	64.7
Biomass	24.3	28.0	46.1	84.7
Other Renewables	1.1	12.3	42.6	87.7
Synfuel Trade	0	0	13.5	46.6
Total	73.6	184.9	319.0	521.3
Cumulative Resource Use in ZJ				
Coal	0.0	1.2	2.7	6.7
Oil	0.0	1.0	3.3	6.3
Gas	0.0	0.3	1.5	4.5
Cumulative CO_2 Emissions in GtC [b]	0.0	53.6	158.5	378.0
Land Use in Million ha [c]				
Forests	527.1	411.7	439.5	482.7
Grasslands	521.2	578.2	614.2	606.0
Cropland	389.9	457.5	401.4	325.7
Energy Biomass	0.0	0.5	96.0	153.6
Other	576.1	566.6	463.3	446.3
Total	2014.4	2014.4	2014.4	2014.4
Anthropogenic Emissions (not standardized) [d]				
CO_2 (as C)[b] in GtC	1.2	3.1	4.2	5.8
Other CO_2 (as C) [e] in GtC	0.3	0.3	-0.1	-0.1
Total CO_2 (as C) in GtC	1.5	3.4	4.1	5.8
CH_4 total (as CH_4) in Mt CH_4	127.1	222.6	265.9	234.9
N_2O total (as N) in Mt N_2O-N	2.3	2.7	2.4	2.1
SO_x total (as S) in MtS	17.3	31.3	24.4	19.0
CO (as CO) in Mt CO	276.5	422.3	550.9	706.0
VOCs in Mt	47.8	75.3	92.2	80.3
NO_x (as N) in MtN	6.2	15.6	23.0	31.2

[a] mex = market exchange rate; ppp= purchasing power parity

[b] CO_2 emissions from fossil fuel and industrial processes (MESSAGE data).

[c] Land-use data taken from AIM B2 emulation run.

[d] Other non-energy emissions categories were calculated based on AIM B2 land-use change estimates and/or AIM B2 non-energy emissions.

[e] CO_2 emissions from land-use changes (AIM B2 emulation run)

Note: Subtotals may not add due to independent rounding.

Preliminary B2 Marker Scenario – ALM (Open process version, date of submission: 11/2/98)

	1990	2020	2050	2100
Population in Million	1192	2263	3289	4139
GNP/GDP (mex) in Trillion US$90 [a]	1.9	5.5	22.8	66.8
GNP/GDP (ppp) in Trillion (1990 prices) [a]	3.8	8.2	23.9	64.9
Final Energy by Fuel in EJ				
Non-commercial	14.2	10.7	4.0	2.7
Solids	1.3	2.6	2.4	4.6
Liquids	17.0	36.7	60.7	93.3
Gas	3.5	9.8	28.3	51.5
Electricity	3.4	10.5	40.0	119.2
Other (e.g. H_2)	0.0	4.3	16.6	34.5
Total	39.4	74.6	152.0	305.8
Primary Energy by Fuel				
Coal	4.7	4.7	6.6	42.9
Oil	20.5	42.4	49.7	13.7
Gas	8.1	27.4	94.5	125.8
Nuclear	0.1	0.4	7.8	38.7
Biomass	14.3	16.6	43.5	151.8
Other Renewables	1.7	8.3	31.7	62.1
Synfuel Trade	0	0	-17.1	2.4
Total	49.4	99.8	216.7	437.4
Cumulative Resource Use in ZJ				
Coal	0.0	0.1	0.3	1.1
Oil	0.0	1.0	2.4	4.2
Gas	0.0	0.4	1.7	8.1
Cumulative CO_2 Emissions in GtC [b]	0.0	28.3	81.0	235.6
Land Use in Million ha [c]				
Forests	1706.0	1316.6	1319.0	1344.3
Grasslands	1635.6	1958.5	2075.9	2068.2
Cropland	357.4	448.4	394.8	311.1
Energy Biomass	0.0	15.0	258.5	378.7
Other	1784.2	1744.7	1435.0	1380.9
Total	5483.2	5483.2	5483.2	5483.2
Anthropogenic Emissions (not standardized)[d]				
CO_2 (as C)[b] in GtC	0.8	1.5	2.5	3.7
Other CO_2 (as C)[e] in GtC	0.8	1.2	-0.2	-0.1
Total CO_2 (as C) in GtC	1.6	2.6	2.3	3.7
CH_4 total (as CH_4) in Mt CH_4	75.9	114.1	153.4	153.6
N_2O total (as N) in Mt N_2O-N	1.0	1.3	0.8	0.8
SO_x total (as S) in MtS	8.7	12.3	17.4	16.8
CO (as CO) in Mt CO	423.7	495.9	472.5	1019.7
VOCs in Mt	62.8	85.7	76.0	60.4
NO_x (as N) in MtN	7.4	10.3	12.0	16.3

[a] mex = market exchange rate; ppp= purchasing power parity

[b] CO_2 emissions from fossil fuel and industrial processes (MESSAGE data).

[c] Land-use data taken from AIM B2 emulation run.

[d] Other non-energy emissions categories were calculated based on AIM B2 land-use change estimates and/or AIM B2 non-energy emissions.

[e] CO_2 emissions from land-use changes (AIM B2 emulation run)

Note: Subtotals may not add due to independent rounding.

VII

Statistical tables

Table VII.1: *Overview tables of the 40 SRES scenarios. Main driving forces and emissions are given in the same format for all scenarios developed by the six modeling teams for the World and the four SRES regions OECD90, REF, ASIA, and ALM in 10-year time steps. (Except for the A2G-IMAGE*[1] *and B2-IMAGE scenarios, for which emissions are in 10-year time steps, and the driving forces are presented for the years 1990, 2020, 2050, and 2100).*

It should be noted that:

- *Subtotals may not add due to independent rounding.*
- *The emissions are all standardized to common 1990 and 2000 values (see Box 5-1).*
- *The chlorofluorocarbons (CFCs) and hydrochlorofluorocarbons (HCFCs) are only available for the World as total. At the regional level only hydrofluorocarbons (HFCs) emissions were estimated. At the World level the row CFC/HFC/HCFC represents an aggregation of all CFCs, HCFCs, and HFCs (using SAR GWPs).*
- *Table footnotes give additional explanations for various scenarios.*
- *During the approval process of the Summary for Policymakers at the 5th Session of Working Group III of the IPCC from 8-11 March 2000 in Katmandu, Nepal, it was decided to combine the A1C and A1G groups into one "fossil intensive" group A1FI in contrast to the non-fossil group A1T, and select two illustrative scenarios from these two A1 groups to facilitate use by modellers and policy makers. Together with the scenario groups A1B, A2, B1, B2, this leads to six scenario groups that constitute the four scenario families. The marker scenarios A1B-AIM, A2, B1 and B2 as well as the illustrative scenarios A1FI-MiniCAM and A1T-MESSAGE are indicated in the Tables. All scenarios are equally sound.*

[1] The IMAGE-results for the A2 and B2 scenarios are based on preliminary model experiments carried out in March 1998. As a consequence of limited resources it has not been possible to rerun these experiments. Hence, the IMAGE team is not able to provide background data and details for these scenario calculations and the population and economic growth assumptions are not fully harmonized, as they have been for the IMAGE A1 and B1 scenarios.

Marker Scenario A1B-AIM World		1990	2000	2010	2020	2030	2040	2050	2060	2070	2080	2090	2100
Population	Million	5262	6117	6805	7493	8182	8439	8704	8538	8375	8030	7528	7056
GNP/GDP (mex)	Trillion US$	20.9	26.7	37.9	56.5	89.1	127.1	181.3	235.1	304.7	377.4	446.6	528.5
GNP/GDP (ppp) Trillion (1990 prices)													
Final Energy	EJ												
Non-commercial		50	50	38	36	26	0	0	0	0	0	0	0
Solids		36	38	50	62	75	75	75	60	49	39	31	25
Liquids		111	124	155	197	231	264	302	334	369	388	387	387
Gas		51	53	89	134	194	239	295	335	380	410	421	432
Electricity		38	49	73	104	143	218	331	438	581	709	798	898
Others													
Total		285	314	405	532	669	819	1002	1180	1388	1550	1643	1741
Primary Energy	EJ												
Coal		93	99	134	163	179	182	186	165	148	126	103	84
Oil		143	167	209	238	239	226	214	188	166	149	136	125
Gas		73	91	147	196	298	372	465	519	578	604	590	576
Nuclear		6	8	16	30	53	81	123	125	127	116	95	78
Biomass		50	48	38	61	85	128	193	247	315	360	368	376
Other Renewables		10	12	15	23	40	82	167	278	464	662	808	987
Total		376	424	559	711	895	1098	1347	1574	1840	2034	2128	2226
Cumulative Resources Use	ZJ												
Coal		0.1	1.1	2.2	3.7	5.4	7.0	9.1	10.5	12.2	13.6	14.7	15.9
Oil		0.1	1.7	3.6	5.8	8.2	10.2	12.7	14.4	16.3	18.0	19.4	20.8
Gas		0.1	0.9	2.1	3.8	6.3	9.3	13.9	18.2	23.9	29.8	35.5	42.2
Cumulative CO2 Emissions	GtC	7.1	82.4	176.7	294.3	429.8	579.0	737.7	899.6	1058.2	1212.8	1360.2	1499.2
Carbon Sequestraction	GtC												
Land Use	Million ha												
Cropland		1459	1466	1462	1457	1454	1448	1442	1436	1429	1424	1422	1420
Grasslands		3389	3404	3429	3446	3458	3478	3498	3525	3552	3568	3572	3576
Energy Biomass		0	0	0	74	158	257	418	484	560	580	536	495
Forest		4296	4237	4173	4164	4164	4177	4190	4194	4199	4202	4203	4204
Others		3805	3842	3886	3807	3715	3554	3400	3299	3201	3173	3213	3253
Total		12949	12949	12949	12949	12949	12949	12949	12949	12949	12949	12949	12949
Anthropogenic Emissions (standardized)													
Fossil Fuel CO2	GtC	5.99	6.90	9.68	12.12	14.01	14.95	16.01	15.70	15.43	14.83	13.94	13.10
Other CO2	GtC	1.11	1.07	1.20	0.52	0.47	0.40	0.37	0.30	0.30	0.35	0.36	0.39
Total CO2	GtC	7.10	7.97	10.88	12.64	14.48	15.35	16.38	16.00	15.73	15.18	14.30	13.49
CH4 total	MtCH4	310	323	373	421	466	458	452	410	373	341	314	289
N2O total	MtN2O-N	6.7	7.0	7.0	7.2	7.3	7.4	7.4	7.3	7.2	7.1	7.1	7.0
SOx total	MtS	70.9	69.0	87.1	100.2	91.0	68.9	64.1	46.9	35.7	30.7	29.1	27.6
CFC/HFC/HCFC	MtC eq.	1672	883	791	337	369	482	566	654	659	654	639	614
PFC	MtC eq.	32	25	31	43	61	77	89	97	106	114	119	115
SF6	MtC eq.	38	40	43	48	66	99	119	127	113	88	84	95
CO	MtCO	879	877	1002	1032	1109	1160	1214	1245	1276	1357	1499	1663
NMVOC	Mt	139	141	178	222	266	272	279	284	289	269	228	193
NOx	MtN	31	32	39	46	50	49	48	46	44	43	41	40

Table VII.1: *Overview tables of the 40 SRES scenarios. Main driving forces and emissions are given in the same format for all scenarios developed by the six modeling teams for the World and the four SRES regions OECD90, REF, ASIA, and ALM in 10-year time steps. (Except for the A2G-IMAGE[1] and B2-IMAGE scenarios, for which emissions are in 10-year time steps, and the driving forces are presented for the years 1990, 2020, 2050, and 2100).*
It should be noted that:

- *Subtotals may not add due to independent rounding.*

Marker Scenario A1B-AIM OECD90		1990	2000	2010	2020	2030	2040	2050	2060	2070	2080	2090	2100
Population	Million	859	919	960	1002	1043	1062	1081	1086	1091	1097	1103	1110
GNP/GDP (mex)	Trillion US$	16.4	20.5	25.3	31.0	38.0	45.4	54.1	64.1	75.9	89.2	103.9	121.1
GNP/GDP (ppp) Trillion (1990 prices)													
Final Energy	EJ												
Non-commercial		6	1	0	0	0	0	0	0	0	0	0	0
Solids		10	9	9	9	8	7	6	5	3	3	2	1
Liquids		64	74	77	82	80	77	75	74	73	71	69	67
Gas		25	30	40	50	63	66	68	72	76	78	78	78
Electricity		22	27	33	40	48	58	70	87	108	129	151	175
Others													
Total		127	142	158	180	199	209	218	239	261	282	301	321
Primary Energy	EJ												
Coal		41	41	41	35	30	29	28	22	17	12	9	7
Oil		76	90	92	89	78	65	54	46	39	34	32	30
Gas		34	47	58	70	83	89	95	99	102	102	99	96
Nuclear		5	6	10	16	23	24	24	23	23	20	18	15
Biomass		6	1	0	4	9	17	29	37	48	54	56	57
Other Renewables		6	6	8	11	16	24	35	55	86	120	152	192
Total		167	191	209	226	239	253	267	293	321	348	372	397
Cumulative Resources Use	ZJ												
Coal		0.0	0.4	0.9	1.2	1.6	1.8	2.1	2.3	2.6	2.7	2.8	2.9
Oil		0.1	0.9	1.8	2.7	3.6	4.2	4.9	5.3	5.8	6.2	6.5	6.8
Gas		0.0	0.4	1.0	1.6	2.4	3.1	4.2	5.0	6.0	7.1	8.0	9.1
Cumulative CO2 Emissions	GtC	2.8	33.0	66.2	101.2	136.4	171.0	204.8	236.9	266.4	294.0	319.7	343.7
Carbon Sequestraction	GtC												
Land Use	Million ha												
Cropland		381	380	377	376	375	372	368	364	359	356	356	355
Grasslands		760	763	771	774	776	787	797	818	839	849	847	846
Energy Biomass		0	0	0	11	24	39	63	73	85	88	81	75
Forest		1050	1053	1052	1053	1056	1062	1068	1061	1054	1050	1049	1048
Others		838	833	828	814	798	765	733	711	690	685	695	705
Total		3029	3029	3029	3029	3029	3029	3029	3029	3029	3029	3029	3029
Anthropogenic Emissions (standardized)													
Fossil Fuel CO2	GtC	2.83	3.20	3.41	3.51	3.48	3.42	3.36	3.06	2.78	2.56	2.40	2.24
Other CO2	GtC	0.00	0.00	0.05	0.03	0.02	0.00	-0.01	0.01	0.06	0.11	0.09	0.07
Total CO2	GtC	2.83	3.20	3.45	3.54	3.50	3.42	3.35	3.06	2.84	2.67	2.48	2.31
CH4 total	MtCH4	73	74	71	69	66	58	52	49	47	46	44	42
N2O total	MtN2O-N	2.6	2.6	2.5	2.6	2.5	2.5	2.4	2.4	2.3	2.3	2.2	2.2
SOx total	MtS	22.7	17.0	9.9	6.9	6.4	6.3	6.3	6.0	5.7	5.3	5.0	4.6
HFC	MtC eq.	19	58	110	108	115	118	122	123	123	124	125	125
PFC	MtC eq.	18	13	13	11	9	8	9	10	12	13	14	16
SF6	MtC eq.	23	23	14	5	6	7	9	11	13	16	18	20
CO	MtCO	179	161	190	205	227	234	241	242	243	247	254	262
NMVOC	Mt	42	36	39	39	38	33	28	24	19	17	16	15
NOx	MtN	13	12	12	12	9	7	6	6	6	5	5	5

Marker Scenario A1B-AIM REF		**1990**	**2000**	**2010**	**2020**	**2030**	**2040**	**2050**	**2060**	**2070**	**2080**	**2090**	**2100**
Population	Million	413	419	424	430	435	429	423	406	391	374	356	339
GNP/GDP (mex)	Trillion US$	1.1	0.8	1.5	2.9	5.3	8.1	12.4	15.6	19.6	24.0	28.6	34.2
GNP/GDP (ppp) Trillion (1990 prices)													
Final Energy	EJ												
Non-commercial		2	4	0	0	0	0	0	0	0	0	0	0
Solids		9	5	5	6	5	5	4	3	2	2	1	1
Liquids		19	10	9	9	8	7	7	7	7	7	7	6
Gas		19	11	18	23	31	38	46	47	49	50	50	50
Electricity		8	8	10	13	15	20	28	33	40	46	50	55
Others													
Total		58	38	43	50	59	70	84	91	99	105	109	113
Primary Energy	EJ												
Coal		18	12	14	14	13	12	11	9	7	5	4	3
Oil		22	14	12	10	7	7	6	5	4	3	3	2
Gas		26	18	26	33	41	48	55	56	56	56	55	54
Nuclear		1	1	2	4	6	7	8	8	8	7	6	5
Biomass		2	4	0	1	2	4	8	10	12	14	14	15
Other Renewables		1	1	1	2	3	6	15	22	32	42	51	61
Total		71	51	56	64	72	86	103	112	122	129	134	139
Cumulative Resources Use	ZJ												
Coal		0.0	0.2	0.3	0.4	0.6	0.7	0.8	0.9	1.0	1.1	1.1	1.1
Oil		0.0	0.2	0.3	0.4	0.5	0.6	0.7	0.7	0.7	0.8	0.8	0.8
Gas		0.0	0.3	0.5	0.8	1.1	1.5	2.1	2.6	3.2	3.7	4.3	4.9
Cumulative CO2 Emissions	GtC	1.3	12.3	23.1	35.0	46.4	57.6	68.4	78.5	87.8	96.5	104.6	112.2
Carbon Sequestraction	GtC												
Land Use	Million ha												
Cropland		268	266	265	265	265	264	263	262	262	261	260	260
Grasslands		341	361	364	366	367	368	370	371	371	372	373	374
Energy Biomass		0	0	0	3	6	10	16	19	22	23	21	20
Forest		966	950	918	904	894	899	905	909	912	916	922	927
Others		701	698	728	738	745	733	722	715	708	703	700	696
Total		2276	2276	2276	2276	2276	2276	2276	2276	2276	2276	2276	2276
Anthropogenic Emissions (standardized)													
Fossil Fuel CO2	GtC	1.30	0.91	1.05	1.11	1.13	1.16	1.18	1.08	0.99	0.91	0.84	0.78
Other CO2	GtC	0.00	0.00	0.20	0.03	0.01	-0.06	-0.13	-0.11	-0.09	-0.07	-0.05	-0.03
Total CO2	GtC	1.30	0.91	1.24	1.14	1.14	1.10	1.05	0.97	0.90	0.84	0.79	0.75
CH4 total	MtCH4	47	39	58	61	60	51	42	41	39	38	36	34
N2O total	MtN2O-N	0.6	0.6	0.6	0.6	0.6	0.6	0.6	0.6	0.6	0.6	0.5	0.5
SOx total	MtS	17.0	11.0	12.2	10.8	7.5	4.3	2.4	2.0	1.7	1.6	1.6	1.6
HFC	MtC eq.	0	4	8	19	29	31	32	33	33	34	33	31
PFC	MtC eq.	7	4	5	8	14	20	21	22	23	24	25	24
SF6	MtC eq.	8	6	8	10	14	18	21	19	15	14	10	11
CO	MtCO	69	41	41	41	42	42	43	42	42	42	43	44
NMVOC	Mt	16	13	15	14	15	15	16	17	17	17	18	18
NOx	MtN	5	3	3	3	3	2	2	2	2	1	1	1

Marker Scenario A1B-AIM ASIA		**1990**	**2000**	**2010**	**2020**	**2030**	**2040**	**2050**	**2060**	**2070**	**2080**	**2090**	**2100**
Population	Million	2798	3261	3556	3851	4147	4183	4220	4016	3822	3541	3194	2882
GNP/GDP (mex)	Trillion US$	1.5	2.7	5.8	12.3	26.2	40.5	62.7	85.4	116.2	147.6	174.9	207.3
GNP/GDP (ppp) Trillion (1990 prices)													
Final Energy	EJ												
Non-commercial		28	25	18	14	4	0	0	0	0	0	0	0
Solids		15	22	34	44	56	56	57	47	39	31	25	20
Liquids		11	17	30	48	65	75	87	99	112	118	115	112
Gas		2	3	8	17	37	51	70	90	115	136	147	160
Electricity		5	8	15	27	44	72	117	161	223	280	319	363
Others													
Total		61	75	105	151	205	261	331	404	495	567	610	655
Primary Energy	EJ												
Coal		30	39	64	87	108	106	104	94	86	73	59	48
Oil		17	27	45	62	68	59	51	44	38	33	30	27
Gas		4	8	17	35	66	95	137	167	203	228	235	243
Nuclear		0	0	2	6	15	26	45	47	50	46	38	32
Biomass		28	24	19	20	19	29	46	58	75	85	87	89
Other Renewables		1	2	2	4	10	24	58	101	177	261	323	399
Total		80	100	149	214	285	354	440	533	646	734	784	838
Cumulative Resources Use	ZJ												
Coal		0.0	0.4	0.9	1.6	2.6	3.5	4.7	5.6	6.5	7.4	8.0	8.7
Oil		0.0	0.2	0.6	1.1	1.8	2.3	3.0	3.4	3.8	4.2	4.5	4.8
Gas		0.0	0.1	0.2	0.4	0.9	1.7	3.0	4.3	6.2	8.4	10.6	13.3
Cumulative CO2 Emissions	GtC	1.5	19.3	45.8	82.8	130.2	185.0	243.1	303.1	363.7	424.2	482.7	538.6
Carbon Sequestraction	GtC												
Land Use	Million ha												
Cropland		438	435	434	433	432	431	431	430	429	428	428	428
Grasslands		608	606	609	611	613	615	617	618	619	621	622	624
Energy Biomass		0	0	0	17	37	61	99	115	133	137	127	117
Forest		535	522	512	524	535	535	535	544	552	556	555	555
Others		583	601	609	579	547	514	483	455	429	421	431	441
Total		2164	2164	2164	2164	2164	2164	2164	2164	2164	2164	2164	2164
Anthropogenic Emissions (standardized)													
Fossil Fuel CO2	GtC	1.15	1.78	2.92	4.11	5.21	5.46	5.73	5.86	5.99	5.89	5.57	5.27
Other CO2	GtC	0.37	0.26	0.33	0.05	0.12	0.17	0.25	0.16	0.11	0.10	0.14	0.19
Total CO2	GtC	1.53	2.03	3.25	4.16	5.32	5.64	5.98	6.02	6.10	5.99	5.71	5.46
CH4 total	MtCH4	113	125	145	171	207	210	214	183	156	138	127	117
N2O total	MtN2O-N	2.3	2.6	2.6	2.7	2.8	2.9	3.0	2.9	2.9	2.9	2.9	2.9
SOx total	MtS	17.7	25.3	42.1	54.2	45.6	19.8	8.4	7.9	7.5	7.1	6.7	6.4
HFC	MtC eq.	0	5	18	45	92	153	224	292	292	285	275	262
PFC	MtC eq.	3	5	8	15	23	30	35	39	43	46	48	46
SF6	MtC eq.	4	7	12	19	28	42	50	55	48	35	33	37
CO	MtCO	235	270	358	360	430	460	492	506	520	555	614	678
NMVOC	Mt	33	37	49	70	94	99	105	112	119	111	90	73
NOx	MtN	7	9	13	16	19	19	19	17	16	15	14	13

Marker Scenario A1B-AIM ALM		**1990**	**2000**	**2010**	**2020**	**2030**	**2040**	**2050**	**2060**	**2070**	**2080**	**2090**	**2100**
Population	Million	1192	1519	1865	2211	2557	2761	2980	3024	3067	3013	2866	2727
GNP/GDP (mex)	Trillion US$	1.9	2.7	5.3	10.3	19.5	31.9	52.0	69.4	92.5	116.6	139.1	165.9
GNP/GDP (ppp) Trillion (1990 prices)													
Final Energy	EJ												
Non-commercial		14	20	20	22	22	0	0	0	0	0	0	0
Solids		1	2	2	4	6	7	7	6	5	4	3	2
Liquids		17	22	38	58	78	102	133	153	176	191	196	202
Gas		4	9	25	43	64	84	112	125	139	147	145	144
Electricity		4	6	15	24	36	65	117	156	210	254	278	304
Others													
Total		40	59	100	151	205	275	369	443	532	596	623	652
Primary Energy	EJ												
Coal		5	7	16	26	29	35	42	40	38	35	30	26
Oil		27	36	60	77	86	94	104	94	85	78	71	65
Gas		9	18	46	58	108	139	178	196	215	217	200	184
Nuclear		0	0	1	4	9	20	45	45	46	42	33	26
Biomass		14	18	19	36	55	78	110	141	181	206	211	215
Other Renewables		2	3	4	6	12	26	59	100	168	238	282	335
Total		57	81	146	208	299	401	538	635	750	822	837	852
Cumulative Resources Use	ZJ												
Coal		0.0	0.1	0.2	0.4	0.7	0.9	1.4	1.7	2.1	2.5	2.8	1.7
Oil		0.0	0.3	0.8	1.5	2.3	3.1	4.2	5.0	6.0	6.9	7.6	6.7
Gas		0.0	0.1	0.5	1.0	1.8	2.9	4.7	6.3	8.4	10.6	12.5	12.5
Cumulative CO2 Emissions	GtC	1.4	17.8	41.6	75.3	116.8	165.4	221.4	281.1	340.3	398.2	453.2	504.6
Carbon Sequestraction	GtC												
Land Use	Million ha												
Cropland		371	385	384	383	382	381	381	380	379	379	378	378
Grasslands		1680	1673	1685	1695	1702	1708	1714	1718	1722	1726	1730	1733
Energy Biomass		0	0	0	42	90	147	240	277	321	332	307	284
Forest		1745	1711	1690	1683	1680	1681	1682	1681	1680	1678	1676	1674
Others		1684	1710	1720	1676	1625	1542	1463	1417	1373	1363	1387	1411
Total		5480	5480	5480	5480	5480	5480	5480	5480	5480	5480	5480	5480
Anthropogenic Emissions (standardized)													
Fossil Fuel CO2	GtC	0.72	1.01	2.30	3.40	4.20	4.91	5.73	5.70	5.67	5.47	5.13	4.81
Other CO2	GtC	0.73	0.82	0.63	0.40	0.32	0.29	0.26	0.25	0.23	0.21	0.19	0.16
Total CO2	GtC	1.45	1.83	2.93	3.80	4.52	5.20	5.99	5.94	5.90	5.68	5.32	4.97
CH4 total	MtCH4	77	85	99	120	134	139	144	137	130	120	107	95
N2O total	MtN2O-N	1.2	1.3	1.3	1.3	1.3	1.4	1.4	1.4	1.4	1.4	1.4	1.4
SOx total	MtS	10.5	12.8	20.0	25.3	28.4	35.5	44.1	28.0	17.8	13.7	12.8	12.1
HFC	MtC eq.	0	2	15	39	81	139	184	205	210	210	206	196
PFC	MtC eq.	4	4	5	9	14	19	23	26	28	30	31	30
SF6	MtC eq.	3	5	10	14	19	32	40	42	37	23	24	26
CO	MtCO	396	404	413	427	410	424	439	455	472	513	588	679
NMVOC	Mt	48	55	75	99	119	124	129	131	134	124	105	88
NOx	MtN	7	8	11	15	19	20	21	21	21	21	21	21

Scenario A1B-ASF World		1990	2000	2010	2020	2030	2040	2050	2060	2070	2080	2090	2100
Population	Million	5264	6117	6827	7537	8039	8526	8704	8527	8444	8022	7282	7056
GNP/GDP (mex)	Trillion US$	20.9	28.9	41.3	60.7	87.9	146.0	174.3	225.7	257.3	367.8	484.9	531.9
GNP/GDP (ppp) Trillion (1990 prices)													
Final Energy	EJ												
Non-commercial													
Solids		56	68	87	105	89	74	59	53	48	43	39	35
Liquids		123	177	243	349	398	447	496	494	492	486	476	465
Gas		51	63	93	154	196	237	278	268	257	229	183	137
Electricity		43	57	84	122	225	327	429	533	637	711	755	798
Others													
Total		273	365	507	730	907	1085	1262	1348	1434	1469	1452	1436
Primary Energy	EJ												
Coal		97	118	154	212	368	523	679	673	667	669	680	691
Oil		141	202	283	381	302	223	143	87	31	3	2	1
Gas		74	87	126	206	278	349	421	391	361	286	166	47
Nuclear		8	14	18	26	72	117	163	191	218	252	292	332
Biomass		0	0	2	27	57	88	119	145	171	198	224	250
Other Renewables		6	9	16	23	44	65	86	173	259	312	329	346
Total		326	430	600	875	1121	1366	1611	1660	1708	1719	1693	1667
Cumulative Resources Use	ZJ												
Coal		0.0	1.2	2.5	4.4	7.3	11.8	17.8	24.6	31.3	37.9	44.7	51.5
Oil		0.0	1.8	4.3	7.6	11.0	13.5	15.3	16.4	16.9	17.0	17.0	17.0
Gas		0.0	0.9	1.9	3.6	6.0	9.2	13.1	17.1	20.8	24.0	26.2	27.0
Cumulative CO2 Emissions	GtC	7.1	82.4	177.9	313.1	495.8	716.8	967.6	1223.9	1461.9	1682.0	1885.3	2072.6
Carbon Sequestration	GtC	-1.8	-1.8	-1.7	-1.7	-1.7	-1.7	-1.6	-1.6	-1.5	-1.5	-1.4	-1.4
Land Use	Million ha												
Cropland													
Grasslands													
Energy Biomass													
Forest													
Others													
Total													
Anthropogenic Emissions (standardized)													
Fossil Fuel CO2	GtC	5.99	6.90	10.01	14.67	19.49	22.60	25.72	24.14	22.55	20.97	19.37	17.78
Other CO2	GtC	1.11	1.07	1.12	1.24	1.13	0.98	0.84	0.58	0.32	0.18	0.16	0.14
Total CO2	GtC	7.10	7.97	11.13	15.91	20.62	23.59	26.56	24.72	22.87	21.15	19.53	17.92
CH4 total	MtCH4	310	323	377	444	518	577	636	616	596	571	540	510
N2O total	MtN2O-N	6.7	7.0	8.1	9.6	10.7	11.2	11.8	11.5	11.2	10.8	10.2	9.7
SOx total	MtS	70.9	69.0	84.7	116.5	110.0	84.7	59.5	54.0	48.6	46.1	46.6	47.0
CFC/HFC/HCFC	MtC eq.	1672	883	791	337	369	482	566	654	659	654	639	614
PFC	MtC eq.	32	25	31	43	61	77	89	97	106	114	119	115
SF6	MtC eq.	38	40	43	48	66	99	119	127	113	88	84	95
CO	MtCO	879	877	1068	1248	1567	1698	1830	1837	1845	1832	1798	1763
NMVOC	Mt	139	141	165	207	248	274	301	346	392	442	497	552
NOx	MtN	31	32	45	66	83	91	100	94	88	84	80	77

Scenario A1B-ASF OECD90		**1990**	**2000**	**2010**	**2020**	**2030**	**2040**	**2050**	**2060**	**2070**	**2080**	**2090**	**2100**
Population	Million	849	919	961	1003	1035	1069	1081	1085	1088	1096	1106	1110
GNP/GDP (mex)	Trillion US$	15.7	20.1	25.2	31.0	37.8	50.1	55.0	65.2	71.0	91.7	118.0	128.4
GNP/GDP (ppp) Trillion (1990 prices)													
Final Energy	EJ												
Non-commercial													
Solids		13	13	14	15	13	12	11	10	8	8	7	7
Liquids		69	88	95	95	87	80	72	68	63	62	65	68
Gas		28	33	36	40	41	41	41	39	36	32	26	20
Electricity		24	29	35	42	52	62	72	86	101	115	129	144
Others													
Total		133	163	179	192	193	195	196	202	209	217	227	238
Primary Energy	EJ												
Coal		32	29	32	39	61	83	106	93	80	78	87	96
Oil		77	97	107	102	72	43	13	8	3	0	0	0
Gas		35	40	44	52	55	59	62	57	51	39	21	4
Nuclear		7	11	12	13	18	24	30	35	39	47	57	68
Biomass		0	0	1	8	12	17	21	26	30	35	40	45
Other Renewables		5	7	9	11	13	15	18	29	41	49	55	60
Total		157	184	205	224	233	241	250	247	244	249	260	272
Cumulative Resources Use		ZJ											
Coal		0.0	0.3	0.6	1.0	1.5	2.2	3.2	4.2	5.0	5.8	6.6	7.6
Oil		0.0	1.0	2.0	3.0	3.9	4.4	4.7	4.8	4.8	4.8	4.8	4.8
Gas		0.0	0.4	0.8	1.3	1.9	2.4	3.0	3.6	4.2	4.6	4.9	5.0
Cumulative CO2 Emissions	GtC	2.8	33.0	66.6	103.4	143.0	183.5	223.3	260.2	292.1	320.1	346.6	372.7
Carbon Sequestration	GtC												
Land Use	Million ha												
Cropland													
Grasslands													
Energy Biomass													
Forest													
Others													
Total													
Anthropogenic Emissions (standardized)													
Fossil Fuel CO2	GtC	2.83	3.20	3.53	3.82	4.09	4.02	3.94	3.44	2.93	2.66	2.63	2.60
Other CO2	GtC	0.00	0.00	0.00	0.00	0.00	0.00	0.00	0.00	0.00	0.00	0.00	0.00
Total CO2	GtC	2.83	3.20	3.53	3.82	4.09	4.02	3.94	3.44	2.93	2.66	2.63	2.60
CH4 total	MtCH4	73	74	77	83	92	100	108	114	121	125	126	128
N2O total	MtN2O-N	2.6	2.6	2.7	3.0	3.0	3.0	2.9	2.8	2.7	2.6	2.6	2.5
SOx total	MtS	22.7	17.0	7.9	8.5	8.2	7.4	6.7	6.6	6.4	6.8	7.5	8.3
HFC	MtC eq.	19	58	110	108	115	118	122	123	123	124	125	125
PFC	MtC eq.	18	13	13	11	9	8	9	10	12	13	14	16
SF6	MtC eq.	23	23	14	5	6	7	9	11	13	16	18	20
CO	MtCO	179	161	169	164	152	124	97	90	83	81	86	90
NMVOC	Mt	42	36	41	45	45	42	38	41	43	48	55	62
NOx	MtN	13	12	14	15	15	14	13	11	10	9	9	9

Scenario A1B-ASF REF		1990	2000	2010	2020	2030	2040	2050	2060	2070	2080	2090	2100
Population	Million	417	419	425	431	432	425	423	406	398	374	347	339
GNP/GDP (mex)	Trillion US$	1.0	1.0	1.4	2.2	3.6	5.9	7.0	9.1	10.3	14.8	20.1	22.2
GNP/GDP (ppp) Trillion (1990 prices)													
Final Energy	EJ												
Non-commercial													
Solids		18	12	12	12	10	9	8	6	5	3	3	2
Liquids		16	12	13	15	16	16	17	16	15	15	15	15
Gas		16	13	18	24	26	29	31	28	26	22	17	12
Electricity		8	8	10	14	17	21	24	27	30	32	32	32
Others													
Total		58	44	53	65	70	74	79	77	76	72	67	61
Primary Energy	EJ												
Coal		23	13	15	18	18	19	19	17	14	13	12	12
Oil		18	13	15	17	18	19	19	12	5	1	0	0
Gas		27	22	29	38	38	39	39	34	30	23	15	6
Nuclear		1	1	2	2	5	7	9	10	11	11	12	13
Biomass		0	0	0	0	0	1	1	7	14	18	21	23
Other Renewables		5	5	6	6	7	8	9	13	16	18	18	18
Total		74	56	66	82	87	92	97	93	89	84	78	73
Cumulative Resources Use	ZJ												
Coal		0.0	0.2	0.3	0.5	0.7	0.9	1.1	1.2	1.4	1.5	1.6	1.8
Oil		0.0	0.2	0.3	0.5	0.7	0.8	1.0	1.2	1.3	1.3	1.3	1.3
Gas		0.0	0.3	0.5	0.9	1.2	1.6	2.0	2.4	2.7	3.0	3.1	3.2
Cumulative CO2 Emissions	GtC	1.3	12.3	22.3	34.1	47.2	60.1	72.2	83.0	92.0	99.5	105.4	110.0
Carbon Sequestration	GtC												
Land Use	Million ha												
Cropland													
Grasslands													
Energy Biomass													
Forest													
Others													
Total													
Anthropogenic Emissions (standardized)													
Fossil Fuel CO2	GtC	1.30	0.91	1.07	1.29	1.34	1.25	1.16	0.99	0.82	0.67	0.53	0.39
Other CO2	GtC	0.00	0.00	0.00	0.00	0.00	0.00	0.00	0.00	0.00	0.00	0.00	0.00
Total CO2	GtC	1.30	0.91	1.07	1.29	1.34	1.25	1.16	0.99	0.82	0.67	0.53	0.39
CH4 total	MtCH4	47	39	45	59	81	102	124	121	118	108	91	73
N2O total	MtN2O-N	0.6	0.6	0.6	0.7	0.7	0.7	0.7	0.6	0.6	0.6	0.5	0.5
SOx total	MtS	17.0	11.0	10.5	10.1	8.8	6.9	5.0	4.0	3.0	2.2	1.7	1.1
HFC	MtC eq.	0	4	8	19	29	31	32	33	33	34	33	31
PFC	MtC eq.	7	4	5	8	14	20	21	22	23	24	25	24
SF6	MtC eq.	8	6	8	10	14	18	21	19	15	14	10	11
CO	MtCO	69	41	42	41	47	48	49	48	47	47	48	49
NMVOC	Mt	16	13	17	25	29	28	28	28	28	31	35	40
NOx	MtN	5	3	3	4	5	5	4	4	3	2	2	2

Scenario A1B-ASF ASIA		**1990**	**2000**	**2010**	**2020**	**2030**	**2040**	**2050**	**2060**	**2070**	**2080**	**2090**	**2100**
Population	Million	2780	3261	3572	3884	4073	4181	4220	4012	3913	3538	3033	2882
GNP/GDP (mex)	Trillion US$	1.7	3.5	7.2	14.1	24.2	47.5	59.5	79.6	92.2	134.9	175.7	191.9
GNP/GDP (ppp) Trillion (1990 prices)													
Final Energy	EJ												
Non-commercial													
Solids		24	42	59	73	57	42	26	23	20	17	15	13
Liquids		15	37	72	133	161	190	218	219	220	217	210	202
Gas		3	8	21	53	72	92	111	106	101	88	69	50
Electricity		6	13	25	43	110	177	243	290	337	363	368	374
Others													
Total		48	100	177	301	400	500	599	638	677	685	662	638
Primary Energy	EJ												
Coal		36	67	92	129	211	293	375	365	356	348	342	336
Oil		19	46	89	144	110	77	43	26	9	0	0	0
Gas		4	11	28	68	107	145	184	168	152	118	66	13
Nuclear		0	1	4	9	38	67	96	108	120	133	147	161
Biomass		0	0	0	11	21	31	41	44	47	53	63	73
Other Renewables		5	5	8	11	24	38	51	95	139	161	162	163
Total		64	130	221	371	511	650	789	806	823	814	780	746
Cumulative Resources Use	ZJ												
Coal		0.0	0.6	1.4	2.5	4.2	6.7	10.1	13.8	17.4	20.9	24.3	27.7
Oil		0.0	0.3	1.0	2.2	3.4	4.3	4.9	5.2	5.4	5.4	5.4	5.4
Gas		0.0	0.1	0.3	0.7	1.6	2.9	4.5	6.3	7.9	9.2	10.1	10.4
Cumulative CO2 Emissions	GtC	1.5	19.3	48.9	102.4	185.2	292.1	417.3	547.7	669.8	783.7	889.5	987.0
Carbon Sequestration	GtC												
Land Use	Million ha												
Cropland													
Grasslands													
Energy Biomass													
Forest													
Others													
Total													
Anthropogenic Emissions (standardized)													
Fossil Fuel CO2	GtC	1.15	1.78	3.55	6.41	9.43	11.34	13.26	12.50	11.73	10.95	10.15	9.35
Other CO2	GtC	0.37	0.26	0.34	0.39	0.33	0.26	0.19	0.13	0.07	0.03	0.01	0.00
Total CO2	GtC	1.53	2.03	3.89	6.80	9.76	11.60	13.45	12.62	11.80	10.98	10.16	9.34
CH4 total	MtCH4	113	125	146	167	190	206	222	208	193	183	179	174
N2O total	MtN2O-N	2.3	2.6	3.2	4.1	4.7	5.0	5.4	5.2	5.0	4.7	4.4	4.0
SOx total	MtS	17.7	25.3	47.2	67.9	58.5	38.4	18.3	15.6	12.9	11.7	12.0	12.3
HFC	MtC eq.	0	5	18	45	92	153	224	292	292	285	275	262
PFC	MtC eq.	3	5	8	15	23	30	35	39	43	46	48	46
SF6	MtC eq.	4	7	12	19	28	42	50	55	48	35	33	37
CO	MtCO	235	270	374	471	629	711	792	802	811	804	778	753
NMVOC	Mt	33	37	43	54	66	75	84	95	105	115	123	132
NOx	MtN	7	9	16	27	36	42	47	44	40	37	35	32

Scenario A1B-ASF ALM		1990	2000	2010	2020	2030	2040	2050	2060	2070	2080	2090	2100
Population	Million	1218	1519	1869	2218	2500	2851	2980	3023	3044	3013	2795	2727
GNP/GDP (mex)	Trillion US$	2.6	4.3	7.4	13.3	22.3	42.5	52.8	71.8	83.8	126.6	171.2	189.4
GNP/GDP (ppp) Trillion (1990 prices)													
Final Energy	EJ												
Non-commercial													
Solids		1	2	3	5	8	11	14	15	15	15	15	14
Liquids		23	40	63	106	134	161	189	191	193	192	186	180
Gas		5	9	18	37	56	76	95	95	94	87	71	55
Electricity		5	8	14	24	46	68	90	130	170	202	225	249
Others													
Total		34	58	98	172	244	316	388	430	473	495	497	499
Primary Energy	EJ												
Coal		5	9	15	26	77	128	179	198	217	230	238	246
Oil		26	45	72	118	101	85	68	42	15	2	1	1
Gas		9	14	25	48	78	107	136	132	128	106	65	24
Nuclear		0	0	1	2	11	19	28	38	48	61	75	90
Biomass		0	0	1	8	24	40	56	68	81	91	100	110
Other Renewables		5	5	7	8	13	17	22	49	77	96	108	120
Total		44	74	121	211	304	396	489	528	566	586	588	590
Cumulative Resources Use	ZJ												
Coal		0.0	0.1	0.2	0.4	0.9	1.9	3.5	5.4	7.5	9.7	12.1	14.5
Oil		0.0	0.4	1.0	1.9	3.0	3.9	4.7	5.2	5.4	5.5	5.5	5.5
Gas		0.0	0.1	0.3	0.7	1.3	2.2	3.5	4.8	6.1	7.3	8.1	8.5
Cumulative CO2 Emissions	GtC	1.4	17.8	40.1	73.3	120.4	181.1	254.7	333.0	407.9	478.7	543.9	602.8
Carbon Sequestration	GtC												
Land Use	Million ha												
Cropland													
Grasslands													
Energy Biomass													
Forest													
Others													
Total													
Anthropogenic Emissions (standardized)													
Fossil Fuel CO2	GtC	0.72	1.01	1.86	3.15	4.63	5.99	7.36	7.21	7.07	6.68	6.06	5.44
Other CO2	GtC	0.73	0.82	0.78	0.85	0.80	0.72	0.65	0.45	0.25	0.15	0.14	0.14
Total CO2	GtC	1.45	1.83	2.63	4.00	5.42	6.71	8.00	7.66	7.32	6.83	6.21	5.58
CH4 total	MtCH4	77	85	109	135	155	168	182	173	164	155	145	135
N2O total	MtN2O-N	1.2	1.3	1.6	1.9	2.3	2.5	2.8	2.8	2.9	2.8	2.8	2.7
SOx total	MtS	10.5	12.8	16.1	27.1	31.5	29.0	26.4	24.9	23.3	22.5	22.4	22.3
HFC	MtC eq.	0	2	15	39	81	139	184	205	210	210	206	196
PFC	MtC eq.	4	4	5	9	14	19	23	26	28	30	31	30
SF6	MtC eq.	3	5	10	14	19	32	40	42	37	23	24	26
CO	MtCO	396	404	483	572	738	815	892	898	904	900	886	872
NMVOC	Mt	48	55	64	82	107	129	151	183	216	249	283	317
NOx	MtN	7	8	12	19	27	31	36	36	35	35	34	33

Scenario A1B-IMAGE World		**1990**	**2000**	**2010**	**2020**	**2030**	**2040**	**2050**	**2060**	**2070**	**2080**	**2090**	**2100**
Population	Million	5280	6122	6892	7618	8196	8547	8708	8671	8484	8142	7663	7047
GNP/GDP (mex)	Trillion US$	21.0	26.7	37.9	55.9	81.2	116.2	163.5	219.4	283.2	365.2	446.4	518.8
	1990 US$/cap [a]	3971.0	4357.0	5503.0	7342.0	9907.0	13598.0	18772.0	25300.0	33383.0	44853.0	58248.0	73621.0
Final Energy	EJ												
Non-commercial		54	59	65	63	59	50	40	38	35	32	29	25
Solids		43	44	45	52	59	66	69	65	59	52	47	40
Liquids		106	119	146	177	198	213	209	204	200	190	175	161
Gas		46	55	73	84	90	100	119	134	145	154	157	156
Electricity		39	55	88	147	220	307	373	410	427	428	408	376
Others		2	4	11	29	50	68	85	93	93	92	91	85
Total		289	336	429	551	677	804	895	944	959	950	907	843
Primary Energy	EJ												
Coal		105	111	129	170	245	298	301	271	237	198	163	129
Oil		129	143	187	242	285	300	348	352	340	320	280	237
Gas		62	73	117	163	188	222	259	278	283	278	264	244
Non-Fossil Electric		8	14	23	39	61	91	128	162	193	218	230	230
Biomass		3	4	12	35	69	102	129	141	138	133	128	116
Other Renewables		61	69	77	76	74	68	59	58	56	54	51	48
Total		368	416	544	725	923	1080	1224	1262	1246	1201	1116	1002
Cumulative Resources Production	ZJ [b]												
Coal		0.0	1.1	2.3	3.7	5.8	8.5	11.6	14.5	17.0	19.2	21.0	22.4
Oil		0.0	1.3	3.0	5.1	7.8	10.7	14.0	17.5	21.0	24.3	27.3	29.9
Gas		0.0	0.6	1.6	3.0	4.8	6.8	9.2	11.9	14.7	17.5	20.3	22.8
Cumulative CO2 Emissions	GtC	7.1	82.4	168.6	275.5	411.1	574.2	755.8	941.6	1123.9	1298.8	1458.9	1601.2
Carbon Sequestration [c]	GtC	2.1	1.7	2.7	4.0	5.3	5.8	5.9	5.7	5.2	5.2	5.2	4.6
Land Use	Million ha												
Cropland [d]		1435	1382	1450	1524	1573	1571	1530	1490	1439	1355	1295	1208
Grasslands [e]		3435	3295	3313	3362	3381	3271	3064	2849	2705	2597	2505	2347
Energy Biomass		8	10	26	82	177	274	374	404	391	384	367	334
Forest		4277	4266	4224	4251	4147	4132	4173	4256	4362	4490	4564	4700
Others		3916	4119	4058	3853	3793	3824	3931	4072	4174	4246	4340	4483
Total		13071	13071	13071	13071	13071	13071	13071	13071	13071	13071	13071	13071
Anthropogenic Emissions (standardized)													
Fossil Fuel CO2	GtC	5.99	6.90	8.50	11.10	14.30	17.60	18.70	18.40	17.80	16.40	14.60	12.90
Other CO2 [f]	GtC	1.11	1.07	0.77	1.01	0.71	0.02	0.02	0.05	0.22	0.56	0.47	0.50
Total CO2	GtC	7.10	7.97	9.27	12.11	15.00	17.61	18.71	18.45	18.02	16.96	15.07	13.39
CH4 total	MtCH4	310	323	360	423	475	499	512	523	525	513	488	452
N2O total	MtN2O-N	6.7	7.0	7.5	8.6	9.3	9.9	10.3	10.3	10.2	9.8	9.7	9.1
SOx total	MtS	70.9	69.0	77.5	90.9	108.5	121.0	119.5	114.8	106.8	95.5	83.0	71.2
CFC/HFC/HCFC	MtC eq.	1672	883	791	337	369	482	566	654	659	654	639	614
PFC	MtC eq.	32	25	31	43	61	77	89	97	106	114	119	115
SF6	MtC eq.	38	40	43	48	66	99	119	127	113	88	84	95
CO	MtCO	879	877	832	978	1054	913	949	1012	1089	1141	1110	1080
NMVOC	Mt	139	141	144	157	163	159	158	154	152	149	142	133
NOx	MtN	31	32	38	48	57	63	63	63	61	57	51	45

a: NOT ppp-corrected.
b: NOT use but production.
c: Net Ecosystem Production (NEP).
d: Arable land for crops excluding energy crops and grass & fodder species.
e: Permanent pasture: FAO category "land for grass & fodder species".
f: Approximate calculation from complex land-use module.

Scenario A1B-IMAGE OECD90		1990	2000	2010	2020	2030	2040	2050	2060	2070	2080	2090	2100
Population	Million	799	849	890	932	965	990	1001	1005	1009	1020	1029	1032
GNP/GDP (mex)	Trillion US$	16.5	20.2	26.9	35.6	44.3	53.0	62.4	71.3	77.9	88.7	101.2	114.6
	1990 US$/cap [a]	20648.9	23840.3	30131.7	38187.4	45863.7	53609.2	62283.6	70840.8	77152.1	87018.1	98251.1	111071.1
Final Energy	EJ												
Non-commercial		5	5	5	5	5	5	5	5	4	4	4	4
Solids		8	7	7	8	8	8	8	8	8	8	8	8
Liquids		57	62	68	71	67	61	53	49	47	46	44	42
Gas		24	27	30	31	32	35	40	43	45	49	52	55
Electricity		24	36	52	66	74	76	76	76	75	76	77	78
Others		2	3	5	10	16	22	26	27	27	26	26	26
Total		120	141	168	191	203	207	208	207	207	209	212	214
Primary Energy	EJ												
Coal		37	47	51	51	51	48	46	42	37	35	32	30
Oil		67	76	84	88	84	67	65	60	57	56	54	50
Gas		28	36	46	52	51	51	53	55	56	58	61	64
Non-Fossil Electric		7	12	19	29	36	40	42	45	46	48	51	54
Biomass		2	3	5	10	16	23	25	28	28	27	28	24
Other Renewables		10	10	10	11	11	11	11	11	11	11	11	11
Total		151	183	216	240	248	240	242	240	236	235	237	233
Cumulative Resources Production	ZJ [b]												
Coal		0.0	0.4	0.8	1.3	1.9	2.6	3.4	4.1	4.8	5.5	6.1	6.7
Oil		0.0	0.3	0.7	1.2	1.7	2.3	3.1	4.3	5.5	6.7	7.9	9.1
Gas		0.0	0.3	0.7	1.1	1.6	2.1	2.5	3.0	3.6	4.1	4.6	5.1
Cumulative CO2 Emissions	GtC	2.8	33.0	65.5	98.2	130.4	162.3	194.4	226.5	258.8	291.4	324.2	357.4
Carbon Sequestration [c]	GtC	0.4	0.4	0.3	0.6	0.9	1.2	1.5	1.6	1.6	1.7	1.7	1.3
Land Use	Million ha												
Cropland [d]		379	379	378	385	382	383	391	394	395	395	399	394
Grasslands [e]		786	676	584	516	487	479	472	467	462	458	459	453
Energy Biomass		3	1	3	9	20	32	43	51	55	52	52	43
Forest		1115	1167	1215	1259	1273	1272	1263	1256	1250	1252	1245	1262
Others		955	1016	1059	1070	1076	1073	1070	1071	1076	1081	1083	1087
Total		3238	3238	3238	3238	3238	3238	3238	3238	3238	3238	3238	3238
Anthropogenic Emissions (standardized)													
Fossil Fuel CO2	GtC	2.83	3.20	3.20	3.20	3.20	3.20	3.20	3.20	3.20	3.20	3.20	3.20
Other CO2 [f]	GtC	0.00	0.00	0.10	0.06	-0.03	0.02	0.00	0.03	0.04	0.07	0.10	0.13
Total CO2	GtC	2.83	3.20	3.30	3.26	3.17	3.22	3.20	3.22	3.24	3.27	3.30	3.33
CH4 total	MtCH4	73	74	78	81	83	85	88	89	89	91	91	86
N2O total	MtN2O-N	2.6	2.6	2.7	2.7	2.7	2.7	2.8	2.8	2.8	2.8	2.9	2.9
SOx total	MtS	22.7	17.0	13.4	12.5	10.9	10.1	9.4	9.1	8.8	8.9	9.0	8.9
HFC	MtC eq.	19	58	110	108	115	118	122	123	123	124	125	125
PFC	MtC eq.	18	13	13	11	9	8	9	10	12	13	14	16
SF6	MtC eq.	23	23	14	5	6	7	9	11	13	16	18	20
CO	MtCO	179	161	177	182	180	180	173	172	169	173	179	177
NMVOC	Mt	42	36	37	37	34	33	30	27	26	25	25	23
NOx	MtN	13	12	13	13	12	10	9	9	8	9	9	8

a: NOT ppp-corrected.
b: NOT use but production.
c: Net Ecosystem Production (NEP).
d: Arable land for crops excluding energy crops and grass & fodder species.
e: Permanent pasture: FAO category "land for grass & fodder species".
f: Approximate calculation from complex land-use module.

Scenario A1B-IMAGE REF		1990	2000	2010	2020	2030	2040	2050	2060	2070	2080	2090	2100
Population	Million	412	429	437	443	445	443	432	419	401	384	365	347
GNP/GDP (mex)	Trillion US$	1.0	0.7	1.0	2.0	3.5	5.9	9.2	13.0	17.2	22.7	28.4	34.3
	1990 US$/cap [a]	2307.2	1657.3	2378.6	4378.7	7868.6	13380.5	21343.7	30936.4	42882.5	59159.7	77741.4	98993.7
Final Energy	EJ												
Non-commercial		3	2	2	2	2	2	2	2	2	1	1	1
Solids		14	10	7	6	5	6	5	5	5	4	4	3
Liquids		21	12	12	15	16	16	15	15	14	13	12	11
Gas		15	11	10	12	12	11	11	10	10	10	9	9
Electricity		9	5	7	15	22	30	34	34	32	31	29	27
Others		0	1	1	1	2	3	4	4	4	4	4	4
Total		62	41	40	50	59	68	71	70	67	63	59	56
Primary Energy	EJ												
Coal		36	19	15	16	20	27	28	25	22	18	15	11
Oil		28	14	16	20	23	26	30	29	27	25	23	20
Gas		25	15	16	23	27	28	25	22	20	17	17	13
Non-Fossil Electric		1	1	1	3	6	9	11	11	12	12	12	12
Biomass		1	1	1	2	3	6	6	6	6	7	7	9
Other Renewables		4	3	3	3	3	4	4	4	4	4	4	4
Total		95	52	52	67	82	99	103	98	90	84	77	70
Cumulative Resources Production	ZJ [b]												
Coal		0.0	0.3	0.5	0.7	1.0	1.4	1.8	2.3	2.7	3.0	3.3	3.4
Oil		0.0	0.2	0.3	0.4	0.5	0.7	1.0	1.2	1.5	1.8	2.1	2.3
Gas		0.0	0.2	0.4	0.6	1.1	1.7	2.4	3.0	3.6	4.3	4.9	5.4
Cumulative CO2 Emissions	GtC	1.3	12.3	21.4	31.0	41.3	53.0	66.7	80.8	94.3	107.7	120.7	133.2
Carbon Sequestration [c]	GtC	0.4	0.2	0.5	0.8	1.2	1.4	1.5	1.5	1.5	1.4	1.4	1.3
Land Use	Million ha												
Cropland [d]		278	223	229	227	224	218	207	199	189	179	169	164
Grasslands [e]		392	307	329	339	333	310	279	258	238	234	226	221
Energy Biomass		1	0	1	6	19	46	84	87	92	83	83	92
Forest		1146	1174	1167	1250	1242	1238	1238	1233	1227	1226	1218	1200
Others		461	574	551	455	460	466	471	501	532	556	582	600
Total		2278	2278	2278	2278	2278	2278	2278	2278	2278	2278	2278	2278
Anthropogenic Emissions (standardized)													
Fossil Fuel CO2	GtC	1.30	0.91	0.91	1.11	1.31	1.61	1.61	1.41	1.31	1.21	1.01	0.91
Other CO2 [f]	GtC	0.00	0.00	-0.01	-0.09	-0.27	-0.30	-0.18	-0.03	0.02	0.15	0.24	0.35
Total CO2	GtC	1.30	0.91	0.90	1.02	1.04	1.31	1.43	1.38	1.33	1.36	1.25	1.26
CH4 total	MtCH4	47	39	41	51	57	58	52	52	51	47	43	40
N2O total	MtN2O-N	0.6	0.6	0.5	0.6	0.6	0.7	0.7	0.7	0.7	0.6	0.6	0.6
SOx total	MtS	17.0	11.0	9.4	9.7	9.6	10.6	10.6	9.9	8.9	7.8	6.8	5.8
HFC	MtC eq.	0	4	8	19	29	31	32	33	33	34	33	31
PFC	MtC eq.	7	4	5	8	14	20	21	22	23	24	25	24
SF6	MtC eq.	8	6	8	10	14	18	21	19	15	14	10	11
CO	MtCO	69	41	34	38	45	52	55	56	56	56	54	52
NMVOC	Mt	16	13	12	12	12	13	12	12	12	12	11	11
NOx	MtN	5	3	3	3	4	4	4	4	4	4	3	3

a: NOT ppp-corrected.
b: NOT use but production.
c: Net Ecosystem Production (NEP).
d: Arable land for crops excluding energy crops and grass & fodder species.
e: Permanent pasture: FAO category "land for grass & fodder species".
f: Approximate calculation from complex land-use module.

Scenario A1B-IMAGE ASIA		1990	2000	2010	2020	2030	2040	2050	2060	2070	2080	2090	2100
Population	Million	2781	3246	3609	3929	4142	4235	4220	4088	3871	3594	3262	2886
GNP/GDP (mex)	Trillion US$	1.4	2.7	4.7	8.7	16.1	28.8	48.2	73.2	103.0	139.3	173.5	200.7
	1990 US$/cap [a]	337.2	531.9	873.7	1553.2	2837.7	5139.5	8819.4	13804.2	19837.3	27431.3	35626.4	44069.4
Final Energy		EJ											
Non-commercial		29	33	33	30	26	20	14	13	11	10	8	7
Solids		20	24	26	29	32	35	33	30	27	23	19	15
Liquids		12	21	29	38	49	59	58	57	55	51	46	42
Gas		2	8	17	23	27	30	39	45	47	49	47	43
Electricity		4	9	20	42	75	119	150	166	173	171	159	139
Others		0	1	3	12	23	33	40	42	40	38	35	31
Total		68	96	127	174	232	295	334	352	352	341	314	277
Primary Energy		EJ											
Coal		28	42	53	71	99	126	125	108	91	70	50	36
Oil		16	27	40	59	77	76	96	98	92	84	71	57
Gas		4	12	29	52	76	87	94	110	98	88	86	65
Non-Fossil Electric		0	1	2	6	14	32	55	75	91	104	107	102
Biomass		0	1	4	14	31	52	61	62	55	53	46	36
Other Renewables		30	34	36	33	29	24	18	17	15	14	13	12
Total		79	116	163	235	326	397	449	468	442	413	373	308
Cumulative Resources Production ZJ [b]													
Coal		0.0	0.4	0.8	1.3	1.9	2.7	3.6	4.5	5.1	5.7	6.1	6.4
Oil		0.0	0.2	0.4	0.7	1.1	1.5	1.8	2.2	2.5	2.8	3.0	3.2
Gas		0.0	0.1	0.2	0.4	0.7	1.1	1.5	2.1	2.5	3.0	3.5	3.8
Cumulative CO2 Emissions	GtC	1.5	19.3	43.6	77.7	125.1	185.8	253.3	319.3	380.0	434.2	479.9	517.1
Carbon Sequestration [c]	GtC	0.2	0.1	0.2	0.3	0.6	0.7	0.5	0.3	0.1	0.4	0.7	0.5
Land Use	Million ha												
Cropland [d]		382	386	396	413	428	430	409	394	372	327	306	267
Grasslands [e]		561	664	692	747	770	778	697	626	601	560	539	473
Energy Biomass		1	1	3	15	53	95	142	141	117	129	126	97
Forest		488	397	369	331	286	252	244	273	335	403	438	515
Others		545	527	515	470	439	421	484	543	551	557	566	624
Total		1976	1976	1976	1976	1976	1976	1976	1976	1976	1976	1976	1976
Anthropogenic Emissions (standardized)													
Fossil Fuel CO2	GtC	1.15	1.78	2.58	3.68	5.08	6.38	6.58	6.18	5.68	4.88	3.98	3.28
Other CO2 [f]	GtC	0.37	0.26	0.24	0.32	0.40	0.29	0.27	0.16	0.14	0.15	0.14	0.06
Total CO2	GtC	1.53	2.03	2.82	4.00	5.48	6.66	6.85	6.33	5.82	5.02	4.12	3.33
CH4 total	MtCH4	113	125	136	157	181	198	206	206	202	191	183	166
N2O total	MtN2O-N	2.3	2.6	2.9	3.4	3.8	4.1	4.1	4.0	3.9	3.6	3.5	3.1
SOx total	MtS	17.7	25.3	29.9	33.6	39.5	42.9	39.4	35.8	32.2	27.3	21.9	17.6
HFC	MtC eq.	0	5	18	45	92	153	224	292	292	285	275	262
PFC	MtC eq.	3	5	8	15	23	30	35	39	43	46	48	46
SF6	MtC eq.	4	7	12	19	28	42	50	55	48	35	33	37
CO	MtCO	235	270	294	332	406	378	403	418	426	421	420	385
NMVOC	Mt	33	37	40	45	50	51	51	49	47	44	42	39
NOx	MtN	7	9	12	16	21	24	23	22	21	18	16	13

a: NOT ppp-corrected.
b: NOT use but production.
c: Net Ecosystem Production (NEP).
d: Arable land for crops excluding energy crops and grass & fodder species.
e: Permanent pasture: FAO category "land for grass & fodder species".
f: Approximate calculation from complex land-use module.

Scenario A1B-IMAGE ALM		**1990**	**2000**	**2010**	**2020**	**2030**	**2040**	**2050**	**2060**	**2070**	**2080**	**2090**	**2100**
Population	Million	1287	1597	1954	2315	2643	2879	3055	3159	3202	3145	3006	2783
GNP/GDP (mex)	Trillion US$	2.1	3.0	5.4	9.8	17.3	28.5	43.7	62.0	85.2	114.4	143.3	169.2
	1990 US$/cap [a]	1653.4	1901.8	2738.2	4210.1	6556.0	9905.2	14317.2	19627.9	26606.2	36388.5	47686.9	60795.4
Final Energy	EJ												
Non-commercial		16	20	25	26	26	24	20	19	18	17	15	13
Solids		1	3	5	9	13	17	22	22	20	17	16	14
Liquids		15	23	37	53	67	77	82	84	85	81	73	65
Gas		4	9	17	18	19	23	30	36	43	47	48	49
Electricity		2	4	9	24	50	81	112	133	147	151	144	132
Others		0	0	1	5	9	11	15	20	22	24	26	24
Total		39	58	94	135	184	233	281	315	334	336	322	296
Primary Energy	EJ												
Coal		3	4	10	32	75	97	102	97	87	75	66	52
Oil		17	27	47	75	101	131	157	165	165	155	132	109
Gas		6	10	26	37	34	56	88	91	109	115	100	102
Non-Fossil Electric		0	0	1	2	5	11	20	31	43	54	60	62
Biomass		0	0	2	9	20	21	36	45	49	45	47	46
Other Renewables		18	22	28	29	30	29	26	26	25	24	23	21
Total		43	64	114	183	266	344	429	456	478	469	428	391
Cumulative Resources Production	ZJ [b]												
Coal		0.0	0.1	0.2	0.4	0.9	1.8	2.7	3.6	4.3	5.0	5.5	5.9
Oil		0.0	0.6	1.6	2.8	4.4	6.3	8.1	9.8	11.5	13.0	14.3	15.4
Gas		0.0	0.1	0.4	0.8	1.3	2.0	2.8	3.8	4.9	6.2	7.3	8.4
Cumulative CO2 Emissions	GtC	1.4	17.8	38.2	68.6	114.4	173.0	241.3	315.0	390.8	465.5	534.1	593.5
Carbon Sequestration [c]	GtC	1.2	1.0	1.6	2.1	2.6	2.5	2.4	2.3	1.9	1.8	1.6	1.3
Land Use	Million ha												
Cropland [d]		397	394	447	499	540	540	523	503	483	455	421	384
Grasslands [e]		1697	1648	1708	1760	1790	1704	1617	1499	1403	1345	1281	1201
Energy Biomass		2	8	19	53	85	101	105	125	127	120	106	102
Forest		1529	1529	1472	1410	1346	1370	1429	1494	1552	1608	1662	1722
Others		1955	2001	1933	1857	1818	1864	1906	1958	2014	2051	2109	2171
Total		5579	5579	5579	5579	5579	5579	5579	5579	5579	5579	5579	5579
Anthropogenic Emissions (standardized)													
Fossil Fuel CO2	GtC	0.72	1.01	1.81	3.11	4.71	6.41	7.31	7.61	7.61	7.11	6.41	5.51
Other CO2 [f]	GtC	0.73	0.82	0.44	0.73	0.60	0.01	-0.08	-0.10	0.02	0.20	-0.01	-0.03
Total CO2	GtC	1.45	1.83	2.25	3.84	5.31	6.42	7.24	7.51	7.63	7.31	6.41	5.48
CH4 total	MtCH4	77	85	105	134	154	158	166	176	183	184	171	160
N2O total	MtN2O-N	1.2	1.3	1.5	2.0	2.3	2.5	2.8	2.9	2.9	2.9	2.8	2.6
SOx total	MtS	10.5	12.8	21.8	32.0	45.5	54.4	57.2	57.0	53.9	48.5	42.4	35.9
HFC	MtC eq.	0	2	15	39	81	139	184	205	210	210	206	196
PFC	MtC eq.	4	4	5	9	14	19	23	26	28	30	31	30
SF6	MtC eq.	3	5	10	14	19	32	40	42	37	23	24	26
CO	MtCO	396	404	326	425	422	302	317	365	437	490	456	465
NMVOC	Mt	48	55	55	63	67	62	65	66	67	68	64	60
NOx	MtN	7	8	10	15	21	24	26	28	28	26	24	21

a: NOT ppp-corrected.
b: NOT use but production.
c: Net Ecosystem Production (NEP).
d: Arable land for crops excluding energy crops and grass & fodder species.
e: Permanent pasture: FAO category "land for grass & fodder species".
f: Approximate calculation from complex land-use module.

Scenario A1B-MARIA World		1990	2000	2010	2020	2030	2040	2050	2060	2070	2080	2090	2100
Population	Million	5262	6117	6888	7617	8048	8207	8704	8536	8372	8028	7527	7056
GNP/GDP (mex)	Trillion US$	19.4	25.8	35.7	51.9	75.4	114.4	179.7	230.1	287.9	355.9	430.8	535.6
GNP/GDP (ppp) Trillion (1990 prices)													
Final Energy	EJ												
Non-commercial													
Solids		48	37	38	27	19	13	26	48	84	140	207	145
Liquids		138	165	200	252	293	350	403	402	486	481	583	651
Gas		56	66	98	139	189	243	303	345	287	303	213	273
Electricity		35	52	63	79	90	116	157	187	218	256	301	363
Others		0	0	0	0	0	0	0	0	0	0	0	0
Total		278	321	399	496	591	723	889	984	1075	1180	1303	1432
Primary Energy	EJ												
Coal		90	69	61	45	33	24	34	54	87	142	209	146
Oil		123	144	168	201	220	250	285	229	263	208	208	226
Gas		71	105	142	188	246	305	354	399	344	363	279	345
Nuclear		22	29	40	61	99	162	240	321	402	505	625	797
Biomass		28	39	56	88	103	121	132	184	230	278	380	429
Other Renewables		9	8	7	7	6	21	68	71	73	73	71	67
Total		343	393	475	589	706	882	1113	1258	1400	1570	1770	2011
Cumulative Resources Use	ZJ												
Coal		0.0	0.9	1.6	2.2	2.6	3.0	3.2	3.5	4.1	5.0	6.4	8.5
Oil		0.0	1.2	2.7	4.4	6.4	8.6	11.1	13.9	16.2	18.8	20.9	23.0
Gas		0.0	0.7	1.8	3.2	5.1	7.5	10.6	14.1	18.1	21.5	25.2	28.0
Cumulative CO2 Emissions	GtC	7.1	82.4	166.3	258.4	358.7	468.5	591.1	720.5	852.4	990.3	1139.9	1301.4
Carbon Sequestration	GtC	0.0	0.0	0.0	0.0	0.0	0.0	0.0	0.0	0.0	0.0	0.0	0.0
Land Use	Million ha												
Cropland		1451	1451	1605	2000	2174	2222	2355	2080	1810	1539	1278	1162
Grasslands		3395	3395	3245	2850	2676	2604	2601	2593	2597	2597	2597	2597
Energy Biomass		0	0	0	0	0	39	43	326	592	872	1572	1932
Forest		4138	4142	4158	4186	4217	4267	4312	4366	4366	4358	3918	3673
Others		4061	4057	4038	4010	3978	3914	3734	3681	3681	3681	3681	3681
Total		13045	13045	13045	13045	13045	13045	13045	13045	13045	13045	13045	13045
Anthropogenic Emissions (standardized)													
Fossil Fuel CO2	GtC	5.99	6.90	7.77	8.69	9.65	10.93	12.66	12.74	13.45	14.07	14.48	14.24
Other CO2	GtC	1.11	1.07	1.03	0.93	0.79	0.58	0.36	0.13	0.05	0.00	1.38	2.20
Total CO2	GtC	7.10	7.97	8.80	9.61	10.44	11.51	13.01	12.87	13.50	14.07	15.86	16.44
CH4 total	MtCH4												
N2O total	MtN2O-N												
SOx total	MtS												
CFC/HFC/HCFC	MtC eq.	1672	883	791	337	369	482	566	654	659	654	639	614
PFC	MtC eq.	32	25	31	43	61	77	89	97	106	114	119	115
SF6	MtC eq.	38	40	43	48	66	99	119	127	113	88	84	95
CO	MtCO												
NMVOC	Mt												
NOx	MtN												

Scenario A1B-MARIA OECD90		1990	2000	2010	2020	2030	2040	2050	2060	2070	2080	2090	2100
Population	Million	859	919	965	1007	1035	1046	1081	1085	1091	1096	1103	1110
GNP/GDP (mex)	Trillion US$	15.6	18.9	22.7	27.0	32.0	38.2	46.0	57.7	71.3	88.4	105.8	123.9
GNP/GDP (ppp) Trillion (1990 prices)													
Final Energy	EJ												
Non-commercial													
Solids		12	8	6	4	3	2	7	15	28	49	65	46
Liquids		71	76	71	68	65	64	59	51	80	73	92	89
Gas		28	32	52	70	82	92	99	111	80	80	56	88
Electricity		21	28	31	36	39	42	47	52	59	67	77	87
Others		0	0	0	0	0	0	0	0	0	0	0	0
Total		132	145	160	177	188	199	212	230	248	269	290	309
Primary Energy	EJ												
Coal		38	28	21	16	13	10	12	19	31	51	67	47
Oil		71	77	71	68	65	64	59	51	80	73	92	87
Gas		34	55	80	105	121	135	133	147	120	124	106	143
Nuclear		18	15	18	23	28	35	43	54	68	85	105	131
Biomass		6	4	3	2	1	1	1	0	1	1	2	3
Other Renewables		5	5	4	4	3	3	19	23	24	24	23	21
Total		171	182	198	217	232	247	267	294	323	357	395	431
Cumulative Resources Use		ZJ											
Coal		0.0	0.4	0.7	0.9	1.0	1.2	1.3	1.4	1.6	1.9	2.4	3.1
Oil		0.0	0.7	1.5	2.2	2.9	3.5	4.2	4.8	5.3	6.1	6.8	7.7
Gas		0.0	0.3	0.9	1.7	2.7	3.9	5.3	6.6	8.1	9.3	10.5	11.6
Cumulative CO2 Emissions	GtC	2.8	33.0	65.6	100.0	136.0	172.5	208.1	243.2	281.5	324.4	372.2	422.8
Carbon Sequestration	GtC	0.0	0.0	0.0	0.0	0.0	0.0	0.0	0.0	0.0	0.0	0.0	0.0
Land Use	Million ha												
Cropland		378	378	378	378	378	378	378	378	378	357	295	237
Grasslands		756	756	756	756	756	756	756	756	756	756	756	756
Energy Biomass		0	0	0	0	0	0	0	0	0	21	82	141
Forest		756	756	756	756	756	778	816	834	834	834	834	834
Others		794	794	794	794	794	772	734	716	716	716	716	716
Total		2684	2684	2684	2684	2684	2684	2684	2684	2684	2684	2684	2684
Anthropogenic Emissions (standardized)													
Fossil Fuel CO2	GtC	2.83	3.20	3.31	3.49	3.60	3.71	3.64	3.86	4.34	4.77	5.30	5.25
Other CO2	GtC	0.00	0.00	0.03	0.05	0.05	-0.04	-0.19	-0.28	-0.27	-0.26	-0.23	-0.20
Total CO2	GtC	2.83	3.20	3.33	3.55	3.65	3.67	3.44	3.58	4.07	4.51	5.07	5.05
CH4 total	MtCH4												
N2O total	MtN2O-N												
SOx total	MtS												
HFC	MtC eq.	19	58	110	108	115	118	122	123	123	124	125	125
PFC	MtC eq.	18	13	13	11	9	8	9	10	12	13	14	16
SF6	MtC eq.	23	23	14	5	6	7	9	11	13	16	18	20
CO	MtCO												
NMVOC	Mt												
NOx	MtN												

Scenario A1B-MARIA REF		**1990**	**2000**	**2010**	**2020**	**2030**	**2040**	**2050**	**2060**	**2070**	**2080**	**2090**	**2100**
Population	Million	413	419	427	433	432	430	423	406	391	374	356	339
GNP/GDP (mex)	Trillion US$	0.9	1.3	1.8	2.6	4.6	8.1	13.7	18.1	22.4	27.1	32.9	40.4
GNP/GDP (ppp) Trillion (1990 prices)													
Final Energy	EJ												
Non-commercial													
Solids		13	9	6	4	3	2	3	6	10	17	27	19
Liquids		17	14	12	11	12	13	12	8	14	16	26	23
Gas		21	24	27	30	41	53	65	70	61	54	38	53
Electricity		7	8	7	7	9	11	13	14	15	16	17	18
Others		0	0	0	0	0	0	0	0	0	0	0	0
Total		58	55	52	53	65	78	93	98	100	102	107	113
Primary Energy	EJ												
Coal		19	13	9	6	4	3	4	6	10	17	27	19
Oil		19	14	13	12	12	13	12	8	14	16	26	23
Gas		27	33	34	38	52	66	80	86	77	69	53	69
Nuclear		3	2	3	4	4	5	7	9	11	13	17	21
Biomass		1	1	0	0	0	0	0	0	0	0	0	0
Other Renewables		1	1	1	1	1	1	4	4	4	4	4	4
Total		69	64	60	61	74	89	106	113	116	120	128	136
Cumulative Resources Use	ZJ												
Coal		0.0	0.2	0.3	0.4	0.5	0.5	0.5	0.6	0.7	0.8	0.9	1.2
Oil		0.0	0.2	0.3	0.5	0.6	0.7	0.8	0.9	1.0	1.2	1.3	1.6
Gas		0.0	0.3	0.6	0.9	1.3	1.8	2.5	3.3	4.2	4.9	5.6	6.2
Cumulative CO2 Emissions	GtC	1.3	12.3	20.9	28.8	37.2	47.3	59.2	71.8	84.6	97.6	111.6	126.2
Carbon Sequestration	GtC	0.0	0.0	0.0	0.0	0.0	0.0	0.0	0.0	0.0	0.0	0.0	0.0
Land Use	Million ha												
Cropland		217	217	217	217	217	217	217	174	139	111	89	71
Grasslands		114	114	114	114	114	114	114	114	114	114	114	114
Energy Biomass		0	0	0	0	0	0	0	43	78	106	128	146
Forest		815	815	815	815	815	815	827	862	862	862	862	862
Others		722	722	722	722	722	722	710	675	675	675	675	675
Total		1868	1868	1868	1868	1868	1868	1868	1868	1868	1868	1868	1868
Anthropogenic Emissions (standardized)													
Fossil Fuel CO2	GtC	1.30	0.91	0.79	0.75	0.93	1.14	1.34	1.43	1.50	1.59	1.82	1.78
Other CO2	GtC	0.00	0.00	0.02	0.01	-0.01	-0.03	-0.07	-0.17	-0.21	-0.27	-0.34	-0.33
Total CO2	GtC	1.30	0.91	0.81	0.76	0.92	1.11	1.27	1.26	1.29	1.32	1.47	1.45
CH4 total	MtCH4												
N2O total	MtN2O-N												
SOx total	MtS												
HFC	MtC eq.	0	4	8	19	29	31	32	33	33	34	33	31
PFC	MtC eq.	7	4	5	8	14	20	21	22	23	24	25	24
SF6	MtC eq.	8	6	8	10	14	18	21	19	15	14	10	11
CO	MtCO												
NMVOC	Mt												
NOx	MtN												

Scenario A1B-MARIA ASIA		1990	2000	2010	2020	2030	2040	2050	2060	2070	2080	2090	2100
Population	Million	2642	3080	3425	3728	3861	3895	4008	3814	3632	3368	3040	2744
GNP/GDP (mex)	Trillion US$	1.2	2.6	5.8	12.5	21.1	36.5	62.9	81.9	104.9	132.1	164.8	210.8
GNP/GDP (ppp) Trillion (1990 prices)													
Final Energy	EJ												
Non-commercial													
Solids		20	18	24	17	12	8	15	25	41	68	104	73
Liquids		26	42	65	96	121	154	181	183	203	199	211	252
Gas		2	3	4	10	17	27	40	53	44	49	35	57
Electricity		4	9	12	17	18	27	40	51	64	79	97	119
Others		0	0	0	0	0	0	0	0	0	0	0	0
Total		52	71	105	140	168	216	276	312	352	395	446	501
Primary Energy	EJ												
Coal		28	24	28	20	14	10	16	26	42	68	104	73
Oil		14	23	36	48	62	87	111	98	98	70	55	59
Gas		3	5	8	14	22	30	42	54	45	50	35	58
Nuclear		1	12	19	35	34	64	78	104	132	169	214	269
Biomass		13	21	31	51	65	71	73	88	106	130	157	193
Other Renewables		1	1	1	1	1	1	20	20	20	20	20	20
Total		61	86	124	168	197	263	340	389	443	507	585	671
Cumulative Resources Use	ZJ												
Coal		0.0	0.3	0.5	0.8	1.0	1.1	1.2	1.4	1.6	2.1	2.7	3.8
Oil		0.0	0.1	0.4	0.7	1.2	1.8	2.7	3.8	4.8	5.8	6.5	7.0
Gas		0.0	0.0	0.1	0.2	0.3	0.5	0.8	1.2	1.8	2.2	2.7	3.1
Cumulative CO2 Emissions	GtC	1.5	19.3	41.5	65.8	91.5	120.9	156.9	197.6	240.3	285.2	332.8	380.3
Carbon Sequestration	GtC	0.0	0.0	0.0	0.0	0.0	0.0	0.0	0.0	0.0	0.0	0.0	0.0
Land Use	Million ha												
Cropland		366	366	380	380	380	380	380	379	374	335	305	266
Grasslands		431	431	421	421	421	382	379	371	375	375	375	375
Energy Biomass		0	0	0	0	0	39	43	52	54	92	122	162
Forest		365	367	381	390	399	406	406	406	406	406	406	406
Others		458	456	440	430	421	414	412	412	412	412	412	412
Total		1621	1621	1621	1621	1621	1621	1621	1621	1621	1621	1621	1621
Anthropogenic Emissions (standardized)													
Fossil Fuel CO2	GtC	1.15	1.78	2.20	2.31	2.55	3.08	3.91	4.08	4.36	4.56	4.96	4.58
Other CO2	GtC	0.37	0.26	0.20	0.16	0.13	0.11	0.09	0.06	0.03	0.01	-0.01	-0.03
Total CO2	GtC	1.53	2.03	2.40	2.46	2.68	3.19	4.00	4.14	4.40	4.57	4.95	4.55
CH4 total	MtCH4												
N2O total	MtN2O-N												
SOx total	MtS												
HFC	MtC eq.	0	5	18	45	92	153	224	292	292	285	275	262
PFC	MtC eq.	3	5	8	15	23	30	35	39	43	46	48	46
SF6	MtC eq.	4	7	12	19	28	42	50	55	48	35	33	37
CO	MtCO												
NMVOC	Mt												
NOx	MtN												

Scenario A1B-MARIA ALM		**1990**	**2000**	**2010**	**2020**	**2030**	**2040**	**2050**	**2060**	**2070**	**2080**	**2090**	**2100**
Population	Million	1348	1699	2071	2449	2720	2836	3192	3231	3259	3190	3029	2864
GNP/GDP (mex)	Trillion US$	1.7	3.0	5.4	9.8	17.6	31.7	57.1	72.4	89.3	108.4	127.4	160.5
GNP/GDP (ppp) Trillion (1990 prices)													
Final Energy	EJ												
Non-commercial													
Solids		3	2	2	1	1	1	1	2	4	7	11	8
Liquids		24	34	51	77	95	119	151	160	190	193	253	288
Gas		5	6	15	29	49	72	99	112	101	120	84	75
Electricity		3	8	12	19	25	37	57	69	81	95	111	139
Others		0	0	0	0	0	0	0	0	0	0	0	0
Total		35	50	81	126	170	229	308	344	376	414	459	510
Primary Energy	EJ												
Coal		6	4	3	2	1	1	2	2	4	7	11	8
Oil		18	30	48	73	80	86	103	72	72	50	35	57
Gas		7	12	19	32	51	74	100	113	102	120	84	75
Nuclear		0	0	0	0	33	58	112	154	192	238	289	377
Biomass		8	13	21	35	36	49	59	96	123	147	221	232
Other Renewables		2	2	1	1	1	15	25	25	24	24	24	23
Total		41	60	93	143	204	282	399	461	517	586	664	772
Cumulative Resources Use	ZJ												
Coal		0.0	0.1	0.1	0.1	0.1	0.2	0.2	0.2	0.2	0.2	0.3	0.4
Oil		0.0	0.2	0.5	1.0	1.7	2.5	3.4	4.4	5.1	5.8	6.3	6.7
Gas		0.0	0.1	0.2	0.4	0.7	1.2	2.0	2.9	4.1	5.1	6.3	7.1
Cumulative CO2 Emissions	GtC	1.4	17.8	38.3	63.8	94.0	127.7	166.9	207.9	246.0	283.1	323.3	372.2
Carbon Sequestration	GtC	0.0	0.0	0.0	0.0	0.0	0.0	0.0	0.0	0.0	0.0	0.0	0.0
Land Use	Million ha												
Cropland		490	490	630	1025	1199	1247	1380	1150	920	736	589	589
Grasslands		2095	2095	1955	1560	1386	1352	1352	1352	1352	1352	1352	1352
Energy Biomass		0	0	0	0	0	0	0	231	460	653	1240	1484
Forest		2202	2204	2207	2226	2248	2268	2263	2263	2263	2255	1815	1571
Others		2086	2084	2081	2063	2040	2006	1878	1878	1878	1878	1878	1878
Total		6873	6873	6873	6873	6873	6873	6873	6873	6873	6873	6873	6873
Anthropogenic Emissions (standardized)													
Fossil Fuel CO2	GtC	0.72	1.01	1.47	2.13	2.57	3.01	3.76	3.37	3.24	3.16	2.41	2.63
Other CO2	GtC	0.73	0.82	0.79	0.71	0.62	0.54	0.54	0.52	0.50	0.51	1.96	2.77
Total CO2	GtC	1.45	1.83	2.26	2.84	3.19	3.54	4.30	3.89	3.74	3.67	4.37	5.40
CH4 total	MtCH4												
N2O total	MtN2O-N												
SOx total	MtS												
HFC	MtC eq.	0	2	15	39	81	139	184	205	210	210	206	196
PFC	MtC eq.	4	4	5	9	14	19	23	26	28	30	31	30
SF6	MtC eq.	3	5	10	14	19	32	40	42	37	23	24	26
CO	MtCO												
NMVOC	Mt												
NOx	MtN												

Scenario A1B-MESSAGE World		1990	2000	2010	2020	2030	2040	2050	2060	2070	2080	2090	2100
Population	Million	5262	6117	6888	7617	8182	8531	8704	8667	8463	8125	7658	7056
GNP/GDP (mex)	Trillion US$	20.9	26.7	37.9	56.5	89.1	135.2	181.3	247.6	313.8	383.3	455.9	528.5
GNP/GDP (ppp) Trillion (1990 prices)		25.7	33.3	47.1	66.6	96.6	138.9	181.0	240.7	304.2	372.2	443.0	513.9
Final Energy	EJ												
Non-commercial		38	28	22	16	10	7	5	0	0	0	0	0
Solids		42	58	59	67	73	70	48	35	29	26	24	15
Liquids		111	125	160	204	254	303	357	396	403	365	308	254
Gas		41	48	67	85	104	135	164	199	210	202	158	96
Electricity		35	47	70	107	164	232	311	414	536	647	745	859
Others		8	10	20	38	58	84	122	169	234	320	422	525
Total		275	316	398	517	662	830	1005	1213	1413	1560	1657	1749
Primary Energy	EJ												
Coal		91	105	120	157	194	227	210	182	192	213	200	192
Oil		128	155	172	198	225	250	281	333	326	253	184	138
Gas		71	85	128	178	250	320	378	445	482	456	398	350
Nuclear		7	9	12	19	34	55	84	126	185	247	306	358
Biomass		46	47	63	87	120	155	204	251	315	394	438	475
Other Renewables		8	13	29	52	78	134	240	371	534	730	945	1172
Total		352	415	524	689	901	1141	1397	1707	2035	2292	2472	2683
Cumulative Resources Use	ZJ												
Coal		0.0	0.9	2.0	3.2	4.8	6.7	9.0	11.1	12.9	14.8	17.0	19.0
Oil		0.0	1.4	3.0	4.7	6.7	8.9	11.4	14.2	17.6	20.8	23.4	25.2
Gas		0.0	0.7	1.6	2.9	4.7	7.1	10.3	14.1	18.6	23.4	28.0	31.9
Cumulative CO2 Emissions	GtC	7.1	82.4	169.1	269.9	390.7	535.1	695.1	866.4	1049.0	1231.7	1398.0	1544.4
Carbon Sequestration	GtC												
Land Use [a]	Million ha												
Cropland		1459	1466	1462	1457	1454	1448	1442	1436	1429	1424	1422	1420
Grasslands		3389	3404	3429	3446	3458	3478	3498	3525	3552	3568	3572	3576
Energy Biomass		0	0	0	74	158	288	418	492	566	581	538	495
Forest		4296	4237	4173	4164	4164	4177	4190	4194	4199	4202	4203	4204
Others		3805	3842	3886	3807	3715	3558	3400	3301	3203	3173	3213	3253
Total		12949	12949	12949	12949	12949	12949	12949	12949	12949	12949	12949	12949
Anthropogenic Emissions (standardized)													
Fossil Fuel CO2 [b]	GtC	5.99	6.90	8.31	10.56	13.21	15.51	16.47	17.84	18.73	17.86	15.46	13.83
Other CO2 [c]	GtC	1.11	1.07	1.04	0.26	0.12	0.05	-0.02	-0.03	-0.03	-0.03	-0.01	0.00
Total CO2	GtC	7.10	7.97	9.36	10.81	13.33	15.56	16.45	17.81	18.70	17.83	15.44	13.83
CH4 total [d]	MtCH4	310	323	362	418	486	517	531	502	477	451	400	347
N2O total [e]	MtN2O-N	6.7	7.0	6.1	6.1	6.2	6.3	6.3	6.2	6.1	6.0	5.9	5.8
SOx total	MtS	70.9	69.0	65.8	62.2	61.8	59.6	49.4	45.4	43.2	40.4	35.9	31.4
CFC/HFC/HCFC	MtC eq.	1672	883	791	337	369	482	566	654	659	654	639	614
PFC	MtC eq.	32	25	31	43	61	77	89	97	106	114	119	115
SF6	MtC eq.	38	40	43	48	66	99	119	127	113	88	84	95
CO	MtCO	879	877	1022	1218	1412	1659	1925	2173	2380	2548	2568	2532
NMVOC	Mt	139	141	164	194	211	235	259	282	276	234	188	137
NOx	MtN	31	32	39	50	60	71	80	90	95	90	81	74

Emissions correlated to land-use change and deforestation were calculated by using AIM A1 marker land-use data.

a: Land-use taken from AIM-A1 marker run.

b:CO2 emissions from fossil fuel and industrial processes (MESSAGE data).

c: CO2 emissions from land-use changes (AIM-A1 marker run).

d: Non-energy related CH4 emissions were taken from AIM-A1 marker run.

e: Non-energy related N2O emissions were taken from AIM-A1 marker run.

Scenario A1B-MESSAGE OECD90		**1990**	**2000**	**2010**	**2020**	**2030**	**2040**	**2050**	**2060**	**2070**	**2080**	**2090**	**2100**
Population	Million	859	919	965	1007	1043	1069	1081	1084	1089	1098	1108	1110
GNP/GDP (mex)	Trillion US$	16.4	20.5	25.3	31.0	38.0	46.1	54.1	65.5	76.9	90.3	105.7	121.1
GNP/GDP (ppp) Trillion (1990 prices)		14.1	17.7	21.8	26.9	33.0	40.1	47.2	57.2	67.3	79.2	92.9	106.6
Final Energy	EJ												
Non-commercial		0	0	0	0	0	0	0	0	0	0	0	0
Solids		13	13	7	3	2	1	0	0	0	0	0	0
Liquids		66	68	73	78	77	75	70	67	55	45	36	27
Gas		22	28	35	40	44	44	43	41	40	36	24	12
Electricity		22	28	38	50	62	70	81	97	117	132	150	172
Others		1	1	4	9	12	17	24	34	50	70	93	109
Total		124	138	157	180	197	207	218	240	262	282	303	320
Primary Energy	EJ												
Coal		38	36	39	42	44	32	20	11	6	5	3	2
Oil		72	82	79	80	75	67	58	55	43	35	28	25
Gas		33	45	60	76	87	96	97	107	118	118	108	113
Nuclear		6	7	9	10	14	20	27	38	48	56	60	59
Biomass		6	9	14	20	27	34	42	44	44	40	41	52
Other Renewables		4	6	12	16	22	35	59	81	114	156	211	246
Total		159	184	210	243	269	283	303	335	372	409	450	497
Cumulative Resources Use	ZJ												
Coal		0.0	0.4	0.7	1.1	1.5	2.0	2.3	2.5	2.6	2.7	2.7	2.7
Oil		0.0	0.8	1.6	2.4	3.2	3.9	4.6	5.2	5.7	6.2	6.5	6.8
Gas		0.0	0.4	0.8	1.4	2.2	3.0	4.0	5.0	6.0	7.2	8.4	9.5
Cumulative CO2 Emissions	GtC	2.8	33.0	66.3	102.5	140.5	176.8	208.3	236.9	264.6	290.9	314.5	336.2
Carbon Sequestration	GtC												
Land Use [a]	Million ha												
Cropland		381	380	377	376	375	372	368	364	359	356	356	355
Grasslands		760	763	771	774	776	787	797	818	839	849	847	846
Energy Biomass		0	0	0	11	24	44	63	74	86	88	81	75
Forest		1050	1053	1052	1053	1056	1062	1068	1061	1054	1050	1049	1048
Others		838	833	828	814	798	765	733	712	691	685	695	705
Total		3029	3029	3029	3029	3029	3029	3029	3029	3029	3029	3029	3029
Anthropogenic Emissions (standardized)													
Fossil Fuel CO2 [b]	GtC	2.83	3.20	3.43	3.76	3.84	3.44	2.94	2.82	2.66	2.49	2.15	2.14
Other CO2 [c]	GtC	0.00	0.00	0.04	0.02	0.00	-0.02	-0.04	0.00	0.05	0.06	0.04	0.01
Total CO2	GtC	2.83	3.20	3.46	3.78	3.83	3.41	2.89	2.83	2.71	2.55	2.18	2.15
CH4 total [d]	MtCH4	73	74	71	71	71	68	65	63	63	63	62	63
N2O total [e]	MtN2O-N	2.6	2.6	2.4	2.4	2.4	2.4	2.4	2.3	2.3	2.3	2.2	2.2
SOx total	MtS	22.7	17.0	9.7	3.7	2.3	1.5	1.1	2.2	2.0	2.7	3.1	4.4
HFC	MtC eq.	19	58	110	108	115	118	122	123	123	124	125	125
PFC	MtC eq.	18	13	13	11	9	8	9	10	12	13	14	16
SF6	MtC eq.	23	23	14	5	6	7	9	11	13	16	18	20
CO	MtCO	179	161	194	218	217	217	214	206	200	186	149	151
NMVOC	Mt	42	36	39	41	37	33	28	27	25	21	14	9
NOx	MtN	13	12	14	16	16	15	14	13	13	12	11	11

Emissions correlated to land-use change and deforestation were calculated by using AIM A1 marker land-use data.

a: Land-use taken from AIM-A1 marker run.

b:CO2 emissions from fossil fuel and industrial processes (MESSAGE data).

c: CO2 emissions from land-use changes (AIM-A1 marker run).

d: Non-energy related CH4 emissions were taken from AIM-A1 marker run.

e: Non-energy related N2O emissions were taken from AIM-A1 marker run.

Scenario A1B-MESSAGE REF		1990	2000	2010	2020	2030	2040	2050	2060	2070	2080	2090	2100
Population	Million	413	419	427	433	435	433	423	409	392	374	357	339
GNP/GDP (mex)	Trillion US$	1.1	0.8	1.5	2.9	5.3	8.8	12.4	16.2	20.0	24.4	29.3	34.2
GNP/GDP (ppp) Trillion (1990 prices)		2.6	2.2	3.1	4.3	6.0	8.8	12.4	16.2	20.0	24.4	29.3	34.2
Final Energy	EJ												
Non-commercial		0	0	0	0	0	0	0	0	0	0	0	0
Solids		9	5	4	3	2	1	0	0	0	0	0	0
Liquids		15	10	10	10	11	10	11	10	10	11	15	21
Gas		14	11	15	17	20	26	29	33	32	25	16	7
Electricity		6	6	7	10	14	20	27	32	38	45	51	56
Others		7	6	7	10	12	15	17	17	19	23	26	28
Total		50	38	42	50	59	72	85	92	99	103	108	112
Primary Energy	EJ												
Coal		19	15	12	14	17	30	32	26	13	8	17	19
Oil		20	15	13	12	12	12	12	13	13	9	7	5
Gas		27	22	30	32	38	42	50	67	81	79	63	53
Nuclear		1	1	1	1	1	2	3	4	5	6	7	8
Biomass		2	1	1	2	3	5	8	10	14	21	32	47
Other Renewables		1	1	2	4	6	9	16	19	26	37	49	55
Total		70	54	58	64	77	99	121	138	151	160	173	187
Cumulative Resources Use	ZJ												
Coal		0.0	0.2	0.3	0.4	0.6	0.7	1.0	1.4	1.7	1.8	1.9	2.1
Oil		0.0	0.2	0.3	0.5	0.6	0.7	0.8	0.9	1.1	1.2	1.3	1.4
Gas		0.0	0.3	0.5	0.8	1.1	1.5	1.9	2.4	3.1	3.9	4.7	5.3
Cumulative CO2 Emissions	GtC	1.3	12.3	22.3	32.7	43.2	56.0	70.8	86.7	102.9	117.8	131.6	144.6
Carbon Sequestration	GtC												
Land Use [a]	Million ha												
Cropland		268	266	265	265	265	264	263	262	262	261	260	260
Grasslands		341	361	364	366	367	368	370	371	371	372	373	374
Energy Biomass		0	0	0	3	6	11	16	19	22	23	21	20
Forest		966	950	918	904	894	899	905	909	912	916	922	927
Others		701	698	728	738	745	733	722	715	709	703	700	696
Total		2276	2276	2276	2276	2276	2276	2276	2276	2276	2276	2276	2276
Anthropogenic Emissions (standardized)													
Fossil Fuel CO2 [b]	GtC	1.30	0.91	0.91	0.96	1.14	1.50	1.67	1.79	1.65	1.49	1.41	1.31
Other CO2 [c]	GtC	0.00	0.00	0.18	0.01	0.00	-0.08	-0.15	-0.12	-0.09	-0.07	-0.06	-0.05
Total CO2	GtC	1.30	0.91	1.09	0.97	1.14	1.42	1.52	1.67	1.56	1.42	1.35	1.25
CH4 total [d]	MtCH4	47	39	44	48	55	58	58	62	60	53	46	40
N2O total [e]	MtN2O-N	0.6	0.6	0.6	0.6	0.6	0.6	0.6	0.6	0.6	0.6	0.6	0.6
SOx total	MtS	17.0	11.0	7.5	4.7	3.0	3.9	3.9	3.2	2.6	2.7	2.7	1.8
HFC	MtC eq.	0	4	8	19	29	31	32	33	33	34	33	31
PFC	MtC eq.	7	4	5	8	14	20	21	22	23	24	25	24
SF6	MtC eq.	8	6	8	10	14	18	21	19	15	14	10	11
CO	MtCO	69	41	53	61	79	112	151	157	152	128	97	65
NMVOC	Mt	16	13	16	20	25	33	41	47	48	42	34	23
NOx	MtN	5	3	3	4	5	7	8	9	10	9	8	8

Emissions correlated to land-use change and deforestation were calculated by using AIM A1 marker land-use data.
a: Land-use taken from AIM-A1 marker run.
b:CO2 emissions from fossil fuel and industrial processes (MESSAGE data).
c: CO2 emissions from land-use changes (AIM-A1 marker run).
d: Non-energy related CH4 emissions were taken from AIM-A1 marker run.
e: Non-energy related N2O emissions were taken from AIM-A1 marker run.

Scenario A1B-MESSAGE ASIA		1990	2000	2010	2020	2030	2040	2050	2060	2070	2080	2090	2100
Population	Million	2798	3261	3620	3937	4147	4238	4220	4085	3867	3589	3258	2882
GNP/GDP (mex)	Trillion US$	1.5	2.7	5.8	12.3	26.2	44.5	62.7	91.9	121.0	150.0	178.6	207.3
GNP/GDP (ppp) Trillion (1990 prices)		5.3	8.2	13.5	21.6	35.2	51.8	67.5	93.3	121.0	150.0	178.6	207.3
Final Energy	EJ												
Non-commercial		24	17	13	8	4	3	2	0	0	0	0	0
Solids		19	36	41	51	59	54	33	26	27	24	22	14
Liquids		13	20	27	47	69	95	122	140	143	134	114	106
Gas		2	3	7	10	14	22	39	64	73	75	62	37
Electricity		4	7	12	24	42	67	97	136	186	231	271	318
Others		1	2	4	10	16	25	37	54	77	107	144	182
Total		62	85	105	150	204	266	330	420	505	570	613	655
Primary Energy	EJ												
Coal		30	48	57	79	99	120	114	104	120	126	105	87
Oil		15	24	28	40	57	71	80	93	77	56	43	41
Gas		3	5	12	22	37	54	81	109	116	112	96	74
Nuclear		0	1	1	4	8	15	26	44	71	102	138	173
Biomass		24	22	27	35	48	64	82	106	133	156	158	163
Other Renewables		1	3	6	14	23	42	75	122	179	239	308	391
Total		74	102	131	193	272	366	459	578	697	791	849	928
Cumulative Resources Use	ZJ												
Coal		0.0	0.3	0.8	1.4	2.2	3.2	4.4	5.5	6.5	7.7	9.0	10.1
Oil		0.0	0.2	0.4	0.7	1.1	1.7	2.4	3.2	4.1	4.9	5.4	5.9
Gas		0.0	0.0	0.1	0.2	0.4	0.8	1.3	2.2	3.2	4.4	5.5	6.5
Cumulative CO2 Emissions	GtC	1.5	19.3	41.7	69.4	106.2	154.8	211.8	273.4	338.5	405.0	466.0	518.3
Carbon Sequestration	GtC												
Land Use [a]	Million ha												
Cropland		438	435	434	433	432	431	431	430	429	428	428	428
Grasslands		608	606	609	611	613	615	617	618	619	621	622	624
Energy Biomass		0	0	0	17	37	68	99	117	134	138	127	117
Forest		535	522	512	524	535	535	535	544	552	556	555	555
Others		583	601	609	579	547	515	483	456	429	421	431	441
Total		2164	2164	2164	2164	2164	2164	2164	2164	2164	2164	2164	2164
Anthropogenic Emissions (standardized)													
Fossil Fuel CO2 [b]	GtC	1.15	1.78	2.22	3.23	4.39	5.49	5.93	6.38	6.78	6.70	5.63	4.85
Other CO2 [c]	GtC	0.37	0.26	0.22	-0.13	-0.12	-0.04	0.03	-0.03	-0.09	-0.10	-0.04	0.01
Total CO2	GtC	1.53	2.03	2.44	3.10	4.27	5.44	5.96	6.35	6.69	6.61	5.59	4.86
CH4 total [d]	MtCH4	113	125	148	180	220	230	237	213	194	179	161	143
N2O total [e]	MtN2O-N	2.3	2.6	2.4	2.4	2.5	2.5	2.5	2.5	2.4	2.4	2.4	2.3
SOx total	MtS	17.7	25.3	27.2	29.7	31.4	30.2	20.9	16.8	15.4	14.3	12.9	11.9
HFC	MtC eq.	0	5	18	45	92	153	224	292	292	285	275	262
PFC	MtC eq.	3	5	8	15	23	30	35	39	43	46	48	46
SF6	MtC eq.	4	7	12	19	28	42	50	55	48	35	33	37
CO	MtCO	235	270	316	413	544	660	776	906	960	1001	968	910
NMVOC	Mt	33	37	40	51	59	66	73	85	79	73	62	42
NOx	MtN	7	9	11	15	19	24	28	32	34	33	30	26

Emissions correlated to land-use change and deforestation were calculated by using AIM A1 marker land-use data.

a: Land-use taken from AIM-A1 marker run.

b:CO2 emissions from fossil fuel and industrial processes (MESSAGE data).

c: CO2 emissions from land-use changes (AIM-A1 marker run).

d: Non-energy related CH4 emissions were taken from AIM-A1 marker run.

e: Non-energy related N2O emissions were taken from AIM-A1 marker run.

Scenario A1B-MESSAGE ALM		1990	2000	2010	2020	2030	2040	2050	2060	2070	2080	2090	2100
Population	Million	1192	1519	1875	2241	2557	2791	2980	3089	3115	3064	2934	2727
GNP/GDP (mex)	Trillion US$	1.9	2.7	5.3	10.3	19.5	35.8	52.0	73.9	95.8	118.6	142.3	165.9
GNP/GDP (ppp) Trillion (1990 prices)		3.8	5.1	8.6	13.8	22.4	38.1	53.9	73.9	95.8	118.6	142.3	165.9
Final Energy	EJ												
Non-commercial		14	11	9	8	5	4	3	0	0	0	0	0
Solids		1	5	7	10	11	14	14	9	3	2	2	2
Liquids		17	27	50	68	98	122	153	180	196	175	144	100
Gas		4	6	11	18	25	43	53	61	66	67	56	40
Electricity		3	6	12	24	47	75	106	150	195	239	274	314
Others		0	1	4	9	17	27	45	63	89	121	159	206
Total		39	55	93	138	203	285	373	463	548	604	634	662
Primary Energy	EJ												
Coal		5	7	14	22	35	46	44	41	53	74	76	83
Oil		21	35	53	67	81	100	131	172	194	154	106	67
Gas		8	14	26	49	88	128	150	163	168	148	131	110
Nuclear		0	0	1	4	10	18	28	40	61	83	101	118
Biomass		14	15	22	30	41	53	72	91	124	176	207	213
Other Renewables		2	3	10	18	27	48	90	148	215	298	377	480
Total		49	74	125	190	283	394	514	656	815	932	999	1071
Cumulative Resources Use	ZJ												
Coal		0.0	0.0	0.1	0.3	0.5	0.8	1.3	1.7	2.1	2.6	3.3	4.0
Oil		0.0	0.3	0.6	1.1	1.8	2.6	3.6	4.9	6.7	8.6	10.1	11.2
Gas		0.0	0.1	0.2	0.5	1.0	1.9	3.1	4.6	6.3	7.9	9.4	10.7
Cumulative CO2 Emissions	GtC	1.4	17.8	38.8	65.4	100.6	147.5	204.2	269.4	343.0	417.9	485.9	545.3
Carbon Sequestration	GtC												
Land Use [a]	Million ha												
Cropland		371	385	384	383	382	381	381	380	379	379	378	378
Grasslands		1680	1673	1685	1695	1702	1708	1714	1718	1722	1726	1730	1733
Energy Biomass		0	0	0	42	90	165	240	282	324	333	308	284
Forest		1745	1711	1690	1683	1680	1681	1682	1681	1680	1678	1676	1674
Others		1684	1710	1720	1676	1625	1544	1463	1419	1374	1364	1387	1411
Total		5480	5480	5480	5480	5480	5480	5480	5480	5480	5480	5480	5480
Anthropogenic Emissions (standardized)													
Fossil Fuel CO2 [b]	GtC	0.72	1.01	1.76	2.61	3.84	5.09	5.93	6.86	7.63	7.18	6.27	5.53
Other CO2 [c]	GtC	0.73	0.82	0.60	0.36	0.25	0.19	0.14	0.12	0.10	0.08	0.06	0.03
Total CO2	GtC	1.45	1.83	2.36	2.96	4.09	5.28	6.07	6.98	7.73	7.26	6.32	5.56
CH4 total [d]	MtCH4	77	85	99	119	141	162	171	165	160	155	131	100
N2O total [e]	MtN2O-N	1.2	1.3	0.7	0.7	0.7	0.7	0.8	0.8	0.7	0.7	0.7	0.6
SOx total	MtS	10.5	12.8	18.5	21.1	22.2	21.0	20.6	20.1	20.2	17.7	14.2	10.4
HFC	MtC eq.	0	2	15	39	81	139	184	205	210	210	206	196
PFC	MtC eq.	4	4	5	9	14	19	23	26	28	30	31	30
SF6	MtC eq.	3	5	10	14	19	32	40	42	37	23	24	26
CO	MtCO	396	404	459	526	572	670	785	904	1069	1233	1353	1405
NMVOC	Mt	48	55	68	82	90	102	117	124	124	98	78	63
NOx	MtN	7	8	11	15	20	25	30	36	38	36	33	30

Emissions correlated to land-use change and deforestation were calculated by using AIM A1 marker land-use data.
a: Land-use taken from AIM-A1 marker run.
b:CO2 emissions from fossil fuel and industrial processes (MESSAGE data).
c: CO2 emissions from land-use changes (AIM-A1 marker run).
d: Non-energy related CH4 emissions were taken from AIM-A1 marker run.
e: Non-energy related N2O emissions were taken from AIM-A1 marker run.

Scenario A1B-MiniCAM World		1990	2000	2010	2020	2030	2040	2050	2060	2070	2080	2090	2100
Population	Million	5293	6100	6874	7618	8122	8484	8703	8623	8430	8126	7621	7137
GNP/GDP (mex)	Trillion US$	20.7	27.4	38.1	52.8	80.2	117.3	164.4	226.8	294.5	367.3	445.1	530.7
GNP/GDP (ppp) Trillion (1990 prices)		na	na	na	na	na	na	na	na	na	na	na	na
Final Energy	EJ												
Non-commercial		0	0	0	0	0	0	0	0	0	0	0	0
Solids		45	57	71	86	98	103	99	73	54	45	46	47
Liquids		121	126	134	145	144	161	197	238	279	318	364	411
Gas		52	63	72	80	81	81	79	90	101	111	73	36
Electricity		35	53	82	123	199	289	392	511	628	745	856	968
Others		0	0	20	59	99	156	231	272	308	341	280	220
Total		253	300	379	493	622	790	998	1184	1371	1559	1620	1682
Primary Energy	EJ												
Coal		88	116	145	174	224	255	265	216	179	156	190	225
Oil		131	136	144	155	141	154	194	244	291	333	387	440
Gas		70	85	137	226	336	437	527	611	678	728	476	224
Nuclear		24	25	33	45	60	75	89	100	112	125	218	311
Biomass		0	5	14	26	55	87	124	148	159	156	187	219
Other Renewables		24	25	28	35	48	97	181	311	448	589	687	785
Total		336	392	500	661	865	1104	1380	1630	1866	2087	2145	2204
Cumulative Resources Use	ZJ												
Coal		0.1	1.2	2.5	4.0	6.2	8.6	11.1	13.4	15.4	17.1	18.9	20.8
Oil		0.1	1.5	2.9	4.4	5.8	7.4	9.0	11.4	14.0	17.1	20.8	24.6
Gas		0.1	0.9	2.1	3.8	6.9	10.8	15.4	21.4	27.8	34.7	40.0	45.2
Cumulative CO2 Emissions	GtC	7.1	82.4	170.3	280.0	417.3	579.7	762.4	956.6	1156.0	1362.1	1559.1	1731.1
Carbon Sequestration	GtC												
Land Use	Million ha												
Cropland		1472	1467	1472	1489	1472	1429	1361	1210	1054	891	769	646
Grasslands		3209	3349	3604	3974	4359	4663	4885	4731	4480	4134	3872	3609
Energy Biomass		0	1	13	35	103	178	261	296	294	255	313	372
Forest		4173	4215	4141	3952	3606	3309	3060	3318	3704	4217	4435	4654
Others		4310	4133	3935	3715	3625	3585	3596	3608	3632	3667	3775	3884
Total		13164	13164	13164	13164	13164	13164	13164	13164	13164	13164	13164	13164
Anthropogenic Emissions (standardized)													
Fossil Fuel CO2	GtC	5.99	6.90	8.51	10.74	13.45	15.88	18.18	19.11	20.07	21.05	19.45	17.93
Other CO2	GtC	1.11	1.07	1.10	1.59	1.68	1.48	1.01	0.53	0.17	-0.07	-1.02	-1.97
Total CO2	GtC	7.10	7.97	9.60	12.33	15.13	17.35	19.19	19.64	20.24	20.98	18.43	15.96
CH4 total	MtCH4	310	323	356	406	467	528	573	592	604	608	572	535
N2O total	MtN2O-N	6.7	7.0	8.0	9.2	10.6	12.3	13.8	14.4	14.8	15.0	15.3	15.6
SOx total	MtS	70.9	69.0	78.8	80.7	77.3	66.1	47.1	32.2	23.9	22.3	25.7	29.2
CFC/HFC/HCFC	MtC eq.	1672	883	791	337	369	482	566	654	659	654	639	614
PFC	MtC eq.	32	25	31	43	61	77	89	97	106	114	119	115
SF6	MtC eq.	38	40	43	48	66	99	119	127	113	88	84	95
CO	MtCO												
NMVOC	Mt												
NOx	MtN												

Scenario A1B-MiniCAM OECD90		1990	2000	2010	2020	2030	2040	2050	2060	2070	2080	2090	2100
Population	Million	838	908	965	1007	1024	1066	1081	1084	1090	1098	1105	1112
GNP/GDP (mex)	Trillion US$	16.3	20.5	25.6	31.5	34.4	44.7	53.6	63.3	74.4	86.7	102.1	118.7
GNP/GDP (ppp) Trillion (1990 prices)		na	na	na	na	na	na	na	na	na	na	na	na
Final Energy	EJ												
Non-commercial		0	0	0	0	0	0	0	0	0	0	0	0
Solids		10	13	12	9	9	7	5	4	3	3	4	5
Liquids		72	73	70	61	51	30	28	30	32	35	43	51
Gas		27	36	41	40	38	30	25	27	30	33	24	14
Electricity		22	28	33	36	37	40	42	51	63	76	102	128
Others		0	0	16	47	59	98	129	140	151	164	141	117
Total		130	151	172	195	195	205	229	252	279	311	313	315
Primary Energy	EJ												
Coal		40	47	45	35	37	42	45	43	38	30	42	54
Oil		76	78	74	64	52	26	22	22	26	35	44	53
Gas		34	47	73	113	120	141	154	169	184	201	143	84
Nuclear		20	16	13	12	11	9	8	9	10	12	26	39
Biomass		0	2	5	10	13	24	33	36	37	37	51	66
Other Renewables		12	11	11	11	11	15	21	32	46	62	82	103
Total		182	201	222	245	245	258	283	310	342	376	388	400
Cumulative Resources Use	ZJ												
Coal		0.0	0.5	0.9	1.4	1.5	2.1	2.6	3.0	3.4	3.8	4.2	4.6
Oil		0.1	0.9	1.6	2.3	2.5	3.1	3.4	3.6	3.8	4.1	4.5	4.9
Gas		0.0	0.5	1.1	2.0	2.6	4.6	6.0	7.7	9.5	11.4	12.9	14.4
Cumulative CO2 Emissions	GtC	2.8	33.0	67.1	104.7	143.5	181.7	220.0	259.2	299.4	340.7	380.2	415.2
Carbon Sequestration	GtC												
Land Use	Million ha												
Cropland		408	411	410	408	402	378	352	309	268	229	196	163
Grasslands		796	821	866	931	963	1044	1076	1048	1007	953	898	843
Energy Biomass		0	1	9	24	34	67	92	95	93	86	120	154
Forest		921	931	922	894	867	794	760	825	904	996	1028	1059
Others		998	959	916	866	857	840	843	846	851	859	882	904
Total		3123	3123	3123	3123	3123	3123	3123	3123	3123	3123	3123	3123
Anthropogenic Emissions (standardized)													
Fossil Fuel CO2	GtC	2.83	3.20	3.48	3.64	3.58	3.57	3.77	3.95	4.14	4.34	4.00	3.67
Other CO2	GtC	0.00	0.00	0.14	0.27	0.28	0.22	0.11	0.02	-0.07	-0.16	-0.28	-0.40
Total CO2	GtC	2.83	3.20	3.62	3.90	3.86	3.79	3.87	3.97	4.07	4.18	3.72	3.28
CH4 total	MtCH4	73	74	83	93	98	117	133	150	160	161	167	173
N2O total	MtN2O-N	2.6	2.6	2.8	3.0	3.1	3.5	3.8	3.9	4.0	4.0	4.0	4.0
SOx total	MtS	22.7	17.0	13.1	2.2	1.0	0.4	0.4	0.9	1.5	2.2	3.3	4.5
HFC	MtC eq.	19	58	110	108	115	118	122	123	123	124	125	125
PFC	MtC eq.	18	13	13	11	9	8	9	10	12	13	14	16
SF6	MtC eq.	23	23	14	5	6	7	9	11	13	16	18	20
CO	MtCO												
NMVOC	Mt												
NOx	MtN												

Scenario A1B-MiniCAM REF		**1990**	**2000**	**2010**	**2020**	**2030**	**2040**	**2050**	**2060**	**2070**	**2080**	**2090**	**2100**
Population	Million	428	425	426	433	434	431	423	408	392	374	357	340
GNP/GDP (mex)	Trillion US$	1.1	1.1	1.4	2.1	3.5	5.1	6.9	10.0	13.6	17.8	21.8	26.3
GNP/GDP (ppp) Trillion (1990 prices)		na	na	na	na	na	na	na	na	na	na	na	na
Final Energy	EJ												
Non-commercial		0	0	0	0	0	0	0	0	0	0	0	0
Solids		13	10	9	9	8	8	8	6	4	4	4	4
Liquids		18	12	9	11	12	14	16	19	21	23	23	24
Gas		19	15	13	15	15	15	15	16	17	18	11	5
Electricity		6	8	13	21	33	44	56	68	79	89	94	98
Others		0	0	0	1	1	2	3	4	4	4	3	1
Total		56	44	45	57	70	84	98	112	125	137	134	132
Primary Energy	EJ												
Coal		18	17	19	22	30	34	34	24	18	17	19	22
Oil		20	13	11	13	10	9	12	18	22	24	25	27
Gas		26	20	22	34	47	53	54	58	60	60	36	12
Nuclear		3	4	6	10	11	12	13	13	14	15	23	31
Biomass		0	1	2	3	7	10	12	14	13	12	12	12
Other Renewables		3	3	4	5	7	14	26	41	56	70	75	80
Total		70	58	64	87	111	132	151	169	184	197	191	184
Cumulative Resources Use	ZJ												
Coal		0.0	0.2	0.4	0.6	0.9	1.2	1.5	1.8	2.0	2.2	2.4	2.5
Oil		0.0	0.2	0.3	0.4	0.5	0.6	0.7	0.9	1.1	1.3	1.6	1.8
Gas		0.0	0.2	0.5	0.7	1.2	1.7	2.2	2.8	3.4	4.0	4.4	4.8
Cumulative CO2 Emissions	GtC	1.3	12.3	21.7	33.1	48.1	66.0	85.3	104.4	122.9	141.1	157.0	168.8
Carbon Sequestration	GtC												
Land Use	Million ha												
Cropland		284	294	304	317	322	321	312	273	230	183	158	134
Grasslands		395	410	454	527	609	676	729	684	616	526	488	451
Energy Biomass		0	0	1	4	17	26	31	32	28	20	18	16
Forest		1007	1016	996	945	861	792	738	818	928	1066	1109	1151
Others		691	657	622	584	569	563	567	570	575	582	604	626
Total		2377	2377	2377	2377	2377	2377	2377	2377	2377	2377	2377	2377
Anthropogenic Emissions (standardized)													
Fossil Fuel CO2	GtC	1.30	0.91	0.94	1.23	1.56	1.77	1.85	1.77	1.73	1.73	1.51	1.31
Other CO2	GtC	0.00	0.00	0.03	0.08	0.13	0.13	0.10	0.10	0.09	0.08	-0.13	-0.34
Total CO2	GtC	1.30	0.91	0.97	1.31	1.69	1.90	1.94	1.87	1.83	1.81	1.38	0.97
CH4 total	MtCH4	47	39	49	68	82	96	108	106	108	112	99	86
N2O total	MtN2O-N	0.6	0.6	0.7	0.9	1.1	1.4	1.6	1.6	1.6	1.5	1.5	1.6
SOx total	MtS	17.0	11.0	10.1	10.1	10.7	9.4	6.4	3.4	1.7	1.4	1.8	2.2
HFC	MtC eq.	0	4	8	19	29	31	32	33	33	34	33	31
PFC	MtC eq.	7	4	5	8	14	20	21	22	23	24	25	24
SF6	MtC eq.	8	6	8	10	14	18	21	19	15	14	10	11
CO	MtCO												
NMVOC	Mt												
NOx	MtN												

Scenario A1B-MiniCAM ASIA		1990	2000	2010	2020	2030	2040	2050	2060	2070	2080	2090	2100
Population	Million	2790	3226	3608	3937	4115	4210	4219	4062	3852	3589	3245	2919
GNP/GDP (mex)	Trillion US$	1.4	3.1	6.5	11.7	23.9	40.4	61.3	86.4	112.3	139.1	166.0	195.3
GNP/GDP (ppp) Trillion (1990 prices)		na	na	na	na	na	na	na	na	na	na	na	na
Final Energy	EJ												
Non-commercial		0	0	0	0	0	0	0	0	0	0	0	0
Solids		20	31	45	62	73	78	76	54	40	32	31	31
Liquids		14	19	27	36	47	61	78	93	106	118	130	143
Gas		2	5	8	11	14	16	17	20	23	25	16	7
Electricity		4	11	25	47	90	141	200	254	305	352	382	412
Others		0	0	2	7	14	26	45	55	62	67	52	36
Total		40	67	108	163	238	322	416	476	535	594	611	629
Primary Energy	EJ												
Coal		26	45	71	104	133	147	147	110	86	73	81	90
Oil		16	21	29	40	47	61	81	101	116	126	140	154
Gas		3	9	24	49	102	149	192	219	235	241	149	57
Nuclear		1	4	9	17	27	37	47	51	55	60	97	135
Biomass		0	2	5	10	22	35	48	58	61	59	64	70
Other Renewables		3	4	5	8	14	40	85	149	212	275	304	332
Total		49	85	144	227	344	468	600	687	765	833	835	838
Cumulative Resources Use	ZJ												
Coal		0.0	0.4	1.1	1.9	3.1	4.5	6.0	7.2	8.1	9.0	9.8	10.6
Oil		0.0	0.2	0.5	0.8	1.3	1.8	2.5	3.4	4.5	5.7	7.1	8.5
Gas		0.0	0.1	0.3	0.6	1.5	2.8	4.4	6.5	8.8	11.1	12.8	14.5
Cumulative CO2 Emissions	GtC	1.5	19.3	44.4	81.5	134.0	201.2	279.3	360.4	439.2	516.5	588.5	651.2
Carbon Sequestration	GtC												
Land Use	Million ha												
Cropland		389	400	410	420	416	405	387	350	312	273	235	198
Grasslands		508	524	555	603	646	681	708	704	691	670	641	612
Energy Biomass		0	0	2	6	28	51	75	85	83	67	73	79
Forest		1168	1144	1102	1041	980	932	896	925	978	1053	1116	1179
Others		664	633	600	565	551	545	547	550	555	562	581	600
Total		2729	2700	2668	2635	2620	2613	2614	2614	2618	2626	2647	2667
Anthropogenic Emissions (standardized)													
Fossil Fuel CO2	GtC	1.15	1.78	2.79	4.16	5.78	7.07	8.03	7.86	7.75	7.70	6.88	6.10
Other CO2	GtC	0.37	0.26	0.20	0.27	0.31	0.29	0.22	0.11	0.03	-0.01	-0.16	-0.30
Total CO2	GtC	1.53	2.03	2.99	4.43	6.08	7.36	8.26	7.97	7.78	7.69	6.72	5.80
CH4 total	MtCH4	113	125	134	146	160	172	180	172	166	163	159	154
N2O total	MtN2O-N	2.3	2.6	2.9	3.3	3.7	4.1	4.5	4.6	4.7	4.9	4.9	5.0
SOx total	MtS	17.7	25.3	37.9	50.0	47.4	38.3	22.7	14.3	9.5	8.2	9.4	10.6
HFC	MtC eq.	0	5	18	45	92	153	224	292	292	285	275	262
PFC	MtC eq.	3	5	8	15	23	30	35	39	43	46	48	46
SF6	MtC eq.	4	7	12	19	28	42	50	55	48	35	33	37
CO	MtCO												
NMVOC	Mt												
NOx	MtN												

Scenario A1B-MiniCAM ALM		1990	2000	2010	2020	2030	2040	2050	2060	2070	2080	2090	2100
Population	Million	1236	1541	1876	2241	2531	2778	2980	3068	3096	3064	2913	2766
GNP/GDP (mex)	Trillion US$	1.9	2.8	4.6	7.4	15.4	27.2	42.6	67.0	94.1	123.7	155.2	190.3
GNP/GDP (ppp) Trillion (1990 prices)		na	na	na	na	na	na	na	na	na	na	na	na
Final Energy	EJ												
Non-commercial		0	0	0	0	0	0	0	0	0	0	0	0
Solids		2	3	4	6	9	10	11	9	7	7	7	8
Liquids		17	22	29	37	43	56	75	98	120	142	167	193
Gas		5	7	10	13	17	20	22	27	31	35	23	10
Electricity		3	6	11	18	38	64	94	137	182	228	279	330
Others		0	0	1	4	13	30	54	74	91	107	86	65
Total		27	38	55	78	120	179	256	344	432	518	562	606
Primary Energy	EJ												
Coal		4	6	9	13	23	31	39	38	38	37	48	59
Oil		20	23	29	38	44	58	80	104	127	149	177	206
Gas		7	10	17	29	61	93	127	165	198	226	148	71
Nuclear		0	2	4	7	11	16	21	27	32	38	72	105
Biomass		0	1	2	3	10	19	31	41	47	49	60	72
Other Renewables		5	6	8	12	16	28	49	89	134	183	226	269
Total		35	48	71	102	165	246	346	464	576	681	731	781
Cumulative Resources Use	ZJ												
Coal		0.0	0.1	0.1	0.2	0.5	0.7	1.1	1.5	1.8	2.2	2.7	3.1
Oil		0.0	0.2	0.5	0.8	1.3	1.8	2.5	3.4	4.6	5.9	7.7	9.4
Gas		0.0	0.1	0.2	0.5	1.0	1.8	2.8	4.4	6.2	8.3	9.9	11.5
Cumulative CO2 Emissions	GtC	1.4	17.8	37.1	60.7	91.7	130.8	177.9	232.6	294.6	363.9	433.4	496.0
Carbon Sequestration	GtC												
Land Use	Million ha												
Cropland		391	363	348	344	338	326	310	278	244	206	179	152
Grasslands		1510	1594	1728	1913	2109	2262	2372	2295	2166	1985	1845	1704
Energy Biomass		0	0	0	1	13	34	64	84	90	82	102	123
Forest		3641	3591	3478	3301	3098	2932	2801	2883	3039	3270	3423	3575
Others		1957	1884	1798	1699	1657	1637	1639	1642	1651	1663	1709	1754
Total		7499	7432	7352	7259	7216	7192	7186	7183	7190	7207	7257	7308
Anthropogenic Emissions (standardized)													
Fossil Fuel CO2	GtC	0.72	1.01	1.30	1.72	2.54	3.47	4.53	5.53	6.45	7.27	7.05	6.84
Other CO2	GtC	0.73	0.82	0.73	0.97	0.97	0.84	0.58	0.30	0.11	0.03	-0.45	-0.93
Total CO2	GtC	1.45	1.83	2.03	2.69	3.51	4.31	5.11	5.83	6.56	7.30	6.60	5.91
CH4 total	MtCH4	77	85	90	100	125	143	152	163	170	171	147	122
N2O total	MtN2O-N	1.2	1.3	1.7	2.1	2.7	3.3	3.9	4.2	4.5	4.6	4.8	5.0
SOx total	MtS	10.5	12.8	14.7	15.4	15.3	15.0	14.5	10.6	8.2	7.4	8.1	8.8
HFC	MtC eq.	0	2	15	39	81	139	184	205	210	210	206	196
PFC	MtC eq.	4	4	5	9	14	19	23	26	28	30	31	30
SF6	MtC eq.	3	5	10	14	19	32	40	42	37	23	24	26
CO	MtCO												
NMVOC	Mt												
NOx	MtN												

Scenario A1C-AIM World		1990	2000	2010	2020	2030	2040	2050	2060	2070	2080	2090	2100
Population	Million	5262	6117	6805	7493	8182	8439	8704	8538	8375	8030	7528	7056
GNP/GDP (mex)	Trillion US$	20.9	26.7	37.9	56.4	89.1	127.2	181.5	235.2	304.7	377.7	447.8	531.0
GNP/GDP (ppp) Trillion (1990 prices)													
Final Energy	EJ												
Non-commercial		50	48	38	34	24	0	0	0	0	0	0	0
Solids		36	42	69	113	164	187	209	192	175	160	146	133
Liquids		111	118	149	187	230	274	319	356	393	430	467	505
Gas		51	51	74	87	103	122	141	156	172	189	207	224
Electricity		38	49	74	104	150	229	307	429	551	651	731	810
Others		0			0			0					0
Total		285	309	404	525	670	823	976	1134	1291	1430	1551	1671
Primary Energy	EJ												
Coal		93	112	195	308	474	593	750	820	898	976	1051	1134
Oil		142	155	193	227	244	232	226	193	165	140	118	99
Gas		72	82	112	124	139	161	191	208	230	252	272	294
Nuclear		6	9	15	23	36	52	83	98	118	135	145	155
Biomass		50	48	38	42	55	78	115	147	187	217	228	240
Other Renewables		10	11	13	17	26	44	78	138	249	372	450	546
Total		374	418	565	742	973	1161	1442	1605	1848	2092	2264	2468
Cumulative Resources Use	ZJ												
Coal		0.1	1.1	2.6	5.1	9.1	13.9	21.3	28.0	36.9	46.7	56.5	68.3
Oil		0.1	1.6	3.4	5.5	7.8	9.9	12.5	14.2	16.2	17.8	19.1	20.4
Gas		0.1	0.8	1.8	3.0	4.3	5.7	7.6	9.4	11.6	14.1	16.7	19.7
Cumulative CO2 Emissions	GtC	7.1	82.4	181.3	320.8	501.7	716.5	965.4	1242.5	1536.3	1848.7	2181.9	2537.7
Carbon Sequestraction	GtC												
Land Use	Million ha												
Cropland		1437	1487	1546	1579	1534	1454	1378	1276	1182	1079	971	873
Grasslands		3290	3287	3477	3874	4235	4271	4307	4231	4157	4056	3930	3808
Energy Biomass		0	0	0	10	227	320	451	535	635	728	807	894
Forest		4249	4120	3975	3808	3742	3778	3813	3896	3981	4071	4166	4264
Others		3966	4048	3944	3671	3205	3097	2994	2984	2974	2995	3048	3102
Total		12942	12942	12942	12942	12942	12942	12942	12942	12942	12942	12942	12942
Anthropogenic Emissions (standardized)													
Fossil Fuel CO2	GtC	5.99	6.90	10.32	14.34	19.28	22.65	26.79	28.47	30.29	32.30	34.48	36.84
Other CO2	GtC	1.11	1.07	1.47	1.78	0.78	0.23	0.12	0.04	-0.03	-0.07	-0.08	-0.09
Total CO2	GtC	7.10	7.97	11.79	16.12	20.06	22.88	26.91	28.51	30.26	32.23	34.41	36.75
CH4 total	MtCH4	310	323	395	479	564	566	568	517	474	442	416	392
N2O total	MtN2O-N	6.7	7.0	7.5	8.2	8.2	8.0	7.8	7.3	6.8	6.5	6.3	6.1
SOx total	MtS	70.9	69.0	99.1	134.2	156.4	133.1	139.0	103.5	82.0	76.3	79.6	83.3
CFC/HFC/HCFC	MtC eq.	1672	883	791	337	369	482	566	654	659	654	639	614
PFC	MtC eq.	32	25	31	43	61	77	89	97	106	114	119	115
SF6	MtC eq.	38	40	43	48	66	99	119	127	113	88	84	95
CO	MtCO	879	877	1101	1320	1402	1506	1619	1710	1806	1937	2110	2298
NMVOC	Mt	139	141	179	220	254	254	256	247	239	221	192	167
NOx	MtN	31	32	41	51	59	61	63	61	59	59	61	63

Scenario A1C-AIM OECD90		**1990**	**2000**	**2010**	**2020**	**2030**	**2040**	**2050**	**2060**	**2070**	**2080**	**2090**	**2100**
Population	Million	859	919	960	1002	1043	1062	1081	1086	1091	1097	1103	1110
GNP/GDP (mex)	Trillion US$	16.4	20.5	25.3	31.0	38.0	45.3	54.1	64.0	75.7	89.0	103.9	121.2
GNP/GDP (ppp) Trillion (1990 prices)													
Final Energy	EJ												
Non-commercial		6	1	0	0	0	0	0	0	0	0	0	0
Solids		10	10	14	23	31	31	30	26	23	21	19	17
Liquids		64	74	80	87	92	94	96	97	97	102	111	121
Gas		25	31	39	39	37	37	36	36	36	37	39	41
Electricity		22	27	35	42	49	59	68	85	103	118	133	148
Others		0			0			0					0
Total		127	143	168	192	210	220	231	245	259	279	303	327
Primary Energy	EJ												
Coal		41	42	60	79	102	118	137	148	161	176	196	218
Oil		76	89	94	97	87	70	56	44	35	30	27	24
Gas		34	45	54	50	46	45	44	42	41	41	43	46
Nuclear		5	8	11	15	18	20	21	22	23	24	26	28
Biomass		6	1	0	2	6	11	22	28	36	42	44	46
Other Renewables		6	6	7	9	12	17	24	38	58	78	91	105
Total		167	191	226	251	271	281	304	323	354	391	426	466
Cumulative Resources Use	ZJ												
Coal		0.0	0.4	1.0	1.7	2.6	3.6	5.0	6.2	7.8	9.6	11.4	13.6
Oil		0.1	0.9	1.8	2.8	3.7	4.4	5.1	5.5	6.0	6.4	6.6	6.9
Gas		0.0	0.4	0.9	1.4	1.9	2.3	2.8	3.2	3.6	4.1	4.5	4.9
Cumulative CO2 Emissions	GtC	2.8	33.0	67.8	108.7	155.0	205.1	258.4	312.9	366.6	420.6	477.3	538.0
Carbon Sequestraction	GtC												
Land Use	Million ha												
Cropland		410	404	389	373	345	329	314	285	259	227	192	163
Grasslands		787	758	751	785	817	825	834	812	791	757	711	668
Energy Biomass		0	0	0	0	47	69	100	112	124	137	149	162
Forest		1056	1065	1102	1131	1168	1174	1179	1217	1257	1304	1360	1418
Others		886	912	897	851	762	737	713	710	706	709	719	729
Total		3140	3140	3140	3140	3140	3140	3140	3140	3140	3140	3140	3140
Anthropogenic Emissions (standardized)													
Fossil Fuel CO2	GtC	2.83	3.20	3.89	4.44	4.94	5.19	5.45	5.42	5.39	5.58	6.00	6.44
Other CO2	GtC	0.00	0.00	-0.12	-0.04	-0.09	-0.02	0.05	-0.01	-0.06	-0.10	-0.13	-0.16
Total CO2	GtC	2.83	3.20	3.78	4.40	4.85	5.16	5.50	5.41	5.33	5.47	5.86	6.28
CH4 total	MtCH4	73	74	71	73	76	73	69	68	66	65	62	60
N2O total	MtN2O-N	2.6	2.6	2.5	2.7	2.6	2.5	2.4	2.2	2.0	1.8	1.7	1.7
SOx total	MtS	22.7	17.0	11.5	9.0	10.0	10.9	12.0	12.5	13.1	13.9	15.0	16.1
HFC	MtC eq.	19	58	110	108	115	118	122	123	123	124	125	125
PFC	MtC eq.	18	13	13	11	9	8	9	10	12	13	14	16
SF6	MtC eq.	23	23	14	5	6	7	9	11	13	16	18	20
CO	MtCO	179	161	186	213	241	256	271	277	284	300	329	359
NMVOC	Mt	42	36	39	40	39	33	29	21	14	11	11	10
NOx	MtN	13	12	12	13	12	10	9	9	9	9	10	11

Scenario A1C-AIM REF		1990	2000	2010	2020	2030	2040	2050	2060	2070	2080	2090	2100
Population	Million	413	419	424	430	435	429	423	406	391	374	356	339
GNP/GDP (mex)	Trillion US$	1.1	0.8	1.5	2.8	5.3	8.1	12.3	15.5	19.4	23.8	28.5	34.1
GNP/GDP (ppp) Trillion (1990 prices)													
Final Energy	EJ												
Non-commercial		2	4	0	0	0	0	0	0	0	0	0	0
Solids		9	7	10	14	18	18	19	17	14	12	11	10
Liquids		19	9	8	8	8	10	12	13	14	15	17	18
Gas		19	11	14	14	14	15	16	17	18	19	21	23
Electricity		8	9	11	12	15	22	29	36	43	49	53	57
Others		0			0			0					0
Total		58	40	43	48	55	65	76	82	89	95	102	109
Primary Energy	EJ												
Coal		18	16	22	27	39	57	84	98	115	131	145	160
Oil		22	12	10	8	6	4	3	2	1	1	1	0
Gas		26	17	18	17	15	15	15	14	14	14	14	15
Nuclear		1	1	2	3	5	7	9	10	10	10	11	11
Biomass		2	4	0	0	1	3	5	7	8	10	10	11
Other Renewables		1	1	1	2	2	4	10	15	24	32	36	40
Total		70	52	54	57	68	90	125	146	173	198	216	237
Cumulative Resources Use	ZJ												
Coal		0.0	0.2	0.4	0.6	0.9	1.4	2.2	3.0	4.1	5.3	6.7	8.3
Oil		0.0	0.2	0.3	0.4	0.5	0.5	0.6	0.6	0.6	0.6	0.6	0.6
Gas		0.0	0.2	0.4	0.6	0.7	0.9	1.0	1.2	1.3	1.5	1.6	1.8
Cumulative CO2 Emissions	GtC	1.3	12.3	22.5	34.2	47.8	64.4	85.6	110.8	139.2	171.2	207.3	248.2
Carbon Sequestraction	GtC												
Land Use	Million ha												
Cropland		279	284	297	299	305	293	282	267	253	233	210	189
Grasslands		346	357	401	468	556	571	586	585	584	571	548	525
Energy Biomass		0	0	0	0	6	11	22	25	29	38	55	80
Forest		960	961	950	950	944	950	955	966	978	993	1010	1028
Others		720	703	658	588	494	477	460	460	460	464	474	483
Total		2305	2305	2305	2305	2305	2305	2305	2305	2305	2305	2305	2305
Anthropogenic Emissions (standardized)													
Fossil Fuel CO2	GtC	1.30	0.91	1.09	1.22	1.45	1.86	2.39	2.69	3.03	3.42	3.88	4.39
Other CO2	GtC	0.00	0.00	0.03	0.01	0.02	-0.01	-0.01	-0.02	-0.03	-0.03	-0.04	-0.05
Total CO2	GtC	1.30	0.91	1.12	1.23	1.48	1.86	2.38	2.67	3.00	3.39	3.84	4.34
CH4 total	MtCH4	47	39	50	62	77	80	84	91	99	103	105	106
N2O total	MtN2O-N	0.6	0.6	0.7	0.7	0.8	0.7	0.7	0.7	0.6	0.5	0.5	0.5
SOx total	MtS	17.0	11.0	14.7	15.6	16.1	12.7	10.0	9.9	9.7	10.6	12.6	14.8
HFC	MtC eq.	0	4	8	19	29	31	32	33	33	34	33	31
PFC	MtC eq.	7	4	5	8	14	20	21	22	23	24	25	24
SF6	MtC eq.	8	6	8	10	14	18	21	19	15	14	10	11
CO	MtCO	69	41	40	42	46	50	54	57	59	63	68	75
NMVOC	Mt	16	13	14	11	11	9	8	7	6	5	5	5
NOx	MtN	5	3	3	4	3	3	3	3	2	2	2	2

Scenario A1C-AIM ASIA		1990	2000	2010	2020	2030	2040	2050	2060	2070	2080	2090	2100
Population	Million	2798	3261	3556	3851	4147	4183	4220	4016	3822	3541	3194	2882
GNP/GDP (mex)	Trillion US$	1.5	2.7	5.8	12.3	26.3	40.8	63.1	85.8	116.7	148.3	176.1	209.1
GNP/GDP (ppp) Trillion (1990 prices)													
Final Energy	EJ												
Non-commercial		28	24	19	14	4	0	0	0	0	0	0	0
Solids		15	23	40	62	90	103	116	107	97	89	81	73
Liquids		11	15	26	40	55	69	84	102	119	135	148	161
Gas		2	3	6	9	18	25	32	41	50	59	67	75
Electricity		5	7	16	29	54	83	113	167	220	264	298	333
Others		0			0			0					0
Total		61	72	106	154	220	283	345	416	487	546	594	642
Primary Energy	EJ												
Coal		30	44	85	144	234	275	324	350	379	407	435	465
Oil		17	21	34	46	56	51	47	40	33	28	22	18
Gas		4	6	11	13	23	31	42	54	68	81	91	101
Nuclear		0	0	1	3	7	12	22	31	43	54	58	64
Biomass		28	24	19	16	11	17	27	34	43	50	52	55
Other Renewables		1	2	2	2	3	6	13	30	69	120	159	210
Total		80	98	151	224	334	393	475	538	636	740	818	913
Cumulative Resources Use	ZJ												
Coal		0.0	0.4	1.0	2.2	4.1	6.3	9.6	12.6	16.3	20.4	24.5	29.4
Oil		0.0	0.2	0.5	0.9	1.4	1.8	2.4	2.8	3.2	3.5	3.7	4.0
Gas		0.0	0.1	0.1	0.3	0.4	0.7	1.1	1.5	2.2	2.9	3.7	4.8
Cumulative CO2 Emissions	GtC	1.5	19.3	47.3	92.2	158.9	243.0	338.5	443.9	556.7	677.4	806.0	942.4
Carbon Sequestraction	GtC												
Land Use	Million ha												
Cropland		390	407	433	458	447	422	398	368	340	313	288	265
Grasslands		521	524	547	595	644	650	656	645	633	622	611	600
Energy Biomass		0	0	0	0	42	61	89	108	130	149	163	179
Forest		527	490	451	416	404	415	427	446	466	482	493	505
Others		576	593	583	546	478	461	444	443	442	446	455	465
Total		2014	2014	2014	2014	2014	2014	2014	2014	2014	2014	2014	2014
Anthropogenic Emissions (standardized)													
Fossil Fuel CO2	GtC	1.15	1.78	3.15	4.99	7.65	8.78	10.08	10.83	11.62	12.41	13.17	13.97
Other CO2	GtC	0.37	0.26	0.40	0.46	0.24	0.15	0.09	0.07	0.05	0.06	0.07	0.09
Total CO2	GtC	1.53	2.03	3.55	5.45	7.89	8.93	10.17	10.89	11.68	12.47	13.24	14.06
CH4 total	MtCH4	113	125	165	211	262	262	261	220	184	162	149	137
N2O total	MtN2O-N	2.3	2.6	2.8	3.1	3.2	3.2	3.1	2.9	2.8	2.7	2.6	2.5
SOx total	MtS	17.7	25.3	47.2	72.2	80.0	39.4	19.2	19.7	20.2	21.1	22.4	23.8
HFC	MtC eq.	0	5	18	45	92	153	224	292	292	285	275	262
PFC	MtC eq.	3	5	8	15	23	30	35	39	43	46	48	46
SF6	MtC eq.	4	7	12	19	28	42	50	55	48	35	33	37
CO	MtCO	235	270	389	509	599	663	734	777	822	881	956	1036
NMVOC	Mt	33	37	54	77	93	97	102	104	105	98	84	71
NOx	MtN	7	9	13	19	25	26	26	25	23	23	23	23

Scenario A1C-AIM ALM		**1990**	**2000**	**2010**	**2020**	**2030**	**2040**	**2050**	**2060**	**2070**	**2080**	**2090**	**2100**
Population	Million	1192	1519	1865	2211	2557	2761	2980	3024	3067	3013	2866	2727
GNP/GDP (mex)	Trillion US$	1.9	2.7	5.3	10.2	19.5	31.9	52.0	69.3	92.4	116.6	139.4	166.6
GNP/GDP (ppp) Trillion (1990 prices)													
Final Energy	EJ												
Non-commercial		14	18	19	20	20	0	0	0	0	0	0	0
Solids		1	2	6	14	25	35	45	43	41	38	35	32
Liquids		17	21	35	52	75	101	127	145	163	179	191	204
Gas		4	7	15	25	34	45	57	62	68	74	80	85
Electricity		4	6	12	21	32	64	97	141	185	220	246	273
Others		0			0			0					0
Total		40	53	87	132	185	255	325	391	457	510	552	594
Primary Energy	EJ												
Coal		5	11	28	58	99	143	205	223	243	261	275	291
Oil		27	33	55	75	95	107	120	107	95	82	69	57
Gas		9	14	29	44	55	71	90	98	107	115	124	132
Nuclear		0	0	1	3	6	14	31	36	42	47	50	53
Biomass		14	18	19	25	36	47	61	78	100	116	122	128
Other Renewables		2	2	3	5	8	16	31	55	98	142	164	191
Total		57	78	135	209	300	397	538	598	686	762	803	852
Cumulative Resources Use	ZJ												
Coal		0.0	0.1	0.3	0.7	1.5	2.6	4.5	6.3	8.7	11.4	13.9	1.7
Oil		0.0	0.3	0.8	1.4	2.3	3.2	4.4	5.3	6.4	7.4	8.1	6.7
Gas		0.0	0.1	0.3	0.7	1.2	1.8	2.7	3.5	4.5	5.7	6.8	12.5
Cumulative CO2 Emissions	GtC	1.4	17.8	43.7	85.6	140.1	204.0	282.9	374.9	473.7	579.5	691.3	809.0
Carbon Sequestraction	GtC												
Land Use	Million ha												
Cropland		357	391	427	449	437	410	384	356	330	305	280	257
Grasslands		1636	1648	1778	2027	2217	2224	2231	2190	2149	2106	2060	2015
Energy Biomass		0	0	0	10	132	178	240	290	351	402	436	473
Forest		1706	1604	1472	1311	1226	1239	1253	1266	1279	1292	1303	1314
Others		1784	1840	1806	1687	1471	1423	1376	1371	1366	1376	1400	1425
Total		5483	5483	5483	5483	5483	5483	5483	5483	5483	5483	5483	5483
Anthropogenic Emissions (standardized)													
Fossil Fuel CO2	GtC	0.72	1.01	2.20	3.69	5.24	6.82	8.87	9.53	10.24	10.89	11.44	12.03
Other CO2	GtC	0.73	0.82	1.15	1.36	0.61	0.11	0.00	0.00	0.00	0.01	0.02	0.04
Total CO2	GtC	1.45	1.83	3.34	5.05	5.85	6.93	8.86	9.53	10.25	10.90	11.47	12.06
CH4 total	MtCH4	77	85	108	132	148	151	154	139	125	112	99	88
N2O total	MtN2O-N	1.2	1.3	1.5	1.8	1.6	1.6	1.6	1.5	1.5	1.5	1.5	1.5
SOx total	MtS	10.5	12.8	22.7	34.4	47.2	67.1	94.8	58.4	36.1	27.7	26.6	25.6
HFC	MtC eq.	0	2	15	39	81	139	184	205	210	210	206	196
PFC	MtC eq.	4	4	5	9	14	19	23	26	28	30	31	30
SF6	MtC eq.	3	5	10	14	19	32	40	42	37	23	24	26
CO	MtCO	396	404	486	556	516	537	560	599	641	693	757	827
NMVOC	Mt	48	55	73	92	111	114	117	115	113	106	93	82
NOx	MtN	7	8	11	16	19	21	24	24	24	25	26	27

Scenario A1C-MESSAGE World		1990	2000	2010	2020	2030	2040	2050	2060	2070	2080	2090	2100
Population	Million	5262	6117	6888	7617	8182	8531	8704	8667	8463	8125	7658	7056
GNP/GDP (mex)	Trillion US$	20.9	26.8	36.8	57.0	91.3	135.4	187.1	254.1	322.9	393.2	469.6	550.0
GNP/GDP (ppp) Trillion (1990 prices)		25.7	33.4	45.7	67.2	98.7	139.0	186.4	246.8	313.2	382.0	456.6	535.0
Final Energy	EJ												
Non-commercial		38	27	19	16	10	7	5	0	0	0	0	0
Solids		42	56	72	82	89	79	72	68	60	55	44	26
Liquids		111	126	158	200	265	353	461	569	641	672	704	685
Gas		41	47	61	76	95	106	110	118	116	119	104	93
Electricity		35	46	62	93	145	209	292	395	493	588	668	726
Others		8	11	20	34	53	70	89	118	143	164	187	213
Total		275	313	393	501	656	824	1031	1268	1453	1599	1706	1743
Primary Energy	EJ												
Coal		91	112	146	194	261	353	463	601	750	872	985	1062
Oil		128	155	172	190	210	209	209	202	170	112	78	56
Gas		71	80	107	149	207	247	283	297	254	207	144	118
Nuclear		7	8	11	21	41	79	127	191	265	348	415	432
Biomass		46	44	52	68	95	133	178	228	287	330	359	376
Other Renewables		8	13	23	37	57	82	117	166	205	234	258	281
Total		352	411	511	659	870	1102	1377	1685	1931	2103	2239	2325
Cumulative Resources Use	ZJ												
Coal		0.0	1.0	2.1	3.5	5.5	8.1	11.6	16.2	22.3	29.8	38.5	48.4
Oil		0.0	1.4	3.0	4.7	6.6	8.7	10.8	12.9	14.9	16.6	17.7	18.5
Gas		0.0	0.7	1.5	2.6	4.1	6.2	8.6	11.5	14.4	17.0	19.0	20.5
Cumulative CO2 Emissions	GtC	7.1	82.4	170.5	274.9	401.9	558.5	747.4	972.8	1231.9	1513.7	1812.7	2127.0
Carbon Sequestration	GtC												
Land Use [a]	Million ha												
Cropland		1459	1466	1462	1457	1454	1448	1442	1436	1429	1424	1422	1420
Grasslands		3389	3404	3429	3446	3458	3478	3498	3525	3552	3568	3572	3576
Energy Biomass		0	0	0	74	158	288	418	492	566	581	538	495
Forest		4296	4237	4173	4164	4164	4177	4190	4194	4199	4202	4203	4204
Others		3805	3842	3886	3807	3715	3558	3400	3301	3203	3173	3213	3253
Total		12949	12949	12949	12949	12949	12949	12949	12949	12949	12949	12949	12949
Anthropogenic Emissions (standardized)													
Fossil Fuel CO2 [b]	GtC	5.99	6.90	8.61	10.97	14.04	17.11	20.64	24.50	27.37	29.05	30.79	32.07
Other CO2 [c]	GtC	1.11	1.07	1.04	0.26	0.12	0.05	-0.02	-0.03	-0.03	-0.03	-0.01	0.00
Total CO2	GtC	7.10	7.97	9.65	11.23	14.17	17.16	20.62	24.47	27.33	29.03	30.78	32.07
CH4 total [d]	MtCH4	310	323	365	427	505	552	606	628	639	652	662	672
N2O total [e]	MtN2O-N	6.7	7.0	6.1	6.1	6.2	6.3	6.3	6.3	6.2	6.2	6.2	6.2
SOx total	MtS	70.9	69.0	77.3	102.4	121.0	126.0	120.2	106.2	97.9	79.6	67.0	49.8
CFC/HFC/HCFC	MtC eq.	1672	883	791	337	369	482	566	654	659	654	639	614
PFC	MtC eq.	32	25	31	43	61	77	89	97	106	114	119	115
SF6	MtC eq.	38	40	43	48	66	99	119	127	113	88	84	95
CO	MtCO	879	877	1005	1165	1415	1805	2277	2758	3217	3547	3772	3766
NMVOC	Mt	139	141	166	194	221	259	310	359	393	408	405	373
NOx	MtN	31	32	39	48	62	77	94	111	124	134	142	151

Emissions correlated to land-use change and deforestation were calculated by using AIM A1 marker land-use data.
a: Land-use taken from AIM-A1 marker run.
b:CO2 emissions from fossil fuel and industrial processes (MESSAGE data).
c: CO2 emissions from land-use changes (AIM-A1 marker run).
d: Non-energy related CH4 emissions were taken from AIM-A1 marker run.
e: Non-energy related N2O emissions were taken from AIM-A1 marker run.

Scenario A1C-MESSAGE OECD90		**1990**	**2000**	**2010**	**2020**	**2030**	**2040**	**2050**	**2060**	**2070**	**2080**	**2090**	**2100**
Population	Million	859	919	965	1007	1043	1069	1081	1084	1089	1098	1108	1110
GNP/GDP (mex)	Trillion US$	16.4	20.6	25.6	31.6	38.7	46.8	55.7	65.7	77.2	90.8	106.6	124.3
GNP/GDP (ppp) Trillion (1990 prices)		14.1	17.8	22.1	27.4	33.7	40.8	48.6	57.4	67.6	79.7	93.6	109.4
Final Energy	EJ												
Non-commercial		0	0	0	0	0	0	0	0	0	0	0	0
Solids		13	11	7	3	2	1	0	0	0	0	0	0
Liquids		66	70	74	74	75	80	87	93	100	98	102	102
Gas		22	28	34	35	39	37	34	29	30	27	21	15
Electricity		22	29	36	44	53	61	75	90	106	124	141	158
Others		1	1	4	8	11	18	24	30	34	37	42	47
Total		124	138	155	164	181	197	220	242	269	285	306	322
Primary Energy	EJ												
Coal		38	46	57	69	80	91	97	102	125	144	183	223
Oil		72	83	81	72	60	46	39	34	29	20	15	12
Gas		33	41	50	60	70	69	75	75	83	75	59	37
Nuclear		6	7	8	10	13	27	41	56	59	64	65	60
Biomass		6	7	6	7	10	14	20	25	28	33	39	44
Other Renewables		4	5	9	12	15	19	26	37	44	48	54	58
Total		159	188	211	228	249	267	299	329	368	383	414	435
Cumulative Resources Use	ZJ												
Coal		0.0	0.4	0.9	1.4	2.1	2.9	3.8	4.9	6.2	7.8	9.6	11.9
Oil		0.0	0.8	1.6	2.4	3.1	3.7	4.2	4.6	4.9	5.2	5.4	5.6
Gas		0.0	0.3	0.8	1.3	1.9	2.6	3.3	4.0	4.8	5.6	6.4	7.0
Cumulative CO2 Emissions	GtC	2.8	33.0	67.0	104.2	143.6	184.0	225.0	267.9	315.8	368.8	427.3	493.4
Carbon Sequestration	GtC												
Land Use [a]	Million ha												
Cropland		381	380	377	376	375	372	368	364	359	356	356	355
Grasslands		760	763	771	774	776	787	797	818	839	849	847	846
Energy Biomass		0	0	0	11	24	44	63	74	86	88	81	75
Forest		1050	1053	1052	1053	1056	1062	1068	1061	1054	1050	1049	1048
Others		838	833	828	814	798	765	733	712	691	685	695	705
Total		3029	3029	3029	3029	3029	3029	3029	3029	3029	3029	3029	3029
Anthropogenic Emissions (standardized)													
Fossil Fuel CO2 [b]	GtC	2.83	3.20	3.57	3.82	4.05	4.04	4.22	4.41	5.10	5.39	6.23	6.94
Other CO2 [c]	GtC	0.00	0.00	0.04	0.02	0.00	-0.02	-0.04	0.00	0.05	0.06	0.04	0.01
Total CO2	GtC	2.83	3.20	3.61	3.83	4.05	4.02	4.18	4.41	5.15	5.45	6.27	6.95
CH4 total [d]	MtCH4	73	74	74	76	80	85	92	101	114	126	147	164
N2O total [e]	MtN2O-N	2.6	2.6	2.4	2.4	2.4	2.4	2.4	2.4	2.4	2.4	2.4	2.4
SOx total	MtS	22.7	17.0	14.9	22.7	25.3	20.2	14.0	7.0	8.2	2.1	3.9	5.6
HFC	MtC eq.	19	58	110	108	115	118	122	123	123	124	125	125
PFC	MtC eq.	18	13	13	11	9	8	9	10	12	13	14	16
SF6	MtC eq.	23	23	14	5	6	7	9	11	13	16	18	20
CO	MtCO	179	161	173	167	162	188	220	241	266	267	271	281
NMVOC	Mt	42	36	38	36	33	34	35	38	44	43	39	35
NOx	MtN	13	12	14	15	15	15	16	16	20	22	25	28

Emissions correlated to land-use change and deforestation were calculated by using AIM A1 marker land-use data.

a: Land-use taken from AIM-A1 marker run.

b:CO2 emissions from fossil fuel and industrial processes (MESSAGE data).

c: CO2 emissions from land-use changes (AIM-A1 marker run).

d: Non-energy related CH4 emissions were taken from AIM-A1 marker run.

e: Non-energy related N2O emissions were taken from AIM-A1 marker run.

Scenario A1C-MESSAGE REF		1990	2000	2010	2020	2030	2040	2050	2060	2070	2080	2090	2100
Population	Million	413	419	427	433	435	433	423	409	392	374	357	339
GNP/GDP (mex)	Trillion US$	1.1	0.8	1.0	2.1	5.4	9.4	12.6	16.2	20.0	24.4	29.2	34.4
GNP/GDP (ppp) Trillion (1990 prices)		2.6	2.2	2.5	3.7	6.0	9.4	12.6	16.2	20.0	24.4	29.2	34.4
Final Energy	EJ												
Non-commercial		0	0	0	0	0	0	0	0	0	0	0	0
Solids		9	5	3	2	2	1	0	0	0	0	0	0
Liquids		15	10	11	15	19	23	25	25	27	32	34	32
Gas		14	10	12	16	19	19	18	20	16	10	4	4
Electricity		6	5	6	8	14	19	23	27	32	37	42	45
Others		7	6	6	8	11	11	13	15	16	15	18	18
Total		50	36	38	48	65	73	80	87	91	95	98	98
Primary Energy	EJ												
Coal		19	13	13	15	21	26	31	37	49	70	104	146
Oil		20	14	13	15	17	19	19	12	9	6	5	4
Gas		27	21	22	30	41	45	47	62	61	48	19	12
Nuclear		1	1	1	1	1	2	5	6	7	7	8	8
Biomass		2	1	1	1	3	4	6	7	10	14	15	19
Other Renewables		1	1	2	3	5	6	8	10	13	15	17	19
Total		70	51	51	65	87	101	115	134	149	161	168	208
Cumulative Resources Use	ZJ												
Coal		0.0	0.2	0.3	0.4	0.6	0.8	1.1	1.4	1.9	2.6	3.6	4.9
Oil		0.0	0.2	0.3	0.5	0.6	0.8	1.0	1.2	1.3	1.4	1.5	1.5
Gas		0.0	0.3	0.5	0.7	1.0	1.4	1.9	2.3	3.0	3.6	4.1	4.3
Cumulative CO2 Emissions	GtC	1.3	12.3	22.3	33.3	46.3	61.8	78.8	98.0	120.6	146.8	177.1	215.4
Carbon Sequestration	GtC												
Land Use [a]	Million ha												
Cropland		268	266	265	265	265	264	263	262	262	261	260	260
Grasslands		341	361	364	366	367	368	370	371	371	372	373	374
Energy Biomass		0	0	0	3	6	11	16	19	22	23	21	20
Forest		966	950	918	904	894	899	905	909	912	916	922	927
Others		701	698	728	738	745	733	722	715	709	703	700	696
Total		2276	2276	2276	2276	2276	2276	2276	2276	2276	2276	2276	2276
Anthropogenic Emissions (standardized)													
Fossil Fuel CO2 [b]	GtC	1.30	0.91	0.90	1.12	1.48	1.70	1.90	2.21	2.52	2.88	3.33	4.44
Other CO2 [c]	GtC	0.00	0.00	0.18	0.01	0.00	-0.08	-0.15	-0.12	-0.09	-0.07	-0.06	-0.05
Total CO2	GtC	1.30	0.91	1.08	1.13	1.47	1.63	1.75	2.10	2.43	2.81	3.27	4.38
CH4 total [d]	MtCH4	47	39	43	51	61	61	62	70	74	81	93	121
N2O total [e]	MtN2O-N	0.6	0.6	0.6	0.6	0.6	0.6	0.6	0.6	0.6	0.6	0.6	0.6
SOx total	MtS	17.0	11.0	11.3	12.8	14.0	14.3	14.4	12.9	16.9	22.5	20.8	12.4
HFC	MtC eq.	0	4	8	19	29	31	32	33	33	34	33	31
PFC	MtC eq.	7	4	5	8	14	20	21	22	23	24	25	24
SF6	MtC eq.	8	6	8	10	14	18	21	19	15	14	10	11
CO	MtCO	69	41	45	59	82	108	129	150	144	136	134	143
NMVOC	Mt	16	13	15	19	25	31	36	43	45	44	37	32
NOx	MtN	5	3	3	4	6	8	9	10	11	12	12	13

Emissions correlated to land-use change and deforestation were calculated by using AIM A1 marker land-use data.
a: Land-use taken from AIM-A1 marker run.
b:CO2 emissions from fossil fuel and industrial processes (MESSAGE data).
c: CO2 emissions from land-use changes (AIM-A1 marker run).
d: Non-energy related CH4 emissions were taken from AIM-A1 marker run.
e: Non-energy related N2O emissions were taken from AIM-A1 marker run.

Scenario A1C-MESSAGE ASIA		**1990**	**2000**	**2010**	**2020**	**2030**	**2040**	**2050**	**2060**	**2070**	**2080**	**2090**	**2100**
Population	Million	2798	3261	3620	3937	4147	4238	4220	4085	3867	3589	3258	2882
GNP/GDP (mex)	Trillion US$	1.5	2.7	5.8	13.5	27.2	44.9	65.3	95.8	126.9	155.5	186.5	218.2
GNP/GDP (ppp) Trillion (1990 prices)		5.3	8.3	13.5	22.8	36.2	52.2	70.0	96.8	126.9	155.5	186.5	218.2
Final Energy	EJ												
Non-commercial		24	16	12	8	5	3	2	0	0	0	0	0
Solids		19	36	53	69	75	62	48	42	39	34	29	19
Liquids		13	23	39	59	93	143	199	259	281	292	295	262
Gas		2	3	6	10	12	18	26	31	34	34	29	23
Electricity		4	7	13	25	47	74	104	145	187	223	249	262
Others		1	3	6	11	18	23	28	37	47	58	67	85
Total		62	88	128	182	249	323	407	513	588	641	668	651
Primary Energy	EJ												
Coal		30	48	68	98	135	191	250	315	377	408	425	386
Oil		15	26	40	54	72	81	80	86	62	40	25	17
Gas		3	5	12	20	30	37	41	40	30	21	15	15
Nuclear		0	1	3	8	20	34	54	83	125	159	175	174
Biomass		24	22	27	34	44	58	77	97	116	134	143	142
Other Renewables		1	3	6	10	18	27	37	54	69	80	89	101
Total		74	105	156	226	319	427	538	676	778	842	871	834
Cumulative Resources Use	ZJ												
Coal		0.0	0.3	0.8	1.5	2.5	3.8	5.7	8.2	11.3	15.3	19.7	24.1
Oil		0.0	0.2	0.5	0.8	1.4	2.1	2.9	3.7	4.6	5.2	5.6	5.8
Gas		0.0	0.0	0.1	0.2	0.4	0.7	1.1	1.5	1.9	2.2	2.4	2.6
Cumulative CO2 Emissions	GtC	1.5	19.3	44.3	78.6	125.2	188.5	269.1	367.5	481.4	604.1	729.9	849.7
Carbon Sequestration	GtC												
Land Use [a]	Million ha												
Cropland		438	435	434	433	432	431	431	430	429	428	428	428
Grasslands		608	606	609	611	613	615	617	618	619	621	622	624
Energy Biomass		0	0	0	17	37	68	99	117	134	138	127	117
Forest		535	522	512	524	535	535	535	544	552	556	555	555
Others		583	601	609	579	547	515	483	456	429	421	431	441
Total		2164	2164	2164	2164	2164	2164	2164	2164	2164	2164	2164	2164
Anthropogenic Emissions (standardized)													
Fossil Fuel CO2 [b]	GtC	1.15	1.78	2.74	4.03	5.54	7.28	8.87	10.80	12.13	12.59	12.70	11.28
Other CO2 [c]	GtC	0.37	0.26	0.22	-0.13	-0.12	-0.04	0.03	-0.03	-0.09	-0.10	-0.04	0.01
Total CO2	GtC	1.53	2.03	2.96	3.90	5.42	7.23	8.90	10.76	12.03	12.50	12.66	11.30
CH4 total [d]	MtCH4	113	125	152	186	231	252	275	274	283	278	277	260
N2O total [e]	MtN2O-N	2.3	2.6	2.4	2.4	2.5	2.6	2.6	2.5	2.5	2.5	2.5	2.5
SOx total	MtS	17.7	25.3	33.9	47.1	60.9	69.5	66.1	60.7	49.5	33.7	22.4	14.9
HFC	MtC eq.	0	5	18	45	92	153	224	292	292	285	275	262
PFC	MtC eq.	3	5	8	15	23	30	35	39	43	46	48	46
SF6	MtC eq.	4	7	12	19	28	42	50	55	48	35	33	37
CO	MtCO	235	270	348	463	621	855	1131	1414	1647	1821	1930	1778
NMVOC	Mt	33	37	49	66	80	101	128	157	180	193	197	168
NOx	MtN	7	9	12	17	24	32	40	48	52	56	60	58

Emissions correlated to land-use change and deforestation were calculated by using AIM A1 marker land-use data.
a: Land-use taken from AIM-A1 marker run.
b:CO2 emissions from fossil fuel and industrial processes (MESSAGE data).
c: CO2 emissions from land-use changes (AIM-A1 marker run).
d: Non-energy related CH4 emissions were taken from AIM-A1 marker run.
e: Non-energy related N2O emissions were taken from AIM-A1 marker run.

Scenario A1C-MESSAGE ALM		1990	2000	2010	2020	2030	2040	2050	2060	2070	2080	2090	2100
Population	Million	1192	1519	1875	2241	2557	2791	2980	3089	3115	3064	2934	2727
GNP/GDP (mex)	Trillion US$	1.9	2.7	4.4	9.8	20.0	34.3	53.5	76.5	98.7	122.5	147.2	173.1
GNP/GDP (ppp) Trillion (1990 prices)		3.8	5.1	7.5	13.3	22.8	36.7	55.3	76.5	98.7	122.5	147.2	173.1
Final Energy	EJ												
Non-commercial		14	10	7	8	5	4	3	0	0	0	0	0
Solids		1	5	9	7	11	16	24	26	21	20	15	7
Liquids		17	23	34	53	78	107	150	193	233	251	274	289
Gas		4	6	10	15	24	31	32	38	37	48	50	51
Electricity		3	5	8	16	31	56	90	133	168	204	236	262
Others		0	2	4	8	12	18	26	36	47	55	60	64
Total		39	51	72	106	161	231	325	426	505	578	634	672
Primary Energy	EJ												
Coal		5	5	7	12	24	46	86	147	199	249	273	307
Oil		21	31	39	50	60	63	71	70	71	47	34	23
Gas		8	13	23	39	66	97	120	121	81	63	52	55
Nuclear		0	0	0	2	7	15	27	46	73	118	166	191
Biomass		14	14	18	26	38	56	75	99	133	150	162	171
Other Renewables		2	4	7	12	19	30	46	65	79	92	99	102
Total		49	67	94	141	214	307	425	547	636	718	786	849
Cumulative Resources Use	ZJ												
Coal		0.0	0.0	0.1	0.2	0.3	0.5	1.0	1.7	2.8	4.1	5.6	7.4
Oil		0.0	0.3	0.6	1.0	1.5	2.1	2.7	3.4	4.1	4.8	5.3	5.6
Gas		0.0	0.1	0.2	0.5	0.8	1.5	2.5	3.6	4.8	5.6	6.1	6.6
Cumulative CO2 Emissions	GtC	1.4	17.8	37.0	58.8	86.7	124.2	174.5	239.4	314.1	394.1	478.4	568.6
Carbon Sequestration	GtC												
Land Use [a]	Million ha												
Cropland		371	385	384	383	382	381	381	380	379	379	378	378
Grasslands		1680	1673	1685	1695	1702	1708	1714	1718	1722	1726	1730	1733
Energy Biomass		0	0	0	42	90	165	240	282	324	333	308	284
Forest		1745	1711	1690	1683	1680	1681	1682	1681	1680	1678	1676	1674
Others		1684	1710	1720	1676	1625	1544	1463	1419	1374	1364	1387	1411
Total		5480	5480	5480	5480	5480	5480	5480	5480	5480	5480	5480	5480
Anthropogenic Emissions (standardized)													
Fossil Fuel CO2 [b]	GtC	0.72	1.01	1.40	2.01	2.97	4.09	5.65	7.09	7.62	8.20	8.53	9.42
Other CO2 [c]	GtC	0.73	0.82	0.60	0.36	0.25	0.19	0.14	0.12	0.10	0.08	0.06	0.03
Total CO2	GtC	1.45	1.83	2.00	2.37	3.22	4.28	5.78	7.20	7.72	8.28	8.59	9.45
CH4 total [d]	MtCH4	77	85	97	114	134	155	178	184	169	167	145	126
N2O total [e]	MtN2O-N	1.2	1.3	0.7	0.7	0.7	0.7	0.7	0.7	0.7	0.7	0.7	0.7
SOx total	MtS	10.5	12.8	14.1	16.7	17.9	19.1	22.7	22.6	20.2	18.2	16.8	13.9
HFC	MtC eq.	0	2	15	39	81	139	184	205	210	210	206	196
PFC	MtC eq.	4	4	5	9	14	19	23	26	28	30	31	30
SF6	MtC eq.	3	5	10	14	19	32	40	42	37	23	24	26
CO	MtCO	396	404	438	476	550	654	798	953	1160	1323	1437	1564
NMVOC	Mt	48	55	65	73	83	94	111	120	124	128	132	138
NOx	MtN	7	8	10	12	16	22	29	37	41	44	46	52

Emissions correlated to land-use change and deforestation were calculated by using AIM A1 marker land-use data.

a: Land-use taken from AIM-A1 marker run.

b:CO2 emissions from fossil fuel and industrial processes (MESSAGE data).

c: CO2 emissions from land-use changes (AIM-A1 marker run).

d: Non-energy related CH4 emissions were taken from AIM-A1 marker run.

e: Non-energy related N2O emissions were taken from AIM-A1 marker run.

Scenario A1C-MiniCAM World		**1990**	**2000**	**2010**	**2020**	**2030**	**2040**	**2050**	**2060**	**2070**	**2080**	**2090**	**2100**
Population	Million	5293	6100	6874	7618	8122	8484	8703	8623	8430	8126	7621	7137
GNP/GDP (mex)	Trillion US$	20.7	27.4	38.1	52.7	79.7	116.4	162.8	224.0	290.2	361.5	437.8	521.8
GNP/GDP (ppp) Trillion (1990 prices)		na	na	na	na	na	na	na	na	na	na	na	na
Final Energy	EJ												
Non-commercial		0	0	0	0	0	0	0	0	0	0	0	0
Solids		45	57	73	94	118	140	162	129	105	89	80	72
Liquids		121	126	137	157	170	205	260	301	338	370	388	405
Gas		52	63	77	96	92	85	76	76	77	80	68	57
Electricity		35	54	86	134	209	299	402	509	608	698	763	829
Others		0	0	0	0	0	0	0	0	0	0	0	0
Total		253	298	374	480	589	729	899	1015	1127	1236	1300	1363
Primary Energy	EJ												
Coal		88	116	152	199	353	543	769	805	821	818	888	958
Oil		131	135	146	163	125	94	68	75	83	93	63	34
Gas		70	85	125	190	216	231	238	251	263	274	215	157
Nuclear		24	25	34	49	82	110	133	154	175	196	204	212
Biomass		0	6	12	19	35	54	78	123	159	186	209	232
Other Renewables		24	24	27	32	44	66	99	160	220	281	338	395
Total		336	391	496	653	854	1098	1384	1567	1721	1847	1918	1988
Cumulative Resources Use	ZJ												
Coal		0.1	1.2	2.6	4.2	7.5	12.1	18.2	26.2	34.3	42.5	51.2	60.0
Oil		0.1	1.5	2.9	4.4	5.7	6.8	7.7	8.4	9.2	10.1	10.8	11.5
Gas		0.1	0.9	2.0	3.5	5.6	7.8	10.1	12.6	15.2	17.8	20.1	22.4
Cumulative CO2 Emissions	GtC	7.1	82.4	170.6	281.6	425.0	608.1	837.3	1095.5	1360.0	1625.5	1888.2	2148.1
Carbon Sequestration	GtC												
Land Use	Million ha												
Cropland		1472	1466	1473	1494	1477	1432	1359	1194	1038	891	776	661
Grasslands		3209	3348	3601	3968	4357	4657	4869	4662	4402	4090	3855	3619
Energy Biomass		0	4	10	20	58	115	190	340	445	505	575	646
Forest		4173	4214	4140	3951	3623	3348	3126	3385	3693	4049	4192	4336
Others		4310	4132	3939	3732	3650	3613	3620	3583	3586	3630	3767	3903
Total		13164	13164	13164	13164	13164	13164	13164	13164	13164	13164	13164	13164
Anthropogenic Emissions (standardized)													
Fossil Fuel CO2	GtC	5.99	6.90	8.58	10.99	14.54	19.10	24.45	25.66	26.42	26.72	27.13	27.70
Other CO2	GtC	1.11	1.07	1.08	1.55	1.60	1.38	0.92	0.61	0.22	-0.27	-1.04	-1.81
Total CO2	GtC	7.10	7.97	9.66	12.54	16.14	20.48	25.37	26.27	26.64	26.46	26.09	25.89
CH4 total	MtCH4	310	323	359	414	482	563	636	651	663	672	682	693
N2O total	MtN2O-N	6.7	7.0	8.0	9.3	10.9	12.7	14.4	14.8	15.1	15.3	15.8	16.2
SOx total	MtS	70.9	69.0	81.1	87.8	97.5	92.3	72.1	48.1	33.7	28.9	27.9	26.9
CFC/HFC/HCFC	MtC eq.	1672	883	791	337	369	482	566	654	659	654	639	614
PFC	MtC eq.	32	25	31	43	61	77	89	97	106	114	119	115
SF6	MtC eq.	38	40	43	48	66	99	119	127	113	88	84	95
CO	MtCO												
NMVOC	Mt												
NOx	MtN												

Scenario A1C-MiniCAM OECD90		**1990**	**2000**	**2010**	**2020**	**2030**	**2040**	**2050**	**2060**	**2070**	**2080**	**2090**	**2100**
Population	Million	838	908	965	1007	1024	1066	1081	1084	1090	1098	1105	1112
GNP/GDP (mex)	Trillion US$	16.3	20.5	25.5	31.5	34.3	44.4	53.1	62.6	73.3	85.3	100.4	116.7
GNP/GDP (ppp) Trillion (1990 prices)		na	na	na	na	na	na	na	na	na	na	na	na
Final Energy	EJ												
Non-commercial		0	0	0	0	0	0	0	0	0	0	0	0
Solids		10	12	14	14	15	17	19	16	13	12	12	12
Liquids		72	73	74	74	69	62	68	70	74	78	82	86
Gas		27	36	45	54	50	40	33	31	31	32	28	24
Electricity		22	28	37	48	54	71	85	95	106	118	130	141
Others		0	0	0	0	0	0	0	0	0	0	0	0
Total		130	150	170	189	187	190	205	212	224	239	251	263
Primary Energy	EJ												
Coal		40	47	49	47	59	98	130	163	203	248	344	441
Oil		76	78	77	75	61	25	11	5	3	2	2	1
Gas		34	47	67	94	92	83	76	73	72	72	51	31
Nuclear		20	16	14	16	17	19	19	21	23	27	30	32
Biomass		0	2	4	5	6	10	14	20	26	30	35	40
Other Renewables		12	11	10	10	11	13	16	23	31	39	50	61
Total		182	200	222	247	244	249	266	306	357	417	512	607
Cumulative Resources Use	ZJ												
Coal		0.0	0.5	1.0	1.5	1.8	3.0	4.1	5.6	7.5	9.6	12.9	16.1
Oil		0.1	0.8	1.6	2.4	2.6	3.3	3.5	3.5	3.6	3.6	3.6	3.6
Gas		0.0	0.5	1.1	1.8	2.3	3.6	4.4	5.1	5.8	6.6	7.1	7.7
Cumulative CO2 Emissions	GtC	2.8	33.0	67.4	106.3	147.8	191.6	239.7	292.3	348.9	410.3	480.1	562.2
Carbon Sequestration	GtC												
Land Use	Million ha												
Cropland		408	410	412	414	408	383	357	310	269	233	203	174
Grasslands		796	820	866	934	967	1050	1084	1043	996	945	901	857
Energy Biomass		0	4	7	10	16	37	53	83	105	119	138	157
Forest		921	931	921	894	869	806	779	845	910	972	995	1018
Others		998	959	917	872	863	847	850	842	843	854	885	917
Total		3123	3123	3123	3123	3123	3123	3123	3123	3123	3123	3123	3123
Anthropogenic Emissions (standardized)													
Fossil Fuel CO2	GtC	2.83	3.20	3.55	3.86	3.93	4.39	4.99	5.39	5.93	6.60	7.88	9.29
Other CO2	GtC	0.00	0.00	0.13	0.25	0.25	0.18	0.08	0.05	-0.04	-0.20	-0.31	-0.43
Total CO2	GtC	2.83	3.20	3.69	4.11	4.18	4.57	5.07	5.44	5.88	6.39	7.57	8.86
CH4 total	MtCH4	73	74	84	97	104	126	145	159	178	201	236	270
N2O total	MtN2O-N	2.6	2.6	2.8	3.0	3.2	3.7	4.0	4.0	4.1	4.1	4.2	4.3
SOx total	MtS	22.7	17.0	14.0	4.8	4.0	3.9	4.2	4.1	4.3	4.8	5.3	5.8
HFC	MtC eq.	19	58	110	108	115	118	122	123	123	124	125	125
PFC	MtC eq.	18	13	13	11	9	8	9	10	12	13	14	16
SF6	MtC eq.	23	23	14	5	6	7	9	11	13	16	18	20
CO	MtCO												
NMVOC	Mt												
NOx	MtN												

Scenario A1C-MiniCAM REF		1990	2000	2010	2020	2030	2040	2050	2060	2070	2080	2090	2100
Population	Million	428	425	426	433	434	431	423	408	392	374	357	340
GNP/GDP (mex)	Trillion US$	1.1	1.1	1.4	2.1	3.5	5.0	6.9	9.9	13.4	17.4	21.4	25.8
GNP/GDP (ppp) Trillion (1990 prices)		na	na	na	na	na	na	na	na	na	na	na	na
Final Energy	EJ												
Non-commercial		0	0	0	0	0	0	0	0	0	0	0	0
Solids		13	10	9	9	10	11	11	9	7	6	5	5
Liquids		18	12	9	10	11	13	14	16	17	18	18	18
Gas		19	15	13	15	15	13	11	11	10	10	8	6
Electricity		6	8	13	21	30	39	48	57	64	71	73	76
Others		0	0	0	0	0	0	0	0	0	0	0	0
Total		56	44	44	55	65	75	85	92	99	105	105	104
Primary Energy	EJ												
Coal		18	17	20	25	68	127	201	175	196	264	228	192
Oil		20	13	11	12	4	0	0	0	0	0	0	0
Gas		26	20	21	30	33	32	28	29	28	27	21	14
Nuclear		3	4	6	10	14	16	17	18	19	20	19	19
Biomass		0	1	1	2	5	7	10	16	20	23	25	27
Other Renewables		3	3	3	4	6	8	11	17	23	29	33	38
Total		70	58	63	84	129	190	267	254	286	362	326	290
Cumulative Resources Use	ZJ												
Coal		0.0	0.2	0.4	0.6	1.2	2.2	3.7	5.5	7.5	9.7	12.0	14.4
Oil		0.0	0.2	0.3	0.4	0.5	0.5	0.5	0.5	0.5	0.5	0.5	0.5
Gas		0.0	0.2	0.5	0.7	1.0	1.3	1.6	1.9	2.2	2.5	2.7	2.9
Cumulative CO2 Emissions	GtC	1.3	12.3	21.7	33.1	50.1	75.8	111.6	151.1	190.0	234.3	279.1	316.2
Carbon Sequestration	GtC												
Land Use	Million ha												
Cropland		284	294	304	316	320	315	303	263	224	186	162	138
Grasslands		395	410	453	525	604	666	710	655	593	524	491	458
Energy Biomass		0	0	2	5	16	29	44	70	86	91	97	102
Forest		1007	1016	996	945	864	800	752	827	912	1005	1031	1057
Others		691	657	622	586	573	567	569	562	563	571	597	622
Total		2377	2377	2377	2377	2377	2377	2377	2377	2377	2377	2377	2377
Anthropogenic Emissions (standardized)													
Fossil Fuel CO2	GtC	1.30	0.91	0.94	1.23	1.97	2.91	4.05	3.67	3.93	4.84	4.20	3.60
Other CO2	GtC	0.00	0.00	0.03	0.08	0.12	0.12	0.09	0.09	0.07	0.02	-0.12	-0.26
Total CO2	GtC	1.30	0.91	0.97	1.31	2.09	3.03	4.14	3.76	4.01	4.87	4.08	3.34
CH4 total	MtCH4	47	39	49	69	94	123	154	141	149	178	160	143
N2O total	MtN2O-N	0.6	0.6	0.7	0.9	1.2	1.4	1.6	1.6	1.6	1.5	1.6	1.6
SOx total	MtS	17.0	11.0	10.4	10.9	14.4	14.6	11.4	6.0	2.8	2.0	2.0	2.0
HFC	MtC eq.	0	4	8	19	29	31	32	33	33	34	33	31
PFC	MtC eq.	7	4	5	8	14	20	21	22	23	24	25	24
SF6	MtC eq.	8	6	8	10	14	18	21	19	15	14	10	11
CO	MtCO												
NMVOC	Mt												
NOx	MtN												

Scenario A1C-MiniCAM ASIA		**1990**	**2000**	**2010**	**2020**	**2030**	**2040**	**2050**	**2060**	**2070**	**2080**	**2090**	**2100**
Population	Million	2790	3226	3608	3937	4115	4210	4219	4062	3852	3589	3245	2919
GNP/GDP (mex)	Trillion US$	1.4	3.1	6.5	11.7	23.7	40.0	60.5	85.0	110.3	136.4	162.6	191.2
GNP/GDP (ppp) Trillion (1990 prices)		na	na	na	na	na	na	na	na	na	na	na	na
Final Energy	EJ												
Non-commercial		0	0	0	0	0	0	0	0	0	0	0	0
Solids		20	31	46	64	81	97	110	86	68	56	49	42
Liquids		14	19	27	37	49	66	88	102	113	122	124	127
Gas		2	5	9	13	13	14	14	15	15	16	13	11
Electricity		4	11	25	47	82	123	170	215	255	288	308	328
Others		0	0	0	0	0	0	0	0	0	0	0	0
Total		40	66	107	160	226	300	381	418	451	482	495	509
Primary Energy	EJ												
Coal		26	45	73	111	177	252	336	362	313	191	193	195
Oil		16	21	29	39	33	27	21	24	28	35	18	2
Gas		3	9	21	40	55	67	78	82	87	92	74	55
Nuclear		1	4	9	17	32	47	62	70	78	84	85	86
Biomass		0	2	5	9	17	26	35	51	64	73	80	86
Other Renewables		3	4	5	6	11	22	37	64	90	115	136	157
Total		49	85	142	223	327	442	568	653	661	590	586	581
Cumulative Resources Use	ZJ												
Coal		0.0	0.4	1.1	1.9	3.6	5.8	8.5	12.1	15.3	18.0	20.0	21.9
Oil		0.0	0.2	0.5	0.8	1.1	1.4	1.7	1.9	2.2	2.5	2.7	2.9
Gas		0.0	0.1	0.3	0.5	1.0	1.6	2.4	3.2	4.0	4.9	5.7	6.5
Cumulative CO2 Emissions	GtC	1.5	19.3	44.5	81.8	136.0	209.6	303.9	410.9	516.5	609.1	690.1	766.4
Carbon Sequestration	GtC												
Land Use	Million ha												
Cropland		389	400	410	420	418	409	392	346	303	263	229	195
Grasslands		508	524	555	602	648	685	714	704	687	662	632	603
Energy Biomass		0	0	2	5	18	32	49	79	99	109	126	143
Forest		1168	1144	1102	1043	986	942	910	936	976	1032	1079	1127
Others		664	632	600	568	555	550	553	547	548	556	579	601
Total		2729	2699	2669	2637	2625	2618	2618	2612	2613	2622	2645	2668
Anthropogenic Emissions (standardized)													
Fossil Fuel CO2	GtC	1.15	1.78	2.80	4.20	6.06	8.09	10.29	10.80	10.15	8.36	8.01	7.68
Other CO2	GtC	0.37	0.26	0.20	0.27	0.29	0.27	0.20	0.12	0.04	-0.03	-0.15	-0.27
Total CO2	GtC	1.53	2.03	3.00	4.47	6.36	8.36	10.49	10.92	10.19	8.33	7.87	7.41
CH4 total	MtCH4	113	125	135	148	164	183	206	217	197	145	147	149
N2O total	MtN2O-N	2.3	2.6	2.9	3.3	3.8	4.3	4.8	4.8	4.9	4.9	5.0	5.1
SOx total	MtS	17.7	25.3	38.8	52.8	57.3	50.9	33.6	21.4	14.0	11.3	10.4	9.4
HFC	MtC eq.	0	5	18	45	92	153	224	292	292	285	275	262
PFC	MtC eq.	3	5	8	15	23	30	35	39	43	46	48	46
SF6	MtC eq.	4	7	12	19	28	42	50	55	48	35	33	37
CO	MtCO												
NMVOC	Mt												
NOx	MtN												

Scenario A1C-MiniCAM ALM		1990	2000	2010	2020	2030	2040	2050	2060	2070	2080	2090	2100
Population	Million	1236	1541	1876	2241	2531	2778	2980	3068	3096	3064	2913	2766
GNP/GDP (mex)	Trillion US$	1.9	2.8	4.6	7.4	15.4	27.0	42.3	66.5	93.2	122.4	153.4	188.0
GNP/GDP (ppp) Trillion (1990 prices)		na	na	na	na	na	na	na	na	na	na	na	na
Final Energy	EJ												
Non-commercial		0	0	0	0	0	0	0	0	0	0	0	0
Solids		2	3	5	7	11	16	21	18	16	15	14	13
Liquids		17	22	28	36	46	64	90	113	134	153	164	175
Gas		5	7	10	14	17	19	18	19	20	22	19	16
Electricity		3	6	11	18	38	65	99	141	182	221	252	283
Others		0	0	0	0	0	0	0	0	0	0	0	0
Total		27	38	54	75	113	164	228	293	353	410	449	487
Primary Energy	EJ												
Coal		4	6	10	15	37	66	102	105	109	116	123	129
Oil		20	23	29	37	41	41	37	46	52	57	43	30
Gas		7	10	16	26	39	49	57	67	76	84	70	56
Nuclear		0	2	5	7	17	26	35	45	55	65	70	76
Biomass		0	1	1	3	6	11	19	35	49	60	70	79
Other Renewables		5	6	8	11	16	23	34	55	76	98	119	140
Total		35	48	69	99	156	218	284	353	418	479	494	510
Cumulative Resources Use	ZJ												
Coal		0.0	0.1	0.1	0.3	0.6	1.1	1.9	3.0	4.0	5.1	6.4	7.6
Oil		0.0	0.2	0.5	0.8	1.2	1.6	2.0	2.5	3.0	3.5	4.0	4.4
Gas		0.0	0.1	0.2	0.4	0.8	1.2	1.7	2.4	3.1	3.9	4.6	5.3
Cumulative CO2 Emissions	GtC	1.4	17.8	37.0	60.3	91.1	131.2	182.1	241.2	304.7	371.8	439.0	503.3
Carbon Sequestration	GtC												
Land Use	Million ha												
Cropland		391	363	348	344	337	324	307	275	243	209	182	154
Grasslands		1510	1594	1727	1907	2105	2256	2361	2260	2126	1960	1830	1701
Energy Biomass		0	0	0	0	2	16	44	108	155	185	215	244
Forest		3641	3591	3480	3307	3114	2957	2834	2900	3020	3192	3321	3450
Others		1957	1883	1800	1707	1668	1649	1649	1631	1631	1649	1706	1762
Total		7499	7432	7354	7265	7226	7202	7195	7175	7175	7195	7254	7312
Anthropogenic Emissions (standardized)													
Fossil Fuel CO2	GtC	0.72	1.01	1.29	1.70	2.57	3.71	5.11	5.80	6.40	6.92	7.03	7.14
Other CO2	GtC	0.73	0.82	0.72	0.95	0.93	0.80	0.55	0.35	0.15	-0.06	-0.46	-0.86
Total CO2	GtC	1.45	1.83	2.01	2.65	3.51	4.51	5.67	6.15	6.55	6.87	6.57	6.28
CH4 total	MtCH4	77	85	90	101	121	131	132	134	139	147	139	130
N2O total	MtN2O-N	1.2	1.3	1.7	2.1	2.7	3.4	4.0	4.3	4.6	4.7	5.0	5.2
SOx total	MtS	10.5	12.8	14.9	16.2	18.8	19.9	19.8	13.6	9.6	7.9	7.2	6.6
HFC	MtC eq.	0	2	15	39	81	139	184	205	210	210	206	196
PFC	MtC eq.	4	4	5	9	14	19	23	26	28	30	31	30
SF6	MtC eq.	3	5	10	14	19	32	40	42	37	23	24	26
CO	MtCO												
NMVOC	Mt												
NOx	MtN												

Scenario A1G-AIM World		**1990**	**2000**	**2010**	**2020**	**2030**	**2040**	**2050**	**2060**	**2070**	**2080**	**2090**	**2100**
Population	Million	5262	6117	6805	7493	8182	8439	8704	8538	8375	8030	7528	7056
GNP/GDP (mex)	Trillion US$	20.9	26.6	37.9	56.6	89.7	128.2	183.4	237.8	308.4	382.3	453.3	537.3
GNP/GDP (ppp) Trillion (1990 prices)													
Final Energy	EJ												
Non-commercial		50	48	37	35	26	0	0	0	0	0	0	0
Solids		36	34	47	65	83	88	93	83	74	66	57	48
Liquids		111	124	159	206	273	342	412	477	543	589	616	643
Gas		51	60	86	115	150	204	258	303	348	379	396	413
Electricity		38	50	71	98	142	220	299	381	464	523	560	596
Others		0			0			0					0
Total		285	315	400	519	674	868	1061	1245	1429	1556	1628	1700
Primary Energy	EJ												
Coal		93	101	140	179	218	212	206	193	180	168	155	144
Oil		143	167	221	290	385	472	582	600	620	640	659	679
Gas		73	99	146	206	285	397	560	659	778	866	910	956
Nuclear		6	10	15	26	45	69	117	140	168	192	208	226
Biomass		50	48	38	34	35	50	75	126	210	277	286	295
Other Renewables		10	11	13	16	24	37	61	95	150	200	225	253
Total		376	435	573	752	993	1237	1601	1813	2106	2342	2442	2554
Cumulative Resources Use	ZJ												
Coal		0.1	1.0	2.3	3.8	5.8	7.7	10.1	11.8	13.7	15.6	17.1	18.8
Oil		0.1	1.7	3.6	6.2	9.6	13.5	19.2	24.2	30.6	37.2	43.4	50.8
Gas		0.1	0.9	2.2	3.9	6.4	9.6	14.8	20.2	27.6	36.0	44.4	54.9
Cumulative CO2 Emissions	GtC	7.1	82.4	177.9	307.7	472.1	669.4	904.7	1168.2	1447.0	1739.1	2039.3	2344.5
Carbon Sequestraction	GtC												
Land Use	Million ha												
Cropland		1437	1487	1546	1580	1554	1469	1389	1266	1154	1052	959	874
Grasslands		3290	3287	3474	3874	4315	4328	4341	4184	4032	3914	3827	3741
Energy Biomass		0	0	0	2	68	154	348	502	723	879	901	923
Forest		4249	4120	3979	3816	3767	3812	3858	3952	4049	4140	4224	4310
Others		3966	4048	3944	3671	3238	3120	3006	2969	2933	2950	3021	3093
Total		12942	12942	12942	12942	12942	12942	12942	12942	12942	12942	12942	12942
Anthropogenic Emissions (standardized)													
Fossil Fuel CO2	GtC	5.99	6.90	9.67	13.09	17.49	21.11	25.58	27.11	28.77	29.88	30.33	30.80
Other CO2	GtC	1.11	1.07	1.45	1.78	0.52	0.33	0.05	-0.03	-0.10	-0.11	-0.07	-0.01
Total CO2	GtC	7.10	7.97	11.11	14.87	18.01	21.44	25.62	27.08	28.67	29.76	30.26	30.79
CH4 total	MtCH4	310	323	393	470	541	525	511	449	396	353	319	289
N2O total	MtN2O-N	6.7	7.0	7.4	7.9	7.8	7.6	7.5	7.0	6.6	6.3	6.1	5.9
SOx total	MtS	70.9	69.0	87.4	100.9	91.4	69.3	64.4	47.0	35.7	30.7	28.9	27.4
CFC/HFC/HCFC	MtC eq.	1672	883	791	337	369	482	566	654	659	654	639	614
PFC	MtC eq.	32	25	31	43	61	77	89	97	106	114	119	115
SF6	MtC eq.	38	40	43	48	66	99	119	127	113	88	84	95
CO	MtCO	879	877	1226	1552	1747	2008	2307	2546	2812	3086	3364	3666
NMVOC	Mt	139	141	181	230	269	273	279	283	288	268	227	192
NOx	MtN	31	32	40	49	52	50	49	47	45	43	41	40

Scenario A1G-AIM OECD90		**1990**	**2000**	**2010**	**2020**	**2030**	**2040**	**2050**	**2060**	**2070**	**2080**	**2090**	**2100**
Population	Million	859	919	960	1002	1043	1062	1081	1086	1091	1097	1103	1110
GNP/GDP (mex)	Trillion US$	16.4	20.4	25.2	31.1	38.3	45.7	54.7	64.7	76.7	90.1	105.2	122.8
GNP/GDP (ppp) Trillion (1990 prices)													
Final Energy	EJ												
Non-commercial		6	1	0	0	0	0	0	0	0	0	0	0
Solids		10	9	9	9	10	9	9	8	7	6	6	5
Liquids		64	75	79	85	91	97	104	107	111	115	120	125
Gas		25	34	39	43	47	53	59	65	71	76	80	83
Electricity		22	28	32	36	42	49	57	67	77	86	96	105
Others		0			0			0					0
Total		127	147	159	173	189	209	229	248	266	284	301	318
Primary Energy	EJ												
Coal		41	42	43	39	33	34	35	32	29	28	27	27
Oil		76	92	98	106	112	121	131	127	124	125	130	136
Gas		34	51	60	68	75	90	108	120	134	147	158	170
Nuclear		5	7	10	14	19	20	20	23	26	30	34	38
Biomass		6	1	0	0	2	6	14	24	40	53	55	57
Other Renewables		6	6	7	8	10	12	14	20	27	35	40	46
Total		167	199	216	233	251	282	322	346	380	416	443	473
Cumulative Resources Use	ZJ												
Coal		0.0	0.4	0.9	1.3	1.6	1.9	2.3	2.6	2.9	3.2	3.5	3.8
Oil		0.1	0.9	1.9	2.9	4.0	5.0	6.4	7.5	8.8	10.1	11.4	12.8
Gas		0.0	0.5	1.0	1.7	2.4	3.2	4.2	5.2	6.5	8.0	9.4	11.2
Cumulative CO2 Emissions	GtC	2.8	33.0	65.5	100.6	139.4	182.6	232.5	286.1	340.0	394.3	449.3	505.0
Carbon Sequestraction	GtC												
Land Use	Million ha												
Cropland		410	404	389	373	350	336	321	286	255	227	202	180
Grasslands		787	758	750	783	834	843	852	817	783	754	728	703
Energy Biomass		0	0	0	0	7	20	56	78	108	132	142	152
Forest		1056	1065	1103	1133	1176	1183	1190	1231	1274	1312	1344	1378
Others		886	912	897	850	772	746	721	716	711	712	720	728
Total		3140	3140	3140	3140	3140	3140	3140	3140	3140	3140	3140	3140
Anthropogenic Emissions (standardized)													
Fossil Fuel CO2	GtC	2.83	3.20	3.42	3.78	4.14	4.68	5.30	5.39	5.48	5.56	5.62	5.68
Other CO2	GtC	0.00	0.00	-0.12	-0.05	-0.12	-0.05	0.04	-0.02	-0.07	-0.10	-0.08	-0.07
Total CO2	GtC	2.83	3.20	3.30	3.73	4.02	4.63	5.34	5.37	5.41	5.46	5.54	5.61
CH4 total	MtCH4	73	74	71	71	71	63	56	51	47	43	41	38
N2O total	MtN2O-N	2.6	2.6	2.5	2.6	2.5	2.4	2.3	2.1	1.9	1.8	1.8	1.8
SOx total	MtS	22.7	17.0	9.9	6.9	6.4	6.3	6.2	5.9	5.6	5.3	4.9	4.6
HFC	MtC eq.	19	58	110	108	115	118	122	123	123	124	125	125
PFC	MtC eq.	18	13	13	11	9	8	9	10	12	13	14	16
SF6	MtC eq.	23	23	14	5	6	7	9	11	13	16	18	20
CO	MtCO	179	161	175	191	213	231	250	260	270	286	307	330
NMVOC	Mt	42	36	37	39	39	34	30	25	21	18	17	16
NOx	MtN	13	12	12	12	9	7	6	6	6	5	5	5

Scenario A1G-AIM REF		**1990**	**2000**	**2010**	**2020**	**2030**	**2040**	**2050**	**2060**	**2070**	**2080**	**2090**	**2100**
Population	Million	413	419	424	430	435	429	423	406	391	374	356	339
GNP/GDP (mex)	Trillion US$	1.1	0.8	1.5	2.9	5.3	8.2	12.5	15.7	19.8	24.2	29.0	34.6
GNP/GDP (ppp) Trillion (1990 prices)													
Final Energy	EJ												
Non-commercial		2	5	0	0	0	0	0	0	0	0	0	0
Solids		9	4	4	5	6	6	6	5	4	3	3	2
Liquids		19	11	11	11	13	14	16	17	18	20	21	22
Gas		19	14	20	23	27	37	46	49	52	53	53	53
Electricity		8	9	11	13	16	20	25	27	29	30	30	30
Others		0			0			0					0
Total		58	43	47	53	61	77	92	97	102	105	106	106
Primary Energy	EJ												
Coal		18	12	14	14	14	14	15	13	11	9	8	7
Oil		22	16	17	16	18	20	23	22	22	22	23	24
Gas		26	22	30	35	41	52	65	69	73	75	76	76
Nuclear		1	1	2	4	5	7	9	10	10	10	11	11
Biomass		2	4	0	0	1	1	3	6	9	12	13	13
Other Renewables		1	1	1	2	2	3	5	7	9	11	12	13
Total		71	57	64	71	80	98	121	126	135	141	142	144
Cumulative Resources Use	ZJ												
Coal		0.0	0.2	0.3	0.4	0.6	0.7	0.9	1.0	1.1	1.2	1.3	1.4
Oil		0.0	0.2	0.4	0.5	0.7	0.9	1.1	1.3	1.5	1.8	2.0	2.2
Gas		0.0	0.3	0.5	0.9	1.2	1.7	2.3	2.9	3.6	4.4	5.1	5.9
Cumulative CO2 Emissions	GtC	1.3	12.3	22.3	33.4	45.4	59.2	75.5	92.9	109.5	125.5	141.3	157.0
Carbon Sequestraction	GtC												
Land Use	Million ha												
Cropland		279	284	296	298	298	287	276	260	245	225	202	181
Grasslands		346	357	400	466	551	563	575	565	555	540	521	502
Energy Biomass		0	0	0	0	7	15	31	38	46	46	38	31
Forest		960	961	951	952	951	957	963	984	1007	1034	1068	1103
Others		720	703	658	588	497	478	460	455	450	456	471	487
Total		2305	2305	2305	2305	2305	2305	2305	2305	2305	2305	2305	2305
Anthropogenic Emissions (standardized)													
Fossil Fuel CO2	GtC	1.30	0.91	1.06	1.14	1.26	1.50	1.79	1.74	1.69	1.67	1.67	1.68
Other CO2	GtC	0.00	0.00	0.03	0.00	0.00	-0.01	-0.02	-0.04	-0.06	-0.08	-0.10	-0.12
Total CO2	GtC	1.30	0.91	1.09	1.14	1.26	1.49	1.78	1.70	1.63	1.58	1.57	1.56
CH4 total	MtCH4	47	39	60	66	68	59	51	49	47	44	42	39
N2O total	MtN2O-N	0.6	0.6	0.6	0.7	0.7	0.6	0.6	0.6	0.5	0.5	0.4	0.4
SOx total	MtS	17.0	11.0	12.2	10.8	7.5	4.3	2.4	2.0	1.7	1.6	1.6	1.6
HFC	MtC eq.	0	4	8	19	29	31	32	33	33	34	33	31
PFC	MtC eq.	7	4	5	8	14	20	21	22	23	24	25	24
SF6	MtC eq.	8	6	8	10	14	18	21	19	15	14	10	11
CO	MtCO	69	41	42	45	50	55	61	64	67	71	76	80
NMVOC	Mt	16	13	15	13	13	14	15	15	16	16	16	16
NOx	MtN	5	3	3	3	3	2	2	2	2	1	1	1

Scenario A1G-AIM ASIA		1990	2000	2010	2020	2030	2040	2050	2060	2070	2080	2090	2100
Population	Million	2798	3261	3556	3851	4147	4183	4220	4016	3822	3541	3194	2882
GNP/GDP (mex)	Trillion US$	1.5	2.7	5.8	12.3	26.3	40.8	63.4	86.3	117.5	149.4	177.4	210.6
GNP/GDP (ppp) Trillion (1990 prices)													
Final Energy	EJ												
Non-commercial		28	23	19	15	4	0	0	0	0	0	0	0
Solids		15	19	32	45	59	63	66	60	54	48	42	35
Liquids		11	15	30	49	77	100	124	155	187	210	224	238
Gas		2	4	9	16	27	42	56	75	94	108	119	129
Electricity		5	7	15	27	46	76	106	143	180	208	227	246
Others		0			0			0					0
Total		61	68	105	153	213	283	352	433	514	574	611	648
Primary Energy	EJ												
Coal		30	38	68	99	128	117	107	100	94	86	78	70
Oil		17	24	45	74	114	144	182	199	217	233	245	257
Gas		4	8	18	35	61	97	156	196	247	289	313	340
Nuclear		0	1	2	5	13	23	43	54	66	78	86	95
Biomass		28	24	19	14	7	11	17	29	48	64	66	68
Other Renewables		1	2	2	3	5	10	20	34	56	78	90	103
Total		80	96	154	230	327	403	525	611	729	827	877	933
Cumulative Resources Use	ZJ												
Coal		0.0	0.4	0.9	1.7	2.9	3.9	5.2	6.1	7.1	8.1	8.9	9.7
Oil		0.0	0.2	0.6	1.2	2.1	3.3	5.1	6.7	8.8	11.2	13.4	16.2
Gas		0.0	0.1	0.2	0.5	0.9	1.7	3.1	4.7	7.0	9.7	12.6	16.2
Cumulative CO2 Emissions	GtC	1.5	19.3	46.7	89.0	147.1	218.4	300.9	394.0	496.5	607.6	724.2	844.6
Carbon Sequestraction	GtC												
Land Use	Million ha												
Cropland		390	407	433	457	459	429	401	363	327	302	284	268
Grasslands		521	524	547	593	643	647	650	631	613	601	597	592
Energy Biomass		0	0	0	2	21	41	81	116	166	198	200	201
Forest		527	490	452	417	410	423	436	451	467	478	485	492
Others		576	593	583	545	482	463	445	438	431	434	448	462
Total		2014	2014	2014	2014	2014	2014	2014	2014	2014	2014	2014	2014
Anthropogenic Emissions (standardized)													
Fossil Fuel CO2	GtC	1.15	1.78	3.04	4.56	6.45	7.53	8.79	9.69	10.69	11.39	11.74	12.09
Other CO2	GtC	0.37	0.26	0.40	0.45	0.17	0.11	0.07	0.07	0.07	0.08	0.11	0.14
Total CO2	GtC	1.53	2.03	3.44	5.02	6.62	7.64	8.86	9.76	10.75	11.47	11.85	12.23
CH4 total	MtCH4	113	125	156	200	248	246	244	202	166	144	130	118
N2O total	MtN2O-N	2.3	2.6	2.8	3.0	3.1	3.0	3.0	2.8	2.7	2.6	2.5	2.4
SOx total	MtS	17.7	25.3	42.2	54.5	45.9	20.0	8.5	8.0	7.5	7.1	6.7	6.3
HFC	MtC eq.	0	5	18	45	92	153	224	292	292	285	275	262
PFC	MtC eq.	3	5	8	15	23	30	35	39	43	46	48	46
SF6	MtC eq.	4	7	12	19	28	42	50	55	48	35	33	37
CO	MtCO	235	270	487	676	836	977	1139	1251	1373	1502	1637	1783
NMVOC	Mt	33	37	55	79	96	100	105	111	117	109	89	71
NOx	MtN	7	9	13	17	20	20	19	18	17	15	14	13

Scenario A1G-AIM ALM		1990	2000	2010	2020	2030	2040	2050	2060	2070	2080	2090	2100
Population	Million	1192	1519	1865	2211	2557	2761	2980	3024	3067	3013	2866	2727
GNP/GDP (mex)	Trillion US$	1.9	2.7	5.3	10.3	19.8	32.3	52.8	70.4	93.9	118.6	141.7	169.4
GNP/GDP (ppp) Trillion (1990 prices)													
Final Energy	EJ												
Non-commercial		14	19	18	20	22	0	0	0	0	0	0	0
Solids		1	2	3	6	9	10	12	11	10	8	7	6
Liquids		17	22	39	60	93	131	169	198	227	244	251	258
Gas		4	9	18	33	49	73	96	114	132	142	145	147
Electricity		4	6	12	21	39	75	111	144	178	199	208	216
Others		0			0			0					0
Total		40	57	90	140	211	299	388	467	546	594	611	628
Primary Energy	EJ												
Coal		5	8	16	27	44	47	50	48	46	44	42	40
Oil		27	35	62	94	142	187	247	252	257	260	261	262
Gas		9	18	38	68	108	158	231	273	323	356	363	371
Nuclear		0	0	1	3	8	19	44	54	65	74	78	82
Biomass		14	18	19	20	26	32	40	67	112	148	152	157
Other Renewables		2	2	3	4	7	12	22	35	57	76	83	92
Total		57	82	138	217	334	455	634	729	861	957	980	1004
Cumulative Resources Use	ZJ												
Coal		0.0	0.1	0.2	0.4	0.8	1.1	1.7	2.1	2.6	3.1	3.5	1.7
Oil		0.0	0.3	0.8	1.6	2.8	4.3	6.7	8.7	11.4	14.1	16.6	6.7
Gas		0.0	0.1	0.4	1.0	1.8	3.1	5.2	7.4	10.5	13.9	17.3	12.5
Cumulative CO2 Emissions	GtC	1.4	17.8	43.4	84.7	140.2	209.2	295.8	395.3	501.0	611.7	724.5	837.9
Carbon Sequestraction	GtC												
Land Use	Million ha												
Cropland		357	391	427	451	446	417	390	357	327	298	271	246
Grasslands		1636	1648	1776	2032	2287	2275	2264	2171	2081	2019	1981	1944
Energy Biomass		0	0	0	0	33	77	180	269	402	501	520	539
Forest		1706	1604	1474	1313	1230	1249	1269	1285	1302	1315	1327	1338
Others		1784	1840	1806	1687	1487	1433	1380	1361	1341	1348	1382	1416
Total		5483	5483	5483	5483	5483	5483	5483	5483	5483	5483	5483	5483
Anthropogenic Emissions (standardized)													
Fossil Fuel CO2	GtC	0.72	1.01	2.15	3.61	5.64	7.40	9.70	10.29	10.92	11.26	11.30	11.35
Other CO2	GtC	0.73	0.82	1.14	1.37	0.47	0.28	-0.05	-0.04	-0.03	-0.01	0.00	0.04
Total CO2	GtC	1.45	1.83	3.29	4.98	6.11	7.68	9.65	10.25	10.89	11.25	11.31	11.39
CH4 total	MtCH4	77	85	106	134	155	157	160	148	136	122	107	93
N2O total	MtN2O-N	1.2	1.3	1.5	1.8	1.6	1.5	1.5	1.5	1.5	1.4	1.4	1.4
SOx total	MtS	10.5	12.8	20.3	25.7	28.7	35.7	44.2	28.0	17.8	13.7	12.8	12.0
HFC	MtC eq.	0	2	15	39	81	139	184	205	210	210	206	196
PFC	MtC eq.	4	4	5	9	14	19	23	26	28	30	31	30
SF6	MtC eq.	3	5	10	14	19	32	40	42	37	23	24	26
CO	MtCO	396	404	522	640	649	745	857	971	1101	1227	1344	1472
NMVOC	Mt	48	55	75	99	121	125	130	132	134	125	106	89
NOx	MtN	7	8	12	17	20	21	21	21	21	21	21	21

Scenario A1G-MESSAGE World		1990	2000	2010	2020	2030	2040	2050	2060	2070	2080	2090	2100
Population	Million	5262	6117	6888	7617	8182	8531	8704	8667	8463	8125	7658	7056
GNP/GDP (mex)	Trillion US$	20.9	26.8	36.8	57.0	91.3	135.4	187.1	254.1	322.9	393.2	469.6	550.0
GNP/GDP (ppp) Trillion (1990 prices)		25.7	33.4	45.7	67.2	98.7	139.0	186.4	246.8	313.2	382.0	456.6	535.0
Final Energy	EJ												
Non-commercial		38	25	20	16	10	7	5	0	0	0	0	0
Solids		42	57	67	75	84	88	74	47	39	25	5	3
Liquids		111	125	157	190	241	292	374	453	491	525	525	510
Gas		41	47	62	76	100	132	161	215	235	253	284	300
Electricity		35	46	64	101	161	242	344	464	573	668	749	800
Others		8	11	18	29	45	55	67	86	106	124	143	152
Total		275	311	387	487	640	815	1024	1265	1444	1595	1705	1765
Primary Energy	EJ												
Coal		91	104	130	167	220	266	272	261	259	214	153	84
Oil		128	155	172	194	235	279	365	451	479	506	461	391
Gas		71	85	119	164	236	328	449	649	796	916	1089	1239
Nuclear		7	8	11	17	30	54	90	128	169	225	282	332
Biomass		46	44	54	76	107	145	193	243	311	381	409	414
Other Renewables		8	13	23	38	61	91	125	165	203	237	264	277
Total		352	409	509	657	889	1163	1495	1898	2217	2479	2658	2737
Cumulative Resources Use	ZJ												
Coal		0.0	0.9	2.0	3.3	4.9	7.1	9.8	12.5	15.1	17.7	19.9	21.4
Oil		0.0	1.4	3.0	4.7	6.7	9.0	11.8	15.4	20.0	24.7	29.8	34.4
Gas		0.0	0.8	1.6	2.8	4.4	6.8	10.1	14.6	21.1	29.0	38.2	49.1
Cumulative CO2 Emissions	GtC	7.1	82.4	170.0	272.2	397.4	556.1	751.2	988.8	1264.4	1562.3	1870.7	2178.0
Carbon Sequestration	GtC												
Land Use [a]	Million ha												
Cropland		1459	1466	1462	1457	1454	1448	1442	1436	1429	1424	1422	1420
Grasslands		3389	3404	3429	3446	3458	3478	3498	3525	3552	3568	3572	3576
Energy Biomass		0	0	0	74	158	288	418	492	566	581	538	495
Forest		4296	4237	4173	4164	4164	4177	4190	4194	4199	4202	4203	4204
Others		3805	3842	3886	3807	3715	3558	3400	3301	3203	3173	3213	3253
Total		12949	12949	12949	12949	12949	12949	12949	12949	12949	12949	12949	12949
Anthropogenic Emissions (standardized)													
Fossil Fuel CO2 [b]	GtC	5.99	6.90	8.49	10.66	14.00	17.56	21.45	26.11	29.08	30.58	31.14	30.31
Other CO2 [c]	GtC	1.11	1.07	1.04	0.26	0.12	0.05	-0.02	-0.03	-0.03	-0.03	-0.01	0.00
Total CO2	GtC	7.10	7.97	9.54	10.91	14.12	17.61	21.42	26.08	29.04	30.55	31.13	30.31
CH4 total [d]	MtCH4	310	323	363	419	496	538	566	555	531	497	474	436
N2O total [e]	MtN2O-N	6.7	7.0	6.1	6.1	6.2	6.3	6.4	6.3	6.3	6.2	6.2	6.2
SOx total	MtS	70.9	69.0	64.3	60.3	63.9	71.9	69.0	59.1	51.2	45.7	43.1	40.5
CFC/HFC/HCFC	MtC eq.	1672	883	791	337	369	482	566	654	659	654	639	614
PFC	MtC eq.	32	25	31	43	61	77	89	97	106	114	119	115
SF6	MtC eq.	38	40	43	48	66	99	119	127	113	88	84	95
CO	MtCO	879	877	986	1123	1344	1616	2026	2399	2717	3033	3199	3261
NMVOC	Mt	139	141	160	178	200	237	297	370	416	459	474	484
NOx	MtN	31	32	38	46	59	72	89	108	118	124	129	133

Emissions correlated to land-use change and deforestation were calculated by using AIM A1 marker land-use data.

a: Land-use taken from AIM-A1 marker run.

b:CO2 emissions from fossil fuel and industrial processes (MESSAGE data).

c: CO2 emissions from land-use changes (AIM-A1 marker run).

d: Non-energy related CH4 emissions were taken from AIM-A1 marker run.

e: Non-energy related N2O emissions were taken from AIM-A1 marker run.

Scenario A1G-MESSAGE OECD90		1990	2000	2010	2020	2030	2040	2050	2060	2070	2080	2090	2100
Population	Million	859	919	965	1007	1043	1069	1081	1084	1089	1098	1108	1110
GNP/GDP (mex)	Trillion US$	16.4	20.6	25.6	31.6	38.7	46.8	55.7	65.7	77.2	90.8	106.6	124.3
GNP/GDP (ppp) Trillion (1990 prices)		14.1	17.8	22.1	27.4	33.7	40.8	48.6	57.4	67.6	79.7	93.6	109.4
Final Energy	EJ												
Non-commercial		0	0	0	0	0	0	0	0	0	0	0	0
Solids		13	11	6	3	1	1	0	0	0	0	0	0
Liquids		66	69	77	75	73	71	79	86	87	94	100	105
Gas		22	28	33	34	38	41	38	33	30	25	22	18
Electricity		22	28	35	45	58	73	89	106	124	139	153	165
Others		1	1	3	5	8	10	12	16	23	27	31	32
Total		124	138	153	162	178	195	218	241	264	284	305	321
Primary Energy	EJ												
Coal		38	38	45	52	57	55	40	26	20	19	17	15
Oil		72	84	83	75	67	64	76	80	82	91	99	99
Gas		33	45	55	66	78	87	101	118	140	139	156	162
Nuclear		6	7	8	10	15	26	41	53	59	68	68	74
Biomass		6	7	7	13	19	24	29	32	36	40	43	47
Other Renewables		4	5	9	12	16	22	28	35	44	49	54	57
Total		159	186	206	227	252	278	315	345	381	406	437	453
Cumulative Resources Use	ZJ												
Coal		0.0	0.4	0.8	1.2	1.7	2.3	2.9	3.3	3.7	4.0	4.2	4.4
Oil		0.0	0.8	1.6	2.5	3.2	3.9	4.5	5.3	6.1	6.9	7.8	8.8
Gas		0.0	0.4	0.8	1.4	2.0	2.8	3.7	4.7	5.9	7.3	8.7	10.3
Cumulative CO2 Emissions	GtC	2.8	33.0	66.6	102.6	140.2	178.6	217.3	256.6	297.5	340.3	385.3	431.5
Carbon Sequestration	GtC												
Land Use [a]	Million ha												
Cropland		381	380	377	376	375	372	368	364	359	356	356	355
Grasslands		760	763	771	774	776	787	797	818	839	849	847	846
Energy Biomass		0	0	0	11	24	44	63	74	86	88	81	75
Forest		1050	1053	1052	1053	1056	1062	1068	1061	1054	1050	1049	1048
Others		838	833	828	814	798	765	733	712	691	685	695	705
Total		3029	3029	3029	3029	3029	3029	3029	3029	3029	3029	3029	3029
Anthropogenic Emissions (standardized)													
Fossil Fuel CO2 [b]	GtC	2.83	3.20	3.49	3.67	3.84	3.87	3.93	3.96	4.17	4.28	4.61	4.58
Other CO2 [c]	GtC	0.00	0.00	0.04	0.02	0.00	-0.02	-0.04	0.00	0.05	0.06	0.04	0.01
Total CO2	GtC	2.83	3.20	3.52	3.69	3.84	3.84	3.89	3.97	4.22	4.34	4.65	4.59
CH4 total [d]	MtCH4	73	74	73	73	77	81	79	75	73	66	62	56
N2O total [e]	MtN2O-N	2.6	2.6	2.4	2.4	2.4	2.4	2.3	2.3	2.3	2.3	2.3	2.3
SOx total	MtS	22.7	17.0	8.9	1.7	0.8	0.5	0.8	1.5	2.1	3.4	4.9	5.8
HFC	MtC eq.	19	58	110	108	115	118	122	123	123	124	125	125
PFC	MtC eq.	18	13	13	11	9	8	9	10	12	13	14	16
SF6	MtC eq.	23	23	14	5	6	7	9	11	13	16	18	20
CO	MtCO	179	161	172	176	173	184	205	214	217	228	236	234
NMVOC	Mt	42	36	37	35	30	29	31	33	36	39	43	44
NOx	MtN	13	12	13	14	14	14	13	13	14	15	17	18

Emissions correlated to land-use change and deforestation were calculated by using AIM A1 marker land-use data.
a: Land-use taken from AIM-A1 marker run.
b:CO2 emissions from fossil fuel and industrial processes (MESSAGE data).
c: CO2 emissions from land-use changes (AIM-A1 marker run).
d: Non-energy related CH4 emissions were taken from AIM-A1 marker run.
e: Non-energy related N2O emissions were taken from AIM-A1 marker run.

Scenario A1G-MESSAGE REF		1990	2000	2010	2020	2030	2040	2050	2060	2070	2080	2090	2100
Population	Million	413	419	427	433	435	433	423	409	392	374	357	339
GNP/GDP (mex)	Trillion US$	1.1	0.8	1.0	2.1	5.4	9.4	12.6	16.2	20.0	24.4	29.2	34.4
GNP/GDP (ppp) Trillion (1990 prices)		2.6	2.2	2.5	3.7	6.0	9.4	12.6	16.2	20.0	24.4	29.2	34.4
Final Energy	EJ												
Non-commercial		0	0	0	0	0	0	0	0	0	0	0	0
Solids		9	4	3	3	3	1	1	0	0	0	0	0
Liquids		15	10	10	12	16	16	22	28	29	28	27	28
Gas		14	10	12	16	19	20	16	12	9	8	7	5
Electricity		6	5	6	9	15	21	27	33	38	42	46	49
Others		7	6	6	8	12	14	13	14	15	16	16	16
Total		50	35	38	47	64	71	79	86	90	94	97	97
Primary Energy	EJ												
Coal		19	13	10	14	20	23	22	25	23	21	21	20
Oil		20	14	13	13	16	16	19	27	30	32	23	19
Gas		27	21	25	29	36	38	40	45	49	60	71	92
Nuclear		1	1	1	1	2	4	7	11	16	20	25	24
Biomass		2	1	1	3	6	10	14	17	28	32	33	33
Other Renewables		1	1	2	3	5	7	9	12	16	19	21	23
Total		70	51	51	63	85	98	112	138	162	183	193	211
Cumulative Resources Use	ZJ												
Coal		0.0	0.2	0.3	0.4	0.5	0.7	1.0	1.2	1.4	1.7	1.9	2.1
Oil		0.0	0.2	0.3	0.5	0.6	0.8	0.9	1.1	1.4	1.7	2.0	2.2
Gas		0.0	0.3	0.5	0.7	1.0	1.4	1.8	2.2	2.6	3.1	3.7	4.4
Cumulative CO2 Emissions	GtC	1.3	12.3	22.0	32.3	44.2	57.9	71.8	87.4	105.3	124.5	144.4	165.2
Carbon Sequestration	GtC												
Land Use [a]	Million ha												
Cropland		268	266	265	265	265	264	263	262	262	261	260	260
Grasslands		341	361	364	366	367	368	370	371	371	372	373	374
Energy Biomass		0	0	0	3	6	11	16	19	22	23	21	20
Forest		966	950	918	904	894	899	905	909	912	916	922	927
Others		701	698	728	738	745	733	722	715	709	703	700	696
Total		2276	2276	2276	2276	2276	2276	2276	2276	2276	2276	2276	2276
Anthropogenic Emissions (standardized)													
Fossil Fuel CO2 [b]	GtC	1.30	0.91	0.85	1.02	1.35	1.46	1.54	1.86	1.94	2.06	2.05	2.24
Other CO2 [c]	GtC	0.00	0.00	0.18	0.01	0.00	-0.08	-0.15	-0.12	-0.09	-0.07	-0.06	-0.05
Total CO2	GtC	1.30	0.91	1.03	1.03	1.34	1.39	1.39	1.74	1.85	1.99	1.98	2.19
CH4 total [d]	MtCH4	47	39	42	50	59	59	53	52	51	51	51	50
N2O total [e]	MtN2O-N	0.6	0.6	0.6	0.6	0.6	0.6	0.6	0.6	0.6	0.6	0.6	0.6
SOx total	MtS	17.0	11.0	7.2	4.8	3.9	4.5	4.1	4.2	3.3	3.3	3.1	3.3
HFC	MtC eq.	0	4	8	19	29	31	32	33	33	34	33	31
PFC	MtC eq.	7	4	5	8	14	20	21	22	23	24	25	24
SF6	MtC eq.	8	6	8	10	14	18	21	19	15	14	10	11
CO	MtCO	69	41	46	52	82	97	110	115	118	110	99	81
NMVOC	Mt	16	13	16	20	26	32	38	55	78	96	93	99
NOx	MtN	5	3	3	4	5	6	6	6	6	5	4	5

Emissions correlated to land-use change and deforestation were calculated by using AIM A1 marker land-use data.

a: Land-use taken from AIM-A1 marker run.

b:CO2 emissions from fossil fuel and industrial processes (MESSAGE data).

c: CO2 emissions from land-use changes (AIM-A1 marker run).

d: Non-energy related CH4 emissions were taken from AIM-A1 marker run.

e: Non-energy related N2O emissions were taken from AIM-A1 marker run.

Scenario A1G-MESSAGE ASIA		1990	2000	2010	2020	2030	2040	2050	2060	2070	2080	2090	2100
Population	Million	2798	3261	3620	3937	4147	4238	4220	4085	3867	3589	3258	2882
GNP/GDP (mex)	Trillion US$	1.5	2.7	5.8	13.5	27.2	44.9	65.3	95.8	126.9	155.5	186.5	218.2
GNP/GDP (ppp) Trillion (1990 prices)		5.3	8.3	13.5	22.8	36.2	52.2	70.0	96.8	126.9	155.5	186.5	218.2
Final Energy	EJ												
Non-commercial		24	15	11	8	5	3	2	0	0	0	0	0
Solids		19	37	52	63	71	74	56	37	28	17	3	2
Liquids		13	22	37	55	83	116	162	188	201	213	211	198
Gas		2	3	6	10	17	29	43	87	105	111	117	119
Electricity		4	7	14	27	48	79	118	167	212	253	285	298
Others		1	2	6	10	16	17	22	31	38	43	52	59
Total		62	87	125	174	239	318	404	510	585	638	668	676
Primary Energy	EJ												
Coal		30	48	67	90	124	165	176	163	145	116	79	42
Oil		15	26	38	55	79	108	152	170	169	172	139	103
Gas		3	5	15	29	49	77	116	212	287	332	381	429
Nuclear		0	1	2	4	8	14	23	36	56	87	121	143
Biomass		24	22	28	37	47	59	80	109	127	142	151	153
Other Renewables		1	3	5	12	21	30	42	56	69	81	91	96
Total		74	105	156	226	327	453	588	746	853	929	962	965
Cumulative Resources Use	ZJ												
Coal		0.0	0.3	0.8	1.5	2.4	3.6	5.3	7.0	8.6	10.0	11.2	12.1
Oil		0.0	0.2	0.4	0.8	1.4	2.2	3.3	4.8	6.5	8.2	9.9	11.3
Gas		0.0	0.0	0.1	0.2	0.5	1.0	1.8	2.9	5.0	7.9	11.2	15.1
Cumulative CO2 Emissions	GtC	1.5	19.3	44.3	78.0	124.8	191.0	277.7	381.2	496.0	615.6	733.8	845.2
Carbon Sequestration	GtC												
Land Use [a]	Million ha												
Cropland		438	435	434	433	432	431	431	430	429	428	428	428
Grasslands		608	606	609	611	613	615	617	618	619	621	622	624
Energy Biomass		0	0	0	17	37	68	99	117	134	138	127	117
Forest		535	522	512	524	535	535	535	544	552	556	555	555
Others		583	601	609	579	547	515	483	456	429	421	431	441
Total		2164	2164	2164	2164	2164	2164	2164	2164	2164	2164	2164	2164
Anthropogenic Emissions (standardized)													
Fossil Fuel CO2 [b]	GtC	1.15	1.78	2.73	3.94	5.65	7.77	9.58	11.13	11.96	12.16	11.61	10.70
Other CO2 [c]	GtC	0.37	0.26	0.22	-0.13	-0.12	-0.04	0.03	-0.03	-0.09	-0.10	-0.04	0.01
Total CO2	GtC	1.53	2.03	2.95	3.81	5.54	7.72	9.60	11.09	11.87	12.07	11.57	10.71
CH4 total [d]	MtCH4	113	125	151	182	226	242	254	238	216	198	183	162
N2O total [e]	MtN2O-N	2.3	2.6	2.4	2.4	2.5	2.6	2.6	2.6	2.6	2.6	2.5	2.5
SOx total	MtS	17.7	25.3	31.2	36.0	40.5	47.4	42.2	31.7	24.2	19.7	17.1	15.1
HFC	MtC eq.	0	5	18	45	92	153	224	292	292	285	275	262
PFC	MtC eq.	3	5	8	15	23	30	35	39	43	46	48	46
SF6	MtC eq.	4	7	12	19	28	42	50	55	48	35	33	37
CO	MtCO	235	270	345	425	554	701	928	1103	1227	1318	1390	1372
NMVOC	Mt	33	37	46	54	63	77	107	130	144	155	155	150
NOx	MtN	7	9	12	16	23	31	40	49	53	55	54	54

Emissions correlated to land-use change and deforestation were calculated by using AIM A1 marker land-use data.

a: Land-use taken from AIM-A1 marker run.

b:CO2 emissions from fossil fuel and industrial processes (MESSAGE data).

c: CO2 emissions from land-use changes (AIM-A1 marker run).

d: Non-energy related CH4 emissions were taken from AIM-A1 marker run.

e: Non-energy related N2O emissions were taken from AIM-A1 marker run.

Scenario A1G-MESSAGE ALM		1990	2000	2010	2020	2030	2040	2050	2060	2070	2080	2090	2100
Population	Million	1192	1519	1875	2241	2557	2791	2980	3089	3115	3064	2934	2727
GNP/GDP (mex)	Trillion US$	1.9	2.7	4.4	9.8	20.0	34.3	53.5	76.5	98.7	122.5	147.2	173.1
GNP/GDP (ppp) Trillion (1990 prices)		3.8	5.1	7.5	13.3	22.8	36.7	55.3	76.5	98.7	122.5	147.2	173.1
Final Energy	EJ												
Non-commercial		14	10	9	8	5	4	3	0	0	0	0	0
Solids		1	5	6	6	9	13	17	10	11	8	2	1
Liquids		17	24	33	49	70	90	111	151	174	190	186	179
Gas		4	6	11	16	26	43	63	83	91	109	138	159
Electricity		3	5	9	19	40	68	110	158	199	234	266	288
Others		0	2	3	6	9	14	20	26	31	38	45	46
Total		39	51	71	104	159	231	324	428	506	579	635	672
Primary Energy	EJ												
Coal		5	5	8	12	19	23	35	47	71	59	36	8
Oil		21	31	39	51	72	91	119	175	197	212	199	170
Gas		8	14	24	41	73	126	192	273	321	385	482	555
Nuclear		0	0	1	2	6	11	18	27	38	51	68	92
Biomass		14	14	17	24	36	53	70	86	121	168	183	182
Other Renewables		2	4	7	12	21	32	47	62	75	88	98	102
Total		49	67	96	142	225	335	481	670	822	961	1065	1109
Cumulative Resources Use	ZJ												
Coal		0.0	0.0	0.1	0.2	0.3	0.5	0.7	1.0	1.5	2.1	2.6	2.9
Oil		0.0	0.3	0.6	1.0	1.5	2.2	3.1	4.3	6.0	8.0	10.1	12.1
Gas		0.0	0.1	0.2	0.5	0.9	1.6	2.9	4.8	7.5	10.7	14.5	19.3
Cumulative CO2 Emissions	GtC	1.4	17.8	37.1	59.2	88.2	128.5	184.5	263.6	365.5	481.8	607.3	736.0
Carbon Sequestration	GtC												
Land Use [a]	Million ha												
Cropland		371	385	384	383	382	381	381	380	379	379	378	378
Grasslands		1680	1673	1685	1695	1702	1708	1714	1718	1722	1726	1730	1733
Energy Biomass		0	0	0	42	90	165	240	282	324	333	308	284
Forest		1745	1711	1690	1683	1680	1681	1682	1681	1680	1678	1676	1674
Others		1684	1710	1720	1676	1625	1544	1463	1419	1374	1364	1387	1411
Total		5480	5480	5480	5480	5480	5480	5480	5480	5480	5480	5480	5480
Anthropogenic Emissions (standardized)													
Fossil Fuel CO2 [b]	GtC	0.72	1.01	1.43	2.03	3.15	4.47	6.41	9.16	11.01	12.07	12.88	12.78
Other CO2 [c]	GtC	0.73	0.82	0.60	0.36	0.25	0.19	0.14	0.12	0.10	0.08	0.06	0.03
Total CO2	GtC	1.45	1.83	2.03	2.39	3.40	4.66	6.54	9.28	11.11	12.15	12.93	12.82
CH4 total [d]	MtCH4	77	85	98	114	134	155	180	190	192	181	179	168
N2O total [e]	MtN2O-N	1.2	1.3	0.7	0.7	0.7	0.7	0.8	0.8	0.8	0.8	0.8	0.8
SOx total	MtS	10.5	12.8	14.0	14.7	15.8	16.5	18.8	18.6	18.7	16.4	14.9	13.3
HFC	MtC eq.	0	2	15	39	81	139	184	205	210	210	206	196
PFC	MtC eq.	4	4	5	9	14	19	23	26	28	30	31	30
SF6	MtC eq.	3	5	10	14	19	32	40	42	37	23	24	26
CO	MtCO	396	404	423	469	535	634	782	967	1154	1378	1473	1573
NMVOC	Mt	48	55	61	70	82	98	122	152	159	169	184	191
NOx	MtN	7	8	9	12	17	22	30	40	45	49	53	56

Emissions correlated to land-use change and deforestation were calculated by using AIM A1 marker land-use data.

a: Land-use taken from AIM-A1 marker run.

b:CO2 emissions from fossil fuel and industrial processes (MESSAGE data).

c: CO2 emissions from land-use changes (AIM-A1 marker run).

d: Non-energy related CH4 emissions were taken from AIM-A1 marker run.

e: Non-energy related N2O emissions were taken from AIM-A1 marker run.

Illustrative Scenario A1FI-MiniCAM, World previously A1G-MiniCAM		1990	2000	2010	2020	2030	2040	2050	2060	2070	2080	2090	2100
Population	Million	5293	6100	6874	7618	8122	8484	8703	8623	8430	8126	7621	7137
GNP/GDP (mex)	Trillion US$	20.7	27.4	38.1	52.8	80.0	117.1	164.0	226.1	293.3	365.7	441.6	525.0
GNP/GDP (ppp) Trillion (1990 prices)		na	na	na	na	na	na	na	na	na	na	na	na
Final Energy	EJ												
Non-commercial		0	0	0	0	0	0	0	0	0	0	0	0
Solids		45	57	73	94	118	140	158	128	108	99	89	79
Liquids		121	126	141	165	182	229	306	366	423	476	455	435
Gas		52	63	81	105	126	141	150	182	212	242	232	223
Electricity		35	54	86	133	207	301	413	525	630	728	782	837
Others		0	0	0	0	0	0	0	0	0	0	0	0
Total		253	300	381	497	634	810	1027	1201	1373	1546	1559	1572
Primary Energy	EJ												
Coal		88	115	150	193	299	393	475	448	432	429	518	607
Oil		131	136	150	173	165	202	283	353	416	471	359	248
Gas		70	85	129	203	268	333	398	494	573	634	606	578
Nuclear		24	26	35	51	79	108	137	155	177	201	217	233
Biomass		0	6	12	18	28	40	52	73	85	89	106	123
Other Renewables		24	24	27	32	42	60	86	128	169	208	246	284
Total		336	392	503	669	882	1135	1431	1650	1850	2032	2052	2073
Cumulative Resources Use	ZJ												
Coal		0.1	1.2	2.6	4.2	7.0	10.4	14.6	19.1	23.6	27.9	32.9	37.9
Oil		0.1	1.5	2.9	4.5	6.2	8.1	10.4	13.8	17.7	22.0	25.8	29.6
Gas		0.1	0.9	2.1	3.6	6.1	9.2	12.7	17.4	22.8	28.7	34.8	40.9
Cumulative CO2 Emissions	GtC	7.1	82.4	171.0	283.3	427.9	608.7	828.0	1075.9	1340.8	1620.6	1906.1	2189.4
Carbon Sequestration	GtC												
Land Use	Million ha												
Cropland		1472	1466	1474	1495	1481	1439	1369	1208	1053	903	786	669
Grasslands		3209	3348	3602	3970	4367	4681	4911	4730	4478	4155	3918	3682
Energy Biomass		0	4	9	16	30	47	68	115	130	116	162	208
Forest		4173	4214	4140	3951	3625	3362	3162	3471	3838	4265	4432	4598
Others		4310	4132	3940	3733	3662	3636	3655	3641	3664	3725	3867	4008
Total		13164	13164	13164	13164	13164	13164	13164	13164	13164	13164	13164	13164
Anthropogenic Emissions (standardized)													
Fossil Fuel CO2	GtC	5.99	6.90	8.65	11.19	14.61	18.66	23.10	25.14	27.12	29.04	29.64	30.32
Other CO2	GtC	1.11	1.07	1.08	1.55	1.57	1.31	0.80	0.55	0.16	-0.36	-1.22	-2.08
Total CO2	GtC	7.10	7.97	9.73	12.73	16.19	19.97	23.90	25.69	27.28	28.68	28.42	28.24
CH4 total	MtCH4	310	323	359	416	489	567	630	655	677	695	715	735
N2O total	MtN2O-N	6.7	7.0	8.0	9.3	10.9	12.8	14.5	15.0	15.4	15.7	16.1	16.6
SOx total	MtS	70.9	69.0	80.8	86.9	96.1	94.0	80.5	56.3	42.6	39.4	39.8	40.1
CFC/HFC/HCFC	MtC eq.	1672	883	791	337	369	482	566	654	659	654	639	614
PFC	MtC eq.	32	25	31	43	61	77	89	97	106	114	119	115
SF6	MtC eq.	38	40	43	48	66	99	119	127	113	88	84	95
CO	MtCO	879	877	1020	1204	1436	1726	2159	2270	2483	2776	2685	2570
NMVOC	Mt	139	141	166	192	214	256	322	361	405	449	435	420
NOx	MtN	31	32	40	50	63	77	95	102	109	115	111	110

Illustrative Scenario A1FI-MiniCAM, OECD90 previously A1G-MiniCAM		**1990**	**2000**	**2010**	**2020**	**2030**	**2040**	**2050**	**2060**	**2070**	**2080**	**2090**	**2100**
Population	Million	838	908	965	1007	1024	1066	1081	1084	1090	1098	1105	1112
GNP/GDP (mex)	Trillion US$	16.3	20.5	25.6	31.5	34.4	44.6	53.5	63.2	74.2	86.4	101.4	117.7
GNP/GDP (ppp) Trillion (1990 prices)		na	na	na	na	na	na	na	na	na	na	na	na
Final Energy	EJ												
Non-commercial		0	0	0	0	0	0	0	0	0	0	0	0
Solids		10	13	14	14	14	15	17	13	12	12	12	12
Liquids		72	73	75	77	73	70	81	87	94	102	97	93
Gas		27	36	47	59	60	62	62	70	80	91	90	89
Electricity		22	28	37	48	53	70	85	95	105	116	126	137
Others		0	0	0	0	0	0	0	0	0	0	0	0
Total		130	151	173	197	199	217	245	265	290	320	325	330
Primary Energy	EJ												
Coal		40	47	48	45	51	70	82	86	93	104	167	231
Oil		76	78	79	79	71	59	70	71	72	74	34	-6
Gas		34	47	69	100	103	112	116	131	146	160	158	157
Nuclear		20	16	14	16	17	18	18	19	22	25	29	33
Biomass		0	2	4	5	5	7	9	11	13	13	17	20
Other Renewables		12	11	10	10	10	12	14	18	23	28	35	42
Total		182	201	225	254	257	279	310	336	368	404	440	476
Cumulative Resources Use	ZJ												
Coal		0.0	0.5	1.0	1.4	1.7	2.6	3.4	4.2	5.1	6.1	7.6	9.2
Oil		0.1	0.9	1.6	2.4	2.8	3.7	4.3	5.1	5.8	6.5	6.9	7.3
Gas		0.0	0.5	1.1	1.9	2.4	4.0	5.2	6.4	7.8	9.3	10.9	12.5
Cumulative CO2 Emissions	GtC	2.8	33.0	67.5	107.1	149.4	193.9	242.3	294.5	349.7	408.5	472.7	544.4
Carbon Sequestration	GtC												
Land Use	Million ha												
Cropland		408	410	412	414	409	385	358	312	272	238	207	177
Grasslands		796	820	866	935	970	1058	1094	1058	1013	962	918	873
Energy Biomass		0	4	6	8	10	18	24	34	38	36	50	63
Forest		921	931	921	894	869	809	787	864	938	1011	1039	1067
Others		998	959	917	872	865	853	858	856	861	876	909	942
Total		3123	3123	3123	3123	3123	3123	3123	3123	3123	3123	3123	3123
Anthropogenic Emissions (standardized)													
Fossil Fuel CO2	GtC	2.83	3.20	3.58	3.95	4.03	4.46	5.02	5.34	5.76	6.30	7.14	8.02
Other CO2	GtC	0.00	0.00	0.13	0.25	0.24	0.16	0.05	0.02	-0.07	-0.24	-0.35	-0.47
Total CO2	GtC	2.83	3.20	3.72	4.19	4.27	4.63	5.07	5.36	5.69	6.06	6.78	7.55
CH4 total	MtCH4	73	74	86	101	108	131	149	172	196	223	246	270
N2O total	MtN2O-N	2.6	2.6	2.8	3.0	3.2	3.7	4.0	4.1	4.2	4.2	4.3	4.4
SOx total	MtS	22.7	17.0	13.9	4.7	4.0	4.2	5.1	5.2	5.7	6.5	7.3	8.0
HFC	MtC eq.	19	58	110	108	115	118	122	123	123	124	125	125
PFC	MtC eq.	18	13	13	11	9	8	9	10	12	13	14	16
SF6	MtC eq.	23	23	14	5	6	7	9	11	13	16	18	20
CO	MtCO	179	161	181	199	187	211	250	269	274	304	307	319
NMVOC	Mt	42	36	39	39	32	33	37	41	45	52	55	60
NOx	MtN	13	12	14	16	15	16	16	16	18	20	23	24

Illustrative Scenario A1FI-MiniCAM, REF previously A1G-MiniCAM		1990	2000	2010	2020	2030	2040	2050	2060	2070	2080	2090	2100
Population	Million	428	425	426	433	434	431	423	408	392	374	357	340
GNP/GDP (mex)	Trillion US$	1.1	1.1	1.4	2.1	3.5	5.1	6.9	10.0	13.6	17.7	21.6	26.0
GNP/GDP (ppp) Trillion (1990 prices)		na	na	na	na	na	na	na	na	na	na	na	na
Final Energy	EJ												
Non-commercial		0	0	0	0	0	0	0	0	0	0	0	0
Solids		13	10	9	9	9	10	11	8	7	6	5	5
Liquids		18	12	9	11	12	14	17	19	21	22	21	19
Gas		19	15	14	17	20	22	22	26	29	32	28	25
Electricity		6	8	13	21	29	38	48	58	66	73	73	74
Others		0	0	0	0	0	0	0	0	0	0	0	0
Total		56	44	45	57	70	84	97	110	122	133	127	122
Primary Energy	EJ												
Coal		18	17	19	24	46	60	68	60	53	45	61	77
Oil		20	14	11	13	6	3	5	10	15	22	19	15
Gas		26	20	22	33	41	47	50	59	65	69	61	54
Nuclear		3	4	6	10	13	16	17	18	19	20	20	20
Biomass		0	1	1	2	3	5	6	8	9	9	10	11
Other Renewables		3	3	3	4	5	7	10	14	18	21	24	27
Total		70	58	63	85	115	138	156	168	178	186	195	205
Cumulative Resources Use	ZJ												
Coal		0.0	0.2	0.4	0.6	1.0	1.5	2.1	2.8	3.3	3.8	4.4	5.0
Oil		0.0	0.2	0.3	0.4	0.5	0.5	0.6	0.7	0.8	1.0	1.2	1.4
Gas		0.0	0.2	0.5	0.7	1.1	1.6	2.0	2.6	3.2	3.9	4.5	5.1
Cumulative CO2 Emissions	GtC	1.3	12.3	21.7	33.2	49.2	70.0	94.2	120.1	146.4	172.3	197.7	222.6
Carbon Sequestration	GtC												
Land Use	Million ha												
Cropland		284	294	304	316	320	316	303	263	223	183	159	135
Grasslands		395	410	454	526	608	673	719	666	601	525	492	458
Energy Biomass		0	0	1	4	8	13	18	25	25	19	23	27
Forest		1007	1016	996	945	865	804	762	850	950	1061	1088	1115
Others		691	657	622	587	575	571	575	573	578	590	616	642
Total		2377	2377	2377	2377	2377	2377	2377	2377	2377	2377	2377	2377
Anthropogenic Emissions (standardized)													
Fossil Fuel CO2	GtC	1.30	0.91	0.94	1.24	1.76	2.17	2.47	2.53	2.54	2.53	2.66	2.79
Other CO2	GtC	0.00	0.00	0.03	0.08	0.12	0.12	0.07	0.10	0.09	0.03	-0.14	-0.32
Total CO2	GtC	1.30	0.91	0.97	1.32	1.88	2.29	2.55	2.63	2.63	2.56	2.52	2.48
CH4 total	MtCH4	47	39	49	68	91	109	122	123	122	120	120	121
N2O total	MtN2O-N	0.6	0.6	0.7	0.9	1.2	1.4	1.6	1.6	1.6	1.5	1.6	1.6
SOx total	MtS	17.0	11.0	10.3	10.5	13.3	13.4	10.7	5.8	3.1	2.4	2.5	2.6
HFC	MtC eq.	0	4	8	19	29	31	32	33	33	34	33	31
PFC	MtC eq.	7	4	5	8	14	20	21	22	23	24	25	24
SF6	MtC eq.	8	6	8	10	14	18	21	19	15	14	10	11
CO	MtCO	69	41	49	63	103	136	166	151	151	130	120	89
NMVOC	Mt	16	13	17	24	32	45	57	72	100	114	113	110
NOx	MtN	5	3	3	4	7	8	9	8	7	6	5	5

Illustrative Scenario A1FI-MiniCAM, ASIA previously A1G-MiniCAM		**1990**	**2000**	**2010**	**2020**	**2030**	**2040**	**2050**	**2060**	**2070**	**2080**	**2090**	**2100**
Population	Million	2790	3226	3608	3937	4115	4210	4219	4062	3852	3589	3245	2919
GNP/GDP (mex)	Trillion US$	1.4	3.1	6.5	11.7	23.8	40.3	61.0	86.0	111.7	138.3	164.3	192.6
GNP/GDP (ppp) Trillion (1990 prices)		na	na	na	na	na	na	na	na	na	na	na	na
Final Energy	EJ												
Non-commercial		0	0	0	0	0	0	0	0	0	0	0	0
Solids		20	31	46	64	84	99	111	88	73	66	57	48
Liquids		14	19	28	39	53	74	103	123	141	156	146	136
Gas		2	5	9	15	20	24	29	37	43	49	46	42
Electricity		4	11	25	47	84	129	182	232	276	314	328	343
Others		0	0	0	0	0	0	0	0	0	0	0	0
Total		40	67	109	165	240	327	424	480	533	584	576	569
Primary Energy	EJ												
Coal		26	45	73	110	165	212	251	221	200	188	194	200
Oil		16	21	30	42	48	66	97	126	148	165	136	107
Gas		3	9	22	43	71	103	137	174	200	216	200	183
Nuclear		1	4	9	18	33	49	67	74	82	91	94	98
Biomass		0	2	5	9	15	20	26	35	41	42	48	54
Other Renewables		3	4	5	6	11	20	33	52	71	88	102	116
Total		49	85	145	228	342	470	611	682	742	790	774	758
Cumulative Resources Use	ZJ												
Coal		0.0	0.4	1.1	1.9	3.4	5.3	7.6	9.8	12.0	13.9	15.8	17.8
Oil		0.0	0.2	0.5	0.8	1.3	1.9	2.7	3.9	5.2	6.8	8.2	9.6
Gas		0.0	0.1	0.3	0.5	1.2	2.1	3.2	4.9	6.7	8.8	10.8	12.8
Cumulative CO2 Emissions	GtC	1.5	19.3	44.6	82.4	137.4	211.6	305.0	409.2	515.3	623.3	728.8	827.9
Carbon Sequestration	GtC												
Land Use	Million ha												
Cropland		389	400	410	420	422	416	403	358	316	275	238	202
Grasslands		508	524	555	602	650	690	721	712	695	669	640	610
Energy Biomass		0	0	1	4	8	13	19	33	37	32	44	55
Forest		1168	1144	1102	1044	988	948	922	960	1016	1088	1144	1199
Others		664	632	600	568	557	555	559	557	562	573	596	620
Total		2729	2699	2669	2637	2627	2622	2624	2621	2625	2637	2662	2686
Anthropogenic Emissions (standardized)													
Fossil Fuel CO2	GtC	1.15	1.78	2.82	4.27	6.17	8.12	10.13	10.42	10.68	10.93	10.42	9.92
Other CO2	GtC	0.37	0.26	0.20	0.27	0.29	0.26	0.17	0.10	0.03	-0.05	-0.19	-0.33
Total CO2	GtC	1.53	2.03	3.02	4.54	6.46	8.38	10.30	10.52	10.71	10.88	10.23	9.59
CH4 total	MtCH4	113	125	135	148	166	186	208	198	189	179	181	182
N2O total	MtN2O-N	2.3	2.6	2.9	3.3	3.8	4.3	4.8	4.9	5.0	5.1	5.2	5.3
SOx total	MtS	17.7	25.3	38.8	52.7	57.1	51.9	37.0	24.3	17.0	15.1	14.6	14.1
HFC	MtC eq.	0	5	18	45	92	153	224	292	292	285	275	262
PFC	MtC eq.	3	5	8	15	23	30	35	39	43	46	48	46
SF6	MtC eq.	4	7	12	19	28	42	50	55	48	35	33	37
CO	MtCO	235	270	371	494	639	777	1030	1067	1133	1226	1242	1187
NMVOC	Mt	33	37	50	63	73	86	119	126	133	144	139	130
NOx	MtN	7	9	13	19	26	34	44	47	48	51	49	47

Illustrative Scenario A1FI-MiniCAM, ALM previously A1G-MiniCAM		**1990**	**2000**	**2010**	**2020**	**2030**	**2040**	**2050**	**2060**	**2070**	**2080**	**2090**	**2100**
Population	Million	1236	1541	1876	2241	2531	2778	2980	3068	3096	3064	2913	2766
GNP/GDP (mex)	Trillion US$	1.9	2.8	4.6	7.4	15.4	27.1	42.5	66.9	93.9	123.4	154.2	188.7
GNP/GDP (ppp) Trillion (1990 prices)		na	na	na	na	na	na	na	na	na	na	na	na
Final Energy	EJ												
Non-commercial		0	0	0	0	0	0	0	0	0	0	0	0
Solids		2	3	5	7	11	15	20	17	16	16	15	14
Liquids		17	22	29	38	49	71	105	138	168	197	192	187
Gas		5	7	10	15	25	32	38	49	60	71	69	66
Electricity		3	6	11	18	37	63	98	141	184	225	254	283
Others		0	0	0	0	0	0	0	0	0	0	0	0
Total		27	38	55	78	121	182	261	346	428	509	530	551
Primary Energy	EJ												
Coal		4	6	10	14	31	50	73	80	87	92	95	99
Oil		20	23	30	39	49	73	111	147	180	210	171	132
Gas		7	10	17	27	49	72	95	131	162	189	187	185
Nuclear		0	2	5	7	16	25	35	44	54	65	73	81
Biomass		0	1	1	2	5	8	12	18	22	25	31	38
Other Renewables		5	6	8	11	15	21	29	44	58	71	85	99
Total		35	48	70	102	165	249	355	464	563	652	643	634
Cumulative Resources Use	ZJ												
Coal		0.0	0.1	0.1	0.3	0.5	1.0	1.5	2.3	3.2	4.0	5.0	5.9
Oil		0.0	0.2	0.5	0.8	1.3	2.0	2.8	4.2	5.9	7.8	9.5	11.3
Gas		0.0	0.1	0.2	0.4	0.9	1.5	2.3	3.5	5.0	6.7	8.6	10.5
Cumulative CO2 Emissions	GtC	1.4	17.8	37.1	60.6	91.9	133.2	186.4	252.2	329.4	416.5	506.9	594.6
Carbon Sequestration	GtC												
Land Use	Million ha												
Cropland		391	363	348	344	335	322	305	274	242	208	181	154
Grasslands		1510	1594	1727	1907	2104	2260	2376	2294	2168	1998	1869	1740
Energy Biomass		0	0	0	0	0	2	6	23	31	29	46	63
Forest		3641	3591	3480	3307	3118	2968	2857	2949	3100	3309	3451	3593
Others		1957	1883	1800	1707	1672	1657	1662	1655	1663	1688	1746	1805
Total		7499	7432	7354	7265	7229	7210	7206	7196	7204	7232	7293	7355
Anthropogenic Emissions (standardized)													
Fossil Fuel CO2	GtC	0.72	1.01	1.30	1.73	2.66	3.91	5.48	6.86	8.13	9.27	9.43	9.58
Other CO2	GtC	0.73	0.82	0.72	0.95	0.92	0.77	0.50	0.32	0.12	-0.09	-0.53	-0.96
Total CO2	GtC	1.45	1.83	2.02	2.68	3.58	4.68	5.98	7.18	8.25	9.18	8.90	8.62
CH4 total	MtCH4	77	85	90	100	124	141	150	162	170	173	168	162
N2O total	MtN2O-N	1.2	1.3	1.7	2.1	2.7	3.4	4.1	4.4	4.6	4.8	5.0	5.3
SOx total	MtS	10.5	12.8	14.9	16.1	18.6	21.5	24.7	17.9	13.8	12.4	12.4	12.4
HFC	MtC eq.	0	2	15	39	81	139	184	205	210	210	206	196
PFC	MtC eq.	4	4	5	9	14	19	23	26	28	30	31	30
SF6	MtC eq.	3	5	10	14	19	32	40	42	37	23	24	26
CO	MtCO	396	404	420	448	506	601	714	783	925	1117	1016	975
NMVOC	Mt	48	55	60	66	76	91	110	121	127	138	128	119
NOx	MtN	7	8	9	11	15	19	25	30	35	38	35	33

Scenario A1T-AIM World		1990	2000	2010	2020	2030	2040	2050	2060	2070	2080	2090	2100
Population	Million	5262	6117	6805	7493	8182	8439	8704	8538	8375	8030	7528	7056
GNP/GDP (mex)	Trillion US$	20.9	26.6	37.7	56.3	89.2	127.4	182.0	235.8	305.5	378.2	447.3	529.0
GNP/GDP (ppp) Trillion (1990 prices)													
Final Energy	EJ												
Non-commercial		50	48	34	27	18	0	0	0	0	0	0	0
Solids		36	36	46	59	76	76	76	58	45	35	29	24
Liquids		111	118	130	148	177	205	239	260	283	299	305	311
Gas		51	58	84	113	146	167	191	200	210	219	227	235
Electricity		38	48	63	82	115	180	282	370	484	573	614	657
Others													
Total		285	307	356	430	532	647	787	901	1032	1128	1176	1226
Primary Energy	EJ												
Coal		93	89	116	142	174	159	146	112	86	68	57	48
Oil		143	158	176	186	183	168	158	141	126	116	110	104
Gas		73	99	142	190	244	284	336	344	355	366	378	390
Nuclear		6	11	17	27	43	68	116	114	111	110	111	112
Biomass		50	48	38	47	81	120	179	215	258	289	302	315
Other Renewables		10	11	14	20	34	70	151	249	409	550	603	663
Total		376	416	502	611	758	868	1086	1174	1345	1500	1561	1632
Cumulative Resources Use	ZJ												
Coal		0.1	1.0	2.0	3.3	4.9	6.3	8.1	9.1	10.2	11.1	11.7	12.4
Oil		0.1	1.6	3.3	5.1	7.0	8.5	10.4	11.6	13.0	14.3	15.4	16.6
Gas		0.1	1.0	2.2	3.8	6.0	8.4	11.8	14.7	18.3	22.1	25.7	29.9
Cumulative CO2 Emissions	GtC	7.1	82.4	171.3	278.0	396.3	515.4	630.1	738.9	839.5	934.0	1025.3	1113.5
Carbon Sequestraction	GtC												
Land Use	Million ha												
Cropland		1437	1486	1548	1573	1533	1434	1341	1251	1167	1076	981	895
Grasslands		3290	3289	3481	3846	4193	4145	4098	4062	4027	3964	3875	3788
Energy Biomass		0	0	0	66	328	494	745	813	888	949	993	1039
Forest		4249	4122	3965	3801	3709	3786	3864	3898	3932	3980	4042	4106
Others		3966	4046	3948	3655	3179	3033	2894	2908	2922	2966	3039	3114
Total		12942	12942	12942	12942	12942	12942	12942	12942	12942	12942	12942	12942
Anthropogenic Emissions (standardized)													
Fossil Fuel CO2	GtC	5.99	6.90	8.37	9.79	11.28	11.32	11.43	10.49	9.66	9.14	8.85	8.58
Other CO2	GtC	1.11	1.07	1.44	1.73	0.87	0.34	-0.15	0.00	-0.03	0.15	0.12	0.08
Total CO2	GtC	7.10	7.97	9.81	11.52	12.15	11.67	11.28	10.49	9.63	9.29	8.97	8.66
CH4 total	MtCH4	310	323	393	466	528	509	492	442	397	358	323	291
N2O total	MtN2O-N	6.7	7.0	7.4	7.8	7.6	7.1	6.7	6.2	5.7	5.4	5.1	4.8
SOx total	MtS	70.9	69.0	87.4	100.9	91.5	69.3	64.3	46.9	35.7	30.7	29.0	27.4
CFC/HFC/HCFC	MtC eq.	1672	883	791	337	369	482	566	654	659	654	639	614
PFC	MtC eq.	32	25	31	43	61	77	89	97	106	114	119	115
SF6	MtC eq.	38	40	43	48	66	99	119	127	113	88	84	95
CO	MtCO	879	877	1030	1160	1161	1202	1244	1294	1347	1402	1460	1520
NMVOC	Mt	139	141	163	188	213	209	206	195	186	167	138	114
NOx	MtN	31	32	40	49	52	50	49	47	45	43	41	40

Scenario A1T-AIM OECD90		1990	2000	2010	2020	2030	2040	2050	2060	2070	2080	2090	2100
Population	Million	859	919	960	1002	1043	1062	1081	1086	1091	1097	1103	1110
GNP/GDP (mex)	Trillion US$	16.4	20.4	25.2	31.0	38.1	45.5	54.3	64.2	76.0	89.3	104.0	121.1
GNP/GDP (ppp) Trillion (1990 prices)													
Final Energy	EJ												
Non-commercial		6	1	0	0	0	0	0	0	0	0	0	0
Solids		10	10	10	10	10	9	8	6	5	4	3	2
Liquids		64	72	68	69	70	71	72	72	72	73	75	76
Gas		25	32	38	44	47	45	43	40	38	36	36	35
Electricity		22	27	30	33	37	46	58	72	90	105	115	126
Others													
Total		127	142	146	156	165	173	181	194	207	219	229	239
Primary Energy	EJ												
Coal		41	36	35	31	27	26	24	18	13	10	8	7
Oil		76	87	82	80	68	52	40	31	24	22	22	23
Gas		34	51	59	65	68	68	67	62	58	55	54	52
Nuclear		5	8	11	14	18	20	22	21	20	19	20	20
Biomass		6	1	0	3	11	19	34	41	49	55	58	60
Other Renewables		6	6	7	9	13	21	32	50	77	102	114	128
Total		167	189	194	203	205	205	220	223	242	264	276	291
Cumulative Resources Use	ZJ												
Coal		0.0	0.4	0.8	1.1	1.4	1.6	1.9	2.1	2.3	2.4	2.5	2.6
Oil		0.1	0.9	1.7	2.6	3.3	3.8	4.4	4.7	5.0	5.2	5.5	5.7
Gas		0.0	0.5	1.0	1.6	2.3	2.9	3.7	4.2	4.8	5.4	6.0	6.5
Cumulative CO2 Emissions	GtC	2.8	33.0	64.3	94.6	123.2	149.5	173.8	195.6	214.1	230.2	245.0	258.9
Carbon Sequestraction	GtC												
Land Use	Million ha												
Cropland		410	402	394	377	353	327	302	281	261	237	208	183
Grasslands		787	762	764	792	827	813	799	794	790	775	750	727
Energy Biomass		0	0	0	19	67	102	156	165	174	182	189	195
Forest		1056	1066	1079	1099	1127	1156	1185	1196	1208	1229	1261	1294
Others		886	910	903	853	765	731	697	702	706	715	727	740
Total		3140	3140	3140	3140	3140	3140	3140	3140	3140	3140	3140	3140
Anthropogenic Emissions (standardized)													
Fossil Fuel CO2	GtC	2.83	3.20	3.14	3.05	2.82	2.61	2.41	2.04	1.72	1.55	1.48	1.41
Other CO2	GtC	0.00	0.00	-0.07	-0.05	-0.09	-0.08	-0.07	-0.03	-0.03	-0.03	-0.04	-0.07
Total CO2	GtC	2.83	3.20	3.06	3.00	2.72	2.53	2.34	2.01	1.70	1.52	1.44	1.34
CH4 total	MtCH4	73	74	72	71	70	61	53	50	47	44	41	39
N2O total	MtN2O-N	2.6	2.6	2.5	2.5	2.4	2.2	2.0	1.9	1.7	1.5	1.4	1.3
SOx total	MtS	22.7	17.0	9.9	6.9	6.4	6.3	6.2	5.9	5.6	5.3	4.9	4.6
HFC	MtC eq.	19	58	110	108	115	118	122	123	123	124	125	125
PFC	MtC eq.	18	13	13	11	9	8	9	10	12	13	14	16
SF6	MtC eq.	23	23	14	5	6	7	9	11	13	16	18	20
CO	MtCO	179	161	170	178	192	199	206	210	215	222	234	245
NMVOC	Mt	42	36	34	33	31	26	21	14	8	5	4	3
NOx	MtN	13	12	12	12	9	7	6	6	6	5	5	5

Scenario A1T-AIM REF		**1990**	**2000**	**2010**	**2020**	**2030**	**2040**	**2050**	**2060**	**2070**	**2080**	**2090**	**2100**
Population	Million	413	419	424	430	435	429	423	406	391	374	356	339
GNP/GDP (mex)	Trillion US$	1.1	0.8	1.5	2.9	5.3	8.1	12.4	15.6	19.6	24.0	28.6	34.1
GNP/GDP (ppp) Trillion (1990 prices)													
Final Energy	EJ												
Non-commercial		2	4	0	0	0	0	0	0	0	0	0	0
Solids		9	4	4	5	5	5	4	3	2	2	2	1
Liquids		19	10	8	7	7	8	9	9	10	10	11	11
Gas		19	13	18	20	22	25	28	26	24	22	21	20
Electricity		8	9	10	11	12	18	26	31	37	40	40	39
Others													
Total		58	40	39	42	46	56	67	70	74	75	73	72
Primary Energy	EJ												
Coal		18	8	8	9	10	10	10	7	5	4	4	4
Oil		22	13	10	8	7	7	7	6	5	5	4	4
Gas		26	22	27	29	30	32	34	31	27	26	25	25
Nuclear		1	2	3	4	5	7	10	9	8	8	9	10
Biomass		2	4	0	1	3	5	8	10	12	13	13	14
Other Renewables		1	1	1	2	2	6	14	21	31	37	34	32
Total		71	51	50	52	57	66	83	83	89	93	91	90
Cumulative Resources Use	ZJ												
Coal		0.0	0.1	0.2	0.3	0.4	0.5	0.6	0.7	0.7	0.8	0.8	0.9
Oil		0.0	0.2	0.3	0.4	0.5	0.5	0.6	0.7	0.7	0.8	0.8	0.9
Gas		0.0	0.3	0.5	0.8	1.1	1.4	1.7	2.0	2.3	2.6	2.8	3.1
Cumulative CO2 Emissions	GtC	1.3	12.3	21.6	30.9	40.3	49.7	58.9	67.4	74.4	80.4	86.2	92.0
Carbon Sequestraction	GtC												
Land Use	Million ha												
Cropland		279	284	295	299	308	300	292	273	255	235	215	196
Grasslands		346	356	398	462	550	565	581	577	574	566	553	541
Energy Biomass		0	0	0	0	6	11	24	30	38	46	52	59
Forest		960	961	954	960	953	958	963	974	985	997	1009	1021
Others		720	703	657	585	488	466	445	448	451	459	473	488
Total		2305	2305	2305	2305	2305	2305	2305	2305	2305	2305	2305	2305
Anthropogenic Emissions (standardized)													
Fossil Fuel CO2	GtC	1.30	0.91	0.92	0.93	0.92	0.93	0.94	0.78	0.66	0.61	0.61	0.60
Other CO2	GtC	0.00	0.00	0.02	-0.01	0.03	0.00	-0.01	-0.02	-0.03	-0.03	-0.03	-0.03
Total CO2	GtC	1.30	0.91	0.94	0.92	0.95	0.93	0.93	0.76	0.63	0.58	0.58	0.58
CH4 total	MtCH4	47	39	59	65	68	59	51	49	48	45	43	41
N2O total	MtN2O-N	0.6	0.6	0.6	0.6	0.7	0.7	0.6	0.6	0.5	0.5	0.4	0.4
SOx total	MtS	17.0	11.0	12.2	10.8	7.5	4.3	2.4	2.0	1.7	1.6	1.6	1.6
HFC	MtC eq.	0	4	8	19	29	31	32	33	33	34	33	31
PFC	MtC eq.	7	4	5	8	14	20	21	22	23	24	25	24
SF6	MtC eq.	8	6	8	10	14	18	21	19	15	14	10	11
CO	MtCO	69	41	39	40	43	46	49	50	51	52	54	55
NMVOC	Mt	16	13	13	11	10	10	9	7	6	5	4	4
NOx	MtN	5	3	3	3	3	3	2	2	2	1	1	1

Scenario A1T-AIM ASIA		1990	2000	2010	2020	2030	2040	2050	2060	2070	2080	2090	2100
Population	Million	2798	3261	3556	3851	4147	4183	4220	4016	3822	3541	3194	2882
GNP/GDP (mex)	Trillion US$	1.5	2.7	5.7	12.2	26.2	40.6	62.9	85.5	116.4	147.7	174.9	207.0
GNP/GDP (ppp) Trillion (1990 prices)													
Final Energy	EJ												
Non-commercial		28	24	17	11	3	0	0	0	0	0	0	0
Solids		15	20	29	39	53	52	51	39	31	24	20	16
Liquids		11	16	23	29	41	49	59	67	77	83	83	83
Gas		2	4	8	16	29	36	45	53	63	73	82	91
Electricity		5	7	13	22	40	67	111	148	197	236	254	272
Others													
Total		61	70	91	118	166	210	265	315	373	417	439	462
Primary Energy	EJ												
Coal		30	38	58	79	108	92	79	62	48	38	31	25
Oil		17	24	35	40	44	37	32	28	25	22	20	18
Gas		4	9	18	33	58	76	99	111	125	138	151	165
Nuclear		0	1	3	5	13	25	47	47	46	46	45	45
Biomass		28	24	19	17	17	26	41	49	59	66	69	73
Other Renewables		1	2	2	4	9	23	58	98	166	227	251	278
Total		80	97	135	178	250	281	357	395	469	537	568	604
Cumulative Resources Use	ZJ												
Coal		0.0	0.4	0.8	1.5	2.5	3.3	4.3	4.9	5.5	6.0	6.3	6.7
Oil		0.0	0.2	0.5	0.9	1.3	1.6	2.1	2.3	2.6	2.8	3.1	3.3
Gas		0.0	0.1	0.2	0.5	0.9	1.5	2.5	3.4	4.6	6.0	7.4	9.1
Cumulative CO2 Emissions	GtC	1.5	19.3	44.4	78.4	121.3	167.9	212.4	255.0	295.8	334.9	372.5	409.0
Carbon Sequestraction	GtC												
Land Use	Million ha												
Cropland		390	408	432	448	432	401	372	349	328	307	286	267
Grasslands		521	523	546	584	620	618	615	611	606	599	589	580
Energy Biomass		0	0	0	21	88	123	172	189	209	224	233	242
Forest		527	490	453	420	406	419	432	437	442	448	456	464
Others		576	593	582	540	469	446	423	426	428	435	448	462
Total		2014	2014	2014	2014	2014	2014	2014	2014	2014	2014	2014	2014
Anthropogenic Emissions (standardized)													
Fossil Fuel CO2	GtC	1.15	1.78	2.59	3.36	4.51	4.40	4.29	4.06	3.85	3.69	3.57	3.45
Other CO2	GtC	0.37	0.26	0.39	0.44	0.27	0.14	0.07	0.10	0.13	0.14	0.13	0.13
Total CO2	GtC	1.53	2.03	2.98	3.81	4.78	4.54	4.36	4.16	3.99	3.83	3.70	3.58
CH4 total	MtCH4	113	125	156	196	238	235	233	197	167	146	131	118
N2O total	MtN2O-N	2.3	2.6	2.7	2.9	3.0	2.8	2.7	2.5	2.4	2.3	2.1	2.0
SOx total	MtS	17.7	25.3	42.2	54.5	45.9	20.0	8.5	8.0	7.5	7.1	6.7	6.3
HFC	MtC eq.	0	5	18	45	92	153	224	292	292	285	275	262
PFC	MtC eq.	3	5	8	15	23	30	35	39	43	46	48	46
SF6	MtC eq.	4	7	12	19	28	42	50	55	48	35	33	37
CO	MtCO	235	270	342	397	421	444	468	484	500	512	522	532
NMVOC	Mt	33	37	48	61	73	73	73	74	76	68	55	43
NOx	MtN	7	9	13	17	20	20	19	18	17	15	14	13

Scenario A1T-AIM ALM		1990	2000	2010	2020	2030	2040	2050	2060	2070	2080	2090	2100
Population	Million	1192	1519	1865	2211	2557	2761	2980	3024	3067	3013	2866	2727
GNP/GDP (mex)	Trillion US$	1.9	2.7	5.3	10.3	19.6	32.1	52.4	69.8	93.0	117.2	139.8	166.8
GNP/GDP (ppp) Trillion (1990 prices)													
Final Energy	EJ												
Non-commercial		14	18	17	16	15	0	0	0	0	0	0	0
Solids		1	2	3	5	9	10	12	9	7	6	5	4
Liquids		17	21	31	43	59	76	99	111	124	133	137	141
Gas		4	8	19	33	48	60	75	80	84	87	88	89
Electricity		4	6	11	17	26	47	87	118	159	191	205	220
Others													
Total		40	55	80	115	156	206	274	321	377	417	435	454
Primary Energy	EJ												
Coal		5	7	14	22	29	31	33	26	20	16	14	12
Oil		27	34	48	58	63	71	79	75	72	67	63	59
Gas		9	18	38	64	88	109	135	140	145	147	148	148
Nuclear		0	0	1	3	6	15	37	37	37	37	36	36
Biomass		14	18	19	27	50	69	96	115	138	154	161	168
Other Renewables		2	2	3	5	9	21	47	80	135	184	203	225
Total		57	80	124	179	246	316	427	472	545	606	625	648
Cumulative Resources Use	ZJ												
Coal		0.0	0.1	0.2	0.3	0.6	0.9	1.2	1.5	1.7	1.9	2.1	1.7
Oil		0.0	0.3	0.7	1.3	1.9	2.5	3.3	4.0	4.7	5.5	6.1	6.7
Gas		0.0	0.1	0.4	0.9	1.7	2.6	3.9	5.1	6.6	8.1	9.5	12.5
Cumulative CO2 Emissions	GtC	1.4	17.8	41.1	74.1	111.5	148.3	184.9	220.9	255.2	288.5	321.5	353.6
Carbon Sequestraction	GtC												
Land Use	Million ha												
Cropland		357	392	426	449	441	406	374	348	323	297	272	248
Grasslands		1636	1647	1773	2009	2196	2149	2103	2080	2057	2024	1982	1941
Energy Biomass		0	0	0	26	167	257	393	428	466	497	520	543
Forest		1706	1605	1478	1322	1223	1253	1284	1291	1297	1305	1316	1327
Others		1784	1840	1806	1677	1456	1391	1328	1333	1338	1357	1390	1424
Total		5483	5483	5483	5483	5483	5483	5483	5483	5483	5483	5483	5483
Anthropogenic Emissions (standardized)													
Fossil Fuel CO2	GtC	0.72	1.01	1.71	2.45	3.02	3.39	3.80	3.60	3.42	3.29	3.20	3.11
Other CO2	GtC	0.73	0.82	1.11	1.35	0.66	0.28	-0.15	-0.05	-0.11	0.06	0.05	0.05
Total CO2	GtC	1.45	1.83	2.82	3.79	3.69	3.67	3.65	3.55	3.31	3.35	3.25	3.16
CH4 total	MtCH4	77	85	106	133	152	154	155	145	136	123	108	94
N2O total	MtN2O-N	1.2	1.3	1.5	1.7	1.5	1.4	1.3	1.2	1.2	1.1	1.1	1.1
SOx total	MtS	10.5	12.8	20.2	25.7	28.8	35.8	44.2	28.0	17.8	13.7	12.8	12.0
HFC	MtC eq.	0	2	15	39	81	139	184	205	210	210	206	196
PFC	MtC eq.	4	4	5	9	14	19	23	26	28	30	31	30
SF6	MtC eq.	3	5	10	14	19	32	40	42	37	23	24	26
CO	MtCO	396	404	479	545	506	513	521	550	582	615	650	688
NMVOC	Mt	48	55	68	84	99	101	103	100	97	88	75	63
NOx	MtN	7	8	12	17	20	21	21	21	21	21	21	21

Illustrative Scenario A1T-MESSAGE World		**1990**	**2000**	**2010**	**2020**	**2030**	**2040**	**2050**	**2060**	**2070**	**2080**	**2090**	**2100**
Population	Million	5262	6117	6888	7617	8182	8531	8704	8667	8463	8125	7658	7056
GNP/GDP (mex)	Trillion US$	20.9	26.8	36.8	57.0	91.3	135.4	187.1	254.1	322.9	393.2	469.6	550.0
GNP/GDP (ppp) Trillion (1990 prices)		25.7	33.4	45.7	67.2	98.7	139.0	186.4	246.8	313.2	382.0	456.6	535.0
Final Energy	EJ												
Non-commercial		38	25	20	16	10	7	5	0	0	0	0	0
Solids		42	60	66	71	72	46	31	27	22	15	2	1
Liquids		111	125	157	193	246	300	344	357	354	346	343	311
Gas		41	48	66	83	107	135	155	180	177	159	132	100
Electricity		35	48	66	100	153	209	275	349	424	483	534	589
Others		8	12	18	33	48	61	83	108	138	179	220	268
Total		275	317	393	495	634	757	893	1020	1115	1182	1231	1270
Primary Energy	EJ												
Coal		91	106	125	151	180	153	119	87	60	53	40	25
Oil		128	155	172	193	223	241	250	236	205	143	113	77
Gas		71	87	124	166	231	288	324	344	324	291	240	196
Nuclear		7	8	11	17	40	78	115	145	175	175	153	114
Biomass		46	46	55	75	104	137	183	235	280	324	353	370
Other Renewables		8	15	25	48	73	122	222	358	544	756	985	1239
Total		352	416	513	649	850	1018	1213	1407	1588	1743	1884	2021
Cumulative Resources Use	ZJ												
Coal		0.0	0.9	2.0	3.2	4.7	6.5	8.1	9.3	10.1	10.7	11.3	11.7
Oil		0.0	1.4	3.0	4.7	6.6	8.9	11.3	13.8	16.1	18.2	19.6	20.8
Gas		0.0	0.8	1.6	2.9	4.5	6.8	9.7	13.0	16.4	19.6	22.6	25.0
Cumulative CO2 Emissions	GtC	7.1	82.4	169.2	267.4	380.5	505.7	630.2	748.4	854.7	944.1	1015.5	1068.4
Carbon Sequestration	GtC												
Land Use [a]	Million ha												
Cropland		1459	1466	1462	1457	1454	1448	1442	1436	1429	1424	1422	1420
Grasslands		3389	3404	3429	3446	3458	3478	3498	3525	3552	3568	3572	3576
Energy Biomass		0	0	0	74	158	288	418	492	566	581	538	495
Forest		4296	4237	4173	4164	4164	4177	4190	4194	4199	4202	4203	4204
Others		3805	3842	3886	3807	3715	3558	3400	3301	3203	3173	3213	3253
Total		12949	12949	12949	12949	12949	12949	12949	12949	12949	12949	12949	12949
Anthropogenic Emissions (standardized)													
Fossil Fuel CO2 [b]	GtC	5.99	6.90	8.33	10.00	12.26	12.60	12.29	11.41	9.91	8.05	6.27	4.31
Other CO2 [c]	GtC	1.11	1.07	1.04	0.26	0.12	0.05	-0.02	-0.03	-0.03	-0.03	-0.01	0.00
Total CO2	GtC	7.10	7.97	9.38	10.26	12.38	12.65	12.26	11.38	9.87	8.02	6.26	4.32
CH4 total [d]	MtCH4	310	323	362	415	483	495	500	459	404	359	317	274
N2O total [e]	MtN2O-N	6.7	7.0	6.1	6.1	6.2	6.2	6.1	6.0	5.7	5.6	5.5	5.4
SOx total	MtS	70.9	69.0	64.7	59.9	59.6	45.9	40.2	34.4	30.1	25.2	23.3	20.2
CFC/HFC/HCFC	MtC eq.	1672	883	791	337	369	482	566	654	659	654	639	614
PFC	MtC eq.	32	25	31	43	61	77	89	97	106	114	119	115
SF6	MtC eq.	38	40	43	48	66	99	119	127	113	88	84	95
CO	MtCO	879	877	1003	1147	1362	1555	1770	1944	2078	2164	2156	2077
NMVOC	Mt	139	141	164	190	212	229	241	242	229	199	167	128
NOx	MtN	31	32	39	46	56	60	61	60	52	43	35	28

Emissions correlated to land-use change and deforestation were calculated by using AIM A1 marker land-use data.
a: Land-use taken from AIM-A1 marker run.
b:CO2 emissions from fossil fuel and industrial processes (MESSAGE data).
c: CO2 emissions from land-use changes (AIM A1T run).
d: Non-energy related CH4 emissions were taken from AIM A1 marker run.
e: Non-energy related N2O emissions were taken from AIM A1T run.

Illustrative Scenario A1T-MESSAGE OECD90		1990	2000	2010	2020	2030	2040	2050	2060	2070	2080	2090	2100
Population	Million	859	919	965	1007	1043	1069	1081	1084	1089	1098	1108	1110
GNP/GDP (mex)	Trillion US$	16.4	20.6	25.6	31.6	38.7	46.8	55.7	65.7	77.2	90.8	106.6	124.3
GNP/GDP (ppp) Trillion (1990 prices)		14.1	17.8	22.1	27.4	33.7	40.8	48.6	57.4	67.6	79.7	93.6	109.4
Final Energy	EJ												
Non-commercial		0	0	0	0	0	0	0	0	0	0	0	0
Solids		13	13	6	3	1	1	0	0	0	0	0	0
Liquids		66	69	74	73	72	70	66	65	62	61	60	55
Gas		22	28	33	34	35	35	32	27	21	19	15	12
Electricity		22	29	36	44	55	65	74	83	95	104	112	118
Others		1	1	3	6	7	10	13	17	22	29	38	48
Total		124	139	152	160	171	180	185	193	201	212	225	233
Primary Energy	EJ												
Coal		38	36	36	32	25	11	4	3	2	1	1	1
Oil		72	83	81	72	63	55	46	39	30	23	20	14
Gas		33	45	57	70	79	78	76	66	59	56	45	34
Nuclear		6	7	8	10	22	43	56	64	67	58	45	28
Biomass		6	9	11	17	21	24	31	37	41	45	52	55
Other Renewables		4	6	10	15	20	27	44	65	98	143	194	245
Total		159	186	202	214	230	238	257	275	297	325	356	377
Cumulative Resources Use	ZJ												
Coal		0.0	0.4	0.7	1.1	1.4	1.7	1.7	1.8	1.8	1.8	1.8	1.8
Oil		0.0	0.8	1.6	2.4	3.1	3.8	4.3	4.8	5.2	5.5	5.7	5.9
Gas		0.0	0.4	0.8	1.4	2.1	2.9	3.6	4.4	5.0	5.6	6.2	6.6
Cumulative CO2 Emissions	GtC	2.8	33.0	65.7	98.4	129.2	155.8	177.3	195.7	211.4	224.7	235.8	244.1
Carbon Sequestration	GtC												
Land Use [a]	Million ha												
Cropland		381	380	377	376	375	372	368	364	359	356	356	355
Grasslands		760	763	771	774	776	787	797	818	839	849	847	846
Energy Biomass		0	0	0	11	24	44	63	74	86	88	81	75
Forest		1050	1053	1052	1053	1056	1062	1068	1061	1054	1050	1049	1048
Others		838	833	828	814	798	765	733	712	691	685	695	705
Total		3029	3029	3029	3029	3029	3029	3029	3029	3029	3029	3029	3029
Anthropogenic Emissions (standardized)													
Fossil Fuel CO2 [b]	GtC	2.83	3.20	3.31	3.18	2.96	2.38	2.00	1.71	1.38	1.17	0.95	0.66
Other CO2 [c]	GtC	0.00	0.00	0.04	0.02	0.00	-0.02	-0.04	0.00	0.05	0.06	0.04	0.01
Total CO2	GtC	2.83	3.20	3.35	3.20	2.96	2.35	1.96	1.72	1.43	1.24	0.98	0.67
CH4 total [d]	MtCH4	73	74	71	68	66	60	56	52	48	47	45	43
N2O total [e]	MtN2O-N	2.6	2.6	2.4	2.4	2.4	2.3	2.3	2.3	2.2	2.2	2.2	2.1
SOx total	MtS	22.7	17.0	8.3	2.1	0.6	-0.1	0.0	0.4	0.8	1.3	1.9	2.3
HFC	MtC eq.	19	58	110	108	115	118	122	123	123	124	125	125
PFC	MtC eq.	18	13	13	11	9	8	9	10	12	13	14	16
SF6	MtC eq.	23	23	14	5	6	7	9	11	13	16	18	20
CO	MtCO	179	161	183	190	184	186	202	210	206	206	208	190
NMVOC	Mt	42	36	38	38	33	30	27	26	23	20	16	10
NOx	MtN	13	12	14	13	13	11	9	7	6	5	4	2

Emissions correlated to land-use change and deforestation were calculated by using AIM A1 marker land-use data.
a: Land-use taken from AIM-A1 marker run.
b:CO2 emissions from fossil fuel and industrial processes (MESSAGE data).
c: CO2 emissions from land-use changes (AIM A1T run).
d: Non-energy related CH4 emissions were taken from AIM A1 marker run.
e: Non-energy related N2O emissions were taken from AIM A1T run.

Illustrative Scenario A1T-MESSAGE REF		1990	2000	2010	2020	2030	2040	2050	2060	2070	2080	2090	2100
Population	Million	413	419	427	433	435	433	423	409	392	374	357	339
GNP/GDP (mex)	Trillion US$	1.1	0.8	1.0	2.1	5.4	9.4	12.6	16.2	20.0	24.4	29.2	34.4
GNP/GDP (ppp) Trillion (1990 prices)		2.6	2.2	2.5	3.7	6.0	9.4	12.6	16.2	20.0	24.4	29.2	34.4
Final Energy	EJ												
Non-commercial		0	0	0	0	0	0	0	0	0	0	0	0
Solids		9	5	4	3	2	0	0	0	0	0	0	0
Liquids		15	11	10	12	14	14	12	13	13	14	16	17
Gas		14	11	14	17	22	26	27	26	23	19	13	9
Electricity		6	6	7	10	14	20	23	26	28	32	34	34
Others		7	6	6	9	13	13	12	12	11	11	12	13
Total		50	38	41	51	65	72	73	76	76	76	75	73
Primary Energy	EJ												
Coal		19	14	11	13	17	16	11	4	2	1	8	8
Oil		20	15	13	14	15	14	12	12	8	6	5	3
Gas		27	23	29	32	40	44	50	63	72	70	46	31
Nuclear		1	1	1	1	4	9	10	9	5	3	3	2
Biomass		2	1	1	1	3	4	7	10	12	14	16	19
Other Renewables		1	1	2	4	6	7	12	14	19	26	42	52
Total		70	54	56	65	84	95	101	112	117	120	119	116
Cumulative Resources Use	ZJ												
Coal		0.0	0.2	0.3	0.4	0.5	0.7	0.9	1.0	1.1	1.1	1.1	1.2
Oil		0.0	0.2	0.3	0.5	0.6	0.8	0.9	1.0	1.1	1.2	1.3	1.3
Gas		0.0	0.3	0.5	0.8	1.1	1.5	1.9	2.4	3.1	3.8	4.5	5.0
Cumulative CO2 Emissions	GtC	1.3	12.3	22.2	32.6	43.7	55.7	66.5	77.0	88.0	98.7	108.1	115.4
Carbon Sequestration	GtC												
Land Use [a]	Million ha												
Cropland		268	266	265	265	265	264	263	262	262	261	260	260
Grasslands		341	361	364	366	367	368	370	371	371	372	373	374
Energy Biomass		0	0	0	3	6	11	16	19	22	23	21	20
Forest		966	950	918	904	894	899	905	909	912	916	922	927
Others		701	698	728	738	745	733	722	715	709	703	700	696
Total		2276	2276	2276	2276	2276	2276	2276	2276	2276	2276	2276	2276
Anthropogenic Emissions (standardized)													
Fossil Fuel CO2 [b]	GtC	1.30	0.91	0.88	1.00	1.22	1.24	1.16	1.20	1.20	1.11	0.91	0.66
Other CO2 [c]	GtC	0.00	0.00	0.18	0.01	0.00	-0.08	-0.15	-0.12	-0.09	-0.07	-0.06	-0.05
Total CO2	GtC	1.30	0.91	1.06	1.01	1.22	1.17	1.01	1.09	1.11	1.03	0.84	0.61
CH4 total [d]	MtCH4	47	39	42	49	58	58	54	52	48	43	36	30
N2O total [e]	MtN2O-N	0.6	0.6	0.6	0.6	0.6	0.6	0.6	0.6	0.6	0.6	0.6	0.6
SOx total	MtS	17.0	11.0	7.5	4.9	5.2	4.5	4.1	1.8	1.2	0.7	3.0	3.1
HFC	MtC eq.	0	4	8	19	29	31	32	33	33	34	33	31
PFC	MtC eq.	7	4	5	8	14	20	21	22	23	24	25	24
SF6	MtC eq.	8	6	8	10	14	18	21	19	15	14	10	11
CO	MtCO	69	41	49	61	82	98	112	121	118	105	90	75
NMVOC	Mt	16	13	16	20	25	31	34	39	37	33	30	26
NOx	MtN	5	3	3	4	5	5	5	5	4	4	3	2

Emissions correlated to land-use change and deforestation were calculated by using AIM A1 marker land-use data.
a: Land-use taken from AIM-A1 marker run.
b:CO2 emissions from fossil fuel and industrial processes (MESSAGE data).
c: CO2 emissions from land-use changes (AIM A1T run).
d: Non-energy related CH4 emissions were taken from AIM A1 marker run.
e: Non-energy related N2O emissions were taken from AIM A1T run.

Illustrative Scenario A1T-MESSAGE ASIA		1990	2000	2010	2020	2030	2040	2050	2060	2070	2080	2090	2100
Population	Million	2798	3261	3620	3937	4147	4238	4220	4085	3867	3589	3258	2882
GNP/GDP (mex)	Trillion US$	1.5	2.7	5.8	13.5	27.2	44.9	65.3	95.8	126.9	155.5	186.5	218.2
GNP/GDP (ppp) Trillion (1990 prices)		5.3	8.3	13.5	22.8	36.2	52.2	70.0	96.8	126.9	155.5	186.5	218.2
Final Energy	EJ												
Non-commercial		24	15	11	8	4	3	2	0	0	0	0	0
Solids		19	38	51	60	63	38	24	21	21	13	1	0
Liquids		13	22	38	56	86	120	152	157	156	149	142	131
Gas		2	3	8	14	22	40	54	73	77	72	62	45
Electricity		4	8	15	30	51	70	98	128	160	183	201	222
Others		1	3	5	11	15	18	25	35	46	59	74	89
Total		62	88	128	178	241	289	356	414	459	476	479	486
Primary Energy	EJ												
Coal		30	51	71	96	122	102	79	57	39	40	26	12
Oil		15	26	39	55	76	91	102	95	90	60	45	31
Gas		3	5	14	28	48	74	98	108	101	88	81	68
Nuclear		0	1	2	4	8	15	27	43	65	74	65	48
Biomass		24	23	27	33	44	59	83	108	123	132	132	142
Other Renewables		1	3	6	15	26	44	79	130	191	261	343	435
Total		74	108	160	231	323	385	467	540	609	655	692	736
Cumulative Resources Use	ZJ												
Coal		0.0	0.3	0.8	1.6	2.5	3.7	4.8	5.6	6.1	6.5	6.9	7.2
Oil		0.0	0.2	0.4	0.8	1.4	2.1	3.1	4.1	5.0	5.9	6.5	7.0
Gas		0.0	0.0	0.1	0.2	0.5	1.0	1.7	2.7	3.8	4.8	5.7	6.5
Cumulative CO2 Emissions	GtC	1.5	19.3	44.3	77.8	122.4	174.6	227.3	276.5	319.5	355.3	383.4	403.2
Carbon Sequestration	GtC												
Land Use [a]	Million ha												
Cropland		438	435	434	433	432	431	431	430	429	428	428	428
Grasslands		608	606	609	611	613	615	617	618	619	621	622	624
Energy Biomass		0	0	0	17	37	68	99	117	134	138	127	117
Forest		535	522	512	524	535	535	535	544	552	556	555	555
Others		583	601	609	579	547	515	483	456	429	421	431	441
Total		2164	2164	2164	2164	2164	2164	2164	2164	2164	2164	2164	2164
Anthropogenic Emissions (standardized)													
Fossil Fuel CO2 [b]	GtC	1.15	1.78	2.73	3.90	5.26	5.36	5.19	4.65	4.06	3.29	2.47	1.52
Other CO2 [c]	GtC	0.37	0.26	0.22	-0.13	-0.12	-0.04	0.03	-0.03	-0.09	-0.10	-0.04	0.01
Total CO2	GtC	1.53	2.03	2.95	3.77	5.14	5.32	5.22	4.62	3.97	3.19	2.43	1.53
CH4 total [d]	MtCH4	113	125	152	185	225	226	227	200	167	147	130	110
N2O total [e]	MtN2O-N	2.3	2.6	2.4	2.4	2.5	2.5	2.5	2.4	2.3	2.3	2.2	2.2
SOx total	MtS	17.7	25.3	32.3	35.4	35.8	22.8	16.3	13.4	11.6	8.8	6.6	5.1
HFC	MtC eq.	0	5	18	45	92	153	224	292	292	285	275	262
PFC	MtC eq.	3	5	8	15	23	30	35	39	43	46	48	46
SF6	MtC eq.	4	7	12	19	28	42	50	55	48	35	33	37
CO	MtCO	235	270	340	422	564	663	784	853	896	914	866	777
NMVOC	Mt	33	37	47	58	70	75	82	82	83	76	63	41
NOx	MtN	7	9	12	17	23	26	28	26	22	18	15	12

Emissions correlated to land-use change and deforestation were calculated by using AIM A1 marker land-use data.
a: Land-use taken from AIM-A1 marker run.
b:CO2 emissions from fossil fuel and industrial processes (MESSAGE data).
c: CO2 emissions from land-use changes (AIM A1T run).
d: Non-energy related CH4 emissions were taken from AIM A1 marker run.
e: Non-energy related N2O emissions were taken from AIM A1T run.

Illustrative Scenario A1T-MESSAGE ALM		**1990**	**2000**	**2010**	**2020**	**2030**	**2040**	**2050**	**2060**	**2070**	**2080**	**2090**	**2100**
Population	Million	1192	1519	1875	2241	2557	2791	2980	3089	3115	3064	2934	2727
GNP/GDP (mex)	Trillion US$	1.9	2.7	4.4	9.8	20.0	34.3	53.5	76.5	98.7	122.5	147.2	173.1
GNP/GDP (ppp) Trillion (1990 prices)		3.8	5.1	7.5	13.3	22.8	36.7	55.3	76.5	98.7	122.5	147.2	173.1
Final Energy	EJ												
Non-commercial		14	10	9	8	5	4	3	0	0	0	0	0
Solids		1	4	5	5	6	8	7	6	2	1	1	1
Liquids		17	24	35	52	75	97	114	122	123	122	125	109
Gas		4	6	11	18	27	35	43	54	56	49	42	36
Electricity		3	5	9	17	33	54	81	111	140	165	188	215
Others		0	2	4	7	12	21	33	44	59	80	96	118
Total		39	52	73	106	157	217	279	336	379	418	452	478
Primary Energy	EJ												
Coal		5	5	8	10	16	24	26	24	18	12	5	4
Oil		21	31	40	53	69	81	90	89	77	54	43	28
Gas		8	14	23	37	65	91	101	107	92	78	69	63
Nuclear		0	0	1	2	6	12	23	30	38	40	40	35
Biomass		14	13	17	24	36	50	62	81	104	134	153	153
Other Renewables		2	4	8	14	21	44	87	150	236	326	406	508
Total		49	68	96	139	213	301	389	480	565	644	717	791
Cumulative Resources Use	ZJ												
Coal		0.0	0.0	0.1	0.2	0.3	0.4	0.7	0.9	1.2	1.4	1.4	1.5
Oil		0.0	0.3	0.6	1.0	1.5	2.2	3.0	3.9	4.8	5.6	6.1	6.5
Gas		0.0	0.1	0.2	0.5	0.8	1.5	2.4	3.4	4.5	5.4	6.2	6.8
Cumulative CO2 Emissions	GtC	1.4	17.8	37.0	58.5	85.3	119.6	159.0	199.2	235.8	265.4	288.2	305.8
Carbon Sequestration	GtC												
Land Use [a]	Million ha												
Cropland		371	385	384	383	382	381	381	380	379	379	378	378
Grasslands		1680	1673	1685	1695	1702	1708	1714	1718	1722	1726	1730	1733
Energy Biomass		0	0	0	42	90	165	240	282	324	333	308	284
Forest		1745	1711	1690	1683	1680	1681	1682	1681	1680	1678	1676	1674
Others		1684	1710	1720	1676	1625	1544	1463	1419	1374	1364	1387	1411
Total		5480	5480	5480	5480	5480	5480	5480	5480	5480	5480	5480	5480
Anthropogenic Emissions (standardized)													
Fossil Fuel CO2 b	GtC	0.72	1.01	1.41	1.92	2.82	3.62	3.94	3.84	3.26	2.48	1.94	1.48
Other CO2 c	GtC	0.73	0.82	0.60	0.36	0.25	0.19	0.14	0.12	0.10	0.08	0.06	0.03
Total CO2	GtC	1.45	1.83	2.01	2.28	3.07	3.81	4.07	3.96	3.36	2.56	2.00	1.51
CH4 total d	MtCH4	77	85	98	114	133	150	162	155	141	123	105	90
N2O total e	MtN2O-N	1.2	1.3	0.7	0.7	0.7	0.7	0.7	0.7	0.6	0.6	0.5	0.5
SOx total	MtS	10.5	12.8	13.7	14.6	15.0	15.7	16.8	15.7	13.5	11.3	8.8	6.8
HFC	MtC eq.	0	2	15	39	81	139	184	205	210	210	206	196
PFC	MtC eq.	4	4	5	9	14	19	23	26	28	30	31	30
SF6	MtC eq.	3	5	10	14	19	32	40	42	37	23	24	26
CO	MtCO	396	404	431	474	533	608	671	760	858	938	992	1035
NMVOC	Mt	48	55	63	74	83	92	98	95	87	69	58	50
NOx	MtN	7	8	10	12	15	18	20	21	19	15	13	12

Emissions correlated to land-use change and deforestation were calculated by using AIM A1 marker land-use data.
a: Land-use taken from AIM-A1 marker run.
b:CO2 emissions from fossil fuel and industrial processes (MESSAGE data).
c: CO2 emissions from land-use changes (AIM A1T run).
d: Non-energy related CH4 emissions were taken from AIM A1 marker run.
e: Non-energy related N2O emissions were taken from AIM A1T run.

Scenario A1T-MARIA World		**1990**	**2000**	**2010**	**2020**	**2030**	**2040**	**2050**	**2060**	**2070**	**2080**	**2090**	**2100**
Population	Million	5262	6117	6888	7617	8048	8207	8704	8536	8372	8028	7527	7056
GNP/GDP (mex)	Trillion US$	19.4	25.9	35.9	52.0	75.1	113.5	177.4	226.2	282.1	347.5	418.0	518.5
GNP/GDP (ppp) Trillion (1990 prices)													
Final Energy	EJ												
Non-commercial													
Solids		48	38	39	27	19	13	9	11	17	26	39	27
Liquids		138	152	185	234	269	310	357	363	429	416	475	512
Gas		56	65	84	110	156	201	250	278	213	232	177	179
Electricity		35	48	56	67	80	96	128	148	160	177	197	215
Others		0	0	0	0	0	0	0	0	0	0	0	0
Total		278	302	364	439	525	621	744	800	819	852	888	934
Primary Energy	EJ												
Coal		90	69	61	44	32	23	16	16	21	29	41	29
Oil		123	132	154	187	190	213	239	190	209	152	169	190
Gas		71	101	134	169	222	266	303	331	268	289	235	239
Nuclear		22	28	33	44	72	117	162	225	252	298	351	404
Biomass		28	31	41	64	100	111	128	180	227	270	313	329
Other Renewables		9	8	7	7	6	17	65	67	69	69	67	65
Total		343	370	431	515	620	747	913	1010	1044	1106	1175	1255
Cumulative Resources Use	ZJ												
Coal		0.0	0.9	1.6	2.2	2.6	3.0	3.2	3.3	3.5	3.7	4.0	4.4
Oil		0.0	1.2	2.5	4.1	6.0	7.9	10.0	12.4	14.3	16.4	17.9	19.6
Gas		0.0	0.7	1.7	3.1	4.8	7.0	9.6	12.7	16.0	18.7	21.5	23.9
Cumulative CO2 Emissions	GtC	7.1	82.4	165.8	256.2	352.3	454.4	563.6	672.1	773.5	868.6	959.2	1049.4
Carbon Sequestration	GtC	0.0	0.0	0.0	0.0	0.0	0.0	0.0	0.0	0.0	0.0	0.0	0.0
Land Use	Million ha												
Cropland		1451	1451	1613	2001	2170	2213	2344	2069	1802	1545	1287	1182
Grasslands		3395	3395	3236	2851	2682	2645	2606	2597	2597	2597	2597	2597
Energy Biomass		0	0	0	0	0	0	40	324	592	848	1107	1212
Forest		4138	4142	4157	4177	4208	4244	4301	4374	4375	4375	4375	4375
Others		4061	4057	4039	4017	3985	3944	3754	3681	3681	3681	3681	3681
Total		13045	13045	13045	13045	13045	13045	13045	13045	13045	13045	13045	13045
Anthropogenic Emissions (standardized)													
Fossil Fuel CO2	GtC	5.99	6.90	7.66	8.41	8.95	9.88	10.80	10.25	9.76	9.15	8.98	9.14
Other CO2	GtC	1.11	1.07	1.04	0.99	0.87	0.70	0.47	0.18	0.09	0.03	-0.04	-0.04
Total CO2	GtC	7.10	7.97	8.70	9.39	9.82	10.58	11.26	10.43	9.85	9.18	8.94	9.10
CH4 total	MtCH4												
N2O total	MtN2O-N												
SOx total	MtS												
CFC/HFC/HCFC	MtC eq.	1672	883	791	337	369	482	566	654	659	654	639	614
PFC	MtC eq.	32	25	31	43	61	77	89	97	106	114	119	115
SF6	MtC eq.	38	40	43	48	66	99	119	127	113	88	84	95
CO	MtCO												
NMVOC	Mt												
NOx	MtN												

Scenario A1T-MARIA OECD90		1990	2000	2010	2020	2030	2040	2050	2060	2070	2080	2090	2100
Population	Million	859	919	965	1007	1035	1046	1081	1085	1091	1096	1103	1110
GNP/GDP (mex)	Trillion US$	15.6	18.9	22.8	27.1	32.0	38.1	45.7	57.0	70.0	86.1	102.5	119.9
GNP/GDP (ppp) Trillion (1990 prices)													
Final Energy	EJ												
Non-commercial													
Solids		12	8	6	4	3	2	1	5	11	19	30	21
Liquids		71	64	57	53	51	49	49	47	74	58	76	71
Gas		28	35	48	57	65	70	75	84	60	78	55	74
Electricity		21	25	27	29	31	32	36	40	45	51	57	63
Others		0	0	0	0	0	0	0	0	0	0	0	0
Total		132	133	139	144	150	154	161	176	190	206	219	229
Primary Energy	EJ												
Coal		38	28	20	15	12	9	6	9	13	21	32	22
Oil		71	64	58	53	51	49	49	47	74	58	76	70
Gas		34	53	73	85	97	105	101	111	91	112	94	114
Nuclear		18	15	12	15	18	23	29	36	45	56	68	83
Biomass		6	4	3	2	1	1	1	1	1	2	3	5
Other Renewables		5	5	4	4	3	3	16	20	21	21	20	18
Total		171	168	170	175	183	190	202	223	245	270	292	313
Cumulative Resources Use	ZJ												
Coal		0.0	0.4	0.7	0.9	1.0	1.1	1.2	1.3	1.4	1.5	1.7	2.0
Oil		0.0	0.7	1.4	1.9	2.5	3.0	3.5	4.0	4.4	5.2	5.8	6.5
Gas		0.0	0.3	0.9	1.6	2.4	3.4	4.5	5.5	6.6	7.5	8.6	9.5
Cumulative CO2 Emissions	GtC	2.8	33.0	65.0	97.2	129.6	162.4	193.6	223.7	255.5	290.1	327.7	367.2
Carbon Sequestration	GtC	0.0	0.0	0.0	0.0	0.0	0.0	0.0	0.0	0.0	0.0	0.0	0.0
Land Use	Million ha												
Cropland		378	378	378	378	378	378	378	378	378	331	266	213
Grasslands		756	756	756	756	756	756	756	756	756	756	756	756
Energy Biomass		0	0	0	0	0	0	0	0	0	47	112	164
Forest		756	756	756	756	756	756	796	834	834	834	834	834
Others		794	794	794	794	794	794	754	716	716	716	716	716
Total		2684	2684	2684	2684	2684	2684	2684	2684	2684	2684	2684	2684
Anthropogenic Emissions (standardized)													
Fossil Fuel CO2	GtC	2.83	3.20	3.19	3.16	3.20	3.21	3.09	3.27	3.62	3.81	4.17	4.12
Other CO2	GtC	0.00	0.00	0.03	0.05	0.08	0.05	-0.10	-0.25	-0.27	-0.24	-0.21	-0.19
Total CO2	GtC	2.83	3.20	3.22	3.21	3.28	3.27	2.99	3.02	3.35	3.57	3.96	3.94
CH4 total	MtCH4												
N2O total	MtN2O-N												
SOx total	MtS												
HFC	MtC eq.	19	58	110	108	115	118	122	123	123	124	125	125
PFC	MtC eq.	18	13	13	11	9	8	9	10	12	13	14	16
SF6	MtC eq.	23	23	14	5	6	7	9	11	13	16	18	20
CO	MtCO												
NMVOC	Mt												
NOx	MtN												

Scenario A1T-MARIA REF		1990	2000	2010	2020	2030	2040	2050	2060	2070	2080	2090	2100
Population	Million	413	419	427	433	432	430	423	406	391	374	356	339
GNP/GDP (mex)	Trillion US$	0.9	1.3	1.8	2.6	4.5	7.9	13.4	17.6	21.7	26.1	31.5	38.7
GNP/GDP (ppp) Trillion (1990 prices)													
Final Energy	EJ												
Non-commercial													
Solids		13	9	6	4	3	2	2	2	3	5	7	5
Liquids		17	13	12	11	12	13	13	12	15	11	17	28
Gas		21	23	26	30	41	52	64	69	66	68	61	53
Electricity		7	8	7	7	9	11	13	14	15	15	16	16
Others		0	0	0	0	0	0	0	0	0	0	0	0
Total		58	53	51	53	65	78	92	98	99	99	101	102
Primary Energy	EJ												
Coal		19	13	9	6	4	3	2	2	3	6	7	5
Oil		19	14	12	11	12	13	14	13	15	11	17	28
Gas		27	31	34	39	53	68	81	88	85	86	79	70
Nuclear		3	2	2	2	3	3	4	5	7	8	10	13
Biomass		1	1	0	0	0	0	0	0	0	0	1	1
Other Renewables		1	1	1	1	1	1	4	4	4	4	4	3
Total		69	63	58	60	73	89	105	112	114	115	118	121
Cumulative Resources Use	ZJ												
Coal		0.0	0.2	0.3	0.4	0.5	0.5	0.5	0.6	0.6	0.6	0.7	0.8
Oil		0.0	0.2	0.3	0.5	0.6	0.7	0.8	1.0	1.1	1.2	1.3	1.5
Gas		0.0	0.3	0.6	0.9	1.3	1.8	2.5	3.3	4.2	5.1	5.9	6.7
Cumulative CO2 Emissions	GtC	1.3	12.3	21.0	29.2	38.3	49.3	62.1	75.6	89.0	102.2	115.0	128.1
Carbon Sequestration	GtC	0.0	0.0	0.0	0.0	0.0	0.0	0.0	0.0	0.0	0.0	0.0	0.0
Land Use	Million ha												
Cropland		217	217	217	217	217	217	217	174	139	111	89	78
Grasslands		114	114	114	114	114	114	114	114	114	114	114	114
Energy Biomass		0	0	0	0	0	0	0	43	78	106	128	139
Forest		815	815	815	815	815	815	827	862	862	862	862	862
Others		722	722	722	722	722	722	710	675	675	675	675	675
Total		1868	1868	1868	1868	1868	1868	1868	1868	1868	1868	1868	1868
Anthropogenic Emissions (standardized)													
Fossil Fuel CO2	GtC	1.30	0.91	0.80	0.80	0.97	1.19	1.38	1.46	1.50	1.48	1.54	1.58
Other CO2	GtC	0.00	0.00	0.02	0.03	0.02	0.01	-0.03	-0.12	-0.15	-0.20	-0.25	-0.24
Total CO2	GtC	1.30	0.91	0.82	0.83	1.00	1.20	1.35	1.34	1.35	1.29	1.28	1.34
CH4 total	MtCH4												
N2O total	MtN2O-N												
SOx total	MtS												
HFC	MtC eq.	0	4	8	19	29	31	32	33	33	34	33	31
PFC	MtC eq.	7	4	5	8	14	20	21	22	23	24	25	24
SF6	MtC eq.	8	6	8	10	14	18	21	19	15	14	10	11
CO	MtCO												
NMVOC	Mt												
NOx	MtN												

Scenario A1T-MARIA ASIA		**1990**	**2000**	**2010**	**2020**	**2030**	**2040**	**2050**	**2060**	**2070**	**2080**	**2090**	**2100**
Population	Million	2642	3080	3425	3728	3861	3895	4008	3814	3632	3368	3040	2744
GNP/GDP (mex)	Trillion US$	1.2	2.6	5.8	12.3	20.8	35.9	61.8	80.3	102.5	128.8	159.8	204.1
GNP/GDP (ppp) Trillion (1990 prices)													
Final Energy	EJ												
Non-commercial													
Solids		20	18	25	17	12	8	6	4	3	2	2	1
Liquids		26	42	65	97	120	146	168	172	190	200	212	214
Gas		2	3	4	8	16	25	35	41	29	24	17	22
Electricity		4	9	11	16	19	24	34	41	46	52	59	64
Others		0	0	0	0	0	0	0	0	0	0	0	0
Total		52	72	106	139	168	203	243	257	267	277	289	301
Primary Energy	EJ												
Coal		28	24	29	20	14	10	7	5	3	2	2	1
Oil		14	23	36	51	51	72	92	72	56	39	45	41
Gas		3	5	8	14	20	28	38	42	30	25	17	22
Nuclear		1	12	18	26	28	50	55	79	89	103	122	133
Biomass		13	21	31	51	80	81	81	104	137	162	168	174
Other Renewables		1	1	1	1	1	1	20	20	20	20	20	20
Total		61	86	124	163	195	243	293	322	335	352	373	391
Cumulative Resources Use	ZJ												
Coal		0.0	0.3	0.5	0.8	1.0	1.2	1.3	1.3	1.4	1.4	1.4	1.4
Oil		0.0	0.1	0.4	0.7	1.3	1.8	2.5	3.4	4.1	4.7	5.1	5.5
Gas		0.0	0.0	0.1	0.2	0.3	0.5	0.8	1.2	1.6	1.9	2.1	2.3
Cumulative CO2 Emissions	GtC	1.5	19.3	41.5	66.1	90.9	117.3	148.0	179.0	205.0	225.8	244.1	261.9
Carbon Sequestration	GtC	0.0	0.0	0.0	0.0	0.0	0.0	0.0	0.0	0.0	0.0	0.0	0.0
Land Use	Million ha												
Cropland		366	366	380	380	380	380	380	377	373	336	305	263
Grasslands		431	431	421	423	423	423	384	375	375	375	375	375
Energy Biomass		0	0	0	0	0	0	40	52	57	93	124	167
Forest		365	367	380	388	397	404	404	404	404	404	404	404
Others		458	456	440	430	421	414	412	412	412	412	412	412
Total		1621	1621	1621	1621	1621	1621	1621	1621	1621	1621	1621	1621
Anthropogenic Emissions (standardized)													
Fossil Fuel CO2	GtC	1.15	1.78	2.20	2.36	2.31	2.73	3.21	2.82	2.27	1.83	1.81	1.80
Other CO2	GtC	0.37	0.26	0.20	0.16	0.12	0.10	0.09	0.07	0.04	0.02	0.00	-0.03
Total CO2	GtC	1.53	2.03	2.40	2.53	2.44	2.83	3.31	2.89	2.32	1.85	1.80	1.77
CH4 total	MtCH4												
N2O total	MtN2O-N												
SOx total	MtS												
HFC	MtC eq.	0	5	18	45	92	153	224	292	292	285	275	262
PFC	MtC eq.	3	5	8	15	23	30	35	39	43	46	48	46
SF6	MtC eq.	4	7	12	19	28	42	50	55	48	35	33	37
CO	MtCO												
NMVOC	Mt												
NOx	MtN												

Scenario A1T-MARIA ALM		1990	2000	2010	2020	2030	2040	2050	2060	2070	2080	2090	2100
Population	Million	1348	1699	2071	2449	2720	2836	3192	3231	3259	3190	3029	2864
GNP/GDP (mex)	Trillion US$	1.7	3.1	5.5	10.0	17.8	31.7	56.5	71.4	87.9	106.5	124.2	155.9
GNP/GDP (ppp) Trillion (1990 prices)													
Final Energy	EJ												
Non-commercial													
Solids		3	2	2	1	1	1	0	0	0	0	0	0
Liquids		24	33	50	72	86	103	127	132	150	148	171	199
Gas		5	4	6	14	35	54	75	84	59	63	44	31
Electricity		3	6	10	15	21	29	45	53	54	59	65	72
Others		0	0	0	0	0	0	0	0	0	0	0	0
Total		35	45	68	103	143	186	247	268	263	270	280	302
Primary Energy	EJ												
Coal		6	4	3	2	1	1	1	0	0	0	0	0
Oil		18	30	48	71	75	79	84	59	63	44	31	50
Gas		7	12	19	32	51	65	83	90	63	65	46	32
Nuclear		0	0	0	0	23	40	74	105	112	130	151	174
Biomass		8	6	7	11	18	28	46	75	88	105	141	149
Other Renewables		2	2	1	1	1	12	25	24	25	24	24	24
Total		41	53	79	117	170	226	313	353	351	369	392	430
Cumulative Resources Use	ZJ												
Coal		0.0	0.1	0.1	0.1	0.1	0.2	0.2	0.2	0.2	0.2	0.2	0.2
Oil		0.0	0.2	0.5	1.0	1.7	2.4	3.2	4.1	4.6	5.3	5.7	6.0
Gas		0.0	0.1	0.2	0.4	0.7	1.2	1.9	2.7	3.6	4.2	4.9	5.3
Cumulative CO2 Emissions	GtC	1.4	17.8	38.3	63.7	93.4	125.4	159.9	193.9	224.0	250.6	272.4	292.1
Carbon Sequestration	GtC	0.0	0.0	0.0	0.0	0.0	0.0	0.0	0.0	0.0	0.0	0.0	0.0
Land Use	Million ha												
Cropland		490	490	639	1026	1195	1238	1370	1141	913	767	627	627
Grasslands		2095	2095	1946	1559	1390	1352	1352	1352	1352	1352	1352	1352
Energy Biomass		0	0	0	0	0	0	0	229	457	603	742	742
Forest		2202	2204	2207	2218	2241	2268	2274	2274	2274	2274	2274	2274
Others		2086	2084	2081	2070	2047	2014	1878	1878	1878	1878	1878	1878
Total		6873	6873	6873	6873	6873	6873	6873	6873	6873	6873	6873	6873
Anthropogenic Emissions (standardized)													
Fossil Fuel CO2	GtC	0.72	1.01	1.47	2.09	2.47	2.74	3.12	2.70	2.37	2.03	1.46	1.64
Other CO2	GtC	0.73	0.82	0.79	0.74	0.64	0.54	0.50	0.48	0.47	0.45	0.43	0.42
Total CO2	GtC	1.45	1.83	2.26	2.83	3.11	3.28	3.62	3.19	2.84	2.48	1.89	2.05
CH4 total	MtCH4												
N2O total	MtN2O-N												
SOx total	MtS												
HFC	MtC eq.	0	2	15	39	81	139	184	205	210	210	206	196
PFC	MtC eq.	4	4	5	9	14	19	23	26	28	30	31	30
SF6	MtC eq.	3	5	10	14	19	32	40	42	37	23	24	26
CO	MtCO												
NMVOC	Mt												
NOx	MtN												

Scenario A1v1-MiniCAM World		1990	2000	2010	2020	2030	2040	2050	2060	2070	2080	2090	2100
Population	Million	5293	6100	6874	7618	8122	8484	8703	8623	8430	8126	7621	7137
GNP/GDP (mex)	Trillion US$	20.7	27.4	38.1	52.8	80.2	117.3	164.2	226.6	294.1	366.9	445.1	531.2
GNP/GDP (ppp) Trillion (1990 prices)		na	na	na	na	na	na	na	na	na	na	na	na
Final Energy	EJ												
Non-commercial		0	0	0	0	0	0	0	0	0	0	0	0
Solids		45	54	64	75	87	96	101	78	63	55	50	46
Liquids		121	121	129	146	152	177	221	256	289	318	334	350
Gas		52	61	75	96	110	117	116	136	156	175	176	177
Electricity		35	51	76	112	161	218	283	338	387	429	459	488
Others		0	0	0	0	0	0	0	0	0	0	0	0
Total		253	286	344	429	510	607	721	808	894	978	1019	1061
Primary Energy	EJ												
Coal		88	108	130	153	206	250	284	249	223	205	192	179
Oil		131	130	137	152	147	167	214	259	299	333	351	369
Gas		70	82	118	179	223	259	287	343	387	417	408	399
Nuclear		24	24	30	41	60	81	102	101	105	116	126	135
Biomass		0	5	11	16	23	30	37	48	54	54	52	50
Other Renewables		24	24	27	31	39	51	68	91	112	131	146	162
Total		336	374	453	573	698	838	992	1091	1179	1255	1274	1293
Cumulative Resources Use	ZJ												
Coal		0.1	1.1	2.3	3.7	5.7	7.9	10.5	13.1	15.5	17.6	19.6	21.5
Oil		0.1	1.4	2.8	4.2	5.7	7.3	9.2	11.7	14.5	17.5	21.0	24.5
Gas		0.1	0.9	1.9	3.3	5.4	7.9	10.5	13.8	17.5	21.4	25.5	29.6
Cumulative CO2 Emissions	GtC	7.1	82.4	168.1	270.7	393.9	535.2	692.8	861.1	1034.4	1210.5	1383.2	1547.3
Carbon Sequestration	GtC												
Land Use	Million ha												
Cropland		1472	1466	1474	1495	1483	1442	1373	1212	1055	903	791	679
Grasslands		3209	3348	3603	3974	4374	4690	4923	4749	4493	4155	3951	3748
Energy Biomass		0	3	6	8	10	11	12	22	21	10	6	3
Forest		4173	4214	4140	3952	3628	3372	3184	3515	3904	4349	4496	4643
Others		4310	4132	3941	3735	3669	3648	3673	3666	3690	3747	3920	4093
Total		13164	13164	13164	13164	13164	13164	13164	13164	13164	13164	13164	13164
Anthropogenic Emissions (standardized)													
Fossil Fuel CO2	GtC	5.99	6.90	8.08	9.81	11.74	13.70	15.80	16.62	17.37	18.06	18.20	18.36
Other CO2	GtC	1.11	1.07	1.08	1.54	1.56	1.27	0.74	0.52	0.16	-0.36	-1.37	-2.37
Total CO2	GtC	7.10	7.97	9.16	11.35	13.29	14.97	16.53	17.14	17.53	17.70	16.83	15.98
CH4 total	MtCH4	310	323	355	405	462	523	567	585	599	611	625	640
N2O total	MtN2O-N	6.7	7.0	8.0	9.3	10.8	12.6	14.3	14.8	15.2	15.3	16.0	16.7
SOx total	MtS	70.9	69.0	74.2	79.5	79.6	72.7	58.9	41.5	31.4	28.6	28.5	28.5
CFC/HFC/HCFC	MtC eq.	1672	883	791	337	369	482	566	654	659	654	639	614
PFC	MtC eq.	32	25	31	43	61	77	89	97	106	114	119	115
SF6	MtC eq.	38	40	43	48	66	99	119	127	113	88	84	95
CO	MtCO												
NMVOC	Mt												
NOx	MtN												

Scenario A1v1-MiniCAM OECD90		1990	2000	2010	2020	2030	2040	2050	2060	2070	2080	2090	2100
Population	Million	838	908	965	1007	1024	1066	1081	1084	1090	1098	1105	1112
GNP/GDP (mex)	Trillion US$	16.3	20.5	25.6	31.6	34.5	44.7	53.6	63.3	74.4	86.7	102.2	119.1
GNP/GDP (ppp) Trillion (1990 prices)		na	na	na	na	na	na	na	na	na	na	na	na
Final Energy	EJ												
Non-commercial		0	0	0	0	0	0	0	0	0	0	0	0
Solids		10	12	12	11	11	11	12	9	7	7	7	7
Liquids		72	70	69	69	63	55	59	61	64	68	72	75
Gas		27	35	44	54	54	53	49	54	60	67	69	71
Electricity		22	27	33	41	44	53	61	64	67	70	74	79
Others		0	0	0	0	0	0	0	0	0	0	0	0
Total		130	144	159	175	173	173	181	188	198	211	221	231
Primary Energy	EJ												
Coal		40	44	43	36	39	48	54	51	48	42	37	32
Oil		76	75	73	70	63	50	52	52	56	62	68	74
Gas		34	45	63	90	91	92	88	95	103	109	110	111
Nuclear		20	15	13	13	14	14	15	13	14	15	17	19
Biomass		0	2	3	4	4	5	6	7	8	8	7	7
Other Renewables		12	11	10	10	10	11	12	14	16	18	21	24
Total		182	192	206	223	221	219	226	234	243	254	261	268
Cumulative Resources Use	ZJ												
Coal		0.0	0.5	0.9	1.3	1.5	2.2	2.7	3.2	3.7	4.1	4.5	4.9
Oil		0.1	0.8	1.6	2.3	2.6	3.4	3.9	4.4	5.0	5.6	6.2	6.9
Gas		0.0	0.5	1.0	1.7	2.2	3.6	4.5	5.4	6.4	7.5	8.5	9.6
Cumulative CO2 Emissions	GtC	2.8	33.0	66.7	103.5	141.5	179.0	216.3	254.0	291.9	329.4	365.9	401.4
Carbon Sequestration	GtC												
Land Use	Million ha												
Cropland		408	410	412	414	409	386	359	313	273	240	211	182
Grasslands		796	820	867	937	972	1062	1100	1064	1019	967	932	896
Energy Biomass		0	3	5	5	6	8	9	13	12	9	6	3
Forest		921	931	922	894	870	812	792	873	951	1026	1053	1080
Others		998	959	917	872	866	856	862	861	867	880	921	962
Total		3123	3123	3123	3123	3123	3123	3123	3123	3123	3123	3123	3123
Anthropogenic Emissions (standardized)													
Fossil Fuel CO2	GtC	2.83	3.20	3.41	3.58	3.55	3.56	3.72	3.78	3.86	3.94	3.99	4.04
Other CO2	GtC	0.00	0.00	0.13	0.24	0.24	0.15	0.03	0.01	-0.08	-0.23	-0.39	-0.55
Total CO2	GtC	2.83	3.20	3.54	3.82	3.79	3.71	3.75	3.79	3.78	3.71	3.60	3.49
CH4 total	MtCH4	73	74	84	97	104	127	145	163	174	179	182	185
N2O total	MtN2O-N	2.6	2.6	2.8	3.0	3.2	3.7	4.0	4.1	4.1	4.2	4.4	4.5
SOx total	MtS	22.7	17.0	12.5	6.2	5.4	5.0	5.2	5.1	5.2	5.5	5.9	6.2
HFC	MtC eq.	19	58	110	108	115	118	122	123	123	124	125	125
PFC	MtC eq.	18	13	13	11	9	8	9	10	12	13	14	16
SF6	MtC eq.	23	23	14	5	6	7	9	11	13	16	18	20
CO	MtCO												
NMVOC	Mt												
NOx	MtN												

Scenario A1v1-MiniCAM REF		1990	2000	2010	2020	2030	2040	2050	2060	2070	2080	2090	2100
Population	Million	428	425	426	433	434	431	423	408	392	374	357	340
GNP/GDP (mex)	Trillion US$	1.1	1.1	1.4	2.1	3.5	5.1	6.9	10.0	13.6	17.7	21.8	26.3
GNP/GDP (ppp) Trillion (1990 prices)		na	na	na	na	na	na	na	na	na	na	na	na
Final Energy	EJ												
Non-commercial		0	0	0	0	0	0	0	0	0	0	0	0
Solids		13	10	8	7	7	6	6	4	3	3	3	2
Liquids		18	11	8	9	9	9	10	10	10	11	11	11
Gas		19	14	13	15	17	16	14	15	16	17	16	15
Electricity		6	8	11	17	21	25	28	30	31	33	33	34
Others		0	0	0	0	0	0	0	0	0	0	0	0
Total		56	44	41	49	54	57	58	59	61	64	63	62
Primary Energy	EJ												
Coal		18	17	17	19	27	32	34	26	21	18	18	17
Oil		20	13	10	11	7	6	8	10	12	13	13	13
Gas		26	20	21	29	32	33	31	32	34	34	32	29
Nuclear		3	4	6	8	10	10	11	9	8	9	9	9
Biomass		0	1	1	2	2	3	3	4	4	4	4	4
Other Renewables		3	3	3	4	5	6	8	9	11	12	13	14
Total		70	57	58	73	83	90	93	91	89	90	87	85
Cumulative Resources Use	ZJ												
Coal		0.0	0.2	0.4	0.5	0.8	1.1	1.4	1.7	1.9	2.1	2.3	2.5
Oil		0.0	0.2	0.3	0.4	0.5	0.6	0.6	0.7	0.8	0.9	1.1	1.2
Gas		0.0	0.2	0.5	0.7	1.0	1.3	1.6	2.0	2.3	2.6	2.9	3.3
Cumulative CO2 Emissions	GtC	1.3	12.3	21.4	31.6	44.0	57.9	72.3	86.2	99.5	112.0	123.0	131.9
Carbon Sequestration	GtC												
Land Use	Million ha												
Cropland		284	294	305	317	321	317	304	263	221	180	157	134
Grasslands		395	410	454	527	612	678	725	671	602	517	491	465
Energy Biomass		0	0	0	1	1	1	0	0	0	0	0	0
Forest		1007	1016	996	945	866	808	769	865	970	1086	1104	1122
Others		691	657	623	587	577	574	579	579	584	594	625	657
Total		2377	2377	2377	2377	2377	2377	2377	2377	2377	2377	2377	2377
Anthropogenic Emissions (standardized)													
Fossil Fuel CO2	GtC	1.30	0.91	0.87	1.05	1.22	1.33	1.36	1.26	1.20	1.18	1.16	1.15
Other CO2	GtC	0.00	0.00	0.03	0.08	0.12	0.12	0.07	0.11	0.09	0.03	-0.16	-0.35
Total CO2	GtC	1.30	0.91	0.91	1.13	1.34	1.44	1.43	1.37	1.29	1.20	1.00	0.80
CH4 total	MtCH4	47	39	48	64	78	89	98	99	101	106	113	120
N2O total	MtN2O-N	0.6	0.6	0.7	0.9	1.1	1.4	1.6	1.6	1.5	1.5	1.5	1.6
SOx total	MtS	17.0	11.0	10.2	10.4	11.8	11.2	8.6	4.6	2.3	1.8	1.9	2.1
HFC	MtC eq.	0	4	8	19	29	31	32	33	33	34	33	31
PFC	MtC eq.	7	4	5	8	14	20	21	22	23	24	25	24
SF6	MtC eq.	8	6	8	10	14	18	21	19	15	14	10	11
CO	MtCO												
NMVOC	Mt												
NOx	MtN												

Scenario A1v1-MiniCAM ASIA		**1990**	**2000**	**2010**	**2020**	**2030**	**2040**	**2050**	**2060**	**2070**	**2080**	**2090**	**2100**
Population	Million	2790	3226	3608	3937	4115	4210	4219	4062	3852	3589	3245	2919
GNP/GDP (mex)	Trillion US$	1.4	3.1	6.5	11.7	23.9	40.4	61.1	86.2	112.1	138.8	165.9	195.4
GNP/GDP (ppp) Trillion (1990 prices)		na	na	na	na	na	na	na	na	na	na	na	na
Final Energy	EJ												
Non-commercial		0	0	0	0	0	0	0	0	0	0	0	0
Solids		20	29	39	51	60	66	68	52	41	35	31	27
Liquids		14	19	25	35	44	57	74	86	96	104	106	108
Gas		2	5	9	13	18	21	23	29	33	37	36	35
Electricity		4	10	22	39	64	93	124	149	169	184	192	200
Others		0	0	0	0	0	0	0	0	0	0	0	0
Total		40	62	95	138	186	237	290	316	339	359	364	370
Primary Energy	EJ												
Coal		26	42	62	87	116	137	149	122	102	92	85	77
Oil		16	20	27	37	42	54	73	92	105	112	114	115
Gas		3	8	20	38	59	79	99	121	135	141	132	124
Nuclear		1	4	8	14	25	37	50	49	50	53	55	57
Biomass		0	2	5	8	12	16	19	25	27	27	25	24
Other Renewables		3	4	5	6	10	16	24	35	45	53	59	65
Total		49	79	126	190	264	339	415	443	464	477	470	462
Cumulative Resources Use	ZJ												
Coal		0.0	0.4	1.0	1.6	2.7	4.0	5.4	6.7	7.8	8.8	9.6	10.5
Oil		0.0	0.2	0.5	0.8	1.2	1.7	2.3	3.2	4.1	5.2	6.3	7.5
Gas		0.0	0.1	0.2	0.5	1.0	1.7	2.6	3.8	5.0	6.4	7.7	9.1
Cumulative CO2 Emissions	GtC	1.5	19.3	43.4	76.7	121.5	177.0	241.6	309.9	377.3	443.5	507.1	566.8
Carbon Sequestration	GtC												
Land Use	Million ha												
Cropland		389	400	410	420	424	419	407	365	321	276	243	211
Grasslands		508	524	555	603	653	694	726	719	702	674	649	624
Energy Biomass		0	0	1	2	3	3	3	6	6	2	1	-1
Forest		1168	1144	1102	1044	990	952	929	973	1034	1113	1172	1231
Others		664	632	601	568	559	557	563	562	567	577	607	637
Total		2729	2699	2669	2638	2628	2624	2627	2625	2630	2642	2672	2702
Anthropogenic Emissions (standardized)													
Fossil Fuel CO2	GtC	1.15	1.78	2.58	3.64	4.77	5.80	6.72	6.69	6.66	6.62	6.39	6.17
Other CO2	GtC	0.37	0.26	0.20	0.27	0.28	0.25	0.16	0.10	0.03	-0.06	-0.23	-0.39
Total CO2	GtC	1.53	2.03	2.77	3.90	5.05	6.04	6.87	6.79	6.68	6.56	6.17	5.78
CH4 total	MtCH4	113	125	134	145	161	176	189	180	174	173	174	174
N2O total	MtN2O-N	2.3	2.6	2.9	3.2	3.7	4.2	4.7	4.9	4.9	5.0	5.2	5.4
SOx total	MtS	17.7	25.3	34.8	45.1	44.1	37.4	25.1	16.4	11.3	9.8	9.5	9.1
HFC	MtC eq.	0	5	18	45	92	153	224	292	292	285	275	262
PFC	MtC eq.	3	5	8	15	23	30	35	39	43	46	48	46
SF6	MtC eq.	4	7	12	19	28	42	50	55	48	35	33	37
CO	MtCO												
NMVOC	Mt												
NOx	MtN												

Scenario A1v1-MiniCAM ALM		1990	2000	2010	2020	2030	2040	2050	2060	2070	2080	2090	2100
Population	Million	1236	1541	1876	2241	2531	2778	2980	3068	3096	3064	2913	2766
GNP/GDP (mex)	Trillion US$	1.9	2.8	4.6	7.4	15.4	27.2	42.5	67.0	94.1	123.7	155.2	190.4
GNP/GDP (ppp) Trillion (1990 prices)		na	na	na	na	na	na	na	na	na	na	na	na
Final Energy	EJ												
Non-commercial		0	0	0	0	0	0	0	0	0	0	0	0
Solids		2	3	4	6	9	12	15	13	11	11	10	10
Liquids		17	21	26	33	40	56	78	99	118	136	146	156
Gas		5	6	9	13	21	26	29	38	46	54	55	56
Electricity		3	6	9	15	29	47	70	96	120	143	159	176
Others		0	0	0	0	0	0	0	0	0	0	0	0
Total		27	35	49	67	99	141	192	245	296	344	371	399
Primary Energy	EJ												
Coal		4	5	8	11	22	34	48	50	52	53	52	52
Oil		20	22	27	34	41	57	82	105	127	146	156	166
Gas		7	9	14	23	40	55	70	95	116	133	134	136
Nuclear		0	2	4	5	12	19	27	29	34	40	45	50
Biomass		0	1	1	2	4	6	9	12	15	15	15	15
Other Renewables		5	6	8	11	15	19	24	33	40	47	53	60
Total		35	45	62	86	133	190	258	324	383	434	456	478
Cumulative Resources Use	ZJ												
Coal		0.0	0.1	0.1	0.2	0.4	0.7	1.1	1.6	2.1	2.6	3.1	3.7
Oil		0.0	0.2	0.5	0.8	1.2	1.7	2.3	3.3	4.5	5.8	7.4	8.9
Gas		0.0	0.1	0.2	0.4	0.8	1.2	1.8	2.7	3.8	5.0	6.3	7.7
Cumulative CO2 Emissions	GtC	1.4	17.8	36.7	58.8	86.9	121.3	162.6	210.9	265.7	325.7	387.2	447.1
Carbon Sequestration	GtC												
Land Use	Million ha												
Cropland		391	363	347	344	334	320	303	272	240	207	180	152
Grasslands		1510	1594	1727	1906	2102	2257	2373	2296	2170	1996	1880	1763
Energy Biomass		0	0	0	0	0	0	0	3	3	0	0	0
Forest		3641	3591	3480	3308	3121	2975	2868	2970	3128	3341	3487	3633
Others		1957	1884	1800	1707	1674	1661	1668	1664	1673	1695	1766	1837
Total		7499	7432	7354	7266	7231	7213	7212	7204	7213	7240	7313	7385
Anthropogenic Emissions (standardized)													
Fossil Fuel CO2	GtC	0.72	1.01	1.22	1.54	2.20	3.02	4.00	4.88	5.66	6.32	6.65	7.00
Other CO2	GtC	0.73	0.82	0.72	0.95	0.91	0.76	0.48	0.31	0.12	-0.09	-0.58	-1.08
Total CO2	GtC	1.45	1.83	1.94	2.49	3.11	3.77	4.48	5.19	5.77	6.23	6.07	5.92
CH4 total	MtCH4	77	85	89	99	118	130	134	144	150	154	157	160
N2O total	MtN2O-N	1.2	1.3	1.6	2.1	2.7	3.3	4.0	4.3	4.5	4.7	4.9	5.2
SOx total	MtS	10.5	12.8	13.7	14.8	15.3	16.1	17.0	12.4	9.6	8.4	8.3	8.1
HFC	MtC eq.	0	2	15	39	81	139	184	205	210	210	206	196
PFC	MtC eq.	4	4	5	9	14	19	23	26	28	30	31	30
SF6	MtC eq.	3	5	10	14	19	32	40	42	37	23	24	26
CO	MtCO												
NMVOC	Mt												
NOx	MtN												

Scenario A1v2-MiniCAM World		1990	2000	2010	2020	2030	2040	2050	2060	2070	2080	2090	2100
Population	Million	5293	5953	6597	7228	7725	8113	8393	8413	8359	8229	8001	7780
GNP/GDP (mex)	Trillion US$	20.7	27.4	36.7	48.4	67.6	91.5	120.2	157.1	197.2	240.8	287.8	339.7
GNP/GDP (ppp) Trillion (1990 prices)		na	na	na	na	na	na	na	na	na	na	na	na
Final Energy	EJ												
Non-commercial		0	0	0	0	0	0	0	0	0	0	0	0
Solids		45	56	68	81	98	110	119	97	82	75	72	68
Liquids		121	123	130	143	145	164	201	232	262	292	315	337
Gas		52	62	77	97	108	113	112	129	148	168	170	172
Electricity		35	52	80	117	163	214	272	325	374	420	460	501
Others		0	0	0	0	0	0	0	0	0	0	0	0
Total		253	293	355	438	513	602	705	783	866	955	1016	1078
Primary Energy	EJ												
Coal		88	112	137	163	220	264	295	258	232	217	210	203
Oil		131	132	139	151	140	154	195	236	273	308	332	356
Gas		70	84	122	183	221	252	277	328	370	405	403	401
Nuclear		24	25	32	44	62	80	98	98	104	115	128	140
Biomass		0	6	11	17	24	30	37	49	54	54	53	51
Other Renewables		24	24	27	31	39	51	66	89	110	129	148	167
Total		336	384	468	589	705	831	968	1057	1144	1227	1273	1318
Cumulative Resources Use	ZJ												
Coal		0.1	1.2	2.4	3.9	6.0	8.4	11.1	13.8	16.2	18.5	20.6	22.7
Oil		0.1	1.5	2.8	4.3	5.7	7.2	8.9	11.2	13.7	16.6	19.8	23.1
Gas		0.1	0.9	2.0	3.4	5.5	7.9	10.5	13.7	17.2	21.0	25.0	29.0
Cumulative CO2 Emissions	GtC	7.1	82.4	168.5	271.7	395.5	536.0	690.4	853.3	1019.8	1189.0	1356.5	1518.5
Carbon Sequestration	GtC												
Land Use	Million ha												
Cropland		1472	1456	1457	1477	1472	1439	1378	1226	1081	943	847	752
Grasslands		3209	3344	3588	3941	4316	4609	4820	4651	4417	4119	3975	3832
Energy Biomass		0	2	5	11	14	18	20	31	31	20	15	10
Forest		4173	4215	4142	3957	3630	3376	3193	3517	3876	4272	4349	4425
Others		4310	4148	3971	3779	3732	3723	3753	3739	3758	3810	3978	4145
Total		13164	13164	13164	13164	13164	13164	13164	13164	13164	13164	13164	13164
Anthropogenic Emissions (standardized)													
Fossil Fuel CO2	GtC	5.99	6.90	8.18	9.91	11.81	13.54	15.39	15.95	16.62	17.39	17.79	18.22
Other CO2	GtC	1.11	1.07	1.05	1.51	1.52	1.24	0.71	0.53	0.18	-0.35	-1.32	-2.29
Total CO2	GtC	7.10	7.97	9.23	11.42	13.34	14.78	16.10	16.49	16.80	17.04	16.47	15.93
CH4 total	MtCH4	310	323	353	400	453	504	541	545	550	556	565	574
N2O total	MtN2O-N	6.7	7.0	8.0	9.2	10.7	12.4	14.0	14.6	15.0	15.3	16.2	17.2
SOx total	MtS	70.9	69.0	76.3	82.8	86.0	81.7	69.9	49.2	35.3	28.2	27.0	25.7
CFC/HFC/HCFC	MtC eq.	1672	883	791	337	369	482	566	654	659	654	639	614
PFC	MtC eq.	32	25	31	43	61	77	89	97	106	114	119	115
SF6	MtC eq.	38	40	43	48	66	99	119	127	113	88	84	95
CO	MtCO												
NMVOC	Mt												
NOx	MtN												

Scenario A1v2-MiniCAM OECD90		1990	2000	2010	2020	2030	2040	2050	2060	2070	2080	2090	2100
Population	Million	838	870	893	906	902	879	851	806	763	722	683	646
GNP/GDP (mex)	Trillion US$	16.3	20.5	24.3	27.8	28.7	31.5	33.5	35.7	37.9	40.2	42.6	45.1
GNP/GDP (ppp) Trillion (1990 prices)		na	na	na	na	na	na	na	na	na	na	na	na
Final Energy	EJ												
Non-commercial		0	0	0	0	0	0	0	0	0	0	0	0
Solids		10	12	12	11	10	10	10	8	6	6	5	5
Liquids		72	71	68	63	57	47	49	49	49	50	50	50
Gas		27	35	42	48	48	43	39	41	44	48	47	46
Electricity		22	27	33	38	40	45	49	49	50	50	51	51
Others		0	0	0	0	0	0	0	0	0	0	0	0
Total		130	145	155	160	155	145	146	147	149	153	152	152
Primary Energy	EJ												
Coal		40	44	43	34	36	41	43	38	35	32	32	32
Oil		76	75	71	65	57	43	44	43	43	44	44	44
Gas		34	45	61	81	80	75	69	72	75	77	74	71
Nuclear		20	15	13	12	12	12	11	10	10	10	11	12
Biomass		0	2	3	4	4	5	5	6	6	6	5	5
Other Renewables		12	11	10	10	10	10	11	12	14	15	16	18
Total		182	193	201	206	199	185	184	182	182	184	182	181
Cumulative Resources Use	ZJ												
Coal		0.0	0.5	0.9	1.3	1.5	2.1	2.5	2.9	3.2	3.6	3.9	4.2
Oil		0.1	0.8	1.6	2.2	2.5	3.3	3.7	4.1	4.6	5.0	5.4	5.9
Gas		0.0	0.5	1.0	1.7	2.1	3.2	4.0	4.7	5.4	6.2	6.9	7.7
Cumulative CO2 Emissions	GtC	2.8	33.0	66.2	101.0	135.8	168.6	199.5	229.3	258.1	285.5	311.0	334.6
Carbon Sequestration	GtC												
Land Use	Million ha												
Cropland		408	407	408	410	406	386	362	318	281	251	226	201
Grasslands		796	820	866	933	966	1049	1082	1048	1008	962	936	911
Energy Biomass		0	2	3	3	3	5	5	7	7	5	4	2
Forest		921	931	922	894	869	811	792	871	943	1009	1022	1035
Others		998	963	924	883	878	873	881	878	883	895	935	974
Total		3123	3123	3123	3123	3123	3123	3123	3123	3123	3123	3123	3123
Anthropogenic Emissions (standardized)													
Fossil Fuel CO2	GtC	2.83	3.20	3.32	3.29	3.19	3.00	3.00	2.92	2.88	2.87	2.82	2.78
Other CO2	GtC	0.00	0.00	0.13	0.24	0.23	0.14	0.03	0.02	-0.06	-0.21	-0.37	-0.52
Total CO2	GtC	2.83	3.20	3.44	3.53	3.42	3.14	3.03	2.94	2.82	2.65	2.45	2.26
CH4 total	MtCH4	73	74	82	92	96	110	120	127	135	142	146	151
N2O total	MtN2O-N	2.6	2.6	2.8	3.0	3.2	3.6	3.9	4.0	4.1	4.2	4.4	4.6
SOx total	MtS	22.7	17.0	13.1	7.9	6.5	5.5	4.8	4.3	4.1	4.2	4.3	4.4
HFC	MtC eq.	19	58	110	108	115	118	122	123	123	124	125	125
PFC	MtC eq.	18	13	13	11	9	8	9	10	12	13	14	16
SF6	MtC eq.	23	23	14	5	6	7	9	11	13	16	18	20
CO	MtCO												
NMVOC	Mt												
NOx	MtN												

Scenario A1v2-MiniCAM REF		**1990**	**2000**	**2010**	**2020**	**2030**	**2040**	**2050**	**2060**	**2070**	**2080**	**2090**	**2100**
Population	Million	428	446	460	470	466	454	434	408	380	352	326	301
GNP/GDP (mex)	Trillion US$	1.1	1.1	1.4	2.2	3.4	4.6	5.9	7.2	8.4	9.4	10.7	12.0
GNP/GDP (ppp) Trillion (1990 prices)		na	na	na	na	na	na	na	na	na	na	na	na
Final Energy	EJ												
Non-commercial		0	0	0	0	0	0	0	0	0	0	0	0
Solids		13	11	10	10	10	10	9	7	5	4	4	3
Liquids		18	13	11	13	13	13	14	14	14	13	13	13
Gas		19	16	16	21	23	23	21	21	22	22	20	19
Electricity		6	9	15	24	29	35	40	41	40	39	40	40
Others		0	0	0	0	0	0	0	0	0	0	0	0
Total		56	48	52	67	75	80	84	83	81	78	77	75
Primary Energy	EJ												
Coal		18	19	22	26	37	43	44	36	29	22	18	13
Oil		20	15	13	15	10	8	10	11	13	14	15	16
Gas		26	22	26	40	45	46	44	45	45	43	39	36
Nuclear		3	4	7	11	14	15	15	13	11	11	11	11
Biomass		0	1	2	3	3	4	5	5	5	4	4	3
Other Renewables		3	3	3	4	5	7	9	11	12	13	14	15
Total		70	64	73	99	113	122	127	122	115	107	100	94
Cumulative Resources Use	ZJ												
Coal		0.0	0.2	0.4	0.7	1.0	1.4	1.8	2.2	2.5	2.8	3.0	3.2
Oil		0.0	0.2	0.3	0.5	0.6	0.7	0.7	0.9	1.0	1.1	1.3	1.4
Gas		0.0	0.3	0.5	0.8	1.3	1.7	2.2	2.6	3.1	3.5	3.9	4.3
Cumulative CO2 Emissions	GtC	1.3	12.3	22.1	34.5	50.3	68.3	87.0	105.0	121.5	135.9	147.5	156.1
Carbon Sequestration	GtC												
Land Use	Million ha												
Cropland		284	291	300	312	318	315	304	265	227	189	170	152
Grasslands		395	410	452	521	601	662	704	652	590	519	505	491
Energy Biomass		0	0	1	3	5	6	6	7	6	2	1	-1
Forest		1007	1016	996	946	867	808	770	863	960	1063	1066	1069
Others		691	660	628	595	588	587	593	591	595	605	635	666
Total		2377	2377	2377	2377	2377	2377	2377	2377	2377	2377	2377	2377
Anthropogenic Emissions (standardized)													
Fossil Fuel CO2	GtC	1.30	0.91	1.01	1.36	1.61	1.75	1.80	1.64	1.48	1.30	1.18	1.07
Other CO2	GtC	0.00	0.00	0.03	0.08	0.12	0.12	0.06	0.10	0.08	0.01	-0.17	-0.35
Total CO2	GtC	1.30	0.91	1.04	1.44	1.73	1.87	1.86	1.74	1.56	1.31	1.01	0.72
CH4 total	MtCH4	47	39	49	67	81	91	97	101	99	93	91	90
N2O total	MtN2O-N	0.6	0.6	0.7	0.9	1.1	1.4	1.6	1.6	1.5	1.5	1.6	1.7
SOx total	MtS	17.0	11.0	11.5	12.6	14.8	14.8	12.6	7.5	4.2	2.7	2.5	2.2
HFC	MtC eq.	0	4	8	19	29	31	32	33	33	34	33	31
PFC	MtC eq.	7	4	5	8	14	20	21	22	23	24	25	24
SF6	MtC eq.	8	6	8	10	14	18	21	19	15	14	10	11
CO	MtCO												
NMVOC	Mt												
NOx	MtN												

Scenario A1v2-MiniCAM ASIA		**1990**	**2000**	**2010**	**2020**	**2030**	**2040**	**2050**	**2060**	**2070**	**2080**	**2090**	**2100**
Population	Million	2790	3165	3517	3847	4078	4237	4322	4242	4123	3965	3779	3597
GNP/GDP (mex)	Trillion US$	1.4	3.1	6.3	11.1	20.7	32.7	46.9	63.7	81.5	100.2	119.0	139.5
GNP/GDP (ppp) Trillion (1990 prices)		na	na	na	na	na	na	na	na	na	na	na	na
Final Energy	EJ												
Non-commercial		0	0	0	0	0	0	0	0	0	0	0	0
Solids		20	30	42	55	70	81	88	71	60	55	52	49
Liquids		14	19	25	34	42	53	69	80	91	101	108	114
Gas		2	5	9	13	17	19	21	27	32	38	38	39
Electricity		4	11	22	40	64	91	122	152	179	203	219	236
Others		0	0	0	0	0	0	0	0	0	0	0	0
Total		40	64	98	142	192	244	299	330	362	397	417	438
Primary Energy	EJ												
Coal		26	43	65	91	125	151	168	141	124	116	112	107
Oil		16	20	27	37	41	51	67	86	102	113	119	125
Gas		3	8	20	37	57	77	96	121	140	153	150	146
Nuclear		1	4	8	14	25	36	48	49	53	58	63	68
Biomass		0	2	5	9	13	16	20	27	31	31	30	29
Other Renewables		3	4	5	6	10	15	23	35	46	56	65	74
Total		49	81	130	195	270	346	422	460	495	527	538	549
Cumulative Resources Use	ZJ												
Coal		0.0	0.4	1.0	1.7	2.9	4.3	5.8	7.3	8.6	9.9	11.0	12.1
Oil		0.0	0.2	0.5	0.8	1.2	1.6	2.2	3.0	4.0	5.0	6.2	7.3
Gas		0.0	0.1	0.2	0.5	1.0	1.7	2.5	3.7	5.0	6.4	7.9	9.4
Cumulative CO_2 Emissions	GtC	1.5	19.3	43.5	77.3	123.0	180.2	247.1	318.5	390.0	462.4	534.5	604.7
Carbon Sequestration	GtC												
Land Use	Million ha												
Cropland		389	397	406	415	420	417	406	366	326	286	259	232
Grasslands		508	523	553	598	642	678	706	698	682	658	640	622
Energy Biomass		0	0	1	4	6	8	9	16	18	13	11	9
Forest		1168	1146	1108	1052	1001	964	943	983	1038	1108	1156	1204
Others		664	635	606	575	569	569	575	573	577	586	615	644
Total		2729	2702	2674	2645	2638	2636	2639	2637	2641	2652	2681	2710
Anthropogenic Emissions (standardized)													
Fossil Fuel CO_2	GtC	1.15	1.78	2.61	3.70	4.90	5.99	6.99	7.03	7.15	7.35	7.33	7.31
Other CO_2	GtC	0.37	0.26	0.19	0.26	0.28	0.25	0.16	0.10	0.03	-0.06	-0.21	-0.37
Total CO_2	GtC	1.53	2.03	2.80	3.96	5.18	6.24	7.15	7.13	7.18	7.30	7.12	6.94
CH_4 total	$MtCH_4$	113	125	133	144	160	174	187	174	169	171	171	172
N_2O total	MtN_2O-N	2.3	2.6	2.9	3.2	3.7	4.2	4.6	4.8	4.9	5.0	5.3	5.5
SO_x total	MtS	17.7	25.3	34.9	44.6	46.3	42.4	32.7	20.9	13.2	9.6	9.0	8.4
HFC	MtC eq.	0	5	18	45	92	153	224	292	292	285	275	262
PFC	MtC eq.	3	5	8	15	23	30	35	39	43	46	48	46
SF_6	MtC eq.	4	7	12	19	28	42	50	55	48	35	33	37
CO	MtCO												
NMVOC	Mt												
NO_x	MtN												

Scenario A1v2-MiniCAM ALM		**1990**	**2000**	**2010**	**2020**	**2030**	**2040**	**2050**	**2060**	**2070**	**2080**	**2090**	**2100**
Population	Million	1236	1472	1727	2004	2282	2543	2785	2957	3092	3190	3213	3236
GNP/GDP (mex)	Trillion US$	1.9	2.8	4.6	7.4	13.9	22.8	33.9	50.5	69.5	91.0	115.5	143.1
GNP/GDP (ppp) Trillion (1990 prices)		na	na	na	na	na	na	na	na	na	na	na	na
Final Energy	EJ												
Non-commercial		0	0	0	0	0	0	0	0	0	0	0	0
Solids		2	3	4	6	8	10	12	11	10	10	11	11
Liquids		17	21	26	33	39	51	70	89	109	128	145	161
Gas		5	7	10	15	22	27	31	40	50	61	64	68
Electricity		3	6	10	15	28	43	62	83	106	128	151	174
Others		0	0	0	0	0	0	0	0	0	0	0	0
Total		27	36	50	69	96	132	175	223	274	327	370	413
Primary Energy	EJ												
Coal		4	6	8	11	20	30	40	42	44	46	48	51
Oil		20	22	27	34	40	53	74	95	116	137	154	171
Gas		7	9	15	25	40	54	67	90	111	132	140	148
Nuclear		0	2	4	6	11	17	23	26	30	36	43	50
Biomass		0	1	1	2	4	5	7	10	12	14	14	14
Other Renewables		5	6	8	11	14	18	23	30	37	44	52	59
Total		35	46	64	89	129	178	235	293	351	409	452	494
Cumulative Resources Use	ZJ												
Coal		0.0	0.1	0.1	0.2	0.4	0.7	1.0	1.4	1.8	2.3	2.8	3.2
Oil		0.0	0.2	0.5	0.8	1.2	1.7	2.3	3.2	4.2	5.4	6.9	8.5
Gas		0.0	0.1	0.2	0.4	0.8	1.3	1.8	2.7	3.7	4.9	6.3	7.6
Cumulative CO2 Emissions	GtC	1.4	17.8	36.7	58.8	86.3	118.9	156.9	200.5	250.1	305.2	363.5	423.0
Carbon Sequestration	GtC												
Land Use	Million ha												
Cropland		391	360	343	340	333	321	306	277	247	216	192	168
Grasslands		1510	1591	1718	1890	2074	2220	2327	2253	2137	1980	1894	1808
Energy Biomass		0	0	0	0	0	0	0	0	0	0	0	0
Forest		3641	3597	3494	3329	3150	3010	2911	3011	3158	3351	3464	3576
Others		1957	1890	1813	1726	1702	1694	1704	1697	1704	1724	1793	1861
Total		7499	7438	7367	7285	7258	7246	7248	7238	7246	7271	7342	7413
Anthropogenic Emissions (standardized)													
Fossil Fuel CO2	GtC	0.72	1.01	1.23	1.56	2.11	2.79	3.60	4.36	5.12	5.88	6.46	7.07
Other CO2	GtC	0.73	0.82	0.71	0.93	0.89	0.73	0.46	0.31	0.13	-0.09	-0.57	-1.05
Total CO2	GtC	1.45	1.83	1.94	2.49	3.00	3.52	4.06	4.67	5.24	5.78	5.88	6.02
CH4 total	MtCH4	77	85	89	98	116	129	137	142	147	151	156	162
N2O total	MtN2O-N	1.2	1.3	1.6	2.1	2.7	3.3	3.9	4.2	4.4	4.6	5.0	5.4
SOx total	MtS	10.5	12.8	13.8	14.8	15.5	16.1	16.7	13.5	10.8	8.6	8.2	7.7
HFC	MtC eq.	0	2	15	39	81	139	184	205	210	210	206	196
PFC	MtC eq.	4	4	5	9	14	19	23	26	28	30	31	30
SF6	MtC eq.	3	5	10	14	19	32	40	42	37	23	24	26
CO	MtCO												
NMVOC	Mt												
NOx	MtN												

Scenario A2-AIM World		**1990**	**2000**	**2010**	**2020**	**2030**	**2040**	**2050**	**2060**	**2070**	**2080**	**2090**	**2100**
Population	Million	5262	6132	7165	8198	9231	10208	11287	12008	12776	13536	14281	15068
GNP/GDP (mex)	Trillion US$	20.1	25.1	31.8	40.2	50.7	63.9	80.7	101.2	126.9	157.7	194.3	239.4
GNP/GDP (ppp) Trillion (1990 prices)													
Final Energy	EJ												
Non-commercial		50	57	69	68	69	62	55	36	24	0	0	0
Solids		36	48	65	79	95	108	121	131	141	150	157	165
Liquids		111	115	135	162	190	204	220	244	270	308	360	421
Gas		51	56	80	101	129	147	167	177	186	201	220	241
Electricity		38	51	69	88	116	145	181	226	283	348	420	507
Others													
Total		285	327	418	499	599	668	744	824	911	1024	1169	1334
Primary Energy	EJ												
Coal		93	119	177	230	301	344	396	477	575	692	837	1016
Oil		143	155	179	182	176	163	152	134	121	107	92	80
Gas		73	96	133	170	216	244	276	292	309	334	368	406
Nuclear		6	9	11	14	16	27	48	65	88	116	150	195
Biomass		50	48	38	66	99	119	143	161	181	194	199	203
Other Renewables		10	11	13	15	20	25	31	45	67	91	112	140
Total		376	438	551	677	829	921	1046	1173	1341	1535	1758	2040
Cumulative Resources Use	ZJ												
Coal		0.1	1.2	2.6	4.7	7.3	10.2	14.3	18.5	23.9	30.4	38.1	47.7
Oil		0.1	1.6	3.3	5.1	6.9	8.3	10.2	11.4	12.7	14.0	14.9	16.0
Gas		0.1	0.9	2.1	3.6	5.5	7.6	10.4	12.9	16.0	19.4	22.8	26.9
Cumulative CO2 Emissions	GtC	7.1	82.4	173.4	281.3	406.7	551.9	714.6	898.1	1105.7	1342.9	1617.6	1937.8
Carbon Sequestraction	GtC												
Land Use	Million ha												
Cropland		1459	1484	1518	1551	1586	1623	1660	1702	1744	1787	1833	1879
Grasslands		3389	3411	3446	3478	3510	3546	3582	3611	3640	3664	3683	3701
Energy Biomass		0	0	0	89	194	252	311	316	322	314	291	268
Forest		4296	4248	4217	4254	4301	4307	4314	4306	4299	4292	4285	4277
Others		3805	3805	3768	3577	3359	3220	3082	3013	2944	2892	2858	2823
Total		12949	12949	12949	12949	12949	12949	12949	12949	12949	12949	12949	12949
Anthropogenic Emissions (standardized)													
Fossil Fuel CO2	GtC	5.99	6.90	9.30	11.29	13.70	15.05	16.60	18.73	21.20	24.41	28.56	33.43
Other CO2	GtC	1.11	1.07	0.92	0.07	0.02	0.26	0.64	0.74	0.87	0.97	1.00	1.04
Total CO2	GtC	7.10	7.97	10.22	11.36	13.72	15.31	17.23	19.47	22.07	25.38	29.56	34.47
CH4 total	MtCH4	310	323	362	385	412	422	434	457	481	505	526	549
N2O total	MtN2O-N	6.7	7.0	7.3	7.4	7.6	7.9	8.1	8.4	8.8	9.1	9.5	9.8
SOx total	MtS	70.9	69.0	86.7	105.0	132.8	135.1	140.7	151.9	165.1	148.0	113.3	92.9
CFC/HFC/HCFC	MtC eq.	1672	883	785	292	258	291	312	384	457	549	662	753
PFC	MtC eq.	32	25	41	51	64	77	92	113	129	148	168	178
SF6	MtC eq.	38	40	50	64	75	89	104	122	129	135	153	165
CO	MtCO	879	877	983	952	1029	1101	1179	1268	1366	1500	1679	1881
NMVOC	Mt	139	141	183	205	230	225	221	208	197	186	177	169
NOx	MtN	31	32	37	42	47	48	50	53	56	60	65	71

Scenario A2-AIM OECD90		1990	2000	2010	2020	2030	2040	2050	2060	2070	2080	2090	2100
Population	Million	859	914	966	1018	1071	1111	1152	1218	1288	1357	1425	1496
GNP/GDP (mex)	Trillion US$	15.3	18.6	21.9	25.5	29.7	34.2	39.4	45.8	53.2	62.3	73.4	86.6
GNP/GDP (ppp) Trillion (1990 prices)													
Final Energy	EJ												
Non-commercial		6	2	1	0	0	0	0	0	0	0	0	0
Solids		10	14	16	18	20	21	21	22	23	23	24	25
Liquids		64	69	72	82	88	87	87	88	89	95	106	118
Gas		25	31	38	44	49	51	52	52	51	53	57	61
Electricity		22	27	32	37	44	50	57	66	76	88	102	118
Others													
Total		127	142	158	181	201	209	217	228	240	259	289	322
Primary Energy	EJ												
Coal		41	46	60	70	80	86	92	106	122	147	187	238
Oil		76	83	84	84	74	61	50	36	26	19	13	9
Gas		34	46	54	62	70	72	75	73	71	73	79	85
Nuclear		5	7	8	9	9	11	14	18	22	28	35	45
Biomass		6	1	0	6	15	20	27	31	35	37	38	39
Other Renewables		6	6	7	8	10	12	13	18	23	28	31	34
Total		167	189	211	238	257	262	272	281	299	331	383	450
Cumulative Resources Use	ZJ												
Coal		0.0	0.5	1.0	1.6	2.4	3.1	4.1	5.1	6.2	7.6	9.4	11.5
Oil		0.1	0.9	1.7	2.5	3.3	3.9	4.6	4.9	5.3	5.5	5.7	5.9
Gas		0.0	0.4	0.9	1.5	2.2	2.8	3.6	4.3	5.0	5.8	6.5	7.4
Cumulative CO2 Emissions	GtC	2.8	33.0	66.7	104.0	144.7	188.0	233.4	280.7	329.4	381.3	439.7	507.3
Carbon Sequestraction	GtC												
Land Use	Million ha												
Cropland		381	384	391	398	405	413	422	432	442	453	465	477
Grasslands		760	765	775	786	797	813	830	841	853	860	864	867
Energy Biomass		0	0	0	17	37	48	59	61	62	60	56	51
Forest		1050	1062	1080	1109	1137	1137	1136	1136	1135	1133	1131	1128
Others		838	818	783	720	652	617	582	560	537	522	513	504
Total		3029	3029	3029	3029	3029	3029	3029	3029	3029	3029	3029	3029
Anthropogenic Emissions (standardized)													
Fossil Fuel CO2	GtC	2.83	3.20	3.62	4.08	4.41	4.48	4.55	4.70	4.86	5.32	6.17	7.15
Other CO2	GtC	0.00	0.00	-0.07	-0.18	-0.18	-0.04	0.10	0.10	0.10	0.10	0.10	0.10
Total CO2	GtC	2.83	3.20	3.55	3.90	4.23	4.44	4.65	4.80	4.95	5.42	6.27	7.26
CH4 total	MtCH4	73	74	74	74	75	69	64	66	67	70	73	76
N2O total	MtN2O-N	2.6	2.6	2.7	2.8	2.7	2.8	2.8	2.8	2.9	2.9	3.1	3.2
SOx total	MtS	22.7	17.0	11.1	8.6	9.2	9.6	10.1	11.0	12.0	13.7	16.1	19.0
HFC	MtC eq.	19	57	107	103	111	116	125	130	135	142	151	160
PFC	MtC eq.	18	13	14	14	14	14	14	13	11	13	15	17
SF6	MtC eq.	23	23	25	28	29	29	28	26	20	12	14	16
CO	MtCO	179	161	189	216	251	265	280	294	307	331	365	404
NMVOC	Mt	42	36	37	38	39	35	32	26	21	20	22	24
NOx	MtN	13	12	12	12	12	10	9	9	10	10	11	13

Scenario A2-AIM REF		1990	2000	2010	2020	2030	2040	2050	2060	2070	2080	2090	2100
Population	Million	413	428	444	460	476	497	519	555	593	630	667	706
GNP/GDP (mex)	Trillion US$	0.9	0.8	1.1	1.5	2.0	2.7	3.7	4.9	6.4	8.4	10.8	13.9
GNP/GDP (ppp) Trillion (1990 prices)													
Final Energy	EJ												
Non-commercial		2	4	1	0	0	0	0	0	0	0	0	0
Solids		9	7	10	11	13	14	15	15	16	16	16	17
Liquids		19	9	8	8	8	8	7	8	9	10	12	14
Gas		19	13	18	22	27	31	36	37	39	42	46	51
Electricity		8	9	12	14	16	20	25	30	37	44	52	61
Others													
Total		58	43	48	55	64	72	82	91	101	113	127	143
Primary Energy	EJ												
Coal		18	17	23	28	33	36	39	49	62	83	118	166
Oil		22	13	11	10	9	8	7	5	3	2	1	1
Gas		26	21	27	32	37	40	44	46	47	51	56	61
Nuclear		1	1	2	2	2	4	7	9	12	15	19	24
Biomass		2	4	0	1	3	5	6	7	8	9	9	9
Other Renewables		1	1	1	1	2	2	3	5	7	10	13	16
Total		71	58	63	75	85	94	106	121	141	170	215	277
Cumulative Resources Use	ZJ												
Coal		0.0	0.2	0.4	0.7	1.0	1.3	1.7	2.1	2.7	3.5	4.6	6.0
Oil		0.0	0.2	0.3	0.4	0.5	0.6	0.6	0.7	0.7	0.8	0.8	0.8
Gas		0.0	0.3	0.5	0.8	1.1	1.5	1.9	2.3	2.8	3.3	3.9	4.5
Cumulative CO2 Emissions	GtC	1.3	12.3	22.5	34.0	46.5	60.7	76.6	94.8	116.2	141.8	173.5	213.4
Carbon Sequestraction	GtC												
Land Use	Million ha												
Cropland		268	270	276	282	288	295	302	309	317	325	333	341
Grasslands		341	362	366	370	373	376	379	382	385	387	390	392
Energy Biomass		0	0	0	4	9	11	14	14	14	14	13	12
Forest		966	950	936	938	941	942	942	936	929	924	920	915
Others		701	694	698	682	665	651	638	634	630	626	621	615
Total		2276	2276	2276	2276	2276	2276	2276	2276	2276	2276	2276	2276
Anthropogenic Emissions (standardized)													
Fossil Fuel CO2	GtC	1.30	0.91	1.11	1.30	1.47	1.59	1.73	1.98	2.27	2.73	3.45	4.36
Other CO2	GtC	0.00	0.00	0.01	-0.14	-0.12	-0.09	-0.05	-0.01	0.04	0.07	0.08	0.10
Total CO2	GtC	1.30	0.91	1.13	1.16	1.35	1.50	1.68	1.97	2.31	2.80	3.54	4.45
CH4 total	MtCH4	47	39	48	47	45	46	46	53	60	69	79	90
N2O total	MtN2O-N	0.6	0.6	0.6	0.7	0.7	0.7	0.8	0.8	0.8	0.8	0.8	0.9
SOx total	MtS	17.0	11.0	14.8	17.4	15.3	10.4	7.0	6.4	5.9	6.8	9.5	13.0
HFC	MtC eq.	0	4	8	13	20	27	31	37	41	44	48	52
PFC	MtC eq.	7	4	8	10	14	17	20	25	30	35	40	42
SF6	MtC eq.	8	6	7	10	12	15	19	24	28	32	36	38
CO	MtCO	69	41	41	43	46	47	48	50	53	57	62	68
NMVOC	Mt	16	13	14	14	14	13	13	13	12	12	12	12
NOx	MtN	5	3	3	4	4	4	4	4	3	3	3	3

Scenario A2-AIM ASIA		1990	2000	2010	2020	2030	2040	2050	2060	2070	2080	2090	2100
Population	Million	2798	3278	3783	4288	4793	5255	5762	6065	6385	6701	7013	7339
GNP/GDP (mex)	Trillion US$	1.4	2.3	3.4	5.1	7.4	10.4	14.5	19.4	25.9	33.7	42.9	54.7
GNP/GDP (ppp) Trillion (1990 prices)													
Final Energy	EJ												
Non-commercial		28	31	37	35	35	30	26	14	7	0	0	0
Solids		15	24	33	41	50	57	67	72	79	84	90	95
Liquids		11	16	22	30	41	47	54	62	71	83	99	118
Gas		2	3	6	9	14	17	21	24	26	28	30	31
Electricity		5	8	14	21	33	42	55	72	94	119	147	181
Others													
Total		61	82	114	136	172	196	223	250	281	320	369	426
Primary Energy	EJ												
Coal		30	46	72	95	130	148	168	203	245	291	338	393
Oil		17	24	34	37	38	36	33	28	24	20	15	11
Gas		4	9	16	23	36	46	58	64	71	78	85	92
Nuclear		0	1	1	2	3	7	15	21	30	40	53	70
Biomass		28	24	19	21	21	26	33	37	42	45	46	47
Other Renewables		1	2	2	2	4	5	7	12	20	29	37	48
Total		80	106	144	180	233	267	314	365	431	502	574	662
Cumulative Resources Use	ZJ												
Coal		0.0	0.4	1.0	1.8	3.0	4.2	6.0	7.7	10.0	12.7	15.8	19.7
Oil		0.0	0.2	0.5	0.9	1.2	1.6	2.0	2.2	2.5	2.7	2.9	3.1
Gas		0.0	0.1	0.2	0.4	0.7	1.0	1.6	2.1	2.8	3.6	4.4	5.3
Cumulative CO2 Emissions	GtC	1.5	19.3	44.2	76.0	116.3	166.1	223.4	289.9	366.7	455.2	556.7	672.7
Carbon Sequestraction	GtC												
Land Use	Million ha												
Cropland		438	440	451	462	473	485	497	509	522	535	549	562
Grasslands		608	607	612	616	621	625	629	633	637	641	644	648
Energy Biomass		0	0	0	21	45	58	72	73	74	72	67	61
Forest		535	523	515	529	542	544	546	546	547	548	548	548
Others		583	593	586	537	484	452	421	402	384	368	356	344
Total		2164	2164	2164	2164	2164	2164	2164	2164	2164	2164	2164	2164
Anthropogenic Emissions (standardized)													
Fossil Fuel CO2	GtC	1.15	1.78	2.68	3.44	4.59	5.16	5.80	6.71	7.77	8.96	10.30	11.83
Other CO2	GtC	0.37	0.26	0.26	-0.03	0.07	0.16	0.37	0.41	0.47	0.51	0.53	0.55
Total CO2	GtC	1.53	2.03	2.94	3.41	4.65	5.31	6.16	7.12	8.24	9.47	10.83	12.38
CH4 total	MtCH4	113	125	141	154	171	181	192	203	214	223	229	236
N2O total	MtN2O-N	2.3	2.6	2.7	2.7	2.8	3.0	3.1	3.2	3.4	3.5	3.7	3.8
SOx total	MtS	17.7	25.3	38.6	52.7	71.2	70.2	69.2	70.0	70.7	56.8	36.3	23.4
HFC	MtC eq.	0	5	11	18	27	38	54	77	100	130	167	204
PFC	MtC eq.	3	5	11	15	20	25	32	42	49	56	64	67
SF6	MtC eq.	4	7	11	16	20	27	34	43	49	55	62	67
CO	MtCO	235	270	304	302	352	383	417	452	491	543	615	696
NMVOC	Mt	33	37	51	60	70	69	69	65	62	57	52	47
NOx	MtN	7	9	11	14	17	19	21	22	22	23	24	25

Scenario A2-AIM ALM		1990	2000	2010	2020	2030	2040	2050	2060	2070	2080	2090	2100
Population	Million	1192	1512	1972	2432	2892	3338	3854	4169	4510	4847	5176	5527
GNP/GDP (mex)	Trillion US$	2.4	3.4	5.4	8.1	11.5	16.3	23.1	30.8	41.1	53.2	66.9	84.2
GNP/GDP (ppp) Trillion (1990 prices)													
Final Energy	EJ												
Non-commercial		14	20	31	33	34	32	29	22	17	0	0	0
Solids		1	3	6	9	13	16	19	21	24	26	27	29
Liquids		17	22	33	42	53	62	72	85	100	119	143	171
Gas		4	9	18	27	39	48	58	64	70	78	87	97
Electricity		4	7	11	16	23	32	44	58	76	96	119	146
Others													
Total		40	60	99	126	163	190	221	253	289	332	384	444
Primary Energy	EJ												
Coal		5	10	23	37	58	75	97	119	145	171	194	220
Oil		27	35	50	52	55	59	62	65	67	66	62	58
Gas		9	20	37	53	74	86	100	109	120	133	149	167
Nuclear		0	0	1	1	2	5	12	17	24	33	43	56
Biomass		14	18	19	37	60	68	76	86	97	104	106	109
Other Renewables		2	2	3	3	4	6	7	11	17	24	31	41
Total		57	85	132	184	254	297	354	406	470	531	586	651
Cumulative Resources Use	ZJ												
Coal		0.0	0.1	0.2	0.5	1.0	1.6	2.6	3.6	4.9	6.5	8.3	1.7
Oil		0.0	0.3	0.8	1.3	1.8	2.3	3.0	3.6	4.2	4.9	5.5	6.7
Gas		0.0	0.2	0.4	0.9	1.5	2.2	3.3	4.2	5.4	6.7	8.1	12.5
Cumulative CO2 Emissions	GtC	1.4	17.8	39.9	67.3	99.3	137.0	181.1	232.7	293.4	364.7	447.8	544.3
Carbon Sequestraction	GtC												
Land Use	Million ha												
Cropland		371	390	400	409	419	430	440	451	462	474	486	498
Grasslands		1680	1677	1693	1706	1719	1732	1744	1755	1765	1775	1785	1794
Energy Biomass		0	0	0	47	103	134	166	169	172	168	155	143
Forest		1745	1712	1686	1678	1680	1685	1689	1689	1688	1687	1686	1685
Others		1684	1700	1701	1639	1558	1499	1440	1416	1392	1376	1368	1359
Total		5480	5480	5480	5480	5480	5480	5480	5480	5480	5480	5480	5480
Anthropogenic Emissions (standardized)													
Fossil Fuel CO2	GtC	0.72	1.01	1.88	2.47	3.23	3.83	4.52	5.34	6.29	7.39	8.64	10.10
Other CO2	GtC	0.73	0.82	0.71	0.42	0.26	0.24	0.22	0.24	0.27	0.30	0.29	0.28
Total CO2	GtC	1.45	1.83	2.59	2.89	3.49	4.06	4.75	5.58	6.57	7.69	8.93	10.38
CH4 total	MtCH4	77	85	99	109	121	126	131	135	140	143	145	148
N2O total	MtN2O-N	1.2	1.3	1.3	1.3	1.3	1.4	1.5	1.6	1.7	1.8	1.9	2.0
SOx total	MtS	10.5	12.8	19.3	23.4	34.2	42.0	51.4	61.5	73.5	67.8	48.3	34.5
HFC	MtC eq.	0	2	19	32	49	69	98	139	181	233	295	336
PFC	MtC eq.	4	4	8	12	16	21	26	34	39	45	50	52
SF6	MtC eq.	3	5	7	10	13	18	23	29	32	37	41	43
CO	MtCO	396	404	449	391	380	405	434	472	516	569	636	714
NMVOC	Mt	48	55	80	94	107	107	107	104	102	97	91	86
NOx	MtN	7	8	10	12	14	15	17	19	21	24	27	30

Marker Scenario A2-ASF World		1990	2000	2010	2020	2030	2040	2050	2060	2070	2080	2090	2100
Population	Million	5282	6170	7188	8206	9170	10715	11296	12139	12587	13828	14743	15068
GNP/GDP (mex)	Trillion US$	20.1	25.2	31.9	40.5	51.2	72.3	81.6	101.9	114.1	159.3	218.4	242.8
GNP/GDP (ppp) Trillion (1990 prices)													
Final Energy	EJ												
Non-commercial													
Solids		52	51	58	65	61	58	55	58	61	63	65	67
Liquids		117	150	187	246	275	304	334	371	408	469	552	635
Gas		49	53	65	92	124	155	187	209	231	246	254	262
Electricity		41	48	63	85	125	164	204	247	290	343	405	468
Others													
Total		257	303	373	488	585	682	779	884	990	1120	1276	1431
Primary Energy	EJ												
Coal		92	90	106	129	184	239	294	415	536	658	781	904
Oil		134	172	220	291	270	249	228	148	69	23	12	0
Gas		71	74	89	126	176	225	275	297	319	330	331	331
Nuclear		8	13	14	17	32	47	62	87	112	147	190	234
Biomass		0	0	6	12	32	52	71	92	112	130	146	162
Other Renewables		8	11	15	20	27	34	42	49	56	65	75	86
Total		313	360	450	595	720	846	971	1088	1204	1353	1535	1717
Cumulative Resources Use	ZJ												
Coal		0.0	1.0	2.0	3.2	4.7	6.9	9.5	13.1	17.9	31.1	38.3	46.8
Oil		0.0	1.7	3.6	6.2	9.0	11.6	13.9	15.7	16.7	17.0	17.2	17.2
Gas		0.0	0.8	1.6	2.7	4.2	6.2	8.7	11.6	14.7	17.9	21.3	24.6
Cumulative CO2 Emissions	GtC	7.1	82.4	170.2	279.3	414.2	568.2	735.7	918.6	1118.8	1339.3	1586.2	1862.4
Carbon Sequestration	GtC	-1.8	-1.6	-1.5	-1.3	-1.1	-0.9	-0.8	-0.6	-0.4	-0.2	0.0	0.2
Land Use	Million ha												
Cropland													
Grasslands													
Energy Biomass													
Forest													
Others													
Total													
Anthropogenic Emissions (standardized)													
Fossil Fuel CO2	GtC	5.99	6.90	8.46	11.01	13.53	15.01	16.49	18.49	20.49	22.97	25.94	28.91
Other CO2	GtC	1.11	1.07	1.12	1.25	1.19	1.06	0.93	0.67	0.40	0.25	0.21	0.18
Total CO2	GtC	7.10	7.97	9.58	12.25	14.72	16.07	17.43	19.16	20.89	23.22	26.15	29.09
CH4 total	MtCH4	310	323	370	424	486	542	598	654	711	770	829	889
N2O total	MtN2O-N	6.7	7.0	8.1	9.6	10.7	11.3	12.0	12.9	13.9	14.8	15.7	16.5
SOx total	MtS	70.9	69.0	74.7	99.5	112.5	109.0	105.4	89.6	73.7	64.7	62.5	60.3
CFC/HFC/HCFC	MtC eq.	1672	883	785	292	258	291	312	384	457	549	662	753
PFC	MtC eq.	32	25	41	51	64	77	92	113	129	148	168	178
SF6	MtC eq.	38	40	50	64	75	89	104	122	129	135	153	165
CO	MtCO	879	877	977	1075	1259	1344	1428	1545	1662	1842	2084	2326
NMVOC	Mt	139	141	155	179	202	214	225	238	251	275	309	342
NOx	MtN	31	32	39	50	61	66	71	75	80	87	98	109

Marker Scenario A2-ASF OECD90		**1990**	**2000**	**2010**	**2020**	**2030**	**2040**	**2050**	**2060**	**2070**	**2080**	**2090**	**2100**
Population	Million	851	923	975	1027	1072	1131	1151	1202	1228	1323	1451	1496
GNP/GDP (mex)	Trillion US$	15.3	18.7	22.3	26.0	30.0	37.1	39.9	46.3	50.0	63.1	80.7	87.6
GNP/GDP (ppp) Trillion (1990 prices)													
Final Energy	EJ												
Non-commercial													
Solids		12	12	12	13	13	14	14	15	15	16	16	16
Liquids		67	85	93	99	97	94	92	95	98	109	128	146
Gas		27	30	29	33	37	41	46	49	52	54	57	60
Electricity		23	27	33	39	46	52	59	69	79	93	112	130
Others													
Total		130	154	167	184	193	202	211	227	244	272	312	352
Primary Energy	EJ												
Coal		33	29	32	37	55	74	92	115	138	164	191	218
Oil		76	94	107	116	94	72	49	31	12	2	1	0
Gas		34	37	37	43	52	61	69	72	74	75	76	77
Nuclear		7	10	11	11	14	17	20	27	34	45	58	72
Biomass		0	0	1	4	10	16	23	23	23	24	26	29
Other Renewables		5	6	7	8	10	11	12	13	15	17	20	23
Total		155	176	194	220	235	250	266	281	296	326	372	418
Cumulative Resources Use	ZJ												
Coal		0.0	0.3	0.6	1.0	1.4	2.1	2.9	4.0	5.3	6.8	8.6	10.6
Oil		0.0	0.9	1.9	3.1	4.1	4.9	5.5	5.9	6.1	6.1	6.1	6.1
Gas		0.0	0.4	0.8	1.2	1.6	2.2	2.9	3.6	4.3	5.0	5.8	6.6
Cumulative CO2 Emissions	GtC	2.8	33.0	66.5	103.8	145.7	190.6	237.2	285.6	335.8	389.1	448.0	513.7
Carbon Sequestration	GtC												
Land Use	Million ha												
Cropland													
Grasslands													
Energy Biomass													
Forest													
Others													
Total													
Anthropogenic Emissions (standardized)													
Fossil Fuel CO2	GtC	2.83	3.20	3.51	3.96	4.42	4.58	4.74	4.93	5.11	5.55	6.23	6.91
Other CO2	GtC	0.00	0.00	0.00	0.00	0.00	0.00	0.00	0.00	0.00	0.00	0.00	0.00
Total CO2	GtC	2.83	3.20	3.51	3.96	4.42	4.58	4.74	4.93	5.11	5.55	6.23	6.91
CH4 total	MtCH4	73	74	78	84	91	98	105	113	121	133	149	166
N2O total	MtN2O-N	2.6	2.6	2.7	3.0	3.1	3.1	3.0	3.2	3.3	3.5	3.7	3.9
SOx total	MtS	22.7	17.0	7.9	8.7	9.3	9.5	9.8	9.3	8.8	9.1	10.5	11.8
HFC	MtC eq.	19	57	107	103	111	116	125	130	135	142	151	160
PFC	MtC eq.	18	13	14	14	14	14	14	13	11	13	15	17
SF6	MtC eq.	23	23	25	28	29	29	28	26	20	12	14	16
CO	MtCO	179	161	168	175	175	158	141	147	154	174	209	243
NMVOC	Mt	42	36	40	44	46	44	42	44	45	50	58	67
NOx	MtN	13	12	14	16	17	17	16	16	16	17	19	21

Marker Scenario A2-ASF REF		**1990**	**2000**	**2010**	**2020**	**2030**	**2040**	**2050**	**2060**	**2070**	**2080**	**2090**	**2100**
Population	Million	418	421	438	454	473	507	519	551	568	622	684	706
GNP/GDP (mex)	Trillion US$	0.9	0.8	1.0	1.4	2.1	3.2	3.7	4.9	5.7	8.5	12.5	14.2
GNP/GDP (ppp) Trillion (1990 prices)													
Final Energy	EJ												
Non-commercial													
Solids		17	10	10	11	11	11	11	11	11	11	11	11
Liquids		15	10	11	13	14	16	17	19	22	25	31	37
Gas		15	11	13	17	22	27	32	36	40	42	44	46
Electricity		8	6	8	11	14	16	19	22	26	31	37	43
Others													
Total		55	38	43	52	61	70	79	89	99	110	123	137
Primary Energy	EJ												
Coal		23	12	13	17	19	21	23	26	30	37	48	59
Oil		17	12	13	15	17	19	21	16	11	7	3	0
Gas		25	19	22	29	33	37	40	43	46	49	51	52
Nuclear		1	1	1	2	3	4	6	8	10	14	18	22
Biomass		0	0	1	0	0	0	0	5	9	12	14	15
Other Renewables		1	1	1	2	2	3	3	4	4	5	6	7
Total		67	45	52	64	74	84	93	102	111	124	139	155
Cumulative Resources Use	ZJ												
Coal		0.0	0.2	0.3	0.5	0.6	0.8	1.1	1.3	1.6	9.1	9.5	10.1
Oil		0.0	0.2	0.3	0.4	0.6	0.8	1.0	1.1	1.3	1.3	1.3	1.3
Gas		0.0	0.2	0.5	0.7	1.0	1.4	1.8	2.2	2.6	3.1	3.6	4.1
Cumulative CO2 Emissions	GtC	1.3	12.3	22.0	33.2	46.3	60.6	75.6	91.3	108.3	126.6	147.1	170.0
Carbon Sequestration	GtC												
Land Use	Million ha												
Cropland													
Grasslands													
Energy Biomass													
Forest													
Others													
Total													
Anthropogenic Emissions (standardized)													
Fossil Fuel CO2	GtC	1.30	0.91	1.03	1.22	1.40	1.46	1.52	1.63	1.75	1.93	2.17	2.41
Other CO2	GtC	0.00	0.00	0.00	0.00	0.00	0.00	0.00	0.00	0.00	0.00	0.00	0.00
Total CO2	GtC	1.30	0.91	1.03	1.22	1.40	1.46	1.52	1.63	1.75	1.93	2.17	2.41
CH4 total	MtCH4	47	39	41	46	57	67	78	92	106	119	131	143
N2O total	MtN2O-N	0.6	0.6	0.6	0.7	0.7	0.8	0.8	0.8	0.9	0.9	0.9	1.0
SOx total	MtS	17.0	11.0	11.1	12.0	12.0	11.1	10.2	8.2	6.2	4.8	4.0	3.2
HFC	MtC eq.	0	4	8	13	20	27	31	37	41	44	48	52
PFC	MtC eq.	7	4	8	10	14	17	20	25	30	35	40	42
SF6	MtC eq.	8	6	7	10	12	15	19	24	28	32	36	38
CO	MtCO	69	41	40	40	47	51	56	63	70	83	101	119
NMVOC	Mt	16	13	15	19	25	28	31	29	27	29	33	37
NOx	MtN	5	3	3	4	5	5	5	5	6	6	7	8

Marker Scenario A2-ASF ASIA		1990	2000	2010	2020	2030	2040	2050	2060	2070	2080	2090	2100
Population	Million	2791	3295	3801	4308	4779	5500	5764	6137	6333	6858	7214	7340
GNP/GDP (mex)	Trillion US$	1.4	2.3	3.5	5.3	7.6	12.6	15.0	20.0	23.1	34.9	50.5	57.1
GNP/GDP (ppp) Trillion (1990 prices)													
Final Energy	EJ												
Non-commercial													
Solids		21	28	33	38	33	27	22	22	23	23	23	23
Liquids		13	24	39	64	77	90	103	118	134	156	185	215
Gas		2	5	9	19	28	36	45	51	56	59	60	60
Electricity		5	9	13	20	42	64	86	104	121	140	160	181
Others													
Total		42	65	95	141	179	218	256	295	334	379	429	479
Primary Energy	EJ												
Coal		32	43	52	62	86	110	134	181	227	272	313	355
Oil		17	30	48	77	70	64	57	35	13	1	1	0
Gas		3	6	13	24	42	60	79	85	91	93	88	84
Nuclear		0	1	2	3	11	19	26	37	47	60	75	90
Biomass		0	0	1	3	10	16	23	22	21	20	20	20
Other Renewables		1	2	3	5	8	12	16	18	21	25	28	32
Total		53	82	119	174	228	281	335	378	421	470	526	581
Cumulative Resources Use	ZJ												
Coal		0.0	0.4	0.9	1.5	2.2	3.2	4.4	6.0	8.1	10.6	13.5	16.9
Oil		0.0	0.3	0.6	1.3	2.0	2.7	3.3	3.7	3.9	4.0	4.0	4.0
Gas		0.0	0.1	0.1	0.3	0.7	1.2	1.9	2.7	3.6	4.5	5.4	6.3
Cumulative CO2 Emissions	GtC	1.5	19.3	43.6	77.2	121.7	175.4	236.4	305.0	381.1	465.1	558.2	660.8
Carbon Sequestration	GtC												
Land Use	Million ha												
Cropland													
Grasslands													
Energy Biomass													
Forest													
Others													
Total													
Anthropogenic Emissions (standardized)													
Fossil Fuel CO2	GtC	1.15	1.78	2.47	3.52	4.64	5.45	6.27	7.08	7.90	8.79	9.75	10.71
Other CO2	GtC	0.37	0.26	0.34	0.39	0.35	0.28	0.22	0.15	0.08	0.04	0.03	0.02
Total CO2	GtC	1.53	2.03	2.81	3.92	4.99	5.73	6.48	7.23	7.98	8.83	9.78	10.74
CH4 total	MtCH4	113	125	144	163	184	205	227	241	255	272	290	308
N2O total	MtN2O-N	2.3	2.6	3.2	4.0	4.6	4.9	5.3	5.8	6.2	6.6	6.9	7.2
SOx total	MtS	17.7	25.3	36.2	51.5	56.6	52.7	48.9	40.1	31.3	25.6	23.1	20.5
HFC	MtC eq.	0	5	11	18	27	38	54	77	100	130	167	204
PFC	MtC eq.	3	5	11	15	20	25	32	42	49	56	64	67
SF6	MtC eq.	4	7	11	16	20	27	34	43	49	55	62	67
CO	MtCO	235	270	321	361	435	479	522	576	630	707	806	906
NMVOC	Mt	33	37	41	46	50	53	56	60	64	69	76	82
NOx	MtN	7	9	12	16	20	23	26	28	30	33	36	40

Marker Scenario A2-ASF ALM		**1990**	**2000**	**2010**	**2020**	**2030**	**2040**	**2050**	**2060**	**2070**	**2080**	**2090**	**2100**
Population	Million	1222	1530	1974	2417	2846	3578	3862	4250	4458	5025	5394	5526
GNP/GDP (mex)	Trillion US$	2.4	3.4	5.1	7.8	11.5	19.3	23.0	30.6	35.4	52.9	74.7	83.8
GNP/GDP (ppp) Trillion (1990 prices)													
Final Energy	EJ												
Non-commercial													
Solids		1	1	2	3	5	7	8	10	12	13	14	16
Liquids		21	31	45	70	87	104	121	138	155	178	208	238
Gas		5	8	13	24	37	50	64	73	83	89	93	96
Electricity		4	6	9	15	23	32	40	52	64	79	97	115
Others													
Total		31	47	69	112	153	193	234	273	313	359	411	464
Primary Energy	EJ												
Coal		4	7	9	14	24	35	45	92	140	185	228	272
Oil		24	36	53	82	88	94	100	67	34	14	7	0
Gas		8	11	17	29	48	68	87	97	107	114	116	118
Nuclear		0	0	0	1	4	7	9	15	20	28	39	50
Biomass		0	0	2	5	12	19	25	43	60	74	86	98
Other Renewables		1	2	3	5	7	9	11	13	16	18	21	25
Total		38	56	85	137	184	231	278	327	376	433	498	563
Cumulative Resources Use	ZJ												
Coal		0.0	0.1	0.1	0.3	0.4	0.7	1.1	1.8	3.0	4.6	6.7	9.2
Oil		0.0	0.3	0.8	1.4	2.3	3.2	4.2	5.0	5.5	5.7	5.8	5.8
Gas		0.0	0.1	0.2	0.5	0.9	1.5	2.2	3.2	4.2	5.3	6.4	7.6
Cumulative CO2 Emissions	GtC	1.4	17.8	38.1	65.1	100.5	141.6	186.5	236.7	293.7	358.5	432.9	517.9
Carbon Sequestration	GtC												
Land Use	Million ha												
Cropland													
Grasslands													
Energy Biomass													
Forest													
Others													
Total													
Anthropogenic Emissions (standardized)													
Fossil Fuel CO2	GtC	0.72	1.01	1.45	2.31	3.08	3.52	3.96	4.84	5.72	6.70	7.79	8.87
Other CO2	GtC	0.73	0.82	0.77	0.85	0.84	0.78	0.71	0.52	0.32	0.21	0.18	0.16
Total CO2	GtC	1.45	1.83	2.23	3.16	3.92	4.30	4.68	5.36	6.04	6.91	7.97	9.03
CH4 total	MtCH4	77	85	107	132	153	171	188	208	229	246	259	272
N2O total	MtN2O-N	1.2	1.3	1.6	1.9	2.2	2.5	2.8	3.1	3.5	3.8	4.1	4.4
SOx total	MtS	10.5	12.8	16.5	24.4	31.7	32.6	33.5	29.0	24.4	22.1	21.9	21.8
HFC	MtC eq.	0	2	19	32	49	69	98	139	181	233	295	336
PFC	MtC eq.	4	4	8	12	16	21	26	34	39	45	50	52
SF6	MtC eq.	3	5	7	10	13	18	23	29	32	37	41	43
CO	MtCO	396	404	448	499	603	656	709	758	808	877	968	1058
NMVOC	Mt	48	55	60	70	81	89	96	106	115	127	142	156
NOx	MtN	7	8	10	14	19	21	24	26	29	32	36	40

Scenario A2G-IMAGE World		1990	2000	2010	2020	2030	2040	2050	2060	2070	2080	2090	2100
Population	Million	5297			8225			11298					14719
GNP/GDP (mex)	Trillion US$	21.0			45.3			111.3					248.5
GNP/GDP (ppp) Trillion (1990 prices)		0.0			0.0			0.0					0.0
Final Energy	EJ												
Non-commercial		50			40			22					16
Solids		40			44			65					99
Liquids		98			145			203					191
Gas		50			108			229					341
Electricity		35			91			262					444
Others		0			0			0					0
Total		272			428			781					1092
Primary Energy	EJ												
Coal		82			101			253					237
Oil		116			175			235					178
Gas		78			187			396					458
Nuclear													
Biomass		1			6			46					71
Non-commercial		6			40			22					16
NTE (Nuclear/Solar) & Hydro		17			37			107					344
Total		344			546			1059					1304
Cumulative Resources Use	ZJ												
Coal		0.1			2.3			7.4					21.6
Oil		0.1			4.3			10.5					21.5
Gas		0.1			3.8			12.2					35.5
Cumulative CO2 Emissions	GtC	7.1	82.4	172.4	282.9	415.3	570.7	749.2	939.9	1131.6	1324.4	1518.4	1713.4
Carbon Sequestration	GtC	0.0			0.0			0.0					0.0
Land Use	Million ha												
Cropland		0			0			0					0
Grasslands		0			0			0					0
Energy Biomass		0			0			0					0
Forest		0			0			0					0
Others		0			0			0					0
Total		0			0			0					0
Anthropogenic Emissions (standardized)													
Fossil Fuel CO2	GtC	5.99	6.90	7.98	9.07	12.10	15.14	18.17	18.05	17.93	17.81	17.69	17.57
Other CO2	GtC	1.11	1.07	2.04	3.01	2.29	1.56	0.84	1.07	1.30	1.53	1.76	1.99
Total CO2	GtC	7.10	7.97	10.02	12.08	14.39	16.70	19.01	19.12	19.23	19.34	19.45	19.56
CH4 total	MtCH4	310	323	408	493	537	581	625	653	681	709	737	765
N2O total	MtN2O-N	6.7	7.0	9.6	12.2	12.7	13.3	13.9	14.5	15.0	15.6	16.2	16.7
SOx total	MtS	70.9	69.0	69.9	70.7	83.8	96.9	110.0	104.9	99.7	94.6	89.4	84.3
CFC/HFC/HCFC	MtC eq.	1672	883	785	292	258	291	312	384	457	549	662	753
PFC	MtC eq.	32	25	41	51	64	77	92	113	129	148	168	178
SF6	MtC eq.	38	40	50	64	75	89	104	122	129	135	153	165
CO	MtCO	879	877	812	748	713	678	642	669	696	723	750	776
NMVOC	Mt	139	141	154	166	165	163	161	167	172	177	183	188
NOx	MtN	31	32	37	42	55	68	82	80	78	77	75	73

Scenario A2G-IMAGE OECD90		**1990**	**2000**	**2010**	**2020**	**2030**	**2040**	**2050**	**2060**	**2070**	**2080**	**2090**	**2100**
Population	Million	801			953			1068					1393
GNP/GDP (mex)	Trillion US$	16.5			30.7			57.7					109.6
GNP/GDP (ppp) Trillion (1990 prices)													
Final Energy	EJ												
Non-commercial		6			4			2					2
Solids		7			7			9					17
Liquids		51			55			56					57
Gas		34			49			66					103
Electricity		21			46			70					111
Others													
Total		119			161			201					289
Primary Energy	EJ												
Coal		24			17			23					36
Oil		55			62			58					50
Gas		51			96			119					151
Nuclear													
Biomass		1			2			6					11
Non-commercial		6			4			2					2
NTE (Nuclear/Solar) & Hydro		10			22			40					80
Total		147			202			248					330
Cumulative Resources Use	ZJ												
Coal		0.0			0.5			1.0					2.5
Oil		0.1			1.1			2.4					4.4
Gas		0.1			2.1			5.1					11.4
Cumulative CO2 Emissions	GtC	2.8	33.0	66.5	103.2	142.6	184.6	229.0	276.0	325.8	378.2	433.4	491.2
Carbon Sequestration	GtC												
Land Use	Million ha												
Cropland													
Grasslands													
Energy Biomass													
Forest													
Others													
Total													
Anthropogenic Emissions (standardized)													
Fossil Fuel CO2	GtC	2.83	3.20	3.40	3.60	3.77	3.95	4.12	4.29	4.45	4.62	4.78	4.95
Other CO2	GtC	0.00	0.00	0.11	0.22	0.30	0.37	0.45	0.55	0.66	0.76	0.87	0.98
Total CO2	GtC	2.83	3.20	3.51	3.82	4.07	4.32	4.57	4.84	5.11	5.38	5.65	5.92
CH4 total	MtCH4	73	74	88	103	107	111	116	121	127	133	139	144
N2O total	MtN2O-N	2.6	2.6	2.8	3.0	3.0	3.1	3.2	3.3	3.4	3.6	3.7	3.8
SOx total	MtS	22.7	17.0	13.4	9.8	10.7	11.6	12.6	13.7	14.8	15.9	17.1	18.2
HFC	MtC eq.	19	57	107	103	111	116	125	130	135	142	151	160
PFC	MtC eq.	18	13	14	14	14	14	14	13	11	13	15	17
SF6	MtC eq.	23	23	25	28	29	29	28	26	20	12	14	16
CO	MtCO	179	161	123	85	87	89	91	96	102	107	113	118
NMVOC	Mt	42	36	35	34	36	37	38	40	43	45	48	50
NOx	MtN	13	12	11	9	10	11	11	12	13	13	14	14

Scenario A2G-IMAGE REF		1990	2000	2010	2020	2030	2040	2050	2060	2070	2080	2090	2100
Population	Million	413			421			391					405
GNP/GDP (mex)	Trillion US$	1.0			1.5			3.5					6.4
GNP/GDP (ppp) Trillion (1990 prices)													
Final Energy	EJ												
Non-commercial		2			1			1					0
Solids		11			5			3					4
Liquids		18			15			12					9
Gas		12			10			12					12
Electricity		7			10			18					21
Others													
Total		50			41			46					46
Primary Energy	EJ												
Coal		26			12			14					12
Oil		26			21			21					12
Gas		21			19			27					21
Nuclear													
Biomass		0			0			1					2
Non-commercial		2			1			1					0
NTE (Nuclear/Solar) & Hydro		2			4			9					14
Total		78			58			71					61
Cumulative Resources Use	ZJ												
Coal		0.0			0.5			0.9					1.6
Oil		0.0			0.5			1.1					1.8
Gas		0.0			0.5			1.2					2.6
Cumulative CO2 Emissions	GtC	1.3	12.3	20.2	25.5	30.0	35.3	41.2	47.5	53.4	59.1	64.4	69.5
Carbon Sequestration	GtC												
Land Use	Million ha												
Cropland													
Grasslands													
Energy Biomass													
Forest													
Others													
Total													
Anthropogenic Emissions (standardized)													
Fossil Fuel CO2	GtC	1.30	0.91	0.65	0.39	0.44	0.49	0.53	0.47	0.41	0.35	0.28	0.22
Other CO2	GtC	0.00	0.00	0.01	0.02	0.05	0.08	0.10	0.14	0.17	0.21	0.24	0.28
Total CO2	GtC	1.30	0.91	0.66	0.41	0.49	0.56	0.64	0.61	0.58	0.55	0.52	0.49
CH4 total	MtCH4	47	39	44	50	49	48	48	47	47	46	45	45
N2O total	MtN2O-N	0.6	0.6	0.6	0.6	0.6	0.6	0.6	0.6	0.6	0.7	0.7	0.7
SOx total	MtS	17.0	11.0	7.5	4.1	4.1	4.2	4.3	4.2	4.0	3.9	3.8	3.7
HFC	MtC eq.	0	4	8	13	20	27	31	37	41	44	48	52
PFC	MtC eq.	7	4	8	10	14	17	20	25	30	35	40	42
SF6	MtC eq.	8	6	7	10	12	15	19	24	28	32	36	38
CO	MtCO	69	41	25	10	11	12	13	12	11	10	10	9
NMVOC	Mt	16	13	12	11	10	10	10	10	10	10	10	10
NOx	MtN	5	3	3	3	4	4	4	4	3	3	2	2

Scenario A2G-IMAGE ASIA		**1990**	**2000**	**2010**	**2020**	**2030**	**2040**	**2050**	**2060**	**2070**	**2080**	**2090**	**2100**
Population	Million	2790			4294			5746					7284
GNP/GDP (mex)	Trillion US$	1.4			5.8			25.6					69.0
GNP/GDP (ppp) Trillion (1990 prices)													
Final Energy	EJ												
Non-commercial		28			25			11					7
Solids		20			26			35					48
Liquids		14			36			62					39
Gas		2			35			110					144
Electricity		4			24			113					170
Others													
Total		68			145			331					409
Primary Energy	EJ												
Coal		29			56			147					114
Oil		17			40			50					20
Gas		3			53			172					170
Nuclear													
Biomass		0			4			28					26
Non-commercial		28			25			11					7
NTE (Nuclear/Solar) & Hydro		2			6			41					141
Total		79			184			449					477
Cumulative Resources Use	ZJ												
Coal		0.0			1.2			4.1					11.9
Oil		0.0			0.8			2.2					3.8
Gas		0.0			0.8			4.0					13.8
Cumulative CO2 Emissions	GtC	1.5	19.3	43.1	73.6	115.8	174.5	249.6	331.3	409.8	485.1	557.2	626.1
Carbon Sequestration	GtC												
Land Use	Million ha												
Cropland													
Grasslands													
Energy Biomass													
Forest													
Others													
Total													
Anthropogenic Emissions (standardized)													
Fossil Fuel CO2	GtC	1.15	1.78	2.48	3.19	4.78	6.36	7.95	7.62	7.29	6.96	6.63	6.30
Other CO2	GtC	0.37	0.26	0.23	0.21	0.27	0.32	0.38	0.39	0.40	0.41	0.42	0.43
Total CO2	GtC	1.53	2.03	2.72	3.40	5.04	6.69	8.33	8.01	7.69	7.37	7.05	6.73
CH4 total	MtCH4	113	125	156	186	216	245	275	285	295	305	316	326
N2O total	MtN2O-N	2.3	2.6	3.0	3.4	3.8	4.2	4.5	4.6	4.8	4.9	5.0	5.1
SOx total	MtS	17.7	25.3	28.4	31.5	35.5	39.6	43.6	38.1	32.6	27.1	21.5	16.0
HFC	MtC eq.	0	5	11	18	27	38	54	77	100	130	167	204
PFC	MtC eq.	3	5	11	15	20	25	32	42	49	56	64	67
SF6	MtC eq.	4	7	11	16	20	27	34	43	49	55	62	67
CO	MtCO	235	270	252	235	269	304	339	346	353	360	368	375
NMVOC	Mt	33	37	40	42	46	51	55	56	56	57	57	58
NOx	MtN	7	9	12	16	23	31	39	37	35	32	30	28

Scenario A2G-IMAGE ALM		1990	2000	2010	2020	2030	2040	2050	2060	2070	2080	2090	2100
Population	Million	1293			2557			4093					5637
GNP/GDP (mex)	Trillion US$	2.1			7.3			24.5					63.5
GNP/GDP (ppp) Trillion (1990 prices)													
Final Energy	EJ												
Non-commercial		14			10			9					7
Solids		2			6			17					30
Liquids		15			40			74					87
Gas		3			14			42					83
Electricity		2			12			61					142
Others													
Total		35			81			203					349
Primary Energy	EJ												
Coal		4			16			69					75
Oil		18			51			107					96
Gas		3			19			78					117
Nuclear													
Biomass		0			1			11					32
Non-commercial		14			10			9					7
NTE (Nuclear/Solar) & Hydro		3			5			18					109
Total		41			103			291					437
Cumulative Resources Use	ZJ												
Coal		0.0			0.2			1.4					5.6
Oil		0.0			1.9			4.9					11.5
Gas		0.0			0.4			1.8					7.6
Cumulative CO2 Emissions	GtC	1.4	17.8	42.7	80.6	126.8	176.4	229.4	285.1	342.6	402.0	463.3	526.5
Carbon Sequestration	GtC												
Land Use	Million ha												
Cropland													
Grasslands													
Energy Biomass													
Forest													
Others													
Total													
Anthropogenic Emissions (standardized)													
Fossil Fuel CO2	GtC	0.72	1.01	1.45	1.89	3.12	4.34	5.56	5.67	5.78	5.89	6.00	6.11
Other CO2	GtC	0.73	0.82	1.69	2.56	1.67	0.79	-0.09	-0.01	0.07	0.15	0.23	0.31
Total CO2	GtC	1.45	1.83	3.14	4.45	4.79	5.13	5.47	5.66	5.85	6.04	6.22	6.41
CH4 total	MtCH4	77	85	120	154	165	176	187	200	212	225	237	250
N2O total	MtN2O-N	1.2	1.3	3.2	5.2	5.3	5.5	5.6	5.9	6.2	6.5	6.8	7.1
SOx total	MtS	10.5	12.8	17.6	22.4	30.4	38.5	46.6	45.9	45.3	44.7	44.0	43.4
HFC	MtC eq.	0	2	19	32	49	69	98	139	181	233	295	336
PFC	MtC eq.	4	4	8	12	16	21	26	34	39	45	50	52
SF6	MtC eq.	3	5	7	10	13	18	23	29	32	37	41	43
CO	MtCO	396	404	412	419	346	273	201	215	230	245	260	274
NMVOC	Mt	48	55	67	79	72	65	58	60	63	66	68	71
NOx	MtN	7	8	11	14	18	23	28	28	28	28	28	29

Scenario A2-MESSAGE World		**1990**	**2000**	**2010**	**2020**	**2030**	**2040**	**2050**	**2060**	**2070**	**2080**	**2090**	**2100**
Population	Million	5262	6170	7188	8206	9170	10715	11296	12139	12587	13828	14743	15068
GNP/GDP (mex)	Trillion US$	20.9	25.2	31.9	40.5	51.2	72.3	81.6	101.9	114.1	159.3	218.4	242.8
GNP/GDP (ppp) Trillion (1990 prices)		25.7	31.2	39.7	50.5	63.3	82.2	106.0	132.2	155.2	180.1	208.0	225.6
Final Energy	EJ												
Non-commercial		38	23	26	24	22	20	18	17	16	16	15	15
Solids		42	39	27	20	13	24	32	30	21	9	7	6
Liquids		111	137	181	235	279	284	296	302	352	391	456	501
Gas		41	51	58	68	85	105	132	162	145	156	143	161
Electricity		35	46	60	82	109	147	184	230	277	337	397	459
Others		8	11	17	28	44	65	88	113	137	158	180	200
Total		275	306	368	457	552	645	750	855	947	1066	1197	1342
Primary Energy	EJ												
Coal		91	83	97	122	157	211	273	342	458	569	735	871
Oil		128	171	208	253	274	244	205	146	109	86	64	47
Gas		71	86	108	135	173	208	245	292	268	285	281	289
Nuclear		7	9	13	17	22	34	47	62	79	97	116	136
Biomass		46	41	42	50	69	97	141	194	245	271	287	299
Other Renewables		8	14	21	34	50	77	103	132	165	196	228	280
Total		352	404	489	610	745	871	1014	1168	1323	1504	1710	1921
Cumulative Resources Use	ZJ												
Coal		0.0	0.9	1.7	2.7	3.9	5.5	7.6	10.3	13.7	18.3	24.0	31.4
Oil		0.0	1.5	3.2	5.3	7.8	10.5	13.0	15.0	16.5	17.6	18.4	19.1
Gas		0.0	0.8	1.6	2.7	4.0	5.8	7.9	10.3	13.2	15.9	18.8	21.6
Cumulative CO2 Emissions	GtC	7.1	82.4	169.6	274.1	398.1	537.8	690.5	855.3	1034.3	1234.7	1466.3	1732.3
Carbon Sequestration	GtC												
Land Use	Million ha												
Cropland													
Grasslands													
Energy Biomass													
Forest													
Others													
Total													
Anthropogenic Emissions (standardized)													
Fossil Fuel CO2 [a]	GtC	5.99	6.90	8.34	10.32	12.28	13.66	15.11	16.46	18.38	21.10	24.90	28.21
Other CO2 [b]	GtC	1.11	1.07	1.11	1.14	1.05	0.96	0.81	0.59	0.38	0.22	0.10	-0.02
Total CO2	GtC	7.10	7.97	9.45	11.46	13.33	14.62	15.91	17.05	18.76	21.32	25.00	28.19
CH4 total [c]	MtCH4	310	323	363	418	489	592	671	743	803	900	993	1069
N2O total [d]	MtN2O-N	6.7	7.0	6.4	6.3	6.3	6.6	6.8	7.1	7.3	7.6	7.9	8.1
SOx total	MtS	70.9	69.0	68.3	79.8	94.2	106.5	103.6	99.0	83.8	74.3	68.1	68.7
CFC/HFC/HCFC	MtC eq.	1672	883	785	292	258	291	312	384	457	549	662	753
PFC	MtC eq.	32	25	41	51	64	77	92	113	129	148	168	178
SF6	MtC eq.	38	40	50	64	75	89	104	122	129	135	153	165
CO	MtCO	879	877	972	1100	1246	1396	1585	1810	2075	2250	2426	2646
NMVOC	Mt	139	141	170	204	229	239	242	241	247	262	281	311
NOx	MtN	31	32	38	47	56	61	66	71	78	87	97	110

Emissions correlated to land-use change and deforestation were calculated by using AIM A2 land-use data.
a:CO2 emissions from fossil fuel and industrial processes (MESSAGE data).
b: CO2 emissions from land-use changes (IS92f).
c: Non-energy related CH4 emissions were taken from AIM-A2 run.
d: Non-energy related N2O emissions were taken from AIM-A2 run.

Scenario A2-MESSAGE OECD90		1990	2000	2010	2020	2030	2040	2050	2060	2070	2080	2090	2100
Population	Million	851	923	975	1027	1072	1131	1151	1202	1228	1323	1451	1496
GNP/GDP (mex)	Trillion US$	15.3	18.7	22.3	26.0	30.0	37.1	39.9	46.3	50.0	63.1	80.7	87.6
GNP/GDP (ppp) Trillion (1990 prices)		14.1	16.0	18.5	20.8	23.1	28.2	30.8	36.2	39.0	49.5	63.8	69.8
Final Energy	EJ												
Non-commercial		0	0	0	0	0	0	0	0	0	0	0	0
Solids		13	9	5	3	1	1	0	0	0	0	0	0
Liquids		66	72	88	102	105	99	88	78	79	80	85	94
Gas		22	26	22	16	19	23	28	33	27	26	26	27
Electricity		22	28	35	43	53	65	75	88	101	117	133	153
Others		1	1	1	4	7	10	14	18	25	28	33	39
Total		124	136	151	168	185	197	205	217	232	251	277	314
Primary Energy	EJ												
Coal		38	40	47	56	71	91	112	126	151	195	232	240
Oil		72	89	99	105	96	83	62	41	31	23	16	11
Gas		33	38	42	43	52	52	58	77	80	72	81	99
Nuclear		6	7	9	11	13	17	19	25	31	36	42	50
Biomass		6	6	6	8	12	16	21	29	32	35	37	41
Other Renewables		4	6	7	11	15	21	27	32	39	43	48	55
Total		159	185	209	233	259	280	300	328	363	404	457	496
Cumulative Resources Use	ZJ												
Coal		0.0	0.4	0.8	1.3	1.8	2.5	3.4	4.6	5.9	7.7	10.1	13.0
Oil		0.0	0.8	1.7	2.7	3.7	4.7	5.5	6.1	6.5	6.8	7.1	7.2
Gas		0.0	0.4	0.7	1.2	1.6	2.1	2.6	3.2	4.0	4.8	5.5	6.3
Cumulative CO2 Emissions	GtC	2.8	33.0	67.0	105.0	146.5	191.1	238.2	287.7	341.8	404.0	476.7	556.4
Carbon Sequestration	GtC												
Land Use	Million ha												
Cropland													
Grasslands													
Energy Biomass													
Forest													
Others													
Total													
Anthropogenic Emissions (standardized)													
Fossil Fuel CO2 [a]	GtC	2.83	3.20	3.61	3.98	4.33	4.58	4.83	5.08	5.74	6.70	7.86	8.06
Other CO2 [b]	GtC	0.00	0.00	0.00	0.00	0.00	0.00	0.00	0.00	0.00	0.00	0.00	0.00
Total CO2	GtC	2.83	3.20	3.61	3.98	4.33	4.58	4.83	5.08	5.74	6.70	7.86	8.06
CH4 total [c]	MtCH4	73	74	72	72	78	89	99	111	126	149	176	175
N2O total [d]	MtN2O-N	2.6	2.6	2.6	2.5	2.5	2.5	2.6	2.6	2.7	2.7	2.8	2.9
SOx total	MtS	22.7	17.0	11.0	5.9	4.0	3.1	2.1	1.9	2.8	3.9	5.1	5.7
HFC	MtC eq.	19	57	107	103	111	116	125	130	135	142	151	160
PFC	MtC eq.	18	13	14	14	14	14	14	13	11	13	15	17
SF6	MtC eq.	23	23	25	28	29	29	28	26	20	12	14	16
CO	MtCO	179	161	170	180	185	200	208	226	223	229	233	243
NMVOC	Mt	42	36	37	36	35	35	30	27	25	24	24	26
NOx	MtN	13	12	14	15	16	17	17	18	19	20	22	25

Emissions correlated to land-use change and deforestation were calculated by using AIM A2 land-use data.

a:CO2 emissions from fossil fuel and industrial processes (MESSAGE data).

b: CO2 emissions from land-use changes (IS92f).

c: Non-energy related CH4 emissions were taken from AIM-A2 run.

d: Non-energy related N2O emissions were taken from AIM-A2 run.

Scenario A2-MESSAGE REF		1990	2000	2010	2020	2030	2040	2050	2060	2070	2080	2090	2100
Population	Million	418	421	438	454	473	507	519	551	568	622	684	706
GNP/GDP (mex)	Trillion US$	0.9	0.8	1.0	1.4	2.1	3.2	3.7	4.9	5.7	8.5	12.5	14.2
GNP/GDP (ppp) Trillion (1990 prices)		2.6	2.6	3.2	4.2	5.0	6.2	9.0	10.9	13.0	12.5	12.8	13.8
Final Energy	EJ												
Non-commercial		0	0	0	0	0	0	0	0	0	0	0	0
Solids		9	4	3	2	1	0	0	0	0	0	0	0
Liquids		15	13	15	18	22	21	21	24	29	33	38	45
Gas		14	14	16	17	18	18	20	19	18	18	19	19
Electricity		6	6	7	9	11	15	19	24	27	31	37	43
Others		7	6	5	7	10	13	14	18	21	22	24	24
Total		50	43	46	53	62	67	74	85	94	104	117	131
Primary Energy	EJ												
Coal		19	11	11	13	17	18	23	34	47	50	64	82
Oil		20	18	18	20	21	18	13	12	9	6	5	2
Gas		27	28	30	33	39	45	52	47	35	39	45	51
Nuclear		1	1	1	0	1	1	2	4	5	7	9	10
Biomass		2	1	1	1	2	4	6	11	18	28	26	24
Other Renewables		1	1	2	3	5	8	10	13	16	19	22	27
Total		70	60	62	70	84	95	107	121	131	148	170	195
Cumulative Resources Use	ZJ												
Coal		0.0	0.2	0.3	0.4	0.5	0.7	0.9	1.2	1.5	2.0	2.5	3.1
Oil		0.0	0.2	0.4	0.6	0.8	1.0	1.2	1.3	1.4	1.5	1.6	1.6
Gas		0.0	0.3	0.6	0.9	1.2	1.6	2.0	2.6	3.0	3.4	3.8	4.2
Cumulative CO_2 Emissions	GtC	1.3	12.3	21.7	31.9	44.0	57.5	72.3	88.5	106.0	124.1	144.9	170.6
Carbon Sequestration	GtC												
Land Use	Million ha												
Cropland													
Grasslands													
Energy Biomass													
Forest													
Others													
Total													
Anthropogenic Emissions (standardized)													
Fossil Fuel CO_2 [a]	GtC	1.30	0.91	0.96	1.09	1.31	1.40	1.54	1.71	1.78	1.86	2.30	2.84
Other CO_2 [b]	GtC	0.00	0.00	0.00	0.00	0.00	0.00	0.00	0.00	0.00	0.00	0.00	0.00
Total CO_2	GtC	1.30	0.91	0.96	1.09	1.31	1.40	1.54	1.71	1.78	1.86	2.30	2.84
CH_4 total [c]	$MtCH_4$	47	39	41	44	51	58	66	69	72	78	94	113
N_2O total [d]	MtN_2O-N	0.6	0.6	0.6	0.6	0.7	0.7	0.7	0.7	0.7	0.8	0.8	0.8
SO_x total	MtS	17.0	11.0	10.9	12.5	15.3	18.2	20.7	23.3	18.2	11.8	6.2	6.4
HFC	MtC eq.	0	4	8	13	20	27	31	37	41	44	48	52
PFC	MtC eq.	7	4	8	10	14	17	20	25	30	35	40	42
SF_6	MtC eq.	8	6	7	10	12	15	19	24	28	32	36	38
CO	MtCO	69	41	43	51	61	68	75	82	100	114	145	183
NMVOC	Mt	16	13	16	19	21	24	23	20	20	21	25	32
NO_x	MtN	5	3	3	4	5	5	6	7	7	7	8	9

Emissions correlated to land-use change and deforestation were calculated by using AIM A2 land-use data.

a: CO_2 emissions from fossil fuel and industrial processes (MESSAGE data).

b: CO_2 emissions from land-use changes (IS92f).

c: Non-energy related CH_4 emissions were taken from AIM-A2 run.

d: Non-energy related N_2O emissions were taken from AIM-A2 run.

Scenario A2-MESSAGE ASIA		1990	2000	2010	2020	2030	2040	2050	2060	2070	2080	2090	2100
Population	Million	2791	3295	3801	4308	4779	5500	5764	6137	6333	6858	7214	7340
GNP/GDP (mex)	Trillion US$	1.4	2.3	3.5	5.3	7.6	12.6	15.0	20.0	23.1	34.9	50.5	57.1
GNP/GDP (ppp) Trillion (1990 prices)		5.3	7.4	10.3	14.3	19.5	27.3	36.8	47.5	57.1	66.1	74.7	79.9
Final Energy	EJ												
Non-commercial		24	16	17	17	15	14	12	11	11	10	10	9
Solids		19	19	15	12	9	21	29	28	18	6	4	2
Liquids		13	26	42	61	85	93	105	104	119	133	155	180
Gas		2	4	8	17	21	19	26	46	43	53	49	45
Electricity		4	6	10	17	25	35	46	58	75	96	114	131
Others		1	4	7	11	17	25	35	44	53	62	71	78
Total		62	75	100	135	172	207	252	291	319	360	402	445
Primary Energy	EJ												
Coal		30	27	33	42	54	82	104	128	187	235	310	390
Oil		15	31	47	67	86	80	71	47	27	16	10	4
Gas		3	7	15	23	27	23	29	50	47	56	49	41
Nuclear		0	1	2	3	5	8	13	17	22	29	34	39
Biomass		24	21	24	29	36	47	63	77	77	79	82	84
Other Renewables		1	3	6	9	14	22	31	42	57	73	81	90
Total		74	90	125	173	222	262	312	361	417	489	566	648
Cumulative Resources Use	ZJ												
Coal		0.0	0.3	0.6	0.9	1.3	1.9	2.7	3.7	4.9	6.9	9.4	12.8
Oil		0.0	0.2	0.5	1.0	1.6	2.5	3.3	4.0	4.5	4.8	4.9	5.0
Gas		0.0	0.0	0.1	0.3	0.5	0.7	1.0	1.2	1.8	2.2	2.8	3.3
Cumulative CO2 Emissions	GtC	1.5	19.3	42.6	72.8	110.7	154.8	203.5	256.9	318.0	390.9	478.9	585.4
Carbon Sequestration	GtC												
Land Use	Million ha												
Cropland													
Grasslands													
Energy Biomass													
Forest													
Others													
Total													
Anthropogenic Emissions (standardized)													
Fossil Fuel CO2 [a]	GtC	1.15	1.78	2.36	3.16	3.93	4.48	4.97	5.48	6.61	7.90	9.72	11.69
Other CO2 [b]	GtC	0.37	0.26	0.26	0.27	0.22	0.17	0.12	0.09	0.05	0.01	-0.03	-0.08
Total CO2	GtC	1.53	2.03	2.62	3.42	4.15	4.66	5.10	5.57	6.66	7.91	9.69	11.62
CH4 total [c]	MtCH4	113	125	153	190	232	283	319	356	394	444	487	523
N2O total [d]	MtN2O-N	2.3	2.6	2.4	2.4	2.5	2.6	2.7	2.9	3.0	3.1	3.2	3.3
SOx total	MtS	17.7	25.3	27.8	36.7	44.3	52.9	47.2	40.7	30.8	27.5	26.8	26.8
HFC	MtC eq.	0	5	11	18	27	38	54	77	100	130	167	204
PFC	MtC eq.	3	5	11	15	20	25	32	42	49	56	64	67
SF6	MtC eq.	4	7	11	16	20	27	34	43	49	55	62	67
CO	MtCO	235	270	335	422	497	538	589	653	746	859	968	1107
NMVOC	Mt	33	37	49	63	72	73	75	82	91	102	113	128
NOx	MtN	7	9	12	16	19	21	23	25	28	33	38	43

Emissions correlated to land-use change and deforestation were calculated by using AIM A2 land-use data.

a:CO2 emissions from fossil fuel and industrial processes (MESSAGE data).

b: CO2 emissions from land-use changes (IS92f).

c: Non-energy related CH4 emissions were taken from AIM-A2 run.

d: Non-energy related N2O emissions were taken from AIM-A2 run.

Scenario A2-MESSAGE ALM		1990	2000	2010	2020	2030	2040	2050	2060	2070	2080	2090	2100
Population	Million	1222	1530	1974	2417	2846	3578	3862	4250	4458	5025	5394	5526
GNP/GDP (mex)	Trillion US$	2.4	3.4	5.1	7.8	11.5	19.3	23.0	30.6	35.4	52.9	74.7	83.8
GNP/GDP (ppp) Trillion (1990 prices)		3.8	5.2	7.7	11.2	15.7	20.4	29.4	37.5	46.1	52.0	56.7	62.2
Final Energy	EJ												
Non-commercial		14	8	9	7	6	6	6	6	6	6	6	6
Solids		1	7	4	3	2	2	3	2	2	3	3	3
Liquids		17	26	36	54	68	71	83	96	125	145	178	183
Gas		4	6	11	19	27	45	58	64	57	59	50	70
Electricity		3	5	8	14	20	33	45	61	75	92	113	133
Others		0	1	3	6	10	18	25	33	39	45	52	59
Total		39	53	71	101	134	174	219	262	303	350	402	454
Primary Energy	EJ												
Coal		5	5	7	11	15	19	34	53	73	90	129	159
Oil		21	34	44	62	71	62	59	46	42	41	34	30
Gas		8	13	22	36	55	89	107	119	107	118	107	98
Nuclear		0	1	1	2	4	7	12	17	21	26	31	37
Biomass		14	13	12	12	18	31	50	78	117	129	141	151
Other Renewables		2	3	7	11	16	26	35	45	53	60	75	107
Total		49	69	93	134	179	234	296	358	412	463	517	582
Cumulative Resources Use	ZJ												
Coal		0.0	0.0	0.1	0.2	0.3	0.4	0.6	0.9	1.4	1.8	2.1	2.5
Oil		0.0	0.3	0.6	1.1	1.7	2.4	3.0	3.6	4.1	4.5	4.9	5.2
Gas		0.0	0.1	0.2	0.4	0.8	1.4	2.2	3.3	4.5	5.5	6.7	7.8
Cumulative CO2 Emissions	GtC	1.4	17.8	38.2	64.3	96.8	134.4	176.5	222.2	268.6	315.7	365.8	419.9
Carbon Sequestration	GtC												
Land Use	Million ha												
Cropland													
Grasslands													
Energy Biomass													
Forest													
Others													
Total													
Anthropogenic Emissions (standardized)													
Fossil Fuel CO2 [a]	GtC	0.72	1.01	1.40	2.10	2.71	3.19	3.77	4.18	4.25	4.65	5.02	5.61
Other CO2 [b]	GtC	0.73	0.82	0.85	0.87	0.83	0.79	0.68	0.51	0.33	0.20	0.13	0.06
Total CO2	GtC	1.45	1.83	2.25	2.97	3.54	3.98	4.45	4.69	4.58	4.85	5.16	5.67
CH4 total [c]	MtCH4	77	85	99	112	128	163	187	207	212	229	235	259
N2O total [d]	MtN2O-N	1.2	1.3	0.8	0.7	0.7	0.7	0.8	0.9	1.0	1.0	1.1	1.1
SOx total	MtS	10.5	12.8	15.6	21.7	27.6	29.4	30.6	30.2	29.1	28.1	27.0	26.9
HFC	MtC eq.	0	2	19	32	49	69	98	139	181	233	295	336
PFC	MtC eq.	4	4	8	12	16	21	26	34	39	45	50	52
SF6	MtC eq.	3	5	7	10	13	18	23	29	32	37	41	43
CO	MtCO	396	404	424	448	503	591	712	848	1007	1048	1081	1113
NMVOC	Mt	48	55	68	86	101	108	113	112	112	115	119	125
NOx	MtN	7	8	10	12	15	18	20	22	24	27	29	32

Emissions correlated to land-use change and deforestation were calculated by using AIM A2 land-use data.
a:CO2 emissions from fossil fuel and industrial processes (MESSAGE data).
b: CO2 emissions from land-use changes (IS92f).
c: Non-energy related CH4 emissions were taken from AIM-A2 run.
d: Non-energy related N2O emissions were taken from AIM-A2 run.

Scenario A2-MiniCAM World		**1990**	**2000**	**2010**	**2020**	**2030**	**2040**	**2050**	**2060**	**2070**	**2080**	**2090**	**2100**
Population	Million	5293	6208	7174	8192	9250	10284	11296	12238	13108	13905	14508	15124
GNP/GDP (mex)	Trillion US$	20.7	27.4	35.4	44.7	55.6	69.5	86.4	108.7	135.2	165.9	204.1	246.6
GNP/GDP (ppp) Trillion (1990 prices)		na	na	na	na	na	na	na	na	na	na	na	na
Final Energy	EJ												
Non-commercial		0	0	0	0	0	0	0	0	0	0	0	0
Solids		45	58	73	88	100	114	131	131	136	145	150	155
Liquids		121	125	131	139	136	150	183	216	257	306	354	403
Gas		52	62	72	80	82	77	67	64	66	71	66	62
Electricity		35	53	78	109	144	190	246	320	410	517	643	769
Others		0	0	0	0	0	0	0	0	0	0	0	0
Total		253	299	353	416	462	532	628	732	869	1039	1214	1388
Primary Energy	EJ												
Coal		88	117	144	170	239	325	429	529	629	729	858	988
Oil		131	135	139	144	111	91	84	69	68	79	72	64
Gas		70	84	113	155	173	181	179	186	201	222	202	183
Nuclear		24	25	30	39	53	71	93	105	123	148	181	214
Biomass		0	6	11	17	25	37	52	77	99	116	139	163
Other Renewables		24	24	27	31	39	52	70	108	153	206	279	353
Total		336	392	465	556	640	757	907	1074	1272	1501	1732	1964
Cumulative Resources Use	ZJ												
Coal		0.1	1.2	2.5	4.0	6.3	9.2	12.8	17.9	23.7	30.3	38.6	46.9
Oil		0.1	1.5	2.8	4.3	5.4	6.5	7.3	8.1	8.8	9.5	10.2	11.0
Gas		0.1	0.9	1.9	3.2	4.9	6.6	8.4	10.3	12.2	14.3	16.4	18.4
Cumulative CO2 Emissions	GtC	7.1	82.4	167.9	267.6	382.8	513.5	664.0	837.0	1033.3	1254.0	1501.5	1779.0
Carbon Sequestration	GtC												
Land Use	Million ha												
Cropland		1472	1481	1531	1620	1689	1729	1739	1676	1629	1598	1585	1573
Grasslands		3209	3348	3554	3828	4074	4274	4427	4373	4348	4351	4411	4471
Energy Biomass		0	4	8	15	38	75	126	201	251	277	336	396
Forest		4173	4214	4137	3942	3685	3503	3396	3591	3717	3772	3636	3500
Others		4310	4118	3934	3759	3678	3584	3476	3323	3220	3166	3195	3224
Total		13164	13164	13164	13164	13164	13164	13164	13164	13164	13164	13164	13164
Anthropogenic Emissions (standardized)													
Fossil Fuel CO2	GtC	5.99	6.90	8.06	9.40	10.81	12.77	15.24	17.62	20.34	23.41	26.28	29.39
Other CO2	GtC	1.11	1.07	1.06	1.44	1.38	1.18	0.91	0.82	0.48	-0.09	-0.09	-0.09
Total CO2	GtC	7.10	7.97	9.12	10.83	12.19	13.96	16.15	18.44	20.83	23.32	26.19	29.30
CH4 total	MtCH4	310	323	348	381	415	452	485	508	540	580	621	663
N2O total	MtN2O-N	6.7	7.0	7.9	8.8	9.7	10.8	11.7	12.1	12.7	13.5	14.6	15.6
SOx total	MtS	70.9	69.0	81.1	82.7	91.4	100.7	110.6	112.1	108.7	100.2	89.7	79.3
CFC/HFC/HCFC	MtC eq.	1672	883	785	292	258	291	312	384	457	549	662	753
PFC	MtC eq.	32	25	41	51	64	77	92	113	129	148	168	178
SF6	MtC eq.	38	40	50	64	75	89	104	122	129	135	153	165
CO	MtCO												
NMVOC	Mt												
NOx	MtN												

Scenario A2-MiniCAM OECD90		1990	2000	2010	2020	2030	2040	2050	2060	2070	2080	2090	2100
Population	Million	838	916	980	1030	1053	1116	1151	1197	1253	1319	1405	1493
GNP/GDP (mex)	Trillion US$	16.3	20.5	24.5	28.3	29.6	34.2	38.4	42.5	47.6	53.8	62.4	71.6
GNP/GDP (ppp) Trillion (1990 prices)		na	na	na	na	na	na	na	na	na	na	na	na
Final Energy	EJ												
Non-commercial		0	0	0	0	0	0	0	0	0	0	0	0
Solids		10	12	13	13	13	14	16	15	14	15	15	16
Liquids		72	72	70	67	62	53	57	57	61	68	76	83
Gas		27	35	42	47	46	39	32	28	27	28	26	24
Electricity		22	28	35	43	47	59	69	74	83	96	112	128
Others		0	0	0	0	0	0	0	0	0	0	0	0
Total		130	147	160	170	167	166	173	174	185	206	229	251
Primary Energy	EJ												
Coal		40	46	47	43	51	75	94	112	131	152	194	236
Oil		76	76	73	68	56	28	18	10	5	5	4	3
Gas		34	46	62	82	82	76	68	63	62	63	52	40
Nuclear		20	15	13	14	14	16	18	18	19	22	27	32
Biomass		0	2	4	4	5	8	11	14	16	18	22	26
Other Renewables		12	11	10	10	10	12	14	18	24	31	42	53
Total		182	196	209	221	218	216	223	236	259	292	341	390
Cumulative Resources Use	ZJ												
Coal		0.0	0.5	0.9	1.4	1.7	2.6	3.4	4.5	5.7	7.1	9.0	10.8
Oil		0.1	0.8	1.6	2.3	2.5	3.2	3.4	3.5	3.6	3.7	3.7	3.8
Gas		0.0	0.5	1.0	1.7	2.1	3.3	4.0	4.7	5.3	5.9	6.5	7.0
Cumulative CO2 Emissions	GtC	2.8	33.0	66.7	103.1	140.5	178.5	218.5	260.7	305.1	352.5	405.0	464.7
Carbon Sequestration	GtC												
Land Use	Million ha												
Cropland		408	414	427	447	452	461	457	440	429	424	419	414
Grasslands		796	819	854	899	916	961	984	979	978	981	989	996
Energy Biomass		0	4	6	8	13	29	41	50	56	60	72	85
Forest		921	930	921	892	875	836	829	877	906	916	894	872
Others		998	955	915	875	866	835	811	777	754	742	749	756
Total		3123	3123	3123	3123	3123	3123	3123	3123	3123	3123	3123	3123
Anthropogenic Emissions (standardized)													
Fossil Fuel CO2	GtC	2.83	3.20	3.42	3.53	3.54	3.71	3.99	4.17	4.51	4.98	5.68	6.44
Other CO2	GtC	0.00	0.00	0.13	0.21	0.20	0.16	0.14	0.14	0.07	-0.07	-0.09	-0.11
Total CO2	GtC	2.83	3.20	3.54	3.74	3.74	3.86	4.12	4.31	4.57	4.91	5.60	6.33
CH4 total	MtCH4	73	74	81	88	90	99	105	114	123	133	148	163
N2O total	MtN2O-N	2.6	2.6	2.7	2.9	3.0	3.3	3.5	3.5	3.6	3.8	4.0	4.2
SOx total	MtS	22.7	17.0	16.2	8.3	7.5	6.6	5.8	5.4	5.4	5.7	6.5	7.3
HFC	MtC eq.	19	57	107	103	111	116	125	130	135	142	151	160
PFC	MtC eq.	18	13	14	14	14	14	14	13	11	13	15	17
SF6	MtC eq.	23	23	25	28	29	29	28	26	20	12	14	16
CO	MtCO												
NMVOC	Mt												
NOx	MtN												

Scenario A2-MiniCAM REF		1990	2000	2010	2020	2030	2040	2050	2060	2070	2080	2090	2100
Population	Million	428	430	439	455	474	495	519	550	585	623	664	707
GNP/GDP (mex)	Trillion US$	1.1	1.1	1.2	1.5	1.8	2.2	2.7	3.7	4.9	6.4	8.3	10.5
GNP/GDP (ppp) Trillion (1990 prices)		na	na	na	na	na	na	na	na	na	na	na	na
Final Energy	EJ												
Non-commercial		0	0	0	0	0	0	0	0	0	0	0	0
Solids		13	10	8	7	7	8	9	9	9	9	9	9
Liquids		18	12	8	7	7	7	8	9	11	14	16	17
Gas		19	15	12	10	10	9	7	7	7	8	7	7
Electricity		6	8	11	14	16	20	24	32	41	51	62	73
Others		0	0	0	0	0	0	0	0	0	0	0	0
Total		56	44	39	39	41	43	48	57	68	81	94	106
Primary Energy	EJ												
Coal		18	17	17	18	34	53	76	123	169	214	290	367
Oil		20	14	10	9	3	0	0	0	0	0	0	0
Gas		26	20	18	20	20	19	17	17	19	21	18	16
Nuclear		3	4	5	6	7	8	10	11	12	14	17	19
Biomass		0	1	1	2	3	4	6	9	12	14	17	19
Other Renewables		3	3	3	4	5	6	8	11	16	21	28	34
Total		70	58	55	59	72	91	116	171	227	283	370	457
Cumulative Resources Use	ZJ												
Coal		0.0	0.2	0.4	0.5	0.8	1.3	1.9	3.0	4.5	6.3	9.1	11.8
Oil		0.0	0.2	0.3	0.4	0.4	0.4	0.4	0.4	0.4	0.4	0.4	0.4
Gas		0.0	0.2	0.4	0.6	0.8	1.0	1.2	1.4	1.6	1.8	1.9	2.1
Cumulative CO2 Emissions	GtC	1.3	12.3	21.1	29.9	40.3	53.1	68.8	89.8	118.1	153.2	197.4	253.9
Carbon Sequestration	GtC												
Land Use	Million ha												
Cropland		284	297	317	345	368	382	388	368	353	345	348	350
Grasslands		395	410	444	498	548	585	611	586	573	572	594	615
Energy Biomass		0	0	1	2	7	14	23	39	49	54	64	74
Forest		1007	1016	995	942	879	836	815	871	906	919	879	839
Others		691	655	621	590	576	559	541	514	496	487	493	498
Total		2377	2377	2377	2377	2377	2377	2377	2377	2377	2377	2377	2377
Anthropogenic Emissions (standardized)													
Fossil Fuel CO2	GtC	1.30	0.91	0.81	0.83	1.05	1.34	1.68	2.42	3.15	3.90	5.03	6.31
Other CO2	GtC	0.00	0.00	0.03	0.09	0.10	0.08	0.03	0.06	0.03	-0.06	-0.03	0.00
Total CO2	GtC	1.30	0.91	0.84	0.92	1.16	1.42	1.72	2.47	3.18	3.84	5.00	6.31
CH4 total	MtCH4	47	39	45	55	65	74	83	99	118	140	171	203
N2O total	MtN2O-N	0.6	0.6	0.7	0.8	1.0	1.1	1.3	1.3	1.3	1.4	1.6	1.7
SOx total	MtS	17.0	11.0	10.4	10.0	11.8	13.6	15.4	16.1	16.4	16.2	14.9	13.6
HFC	MtC eq.	0	4	8	13	20	27	31	37	41	44	48	52
PFC	MtC eq.	7	4	8	10	14	17	20	25	30	35	40	42
SF6	MtC eq.	8	6	7	10	12	15	19	24	28	32	36	38
CO	MtCO												
NMVOC	Mt												
NOx	MtN												

Scenario A2-MiniCAM ASIA		**1990**	**2000**	**2010**	**2020**	**2030**	**2040**	**2050**	**2060**	**2070**	**2080**	**2090**	**2100**
Population	Million	2790	3296	3802	4308	4817	5302	5763	6180	6558	6898	7133	7372
GNP/GDP (mex)	Trillion US$	1.4	3.1	5.5	8.7	13.0	18.3	24.6	33.0	43.1	54.8	69.1	85.2
GNP/GDP (ppp) Trillion (1990 prices)		na	na	na	na	na	na	na	na	na	na	na	na
Final Energy	EJ												
Non-commercial		0	0	0	0	0	0	0	0	0	0	0	0
Solids		20	33	46	60	68	78	88	88	92	99	102	106
Liquids		14	20	26	32	35	42	54	66	80	96	112	128
Gas		2	5	8	11	11	10	10	10	11	12	12	11
Electricity		4	11	22	37	49	67	88	120	159	204	259	314
Others		0	0	0	0	0	0	0	0	0	0	0	0
Total		40	70	103	139	163	197	240	284	341	412	485	559
Primary Energy	EJ												
Coal		26	48	71	96	121	154	192	225	252	273	273	272
Oil		16	22	28	34	28	24	22	16	15	17	14	10
Gas		3	9	18	31	38	43	48	52	58	67	63	58
Nuclear		1	4	8	13	19	27	38	43	50	61	75	89
Biomass		0	2	5	8	12	17	23	33	41	48	57	65
Other Renewables		3	4	5	6	9	15	23	38	57	80	111	142
Total		49	89	135	187	227	280	345	407	474	546	591	637
Cumulative Resources Use	ZJ												
Coal		0.0	0.5	1.1	1.9	3.0	4.4	6.1	8.3	10.6	13.2	16.0	18.7
Oil		0.0	0.2	0.5	0.8	1.1	1.3	1.6	1.7	1.9	2.1	2.2	2.3
Gas		0.0	0.1	0.2	0.5	0.8	1.2	1.7	2.2	2.7	3.4	4.0	4.6
Cumulative CO2 Emissions	GtC	1.5	19.3	43.5	76.3	117.2	165.4	222.2	287.9	362.0	444.8	534.1	627.8
Carbon Sequestration	GtC												
Land Use	Million ha												
Cropland		389	403	421	444	461	472	476	462	452	445	439	433
Grasslands		508	523	546	578	605	630	653	659	665	670	673	676
Energy Biomass		0	0	1	4	12	22	33	47	56	60	71	83
Forest		1168	1141	1100	1043	991	944	904	899	896	895	885	876
Others		664	630	599	570	557	542	525	501	484	476	481	486
Total		2729	2697	2668	2640	2626	2610	2592	2568	2553	2545	2550	2554
Anthropogenic Emissions (standardized)													
Fossil Fuel CO2	GtC	1.15	1.78	2.60	3.50	4.14	4.96	5.97	6.85	7.77	8.72	9.14	9.57
Other CO2	GtC	0.37	0.26	0.20	0.27	0.28	0.25	0.19	0.13	0.06	0.00	0.01	0.01
Total CO2	GtC	1.53	2.03	2.79	3.77	4.42	5.21	6.16	6.98	7.84	8.72	9.15	9.59
CH4 total	MtCH4	113	125	133	143	152	163	176	180	184	186	179	172
N2O total	MtN2O-N	2.3	2.6	2.8	3.1	3.4	3.7	3.9	4.1	4.3	4.5	4.8	5.0
SOx total	MtS	17.7	25.3	36.8	46.0	51.1	55.8	60.2	59.5	56.1	50.1	43.5	36.9
HFC	MtC eq.	0	5	11	18	27	38	54	77	100	130	167	204
PFC	MtC eq.	3	5	11	15	20	25	32	42	49	56	64	67
SF6	MtC eq.	4	7	11	16	20	27	34	43	49	55	62	67
CO	MtCO												
NMVOC	Mt												
NOx	MtN												

Scenario A2-MiniCAM ALM		1990	2000	2010	2020	2030	2040	2050	2060	2070	2080	2090	2100
Population	Million	1236	1566	1953	2399	2884	3372	3862	4311	4712	5065	5306	5552
GNP/GDP (mex)	Trillion US$	1.9	2.8	4.3	6.2	9.9	14.7	20.7	29.5	39.5	50.9	64.3	79.3
GNP/GDP (ppp) Trillion (1990 prices)		na	na	na	na	na	na	na	na	na	na	na	na
Final Energy	EJ												
Non-commercial		0	0	0	0	0	0	0	0	0	0	0	0
Solids		2	3	5	7	11	15	19	20	22	23	24	25
Liquids		17	22	27	33	37	48	65	83	104	128	151	174
Gas		5	7	10	13	17	19	18	19	20	23	21	20
Electricity		3	6	10	16	28	45	65	94	128	166	210	254
Others		0	0	0	0	0	0	0	0	0	0	0	0
Total		27	38	52	68	93	126	167	216	274	340	406	472
Primary Energy	EJ												
Coal		4	6	9	13	26	44	67	69	77	91	101	112
Oil		20	23	28	33	35	38	44	43	48	57	54	51
Gas		7	10	15	22	34	42	46	53	62	72	70	68
Nuclear		0	2	4	6	12	19	27	33	41	50	62	74
Biomass		0	1	1	2	4	8	12	22	30	36	44	53
Other Renewables		5	6	8	11	15	20	26	40	56	74	98	123
Total		35	48	66	89	126	171	223	261	313	380	430	480
Cumulative Resources Use	ZJ												
Coal		0.0	0.1	0.1	0.2	0.5	0.9	1.4	2.1	2.8	3.6	4.6	5.6
Oil		0.0	0.2	0.5	0.8	1.1	1.5	1.9	2.4	2.8	3.3	3.9	4.4
Gas		0.0	0.1	0.2	0.4	0.7	1.1	1.5	2.0	2.6	3.3	4.0	4.7
Cumulative CO2 Emissions	GtC	1.4	17.8	36.6	58.3	84.8	116.4	154.5	198.6	248.1	303.5	365.0	432.5
Carbon Sequestration	GtC												
Land Use	Million ha												
Cropland		391	368	365	384	403	414	418	405	394	384	379	375
Grasslands		1510	1596	1710	1853	1989	2097	2179	2149	2132	2128	2156	2184
Energy Biomass		0	0	0	0	0	10	28	65	90	104	129	153
Forest		3641	3585	3478	3318	3161	3028	2920	2930	2936	2939	2890	2841
Others		1957	1878	1800	1723	1688	1647	1599	1532	1486	1461	1473	1484
Total		7499	7427	7353	7279	7241	7196	7145	7081	7038	7016	7027	7037
Anthropogenic Emissions (standardized)													
Fossil Fuel CO2	GtC	0.72	1.01	1.24	1.54	2.08	2.77	3.61	4.18	4.91	5.81	6.42	7.07
Other CO2	GtC	0.73	0.82	0.70	0.87	0.80	0.69	0.54	0.49	0.32	0.04	0.02	0.00
Total CO2	GtC	1.45	1.83	1.94	2.41	2.88	3.46	4.15	4.67	5.24	5.84	6.44	7.07
CH4 total	MtCH4	77	85	89	96	108	116	120	114	115	121	123	125
N2O total	MtN2O-N	1.2	1.3	1.6	1.9	2.3	2.7	3.1	3.3	3.5	3.8	4.2	4.6
SOx total	MtS	10.5	12.8	14.7	15.4	18.1	21.7	26.1	28.1	27.8	25.2	21.9	18.5
HFC	MtC eq.	0	2	19	32	49	69	98	139	181	233	295	336
PFC	MtC eq.	4	4	8	12	16	21	26	34	39	45	50	52
SF6	MtC eq.	3	5	7	10	13	18	23	29	32	37	41	43
CO	MtCO												
NMVOC	Mt												
NOx	MtN												

Scenario A2-A1-MiniCAM World		1990	2000	2010	2020	2030	2040	2050	2060	2070	2080	2090	2100
Population	Million	5293	6007	6762	7558	8332	9054	9723	10217	10696	11161	11620	12090
GNP/GDP (mex)	Trillion US$	20.7	27.4	33.0	37.6	41.8	48.8	58.6	76.4	98.2	123.8	158.1	197.3
GNP/GDP (ppp) Trillion (1990 prices)		na	na	na	na	na	na	na	na	na	na	na	na
Final Energy	EJ												
Non-commercial		0	0	0	0	0	0	0	0	0	0	0	0
Solids		45	60	70	73	80	90	103	102	106	116	125	135
Liquids		121	128	130	128	116	124	151	198	253	314	381	447
Gas		52	64	73	78	81	91	110	128	155	191	208	225
Electricity		35	55	74	90	103	128	164	237	317	406	523	640
Others		0	0	0	0	0	0	0	0	0	0	0	0
Total		253	308	346	369	380	433	528	665	831	1027	1237	1447
Primary Energy	EJ												
Coal		88	121	136	133	148	167	189	196	213	239	266	294
Oil		131	138	138	133	114	118	145	200	260	327	395	464
Gas		70	87	111	142	149	171	209	263	328	405	444	483
Nuclear		24	27	30	32	37	47	64	95	123	149	214	278
Biomass		0	6	10	13	16	18	22	34	44	51	55	59
Other Renewables		24	24	27	31	36	43	51	70	92	118	151	183
Total		336	403	452	485	499	564	679	857	1061	1289	1525	1761
Cumulative Resources Use	ZJ												
Coal		0.1	1.2	2.5	3.8	5.3	6.9	8.6	10.6	12.6	14.8	17.5	20.1
Oil		0.1	1.5	2.9	4.2	5.4	6.6	7.9	9.8	12.1	14.9	18.7	22.5
Gas		0.1	0.9	1.9	3.1	4.6	6.3	8.1	10.6	13.6	17.1	21.5	25.8
Cumulative CO2 Emissions	GtC	7.1	82.4	164.2	248.7	333.8	422.5	523.9	641.8	778.8	943.5	1135.8	1352.2
Carbon Sequestration	GtC												
Land Use	Million ha												
Cropland		1472	1461	1463	1478	1423	1359	1285	1214	1174	1165	1107	1050
Grasslands		3209	3343	3464	3572	3545	3524	3508	3557	3703	3948	4091	4234
Energy Biomass		0	4	8	14	16	17	18	45	65	75	71	67
Forest		4173	4214	4155	3995	4092	4252	4475	4571	4436	4069	3925	3782
Others		4310	4142	4074	4106	4088	4012	3878	3777	3787	3908	3970	4032
Total		13164	13164	13164	13164	13164	13164	13164	13164	13164	13164	13164	13164
Anthropogenic Emissions (standardized)													
Fossil Fuel CO2	GtC	5.99	6.90	7.62	7.89	8.13	8.89	10.46	12.40	14.83	17.77	20.28	23.00
Other CO2	GtC	1.11	1.07	0.76	0.62	0.37	0.35	0.58	0.13	0.04	0.31	0.10	-0.11
Total CO2	GtC	7.10	7.97	8.39	8.51	8.50	9.24	11.05	12.52	14.87	18.08	20.38	22.89
CH4 total	MtCH4	310	323	339	354	363	379	402	428	473	537	582	627
N2O total	MtN2O-N	6.7	7.0	7.8	8.6	9.1	9.9	10.6	11.5	13.0	15.1	17.2	19.3
SOx total	MtS	70.9	69.0	73.9	65.7	66.6	70.8	78.1	81.4	84.0	86.2	85.4	84.5
CFC/HFC/HCFC	MtC eq.	1672	883	785	292	258	291	312	384	457	549	662	753
PFC	MtC eq.	32	25	41	51	64	77	92	113	129	148	168	178
SF6	MtC eq.	38	40	50	64	75	89	104	122	129	135	153	165
CO	MtCO												
NMVOC	Mt												
NOx	MtN												

Scenario A2-A1-MiniCAM OECD90		1990	2000	2010	2020	2030	2040	2050	2060	2070	2080	2090	2100
Population	Million	838	874	903	924	926	924	913	895	883	877	882	886
GNP/GDP (mex)	Trillion US$	16.3	20.5	23.5	25.5	26.1	27.9	29.5	31.4	33.7	36.4	40.1	44.1
GNP/GDP (ppp) Trillion (1990 prices)		na	na	na	na	na	na	na	na	na	na	na	na
Final Energy	EJ												
Non-commercial		0	0	0	0	0	0	0	0	0	0	0	0
Solids		10	12	13	12	12	12	11	9	8	8	9	9
Liquids		72	73	71	66	61	51	54	57	60	64	69	73
Gas		27	36	43	47	47	48	52	52	55	61	62	64
Electricity		22	28	34	40	42	48	53	57	61	64	72	80
Others		0	0	0	0	0	0	0	0	0	0	0	0
Total		130	149	161	166	161	159	170	175	185	198	212	226
Primary Energy	EJ												
Coal		40	47	46	38	39	42	42	39	37	35	38	40
Oil		76	77	75	68	60	47	49	53	57	61	65	69
Gas		34	46	62	80	80	81	86	87	90	97	98	99
Nuclear		20	16	14	13	13	14	15	16	18	19	24	30
Biomass		0	2	4	4	4	5	5	6	7	7	7	7
Other Renewables		12	11	10	10	10	10	11	12	14	17	19	22
Total		182	199	210	213	207	198	207	213	223	236	251	267
Cumulative Resources Use	ZJ												
Coal		0.0	0.5	0.9	1.4	1.6	2.2	2.6	3.0	3.4	3.7	4.1	4.5
Oil		0.1	0.8	1.6	2.3	2.6	3.4	3.9	4.4	4.9	5.5	6.2	6.8
Gas		0.0	0.5	1.0	1.7	2.1	3.3	4.1	5.0	5.9	6.8	7.8	8.8
Cumulative CO2 Emissions	GtC	2.8	33.0	65.9	99.3	131.6	162.9	194.5	226.6	258.3	291.1	325.5	361.1
Carbon Sequestration	GtC												
Land Use	Million ha												
Cropland		408	408	408	406	395	365	349	329	317	311	285	259
Grasslands		796	819	838	854	849	840	842	848	873	915	930	944
Energy Biomass		0	4	6	7	8	9	9	15	18	18	16	15
Forest		921	931	924	902	919	976	1020	1048	1029	963	962	961
Others		998	961	946	954	952	933	903	882	886	915	930	945
Total		3123	3123	3123	3123	3123	3123	3123	3123	3123	3123	3123	3123
Anthropogenic Emissions (standardized)													
Fossil Fuel CO2	GtC	2.83	3.20	3.37	3.33	3.23	3.10	3.22	3.21	3.28	3.43	3.58	3.74
Other CO2	GtC	0.00	0.00	0.03	-0.05	-0.05	-0.03	0.03	-0.06	-0.09	-0.05	-0.08	-0.11
Total CO2	GtC	2.83	3.20	3.40	3.27	3.19	3.08	3.25	3.15	3.19	3.37	3.50	3.63
CH4 total	MtCH4	73	74	81	85	85	87	92	100	109	120	131	143
N2O total	MtN2O-N	2.6	2.6	2.7	2.9	2.9	3.1	3.3	3.4	3.7	4.2	4.5	4.9
SOx total	MtS	22.7	17.0	15.0	9.2	7.8	6.6	5.6	5.0	5.0	5.6	6.3	7.0
HFC	MtC eq.	19	57	107	103	111	116	125	130	135	142	151	160
PFC	MtC eq.	18	13	14	14	14	14	14	13	11	13	15	17
SF6	MtC eq.	23	23	25	28	29	29	28	26	20	12	14	16
CO	MtCO												
NMVOC	Mt												
NOx	MtN												

Scenario A2-A1-MiniCAM REF		1990	2000	2010	2020	2030	2040	2050	2060	2070	2080	2090	2100
Population	Million	428	446	462	475	480	480	475	470	466	464	470	475
GNP/GDP (mex)	Trillion US$	1.1	1.1	1.1	1.2	1.3	1.4	1.6	2.2	3.0	3.9	5.1	6.4
GNP/GDP (ppp) Trillion (1990 prices)		na	na	na	na	na	na	na	na	na	na	na	na
Final Energy	EJ												
Non-commercial		0	0	0	0	0	0	0	0	0	0	0	0
Solids		13	11	10	8	7	6	6	6	5	5	6	6
Liquids		18	12	9	8	6	6	6	8	9	11	13	15
Gas		19	16	13	11	10	10	11	12	15	18	19	20
Electricity		6	9	11	13	13	13	14	20	26	32	39	47
Others		0	0	0	0	0	0	0	0	0	0	0	0
Total		56	48	43	40	36	35	37	45	55	66	77	88
Primary Energy	EJ												
Coal		18	19	19	17	18	19	20	18	18	20	22	24
Oil		20	15	11	9	6	5	6	8	11	13	15	17
Gas		26	21	20	21	19	18	18	23	28	33	35	37
Nuclear		3	4	6	6	6	6	6	8	10	11	15	19
Biomass		0	1	1	2	2	2	2	3	3	4	4	4
Other Renewables		3	3	3	4	5	5	6	8	9	11	13	16
Total		70	64	60	58	55	54	57	67	79	92	105	118
Cumulative Resources Use	ZJ												
Coal		0.0	0.2	0.4	0.6	0.8	0.9	1.1	1.3	1.5	1.7	1.9	2.1
Oil		0.0	0.2	0.3	0.4	0.5	0.5	0.6	0.7	0.8	0.9	1.0	1.2
Gas		0.0	0.2	0.5	0.7	0.9	1.0	1.2	1.4	1.7	2.0	2.3	2.7
Cumulative CO2 Emissions	GtC	1.3	12.3	21.0	29.3	36.8	43.6	50.4	57.5	65.2	75.0	86.5	98.8
Carbon Sequestration	GtC												
Land Use	Million ha												
Cropland		284	293	301	310	298	281	260	247	244	251	242	232
Grasslands		395	410	430	458	450	441	428	436	473	539	571	604
Energy Biomass		0	0	1	2	2	2	2	6	9	9	7	5
Forest		1007	1016	999	956	978	1018	1075	1093	1054	958	925	891
Others		691	659	646	652	649	636	612	595	598	621	633	644
Total		2377	2377	2377	2377	2377	2377	2377	2377	2377	2377	2377	2377
Anthropogenic Emissions (standardized)													
Fossil Fuel CO2	GtC	1.30	0.91	0.81	0.74	0.67	0.65	0.66	0.75	0.88	1.04	1.16	1.29
Other CO2	GtC	0.00	0.00	0.03	0.07	0.03	0.02	0.04	-0.04	-0.03	0.07	0.03	-0.02
Total CO2	GtC	1.30	0.91	0.83	0.81	0.70	0.66	0.70	0.71	0.84	1.11	1.19	1.27
CH4 total	MtCH4	47	39	43	49	51	55	61	67	78	94	107	120
N2O total	MtN2O-N	0.6	0.6	0.7	0.8	0.9	0.9	1.0	1.1	1.3	1.6	1.9	2.2
SOx total	MtS	17.0	11.0	10.3	8.9	8.8	8.8	8.8	9.3	9.8	10.1	10.3	10.5
HFC	MtC eq.	0	4	8	13	20	27	31	37	41	44	48	52
PFC	MtC eq.	7	4	8	10	14	17	20	25	30	35	40	42
SF6	MtC eq.	8	6	7	10	12	15	19	24	28	32	36	38
CO	MtCO												
NMVOC	Mt												
NOx	MtN												

Scenario A2-A1-MiniCAM ASIA		1990	2000	2010	2020	2030	2040	2050	2060	2070	2080	2090	2100
Population	Million	2790	3193	3603	4022	4396	4723	5004	5179	5343	5496	5663	5833
GNP/GDP (mex)	Trillion US$	1.4	3.1	4.5	5.6	6.9	9.6	13.6	22.0	31.9	43.4	58.1	75.1
GNP/GDP (ppp) Trillion (1990 prices)		na	na	na	na	na	na	na	na	na	na	na	na
Final Energy	EJ												
Non-commercial		0	0	0	0	0	0	0	0	0	0	0	0
Solids		20	34	43	48	54	63	76	77	80	88	94	99
Liquids		14	21	24	24	24	30	41	62	84	107	132	156
Gas		2	6	7	8	8	10	15	21	29	39	43	47
Electricity		4	12	19	24	28	40	59	100	145	193	250	306
Others		0	0	0	0	0	0	0	0	0	0	0	0
Total		40	72	93	104	114	143	191	260	338	426	517	608
Primary Energy	EJ												
Coal		26	49	63	68	75	87	105	112	124	141	154	168
Oil		16	22	26	26	24	28	39	64	90	115	140	165
Gas		3	10	16	21	24	34	51	79	109	140	154	168
Nuclear		1	4	7	8	11	17	27	45	60	75	105	135
Biomass		0	2	4	6	7	9	11	18	24	29	31	34
Other Renewables		3	4	5	6	8	11	15	25	37	51	67	83
Total		49	92	121	136	148	186	249	344	444	550	652	753
Cumulative Resources Use	ZJ												
Coal		0.0	0.5	1.0	1.7	2.4	3.2	4.2	5.3	6.5	7.8	9.3	10.8
Oil		0.0	0.2	0.5	0.7	1.0	1.2	1.6	2.1	2.9	3.9	5.3	6.6
Gas		0.0	0.1	0.2	0.4	0.6	0.9	1.3	2.1	3.0	4.2	5.7	7.2
Cumulative CO2 Emissions	GtC	1.5	19.3	41.7	67.2	94.3	124.5	161.7	208.2	265.2	335.0	416.5	507.9
Carbon Sequestration	GtC												
Land Use	Million ha												
Cropland		389	399	406	411	403	390	373	351	339	336	327	319
Grasslands		508	523	535	545	548	554	562	570	583	601	621	641
Energy Biomass		0	0	1	4	5	6	7	21	32	39	39	38
Forest		1168	1145	1124	1104	1108	1114	1122	1125	1114	1090	1075	1061
Others		664	634	622	628	625	613	592	575	576	597	609	621
Total		2729	2701	2689	2693	2689	2677	2657	2642	2644	2663	2672	2680
Anthropogenic Emissions (standardized)													
Fossil Fuel CO2	GtC	1.15	1.78	2.27	2.48	2.66	3.19	4.08	5.10	6.27	7.59	8.59	9.65
Other CO2	GtC	0.37	0.26	0.17	0.18	0.10	0.07	0.09	0.03	0.02	0.08	0.04	0.00
Total CO2	GtC	1.53	2.03	2.44	2.66	2.76	3.27	4.17	5.13	6.29	7.67	8.63	9.64
CH4 total	MtCH4	113	125	129	128	131	133	136	139	146	159	168	176
N2O total	MtN2O-N	2.3	2.6	2.8	3.0	3.2	3.4	3.7	3.9	4.3	4.9	5.5	6.1
SOx total	MtS	17.7	25.3	31.5	30.8	32.3	36.0	41.9	43.4	43.8	43.1	40.1	37.1
HFC	MtC eq.	0	5	11	18	27	38	54	77	100	130	167	204
PFC	MtC eq.	3	5	11	15	20	25	32	42	49	56	64	67
SF6	MtC eq.	4	7	11	16	20	27	34	43	49	55	62	67
CO	MtCO												
NMVOC	Mt												
NOx	MtN												

Scenario A2-A1-MiniCAM ALM		1990	2000	2010	2020	2030	2040	2050	2060	2070	2080	2090	2100
Population	Million	1236	1494	1795	2136	2528	2926	3331	3673	4004	4323	4605	4896
GNP/GDP (mex)	Trillion US$	1.9	2.8	3.9	5.2	7.0	9.9	13.9	20.8	29.6	40.2	54.8	71.7
GNP/GDP (ppp) Trillion (1990 prices)		na	na	na	na	na	na	na	na	na	na	na	na
Final Energy	EJ												
Non-commercial		0	0	0	0	0	0	0	0	0	0	0	0
Solids		2	3	4	6	8	9	10	11	12	15	17	20
Liquids		17	22	26	30	30	37	50	72	99	132	168	203
Gas		5	7	9	12	16	23	32	42	56	74	84	94
Electricity		3	6	9	13	19	27	38	60	86	117	162	207
Others		0	0	0	0	0	0	0	0	0	0	0	0
Total		27	38	49	60	73	96	130	184	253	337	431	525
Primary Energy	EJ												
Coal		4	6	8	10	14	18	22	27	34	42	52	61
Oil		20	23	27	30	31	38	51	75	104	138	175	212
Gas		7	10	14	20	27	38	53	75	102	135	157	178
Nuclear		0	2	4	5	7	11	16	25	35	45	69	93
Biomass		0	1	1	2	3	3	4	7	9	11	13	15
Other Renewables		5	6	8	11	14	16	19	25	32	40	51	63
Total		35	48	63	78	95	125	166	233	315	411	517	622
Cumulative Resources Use	ZJ												
Coal		0.0	0.1	0.1	0.2	0.4	0.5	0.7	1.0	1.3	1.7	2.1	2.6
Oil		0.0	0.2	0.5	0.8	1.1	1.4	1.9	2.6	3.5	4.6	6.3	8.0
Gas		0.0	0.1	0.2	0.4	0.6	1.0	1.4	2.1	3.0	4.1	5.7	7.2
Cumulative CO2 Emissions	GtC	1.4	17.8	35.5	52.9	71.0	91.5	117.3	149.6	190.0	242.3	307.3	384.4
Carbon Sequestration	GtC												
Land Use	Million ha												
Cropland		391	362	348	350	339	323	303	287	275	267	253	239
Grasslands		1510	1592	1660	1716	1703	1690	1676	1702	1775	1892	1969	2045
Energy Biomass		0	0	0	0	0	0	0	3	6	9	10	10
Forest		3641	3595	3539	3474	3495	3525	3565	3560	3498	3379	3306	3232
Others		1957	1887	1859	1872	1864	1830	1771	1725	1727	1775	1799	1822
Total		7499	7436	7407	7413	7401	7368	7315	7277	7279	7323	7336	7349
Anthropogenic Emissions (standardized)													
Fossil Fuel CO2	GtC	0.72	1.01	1.17	1.35	1.57	1.95	2.50	3.33	4.41	5.71	6.96	8.33
Other CO2	GtC	0.73	0.82	0.54	0.42	0.28	0.29	0.43	0.21	0.13	0.21	0.11	0.02
Total CO2	GtC	1.45	1.83	1.71	1.77	1.85	2.24	2.92	3.54	4.54	5.92	7.07	8.35
CH4 total	MtCH4	77	85	87	91	96	104	113	123	140	164	176	187
N2O total	MtN2O-N	1.2	1.3	1.6	1.9	2.1	2.4	2.7	3.1	3.7	4.5	5.3	6.1
SOx total	MtS	10.5	12.8	14.1	13.7	14.7	16.4	18.9	20.6	22.4	24.4	25.7	27.0
HFC	MtC eq.	0	2	19	32	49	69	98	139	181	233	295	336
PFC	MtC eq.	4	4	8	12	16	21	26	34	39	45	50	52
SF6	MtC eq.	3	5	7	10	13	18	23	29	32	37	41	43
CO	MtCO												
NMVOC	Mt												
NOx	MtN												

Scenario B1-AIM World		1990	2000	2010	2020	2030	2040	2050	2060	2070	2080	2090	2100
Population	Million	5204	6056	6741	7426	8112	8368	8631	8465	8301	7957	7453	6982
GNP/GDP (mex)	Trillion US$	20.9	27.4	37.3	51.9	73.9	99.7	134.5	165.9	204.8	245.7	286.6	334.3
GNP/GDP (ppp) Trillion (1990 prices)													
Final Energy	EJ												
Non-commercial		50	46	30	24	17	0	0	0	0	0	0	0
Solids		36	39	47	56	62	65	67	52	41	31	24	18
Liquids		111	119	130	152	173	186	200	202	204	204	200	196
Gas		51	59	82	104	131	149	170	161	152	137	117	100
Electricity		38	50	63	81	103	138	186	228	279	314	324	334
Others													
Total		285	311	353	417	485	549	623	652	683	688	668	648
Primary Energy	EJ												
Coal		93	93	126	162	195	198	204	168	138	113	92	75
Oil		143	161	176	177	170	152	137	125	115	108	106	103
Gas		73	104	143	183	229	264	308	288	270	241	204	172
Nuclear		6	10	11	14	15	23	38	36	33	30	27	25
Biomass		50	48	38	56	77	94	117	133	150	162	166	171
Other Renewables		10	11	14	19	29	49	87	138	220	290	316	344
Total		376	427	508	610	715	781	892	887	926	944	911	890
Cumulative Resources Use	ZJ												
Coal		0.1	1.0	2.1	3.5	5.3	7.0	9.3	10.8	12.5	13.9	14.9	16.0
Oil		0.1	1.7	3.4	5.1	6.9	8.2	9.9	11.0	12.3	13.5	14.5	15.7
Gas		0.1	1.0	2.2	3.8	5.9	8.1	11.3	13.7	16.6	19.3	21.5	23.8
Cumulative CO2 Emissions	GtC	7.1	82.4	169.3	267.6	375.1	489.7	609.9	725.8	828.4	917.8	994.1	1060.1
Carbon Sequestraction	GtC												
Land Use	Million ha												
Cropland		1459	1466	1464	1461	1458	1452	1447	1442	1437	1433	1431	1429
Grasslands		3389	3407	3432	3453	3474	3510	3547	3591	3635	3668	3688	3709
Energy Biomass		0	0	0	59	137	187	254	260	266	260	242	225
Forest		4296	4255	4231	4264	4335	4404	4475	4522	4570	4619	4668	4719
Others		3805	3820	3822	3711	3545	3382	3226	3131	3039	2968	2917	2868
Total		12949	12949	12949	12949	12949	12949	12949	12949	12949	12949	12949	12949
Anthropogenic Emissions (standardized)													
Fossil Fuel CO2	GtC	5.99	6.90	8.47	10.05	11.50	12.00	12.59	11.12	9.86	8.29	7.27	6.40
Other CO2	GtC	1.11	1.07	0.93	0.23	-0.29	-0.31	-0.24	-0.28	-0.19	-0.09	-0.21	-0.26
Total CO2	GtC	7.10	7.97	9.39	10.28	11.22	11.69	12.35	10.84	9.67	8.20	7.06	6.15
CH4 total	MtCH4	310	323	358	391	426	437	449	429	410	364	300	248
N2O total	MtN2O-N	6.7	7.0	6.9	6.9	6.8	6.8	6.8	6.7	6.6	6.5	6.3	6.2
SOx total	MtS	70.9	69.0	76.3	89.2	92.9	60.7	42.3	33.1	26.3	22.8	21.4	20.1
CFC/HFC/HCFC	MtC eq.	1672	883	784	291	257	298	338	337	333	327	315	299
PFC	MtC eq.	32	25	29	32	33	37	42	47	45	43	46	45
SF6	MtC eq.	38	40	36	37	47	58	68	71	62	46	43	43
CO	MtCO	879	877	968	981	1076	1146	1223	1241	1260	1278	1296	1314
NMVOC	Mt	139	141	157	177	176	171	167	141	119	97	76	58
NOx	MtN	31	32	35	38	41	41	40	36	31	28	26	24

Scenario B1-AIM OECD90		**1990**	**2000**	**2010**	**2020**	**2030**	**2040**	**2050**	**2060**	**2070**	**2080**	**2090**	**2100**
Population	Million	801	857	896	935	973	991	1008	1013	1018	1023	1029	1035
GNP/GDP (mex)	Trillion US$	16.4	20.7	25.5	31.3	38.5	44.9	52.4	59.1	66.7	74.0	80.5	87.7
GNP/GDP (ppp) Trillion (1990 prices)													
Final Energy	EJ												
Non-commercial		6	2	0	0	0	0	0	0	0	0	0	0
Solids		10	11	10	10	9	8	7	6	4	3	2	2
Liquids		64	70	67	70	68	66	64	60	56	53	50	47
Gas		25	31	37	40	43	41	40	35	31	27	23	19
Electricity		22	27	29	31	33	38	43	50	59	64	66	68
Others													
Total		127	141	144	151	153	154	154	153	152	148	142	137
Primary Energy	EJ												
Coal		41	34	37	38	38	36	33	26	21	17	14	11
Oil		76	88	85	81	69	56	46	39	33	29	28	26
Gas		34	51	58	61	63	62	62	54	47	40	34	28
Nuclear		5	7	8	8	7	8	8	8	7	6	5	5
Biomass		6	1	0	4	9	14	20	23	26	28	28	29
Other Renewables		6	6	7	9	13	17	22	32	48	60	65	70
Total		167	187	195	201	199	192	192	182	181	180	174	170
Cumulative Resources Use	ZJ												
Coal		0.0	0.4	0.8	1.1	1.5	1.8	2.2	2.5	2.7	3.0	3.1	3.3
Oil		0.1	0.9	1.8	2.6	3.4	3.9	4.5	4.9	5.3	5.6	5.9	6.2
Gas		0.0	0.5	1.0	1.6	2.2	2.8	3.5	4.0	4.5	5.0	5.3	5.7
Cumulative CO2 Emissions	GtC	2.8	33.0	64.8	96.5	127.3	156.7	184.3	209.3	230.8	249.1	264.6	277.7
Carbon Sequestraction	GtC												
Land Use	Million ha												
Cropland		381	380	378	376	374	370	366	361	357	354	353	352
Grasslands		760	765	772	779	787	809	832	862	893	912	918	924
Energy Biomass		0	0	0	10	23	32	44	45	46	45	41	39
Forest		1050	1059	1074	1094	1117	1133	1149	1154	1159	1167	1179	1192
Others		838	826	805	770	728	682	639	605	574	551	536	522
Total		3029	3029	3029	3029	3029	3029	3029	3029	3029	3029	3029	3029
Anthropogenic Emissions (standardized)													
Fossil Fuel CO2	GtC	2.83	3.20	3.24	3.27	3.16	2.96	2.77	2.34	1.97	1.68	1.45	1.25
Other CO2	GtC	0.00	0.00	-0.06	-0.12	-0.14	-0.11	-0.08	-0.03	0.01	0.00	-0.03	-0.06
Total CO2	GtC	2.83	3.20	3.18	3.15	3.02	2.85	2.69	2.31	1.98	1.68	1.42	1.19
CH4 total	MtCH4	73	74	71	69	66	60	54	50	46	43	41	38
N2O total	MtN2O-N	2.6	2.6	2.5	2.5	2.5	2.4	2.3	2.2	2.2	2.1	2.0	2.0
SOx total	MtS	22.7	17.0	10.4	7.8	7.5	7.2	6.9	6.2	5.5	4.9	4.3	3.8
HFC	MtC eq.	19	58	108	103	109	112	116	117	117	119	120	120
PFC	MtC eq.	18	13	12	10	8	7	7	7	7	7	7	6
SF6	MtC eq.	24	23	16	5	6	6	7	7	8	8	8	8
CO	MtCO	179	161	172	183	200	205	210	209	209	209	210	212
NMVOC	Mt	42	36	34	31	28	23	18	11	5	1	-1	-3
NOx	MtN	13	12	11	10	9	7	6	6	5	5	4	4

Scenario B1-AIM REF		1990	2000	2010	2020	2030	2040	2050	2060	2070	2080	2090	2100
Population	Million	413	419	424	430	435	429	423	406	391	374	356	339
GNP/GDP (mex)	Trillion US$	1.1	0.8	1.2	1.8	2.5	3.9	5.9	7.8	10.2	12.9	15.7	19.1
GNP/GDP (ppp) Trillion (1990 prices)													
Final Energy	EJ												
Non-commercial		2	5	0	0	0	0	0	0	0	0	0	0
Solids		9	6	5	5	4	4	4	3	2	2	1	1
Liquids		19	9	7	5	4	4	4	4	3	3	3	2
Gas		19	13	16	17	18	21	26	23	21	17	13	10
Electricity		8	9	9	9	9	12	17	21	25	27	27	27
Others													
Total		58	42	37	36	35	42	51	51	52	50	45	40
Primary Energy	EJ												
Coal		18	13	13	12	11	12	13	11	9	7	6	5
Oil		22	12	9	7	6	5	4	3	2	1	1	1
Gas		26	22	25	26	25	29	33	30	26	22	17	13
Nuclear		1	1	1	2	1	2	3	3	3	3	2	2
Biomass		2	4	0	3	8	12	17	19	22	24	24	25
Other Renewables		1	1	1	2	2	4	8	13	20	25	26	28
Total		71	54	50	51	53	63	79	78	82	82	76	73
Cumulative Resources Use	ZJ												
Coal		0.0	0.2	0.3	0.4	0.5	0.7	0.8	0.9	1.0	1.1	1.1	1.2
Oil		0.0	0.2	0.3	0.4	0.4	0.5	0.5	0.5	0.6	0.6	0.6	0.6
Gas		0.0	0.3	0.5	0.8	1.0	1.3	1.6	1.9	2.2	2.4	2.6	2.8
Cumulative CO2 Emissions	GtC	1.3	12.3	21.5	29.6	35.8	40.9	45.3	48.6	51.0	51.3	50.1	49.0
Carbon Sequestraction	GtC												
Land Use	Million ha												
Cropland		268	266	266	266	265	265	264	264	263	263	263	262
Grasslands		341	362	364	366	368	370	372	374	376	378	380	382
Energy Biomass		0	0	0	9	20	27	37	38	39	38	35	33
Forest		966	951	934	933	934	953	973	983	993	1005	1018	1031
Others		701	696	711	702	688	658	630	617	604	592	580	568
Total		2276	2276	2276	2276	2276	2276	2276	2276	2276	2276	2276	2276
Anthropogenic Emissions (standardized)													
Fossil Fuel CO2	GtC	1.30	0.91	0.88	0.78	0.65	0.66	0.66	0.48	0.35	-0.11	-0.05	0.02
Other CO2	GtC	0.00	0.00	0.05	-0.10	-0.10	-0.19	-0.26	-0.20	-0.15	-0.02	-0.07	-0.11
Total CO2	GtC	1.30	0.91	0.93	0.68	0.56	0.47	0.40	0.28	0.20	-0.14	-0.12	-0.10
CH4 total	MtCH4	47	39	48	50	49	46	43	34	26	20	16	12
N2O total	MtN2O-N	0.6	0.6	0.6	0.6	0.6	0.6	0.6	0.6	0.6	0.5	0.5	0.5
SOx total	MtS	17.0	11.0	10.6	9.2	6.9	4.3	2.7	2.2	1.9	1.7	1.7	1.6
HFC	MtC eq.	0	4	9	15	20	24	26	27	27	27	27	26
PFC	MtC eq.	7	4	6	6	7	8	9	10	9	8	8	8
SF6	MtC eq.	8	6	4	7	7	8	9	9	7	6	4	4
CO	MtCO	69	41	36	34	33	33	33	32	32	31	30	29
NMVOC	Mt	16	13	12	11	10	9	9	7	6	4	3	1
NOx	MtN	5	3	3	3	2	2	2	2	1	1	1	1

Scenario B1-AIM ASIA		**1990**	**2000**	**2010**	**2020**	**2030**	**2040**	**2050**	**2060**	**2070**	**2080**	**2090**	**2100**
Population	Million	2798	3261	3556	3851	4147	4183	4220	4016	3822	3541	3194	2882
GNP/GDP (mex)	Trillion US$	1.5	2.9	5.2	9.2	16.2	24.0	35.8	46.0	59.2	73.5	88.2	105.9
GNP/GDP (ppp) Trillion (1990 prices)													
Final Energy	EJ												
Non-commercial		28	22	20	16	8	0	0	0	0	0	0	0
Solids		15	20	30	39	46	49	52	40	31	24	18	14
Liquids		11	15	26	35	45	51	58	59	60	60	58	57
Gas		2	4	10	18	24	28	33	32	31	28	24	20
Electricity		5	7	15	25	38	54	78	96	119	133	135	137
Others													
Total		61	69	101	134	162	189	220	232	244	246	236	227
Primary Energy	EJ												
Coal		30	39	66	95	122	119	116	94	76	61	49	38
Oil		17	29	39	42	44	40	35	31	28	25	24	23
Gas		4	9	21	37	52	67	87	81	76	67	55	45
Nuclear		0	1	2	3	4	8	16	15	14	13	12	10
Biomass		28	24	19	18	15	19	24	27	31	33	34	35
Other Renewables		1	2	2	4	7	16	34	56	92	122	131	141
Total		80	103	148	199	244	269	314	306	317	321	304	293
Cumulative Resources Use	ZJ												
Coal		0.0	0.4	0.9	1.7	2.8	3.8	5.2	6.0	7.0	7.7	8.3	8.8
Oil		0.0	0.2	0.6	1.0	1.4	1.8	2.2	2.5	2.8	3.1	3.3	3.6
Gas		0.0	0.1	0.2	0.5	0.9	1.5	2.3	3.0	3.8	4.6	5.2	5.8
Cumulative CO2 Emissions	GtC	1.5	19.3	45.3	81.0	126.0	177.6	232.2	284.2	328.8	367.4	400.5	429.0
Carbon Sequestraction	GtC												
Land Use	Million ha												
Cropland		438	435	435	435	434	434	434	434	433	433	433	433
Grasslands		608	606	609	612	614	617	619	622	625	628	631	634
Energy Biomass		0	0	0	12	28	39	52	54	55	53	50	46
Forest		535	525	518	535	551	557	563	577	593	604	610	616
Others		583	598	602	571	536	516	496	477	458	446	441	436
Total		2164	2164	2164	2164	2164	2164	2164	2164	2164	2164	2164	2164
Anthropogenic Emissions (standardized)													
Fossil Fuel CO2	GtC	1.15	1.78	2.87	3.99	4.97	5.19	5.42	4.74	4.15	3.57	3.03	2.57
Other CO2	GtC	0.37	0.26	0.29	0.00	0.05	0.10	0.22	0.03	0.01	0.00	0.02	0.08
Total CO2	GtC	1.53	2.03	3.16	3.99	5.02	5.29	5.63	4.77	4.15	3.58	3.05	2.66
CH4 total	MtCH4	113	125	142	160	183	196	209	205	200	177	143	114
N2O total	MtN2O-N	2.3	2.6	2.6	2.6	2.6	2.7	2.7	2.7	2.6	2.6	2.6	2.5
SOx total	MtS	17.7	25.3	40.0	56.5	58.2	31.2	16.7	11.4	7.8	6.0	5.3	4.7
HFC	MtC eq.	0	4	11	20	34	56	93	90	85	79	72	64
PFC	MtC eq.	3	5	7	9	11	14	17	20	18	18	18	18
SF6	MtC eq.	4	7	11	17	23	27	30	32	26	19	16	16
CO	MtCO	235	270	335	366	438	473	510	511	511	510	507	504
NMVOC	Mt	33	37	51	63	59	60	61	51	42	33	24	17
NOx	MtN	7	9	12	15	18	19	19	16	13	11	9	8

Scenario B1-AIM ALM		1990	2000	2010	2020	2030	2040	2050	2060	2070	2080	2090	2100
Population	Million	1192	1519	1865	2211	2557	2761	2980	3024	3067	3013	2866	2726
GNP/GDP (mex)	Trillion US$	1.9	3.0	5.4	9.6	16.7	26.0	40.4	52.5	68.2	85.1	101.7	121.7
GNP/GDP (ppp) Trillion (1990 prices)													
Final Energy	EJ												
Non-commercial		14	17	9	8	8	0	0	0	0	0	0	0
Solids		1	2	2	2	3	4	4	3	3	2	2	2
Liquids		17	24	31	41	55	64	75	79	84	88	89	90
Gas		4	10	19	29	46	57	71	70	69	65	57	50
Electricity		4	7	11	16	23	33	48	61	77	89	95	102
Others													
Total		40	60	71	97	135	163	198	215	235	245	244	244
Primary Energy	EJ												
Coal		5	7	11	16	24	31	41	36	32	28	24	21
Oil		27	33	44	47	51	51	52	52	52	52	53	54
Gas		9	22	39	59	89	106	126	123	121	112	98	86
Nuclear		0	0	1	1	2	5	10	9	9	9	8	8
Biomass		14	18	19	31	45	50	56	63	72	77	79	81
Other Renewables		2	2	3	5	7	13	23	37	61	82	93	105
Total		57	83	115	159	219	257	308	322	346	360	356	355
Cumulative Resources Use	ZJ												
Coal		0.0	0.1	0.2	0.3	0.5	0.7	1.1	1.4	1.8	2.1	2.4	1.7
Oil		0.0	0.3	0.7	1.2	1.7	2.1	2.7	3.1	3.7	4.2	4.7	6.7
Gas		0.0	0.2	0.5	1.0	1.7	2.6	3.8	4.9	6.2	7.4	8.4	12.5
Cumulative CO2 Emissions	GtC	1.4	17.8	37.6	60.5	86.0	114.5	148.1	183.7	217.8	249.9	278.9	304.4
Carbon Sequestraction	GtC												
Land Use	Million ha												
Cropland		371	385	385	385	384	384	383	383	383	383	382	382
Grasslands		1680	1675	1686	1696	1705	1714	1723	1732	1741	1750	1760	1769
Energy Biomass		0	0	0	29	66	89	121	124	127	124	115	107
Forest		1745	1720	1704	1702	1732	1761	1790	1808	1825	1843	1861	1880
Others		1684	1700	1704	1669	1593	1526	1461	1432	1403	1379	1360	1342
Total		5480	5480	5480	5480	5480	5480	5480	5480	5480	5480	5480	5480
Anthropogenic Emissions (standardized)													
Fossil Fuel CO2	GtC	0.72	1.01	1.48	2.02	2.73	3.20	3.75	3.56	3.39	3.14	2.84	2.56
Other CO2	GtC	0.73	0.82	0.65	0.44	-0.10	-0.11	-0.11	-0.08	-0.05	-0.07	-0.12	-0.17
Total CO2	GtC	1.45	1.83	2.13	2.46	2.63	3.08	3.63	3.48	3.34	3.08	2.72	2.39
CH4 total	MtCH4	77	85	97	112	128	135	143	141	138	124	101	84
N2O total	MtN2O-N	1.2	1.3	1.3	1.2	1.1	1.2	1.2	1.2	1.2	1.2	1.2	1.2
SOx total	MtS	10.5	12.8	12.2	12.6	17.3	15.0	13.1	10.3	8.1	7.2	7.1	7.0
HFC	MtC eq.	0	2	16	28	42	64	98	102	102	101	96	89
PFC	MtC eq.	4	4	5	6	7	8	9	11	11	11	12	12
SF6	MtC eq.	2	5	5	8	12	17	22	24	21	14	14	14
CO	MtCO	396	404	425	398	405	436	470	489	508	528	548	569
NMVOC	Mt	48	55	60	72	80	79	79	73	67	59	50	42
NOx	MtN	7	8	9	10	12	12	13	13	12	12	11	11

Scenario B1-ASF World		**1990**	**2000**	**2010**	**2020**	**2030**	**2040**	**2050**	**2060**	**2070**	**2080**	**2090**	**2100**
Population	Million	5264	6117	6827	7537	8039	8526	8704	8527	8444	8022	7282	7056
GNP/GDP (mex)	Trillion US$	20.6	27.6	38.0	53.5	73.8	114.9	134.3	167.2	186.9	252.2	314.9	339.3
GNP/GDP (ppp) Trillion (1990 prices)													
Final Energy	EJ												
Non-commercial													
Solids		55	66	80	89	76	62	49	42	35	30	28	25
Liquids		121	170	228	304	308	311	314	280	247	221	203	186
Gas		50	58	82	124	149	174	198	185	171	157	143	129
Electricity		42	55	77	104	149	195	241	283	325	343	339	334
Others													
Total		269	350	467	621	682	742	802	790	778	752	712	673
Primary Energy	EJ												
Coal		97	119	149	191	262	332	403	354	305	259	215	172
Oil		139	198	272	354	289	225	160	103	46	14	8	2
Gas		74	83	118	179	231	283	335	303	272	242	213	184
Nuclear		7	10	8	7	9	11	13	11	10	9	8	8
Biomass		0	0	0	14	43	73	102	125	148	167	183	198
Other Renewables		10	14	23	30	46	61	77	135	192	223	226	229
Total		326	424	570	774	879	984	1090	1031	973	914	853	791
Cumulative Resources Use	ZJ												
Coal		0.0	1.2	2.5	4.2	6.5	9.5	13.2	16.9	20.2	23.0	25.3	27.2
Oil		0.0	1.8	4.2	7.3	10.5	13.0	14.9	16.1	16.8	17.0	17.1	17.2
Gas		0.0	0.9	1.9	3.3	5.4	8.0	11.1	14.3	17.1	19.7	21.9	23.9
Cumulative CO2 Emissions	GtC	7.1	82.4	176.1	302.2	458.8	631.1	810.7	978.8	1116.5	1227.0	1317.0	1389.7
Carbon Sequestration	GtC	-1.8	-2.0	-2.1	-2.3	-2.5	-2.7	-2.9	-3.1	-3.2	-3.3	-3.4	-3.4
Land Use	Million ha												
Cropland													
Grasslands													
Energy Biomass													
Forest													
Others													
Total													
Anthropogenic Emissions (standardized)													
Fossil Fuel CO2	GtC	5.99	6.90	9.65	13.22	15.72	16.61	17.50	14.71	11.93	9.68	7.98	6.27
Other CO2	GtC	1.11	1.07	1.12	1.24	1.13	0.98	0.84	0.58	0.32	0.18	0.16	0.14
Total CO2	GtC	7.10	7.97	10.76	14.46	16.85	17.59	18.33	15.29	12.24	9.86	8.13	6.41
CH4 total	MtCH4	310	323	373	430	479	513	546	513	481	447	412	377
N2O total	MtN2O-N	6.7	7.0	8.1	9.5	10.5	11.0	11.5	11.2	10.8	10.4	9.9	9.3
SOx total	MtS	70.9	69.0	83.6	112.3	104.7	76.4	48.1	38.1	28.1	23.4	24.0	24.6
CFC/HFC/HCFC	MtC eq.	1672	883	784	291	257	298	338	337	333	327	315	299
PFC	MtC eq.	32	25	29	32	33	37	42	47	45	43	46	45
SF6	MtC eq.	38	40	36	37	47	58	68	71	62	46	43	43
CO	MtCO	879	877	1039	1162	1338	1316	1293	1199	1105	1029	971	913
NMVOC	Mt	139	141	162	193	218	228	237	255	272	295	322	349
NOx	MtN	31	32	44	59	70	71	72	62	52	45	40	35

Scenario B1-ASF OECD90	1990	2000	2010	2020	2030	2040	2050	2060	2070	2080	2090	2100	
Population	Million	849	919	961	1003	1035	1069	1081	1085	1088	1096	1106	1110
GNP/GDP (mex)	Trillion US$	15.5	19.2	23.4	27.7	32.4	40.4	43.5	49.7	53.0	64.7	78.7	84.1
GNP/GDP (ppp) Trillion (1990 prices)													
Final Energy	EJ												
Non-commercial													
Solids		12	13	12	12	12	11	11	9	8	7	7	6
Liquids		68	86	91	88	77	66	55	48	41	37	37	36
Gas		27	29	31	35	36	37	38	35	33	31	31	31
Electricity		24	27	33	37	40	43	46	54	63	69	74	79
Others													
Total		131	155	167	172	165	157	150	147	144	144	148	152
Primary Energy	EJ												
Coal		33	32	34	39	52	66	79	66	52	44	40	36
Oil		77	97	106	102	75	47	20	12	5	1	1	0
Gas		35	38	42	50	54	59	63	57	51	47	45	43
Nuclear		6	8	6	4	4	3	3	3	3	2	2	3
Biomass		0	0	0	4	10	15	21	24	27	30	34	38
Other Renewables		6	7	10	11	13	15	16	27	38	46	50	55
Total		156	181	198	211	208	205	202	189	176	171	173	175
Cumulative Resources Use	ZJ												
Coal		0.0	0.4	0.7	1.1	1.5	2.1	2.8	3.6	4.1	4.6	5.0	5.4
Oil		0.0	1.0	2.0	3.0	3.9	4.5	4.8	4.9	5.0	5.0	5.0	5.0
Gas		0.0	0.4	0.8	1.3	1.8	2.3	3.0	3.6	4.1	4.6	5.1	5.5
Cumulative CO2 Emissions	GtC	2.8	33.0	66.5	102.6	140.3	177.6	212.9	244.3	269.7	290.4	308.8	326.1
Carbon Sequestration	GtC												
Land Use	Million ha												
Cropland													
Grasslands													
Energy Biomass													
Forest													
Others													
Total													
Anthropogenic Emissions (standardized)													
Fossil Fuel CO2	GtC	2.83	3.20	3.51	3.72	3.82	3.63	3.43	2.84	2.24	1.89	1.78	1.68
Other CO2	GtC	0.00	0.00	0.00	0.00	0.00	0.00	0.00	0.00	0.00	0.00	0.00	0.00
Total CO2	GtC	2.83	3.20	3.51	3.72	3.82	3.63	3.43	2.84	2.24	1.89	1.78	1.68
CH4 total	MtCH4	73	74	77	81	86	90	93	90	88	85	81	78
N2O total	MtN2O-N	2.6	2.6	2.7	2.9	3.0	2.9	2.8	2.7	2.6	2.5	2.4	2.3
SOx total	MtS	22.7	17.0	7.6	7.5	7.0	6.3	5.6	5.1	4.6	4.8	5.5	6.3
HFC	MtC eq.	19	58	108	103	109	112	116	117	117	119	120	120
PFC	MtC eq.	18	13	12	10	8	7	7	7	7	7	7	6
SF6	MtC eq.	24	23	16	5	6	6	7	7	8	8	8	8
CO	MtCO	179	161	165	154	135	100	66	54	42	35	34	32
NMVOC	Mt	42	36	40	43	41	36	32	32	31	33	37	40
NOx	MtN	13	12	13	14	14	13	11	9	7	6	6	5

Scenario B1-ASF REF		1990	2000	2010	2020	2030	2040	2050	2060	2070	2080	2090	2100
Population	Million	417	419	425	431	432	425	423	406	398	374	347	339
GNP/GDP (mex)	Trillion US$	1.0	0.9	1.3	1.9	2.9	4.6	5.3	6.6	7.4	10.0	12.8	14.0
GNP/GDP (ppp) Trillion (1990 prices)													
Final Energy	EJ												
Non-commercial													
Solids		18	11	11	9	8	6	5	3	2	1	1	1
Liquids		16	11	12	13	12	11	11	9	8	7	7	7
Gas		16	12	15	18	17	17	16	14	11	9	8	6
Electricity		8	7	9	11	10	10	10	10	10	10	9	8
Others													
Total		57	42	47	51	48	44	41	36	32	28	25	21
Primary Energy	EJ												
Coal		23	13	14	14	14	13	12	9	5	3	2	2
Oil		18	13	14	15	14	13	13	9	6	4	2	1
Gas		26	22	26	30	27	24	21	17	14	11	9	7
Nuclear		1	1	1	0	1	1	1	0	0	0	0	0
Biomass		0	0	0	0	0	0	1	2	4	6	8	9
Other Renewables		1	2	3	3	3	3	3	5	6	7	6	6
Total		69	50	57	64	59	55	50	43	36	31	28	25
Cumulative Resources Use	ZJ												
Coal		0.0	0.2	0.3	0.5	0.6	0.7	0.9	1.0	1.0	1.1	1.1	1.1
Oil		0.0	0.2	0.3	0.4	0.6	0.7	0.9	1.0	1.0	1.1	1.1	1.1
Gas		0.0	0.3	0.5	0.8	1.1	1.3	1.6	1.7	1.9	2.0	2.1	2.2
Cumulative CO2 Emissions	GtC	1.3	12.3	21.9	32.5	42.8	51.8	59.0	64.2	67.4	68.9	69.3	68.9
Carbon Sequestration	GtC												
Land Use	Million ha												
Cropland													
Grasslands													
Energy Biomass													
Forest													
Others													
Total													
Anthropogenic Emissions (standardized)													
Fossil Fuel CO2	GtC	1.30	0.91	1.01	1.09	0.98	0.81	0.63	0.42	0.22	0.07	0.00	-0.08
Other CO2	GtC	0.00	0.00	0.00	0.00	0.00	0.00	0.00	0.00	0.00	0.00	0.00	0.00
Total CO2	GtC	1.30	0.91	1.01	1.09	0.98	0.81	0.63	0.42	0.22	0.07	0.00	-0.08
CH4 total	MtCH4	47	39	43	53	67	78	89	77	65	54	46	37
N2O total	MtN2O-N	0.6	0.6	0.6	0.6	0.7	0.7	0.7	0.6	0.6	0.5	0.5	0.5
SOx total	MtS	17.0	11.0	10.4	9.6	8.0	5.7	3.4	2.5	1.6	1.0	0.7	0.4
HFC	MtC eq.	0	4	9	15	20	24	26	27	27	27	27	26
PFC	MtC eq.	7	4	6	6	7	8	9	10	9	8	8	8
SF6	MtC eq.	8	6	4	7	7	8	9	9	7	6	4	4
CO	MtCO	69	41	41	38	41	38	35	31	28	25	24	23
NMVOC	Mt	16	13	16	22	27	26	25	24	22	22	25	27
NOx	MtN	5	3	3	4	3	3	2	2	1	1	1	0

Scenario B1-ASF ASIA		1990	2000	2010	2020	2030	2040	2050	2060	2070	2080	2090	2100
Population	Million	2780	3261	3572	3884	4073	4181	4220	4012	3913	3538	3033	2882
GNP/GDP (mex)	Trillion US$	1.6	3.4	6.7	12.5	20.5	37.9	46.6	60.0	68.2	94.3	115.9	124.2
GNP/GDP (ppp) Trillion (1990 prices)													
Final Energy	EJ												
Non-commercial													
Solids		24	41	55	64	50	36	22	18	15	12	10	8
Liquids		15	36	67	114	124	134	144	130	116	104	93	83
Gas		3	8	19	41	55	70	85	78	71	63	56	48
Electricity		6	13	23	36	68	100	133	149	165	167	155	143
Others													
Total		47	97	163	255	298	340	383	375	366	346	314	282
Primary Energy	EJ												
Coal		36	65	88	116	154	193	232	201	170	141	112	83
Oil		19	45	85	134	110	86	62	38	13	1	0	0
Gas		4	10	27	56	89	121	154	137	120	103	86	69
Nuclear		0	1	1	2	4	6	8	7	6	5	4	4
Biomass		0	0	0	5	18	30	43	54	65	72	76	80
Other Renewables		1	3	6	9	19	29	39	67	95	107	102	98
Total		61	124	207	323	395	466	537	504	470	429	381	333
Cumulative Resources Use	ZJ												
Coal		0.0	0.5	1.3	2.3	3.7	5.5	7.6	9.7	11.6	13.1	14.4	15.3
Oil		0.0	0.3	1.0	2.1	3.3	4.3	5.0	5.5	5.7	5.7	5.7	5.7
Gas		0.0	0.1	0.3	0.7	1.4	2.5	3.8	5.3	6.6	7.7	8.6	9.4
Cumulative CO2 Emissions	GtC	1.5	19.3	48.1	96.8	166.3	249.5	341.8	431.8	508.5	572.7	626.1	669.6
Carbon Sequestration	GtC												
Land Use	Million ha												
Cropland													
Grasslands													
Energy Biomass													
Forest													
Others													
Total													
Anthropogenic Emissions (standardized)													
Fossil Fuel CO2	GtC	1.15	1.78	3.38	5.63	7.54	8.51	9.48	8.21	6.93	5.81	4.83	3.86
Other CO2	GtC	0.37	0.26	0.34	0.39	0.33	0.26	0.19	0.13	0.07	0.03	0.01	0.00
Total CO2	GtC	1.53	2.03	3.72	6.02	7.87	8.77	9.67	8.34	7.00	5.84	4.85	3.86
CH4 total	MtCH4	113	125	145	163	179	190	202	187	173	158	142	126
N2O total	MtN2O-N	2.3	2.6	3.2	4.0	4.6	5.0	5.3	5.1	4.9	4.6	4.3	3.9
SOx total	MtS	17.7	25.3	47.3	66.7	58.2	37.6	17.0	12.1	7.2	4.9	5.2	5.4
HFC	MtC eq.	0	4	11	20	34	56	93	90	85	79	72	64
PFC	MtC eq.	3	5	7	9	11	14	17	20	18	18	18	18
SF6	MtC eq.	4	7	11	17	23	27	30	32	26	19	16	16
CO	MtCO	235	270	363	438	537	549	561	521	482	445	411	376
NMVOC	Mt	33	37	43	51	58	63	67	72	76	80	84	88
NOx	MtN	7	9	16	25	31	33	35	30	25	20	17	14

Scenario B1-ASF ALM		**1990**	**2000**	**2010**	**2020**	**2030**	**2040**	**2050**	**2060**	**2070**	**2080**	**2090**	**2100**
Population	Million	1218	1519	1869	2218	2500	2851	2980	3023	3044	3013	2795	2727
GNP/GDP (mex)	Trillion US$	2.5	4.0	6.7	11.3	18.0	32.0	38.9	50.9	58.2	83.3	107.4	117.1
GNP/GDP (ppp) Trillion (1990 prices)													
Final Energy	EJ												
Non-commercial													
Solids		1	2	3	4	7	9	11	11	11	10	10	10
Liquids		22	38	58	89	94	100	105	94	82	73	66	60
Gas		5	9	17	31	40	50	60	58	56	53	49	44
Electricity		4	7	12	20	31	42	54	71	88	98	101	103
Others													
Total		33	56	89	144	172	201	229	233	237	234	226	217
Primary Energy	EJ												
Coal		5	9	14	21	41	61	81	79	77	71	61	51
Oil		25	43	67	102	90	77	65	43	21	8	5	1
Gas		8	13	23	42	60	79	97	92	87	80	72	64
Nuclear		0	0	0	0	1	1	1	1	1	1	1	1
Biomass		0	0	0	4	15	26	38	45	53	59	65	71
Other Renewables		2	3	5	7	10	14	18	35	53	63	67	71
Total		40	68	108	177	218	259	300	295	291	283	271	259
Cumulative Resources Use	ZJ												
Coal		0.0	0.1	0.2	0.4	0.7	1.2	1.9	2.7	3.5	4.2	4.9	5.4
Oil		0.0	0.4	0.9	1.8	2.7	3.5	4.2	4.8	5.1	5.2	5.3	5.3
Gas		0.0	0.1	0.3	0.6	1.1	1.8	2.7	3.7	4.6	5.4	6.2	6.8
Cumulative CO2 Emissions	GtC	1.4	17.8	39.6	70.3	109.4	152.2	197.1	238.5	270.9	295.1	312.9	325.1
Carbon Sequestration	GtC												
Land Use	Million ha												
Cropland													
Grasslands													
Energy Biomass													
Forest													
Others													
Total													
Anthropogenic Emissions (standardized)													
Fossil Fuel CO2	GtC	0.72	1.01	1.74	2.79	3.37	3.66	3.95	3.24	2.53	1.91	1.36	0.81
Other CO2	GtC	0.73	0.82	0.78	0.85	0.80	0.72	0.65	0.45	0.25	0.15	0.14	0.14
Total CO2	GtC	1.45	1.83	2.52	3.64	4.17	4.38	4.59	3.69	2.78	2.06	1.50	0.95
CH4 total	MtCH4	77	85	108	132	147	155	163	159	155	149	142	136
N2O total	MtN2O-N	1.2	1.3	1.6	1.9	2.2	2.5	2.7	2.7	2.8	2.7	2.7	2.6
SOx total	MtS	10.5	12.8	15.2	25.4	28.5	23.8	19.1	15.4	11.7	9.8	9.6	9.4
HFC	MtC eq.	0	2	16	28	42	64	98	102	102	101	96	89
PFC	MtC eq.	4	4	5	6	7	8	9	11	11	11	12	12
SF6	MtC eq.	2	5	5	8	12	17	22	24	21	14	14	14
CO	MtCO	396	404	470	532	626	629	631	592	553	523	503	482
NMVOC	Mt	48	55	63	77	92	103	113	128	143	159	176	193
NOx	MtN	7	8	11	17	21	22	24	21	19	18	16	15

Marker Scenario B1-IMAGE World		1990	2000	2010	2020	2030	2040	2050	2060	2070	2080	2090	2100
Population	Million	5280	6122	6892	7618	8196	8547	8708	8671	8484	8142	7663	7047
GNP/GDP (mex)	Trillion US$	21.0	26.8	37.3	52.6	73.1	100.7	135.6	171.7	208.5	249.7	290.1	328.4
	1990 US$/cap [a]	3971.0	4372.0	5416.0	6900.0	8916.0	11783.0	15569.0	19803.0	24577.0	30667.0	37856.0	46598.0
Final Energy	EJ												
Non-commercial		54	57	60	53	46	37	27	24	22	19	16	14
Solids		43	43	41	42	43	44	44	40	35	30	25	21
Liquids		106	117	137	152	156	151	135	115	102	93	84	75
Gas		46	52	66	70	71	74	79	82	83	82	80	76
Electricity		39	54	83	122	168	223	260	270	266	255	240	220
Others		2	4	10	23	38	52	64	70	69	65	59	52
Total		289	327	398	462	523	581	608	601	576	544	505	458
Primary Energy	EJ												
Coal		105	109	120	134	163	181	167	133	101	76	58	44
Oil		129	141	176	206	230	236	228	199	167	143	119	99
Gas		62	71	108	138	153	166	173	168	154	136	121	103
Non-Fossil Electric		8	14	22	33	49	73	105	132	151	163	168	165
Biomass		3	4	11	29	54	77	95	102	99	90	79	67
Other Renewables		61	68	71	66	61	54	46	44	42	41	39	36
Total		368	407	508	606	710	788	813	778	715	650	584	514
Cumulative Resources Production	ZJ [b]												
Coal		0.0	1.1	2.2	3.5	4.9	6.7	8.5	10.0	11.1	12.0	12.7	13.2
Oil		0.0	1.3	2.9	4.8	7.0	9.3	11.6	13.8	15.6	17.2	18.5	19.6
Gas		0.0	0.6	1.5	2.8	4.2	5.8	7.5	9.2	10.8	12.3	13.6	14.7
Cumulative CO2 Emissions	GtC	7.1	82.4	168.7	268.2	376.9	491.0	606.1	711.2	800.8	875.1	935.2	983.0
Carbon Sequestration [c]	GtC	2.1	1.7	2.8	3.8	4.5	4.3	4.2	3.9	3.6	3.5	3.7	3.5
Land Use	Million ha												
Cropland [d]		1436	1371	1445	1499	1516	1486	1429	1360	1288	1209	1124	1038
Grasslands [e]		3435	3316	3424	3404	3292	3052	2785	2500	2293	2142	2023	1899
Energy Biomass		8	10	24	68	134	209	268	294	287	262	227	194
Forest		4277	4261	4179	4258	4274	4375	4551	4711	4897	5130	5349	5543
Others		3916	4114	3999	3843	3853	3950	4038	4207	4305	4327	4348	4398
Total		13071	13071	13071	13071	13071	13071	13071	13071	13071	13071	13071	13071
Anthropogenic Emissions (standardized)													
Fossil Fuel CO2	GtC	5.99	6.90	8.50	10.00	11.20	12.20	11.70	10.20	8.60	7.30	6.10	5.20
Other CO2 [f]	GtC	1.11	1.07	0.78	0.63	-0.09	-0.48	-0.41	-0.46	-0.42	-0.60	-0.78	-0.97
Total CO2	GtC	7.10	7.97	9.28	10.63	11.11	11.72	11.29	9.74	8.18	6.70	5.32	4.23
CH4 total	MtCH4	310	323	349	377	385	381	359	342	324	293	266	236
N2O total	MtN2O-N	6.7	7.0	7.5	8.1	8.2	8.3	8.3	7.7	7.4	7.0	6.4	5.7
SOx total	MtS	70.9	69.0	73.9	74.6	78.2	78.5	68.9	55.8	44.3	36.1	29.8	24.9
CFC/HFC/HCFC	MtC eq.	1672	883	784	291	257	298	338	337	333	327	315	299
PFC	MtC eq.	32	25	29	32	33	37	42	47	45	43	46	45
SF6	MtC eq.	38	40	36	37	47	58	68	71	62	46	43	43
CO	MtCO	879	877	789	751	603	531	471	459	456	426	399	363
NMVOC	Mt	139	141	141	140	131	123	116	111	103	99	96	87
NOx	MtN	31	32	36	40	42	43	39	34	30	26	22	19

a: NOT ppp-corrected.
b: NOT use but production
c: Net Ecosystem Production (NEP).
d: Arable land for crops excluding energy crops and grass & fodder species.
e: Permanent pasture: FAO category "land for grass & fodder species".
f: Approximate calculation from complex land-use module.

Marker Scenario B1-IMAGE OECD90		1990	2000	2010	2020	2030	2040	2050	2060	2070	2080	2090	2100
Population	Million	799	849	890	932	965	990	1001	1005	1009	1020	1029	1032
GNP/GDP (mex)	Trillion US$	16.5	20.2	25.9	32.4	38.2	43.9	49.9	55.4	59.8	66.3	73.8	82.3
	1990 US$/cap [a]	20648.9	23841.8	29100.0	34761.7	39587.5	44382.4	49810.4	55070.9	59257.3	65005.4	71721.0	79765.3
Final Energy	EJ												
Non-commercial		5	5	5	4	4	4	3	3	3	3	2	2
Solids		8	7	7	6	6	6	6	5	5	5	4	4
Liquids		57	60	60	58	51	41	32	25	20	19	17	16
Gas		24	27	28	27	27	27	27	28	28	28	28	28
Electricity		24	36	47	54	58	58	57	55	52	51	50	49
Others		2	3	5	8	13	18	21	22	22	20	19	17
Total		120	137	152	159	159	153	146	138	130	125	121	116
Primary Energy	EJ												
Coal		37	46	46	40	36	30	24	19	15	13	11	10
Oil		67	73	74	72	64	50	39	31	25	22	20	19
Gas		28	36	42	44	41	38	37	35	34	33	32	31
Non-Fossil Electric		7	12	18	24	29	32	35	37	37	37	37	37
Biomass		2	3	5	8	14	18	21	22	23	21	20	19
Other Renewables		10	10	10	10	10	10	10	10	10	10	10	9
Total		151	178	195	199	193	178	166	154	144	136	131	126
Cumulative Resources Production	ZJ [b]												
Coal		0.0	0.4	0.8	1.2	1.7	2.1	2.5	2.9	3.1	3.4	3.6	3.8
Oil		0.0	0.3	0.7	1.1	1.5	1.9	2.3	2.9	3.5	4.0	4.5	4.9
Gas		0.0	0.3	0.6	1.1	1.5	1.9	2.2	2.5	2.8	3.0	3.2	3.4
Cumulative CO2 Emissions	GtC	2.8	33.0	66.0	99.3	129.4	155.0	176.1	193.2	207.5	220.0	231.1	241.5
Carbon Sequestration [c]	GtC	0.4	0.4	0.3	0.6	0.8	1.1	1.2	1.4	1.4	1.3	1.3	1.2
Land Use	Million ha												
Cropland [d]		379	354	344	349	348	349	352	351	347	345	343	339
Grasslands [e]		785	704	619	553	498	459	428	418	408	402	396	389
Energy Biomass		3	1	2	8	16	27	36	42	42	37	31	26
Forest		1115	1165	1215	1259	1299	1328	1351	1361	1378	1401	1423	1451
Others		956	1015	1058	1069	1077	1076	1072	1065	1063	1054	1045	1033
Total		3238	3238	3238	3238	3238	3238	3238	3238	3238	3238	3238	3238
Anthropogenic Emissions (standardized)													
Fossil Fuel CO2	GtC	2.83	3.20	3.30	3.20	2.80	2.40	2.00	1.60	1.40	1.30	1.20	1.10
Other CO2 [f]	GtC	0.00	0.00	0.10	0.06	-0.02	-0.07	-0.09	-0.08	-0.06	-0.14	-0.12	-0.11
Total CO2	GtC	2.83	3.20	3.40	3.26	2.78	2.33	1.90	1.52	1.34	1.16	1.07	0.99
CH4 total	MtCH4	73	74	75	72	68	63	56	52	50	46	42	40
N2O total	MtN2O-N	2.6	2.6	2.6	2.6	2.5	2.4	2.4	2.3	2.3	2.2	2.2	2.0
SOx total	MtS	22.7	17.0	11.8	7.9	5.4	3.6	2.5	2.0	2.0	2.1	2.3	2.6
HFC	MtC eq.	19	58	108	103	109	112	116	117	117	119	120	120
PFC	MtC eq.	18	13	12	10	8	7	7	7	7	7	7	6
SF6	MtC eq.	24	23	16	5	6	6	7	7	8	8	8	8
CO	MtCO	179	161	153	137	116	100	85	72	64	63	59	56
NMVOC	Mt	42	36	35	33	29	24	21	18	16	16	15	13
NOx	MtN	13	12	12	10	8	6	5	4	3	3	3	2

a: NOT ppp-corrected.
b: NOT use but production
c: Net Ecosystem Production (NEP).
d: Arable land for crops excluding energy crops and grass & fodder species.
e: Permanent pasture: FAO category "land for grass & fodder species".
f: Approximate calculation from complex land-use module.

Marker Scenario B1-IMAGE REF		1990	2000	2010	2020	2030	2040	2050	2060	2070	2080	2090	2100
Population	Million	412	429	437	443	445	443	432	419	401	384	365	347
GNP/GDP (mex)	Trillion US$	1.0	0.7	1.0	1.7	2.8	4.3	6.2	8.2	10.3	12.8	15.4	18.1
	1990 US$/cap [a]	2307.2	1675.8	2335.2	3864.4	6258.7	9702.5	14328.5	19522.7	25650.8	33306.4	42087.4	52209.7
Final Energy	EJ												
Non-commercial		3	2	2	2	1	1	1	1	1	1	1	1
Solids		14	10	6	5	4	4	4	3	3	3	2	2
Liquids		21	12	11	12	12	11	10	8	7	6	6	5
Gas		15	11	10	10	10	9	9	8	8	7	6	6
Electricity		9	5	7	11	14	18	20	20	19	18	17	16
Others		0	1	1	1	2	2	3	3	3	3	3	3
Total		62	40	37	40	42	46	46	45	41	38	35	32
Primary Energy	EJ												
Coal		36	18	13	12	12	14	14	12	9	8	6	5
Oil		28	13	14	16	16	18	18	16	13	11	10	8
Gas		25	15	15	18	19	20	18	17	15	13	11	10
Non-Fossil Electric		1	1	1	2	4	5	6	7	8	8	8	8
Biomass		1	1	1	2	2	4	5	5	5	5	5	5
Other Renewables		4	3	3	3	3	3	3	3	3	3	3	3
Total		95	52	48	52	56	63	64	60	54	48	43	39
Cumulative Resources Production	ZJ [b]												
Coal		0.0	0.3	0.5	0.7	0.9	1.1	1.3	1.5	1.7	1.8	2.0	2.0
Oil		0.0	0.1	0.2	0.4	0.5	0.6	0.7	0.9	1.0	1.1	1.2	1.3
Gas		0.0	0.2	0.3	0.6	1.0	1.4	1.9	2.3	2.8	3.2	3.5	3.8
Cumulative CO2 Emissions	GtC	1.3	12.3	20.9	29.0	36.0	41.8	47.4	52.3	55.4	57.2	58.3	59.2
Carbon Sequestration [c]	GtC	0.4	0.2	0.5	0.8	1.1	1.2	1.3	1.4	1.4	1.4	1.3	1.2
Land Use	Million ha												
Cropland [d]		278	217	219	209	190	173	154	141	128	117	106	98
Grasslands [e]		392	300	321	319	286	248	211	187	170	162	155	148
Energy Biomass		1	0	1	4	12	28	41	51	57	58	50	46
Forest		1146	1177	1175	1282	1295	1310	1350	1378	1410	1443	1459	1480
Others		461	583	562	463	495	519	521	521	512	498	507	506
Total		2278	2278	2278	2278	2278	2278	2278	2278	2278	2278	2278	2278
Anthropogenic Emissions (standardized)													
Fossil Fuel CO2	GtC	1.30	0.91	0.81	0.91	0.91	0.91	0.91	0.81	0.61	0.51	0.41	0.41
Other CO2 [f]	GtC	0.00	0.00	-0.01	-0.10	-0.31	-0.35	-0.36	-0.38	-0.41	-0.36	-0.34	-0.29
Total CO2	GtC	1.30	0.91	0.80	0.81	0.60	0.56	0.55	0.43	0.20	0.15	0.06	0.12
CH4 total	MtCH4	47	39	38	42	44	40	34	33	30	27	24	21
N2O total	MtN2O-N	0.6	0.6	0.6	0.6	0.6	0.6	0.5	0.4	0.4	0.4	0.3	0.3
SOx total	MtS	17.0	11.0	9.0	7.7	6.9	7.0	6.5	5.6	4.6	3.9	3.0	2.5
HFC	MtC eq.	0	4	9	15	20	24	26	27	27	27	27	26
PFC	MtC eq.	7	4	6	6	7	8	9	10	9	8	8	8
SF6	MtC eq.	8	6	4	7	7	8	9	9	7	6	4	4
CO	MtCO	69	41	28	23	20	19	19	17	15	13	12	9
NMVOC	Mt	16	13	12	12	11	11	11	10	9	9	9	9
NOx	MtN	5	3	3	3	3	3	3	2	2	2	2	1

a: NOT ppp-corrected.
b: NOT use but production
c: Net Ecosystem Production (NEP).
d: Arable land for crops excluding energy crops and grass & fodder species.
e: Permanent pasture: FAO category “land for grass & fodder species”.
f: Approximate calculation from complex land-use module.

Marker Scenario B1-IMAGE ASIA		1990	2000	2010	2020	2030	2040	2050	2060	2070	2080	2090	2100
Population	Million	2781	3246	3609	3929	4142	4235	4220	4088	3871	3594	3262	2886
GNP/GDP (mex)	Trillion US$	1.4	2.7	4.8	8.7	15.1	24.9	37.9	51.4	64.8	78.7	91.7	103.1
	1990 US$/cap [a]	502.0	836.8	1336.1	2206.5	3651.1	5885.3	8980.9	12579.4	16732.0	21889.1	28118.6	35726.2
Final Energy	EJ												
Non-commercial		29	31	30	24	19	14	9	8	6	5	5	4
Solids		20	23	23	23	23	22	20	17	14	12	10	8
Liquids		12	22	29	34	38	39	35	30	26	24	21	19
Gas		2	6	13	17	19	20	22	23	23	23	21	20
Electricity		4	9	19	36	56	80	93	95	93	88	84	77
Others		0	1	3	9	16	23	27	28	26	24	21	18
Total		68	92	117	144	171	198	205	200	189	175	162	144
Primary Energy	EJ												
Coal		28	41	51	58	67	74	64	48	35	24	17	11
Oil		16	27	39	51	63	61	56	49	40	33	27	21
Gas		4	11	25	42	56	63	59	52	45	38	31	25
Non-Fossil Electric		0	1	2	5	11	24	40	52	60	65	68	67
Biomass		0	1	3	11	23	34	40	38	37	31	26	22
Other Renewables		30	33	32	27	23	18	13	12	11	10	9	8
Total		79	113	153	195	244	273	272	251	227	200	178	154
Cumulative Resources Production	ZJ [b]												
Coal		0.0	0.4	0.8	1.2	1.7	2.2	2.7	3.2	3.5	3.7	3.8	3.9
Oil		0.0	0.2	0.4	0.7	1.1	1.4	1.6	1.8	2.0	2.1	2.2	2.3
Gas		0.0	0.1	0.2	0.4	0.6	0.8	1.2	1.5	1.8	2.1	2.4	2.6
Cumulative CO2 Emissions	GtC	1.5	19.3	43.6	74.7	111.6	152.8	193.4	228.9	258.2	281.7	299.6	311.7
Carbon Sequestration [c]	GtC	0.2	0.1	0.2	0.3	0.4	0.5	0.5	0.2	-0.2	-0.1	0.2	0.4
Land Use	Million ha												
Cropland [d]		382	392	405	419	422	411	393	362	327	288	252	216
Grasslands [e]		561	652	734	748	753	753	723	600	531	471	431	386
Energy Biomass		1	1	3	12	32	71	94	110	95	84	63	50
Forest		488	403	348	330	316	300	302	334	383	489	599	671
Others		544	528	486	468	454	441	464	570	640	645	631	653
Total		1976	1976	1976	1976	1976	1976	1976	1976	1976	1976	1976	1976
Anthropogenic Emissions (standardized)													
Fossil Fuel CO2	GtC	1.15	1.78	2.58	3.18	3.78	4.08	3.68	3.08	2.48	1.98	1.58	1.28
Other CO2 [f]	GtC	0.37	0.26	0.24	0.22	0.21	0.19	0.18	0.16	0.15	0.11	-0.09	-0.35
Total CO2	GtC	1.53	2.03	2.82	3.40	3.98	4.27	3.85	3.24	2.62	2.09	1.49	0.93
CH4 total	MtCH4	113	125	136	148	154	161	157	149	141	128	118	105
N2O total	MtN2O-N	2.3	2.6	2.9	3.2	3.3	3.4	3.3	3.0	2.7	2.5	2.1	1.9
SOx total	MtS	17.7	25.3	29.0	29.1	29.2	27.6	21.4	15.3	10.7	7.7	5.8	4.2
HFC	MtC eq.	0	4	11	20	34	56	93	90	85	79	72	64
PFC	MtC eq.	3	5	7	9	11	14	17	20	18	18	18	18
SF6	MtC eq.	4	7	11	17	23	27	30	32	26	19	16	16
CO	MtCO	235	270	286	289	280	285	245	238	233	228	222	214
NMVOC	Mt	33	37	40	40	40	40	37	36	33	32	32	29
NOx	MtN	7	9	11	13	15	15	13	11	9	8	7	5

a: NOT ppp-corrected.
b: NOT use but production
c: Net Ecosystem Production (NEP).
d: Arable land for crops excluding energy crops and grass & fodder species.
e: Permanent pasture: FAO category "land for grass & fodder species".
f: Approximate calculation from complex land-use module.

Marker Scenario B1-IMAGE ALM		1990	2000	2010	2020	2030	2040	2050	2060	2070	2080	2090	2100
Population	Million	1287	1597	1954	2315	2643	2879	3055	3159	3202	3145	3006	2783
GNP/GDP (mex)	Trillion US$	2.1	3.1	5.6	9.8	16.9	27.6	41.6	56.7	73.6	92.0	109.1	124.9
	1990 US$/cap [a]	1653.4	1933.0	2839.5	4243.2	6406.6	9580.8	13623.6	17954.7	22996.3	29243.6	36311.8	44869.7
Final Energy	EJ												
Non-commercial		16	19	24	23	21	18	14	13	11	10	9	7
Solids		1	3	5	7	10	12	14	14	12	10	8	7
Liquids		15	23	36	48	55	60	58	53	48	45	40	35
Gas		4	8	15	16	16	17	21	23	25	25	25	23
Electricity		2	4	10	21	41	67	90	100	101	97	89	79
Others		0	0	1	4	8	9	13	17	18	18	17	15
Total		39	57	91	119	150	183	211	219	216	205	187	165
Primary Energy	EJ												
Coal		3	4	10	24	48	64	65	54	42	32	24	18
Oil		17	27	48	68	87	107	114	103	89	77	62	51
Gas		6	10	25	33	36	45	59	64	60	52	47	37
Non-Fossil Electric		0	0	1	2	5	12	24	36	46	53	55	53
Biomass		0	0	2	7	14	23	29	36	33	33	28	22
Other Renewables		18	22	27	26	25	23	20	19	19	18	16	15
Total		43	64	112	160	216	273	312	313	290	265	232	196
Cumulative Resources Production	ZJ [b]												
Coal		0.0	0.1	0.2	0.3	0.7	1.3	1.9	2.4	2.8	3.1	3.3	3.4
Oil		0.0	0.6	1.5	2.6	4.0	5.5	6.9	8.2	9.1	9.9	10.6	11.1
Gas		0.0	0.1	0.4	0.7	1.2	1.7	2.3	2.9	3.5	4.0	4.5	4.9
Cumulative CO2 Emissions	GtC	1.4	17.8	38.3	65.3	99.8	141.4	189.1	236.8	279.7	316.2	346.2	370.6
Carbon Sequestration [c]	GtC	1.2	1.1	1.7	2.1	2.1	1.7	1.3	1.1	1.0	0.9	0.8	0.7
Land Use	Million ha												
Cropland [d]		397	408	477	521	558	553	530	506	486	460	423	385
Grasslands [e]		1697	1660	1750	1784	1755	1592	1424	1295	1184	1108	1041	975
Energy Biomass		2	7	18	44	74	84	96	90	94	84	83	74
Forest		1528	1517	1441	1386	1365	1437	1548	1639	1726	1798	1868	1940
Others		1955	1987	1893	1844	1827	1915	1981	2050	2090	2130	2165	2205
Total		5579	5579	5579	5579	5579	5579	5579	5579	5579	5579	5579	5579
Anthropogenic Emissions (standardized)													
Fossil Fuel CO2	GtC	0.72	1.01	1.81	2.71	3.71	4.81	5.11	4.71	4.11	3.51	2.91	2.41
Other CO2 [f]	GtC	0.73	0.82	0.45	0.45	0.04	-0.25	-0.13	-0.15	-0.10	-0.21	-0.22	-0.22
Total CO2	GtC	1.45	1.83	2.26	3.16	3.75	4.56	4.98	4.56	4.01	3.30	2.69	2.19
CH4 total	MtCH4	77	85	100	115	119	117	112	108	103	92	82	70
N2O total	MtN2O-N	1.2	1.3	1.5	1.8	1.9	2.0	2.2	2.1	2.1	2.0	1.9	1.6
SOx total	MtS	10.5	12.8	21.2	26.8	33.7	37.4	35.6	29.9	24.0	19.5	15.7	12.6
HFC	MtC eq.	0	2	16	28	42	64	98	102	102	101	96	89
PFC	MtC eq.	4	4	5	6	7	8	9	11	11	11	12	12
SF6	MtC eq.	2	5	5	8	12	17	22	24	21	14	14	14
CO	MtCO	396	404	321	301	186	126	121	131	143	121	105	83
NMVOC	Mt	48	55	54	55	51	48	47	47	45	42	40	36
NOx	MtN	7	8	10	13	16	18	18	17	15	13	11	10

a: NOT ppp-corrected.
b: NOT use but production
c: Net Ecosystem Production (NEP).
d: Arable land for crops excluding energy crops and grass & fodder species.
e: Permanent pasture: FAO category "land for grass & fodder species".
f: Approximate calculation from complex land-use module.

Scenario B1-MARIA World		**1990**	**2000**	**2010**	**2020**	**2030**	**2040**	**2050**	**2060**	**2070**	**2080**	**2090**	**2100**
Population	Million	5262	6117	6888	7617	8048	8207	8704	8536	8372	8028	7527	7056
GNP/GDP (mex)	Trillion US$	19.4	25.5	34.2	46.0	60.0	80.7	110.2	135.8	169.8	214.5	275.5	348.5
GNP/GDP (ppp) Trillion (1990 prices)													
Final Energy	EJ												
Non-commercial													
Solids		48	34	24	17	12	8	6	4	3	2	1	1
Liquids		138	148	167	188	215	247	288	349	368	400	410	448
Gas		56	54	79	111	142	171	197	175	173	165	185	177
Electricity		35	46	53	61	70	80	98	105	115	131	150	164
Others		0	0	0	0	0	0	0	0	0	0	0	0
Total		278	282	323	377	438	507	590	633	658	697	747	791
Primary Energy	EJ												
Coal		90	64	46	34	25	19	14	12	9	7	5	5
Oil		123	134	146	158	177	192	186	177	156	158	147	170
Gas		71	91	129	169	205	237	251	228	223	212	231	219
Nuclear		22	18	17	15	16	26	43	70	114	173	237	283
Biomass		28	30	37	55	75	104	136	198	231	257	276	291
Other Renewables		9	8	7	7	6	5	57	63	64	64	63	62
Total		343	345	382	438	504	583	687	747	797	873	959	1030
Cumulative Resources Use	ZJ												
Coal		0.0	0.9	1.5	2.0	2.4	2.6	2.8	2.9	3.0	3.1	3.2	3.3
Oil		0.0	1.2	2.6	4.0	5.6	7.4	9.3	11.2	12.9	14.5	16.1	17.5
Gas		0.0	0.7	1.6	2.9	4.6	6.6	9.0	11.5	13.8	16.0	18.2	20.5
Cumulative CO2 Emissions	GtC	7.1	82.4	163.5	248.3	338.4	434.2	531.6	623.8	708.1	787.8	866.6	947.1
Carbon Sequestration	GtC	0.0	0.0	0.0	0.0	0.0	0.0	0.0	0.0	0.0	0.0	0.0	0.0
Land Use	Million ha												
Cropland		1451	1451	1454	1689	1809	1821	1912	1713	1527	1363	1240	1145
Grasslands		3395	3395	3392	3160	3040	3029	2773	2609	2609	2609	2609	2609
Energy Biomass		0	0	0	0	0	0	164	528	713	877	1000	1095
Forest		4138	4142	4170	4199	4231	4265	4310	4361	4408	4412	4412	4412
Others		4061	4057	4029	3997	3965	3930	3886	3835	3788	3784	3784	3784
Total		13045	13045	13045	13045	13045	13045	13045	13045	13045	13045	13045	13045
Anthropogenic Emissions (standardized)													
Fossil Fuel CO2	GtC	5.99	6.90	7.25	7.80	8.49	9.13	9.11	8.49	7.94	7.77	7.79	8.06
Other CO2	GtC	1.11	1.07	0.99	0.91	0.82	0.72	0.53	0.31	0.12	0.11	0.12	0.13
Total CO2	GtC	7.10	7.97	8.25	8.71	9.31	9.85	9.64	8.80	8.05	7.88	7.90	8.19
CH4 total	MtCH4												
N2O total	MtN2O-N												
SOx total	MtS												
CFC/HFC/HCFC	MtC eq.	1672	883	784	291	257	298	338	337	333	327	315	299
PFC	MtC eq.	32	25	29	32	33	37	42	47	45	43	46	45
SF6	MtC eq.	38	40	36	37	47	58	68	71	62	46	43	43
CO	MtCO												
NMVOC	Mt												
NOx	MtN												

Scenario B1-MARIA OECD90		1990	2000	2010	2020	2030	2040	2050	2060	2070	2080	2090	2100
Population	Million	859	919	965	1007	1035	1046	1081	1085	1091	1096	1103	1110
GNP/GDP (mex)	Trillion US$	15.6	19.6	24.7	31.0	36.4	43.3	50.6	55.3	60.3	66.5	72.2	78.3
GNP/GDP (ppp) Trillion (1990 prices)													
Final Energy	EJ												
Non-commercial													
Solids		12	8	6	4	3	2	1	1	1	0	0	0
Liquids		71	74	66	60	59	58	57	78	72	66	62	59
Gas		28	27	44	58	66	71	77	59	68	76	81	85
Electricity		21	26	29	32	34	37	39	41	42	44	46	47
Others		0	0	0	0	0	0	0	0	0	0	0	0
Total		132	136	145	155	162	168	175	179	182	186	189	191
Primary Energy	EJ												
Coal		38	28	21	16	12	10	8	7	6	5	4	4
Oil		71	74	66	60	59	58	57	78	72	66	61	57
Gas		34	47	72	94	107	117	115	99	106	112	116	117
Nuclear		18	15	12	11	13	15	19	21	26	32	40	48
Biomass		6	4	3	2	1	1	1	1	2	3	4	7
Other Renewables		5	5	4	4	3	3	14	14	16	17	16	16
Total		171	171	178	187	196	204	214	221	227	235	242	248
Cumulative Resources Use	ZJ												
Coal		0.0	0.4	0.7	0.9	1.0	1.1	1.2	1.3	1.4	1.5	1.5	1.6
Oil		0.0	0.7	1.5	2.1	2.7	3.3	3.9	4.5	5.2	6.0	6.6	7.2
Gas		0.0	0.3	0.8	1.5	2.5	3.5	4.7	5.8	6.8	7.9	9.0	10.2
Cumulative CO2 Emissions	GtC	2.8	33.0	65.3	98.7	133.3	168.9	204.2	239.2	274.5	309.6	344.4	378.9
Carbon Sequestration	GtC	0.0	0.0	0.0	0.0	0.0	0.0	0.0	0.0	0.0	0.0	0.0	0.0
Land Use	Million ha												
Cropland		378	378	378	378	378	378	378	378	378	303	243	195
Grasslands		756	756	756	756	756	756	756	756	756	756	756	756
Energy Biomass		0	0	0	0	0	0	0	0	0	75	134	182
Forest		756	756	756	756	756	756	768	780	780	780	780	780
Others		794	794	794	794	794	794	782	770	770	770	770	770
Total		2684	2684	2684	2684	2684	2684	2684	2684	2684	2684	2684	2684
Anthropogenic Emissions (standardized)													
Fossil Fuel CO2	GtC	2.83	3.20	3.25	3.35	3.43	3.50	3.40	3.55	3.51	3.46	3.40	3.33
Other CO2	GtC	0.00	0.00	0.03	0.05	0.08	0.11	0.05	-0.01	0.01	0.04	0.07	0.09
Total CO2	GtC	2.83	3.20	3.27	3.40	3.51	3.61	3.46	3.55	3.52	3.50	3.46	3.43
CH4 total	MtCH4												
N2O total	MtN2O-N												
SOx total	MtS												
HFC	MtC eq.	19	58	108	103	109	112	116	117	117	119	120	120
PFC	MtC eq.	18	13	12	10	8	7	7	7	7	7	7	6
SF6	MtC eq.	24	23	16	5	6	6	7	7	8	8	8	8
CO	MtCO												
NMVOC	Mt												
NOx	MtN												

Scenario B1-MARIA REF		**1990**	**2000**	**2010**	**2020**	**2030**	**2040**	**2050**	**2060**	**2070**	**2080**	**2090**	**2100**
Population	Million	413	419	427	433	432	430	423	406	391	374	356	339
GNP/GDP (mex)	Trillion US$	0.9	1.1	1.5	1.9	2.6	3.6	5.2	6.2	7.9	10.3	13.0	16.5
GNP/GDP (ppp) Trillion (1990 prices)													
Final Energy	EJ												
Non-commercial													
Solids		13	9	6	4	3	2	2	1	1	1	0	0
Liquids		17	12	11	10	10	10	10	10	9	9	12	20
Gas		21	20	23	25	30	35	40	41	43	46	45	39
Electricity		7	7	7	6	7	7	8	8	9	9	9	10
Others		0	0	0	0	0	0	0	0	0	0	0	0
Total		58	48	47	45	49	54	60	60	62	65	67	68
Primary Energy	EJ												
Coal		19	13	9	6	4	3	2	2	1	1	1	0
Oil		19	13	12	10	10	10	10	10	9	9	12	20
Gas		27	27	30	32	39	45	50	51	53	55	54	47
Nuclear		3	2	2	1	2	2	3	3	4	5	6	8
Biomass		1	1	0	0	0	0	0	0	0	0	1	1
Other Renewables		1	1	1	1	1	1	3	3	3	3	3	3
Total		69	57	54	51	56	61	68	68	71	74	77	80
Cumulative Resources Use	ZJ												
Coal		0.0	0.2	0.3	0.4	0.5	0.5	0.5	0.6	0.6	0.6	0.6	0.6
Oil		0.0	0.2	0.3	0.4	0.5	0.6	0.7	0.8	0.9	1.0	1.1	1.2
Gas		0.0	0.3	0.5	0.8	1.2	1.6	2.0	2.5	3.0	3.5	4.1	4.6
Cumulative CO2 Emissions	GtC	1.3	12.3	21.1	29.2	37.4	46.2	55.7	65.3	74.7	84.2	94.3	104.9
Carbon Sequestration	GtC	0.0	0.0	0.0	0.0	0.0	0.0	0.0	0.0	0.0	0.0	0.0	0.0
Land Use	Million ha												
Cropland		217	217	217	217	217	217	217	174	139	111	89	79
Grasslands		114	114	114	114	114	114	114	114	114	114	114	114
Energy Biomass		0	0	0	0	0	0	0	43	78	106	128	138
Forest		815	815	815	815	815	815	815	815	815	815	815	815
Others		722	722	722	722	722	722	722	722	722	722	722	722
Total		1868	1868	1868	1868	1868	1868	1868	1868	1868	1868	1868	1868
Anthropogenic Emissions (standardized)													
Fossil Fuel CO2	GtC	1.30	0.91	0.82	0.76	0.80	0.86	0.91	0.90	0.91	0.94	0.97	1.02
Other CO2	GtC	0.00	0.00	0.02	0.03	0.05	0.06	0.06	0.04	0.02	0.04	0.05	0.07
Total CO2	GtC	1.30	0.91	0.83	0.79	0.85	0.93	0.97	0.94	0.94	0.98	1.03	1.09
CH4 total	MtCH4												
N2O total	MtN2O-N												
SOx total	MtS												
HFC	MtC eq.	0	4	9	15	20	24	26	27	27	27	27	26
PFC	MtC eq.	7	4	6	6	7	8	9	10	9	8	8	8
SF6	MtC eq.	8	6	4	7	7	8	9	9	7	6	4	4
CO	MtCO												
NMVOC	Mt												
NOx	MtN												

Scenario B1-MARIA ASIA		1990	2000	2010	2020	2030	2040	2050	2060	2070	2080	2090	2100
Population	Million	2642	3080	3425	3728	3861	3895	4008	3814	3632	3368	3040	2744
GNP/GDP (mex)	Trillion US$	1.2	2.0	3.4	5.9	10.1	17.0	28.8	39.0	52.6	70.5	97.3	126.4
GNP/GDP (ppp) Trillion (1990 prices)													
Final Energy	EJ												
Non-commercial													
Solids		20	14	10	7	5	3	2	2	1	1	1	0
Liquids		26	33	49	68	86	107	131	152	165	178	186	187
Gas		2	3	4	8	14	21	24	17	12	8	11	19
Electricity		4	7	8	11	13	16	23	24	29	35	42	45
Others		0	0	0	0	0	0	0	0	0	0	0	0
Total		52	57	72	93	118	147	180	194	207	221	240	251
Primary Energy	EJ												
Coal		28	20	14	10	7	5	3	2	2	1	1	1
Oil		14	22	30	39	49	56	61	49	46	45	45	46
Gas		3	5	7	11	17	23	25	18	13	9	12	19
Nuclear		1	1	3	2	2	6	13	18	38	61	79	87
Biomass		13	20	27	41	55	72	85	114	127	138	145	144
Other Renewables		1	1	1	1	1	1	17	20	20	20	20	20
Total		61	68	81	104	129	163	204	222	245	274	301	315
Cumulative Resources Use	ZJ												
Coal		0.0	0.3	0.5	0.6	0.7	0.8	0.8	0.9	0.9	0.9	0.9	0.9
Oil		0.0	0.1	0.4	0.7	1.1	1.5	2.1	2.7	3.2	3.7	4.1	4.6
Gas		0.0	0.0	0.1	0.2	0.3	0.4	0.7	0.9	1.1	1.2	1.3	1.4
Cumulative CO2 Emissions	GtC	1.5	19.3	39.6	60.2	82.0	105.5	130.2	153.4	174.0	193.1	211.7	230.9
Carbon Sequestration	GtC	0.0	0.0	0.0	0.0	0.0	0.0	0.0	0.0	0.0	0.0	0.0	0.0
Land Use	Million ha												
Cropland		366	366	369	380	380	380	379	375	362	326	285	248
Grasslands		431	431	428	420	420	420	422	387	387	387	387	387
Energy Biomass		0	0	0	0	0	0	0	39	51	88	129	166
Forest		365	367	380	390	399	407	407	407	407	407	407	407
Others		458	456	444	430	421	414	414	414	414	414	414	414
Total		1621	1621	1621	1621	1621	1621	1621	1621	1621	1621	1621	1621
Anthropogenic Emissions (standardized)													
Fossil Fuel CO2	GtC	1.15	1.78	1.81	1.95	2.16	2.35	2.45	2.05	1.91	1.82	1.84	1.96
Other CO2	GtC	0.37	0.26	0.20	0.15	0.11	0.08	0.07	0.08	0.06	0.04	0.02	0.01
Total CO2	GtC	1.53	2.03	2.01	2.10	2.27	2.42	2.52	2.13	1.97	1.86	1.86	1.97
CH4 total	MtCH4												
N2O total	MtN2O-N												
SOx total	MtS												
HFC	MtC eq.	0	4	11	20	34	56	93	90	85	79	72	64
PFC	MtC eq.	3	5	7	9	11	14	17	20	18	18	18	18
SF6	MtC eq.	4	7	11	17	23	27	30	32	26	19	16	16
CO	MtCO												
NMVOC	Mt												
NOx	MtN												

Scenario B1-MARIA ALM		1990	2000	2010	2020	2030	2040	2050	2060	2070	2080	2090	2100
Population	Million	1348	1699	2071	2449	2720	2836	3192	3231	3259	3190	3029	2864
GNP/GDP (mex)	Trillion US$	1.7	2.7	4.6	7.2	10.9	16.8	25.6	35.3	49.0	67.1	93.0	127.4
GNP/GDP (ppp) Trillion (1990 prices)													
Final Energy	EJ												
Non-commercial													
Solids		3	2	2	1	1	1	0	0	0	0	0	0
Liquids		24	29	41	51	61	73	90	110	122	148	150	182
Gas		5	4	7	19	32	44	56	57	49	35	48	36
Electricity		3	6	9	13	16	20	28	32	35	43	53	62
Others		0	0	0	0	0	0	0	0	0	0	0	0
Total		35	40	60	84	110	138	175	199	207	225	252	280
Primary Energy	EJ												
Coal		6	4	3	2	1	1	1	0	0	0	0	0
Oil		18	25	38	49	59	68	58	40	28	38	29	47
Gas		7	12	19	32	42	51	61	61	52	36	50	37
Nuclear		0	0	0	0	0	3	9	27	47	76	111	140
Biomass		8	6	7	12	19	31	51	83	102	116	126	139
Other Renewables		2	2	1	1	1	1	22	25	25	24	24	24
Total		41	48	69	95	123	155	202	236	254	290	339	386
Cumulative Resources Use	ZJ												
Coal		0.0	0.1	0.1	0.1	0.1	0.2	0.2	0.2	0.2	0.2	0.2	0.2
Oil		0.0	0.2	0.4	0.8	1.3	1.9	2.6	3.2	3.6	3.8	4.2	4.5
Gas		0.0	0.1	0.2	0.4	0.7	1.1	1.6	2.3	2.9	3.4	3.7	4.2
Cumulative CO2 Emissions	GtC	1.4	17.8	37.6	60.3	85.8	113.7	141.5	165.9	185.0	200.8	216.3	232.5
Carbon Sequestration	GtC	0.0	0.0	0.0	0.0	0.0	0.0	0.0	0.0	0.0	0.0	0.0	0.0
Land Use	Million ha												
Cropland		490	490	490	714	834	846	939	787	648	623	623	623
Grasslands		2095	2095	2095	1870	1750	1739	1482	1352	1352	1352	1352	1352
Energy Biomass		0	0	0	0	0	0	164	445	584	609	609	609
Forest		2202	2204	2220	2239	2261	2288	2321	2360	2407	2411	2411	2411
Others		2086	2084	2068	2050	2027	2000	1967	1928	1881	1878	1878	1878
Total		6873	6873	6873	6873	6873	6873	6873	6873	6873	6873	6873	6873
Anthropogenic Emissions (standardized)													
Fossil Fuel CO2	GtC	0.72	1.01	1.37	1.75	2.11	2.41	2.35	1.99	1.60	1.55	1.58	1.74
Other CO2	GtC	0.73	0.82	0.75	0.67	0.58	0.47	0.34	0.19	0.02	-0.01	-0.02	-0.04
Total CO2	GtC	1.45	1.83	2.12	2.42	2.69	2.88	2.69	2.18	1.62	1.54	1.55	1.70
CH4 total	MtCH4												
N2O total	MtN2O-N												
SOx total	MtS												
HFC	MtC eq.	0	2	16	28	42	64	98	102	102	101	96	89
PFC	MtC eq.	4	4	5	6	7	8	9	11	11	11	12	12
SF6	MtC eq.	2	5	5	8	12	17	22	24	21	14	14	14
CO	MtCO												
NMVOC	Mt												
NOx	MtN												

Scenario B1-MESSAGE World		1990	2000	2010	2020	2030	2040	2050	2060	2070	2080	2090	2100
Population	Million	5262	6117	6888	7617	8182	8531	8704	8667	8463	8125	7658	7056
GNP/GDP (mex)	Trillion US$	20.9	26.8	36.2	52.1	73.1	100.7	135.6	171.6	208.5	249.8	290.0	328.4
GNP/GDP (ppp) Trillion (1990 prices)		25.7	33.3	44.6	61.6	82.2	108.4	140.0	171.8	204.1	242.5	281.3	318.8
Final Energy	EJ												
Non-commercial		38	27	22	16	10	8	6	0	0	0	0	0
Solids		42	52	61	59	55	39	23	18	9	3	2	1
Liquids		111	123	155	185	214	239	243	220	190	153	131	114
Gas		41	47	61	71	74	73	63	53	47	45	39	31
Electricity		35	44	61	87	117	148	172	194	204	213	209	200
Others		8	10	18	33	52	75	98	123	136	144	138	124
Total		275	303	379	450	522	581	604	608	585	556	520	469
Primary Energy	EJ												
Coal		91	91	109	110	97	71	37	21	15	16	18	22
Oil		128	155	172	189	203	198	192	161	132	95	64	46
Gas		71	84	119	161	221	278	297	302	281	267	232	215
Nuclear		7	8	11	15	20	27	36	42	47	48	46	41
Biomass		46	45	55	70	89	111	121	129	150	174	215	235
Other Renewables		8	13	23	41	62	103	156	211	239	248	239	197
Total		352	395	488	586	692	788	837	865	865	848	813	755
Cumulative Resources Use	ZJ												
Coal		0.0	0.9	1.8	2.9	4.0	5.0	5.7	6.1	6.3	6.4	6.6	6.8
Oil		0.0	1.4	3.0	4.7	6.6	8.6	10.6	12.5	14.1	15.4	16.4	17.0
Gas		0.0	0.7	1.6	2.8	4.4	6.6	9.4	12.3	15.3	18.2	20.8	23.2
Cumulative CO2 Emissions	GtC	7.1	82.4	167.5	258.6	350.7	443.3	532.4	613.9	686.3	749.0	801.1	844.4
Carbon Sequestration	GtC												
Land Use a	Million ha												
Cropland		1459	1466	1464	1461	1458	1452	1447	1442	1437	1433	1431	1429
Grasslands		3389	3407	3432	3453	3474	3510	3547	3591	3635	3668	3688	3709
Energy Biomass		0	0	0	59	137	196	254	260	266	261	243	225
Forest		4296	4255	4231	4264	4335	4405	4475	4522	4570	4619	4669	4719
Others		3805	3820	3822	3711	3545	3386	3226	3133	3040	2968	2918	2868
Total		12949	12949	12949	12949	12949	12949	12949	12949	12949	12949	12949	12949
Anthropogenic Emissions (standardized)													
Fossil Fuel CO2 [b]	GtC	5.99	6.90	8.26	9.19	9.93	9.91	9.24	8.36	7.32	6.36	5.24	4.68
Other CO2 [c]	GtC	1.11	1.07	0.79	-0.03	-0.65	-0.66	-0.67	-0.62	-0.57	-0.57	-0.61	-0.65
Total CO2	GtC	7.10	7.97	9.05	9.16	9.27	9.25	8.57	7.74	6.75	5.79	4.63	4.04
CH4 total [d]	MtCH4	310	323	356	386	421	442	453	436	424	392	334	279
N2O total [e]	MtN2O-N	6.7	7.0	6.0	5.8	5.7	5.6	5.6	5.5	5.4	5.4	5.4	5.3
SOx total	MtS	70.9	69.0	65.6	55.2	47.5	37.9	29.9	24.7	20.2	16.5	13.9	12.2
CFC/HFC/HCFC	MtC eq.	1672	883	784	291	257	298	338	337	333	327	315	299
PFC	MtC eq.	32	25	29	32	33	37	42	47	45	43	46	45
SF6	MtC eq.	38	40	36	37	47	58	68	71	62	46	43	43
CO	MtCO	879	877	1001	1120	1233	1326	1331	1268	1288	1307	1409	1436
NMVOC	Mt	139	141	164	180	183	180	175	154	141	121	93	70
NOx	MtN	31	32	38	44	47	46	40	33	27	23	20	17

Emissions correlated to land-use change and deforestation were calculated by using AIM B1 land-use data.
a: Land-use taken from AIM-B1 emulation run.
b:CO2 emissions from fossil fuel and industrial processes (MESSAGE data).
c: CO2 emissions from land-use changes (AIM-B1 emulation run).
d: Non-energy related CH4 emissions were taken from AIM-B1 emulation run.
e: Non-energy related N2O emissions were taken from AIM-B1 emulation run.

Scenario B1-MESSAGE OECD90		1990	2000	2010	2020	2030	2040	2050	2060	2070	2080	2090	2100
Population	Million	859	919	965	1007	1043	1069	1081	1084	1089	1098	1108	1110
GNP/GDP (mex)	Trillion US$	16.4	20.6	26.0	32.4	38.3	43.9	49.9	55.4	59.8	66.3	73.9	82.3
GNP/GDP (ppp) Trillion (1990 prices)		14.1	17.7	22.4	28.1	33.3	38.3	43.6	48.5	52.5	58.4	65.1	72.7
Final Energy	EJ												
Non-commercial		0	0	0	0	0	0	0	0	0	0	0	0
Solids		13	10	6	3	1	0	0	0	0	0	0	0
Liquids		66	69	75	73	69	65	61	57	52	44	42	39
Gas		22	28	34	35	33	29	21	14	10	9	7	5
Electricity		22	28	35	43	49	53	54	56	57	60	61	62
Others		1	1	2	5	7	8	9	12	13	13	12	11
Total		124	136	152	159	159	154	146	139	131	126	123	117
Primary Energy	EJ												
Coal		38	33	36	37	31	14	5	4	2	1	1	0
Oil		72	85	83	73	62	53	44	36	29	21	15	12
Gas		33	45	57	68	80	87	84	75	73	77	76	78
Nuclear		6	7	8	10	11	14	21	25	27	24	20	15
Biomass		6	8	10	13	15	18	20	26	31	39	50	53
Other Renewables		4	5	8	13	17	22	24	27	29	31	32	32
Total		159	183	202	213	215	207	198	192	191	192	193	190
Cumulative Resources Use	ZJ												
Coal		0.0	0.4	0.7	1.1	1.4	1.7	1.9	1.9	2.0	2.0	2.0	2.0
Oil		0.0	0.8	1.6	2.5	3.2	3.8	4.3	4.8	5.1	5.4	5.6	5.8
Gas		0.0	0.4	0.8	1.4	2.0	2.9	3.7	4.5	5.3	6.0	6.8	7.6
Cumulative CO2 Emissions	GtC	2.8	33.0	65.6	98.3	129.2	156.4	178.9	198.0	214.7	229.9	243.6	255.9
Carbon Sequestration	GtC												
Land Use [a]	Million ha												
Cropland		381	380	378	376	374	370	366	361	357	354	353	352
Grasslands		760	765	772	779	787	809	832	862	893	912	918	924
Energy Biomass		0	0	0	10	23	33	44	45	46	45	42	39
Forest		1050	1059	1074	1094	1117	1133	1149	1154	1159	1167	1179	1192
Others		838	826	805	770	728	684	639	607	574	551	537	522
Total		3029	3029	3029	3029	3029	3029	3029	3029	3029	3029	3029	3029
Anthropogenic Emissions (standardized)													
Fossil Fuel CO2 [b]	GtC	2.83	3.20	3.40	3.35	3.13	2.60	2.16	1.84	1.63	1.50	1.35	1.31
Other CO2 [c]	GtC	0.00	0.00	-0.07	-0.13	-0.16	-0.14	-0.12	-0.08	-0.04	-0.04	-0.08	-0.12
Total CO2	GtC	2.83	3.20	3.32	3.22	2.97	2.46	2.04	1.76	1.58	1.46	1.27	1.19
CH4 total [d]	MtCH4	73	74	73	70	67	62	57	54	52	51	50	51
N2O total [e]	MtN2O-N	2.6	2.6	2.4	2.4	2.3	2.3	2.3	2.2	2.2	2.1	2.1	2.1
SOx total	MtS	22.7	17.0	8.9	2.3	0.6	-0.3	-0.6	-0.1	0.3	0.6	0.9	1.5
HFC	MtC eq.	19	58	108	103	109	112	116	117	117	119	120	120
PFC	MtC eq.	18	13	12	10	8	7	7	7	7	7	7	6
SF6	MtC eq.	24	23	16	5	6	6	7	7	8	8	8	8
CO	MtCO	179	161	177	181	160	145	125	123	121	122	139	134
NMVOC	Mt	42	36	38	36	31	27	23	21	21	20	14	9
NOx	MtN	13	12	14	14	13	10	7	5	3	2	1	1

Emissions correlated to land-use change and deforestation were calculated by using AIM B1 land-use data.

a: Land-use taken from AIM-B1 emulation run.

b:CO2 emissions from fossil fuel and industrial processes (MESSAGE data).

c: CO2 emissions from land-use changes (AIM-B1 emulation run).

d: Non-energy related CH4 emissions were taken from AIM-B1 emulation run.

e: Non-energy related N2O emissions were taken from AIM-B1 emulation run.

Scenario B1-MESSAGE REF		1990	2000	2010	2020	2030	2040	2050	2060	2070	2080	2090	2100
Population	Million	413	419	427	433	435	433	423	409	392	374	357	339
GNP/GDP (mex)	Trillion US$	1.1	0.8	1.1	1.7	2.8	4.3	6.2	8.2	10.3	12.8	15.3	18.1
GNP/GDP (ppp) Trillion (1990 prices)		2.6	2.2	2.6	3.3	4.3	5.3	6.4	8.2	10.3	12.8	15.3	18.1
Final Energy	EJ												
Non-commercial		0	0	0	0	0	0	0	0	0	0	0	0
Solids		9	5	3	2	1	0	0	0	0	0	0	0
Liquids		15	10	10	13	15	17	18	16	13	10	9	8
Gas		14	10	11	10	9	8	6	4	4	4	3	2
Electricity		6	5	6	7	9	12	14	15	15	15	15	15
Others		7	6	6	7	9	9	9	10	10	10	9	7
Total		50	36	37	40	43	46	47	45	41	39	35	32
Primary Energy	EJ												
Coal		19	12	9	6	4	3	2	1	1	1	1	2
Oil		20	14	13	15	17	16	16	15	11	9	6	5
Gas		27	21	25	28	33	40	43	49	53	48	40	31
Nuclear		1	1	1	0	0	0	0	0	0	1	1	1
Biomass		2	1	1	1	2	3	3	3	3	3	6	9
Other Renewables		1	1	1	3	4	5	6	6	6	7	9	10
Total		70	51	49	53	59	67	70	74	74	69	63	57
Cumulative Resources Use	ZJ												
Coal		0.0	0.2	0.3	0.4	0.4	0.5	0.5	0.5	0.5	0.5	0.6	0.6
Oil		0.0	0.2	0.3	0.5	0.6	0.8	0.9	1.1	1.3	1.4	1.5	1.5
Gas		0.0	0.3	0.5	0.7	1.0	1.3	1.8	2.3	2.8	3.4	4.0	4.4
Cumulative CO2 Emissions	GtC	1.3	12.3	21.3	29.5	36.9	44.2	51.3	58.7	67.0	75.0	81.7	86.8
Carbon Sequestration	GtC												
Land Use [a]	Million ha												
Cropland		268	266	266	266	265	265	264	264	263	263	263	262
Grasslands		341	362	364	366	368	370	372	374	376	378	380	382
Energy Biomass		0	0	0	9	20	29	37	38	39	38	35	33
Forest		966	951	934	933	934	954	973	983	993	1005	1018	1031
Others		701	696	711	702	688	659	630	617	604	592	580	568
Total		2276	2276	2276	2276	2276	2276	2276	2276	2276	2276	2276	2276
Anthropogenic Emissions (standardized)													
Fossil Fuel CO2 [b]	GtC	1.30	0.91	0.85	0.85	0.87	0.92	0.97	1.03	1.02	0.90	0.73	0.56
Other CO2 [c]	GtC	0.00	0.00	0.04	-0.11	-0.12	-0.20	-0.28	-0.23	-0.17	-0.15	-0.14	-0.14
Total CO2	GtC	1.30	0.91	0.89	0.73	0.75	0.72	0.69	0.81	0.85	0.75	0.59	0.42
CH4 total [d]	MtCH4	47	39	40	40	41	43	43	39	37	33	30	26
N2O total [e]	MtN2O-N	0.6	0.6	0.6	0.6	0.6	0.6	0.6	0.6	0.6	0.6	0.6	0.6
SOx total	MtS	17.0	11.0	7.1	4.6	3.0	2.0	1.5	1.4	1.3	1.4	1.5	1.6
HFC	MtC eq.	0	4	9	15	20	24	26	27	27	27	27	26
PFC	MtC eq.	7	4	6	6	7	8	9	10	9	8	8	8
SF6	MtC eq.	8	6	4	7	7	8	9	9	7	6	4	4
CO	MtCO	69	41	45	46	51	60	65	60	50	38	34	30
NMVOC	Mt	16	13	15	17	20	23	27	29	31	28	21	14
NOx	MtN	5	3	3	3	4	4	4	4	3	2	1	1

Emissions correlated to land-use change and deforestation were calculated by using AIM B1 land-use data.
a: Land-use taken from AIM-B1 emulation run.
b:CO2 emissions from fossil fuel and industrial processes (MESSAGE data).
c: CO2 emissions from land-use changes (AIM-B1 emulation run).
d: Non-energy related CH4 emissions were taken from AIM-B1 emulation run.
e: Non-energy related N2O emissions were taken from AIM-B1 emulation run.

Scenario B1-MESSAGE ASIA		**1990**	**2000**	**2010**	**2020**	**2030**	**2040**	**2050**	**2060**	**2070**	**2080**	**2090**	**2100**
Population	Million	2798	3261	3620	3937	4147	4238	4220	4085	3867	3589	3258	2882
GNP/GDP (mex)	Trillion US$	1.5	2.7	4.8	8.7	15.1	24.9	37.9	51.4	64.8	78.7	91.7	103.1
GNP/GDP (ppp) Trillion (1990 prices)		5.3	8.2	12.0	17.3	24.6	34.1	46.1	57.4	67.7	79.3	91.7	103.1
Final Energy	EJ												
Non-commercial		24	16	12	8	5	3	2	0	0	0	0	0
Solids		19	33	47	48	46	32	19	14	6	2	1	0
Liquids		13	22	35	50	65	82	87	78	68	55	45	38
Gas		2	3	7	11	11	14	11	10	10	10	10	10
Electricity		4	6	11	19	30	42	49	57	60	64	63	58
Others		1	2	6	11	18	27	34	45	49	52	50	45
Total		62	82	117	146	175	199	202	204	193	182	170	150
Primary Energy	EJ												
Coal		30	42	57	58	51	44	24	13	8	8	9	10
Oil		15	26	38	52	66	68	72	61	52	37	25	17
Gas		3	5	13	24	36	51	53	53	50	46	34	27
Nuclear		0	1	1	3	5	9	11	13	14	17	17	17
Biomass		24	22	26	30	39	49	54	61	63	67	72	73
Other Renewables		1	3	6	12	20	35	54	77	83	85	87	74
Total		74	98	141	178	217	256	268	278	270	260	244	218
Cumulative Resources Use	ZJ												
Coal		0.0	0.3	0.7	1.3	1.9	2.4	2.8	3.1	3.2	3.3	3.4	3.5
Oil		0.0	0.2	0.4	0.8	1.3	2.0	2.7	3.4	4.0	4.5	4.9	5.1
Gas		0.0	0.0	0.1	0.2	0.5	0.8	1.3	1.8	2.2	2.6	3.0	3.3
Cumulative CO2 Emissions	GtC	1.5	19.3	43.3	71.2	100.9	132.5	163.1	189.1	210.1	227.1	240.4	251.0
Carbon Sequestration	GtC												
Land Use [a]	Million ha												
Cropland		438	435	435	435	434	434	434	434	433	433	433	433
Grasslands		608	606	609	612	614	617	619	622	625	628	631	634
Energy Biomass		0	0	0	12	28	40	52	54	55	54	50	46
Forest		535	525	518	535	551	557	563	578	593	604	610	616
Others		583	598	602	571	536	516	496	477	458	446	441	436
Total		2164	2164	2164	2164	2164	2164	2164	2164	2164	2164	2164	2164
Anthropogenic Emissions (standardized)													
Fossil Fuel CO2 [b]	GtC	1.15	1.78	2.56	3.01	3.27	3.33	2.93	2.41	2.06	1.69	1.29	1.06
Other CO2 [c]	GtC	0.37	0.26	0.20	-0.17	-0.18	-0.10	-0.03	-0.10	-0.17	-0.18	-0.14	-0.10
Total CO2	GtC	1.53	2.03	2.75	2.84	3.09	3.23	2.90	2.31	1.89	1.51	1.15	0.96
CH4 total [d]	MtCH4	113	125	145	164	183	196	201	194	189	172	141	109
N2O total [e]	MtN2O-N	2.3	2.6	2.3	2.3	2.3	2.3	2.2	2.2	2.2	2.2	2.2	2.2
SOx total	MtS	17.7	25.3	31.8	29.9	26.3	19.3	13.1	9.3	6.7	4.4	3.1	2.1
HFC	MtC eq.	0	4	11	20	34	56	93	90	85	79	72	64
PFC	MtC eq.	3	5	7	9	11	14	17	20	18	18	18	18
SF6	MtC eq.	4	7	11	17	23	27	30	32	26	19	16	16
CO	MtCO	235	270	345	421	511	597	632	621	603	587	583	559
NMVOC	Mt	33	37	48	56	58	59	57	46	40	32	25	20
NOx	MtN	7	9	12	14	16	17	15	12	10	8	7	6

Emissions correlated to land-use change and deforestation were calculated by using AIM B1 land-use data.
a: Land-use taken from AIM-B1 emulation run.
b:CO2 emissions from fossil fuel and industrial processes (MESSAGE data).
c: CO2 emissions from land-use changes (AIM-B1 emulation run).
d: Non-energy related CH4 emissions were taken from AIM-B1 emulation run.
e: Non-energy related N2O emissions were taken from AIM-B1 emulation run.

Scenario B1-MESSAGE ALM		**1990**	**2000**	**2010**	**2020**	**2030**	**2040**	**2050**	**2060**	**2070**	**2080**	**2090**	**2100**
Population	Million	1192	1519	1875	2241	2557	2791	2980	3089	3115	3064	2934	2727
GNP/GDP (mex)	Trillion US$	1.9	2.7	4.4	9.3	17.0	27.6	41.6	56.7	73.6	92.0	109.1	124.8
GNP/GDP (ppp) Trillion (1990 prices)		3.8	5.1	7.6	12.8	20.1	30.6	44.0	57.8	73.6	92.0	109.1	124.8
Final Energy	EJ												
Non-commercial		14	11	11	8	6	4	3	0	0	0	0	0
Solids		1	4	5	6	7	7	4	4	3	1	1	1
Liquids		17	22	34	48	64	75	77	69	57	44	35	29
Gas		4	6	10	15	22	22	25	25	23	22	18	14
Electricity		3	5	9	17	29	42	54	66	72	74	70	65
Others		0	1	5	10	18	32	46	56	65	69	67	61
Total		39	49	73	105	145	182	210	220	219	209	192	170
Primary Energy	EJ												
Coal		5	4	8	10	12	10	6	3	4	6	7	9
Oil		21	30	39	49	59	62	59	50	40	29	18	12
Gas		8	12	24	42	73	100	116	125	106	96	81	79
Nuclear		0	0	1	2	4	4	4	4	6	6	7	8
Biomass		14	14	18	25	33	42	44	39	54	65	88	101
Other Renewables		2	4	7	13	22	42	72	100	121	125	111	81
Total		49	65	96	141	201	259	301	321	330	327	313	290
Cumulative Resources Use	ZJ												
Coal		0.0	0.0	0.1	0.2	0.3	0.4	0.5	0.5	0.6	0.6	0.7	0.7
Oil		0.0	0.3	0.6	0.9	1.4	2.0	2.6	3.2	3.7	4.1	4.4	4.6
Gas		0.0	0.1	0.2	0.5	0.9	1.6	2.6	3.8	5.0	6.1	7.0	7.8
Cumulative CO2 Emissions	GtC	1.4	17.8	37.4	59.6	83.7	110.2	139.1	168.0	194.5	217.0	235.4	250.8
Carbon Sequestration	GtC												
Land Use [a]	Million ha												
Cropland		371	385	385	385	384	384	383	383	383	383	382	382
Grasslands		1680	1675	1686	1696	1705	1714	1723	1732	1741	1750	1760	1769
Energy Biomass		0	0	0	29	66	93	121	124	127	124	116	107
Forest		1745	1720	1704	1702	1732	1761	1790	1808	1825	1843	1862	1880
Others		1684	1700	1704	1669	1593	1527	1461	1432	1403	1379	1361	1342
Total		5480	5480	5480	5480	5480	5480	5480	5480	5480	5480	5480	5480
Anthropogenic Emissions (standardized)													
Fossil Fuel CO2 [b]	GtC	0.72	1.01	1.46	1.98	2.66	3.05	3.18	3.08	2.61	2.27	1.86	1.75
Other CO2 [c]	GtC	0.73	0.82	0.62	0.38	-0.19	-0.22	-0.25	-0.22	-0.19	-0.20	-0.24	-0.29
Total CO2	GtC	1.45	1.83	2.08	2.36	2.46	2.83	2.94	2.86	2.43	2.07	1.62	1.46
CH4 total [d]	MtCH4	77	85	98	112	130	141	151	149	146	135	114	93
N2O total [e]	MtN2O-N	1.2	1.3	0.7	0.6	0.5	0.5	0.5	0.5	0.5	0.5	0.5	0.5
SOx total	MtS	10.5	12.8	14.8	15.4	14.8	13.8	12.9	11.1	8.9	7.1	5.4	4.1
HFC	MtC eq.	0	2	16	28	42	64	98	102	102	101	96	89
PFC	MtC eq.	4	4	5	6	7	8	9	11	11	11	12	12
SF6	MtC eq.	2	5	5	8	12	17	22	24	21	14	14	14
CO	MtCO	396	404	434	472	510	523	509	465	513	559	654	714
NMVOC	Mt	48	55	63	70	74	71	68	58	50	42	33	27
NOx	MtN	7	8	10	12	14	15	14	12	11	11	10	9

Emissions correlated to land-use change and deforestation were calculated by using AIM B1 land-use data.

a: Land-use taken from AIM-B1 emulation run.

b:CO2 emissions from fossil fuel and industrial processes (MESSAGE data).

c: CO2 emissions from land-use changes (AIM-B1 emulation run).

d: Non-energy related CH4 emissions were taken from AIM-B1 emulation run.

e: Non-energy related N2O emissions were taken from AIM-B1 emulation run.

Scenario B1-MiniCAM World		1990	2000	2010	2020	2030	2040	2050	2060	2070	2080	2090	2100
Population	Million	5293	6100	6874	7618	8122	8484	8703	8623	8430	8126	7621	7137
GNP/GDP (mex)	Trillion US$	20.7	27.4	37.5	51.0	73.0	100.7	134.1	172.7	211.9	251.8	289.9	330.8
GNP/GDP (ppp) Trillion (1990 prices)		na	na	na	na	na	na	na	na	na	na	na	na
Final Energy	EJ												
Non-commercial		0	0	0	0	0	0	0	0	0	0	0	0
Solids		45	50	55	60	66	66	61	43	32	25	24	22
Liquids		121	112	109	113	113	123	142	152	162	172	176	181
Gas		52	57	66	82	86	86	81	78	78	79	81	84
Electricity		35	46	63	85	113	147	185	219	247	270	272	273
Others		0	0	0	0	0	0	0	0	0	0	0	0
Total		253	264	294	341	378	421	469	493	518	546	553	561
Primary Energy	EJ												
Coal		88	98	107	113	131	136	127	93	68	51	49	47
Oil		131	120	116	119	115	123	144	156	167	177	183	188
Gas		70	76	100	144	162	169	165	160	155	148	149	151
Nuclear		24	22	24	29	41	49	54	51	46	38	40	41
Biomass		0	5	9	13	17	20	23	26	27	24	23	22
Other Renewables		24	24	28	33	44	76	129	185	234	277	257	238
Total		336	345	384	451	509	573	642	672	696	716	701	687
Cumulative Resources Use	ZJ												
Coal		0.1	1.1	2.1	3.2	4.4	5.7	7.1	8.1	8.9	9.5	10.0	10.5
Oil		0.1	1.4	2.6	3.7	4.9	6.1	7.4	8.9	10.6	12.3	14.1	15.9
Gas		0.1	0.8	1.8	2.9	4.5	6.1	7.8	9.4	11.0	12.5	14.0	15.5
Cumulative CO2 Emissions	GtC	7.1	82.4	164.2	253.8	352.1	452.4	549.4	641.3	726.4	801.1	865.9	922.6
Carbon Sequestration	GtC												
Land Use	Million ha												
Cropland		1472	1467	1456	1439	1377	1287	1168	989	833	699	598	496
Grasslands		3209	3349	3590	3933	4240	4443	4544	4268	3976	3668	3522	3375
Energy Biomass		0	1	2	1	1	0	0	0	0	0	0	0
Forest		4173	4214	4144	3963	3737	3611	3584	4038	4428	4754	4796	4837
Others		4310	4133	3972	3828	3810	3823	3868	3870	3928	4043	4250	4458
Total		13164	13164	13164	13164	13164	13164	13164	13164	13164	13164	13165	13166
Anthropogenic Emissions (standardized)													
Fossil Fuel CO2	GtC	5.99	6.90	7.39	8.23	8.99	9.23	9.30	8.58	8.06	7.73	7.83	7.92
Other CO2	GtC	1.11	1.07	1.00	1.30	1.14	0.72	0.15	0.36	0.02	-0.86	-1.75	-2.64
Total CO2	GtC	7.10	7.97	8.38	9.53	10.13	9.94	9.45	8.94	8.08	6.87	6.07	5.28
CH4 total	MtCH4	310	323	351	399	439	483	512	506	505	510	535	561
N2O total	MtN2O-N	6.7	7.0	8.1	9.5	11.1	13.0	14.8	15.4	16.1	16.8	18.5	20.2
SOx total	MtS	70.9	69.0	69.0	69.0	65.6	57.1	43.7	30.4	21.6	17.3	16.9	16.6
CFC/HFC/HCFC	MtC eq.	1672	883	784	291	257	298	338	337	333	327	315	299
PFC	MtC eq.	32	25	29	32	33	37	42	47	45	43	46	45
SF6	MtC eq.	38	40	36	37	47	58	68	71	62	46	43	43
CO	MtCO												
NMVOC	Mt												
NOx	MtN												

Scenario B1-MiniCAM OECD90		**1990**	**2000**	**2010**	**2020**	**2030**	**2040**	**2050**	**2060**	**2070**	**2080**	**2090**	**2100**
Population	Million	838	908	965	1007	1024	1066	1081	1084	1090	1098	1105	1112
GNP/GDP (mex)	Trillion US$	16.3	20.5	25.1	30.1	32.0	38.8	44.6	50.3	56.5	63.2	71.0	79.3
GNP/GDP (ppp) Trillion (1990 prices)		na	na	na	na	na	na	na	na	na	na	na	na
Final Energy	EJ												
Non-commercial		0	0	0	0	0	0	0	0	0	0	0	0
Solids		10	11	10	7	7	6	6	5	4	3	3	3
Liquids		72	63	55	48	43	34	35	35	36	38	39	41
Gas		27	31	36	41	39	35	32	30	30	32	33	35
Electricity		22	24	25	27	28	31	35	38	42	46	47	47
Others		0	0	0	0	0	0	0	0	0	0	0	0
Total		130	128	126	123	117	107	108	108	112	118	122	126
Primary Energy	EJ												
Coal		40	38	32	22	22	23	23	18	13	9	10	11
Oil		76	67	58	49	43	32	32	33	35	38	40	41
Gas		34	40	50	64	61	54	49	46	45	45	47	49
Nuclear		20	13	9	8	8	8	7	7	6	6	6	7
Biomass		0	2	3	3	3	3	3	3	4	3	3	3
Other Renewables		12	11	10	10	10	14	20	27	34	42	41	39
Total		182	171	162	155	148	133	134	134	137	143	147	150
Cumulative Resources Use		ZJ											
Coal		0.0	0.4	0.8	1.1	1.2	1.5	1.7	1.9	2.1	2.2	2.3	2.4
Oil		0.1	0.8	1.4	1.9	2.1	2.7	3.0	3.3	3.7	4.0	4.4	4.8
Gas		0.0	0.4	0.9	1.4	1.7	2.6	3.1	3.6	4.0	4.5	4.9	5.4
Cumulative CO2 Emissions	GtC	2.8	33.0	64.7	95.3	124.5	150.8	174.4	196.7	217.6	236.7	254.0	269.7
Carbon Sequestration	GtC												
Land Use	Million ha												
Cropland		408	410	407	399	386	343	305	259	221	193	158	124
Grasslands		796	821	867	933	960	1022	1037	988	939	891	850	809
Energy Biomass		0	1	1	0	0	0	0	0	0	0	0	0
Forest		921	931	923	897	883	862	873	968	1040	1090	1117	1145
Others		998	959	925	894	893	896	907	908	922	949	998	1046
Total		3123	3123	3123	3123	3123	3123	3123	3123	3123	3123	3123	3124
Anthropogenic Emissions (standardized)													
Fossil Fuel CO2	GtC	2.83	3.20	3.04	2.83	2.71	2.43	2.37	2.23	2.13	2.08	2.16	2.24
Other CO2	GtC	0.00	0.00	0.10	0.16	0.13	0.00	-0.09	-0.05	-0.11	-0.28	-0.51	-0.74
Total CO2	GtC	2.83	3.20	3.14	2.99	2.84	2.43	2.28	2.17	2.02	1.80	1.65	1.50
CH4 total	MtCH4	73	74	82	93	100	119	132	140	143	143	155	167
N2O total	MtN2O-N	2.6	2.6	2.8	3.1	3.2	3.8	4.1	4.3	4.4	4.7	4.9	5.1
SOx total	MtS	22.7	17.0	12.2	6.8	5.8	4.9	4.2	3.7	3.4	3.4	3.6	3.8
HFC	MtC eq.	19	58	108	103	109	112	116	117	117	119	120	120
PFC	MtC eq.	18	13	12	10	8	7	7	7	7	7	7	6
SF6	MtC eq.	24	23	16	5	6	6	7	7	8	8	8	8
CO	MtCO												
NMVOC	Mt												
NOx	MtN												

Scenario B1-MiniCAM REF		1990	2000	2010	2020	2030	2040	2050	2060	2070	2080	2090	2100
Population	Million	428	425	426	433	434	431	423	408	392	374	357	340
GNP/GDP (mex)	Trillion US$	1.1	1.1	1.4	2.0	3.0	4.1	5.3	7.1	9.2	11.5	13.4	15.6
GNP/GDP (ppp) Trillion (1990 prices)		na	na	na	na	na	na	na	na	na	na	na	na
Final Energy	EJ												
Non-commercial		0	0	0	0	0	0	0	0	0	0	0	0
Solids		13	10	8	7	6	5	4	3	2	2	1	1
Liquids		18	12	8	8	8	8	7	7	7	6	6	6
Gas		19	15	14	16	16	14	12	10	9	8	8	7
Electricity		6	8	11	15	17	19	21	22	23	22	21	20
Others		0	0	0	0	0	0	0	0	0	0	0	0
Total		56	44	40	46	47	46	45	42	40	38	36	34
Primary Energy	EJ												
Coal		18	17	16	15	16	16	15	9	6	5	4	3
Oil		20	13	10	10	9	8	8	8	8	7	7	6
Gas		26	20	20	28	28	26	21	18	15	13	12	11
Nuclear		3	4	5	7	7	7	6	5	4	3	3	3
Biomass		0	1	1	2	2	2	2	2	2	2	2	2
Other Renewables		3	3	4	5	6	10	16	20	23	24	21	19
Total		70	57	56	66	68	69	69	63	58	54	49	44
Cumulative Resources Use	ZJ												
Coal		0.0	0.2	0.3	0.5	0.7	0.8	1.0	1.1	1.2	1.2	1.3	1.3
Oil		0.0	0.2	0.3	0.4	0.5	0.6	0.6	0.7	0.8	0.9	1.0	1.0
Gas		0.0	0.2	0.5	0.7	1.0	1.2	1.5	1.7	1.8	2.0	2.1	2.2
Cumulative CO2 Emissions	GtC	1.3	12.3	21.2	30.6	40.8	50.6	59.2	66.6	72.9	77.4	80.0	81.4
Carbon Sequestration	GtC												
Land Use	Million ha												
Cropland		284	294	300	304	295	278	251	205	166	133	117	101
Grasslands		395	410	452	522	587	627	643	564	493	430	428	426
Energy Biomass		0	0	0	0	0	0	0	0	0	0	0	0
Forest		1007	1016	997	948	893	867	869	994	1093	1167	1147	1127
Others		691	657	628	604	602	606	614	615	626	647	685	724
Total		2377	2377	2377	2377	2377	2377	2377	2377	2377	2377	2377	2377
Anthropogenic Emissions (standardized)													
Fossil Fuel CO2	GtC	1.30	0.91	0.83	0.93	0.94	0.89	0.78	0.60	0.48	0.40	0.38	0.35
Other CO2	GtC	0.00	0.00	0.03	0.08	0.09	0.06	-0.01	0.10	0.08	-0.07	-0.17	-0.28
Total CO2	GtC	1.30	0.91	0.86	1.01	1.03	0.95	0.77	0.71	0.56	0.33	0.20	0.07
CH4 total	MtCH4	47	39	46	60	70	78	84	78	78	84	90	97
N2O total	MtN2O-N	0.6	0.6	0.7	0.9	1.2	1.4	1.6	1.6	1.6	1.6	1.8	2.0
SOx total	MtS	17.0	11.0	9.6	8.9	9.5	8.6	6.5	3.6	1.8	1.0	1.2	1.3
HFC	MtC eq.	0	4	9	15	20	24	26	27	27	27	27	26
PFC	MtC eq.	7	4	6	6	7	8	9	10	9	8	8	8
SF6	MtC eq.	8	6	4	7	7	8	9	9	7	6	4	4
CO	MtCO												
NMVOC	Mt												
NOx	MtN												

Scenario B1-MiniCAM ASIA		1990	2000	2010	2020	2030	2040	2050	2060	2070	2080	2090	2100
Population	Million	2790	3226	3608	3937	4115	4210	4219	4062	3852	3589	3245	2919
GNP/GDP (mex)	Trillion US$	1.4	3.1	6.4	11.5	21.7	34.1	48.6	63.3	77.6	91.5	103.9	117.1
GNP/GDP (ppp) Trillion (1990 prices)		na	na	na	na	na	na	na	na	na	na	na	na
Final Energy	EJ												
Non-commercial		0	0	0	0	0	0	0	0	0	0	0	0
Solids		20	27	34	42	46	46	41	29	20	16	14	13
Liquids		14	18	23	29	34	40	47	50	53	55	55	55
Gas		2	5	8	13	15	16	16	16	15	15	15	15
Electricity		4	10	19	32	47	63	81	95	105	113	111	109
Others		0	0	0	0	0	0	0	0	0	0	0	0
Total		40	59	84	115	141	165	185	189	194	200	196	193
Primary Energy	EJ												
Coal		26	39	53	68	79	79	68	47	33	25	22	20
Oil		16	19	25	32	35	41	49	53	56	57	58	58
Gas		3	8	18	33	44	51	54	52	48	44	43	42
Nuclear		1	3	7	11	18	23	26	24	21	17	17	17
Biomass		0	2	4	7	9	11	12	14	14	13	12	11
Other Renewables		3	4	5	7	12	28	54	79	100	117	105	94
Total		49	75	111	157	197	232	263	269	272	272	257	242
Cumulative Resources Use	ZJ												
Coal		0.0	0.4	0.9	1.4	2.2	3.0	3.7	4.2	4.6	4.9	5.2	5.4
Oil		0.0	0.2	0.4	0.7	1.0	1.4	1.9	2.4	2.9	3.5	4.1	4.6
Gas		0.0	0.1	0.2	0.4	0.9	1.3	1.9	2.4	2.9	3.3	3.8	4.2
Cumulative CO2 Emissions	GtC	1.5	19.3	42.2	71.6	107.2	145.8	184.3	219.3	249.8	276.1	299.3	320.4
Carbon Sequestration	GtC												
Land Use	Million ha												
Cropland		389	400	406	408	399	381	354	302	255	213	186	160
Grasslands		508	524	554	600	641	671	691	673	648	615	593	572
Energy Biomass		0	0	0	1	1	0	0	0	0	0	0	0
Forest		1168	1144	1108	1060	1027	1012	1017	1086	1159	1236	1283	1330
Others		664	633	606	584	583	587	596	597	608	628	665	701
Total		2729	2700	2674	2652	2650	2651	2657	2659	2671	2692	2728	2763
Anthropogenic Emissions (standardized)													
Fossil Fuel CO2	GtC	1.15	1.78	2.36	3.08	3.58	3.78	3.68	3.20	2.84	2.59	2.52	2.46
Other CO2	GtC	0.37	0.26	0.19	0.24	0.22	0.16	0.07	0.07	-0.01	-0.16	-0.30	-0.45
Total CO2	GtC	1.53	2.03	2.55	3.32	3.79	3.94	3.75	3.27	2.83	2.42	2.22	2.01
CH4 total	MtCH4	113	125	133	144	154	161	165	158	153	148	149	150
N2O total	MtN2O-N	2.3	2.6	2.9	3.3	3.8	4.3	4.8	5.0	5.1	5.3	5.8	6.4
SOx total	MtS	17.7	25.3	31.2	36.9	34.1	27.8	18.0	11.1	6.7	4.7	4.4	4.1
HFC	MtC eq.	0	4	11	20	34	56	93	90	85	79	72	64
PFC	MtC eq.	3	5	7	9	11	14	17	20	18	18	18	18
SF6	MtC eq.	4	7	11	17	23	27	30	32	26	19	16	16
CO	MtCO												
NMVOC	Mt												
NOx	MtN												

Scenario B1-MiniCAM ALM		1990	2000	2010	2020	2030	2040	2050	2060	2070	2080	2090	2100
Population	Million	1236	1541	1876	2241	2531	2778	2980	3068	3096	3064	2913	2766
GNP/GDP (mex)	Trillion US$	1.9	2.8	4.6	7.4	14.4	23.7	35.6	52.0	68.6	85.6	101.5	118.8
GNP/GDP (ppp) Trillion (1990 prices)		na	na	na	na	na	na	na	na	na	na	na	na
Final Energy	EJ												
Non-commercial		0	0	0	0	0	0	0	0	0	0	0	0
Solids		2	3	4	5	7	8	9	7	6	5	5	5
Liquids		17	19	23	28	33	41	52	60	67	73	76	80
Gas		5	6	9	13	18	21	22	22	23	24	25	27
Electricity		3	5	8	12	21	33	48	63	77	89	93	96
Others		0	0	0	0	0	0	0	0	0	0	0	0
Total		27	33	43	57	79	103	131	153	172	190	199	207
Primary Energy	EJ												
Coal		4	5	6	8	14	18	21	18	15	13	13	13
Oil		20	20	23	29	33	42	54	62	69	75	79	83
Gas		7	8	13	20	30	38	42	44	46	46	48	49
Nuclear		0	2	3	4	8	11	14	15	14	13	14	14
Biomass		0	0	1	2	3	4	5	7	7	7	6	6
Other Renewables		5	6	8	11	15	24	39	59	77	94	89	85
Total		35	42	55	73	103	138	176	205	229	247	249	251
Cumulative Resources Use	ZJ												
Coal		0.0	0.1	0.1	0.2	0.3	0.5	0.7	0.8	1.0	1.2	1.3	1.4
Oil		0.0	0.2	0.4	0.7	1.0	1.4	1.9	2.5	3.1	3.8	4.6	5.4
Gas		0.0	0.1	0.2	0.3	0.6	1.0	1.4	1.8	2.2	2.7	3.2	3.6
Cumulative CO2 Emissions	GtC	1.4	17.8	36.1	56.3	79.6	105.1	131.5	158.7	186.1	211.0	232.6	251.1
Carbon Sequestration	GtC												
Land Use	Million ha												
Cropland		391	363	342	329	309	285	258	223	190	160	136	111
Grasslands		1510	1594	1717	1879	2025	2123	2174	2043	1897	1733	1651	1568
Energy Biomass		0	0	0	0	0	0	0	0	0	0	0	0
Forest		3641	3591	3493	3348	3218	3137	3105	3271	3453	3650	3756	3862
Others		1957	1884	1813	1746	1733	1734	1750	1749	1771	1818	1902	1987
Total		7499	7432	7366	7301	7285	7280	7286	7286	7311	7361	7445	7528
Anthropogenic Emissions (standardized)													
Fossil Fuel CO2	GtC	0.72	1.01	1.15	1.38	1.77	2.13	2.48	2.55	2.62	2.67	2.77	2.87
Other CO2	GtC	0.73	0.82	0.68	0.82	0.71	0.49	0.18	0.24	0.06	-0.35	-0.76	-1.17
Total CO2	GtC	1.45	1.83	1.83	2.20	2.47	2.62	2.65	2.79	2.67	2.31	2.01	1.70
CH4 total	MtCH4	77	85	90	102	116	126	131	131	132	135	141	147
N2O total	MtN2O-N	1.2	1.3	1.7	2.2	2.9	3.6	4.2	4.6	5.0	5.3	6.0	6.6
SOx total	MtS	10.5	12.8	13.0	13.4	13.3	12.8	12.0	9.0	6.7	5.2	4.8	4.4
HFC	MtC eq.	0	2	16	28	42	64	98	102	102	101	96	89
PFC	MtC eq.	4	4	5	6	7	8	9	11	11	11	12	12
SF6	MtC eq.	2	5	5	8	12	17	22	24	21	14	14	14
CO	MtCO												
NMVOC	Mt												
NOx	MtN												

Scenario B1T-MESSAGE World		1990	2000	2010	2020	2030	2040	2050	2060	2070	2080	2090	2100
Population	Million	5262	6117	6888	7617	8182	8531	8704	8667	8463	8125	7658	7056
GNP/GDP (mex)	Trillion US$	20.9	26.8	36.2	52.1	73.1	100.7	135.6	171.6	208.5	249.8	290.0	328.4
GNP/GDP (ppp) Trillion (1990 prices)		25.7	33.3	44.6	61.6	82.2	108.4	140.0	171.8	204.1	242.5	281.3	318.8
Final Energy	EJ												
Non-commercial		38	27	22	16	10	8	6	0	0	0	0	0
Solids		42	51	58	52	47	38	21	16	6	2	1	1
Liquids		111	123	158	192	224	250	261	244	215	191	156	131
Gas		41	47	63	73	87	101	97	92	76	52	41	33
Electricity		35	44	62	89	120	151	174	196	204	208	203	193
Others		8	10	16	27	36	47	57	68	84	96	103	95
Total		275	303	379	449	524	593	616	616	586	548	503	452
Primary Energy	EJ												
Coal		91	91	109	109	98	77	39	19	7	4	2	2
Oil		128	155	172	189	203	197	190	159	124	96	68	48
Gas		71	84	119	159	212	258	258	258	234	210	183	166
Nuclear		7	8	11	16	27	43	48	42	24	17	18	21
Biomass		46	45	53	68	86	107	123	138	148	148	138	128
Other Renewables		8	13	24	43	62	97	161	230	303	347	372	348
Total		352	395	488	583	687	779	819	846	840	822	780	714
Cumulative Resources Use	ZJ												
Coal		0.0	0.9	1.8	2.9	4.0	5.0	5.8	6.1	6.3	6.4	6.4	6.5
Oil		0.0	1.4	3.0	4.7	6.6	8.6	10.6	12.5	14.1	15.3	16.3	16.9
Gas		0.0	0.7	1.6	2.8	4.3	6.5	9.0	11.6	14.2	16.5	18.6	20.5
Cumulative CO2 Emissions	GtC	7.1	82.4	167.6	258.2	349.2	439.5	523.3	595.5	655.3	703.8	742.3	772.5
Carbon Sequestration	GtC												
Land Use [a]	Million ha												
Cropland		1459	1466	1464	1461	1458	1452	1447	1442	1437	1433	1431	1429
Grasslands		3389	3407	3432	3453	3474	3510	3547	3591	3635	3668	3688	3709
Energy Biomass		0	0	0	59	137	196	254	260	266	261	243	225
Forest		4296	4255	4231	4264	4335	4405	4475	4522	4570	4619	4669	4719
Others		3805	3820	3822	3711	3545	3386	3226	3133	3040	2968	2918	2868
Total		12949	12949	12949	12949	12949	12949	12949	12949	12949	12949	12949	12949
Anthropogenic Emissions (standardized)													
Fossil Fuel CO2 [b]	GtC	5.99	6.90	8.26	9.11	9.77	9.61	8.48	7.26	5.92	4.92	3.96	3.33
Other CO2 [c]	GtC	1.11	1.07	0.79	-0.03	-0.65	-0.66	-0.67	-0.62	-0.57	-0.57	-0.61	-0.65
Total CO2	GtC	7.10	7.97	9.05	9.08	9.11	8.95	7.81	6.63	5.35	4.35	3.35	2.68
CH4 total [d]	MtCH4	310	323	357	387	426	455	465	446	427	385	326	268
N2O total [e]	MtN2O-N	6.7	7.0	6.0	5.8	5.7	5.7	5.6	5.6	5.5	5.4	5.4	5.3
SOx total	MtS	70.9	69.0	64.7	52.4	44.5	36.9	29.1	24.3	19.2	15.9	13.3	11.4
CFC/HFC/HCFC	MtC eq.	1672	883	784	291	257	298	338	337	333	327	315	299
PFC	MtC eq.	32	25	29	32	33	37	42	47	45	43	46	45
SF6	MtC eq.	38	40	36	37	47	58	68	71	62	46	43	43
CO	MtCO	879	877	1000	1129	1271	1414	1460	1426	1333	1192	1063	966
NMVOC	Mt	139	141	164	183	192	199	194	171	143	114	87	68
NOx	MtN	31	32	39	45	51	53	49	44	36	28	21	16

Emissions correlated to land-use change and deforestation were calculated by using AIM B1 land-use data.
a: Land-use taken from AIM-B1 emulation run.
b:CO2 emissions from fossil fuel and industrial processes (MESSAGE data).
c: CO2 emissions from land-use changes (AIM-B1 emulation run).
d: Non-energy related CH4 emissions were taken from AIM-B1 emulation run.
e: Non-energy related N2O emissions were taken from AIM-B1 emulation run.

Scenario B1T-MESSAGE OECD90		**1990**	**2000**	**2010**	**2020**	**2030**	**2040**	**2050**	**2060**	**2070**	**2080**	**2090**	**2100**
Population	Million	859	919	965	1007	1043	1069	1081	1084	1089	1098	1108	1110
GNP/GDP (mex)	Trillion US$	16.4	20.6	26.0	32.4	38.3	43.9	49.9	55.4	59.8	66.3	73.9	82.3
GNP/GDP (ppp) Trillion (1990 prices)		14.1	17.7	22.4	28.1	33.3	38.3	43.6	48.5	52.5	58.4	65.1	72.7
Final Energy	EJ												
Non-commercial		0	0	0	0	0	0	0	0	0	0	0	0
Solids		13	10	6	3	1	0	0	0	0	0	0	0
Liquids		66	69	74	72	64	58	55	52	48	45	41	37
Gas		22	28	34	35	38	36	28	21	14	11	9	7
Electricity		22	28	36	44	50	53	56	57	59	61	63	63
Others		1	1	2	5	5	5	6	7	8	8	7	7
Total		124	136	151	158	158	153	145	137	129	124	120	114
Primary Energy	EJ												
Coal		38	33	35	35	30	14	5	3	1	1	0	0
Oil		72	84	81	71	56	45	36	30	24	18	14	10
Gas		33	45	58	67	76	75	73	70	74	71	62	53
Nuclear		6	7	8	10	15	27	30	23	8	3	6	8
Biomass		6	8	10	14	16	17	19	23	27	33	37	42
Other Renewables		4	5	9	14	17	22	31	41	55	61	64	64
Total		159	182	200	211	211	200	194	189	188	186	182	176
Cumulative Resources Use	ZJ												
Coal		0.0	0.4	0.7	1.1	1.4	1.7	1.8	1.8	1.8	1.8	1.8	1.8
Oil		0.0	0.8	1.6	2.4	3.1	3.7	4.2	4.5	4.8	5.1	5.2	5.4
Gas		0.0	0.4	0.8	1.4	2.1	2.8	3.5	4.2	4.9	5.6	6.2	6.8
Cumulative CO2 Emissions	GtC	2.8	33.0	65.4	97.6	127.3	151.5	170.3	186.4	201.5	215.6	227.4	236.5
Carbon Sequestration	GtC												
Land Use [a]	Million ha												
Cropland		381	380	378	376	374	370	366	361	357	354	353	352
Grasslands		760	765	772	779	787	809	832	862	893	912	918	924
Energy Biomass		0	0	0	10	23	33	44	45	46	45	42	39
Forest		1050	1059	1074	1094	1117	1133	1149	1154	1159	1167	1179	1192
Others		838	826	805	770	728	684	639	607	574	551	537	522
Total		3029	3029	3029	3029	3029	3029	3029	3029	3029	3029	3029	3029
Anthropogenic Emissions (standardized)													
Fossil Fuel CO2 [b]	GtC	2.83	3.20	3.37	3.28	2.93	2.22	1.80	1.61	1.54	1.36	1.12	0.90
Other CO2 [c]	GtC	0.00	0.00	-0.07	-0.13	-0.16	-0.14	-0.12	-0.08	-0.04	-0.04	-0.08	-0.12
Total CO2	GtC	2.83	3.20	3.29	3.15	2.77	2.08	1.69	1.53	1.50	1.32	1.04	0.78
CH4 total [d]	MtCH4	73	74	72	70	67	62	58	55	52	51	50	49
N2O total [e]	MtN2O-N	2.6	2.6	2.4	2.4	2.3	2.3	2.3	2.2	2.2	2.2	2.1	2.1
SOx total	MtS	22.7	17.0	8.9	2.4	0.4	-0.5	-0.8	-0.2	0.1	0.5	1.1	1.6
HFC	MtC eq.	19	58	108	103	109	112	116	117	117	119	120	120
PFC	MtC eq.	18	13	12	10	8	7	7	7	7	7	7	6
SF6	MtC eq.	24	23	16	5	6	6	7	7	8	8	8	8
CO	MtCO	179	161	177	183	167	151	137	129	119	113	104	103
NMVOC	Mt	42	36	38	37	32	27	24	22	20	18	14	10
NOx	MtN	13	12	14	14	13	11	9	7	6	4	2	1

Emissions correlated to land-use change and deforestation were calculated by using AIM B1 land-use data.

a: Land-use taken from AIM-B1 emulation run.

b:CO2 emissions from fossil fuel and industrial processes (MESSAGE data).

c: CO2 emissions from land-use changes (AIM-B1 emulation run).

d: Non-energy related CH4 emissions were taken from AIM-B1 emulation run.

e: Non-energy related N2O emissions were taken from AIM-B1 emulation run.

Scenario B1T-MESSAGE REF		1990	2000	2010	2020	2030	2040	2050	2060	2070	2080	2090	2100
Population	Million	413	419	427	433	435	433	423	409	392	374	357	339
GNP/GDP (mex)	Trillion US$	1.1	0.8	1.1	1.7	2.8	4.3	6.2	8.2	10.3	12.8	15.3	18.1
GNP/GDP (ppp) Trillion (1990 prices)		2.6	2.2	2.6	3.3	4.3	5.3	6.4	8.2	10.3	12.8	15.3	18.1
Final Energy	EJ												
Non-commercial		0	0	0	0	0	0	0	0	0	0	0	0
Solids		9	5	3	2	1	0	0	0	0	0	0	0
Liquids		15	10	10	13	14	16	18	17	14	13	12	10
Gas		14	10	12	12	12	12	9	7	5	4	3	2
Electricity		6	5	6	7	9	12	14	15	15	15	15	15
Others		7	6	6	7	7	7	7	7	6	6	5	5
Total		50	36	37	40	43	47	48	45	41	38	35	32
Primary Energy	EJ												
Coal		19	12	9	6	3	1	-1	0	0	0	0	0
Oil		20	14	13	15	16	16	18	16	11	9	7	5
Gas		27	21	25	28	33	39	39	38	39	39	34	30
Nuclear		1	1	1	1	1	1	1	1	0	0	0	0
Biomass		2	1	1	1	1	2	4	6	7	8	9	9
Other Renewables		1	1	1	3	4	5	6	8	8	9	9	10
Total		70	51	50	53	58	65	67	68	66	64	59	54
Cumulative Resources Use	ZJ												
Coal		0.0	0.2	0.3	0.4	0.4	0.5	0.5	0.5	0.5	0.5	0.5	0.5
Oil		0.0	0.2	0.3	0.5	0.6	0.8	0.9	1.1	1.3	1.4	1.5	1.5
Gas		0.0	0.3	0.5	0.7	1.0	1.3	1.7	2.1	2.5	3.0	3.4	3.8
Cumulative CO2 Emissions	GtC	1.3	12.3	21.4	29.5	36.8	43.8	50.0	55.9	62.0	67.9	73.3	77.6
Carbon Sequestration	GtC												
Land Use [a]	Million ha												
Cropland		268	266	266	266	265	265	264	264	263	263	263	262
Grasslands		341	362	364	366	368	370	372	374	376	378	380	382
Energy Biomass		0	0	0	9	20	29	37	38	39	38	35	33
Forest		966	951	934	933	934	954	973	983	993	1005	1018	1031
Others		701	696	711	702	688	659	630	617	604	592	580	568
Total		2276	2276	2276	2276	2276	2276	2276	2276	2276	2276	2276	2276
Anthropogenic Emissions (standardized)													
Fossil Fuel CO2 [b]	GtC	1.30	0.91	0.86	0.84	0.86	0.86	0.86	0.83	0.78	0.73	0.62	0.52
Other CO2 [c]	GtC	0.00	0.00	0.04	-0.11	-0.12	-0.20	-0.28	-0.23	-0.17	-0.15	-0.14	-0.14
Total CO2	GtC	1.30	0.91	0.90	0.72	0.74	0.66	0.59	0.60	0.61	0.58	0.48	0.38
CH4 total [d]	MtCH4	47	39	40	41	44	46	45	40	35	31	28	25
N2O total [e]	MtN2O-N	0.6	0.6	0.6	0.6	0.6	0.6	0.6	0.6	0.6	0.6	0.6	0.6
SOx total	MtS	17.0	11.0	7.2	4.7	3.0	2.1	1.3	1.1	0.8	0.6	0.5	0.4
HFC	MtC eq.	0	4	9	15	20	24	26	27	27	27	27	26
PFC	MtC eq.	7	4	6	6	7	8	9	10	9	8	8	8
SF6	MtC eq.	8	6	4	7	7	8	9	9	7	6	4	4
CO	MtCO	69	41	46	50	58	72	81	72	54	48	35	25
NMVOC	Mt	16	13	15	18	21	25	28	27	26	25	19	14
NOx	MtN	5	3	3	3	4	4	4	4	3	2	2	1

Emissions correlated to land-use change and deforestation were calculated by using AIM B1 land-use data.
a: Land-use taken from AIM-B1 emulation run.
b:CO2 emissions from fossil fuel and industrial processes (MESSAGE data).
c: CO2 emissions from land-use changes (AIM-B1 emulation run).
d: Non-energy related CH4 emissions were taken from AIM-B1 emulation run.
e: Non-energy related N2O emissions were taken from AIM-B1 emulation run.

Scenario B1T-MESSAGE ASIA		1990	2000	2010	2020	2030	2040	2050	2060	2070	2080	2090	2100
Population	Million	2798	3261	3620	3937	4147	4238	4220	4085	3867	3589	3258	2882
GNP/GDP (mex)	Trillion US$	1.5	2.7	4.8	8.7	15.1	24.9	37.9	51.4	64.8	78.7	91.7	103.1
GNP/GDP (ppp) Trillion (1990 prices)		5.3	8.2	12.0	17.3	24.6	34.1	46.1	57.4	67.7	79.3	91.7	103.1
Final Energy	EJ												
Non-commercial		24	16	12	8	5	3	2	0	0	0	0	0
Solids		19	32	45	42	39	31	17	13	5	1	0	0
Liquids		13	22	37	55	73	94	99	92	82	74	62	50
Gas		2	3	7	11	13	17	20	25	23	15	10	8
Electricity		4	6	12	21	32	44	49	57	59	60	58	53
Others		1	2	5	9	12	15	16	20	24	27	30	28
Total		62	82	117	145	173	203	204	206	193	176	159	139
Primary Energy	EJ												
Coal		30	42	57	58	49	44	24	12	4	2	1	0
Oil		15	26	39	53	67	71	71	58	47	37	26	18
Gas		3	5	13	24	35	48	48	48	45	38	31	24
Nuclear		0	1	2	3	7	9	10	11	9	8	7	7
Biomass		24	22	25	30	39	50	53	61	61	62	59	50
Other Renewables		1	3	6	12	20	33	53	75	92	101	113	110
Total		74	98	142	179	217	255	258	266	259	249	237	208
Cumulative Resources Use	ZJ												
Coal		0.0	0.3	0.7	1.3	1.9	2.4	2.8	3.1	3.2	3.2	3.3	3.3
Oil		0.0	0.2	0.4	0.8	1.4	2.0	2.7	3.4	4.0	4.5	4.9	5.1
Gas		0.0	0.0	0.1	0.2	0.5	0.8	1.3	1.8	2.3	2.7	3.1	3.4
Cumulative CO2 Emissions	GtC	1.5	19.3	43.4	71.5	100.9	131.8	161.0	184.6	202.3	215.4	225.4	232.9
Carbon Sequestration	GtC												
Land Use [a]	Million ha												
Cropland		438	435	435	435	434	434	434	434	433	433	433	433
Grasslands		608	606	609	612	614	617	619	622	625	628	631	634
Energy Biomass		0	0	0	12	28	40	52	54	55	54	50	46
Forest		535	525	518	535	551	557	563	578	593	604	610	616
Others		583	598	602	571	536	516	496	477	458	446	441	436
Total		2164	2164	2164	2164	2164	2164	2164	2164	2164	2164	2164	2164
Anthropogenic Emissions (standardized)													
Fossil Fuel CO2 [b]	GtC	1.15	1.78	2.58	3.01	3.21	3.26	2.72	2.13	1.66	1.32	1.00	0.75
Other CO2 [c]	GtC	0.37	0.26	0.20	-0.17	-0.18	-0.10	-0.03	-0.10	-0.17	-0.18	-0.14	-0.10
Total CO2	GtC	1.53	2.03	2.78	2.84	3.04	3.16	2.69	2.03	1.49	1.14	0.86	0.65
CH4 total [d]	MtCH4	113	125	146	163	183	196	203	197	192	172	138	105
N2O total [e]	MtN2O-N	2.3	2.6	2.3	2.3	2.3	2.3	2.2	2.2	2.2	2.2	2.2	2.2
SOx total	MtS	17.7	25.3	30.9	27.3	22.9	18.2	12.4	9.0	6.3	4.6	3.2	2.2
HFC	MtC eq.	0	4	11	20	34	56	93	90	85	79	72	64
PFC	MtC eq.	3	5	7	9	11	14	17	20	18	18	18	18
SF6	MtC eq.	4	7	11	17	23	27	30	32	26	19	16	16
CO	MtCO	235	270	343	422	523	624	654	661	622	558	512	451
NMVOC	Mt	33	37	48	56	60	64	61	53	43	30	21	16
NOx	MtN	7	9	12	15	18	19	17	15	13	10	8	6

Emissions correlated to land-use change and deforestation were calculated by using AIM B1 land-use data.
a: Land-use taken from AIM-B1 emulation run.
b:CO2 emissions from fossil fuel and industrial processes (MESSAGE data).
c: CO2 emissions from land-use changes (AIM-B1 emulation run).
d: Non-energy related CH4 emissions were taken from AIM-B1 emulation run.
e: Non-energy related N2O emissions were taken from AIM-B1 emulation run.

Scenario B1T-MESSAGE ALM		1990	2000	2010	2020	2030	2040	2050	2060	2070	2080	2090	2100
Population	Million	1192	1519	1875	2241	2557	2791	2980	3089	3115	3064	2934	2727
GNP/GDP (mex)	Trillion US$	1.9	2.7	4.4	9.3	17.0	27.6	41.6	56.7	73.6	92.0	109.1	124.8
GNP/GDP (ppp) Trillion (1990 prices)		3.8	5.1	7.6	12.8	20.1	30.6	44.0	57.8	73.6	92.0	109.1	124.8
Final Energy	EJ												
Non-commercial		14	11	11	8	6	4	3	0	0	0	0	0
Solids		1	4	4	5	7	6	4	3	1	1	1	1
Liquids		17	22	37	53	72	82	90	83	72	59	42	33
Gas		4	6	10	15	24	36	40	39	34	22	19	16
Electricity		3	5	9	18	29	42	55	68	71	72	68	62
Others		0	1	3	7	12	19	27	34	46	56	61	56
Total		39	49	73	106	149	190	219	228	224	210	190	168
Primary Energy	EJ												
Coal		5	4	8	11	15	18	11	5	2	2	1	2
Oil		21	30	40	50	64	65	66	55	42	32	22	15
Gas		8	12	23	39	68	96	98	101	76	62	56	60
Nuclear		0	0	1	2	4	5	7	8	8	6	5	7
Biomass		14	14	17	24	30	37	47	48	53	46	32	27
Other Renewables		2	4	7	14	21	37	70	106	148	176	186	165
Total		49	65	96	140	201	259	300	322	328	322	302	275
Cumulative Resources Use	ZJ												
Coal		0.0	0.0	0.1	0.2	0.3	0.4	0.6	0.7	0.8	0.8	0.8	0.8
Oil		0.0	0.3	0.6	1.0	1.5	2.1	2.7	3.4	4.0	4.4	4.7	4.9
Gas		0.0	0.1	0.2	0.4	0.8	1.5	2.5	3.5	4.5	5.3	5.9	6.5
Cumulative CO2 Emissions	GtC	1.4	17.8	37.4	59.6	84.3	112.4	141.9	168.5	189.6	204.9	216.3	225.5
Carbon Sequestration	GtC												
Land Use [a]	Million ha												
Cropland		371	385	385	385	384	384	383	383	383	383	382	382
Grasslands		1680	1675	1686	1696	1705	1714	1723	1732	1741	1750	1760	1769
Energy Biomass		0	0	0	29	66	93	121	124	127	124	116	107
Forest		1745	1720	1704	1702	1732	1761	1790	1808	1825	1843	1862	1880
Others		1684	1700	1704	1669	1593	1527	1461	1432	1403	1379	1361	1342
Total		5480	5480	5480	5480	5480	5480	5480	5480	5480	5480	5480	5480
Anthropogenic Emissions (standardized)													
Fossil Fuel CO2 [b]	GtC	0.72	1.01	1.46	1.98	2.76	3.27	3.09	2.69	1.94	1.50	1.22	1.16
Other CO2 [c]	GtC	0.73	0.82	0.62	0.38	-0.19	-0.22	-0.25	-0.22	-0.19	-0.20	-0.24	-0.29
Total CO2	GtC	1.45	1.83	2.08	2.36	2.57	3.05	2.85	2.47	1.75	1.31	0.97	0.87
CH4 total [d]	MtCH4	77	85	98	112	133	151	159	154	148	132	110	89
N2O total [e]	MtN2O-N	1.2	1.3	0.7	0.6	0.5	0.5	0.5	0.5	0.5	0.5	0.5	0.5
SOx total	MtS	10.5	12.8	14.8	15.0	15.2	14.1	13.1	11.4	9.0	7.1	5.4	4.1
HFC	MtC eq.	0	2	16	28	42	64	98	102	102	101	96	89
PFC	MtC eq.	4	4	5	6	7	8	9	11	11	11	12	12
SF6	MtC eq.	2	5	5	8	12	17	22	24	21	14	14	14
CO	MtCO	396	404	434	475	522	568	588	564	538	473	412	387
NMVOC	Mt	48	55	64	72	80	83	81	68	54	41	33	28
NOx	MtN	7	8	10	13	16	19	19	18	14	11	10	8

Emissions correlated to land-use change and deforestation were calculated by using AIM B1 land-use data.
a: Land-use taken from AIM-B1 emulation run.
b:CO2 emissions from fossil fuel and industrial processes (MESSAGE data).
c: CO2 emissions from land-use changes (AIM-B1 emulation run).
d: Non-energy related CH4 emissions were taken from AIM-B1 emulation run.
e: Non-energy related N2O emissions were taken from AIM-B1 emulation run.

Scenario B1High-MESSAGE World		1990	2000	2010	2020	2030	2040	2050	2060	2070	2080	2090	2100
Population	Million	5262	6117	6888	7617	8182	8531	8704	8667	8463	8125	7658	7056
GNP/GDP (mex)	Trillion US$	20.9	26.8	36.7	56.5	88.5	126.0	166.0	212.3	253.8	290.1	322.7	350.3
GNP/GDP (ppp) Trillion (1990 prices)		25.7	33.3	45.7	66.9	96.0	130.3	166.8	208.9	246.1	281.7	313.4	340.2
Final Energy	EJ												
Non-commercial		38	25	20	16	10	8	6	0	0	0	0	0
Solids		42	52	54	58	58	44	29	24	14	11	2	1
Liquids		111	123	152	180	212	238	259	262	242	208	179	161
Gas		41	47	58	66	71	75	74	67	62	58	53	45
Electricity		35	44	57	83	116	152	186	231	264	291	307	313
Others		8	10	17	32	52	76	104	139	173	201	218	213
Total		275	302	359	434	519	593	659	723	755	768	758	732
Primary Energy	EJ												
Coal		91	91	100	111	103	87	64	50	43	47	60	60
Oil		128	155	172	188	204	197	191	162	139	100	70	53
Gas		71	83	110	149	215	273	306	324	304	287	263	244
Nuclear		7	8	10	14	20	30	45	66	79	81	80	82
Biomass		46	44	47	60	79	101	123	159	203	245	280	316
Other Renewables		8	13	21	39	63	112	183	272	348	406	431	404
Total		352	393	460	560	682	798	911	1033	1117	1167	1184	1157
Cumulative Resources Use	ZJ												
Coal		0.0	0.9	1.8	2.8	3.9	5.0	5.8	6.5	7.0	7.4	7.9	8.5
Oil		0.0	1.4	3.0	4.7	6.6	8.6	10.6	12.5	14.1	15.5	16.5	17.2
Gas		0.0	0.7	1.6	2.7	4.1	6.3	9.0	12.1	15.3	18.4	21.2	23.9
Cumulative CO2 Emissions	GtC	7.1	82.4	165.7	254.0	345.4	440.0	535.2	627.2	712.0	787.5	854.7	915.0
Carbon Sequestration	GtC												
Land Use	Million ha												
Cropland		1459	1466	1464	1461	1458	1452	1447	1442	1437	1433	1431	1429
Grasslands		3389	3407	3432	3453	3474	3510	3547	3591	3635	3668	3688	3709
Energy Biomass		0	0	0	59	137	196	254	260	266	261	243	225
Forest		4296	4255	4231	4264	4335	4405	4475	4522	4570	4619	4669	4719
Others		3805	3820	3822	3711	3545	3386	3226	3133	3040	2968	2918	2868
Total		12949	12949	12949	12949	12949	12949	12949	12949	12949	12949	12949	12949
Anthropogenic Emissions (standardized)													
Fossil Fuel CO2	GtC	5.99	6.90	7.90	8.99	9.97	10.27	10.11	9.57	8.60	7.63	6.99	6.33
Other CO2	GtC	1.11	1.07	0.79	-0.03	-0.65	-0.66	-0.67	-0.62	-0.57	-0.57	-0.61	-0.65
Total CO2	GtC	7.10	7.97	8.69	8.96	9.32	9.61	9.44	8.94	8.03	7.06	6.38	5.68
CH4 total	MtCH4	310	323	351	385	423	449	468	452	441	411	359	302
N2O total	MtN2O-N	6.7	7.0	6.0	5.8	5.7	5.7	5.6	5.5	5.5	5.4	5.4	5.4
SOx total	MtS	70.9	69.0	62.0	53.8	48.1	39.7	32.4	26.9	24.1	20.6	17.8	15.5
CFC/HFC/HCFC	MtC eq.	1672	883	784	291	257	298	338	337	333	327	315	299
PFC	MtC eq.	32	25	29	32	33	37	42	47	45	43	46	45
SF6	MtC eq.	38	40	36	37	47	58	68	71	62	46	43	43
CO	MtCO	879	877	951	1065	1193	1323	1470	1549	1658	1731	1793	1871
NMVOC	Mt	139	141	157	177	184	190	199	185	173	152	125	103
NOx	MtN	31	32	37	42	47	47	45	40	35	29	26	23

Emissions correlated to land-use change and deforestation were calculated by using AIM B1 land-use data.
a: Land-use taken from AIM-B1 emulation run.
b:CO2 emissions from fossil fuel and industrial processes (MESSAGE data).
c: CO2 emissions from land-use changes (AIM-B1 emulation run).
d: Non-energy related CH4 emissions were taken from AIM-B1 emulation run.
e: Non-energy related N2O emissions were taken from AIM-B1 emulation run.

Scenario B1High-MESSAGE OECD90		1990	2000	2010	2020	2030	2040	2050	2060	2070	2080	2090	2100
Population	Million	859	919	965	1007	1043	1069	1081	1084	1089	1098	1108	1110
GNP/GDP (mex)	Trillion US$	16.4	20.6	25.5	31.1	37.2	43.5	49.8	56.2	63.0	70.5	78.6	86.9
GNP/GDP (ppp) Trillion (1990 prices)		14.1	17.7	22.1	27.0	32.4	37.9	43.5	49.2	55.3	62.1	69.3	76.8
Final Energy	EJ												
Non-commercial		0	0	0	0	0	0	0	0	0	0	0	0
Solids		13	10	5	2	1	0	0	0	0	0	0	0
Liquids		66	69	75	75	73	70	71	68	66	62	58	55
Gas		22	28	31	30	31	30	25	20	14	12	10	7
Electricity		22	28	34	42	50	58	64	70	76	83	88	92
Others		1	1	1	5	7	9	11	15	17	19	19	18
Total		124	136	146	155	162	168	171	173	174	175	175	173
Primary Energy	EJ												
Coal		38	33	34	35	30	14	7	7	4	1	1	1
Oil		72	84	83	75	64	55	48	38	31	24	18	14
Gas		33	45	53	62	81	91	95	86	90	104	114	116
Nuclear		6	7	8	9	11	16	27	42	50	47	38	32
Biomass		6	8	8	13	17	20	23	31	40	50	62	68
Other Renewables		4	5	8	13	17	27	34	37	37	39	41	43
Total		159	182	194	207	220	224	233	240	252	264	275	274
Cumulative Resources Use	ZJ												
Coal		0.0	0.4	0.7	1.0	1.4	1.7	1.8	1.9	2.0	2.0	2.0	2.0
Oil		0.0	0.8	1.6	2.5	3.2	3.8	4.4	4.9	5.3	5.6	5.8	6.0
Gas		0.0	0.4	0.8	1.3	2.0	2.8	3.7	4.6	5.5	6.5	7.5	8.7
Cumulative CO2 Emissions	GtC	2.8	33.0	65.2	97.1	127.7	155.5	180.2	202.2	222.4	241.9	261.3	280.1
Carbon Sequestration	GtC												
Land Use	Million ha												
Cropland		381	380	378	376	374	370	366	361	357	354	353	352
Grasslands		760	765	772	779	787	809	832	862	893	912	918	924
Energy Biomass		0	0	0	10	23	33	44	45	46	45	42	39
Forest		1050	1059	1074	1094	1117	1133	1149	1154	1159	1167	1179	1192
Others		838	826	805	770	728	684	639	607	574	551	537	522
Total		3029	3029	3029	3029	3029	3029	3029	3029	3029	3029	3029	3029
Anthropogenic Emissions (standardized)													
Fossil Fuel CO2	GtC	2.83	3.20	3.32	3.27	3.14	2.73	2.46	2.16	2.00	1.99	2.01	1.96
Other CO2	GtC	0.00	0.00	-0.07	-0.13	-0.16	-0.14	-0.12	-0.08	-0.04	-0.04	-0.08	-0.12
Total CO2	GtC	2.83	3.20	3.25	3.14	2.98	2.59	2.34	2.08	1.96	1.94	1.93	1.84
CH4 total	MtCH4	73	74	71	69	66	62	59	56	53	53	53	53
N2O total	MtN2O-N	2.6	2.6	2.4	2.3	2.3	2.3	2.3	2.2	2.2	2.2	2.1	2.1
SOx total	MtS	22.7	17.0	9.1	2.2	0.6	-0.1	-0.3	0.0	0.5	0.9	1.5	1.9
HFC	MtC eq.	19	58	108	103	109	112	116	117	117	119	120	120
PFC	MtC eq.	18	13	12	10	8	7	7	7	7	7	7	6
SF6	MtC eq.	24	23	16	5	6	6	7	7	8	8	8	8
CO	MtCO	179	161	166	170	163	162	157	164	171	177	180	166
NMVOC	Mt	42	36	36	35	32	31	30	29	29	29	24	18
NOx	MtN	13	12	13	13	13	11	9	7	6	4	3	2

Emissions correlated to land-use change and deforestation were calculated by using AIM B1 land-use data.
a: Land-use taken from AIM-B1 emulation run.
b:CO2 emissions from fossil fuel and industrial processes (MESSAGE data).
c: CO2 emissions from land-use changes (AIM-B1 emulation run).
d: Non-energy related CH4 emissions were taken from AIM-B1 emulation run.
e: Non-energy related N2O emissions were taken from AIM-B1 emulation run.

Scenario B1High-MESSAGE REF		1990	2000	2010	2020	2030	2040	2050	2060	2070	2080	2090	2100
Population	Million	413	419	427	433	435	433	423	409	392	374	357	339
GNP/GDP (mex)	Trillion US$	1.1	0.8	1.0	2.0	4.9	7.9	9.9	11.9	13.8	15.6	17.4	19.0
GNP/GDP (ppp) Trillion (1990 prices)		2.6	2.2	2.5	3.6	5.7	7.9	9.9	11.9	13.8	15.6	17.4	19.0
Final Energy	EJ												
Non-commercial		0	0	0	0	0	0	0	0	0	0	0	0
Solids		9	5	3	2	1	0	0	0	0	0	0	0
Liquids		15	10	11	14	17	17	15	14	14	12	11	10
Gas		14	10	11	11	11	10	8	5	4	4	3	2
Electricity		6	5	6	7	10	12	14	15	16	17	18	18
Others		7	6	6	7	9	9	9	10	11	11	10	8
Total		50	36	37	42	48	48	47	45	45	43	41	38
Primary Energy	EJ												
Coal		19	13	9	6	4	3	3	3	4	5	5	4
Oil		20	14	13	16	18	16	13	11	10	8	8	7
Gas		27	20	23	28	35	40	44	49	50	46	37	27
Nuclear		1	1	1	0	0	0	0	0	1	1	2	2
Biomass		2	1	1	1	2	3	3	3	5	8	13	18
Other Renewables		1	1	2	3	4	5	6	8	9	11	12	13
Total		70	50	48	55	64	68	70	74	78	77	76	71
Cumulative Resources Use	ZJ												
Coal		0.0	0.2	0.3	0.4	0.4	0.5	0.5	0.6	0.6	0.6	0.7	0.7
Oil		0.0	0.2	0.3	0.5	0.6	0.8	1.0	1.1	1.2	1.3	1.4	1.5
Gas		0.0	0.3	0.5	0.7	1.0	1.3	1.8	2.2	2.8	3.4	3.9	4.4
Cumulative CO2 Emissions	GtC	1.3	12.3	21.3	29.4	37.4	45.3	52.5	59.7	67.7	75.8	83.0	88.7
Carbon Sequestration	GtC												
Land Use	Million ha												
Cropland		268	266	266	266	265	265	264	264	263	263	263	262
Grasslands		341	362	364	366	368	370	372	374	376	378	380	382
Energy Biomass		0	0	0	9	20	29	37	38	39	38	35	33
Forest		966	951	934	933	934	954	973	983	993	1005	1018	1031
Others		701	696	711	702	688	659	630	617	604	592	580	568
Total		2276	2276	2276	2276	2276	2276	2276	2276	2276	2276	2276	2276
Anthropogenic Emissions (standardized)													
Fossil Fuel CO2	GtC	1.30	0.91	0.84	0.87	0.96	0.95	0.96	0.98	1.01	0.93	0.79	0.62
Other CO2	GtC	0.00	0.00	0.04	-0.11	-0.12	-0.20	-0.28	-0.23	-0.17	-0.15	-0.14	-0.14
Total CO2	GtC	1.30	0.91	0.88	0.76	0.84	0.75	0.69	0.76	0.84	0.78	0.65	0.48
CH4 total	MtCH4	47	39	39	41	43	45	46	41	38	36	32	27
N2O total	MtN2O-N	0.6	0.6	0.6	0.6	0.6	0.6	0.6	0.6	0.6	0.6	0.6	0.6
SOx total	MtS	17.0	11.0	7.1	4.6	3.0	1.7	2.0	1.9	2.7	3.2	3.4	2.8
HFC	MtC eq.	0	4	9	15	20	24	26	27	27	27	27	26
PFC	MtC eq.	7	4	6	6	7	8	9	10	9	8	8	8
SF6	MtC eq.	8	6	4	7	7	8	9	9	7	6	4	4
CO	MtCO	69	41	44	49	58	63	60	53	49	40	35	33
NMVOC	Mt	16	13	15	18	21	23	25	27	30	30	23	15
NOx	MtN	5	3	3	4	4	4	4	3	2	2	1	1

Emissions correlated to land-use change and deforestation were calculated by using AIM B1 land-use data.
a: Land-use taken from AIM-B1 emulation run.
b:CO2 emissions from fossil fuel and industrial processes (MESSAGE data).
c: CO2 emissions from land-use changes (AIM-B1 emulation run).
d: Non-energy related CH4 emissions were taken from AIM-B1 emulation run.
e: Non-energy related N2O emissions were taken from AIM-B1 emulation run.

Scenario B1High-MESSAGE ASIA		**1990**	**2000**	**2010**	**2020**	**2030**	**2040**	**2050**	**2060**	**2070**	**2080**	**2090**	**2100**
Population	Million	2798	3261	3620	3937	4147	4238	4220	4085	3867	3589	3258	2882
GNP/GDP (mex)	Trillion US$	1.5	2.7	5.9	14.0	28.0	44.1	60.5	81.5	99.6	112.8	123.9	131.7
GNP/GDP (ppp) Trillion (1990 prices)		5.3	8.3	13.7	23.5	36.7	51.2	65.4	84.4	99.6	112.8	123.9	131.7
Final Energy	EJ												
Non-commercial		24	15	12	8	5	3	2	0	0	0	0	0
Solids		19	33	42	49	52	39	26	21	13	10	1	0
Liquids		13	22	37	55	74	94	111	113	106	85	68	58
Gas		2	3	7	11	13	16	18	17	17	16	17	16
Electricity		4	6	11	21	35	49	63	83	96	104	107	106
Others		1	2	6	11	21	31	44	59	75	86	95	91
Total		62	82	114	155	198	234	264	293	306	302	288	271
Primary Energy	EJ												
Coal		30	41	52	64	64	63	46	32	23	24	34	29
Oil		15	26	40	57	72	74	77	69	66	48	31	23
Gas		3	5	14	24	39	55	68	74	68	52	44	34
Nuclear		0	1	1	3	7	11	16	19	22	24	27	32
Biomass		24	22	25	31	41	53	64	75	82	95	98	99
Other Renewables		1	3	5	12	25	44	75	124	159	183	183	169
Total		74	98	138	191	247	300	346	393	419	425	417	386
Cumulative Resources Use	ZJ												
Coal		0.0	0.3	0.7	1.3	1.9	2.5	3.2	3.6	3.9	4.2	4.4	4.8
Oil		0.0	0.2	0.4	0.9	1.4	2.1	2.9	3.7	4.3	5.0	5.5	5.8
Gas		0.0	0.0	0.1	0.2	0.5	0.8	1.3	1.9	2.4	2.9	3.3	3.5
Cumulative CO2 Emissions	GtC	1.5	19.3	42.9	72.0	105.6	143.1	181.7	217.4	248.4	274.1	296.0	314.9
Carbon Sequestration	GtC												
Land Use	Million ha												
Cropland		438	435	435	435	434	434	434	434	433	433	433	433
Grasslands		608	606	609	612	614	617	619	622	625	628	631	634
Energy Biomass		0	0	0	12	28	40	52	54	55	54	50	46
Forest		535	525	518	535	551	557	563	578	593	604	610	616
Others		583	598	602	571	536	516	496	477	458	446	441	436
Total		2164	2164	2164	2164	2164	2164	2164	2164	2164	2164	2164	2164
Anthropogenic Emissions (standardized)													
Fossil Fuel CO2	GtC	1.15	1.78	2.49	3.29	3.78	4.01	3.84	3.44	3.04	2.46	2.23	1.79
Other CO2	GtC	0.37	0.26	0.20	-0.17	-0.18	-0.10	-0.03	-0.10	-0.17	-0.18	-0.14	-0.10
Total CO2	GtC	1.53	2.03	2.69	3.12	3.60	3.90	3.81	3.33	2.87	2.28	2.09	1.70
CH4 total	MtCH4	113	125	144	166	188	203	212	203	197	180	152	118
N2O total	MtN2O-N	2.3	2.6	2.3	2.3	2.3	2.3	2.2	2.2	2.2	2.2	2.2	2.2
SOx total	MtS	17.7	25.3	30.2	31.5	29.6	23.3	16.1	11.5	9.4	6.3	4.0	3.0
HFC	MtC eq.	0	4	11	20	34	56	93	90	85	79	72	64
PFC	MtC eq.	3	5	7	9	11	14	17	20	18	18	18	18
SF6	MtC eq.	4	7	11	17	23	27	30	32	26	19	16	16
CO	MtCO	235	270	338	432	543	645	764	792	800	775	773	751
NMVOC	Mt	33	37	47	60	65	68	75	69	61	48	40	34
NOx	MtN	7	9	12	16	18	20	19	17	14	12	10	9

Emissions correlated to land-use change and deforestation were calculated by using AIM B1 land-use data.
a: Land-use taken from AIM-B1 emulation run.
b:CO2 emissions from fossil fuel and industrial processes (MESSAGE data).
c: CO2 emissions from land-use changes (AIM-B1 emulation run).
d: Non-energy related CH4 emissions were taken from AIM-B1 emulation run.
e: Non-energy related N2O emissions were taken from AIM-B1 emulation run.

Scenario B1High-MESSAGE ALM		1990	2000	2010	2020	2030	2040	2050	2060	2070	2080	2090	2100
Population	Million	1192	1519	1875	2241	2557	2791	2980	3089	3115	3064	2934	2727
GNP/GDP (mex)	Trillion US$	1.9	2.7	4.3	9.3	18.3	30.5	45.7	62.7	77.5	91.1	102.9	112.7
GNP/GDP (ppp) Trillion (1990 prices)		3.8	5.1	7.4	12.8	21.2	33.3	47.9	63.4	77.5	91.1	102.9	112.7
Final Energy	EJ												
Non-commercial		14	10	9	8	6	4	3	0	0	0	0	0
Solids		1	4	4	4	4	4	3	3	1	1	1	1
Liquids		17	22	28	36	49	57	63	66	57	49	43	39
Gas		4	6	9	13	17	18	22	24	27	26	23	19
Electricity		3	5	7	13	21	33	46	63	76	87	94	96
Others		0	1	4	8	15	27	40	55	70	85	94	96
Total		39	48	61	82	111	143	178	211	231	248	254	251
Primary Energy	EJ												
Coal		5	4	5	5	5	6	7	10	13	17	21	26
Oil		21	30	35	41	49	52	52	44	31	20	13	8
Gas		8	12	20	34	60	87	100	114	96	86	68	67
Nuclear		0	0	0	1	2	2	3	5	7	10	13	16
Biomass		14	13	13	16	19	25	33	50	77	94	108	131
Other Renewables		2	4	7	11	17	35	67	103	143	174	194	179
Total		49	63	80	108	152	206	262	326	368	400	417	426
Cumulative Resources Use	ZJ												
Coal		0.0	0.0	0.1	0.1	0.2	0.2	0.3	0.4	0.5	0.6	0.8	1.0
Oil		0.0	0.3	0.6	0.9	1.3	1.8	2.3	2.8	3.3	3.6	3.8	3.9
Gas		0.0	0.1	0.2	0.4	0.8	1.4	2.3	3.3	4.6	5.6	6.5	7.3
Cumulative CO2 Emissions	GtC	1.4	17.8	36.3	55.4	74.7	96.0	120.9	147.8	173.5	195.6	214.4	231.3
Carbon Sequestration	GtC												
Land Use	Million ha												
Cropland		371	385	385	385	384	384	383	383	383	383	382	382
Grasslands		1680	1675	1686	1696	1705	1714	1723	1732	1741	1750	1760	1769
Energy Biomass		0	0	0	29	66	93	121	124	127	124	116	107
Forest		1745	1720	1704	1702	1732	1761	1790	1808	1825	1843	1862	1880
Others		1684	1700	1704	1669	1593	1527	1461	1432	1403	1379	1361	1342
Total		5480	5480	5480	5480	5480	5480	5480	5480	5480	5480	5480	5480
Anthropogenic Emissions (standardized)													
Fossil Fuel CO2	GtC	0.72	1.01	1.25	1.56	2.10	2.58	2.85	2.99	2.55	2.26	1.96	1.96
Other CO2	GtC	0.73	0.82	0.62	0.38	-0.19	-0.22	-0.25	-0.22	-0.19	-0.20	-0.24	-0.29
Total CO2	GtC	1.45	1.83	1.87	1.95	1.90	2.36	2.60	2.77	2.36	2.06	1.71	1.67
CH4 total	MtCH4	77	85	97	109	125	138	151	153	153	143	122	104
N2O total	MtN2O-N	1.2	1.3	0.7	0.6	0.5	0.5	0.5	0.5	0.5	0.5	0.5	0.5
SOx total	MtS	10.5	12.8	12.7	12.5	12.0	11.8	11.6	10.4	8.5	7.2	5.9	4.8
HFC	MtC eq.	0	2	16	28	42	64	98	102	102	101	96	89
PFC	MtC eq.	4	4	5	6	7	8	9	11	11	11	12	12
SF6	MtC eq.	2	5	5	8	12	17	22	24	21	14	14	14
CO	MtCO	396	404	403	414	429	453	489	540	638	738	806	922
NMVOC	Mt	48	55	59	64	67	67	68	60	53	45	39	36
NOx	MtN	7	8	9	10	12	12	13	13	12	12	11	11

Emissions correlated to land-use change and deforestation were calculated by using AIM B1 land-use data.

a: Land-use taken from AIM-B1 emulation run.

b:CO2 emissions from fossil fuel and industrial processes (MESSAGE data).

c: CO2 emissions from land-use changes (AIM-B1 emulation run).

d: Non-energy related CH4 emissions were taken from AIM-B1 emulation run.

e: Non-energy related N2O emissions were taken from AIM-B1 emulation run.

Scenario B1High-MiniCAM World		**1990**	**2000**	**2010**	**2020**	**2030**	**2040**	**2050**	**2060**	**2070**	**2080**	**2090**	**2100**
Population	Million	5293	6100	6874	7618	8122	8484	8703	8623	8430	8126	7621	7137
GNP/GDP (mex)	Trillion US$	20.7	27.4	37.5	51.0	73.0	100.7	134.0	172.6	211.9	251.7	289.7	330.6
GNP/GDP (ppp) Trillion (1990 prices)		na	na	na	na	na	na	na	na	na	na	na	na
Final Energy	EJ												
Non-commercial		0	0	0	0	0	0	0	0	0	0	0	0
Solids		45	54	62	69	71	70	65	47	34	28	28	27
Liquids		121	118	117	119	113	118	132	152	170	187	197	208
Gas		52	59	64	67	64	58	50	55	59	63	65	67
Electricity		35	49	69	97	143	189	235	281	322	360	370	381
Others		0	0	16	47	73	107	148	167	182	194	201	209
Total		253	279	328	400	465	542	630	701	768	832	861	891
Primary Energy	EJ												
Coal		88	106	122	134	146	153	155	119	94	80	80	81
Oil		131	127	125	127	116	118	133	158	179	197	210	222
Gas		70	79	118	186	261	310	334	371	399	419	429	438
Nuclear		24	23	26	31	35	41	48	49	46	41	43	46
Biomass		0	5	11	20	36	54	74	81	81	75	75	74
Other Renewables		24	24	28	33	41	68	113	169	224	278	265	253
Total		336	365	430	532	635	743	857	946	1023	1089	1101	1114
Cumulative Resources Use	ZJ												
Coal		0.1	1.1	2.3	3.5	5.0	6.4	8.0	9.2	10.3	11.2	12.0	12.8
Oil		0.1	1.4	2.7	3.9	5.1	6.3	7.5	9.1	10.8	12.6	14.7	16.7
Gas		0.1	0.8	1.9	3.3	5.8	8.6	11.7	15.4	19.2	23.3	27.5	31.8
Cumulative CO2 Emissions	GtC	7.1	82.4	166.4	262.9	373.9	492.9	614.2	736.9	859.7	979.5	1093.6	1201.1
Carbon Sequestration	GtC												
Land Use	Million ha												
Cropland		1472	1467	1455	1435	1373	1284	1167	992	833	691	592	493
Grasslands		3209	3349	3591	3935	4239	4439	4537	4270	3979	3665	3479	3293
Energy Biomass		0	0	6	18	47	83	125	125	109	79	70	61
Forest		4173	4215	4144	3963	3720	3568	3506	3931	4339	4731	4805	4879
Others		4310	4134	3968	3813	3785	3790	3829	3847	3904	3999	4219	4438
Total		13164	13164	13164	13164	13164	13164	13164	13164	13164	13164	13164	13164
Anthropogenic Emissions (standardized)													
Fossil Fuel CO2	GtC	5.99	6.90	7.81	9.15	10.49	11.28	11.93	12.00	12.15	12.40	12.81	13.22
Other CO2	GtC	1.11	1.07	1.01	1.34	1.22	0.82	0.25	0.36	0.05	-0.66	-1.73	-2.79
Total CO2	GtC	7.10	7.97	8.82	10.48	11.70	12.09	12.18	12.35	12.21	11.74	11.08	10.43
CH4 total	MtCH4	310	323	352	398	443	489	521	524	529	536	558	579
N2O total	MtN2O-N	6.7	7.0	8.1	9.5	11.0	12.9	14.6	15.3	16.1	16.8	18.2	19.7
SOx total	MtS	70.9	69.0	73.4	75.6	67.0	56.1	42.8	29.8	21.4	17.6	17.7	17.9
CFC/HFC/HCFC	MtC eq.	1672	883	784	291	257	298	338	337	333	327	315	299
PFC	MtC eq.	32	25	29	32	33	37	42	47	45	43	46	45
SF6	MtC eq.	38	40	36	37	47	58	68	71	62	46	43	43
CO	MtCO												
NMVOC	Mt												
NOx	MtN												

Scenario B1High-MiniCAM OECD90		1990	2000	2010	2020	2030	2040	2050	2060	2070	2080	2090	2100
Population	Million	838	908	965	1007	1024	1066	1081	1084	1090	1098	1105	1112
GNP/GDP (mex)	Trillion US$	16.3	20.5	25.1	30.1	32.0	38.8	44.6	50.2	56.4	63.1	71.0	79.3
GNP/GDP (ppp) Trillion (1990 prices)		na	na	na	na	na	na	na	na	na	na	na	na
Final Energy	EJ												
Non-commercial		0	0	0	0	0	0	0	0	0	0	0	0
Solids		10	12	10	7	6	4	4	3	2	2	2	2
Liquids		72	66	57	45	37	20	16	17	18	19	20	21
Gas		27	33	34	32	29	20	14	15	16	17	18	19
Electricity		22	25	27	26	25	24	23	26	29	32	33	34
Others		0	0	12	37	43	61	73	75	77	80	83	85
Total		130	136	141	146	140	129	131	135	141	150	155	160
Primary Energy	EJ												
Coal		40	41	36	23	23	22	24	20	16	11	12	13
Oil		76	71	61	47	38	18	14	15	17	20	22	23
Gas		34	42	60	87	89	91	89	92	97	101	104	107
Nuclear		20	14	10	7	6	4	4	4	4	3	4	4
Biomass		0	2	4	7	8	13	17	17	16	15	15	15
Other Renewables		12	11	11	10	10	10	13	17	21	26	26	25
Total		182	181	180	181	173	159	160	165	171	177	183	188
Cumulative Resources Use	ZJ												
Coal		0.0	0.5	0.8	1.1	1.2	1.6	1.8	2.0	2.2	2.3	2.5	2.6
Oil		0.1	0.8	1.4	2.0	2.2	2.6	2.7	2.9	3.1	3.2	3.5	3.7
Gas		0.0	0.4	1.0	1.7	2.1	3.5	4.4	5.3	6.2	7.2	8.2	9.3
Cumulative CO2 Emissions	GtC	2.8	33.0	65.3	97.4	128.3	156.1	181.0	204.9	228.5	251.3	272.7	292.5
Carbon Sequestration	GtC												
Land Use	Million ha												
Cropland		408	411	406	394	382	340	303	258	219	187	156	125
Grasslands		796	822	866	929	955	1014	1028	983	937	890	841	793
Energy Biomass		0	0	4	13	17	30	42	38	34	28	27	26
Forest		921	931	923	897	881	850	853	941	1017	1079	1110	1140
Others		998	959	924	890	888	888	897	902	916	938	989	1039
Total		3123	3123	3123	3123	3123	3123	3123	3123	3123	3123	3123	3123
Anthropogenic Emissions (standardized)													
Fossil Fuel CO2	GtC	2.83	3.20	3.15	2.99	2.85	2.54	2.46	2.44	2.44	2.47	2.56	2.65
Other CO2	GtC	0.00	0.00	0.11	0.18	0.15	0.03	-0.07	-0.05	-0.11	-0.24	-0.50	-0.75
Total CO2	GtC	2.83	3.20	3.26	3.17	3.01	2.57	2.39	2.39	2.33	2.22	2.06	1.90
CH4 total	MtCH4	73	74	82	91	96	114	127	137	143	145	153	161
N2O total	MtN2O-N	2.6	2.6	2.8	3.0	3.2	3.7	4.0	4.2	4.4	4.6	4.8	5.0
SOx total	MtS	22.7	17.0	12.1	5.1	3.6	2.6	2.2	2.1	2.2	2.4	2.7	3.0
HFC	MtC eq.	19	58	108	103	109	112	116	117	117	119	120	120
PFC	MtC eq.	18	13	12	10	8	7	7	7	7	7	7	6
SF6	MtC eq.	24	23	16	5	6	6	7	7	8	8	8	8
CO	MtCO												
NMVOC	Mt												
NOx	MtN												

Scenario B1High-MiniCAM REF		**1990**	**2000**	**2010**	**2020**	**2030**	**2040**	**2050**	**2060**	**2070**	**2080**	**2090**	**2100**
Population	Million	428	425	426	433	434	431	423	408	392	374	357	340
GNP/GDP (mex)	Trillion US$	1.1	1.1	1.4	2.0	3.0	4.1	5.3	7.1	9.2	11.4	13.4	15.5
GNP/GDP (ppp) Trillion (1990 prices)		na	na	na	na	na	na	na	na	na	na	na	na
Final Energy	EJ												
Non-commercial		0	0	0	0	0	0	0	0	0	0	0	0
Solids		13	10	8	7	5	4	4	2	2	1	1	1
Liquids		18	11	8	8	8	7	7	7	6	6	6	6
Gas		19	15	13	13	11	9	7	6	6	5	5	5
Electricity		6	8	11	16	20	23	24	24	23	23	22	21
Others		0	0	0	1	1	1	2	1	1	1	1	1
Total		56	44	40	44	45	44	43	41	39	36	35	33
Primary Energy	EJ												
Coal		18	17	16	16	16	16	17	9	6	5	5	5
Oil		20	13	10	10	8	7	7	7	7	7	7	6
Gas		26	20	20	28	31	30	24	22	19	17	16	15
Nuclear		3	4	5	6	6	5	5	4	3	2	2	2
Biomass		0	1	1	2	4	4	5	4	4	3	3	3
Other Renewables		3	3	4	5	6	8	13	16	18	19	17	16
Total		70	57	56	67	69	70	69	62	57	53	50	46
Cumulative Resources Use	ZJ												
Coal		0.0	0.2	0.4	0.5	0.7	0.8	1.0	1.1	1.2	1.2	1.3	1.3
Oil		0.0	0.2	0.3	0.4	0.5	0.6	0.6	0.7	0.8	0.8	0.9	1.0
Gas		0.0	0.2	0.5	0.7	1.0	1.3	1.6	1.8	2.0	2.2	2.3	2.5
Cumulative CO2 Emissions	GtC	1.3	12.3	21.2	30.7	41.0	51.1	60.1	67.9	74.8	80.0	83.5	85.2
Carbon Sequestration	GtC												
Land Use	Million ha												
Cropland		284	293	300	304	297	282	259	211	169	131	115	99
Grasslands		395	410	453	523	588	631	652	576	502	428	417	406
Energy Biomass		0	0	1	2	5	7	7	3	1	0	0	0
Forest		1007	1016	997	948	889	857	851	973	1081	1177	1163	1150
Others		691	657	628	602	598	600	609	613	624	642	682	722
Total		2377	2377	2377	2377	2377	2377	2377	2377	2377	2377	2377	2377
Anthropogenic Emissions (standardized)													
Fossil Fuel CO2	GtC	1.30	0.91	0.84	0.95	0.95	0.91	0.82	0.65	0.53	0.46	0.44	0.43
Other CO2	GtC	0.00	0.00	0.03	0.08	0.09	0.07	0.00	0.11	0.09	-0.04	-0.19	-0.34
Total CO2	GtC	1.30	0.91	0.87	1.02	1.04	0.97	0.82	0.75	0.62	0.42	0.26	0.09
CH4 total	MtCH4	47	39	47	63	72	82	92	86	86	90	100	111
N2O total	MtN2O-N	0.6	0.6	0.7	0.9	1.2	1.4	1.6	1.6	1.6	1.5	1.7	1.9
SOx total	MtS	17.0	11.0	9.9	9.5	9.1	8.1	6.5	3.8	2.1	1.3	1.4	1.6
HFC	MtC eq.	0	4	9	15	20	24	26	27	27	27	27	26
PFC	MtC eq.	7	4	6	6	7	8	9	10	9	8	8	8
SF6	MtC eq.	8	6	4	7	7	8	9	9	7	6	4	4
CO	MtCO												
NMVOC	Mt												
NOx	MtN												

Scenario B1High-MiniCAM ASIA		**1990**	**2000**	**2010**	**2020**	**2030**	**2040**	**2050**	**2060**	**2070**	**2080**	**2090**	**2100**
Population	Million	2790	3226	3608	3937	4115	4210	4219	4062	3852	3589	3245	2919
GNP/GDP (mex)	Trillion US$	1.4	3.1	6.4	11.5	21.7	34.1	48.6	63.3	77.6	91.5	103.9	117.1
GNP/GDP (ppp) Trillion (1990 prices)		na	na	na	na	na	na	na	na	na	na	na	na
Final Energy	EJ												
Non-commercial		0	0	0	0	0	0	0	0	0	0	0	0
Solids		20	29	40	51	53	53	49	35	25	20	19	18
Liquids		14	19	25	33	39	47	54	60	66	70	73	75
Gas		2	5	8	11	13	13	13	14	15	16	16	16
Electricity		4	11	22	40	68	96	123	143	161	176	179	182
Others		0	0	2	6	12	21	33	38	41	43	43	44
Total		40	64	97	140	185	229	271	290	308	326	330	335
Primary Energy	EJ												
Coal		26	43	62	84	92	93	87	65	49	41	40	39
Oil		16	20	27	36	41	47	55	63	70	74	78	81
Gas		3	9	23	45	87	115	129	140	146	147	147	147
Nuclear		1	4	8	13	17	21	27	26	24	21	21	22
Biomass		0	2	5	9	16	24	32	35	34	31	30	29
Other Renewables		3	4	5	7	11	26	52	80	107	132	124	116
Total		49	81	129	194	264	327	381	408	430	446	440	434
Cumulative Resources Use	ZJ												
Coal		0.0	0.4	1.0	1.6	2.6	3.5	4.4	5.1	5.6	6.1	6.5	6.9
Oil		0.0	0.2	0.5	0.8	1.1	1.6	2.1	2.7	3.4	4.1	4.9	5.6
Gas		0.0	0.1	0.3	0.6	1.3	2.3	3.5	4.9	6.3	7.8	9.2	10.7
Cumulative CO2 Emissions	GtC	1.5	19.3	43.3	76.5	119.4	169.3	222.3	274.3	323.6	370.4	415.4	459.2
Carbon Sequestration	GtC												
Land Use	Million ha												
Cropland		389	399	406	408	395	373	343	296	254	215	186	158
Grasslands		508	524	554	599	636	663	680	666	646	620	593	565
Energy Biomass		0	0	1	4	16	30	44	44	36	22	16	10
Forest		1168	1144	1107	1058	1020	999	995	1056	1126	1207	1268	1329
Others		664	633	605	581	578	580	587	591	602	620	659	697
Total		2729	2700	2674	2650	2645	2645	2650	2653	2664	2683	2721	2759
Anthropogenic Emissions (standardized)													
Fossil Fuel CO2	GtC	1.15	1.78	2.58	3.61	4.51	5.06	5.27	4.99	4.79	4.68	4.71	4.75
Other CO2	GtC	0.37	0.26	0.19	0.24	0.23	0.18	0.08	0.07	0.01	-0.12	-0.27	-0.43
Total CO2	GtC	1.53	2.03	2.77	3.86	4.74	5.24	5.36	5.06	4.79	4.56	4.44	4.31
CH4 total	MtCH4	113	125	133	144	155	160	161	157	155	154	151	148
N2O total	MtN2O-N	2.3	2.6	2.9	3.3	3.8	4.2	4.7	4.9	5.1	5.4	5.8	6.3
SOx total	MtS	17.7	25.3	34.7	43.5	37.6	29.4	19.1	11.7	7.1	5.3	5.3	5.3
HFC	MtC eq.	0	4	11	20	34	56	93	90	85	79	72	64
PFC	MtC eq.	3	5	7	9	11	14	17	20	18	18	18	18
SF6	MtC eq.	4	7	11	17	23	27	30	32	26	19	16	16
CO	MtCO												
NMVOC	Mt												
NOx	MtN												

Scenario B1High-MiniCAM ALM		1990	2000	2010	2020	2030	2040	2050	2060	2070	2080	2090	2100
Population	Million	1236	1541	1876	2241	2531	2778	2980	3068	3096	3064	2913	2766
GNP/GDP (mex)	Trillion US$	1.9	2.8	4.6	7.4	14.3	23.7	35.6	52.0	68.6	85.6	101.5	118.7
GNP/GDP (ppp) Trillion (1990 prices)		na	na	na	na	na	na	na	na	na	na	na	na
Final Energy	EJ												
Non-commercial		0	0	0	0	0	0	0	0	0	0	0	0
Solids		2	3	4	6	7	8	9	7	6	5	5	5
Liquids		17	21	26	33	37	44	55	68	80	91	99	106
Gas		5	7	9	12	15	17	16	19	22	25	26	27
Electricity		3	6	10	15	30	47	65	88	109	129	137	145
Others		0	0	1	4	11	24	40	52	63	70	75	79
Total		27	36	50	70	101	139	185	235	280	321	341	362
Primary Energy	EJ												
Coal		4	6	8	11	16	22	28	26	24	22	23	25
Oil		20	22	27	34	38	46	57	72	85	96	104	112
Gas		7	9	16	27	52	74	93	117	137	153	161	169
Nuclear		0	2	3	5	7	10	13	15	15	15	16	17
Biomass		0	1	2	3	7	13	20	25	27	26	27	27
Other Renewables		5	6	8	12	15	23	36	57	78	101	98	96
Total		35	46	65	91	136	188	246	311	366	413	429	446
Cumulative Resources Use	ZJ												
Coal		0.0	0.1	0.1	0.2	0.4	0.6	0.8	1.1	1.3	1.5	1.8	2.0
Oil		0.0	0.2	0.5	0.8	1.2	1.6	2.1	2.8	3.6	4.4	5.5	6.5
Gas		0.0	0.1	0.2	0.4	0.9	1.5	2.3	3.5	4.7	6.1	7.7	9.3
Cumulative CO2 Emissions	GtC	1.4	17.8	36.6	58.4	85.1	116.3	150.9	189.8	232.8	277.7	322.0	364.2
Carbon Sequestration	GtC												
Land Use	Million ha												
Cropland		391	363	342	329	311	289	263	226	191	157	134	111
Grasslands		1510	1594	1719	1884	2033	2131	2177	2045	1895	1727	1628	1528
Energy Biomass		0	0	0	0	5	16	33	39	38	30	27	25
Forest		3641	3592	3492	3343	3203	3111	3065	3227	3415	3629	3753	3876
Others		1957	1884	1812	1740	1724	1722	1736	1740	1762	1799	1890	1980
Total		7499	7432	7365	7296	7276	7269	7273	7277	7300	7343	7431	7520
Anthropogenic Emissions (standardized)													
Fossil Fuel CO2	GtC	0.72	1.01	1.24	1.59	2.18	2.78	3.38	3.92	4.40	4.79	5.09	5.40
Other CO2	GtC	0.73	0.82	0.68	0.84	0.74	0.54	0.23	0.23	0.06	-0.27	-0.77	-1.27
Total CO2	GtC	1.45	1.83	1.93	2.43	2.92	3.31	3.62	4.15	4.46	4.53	4.32	4.13
CH4 total	MtCH4	77	85	90	101	120	134	142	143	145	147	154	160
N2O total	MtN2O-N	1.2	1.3	1.7	2.2	2.9	3.6	4.3	4.6	5.0	5.3	5.9	6.5
SOx total	MtS	10.5	12.8	13.7	14.5	13.7	12.9	12.1	9.1	7.0	5.6	5.3	5.0
HFC	MtC eq.	0	2	16	28	42	64	98	102	102	101	96	89
PFC	MtC eq.	4	4	5	6	7	8	9	11	11	11	12	12
SF6	MtC eq.	2	5	5	8	12	17	22	24	21	14	14	14
CO	MtCO												
NMVOC	Mt												
NOx	MtN												

Scenario B2-AIM World		**1990**	**2000**	**2010**	**2020**	**2030**	**2040**	**2050**	**2060**	**2070**	**2080**	**2090**	**2100**
Population	Million	5262	6091	6851	7612	8372	8855	9367	9638	9917	10129	10271	10414
GNP/GDP (mex)	Trillion US$	20.9	28.2	36.4	48.4	66.9	86.3	111.2	133.7	160.7	186.9	210.5	237.1
GNP/GDP (ppp) Trillion (1990 prices)													
Final Energy	EJ												
Non-commercial		50	38	34	30	26	18	13	0	0	0	0	0
Solids		36	43	54	66	77	82	88	84	80	76	72	69
Liquids		111	118	132	146	168	191	217	225	234	241	246	251
Gas		51	54	74	97	123	148	179	193	209	221	228	236
Electricity		38	50	65	81	105	135	175	215	266	311	344	381
Others													
Total		285	304	359	420	499	578	671	730	795	850	892	936
Primary Energy	EJ												
Coal		93	105	141	180	214	242	278	272	268	262	254	247
Oil		143	160	175	185	188	178	173	154	139	126	113	102
Gas		73	93	123	153	187	226	277	297	320	337	344	352
Nuclear		6	9	13	19	29	35	46	53	63	72	79	88
Biomass		50	48	38	38	53	79	120	144	173	198	215	233
Other Renewables		10	11	13	16	22	32	48	79	132	186	222	264
Total		376	427	502	590	693	793	941	1000	1095	1181	1227	1285
Cumulative Resources Use	ZJ												
Coal		0.1	1.1	2.3	3.9	5.9	8.0	10.8	13.1	15.9	18.7	21.2	24.0
Oil		0.1	1.6	3.3	5.1	7.0	8.6	10.6	12.0	13.5	14.9	16.1	17.3
Gas		0.1	0.9	2.0	3.4	5.1	7.0	9.7	12.2	15.4	18.8	22.0	25.9
Cumulative CO2 Emissions	GtC	7.1	82.4	169.4	268.9	380.8	507.1	649.8	801.2	951.5	1100.3	1246.1	1388.4
Carbon Sequestraction	GtC												
Land Use	Million ha												
Cropland		1459	1481	1510	1540	1570	1598	1626	1656	1686	1718	1751	1784
Grasslands		3389	3409	3436	3454	3473	3508	3543	3577	3610	3641	3668	3696
Energy Biomass		0	0	0	14	92	163	288	312	337	341	324	307
Forest		4296	4252	4225	4251	4292	4322	4353	4380	4407	4441	4481	4522
Others		3805	3806	3778	3690	3522	3325	3139	3019	2904	2805	2721	2639
Total		12949	12949	12949	12949	12949	12949	12949	12949	12949	12949	12949	12949
Anthropogenic Emissions (standardized)													
Fossil Fuel CO2	GtC	5.99	6.90	8.51	10.21	11.83	13.24	14.96	14.82	14.75	14.56	14.24	13.93
Other CO2	GtC	1.11	1.07	0.92	0.26	0.07	0.12	0.24	0.24	0.26	0.19	0.17	0.10
Total CO2	GtC	7.10	7.97	9.43	10.47	11.90	13.36	15.20	15.07	15.01	14.76	14.41	14.04
CH4 total	MtCH4	310	323	358	391	429	453	482	485	489	485	475	465
N2O total	MtN2O-N	6.7	7.0	7.1	7.1	7.3	7.4	7.5	7.6	7.7	7.8	7.9	8.0
SOx total	MtS	70.9	69.0	78.3	77.6	69.8	51.9	44.2	38.4	34.2	32.8	33.3	33.8
CFC/HFC/HCFC	MtC eq.	1672	883	786	299	272	315	346	413	483	548	603	649
PFC	MtC eq.	32	25	42	55	70	88	107	120	128	130	127	121
SF6	MtC eq.	38	40	48	55	60	76	79	80	74	63	65	69
CO	MtCO	879	877	953	941	1002	1086	1180	1224	1272	1332	1407	1487
NMVOC	Mt	139	141	161	179	186	191	197	181	167	153	141	130
NOx	MtN	31	32	35	38	39	40	42	40	38	37	36	34

Scenario B2-AIM OECD90		1990	2000	2010	2020	2030	2040	2050	2060	2070	2080	2090	2100
Population	Million	859	916	942	968	994	985	976	964	952	943	935	928
GNP/GDP (mex)	Trillion US$	16.4	21.1	24.5	28.3	32.8	35.3	37.9	41.0	44.3	48.0	52.1	56.7
GNP/GDP (ppp) Trillion (1990 prices)													
Final Energy	EJ												
Non-commercial		6	0	0	0	0	0	0	0	0	0	0	0
Solids		10	12	12	12	11	10	9	8	7	6	5	4
Liquids		64	71	70	71	72	70	67	62	58	55	54	52
Gas		25	30	36	41	45	46	47	45	44	43	43	43
Electricity		22	27	31	34	38	40	43	48	53	59	63	68
Others													
Total		127	141	148	157	166	166	166	164	163	163	165	168
Primary Energy	EJ												
Coal		41	39	44	46	46	48	49	43	38	34	32	30
Oil		76	88	84	82	73	58	47	38	31	27	25	24
Gas		34	48	55	59	62	64	66	63	60	58	58	57
Nuclear		5	7	8	11	13	12	11	11	12	13	14	15
Biomass		6	1	0	1	6	11	23	27	33	37	40	44
Other Renewables		6	6	7	8	10	12	14	20	29	37	42	49
Total		167	189	198	206	210	204	208	202	202	206	212	219
Cumulative Resources Use	ZJ												
Coal		0.0	0.4	0.9	1.3	1.8	2.2	2.7	3.1	3.5	3.9	4.3	4.6
Oil		0.1	0.9	1.8	2.6	3.4	3.9	4.6	4.9	5.3	5.6	5.9	6.1
Gas		0.0	0.4	1.0	1.5	2.1	2.7	3.4	4.0	4.6	5.2	5.8	6.4
Cumulative CO2 Emissions	GtC	2.8	33.0	65.4	98.7	132.1	164.8	196.8	226.5	252.7	275.9	297.2	317.0
Carbon Sequestraction	GtC												
Land Use	Million ha												
Cropland		381	384	391	398	406	411	417	423	429	435	442	449
Grasslands		760	764	770	773	776	787	798	811	826	838	849	860
Energy Biomass		0	0	0	2	17	31	54	59	63	64	61	58
Forest		1050	1059	1069	1082	1095	1104	1113	1124	1136	1151	1171	1190
Others		838	822	799	774	734	689	647	609	574	538	504	471
Total		3029	3029	3029	3029	3029	3029	3029	3029	3029	3029	3029	3029
Anthropogenic Emissions (standardized)													
Fossil Fuel CO2	GtC	2.83	3.20	3.32	3.41	3.37	3.27	3.18	2.81	2.49	2.28	2.16	2.04
Other CO2	GtC	0.00	0.00	-0.02	-0.05	-0.06	-0.04	-0.02	-0.03	-0.05	-0.07	-0.11	-0.14
Total CO2	GtC	2.83	3.20	3.30	3.36	3.31	3.24	3.16	2.78	2.45	2.21	2.05	1.90
CH4 total	MtCH4	73	74	71	67	64	59	54	52	51	49	48	47
N2O total	MtN2O-N	2.6	2.6	2.6	2.6	2.6	2.6	2.6	2.5	2.5	2.5	2.5	2.5
SOx total	MtS	22.7	17.0	10.3	7.4	7.2	7.3	7.4	7.0	6.6	6.4	6.2	6.0
HFC	MtC eq.	19	57	105	99	102	101	102	101	100	98	98	97
PFC	MtC eq.	18	13	14	13	12	11	10	8	7	7	7	7
SF6	MtC eq.	24	23	25	23	17	13	13	11	10	11	10	10
CO	MtCO	179	161	174	186	203	210	216	215	214	216	221	227
NMVOC	Mt	42	36	34	32	29	25	21	15	10	7	6	6
NOx	MtN	13	12	11	10	9	8	6	6	5	5	5	5

Scenario B2-AIM REF		1990	2000	2010	2020	2030	2040	2050	2060	2070	2080	2090	2100
Population	Million	413	415	416	416	416	411	406	398	390	385	382	379
GNP/GDP (mex)	Trillion US$	1.1	1.0	1.4	1.9	2.7	4.2	6.5	8.0	10.0	11.7	12.9	14.3
GNP/GDP (ppp) Trillion (1990 prices)													
Final Energy	EJ												
Non-commercial		2	3	0	0	0	0	0	0	0	0	0	0
Solids		9	7	7	7	6	6	5	4	4	3	3	2
Liquids		19	10	9	8	8	9	10	10	10	10	10	9
Gas		19	13	18	22	25	30	36	36	37	37	35	34
Electricity		8	9	11	12	14	17	20	24	28	31	32	33
Others													
Total		58	42	45	49	53	61	71	75	80	81	80	79
Primary Energy	EJ												
Coal		18	15	17	19	19	19	20	19	19	18	16	15
Oil		22	15	13	11	10	9	8	7	6	5	4	3
Gas		26	22	28	31	33	38	44	44	44	43	41	39
Nuclear		1	1	2	3	4	5	6	6	7	7	7	8
Biomass		2	4	0	0	2	4	8	9	11	12	14	15
Other Renewables		1	1	1	2	2	3	6	9	14	19	21	23
Total		71	58	61	65	69	78	91	95	102	105	103	102
Cumulative Resources Use	ZJ												
Coal		0.0	0.2	0.3	0.5	0.7	0.9	1.1	1.3	1.5	1.7	1.8	2.0
Oil		0.0	0.2	0.3	0.4	0.5	0.6	0.7	0.8	0.9	0.9	1.0	1.0
Gas		0.0	0.3	0.5	0.8	1.1	1.5	1.9	2.3	2.7	3.2	3.6	4.0
Cumulative CO2 Emissions	GtC	1.3	12.3	21.9	31.4	40.5	50.3	61.1	72.3	83.2	93.4	102.6	111.1
Carbon Sequestraction	GtC												
Land Use	Million ha												
Cropland		268	269	274	280	285	291	296	302	308	314	320	327
Grasslands		341	362	366	368	370	373	377	380	383	386	389	392
Energy Biomass		0	0	0	1	6	10	18	20	21	22	20	19
Forest		966	951	939	944	947	944	941	940	939	941	945	949
Others		701	693	697	683	668	656	644	634	624	613	601	589
Total		2276	2276	2276	2276	2276	2276	2276	2276	2276	2276	2276	2276
Anthropogenic Emissions (standardized)													
Fossil Fuel CO2	GtC	1.30	0.91	1.01	1.05	1.06	1.11	1.17	1.13	1.09	1.02	0.91	0.82
Other CO2	GtC	0.00	0.00	0.00	-0.16	-0.12	-0.09	-0.03	-0.02	-0.02	-0.07	-0.02	-0.02
Total CO2	GtC	1.30	0.91	1.01	0.89	0.94	1.02	1.14	1.10	1.08	0.95	0.89	0.80
CH4 total	MtCH4	47	39	47	48	47	48	48	44	40	36	32	28
N2O total	MtN2O-N	0.6	0.6	0.6	0.6	0.6	0.6	0.6	0.6	0.6	0.7	0.7	0.7
SOx total	MtS	17.0	11.0	11.7	9.7	6.9	3.4	1.6	1.8	2.1	2.3	2.5	2.6
HFC	MtC eq.	0	4	9	15	21	25	25	26	26	26	27	27
PFC	MtC eq.	7	4	8	10	14	19	25	28	30	30	28	27
SF6	MtC eq.	7	6	5	7	9	12	15	16	15	14	14	14
CO	MtCO	69	41	40	40	42	45	48	50	51	53	55	57
NMVOC	Mt	16	13	13	13	12	12	12	11	10	9	9	8
NOx	MtN	5	3	3	3	3	3	2	2	2	2	1	1

Scenario B2-AIM ASIA		**1990**	**2000**	**2010**	**2020**	**2030**	**2040**	**2050**	**2060**	**2070**	**2080**	**2090**	**2100**
Population	Million	2798	3248	3603	3958	4312	4500	4696	4768	4842	4897	4932	4968
GNP/GDP (mex)	Trillion US$	1.5	3.5	6.5	12.1	22.7	31.8	44.5	54.4	66.4	78.3	89.2	101.6
GNP/GDP (ppp) Trillion (1990 prices)													
Final Energy	EJ												
Non-commercial		28	18	17	14	8	0	0	0	0	0	0	0
Solids		15	23	33	44	54	59	65	63	60	58	56	54
Liquids		11	17	27	37	51	62	74	76	79	81	82	83
Gas		2	4	8	17	31	41	55	62	71	79	85	91
Electricity		5	8	15	25	40	54	75	94	118	140	157	176
Others													
Total		61	69	100	137	184	222	268	298	330	358	381	404
Primary Energy	EJ												
Coal		30	44	69	98	125	142	161	157	154	151	147	144
Oil		17	26	41	53	63	61	59	48	39	33	29	25
Gas		4	8	17	33	56	76	103	115	128	138	145	153
Nuclear		0	1	2	4	9	14	20	24	29	33	37	41
Biomass		28	24	19	15	11	17	27	33	39	45	48	53
Other Renewables		1	2	2	3	6	10	18	31	56	82	99	120
Total		80	104	150	206	269	319	388	408	444	481	506	536
Cumulative Resources Use	ZJ												
Coal		0.0	0.4	1.0	1.8	2.9	4.1	5.8	7.1	8.7	10.3	11.8	13.4
Oil		0.0	0.2	0.6	1.0	1.6	2.1	2.8	3.2	3.7	4.1	4.4	4.7
Gas		0.0	0.1	0.2	0.4	0.9	1.5	2.5	3.4	4.6	6.0	7.4	9.0
Cumulative CO2 Emissions	GtC	1.5	19.3	45.1	80.6	127.3	185.2	253.4	327.3	401.4	475.5	549.2	622.4
Carbon Sequestraction	GtC												
Land Use	Million ha												
Cropland		438	439	448	457	466	475	484	494	503	514	524	535
Grasslands		608	607	611	615	618	623	627	631	636	640	644	647
Energy Biomass		0	0	0	3	21	37	65	70	76	77	73	69
Forest		535	525	518	533	548	553	558	563	567	571	575	580
Others		583	593	587	557	512	469	430	405	381	362	347	333
Total		2164	2164	2164	2164	2164	2164	2164	2164	2164	2164	2164	2164
Anthropogenic Emissions (standardized)													
Fossil Fuel CO2	GtC	1.15	1.78	2.82	4.00	5.26	6.13	7.13	7.15	7.17	7.15	7.09	7.03
Other CO2	GtC	0.37	0.26	0.29	0.01	0.06	0.13	0.25	0.25	0.24	0.25	0.25	0.26
Total CO2	GtC	1.53	2.03	3.11	4.00	5.33	6.26	7.38	7.40	7.42	7.40	7.35	7.29
CH4 total	MtCH4	113	125	148	176	209	224	240	241	243	239	230	221
N2O total	MtN2O-N	2.3	2.6	2.7	2.7	2.9	3.0	3.1	3.1	3.2	3.3	3.3	3.4
SOx total	MtS	17.7	25.3	40.1	44.9	36.2	19.0	9.8	10.0	10.3	10.5	10.6	10.8
HFC	MtC eq.	0	5	21	40	66	95	130	164	199	233	267	302
PFC	MtC eq.	3	5	14	22	30	38	46	51	54	54	53	51
SF6	MtC eq.	4	7	12	18	25	32	36	36	33	27	29	30
CO	MtCO	235	270	332	356	412	455	503	521	538	563	594	627
NMVOC	Mt	33	37	50	66	72	77	83	78	74	69	62	56
NOx	MtN	7	9	12	15	18	18	19	17	16	15	14	14

Scenario B2-AIM ALM		1990	2000	2010	2020	2030	2040	2050	2060	2070	2080	2090	2100
Population	Million	1192	1511	1891	2270	2649	2952	3289	3502	3729	3904	4020	4139
GNP/GDP (mex)	Trillion US$	1.9	2.7	4.1	6.0	8.6	13.9	22.3	29.7	39.6	48.9	56.2	64.4
GNP/GDP (ppp) Trillion (1990 prices)													
Final Energy	EJ												
Non-commercial		14	17	18	16	18	15	13	0	0	0	0	0
Solids		1	2	3	4	6	7	9	9	9	9	9	8
Liquids		17	20	25	29	36	49	66	75	86	95	100	106
Gas		4	7	12	17	22	30	42	49	56	62	65	68
Electricity		4	6	8	10	14	23	36	49	65	81	91	103
Others													
Total		40	52	66	77	96	126	166	191	221	246	265	285
Primary Energy	EJ												
Coal		5	7	11	17	24	34	48	52	57	59	58	58
Oil		27	32	37	39	43	50	59	61	63	61	55	49
Gas		9	15	23	31	37	49	64	76	89	97	100	103
Nuclear		0	0	0	1	2	4	9	12	15	19	21	24
Biomass		14	18	19	22	35	47	63	76	91	104	112	122
Other Renewables		2	2	3	3	5	7	11	19	33	49	59	72
Total		57	75	94	113	145	191	254	295	348	388	406	428
Cumulative Resources Use	ZJ												
Coal		0.0	0.1	0.2	0.3	0.5	0.8	1.2	1.6	2.2	2.8	3.3	1.7
Oil		0.0	0.3	0.7	1.1	1.5	1.9	2.5	3.0	3.7	4.3	4.8	6.7
Gas		0.0	0.1	0.3	0.6	0.9	1.3	1.9	2.6	3.4	4.3	5.3	12.5
Cumulative CO2 Emissions	GtC	1.4	17.8	37.0	58.2	80.9	106.7	138.5	175.0	214.2	255.6	297.1	337.9
Carbon Sequestraction	GtC												
Land Use	Million ha												
Cropland		371	389	397	405	413	421	429	437	446	455	464	474
Grasslands		1680	1676	1689	1699	1709	1725	1741	1754	1766	1777	1787	1797
Energy Biomass		0	0	0	7	48	85	151	163	176	179	170	161
Forest		1745	1717	1699	1692	1701	1721	1740	1752	1765	1777	1790	1803
Others		1684	1698	1695	1676	1608	1510	1419	1371	1325	1291	1268	1246
Total		5480	5480	5480	5480	5480	5480	5480	5480	5480	5480	5480	5480
Anthropogenic Emissions (standardized)													
Fossil Fuel CO2	GtC	0.72	1.01	1.36	1.76	2.14	2.73	3.48	3.73	3.99	4.11	4.07	4.04
Other CO2	GtC	0.73	0.82	0.65	0.46	0.18	0.12	0.03	0.05	0.08	0.08	0.04	0.00
Total CO2	GtC	1.45	1.83	2.01	2.22	2.32	2.84	3.52	3.78	4.07	4.20	4.12	4.04
CH4 total	MtCH4	77	85	93	100	108	123	139	147	156	162	165	168
N2O total	MtN2O-N	1.2	1.3	1.3	1.2	1.2	1.2	1.3	1.3	1.4	1.4	1.5	1.5
SOx total	MtS	10.5	12.8	13.2	12.6	16.4	19.2	22.4	16.6	12.3	10.7	11.0	11.3
HFC	MtC eq.	0	2	12	20	32	54	84	120	157	190	211	223
PFC	MtC eq.	4	4	7	10	14	20	27	33	37	39	39	37
SF6	MtC eq.	3	5	5	7	9	20	16	17	15	11	13	14
CO	MtCO	396	404	408	359	345	376	413	439	468	501	536	576
NMVOC	Mt	48	55	64	69	72	77	81	77	73	68	64	60
NOx	MtN	7	8	9	9	10	12	14	15	16	16	15	15

Scenario B2-ASF World		**1990**	**2000**	**2010**	**2020**	**2030**	**2040**	**2050**	**2060**	**2070**	**2080**	**2090**	**2100**
Population	Million	5256	6091	6870	7650	8277	9072	9367	9632	9771	10132	10341	10414
GNP/GDP (mex)	Trillion US$	20.3	26.1	33.8	43.3	54.8	76.3	85.7	104.8	116.3	158.9	214.6	237.9
GNP/GDP (ppp) Trillion (1990 prices)													
Final Energy	EJ												
Non-commercial													
Solids		53	55	66	71	64	57	49	49	48	48	48	48
Liquids		118	156	194	254	271	288	305	317	329	353	391	428
Gas		49	55	70	98	121	144	167	177	187	190	186	183
Electricity		41	50	67	89	116	144	171	185	199	216	234	253
Others													
Total		261	316	397	512	572	632	692	727	763	807	859	912
Primary Energy	EJ												
Coal		94	99	124	146	197	247	297	363	430	489	541	594
Oil		136	181	232	305	273	240	207	136	65	24	13	2
Gas		73	78	99	140	180	220	260	268	276	273	258	243
Nuclear		8	11	10	9	12	15	17	19	21	22	24	25
Biomass		0	0	4	8	28	49	69	88	108	125	141	156
Other Renewables		8	11	16	24	31	39	46	52	58	70	86	103
Total		319	380	486	633	721	809	897	927	957	1002	1062	1122
Cumulative Resources Use	ZJ												
Coal		0.0	1.0	2.2	3.5	5.3	7.5	10.2	13.5	17.5	22.1	27.3	33.0
Oil		0.0	1.7	3.8	6.5	9.4	11.9	14.1	15.8	16.7	17.1	17.2	17.3
Gas		0.0	0.8	1.7	2.9	4.5	6.5	8.9	11.6	14.3	17.1	19.7	22.2
Cumulative CO2 Emissions	GtC	7.1	82.4	172.2	285.7	423.0	574.4	733.3	897.2	1063.8	1234.4	1412.4	1599.2
Carbon Sequestration	GtC	-1.8	-1.8	-1.8	-1.9	-1.9	-1.6	-2.0	-1.6	-2.0	-1.9	-1.5	-1.8
Land Use	Million ha												
Cropland													
Grasslands													
Energy Biomass													
Forest													
Others													
Total													
Anthropogenic Emissions (standardized)													
Fossil Fuel CO2	GtC	5.99	6.90	8.85	11.48	13.60	14.51	15.42	15.93	16.43	17.13	18.03	18.93
Other CO2	GtC	1.11	1.07	1.12	1.25	1.15	1.00	0.85	0.60	0.35	0.22	0.21	0.20
Total CO2	GtC	7.10	7.97	9.97	12.73	14.75	15.51	16.27	16.52	16.78	17.35	18.24	19.13
CH4 total	MtCH4	310	323	367	414	459	493	527	545	563	579	593	607
N2O total	MtN2O-N	6.7	7.0	8.0	9.5	10.5	11.0	11.5	11.7	11.8	11.9	12.0	12.0
SOx total	MtS	70.9	69.0	77.2	101.3	108.1	97.4	86.8	68.1	49.4	38.7	36.0	33.3
CFC/HFC/HCFC	MtC eq.	1672	883	786	299	272	315	346	413	483	548	603	649
PFC	MtC eq.	32	25	42	55	70	88	107	120	128	130	127	121
SF6	MtC eq.	38	40	48	55	60	76	79	80	74	63	65	69
CO	MtCO	879	877	984	1077	1233	1275	1318	1361	1405	1485	1600	1716
NMVOC	Mt	139	141	157	180	199	205	211	221	230	248	276	304
NOx	MtN	31	32	41	52	61	64	66	66	66	68	72	77

Scenario B2-ASF OECD90		1990	2000	2010	2020	2030	2040	2050	2060	2070	2080	2090	2100
Population	Million	848	916	947	978	990	979	976	963	957	941	930	928
GNP/GDP (mex)	Trillion US$	15.4	19.1	23.1	27.0	30.9	37.0	39.4	43.8	46.2	54.3	64.2	68.0
GNP/GDP (ppp) Trillion (1990 prices)													
Final Energy	EJ												
Non-commercial													
Solids		12	12	12	12	13	13	13	13	12	12	12	12
Liquids		68	86	92	97	91	85	78	75	72	73	77	82
Gas		27	30	32	35	38	40	43	45	47	48	46	45
Electricity		23	27	33	39	41	43	45	47	49	53	58	62
Others													
Total		130	156	168	183	182	181	180	180	181	185	194	202
Primary Energy	EJ												
Coal		33	30	34	39	54	68	83	92	102	109	116	122
Oil		77	97	108	117	94	71	47	29	11	2	1	0
Gas		35	38	41	48	55	61	68	69	69	68	64	60
Nuclear		6	9	8	6	6	6	6	6	7	7	8	8
Biomass		0	0	1	2	8	13	18	19	21	23	25	28
Other Renewables		5	6	8	10	11	12	13	14	15	17	21	25
Total		156	180	200	223	227	230	234	229	225	226	235	244
Cumulative Resources Use	ZJ												
Coal		0.0	0.3	0.7	1.0	1.5	2.1	2.9	3.8	4.7	5.8	6.9	8.1
Oil		0.0	1.0	2.0	3.1	4.2	5.0	5.5	5.9	6.1	6.1	6.1	6.1
Gas		0.0	0.4	0.8	1.2	1.8	2.3	3.0	3.7	4.4	5.0	5.7	6.3
Cumulative CO2 Emissions	GtC	2.8	33.0	66.8	104.6	146.1	188.9	231.3	272.9	313.2	352.7	392.5	433.1
Carbon Sequestration	GtC												
Land Use	Million ha												
Cropland													
Grasslands													
Energy Biomass													
Forest													
Others													
Total													
Anthropogenic Emissions (standardized)													
Fossil Fuel CO2	GtC	2.83	3.20	3.57	3.99	4.30	4.26	4.22	4.09	3.97	3.94	4.02	4.10
Other CO2	GtC	0.00	0.00	0.00	0.00	0.00	0.00	0.00	0.00	0.00	0.00	0.00	0.00
Total CO2	GtC	2.83	3.20	3.57	3.99	4.30	4.26	4.22	4.09	3.97	3.94	4.02	4.10
CH4 total	MtCH4	73	74	77	81	85	87	89	90	91	93	96	99
N2O total	MtN2O-N	2.6	2.6	2.7	3.0	3.1	3.0	2.9	2.8	2.8	2.7	2.6	2.5
SOx total	MtS	22.7	17.0	8.0	8.5	8.7	8.5	8.3	7.5	6.8	6.7	7.4	8.1
HFC	MtC eq.	19	57	105	99	102	101	102	101	100	98	98	97
PFC	MtC eq.	18	13	14	13	12	11	10	8	7	7	7	7
SF6	MtC eq.	24	23	25	23	17	13	13	11	10	11	10	10
CO	MtCO	179	161	167	170	161	135	110	105	100	102	110	119
NMVOC	Mt	42	36	40	43	43	39	36	35	34	35	37	40
NOx	MtN	13	12	14	16	16	15	14	13	12	12	12	13

Scenario B2-ASF REF		1990	2000	2010	2020	2030	2040	2050	2060	2070	2080	2090	2100
Population	Million	416	415	416	417	415	408	406	398	394	385	380	379
GNP/GDP (mex)	Trillion US$	1.0	0.9	1.1	1.6	2.3	3.5	4.1	5.1	5.7	7.9	10.8	11.9
GNP/GDP (ppp) Trillion (1990 prices)													
Final Energy	EJ												
Non-commercial													
Solids		18	11	11	10	9	8	7	6	5	4	4	3
Liquids		16	11	11	13	14	14	15	15	15	16	18	19
Gas		15	11	14	17	19	21	23	22	21	20	18	16
Electricity		8	7	9	11	11	11	11	11	11	10	10	10
Others													
Total		56	39	44	51	52	54	55	53	52	50	50	49
Primary Energy	EJ												
Coal		23	12	14	15	16	16	16	14	11	12	15	19
Oil		18	12	13	16	16	17	17	14	10	6	3	0
Gas		26	21	24	30	30	29	29	27	25	23	21	18
Nuclear		1	1	1	1	1	1	1	1	1	1	1	1
Biomass		0	0	1	0	0	0	0	5	9	12	13	15
Other Renewables		1	1	2	2	2	3	3	3	3	3	4	4
Total		68	48	55	64	65	66	67	63	59	57	57	58
Cumulative Resources Use	ZJ												
Coal		0.0	0.2	0.3	0.5	0.6	0.8	0.9	1.1	1.2	1.3	1.5	1.6
Oil		0.0	0.2	0.3	0.4	0.6	0.8	0.9	1.1	1.2	1.3	1.3	1.3
Gas		0.0	0.3	0.5	0.8	1.1	1.3	1.6	1.9	2.2	2.4	2.6	2.8
Cumulative CO2 Emissions	GtC	1.3	12.3	22.0	33.0	45.0	56.7	67.7	77.8	87.0	95.3	103.4	111.5
Carbon Sequestration	GtC												
Land Use	Million ha												
Cropland													
Grasslands													
Energy Biomass													
Forest													
Others													
Total													
Anthropogenic Emissions (standardized)													
Fossil Fuel CO2	GtC	1.30	0.91	1.03	1.17	1.21	1.14	1.06	0.96	0.86	0.81	0.81	0.81
Other CO2	GtC	0.00	0.00	0.00	0.00	0.00	0.00	0.00	0.00	0.00	0.00	0.00	0.00
Total CO2	GtC	1.30	0.91	1.03	1.17	1.21	1.14	1.06	0.96	0.86	0.81	0.81	0.81
CH4 total	MtCH4	47	39	41	46	55	63	71	76	82	87	91	95
N2O total	MtN2O-N	0.6	0.6	0.6	0.7	0.7	0.7	0.7	0.7	0.6	0.6	0.6	0.6
SOx total	MtS	17.0	11.0	10.9	10.9	9.6	7.5	5.3	4.5	3.6	2.8	1.9	1.0
HFC	MtC eq.	0	4	9	15	21	25	25	26	26	26	27	27
PFC	MtC eq.	7	4	8	10	14	19	25	28	30	30	28	27
SF6	MtC eq.	7	6	5	7	9	12	15	16	15	14	14	14
CO	MtCO	69	41	40	39	45	46	48	49	51	55	61	67
NMVOC	Mt	16	13	15	20	25	26	27	25	22	22	25	27
NOx	MtN	5	3	3	4	4	4	4	3	3	3	3	3

Scenario B2-ASF ASIA		**1990**	**2000**	**2010**	**2020**	**2030**	**2040**	**2050**	**2060**	**2070**	**2080**	**2090**	**2100**
Population	Million	2776	3248	3617	3986	4270	4584	4696	4769	4806	4899	4951	4968
GNP/GDP (mex)	Trillion US$	1.5	2.5	4.1	6.6	9.7	16.5	19.7	26.3	30.4	46.2	67.6	76.7
GNP/GDP (ppp) Trillion (1990 prices)													
Final Energy	EJ												
Non-commercial													
Solids		22	31	41	46	38	29	21	20	19	19	18	17
Liquids		14	26	43	72	83	95	106	115	124	138	156	174
Gas		2	5	10	22	32	42	51	55	60	61	59	57
Electricity		6	10	16	24	43	62	81	86	92	97	103	108
Others													
Total		43	72	110	163	195	227	259	277	295	314	335	356
Primary Energy	EJ												
Coal		33	49	65	77	101	126	151	184	218	245	265	286
Oil		17	34	55	88	79	71	63	38	14	1	1	0
Gas		4	7	15	29	50	71	92	96	100	97	89	80
Nuclear		0	1	1	2	4	6	9	9	10	11	11	11
Biomass		0	0	1	3	7	12	17	18	20	22	24	26
Other Renewables		1	2	4	6	11	15	19	22	24	29	35	42
Total		55	92	140	205	253	302	351	368	386	404	425	445
Cumulative Resources Use	ZJ												
Coal		0.0	0.4	1.0	1.7	2.6	3.8	5.2	6.9	8.9	11.2	13.8	16.5
Oil		0.0	0.3	0.7	1.4	2.3	3.0	3.7	4.2	4.4	4.4	4.4	4.4
Gas		0.0	0.1	0.2	0.4	0.8	1.4	2.2	3.2	4.1	5.1	6.1	6.9
Cumulative CO2 Emissions	GtC	1.5	19.3	44.9	81.9	130.7	188.3	252.6	322.3	395.8	473.1	554.2	639.1
Carbon Sequestration	GtC												
Land Use	Million ha												
Cropland													
Grasslands													
Energy Biomass													
Forest													
Others													
Total													
Anthropogenic Emissions (standardized)													
Fossil Fuel CO2	GtC	1.15	1.78	2.73	3.94	5.08	5.84	6.59	7.03	7.47	7.89	8.27	8.66
Other CO2	GtC	0.37	0.26	0.35	0.40	0.34	0.26	0.18	0.13	0.07	0.04	0.02	0.01
Total CO2	GtC	1.53	2.03	3.07	4.34	5.42	6.10	6.77	7.16	7.54	7.92	8.30	8.67
CH4 total	MtCH4	113	125	144	159	176	191	206	212	217	220	221	221
N2O total	MtN2O-N	2.3	2.6	3.2	4.0	4.5	4.9	5.2	5.3	5.4	5.5	5.6	5.6
SOx total	MtS	17.7	25.3	39.6	56.4	59.3	52.1	45.0	31.7	18.4	11.0	9.3	7.7
HFC	MtC eq.	0	5	21	40	66	95	130	164	199	233	267	302
PFC	MtC eq.	3	5	14	22	30	38	46	51	54	54	53	51
SF6	MtC eq.	4	7	12	18	25	32	36	36	33	27	29	30
CO	MtCO	235	270	326	369	438	470	502	533	565	611	672	734
NMVOC	Mt	33	37	41	46	50	52	54	57	60	64	70	76
NOx	MtN	7	9	13	18	22	24	27	27	28	29	30	32

Scenario B2-ASF ALM		**1990**	**2000**	**2010**	**2020**	**2030**	**2040**	**2050**	**2060**	**2070**	**2080**	**2090**	**2100**
Population	Million	1216	1511	1890	2269	2603	3100	3289	3502	3615	3908	4080	4139
GNP/GDP (mex)	Trillion US$	2.4	3.6	5.4	8.2	11.8	19.2	22.6	29.7	34.0	50.5	72.1	81.2
GNP/GDP (ppp) Trillion (1990 prices)													
Final Energy	EJ												
Non-commercial													
Solids		1	1	2	3	5	7	9	10	11	13	14	15
Liquids		21	33	47	72	83	94	106	112	117	127	139	152
Gas		5	8	14	25	33	41	49	54	59	62	63	65
Electricity		4	6	10	16	22	28	35	41	48	55	64	73
Others													
Total		31	49	74	115	143	171	198	217	235	257	281	305
Primary Energy	EJ												
Coal		4	7	10	15	26	37	48	73	99	123	145	167
Oil		25	38	56	85	83	81	79	54	30	14	8	1
Gas		8	12	19	32	45	58	71	77	82	84	84	84
Nuclear		0	0	0	1	1	1	2	2	3	3	4	4
Biomass		0	0	2	3	13	23	34	46	58	69	79	88
Other Renewables		1	2	4	5	7	9	12	14	16	20	26	31
Total		39	60	91	141	176	210	245	267	288	314	345	375
Cumulative Resources Use	ZJ												
Coal		0.0	0.1	0.2	0.3	0.5	0.8	1.2	1.8	2.7	3.8	5.2	6.7
Oil		0.0	0.3	0.8	1.5	2.4	3.2	4.0	4.6	5.0	5.2	5.3	5.4
Gas		0.0	0.1	0.3	0.5	0.9	1.4	2.1	2.8	3.6	4.5	5.3	6.1
Cumulative CO2 Emissions	GtC	1.4	17.8	38.5	66.1	101.3	140.4	181.6	224.2	267.8	313.2	362.2	415.5
Carbon Sequestration	GtC												
Land Use	Million ha												
Cropland													
Grasslands													
Energy Biomass													
Forest													
Others													
Total													
Anthropogenic Emissions (standardized)													
Fossil Fuel CO2	GtC	0.72	1.01	1.52	2.38	3.00	3.28	3.55	3.84	4.13	4.50	4.93	5.36
Other CO2	GtC	0.73	0.82	0.78	0.85	0.82	0.74	0.66	0.47	0.28	0.18	0.19	0.19
Total CO2	GtC	1.45	1.83	2.30	3.22	3.82	4.02	4.21	4.31	4.41	4.68	5.12	5.55
CH4 total	MtCH4	77	85	105	128	143	151	160	166	173	179	185	191
N2O total	MtN2O-N	1.2	1.3	1.6	1.9	2.2	2.5	2.7	2.9	3.0	3.1	3.2	3.3
SOx total	MtS	10.5	12.8	15.6	22.5	27.5	26.3	25.2	21.4	17.5	15.2	14.4	13.5
HFC	MtC eq.	0	2	12	20	32	54	84	120	157	190	211	223
PFC	MtC eq.	4	4	7	10	14	20	27	33	37	39	39	37
SF6	MtC eq.	3	5	5	7	9	20	16	17	15	11	13	14
CO	MtCO	396	404	451	499	590	624	658	674	689	717	757	797
NMVOC	Mt	48	55	61	71	81	88	95	104	114	127	143	160
NOx	MtN	7	8	10	14	18	20	22	23	24	25	27	29

Scenario B2-IMAGE World		**1990**	**2000**	**2010**	**2020**	**2030**	**2040**	**2050**	**2060**	**2070**	**2080**	**2090**	**2100**
Population	Million	5297			7869			9875					10360
GNP/GDP (mex)	Trillion US$	21.0			41.2			75.7					198.7
GNP/GDP (ppp) Trillion (1990 prices)		0.0			0.0			0.0					0.0
Final Energy	EJ												
Non-commercial		50			35			15					9
Solids		40			36			46					73
Liquids		98			106			124					128
Gas		50			102			143					160
Electricity		35			99			177					355
Others		0			0			0					0
Total		272			377			504					726
Primary Energy	EJ												
Coal		82			113			197					168
Oil		116			130			133					86
Gas		78			182			212					187
Nuclear													
Biomass		1			6			34					76
Non-commercial		6			35			15					9
NTE (Nuclear/Solar) & Hydro		17			41			90					322
Total		344			506			679					846
Cumulative Resources Use	ZJ												
Coal		0.1			3.1			8.0					17.9
Oil		0.1			3.3			7.3					12.3
Gas		0.1			3.5			9.6					19.6
Cumulative CO2 Emissions	GtC	7.1	82.4	168.2	266.2	373.5	487.3	607.7	729.7	848.5	964.0	1076.4	1185.5
Carbon Sequestration	GtC	0.0			0.0			0.0					0.0
Land Use	Million ha												
Cropland		0			0			0					0
Grasslands		0			0			0					0
Energy Biomass		0			0			0					0
Forest		0			0			0					0
Others		0			0			0					0
Total		0			0			0					0
Anthropogenic Emissions (standardized)													
Fossil Fuel CO2	GtC	5.99	6.90	7.68	8.47	9.39	10.31	11.23	10.84	10.46	10.07	9.68	9.30
Other CO2	GtC	1.11	1.07	1.50	1.94	1.67	1.40	1.13	1.20	1.26	1.33	1.39	1.46
Total CO2	GtC	7.10	7.97	9.19	10.40	11.06	11.71	12.36	12.04	11.72	11.40	11.07	10.75
CH4 total	MtCH4	310	323	396	469	491	514	536	540	544	548	551	555
N2O total	MtN2O-N	6.7	7.0	9.3	11.5	11.8	12.2	12.5	12.5	12.6	12.6	12.7	12.7
SOx total	MtS	70.9	69.0	52.6	47.7	42.4	44.8	41.7	42.5	43.3	41.7	37.9	34.0
CFC/HFC/HCFC	MtC eq.	1672	883	786	299	272	315	346	413	483	548	603	649
PFC	MtC eq.	32	25	42	55	70	88	107	120	128	130	127	121
SF6	MtC eq.	38	40	48	55	60	76	79	80	74	63	65	69
CO	MtCO	879	877	755	632	615	597	580	596	612	629	645	661
NMVOC	Mt	139	141	147	152	151	149	147	149	151	154	156	158
NOx	MtN	31	32	36	39	43	47	51	49	48	46	44	43

Scenario B2-IMAGE OECD90		1990	2000	2010	2020	2030	2040	2050	2060	2070	2080	2090	2100
Population	Million	801			993			1022					1005
GNP/GDP (mex)	Trillion US$	16.5			27.9			39.2					65.7
GNP/GDP (ppp) Trillion (1990 prices)													
Final Energy	EJ												
Non-commercial		6			5			2					1
Solids		7			6			6					8
Liquids		51			40			38					25
Gas		34			53			64					78
Electricity		21			48			57					68
Others													
Total		119			152			166					179
Primary Energy	EJ												
Coal		24			22			27					17
Oil		55			45			39					20
Gas		51			96			97					88
Nuclear													
Biomass		1			2			6					10
Non-commercial		6			5			2					1
NTE (Nuclear/Solar) & Hydro		10			25			35					62
Total		147			195			206					197
Cumulative Resources Use		ZJ											
Coal		0.0			0.9			1.6					2.6
Oil		0.1			0.8			1.7					2.5
Gas		0.1			1.8			4.6					9.1
Cumulative CO2 Emissions	GtC	2.8	33.0	66.1	101.4	138.2	175.6	213.5	251.2	288.0	323.9	358.9	392.9
Carbon Sequestration	GtC												
Land Use	Million ha												
Cropland													
Grasslands													
Energy Biomass													
Forest													
Others													
Total													
Anthropogenic Emissions (standardized)													
Fossil Fuel CO2	GtC	2.83	3.20	3.30	3.40	3.40	3.40	3.40	3.25	3.10	2.96	2.81	2.66
Other CO2	GtC	0.00	0.00	0.13	0.25	0.31	0.36	0.42	0.47	0.53	0.58	0.64	0.69
Total CO2	GtC	2.83	3.20	3.42	3.65	3.71	3.76	3.82	3.73	3.63	3.54	3.45	3.35
CH4 total	MtCH4	73	74	87	101	101	102	103	102	100	99	98	97
N2O total	MtN2O-N	2.6	2.6	2.7	2.9	2.9	2.9	2.9	2.9	2.9	2.9	2.9	2.9
SOx total	MtS	22.7	17.0	10.2	7.5	6.5	5.6	4.8	4.3	3.8	3.4	3.1	2.8
HFC	MtC eq.	19	57	105	99	102	101	102	101	100	98	98	97
PFC	MtC eq.	18	13	14	13	12	11	10	8	7	7	7	7
SF6	MtC eq.	24	23	25	23	17	13	13	11	10	11	10	10
CO	MtCO	179	161	119	77	79	80	82	83	84	84	85	86
NMVOC	Mt	42	36	36	36	36	36	35	35	35	34	34	33
NOx	MtN	13	12	11	9	9	9	9	8	8	8	8	7

Scenario B2-IMAGE REF		**1990**	**2000**	**2010**	**2020**	**2030**	**2040**	**2050**	**2060**	**2070**	**2080**	**2090**	**2100**
Population	Million	413			432			435					407
GNP/GDP (mex)	Trillion US$	1.0			2.1			3.5					5.7
GNP/GDP (ppp) Trillion (1990 prices)													
Final Energy	EJ												
Non-commercial		2			1			0					0
Solids		11			4			3					3
Liquids		18			15			10					7
Gas		12			9			9					9
Electricity		7			13			14					13
Others													
Total		50			41			36					31
Primary Energy	EJ												
Coal		26			16			13					6
Oil		26			21			13					8
Gas		21			19			16					13
Nuclear													
Biomass		0			0			3					3
Non-commercial		2			1			0					0
NTE (Nuclear/Solar) & Hydro		2			5			8					10
Total		78			63			53					40
Cumulative Resources Use	ZJ												
Coal		0.0			0.6			1.1					1.6
Oil		0.0			0.6			1.0					1.5
Gas		0.0			0.5			1.1					2.0
Cumulative CO2 Emissions	GtC	1.3	12.3	20.4	26.2	30.6	34.1	36.8	38.9	40.7	42.3	43.6	44.6
Carbon Sequestration	GtC												
Land Use	Million ha												
Cropland													
Grasslands													
Energy Biomass													
Forest													
Others													
Total													
Anthropogenic Emissions (standardized)													
Fossil Fuel CO2	GtC	1.30	0.91	0.68	0.46	0.34	0.23	0.12	0.06	-0.01	-0.07	-0.13	-0.20
Other CO2	GtC	0.00	0.00	0.01	0.02	0.05	0.08	0.10	0.14	0.18	0.21	0.25	0.29
Total CO2	GtC	1.30	0.91	0.69	0.48	0.39	0.31	0.22	0.20	0.17	0.14	0.12	0.09
CH4 total	MtCH4	47	39	45	50	48	45	42	41	39	38	36	34
N2O total	MtN2O-N	0.6	0.6	0.6	0.6	0.6	0.5	0.5	0.5	0.5	0.5	0.5	0.5
SOx total	MtS	17.0	11.0	5.0	3.6	2.7	1.8	1.5	1.2	0.9	0.7	0.4	0.2
HFC	MtC eq.	0	4	9	15	21	25	25	26	26	26	27	27
PFC	MtC eq.	7	4	8	10	14	19	25	28	30	30	28	27
SF6	MtC eq.	7	6	5	7	9	12	15	16	15	14	14	14
CO	MtCO	69	41	26	10	8	6	3	2	1	-1	-2	-3
NMVOC	Mt	16	13	12	10	10	9	9	9	8	8	8	8
NOx	MtN	5	3	4	4	3	3	2	1	1	1	0	0

Scenario B2-IMAGE ASIA		**1990**	**2000**	**2010**	**2020**	**2030**	**2040**	**2050**	**2060**	**2070**	**2080**	**2090**	**2100**
Population	Million	2790			4123			4956					4783
GNP/GDP (mex)	Trillion US$	1.4			5.7			17.7					70.6
GNP/GDP (ppp) Trillion (1990 prices)													
Final Energy	EJ												
Non-commercial		28			21			7					3
Solids		20			22			26					39
Liquids		14			26			32					28
Gas		2			34			59					48
Electricity		4			31			79					164
Others													
Total		68			134			203					282
Primary Energy	EJ												
Coal		29			63			120					64
Oil		17			32			22					3
Gas		3			60			82					48
Nuclear													
Biomass		0			3			19					28
Non-commercial		28			21			7					3
NTE (Nuclear/Solar) & Hydro		2			7			34					163
Total		79			185			284					309
Cumulative Resources Use	ZJ												
Coal		0.0			1.4			4.3					9.4
Oil		0.0			0.7			1.5					1.9
Gas		0.0			0.9			3.1					5.9
Cumulative CO2 Emissions	GtC	1.5	19.3	43.2	73.9	111.3	154.6	204.1	254.3	299.7	340.6	376.7	408.1
Carbon Sequestration	GtC												
Land Use	Million ha												
Cropland													
Grasslands													
Energy Biomass													
Forest													
Others													
Total													
Anthropogenic Emissions (standardized)													
Fossil Fuel CO2	GtC	1.15	1.78	2.51	3.24	3.80	4.36	4.92	4.46	4.00	3.54	3.08	2.62
Other CO2	GtC	0.37	0.26	0.22	0.19	0.24	0.28	0.33	0.32	0.31	0.31	0.30	0.29
Total CO2	GtC	1.53	2.03	2.73	3.43	4.03	4.64	5.25	4.78	4.31	3.85	3.38	2.91
CH4 total	MtCH4	113	125	153	180	194	209	224	225	226	227	228	229
N2O total	MtN2O-N	2.3	2.6	3.0	3.4	3.6	3.8	4.0	4.0	4.0	4.0	4.0	4.0
SOx total	MtS	17.7	25.3	20.5	18.1	17.6	17.1	15.3	15.2	15.0	14.0	12.1	10.2
HFC	MtC eq.	0	5	21	40	66	95	130	164	199	233	267	302
PFC	MtC eq.	3	5	14	22	30	38	46	51	54	54	53	51
SF6	MtC eq.	4	7	12	18	25	32	36	36	33	27	29	30
CO	MtCO	235	270	247	225	237	249	262	262	263	263	264	264
NMVOC	Mt	33	37	39	41	43	45	46	46	46	46	46	46
NOx	MtN	7	9	12	16	19	22	25	22	20	17	15	13

Scenario B2-IMAGE ALM		1990	2000	2010	2020	2030	2040	2050	2060	2070	2080	2090	2100
Population	Million	1293			2321			3462					4165
GNP/GDP (mex)	Trillion US$	2.1			5.5			15.3					56.7
GNP/GDP (ppp) Trillion (1990 prices)													
Final Energy	EJ												
Non-commercial		14			8			6					4
Solids		2			4			10					24
Liquids		15			25			45					68
Gas		3			6			11					25
Electricity		2			7			26					111
Others													
Total		35			51			99					233
Primary Energy	EJ												
Coal		4			12			38					80
Oil		18			31			59					55
Gas		3			8			17					38
Nuclear													
Biomass		0			1			5					36
Non-commercial		14			8			6					4
NTE (Nuclear/Solar) & Hydro		3			4			12					86
Total		41			64			137					300
Cumulative Resources Use	ZJ												
Coal		0.0			0.2			1.0					4.4
Oil		0.0			1.2			3.0					6.4
Gas		0.0			0.3			0.8					2.5
Cumulative CO2 Emissions	GtC	1.4	17.8	38.7	64.6	93.4	123.0	153.3	185.3	220.0	257.3	297.3	339.9
Carbon Sequestration	GtC												
Land Use	Million ha												
Cropland													
Grasslands													
Energy Biomass													
Forest													
Others													
Total													
Anthropogenic Emissions (standardized)													
Fossil Fuel CO2	GtC	0.72	1.01	1.19	1.37	1.84	2.32	2.79	3.07	3.36	3.64	3.93	4.21
Other CO2	GtC	0.73	0.82	1.15	1.47	1.08	0.68	0.28	0.26	0.24	0.22	0.20	0.19
Total CO2	GtC	1.45	1.83	2.34	2.85	2.92	2.99	3.07	3.33	3.60	3.86	4.13	4.40
CH4 total	MtCH4	77	85	111	138	148	158	168	173	179	184	189	195
N2O total	MtN2O-N	1.2	1.3	3.0	4.7	4.8	4.9	5.1	5.1	5.2	5.2	5.3	5.4
SOx total	MtS	10.5	12.8	14.1	15.6	12.8	17.2	17.1	18.8	20.5	20.7	19.2	17.8
HFC	MtC eq.	0	2	12	20	32	54	84	120	157	190	211	223
PFC	MtC eq.	4	4	7	10	14	20	27	33	37	39	39	37
SF6	MtC eq.	3	5	5	7	9	20	16	17	15	11	13	14
CO	MtCO	396	404	362	320	291	262	233	249	265	282	298	315
NMVOC	Mt	48	55	60	64	62	59	56	59	62	65	68	71
NOx	MtN	7	8	9	10	12	14	15	17	18	20	21	23

Scenario B2-MARIA World		1990	2000	2010	2020	2030	2040	2050	2060	2070	2080	2090	2100
Population	Million	5262	6091	6891	7672	8372	8930	9367	9704	9960	10159	10306	10414
GNP/GDP (mex)	Trillion US$	19.5	27.2	37.0	49.7	64.0	83.1	108.0	130.5	156.3	182.0	206.2	232.9
GNP/GDP (ppp) Trillion (1990 prices)													
Final Energy	EJ												
Non-commercial													
Solids		48	40	45	55	80	124	174	234	287	328	333	264
Liquids		138	159	180	204	229	261	292	308	301	317	348	375
Gas		56	56	87	119	141	144	132	118	115	96	86	148
Electricity		35	49	60	71	84	95	116	125	141	152	158	159
Others		0	0	0	0	0	0	0	0	0	0	0	0
Total		278	305	371	449	535	625	714	784	844	893	925	946
Primary Energy	EJ												
Coal		90	71	68	72	93	135	182	239	291	331	335	265
Oil		123	140	150	167	172	181	197	189	152	131	141	185
Gas		71	99	147	190	222	248	217	196	188	164	147	202
Nuclear		22	21	28	26	27	44	69	112	180	228	264	233
Biomass		28	32	40	62	97	107	114	132	159	192	212	225
Other Renewables		9	8	7	7	6	6	58	62	61	60	59	58
Total		343	372	441	523	617	721	837	931	1031	1107	1159	1169
Cumulative Resources Use	ZJ												
Coal		0.0	0.9	1.6	2.3	3.0	3.9	5.3	7.1	9.5	12.4	15.7	19.1
Oil		0.0	1.2	2.6	4.1	5.8	7.5	9.3	11.3	13.2	14.7	16.0	17.4
Gas		0.0	0.7	1.7	3.2	5.1	7.3	9.8	11.9	13.9	15.8	17.4	18.9
Cumulative CO2 Emissions	GtC	7.1	82.4	166.5	260.4	365.3	484.0	616.0	757.3	904.8	1055.4	1207.2	1358.9
Carbon Sequestration	GtC	0.0	0.0	0.0	0.0	0.0	0.0	0.0	0.0	0.0	0.0	0.0	0.0
Land Use	Million ha												
Cropland		1451	1451	1457	1743	1937	2049	2079	1999	1868	1701	1578	1500
Grasslands		3395	3395	3392	3114	2944	2904	2904	2904	2904	2904	2904	2904
Energy Biomass		0	0	0	0	0	0	0	81	219	396	519	597
Forest		4138	4138	4138	4138	4138	4122	4124	4138	4138	4138	4138	4138
Others		4061	4061	4058	4051	4027	3970	3937	3924	3916	3907	3907	3907
Total		13045	13045	13045	13045	13045	13045	13045	13045	13045	13045	13045	13045
Anthropogenic Emissions (standardized)													
Fossil Fuel CO2	GtC	5.99	6.90	7.75	8.85	10.00	11.66	12.74	13.72	14.19	14.46	14.49	14.42
Other CO2	GtC	1.11	1.07	1.09	1.08	1.05	1.04	0.96	0.83	0.77	0.70	0.71	0.73
Total CO2	GtC	7.10	7.97	8.84	9.93	11.05	12.70	13.70	14.55	14.96	15.16	15.20	15.15
CH4 total	MtCH4												
N2O total	MtN2O-N												
SOx total	MtS												
CFC/HFC/HCFC	MtC eq.	1672	883	786	299	272	315	346	413	483	548	603	649
PFC	MtC eq.	32	25	42	55	70	88	107	120	128	130	127	121
SF6	MtC eq.	38	40	48	55	60	76	79	80	74	63	65	69
CO	MtCO												
NMVOC	Mt												
NOx	MtN												

Scenario B2-MARIA OECD90		1990	2000	2010	2020	2030	2040	2050	2060	2070	2080	2090	2100
Population	Million	859	917	953	982	994	988	976	965	952	941	934	928
GNP/GDP (mex)	Trillion US$	15.6	20.7	26.1	30.2	33.4	37.1	41.1	43.8	47.9	52.9	57.7	62.7
GNP/GDP (ppp) Trillion (1990 prices)													
Final Energy	EJ												
Non-commercial													
Solids		12	8	6	4	13	29	50	65	76	83	89	75
Liquids		71	80	74	67	62	58	55	52	51	50	50	50
Gas		28	26	45	61	62	53	37	26	18	13	9	20
Electricity		21	28	32	35	37	38	40	41	43	44	45	46
Others		0	0	0	0	0	0	0	0	0	0	0	0
Total		132	142	157	167	174	178	182	184	188	190	194	192
Primary Energy	EJ												
Coal		38	28	20	15	22	36	56	69	79	85	90	76
Oil		71	80	74	67	62	58	55	52	51	50	50	50
Gas		34	49	79	102	107	100	75	63	55	49	43	51
Nuclear		18	15	12	13	16	20	24	30	38	45	54	65
Biomass		6	4	3	2	1	1	1	0	0	0	1	1
Other Renewables		5	5	4	4	3	3	15	14	14	14	14	13
Total		171	180	193	202	212	218	225	230	237	244	252	255
Cumulative Resources Use	ZJ												
Coal		0.0	0.4	0.7	0.9	1.0	1.2	1.6	2.2	2.8	3.6	4.5	5.4
Oil		0.0	0.7	1.5	2.3	2.9	3.5	4.1	4.7	5.2	5.7	6.2	6.7
Gas		0.0	0.3	0.8	1.6	2.6	3.7	4.7	5.5	6.1	6.6	7.1	7.6
Cumulative CO2 Emissions	GtC	2.8	33.0	65.9	100.1	135.8	173.1	211.3	250.1	290.3	331.4	373.3	414.3
Carbon Sequestration	GtC	0.0	0.0	0.0	0.0	0.0	0.0	0.0	0.0	0.0	0.0	0.0	0.0
Land Use	Million ha												
Cropland		378	378	378	378	378	378	378	378	378	378	303	243
Grasslands		756	756	756	756	756	756	756	756	756	756	756	756
Energy Biomass		0	0	0	0	0	0	0	0	0	0	75	134
Forest		756	756	756	756	756	756	756	756	756	756	756	756
Others		794	794	794	794	794	794	794	794	794	794	794	794
Total		2684	2684	2684	2684	2684	2684	2684	2684	2684	2684	2684	2684
Anthropogenic Emissions (standardized)													
Fossil Fuel CO2	GtC	2.83	3.20	3.36	3.42	3.59	3.76	3.81	3.93	4.04	4.09	4.13	3.86
Other CO2	GtC	0.00	0.00	0.03	0.05	0.07	0.04	0.02	0.01	0.04	0.06	0.09	0.12
Total CO2	GtC	2.83	3.20	3.38	3.47	3.66	3.80	3.83	3.95	4.07	4.16	4.22	3.98
CH4 total	MtCH4												
N2O total	MtN2O-N												
SOx total	MtS												
HFC	MtC eq.	19	57	105	99	102	101	102	101	100	98	98	97
PFC	MtC eq.	18	13	14	13	12	11	10	8	7	7	7	7
SF6	MtC eq.	24	23	25	23	17	13	13	11	10	11	10	10
CO	MtCO												
NMVOC	Mt												
NOx	MtN												

Scenario B2-MARIA REF		1990	2000	2010	2020	2030	2040	2050	2060	2070	2080	2090	2100
Population	Million	413	415	417	418	416	411	406	396	389	384	381	379
GNP/GDP (mex)	Trillion US$	0.9	1.2	1.4	1.7	2.6	4.1	6.7	8.1	10.0	11.5	12.9	14.5
GNP/GDP (ppp) Trillion (1990 prices)													
Final Energy	EJ												
Non-commercial													
Solids		13	9	6	4	9	16	27	45	58	66	72	68
Liquids		17	13	12	11	12	13	14	14	14	14	14	14
Gas		21	24	30	36	40	43	44	31	22	15	11	12
Electricity		7	8	8	9	10	12	15	16	17	17	17	17
Others		0	0	0	0	0	0	0	0	0	0	0	0
Total		58	54	56	60	71	84	100	105	110	112	113	111
Primary Energy	EJ												
Coal		19	13	9	6	10	17	28	45	58	67	72	68
Oil		19	14	12	11	12	13	14	14	14	14	14	14
Gas		27	33	40	47	55	62	65	53	44	37	31	32
Nuclear		3	2	2	2	3	3	4	5	7	8	10	13
Biomass		1	1	0	0	0	0	0	0	0	0	0	0
Other Renewables		1	1	1	1	1	1	3	3	3	3	3	3
Total		69	64	64	68	81	96	114	120	126	129	130	130
Cumulative Resources Use	ZJ												
Coal		0.0	0.2	0.3	0.4	0.5	0.6	0.7	1.0	1.5	2.1	2.7	3.4
Oil		0.0	0.2	0.3	0.4	0.6	0.7	0.8	1.0	1.1	1.2	1.4	1.5
Gas		0.0	0.3	0.6	1.0	1.5	2.0	2.6	3.3	3.8	4.2	4.6	4.9
Cumulative CO2 Emissions	GtC	1.3	12.3	21.4	30.4	40.6	53.3	69.0	87.4	107.7	129.2	151.1	173.0
Carbon Sequestration	GtC	0.0	0.0	0.0	0.0	0.0	0.0	0.0	0.0	0.0	0.0	0.0	0.0
Land Use	Million ha												
Cropland		217	217	217	217	217	217	217	174	139	111	89	89
Grasslands		114	114	114	114	114	114	114	114	114	114	114	114
Energy Biomass		0	0	0	0	0	0	0	43	78	106	128	128
Forest		815	815	815	815	815	815	815	815	815	815	815	815
Others		722	722	722	722	722	722	722	722	722	722	722	722
Total		1868	1868	1868	1868	1868	1868	1868	1868	1868	1868	1868	1868
Anthropogenic Emissions (standardized)													
Fossil Fuel CO2	GtC	1.30	0.91	0.88	0.90	1.14	1.44	1.78	2.04	2.25	2.36	2.40	2.32
Other CO2	GtC	0.00	0.00	0.02	0.01	-0.01	-0.03	-0.05	-0.09	-0.13	-0.19	-0.18	-0.16
Total CO2	GtC	1.30	0.91	0.90	0.91	1.13	1.41	1.73	1.95	2.12	2.16	2.22	2.16
CH4 total	MtCH4												
N2O total	MtN2O-N												
SOx total	MtS												
HFC	MtC eq.	0	4	9	15	21	25	25	26	26	26	27	27
PFC	MtC eq.	7	4	8	10	14	19	25	28	30	30	28	27
SF6	MtC eq.	7	6	5	7	9	12	15	16	15	14	14	14
CO	MtCO												
NMVOC	Mt												
NOx	MtN												

Scenario B2-MARIA ASIA		**1990**	**2000**	**2010**	**2020**	**2030**	**2040**	**2050**	**2060**	**2070**	**2080**	**2090**	**2100**
Population	Million	2642	3062	3450	3796	4089	4316	4479	4613	4720	4813	4851	4888
GNP/GDP (mex)	Trillion US$	1.2	2.9	6.3	13.0	19.9	28.1	37.8	46.9	57.0	67.2	77.1	87.9
GNP/GDP (ppp) Trillion (1990 prices)													
Final Energy	EJ												
Non-commercial													
Solids		20	20	31	43	51	68	79	93	102	111	96	67
Liquids		26	39	66	93	109	125	139	148	146	148	162	176
Gas		2	2	4	7	15	15	14	15	23	27	38	64
Electricity		4	8	12	18	24	27	35	37	43	47	49	48
Others		0	0	0	0	0	0	0	0	0	0	0	0
Total		52	70	113	160	198	236	266	293	314	333	345	355
Primary Energy	EJ												
Coal		28	26	36	46	54	70	80	94	102	111	96	67
Oil		14	23	38	57	54	58	69	72	52	43	42	65
Gas		3	5	8	14	21	29	23	22	27	30	40	65
Nuclear		1	5	14	11	9	21	35	49	76	94	105	81
Biomass		13	22	33	53	85	89	85	86	101	110	124	128
Other Renewables		1	1	1	1	1	1	17	20	20	20	20	19
Total		61	82	130	181	222	266	309	343	379	408	426	427
Cumulative Resources Use	ZJ												
Coal		0.0	0.3	0.5	0.9	1.4	1.9	2.6	3.4	4.3	5.3	6.5	7.4
Oil		0.0	0.1	0.4	0.8	1.3	1.9	2.4	3.1	3.9	4.4	4.8	5.2
Gas		0.0	0.0	0.1	0.2	0.3	0.5	0.8	1.0	1.2	1.5	1.8	2.2
Cumulative CO2 Emissions	GtC	1.5	19.3	42.5	72.1	106.3	144.9	188.4	235.8	284.6	333.2	380.8	427.5
Carbon Sequestration	GtC	0.0	0.0	0.0	0.0	0.0	0.0	0.0	0.0	0.0	0.0	0.0	0.0
Land Use	Million ha												
Cropland		366	366	373	380	380	380	380	380	380	380	361	343
Grasslands		431	431	428	425	426	426	426	426	426	426	426	426
Energy Biomass		0	0	0	0	0	0	0	0	0	0	19	37
Forest		365	365	365	365	365	365	365	365	365	365	365	365
Others		458	458	455	451	449	449	449	449	449	449	449	449
Total		1621	1621	1621	1621	1621	1621	1621	1621	1621	1621	1621	1621
Anthropogenic Emissions (standardized)													
Fossil Fuel CO2	GtC	1.15	1.78	2.36	3.07	3.33	3.94	4.34	4.75	4.66	4.76	4.50	4.61
Other CO2	GtC	0.37	0.26	0.25	0.23	0.22	0.21	0.21	0.19	0.16	0.14	0.12	0.11
Total CO2	GtC	1.53	2.03	2.61	3.31	3.55	4.15	4.55	4.94	4.82	4.90	4.62	4.72
CH4 total	MtCH4												
N2O total	MtN2O-N												
SOx total	MtS												
HFC	MtC eq.	0	5	21	40	66	95	130	164	199	233	267	302
PFC	MtC eq.	3	5	14	22	30	38	46	51	54	54	53	51
SF6	MtC eq.	4	7	12	18	25	32	36	36	33	27	29	30
CO	MtCO												
NMVOC	Mt												
NOx	MtN												

Scenario B2-MARIA ALM		1990	2000	2010	2020	2030	2040	2050	2060	2070	2080	2090	2100
Population	Million	1348	1697	2071	2476	2873	3214	3506	3730	3900	4021	4141	4220
GNP/GDP (mex)	Trillion US$	1.7	2.4	3.2	4.9	8.1	13.8	22.4	31.6	41.4	50.4	58.5	67.8
GNP/GDP (ppp) Trillion (1990 prices)													
Final Energy	EJ												
Non-commercial													
Solids		3	2	2	4	7	11	19	31	51	68	76	54
Liquids		24	27	28	34	47	65	85	94	91	105	122	135
Gas		5	4	7	15	24	32	37	46	52	41	29	51
Electricity		3	6	8	10	14	18	26	31	39	44	47	48
Others		0	0	0	0	0	0	0	0	0	0	0	0
Total		35	38	45	63	91	127	167	202	233	258	274	288
Primary Energy	EJ												
Coal		6	4	3	4	7	12	19	31	51	68	76	54
Oil		18	23	26	32	44	53	60	50	35	25	35	57
Gas		7	12	19	28	40	58	55	58	61	47	33	54
Nuclear		0	0	0	0	0	0	5	27	59	81	95	74
Biomass		8	6	4	7	11	17	28	46	57	82	88	96
Other Renewables		2	2	1	1	1	2	23	25	24	24	23	23
Total		41	46	53	72	103	141	190	238	288	326	350	357
Cumulative Resources Use	ZJ												
Coal		0.0	0.1	0.1	0.1	0.2	0.2	0.4	0.6	0.9	1.4	2.1	2.8
Oil		0.0	0.2	0.4	0.7	1.0	1.4	1.9	2.5	3.0	3.4	3.6	4.0
Gas		0.0	0.1	0.2	0.4	0.7	1.1	1.6	2.2	2.8	3.4	3.9	4.2
Cumulative CO2 Emissions	GtC	1.4	17.8	36.8	57.8	82.6	112.8	147.4	184.0	222.2	261.6	302.0	344.1
Carbon Sequestration	GtC	0.0	0.0	0.0	0.0	0.0	0.0	0.0	0.0	0.0	0.0	0.0	0.0
Land Use	Million ha												
Cropland		490	490	490	768	963	1075	1105	1067	972	832	825	825
Grasslands		2095	2095	2095	1820	1648	1609	1609	1609	1609	1609	1609	1609
Energy Biomass		0	0	0	0	0	0	0	37	141	290	297	297
Forest		2202	2202	2202	2202	2202	2186	2188	2202	2202	2202	2202	2202
Others		2086	2086	2086	2083	2060	2004	1971	1958	1950	1941	1941	1941
Total		6873	6873	6873	6873	6873	6873	6873	6873	6873	6873	6873	6873
Anthropogenic Emissions (standardized)													
Fossil Fuel CO2	GtC	0.72	1.01	1.16	1.46	1.95	2.52	2.81	2.99	3.24	3.25	3.46	3.63
Other CO2	GtC	0.73	0.82	0.80	0.79	0.77	0.81	0.78	0.72	0.70	0.69	0.67	0.66
Total CO2	GtC	1.45	1.83	1.96	2.24	2.72	3.33	3.60	3.71	3.94	3.94	4.14	4.29
CH4 total	MtCH4												
N2O total	MtN2O-N												
SOx total	MtS												
HFC	MtC eq.	0	2	12	20	32	54	84	120	157	190	211	223
PFC	MtC eq.	4	4	7	10	14	20	27	33	37	39	39	37
SF6	MtC eq.	3	5	5	7	9	20	16	17	15	11	13	14
CO	MtCO												
NMVOC	Mt												
NOx	MtN												

Marker Scenario B2-MESSAGE World		**1990**	**2000**	**2010**	**2020**	**2030**	**2040**	**2050**	**2060**	**2070**	**2080**	**2090**	**2100**
Population	Million	5262	6091	6891	7672	8372	8930	9367	9704	9960	10158	10306	10414
GNP/GDP (mex)	Trillion US$	20.9	28.3	38.6	50.7	66.0	85.5	109.5	134.8	161.5	186.3	210.3	234.9
GNP/GDP (ppp) Trillion (1990 prices)		25.7	34.8	46.9	60.2	75.5	93.2	113.9	136.8	160.7	183.8	207.4	231.8
Final Energy	EJ												
Non-commercial		38	27	27	24	18	14	11	10	9	8	7	7
Solids		42	47	40	36	29	30	19	17	16	16	13	7
Liquids		111	134	167	200	236	255	268	273	267	270	289	294
Gas		41	46	50	59	70	86	105	119	130	130	118	111
Electricity		35	47	62	85	113	150	188	227	272	321	366	409
Others		8	11	17	25	41	51	63	78	90	104	116	123
Total		275	311	362	429	507	586	654	723	783	848	909	951
Primary Energy	EJ												
Coal		91	91	98	98	96	93	86	91	119	170	231	300
Oil		128	168	195	214	240	238	227	201	146	101	72	52
Gas		71	84	107	150	194	251	297	356	390	402	385	336
Nuclear		7	8	11	16	23	32	48	61	83	99	120	142
Biomass		46	43	46	53	61	79	105	136	184	236	280	315
Other Renewables		8	14	22	34	54	80	107	131	153	176	197	212
Total		352	408	479	566	667	773	869	976	1074	1184	1285	1357
Cumulative Resources Use	ZJ												
Coal		0.0	0.9	1.8	2.8	3.8	4.7	5.7	6.5	7.4	8.6	10.3	12.6
Oil		0.0	1.4	3.1	5.1	7.2	9.6	12.0	14.3	16.3	17.7	18.7	19.5
Gas		0.0	0.7	1.6	2.7	4.2	6.1	8.6	11.6	15.1	19.0	23.1	26.9
Cumulative CO2 Emissions	GtC	7.1	82.4	166.2	255.4	350.1	453.1	561.5	674.1	789.6	908.5	1033.2	1163.8
Carbon Sequestration	GtC												
Land Use [a]	Million ha												
Cropland		1459	1481	1510	1540	1570	1598	1626	1656	1687	1718	1751	1784
Grasslands		3389	3409	3436	3454	3473	3508	3543	3577	3611	3641	3668	3696
Energy Biomass		0	0	0	14	92	190	288	313	338	342	325	307
Forest		4296	4252	4225	4251	4292	4323	4353	4380	4407	4441	4482	4522
Others		3805	3806	3778	3690	3522	3330	3139	3023	2907	2807	2723	2639
Total		12949	12949	12949	12949	12949	12949	12949	12949	12949	12949	12949	12949
Anthropogenic Emissions (standardized)													
Fossil Fuel CO2 [b]	GtC	5.99	6.90	7.99	9.02	10.15	10.93	11.23	11.74	11.87	12.46	13.20	13.82
Other CO2 [c]	GtC	1.11	1.07	0.80	0.03	-0.25	-0.24	-0.23	-0.24	-0.25	-0.31	-0.41	-0.50
Total CO2	GtC	7.10	7.97	8.78	9.05	9.90	10.69	11.01	11.49	11.62	12.15	12.79	13.32
CH4 total [d]	MtCH4	310	323	349	384	426	466	504	522	544	566	579	597
N2O total [e]	MtN2O-N	6.7	7.0	6.2	6.1	6.1	6.2	6.3	6.4	6.6	6.7	6.8	6.9
SOx total	MtS	70.9	69.0	65.9	61.3	60.3	59.0	55.7	53.8	50.9	50.0	49.0	47.9
CFC/HFC/HCFC	MtC eq.	1672	883	786	299	272	315	346	413	483	548	603	649
PFC	MtC eq.	32	25	42	55	70	88	107	120	128	130	127	121
SF6	MtC eq.	38	40	48	55	60	76	79	80	74	63	65	69
CO	MtCO	879	877	935	1022	1111	1220	1319	1423	1570	1742	1886	2002
NMVOC	Mt	139	141	159	180	199	214	217	214	202	192	178	170
NOx	MtN	31	32	37	43	49	53	54	56	56	59	61	61

Emissions correlated to land-use change and deforestation were calculated by using AIM B2 land-use data.
a: Land-use taken from AIM-B2 run.
b:CO2 emissions from fossil fuel and industrial processes (MESSAGE data).
c: CO2 emissions from land-use changes (AIM-B2 run).
d: Non-energy related CH4 emissions were taken from AIM-B2 run.
e: Non-energy related N2O emissions were taken from AIM-B2 run.

Marker Scenario B2-MESSAGE OECD90		1990	2000	2010	2020	2030	2040	2050	2060	2070	2080	2090	2100
Population	Million	859	916	953	982	994	988	976	965	951	941	934	928
GNP/GDP (mex)	Trillion US$	16.4	21.1	26.5	30.3	33.1	35.8	38.3	40.9	44.4	47.9	52.0	56.6
GNP/GDP (ppp) Trillion (1990 prices)		14.1	18.3	23.0	26.3	28.8	31.3	33.5	35.9	39.2	42.4	46.1	50.4
Final Energy	EJ												
Non-commercial		0	0	0	0	0	0	0	0	0	0	0	0
Solids		13	9	4	3	1	0	0	0	0	0	0	0
Liquids		66	71	79	86	92	88	76	68	64	60	60	58
Gas		22	24	26	25	27	31	35	38	34	29	24	17
Electricity		22	29	36	44	50	56	61	65	71	76	83	89
Others		1	1	2	4	7	8	10	12	14	16	17	18
Total		124	133	146	162	177	183	182	183	183	182	184	182
Primary Energy	EJ												
Coal		38	34	36	39	33	24	20	19	23	26	29	34
Oil		72	86	91	90	90	81	65	54	42	30	23	17
Gas		33	41	50	61	71	86	99	114	113	123	127	121
Nuclear		6	7	9	11	13	15	17	16	21	21	25	29
Biomass		6	7	6	7	8	9	12	15	22	28	31	33
Other Renewables		4	5	8	11	15	20	25	29	33	36	39	41
Total		159	180	200	219	230	234	236	248	253	265	273	274
Cumulative Resources Use	ZJ												
Coal		0.0	0.4	0.7	1.1	1.5	1.8	2.0	2.2	2.4	2.6	2.9	3.3
Oil		0.0	0.8	1.7	2.6	3.5	4.4	5.2	5.8	6.4	6.8	7.1	7.3
Gas		0.0	0.3	0.8	1.3	1.9	2.6	3.4	4.3	5.4	6.6	7.8	9.2
Cumulative CO2 Emissions	GtC	2.8	33.0	66.3	101.9	138.3	173.6	206.9	239.0	269.9	299.8	329.6	359.1
Carbon Sequestration	GtC												
Land Use [a]	Million ha												
Cropland		381	384	391	398	406	411	417	423	429	435	442	449
Grasslands		760	764	770	773	776	787	798	812	826	838	849	860
Energy Biomass		0	0	0	2	17	36	54	59	64	64	61	58
Forest		1050	1059	1069	1082	1095	1104	1113	1125	1136	1151	1171	1190
Others		838	822	799	774	734	691	647	611	575	540	506	471
Total		3029	3029	3029	3029	3029	3029	3029	3029	3029	3029	3029	3029
Anthropogenic Emissions (standardized)													
Fossil Fuel CO2 b	GtC	2.83	3.20	3.50	3.71	3.70	3.51	3.26	3.25	3.08	3.10	3.14	3.10
Other CO2 c	GtC	0.00	0.00	-0.02	-0.06	-0.08	-0.06	-0.05	-0.06	-0.08	-0.11	-0.15	-0.19
Total CO2	GtC	2.83	3.20	3.48	3.64	3.62	3.45	3.22	3.19	2.99	2.99	2.98	2.91
CH4 total d	MtCH4	73	74	72	71	68	67	69	69	70	72	75	78
N2O total e	MtN2O-N	2.6	2.6	2.5	2.4	2.5	2.5	2.5	2.5	2.5	2.6	2.6	2.6
SOx total	MtS	22.7	17.0	11.3	6.7	5.7	4.9	4.1	3.7	3.4	3.3	3.3	3.5
HFC	MtC eq.	19	57	105	99	102	101	102	101	100	98	98	97
PFC	MtC eq.	18	13	14	13	12	11	10	8	7	7	7	7
SF6	MtC eq.	24	23	25	23	17	13	13	11	10	11	10	10
CO	MtCO	179	161	163	173	180	187	185	185	195	213	207	197
NMVOC	Mt	42	36	37	39	42	45	43	41	38	36	33	30
NOx	MtN	13	12	13	15	16	16	15	15	14	13	12	11

Emissions correlated to land-use change and deforestation were calculated by using AIM B2 land-use data.

a: Land-use taken from AIM-B2 run.

b:CO2 emissions from fossil fuel and industrial processes (MESSAGE data).

c: CO2 emissions from land-use changes (AIM-B2 run).

d: Non-energy related CH4 emissions were taken from AIM-B2 run.

e: Non-energy related N2O emissions were taken from AIM-B2 run.

Marker Scenario B2-MESSAGE REF		**1990**	**2000**	**2010**	**2020**	**2030**	**2040**	**2050**	**2060**	**2070**	**2080**	**2090**	**2100**
Population	Million	413	415	417	418	416	411	406	396	389	384	381	379
GNP/GDP (mex)	Trillion US$	1.1	1.0	1.2	1.8	2.8	4.5	6.6	8.6	10.5	11.9	13.2	14.5
GNP/GDP (ppp) Trillion (1990 prices)		2.6	2.4	2.7	3.3	4.3	5.6	7.2	9.5	11.6	13.3	14.8	16.2
Final Energy	EJ												
Non-commercial		0	0	0	0	0	0	0	0	0	0	0	0
Solids		9	5	3	2	1	1	0	0	0	0	0	0
Liquids		15	14	17	17	17	19	19	18	18	19	20	22
Gas		14	14	12	15	18	19	19	18	16	15	12	10
Electricity		6	6	6	7	9	13	18	21	24	27	30	32
Others		7	6	6	7	10	11	12	14	15	16	16	15
Total		50	45	44	49	55	63	68	71	73	76	78	79
Primary Energy	EJ												
Coal		19	15	11	7	9	10	12	8	11	15	21	29
Oil		20	18	20	19	18	18	20	16	7	5	4	0
Gas		27	26	24	31	38	49	51	59	66	63	52	43
Nuclear		1	1	1	1	1	1	2	3	4	6	7	9
Biomass		2	1	1	1	2	2	4	6	10	13	16	22
Other Renewables		1	1	2	3	4	6	8	10	12	16	19	21
Total		70	62	57	62	73	86	97	102	111	117	119	125
Cumulative Resources Use	ZJ												
Coal		0.0	0.2	0.3	0.4	0.5	0.6	0.7	0.8	0.9	1.1	1.3	1.5
Oil		0.0	0.2	0.4	0.6	0.8	1.0	1.1	1.3	1.5	1.6	1.6	1.6
Gas		0.0	0.3	0.5	0.8	1.1	1.5	2.0	2.5	3.2	3.9	4.6	5.2
Cumulative CO2 Emissions	GtC	1.3	12.3	20.8	28.0	35.1	44.2	55.3	67.0	78.7	90.7	102.5	114.1
Carbon Sequestration	GtC												
Land Use [a]	Million ha												
Cropland		268	269	274	280	285	291	296	302	308	314	320	327
Grasslands		341	362	366	368	370	373	377	380	383	386	389	392
Energy Biomass		0	0	0	1	6	12	18	20	21	22	20	19
Forest		966	951	939	944	947	944	941	940	939	941	945	949
Others		701	693	697	683	668	656	644	634	624	613	601	589
Total		2276	2276	2276	2276	2276	2276	2276	2276	2276	2276	2276	2276
Anthropogenic Emissions (standardized)													
Fossil Fuel CO2 [b]	GtC	1.30	0.91	0.80	0.81	0.94	1.11	1.24	1.18	1.22	1.24	1.21	1.18
Other CO2 [c]	GtC	0.00	0.00	-0.01	-0.18	-0.14	-0.09	-0.04	-0.04	-0.03	-0.03	-0.04	-0.04
Total CO2	GtC	1.30	0.91	0.79	0.63	0.80	1.02	1.20	1.15	1.19	1.21	1.17	1.14
CH4 total [d]	MtCH4	47	39	36	40	45	50	53	50	47	47	46	47
N2O total [e]	MtN2O-N	0.6	0.6	0.6	0.6	0.6	0.6	0.6	0.7	0.7	0.7	0.7	0.7
SOx total	MtS	17.0	11.0	7.1	3.5	1.6	1.3	2.9	2.8	2.9	3.1	3.4	3.6
HFC	MtC eq.	0	4	9	15	21	25	25	26	26	26	27	27
PFC	MtC eq.	7	4	8	10	14	19	25	28	30	30	28	27
SF6	MtC eq.	7	6	5	7	9	12	15	16	15	14	14	14
CO	MtCO	69	41	43	48	59	69	74	73	68	72	70	79
NMVOC	Mt	16	13	14	16	21	28	32	31	30	30	27	26
NOx	MtN	5	3	3	3	4	5	5	5	5	5	4	4

Emissions correlated to land-use change and deforestation were calculated by using AIM B2 land-use data.
a: Land-use taken from AIM-B2 run.
b:CO2 emissions from fossil fuel and industrial processes (MESSAGE data).
c: CO2 emissions from land-use changes (AIM-B2 run).
d: Non-energy related CH4 emissions were taken from AIM-B2 run.
e: Non-energy related N2O emissions were taken from AIM-B2 run.

Marker Scenario B2-MESSAGE ASIA		**1990**	**2000**	**2010**	**2020**	**2030**	**2040**	**2050**	**2060**	**2070**	**2080**	**2090**	**2100**
Population	Million	2798	3248	3649	4008	4312	4538	4696	4790	4856	4902	4938	4968
GNP/GDP (mex)	Trillion US$	1.5	3.5	7.2	13.2	21.3	30.7	41.8	52.7	64.1	75.0	85.8	97.1
GNP/GDP (ppp) Trillion (1990 prices)		5.3	9.3	15.1	22.4	30.7	39.3	49.3	59.0	68.7	78.5	89.2	100.4
Final Energy	EJ												
Non-commercial		24	16	16	13	11	9	7	6	5	5	5	4
Solids		19	30	29	28	24	25	17	14	12	12	9	2
Liquids		13	26	41	60	81	96	112	117	109	107	114	122
Gas		2	3	5	9	13	18	22	30	38	37	34	33
Electricity		4	8	14	24	37	53	70	87	108	131	151	169
Others		1	3	7	10	16	20	25	31	35	43	51	55
Total		62	86	111	144	181	220	252	284	308	335	363	385
Primary Energy	EJ												
Coal		30	38	47	47	48	53	48	56	71	105	144	180
Oil		15	32	45	63	82	88	93	82	61	41	27	21
Gas		3	6	16	31	43	55	61	72	79	66	51	39
Nuclear		0	1	2	4	7	13	21	29	38	47	56	65
Biomass		24	22	25	28	32	41	54	68	88	109	123	129
Other Renewables		1	4	7	12	21	31	43	52	61	71	81	88
Total		74	103	141	185	233	281	319	359	397	439	482	521
Cumulative Resources Use	ZJ												
Coal		0.0	0.3	0.7	1.2	1.7	2.1	2.7	3.1	3.7	4.3	5.3	6.7
Oil		0.0	0.2	0.5	1.0	1.6	2.4	3.3	4.2	5.0	5.6	6.1	6.3
Gas		0.0	0.0	0.1	0.3	0.6	1.0	1.5	2.1	2.7	3.4	4.0	4.5
Cumulative CO2 Emissions	GtC	1.5	19.3	42.6	70.0	101.7	138.9	179.3	220.9	263.3	307.3	355.0	408.2
Carbon Sequestration	GtC												
Land Use [a]	Million ha												
Cropland		438	439	448	457	466	475	484	494	504	514	524	535
Grasslands		608	607	611	615	618	623	627	632	636	640	644	647
Energy Biomass		0	0	0	3	21	43	65	70	76	77	73	69
Forest		535	525	518	533	548	553	558	563	567	571	575	580
Others		583	593	587	557	512	471	430	406	382	362	348	333
Total		2164	2164	2164	2164	2164	2164	2164	2164	2164	2164	2164	2164
Anthropogenic Emissions (standardized)													
Fossil Fuel CO2 [b]	GtC	1.15	1.78	2.42	3.02	3.62	4.07	4.12	4.26	4.30	4.57	5.07	5.69
Other CO2 [c]	GtC	0.37	0.26	0.20	-0.15	-0.16	-0.09	-0.03	-0.03	-0.04	-0.05	-0.05	-0.06
Total CO2	GtC	1.53	2.03	2.62	2.87	3.46	3.98	4.10	4.22	4.26	4.53	5.01	5.63
CH4 total [d]	MtCH4	113	125	146	171	201	219	234	241	251	260	265	272
N2O total [e]	MtN2O-N	2.3	2.6	2.4	2.4	2.5	2.5	2.6	2.6	2.7	2.8	2.8	2.9
SOx total	MtS	17.7	25.3	30.1	32.9	33.3	32.0	26.4	24.3	22.1	21.5	21.3	20.6
HFC	MtC eq.	0	5	21	40	66	95	130	164	199	233	267	302
PFC	MtC eq.	3	5	14	22	30	38	46	51	54	54	53	51
SF6	MtC eq.	4	7	12	18	25	32	36	36	33	27	29	30
CO	MtCO	235	270	314	375	428	472	518	549	579	598	617	651
NMVOC	Mt	33	37	44	53	59	60	59	56	52	45	41	39
NOx	MtN	7	9	12	15	18	20	19	20	20	22	24	26

Emissions correlated to land-use change and deforestation were calculated by using AIM B2 land-use data.
a: Land-use taken from AIM-B2 run.
b:CO2 emissions from fossil fuel and industrial processes (MESSAGE data).
c: CO2 emissions from land-use changes (AIM-B2 run).
d: Non-energy related CH4 emissions were taken from AIM-B2 run.
e: Non-energy related N2O emissions were taken from AIM-B2 run.

Marker Scenario B2-MESSAGE ALM		1990	2000	2010	2020	2030	2040	2050	2060	2070	2080	2090	2100
Population	Million	1192	1511	1872	2263	2649	2992	3289	3554	3764	3931	4053	4139
GNP/GDP (mex)	Trillion US$	1.9	2.7	3.7	5.5	8.8	14.6	22.8	32.6	42.6	51.4	59.3	66.8
GNP/GDP (ppp) Trillion (1990 prices)		3.8	4.9	6.2	8.2	11.7	17.0	23.9	32.4	41.2	49.6	57.2	64.9
Final Energy	EJ												
Non-commercial		14	10	11	11	7	5	4	4	3	3	3	3
Solids		1	4	3	3	3	3	2	3	4	4	4	5
Liquids		17	23	30	37	46	53	61	71	76	83	95	93
Gas		4	5	8	10	14	19	28	34	42	48	48	52
Electricity		3	5	7	11	16	28	40	54	70	86	102	119
Others		0	1	3	4	8	12	17	21	25	30	32	35
Total		39	47	61	75	95	120	152	186	219	255	284	306
Primary Energy	EJ												
Coal		5	5	4	5	6	6	6	7	14	24	37	58
Oil		21	32	40	42	50	51	50	50	36	25	18	14
Gas		8	11	18	27	41	62	87	111	132	150	155	133
Nuclear		0	0	0	0	1	3	8	13	20	26	32	39
Biomass		14	13	14	17	20	28	36	46	64	86	111	132
Other Renewables		2	3	6	8	13	23	32	40	47	53	58	62
Total		49	63	81	100	131	172	217	267	314	364	411	437
Cumulative Resources Use	ZJ												
Coal		0.0	0.0	0.1	0.1	0.2	0.2	0.3	0.4	0.5	0.6	0.8	1.1
Oil		0.0	0.3	0.6	1.0	1.4	1.9	2.4	2.9	3.4	3.8	4.0	4.2
Gas		0.0	0.1	0.2	0.4	0.6	1.1	1.7	2.7	3.9	5.2	6.6	8.1
Cumulative CO2 Emissions	GtC	1.4	17.8	36.5	55.5	75.0	96.3	120.0	147.2	177.7	210.7	246.0	282.4
Carbon Sequestration	GtC												
Land Use [a]	Million ha												
Cropland		371	389	397	405	413	421	429	438	446	455	464	474
Grasslands		1680	1676	1689	1699	1709	1725	1741	1754	1766	1777	1787	1797
Energy Biomass		0	0	0	7	48	100	151	164	177	179	170	161
Forest		1745	1717	1699	1692	1701	1721	1740	1752	1765	1777	1790	1803
Others		1684	1698	1695	1676	1608	1513	1419	1372	1326	1292	1269	1246
Total		5480	5480	5480	5480	5480	5480	5480	5480	5480	5480	5480	5480
Anthropogenic Emissions (standardized)													
Fossil Fuel CO2 [b]	GtC	0.72	1.01	1.26	1.48	1.88	2.24	2.60	3.04	3.27	3.55	3.79	3.84
Other CO2 [c]	GtC	0.73	0.82	0.63	0.42	0.12	0.01	-0.11	-0.11	-0.10	-0.12	-0.16	-0.20
Total CO2	GtC	1.45	1.83	1.90	1.90	2.01	2.25	2.49	2.94	3.17	3.43	3.63	3.64
CH4 total [d]	MtCH4	77	85	94	102	112	129	149	162	175	187	193	200
N2O total [e]	MtN2O-N	1.2	1.3	0.7	0.6	0.5	0.6	0.6	0.6	0.7	0.7	0.7	0.8
SOx total	MtS	10.5	12.8	14.4	15.2	16.7	17.8	19.4	20.0	19.6	19.1	18.0	17.2
HFC	MtC eq.	0	2	12	20	32	54	84	120	157	190	211	223
PFC	MtC eq.	4	4	7	10	14	20	27	33	37	39	39	37
SF6	MtC eq.	3	5	5	7	9	20	16	17	15	11	13	14
CO	MtCO	396	404	416	427	445	492	542	616	728	860	992	1075
NMVOC	Mt	48	55	65	72	78	81	83	86	82	80	77	75
NOx	MtN	7	8	9	10	11	13	15	16	17	19	20	20

Emissions correlated to land-use change and deforestation were calculated by using AIM B2 land-use data.
a: Land-use taken from AIM-B2 run.
b:CO2 emissions from fossil fuel and industrial processes (MESSAGE data).
c: CO2 emissions from land-use changes (AIM-B2 run).
d: Non-energy related CH4 emissions were taken from AIM-B2 run.
e: Non-energy related N2O emissions were taken from AIM-B2 run.

Scenario B2-MiniCAM World		**1990**	**2000**	**2010**	**2020**	**2030**	**2040**	**2050**	**2060**	**2070**	**2080**	**2090**	**2100**
Population	Million	5293	6147	7009	7880	8640	9304	9874	10216	10453	10585	10501	10418
GNP/GDP (mex)	Trillion US$	20.7	27.4	36.3	47.4	62.1	80.3	102.0	127.9	156.0	186.3	219.2	255.1
GNP/GDP (ppp) Trillion (1990 prices)		na	na	na	na	na	na	na	na	na	na	na	na
Final Energy	EJ												
Non-commercial		0	0	0	0	0	0	0	0	0	0	0	0
Solids		45	55	67	79	90	97	103	85	75	73	73	72
Liquids		121	121	127	140	138	156	192	228	264	300	327	355
Gas		52	60	73	91	101	107	109	118	132	152	157	162
Electricity		35	51	73	102	137	181	234	292	352	413	470	527
Others		0	0	0	0	0	0	0	0	0	0	0	0
Total		253	286	339	411	465	541	637	723	823	937	1027	1117
Primary Energy	EJ												
Coal		88	110	128	143	176	201	218	191	175	170	169	167
Oil		131	130	134	145	134	147	185	228	270	309	338	367
Gas		70	81	111	159	186	209	229	260	292	325	330	335
Nuclear		24	24	28	34	46	62	82	93	105	120	150	181
Biomass		0	5	10	15	19	24	28	38	43	45	44	43
Other Renewables		24	24	32	47	65	88	116	155	195	234	255	275
Total		336	375	443	542	626	731	858	965	1080	1203	1286	1370
Cumulative Resources Use	ZJ												
Coal		0.1	1.1	2.3	3.7	5.3	7.2	9.3	11.2	13.1	14.8	16.5	18.2
Oil		0.1	1.4	2.8	4.1	5.5	7.0	8.6	10.8	13.3	16.1	19.4	22.7
Gas		0.1	0.9	1.9	3.1	4.9	6.9	9.1	11.6	14.4	17.4	20.7	24.0
Cumulative CO2 Emissions	GtC	7.1	82.4	166.7	263.6	373.5	492.4	619.7	754.8	895.8	1040.8	1187.8	1335.6
Carbon Sequestration	GtC												
Land Use	Million ha												
Cropland		1472	1473	1491	1526	1525	1493	1431	1295	1175	1070	988	906
Grasslands		3209	3347	3572	3885	4175	4393	4540	4391	4245	4101	4067	4033
Energy Biomass		0	2	4	6	6	7	8	22	25	16	10	4
Forest		4173	4214	4140	3952	3705	3547	3480	3818	4092	4304	4278	4252
Others		4310	4128	3956	3795	3753	3724	3706	3639	3628	3673	3821	3970
Total		13164	13164	13164	13164	13164	13164	13164	13164	13164	13164	13164	13164
Anthropogenic Emissions (standardized)													
Fossil Fuel CO2	GtC	5.99	6.90	7.86	9.11	10.24	11.35	12.73	13.29	14.11	15.20	15.86	16.54
Other CO2	GtC	1.11	1.07	1.03	1.38	1.27	0.92	0.45	0.56	0.23	-0.53	-1.13	-1.72
Total CO2	GtC	7.10	7.97	8.88	10.48	11.51	12.27	13.18	13.85	14.34	14.67	14.73	14.82
CH4 total	MtCH4	310	323	350	391	430	469	498	499	509	528	548	569
N2O total	MtN2O-N	6.7	7.0	8.0	9.1	10.4	11.9	13.2	13.7	14.4	15.3	16.7	18.1
SOx total	MtS	70.9	69.0	76.6	80.6	85.2	87.1	86.4	71.9	59.0	47.7	41.1	34.6
CFC/HFC/HCFC	MtC eq.	1672	883	786	299	272	315	346	413	483	548	603	649
PFC	MtC eq.	32	25	42	55	70	88	107	120	128	130	127	121
SF6	MtC eq.	38	40	48	55	60	76	79	80	74	63	65	69
CO	MtCO												
NMVOC	Mt												
NOx	MtN												

Scenario B2-MiniCAM OECD90		1990	2000	2010	2020	2030	2040	2050	2060	2070	2080	2090	2100
Population	Million	838	907	958	993	1003	1024	1023	1013	1006	1002	1003	1004
GNP/GDP (mex)	Trillion US$	16.3	20.5	24.5	28.5	29.5	33.2	36.3	38.4	40.9	43.7	47.4	51.3
GNP/GDP (ppp) Trillion (1990 prices)		na	na	na	na	na	na	na	na	na	na	na	na
Final Energy	EJ												
Non-commercial		0	0	0	0	0	0	0	0	0	0	0	0
Solids		10	12	12	11	11	11	11	8	7	6	6	6
Liquids		72	69	66	63	57	47	49	49	50	52	53	55
Gas		27	34	42	50	50	48	45	44	45	49	49	49
Electricity		22	26	31	36	38	44	50	52	54	56	59	63
Others		0	0	0	0	0	0	0	0	0	0	0	0
Total		130	141	151	160	156	149	154	153	156	163	168	173
Primary Energy	EJ												
Coal		40	43	40	31	32	35	37	33	30	29	28	27
Oil		76	73	69	64	56	42	44	44	45	47	49	51
Gas		34	43	59	79	79	76	73	72	72	75	74	73
Nuclear		20	15	12	11	11	11	12	12	12	13	16	20
Biomass		0	2	3	3	4	4	4	5	5	5	5	5
Other Renewables		12	11	12	15	15	18	19	22	25	28	30	32
Total		182	188	194	203	197	187	190	189	191	197	201	206
Cumulative Resources Use	ZJ												
Coal		0.0	0.5	0.9	1.2	1.4	1.9	2.3	2.6	2.9	3.2	3.5	3.8
Oil		0.1	0.8	1.5	2.2	2.5	3.2	3.6	4.1	4.5	5.0	5.4	5.9
Gas		0.0	0.4	1.0	1.6	2.0	3.2	3.9	4.6	5.4	6.1	6.8	7.6
Cumulative CO2 Emissions	GtC	2.8	33.0	65.9	100.1	134.0	165.9	196.2	226.0	254.8	282.2	308.3	333.2
Carbon Sequestration	GtC												
Land Use	Million ha												
Cropland		408	412	417	423	419	399	375	339	309	286	261	236
Grasslands		796	820	861	918	942	1002	1025	998	974	955	945	935
Energy Biomass		0	2	3	3	3	4	4	8	8	6	3	1
Forest		921	931	922	894	878	847	850	926	980	1014	1016	1018
Others		998	958	920	885	881	871	868	853	851	862	897	932
Total		3123	3123	3123	3123	3123	3123	3123	3123	3123	3123	3123	3123
Anthropogenic Emissions (standardized)													
Fossil Fuel CO2	GtC	2.83	3.20	3.28	3.26	3.16	2.98	3.02	2.90	2.86	2.88	2.90	2.91
Other CO2	GtC	0.00	0.00	0.12	0.19	0.17	0.07	0.00	0.03	-0.03	-0.21	-0.35	-0.49
Total CO2	GtC	2.83	3.20	3.39	3.45	3.33	3.05	3.02	2.94	2.82	2.68	2.55	2.42
CH4 total	MtCH4	73	74	82	92	95	106	114	120	128	138	148	158
N2O total	MtN2O-N	2.6	2.6	2.8	3.0	3.1	3.5	3.8	3.9	4.0	4.2	4.5	4.8
SOx total	MtS	22.7	17.0	13.2	7.0	6.2	5.6	5.0	4.5	4.3	4.5	4.9	5.3
HFC	MtC eq.	19	57	105	99	102	101	102	101	100	98	98	97
PFC	MtC eq.	18	13	14	13	12	11	10	8	7	7	7	7
SF6	MtC eq.	24	23	25	23	17	13	13	11	10	11	10	10
CO	MtCO												
NMVOC	Mt												
NOx	MtN												

Scenario B2-MiniCAM REF		1990	2000	2010	2020	2030	2040	2050	2060	2070	2080	2090	2100
Population	Million	428	425	427	435	438	438	435	429	423	418	412	407
GNP/GDP (mex)	Trillion US$	1.1	1.1	1.3	1.8	2.3	2.9	3.4	4.5	5.7	7.0	8.3	9.8
GNP/GDP (ppp) Trillion (1990 prices)		na	na	na	na	na	na	na	na	na	na	na	na
Final Energy	EJ												
Non-commercial		0	0	0	0	0	0	0	0	0	0	0	0
Solids		13	10	8	7	6	5	5	4	3	3	3	3
Liquids		18	12	8	8	7	7	7	7	7	8	8	9
Gas		19	15	13	13	13	12	11	10	11	12	12	12
Electricity		6	8	10	14	15	17	19	20	22	25	27	29
Others		0	0	0	0	0	0	0	0	0	0	0	0
Total		56	44	39	42	42	41	41	41	43	47	50	52
Primary Energy	EJ												
Coal		18	17	16	15	20	22	22	16	13	12	10	9
Oil		20	13	10	9	6	4	5	6	8	9	9	10
Gas		26	20	19	23	23	22	19	19	20	21	21	20
Nuclear		3	4	5	6	6	6	7	6	6	7	8	9
Biomass		0	1	1	2	2	2	2	2	3	3	2	2
Other Renewables		3	3	4	7	8	10	11	13	15	16	17	17
Total		70	57	54	61	64	66	66	63	63	67	67	68
Cumulative Resources Use	ZJ												
Coal		0.0	0.2	0.4	0.5	0.7	0.9	1.1	1.3	1.4	1.6	1.7	1.8
Oil		0.0	0.2	0.3	0.4	0.4	0.5	0.5	0.6	0.7	0.8	0.8	0.9
Gas		0.0	0.2	0.4	0.6	0.9	1.1	1.3	1.5	1.7	1.9	2.1	2.3
Cumulative CO2 Emissions	GtC	1.3	12.3	21.1	29.8	39.3	48.8	57.8	66.2	74.2	81.3	87.5	93.1
Carbon Sequestration	GtC												
Land Use	Million ha												
Cropland		284	295	308	324	330	326	313	277	245	217	203	189
Grasslands		395	410	448	511	572	615	640	592	553	520	529	539
Energy Biomass		0	0	0	0	0	0	0	0	0	0	0	0
Forest		1007	1016	996	945	885	850	841	935	1008	1060	1038	1017
Others		691	656	625	597	591	587	584	573	571	580	607	633
Total		2377	2377	2377	2377	2377	2377	2377	2377	2377	2377	2377	2377
Anthropogenic Emissions (standardized)													
Fossil Fuel CO2	GtC	1.30	0.91	0.80	0.84	0.87	0.87	0.85	0.74	0.70	0.72	0.71	0.70
Other CO2	GtC	0.00	0.00	0.03	0.08	0.10	0.07	0.01	0.08	0.06	-0.06	-0.12	-0.18
Total CO2	GtC	1.30	0.91	0.84	0.92	0.97	0.94	0.85	0.83	0.76	0.66	0.59	0.52
CH4 total	MtCH4	47	39	46	60	69	76	81	80	81	84	87	91
N2O total	MtN2O-N	0.6	0.6	0.7	0.9	1.1	1.2	1.4	1.4	1.5	1.5	1.7	1.9
SOx total	MtS	17.0	11.0	10.2	9.6	10.5	11.0	11.1	8.8	6.7	5.0	4.1	3.2
HFC	MtC eq.	0	4	9	15	21	25	25	26	26	26	27	27
PFC	MtC eq.	7	4	8	10	14	19	25	28	30	30	28	27
SF6	MtC eq.	7	6	5	7	9	12	15	16	15	14	14	14
CO	MtCO												
NMVOC	Mt												
NOx	MtN												

Scenario B2-MiniCAM ASIA		1990	2000	2010	2020	2030	2040	2050	2060	2070	2080	2090	2100
Population	Million	2790	3261	3707	4127	4463	4739	4953	5037	5070	5052	4932	4814
GNP/GDP (mex)	Trillion US$	1.4	3.1	6.0	10.3	17.5	25.9	35.6	46.6	58.4	71.0	84.5	99.3
GNP/GDP (ppp) Trillion (1990 prices)		na	na	na	na	na	na	na	na	na	na	na	na
Final Energy	EJ												
Non-commercial		0	0	0	0	0	0	0	0	0	0	0	0
Solids		20	30	42	55	64	69	72	59	52	50	50	50
Liquids		14	19	26	35	41	50	64	78	91	103	111	119
Gas		2	5	9	13	17	19	22	25	30	35	35	36
Electricity		4	11	22	37	55	77	102	131	160	189	215	241
Others		0	0	0	0	0	0	0	0	0	0	0	0
Total		40	66	99	140	176	216	260	293	332	377	411	445
Primary Energy	EJ												
Coal		26	44	64	86	104	116	121	103	92	88	88	87
Oil		16	21	28	37	39	48	62	80	95	109	117	126
Gas		3	9	19	35	48	62	75	90	103	114	114	114
Nuclear		1	4	8	12	20	29	40	45	51	57	71	84
Biomass		0	2	5	8	10	13	15	20	23	24	24	24
Other Renewables		3	4	7	12	21	33	48	67	86	105	114	123
Total		49	83	130	190	243	300	361	405	450	498	528	558
Cumulative Resources Use	ZJ												
Coal		0.0	0.4	1.0	1.7	2.7	3.8	5.0	6.0	7.0	7.9	8.8	9.7
Oil		0.0	0.2	0.5	0.8	1.2	1.6	2.1	2.9	3.8	4.8	5.9	7.1
Gas		0.0	0.1	0.2	0.5	0.9	1.5	2.1	3.0	4.0	5.0	6.2	7.3
Cumulative CO2 Emissions	GtC	1.5	19.3	43.2	75.6	116.4	163.7	216.2	271.4	327.4	385.2	444.2	503.6
Carbon Sequestration	GtC												
Land Use	Million ha												
Cropland		389	401	413	425	430	428	417	381	349	321	300	279
Grasslands		508	523	550	589	625	655	679	674	666	655	649	643
Energy Biomass		0	0	1	3	3	3	3	10	12	8	6	4
Forest		1168	1143	1104	1052	1010	981	965	1000	1039	1082	1110	1139
Others		664	632	603	577	572	568	566	555	554	562	588	614
Total		2729	2699	2671	2646	2639	2634	2630	2620	2619	2628	2653	2678
Anthropogenic Emissions (standardized)													
Fossil Fuel CO2	GtC	1.15	1.78	2.55	3.48	4.18	4.82	5.39	5.45	5.64	5.96	6.11	6.26
Other CO2	GtC	0.37	0.26	0.19	0.25	0.25	0.20	0.10	0.09	0.03	-0.08	-0.19	-0.29
Total CO2	GtC	1.53	2.03	2.74	3.73	4.43	5.02	5.49	5.54	5.67	5.88	5.92	5.96
CH4 total	MtCH4	113	125	133	142	154	163	171	164	160	160	163	167
N2O total	MtN2O-N	2.3	2.6	2.9	3.2	3.6	4.0	4.4	4.5	4.7	5.0	5.4	5.8
SOx total	MtS	17.7	25.3	35.9	45.6	48.5	48.2	44.7	34.6	26.2	19.4	16.0	12.5
HFC	MtC eq.	0	5	21	40	66	95	130	164	199	233	267	302
PFC	MtC eq.	3	5	14	22	30	38	46	51	54	54	53	51
SF6	MtC eq.	4	7	12	18	25	32	36	36	33	27	29	30
CO	MtCO												
NMVOC	Mt												
NOx	MtN												

Scenario B2-MiniCAM ALM		1990	2000	2010	2020	2030	2040	2050	2060	2070	2080	2090	2100
Population	Million	1236	1554	1917	2326	2725	3104	3463	3737	3953	4113	4153	4192
GNP/GDP (mex)	Trillion US$	1.9	2.8	4.4	6.8	11.8	18.4	26.8	38.5	51.1	64.6	78.9	94.7
GNP/GDP (ppp) Trillion (1990 prices)		na	na	na	na	na	na	na	na	na	na	na	na
Final Energy	EJ												
Non-commercial		0	0	0	0	0	0	0	0	0	0	0	0
Solids		2	3	4	6	9	12	15	14	14	14	14	14
Liquids		17	21	27	34	40	52	73	94	116	138	155	173
Gas		5	7	10	14	21	27	31	38	46	56	61	65
Electricity		3	6	9	15	27	43	63	89	116	143	169	195
Others		0	0	0	0	0	0	0	0	0	0	0	0
Total		27	36	50	69	97	134	182	235	292	351	399	448
Primary Energy	EJ												
Coal		4	6	8	11	19	28	37	39	40	42	43	45
Oil		20	23	28	35	40	53	75	98	122	145	163	181
Gas		7	9	14	22	36	50	62	80	98	115	122	129
Nuclear		0	2	4	5	9	16	23	29	36	43	55	68
Biomass		0	1	1	2	3	5	7	10	12	13	13	13
Other Renewables		5	6	9	13	20	28	37	53	68	84	94	103
Total		35	46	64	88	128	179	241	309	376	441	490	538
Cumulative Resources Use	ZJ												
Coal		0.0	0.1	0.1	0.2	0.4	0.6	0.9	1.3	1.7	2.1	2.6	3.0
Oil		0.0	0.2	0.5	0.8	1.2	1.7	2.3	3.2	4.3	5.6	7.2	8.8
Gas		0.0	0.1	0.2	0.4	0.7	1.2	1.7	2.5	3.3	4.4	5.6	6.8
Cumulative CO2 Emissions	GtC	1.4	17.8	36.5	58.0	83.8	114.1	149.5	191.3	239.4	292.1	347.8	405.7
Carbon Sequestration	GtC												
Land Use	Million ha												
Cropland		391	365	353	355	351	341	326	299	272	246	224	202
Grasslands		1510	1594	1713	1867	2012	2121	2197	2127	2052	1970	1943	1916
Energy Biomass		0	0	0	0	0	0	0	4	4	2	1	-1
Forest		3641	3589	3487	3334	3190	3085	3018	3114	3218	3330	3379	3428
Others		1957	1882	1808	1735	1714	1698	1688	1658	1651	1669	1730	1791
Total		7499	7430	7361	7291	7266	7246	7229	7201	7197	7217	7276	7336
Anthropogenic Emissions (standardized)													
Fossil Fuel CO2	GtC	0.72	1.01	1.22	1.53	2.03	2.68	3.47	4.19	4.92	5.64	6.14	6.67
Other CO2	GtC	0.73	0.82	0.69	0.85	0.75	0.58	0.34	0.35	0.17	-0.18	-0.47	-0.76
Total CO2	GtC	1.45	1.83	1.91	2.38	2.78	3.26	3.82	4.54	5.09	5.46	5.67	5.91
CH4 total	MtCH4	77	85	89	97	112	123	131	135	140	146	149	153
N2O total	MtN2O-N	1.2	1.3	1.6	2.0	2.6	3.1	3.7	3.9	4.3	4.6	5.1	5.7
SOx total	MtS	10.5	12.8	14.2	15.3	16.9	19.3	22.6	21.0	18.7	15.8	13.2	10.5
HFC	MtC eq.	0	2	12	20	32	54	84	120	157	190	211	223
PFC	MtC eq.	4	4	7	10	14	20	27	33	37	39	39	37
SF6	MtC eq.	3	5	5	7	9	20	16	17	15	11	13	14
CO	MtCO												
NMVOC	Mt												
NOx	MtN												

Scenario B2C-MARIA World		**1990**	**2000**	**2010**	**2020**	**2030**	**2040**	**2050**	**2060**	**2070**	**2080**	**2090**	**2100**
Population	Million	5262	6091	6891	7672	8372	8930	9367	9704	9960	10159	10306	10414
GNP/GDP (mex)	Trillion US$	19.5	27.2	37.0	49.8	64.1	83.2	108.1	130.5	156.2	181.5	205.4	231.7
GNP/GDP (ppp) Trillion (1990 prices)													
Final Energy	EJ												
Non-commercial													
Solids		48	48	67	107	161	221	277	322	356	381	394	395
Liquids		138	148	158	187	217	251	248	284	291	325	354	377
Gas		56	61	88	87	75	60	79	58	67	47	33	23
Electricity		35	49	60	71	84	97	114	124	134	143	147	153
Others		0	0	0	0	0	0	0	0	0	0	0	0
Total		278	305	373	451	537	629	718	788	847	895	928	949
Primary Energy	EJ												
Coal		90	79	90	124	175	232	287	331	363	387	399	400
Oil		123	134	141	158	191	197	162	178	163	172	185	206
Gas		71	105	151	170	183	193	198	197	223	223	223	228
Nuclear		22	19	16	13	11	10	8	7	12	10	9	8
Biomass		28	27	34	47	47	76	101	117	136	158	172	173
Other Renewables		9	8	7	7	6	5	58	63	66	66	64	62
Total		343	372	438	519	613	714	815	893	962	1015	1052	1077
Cumulative Resources Use	ZJ												
Coal		0.0	0.9	1.7	2.6	3.8	5.6	7.9	10.8	14.1	17.7	21.6	25.6
Oil		0.0	1.2	2.6	4.0	5.6	7.5	9.4	11.1	12.8	14.5	16.2	18.0
Gas		0.0	0.7	1.8	3.3	5.0	6.8	8.7	10.7	12.7	14.9	17.2	19.4
Cumulative CO2 Emissions	GtC	7.1	82.4	167.9	266.9	384.5	522.5	673.8	835.5	1008.2	1189.8	1378.5	1572.8
Carbon Sequestration	GtC	0.0	0.0	0.0	0.0	0.0	0.0	0.0	0.0	0.0	0.0	0.0	0.0
Land Use	Million ha												
Cropland		1451	1451	1458	1744	1940	2051	2079	2013	1906	1777	1691	1691
Grasslands		3395	3395	3392	3113	2925	2904	2904	2904	2904	2904	2904	2904
Energy Biomass		0	0	0	0	0	0	0	66	173	302	388	388
Forest		4138	4138	4138	4138	4138	4100	4099	4131	4138	4138	4138	4138
Others		4061	4061	4058	4051	4043	3990	3963	3931	3924	3924	3924	3924
Total		13045	13045	13045	13045	13045	13045	13045	13045	13045	13045	13045	13045
Anthropogenic Emissions (standardized)													
Fossil Fuel CO2	GtC	5.99	6.90	8.03	9.56	11.73	13.49	14.28	15.73	16.65	17.46	18.04	18.57
Other CO2	GtC	1.11	1.07	1.09	1.11	1.12	1.26	1.23	1.09	1.09	1.11	1.12	1.14
Total CO2	GtC	7.10	7.97	9.12	10.67	12.85	14.76	15.51	16.81	17.74	18.57	19.16	19.71
CH4 total	MtCH4												
N2O total	MtN2O-N												
SOx total	MtS												
CFC/HFC/HCFC	MtC eq.	1672	883	786	299	272	315	346	413	483	548	603	649
PFC	MtC eq.	32	25	42	55	70	88	107	120	128	130	127	121
SF6	MtC eq.	38	40	48	55	60	76	79	80	74	63	65	69
CO	MtCO												
NMVOC	Mt												
NOx	MtN												

Scenario B2C-MARIA OECD90		1990	2000	2010	2020	2030	2040	2050	2060	2070	2080	2090	2100
Population	Million	859	917	953	982	994	988	976	965	952	941	934	928
GNP/GDP (mex)	Trillion US$	15.6	20.7	26.1	30.2	33.4	37.1	41.2	43.8	47.9	52.8	57.5	62.5
GNP/GDP (ppp) Trillion (1990 prices)													
Final Energy	EJ												
Non-commercial													
Solids		12	8	6	21	44	62	73	81	88	92	96	96
Liquids		71	74	68	63	59	56	53	51	50	50	50	50
Gas		28	32	51	48	34	24	17	12	8	6	4	3
Electricity		21	28	32	35	38	39	40	41	42	43	44	45
Others		0	0	0	0	0	0	0	0	0	0	0	0
Total		132	142	157	167	175	180	183	185	189	191	194	194
Primary Energy	EJ												
Coal		38	28	21	33	53	70	81	88	94	98	101	100
Oil		71	74	68	63	59	55	50	47	42	36	27	34
Gas		34	55	85	91	84	78	63	61	58	59	61	63
Nuclear		18	15	12	11	9	8	7	6	6	5	5	5
Biomass		6	4	3	2	1	2	4	5	9	14	23	16
Other Renewables		5	5	4	4	3	3	14	14	17	16	15	15
Total		171	180	193	203	210	216	219	221	225	229	232	233
Cumulative Resources Use	ZJ												
Coal		0.0	0.4	0.7	0.9	1.2	1.7	2.4	3.2	4.1	5.1	6.0	7.0
Oil		0.0	0.7	1.5	2.1	2.8	3.4	3.9	4.4	4.9	5.3	5.7	5.9
Gas		0.0	0.3	0.9	1.7	2.6	3.5	4.2	4.9	5.5	6.1	6.7	7.3
Cumulative CO2 Emissions	GtC	2.8	33.0	65.8	101.2	140.0	182.2	225.4	268.7	312.5	356.6	400.6	445.3
Carbon Sequestration	GtC	0.0	0.0	0.0	0.0	0.0	0.0	0.0	0.0	0.0	0.0	0.0	0.0
Land Use	Million ha												
Cropland		378	378	378	378	378	378	378	354	312	251	202	202
Grasslands		756	756	756	756	756	756	756	756	756	756	756	756
Energy Biomass		0	0	0	0	0	0	0	23	65	127	176	176
Forest		756	756	756	756	756	756	756	756	756	756	756	756
Others		794	794	794	794	794	794	794	794	794	794	794	794
Total		2684	2684	2684	2684	2684	2684	2684	2684	2684	2684	2684	2684
Anthropogenic Emissions (standardized)													
Fossil Fuel CO2	GtC	2.83	3.20	3.35	3.64	3.99	4.25	4.21	4.29	4.32	4.31	4.24	4.39
Other CO2	GtC	0.00	0.00	0.03	0.05	0.08	0.11	0.08	0.06	0.09	0.12	0.14	0.17
Total CO2	GtC	2.83	3.20	3.38	3.69	4.07	4.36	4.29	4.36	4.40	4.42	4.39	4.56
CH4 total	MtCH4												
N2O total	MtN2O-N												
SOx total	MtS												
HFC	MtC eq.	19	57	105	99	102	101	102	101	100	98	98	97
PFC	MtC eq.	18	13	14	13	12	11	10	8	7	7	7	7
SF6	MtC eq.	24	23	25	23	17	13	13	11	10	11	10	10
CO	MtCO												
NMVOC	Mt												
NOx	MtN												

Scenario B2C-MARIA REF		1990	2000	2010	2020	2030	2040	2050	2060	2070	2080	2090	2100
Population	Million	413	415	417	418	416	411	406	396	389	384	381	379
GNP/GDP (mex)	Trillion US$	0.9	1.2	1.4	1.7	2.6	4.2	6.8	8.2	10.0	11.5	12.9	14.5
GNP/GDP (ppp) Trillion (1990 prices)													
Final Energy	EJ												
Non-commercial													
Solids		13	9	7	15	26	43	61	68	75	78	80	81
Liquids		17	13	12	11	12	12	9	7	5	7	9	11
Gas		21	24	29	26	24	17	17	15	15	10	7	5
Electricity		7	8	8	9	10	12	15	16	16	16	16	17
Others		0	0	0	0	0	0	0	0	0	0	0	0
Total		58	54	56	60	72	85	101	106	111	113	113	113
Primary Energy	EJ												
Coal		19	13	10	17	27	44	61	69	75	79	81	81
Oil		19	14	12	11	12	12	9	6	4	7	9	11
Gas		27	33	39	38	40	38	40	40	40	36	33	32
Nuclear		3	2	2	1	1	1	1	0	0	0	0	0
Biomass		1	1	0	0	0	0	1	1	1	0	0	0
Other Renewables		1	1	1	1	1	1	3	3	3	3	3	3
Total		69	64	65	68	81	96	114	119	124	126	126	126
Cumulative Resources Use	ZJ												
Coal		0.0	0.2	0.3	0.4	0.6	0.9	1.3	1.9	2.6	3.3	4.1	4.9
Oil		0.0	0.2	0.3	0.4	0.6	0.7	0.8	0.9	1.0	1.0	1.1	1.2
Gas		0.0	0.3	0.6	1.0	1.4	1.8	2.1	2.5	2.9	3.3	3.7	4.0
Cumulative CO2 Emissions	GtC	1.3	12.3	21.4	31.2	43.4	59.3	79.5	102.7	127.5	153.5	180.2	207.5
Carbon Sequestration	GtC	0.0	0.0	0.0	0.0	0.0	0.0	0.0	0.0	0.0	0.0	0.0	0.0
Land Use	Million ha												
Cropland		217	217	217	217	217	217	217	174	139	111	103	103
Grasslands		114	114	114	114	114	114	114	114	114	114	114	114
Energy Biomass		0	0	0	0	0	0	0	43	78	106	114	114
Forest		815	815	815	815	815	815	815	815	815	815	815	815
Others		722	722	722	722	722	722	722	722	722	722	722	722
Total		1868	1868	1868	1868	1868	1868	1868	1868	1868	1868	1868	1868
Anthropogenic Emissions (standardized)													
Fossil Fuel CO2	GtC	1.30	0.91	0.89	1.02	1.33	1.74	2.16	2.30	2.44	2.52	2.56	2.58
Other CO2	GtC	0.00	0.00	0.02	0.03	0.05	0.06	0.08	0.10	0.11	0.13	0.15	0.16
Total CO2	GtC	1.30	0.91	0.91	1.05	1.38	1.81	2.24	2.40	2.55	2.65	2.70	2.74
CH4 total	MtCH4												
N2O total	MtN2O-N												
SOx total	MtS												
HFC	MtC eq.	0	4	9	15	21	25	25	26	26	26	27	27
PFC	MtC eq.	7	4	8	10	14	19	25	28	30	30	28	27
SF6	MtC eq.	7	6	5	7	9	12	15	16	15	14	14	14
CO	MtCO												
NMVOC	Mt												
NOx	MtN												

Scenario B2C-MARIA ASIA		1990	2000	2010	2020	2030	2040	2050	2060	2070	2080	2090	2100
Population	Million	2642	3062	3450	3796	4089	4316	4479	4613	4720	4813	4851	4888
GNP/GDP (mex)	Trillion US$	1.2	2.9	6.3	12.9	19.9	28.1	37.8	46.8	56.9	67.0	76.7	87.3
GNP/GDP (ppp) Trillion (1990 prices)													
Final Energy	EJ												
Non-commercial													
Solids		20	28	48	61	75	89	98	109	116	123	126	126
Liquids		26	34	54	82	100	119	135	148	158	167	176	183
Gas		2	2	1	1	1	1	1	0	0	0	0	0
Electricity		4	8	12	16	23	27	34	36	40	42	44	46
Others		0	0	0	0	0	0	0	0	0	0	0	0
Total		52	71	115	161	199	237	267	294	314	333	346	355
Primary Energy	EJ												
Coal		28	34	52	64	77	91	99	110	116	123	126	126
Oil		14	23	38	55	76	70	58	58	56	56	74	101
Gas		3	5	8	14	21	28	29	35	41	49	54	59
Nuclear		1	2	1	1	1	1	0	0	6	4	3	3
Biomass		13	17	26	42	43	70	91	100	110	116	105	85
Other Renewables		1	1	1	1	1	1	17	20	20	21	20	20
Total		61	82	127	176	219	260	295	324	349	369	383	393
Cumulative Resources Use	ZJ												
Coal		0.0	0.3	0.6	1.1	1.8	2.6	3.5	4.4	5.5	6.7	7.9	9.2
Oil		0.0	0.1	0.4	0.8	1.3	2.1	2.8	3.3	3.9	4.5	5.0	5.8
Gas		0.0	0.0	0.1	0.2	0.3	0.5	0.8	1.1	1.4	1.8	2.3	2.9
Cumulative CO2 Emissions	GtC	1.5	19.3	43.8	75.8	115.8	161.7	209.1	258.1	310.0	364.5	422.9	486.9
Carbon Sequestration	GtC	0.0	0.0	0.0	0.0	0.0	0.0	0.0	0.0	0.0	0.0	0.0	0.0
Land Use	Million ha												
Cropland		366	366	373	380	380	380	380	380	380	379	350	350
Grasslands		431	431	428	425	426	426	426	426	426	426	426	426
Energy Biomass		0	0	0	0	0	0	0	0	0	1	30	30
Forest		365	365	365	365	365	365	365	365	365	365	365	365
Others		458	458	455	451	449	449	449	449	449	449	449	449
Total		1621	1621	1621	1621	1621	1621	1621	1621	1621	1621	1621	1621
Anthropogenic Emissions (standardized)													
Fossil Fuel CO2	GtC	1.15	1.78	2.61	3.33	4.21	4.54	4.53	4.90	5.11	5.43	5.94	6.54
Other CO2	GtC	0.37	0.26	0.25	0.23	0.22	0.21	0.19	0.18	0.18	0.17	0.15	0.14
Total CO2	GtC	1.53	2.03	2.85	3.56	4.43	4.75	4.72	5.08	5.29	5.61	6.09	6.69
CH4 total	MtCH4												
N2O total	MtN2O-N												
SOx total	MtS												
HFC	MtC eq.	0	5	21	40	66	95	130	164	199	233	267	302
PFC	MtC eq.	3	5	14	22	30	38	46	51	54	54	53	51
SF6	MtC eq.	4	7	12	18	25	32	36	36	33	27	29	30
CO	MtCO												
NMVOC	Mt												
NOx	MtN												

Scenario B2C-MARIA ALM		1990	2000	2010	2020	2030	2040	2050	2060	2070	2080	2090	2100
Population	Million	1348	1697	2071	2476	2873	3214	3506	3730	3900	4021	4141	4220
GNP/GDP (mex)	Trillion US$	1.7	2.4	3.2	4.9	8.1	13.9	22.3	31.7	41.4	50.2	58.2	67.3
GNP/GDP (ppp) Trillion (1990 prices)													
Final Energy	EJ												
Non-commercial													
Solids		3	2	5	10	17	28	46	64	77	87	92	94
Liquids		24	26	25	31	45	63	51	77	77	100	118	133
Gas		5	4	7	12	16	18	45	31	43	30	21	15
Electricity		3	6	8	10	14	18	25	31	36	41	43	46
Others		0	0	0	0	0	0	0	0	0	0	0	0
Total		35	38	45	63	91	127	167	203	234	258	275	287
Primary Energy	EJ												
Coal		6	4	7	11	17	28	46	64	77	87	92	94
Oil		18	23	22	30	44	60	45	68	61	73	74	61
Gas		7	12	19	28	39	50	67	62	84	78	75	74
Nuclear		0	0	0	0	0	0	0	0	0	0	0	0
Biomass		8	6	4	3	2	4	6	10	17	27	44	72
Other Renewables		2	2	1	1	1	1	24	25	26	26	25	24
Total		41	46	53	73	103	142	188	228	264	291	310	325
Cumulative Resources Use	ZJ												
Coal		0.0	0.1	0.1	0.2	0.3	0.4	0.7	1.2	1.8	2.6	3.5	4.4
Oil		0.0	0.2	0.4	0.6	0.9	1.4	2.0	2.4	3.1	3.7	4.4	5.2
Gas		0.0	0.1	0.2	0.4	0.7	1.1	1.6	2.2	2.8	3.7	4.5	5.2
Cumulative CO2 Emissions	GtC	1.4	17.8	36.9	58.6	85.2	119.3	159.8	206.0	258.3	315.3	374.6	433.2
Carbon Sequestration	GtC	0.0	0.0	0.0	0.0	0.0	0.0	0.0	0.0	0.0	0.0	0.0	0.0
Land Use	Million ha												
Cropland		490	490	490	769	965	1076	1104	1104	1075	1036	1036	1036
Grasslands		2095	2095	2095	1818	1630	1609	1609	1609	1609	1609	1609	1609
Energy Biomass		0	0	0	0	0	0	0	0	30	68	68	68
Forest		2202	2202	2202	2202	2202	2164	2163	2195	2202	2202	2202	2202
Others		2086	2086	2086	2084	2076	2024	1997	1965	1958	1958	1958	1958
Total		6873	6873	6873	6873	6873	6873	6873	6873	6873	6873	6873	6873
Anthropogenic Emissions (standardized)													
Fossil Fuel CO2	GtC	0.72	1.01	1.18	1.57	2.19	2.96	3.39	4.23	4.78	5.20	5.30	5.07
Other CO2	GtC	0.73	0.82	0.80	0.79	0.77	0.88	0.87	0.75	0.71	0.69	0.68	0.66
Total CO2	GtC	1.45	1.83	1.98	2.36	2.96	3.84	4.26	4.98	5.49	5.90	5.98	5.73
CH4 total	MtCH4												
N2O total	MtN2O-N												
SOx total	MtS												
HFC	MtC eq.	0	2	12	20	32	54	84	120	157	190	211	223
PFC	MtC eq.	4	4	7	10	14	20	27	33	37	39	39	37
SF6	MtC eq.	3	5	5	7	9	20	16	17	15	11	13	14
CO	MtCO												
NMVOC	Mt												
NOx	MtN												

Scenario B2High-MiniCAM World		**1990**	**2000**	**2010**	**2020**	**2030**	**2040**	**2050**	**2060**	**2070**	**2080**	**2090**	**2100**
Population	Million	5293	6147	7009	7880	8640	9304	9874	10216	10453	10585	10501	10418
GNP/GDP (mex)	Trillion US$	20.7	27.4	36.3	47.3	61.8	79.8	101.1	126.5	154.0	183.6	215.7	250.7
GNP/GDP (ppp) Trillion (1990 prices)		na	na	na	na	na	na	na	na	na	na	na	na
Final Energy	EJ												
Non-commercial		0	0	0	0	0	0	0	0	0	0	0	0
Solids		45	57	72	89	104	118	131	119	111	108	105	102
Liquids		121	124	131	143	143	159	193	225	258	292	318	344
Gas		52	62	72	84	81	74	62	58	57	60	52	44
Electricity		35	53	80	116	159	209	268	339	415	497	574	651
Others		0	0	0	0	0	0	0	0	0	0	0	0
Total		253	295	355	433	486	560	654	740	841	957	1049	1141
Primary Energy	EJ												
Coal		88	115	145	178	263	362	474	540	592	630	700	771
Oil		131	133	139	149	114	91	80	74	77	88	72	57
Gas		70	83	115	165	180	185	181	183	191	206	173	140
Nuclear		24	25	31	42	60	79	100	108	121	142	161	180
Biomass		0	6	12	17	27	40	56	86	110	128	150	172
Other Renewables		24	24	27	32	40	55	75	115	158	202	254	305
Total		336	386	468	582	684	812	966	1105	1248	1395	1510	1625
Cumulative Resources Use	ZJ												
Coal		0.1	1.2	2.5	4.1	6.5	9.7	13.7	19.0	24.6	30.6	37.5	44.4
Oil		0.1	1.5	2.8	4.3	5.5	6.5	7.4	8.1	8.9	9.7	10.5	11.2
Gas		0.1	0.9	1.9	3.2	5.0	6.8	8.6	10.5	12.4	14.3	16.1	17.9
Cumulative CO2 Emissions	GtC	7.1	82.4	168.5	271.1	392.9	532.5	691.9	870.2	1062.5	1263.9	1471.7	1686.1
Carbon Sequestration	GtC												
Land Use	Million ha												
Cropland		1472	1473	1491	1526	1522	1488	1423	1284	1163	1060	975	890
Grasslands		3209	3347	3570	3880	4164	4370	4501	4326	4168	4027	3957	3886
Energy Biomass		0	3	8	13	36	74	128	233	308	352	430	507
Forest		4173	4214	4140	3952	3701	3531	3441	3733	3962	4129	4093	4058
Others		4310	4127	3955	3794	3742	3701	3672	3588	3562	3596	3710	3824
Total		13164	13164	13164	13164	13164	13164	13164	13164	13164	13164	13164	13164
Anthropogenic Emissions (standardized)													
Fossil Fuel CO2	GtC	5.99	6.90	8.20	9.92	11.77	13.87	16.44	18.03	19.52	20.92	21.97	23.10
Other CO2	GtC	1.11	1.07	1.03	1.38	1.29	0.99	0.57	0.62	0.28	-0.44	-0.88	-1.32
Total CO2	GtC	7.10	7.97	9.23	11.30	13.06	14.86	17.02	18.65	19.80	20.47	21.09	21.78
CH4 total	MtCH4	310	323	353	397	439	483	520	533	549	567	590	613
N2O total	MtN2O-N	6.7	7.0	8.0	9.2	10.4	11.9	13.2	13.7	14.3	15.2	16.3	17.5
SOx total	MtS	70.9	69.0	81.6	86.9	97.7	104.4	107.0	86.4	69.1	55.0	45.6	36.2
CFC/HFC/HCFC	MtC eq.	1672	883	786	299	272	315	346	413	483	548	603	649
PFC	MtC eq.	32	25	42	55	70	88	107	120	128	130	127	121
SF6	MtC eq.	38	40	48	55	60	76	79	80	74	63	65	69
CO	MtCO												
NMVOC	Mt												
NOx	MtN												

Scenario B2High-MiniCAM OECD90		1990	2000	2010	2020	2030	2040	2050	2060	2070	2080	2090	2100
Population	Million	838	907	958	993	1003	1024	1023	1013	1006	1002	1003	1004
GNP/GDP (mex)	Trillion US$	16.3	20.5	24.5	28.4	29.4	32.9	35.9	38.0	40.4	43.0	46.6	50.3
GNP/GDP (ppp) Trillion (1990 prices)		na	na	na	na	na	na	na	na	na	na	na	na
Final Energy	EJ												
Non-commercial		0	0	0	0	0	0	0	0	0	0	0	0
Solids		10	12	13	13	13	13	14	12	11	10	10	9
Liquids		72	71	69	65	59	47	47	46	45	46	47	47
Gas		27	35	42	47	44	34	26	22	20	20	17	14
Electricity		22	28	34	42	44	52	58	61	64	67	72	76
Others		0	0	0	0	0	0	0	0	0	0	0	0
Total		130	146	158	166	160	146	146	140	139	143	145	146
Primary Energy	EJ												
Coal		40	45	46	41	47	67	82	93	104	113	136	159
Oil		76	76	72	66	53	23	13	7	4	3	3	2
Gas		34	45	61	82	79	67	57	50	46	44	36	27
Nuclear		20	15	13	13	14	15	15	14	14	15	17	18
Biomass		0	2	4	4	5	8	10	14	16	18	21	24
Other Renewables		12	11	10	10	10	11	13	16	20	24	29	34
Total		182	195	206	216	208	190	189	194	203	217	241	265
Cumulative Resources Use	ZJ												
Coal		0.0	0.5	0.9	1.4	1.6	2.5	3.2	4.1	5.1	6.1	7.5	8.8
Oil		0.1	0.8	1.6	2.3	2.5	3.1	3.3	3.3	3.4	3.4	3.5	3.5
Gas		0.0	0.5	1.0	1.7	2.1	3.2	3.8	4.3	4.8	5.3	5.6	6.0
Cumulative CO2 Emissions	GtC	2.8	33.0	66.5	102.2	138.3	173.0	207.2	241.8	276.3	310.1	343.9	378.9
Carbon Sequestration	GtC												
Land Use	Million ha												
Cropland		408	412	416	422	417	397	374	337	308	285	260	235
Grasslands		796	819	859	915	938	993	1013	984	960	940	924	908
Energy Biomass		0	3	6	7	11	24	36	53	65	71	84	97
Forest		921	931	922	894	877	843	840	907	954	980	981	981
Others		998	958	920	885	879	866	860	842	837	846	874	902
Total		3123	3123	3123	3123	3123	3123	3123	3123	3123	3123	3123	3123
Anthropogenic Emissions (standardized)													
Fossil Fuel CO2	GtC	2.83	3.20	3.39	3.46	3.39	3.30	3.42	3.41	3.44	3.52	3.73	3.95
Other CO2	GtC	0.00	0.00	0.12	0.19	0.17	0.09	0.03	0.05	-0.02	-0.18	-0.29	-0.40
Total CO2	GtC	2.83	3.20	3.50	3.65	3.56	3.39	3.45	3.47	3.43	3.33	3.44	3.55
CH4 total	MtCH4	73	74	82	91	95	105	113	117	123	130	141	152
N2O total	MtN2O-N	2.6	2.6	2.8	3.0	3.1	3.5	3.8	3.8	3.9	4.1	4.4	4.6
SOx total	MtS	22.7	17.0	14.8	6.7	5.5	4.8	4.4	3.5	3.2	3.7	4.3	5.0
HFC	MtC eq.	19	57	105	99	102	101	102	101	100	98	98	97
PFC	MtC eq.	18	13	14	13	12	11	10	8	7	7	7	7
SF6	MtC eq.	24	23	25	23	17	13	13	11	10	11	10	10
CO	MtCO												
NMVOC	Mt												
NOx	MtN												

Scenario B2High-MiniCAM REF		**1990**	**2000**	**2010**	**2020**	**2030**	**2040**	**2050**	**2060**	**2070**	**2080**	**2090**	**2100**
Population	Million	428	425	427	435	438	438	435	429	423	418	412	407
GNP/GDP (mex)	Trillion US$	1.1	1.1	1.3	1.8	2.3	2.8	3.4	4.4	5.6	6.9	8.2	9.6
GNP/GDP (ppp) Trillion (1990 prices)		na	na	na	na	na	na	na	na	na	na	na	na
Final Energy	EJ												
Non-commercial		0	0	0	0	0	0	0	0	0	0	0	0
Solids		13	10	8	7	6	6	6	5	5	4	4	4
Liquids		18	11	8	8	7	6	6	7	7	8	8	8
Gas		19	14	12	11	10	8	5	5	5	5	4	3
Electricity		6	8	11	15	16	18	19	22	26	29	31	33
Others		0	0	0	0	0	0	0	0	0	0	0	0
Total		56	44	38	41	39	38	37	39	42	45	46	47
Primary Energy	EJ												
Coal		18	17	17	18	38	61	86	113	147	188	236	283
Oil		20	13	10	9	3	0	0	0	0	0	0	0
Gas		26	20	18	22	20	17	13	12	11	12	9	6
Nuclear		3	4	5	7	7	7	7	7	7	8	8	8
Biomass		0	1	1	2	3	3	5	7	9	10	12	13
Other Renewables		3	3	3	4	5	6	7	9	12	14	16	18
Total		70	57	55	62	76	94	117	149	187	231	280	328
Cumulative Resources Use	ZJ												
Coal		0.0	0.2	0.4	0.5	0.9	1.4	2.1	3.1	4.5	6.1	8.3	10.6
Oil		0.0	0.2	0.3	0.4	0.4	0.4	0.4	0.4	0.4	0.4	0.4	0.4
Gas		0.0	0.2	0.4	0.6	0.8	1.0	1.2	1.3	1.4	1.5	1.6	1.7
Cumulative CO2 Emissions	GtC	1.3	12.3	21.2	30.3	41.4	54.8	70.7	90.0	113.6	141.8	175.3	215.0
Carbon Sequestration	GtC												
Land Use	Million ha												
Cropland		284	295	308	324	331	328	317	281	250	225	211	198
Grasslands		395	410	448	510	569	611	636	588	551	524	529	534
Energy Biomass		0	0	1	2	5	9	15	28	36	38	43	47
Forest		1007	1016	996	945	884	846	831	915	979	1023	1005	987
Others		691	656	625	597	589	583	578	564	561	568	590	612
Total		2377	2377	2377	2377	2377	2377	2377	2377	2377	2377	2377	2377
Anthropogenic Emissions (standardized)													
Fossil Fuel CO2	GtC	1.30	0.91	0.83	0.90	1.13	1.39	1.69	2.06	2.53	3.11	3.75	4.45
Other CO2	GtC	0.00	0.00	0.03	0.08	0.10	0.08	0.02	0.08	0.05	-0.06	-0.11	-0.16
Total CO2	GtC	1.30	0.91	0.86	0.98	1.22	1.47	1.71	2.14	2.59	3.06	3.64	4.29
CH4 total	MtCH4	47	39	47	61	75	88	99	108	123	144	162	179
N2O total	MtN2O-N	0.6	0.6	0.7	0.9	1.1	1.3	1.4	1.4	1.5	1.5	1.7	1.9
SOx total	MtS	17.0	11.0	10.3	10.3	12.6	14.3	15.3	12.8	10.3	7.7	5.9	4.1
HFC	MtC eq.	0	4	9	15	21	25	25	26	26	26	27	27
PFC	MtC eq.	7	4	8	10	14	19	25	28	30	30	28	27
SF6	MtC eq.	7	6	5	7	9	12	15	16	15	14	14	14
CO	MtCO												
NMVOC	Mt												
NOx	MtN												

Scenario B2High-MiniCAM ASIA		1990	2000	2010	2020	2030	2040	2050	2060	2070	2080	2090	2100
Population	Million	2790	3261	3707	4127	4463	4739	4953	5037	5070	5052	4932	4814
GNP/GDP (mex)	Trillion US$	1.4	3.1	6.0	10.3	17.4	25.6	35.2	45.9	57.4	69.7	82.9	97.2
GNP/GDP (ppp) Trillion (1990 prices)		na	na	na	na	na	na	na	na	na	na	na	na
Final Energy	EJ												
Non-commercial		0	0	0	0	0	0	0	0	0	0	0	0
Solids		20	32	46	62	74	83	91	81	75	73	70	68
Liquids		14	20	27	36	43	53	67	79	91	104	112	119
Gas		2	5	9	13	13	13	12	12	13	14	12	11
Electricity		4	11	24	43	64	88	115	147	182	220	253	287
Others		0	0	0	0	0	0	0	0	0	0	0	0
Total		40	68	106	153	193	237	284	319	361	410	447	484
Primary Energy	EJ												
Coal		26	47	72	104	142	182	224	251	254	234	226	218
Oil		16	22	29	39	33	28	25	21	21	25	19	12
Gas		3	9	20	37	48	56	61	63	68	75	62	49
Nuclear		1	4	9	15	25	36	47	50	56	65	73	81
Biomass		0	2	5	9	14	20	27	38	48	55	63	71
Other Renewables		3	4	5	6	10	17	27	45	65	86	109	132
Total		49	87	141	210	272	339	411	470	513	540	551	562
Cumulative Resources Use	ZJ												
Coal		0.0	0.5	1.1	1.9	3.2	4.9	6.8	9.3	11.8	14.2	16.5	18.8
Oil		0.0	0.2	0.5	0.8	1.1	1.4	1.7	1.9	2.2	2.4	2.6	2.8
Gas		0.0	0.1	0.2	0.5	1.0	1.5	2.0	2.7	3.3	4.0	4.7	5.3
Cumulative CO2 Emissions	GtC	1.5	19.3	44.1	79.6	126.6	184.0	251.8	327.7	408.1	489.7	569.9	648.2
Carbon Sequestration	GtC												
Land Use	Million ha												
Cropland		389	401	413	425	426	418	403	365	332	304	278	252
Grasslands		508	523	550	588	621	647	668	662	654	642	628	615
Energy Biomass		0	0	2	5	15	28	43	67	84	95	117	139
Forest		1168	1143	1104	1052	1007	973	951	972	997	1026	1043	1060
Others		664	632	603	577	569	563	559	545	541	547	565	584
Total		2729	2699	2671	2646	2637	2630	2624	2611	2608	2614	2632	2650
Anthropogenic Emissions (standardized)													
Fossil Fuel CO2	GtC	1.15	1.78	2.72	3.92	4.97	6.05	7.14	7.80	8.14	8.18	8.04	7.90
Other CO2	GtC	0.37	0.26	0.19	0.26	0.26	0.22	0.14	0.11	0.04	-0.05	-0.11	-0.17
Total CO2	GtC	1.53	2.03	2.92	4.18	5.23	6.26	7.28	7.90	8.19	8.12	7.93	7.73
CH4 total	MtCH4	113	125	134	145	155	166	178	183	179	166	160	154
N2O total	MtN2O-N	2.3	2.6	2.9	3.2	3.6	4.0	4.4	4.5	4.7	4.9	5.2	5.4
SOx total	MtS	17.7	25.3	38.4	50.6	57.0	59.1	56.9	43.7	32.8	24.3	19.4	14.4
HFC	MtC eq.	0	5	21	40	66	95	130	164	199	233	267	302
PFC	MtC eq.	3	5	14	22	30	38	46	51	54	54	53	51
SF6	MtC eq.	4	7	12	18	25	32	36	36	33	27	29	30
CO	MtCO												
NMVOC	Mt												
NOx	MtN												

Scenario B2High-MiniCAM ALM		1990	2000	2010	2020	2030	2040	2050	2060	2070	2080	2090	2100
Population	Million	1236	1554	1917	2326	2725	3104	3463	3737	3953	4113	4153	4192
GNP/GDP (mex)	Trillion US$	1.9	2.8	4.4	6.8	11.8	18.4	26.6	38.2	50.6	63.9	78.0	93.6
GNP/GDP (ppp) Trillion (1990 prices)		na	na	na	na	na	na	na	na	na	na	na	na
Final Energy	EJ												
Non-commercial		0	0	0	0	0	0	0	0	0	0	0	0
Solids		2	3	5	7	11	15	20	20	20	21	21	21
Liquids		17	22	27	35	41	54	73	94	114	135	152	170
Gas		5	7	10	13	17	19	19	19	20	22	20	17
Electricity		3	6	10	17	32	51	76	109	144	181	218	255
Others		0	0	0	0	0	0	0	0	0	0	0	0
Total		27	37	52	72	101	139	188	242	299	359	411	463
Primary Energy	EJ												
Coal		4	6	10	14	30	53	83	82	86	96	103	110
Oil		20	23	28	36	37	40	42	46	52	59	51	43
Gas		7	10	15	24	36	45	50	58	66	75	66	58
Nuclear		0	2	4	7	14	22	31	37	44	54	64	73
Biomass		0	1	1	3	5	9	15	27	37	44	55	65
Other Renewables		5	6	8	11	15	21	28	44	61	79	100	122
Total		35	47	67	95	137	189	249	293	345	407	439	471
Cumulative Resources Use	ZJ												
Coal		0.0	0.1	0.1	0.3	0.5	1.0	1.6	2.4	3.3	4.2	5.2	6.2
Oil		0.0	0.2	0.5	0.8	1.2	1.6	2.0	2.4	2.9	3.5	4.0	4.5
Gas		0.0	0.1	0.2	0.4	0.7	1.2	1.6	2.2	2.8	3.5	4.2	4.9
Cumulative CO2 Emissions	GtC	1.4	17.8	36.7	59.0	86.7	120.6	162.2	210.8	264.5	322.3	382.5	444.0
Carbon Sequestration	GtC												
Land Use	Million ha												
Cropland		391	365	353	355	353	344	330	301	274	247	226	205
Grasslands		1510	1594	1713	1867	2013	2118	2183	2091	2004	1921	1875	1829
Energy Biomass		0	0	0	0	1	12	34	85	123	148	186	223
Forest		3641	3589	3487	3333	3186	3074	2996	3067	3147	3234	3263	3292
Others		1957	1882	1808	1735	1710	1690	1675	1636	1623	1635	1681	1727
Total		7499	7430	7361	7290	7263	7239	7217	7182	7171	7185	7230	7275
Anthropogenic Emissions (standardized)													
Fossil Fuel CO2	GtC	0.72	1.01	1.27	1.64	2.28	3.13	4.18	4.76	5.40	6.11	6.45	6.81
Other CO2	GtC	0.73	0.82	0.69	0.85	0.77	0.61	0.39	0.38	0.20	-0.15	-0.37	-0.59
Total CO2	GtC	1.45	1.83	1.96	2.49	3.05	3.74	4.58	5.14	5.60	5.96	6.08	6.22
CH4 total	MtCH4	77	85	90	99	114	124	129	125	124	127	127	128
N2O total	MtN2O-N	1.2	1.3	1.6	2.1	2.6	3.1	3.7	4.0	4.3	4.6	5.1	5.6
SOx total	MtS	10.5	12.8	15.1	16.2	19.5	23.3	27.5	23.4	19.7	16.4	13.1	9.7
HFC	MtC eq.	0	2	12	20	32	54	84	120	157	190	211	223
PFC	MtC eq.	4	4	7	10	14	20	27	33	37	39	39	37
SF6	MtC eq.	3	5	5	7	9	20	16	17	15	11	13	14
CO	MtCO												
NMVOC	Mt												
NOx	MtN												

VIII

Acronyms and Abbreviations

Acronyms and Abbreviations

A1	SRES scenario family A1
A1B	Scenariogroup within the A1 scenario family (balanced energy supply mix)
A1C	Scenario group within A1 scenario family (emphasis on coal)
A1FI	Scenariogroup within the A1 scenario family (fossil-intensive, combination of A1C and A1G)
A1G	Scenario group within A1 scenario family (emphasis on oil and gas)
A1T	Scenario group within A1 scenario family (emphasis on non-fossils)
A2	SRES scenario family A2
AAGR	-Average Annual Growth Rate
AEEI	Autonomous Energy Efficiency Improvement (rate)
AFEAS	Alternative Fluorocarbons Environmental Acceptability Study
AFR	Sub-Saharan Africa (see Appendix III)
AGC/MLO	Atmospheric General Circulation/Mixed Layer Ocean model
AIDS	Acquired Immune Deficiency Syndrome
AIM	Asian Integrated Model (see Appendix IV)
ALM	SRES region – Africa, Latin America and Middle East (see Appendix III)
A/O GCM	Atmosphere/Ocean General Circulation Model
AOS	Atmosphere-Ocean System (IMAGE model)
ASF	Atmospheric Stabilization Framework model (see Appendix IV)
ASIA	SRES region – Asia excluding the Middle East (see Appendix III)
B1	SRES scenario family B1
B2	SRES scenario family B2
BGF	Biomass based gaseous fuels (IMAGE model)
BLF	Biomass based liquid fuels (IMAGE model)
BLS	Basic Linked System of National Agricultural Models (see Appendix IV)
BP	British Petroleum
CCGT	combined cycle gas turbine
CES	constant elasticity of substitution
CETA	Carbon Emissions Trajectory Assessment model
CIESIN	Center for International Earth Science Information Network
CO2DB	Carbon Dioxide Database (see Appendix IV)
COP	Conference of the Parties
CORINAIR	Coordination of Information on the Environment - Air
CPA	Centrally Planned Asia and China (see Appendix III)
CPB	Central Planning Bureau (the Netherlands)
CRP	Current Reduction Plans
DEV	Developing countries
DICE	Dynamic Integrated Climate Economy model
DMSP	Defense Meteorological Satellite Program
DOE	Department of Energy
EBC/UDO	Energy Balance Climate/Upwelling Diffusion Ocean model
ECN	Netherlands Energy Research Foundation
EEA	European Environmental Agency
EEU	Central and Eastern Europe (see Appendix III)
EIA	Energy Information Administration (US)
EIS	Energy-Industry System (IMAGE model)
EJ	Exajoules (10^{18}J)
ENDS	Environmental Data Services
EMEP	European Monitoring and Evaluation Programme (for air pollutants)
EMF	Energy Modeling Forum (Stanford University)
EPA	Environmental Protection Agency (US)
ERB	Edmonds, Reilly and Barns model
ESD	Emissions Scenario Database
ETSAP	Energy Technology Systems Analysis Programme (IEA)
EU	European Union
FAO	UN Food and Agriculture Organization
FBC	Fluidized Bed Combustion
FETC	Federal Energy Technology Center (US)
FGD	Flue Gas Desulfurization
FSU	Former Soviet Union (see Appendix III)
FUND	Climate Framework for Uncertainty Negotiation and Distribution model
GCM	General Circulation Model
GCAM	Global Change Assessment Model
GDI	Gender related Development Index (UNDP)
GDP	Gross Domestic Product
GEM	Gender Empowerment Measure (UNDP)
GHG	Greenhouse Gas
GNP	Gross National Product
GRP	Gross Regional Product
GWP	Global Warming Potential
GtC	Gigaton of carbon (1 GtC = 10^{15} gC = 1 PgC ~ 3.7 Gt carbon dioxide)
GWP	Global Warming Potential
HABITAT	United Nations Centre for Human Settlements
HDI	Human Development Index (UNDP)
HIV	Human Immunodeficiency Virus
HS	Harmonized Scenarios
IA	Integrated Assessment
ICAM	Integrated Climate Assessment Model
ICAO	International Civil Aviation Organization
ID	Identification
IEA	International Energy Agency
IEA CIAB	IEA Coal Industry Advisory Board
IEW	International Energy Workshop (IIASA – Stanford University)
IGCC	Integrated Gasification Combined Cycle
IGU	International Gas Union
IIASA	International Institute for Applied Systems Analysis

IMAGE	Integrated Model to Assess the Greenhouse Effect
IND	Industrial(ized) countries
IPCC	Intergovernmental Panel on Climate Change
IS92	IPCC scenarios 1992
IU	Inverted U-curve
J	Joule
ISO	International Standards Organization
km	Kilometers
kt	Kilotons
LAM	Latin America and Caribbean (see Appendix III)
LBNL	Lawrence Berkeley National Laboratory (US)
LNG	Liquid Natural Gas
LPG	Liquefied Petroleum Gas
MACRO	Macroeconomic model (see Appendix IV)
MAGICC	Model for the Assessment of Greenhouse gas-Induced Climate Change
MARIA	Multiregional Approach for Resource and Industry Allocation model (see Appendix IV)
MEA	Middle East and North Africa (see Appendix III)
MERGE	Model for Evaluating the Regional and Global Effects of GHG Reduction Policies
MESSAGE	Model for Energy Supply Strategy Alternatives and their General Environmental Impact (see Appendix IV)
MFR	Maximum Feasible Reduction
MiniCAM	Mini Climate Assessment Model (see Appendix IV)
MJ	Megajoule (10^6 J)
Mt	Megaton = 10^6 tons = Tg
MtS	Megaton (elemental) Sulfur
MWe	Megawatts of electricity (electric capacity)
NAM	North America (see Appendix III)
NAPAP	National Acidic Precipitation Assessment Program (US)
NIES	National Institute for Environmental Studies (Japan)
NIS	Newly Independent States (of the Former Soviet Union)
NGO	Non-Governmental Organization
NMVOC	Non-Methane Volatile Organic Compounds (hydrocarbons)
NTE	Non-Thermal Electric
ODS	Ozone Depleting Substances
ODT	Oven Dry Tons
OECD	Organisation for Economic Cooperation and Development
OECD90	SRES region – OECD member states as at 1990 (see Appendix III)
OLS	Operational Linescan System
OS	Other Scenarios
PAGE	Policy Analysis for the Greenhouse Effect model
PAO	Pacific OECD (see Appendix III)
PAS	Other Pacific Asia (see Appendix III)
PgC	Petagrams of carbon (1 PgC = 1 GtC)
PIEEI	Price Induced Energy Efficiency Improvement
PNLL	Pacific Northwest National Laboratory (US)
ppbv	Parts per billion by volume (10^9)
ppmv	Parts per million by volume (10^6)
PPP	Purchasing Power Parity
pptv	Parts per trillion by volume (10^{12})
RAINS	Regional Acidification INformation and Simulation model (see Appendix IV)
R&D	Research and Development
RCW	Rapidly Changing World scenario
RD&D	Research, Development and Demonstration
REF	SRES region – Central and Eastern Europe and Newly Independent States of the former Soviet Union (see Appendix III)
RIVM	Netherlands National Institute of Public Health and the Environment
SA90	1990 Scientific Assessment of the IPCC
SAR	Second Assessment Report of the IPCC
SAS	South Asia (see Appendix III)
SC	Sulfur Control (scenario)
SCENGEN	Scenario Generator model (University of East Anglia)
SCW	Slowly Changing World scenario
SG	Scenario Generator model (IIASA) (see Appendix IV)
SPM	Summary for Policymakers
SRES	Special Report on Emissions Scenarios
SRTT	Special Report on methodological and technological issues in Technology Transfer
t	Ton
TAR	Third Assessment Report of the IPCC
TE	Thermal Electric
TES	Terrestrial Environment System (IMAGE model)
TFR	Total (average) Fertility Rate
TgC	Teragrams of Carbon (1 TgC = 1 MtC)
TGCIA	Task Group on Climate scenarios for Impact Assessment
TgN	Teragrams (million tons) of (elemental) Nitrogen
TgS	Teragrams (million tons) of (elemental) Sulfur
TIMER	Targets IMAGE Energy Regional simulation model
TSU	Technical Support Unit
UN	United Nations
UNAIDS	Joint United Nations Programme on HIV/AIDS
UNCED	United Nations Conference on Environment and Development
UNDP	United Nations Development Programme
UNEP	United Nations Environment Programme
UNESCO	United Nations Educational, Scientific, and Cultural Organization
UNFCCC	United Nations Framework Convention on Climate Change
UNFPA	United Nations Population Fund

USCB United States Census Bureau
US DOC United States Department of Commerce
US DOE United States Department of Energy
US EPA United States Environmental Protection Agency
VNIR Visible and Near-InfraRed light
VOC Volatile Organic Compounds (hydrocarbons)
WB World Bank
WBCSD World Business Council for Sustainable Development
WEC World Energy Council
WHO World Health Organization
WG Working Group
WMO World Meteorological Organisation
WRI World Resources Institute
ZJ Zetajoule (10^{21}J)

IX

Chemical Symbols

Chemical Symbols

C	Carbon
CF_4	Tetrafluoromethane – CFC-14
C_2F_6	Hexafluoroethane – CFC-116
CFC	Chlorofluorocarbon
CFC-14	Tetrafluoromethane – CF_4
CFC-116	Hexafluoroethylene – C_2F_6
$(CH_2)_4(COOH)_2$	Adipic Acid
CH_4	Methane
CH_3OH	Methanol
CO	Carbon Monoxide
CO_2	Carbon Dioxide
H	Atomic Hydrogen
H_2	Molecular Hydrogen
HCFC	Hydrochlorofluorocarbon
HCO_3	Bicarbonate Ion
HFC	Hydrofluorocarbon (Hydrogenated FluoroCarbons)
HFC-134a	CH_2FCF_3
HNO_3	Nitric Acid
MeOH	Methanol
N	Atomic Nitrogen
N_2	Molecular Nitrogen
NH_3	Ammonia
NMVOC	Non-Methane Volatile Organic Compounds (hydrocarbons)
N_2O	Nitrous Oxide
NO	Nitric Oxide
NO_x	Nitrogen Oxides (the sum of NO and NO_2)
O	Atomic Oxygen
O_2	Molecular Oxygen
O_3	Ozone
PFC	Perfluorocarbon
S	Atomic Sulfur
SF_6	Sulfur Hexafluoride
SO_x	Sulfur Oxides
SO_2	Sulfur Dioxide

X

Units

Units

Table X-1: *SI (Systeme Internationale) Units and Fractions and Multiples Having Special Names*

Physical Quantity	Name of Unit	Symbol
amount of substance	mole	mol
area	hectare	ha
energy	joule	J (kg m^2 s^{-2})
force	newton	N (kg m s^{-2})
frequency	hertz	Hz (s^{-1} = cycles per second)
length	metre (meter in this report)	m
length	micron	μm (10^{-6} m)
power	watt	W (kg m^2 s^{-3} = J s^{-1})
pressure	pascal	Pa (kg $m^{-1}s^{-2}$ = N m^{-2}
temperature	kelvin	K
time	second	s
weight (mass)	kilogram	kg
weight	tonne (ton in this report)	t (10^3 kg)
weight	gram	g (10^{-3} kg)

Fraction	Prefix	Symbol
10^{-12}	pico	p
10^{-9}	nano	n
10^{-6}	micro	μ
10^{-3}	milli	m
10^{-2}	cent	c
10^{-1}	deci	d
10	deca	da
10^2	hecto	h
10^3	kilo	k
10^6	mega	M
10^9	giga	G
10^{12}	tera	T
10^{15}	peta	P
10^{18}	exa	E
10^{21}	zetta	Z

Table X-2: *Non-SI Units*

°C	degrees Celsius (0°C = ~273K); Temperature differences are given in °C rather than the more correct form of "Celsius degrees"
Btu	British Thermal Unit (1.055 kJ)
kWh	kilowatt-hour (3.6 MJ)
MW_e	megawatts of electricity (electrical capacity)
ppmv	parts per million by volume (10^6)
ppbv	parts per billion by volume (10^9)
pptv	parts per trillion by volume (10^{12})
tce	tons of coal equivalent (29.31 GJ)
toe	tons of oil equivalent (41.87 GJ)
TWh	terawatt-hour (3.6 PJ)

XI

Glossary of Terms

Glossary of Terms

Afforestation
The act or process of establishing a forest, especially on land not previously forested.

Alternative Energy
Energy derived from non-fossil fuel sources.

Anthropogenic Emissions
Emissions of greenhouse gases (GHGs) associated with human activities. These include burning of fossil fuels for energy, deforestation, and land-use changes.

Annex I Countries
Annex I to the Climate Convention (UNFCCC) lists all the countries in the Organization of Economic Cooperation and Development (OECD), plus countries with economies in transition, Central, and Eastern Europe (excluding the former Yugoslavia and Albania). By default the other countries are referred to as Non-Annex I countries. Under Article 4.2 (a&b) of the Convention, Annex I countries commit themselves specifically to the aim of returning individually or jointly to their 1990 levels of GHG emissions by the year 2000.

Annex II Countries
Annex II to the Climate Convention lists all countries in the OECD. Under Article 4.2 (g) of the Convention, these countries are expected to provide financial resources to assist developing countries comply with their obligations such as preparing national reports. Annex II countries are also expected to promote the transfer of environmentally sound technologies to developing countries.

Annex B Countries
Annex B in the Kyoto Protocol lists those developed countries that have agreed to a target for their GHG emissions, including those in the OECD, Central and Eastern Europe, and the Russian Federation. Not quite the same but similar to Annex I, which also includes Turkey and Belarus, while Annex B includes Croatia, Monaco, Liechtenstein, and Slovenia.

Baseline
A projected level of future emissions against which reductions by project activities could be determined.

Base Year
A common year for calculating emission inventories or to begin model simulations for future scenarios.

Biofuel
A fuel produced from dry organic matter or combustible oils produced by plants. Examples of biofuel include alcohol (from fermented sugar), black liquor from the paper manufacturing process, wood, and soybean oil.

Biomass
The total dry organic matter or stored energy content of living organisms. Biomass can be used for fuel directly by burning it (e.g., wood), indirectly by fermentation to an alcohol (e.g., sugar), or by extraction of combustible oils (e.g., soybeans).

Carbon Cycle
The natural processes that influence the exchange of carbon (in the form of carbon dioxide (CO_2), carbonates and organic compounds, etc.) among the atmosphere, ocean, and terrestrial systems. Major components include photosynthesis, respiration, and decay between atmospheric and terrestrial systems (approximately 100 billion tons/year (gigatons)); thermodynamic invasion and evasion between the ocean and atmosphere, operation of the carbon pump and mixing in the deep ocean (approx. 90 billion tons/year). Deforestation and fossil fuel burning releases approximately 7 Gt into the atmosphere annually. The total carbon in the reservoirs is approximately 2000 Gt in land biota, soil, and detritus, 750 Gt in the atmosphere, and 38,000 Gt in the oceans. (Figures from IPCC WGI Scientific Assessment 1990.)

Carbon Dioxide
A naturally occurring gas, CO_2 is also a by-product of burning fossil fuels and biomass, as well as land-use changes and other industrial processes. It is the principal anthropogenic GHG that affects the earth's temperature. It is the reference gas against which other GHGs are measured and therefore has a "Global Warming Potential" (GWP) of 1.

Carbon Sequestration
The long-term storage of carbon or CO_2 in the forests, soils, ocean, or underground in depleted oil and gas reservoirs, coal seams, and saline aquifers. Examples include the separation and disposal of CO_2 from flue gases or processing fossil fuels to produce H_2- and CO_2-rich fractions, and the direct removal of CO_2 from the atmosphere through land use change, afforestation, reforestation, ocean fertilization, and agricultural practices to enhance soil carbon.

Carbon Sinks
Natural or man-made systems that absorb CO_2 from the atmosphere and store them. Trees, plants, and the oceans all absorb CO_2 and, therefore, are carbon sinks.

CFCs
See "Chlorofluorocarbons".

CH_4
See "Methane".

Chlorofluorocarbons
Chlorofluorocarbons (CFCs) are GHGs covered under the 1987 Montreal Protocol and used for refrigeration, air conditioning, packaging, insulation, solvents, or aerosol propellants. As they are not destroyed in the lower atmosphere, CFCs drift into the upper atmosphere where, given suitable conditions, they break down ozone. These gases are being replaced by other compounds, including hydrochlorofluorocarbons (HCFCs) and hydrofluorocarbons (HFCs), which are GHGs covered under the Kyoto Protocol.

Climate Change *(UNFCCC definition)*
A change of climate which is attributed directly or indirectly to human activity that alters the composition of the global atmosphere and that is in addition to natural climate variability over comparable time periods.

Climate Convention
See "UN Framework Convention on Climate Change".

Climate Models
Large and complex computer programs used to mathematically simulate global climate. They are based on mathematical equations derived from our knowledge of the physics that governs the earth–atmosphere system.

Co-generation
The use of waste heat from electric generation, such as exhaust from gas turbines, for either industrial purposes or district heating.

Commercialization
Sequence of actions necessary to achieve market entry and general market competitiveness of new innovative technologies, processes, and products.

Conference of the Parties (COP)
The supreme body of the UN Framework Convention on Climate Change (UNFCCC), comprises countries that have ratified or acceded to the Framework Convention on Climate Change. The first session of the Conference of the Parties (COP-1) was held in Berlin in 1995, COP-2 in Geneva 1996, COP-3 in Kyoto 1997, and COP-4 in Buenos Aires. COP-5 will be held in Bonn.

CO_2

See "Carbon Dioxide".

Cost-effective
A criterion that specifies that a technology or measure delivers a good or service at equal or lower cost than current practice, or the lowest cost alternative for the achievement of a given target.

Decarbonization
A decrease in the specific carbon content of primary energy or of fuels.

Deforestation
The removal of forest stands by cutting and burning to provide land for agricultural purposes, residential or industrial building sites, roads, etc., or by harvesting the trees for building materials or fuel.

Demand-Side Management
Policies and programs designed to reduce consumer demand for electricity and other energy sources while maintaining (or even increasing) the services the energy use renders. It helps to reduce the need for constructing new power facilities.

Dematerialization
A decrease in the material intensity of economic activity in general, or of individual production processes and end-use applications.

Economic Potential
The portion of technical potential for GHG emissions reductions or energy efficiency improvements that could be achieved cost-effectively in the absence of market barriers. The achievement of market potential requires additional policies and measures to break down market barriers.

Emissions
The release of GHGs and/or their precursors into the atmosphere over a specified area and period of time.

Emissions Category
The SRES Scenarios are grouped into four categories of cumulative CO_2 emissions (all sources) between 1990 and 2100: low, medium-low, medium-high, and high emissions. Each category contains scenarios with a range of different driving forces yet similar cumulative emissions. See also "(Scenario) Category."

Emission Standard
A level of emission that under law may not be exceeded.

Energy Intensity
This is the ratio of energy consumption to economic or physical output. At the national level, energy intensity is the ratio of total domestic primary energy consumption or final energy consumption to gross domestic product or physical output.

FCCC
See 'UN Framework Convention on Climate Change'.

Final Energy
Energy supplied that is available to the consumer to be converted into useful energy (e.g. electricity at the wall outlet).

Fossil Fuels
Carbon-based fuels, including coal, oil, and natural gas and their derived fuels such as gasoline, synthesis gas from coal, etc.

Fuel Switching
Policy designed to reduce CO_2 emissions by requiring electric utilities or consumers to switch from high-carbon to low-carbon fuels (e.g. from coal to gas).

GHGs
See "Greenhouse Gases".

GHG Reduction Potential
Possible reductions in emissions of greenhouse gasses (quantified in terms of absolute reductions or in percentages of baseline emissions) that could be achieved through the use of technologies and measures.

Global Warming
The hypothesis that the earth's temperature is being increased, in part, because of emissions of GHGs associated with human activities, such as burning fossil fuels, biomass burning, cement manufacture, cow and sheep rearing, deforestation, and other land-use changes.

Global Warming Potential
A measurement technique to define the relative contribution of each GHG to atmospheric warming. A GWP can only be calculated for specified time horizons (e.g. 20 to 500 years) and for given GHG concentration levels (e.g. current). Both direct and indirect effects are considered. (Indirect effects include changes in atmospheric chemistry such as ozone formation and changes in stratospheric water vapor.) CO_2 has been assigned a GWP of 1, against which all other GHGs are compared. For example, methane (CH_4) has a GWP that is currently estimated to be about 21 times greater than that of CO_2 over a 100 year time horizon, and thus CH_4 has a GWP of 21. (Note that in the economic literature GWP usually denotes gross world product, referrred to as global GDP in this report.)

Greenhouse Effect
The trapping of heat by an envelope of naturally occurring heat-retaining gases (water vapour, CO_2, nitrous oxide (N_2O), CH_4, and ozone) that keeps the earth about 30°C (60°F) warmer than if these gases did not exist.

Greenhouse Gases
Gases in the earth's atmosphere that absorb and re-emit infrared radiation. These gases occur through both natural and human-influenced processes. The major GHG is water vapour. Other GHGs include CO_2, N_2O, CH_4, ozone, and CFCs.

Gridding
The provision of emission or socio-economic activity data in spatially highly explicit form.

GWP
See "Global Warming Potential".

HFCs
See "Hydrofluorocarbons".

Harmonization
A procedure to ease comparability of model results by adopting common (exogenous) input assumptions. Through harmonization, differences in emissions outcomes resulting from differences in model input assumptions (e.g. exogenous population growth) can be separated from differences that arise from different internal model parametrizations (e.g. of the dynamics of technological change). The scenarios reported here can be classified into three categories: "fully harmonized" scenarios share population, GDP, and final energy use assumptions at the level of the four SRES regions (and hence also at the global level) between 1990 and 2100 within prespecified bounds. "Globally harmonized" scenarios share global population and GDP assumptions at the global level for the 1990 to 2100 period within prespecified bounds (deviations in one 10-year interval are not considered). "Other scenarios" have adopted alternative assumptions for population and GDP than the ones suggested for scenario harmonization.

Hydrofluorocarbons
HFCs are among the six GHGs to be curbed under the Kyoto Protocol. They are produced commercially as a substitute for CFCs. HFCs are used largely in refrigeration and semi-conductor manufacturing. Their GWPs range from 1300 to 11,700 times that of CO_2 (over a 100 year time horizon), depending on the HFC.

Illustrative Scenario
A scenario that is illustrative for each of the six scenario groups reflected in the Summary for Policymakers of this report. They include four revised "scenario markers" for the scenario groups A1B, A2, B1 and B2, and two additional scenarios for the A1FI and AIT groups. All scenario groups are equally sound. See also "(Scenario) Groups" and "(Scenario) Markers".

Intergovernmental Organization (IGO)
Organizations constituted of governments. Examples include the World Bank, the OECD, and the International Civil Aviation Organization. The UNFCCC allows accreditation of these IGOs to attend the negotiating sessions.

International Energy Agency (IEA)
Paris-based organization formed in 1973 by the major oil-consuming nations to manage future oil supply shortfalls.

International Institute for Applied Systems Analysis (IIASA)
Non-governmental, international, interdisciplinary research institute located in Laxenburg, Austria. IIASA is supported by the Academy of Sciences and similar learned societies from 15 countries. Its research focuses on the human dimensions of global change.

Kyoto Mechanisms *(formerly known as Flexibility Mechanisms)*
Economic mechanisms based on market principles that Parties to the Kyoto Protocol can use in an attempt to lessen the potential economic impacts of GHG emission-reduction requirements. They include *Joint Implementation* (Article 6), the *Clean Development Mechanisms* (Article 12), and *Emissions Trading* (Article 17).

Kyoto Protocol
The Protocol, drafted during the Berlin Mandate process, that, on entry into force, would require countries listed in its Annex B (developed nations) to meet differentiated reduction targets for their GHG emissions relative to 1990 levels by 2008–2012. It was adopted by all Parties to the Climate Convention in Kyoto, Japan, in December 1997.

Marker (Scenario)
See "(Scenario) Marker".

Market Penetration
The share of a given market that is provided by a particular good or service at a given time.

Market Potential (or Currently Realizable Potential)
The portion of the economic potential for GHG emissions reductions or energy efficiency improvements that could be achieved under existing market conditions, assuming no new policies and measures.

Measures
Actions that can be taken by a government or a group of governments, often in conjunction with the private sector, to accelerate the use of technologies or other practices that reduce GHG emissions.

Methane
One of the six GHGs to be mitigated under the Kyoto Protocol, it has a relatively short atmospheric lifetime of 10 ± 2 years. Primary sources of CH_4 are landfills, coal mines, paddy fields, natural gas systems, and livestock (e.g., cows and sheep). It has a GWP of 21 (100 year time horizon).

Model
A formal representation of a system that allows quantification of relevant system variables and simulation of systems' behavior, e.g. the implications on future GHG emissions of alternative demographic, economic and technological developments (scenarios).

Montreal Protocol
International agreement under the UN which entered into force in January 1989 to phase out the use of ozone-depleting compounds such as CFCs, methyl chloroform, carbon tetrachloride, and many others.

NGO
See "Non-Governmental Organization".

Nitrous Oxide
One of the six GHGs to be curbed under the Kyoto Protocol, N_2O is generated by burning fossil fuels and the manufacture of fertilizer. It has a GWP 310 times that of CO_2 (100 year time horizon).

Non-Annex I Parties
The countries that have ratified or acceded to the UNFCCC that are not included in Annex I of the Convention.

Non-Annex B Parties
The countries that are not included in the Annex B list of developed nations in the Kyoto Protocol.

Non-Governmental Organization/Observer
Non-Governmental Organization (NGOs) include registered non-profit organizations and associations from business and industry, environmental groups, cities and municipalities, academics, and social and activist organizations.

No Regrets
Actions that result in GHG limitations and abatement, and that also make good environmental and economic sense in their own right.

Ozone
Ozone (O_3) in the troposphere, or lower part of the atmosphere, can be a constituent of smog and acts as a GHG. It is created naturally and also by reactions in the atmosphere that involve gases resulting from human activities, including nitrogen oxides (NO_x), from motor vehicles and power plants. The Montreal Protocol seeks to control chemicals that destroy ozone in the stratosphere (upper part of the atmosphere), where the ozone absorbs ultra-violet radiation.

PAMs
See "Policies and Measures".

Perfluorocarbons
Among the six GHGs to be abated under the Kyoto Protocol. Perfluorocarbons (PFCs) are a by-product of aluminum smelting and uranium enrichment. They also are the replacement for CFCs in manufacturing semiconductors. The GWP of PFCs is 6500–9200 times that of CO_2 (100 year time horizon).

PFCs
See 'Perfluorocarbons'.

Policies and Measures
In UNFCCC parlance, **policies** are actions that can be taken and/or mandated by a government – often in conjunction with business and industry within its own country, as well as with other countries – to accelerate the application and use of successful measures to curb GHG emissions. **Measures** are technologies, processes, and practices used to implement policies that, if employed, would reduce GHG emissions below anticipated future levels. Examples might include carbon or other energy taxes, standardized fuel efficiency

standards for automobiles, etc. "Common and co-ordinated" or "harmonized" policies refer to those adopted jointly by Parties. (This could be by region, such as the European Union (EU), or by countries that comprise a given classification, for example, all Annex I nations.)

Precautionary Principle
From the UN Framework Convention on Climate Change (Article 3): *Parties should take precautionary measures to anticipate, prevent or minimize the causes of climate change and mitigate its adverse effects. Where there are threats of serious or irreversible damage, lack of full scientific certainty should not be used as a reason for postponing such measures taking into account that policies and measures to deal with climate change should be cost-effective so as to ensure global benefits at the lowest possible cost.*

Primary Energy
Energy embodied in natural resources (e.g., coal, crude oil, sunlight, uranium) that has not undergone any anthropogenic conversion or transformation.

Quantified Emissions Limitations and Reductions Objectives
Abbreviated to QELROs, these are the GHG emissions reduction commitments made by developed countries listed in Annex B of the Protocol. (See also "Targets and Timetables".)

Regulatory Measures
Rules or codes enacted by governments that mandate product specifications or process performance characteristics.

Renewables
Energy sources that are, within a short timeframe relative to the earth's natural cycles, sustainable, and include non-carbon technologies such as solar energy, hydropower, and wind as well as carbon-neutral technologies such as biomass.

Research, Development, and Demonstration
Scientific/technical research and development of new production processes or products, coupled with analysis and measures that provide information to potential users regarding the application of the new product or process; demonstration tests, and feasibility of applying these products processes via pilot plants and other pre-commercial applications.

Scenario
A plausible description of how the future may develop, based on a coherent and internally consistent set of assumptions ("scenario logic") about key relationships and driving forces (e.g., rate of technology change, prices). Note that scenarios are neither predictions nor forecasts.

(Scenario) Category
The SRES Scenarios are grouped into four categories of cumulative CO_2 emissions (all sources) between 1990 and 2100: low, medium-low, medium-high, and high emissions. Each category contains scenarios with a range of different driving forces yet similar cumulative emissions. See also "Emissions Category".

(Scenario) Family
Scenarios that have a similar demographic, societal, economic and technical-change storyline. Four scenario families comprise the SRES scenario set: A1, A2, B1 and B2.

(Scenario) Group
Scenarios within a family that reflect a consistent variation of the storyline. The A1 scenario family includes four groups designated as A1T, A1C, A1G and A1B that explore alternative structures of future energy systems. In the Summary for Policymakers, the A1C and A1G groups have been combined into one "Fossil Intensive" A1FI scenario group. The other three scenario families consist of one group each. The SRES scenario set reflected in the SPM thus consists of six distinct scenario groups, all of which are equally sound and together capture the range of uncertainties associated with driving forces and emissions.

(Scenario) Marker
A scenario that was originally posted in draft form on the SRES website to represent a given scenario family. The choice of markers was based on which of the initial quantifications best reflected the storyline, and the features of specific models. Markers are no more likely than other scenarios, but are considered by the SRES writing team as illustrative of a particular storyline. They are included in revised form in this report. These scenarios have received the closest scrutiny of the entire writing team and via the SRES open process. Scenarios have also been selected to illustrate the other two scenario groups (see also "Scenario Group" and "Illustrative Scenario".

(Scenario) Set
A set of scenarios developed using a particular methodologic approach. The SRES scenario set comprises 40 scenarios grouped into four scenario families, seven (six in the SPM) scenario groups and four (cumulative CO_2) emissions categories.

(Scenario) Storyline
A narrative description of a scenario (or a family of scenarios) highlighting the main scenario characteristics, relationships between key driving forces and the dynamics of their evolution.

Standardization
Adopting standardized numerical values to improve model and scenario comparability. In this report, emissions are standardized for the two reporting years 1990 and 2100 across all models and scenarios, and individual scenario differences thereafter are corrected for differences between original model outputs and standardized values ("offsets"). (Base year differences reflect scientific uncertainty in source/sink strengths for many GHGs as well as differences in model calibration and simulation time horizons, e.g. for some models 1990 is a projected year as simulations begin by an earlier base year.

SF_6
See "Sulfur Hexafluoride".

Sinks *(UNFCCC Definition)*
Any process or activity or mechanism that removes a GHG, aerosol, or precursor of a GHG into the atmosphere.

Source *(UNFCCC Definition)*
Any process or activity that releases a GHG, aerosol, or precursor of a GHG into the atmosphere.

Standards/Performance Criteria
Set of rules or codes that mandate or define product performance (e.g., grades, dimensions, characteristics, test methods, rules for use).

Structural Change
Changes, for example, in the relative share of GDP produced by the industrial, agricultural, or services sectors of an economy; or (more generally) systems transformations whereby some components are either replaced or potentially substituted by other ones.

Sulfur Hexafluoride
One of the six GHGs to be curbed under the Kyoto Protocol. Sulfur hexafluoride (SF_6) is largely used in heavy industry to insulate high-voltage equipment and to assist in the manufacturing of cable-cooling systems. Its GWP is 23,900 times that of CO_2 (100 year time horizon).

Targets and Timetables (see also QELROs)
A target is the reduction of a specific percentage of GHG emissions (e.g., 6%, 7%) from a baseline date (e.g., "below 1990 levels") to be achieved by a set date, or timetable (e.g., 2008–2012). For example, under the Kyoto Protocol's formula, the EU has agreed to reduce its GHG emissions by 8% below 1990 levels by the 2008–2012 commitment period. These targets and timetables are, in effect, a cap on the total amount of GHG emissions that can be emitted by a country or region in a given time period.

Technical Potential
The amount by which it is possible to reduce GHG emissions or improve energy efficiency by using a technology or practice in all applications in which it could technically be adopted, without consideration of its costs or practical feasibility.

Technology
A systems of means towards particular ends that includes both hardware and social information, e.g. a piece of equipment or a technique for performing a particular activity.

Trace Gas
A minor constituent of the atmosphere. The most important trace gases that contribute to the greenhouse effect are CO_2, ozone, CH_4, N_2O, ammonia, nitric acid, ethylene, sulfur dioxide (SO_2), nitric oxide, CFCs, HFCs HCFCs, SF_6, methyl chloride, carbon monoxide, and carbon tetrachloride.

UN Framework Convention on Climate Change (UNFCCC)
A treaty signed at the 1992 Earth Summit in Rio de Janeiro by more than 150 countries. Its ultimate objective is the "stabilization of greenhouse gas concentrations in the atmosphere at a level that would prevent dangerous anthropogenic (human-induced) interference with the climate system". While no legally binding level of emissions is set, the treaty states an aim by Annex I countries to return these emissions to 1990 levels by the year 2000. The treaty took effect in March 1994 upon the ratification of more than 50 countries; a total of some 160 nations have now ratified. In March 1995, the UNFCCC held the first session of the COP, the supreme body of the Convention, in Berlin. Its Secretariat is based in Bonn, Germany. In the biennium 2000–01, its approved budget and staffing level are approximately US$12 million annually with approximately 80 personnel.

Values
Values are based on individual preferences, and the total value of any resource is the sum of the values of the different individuals involved in the use of the resource. The values that are the foundation of the estimation of costs are measured in terms of the willingness to pay (WTP) by individuals to receive the resource or by the willingness of individuals to accept payment (WTA) to part with the resource.

Voluntary Measures
Measures to reduce GHG emissions that are adopted by firms or other actors in the absence of government mandates. Voluntary measures help make climate-friendly products or processes more readily available or encourage consumers to incorporate environmental values in their market choices.

XII

List of Major IPCC Reports

Climate Change—The IPCC Scientific Assessment
The 1990 Report of the IPCC Scientific Assessment Working Group (also in Chinese, French, Russian, and Spanish)

Climate Change—The IPCC Impacts Assessment
The 1990 Report of the IPCC Impacts Assessment Working Group (also in Chinese, French, Russian, and Spanish)

Climate Change—The IPCC Response Strategies
The 1990 Report of the IPCC Response Strategies Working Group (also in Chinese, French, Russian, and Spanish)

Emissions Scenarios
Prepared for the IPCC Response Strategies Working Group, 1990

Assessment of the Vulnerability of Coastal Areas to Sea Level Rise–A Common Methodology
1991 (also in Arabic and French)

Climate Change 1992—The Supplementary Report to the IPCC Scientific Assessment
The 1992 Report of the IPCC Scientific Assessment Working Group

Climate Change 1992—The Supplementary Report to the IPCC Impacts Assessment
The 1992 Report of the IPCC Impacts Assessment Working Group

Climate Change: The IPCC 1990 and 1992 Assessments
IPCC First Assessment Report Overview and Policymaker Summaries, and 1992 IPCC Supplement

Global Climate Change and the Rising Challenge of the Sea
Coastal Zone Management Subgroup of the IPCC Response Strategies Working Group, 1992

Report of the IPCC Country Studies Workshop
1992

Preliminary Guidelines for Assessing Impacts of Climate Change
1992

IPCC Guidelines for National Greenhouse Gas Inventories
Three volumes, 1994 (also in French, Russian, and Spanish)

IPCC Technical Guidelines for Assessing Climate Change Impacts and Adaptations
1995 (also in Arabic, Chinese, French, Russian, and Spanish)

Climate Change 1994—Radiative Forcing of Climate Change and an Evaluation of the IPCC IS92 Emission Scenarios
1995

Climate Change 1995—The Science of Climate Change – Contribution of Working Group I to the Second Assessment Report
1996

Climate Change 1995—Impacts, Adaptations, and Mitigation of Climate Change: Scientific-Technical Analyses – Contribution of Working Group II to the Second Assessment Report
1996

Climate Change 1995—Economic and Social Dimensions of Climate Change – Contribution of Working Group III to the Second Assessment Report
1996

Climate Change 1995—IPCC Second Assessment Synthesis of Scientific-Technical Information Relevant to Interpreting Article 2 of the UN Framework Convention on Climate Change
1996 (also in Arabic, Chinese, French, Russian, and Spanish)

Technologies, Policies, and Measures for Mitigating Climate Change – IPCC Technical Paper I
1996 (also in French and Spanish)

An Introduction to Simple Climate Models used in the IPCC Second Assessment Report – IPCC Technical Paper II
1997 (also in French and Spanish)

Stabilization of Atmospheric Greenhouse Gases: Physical, Biological and Socio-economic Implications – IPCC Technical Paper III
1997 (also in French and Spanish)

Implications of Proposed CO_2 Emissions Limitations – IPCC Technical Paper IV
1997 (also in French and Spanish)

The Regional Impacts of Climate Change: An Assessment of Vulnerability – IPCC Special Report
1998

Aviation and the Global Atmosphere - IPCC Special Report 1999

Land Use, Land Use Changes and Forestry - IPCC Special Report 2000

Methodological and Technological Issues in Technology Transfer - IPCC Special Report 2000

ENQUIRIES: IPCC Secretariat, c/o World Meteorological Organization, 7 bis, Avenue de la Paix, Case Postale 2300, 1211 Geneva 2, Switzerland